Human Form, Human Function

Essentials of Anatomy & Physiology

Thomas H. McConnell, III, MD, FCAP

Clinical Professor
Department of Pathology
University of Texas Southwestern Medical Center
Dallas, Texas

Kerry L. Hull, PhD

Professor, Department of Biology
Bishop's University
Sherbrooke, Quebec
Adjunct Professor, Department of Physiology
Faculty of Medicine
University of Sherbrooke
Sherbrooke, Quebec
Canada

 Wolters Kluwer | Lippincott Williams & Wilkins
Health

Philadelphia · Baltimore · New York · London
Buenos Aires · Hong Kong · Sydney · Tokyo

Acquisitions Editor: David B. Troy
Product Manager: Jennifer Ajello
Development Editor: Laura Bonazzoli
Marketing Manager: Allison Powell
Design Coordinator: Teresa Mallon
Artist: Imagineering Media Services, Inc.
Compositor: Aptara, Inc.

First Edition

9 8 7 6 5 4 3 2

Library of Congress Cataloging-in-Publication Data

McConnell, Thomas H.
  Human form, human function : essentials of anatomy & physiology /
Thomas H. McConnell, III, Kerry L. Hull. – 1st ed.
    p. ; cm.
  ISBN 978-0-7817-8020-9
  1. Human physiology–Textbooks.   2. Human anatomy–Textbooks.   I.
Hull, Kerry L.   II. Title.
  [DNLM: 1. Physiological Phenomena.   2. Anatomy. QT 4 M478h 2011]
  QP34.5.M467 2011
  612–dc22

2010025886

DISCLAIMER

Care has been taken to confirm the accuracy of the information present and to describe generally accepted practices. However, the authors, editors, and publisher are not responsible for errors or omissions or for any consequences from application of the information in this book and make no warranty, expressed or implied, with respect to the currency, completeness, or accuracy of the contents of the publication. Application of this information in a particular situation remains the professional responsibility of the practitioner; the clinical treatments described and recommended may not be considered absolute and universal recommendations.

The authors, editors, and publisher have exerted every effort to ensure that drug selection and dosage set forth in this text are in accordance with the current recommendations and practice at the time of publication. However, in view of ongoing research, changes in government regulations, and the constant flow of information relating to drug therapy and drug reactions, the reader is urged to check the package insert for each drug for any change in indications and dosage and for added warnings and precautions. This is particularly important when the recommended agent is a new or infrequently employed drug.

Some drugs and medical devices presented in this publication have Food and Drug Administration (FDA) clearance for limited use in restricted research settings. It is the responsibility of the health care provider to ascertain the FDA status of each drug or device planned for use in their clinical practice.

Dedication

For Marianne
Thomas H. McConnell, III

For Norm
Kerry L. Hull

Reviewers

Brian Archer, BA
Program Manager
Department of Therapeutic Massage
Heritage College
Denver, Colorado

Benoit A. Bacon, PhD
Associate Professor
Department of Psychology
Bishop's University
Sherbrooke, QC
Canada

Kristen Bebeau, RN, BSN, MA, PHN
Nursing Faculty
Department of Nursing
Hannepin Technical College
Brooklyn Park, Minnesota

Janet Brodsky, MS
Program Chair
Life Sciences
Ivy Tech Community College
Lafayette, Indiana

Wendy Cain, MS Biology
Assistant Professor of Microbiology
and Anatomy and Physiology
Department of Science
Ivy Tech Community College of
Indiana
Columbus, Indiana

Mauro daFonte, MD
Faculty Member
Department of Health Science and
Math & Science
GateWay Community College
Phoenix, Arizona

Clair Eckersell, PhD
Professor of Biology
Brigham Young University
Rexburg, Idaho

John Fishback, MS
Biology Instructor
Department of Biology
Ozarks Technical Community
College
Springfield, Missouri

Anthony Gendill, BS, BA, DC
Medical Instructor
Department of Medicine
Institute of Business and Medical
Careers
Fort Collins, Colorado

Gary Heisermann, PhD
Assistant Professor
Department of Biology
Salem State College
Salem, Massachusetts

Ann Henninger, PhD
Professor of Biology
Department of Biology
Wartburg College
Waverly, Iowa

Krista Hoekstra, BSN, MA
Director of Nursing
Nursing Instructor
Department of Practical Nursing
Hennepin Technical College
Brooklyn Park, Minnesota

Stephanie Irwin, MS
Professor
Department of Science
Front Range Community College
Fort Collins, Colorado

Denise Johnson, MSc
Assistant Professor of Biology
Department of Biology
Sacramento City College
Sacramento, California

Brian Kipp, PhD
Assistant Professor
Biomedical Science
Grand Valley State University
Allendale, Michigan

Kathleen Olewinski, MS
Program Director
Department of Allied Health
Bryant & Stratton College
Franklin, Wisconsin

Matthew Petersen, PhD
Professor of Anatomy & Physiology
Department of General Studies
Northeast Wisconsin Technical
College
Green Bay, Wisconsin

Michelle Roybal, AA
Program Supervisor
Department of Medical Assisting
Intellitec College
Grand Junction, Colorado

Richard Sims, BS, MS
Instructor-Biology
Department of Biology
Jones County Junior College
Ellisville, Mississippi

Bonnie Tarricone, PhD
Director of Advanced Human
Physiology
Department of Health and Human
Services
Ivy Tech Community College
Indianapolis, Indiana

Shannon Thomas, MEd, MS
Instructor of Biology
Department of Biology
McLennan Community College
Waco, Texas

Sales Advisory Team

Doug Dobbs
Instructional Services Consultant

Jennifer Bookwalter
Inside Sales

Brett Martin
Midwest Region

John Willett
Northeast Mid-Atlantic Region

Steve White
Southeast Region

Jane Boles
West Coast Region

Kimberly Morin
Career: Northeast Region

Lucky Cater
Career: West Coast Region

Ann-Marie Spindler
Canadian Great Lakes Region

Preface

Form and function are inseparable in *every* aspect of life, not just in the life sciences: the laws of aerodynamics dictate the shape of an airplane, just as the shape of a claw hammer is dictated by its use to either pound in nails or pull them out. We chose the name of this textbook, *Human Form, Human Function,* to emphasize that human body anatomy (form) and physiology (function) are similarly inseparable. For example, it is no accident that when you fold your elbow it brings your hand near your mouth. The length of your upper and lower arm along with the joint in between, function together to make it easy to feed yourself. Also, on the molecular level, form and function are coupled—the function of protein molecules relies on the intricate folds of their structure; a misfolded protein does not function properly.

Approach

Just as form and function are inseparable, so are the workings of the various organs and organ systems of the body, which cooperate in a mutually beneficial way to sustain normal body function. For example, in the broadest sense the lungs support the intestines and the intestines support the lungs. In our discussions, we emphasize these points by returning again and again to the relationship of form to function and the mutual interdependence of body organs and organ systems.

Apart from our emphasis on the interlocking nature of human anatomy and physiology, we also emphasize that injury and disease are nothing more than abnormal form coupled with abnormal function. To this end, the narrative is laced with medical content and features that will deepen your comprehension of the interdependent workings of the body in its healthy state. Let's look a bit more closely at how form, function, and integrated physiology are presented.

The Body as an Integrated Organism

Although many body functions might seem to be performed by a single body system, just the opposite is true. For example, it would be easy to assume that the lungs are solely responsible for respiration and the intestinal tract for digestion, but in fact, neither the lungs nor the bowels can live without the other. They are bound together by certain integrative functions, such as cell-to-cell communication, the uptake of nutrients and oxygen, and so on. Many textbooks attempt to convey this integration through a single end-of-chapter discussion, diagram, or table. In contrast, we believe that the integrated nature of our functionality is fundamental, and must be conveyed continuously throughout the content discussion. Every aspect of *Human Form, Human Function* reflects this belief. Let's look at how we have laid it out for you:

Instead of dividing chapters into anatomy and physiology sections, we weave together form and function. We begin our exploration of the body by introducing you to certain chemical and cellular principles that operate in every cell and every tissue, despite where located, and which support each of the forms and functions you'll discover as the chapters unfold. As we go, we help you to build on your understanding of these fundamental principles by applying them to the body in health and in disease. To this end we have written two unique chapters in which integration is the focus:

- **Chapter 4, Communication: Chemical and Electrical Signaling.** We present this subject much earlier and at a more foundational level than in most textbooks. In most other textbooks, signaling is discussed with the endocrine and nervous systems, in which its importance in the life of other systems gets lost in the dizzying complexities of endocrine and neural function. To the contrary, we believe that an early understanding of cell signaling will help you make sense of the physiology of every body system.
- **Chapter 18, Life.** This closing chapter integrates your new knowledge of form and function from all of the previous chapters and applies it to the discussion on aging, the stages of life, genetics and inherited traits and their importance, good and bad stress, exercise and diet; and other aspects of healthy living.

Several chapters also contain dedicated sections on integration. For example:

- **Chapter 6, Bones and Joints,** includes a section on the role of nutrients and hormones in bone health.
- **Chapter 8, The Nervous System,** includes an integrated view of neural form and function.

- **Chapter 15, Metabolism and Endocrine Control,** discusses energy generation, energy balance, body temperature regulation, growth; in short, processes that affect the health and functioning of every body cell.

Your study of these chapters will give you a deeper appreciation of the integrated nature of our body systems.

The Body in Health and Disease

We believe that normal, healthy form and function is best understood and remembered when contrasted with the abnormal forms and functions that are characteristic of certain diseases or conditions.

- Each chapter opens with a **Case Study,** the presentation of the natural history of an episode of disease. Each of these cases, all of them drawn from real patients, was selected to illustrate the most important concepts discussed in the chapter. For example, the case study in Chapter 5, which discusses the skin, tells the story of a young man who sustained severe burns over much of his body. It is a case based on Dr. McConnell's actual hospital experience, as are many of the other cases.
- As the chapter narrative progresses, we return to the chapter case in a feature called **Case Notes.** The critical thinking questions in this feature prompt you to apply to the case the material you have just learned.
- Finally, near the end of each chapter is the **Case Discussion** in which we revisit the case study and apply the chapter concepts to the case in some detail. These case discussions are accompanied by an illustration to help you visualize the interrelationships among various aspects of the case, such as the breakdown in normal functions and the effect of medical treatment.
- **Clinical Snapshot** boxes are included in most chapters. These allow for a fuller discussion of abnormalities of form and function than can be included within the narrative. For example, a Clinical Snapshot in Chapter 5 discusses skin cancer.

Apart from these particular features, medical discussion of abnormal form and function are regularly woven into the narrative.

Organization and Structure

Just as body systems work together as a unified whole, so the 18 chapters in *Human Form, Human Function* flow smoothly from one to another with no artificial division into units. This choice reflects and underscores our theme of integration. We believe that units, although useful in highlighting similarities, can also distract from an understanding of the inherent anatomical and physiological interrelationships among body systems. Our integrative approach carries through every chapter of the text.

- **Chapter 1** introduces you to the basic features of living organisms and the language of anatomy and physiology.
- **Chapters 2–4** present the chemical and cellular structures and functions that will inform your subsequent learning about the integrated body.
- **Chapters 5–9** discuss the form and function of skin, bones, muscles and tendons, nerves, and our sensory apparatus.
- **Chapters 10–11** explore circulation of blood.
- **Chapter 12** is devoted to the immune system and its role in defending the body against microbes and other threats; also discussed is the lymphatic system, an important feature of the immune system, but part of which plays a role in digestion.
- **Chapter 13** is devoted to respiration: the inhalation and absorption of oxygen by the lungs and its use by cells; and the generation off carbon dioxide by cells and its exhalation by the lungs.
- **Chapters 14–16** explore what happens to the foods and fluids we ingest, as well as how the body removes toxins, generates energy and builds essential substances, and regulates temperature, growth, and other processes.
- **Chapter 17** discusses human reproduction, including a brief look at pregnancy and breastfeeding.
- **Chapter 18** is intended to weave together your understanding of the structures and functions of the human body into a seamless whole as you consider how the body grows and ages, and how good health is maintained.

Style

We deliberately use a narrative technique that is commonly described as "casual" or "conversational." Our classroom and professional experience teaches us that brevity, manner, and style are the essence of effective communication. We find that our style makes reading easier, and enhances understanding and recall of important points without sacrificing scientific relevance. A good story and a dollop of humor help, too.

However, the science we deliver was not casually assembled in a coffee shop discussion. We have scrubbed every sentence to be sure that our casual style conveys rigorous scientific truth.

Terminology

A word about terminology is necessary. One of us is a physician, a pathologist (Dr. McConnell); the other (Dr. Hull) is a physiologist. In the course of our collaboration it became clear that medicine and physiology sometimes have different names for the same thing. In virtually every such instance the preferred term in medicine is an eponym and the preferred term in physiology is a generic scientific one. For example, extending from the uterus to the ovary is a tube. In medicine it is universally called the *fallopian tube* after the Italian anatomist and physician, Gabrielle Fallopius (1523–1562), who first described it. On the other hand, physiologists usually prefer to call it the *uterine tube*, because it is a duct within which ova travel from the ovary to the uterus. Each approach has its virtue. Eponyms, once learned, are singularly distinctive and very unlikely to be forgotten or misunderstood. Generic terms are not so unique, but the parts of their names offer a clue to their function.

However, in most instances the terms in medicine and physiology are identical generic scientific ones. In the few instances in which there is a choice we usually prefer the medical term, reasoning that few of our readers will become physiologists, but all of them will be patients. Even so, when we use an eponym we include the generic term in parenthesis.

Additional Resources

Human Form, Human Function: Essentials of Anatomy & Physiology includes additional resources for both instructors and students that are available on the book's companion Web site at http://thePoint.lww.com/McConnellandHull.

Students

Students who have purchased *Human Form, Human Function* have access to the following additional resources:

- Podcasts provide an audio summary on certain topics that may be more challenging for students.
- Interactive Online Activities—approximately 1,800 additional questions and activities for student practice/quizzing.
- Online Dissection Atlas—fetal pig, human, and cat dissection images are presented in an online library so students get a "real" view of anatomy.
- Animations of various processes are included to help students understand the body's inner workings.
- Additional Clinical Snapshot, History of Science, and Basic Form, Basic Function boxes provide learning opportunity and interest for the student.

Instructors

Approved adopting instructors will be given access to all of the student resources and these additional resources:

- Online Instructor's Manual
- PowerPoint Slides with Lecture Notes and images
- Online Activities—approximately 1,800 additional questions and activities for student practice/quizzing
- Image Bank includes all images in the text plus table images and with labels on/off feature.
- Answers to textbook chapter questions and mini quizzes (Pop Quiz, Case Notes, Figure legend questions) are included online
- Test Bank with more than 500 questions in different formats (multiple choice, true-false, fill-in-the-blank, and matching)
- WebCT and Blackboard-ready Cartridge, which allow you to integrate the ancillary materials into learning management systems

In addition, purchasers of the text can access the searchable Full Text On-line by going to the *Human Form, Human Function: Essentials of Anatomy & Physiology* Web site at **http://thePoint.lww.com/McConnellandHull.** See the inside front cover of this text for more details, including the passcode you will need to gain access to the Web site.

So here it is. We hope you like it. Judge for yourself.

Thomas H. McConnell

Kerry L. Hull

Acknowledgments

Producing *Human Form, Human Function: Essentials of Anatomy & Physiology* required the coordinated efforts of many dedicated professionals. We would first like to thank Dana Knighten, Development Editor, for her insight in pairing a medical doctor and a physiologist to write an integrated anatomy and physiology text. Without this creative inspiration, this text would not have taken any form. Additionally, we are offering our most enthusiastic applause to Laura Bonazzoli, Development Editor, who edited the raw manuscript. It is an understatement to say that her insights and suggestions were invaluable. Laura also had the herculean task of organizing reviewer remarks, and knowing whom among them offered the best advice. Her attention to detail was thorough, and her effort shines through in the final product. Thank you, Laura. In the age of the Internet a textbook is a stampede of emails and file exchanges. Herding us from one milepost to the next was Jennifer Ajello, Product Manager, who somehow managed to keep us together and on time by directing our priorities and coaxing us to meet one deadline after another. Thank you, Jenn. Also, presiding benignly above the fray was David Troy, Acquisitions Editor, upon whose initiative the entire process began. Thank you, David.

We would also like to acknowledge Imagineering for their exceptional job turning rough sketches into artistically pleasing and scientifically precise artwork. A special thanks also goes to the many reviewers for their insightful comments and excellent suggestions for improvement. Thanks to Aleksandar Markovic and Jean Porter for their aid in preparing the art manuscript.

And finally, a special thanks to our families: the McConnell family and the extended Hull-Jones family (Norman, Lauren, and Evan, and Bill, Scotty, Carmen and John) for their inspiration and support over this long and sometimes arduous process.

Thomas H. McConnell, MD, FCAP
Kerry L. Hull, PhD

User's Guide

Human Form, Human Function uses more than a dozen different features to help you understand and retain the text material. Let's take an imaginary tour of these features:

Chapter Opener Features

Need to Know

Except for Chapters 1 and 2, each chapter begins with a list of topics you will need to have mastered before delving into the new material. For example, the idea of "barriers" is first presented in Chapter 1 but is essential to an understanding of the importance of the skin. Thus, the *Need to Know* list in the skin chapter (Chapter 5) will prompt you to make sure you have mastered this discussion from Chapter 1 before beginning your study of Chapter 5.

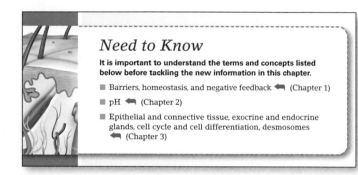

Need to Know

It is important to understand the terms and concepts listed below before tackling the new information in this chapter.

■ Barriers, homeostasis, and negative feedback ◄ (Chapter 1)
■ pH ◄ (Chapter 2)
■ Epithelial and connective tissue, exocrine and endocrine glands, cell cycle and cell differentiation, desmosomes ◄ (Chapter 3)

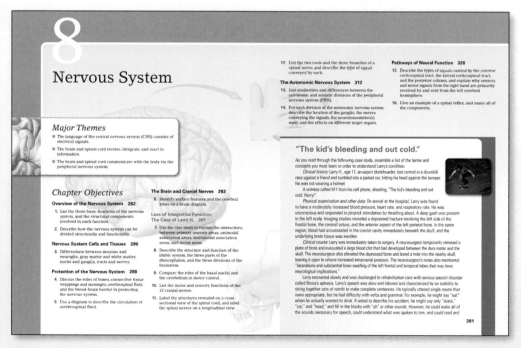

Case Study

As noted in the Preface, a clinical case study opens each chapter. Each case is chosen for its usefulness in illustrating the principles discussed in the chapter.

Quotation

Brief quotations help to illustrate the main idea of each chapter in an entertaining and informative way.

Be careful about reading health books. You may die of a misprint.

Mark Twain (Samuel Langhorne Clemens), American novelist and humorist, 1835–1910

Major Themes

Think of Major Themes as your "heads up" feature. It provides you with a quick overview of the most important principles that govern the forms and functions we are discussing.

Major Themes

- Cell-to-cell communication is critical for homeostasis and life.
- Communication requires a *sender*, a *signal*, a *medium* to carry the signal, and a *receiver* to accept the signal.
- Signals must be "translated" from a code into action.
- The effect of the signal is determined by the receiver, not merely by the signal.
- There are two kinds of physiological signals: chemical and electrical.

Intrachapter Features

Case Notes

As discussed in the Preface, *Case Notes* are critical-thinking questions that challenge you to apply lessons discussed in a previous section to the case study at the beginning of the chapter. Answers to these questions are provided on the student Web site at **http://thePoint.lww.com/McConnellandHull**

Remember This

Within narrative sections, Remember This statements emphasize the most important ideas, or suggest a game or other device to help you memorize them. They may also help clarify concepts or pull together related ideas, as in, "Hair cells and nail cells both grow from a matrix of stem cells."

Remember This! **An unstressed bone is an unhealthy bone.**

Local conditions that affect growth or repair are also important. For example, poor blood flow to a body part, such as a lower limb, can cause affected bone to grow or repair slowly. Diabetes ➡ (Chapter 15) and athero-sclerosis ➡ (Chapter 11) are notable causes of low blood flow to the feet and lower limbs. Bone health can also be affected by other local conditions, such as infection or inflammation.

Case Notes

6.10. Maggie must repair her fragile bones. Remember that she is 57 years old. Which hormone could she take to improve her bone health—estrogen or testosterone?

6.11. Has Maggie's intensive exercise regime helped or harmed her bones?

A Healing Fracture Is a Model of Bone Repair

Healthy bones require the stress of daily life, which stimulates osteoblast and osteoclast activity to rebuild and refresh bone. Lack of stress, as with space flight or prolonged bed rest, causes bones to dissolve partially and leads to weak, brittle bones. However, force from an unusual direction or force that is unusually strong can stress a bone to the point of breaking. A **fracture** is a broken bone. A **pathological fracture** is a break that occurs when the bone is under normal, daily stress. Pathological fractures occur in bones with a

and white blood cells. Repair progresses with the appearance of *woven bone*, which ossifies by intra-membranous ossification. This rich mixture of granulation tissue, cartilage, and woven bone is called a **soft callus**, which loosely unites the ends of the broken bone but cannot bear weight. Repair continues as osteoclasts reabsorb dead bone and smooth off the edges of the fracture.

3. After a couple of weeks, the soft callus matures into a **bony callus**, as more bone is deposited and the woven bone and cartilage are increasingly replaced by spongy bone. The bony callus binds the bone ends more securely and is capable of limited weight bearing.

4. Finally, excess spongy bone and callus are resorbed and replaced by dense, compact bone, which remodels into the previous anatomical outline under the influence of local mechanical forces.

Case Notes

6.12. Maggie's bone repair was very slow. Two weeks after her accident, a soft callus had formed around the broken bones. What did this soft callus contain?

6.13. Why did Maggie's physician consider her ankle fracture pathological?

Calcium Homeostasis Is Critical to Body Functioning

Recall from ⬅ Chapter 1 that homeostasis is "the body's collective communication and control effort to maintain constant, healthy internal conditions." The blood calcium concentration, like the concentration of many other ions, is regulated by a negative feedback loop—if it rises too high, the body reacts to lower it, and

Pop Quiz

At the close of narrative sections, Pop Quiz questions test you on the topics just covered. You can use them to determine whether or not you have fully understood prior material, before going on to new material.

 10.12 Rank the following leukocytes from the most abundant to the least abundant: basophils, eosinophils, lymphocytes, neutrophils, monocytes.

10.13 Which of the above leukocytes are granulocytes?

10.14 What is the clinical term for a low blood leukocyte count?

10.15 Are neutrophils the major players in acute inflammation or chronic inflammation?

10.16 An adult patient has large numbers of malignant lymphocytes in her blood. What would you call the patient's disease—lymphoma or leukemia?

 THE HISTORY OF SCIENCE

Does Blood Flow In and Out Like the Tide?

Nowadays students learn in elementary school that blood flows in an endless loop—from the heart to tissues and back. The ancients, however, had a different view, proposed by Galen, a Greek physician, in the second century A.D. Galen correctly observed that arterial and venous blood are different. He proposed that venous blood originates in the liver and arterial blood in the heart, and that blood moves outward to other organs—which consume it—from these manufacturing sites. This misunderstanding held sway for nearly 1500 years until English physician William Harvey (1578–1657) reasoned otherwise.

Harvey used both mathematics and scientific experimentation to show that blood could not be manufactured and consumed as Galen postulated. First, he estimated the number of heartbeats in a day to be 48,000, although it's actually closer to 100,000, and that the volume moved with each beat was 5 mL, though it's actually closer to 80 mL. Even using these low estimates, Harvey proved that if the liver and heart were producing blood, they would have to manufacture over 500 pounds of new blood each day. Clearly Galen had it wrong.

Harvey also did carefully constructed scientific experiments. As you can see in the figure, which he used to illustrate his experiment, he tied a tourniquet on the upper arm tightly enough to stop blood flow into the arm and observed that the limb below the stricture grew pale and cool. Upon release of the tourniquet, the arm grew red and warm.

Harvey coupled this observation with what he saw happening in superficial veins of the forearm. He noticed little bumps in the veins and concluded correctly that they were the valves that Fabricius, his teacher in Italy, had discovered earlier. Harvey tried to massage blood toward the hand by running his finger

Harvey's experiment. This experiment showed that blood flows only toward the heart. If blood flow from the extremity (toward the heart) is blocked, the vein cannot refill by blood flowing from the heart.

downward along the vein, but to no avail. However, the same technique applied upward readily emptied the vein of its contents. He did the same experiment in neck veins, but found the opposite result—he could massage blood downward but not upward. Harvey concluded correctly that veins from every part of the body moved blood to the heart and that the heart pumped it back out. But he was never able to understand how blood passed from arteries to veins, via a system we now know as the vast capillary network of the body.

Harvey announced his discoveries in 1616, but they were not published until 1628, when he printed *Exercitatio Anatomica de Motu Cordis et Sanguinis in Animalibus* (*An Anatomical Exercise on the Motion of the Heart and Blood in Animals*).

Boxed Essays

These "sidebar" boxes present intriguing information that complements the topics in the chapter. Though not essential to your understanding of the chapter's principles, they will deepen your appreciation for the human body and help you apply your learning.

● **History of Science.** Describe discoveries, social issues, or personalities important in the history of anatomy and physiology. For example, a History of Science box in Chapter 1 discusses how medical students of earlier eras sometimes resorted to grave-robbing to obtain cadavers to study.

● **Basic Form, Basic Function.** Explore intriguing aspects of human form or human function. For example, a Fundamentals of Form and Function box in Chapter 8 explores the question of whether or not we really have a right brain and a left brain.

BASIC FORM, BASIC FUNCTION

The Developing Brain

The relationship between the cerebral ventricles and the different brain regions is much easier to understand if we consider how the brain develops embryonically. The upper figure resembles the human brain during embryonic development as well as the brain of other quadripedal mammals.

The embryonic CNS consists of a long hollow tube of tissue. The hollow interior of the tube forms the CSF-containing ventricular system. The most *rostral* (= "toward the nose") portion of the tube is enlarged into two blobs, the two cerebral hemispheres, each containing a fluid-filled lateral ventricle. If we could float on a raft through the ventricular system, we would proceed from the lateral ventricles into the third ventricle, which is surrounded by the diencephalon. Next would be the cerebral aqueduct, surrounded by the midbrain. Continuing in a *caudal* (= "toward the tail") direction through the fourth ventricle, we would find the cerebellum located dorsally and the upper brainstem ventrally. Finally, our raft would exit the brain and enter the central canal of the spinal cord.

As the human brain develops, an amazing event occurs—the long neural tube gets a kink in it. The cerebrum grows and essentially flops forward. The skull bones restrict the growth of the cerebral hemispheres toward the crown of the head, so they curve outward (towards the ear) and inferiorly. The result? The brain we see in part B of this figure, and the ventricles that we see in Figure 8.7.

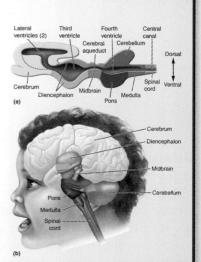

Brain development. A. Basic plan of the nervous system, as in an early embryo. The ventricular system is shown in dark blue. **B.** The human brain.

CLINICAL SNAPSHOT

Congestive Heart Failure: Bob's "Big Heart"

At the memorial service for Bob, many people recalled his generosity, his kindness, his "big heart." Although they were speaking metaphorically, it's also true that Bob's heart was abnormally enlarged at the time of his death. Why?

Recall that the normal heart pumps all of the blood delivered to it, thanks to the relationship between preload and stroke volume. This relationship, known as the Frank–Starling curve, relies on healthy heart muscle. But what happens when, as with Bob, coronary artery disease interferes with heart muscle's blood supply and damages cardiac muscle? The result can be, as it was with Bob, congestive heart failure (CHF), a condition in which *the heart is unable to eject the volume of blood delivered to it* and becomes engorged with blood—the ventricle dilates and cardiac fibers are stretched. In such cases, the engorgement can cause the heart to enlarge so significantly that the abnormal size is detectable by chest x-ray.

Understanding heart failure requires understanding the Frank–Starling curve. As the heart is unable to eject the entire new load of blood delivered to it, preload increases, and the stretched myocardial muscle fibers compensate by contracting more forcefully, just as a stretched spring pulls ever harder as it lengthens. The resulting increase in contraction strength (and thus stroke volume) *compensates* for the impaired functioning of the heart and maintains normal cardiac output. Therefore this initial stage is called **compensated heart failure;** it usually produces no symptoms, because cardiac output is maintained.

However, sometimes the compensation is not enough and blood continues to accumulate. Eventually, the muscle fibers become stretched too far, and further stretching results in *weaker*, not stronger, contractions—just as an overstretched spring loses its power. The force of contraction weakens, stroke volume and cardiac output fall, and the heart enters a vicious cycle: with each beat, more blood accumulates, preload continues to increase, contractile power decreases, stroke volume decreases, cardiac output falls, and even more blood accumulates. This stage, called **uncompensated heart failure,** usually produces symptoms.

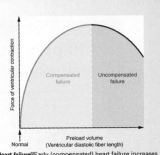

Heart failure Early (compensated) heart failure increases the force of contraction; late (uncompensated) failure reduces it.

Bob's heart failure was due to two mechanisms. Initially it was muscle damage—there were fewer muscle fibers to eject blood. Therefore more blood accumulated with each beat. Second, as blood accumulated and preload increased, the remaining healthy ventricular fibers were stretched. We know from the Frank–Starling curve that the first bit of stretching helped; but as more blood accumulated, it stretched the fibers too far, which had a deleterious effect on the pumping power that remained. With each heartbeat, the volume of blood in the ventricle increased bit by bit. And, as in the traffic jam caused by one stalled car, the congestion "backed up" in Bob's vascular system, first in the left atrium and then in the pulmonary veins and capillaries. This increased pressure caused fluid to ooze from lung capillaries into his lungs, impairing gas exchange. His cardiac output continued to fall, which further starved his tissues for oxygen. Ultimately, Bob's "big heart" caused his demise.

● **Clinical Snapshot.** Explore a disease, injury, method of diagnosis, or medical treatment related to the chapter topic. For example, a Clinical Snapshot box in Chapter 2 (chemistry) discusses medical uses of radioactivity.

End-of-Chapter Features

Word Parts

Learning the language of human form and function is much easier if you master a small number of word parts relevant to each body system. Following the narrative for each chapter is a Word Parts table that identifies prefixes, suffixes, and word roots relevant to the chapter content. For example, in Chapter 3 (cells), you'll learn that the root *cyto-* means cell, and so "cytology" is the study of cells.

Word Parts		
Latin/Greek Word Parts	**English Equivalents**	**Examples**
Ant-/i-	Against	Antagonist: works against the agonist
-crine	Secretion	Endocrine: secreted within the body
De-	Remove	Depolarize: remove the polarization
Endo-	Within	Endogenous: generated inside the body
Exo-	Outside	Exogenous: generated outside the body
Neur-/o-	Neuron, nerve, nervous tissue	Neurocrine: secreted from a neuron
Para-	Beside	Paracrine: secreted to cells beside
Post-	After	Postsynaptic cell: cell after the synapse
Pre-	Before	Presynaptic cell: cell before the synapse
Re-	Again, back	Repolarize: polarize again

Review Questions

To help you check your learning and prepare for exams, we include a variety of review questions at the end of each chapter. These end-of-chapter review questions are organized into three groups by type:

- *Recall* questions check your retention of the facts you learned in the chapter. These are usually multiple-choice and matching questions.

- *Conceptual Understanding* questions test your comprehension of the underlying ideas.

- *Application* questions require you to apply your learning of facts and concepts to new situations. Both the Conceptual Understanding and Application questions are usually short-answer questions.

Answers to the Review Questions are available on the text's companion Web site, **http://thePoint.lww.com/ McConnellandHull,** so that you can check your progress.

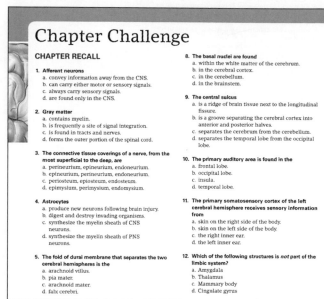

Chapter Challenge

CHAPTER RECALL

1. Afferent neurons
  a. convey information away from the CNS.
  b. can carry either motor or sensory signals.
  c. always carry sensory signals.
  d. are found only in the CNS.

2. Gray matter
  a. contains myelin.
  b. is frequently a site of signal integration.
  c. is found in tracts and nerves.
  d. forms the outer portion of the spinal cord.

3. The connective tissue coverings of a nerve, from the most superficial to the deep, are
  a. perineurium, epineurium, endoneurium.
  b. epineurium, perineurium, endoneurium.
  c. periosteum, epiosteum, endosteum.
  d. epimysium, perimysium, endomysium.

4. Astrocytes
  a. produce new neurons following brain injury.
  b. digest and destroy invading organisms.
  c. synthesize the myelin sheath of CNS neurons.
  d. synthesize the myelin sheath of PNS neurons.

5. The fold of dural membrane that separates the two cerebral hemispheres is the
  a. arachnoid villus.
  b. pia mater.
  c. arachnoid mater.
  d. falx cerebri.

8. The basal nuclei are found
  a. within the white matter of the cerebrum.
  b. in the cerebral cortex.
  c. in the cerebellum.
  d. in the brainstem.

9. The central sulcus
  a. is a ridge of brain tissue next to the longitudinal fissure.
  b. is a groove separating the cerebral cortex into anterior and posterior halves.
  c. separates the cerebrum from the cerebellum.
  d. separates the temporal lobe from the occipital lobe.

10. The primary auditory area is found in the
  a. frontal lobe.
  b. occipital lobe.
  c. insula.
  d. temporal lobe.

11. The primary somatosensory cortex of the left cerebral hemisphere receives sensory information from
  a. skin on the right side of the body.
  b. skin on the left side of the body.
  c. the right inner ear.
  d. the left inner ear.

12. Which of the following structures is *not* part of the limbic system?
  a. Amygdala
  b. Thalamus
  c. Mammary body
  d. Cingulate gyrus

Art Program

As experienced teachers, we know that "a picture is worth a thousand words." Throughout this textbook, you'll find expertly rendered anatomical illustrations, innovative process diagrams, flowcharts, micrographs, and clinical photographs with or without accompanying line drawings. These amply support the narrative, and will help you to visualize the structures and functions of the integrated body.

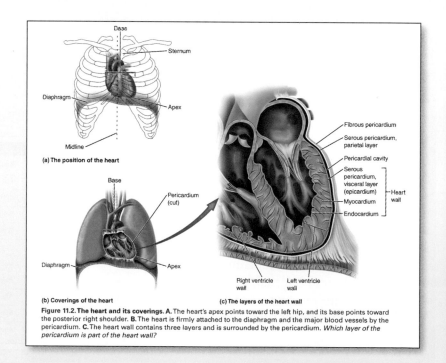

(a) The position of the heart

(b) Coverings of the heart

(c) The layers of the heart wall

Figure 11.2. The heart and its coverings. A. The heart's apex points toward the left hip, and its base points toward the posterior right shoulder. **B.** The heart is firmly attached to the diaphragm and the major blood vessels by the pericardium. **C.** The heart wall contains three layers and is surrounded by the pericardium. *Which layer of the pericardium is part of the heart wall?*

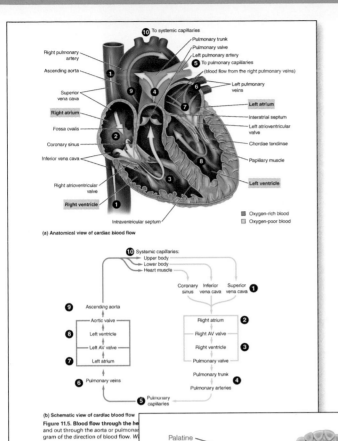

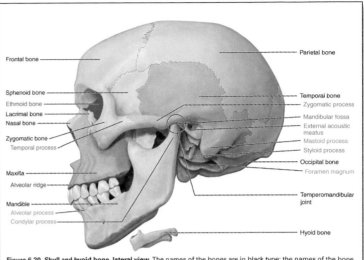

Figure 6.20. **Skull and hyoid bone, lateral view.** The names of the bones are in black type; the names of the bone features are in the same color as the bone under the relevant bone. *Which bone contains the condylar process?*

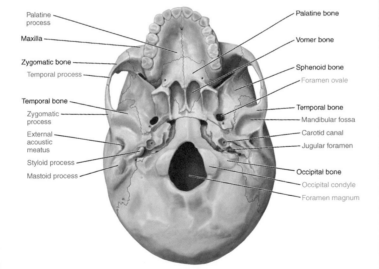

Figure 6.21. **Skull: inferior view.** *Which bone contains the external acoustic meatus?*

Companion Web Site

On our web site you will find additional diagrams, animations, podcasts, test questions, online instructor's manual, online dissection manual, PowerPoint slides, online activities, and much more...

Summary

We wrote this book to be read and enjoyed. Our intention is to make your studying easier and more pleasurable, and in doing so, to enhance your retention. Every sentence and every illustration is crafted with you, the student, in mind. Most of all, we want to inform you in a way that might make you say, "Wow! How interesting is that!"

We hope you'll enjoy your tour of discovery through the integrated body. May it be the beginning of a life-long interest in the human condition.

Thomas H. McConnell
Kerry L. Hull

Contents

Chapter 4
Communication: Chemical and Electrical Signaling   110

Case Study: "I must be getting the flu."   111

Chapter 8
Nervous System 280

Chapter 9
Sensation: The Somatic and Special Senses 328

Chapter 10
Blood 374

Chapter 11
The Cardiovascular
System 408

**Case Study: "He's been having a
lot of chest pain."   409**

Chapter 16
The Urinary System and Body Fluids 632

Case Study: "He's having one of those acid attacks." 633

Human Form, Human Function

Essentials of Anatomy & Physiology

1

Form, Function, and Life

Major Themes

- Life is defined by specific characteristics.
- Form and function are intimately related at every level of life.
- Life is sustained by specific things in the external environment.
- Life is maintained by strict, automatic reactions that adjust the internal environment.
- Precise communication is a key attribute of science and requires specialized descriptive language.

Chapter Objectives

Form, Function, and Life   4

1. Explain the difference between anatomy and physiology and offer an example of each.

2. List four characteristics of living things.

The Building Blocks of Life   7

3. Within a specific organ system, list the building blocks of life, starting with the smallest.

Life and the External Environment   7

4. Identify five features of the external environment that are necessary for life.

Life and Gradients   10

5. Explain the importance of pressure and concentration gradients in the body.

Homeostasis: Maintaining a Healthy Internal Environment   11

6. Define homeostasis.

7. Name at least one contribution to homeostasis made by the 11 body systems.

Case Study: "Honey, I forgot to duck."

As you read through the following case study, assemble a list of the terms and concepts you must learn in order to understand President Reagan's case.

Clinical History: On March 30, 1981, two months after his first inauguration, Ronald Reagan, 40th president of the United States, was shot in the thorax. James Hinckley, a deranged drifter, thought assassinating Reagan would prove his love for actress Jodie Foster.

President Reagan.

After delivering a short speech to a labor group at a Washington, DC hotel, Reagan approached his limousine, where Hinckley had muscled his way to the front of a waiting crowd of admirers and reporters. At close range, Hinckley fired six shots from a .22-caliber pistol, wounding three of the presidential staff but missing a direct hit on the president. However, one bullet ricocheted off the president's bullet-proof limousine and struck Reagan in the left posterior chest, near his axilla. The bullet traveled inferiorly, anteriorly, and medially, coming to rest in his lung about 2 cm from his heart.

Reagan immediately began coughing up blood and complained of shortness of breath as he was rushed to a nearby hospital. He walked part way to the emergency room but became faint, collapsed to one knee, and was carried the remaining distance.

Physical examination and other data: Reagan was pale, short of breath, and complaining of chest pain. Systolic blood pressure was 80 (normal 120), heart rate was 80 beats per minute (normal 70), and respirations were 30 per minute (normal 14). A small wound was found in the skin of the posterior chest near his left axilla at the level of the fourth rib. A chest x-ray revealed blood in the chest and a small metal object in the inferior left chest.

Clinical course: Oxygen was provided immediately and a solution containing fluid and minerals was administered intravenously while awaiting blood from the blood bank. A suction tube was inserted into the left chest between the rib cage and lung. By the end of the first hour, 2 L of blood had been suctioned from the chest; about 500 mL of concentrated red blood cells and 3,000 mL of fluids had been administered intravenously. Systolic blood pressure was 160 (80 on admission), heart rate was 90 (80 on admission), and respirations were 25 (30 on admission).

Reagan continued to bleed through the suction tube in his chest, and his physicians decided to open his chest anteriorly between the fifth and sixth ribs. When informed he was to have chest surgery, he quipped to the surgeon, "I hope you're all Republicans." On entering the chest, the surgeon evacuated 500 mL of blood, bringing the total blood loss to 3,000 mL. The heart and nearby organs were normal, but a pulsing jet of dark blood was discovered coming from a small lung artery. The vessel was tied off and the bleeding ceased. In the left lung near the heart, a "dime-sized" flattened bullet was found and removed. While in the operating room, Reagan received an intravenous infusion of about 5,000 mL of fluids and blood.

Reagan, who was 69 years old at the time of the injury, convalesced slowly but never lost his famous sense of humor. During recovery, he explained his wound to First Lady Nancy Reagan by saying, "Honey, I forgot to duck," a line used by Jack Dempsey to his wife after he lost the world heavyweight boxing championship to Gene Tunney in 1927. Reagan was discharged on the 12th hospital day but did not return to a full schedule for another 6 weeks.

After you have read Chapter 1, you will be able to explain Reagan's symptoms and signs and how his body's responses and medical treatments restored him to health.

In this and every chapter of this book, form and function will be the underlying theme: the shapes and structures of all things—human hearts, forks, or staplers—are intimately related to their function. In our discussions, anatomy is the form of things; physiology is the function (activity) of things. When form and function combine to produce biologic activity, we call it *life*. When form and function become abnormal, we call it *illness*. When biologic activity ceases, we call it *death*.

In this chapter, we can think of ourselves viewing the landscape of anatomy and physiology from the window of a jet plane at 30,000 feet on a clear day—the big picture is easy to see, but there is too much detail to grasp in a quick overflight. A deeper understanding requires a closer view, which we will take in later chapters. For the moment we are interested only in the big picture; do not expect to master now all of the topics mentioned.

Be careful about reading health books. You may die of a misprint.

Mark Twain (Samuel Langhorne Clemens), American novelist and humorist, 1835–1910

Form, Function, and Life

The goal of human **anatomy** (from Greek *anatome*, to cut up or dissect) is to understand the structure (form) of our body and how it relates to our functions. *Gross (macroscopic) anatomy* examines structures that can be observed with the naked eye. By contrast, *microscopic anatomy* is the study of structures too small to be seen with the unaided eye and requires a microscope. For centuries, anatomists and medical students have dissected cadavers in their study of human anatomy. As discussed in the accompanying History of Science box, *A Grave Offense*, the need for cadavers once led to body snatching and grave robbing—even murder.

The goal of human **physiology** (from Greek *physis*, for nature, and *logia*, to study) is to explain the internal mechanisms of the human body; that is, the gross and microscopic workings of things that give us life. Thinkers have always been fascinated with how the body works;

Aristotle, for instance, postulated that the purpose of the brain was to secrete mucus (perhaps he had frequent head colds). Now we have the sophisticated equipment that enables us to get closer to the truth, from understanding why diabetics have breath smelling of apples to explaining the visual blind spot.

Case Note

1.1. Physicians noted a small wound in the skin of Reagan's posterior chest. Was this a gross anatomic observation or a microscopic one?

Human Beings Share Four Characteristics of All Living Organisms

Like all living things, human beings are **organisms;** that is, complete life forms that can function independently

THE HISTORY OF SCIENCE

A Grave Offense

On a freezing night in January 1824, seventeen-year-old Bathsheba Smith's body was discovered missing from her freshly dug grave in West Haven, Connecticut. Citizens immediately suspected that students from nearby Yale Medical School had snatched the body for their anatomic studies. The next morning the town constable found the body carefully wrapped and concealed beneath paving stones in the school's cellar floor. News of the event spread and by nightfall a feverish mob besieged the medical school for 2 days, hurling stones and burning clods of coal, intent on destruction. Police and militia finally dispersed the mob, and a court later convicted an anatomy lab assistant of the crime.

Grave robbing was a common practice in other nations, too. In England, Ireland, and Scotland grave robbers, known as "resurrectionists," found that grave robbing alone was insufficient to meet the demand for fresh corpses, and some resorted to murder. In Scotland in 1829, Irishman William Burke confessed to the murder of 16 people and was hanged. He had sold each of the fresh corpses to an anatomist at Edinburgh University, who had purchased them without question.

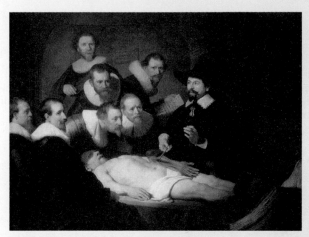

This painting depicts a dissection class in a 17th-century anatomy school.

Grave robbing ended when laws were passed making it unnecessary. These laws made it legal for people to will their bodies to science and for anatomists to dissect the bodies of executed criminals and unknown persons with no one to claim their remains.

of other life forms. Organisms are characterized by four qualities, depicted in Figure 1.1:

- Organisms are *organized*; that is, every component has its place and must remain in place for the organism to thrive. Organization is maintained by barriers that create *boundaries* (Fig. 1.1A). Barriers keep the *internal environment* of living things organized and separated from the *external environment* in which they live. For example, human skin is a tough barrier that shields us from the environment (bacteria, heat and cold, and so on). Skin also keeps fluid loss or gain to a minimum: patients suffering from extensive burns lose large amounts of fluid because the skin is burned away, and swimmers gain a negligible amount of water, which accumulates only in their skin cells. A microscopic barrier is the *cell membrane,* which encloses the contents of a **cell,** the basic unit of life.
- Organisms have a **metabolism;** that is, their cells engage in chemical reactions that build and break down substances they need. For example, they take

simple chemicals present in the food we eat and assemble them into complex substances. This process is called synthesis, growth, or *anabolism*. If you think about building a tower from a variety of simple blocks, you'll have a good idea of anabolism (Fig. 1.1B). Organisms also break down substances into simpler ones, a process known as *catabolism*. Breaking that tower into its component blocks is somewhat like catabolism.
- Organisms *adapt* to the environment; that is, they adjust their form, function, and/or behavior to changes in their environment in order to stay alive and healthy. Human skin, for instance, darkens in the summertime when exposed to sunlight, a reaction designed to protect skin cells from harmful ultraviolet rays (Fig. 1.1C).
- Organisms *reproduce*; that is, a living thing is capable of producing another independent living thing like itself. Humans, for example, reproduce organisms (children) like themselves (Fig. 1.1D).

All of the above are driven by the rules of physics, chemistry, and behavior, so that in a certain sense we are

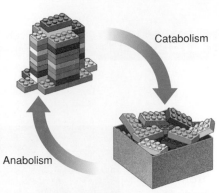

(a) Organization; Organisms use barriers to separate internal and external environments.

(b) Metabolism; Organisms convert simple substances into complex ones, and vice versa.

(c) Adaptation; Organisms change when environmental factors (such as weather) change.

(d) Reproduction; Organisms can produce offspring.

Figure 1.1. Characteristics of life. All living things share these four functions. *We shiver when it is cold in order to generate more heat. Which function of life does this represent?*

like robots reacting to circumstances in our environment: when cold we seek warmth; when hungry we seek food; when tired we rest; when drowsy we sleep. At the cellular level, our life is, so to speak, on autopilot: chemical reactions speed or slow, begin or end, in response to the local cellular environment in an infinitely complex but orderly and automatic chain of events, one event feeding on another, the sum of which is *life*.

Case Notes

1.2. Reagan's bullet wound was closely monitored for infection. Why is broken skin prone to infection?

1.3. In the emergency room, Reagan's heart rate was higher than normal in an attempt to compensate for his blood loss. Which of the four characteristics of living things does this change represent?

Form and Function Are Inseparable

Form and function are inseparable in the creation and maintenance of life: every form has a function, and every function has a form—whether it is the design of houses we construct for shelter or the hands we use to construct

them. For example, notice your thumb: it is rotated 90 degrees from your other fingers so that its natural motion runs back and forth across the flat of your palm (Fig. 1.2A). This makes it easy to touch the tip of your thumb to the tip of each of your other four fingers, whose natural motion runs crossways to that of your thumb. If this doesn't seem extraordinary, hold your thumb against the side of your hand and then try to tear a piece of tape, turn the page of a book, tie your shoes, or grip a pen to write a sentence. Among animals, only humans have a fully opposable (grasping, prehensile) thumb. This anatomic characteristic enables us to make and use the things that our advanced brains design, from paper clips to computers. The form of the monkey hand, in which the thumb folds downward, parallel to the fingers, is much less versatile and better adapted to maintaining a strong grip on a branch (Fig. 1.2B).

1.1 What is the difference between anabolism and catabolism?

1.2 Is the dissection of an eyeball more closely related to anatomy or to physiology?

FORM   **FUNCTION**

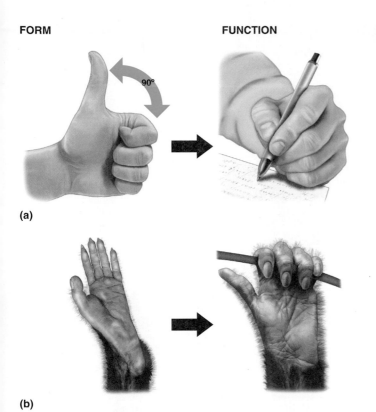

(a)

(b)

Figure 1.2. Form and function. A. The human hand, with thumb rotated at 90 degrees, is capable of a pincer grip. **B.** The monkey's hand, with thumb parallel to other fingers, allows only a less precise (but stronger) palmar grip. *True or false: The structure of the monkey's hand refers to is its physiology.*

Organs making up the digestive system include the esophagus, stomach, small and large intestines, liver, gallbladder, and pancreas. Chapters 4 to 21 discuss the form and function of the different body systems.

In turn, organs are composed of **tissues.** Close examination of a particular organ, the small intestine, for example, reveals that it is composed of several tissues, among them layers of muscle tissue that form its tubular wall and squeeze intestinal contents down the tube. Other tissues in the small intestine include a lacy network of *nerves* in the wall that control the muscular contractions, a specialized layer called *epithelium* that secretes mucus and absorbs nutrients, and a layer of *connective tissue* that holds everything together.

Each tissue is made up of distinctive **cells,** such as muscle cells, epithelial cells, or connective tissue cells. Cells are the fundamental building blocks of all life, which we discuss in detail in ➡ Chapter 3.

The cell and its parts are composed of **molecules,** which in turn are formed of **atoms.** While cells are the building blocks of *life*, atoms are the fundamental building blocks of *everything*—living and nonliving. We will discuss the role of atoms and molecules in physiology in detail in ➡ Chapter 2.

 1.3 Name the six levels of the organization of life.

The Building Blocks of Life

We can study form and function on a large or small scale. For example, we can study the entire body performing a somersault, the behavior of individual muscles as they contract, or the microscopic action of tiny muscle fibers deep within a muscle. These differing scales of study are often described as *levels of organization*, which are shown in Figure 1.3. A human is an *organism*; that is, it can function independently and is composed of multiple smaller parts that constitute a whole. In the same way, a bacterium or a tree is an organism. Each is composed of an organized set of smaller parts that are sustained by the physiologic functions that give the entire organism life.

These physiologic functions are provided by one or more subdivisions of the organism, the largest of which is a collection of organs called a **system,** or *organ system*. Consider a function—say, digestion. Digestion is performed by a subdivision aptly named the *digestive system*, which, like all body systems, is composed of a number of **organs** that work together to perform a function.

Life and the External Environment

Life has evolved for millions of years in an external environment that provides certain essentials. If any one of these essentials is absent, life ceases. Consider an astronaut floating through space in a specialized suit (Fig. 1.4). The space suit provides—or preserves—all of the elements that we normally get from our environment. These include pressure, oxygen, heat, water, and nutrients.

Pressure Is Necessary for Life

Pressure is a force exerted by a solid, gas, or fluid. Although all are important to our functioning, the pressures of gases and fluids are most critical.

Gases and fluids can exert two types of pressure: *Static pressure* is exerted by the weight of a gas or fluid pressing down on a point within it. For example, imagine a fly

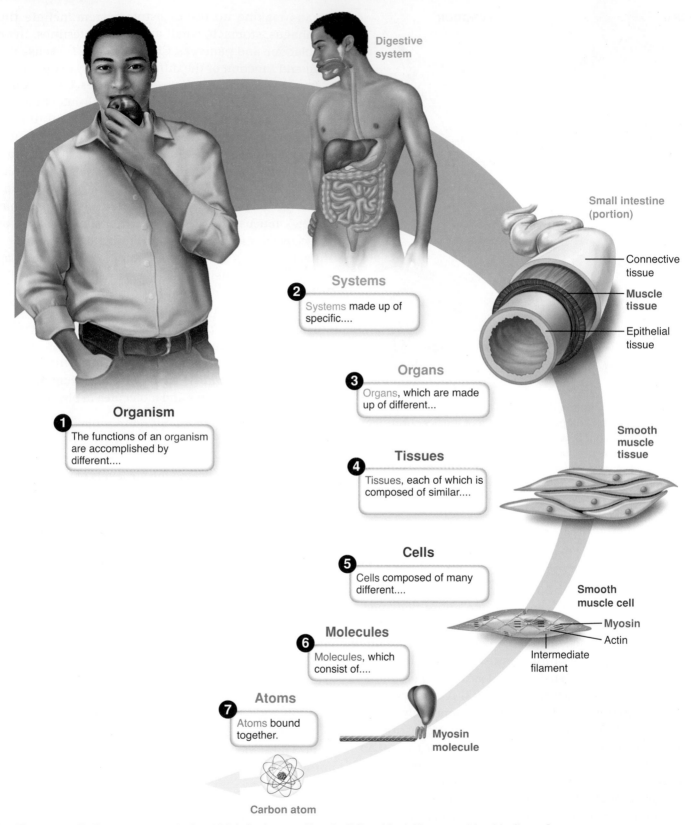

Digestive system

Systems

2 Systems made up of specific....

Organs

3 Organs, which are made up of different...

Organism

1 The functions of an organism are accomplished by different....

Tissues

4 Tissues, each of which is composed of similar....

Cells

5 Cells composed of many different....

Molecules

6 Molecules, which consist of....

Atoms

7 Atoms bound together.

Small intestine (portion)

Connective tissue

Muscle tissue

Epithelial tissue

Smooth muscle tissue

Smooth muscle cell

Myosin

Actin

Intermediate filament

Myosin molecule

Carbon atom

Figure 1.3. Building blocks of life. *Which is the smallest building block illustrated in this figure?*

Figure 1.4. Environment and life. Astronauts in outer space must carry their life-sustaining environment with them. *Which necessities of life are illustrated in this figure?*

at the bottom of a jar. The greater the amount of fluid in the jar, the greater the static pressure exerted on the body of the fly. *Dynamic pressure* is the additional force created by moving gas or fluid. When you're knocked over by an ocean wave, that's dynamic pressure.

Atmospheric Pressure Is Exerted by the Gases in Air

Although it seems weightless, air has substance, as any child knows who has held a hand out of a moving car. Air is a mixture of several gases, each of which has a unique weight. *Atmospheric pressure* is a static pressure and represents the total weight of the air in the atmosphere above us, which presses constantly upon our bodies. Atmospheric pressure at high altitudes is lower than at sea level because there is less air above.

Atmospheric pressure keeps the gases within our bodies, mainly oxygen and nitrogen, dissolved in our body fluids for transport to our tissues. For example, if an astronaut's suit were to lose pressure, gas molecules dissolved in the astronaut's blood and other bodily fluids would suddenly form bubbles, which would explode body cells or obstruct blood vessels and blood flow. To understand why a loss of pressure would cause gas bubbles to form, think of a bottle of carbonated soda. High pressure inside the bottle keeps the carbon dioxide gas dissolved in the beverage. When the bottle is opened, pressure inside the bottle falls and bubbles of carbon dioxide gas are released from the fluid.

As we noted earlier, moving air creates an additional force called dynamic pressure, which you experience whenever you try to walk into a strong wind.

Fluids Exert Pressure

Fluids also exert static and dynamic pressure. Of the two, dynamic pressure is much more important in physiology.

Consider the water pressure used to water flowers from a garden hose. You can increase the water pressure (say, to reach more distant flowers) by two methods. First, you can increase the water flow rate by further opening the faucet. Alternatively, you can decrease the size (the diameter) of the hose outlet by partially covering it with your thumb. Pressure builds behind your thumb and water sprays a greater distance.

Body fluids behave similarly. A particularly vital form of fluid pressure is *blood pressure*, which is created by the pumping action of the heart as it forces blood through a vast network of tubular blood vessels. Blood pressure is necessary for life—the flow of blood not only moves blood to the various organs, carrying oxygen, nutrients, and other vital substances, but it also provides the underlying pressure that assists molecules in their movement out of blood vessels and into tissues and cells.

The factors determining blood pressure are a little different from those determining water pressure in a garden hose, because blood circulates through the body in a closed loop; that is, the fluid has no escape into the surrounding environment. So let's imagine that we put a cap on the end of our garden hose. Now, as we open the faucet to increase the amount (the volume) of water trying to flow into the hose, the pressure within the hose increases. In the same way, increased blood volume increases blood pressure. And any reduction in blood volume (say, by severe bleeding) reduces blood pressure, just as releasing water from a hose reduces pressure in the hose. Other factors we discussed for the garden hose are also important in the determination of blood pressure: The flow rate of blood is largely determined by the heart's activity—if the heart beats faster or more strongly, blood pressure increases as more blood volume per minute is forced through the system. The body also controls blood pressure by altering the diameter of blood vessels. Narrowing blood vessel diameter increases blood pressure, and vice versa. We will learn much more about the control of blood pressure in ➡ Chapter 13.

Oxygen Is Necessary for Life

A gas that forms about 20% of atmospheric air, oxygen is avidly absorbed by blood passing through the lungs. Oxygen is the key that unlocks energy from the chemical nutrients in food. Without oxygen, cells cannot obtain the energy they need to remain alive, and brain cells and all other cells starve and die. If the oxygen level in our astronaut's suit were to fall too low, he would become dizzy, lose consciousness, and eventually die.

Heat Is Necessary for Life

In the very cold environment of outer space, our astronaut's space suit must be heated. Heat is generated by the body's chemical reactions (metabolism) and is naturally shed by radiating away in the same way that an electric heater radiates heat into the air. However, just the right amount of heat must be lost because losing or keeping too much heat can lower or raise body temperature dangerously.

Although body temperature must be maintained within extremely narrow limits, livable environmental temperature can vary considerably. In warm environments or if the body is producing a large amount of heat from, for example, hard exercise, we adapt by sweating: the sweat evaporates with cooling effect. In cold environments, we adapt by shivering: rapid, repetitive, minute muscle contractions that generate considerable body heat.

Case Note

1.4. Which of the following elements necessary to life were immediately compromised by the president's injury: oxygen, heat, or nutrients?

Nutrients and Water Are Necessary for Life

Food supplies the chemicals necessary for life. Some food chemicals (nutrients) are burned for energy; others are used as building blocks for body parts. Since most nutrients can be stored, our astronaut could survive without food for many days. However, he cannot survive long without water. The other chemicals of life are dissolved in it, the molecular reactions of life take place in it, it smoothes the movements of our various body parts, it carries nutrients to cells, it carries wastes away, and it transports the messenger and regulatory molecules that govern every aspect of cell life.

1.4 Life requires nutrients, heat, oxygen, and what else?

1.5 Which change would increase blood pressure: increasing vessel diameter or increasing heart rate?

1.6 Oxygen is what percent of air?

1.7 An ambient air temperature of 95°F feels uncomfortably hot, even though normal body temperature is about 97°F. Can you think why?

Life and Gradients

To understand how the human body functions, you must understand gradients. Simply put, a **gradient** is the difference in the quantity or concentration of a physical value between two areas (Fig. 1.5). For example, there is an *altitude gradient* between the peak of a roof and the roof edge because the peak has more altitude than the edge (Fig. 1.5A). Water from snow melting on a steep roof will flow downward following this gradient. Indeed, unless they are inhibited from doing so, substances always move down their gradients from high to low. Life depends on maintaining gradients—death eliminates the gradients upon which life depends. For example, there is a temperature gradient between the body and the environment, which is usually hotter or colder than body temperature. The body works to maintain body temperature in the range of 96.8°F (36°C) to 98.6°F (37°C) regardless of outside temperature. But with death the temperature gradient between the body and the environment disappears and the body assumes the temperature of its environment.

Turning back to our astronaut, a *pressure gradient* exists between the inside of the suit, where pressure is maintained equal to that on the earth's surface, and the outside of the suit, where pressure is zero. Without the barrier formed by suit material and helmet, air would rush from the area of high pressure (inside the suit) to the area of low pressure (outside); that is, air would move down the pressure gradient.

Many different gradients exist within the body, and they are important in physiology. Physiologic gradients usually occur between the inside and outside of a cell or between one body compartment and another. For example, the pressure inside blood vessels is higher than the pressure outside of them. This simple fact explains why blood flows out of a wound—the blood is flowing down a pressure gradient from an area of high pressure inside

(a) Altitude gradient

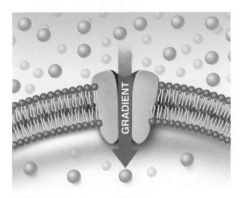

Lower pressure

Higher pressure

(b) Pressure gradient

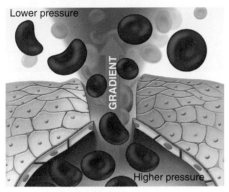

(c) Concentration gradient

Figure 1.5. Gradients. A. Altitude gradient of a roof. **B.** Pressure gradient in a blood vessel. **C.** Concentration gradient in a cell. *In part B, where is the pressure greatest—inside or outside the blood vessel?*

of blood vessels to an area of lower pressure outside (Fig. 1.5B). Pressure gradients also drive blood flow—the pressure in arteries is lower than that of the heart, so blood flows from heart to arteries.

Case Note

1.5. There was a "pulsing jet of blood" from the president's pulmonary artery. The blood was moving down which sort of gradient?

Physiologic gradients can also involve varying concentrations of chemicals. For example, the concentration of the element sodium is higher outside of cells than inside, so it can be said that for sodium, a *concentration gradient* exists across the cell membrane (Fig. 1.5C). If the cell membrane were freely permeable, sodium would flow unopposed down its concentration gradient into the cell. As we will see later, functions such as the production of urine and the conduction of nerve impulses depend upon concentration gradients. The accompanying Basic Form, Basic Function box, titled *You're Getting Warmer,* discusses a gradient important in human reproduction.

 1.8 Define the term *gradient* and give three examples.

Homeostasis: Maintaining a Healthy Internal Environment

The body is an integrated community of about 100 trillion cells, which, like an immense city, must have communication and control mechanisms to maintain order so that food and water are delivered, waste is removed, traffic flows smoothly, and messages are distributed reliably. The body's collective communication and control effort to maintain internal conditions within a narrow, stable physiologic range is called **homeostasis.** The word is derived from the Greek words *homios,* meaning "of the same kind," and *stasis,* meaning "standing still." In biology, the terms are combined to refer to the body's automatic tendency to maintain the "standing still" or normal state. Homeostasis requires the effort of every organ, every tissue, and every cell.

All Body Systems Participate in Homeostasis

All tissues and organs participate in homeostasis. For example, the lungs provide oxygen to enable cells to burn nutrients provided by the digestive system. The cardiovascular system ensures the circulation of essential oxygen and nutrients by increasing heart rate if blood pressure falls (as it did in President Reagan's case). What's more, the respiratory system maintains satisfactory blood oxygen levels by increased breathing rate during exercise; and, in the big picture, the musculoskeletal

BASIC FORM, BASIC FUNCTION

You're Getting Warmer. . . .

Have you wondered how spermatozoa (male sperm cells) find the egg in the female reproductive tract without a map? The chances of a random meeting between sperm and egg are minute, so blind luck is not enough. Scientists now believe that gradients guide sperm to their target.

Two types of gradients exist near the egg: a temperature gradient and a chemical gradient. The area around the egg is warmer than regions further away, so sperm swim up the thermal gradient. Once they arrive in the vicinity of the egg, the chemical gradient comes into play. The egg emits several chemicals (progesterone may be one) that the sperm can detect. These chemicals are more concentrated near the egg, so sperm swim up the chemical gradient. Incidentally, the human nose can also detect some of these chemicals. Perhaps the attraction between sperm and egg is a microcosm of the attraction between male and female?

Sperm use gradients to find the egg.

system helps humans to maintain themselves by enabling them to move about to find water and food.

We trust you are beginning to see that the body is a symphony of cooperation at every level of organization, from molecules to muscles. Just as we cannot appreciate a symphony by listening only for the sound of a single instrument, we cannot appreciate or understand human anatomy and physiology by focusing on the function of a single cell, organ, or system. Enjoying a symphony and understanding physiology require a broader view. As you learn about the different sections in our bodily orchestra by reviewing Figure 1.6, take note of the different ways in which they work together. For instance, the brain, conductor of our orchestra, is classified under the nervous system but it is intimately involved in the functioning of every system. Similarly, the heart is classified under the cardiovascular system, but every organ in the body depends on the heart's pumping action to maintain its blood supply.

Case Note

1.6. The bullet that struck the president severed an artery and lodged in his left lung. To which system(s) do these structures belong?

Negative Feedback Is the Key to Homeostasis

All physiologic conditions, from body temperature to blood oxygen level, have a *set point*—that is, a value at which the condition must be maintained for optimal health. For example, for body temperature, the set point is 98.6°F. Every physiologic condition is associated with a sensor that detects upward or downward deviations from the set point and signals the need for an *opposing* (negative) change. The entire sequence of homeostatic events, from the sensor to the response, is called *negative feedback*. Thus, **negative feedback** is a process that reflexively keeps systems tightly regulated near their set points and thus promotes stability.

Home thermostats work on a similar principle (Fig. 1.7, inner circle). Any change in the temperature of the air (the *condition*) is detected by a thermometer (the *sensor*) in the thermostat. The thermometer relays a signal—for example, that the air temperature has climbed above the set point—to a computer chip within the thermostat. This chip acts as an *integrating center*, receiving the data about air temperature and relaying a signal to the air conditioner (the *effector*), switching it on. The result is cooler air.

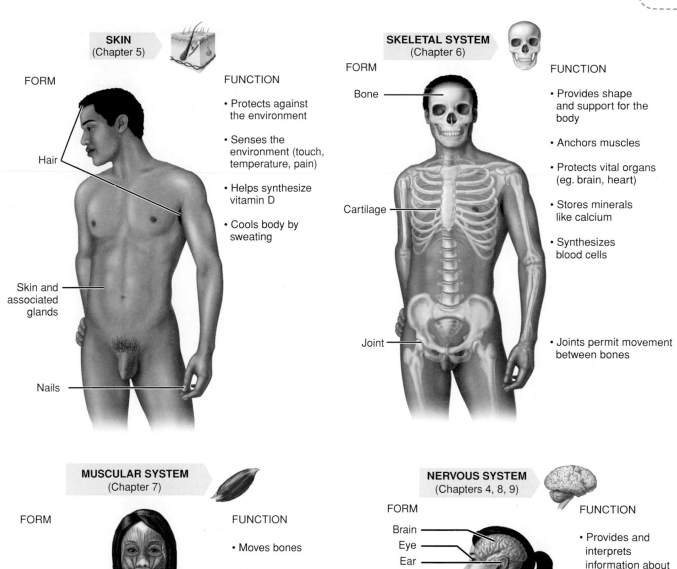

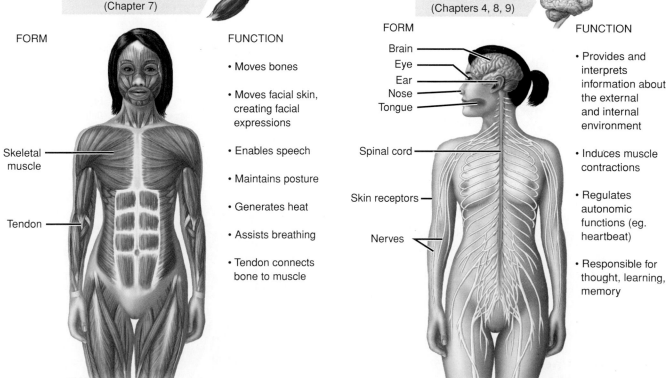

Figure 1.6. Form and function of body systems. *Find two body systems involved in moving oxygen from the atmosphere into body cells.* (*continued*)

CARDIOVASCULAR SYSTEM
(Chapters 10–11)

FORM

Blood

Heart

Capillary

Vein

Artery

FUNCTION

- Transports oxygen and nutrients to cells

- Transports carbon dioxide and wastes away from cells

- Transports hormones

- Blood cells fight infection

- Clotting components in blood prevent excessive blood loss

LYMPHATIC / IMMUNE SYSTEM
(Chapter 14)

FORM

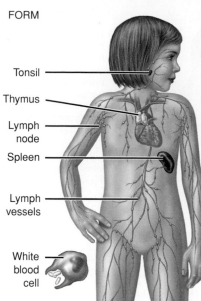

Tonsil

Thymus

Lymph node

Spleen

Lymph vessels

White blood cell

FUNCTION

- Defends against infection and cancer

- Filters and returns fluid from extracellular spaces to blood

- Transports digested fats to the bloodstream

RESPIRATORY SYSTEM
(Chapter 13)

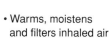

FORM

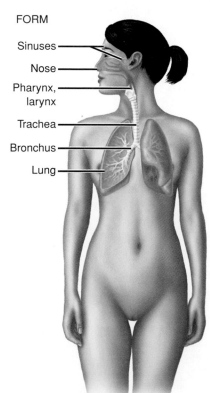

Sinuses

Nose

Pharynx, larynx

Trachea

Bronchus

Lung

FUNCTION

- Warms, moistens and filters inhaled air

- Channels air to lungs

- Extracts oxygen from inhaled air

- Discharges carbon dioxide in exhaled air

- Helps regulate acid-base balance

- Helps produce sounds

DIGESTIVE SYSTEM
(Chapter 14)

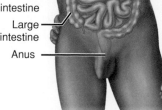

FORM

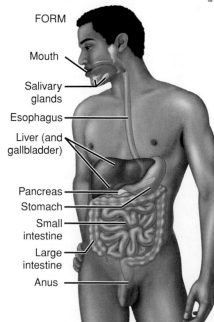

Mouth

Salivary glands

Esophagus

Liver (and gallbladder)

Pancreas

Stomach

Small intestine

Large intestine

Anus

FUNCTION

- Secretes digestive enzymes that break down nutrients

- Secretes mucus to protect digestive tract cells and lubricate food

- Moves food and water through the digestive tract

- Absorbs nutrients

- Eliminates solid waste

Figure 1.6. (*Continued*)

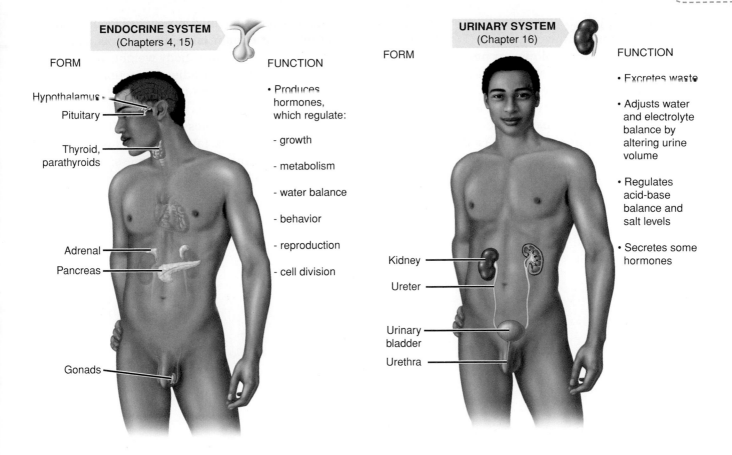

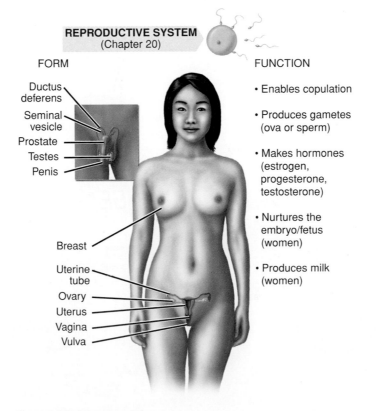

Figure 1.6. (*Continued*)

Likewise, on a hot day, sensors in our skin sense that our body temperature is rising (Fig. 1.7, outer circle). They send a signal along our nerves to an integrating center in the brain, which relays the signal to effectors (for example, sweat glands and blood vessels), which in turn take actions that lower body temperature. For example, increased sweating causes heat loss by water evaporation, and increased blood flow to skin causes heat loss by radiation. As a result, we cool down.

Other factors maintained by negative feedback include blood levels of certain minerals, such as sodium and potassium; blood pressure; body weight; and thousands more. This marvel of self-adjusting action and reaction continues to maintain the body in good health. When one or more systems loses their ability to control their share of function, sickness results. Most illness can be understood best as failed homeostasis—moderate homeostatic dysfunction leads to illness; severe homeostatic dysfunction can cause death.

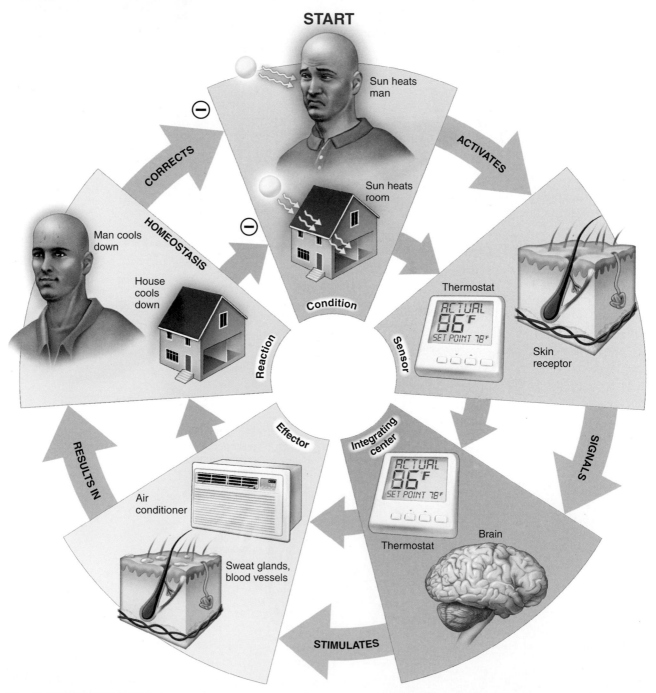

Figure 1.7. Negative feedback restores homeostasis. Negative feedback is used to maintain a constant temperature in a room (inner circle) and in a person (outer circle). *Name the effectors that respond to a change in body temperature.*

Remember This! **Negative feedback keeps systems tightly regulated near their set points and thus promotes stability.**

Case Discussion

President Reagan's Case Shows Homeostasis in Action

 President Reagan's gunshot wound is, well, a textbook example of homeostasis in action. When the bullet tore open blood vessels in his chest, it destabilized two key homeostatic conditions: blood pressure and the oxygen content of blood. Here's how. Blood, which is under pressure in blood vessels, flowed down a pressure gradient and out of the vessels. In the president's case, the blood did not exit the body; instead, it became trapped between his rib cage and lung. This blood was worse than useless: it was unable to carry vital oxygen and nutrients throughout the body, and its accumulation between the rib cage and lung squeezed air space out of the lung. The president, therefore, had less lung space to use to deliver oxygen to the circulating blood. The president's body reacted to correct the situation; in the grand scheme of things, the physicians were just very important homeostatic assistants. Figure 1.8 illustrates the homeostatic mechanisms and medical interventions that helped President Reagan survive.

The Problem: Low Tissue Oxygen. All cells need oxygen. President Reagan felt faint because too little oxygen was reaching his brain (and other tissues). This reduction in oxygen delivery was caused by two factors:

- The blood loss reduced his blood pressure as well as the amount of blood available to deliver the oxygen.
- The impaired lung function reduced the amount of oxygen delivered to the blood.

The Solution (Part 1): Homeostatic Mechanisms. Reagan's body attempted to increase oxygen delivery to the tissues and restore homeostasis. We can see several negative feedback loops at work. The reduction in blood pressure triggered numerous responses. Recall from our discussion of pressure that increased blood flow and decreased vessel diameter both increase blood pressure. So Reagan's heart rate increased in order to increase the rate of blood flow. His peripheral vessels constricted, which made his skin pale. Both responses subsequently increased blood pressure, although not all the way to normal. The low levels of blood oxygen triggered an increase in breathing rate, which subsequently increased blood oxygen levels somewhat. In each instance, abnormally low blood pressure and oxygen were raised back toward higher, healthier levels by homeostatic mechanisms. That is to say, the magnitude of the degree of change (the drop in blood pressure and oxygen content) was reduced. However, these homeostatic mechanisms were not sufficient to restore the president to a state of good health, and medical treatment was required.

The Solution (Part 2): Medically Assisted Homeostasis. The president's treatment addressed both problems: oxygen delivery into the blood and blood delivery to the tissues. First, air is only 20% oxygen, so extra oxygen was administered by mask in order to maximize its delivery to the blood. Second, the president was immediately given a large volume of intravenous solution in order to expand the amount of fluid in his blood vessels. Although these solutions cannot transport oxygen the way blood does, they can temporarily increase blood volume and blood pressure until blood for transfusion can be obtained. And third, the president was given an infusion of concentrated red blood cells to improve the oxygen-carrying capacity of his blood.

The results of treatment were immediate and dramatic: blood pressure rose, although a high heart rate remained necessary to sustain it. The president's breathing rate eased, falling back toward normal, in part because of supplemental oxygen and in part as a response to improved blood flow due to infusions of fluids and blood.

Finally, however, physicians had to perform surgery to stop the blood loss, so that all of the other treatments would no longer be necessary.

Case Notes

1.7. Why was Mr. Reagan pale?

1.8. Which intervention was mainly responsible for increasing Mr. Reagan's blood pressure—concentrated blood cells or intravenous fluid?

Positive Feedback Accelerates Processes to an End Point

Positive feedback is the opposite of negative feedback: negative feedback "pushes back" against a detected change; **positive feedback** pushes the detected change ever more strongly in the same direction until the process is completed. Positive feedback is like a rock slide—a few rocks falling down a hill dislodge more rocks, which dislodge even more rocks, until an avalanche

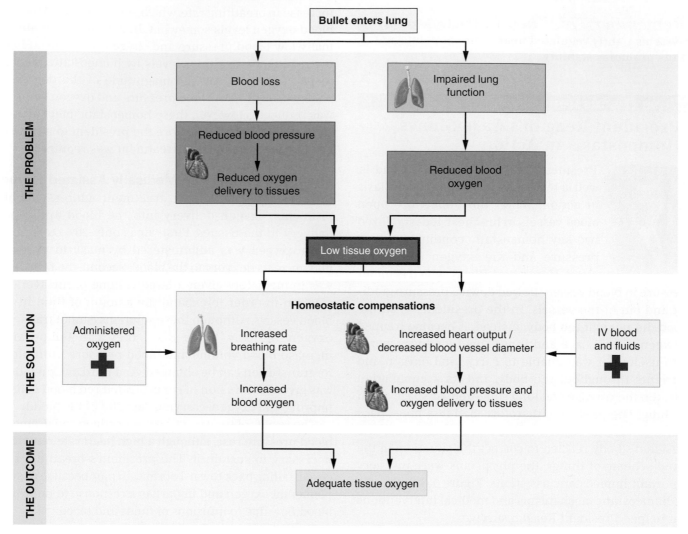

Figure 1.8. Homeostasis and President Reagan. The disruptions in homeostasis caused by the bullet are shown in red. Homeostatic adjustments and medical interventions are shown in yellow. *Why was Reagan's blood pressure low?*

results that is capable of burying a village. Ultimately, the avalanche stops and the process ends.

> *Remember This!* **You can demonstrate positive feedback using dominoes. Arrange the dominoes vertically in a long chain and push the first one over. The dominoes will fall one upon the other until all have fallen.**

Negative feedback is central to the moment-by-moment maintenance of many vital body processes. In contrast, positive feedback plays a role in only a very few normal physiologic events. An example is childbirth (Fig. 1.9). Positive feedback loops share the same elements as negative feedback loops: a *condition* is detected by a *sensor,* which signals an *integrating center,* which, in turn, signals an *effector* to induce an action. The dif-

ference is that this action boosts rather than diminishes the condition. In childbirth, for instance, the cervix (the opening of the uterus [womb] at the upper end of the vagina) is stretched by the baby's head. This stretching sends signals to the nervous system, which in turn signals the upper end of the uterus to contract even more strongly. As a result, the baby's head is pushed further into the cervix, which stretches the cervix even more, and the cycle continues until the baby is born. Events governed by positive feedback do not stabilize; they proceed until an end point is reached (in this instance, birth) or the reaction is exhausted (that is, when all of the rocks have fallen down the hill).

> *Remember This!* **Try to create your own feedback diagrams, using our examples (landslides, dominoes) or your own.**

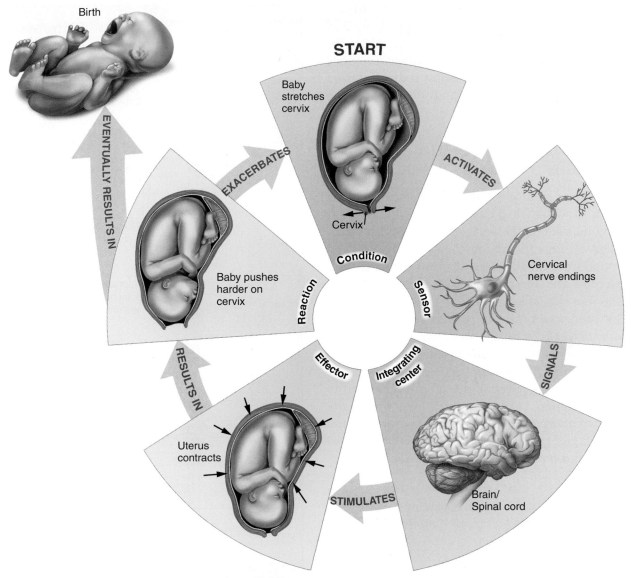

Figure 1.9. Positive feedback reinforces a condition. Positive feedback loops often terminate in crucial events, such as childbirth or death. *Would uterine contractions increase or decrease the activation of the cervical nerve endings (the sensors)?*

Negative and positive feedback are thus similar in that they involve a response to change: negative feedback opposes change; positive feedback enhances change.

Pop Quiz

1.9 Define homeostasis.

1.10 List the main contribution of the respiratory system to homeostasis.

1.11 Is body temperature regulated by negative feedback or by positive feedback?

The Language of Form and Function

It has been rightfully said that acquiring new knowledge is largely a task of learning new language. For example, imagine that you are assisting in Ronald Reagan's surgery and another surgeon asks you to "make an incision next to the left nipple and cut down for 2 cm." Would you be able to make an accurate incision? Of course not. You need to know in which direction from the nipple to make the incision. What's more, *down* could mean deeper into the chest or along the surface of the chest toward the feet. Instead, the surgeon would have given you precise instructions by saying, "Make a superficial 2 cm horizontal incision beginning 2 cm lateral to the nipple and superior to the fifth rib and extend it to the posterior axillary line." The surgeon's use of standard anatomic language ensures exactness. By using the specialized language of anatomy, we can accurately describe the site of an injury, tumor, or incision.

Medical Terms Are Assembled from Smaller Subunits

The language of anatomy and physiology contains many Latin- and Greek-based terms because scientists of previous centuries were educated in these languages. These terms include prefixes, word roots, and suffixes that are assembled in different combinations. The word roots are often the Latin word for a system or organ. The root can be modified by a prefix (which precedes the root) and/or suffix (which comes after the root). Consider the term *pericarditis*. This imposing term for inflammation of the sac surrounding the heart can be broken down into its origins: *peri-* = "around" + *cardia* = "heart" + *ites* = "pertaining to," but now used to denote inflammation. Taken together, these word parts mean *inflammation around the heart*. The table at the end of this chapter contains a number of important word parts; more word parts are introduced in each chapter.

> *Remember This!* **You can remember the order of the word parts by the mnemonic PRS (prefix, root, suffix).**

Anatomic Terms Describe Directions and Body Planes

Just as the geographic terms *north, east, south, and west* depend on reference to the North Pole, anatomic terms refer to the body posed in **standard anatomic position** (Fig. 1.10), in which the body is:

- Standing erect; head upright
- Facing forward
- Arms at sides
- Hands rotated with thumb outside, palm forward
- Toes forward; feet parallel

Directional Terms Describe the Relative Positions of Body Parts

Directional terms describe the position of a part relative to another part or a subdivision of a part. We can describe,

TERM	ILLUSTRATION	DEFINITION	EXAMPLES
Anterior (ventral)	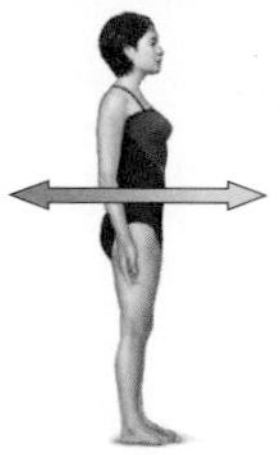	Toward the front (ventral: toward the belly)	The toes are anterior to the heel; the navel is anterior to the sacrum
Posterior (dorsal)		Toward the back	The vertebrae are posterior to the breastbone; the fingernails are on the dorsal surface of the hand
Superior	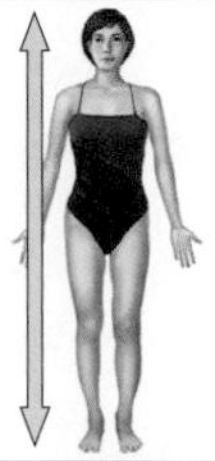	Toward the head	The eyes are superior to the mouth
Inferior		Toward the feet	The navel is inferior to the chin
Proximal	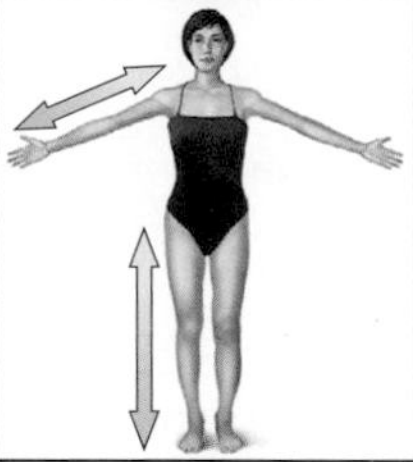	Nearer to the point of attachment	The elbow is proximal to the wrist
Distal		Farther from the point of attachment	The ankle is distal to the knee
Medial	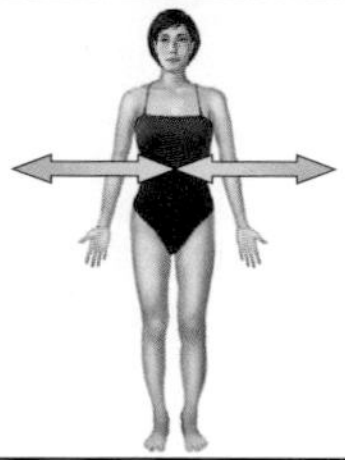	Toward the midline (the midsagittal plane)	The navel is medial to the hipbone
Lateral		Away from the midline (the midsagittal plane)	The ears are lateral to the nose
Superficial	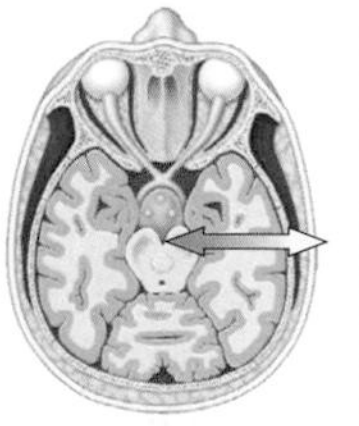	Toward the outside of the body	The hair is superficial to the skull
Deep		Toward the inside of the body	The brain is deep to the skull

Figure 1.10. Directional terms. Directional terms refer to the body in the anatomic position. *Which part is more lateral—the ears or the nose?*

for instance, the inferior (lower) portion of the sternum (breastbone) or make an incision superior to (above) the fifth rib. These terms are identified in Figure 1.10.

Case Note

1.9. The bullet entered and traveled inferiorly, anteriorly, and medially through the president's chest. Did the bullet move:

a. Toward or away from the head?
b. Toward or away from his breastbone?

Body Planes Are Imaginary Flat Surfaces That Divide the Body

Planes are imaginary flat surfaces. Where planes pass through and intersect solid objects, they reveal two-dimensional shapes called **sections**. For example, a plane passing through a sphere reveals a cross-section that is a circle. Some aspects of internal structures can be better understood by viewing two-dimensional images of the shapes that appear in sections along *standard body planes* (Fig. 1.11). These standard planes, which intersect one another at 90-degree angles like the three sides of the corner of a box, are as follows:

- A **frontal plane** is any plane that runs vertically (straight up and down) from superior (above) to inferior (below) and divides structures into anterior and posterior parts. It is also called a *coronal* plane, particularly in reference to the brain. Cuts along this plane are called *frontal* or *coronal* sections.
- A **sagittal plane** is a second type of vertical plane, one which divides structures into right and left parts.

It runs from superior to inferior and is perpendicular to a frontal plane. Cuts along this plane are called *sagittal* sections; cuts right down the middle of the body are called *midsagittal* sections.
- The **transverse plane** or *horizontal plane* is one parallel to the horizon and divides structures into superior and inferior parts. It runs from anterior (front) to posterior (rear), and is perpendicular to both frontal and sagittal planes. Cuts along this plane are called *transverse* sections.
- Any plane not perpendicular to a frontal, sagittal, or transverse (horizontal) plane is an *oblique* plane.

Recall that body planes, like all anatomic directions, are in reference to a body in the anatomic position.

Case Note

1.10. An x-ray was taken of Reagan's chest that showed his lungs side by side. Was this x-ray taken in the frontal or sagittal plane?

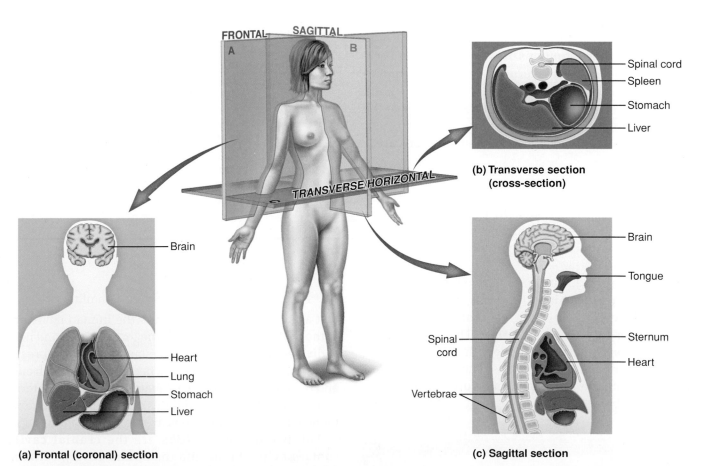

(a) Frontal (coronal) section

(b) Transverse section (cross-section)

(c) Sagittal section

Figure 1.11. Planes and sections. Planes divide the body in specific ways. Cuts along these planes result in different sections. *Which plane divides the body into right and left halves?*

Anatomic Terms Precisely Describe the Body

In everyday language, words used to describe the body are often inaccurate. A person complaining of stomachache, for instance, usually indicates discomfort low in the abdomen, whereas the stomach actually resides high in the abdomen, partly beneath the ribs. Among health care professionals, such inaccuracy can be fatal. Therefore, terms for anatomic landmarks and body regions are defined precisely.

> **Remember This!** Use touch, speaking, and hearing to make memorization easier. For instance, identify each body region aloud on your own body. Compare the location of these regions using relative anatomic terms.

Body Regions Are Defined by Surface Anatomy

Body regions are based on features visible on an intact, unclothed body; since we rarely see beneath the skin, they are referred to as *surface anatomy*. As you learn the new terminology of surface anatomy, you will have to relearn the correct definitions of everyday terms. In correct anatomic language, the term *leg*, for instance, applies only to that part of the limb below (inferior to) the knee, and the entire limb extending from the shoulder to the hand is the upper limb, not the arm.

The regions of the body and their subdivisions are:

- *Head*: skull and face
- *Neck*
- *Trunk or torso*: chest, abdomen, pelvis
- *Upper limb*: shoulder, armpit, arm, forearm, wrist, hand
- *Lower limb*: buttock, thigh, leg, ankle, foot

Many body parts have names derived from Latin. For instance, in Latin the chin is the *mentis*; hence the chin is the mental region. Or, knowing that the word root *brachium* refers to *embrace* and *arm* will help you remember the location of the brachial artery, the brachialis muscle, and the brachial nerve plexus. The most important anatomic landmarks are shown on Figure 1.12 and summarized in Table 1.1.

Case Note

1.11. The president was wounded near his armpit. What is the medical term for *armpit*?

Table 1.1 Some Anatomic Landmarks

Region	Common Name	Adjective
Head	Head	Cephalic
	Eye	Orbital/ocular
	Cheek	Buccal
	Ear	Otic
	Nose	Nasal
	Mouth	Oral
	Chin	Mental
Neck	Neck	Cervical
Trunk	Chest	Thoracic
	Breast	Mammary
	Abdominal/belly	Abdominal
	Navel	Umbilical
	Pelvic	Pelvic
	Loin	Lumbar
	Groin	Inguinal
	Pubic	Pubic
Upper limb	Armpit	Axillary
	Arm	Brachial
	Front of elbow	Antecubital
	Forearm	Antebrachial
	Back of elbow	Olecranal
	Wrist	Carpal
	Hand	Manual
Lower limb	Buttock	Gluteal
	Thigh	Femoral
	Back of knee	Popliteal
	Kneecap	Patellar
	Leg	Crural
	Calf	Sural
	Ankle	Tarsal
	Foot	Pedal
	Heel	Calcaneal
	Sole of foot	Plantar
	Toes	Digital/phalangeal

Body Cavities Are Spaces within the Body

Just as surface landmarks define the human landscape without, the landscape within the torso (trunk) and head is divided into regions. Surface landmarks lack precise margins; for example, *exactly* where the upper limb ends and the torso begins can be debated. In contrast, interior divisions have sharp boundaries defined by specific anatomic margins. Body cavities are shown in Figure 1.13.

The two **dorsal cavities** are the **cranial cavity**, containing the brain, and the **spinal cavity**, containing the spinal cord. Both are formed of bone (skull and spinal column) and lined by membranes (the *meninges*, ➡ Chapter 8).

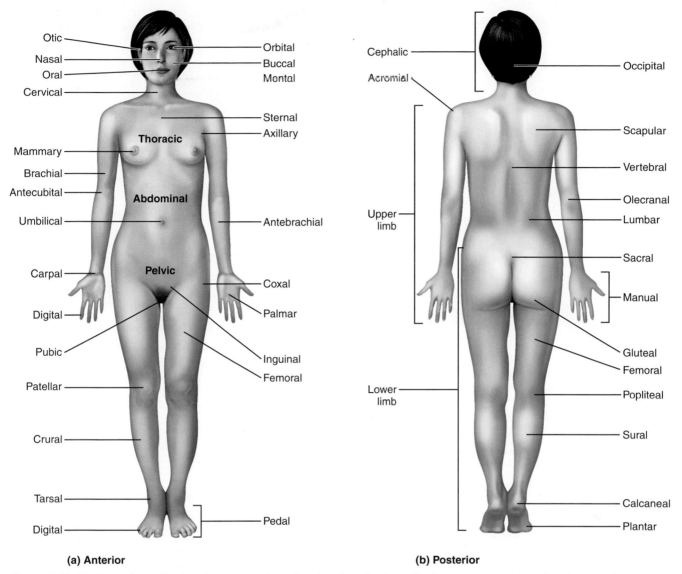

Otic
Nasal
Oral
Cervical
Orbital
Buccal
Montal

Thoracic

Sternal
Axillary

Mammary
Brachial
Antecubital

Abdominal

Umbilical
Antebrachial

Carpal
Pelvic

Digital
Coxal
Palmar

Pubic
Inguinal
Femoral

Patellar

Crural

Tarsal
Digital
Pedal

(a) Anterior

Cephalic
Acromial

Occipital

Scapular

Vertebral

Olecranal

Upper
limb
Lumbar

Sacral

Manual

Gluteal
Femoral

Lower
limb
Popliteal

Sural

Calcaneal
Plantar

(b) Posterior

Figure 1.12. Body regions. Regional terms and surface landmarks describe important areas and points on the surface of a body in the anatomic position. *What is the adjective describing the entire foot?*

The **ventral cavity** is divided by a horizontal (transverse) muscle, the diaphragm. The **thoracic cavity** (chest cavity) lies above the diaphragm; it contains the heart, lungs, and large vessels. The **abdominopelvic cavity** lies below the diaphragm. An imaginary line running along the top of the pelvis separates this large cavity into the superior **abdominal cavity** from the inferior **pelvic cavity.** The abdominal cavity contains most of the digestive organs and abdominal glands, such as the stomach, liver, gallbladder, pancreas, and most of the intestines. The pelvic cavity contains part of the large intestine as well as the reproductive organs.

Double-layered membranes surround many organs of the thoracic and abdominopelvic cavity. The **pericardium** covers the heart, the **pleurae** cover the lungs, and the **peritoneum** covers many abdominal and pelvic organs. One layer of each membrane, the **visceral** layer, attaches to the organs, and the other, the **parietal** layer, attaches to the cavity wall, leaving a potential space between them.

You'll understand the anatomy of the thoracic and abdominal membranes better if you first understand how each is formed in the developing fetus. The organs in each of these cavities develop in a way that can be likened to a fist or hand being pushed into a balloon from one side. For example, as the heart forms, it pushes its way into a balloonlike fetal sac in the thoracic cavity (Fig. 1.14). The near wall of the balloon forms the surface covering of the heart (the *visceral pericardium*). The balloon's interior becomes the *pericardial cavity*, which contains a small amount of lubricant fluid, and the far wall becomes the *parietal pericardium*, a tough, membranous sac that contains the heart. Note that visceral membranes cover organs (from the Latin *viscus*, "organ") and the parietal membranes line the body wall (from the Latin *parietalis*, "wall").

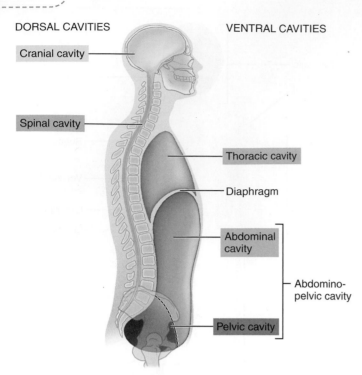

DORSAL CAVITIES   VENTRAL CAVITIES

- Cranial cavity
- Spinal cavity
- Thoracic cavity
- Diaphragm
- Abdominal cavity
- Abdomino-pelvic cavity
- Pelvic cavity

Figure 1.13. Body cavities. Body organs exist in hollow spaces lined by membranes. *Which cavity is superior—pelvic or abdominal?*

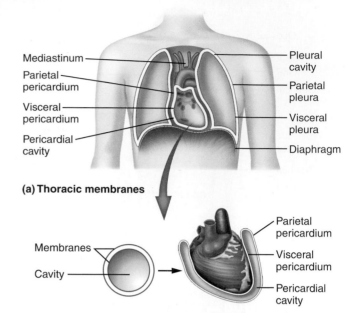

- Mediastinum
- Parietal pericardium
- Visceral pericardium
- Pericardial cavity
- Pleural cavity
- Parietal pleura
- Visceral pleura
- Diaphragm

(a) Thoracic membranes

- Membranes
- Cavity
- Parietal pericardium
- Visceral pericardium
- Pericardial cavity

(b) Formation of the pericardial membranes

Figure 1.14. Thoracic membranes and cavities. A. A very thin cavity separates the two layers of the pericardium and the two pleurae. **B.** The double-layered thoracic membranes formed when the heart and lungs pushed their way into fetal cavities. *Name the membrane in contact with the heart wall.*

In like manner, the lungs push their way into a second fetal sac in the thoracic cavity. The walls of the balloon become the parietal and visceral pleurae of each lung and the balloon's interior becomes the **pleural cavity.**

Between the two pleural cavities, superior to the diaphragm, inferior to the neck, posterior to the anterior chest wall, and anterior to the spine is a space called the **mediastinum.** It contains the heart and pericardial sac, the great arteries and veins near the heart, the lower trachea (windpipe), fat, lymph nodes, nerves, and the large bronchi (hollow tubes that carry air to and from the lungs).

In the abdominopelvic cavity, the liver, spleen, and intestines (and in females the ovaries, uterine tubes, and superior end of the uterus) push their way into a balloonlike fetal sac. The wall of this balloon that covers the organs is the *visceral peritoneum,* and the wall of the balloon lining the remainder of the space becomes the *parietal* peritoneum. The interior of the balloon becomes the **peritoneal space.**

Abdominal Regions Define Parts of the Abdominal Cavity

The abdominal cavity is large and contains many organs. To facilitate discussion, health care professionals often refer to the location of organs by using one of two systems for dividing the abdomen into subsections. In the simpler of the two systems, the abdomen is considered to have four *quadrants* (Fig. 1.15A) created by drawing vertical and horizontal lines that cross at the navel (umbilicus). The liver occupies most of the right upper quadrant, and the appendix is located in the right lower quadrant. The more detailed system divides the abdomen into nine *regions.* These are created by drawing one horizontal line just inferior to the ribs, another just superior to the tops of the hip bones, and one vertical line descending from the middle of each clavicle, just medial to each nipple (Fig. 1.15B).

Case Notes

1.12. The bullet came to rest in President Reagan's lung. In which body cavity are the lungs?

1.13. The bullet passed through two membranes covering his lung. Which membrane was encountered first?

1.12 The suffix *-megaly* means "enlargement." What is the definition of *cardiomegaly*?

1.13 Which structure is more distal—the wrist or the shoulder?

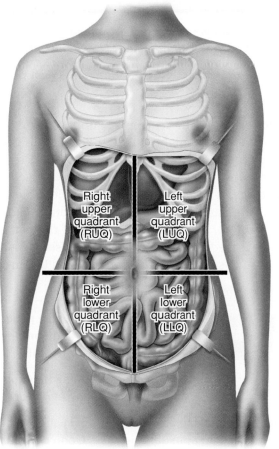

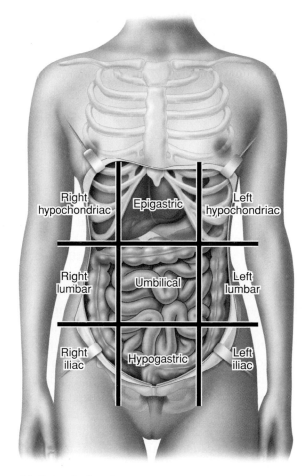

(a) **Abdominal quadrants**

(b) **Abdominal regions**

Figure 1.15. Abdominal regions. A. Abdominal quadrants. **B.** Abdominal regions. *Which region is inferior—the iliac or hypochondriac region?*

1.14 Throughout human history, a common method of execution has been to chop off the victim's head. Recalling that planes relate to a body in the anatomic position, along what type of plane would this cut have been made?

1.15 Which of the nine abdominal regions surrounds the navel?

The Language of Disease

All disease is due to some kind of injury. We can define **disease,** therefore, as an unhealthful state of abnormal form and function due to the effects of injury. The term *injury* is used in the broadest possible sense to include damage to cells and tissues from any cause, including physical trauma, inflammation, radiation, drugs, diet, personal habits, and many others.

All disease is either acute or chronic. **Acute** disease arises rapidly, lasts a short time, and is accompanied by distinct symptoms. For example, a typical infection in a child's ear begins suddenly, lasts a few days, and is accompanied by ear pain and fever. **Chronic** disease usually begins slowly, with signs and symptoms that are difficult to interpret, persists for a long time, and generally cannot be prevented by vaccines or cured by medication. For example, arthritis is a disease that begins with vague stiffness or aches in certain joints, progresses slowly, cannot be cured (but can be treated), and typically lasts a lifetime.

Remember This! **Both acute and chronic diseases can range in severity from mild to life-threatening.**

Signs are direct, measurable observations by an examiner (nurse, physician assistant, physician, etc.), such as body temperature and heart rate. **Symptoms** are complaints reported by the patient or by someone else on behalf of the patient and are a part of the medical history. For example, rapid heart rate reported by the patient is a symptom, whereas rapid heart rate observed by an examiner is a sign. Medical tests such as x-rays or blood tests are used to gather other signs, such as a low red blood cell count. A collection of clinical signs and symptoms is a **syndrome.** For example, overexposure to bright sunlight produces the syndrome of sunburn: skin that is red (sign), painful (symptom), and swollen (sign).

Pathology is the study of changes in bodily structure and function that occur as a result of disease. The purpose of pathology is to discover the **etiology** (cause) of the injury (disease), understand disease **pathogenesis** (natural history and development), explain the **pathophysiology** (the abnormal manner in which the incorrect function is expressed), and describe the **lesion** (the structural [anatomic] abnormality produced by injury). If the etiology is unknown, the disease is said to be **idiopathic** (from Greek *idio,* for "personal," and *pathic,* for "suffering"; thus, suffering of a personal nature, not of known cause). If the disease is a byproduct of medical diagnosis or treatment, it is said to be **iatrogenic** (from Greek *iatros,* for "physician," and *geneos,* for "origin" or "creation").

Case Notes

1.14. Reagan was pale and complaining of chest pain. Is pallor a symptom or a sign? How about chest pain?

1.15. What is the etiology of Reagan's chest pain?

For example, the etiology of sunburn is excess exposure to bright sunlight. The pathogenesis of sunburn is absorption of high-energy ultraviolet (UV) rays, which injure skin. The pathophysiology is characterized by blood vessel dilation and increased blood flow, both of which are part of the reaction to the injury. The lesion is red, swollen, hot, painful skin. And if the patient were taking a medicine that causes increased sensitivity of the skin to sunlight, the sunburn would be iatrogenic.

1.16 After eating sushi, a patient experiences nausea and vomiting. Is this disorder acute or chronic?

1.17 Is a lesion on the hand observed by a nurse a sign or a symptom?

1.18 Upon taking a prescribed medication, a patient experiences an unusual skin rash. Upon discontinuation of the medication, the rash clears up. Is the rash more likely idiopathic or iatrogenic?

Word Parts

Latin/Greek Roots	English Equivalents	Examples
Cardi-/cardio-	Heart	Cardiac: of the heart
Cyt-/cyto-	Cell	Cytology: the study of cells
Derm-/dermo-, dermat-/dermato-	Skin	Dermal: of the skin
Geri-/gero-, geront-/geronto-	Aged	Geriatric: of the aged
Hem-/hemo-, hemat-/hemato-	Blood	Hematology: the study of blood
Hepat-/hepato-	Liver	Hepatic: of the liver
Leuk-/leuko-	White	Leukocyte: a white blood cell
Neur-/neuro-	Nerve cell	Neuritis: inflammation of a nerve
Oste-/osteo-	Bone	Osteocyte: a bone cell
Path-/patho-	Disease	Pathology: the study of disease
Psych-/psycho-	Mind	Psychology: study of the mind

Word Parts (continued)

Latin/Greek Prefixes	English Equivalents	Examples
A-/an-	Without	Anuria: without urine production
Ab-	Away from, outside of, beyond	Abnormal: away from normal; not normal
Ad-	Toward, near to	Adduct: to move toward
Ante-/pre-	Before	Antemortem: before death
Brady-	Slow	Bradycardia: slow heartbeat
Dys-	Abnormal, difficult	Dystocia: difficult childbirth
Ec-, ecto-	Outside	Ectoderm: the outside layer of skin
En-, endo-	Inside	Endocardium: the internal lining of the heart
Epi-	Upon, over, outside	Epineural: outside or upon a nerve
Hyper-	Above normal	Hyperventilation: breathing very rapidly
Hypo-	Below normal	Hypoventilation: breathing very slowly
Inter-	Between	Intervertebral: between the vertebrae
Intra-	Inside	Intracardiac: within the heart
Macro-	Big	Macrocyte: a large cell
Micro-	Small	Microcyte: a small cell
Peri-	Around	Pericardium: around the heart
Tachy-	Abnormally fast	Tachycardia: rapid heart rate

Latin/Greek Suffixes	English Equivalents	Examples
-algia	Pain	Neuralgia: pain from a nerve
-cyte	Cell	Hepatocyte: a liver cell
-ectomy	Surgical removal	Appendectomy: removal of the appendix
-emia	Full of	Leukemia: full of white (blood cells)
-gen, -genic, -genesis	Cause of	Pathogen: the thing causing a disease
-itis	Inflammation	Appendicitis: inflammation of the appendix
-logist, -logy, -ics, -ian, -ist	Study/practice of (or one who studies/practices)	Cardiologist: one who studies the heart
-lysis	Disintegration	Hemolysis: disintegration or destruction of blood cells
-megaly	Enlargement	Hepatomegaly: an enlarged liver

(continued)

Word Parts (continued)

Latin/Greek Suffixes	English Equivalents	Examples
-oma	Tumor	Hepatoma: a tumor of the liver
-pathy	Disease	Neuropathy: disease of a nerve
-penia	Decrease/lack of	Leukopenia: lack of white (blood cells)
-plasia	Growth	Dysplasia: abnormal growth
-stasis	Unchanging, stopped	Hemostasis: stop bleeding

Chapter Challenge

CHAPTER RECALL

1. **Epithelium is an example of a(n)**
   a. Atom
   b. Tissue
   c. System
   d. Molecule

2. **The smallest level of organization is the**
   a. Tissue
   b. Molecule
   c. Atom
   d. Organ

3. **The cavity that contains the ovaries is the**
   a. Thoracic
   b. Dorsal
   c. Abdominal
   d. Pelvic

4. **The mediastinum contains the**
   a. Lungs
   b. Heart
   c. Kidneys
   d. Brain

5. **The center of the abdomen is called the**
   a. Umbilical region
   b. Left inguinal region
   c. Femoral region
   d. Hypogastric region

6. **The olecranal region is found at the**
   a. Back of the elbow
   b. Back of the knee
   c. Base of the skull
   d. Bottom of the foot

7. **Dorsal is to ventral as distal is to**
   a. Inferior
   b. Proximal
   c. Medial
   d. Superficial

8. **The plane that divides the body into superior and inferior parts is called the**
   a. Frontal
   b. Coronal
   c. Transverse
   d. Sagittal

9. **The cause of a disease is most accurately defined as its**
   a. Pathophysiology
   b. Etiology
   c. Anatomy
   d. Pathology

10. **A disease of unknown origin is described as**
    a. Iatrogenic
    b. Idiopathic
    c. Epidemic
    d. A lesion

11. **A cut bleeds because**
    a. Blood moves down a concentration gradient, from high concentration to low.
    b. Blood moves down a concentration gradient, from low concentration to high.
    c. Blood moves down a pressure gradient, from high pressure to low.
    d. Blood moves down a pressure gradient, from low pressure to high.

Match each body function or component with the appropriate body system:

12. **Tonsils**

13. **Bronchi**

14. **Induces muscle contractions**

15. **Delivery of oxygen and nutrients to cells**

16. **Sperm/ovum production**

a. Nervous system
b. Lymphatic/immune system
c. Respiratory system
d. Reproductive system
e. Cardiovascular system

CONCEPTUAL UNDERSTANDING

17. **Using examples from the respiratory system, describe the six levels of organization from largest to smallest.**

18. **Does positive feedback maintain homeostasis? Defend your answer, referring to the definition of homeostasis.**

19. **Compare and contrast the terms *signs* and *symptoms*. List one similarity and one difference.**

APPLICATION

20. **In the standard anatomic position,**
    a. The palms of the hands face forward.
    b. The head is inferior to the diaphragm.
    c. The distal portion of the upper limb is superior to the cranium.
    d. The umbilicus is distal to the epigastric region.

Based on your knowledge of anatomic landmarks, give the most likely site for the following lesions, diseases, and structures:

21. **Plantar wart**

22. **Patellar tendon**

23. **Otitis**

24. **Femoral artery**

25. **Consider the four qualities of life: organization, metabolism, adaptation, and reproduction. How do these four qualities distinguish living from nonliving things? Can you think of nonliving things as possessing some of these qualities?**

26. **Your new apartment has a heater controlled by a thermostat. What would happen on a hot day if your thermostat worked according to the principles of positive feedback?**

27. **A new café has recently opened on campus that sells fantastic iced coffees. On a hot day, 37 customers are inside waiting in line and 7 people are sitting outside on the terrace drinking their coffees. The terrace is the same size as the café.**
    a. Michele had been sitting outside. She enters the café to use the washroom. Is she moving up or down the concentration gradient of people?
    b. If an equal number of patrons can be found inside and outside the café, does a gradient still exist?

You can find the answers to these questions on the student Web site at
http://thepoint.lww.com/McConnellandHull

2

Chemistry in Context: The Molecules of Life

Major Themes

- All matter is composed of atoms.
- Atoms are bound together by chemical bonds to form molecules.
- Chemical bonds store energy.
- The electrical charge of atoms and molecules is important in some chemical reactions.
- Chemical reactions create or break chemical bonds.
- Water is an important chemical in some chemical reactions.
- The body strictly controls the acidity of body tissues and fluids.
- Organic molecules are large, complex, and made only by living things.

Chapter Objectives

The Elements of Life   32

1. Give some examples of trace elements and bulk elements.

The Form and Function of Atoms   33

2. Use the periodic table to determine the atomic number and mass of an atom and then calculate the number of protons, electrons, and neutrons.

3. For a given atom, predict the effect of changing the number of protons, electrons, and neutrons.

Chemical Bonds   40

4. Use the electron shell model to explain the formation of ions and ionic bonds, nonpolar covalent bonds, and polar covalent bonds.

5. Explain why only polar molecules and ions can dissolve in water.

6. Explain the importance of functional groups, and list five examples.

7. Explain why an increased concentration of hydrogen or bicarbonate ions results in increased carbon dioxide production, using chemical symbols to illustrate the chemical reaction.

The Chemistry of Living Things 48

8. Explain the difference between organic and inorganic molecules.

9. List five important properties of water.

10. Relate the pH scale to changes in hydrogen ion concentrations.

11. Illustrate how different macromolecules (carbohydrates, lipids, proteins, and nucleotides) are assembled from different subunits.

12. Explain how water is involved in dehydration synthesis and hydrolysis.

13. Explain why lipids are hydrophobic but carbohydrates are hydrophilic.

14. Describe the functions of nucleic acid and adenosine triphosphate (ATP).

Chemistry in Context: The Case of Joe G. 51

15. Use the case study to illustrate the importance of buffers.

Case Study: "They made me drink pure lemon juice."

As you read through the following case study, assemble a list of the terms and concepts you must learn in order to understand Joe's case.

Clinical history: Joe G. is an 18-year-old college freshman in otherwise good health who was deposited in the emergency room by friends who left without providing any information. On questioning by an emergency physician, Joe said, "They made me drink pure lemon juice" and described a fraternity hazing ritual in which he drank most of the contents of three large bottles of commercial lemon juice. He said he vomited once, "but not much came up."

Physical examination and other data: He was breathing rapidly (24 respirations per minute) and seemed drowsy. He complained of nausea and shortness of breath, but no other physical abnormalities were present. A blood specimen was obtained and analyzed for acidity (pH), bicarbonate, oxygen, and carbon dioxide.

- pH   7.26   Normal, 7.40
- Bicarbonate (HCO_3^-)   21   Normal, 24
- Oxygen   96   Normal, over 80
- Carbon dioxide (CO_2)   33   Normal, 40

Clinical course: A diagnosis of acidosis was made and Joe was given an intravenous solution containing bicarbonate and electrolytes, which was continued overnight. The next morning his respirations were 14 per minute, his laboratory tests were normal, and he was discharged.

Nutrition Facts	
Serving Size 1 cup 244g (244 g)	

Amount Per Serving	
Calories 51	Calories from Fat 6

	% Daily Value*
Total Fat 1g	1%
Saturated Fat 0g	0%
Trans Fat	
Cholesterol 0mg	0%
Sodium 51mg	2%
Total Carbohydrate 16g	5%
Dietary Fiber 1g	4%
Sugars 6g	
Protein 1g	

Vitamin A	1%	•	Vitamin C	101%
Calcium	3%	•	Iron	2%

*Percent Daily Values are based on a 2,000 calorie diet. Your daily values may be higher or lower depending on your calorie needs.

© www.NutritionData.com

Nutrition facts panel from a bottle of lemon juice.

In ← Chapter 1, we thought of ourselves as "viewing the landscape of anatomy and physiology from the window of a jet plane." In this chapter, we can think of ourselves on hands and knees, peering through a magnifying glass, studying "the smallest things" our eyes can see: the rocks, roots, and soil of which the landscape is formed. The view from high above and the view at very close range are necessary for us to understand what we will see in the remainder of our journey, which will be, so to speak, on foot, hiking through the hills and dales of human form and function.

Recall that in ← Chapter 1, we introduced you to the levels of organization in the body. Now we are going to discuss the organization of the smallest things: molecules and the atoms of which they are made.

Nothing exists but atoms and empty space.

Democritus, Greek philosopher, circa 470 to 380 BC, who gave us the word *atom*, which in Greek means "indivisible"; that is, the smallest thing that cannot be divided into something even smaller.

The Elements of Life

What are we made of? Although an old nursery rhyme suggests that girls are made of sugar and spice and boys of puppy dogs' tails, we know that all humans are composed of the same fundamental chemicals in roughly the same proportions (Fig. 2.1). These fundamental chemicals, such as oxygen, hydrogen, and nitrogen, are examples of *elements*. Not only human beings but all matter is composed of elements. An **element** is a sub-stance that cannot be reduced to a simpler substance by normal forces such as heat, electricity, magnetism, or a chemical reaction. For example, the element iron always remains iron in chemical reactions; it does not change into another metal such as copper or lead; and elemental iron is the same substance whether it occurs in human blood or in a cast-iron skillet.

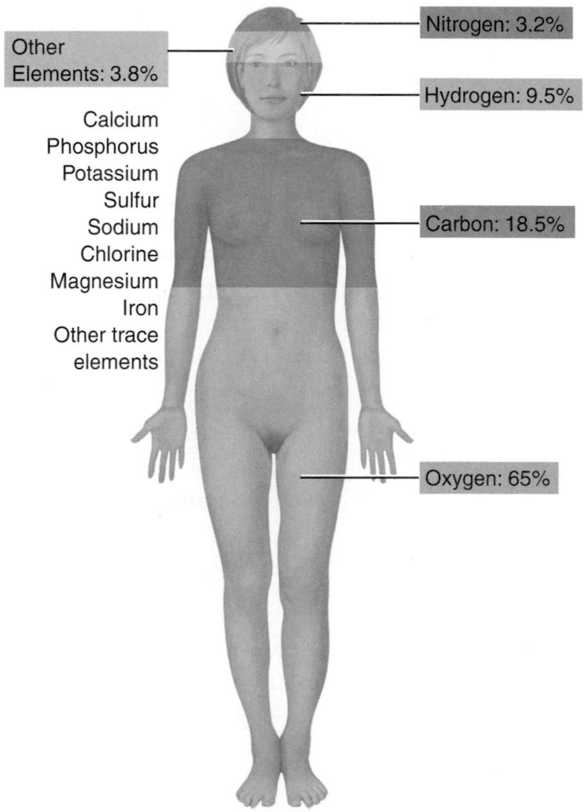

> *Remember This!* **Many breakfast cereals are supplemented with iron. You can run an experiment to see whether your favorite cereal contains iron. First crush it up, then stir it with a magnet. If iron is present, bits of cereal should stick to the magnet.**

Over 90 elements are found in nature; about 30 of these are essential for life. Figure 2.2 presents the **periodic table,** a chart illustrating the essential characteristics of all elements known to exist. Each element is listed by its abbreviation (H for hydrogen, for instance). The elements are numbered from the smallest (1, H, hydrogen) to the very large (88, Ra, radium). Elements in the same column tend to share chemical properties. A bit further on, we discuss the logic of the different numbers and the arrangements of the elements into rows.

The four most abundant elements, which constitute 99% of the mass of body tissues, are oxygen (O), carbon (C), hydrogen (H), and nitrogen (N). These elements and a few others are the *bulk elements* (Fig. 2.2), which must be obtained daily in the diet in large quantities. *Trace elements* (*trace minerals, micronutrients*) are chemical elements required for life but only in very small amounts, ordinarily far less than 0.10 g per day. They, too, are obtained through the diet and include cobalt (Co), copper (Cu), chromium (Cr), iron (Fe), iodine (I), selenium (Se), and others. Iron, for instance, is an integral part of hemoglobin, the chemical that carries oxygen in blood.

Other Elements: 3.8%
Calcium
Phosphorus
Potassium
Sulfur
Sodium
Chlorine
Magnesium
Iron
Other trace elements

Nitrogen: 3.2%
Hydrogen: 9.5%
Carbon: 18.5%
Oxygen: 65%

Figure 2.1. Elements of the body. The percentage of body mass made up by each element is shown on the figure. *What is the third most abundant element in the human body?*

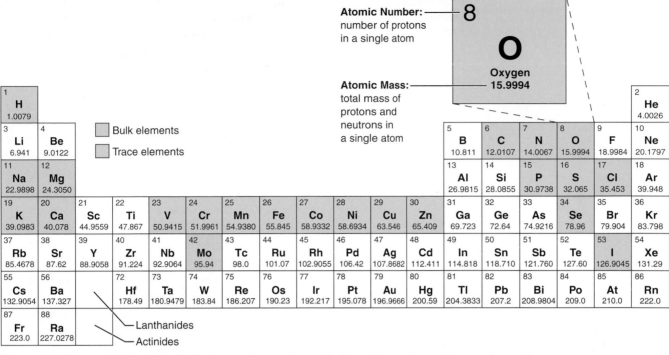

Figure 2.2. The periodic table. Bulk elements are shown in blue; trace elements are shown in green.
Is chromium (Cr) a bulk element or a trace element?

Case Note

2.1. As shown on the Nutrition Facts panel in the case study box, lemon juice contains significant amounts of calcium (Ca) and iron (Fe). Which of these is a bulk element and which is a trace element?

People consuming a varied diet will meet their need for most bulk elements because they are found in a wide range of foods. In contrast, trace elements are found only in specific foods, so inadequate intake is more common. When people do not meet their daily need for a particular trace element, a deficiency may develop. Iron deficiency is the most common dietary deficiency in the western world.

To avoid such deficiencies, many basic foods, such as cereals and bread, are supplemented with trace elements such as iron.

2.1   Name the four most abundant elements in the human body.

2.2   True or false. Iron is a trace element.

2.3   True or false. The daily requirement for each trace element is about 1 g.

The Form and Function of Atoms

The average 70-kg (155-lb) human contains 12.6 kg (28 lb) of the element carbon. The carbon is, of course, not gathered into a large chunk—it is divided into approximately $7.0 * 10^{26}$ (roughly a billion billion billion) carbon *atoms*. An **atom** is the smallest particle of an element that behaves like the element; for example, a carbon atom is the smallest particle of the element carbon that still acts like carbon.

Atoms Are Composed of Subatomic Particles

For over 2,000 years, science has been looking for "the smallest thing" that makes up matter, and in the early 19th century the existence of atoms was proved. However, experiments soon revealed that atoms have an internal structure composed of even smaller *subatomic particles—neutrons, protons, and electrons* (Fig. 2.3A). The behavior of these subatomic particles is key to all physiological functions, from the beating of the heart to the formation of urine. Later discoveries proved that subatomic particles are themselves composed of even tinier *elementary particles* called *quarks*, which now

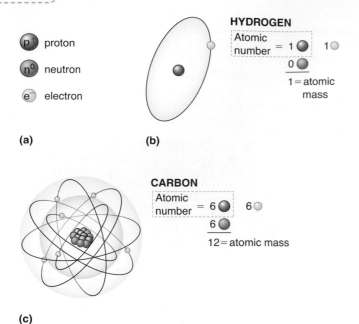

Figure 2.3. Structure of atoms. A. Atoms are composed of protons, neutrons, and electrons. All protons are identical, as are all neutrons, and all electrons. **B.** A hydrogen atom consists of a single electron orbiting a single proton. **C.** The heavier carbon atom contains more protons and electrons, as well as neutrons. Note that the illustrations are NOT to scale—the electrons are much smaller and further away from the nucleus than shown here. *Which atom contains an equal number of protons and neutrons—carbon or hydrogen?*

seem to be the "smallest thing." There are six different types, or *flavors*, of quarks, which combine in different proportions to form subatomic particles.

It is convenient to think of atoms and their subatomic particles as miniature solar systems composed of a central nucleus (the sun), about which other subatomic particles (electrons) orbit like planets (Fig. 2.3). The nucleus of an atom sits in its center and is composed of *protons* and *neutrons*, which are heavy. Surrounding the nucleus are *electrons*, which are very lightweight. Because the nucleus is composed of two heavy particles (protons and neutrons) and electrons are so light, an atom can also be likened to a bowling ball of heavy protons and neutrons with a few gnatlike electrons buzzing around it at a great distance. It is important to realize that this "planetary model" is merely a useful way of thinking about the behavior of atoms because it explains *some* atomic behavior, the behavior that concerns us in this chapter. The *actual* behavior of atoms is very unusual, so unusual that it led Sir Arthur Eddington (1882–1944), an English astronomer, to say that the behavior of atoms and subatomic particles is "…not only stranger than we imagine, it is stranger than we *can* imagine."

Protons and electrons have an *electric charge*; that is, they possess electrical **energy** or the ability to perform work (the accompanying Basic Form, Basic Function box, titled *Energy and Life,* offers more information about the nature of energy). Electric charge is easier to demonstrate than to define—suffice it to say that two opposing charges exist in nature: one called *negative* (−), and its opposite, *positive* (+). By convention, the electron's charge is considered negative, and the proton's charge is positive. The charge of a single electron is designated −1, whereas the charge of a proton is designated +1. The neutron does not possess electrical energy; its charge is 0. If you've ever experimented with magnets, you know that opposite charges attract one another; that is, negative particles are attracted to positive ones and vice versa. In contrast, like charges repel one another: negative particles are repelled by other negative particles; and positives repel positives.

Although the three types of subatomic particles differ in their location, weight, and charge, every particular subatomic particle of the same type is identical: that is, all protons are alike, all electrons are alike, and all neutrons are alike. Therefore, the properties of an atom depend on the number of protons, neutrons, and electrons it contains. For example, hydrogen (H) is a very lightweight gas, and carbon (symbolized by C) is the hard substance that makes up pencil lead. The physical and chemical differences between them are due to their differing numbers of protons, neutrons, and electrons: hydrogen has a nucleus of one proton and no neutrons and is orbited by a single electron (Fig. 2.3B); carbon, on the other hand, is composed of six protons, six neutrons, and six electrons (Fig. 2.3C).

Differences in the number of protons determine the physical and chemical character of the atom and the element it represents. All atoms of a particular element have the same number of protons. The number of protons in the nucleus defines the element: its **atomic number.** Changing the number of protons changes one element into another. For example, by definition an iron atom is one that has a nucleus that contains 26 protons; the number of neutrons and electrons can vary somewhat, but it remains iron only as long as it contains 26 protons. Every atom in iron, therefore, contains 26 protons, and the *atomic number* of iron is 26. Were we to take away one proton, the atoms would become manganese (Mn), a metal defined as having 25 protons (or atomic number 25); or, if we were to add one proton, we would produce cobalt (Co), which by definition is an atom with 27 protons (atomic number 27). Manganese and cobalt have physical and chemical characteristics that are distinctly different from those of iron.

BASIC FORM, BASIC FUNCTION

Energy and Life

The Greek philosopher Democritus said, *"Nothing exists but atoms and empty space."* But that's not exactly true—**energy** occupies "empty space" but has no mass (or weight). Although it is not a physical thing like an atom or an axe, it is as real as sunlight. It is *not* nothing.

Energy is observable only by the *effect* it has on matter. For example, light waves are a form of energy that influences every aspect of life on earth. Sunlight heats the earth; without it earth would be a ball of ice. Sugar cane plants capture sunlight and use it to power the synthesis of sugar, which is then consumed by humans and other animals and converted back into energy that powers chemical reactions and maintains body temperature.

Thus energy and matter are intimately related. Energy can be converted into matter, as in the example of sugar cane sugar, and matter can be converted into energy, as the heat from a fire proves. As the wood is transformed into ashes and smoke, the energy contained in the wood is released as light and heat. After the fire has burned out, the total mass of the wood placed into the fire is equal to the mass of ashes and smoke plus the energy released as heat and light.

Energy can put matter into motion or can stop its motion. It can also be emitted as waves of sound or light that travel through space. Energy exists in several forms:

- *Radiant energy* is emitted as *atomic particles* or as *waves* that travel through space. Ordinary sunlight is a form of radiant energy, for example.
- *Mechanical energy* is stored in mechanical systems. For example, the tightly coiled spring that powers a wind-up toy contains *potential energy*, which is converted to *kinetic energy* (the energy of a moving object) as the toy moves.
- *Electric energy* is the energy associated with the movement of electrons. For example, electric currents and lightning bolts are flows of electrons.
- *Chemical energy* is stored in the bonds that hold molecules together and is released as the bonds are broken and parts of the molecule move apart.

Radiant energy   Mechanical energy   Electrical energy   Chemical energy

(a) Energy takes many forms.

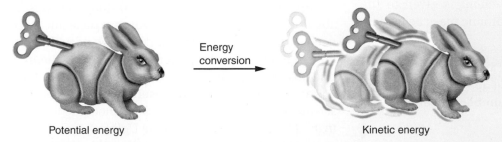

Potential energy   Kinetic energy

(b) Energy can be converted between different forms.

Energy. A. Energy can take many forms, including radiant, mechanical, electric, and chemical. **B.** Potential energy can be converted into kinetic energy.

(continued)

Energy and Life *(Continued)*

When energy is stored, it is called **potential energy.** For example, the sun is a ball of hydrogen and helium, each of which is a source of potential energy. It uses some of this energy every day as it burns hydrogen to produce light. As just noted, a wind-up toy contains potential mechanical energy in its coiled spring; a battery contains potential electrical energy in the chemicals and metals it contains; and glucose contains potential chemical energy in the bonds that hold the atoms together. When potential energy is used to put something in motion, the energy is called **kinetic energy.**

Energy of one form can be *converted* into another, and *all energy conversions liberate heat.* For example, a lightbulb converts electrical energy into radiant energy (light waves), and in doing so the bulb becomes hot. The same thing happens in the body: when dietary nutrients, such as sugar and fat, are converted into the mechanical action of running, heat is liberated, which raises body temperature.

Remember This! **No two elements have the same atomic number.**

The total number of protons and neutrons in the nucleus determines the weight of an atom—its **atomic mass.** Electrons are so extremely light that they are not considered. Atoms with a high atomic mass are heavy; those with a low mass are light. In notations of elements, a number to the upper left of the symbol indicates the atomic mass of that element, not its atomic number. For example, normal carbon (C) contains six protons and six neutrons and has an atomic mass of 12, which is indicated as ^{12}C (Fig. 2.3C). Hydrogen (^{1}H) is the smallest and lightest element: it exists as a gas and consists of one proton, no neutrons, and one electron (Fig. 2.3B). Toward the other end of the scale, uranium (^{238}U) is a very heavy metal and each of its atoms consists of 92 protons, 146 neutrons, and 92 electrons. Note that the number of protons does not necessarily equal the number of neutrons.

You can discover the atomic number and mass for each element using a periodic table (Fig. 2.2). Note that the atomic number for each element is given to the upper left of its symbol and the atomic mass below it.

Case Note

2.2. **The oxygen concentration in Joe's blood was measured. What is the atomic number and atomic mass of one oxygen atom?**

Isotopes Differ in the Number of Neutrons in the Nucleus

Atoms with the same number of protons but different numbers of neutrons are known as **isotopes.** For exam-ple, the most common form of carbon contains six protons and six neutrons and has an atomic mass of 12 (^{12}C). If, however, we were to add two neutrons to the nucleus, we would create a *heavy* isotope of carbon (^{14}C) because it contains eight, not six, neutrons. The isotope is said to be heavy because it contains more than the normal number of neutrons. Heavy isotopes are not very stable; they tend to break apart into other, more stable atoms. This process releases *ionizing radiation,* which is particles or energy waves that collide with other atoms and destabilize them. Ionizing radiation can kill cells or damage their DNA in a way that causes cancer.

That isotopes release ionizing particles makes them valuable in medicine and physiology because laboratory instruments can detect the radiation they emit. Typically, the isotope is built into a custom-made molecule that is identical to a natural one except that it emits radiation. After entering the body, the radioactive molecule participates in chemical reactions alongside the natural molecules and emits radiation as it does so, which is detected with a laboratory instrument.

Suppose, for example, we want to study the intestinal absorption of a certain dietary nutrient molecule. All we need to do is to make some of the molecule with an isotope in it and have the patient consume it. If the intestine absorbs it, radioactivity will show up in the blood. For another example of the usefulness of radioactivity, see the following page Clinical Snapshot, titled *Radioactivity: The Good, the Bad, and the Ugly.*

What's more, the radiation produced by isotopes can be used to treat cancer. For example, certain cancers may need a particular molecule to survive. Knowing this, medical scientists can insert an isotopic atom into the molecule and administer it to a patient. As cancer cells use the molecule, they are killed by the radiation.

CLINICAL SNAPSHOT

Radioactivity: The Good, the Bad, and the Ugly

"Radioactive." Scary word. Stay away. Handle with extreme caution. Visions of atomic bombs, deadly poisons, and pollution. And all of that is true. For evidence we need look no further than the atomic bombs dropped on Japan in 1945. The blasts in Hiroshima and Nagasaki killed tens of thousands of people immediately and others died of radiation poisoning in the following days, months, and years. Those who survived the shock and heat of the explosion were irradiated by ionizing rays from the blast or were later slowly poisoned by rays from radioactive compounds deposited in the environment. The radiation damaged DNA and caused a great increase in birth defects and cancers in the exposed population that survived. That's "the bad and the ugly" part of radioactivity.

Less well known is the good—in certain well-controlled circumstances radioactive molecules can be used in medical diagnosis and treatment. For example, to make its hormones, the thyroid gland must use tiny amounts of iodine (a trace element). Glands that make too much thyroid hormone (a medical condition called hyperthyroidism) require a lot of iodine. One way to see if the thyroid gland is making too much hormone is to give the patient a tiny dose of radioactive iodine (too little to cause health risk) and monitor the radioactivity that accumulates in the thyroid gland. Patients with hyperthyroidism have very "hot" (radioactive) glands.

Another interesting example is a positron emission tomography (PET) scan using a camera sensitive to radioactive rays instead of rays of light. PET scan technique depends upon the fact that cells burn glucose for energy, and cells that are more active burn more glucose than those that are less active. In the PET procedure, a patient is injected with a tiny dose of radioactive glucose, and images are obtained of

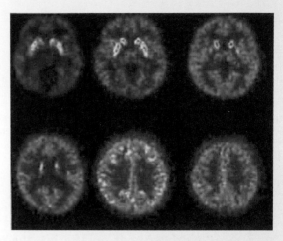

PET scan of the brain. The bright yellow areas use more glucose than normal and may indicate tumors.

various sections of tissue. Physicians can then study which tissues are burning the most glucose—that is, which are metabolically most active. These tissues take up more of the radioactive glucose and, therefore, emit more radioactive waves, which show up as "hot spots" to the PET camera. PET scans can be used to detect cancer because cancer cells multiply rapidly, an activity that requires a lot of glucose.

But that's not the end of the good news about radioactivity. Once a thyroid tumor has been diagnosed, radioactive iodine can be used to treat the tumor. In this instance the patient can be injected with a larger dose of radioactive iodine. As the tumor accumulates radioactive iodine, the concentration of radioactivity becomes high enough to kill cancer cells. For some patients with thyroid cancer, this is the only treatment required.

Electrons Are Arranged in Electron Shells around the Atomic Nucleus

Recall that we compared atoms to a solar system: electrons follow particular paths about the nucleus, just as planets follow particular orbits around the sun. Just as some planets orbit nearer than others to the sun, some electrons orbit close to the nucleus and others orbit at various distances farther away. The paths along which the electrons move can be thought of as tracing out a set of hollow spheres—called *electron shells*—the smaller ones nested inside the larger ones like a child's set of nesting balls. The nucleus sits in the center of these spheres.

If we arrange atoms by size, we find that the two smallest—hydrogen and helium—have only one electron shell. As atoms grow larger—having nuclei with more neutrons and protons—additional electron shells are necessary to hold the electrons. Atoms can have up

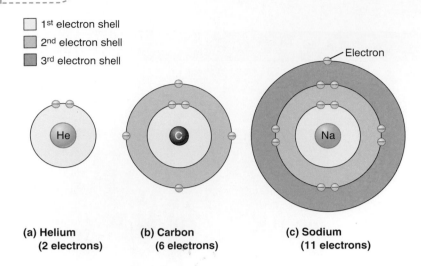

□ 1st electron shell
■ 2nd electron shell
■ 3rd electron shell

(a) Helium
(2 electrons)

(b) Carbon
(6 electrons)

(c) Sodium
(11 electrons)

Figure 2.4. Electron shells. Electrons are organized in energy levels called electron shells. **A.** Helium has a single shell. **B.** Carbon has two shells. **C.** Sodium has three shells. *Which of these atoms has a full outer shell?*

to five shells, each with a greater capacity than the previous one. The innermost (smallest) shell has a capacity of two electrons; that is, it can contain either one or two electrons. Again, hydrogen and helium are the only atoms with just one shell: hydrogen has one electron in the shell and helium has two (Fig. 2.4A). All other atoms have more electrons and, therefore, need more shells. The second shell has a capacity of eight electrons; the third shell has a larger capacity but is quite stable if it also contains eight electrons.

Two examples, using carbon and sodium, will illustrate the point. Carbon needs two shells to hold its six electrons: two electrons in the inner shell and four electrons in a second, outer shell (Fig. 2.4B). On the other hand, sodium, with 11 electrons, has 2 electrons in the inner shell (which fill it up), 8 in a middle shell (which fill it up), and 1 in a third shell, which leaves room for more electrons to join (Fig. 2.4C). Atoms are stable when they have eight electrons in the second shell and, if necessary, the third shell. Since both carbon and sodium have fewer than eight electrons in their outer shells, they are unstable and more readily participate in chemical reactions (see below).

Referring back to the periodic table (Fig. 2.2), the elements are arranged into rows based on the number of electron shells—hydrogen (H) and helium (He) have one shell, while potassium (K) and calcium (Ca) have four shells.

Electrically Charged Atoms Are Called Ions

Remember that protons are positively charged and electrons are negatively charged. If the number of protons and electrons in an atom is equal, the positive and negative forces exactly cancel one another and the atom has no *net* electrical charge. However, if the number

of protons and electrons is not equal, the atom may have a *net* positive or negative electrical charge, in which case the atom is referred to as a positive or negative **ion.** Elements at the far left and right of the periodic table tend to form ions, while those closer to the middle do not.

If electrons outnumber protons, the ion's net charge is negative, and if protons outnumber electrons, the ion's charge is positive. For example, a hydrogen atom missing its electron is a *hydrogen ion* (H^+, which is actually just the remaining proton) (Fig. 2.5A). A positively charged ion is called a **cation,** and a negatively charged ion is called an **anion.** If the number of protons outnumbers the number of electrons by two, as is the case with magnesium (Mg) ions, the magnesium atom has a net charge of +2, and the notation is written Mg^{++} or Mg^{2+}. Likewise, if electrons outnumber protons, as they do in negatively charged ions, the same principle applies. For example, an ion of the element chlorine (Cl) has one more electron than the number of protons and is written as Cl^-.

Ions are attracted to and tend to interact with other atoms that have an opposite charge. This characteristic of ions renders them very important in body processes— they can give us the power to move mountains! Ions will move a short distance to meet up with their opposing charge: this tiny movement enables muscle contractions that can lift a shovel full of dirt (or push the start button of a bulldozer).

Case Note

2.3. Joe's treatment involved the administration of bicarbonate (HCO_3^-). Based on the negative sign in the formula, is bicarbonate an anion or a cation?

Atoms that have one or two electrons in their outermost shell are likely to *lose* them, which can empty the shell,

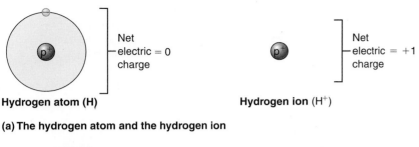

Hydrogen atom (H)

Net
electric = 0
charge

Hydrogen ion (H⁺)

Net
electric = +1
charge

(a) The hydrogen atom and the hydrogen ion

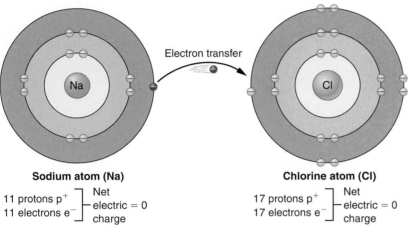

Electron transfer

Sodium atom (Na)

11 protons p⁺
11 electrons e⁻

Net
electric = 0
charge

Chlorine atom (Cl)

17 protons p⁺
17 electrons e⁻

Net
electric = 0
charge

(b) Sodium and chlorine atoms

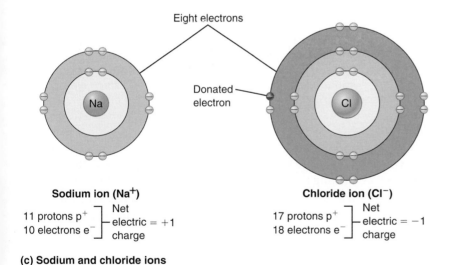

Eight electrons

Donated
electron

Sodium ion (Na⁺)

11 protons p⁺
10 electrons e⁻

Net
electric = +1
charge

Chloride ion (Cl⁻)

17 protons p⁺
18 electrons e⁻

Net
electric = −1
charge

(c) Sodium and chloride ions

Figure 2.5. Formation of ions. A. The hydrogen atom has a single electron in its outer shell. It loses this electron to another substance (not shown) to become a positively charged hydrogen ion (H⁺). **B.** Sodium has one electron in its outer shell, and chlorine has seven. Sodium readily donates a single electron to chlorine. **C.** This reaction produces sodium (Na⁺) and chloride (Cl⁻) ions. Both ions now have eight electrons in their outer shell. *How many electrons are in the middle shell of the chlorine ion?*

leaving the underlying full shell as the outermost one and resulting in the formation of a positive ion. Atoms with six or seven electrons in the outermost shell tend to *gain* electrons to fill the shell, which stabilizes the shell (brings it near or to eight electrons) and forms a negative ion.

> *Remember This!* **For a particular atom, changing the number of electrons results in an ion, changing the number of neutrons results in an isotope, and changing the number of protons results in a different element.**

Atoms gain or lose electrons by interacting with other atoms. Let's turn our attention back to the chlorine atom. Chlorine has only seven electrons in its third and outermost shell; so, in order to reach the stability of eight electrons, it is eager to "steal" an electron from another atom (Fig. 2.5B). On the other hand, sodium has only a single electron in its outer shell and is eager to lose it so that the underlying stable shell of eight electrons will become outermost. With the transfer of one electron from sodium to chlorine, both atoms attain a stable outermost electron shell containing eight electrons. Sodium is now a positively charged ion (Na⁺) with 10 electrons

(2 in the innermost shell and 8 in the second, outermost shell) and 11 protons (Fig. 2.6C). And chlorine is now a negatively charged ion (Cl⁻) with 18 electrons (2 in the innermost shell, 8 in the second shell, and a stable 8 in the third, outermost shell) and 17 protons. We refer to this form of chlorine by a different name—*chloride*—to signify the ionic form of the element.

2.4 Which subatomic particles do not have an electric charge?

2.5 The atomic number of neon is 10 and its atomic mass is 20. How many protons and neutrons are present in a neon atom?

2.6 True or false: Anions have more protons than electrons.

2.7 Aluminum atoms have 13 electrons. How many electrons will be in the outer shell of an aluminum atom?

2.8 A dangerous form of oxygen is written as O22. Remember that oxygen has eight protons; how many electrons does this form of oxygen have?

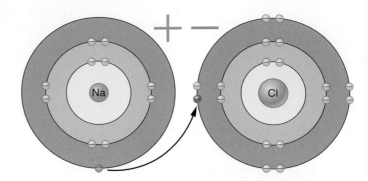

(a) Na⁺ and Cl⁻ attract each other, forming an ionic bond.

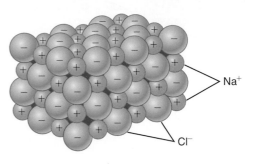

(b) A salt crystal results from ionic bonds between many Na⁺ and Cl⁻ ions.

Figure 2.6. Ionic bonds. A. Positively charged sodium (Na⁺) ions form ionic bonds with negatively charged chloride (Cl⁻) ions. Note that both ions have stable outer electron shells. **B.** Many sodium and chloride ions are ionically bonded to form a salt (NaCl) crystal. Each ball represents a sodium (orange) or chloride (green) ion. *Which element loses a shell in the formation of NaCl, Na or Cl?*

Chemical Bonds

People do not live alone. Each person forms bonds of varying strength and duration with other people, ranging, say, from a 50-year marriage to a 5-minute conversation on the bus. In a similar way, atoms form bonds that are strong and durable or weak and transient.

Ionic Bonds Form between Ions

Remember that ions are formed when one atom *donates* an electron and another accepts it. However, just like a child who lends a treasured book to her brother, the electron donor stays close to the electron acceptor and does not completely lose control of the donated electron. Moreover, as a result of the donation, the receiving atom becomes negatively charged, the donor positively charged, and the two are strongly attracted to each other. The interaction between these opposing charges is called an *ionic bond.* For example, sodium (Na) transfers an electron to chlorine (Cl), forming Na⁺ and Cl⁻ ions. The Na⁺ and Cl⁻ ions form an ionic bond because of their opposing charges. The resulting substance is sodium chloride (NaCl, or table salt) (Fig. 2.6A).

NaCl is an example of a **compound,** a substance containing at least two *different* elements linked by a chemical bond. And any substance that is created by an ionic bond, not just NaCl, is a **salt.** Salts have a crystalline structure, created by arrays of alternating cations and anions interacting by ionic bonds (Fig. 2.6B).

Salts separate into their ionic parts when dissolved in water; that is, they *ionize.* For example, when NaCl is dissolved in water, it divides into Na⁺ and Cl⁻ ions. Water containing ions conducts electricity easily, whereas pure water without dissolved ionic compounds does not. In fact, salts are also called **electrolytes** because they conduct electricity when dissolved in water.

Case Note

2.4. Joe was given an intravenous solution containing electrolytes. What is another name for electrolytes?

Molecules Are Atoms Joined by Covalent Bonds

Covalent bonds are bonds formed when atoms *share* electrons. Much like two children content with sharing a book, covalently bonded atoms share one or more electrons to fill their outer shell. The shared nature of covalent bonds renders them stronger than ionic bonds. For example, covalent bonds (unlike ionic bonds) remain intact when a covalently bonded compound is dissolved in water.

Two or more atoms linked by covalent bonds create a **molecule.** Chemical formulas are used to illustrate the structure of molecules. These formulas consist of symbols for the elements (H for hydrogen, C for carbon, and so on) together with a subscript number that indicates the number of atoms involved. For example, the symbols in the formula for water, H_2O, indicate two atoms of hydrogen combined with one atom of oxygen. And the formula for the main sugar in blood (*glucose*) is $C_6H_{12}O_6$, which indicates that a molecule of glucose is composed of 6 atoms of carbon, 12 atoms of hydrogen, and 6 atoms of oxygen.

> *Remember This!* **Two or more atoms linked together by covalent bonds form a molecule. The atoms can be the same, such as two hydrogens, or different, such as hydrogen and oxygen. Two or more atoms of different elements linked together by any chemical bond form a compound. Thus, H_2O is both a molecule and a compound.**

Even in the invisible world of atoms and molecules, form and function go together. Because the molecules we are studying are described on flat pieces of paper, it is easy to forget that molecules are three-dimensional objects; moreover, the function of a molecule depends not only on the atoms it contains but also on its three-dimensional shape (form). For example, glucose, which makes up part of table sugar, can assume two slightly different shapes: d-glucose and l-glucose. Even though these two forms contain identical sets of atoms, only one shape, d-glucose, can be burned for energy—l-glucose just doesn't "fit" the body's chemistry. Therefore, only d-glucose goes into sugar cookies.

In Covalent Bonds, Atoms Share Electrons

The formation of hydrogen gas (H_2) from two hydrogen atoms is a good example of covalent bonding. Recall that an atom of hydrogen is formed of a nucleus that contains a single proton, which is orbited by a single electron (Fig. 2.7A, left side). When two hydrogen atoms (H) unite, the electron of each atom orbits *both* hydrogen nuclei, each electron flitting back and forth from one atom to the other (Fig. 2.7A, right side). This electron sharing is so complete that each atom of hydrogen acts as if its outer (and only) electron shell is filled to its stable capacity of two electrons.

In the above example, the covalent bond is composed of a single electron from each hydrogen atom. However, atoms can bond by each contributing two

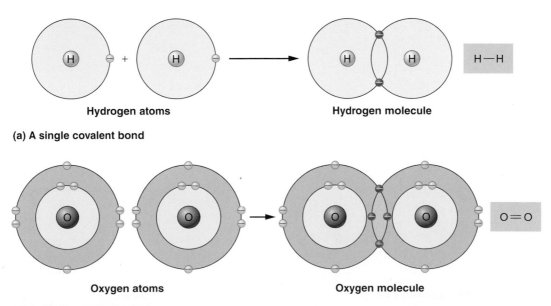

Hydrogen atoms   **Hydrogen molecule**

(a) A single covalent bond

Oxygen atoms   **Oxygen molecule**

(b) A double covalent bond

Figure 2.7. Covalent bonds. A. Two hydrogen atoms form a single covalent bond when they share their single electrons. **B.** Two oxygen atoms form a double covalent bond by sharing two electrons each. *How many electrons are circling the hydrogen gas molecule?*

electrons instead of one, as when two oxygen atoms combine to form oxygen gas (O_2) (Fig. 2.7B). The resulting bond, which is quite sensibly called a *double bond*, is a very strong covalent bond, which stores a lot of energy.

In both of these examples, the electrons are equally shared between the two atoms. Figure 2.7 also provides a simple way of illustrating covalent bonds—drawing lines between the symbols for the bound atoms. A single line indicates that the atoms form a single bond (H–H in Fig. 2.7A); a double line indicates a double bond (O=O in Fig. 2.7B).

Polar Covalent Bonds Can Form Polar Molecules

We've said that covalent bonds can also form between atoms of different elements. Water, for instance, consists of one hydrogen atom and two oxygen atoms bound by covalent bonds (H_2O) (Fig. 2.8). However, different atoms do not share as nicely as identical atoms. Whichever one has a heavier nucleus (with more protons) exerts a stronger pull on the shared electrons than the atom with the lighter nucleus (fewer protons). The electrons, therefore, spend more time hanging out with the heavier atom. This causes the heavy end of the bond to be slightly negative and the lighter end to be slightly positive. Because the ends of the bond have opposing positive and negative charges, such a bond is called a *polar covalent bond*. That is, the bond has a negative pole and a positive pole, much like a magnet.

Nitrogen and oxygen, but not carbon, frequently form polar covalent bonds with hydrogen atoms. For instance,

oxygen is heavier (has more protons) than hydrogen. Thus, when oxygen and hydrogen bond, the shared electron gravitates toward the oxygen, making that end of the bond more negative.

Because of the arrangement of its polar bonds, the water molecule as a whole is also polar. The oxygen side of the molecule is slightly negative and the hydrogen side slightly positive. The key word here is *slightly*: The positive and negative charges of polarized molecules are much weaker than the positive and negative charges associated with ions. For this reason, chemists designate these weak polarities with special symbols, δ^+ and δ^- (Fig. 2.8, right side).

Because a polar molecule has two poles—that is, regions of opposite charges separated by a short distance—it is also called a *dipole*. Molecules containing oxygen and nitrogen are frequently dipoles. By contrast, charges are evenly distributed in *nonpolar molecules*, which either are composed of a single element, such O_2, or, like most fats, contain only carbon and hydrogen.

Strong Covalent Bonds Create Functional Groups

Some covalent bonds are so strong that the participating atoms behave like a single atom, much like a close-knit group of friends. These atomic gangs are called *functional groups*. The hydroxyl functional group (OH), for instance, rarely dissociates into oxygen and hydrogen atoms. Five functional groups that play important roles in the chemistry of life are identified in Table 2.1. Functional groups form covalent bonds with other atoms and functional groups, creating large, complex

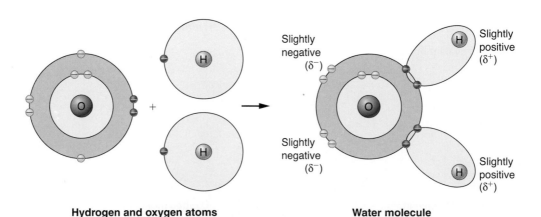

Hydrogen and oxygen atoms   **Water molecule**

Figure 2.8. Polar covalent bonds. Water is formed by one oxygen atom and two hydrogen atoms bound by polar covalent bonds. The slightly positive end of the water molecule is shown by the symbol δ^+; the negative end by the symbol δ^-. *Which side of the water molecule is slightly positive: the hydrogen end or the oxygen end?*

Table 2.1 Functional Groups

FUNCTIONAL GROUP	FORMULA	EXAMPLES
Hydroxyl	$-OH$	
Carboxyl	$-COOH$	
Amino	$-NH_2$	
Methyl	$-CH_3$	
Phosphate	PO_4^{-2}	

Threonine (amino acid)

Adenine neucleotide (DNA)

molecules. Citric acid, the major component in lemon juice, is one such example. Its chemical structure is shown below.

Case Note

2.5. How many functional groups can you find in citric acid?

Some Molecules Are Ions

Recall that salts dissolve in water to form ions. Some salts contain more than two atoms; $CaPO_4$, for instance, is a salt that makes bones hard. When $CaPO_4$ dissolves in water, the PO_4 functional group behaves like a single atom. It receives two electrons from calcium, resulting in Ca^{2+} and PO_4^{2-}. The atoms in PO_4^{2-} are covalently bonded, forming a molecule, but the entire molecule has a negative charge. PO_4^{2-} is thus a *molecular ion* and can participate in ionic bonds with oppositely charged ions.

Highly polar molecules become ions by interacting with water molecules. In acetic acid (CH_3COOH), for instance, the H of the COOH end is virtually a naked proton because its electron spends all of its time at the other end of the molecule. When dissolved in water, this neglected proton forms a covalent bond with a water molecule, like a neglected cat bonding with a neighboring family, resulting in two ions, H_3O^+ and CH_3COO^-. Other molecules, such as ammonium (NH_3), perform the opposite feat, grabbing a hydrogen ion (i.e., a proton) from water to form OH^- and NH_4^+. As discussed later in the text, proton donors are called *acids* and proton grabbers are called *bases*.

> ***Remember This!*** In ions, the number of protons and electrons is unequal. Dipoles contain equal numbers of protons and electrons, but the electrons spend more time near some protons than others.

Case Notes

2.6. A laboratory test measured the amount of three molecules in Joe's blood: bicarbonate (HCO_3^-), carbon dioxide (CO_2), and oxygen (O_2). Which of these molecules are compounds?

2.7. Which of these measured molecules would participate in an ionic bond: bicarbonate or oxygen?

Molecules Interact through Intermolecular Bonds

In the molecular world, opposites attract. For example:

- An ionic bond can form between a molecular cation (+) and a molecular anion (–).
- The negative end of a polar molecule can interact with a molecular cation, or the positive end of a polar molecule can interact with a molecular anion. Bonds between polar molecules and ions are called *dipole–ion bonds*.
- The negative end of a polar molecule can form a weak bond with the positive end of a different polar molecule, and vice versa. Interactions between two polar molecules are called *dipole–dipole bonds*.

Now let's take a quick look at the most physiologically important type of dipole–dipole bond, the *hydrogen bond*.

Water Molecules Form Hydrogen Bonds

Hydrogen bonds are weak dipole–dipole bonds between the weak positive charge on the hydrogen atoms and the

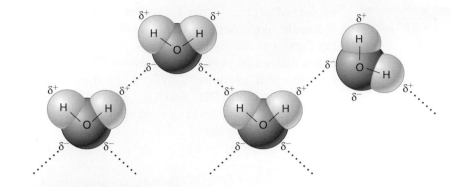

(a) Hydrogen bonds between water molecules

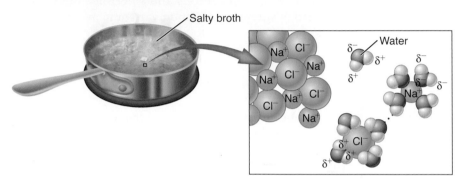

(b) Salt dissociates in water

Figure 2.9. Intermolecular bonds. **A.** Hydrogen bonds form between polar water molecules. **B.** Table salt (NaCl) dissolves in water because polar water molecules form dipole–ion bonds with Na^+ and Cl^-. *What sort of bond is represented by the lines between the H and O symbols?*

weak negative charges of other polar molecules. In Figure 2.9A, for example, multiple water molecules form hydrogen bonds with one another. The negative oxygen end of one water molecule interacts with the positive hydrogen end of another water molecule.

Individual hydrogen bonds are much weaker than covalent or ionic bonds and are easily disrupted. Liquid water, for instance, is held together by hydrogen bonds. However, once water is heated above 100°F, the increased thermal energy of individual water molecules overcomes the hydrogen bonds and water evaporates. In contrast, when water reaches the freezing point, the bonds lock the sluggish water molecules solidly into place—and voila! Ice.

The important roles played by hydrogen bonds in the three-dimensional shape of proteins and DNA are discussed further on.

Charged Molecules Are Soluble in Water

The mixture of solute and solvent is called a **solution.** In a solution, the minor component is the **solute;** the major component is the **solvent.** *Solubility* is the ability of the solute to dissolve in the solvent.

The ability of charged molecules to form intermolecular bonds with each other determines their solubility. In the body, the most abundant polar solvent is water. Charged solutes are soluble in water because they can form dipole–dipole or dipole–ion bonds with water molecules. For instance, Figure 2.9B shows water forming dipole–ion bonds to dissolve NaCl into Na^+ and Cl^-. The slightly negative oxygen ends of water molecules interact with Na^+; the slightly positive hydrogen ends with Cl^-.

Polar and ionic molecules are said to be **hydrophilic,** or water-loving, because they are soluble in any watery substance, such as blood or cytosol. They are not soluble in nonpolar solvents because their charged solute molecules cannot form bonds with the nonpolar solvent molecules. For our purposes the most important nonpolar solvents are *lipids,* a class of molecules that includes oils and fats, which will be discussed later in the text.

Nonpolar molecules such as fats or oils cannot form dipole–dipole bonds with water and are thus not soluble in watery fluids like blood. Instead, they are soluble in other nonpolar substances and are called **hydrophobic,** or water-hating. In fact, it's the water that "hates" the lipid—water repels any nonpolar molecule even as it forms more bonds with other water molecules. This repulsion forces hydrophobic molecules to cluster together in little droplets, much as the icy Antarctic winds force penguins to huddle together. An everyday example of polar–nonpolar interactions can be found at the dinner table in a simple salad dressing of hydrophobic olive oil and hydrophilic vinegar. If left to stand for a few minutes, the olive oil and vinegar separate into two layers because the polar, watery vinegar solution drives away the nonpolar olive oil molecules.

Polarity and solubility are of immense practical importance in the administration of drugs–some drugs are polar and others are nonpolar. The polarity and thus the solubility of a drug determines how it can be administered. See the nearby Clinical Snapshot, titled *Drug Delivery Methods: Pill, Patch, Puff, Pump, or Poke?* to find out more about drug delivery methods.

Case Notes

2.8. A major solute in Joe's intravenous solution was bicarbonate ions. Do you think bicarbonate ions are hydrophilic or hydrophobic?

2.9. What sort of bond can bicarbonate ions form with water, if any?

2.10. Which solvent do you think was used for the intravenous solution: water or oil?

Remember This! Like dissolves like. Polar solvents dissolve polar solutes, and nonpolar solvents dissolve nonpolar solutes.

 CLINICAL SNAPSHOT

Drug Delivery Methods: Pill, Patch, Puff, Pump, or Poke?

No one likes injections, but for most people with diabetes, multiple daily "pokes," or injections of insulin, remain a way of life. Insulin enables the body to metabolize glucose (blood sugar) for energy. Without glucose to burn for energy, human life is not possible. There are good reasons diabetic patients must inject insulin instead of swallowing it or smearing it on their skin. And the reasons are a matter of *chemistry*.

First, insulin cannot be effectively administered in a pill because stomach acid and intestinal digestive juices are designed to break down protein, and insulin, a protein, is destroyed by passage through the intestines. Second, insulin cannot be administered as a skin cream or patch because insulin, like most proteins, is hydrophilic (water-loving) and skin is hydrophobic (water-hating). On the other hand, hydrophobic drugs can be administered by placing them on skin, often as a patch; nicotine and estrogen are examples.

So, until recently, the only effective way to get insulin into the bloodstream has been by injection. Fortunately, advances in chemistry have recently led to some less painful methods of administration. Insulin molecules can be packaged into microscopic particles that can be inhaled deep into the lungs, where they can pass readily into the bloodstream. Another method is an insulin infusion pump, which can be worn in a small

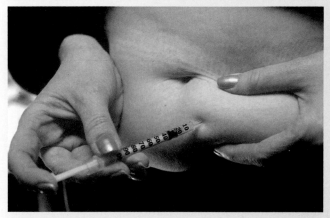

Insulin delivery. With the development of new insulin delivery methods, painful insulin injections may soon be a thing of the past.

case like a cell phone and, periodically throughout the day, pumps insulin through a thin tube into a "port" in a blood vessel. What's more, under development is an artificial pancreas, a small device that senses blood sugar and releases exactly the right amount of insulin to control blood glucose. Soon, it seems, insulin injections will be a thing of the past for people with diabetes.

Enzymes Promote Chemical Reactions

Life is maintained by an infinite number of chemical reactions, the sum of which is an organism's **metabolism.** Metabolic chemical reactions are usually described using symbols representing the chemical composition of molecules, which are linked by an arrow that indicates the way in which the reaction proceeds. For example, the following formula indicates how water (H_2O) and carbon dioxide (CO_2) combine to form *carbonic acid* (H_2CO_3):

$$CO_2 + H_2O \rightarrow H_2CO_3$$

Enzymes are specialized proteins that facilitate chemical reactions. Enzymes can be matchmakers, helping two molecules combine into one, or cleavers, chopping large molecules into smaller ones. Others are more subtle, adding an atom here or subtracting one there. Much like a kitchen knife is not changed after it has cut many carrots, an enzyme is not affected by the chemical reaction it facilitates and can enable many different reactions. Without the appropriate enzyme, the chemical reactions necessary for life would be so slow that life could not exist. The reaction above, for instance, is enabled by the enzyme *carbonic anhydrase*. The suffix *-ase* identifies most enzymes.

Recall that functional groups are particular atomic combinations that tend to stay together. Each functional group reacts in a characteristic way, usually with a functional group of a different kind on another molecule, regardless of the other parts of the molecules. Amino groups, for instance, usually react with carboxyl groups—a reaction important in the formation of proteins (see below).

Chemical Reactions Create or Break Chemical Bonds

Chemical bonds store energy, and it is not far-fetched to think of them as tiny sticks of dynamite. Just as dynamite is a store of energy capable of releasing an explosion of energy, foods are stores of energy that is released as their chemical bonds are broken in metabolic reactions. (Recall the box Basic Form, Basic Function: *Energy and Life* on page 35–36.)

For example, apples are sweet because the apple tree synthesizes *fructose,* or fruit sugar. The synthesis of fructose requires energy from the sun (as rays of light) and atoms of carbon, oxygen, and hydrogen (Fig. 2.10A). It gets these atoms in the form of molecules found in the air, water, and soil. For instance, water (H_2O) is a source of hydrogen and oxygen atoms and CO_2 a source of

oxygen and carbon. Fructose thus stores the sun's energy in chemical form.

When we consume an apple, fructose enters our body cells. Enzymes break the chemical bonds joining the atoms, releasing the bond energy to power cellular activity. The sugar's atoms are reorganized into smaller molecules of water and carbon dioxide (Fig. 2.10B). In this way, the energy from sunlight captured and stored in sugar is released to power animal life.

According to whether they make or break chemical bonds, metabolic reactions can be classified in one of three ways:

- *Synthesis* reactions (Fig. 2.10A) are *anabolic* reactions that use chemical bonds to combine atoms or small molecules to produce new, larger molecules. Synthesis *consumes energy and stores it* and is essential in building cell and body parts. For example, growing children depend on synthesis reactions to build new bone and muscle, and these reactions are powered by energy obtained from food. Synthesis reactions can be characterized by the formula:

$$Energy + A + B \rightarrow AB$$

- *Decomposition* reactions (Fig. 2.10B) are *catabolic* reactions that break molecules into smaller parts. Decomposition releases stored energy and is essential in the maintenance of life. For example, growing children require energy to build new tissue as well as to run and jump. To obtain this energy, their cells decompose molecules of dietary nutrients into their component parts. Decomposition reactions can be characterized by the formula:

$$CD \rightarrow C + D + energy$$

- *Exchange reactions* are metabolic reactions that involve both synthesis and decomposition, and chemical bonds are both made and broken. Energy production and consumption vary. In their simplest form, exchange reactions can be characterized by the formula:

$$EF + G \rightarrow EG + F$$

> *Remember This!* **Anabolism is synthesis. Catabolism is breakdown.**

Chemical Reactions Can Proceed in Both Directions

Any chemical reaction can proceed in either direction; however, most reactions have a preferred direction that

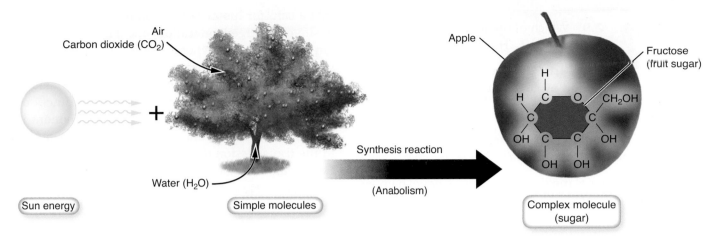

(a) Synthesis reaction (Anabolism)

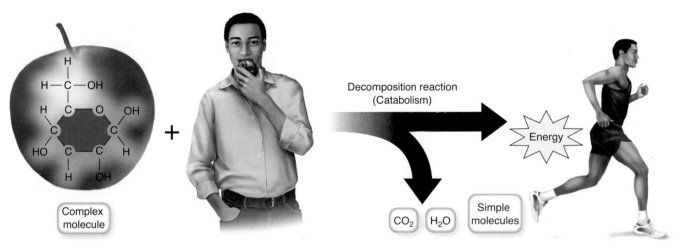

(b) Decomposition reaction (Catabolism)

Figure 2.10. Chemical reactions. A. Anabolic (synthesis) reactions combine atoms (or simple molecules) into larger ones, incorporating energy in the form of chemical bonds. **B.** Catabolic (decomposition) reactions release this energy by breaking the bonds. *Which type of reaction would be used to break a substance into its component parts?*

requires less energy, like pushing a stone down a slope. Reactions proceeding in the opposite direction are unusual because they require a great deal of energy, like rolling a stone uphill. However, some reactions can readily go back and forth, depending on the concentrations of the molecules involved. Such back-and-forth reactions are displayed by formulas that contain arrows pointing in both directions. Consider the reaction between hemoglobin (Hb, a molecule in red blood cells) and oxygen (O_2):

$$Hb + O_2 \underset{\text{tissues}}{\overset{\text{lungs}}{\rightleftharpoons}} HbO_2$$

An increase in the concentration of either hemoglobin (Hb) or oxygen will push the reaction toward the right, increasing the amount of HbO_2 formed. Conversely, a decrease in the concentration of either hemoglobin or oxygen will drive the reaction to the left, releasing oxygen from the hemoglobin. As explained in ➡ Chapter 13, this principle drives the reaction from left to right in the lungs, where oxygen concentration is high, and from right to left in tissues, where oxygen concentration is low because working cells have used up available oxygen.

Case Note

2.11. The concentration of oxygen was abnormally high in Joe's blood. Would he have more HbO_2 than usual, or less? Explain why.

2.9   Which of the following is a compound: H_2, H_2O, O_2?

2.10   True or false: Electrolytes are created by covalent bonds.

2.11   Is NaCl a molecule?

2.12   Which contains more hydrogen: a molecule of water or a molecule of glucose?

2.13   Which type of molecule contains an equal number of protons and electrons: a dipole or a molecular ion?

2.14   The glucose molecule is polar. Is it soluble in water or in fat? Explain why.

2.15   Hydrochloric acid (HCl) forms between the hydrogen cation and the chloride anion. Which type of bond holds HCl together—an ionic bond or a hydrogen bond?

2.16   Some women get injections of estrogen to prevent conception. Estrogen is a nonpolar molecule; should it be dissolved in water or oil?

2.17   Which atomic particles participate in chemical bonds?

The Chemistry of Living Things

To this point we have discussed processes and characteristics common to substances in all types of matter, from rocks to ears. Next we narrow our focus to the characteristics of substances found in the human body. These substances can be divided into two large categories: *inorganic* and *organic*. **Inorganic** substances are found in living and nonliving things, and contain no carbon atoms (an exception is carbon dioxide, CO_2). Examples include water (formed of hydrogen and oxygen) and table salt (formed of sodium and chloride). **Organic** substances (organic matter) contain carbon and originate in living things; that is, they are created by the chemical reactions (metabolism) of living things. Examples of organic molecules include carbohydrates, such as the starch found in wheat, and the proteins found in muscle.

Important Inorganic Substances Are Salts, Water, Acids, and Bases

The molecules and compounds of inorganic matter are smaller and less complex than organic ones. Remember

that a salt, for instance, is a compound that contains a positively charged ion (cation) and a negatively charged ion (anion) bound by ionic bonds (Fig. 2.6). Examples include calcium phosphate ($CaPO_4$), which provides the strong framework of bones and teeth. Many salts perform their most important physiological functions in water after they dissociate into their component ions. Both calcium and sodium ions, for instance, are important in the transmission of nerve impulses ➡ (Chapter 8). Other important inorganic molecules include water, acids, and bases.

Water

Jupiter has its moons, Saturn its rings, and Mars its red hue, reflecting the abundance of iron in its soil and dust. Earth is called "the blue planet," a title that reflects its abundance of water. Life is not possible without *water*: It is the main ingredient in every cell, every tissue but bone, and every fluid in the body. For example, blood is about 90% water. The *physical properties* of water are important:

- It is a *solvent*. Recall that a solvent is a liquid in which a solute can be dissolved to form a solution. With the exception of fats and oils, most of the chemicals in living things are soluble in water, and with good reason: molecules cannot react unless they are in solution with one another. That is, the chemical reactions of life (metabolism) occur between molecules dissolved in water.
- It is a *lubricant*. Water is the main ingredient in fluids that smooth the movement of muscles, tendons, bones, and other tissues against one another.
- It is a *cushion*. Water insulates tissues and cells from the bumps and blows of daily life. For example, water is the main ingredient in the amniotic fluid in which the developing fetus floats in the mother's body; on a smaller scale, it is the main ingredient in the fluid (*interstitial fluid*) that surrounds every cell in the body.
- It is a *heat sink*. A heat sink is a substance that absorbs heat without much change in temperature. Metabolic reactions release large amounts of heat and body water absorbs it without much change; otherwise, body temperature would vary widely. This ability to absorb heat makes it easier for homeostatic mechanisms to maintain body temperature very close to the body's set point of 98.6°F.

Additionally, water is an important *chemical* in metabolic reactions. As discussed below, water participates in chemical reactions in one of two ways: It is either created or consumed.

Dehydration Synthesis

In some chemical reactions, water is *created* with the combination of a hydrogen ion (H$^+$) taken from one molecule and a hydroxyl ion (OH$^-$) from another molecule. This process is termed *dehydration synthesis*, because the two substances are "dehydrated" to create a new water molecule.

Our body cells use repeated cycles of dehydration synthesis to assemble the large molecules they need, such as new proteins. Called **polymers,** these large molecules are essentially long chains of similar or identical molecular subunits called **monomers,** which can be likened to beads strung into a necklace—the bead is the monomer; the necklace is the polymer.

The process of dehydration is analogous to assembling beads into a necklace. Imagine a partially assembled chain that has been capped to prevent the beads from falling off. Each loose bead similarly has a small protective cap. Removing the protective caps and sticking the new bead onto the chain lengthens the necklace.

In a similar fashion, monomers are added to short polymers to create longer polymers (Fig. 2.11A). Following removal of the OH and H "caps" on the polymer and monomer (respectively), the monomer is joined to the end of the short polymer. In doing so, this reaction stores energy in the bond between the newly joined pieces, and the OH$^-$ and H$^+$ combine to form water. Dehydration synthesis is anabolic, because it creates a larger molecule from smaller ones.

Hydrolysis Reactions

Water is *required* and *consumed* for the breakdown of a polymer into shorter polymers or into monomers. Contrary to dehydration synthesis, this process is analogous to removing a bead from a necklace. In this process, a water molecule is inserted into a polymer, breaking its bonds. The water molecule separates into a hydrogen ion (H$^+$) and a hydroxyl ion (OH$^-$), and one part of the broken polymer receives the hydrogen ion (H$^+$) while the other part receives the hydroxyl ion (OH$^-$) (Fig. 2.11B). This reaction liberates energy stored in the chemical bond. This process is called **hydrolysis** because water is broken (*hydro-*, water; and *-lysis*, break) into its constituent parts.

Acids and Bases

An **acid** is a chemical that releases (increases) hydrogen ions (H$^+$) when dissolved in water. Acids are important because they play an important role in metabolism: the hydrogen ions (H$^+$) they release are eager to interact with other chemicals. Acids that ionize completely and release large amounts of H$^+$ ion are strong acids; those that ionize only partially and release fewer H$^+$ ions are weak acids.

For example, *hydrochloric acid* (HCl), the chemical that gives stomach juice its acidity, is a strong acid because it ionizes completely:

$$HCl \rightarrow H^+ + Cl^-$$

In contrast, *carbonic acid* is a weak acid created as cells that produce *carbon dioxide* (CO$_2$), a waste product of energy metabolism. The CO$_2$ mixes with water to create carbonic acid in the following way:

$$CO_2 + H_2O \rightarrow H_2CO_3$$

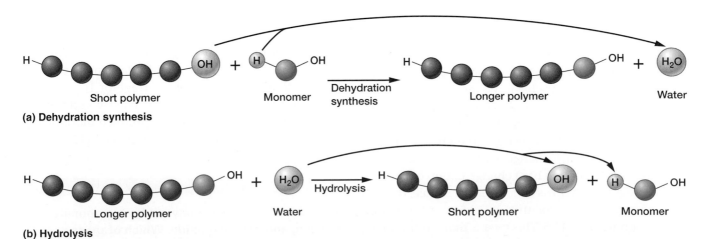

(a) Dehydration synthesis

(b) Hydrolysis

Figure 2.11. Water reactions. Dehydration synthesis generates a water molecule as it assembles polymers from monomers. Hydrolysis breaks a water molecule to remove monomers from a polymer. *During dehydration synthesis, is the monomer added to the hydroxy (OH) end or the hydrogen (H) end of the polymer?*

Carbonic acid (which is used in carbonated beverages to give them their tart taste and fizz) is weak because it does not completely ionize, as indicated by the following formula:

$$H_2CO_3 \rightarrow H^+ + HCO_3^- + H_2CO_3$$

That is, only some of the H_2CO_3 molecules will dissociate, while other molecules will remain intact.

A **base** (*alkali, caustic*) is a chemical that decreases (absorbs) hydrogen ions when dissolved in water. The strongest bases release hydroxyl (OH^-) groups, which combine with H^+ ions to produce water (H_2O). For example, *sodium hydroxide* (NaOH, the main ingredient in household liquid treatments for clogged drains) is a strong base:

$$NaOH \rightarrow Na^+ + OH^-$$

Weak bases, like weaker acids, do not completely dissociate in water and do not attract hydrogen ions with OH^-. The *bicarbonate ion* (HCO_3^-), the most abundant base in blood, is a weak base that attracts hydrogen with the HCO_3^- ion instead of the OH^- ion.

Case Note

2.12. The lemon juice Joe consumed was rich in citric acid. Citric acid does not completely ionize in water. Is it a weak acid or a strong acid?

The pH Scale Is a Measure of Acidity

A convenient way of measuring the strength of an acid or base is the **pH** scale, a series of numbers from 0 to 14, in which 0 is purely acidic (all H^+ ions) and 14 is purely basic (no H^+ ions, all OH^- ions). As the concentration of H^+ ions rises, the concentration of OH^- ions falls, and the solution becomes more acidic. As the concentration of OH^- ions rises, the concentration of H^+ ions falls, and the solution becomes more basic. Water is neither acidic nor basic; it is neutral and has a pH of 7.0. That is, it has an equal concentration of H^+ and OH^- ions.

Small changes in pH represent large changes in H^+ and OH^- concentration. A change of pH by one unit on the scale represents a tenfold ($10\times$) change in the concentration of hydrogen (or hydroxyl) ions. For example, the pH of milk of magnesia (a laxative) is 10.5, and the pH of household ammonia is 11.5. This means that the ammonia has 10 times more hydroxyl ions (and 1/10 as many hydrogen ions) as milk of magnesia. The pH value of other fluids

and everyday items is depicted in Figure 2.12. Notice that human blood is slightly basic, about pH 7.4, which reflects its concentration of bicarbonate (HCO_3^-) ions. As discussed immediately below, bicarbonate is an abundant base whose job is to absorb the acidic hydrogen-ion waste products of the body's energy metabolism.

Case Note

2.13. The pH of Joe's blood was 7.26. Does his blood contain more hydrogen ions or fewer hydrogen ions than normal?

Buffers Protect against Changes in pH

Human cells require a pH near 7.4; variation of just a few tenths of a unit can be life-threatening. To maintain pH at this homeostatic set point, the body utilizes *buffers* that resist changes away from 7.4 in either direction. Specifically, a **buffer** is a chemical that converts a stronger acid into a weaker acid (or a stronger base into a weaker base) that can be safely expelled from the body. Because energy metabolism produces an excess of acid, the body needs a way to neutralize hydrogen ions. One chemical that fills this role is the *bicarbonate* ion (HCO_3^-), which we encountered above. As hydrogen ions (H^+) are produced by metabolic reactions, bicarbonate neutralizes them in the following way:

$$H^+ + HCO_3^- \rightleftharpoons H_2CO_3 \rightleftharpoons H_2O + CO_2$$

Normally, the CO_2 produced by the reaction is blown off by the lungs ➡ (Chapter 13) and the small amount of water created by the reaction is absorbed into body water.

As the arrows in the formula indicate, the chemical reaction can go either way, absorbing or producing H^+ ions as needed to keep blood pH stable near 7.4. Repeated vomiting, for example, can cause the loss of gastric acid, resulting in a shortage of H^+ ions in the body. In such a case the reaction above would run "backward" (right to left) in order to produce more blood H^+ ions to make up for the loss of stomach acid.

Case Note

2.14. Joe's blood was analyzed for bicarbonate, oxygen, and carbon dioxide. Which of these molecules is a chemical buffer?

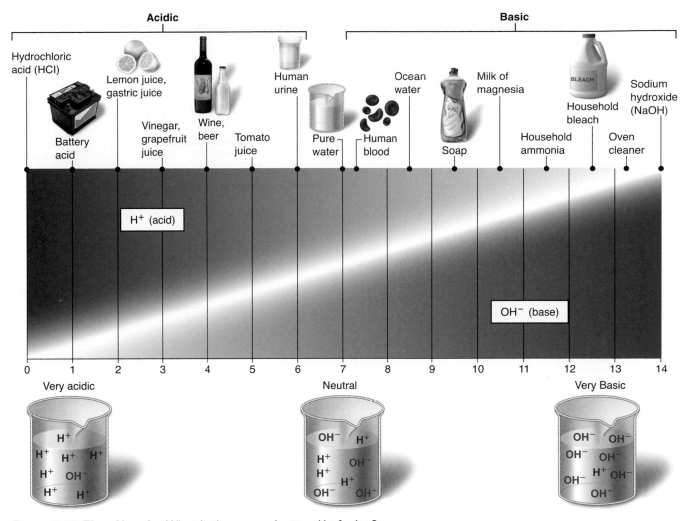

Figure 2.12. The pH scale. *What is the approximate pH of wine?*

Case Discussion

Chemistry in Context: The Case of Joe G.

Recall that Joe was sick because he was forced to drink a large amount of lemon juice, which contains a lot of acid.

The Problem: Excess Acidity. Joe had an acid–base problem (Fig. 2.13). In a short time he ingested far, far more lemon juice, a fairly strong acid, than the body could handle safely. In short, although lemon juice is a necessary ingredient in lemonade, it can be *toxic* (poisonous) in large amounts. Strictly defined, a **toxin** is a substance of plant or animal origin that causes disease when present at low concentration in the body. However, *any* substance, even water, can be injurious (toxic) in unusually large amounts, as was lemon juice in this instance.

Refer to ➡ Chapter 16 for a discussion of water intoxication (poisoning).

Lemon juice is very acidic (pH 2); that is, it has a much higher concentration of hydrogen ions (H^+) than hydroxyl ions (OH^-). Joe's intestine absorbed these H^+ ions into his blood, so that it became more acidic than normal: the pH of his blood fell to 7.26 instead of remaining at 7.4, where it should be. Blood pH this low is not compatible with good health—Joe is breathing unusually fast and his brain function is affected: he is drowsy and nauseated.

The Solution (Part 1): Homeostatic Mechanisms. Joe must get rid of the excess acid he has consumed, which can be accomplished in one of two ways: through the kidneys, which work relatively slowly, or the lungs, which work much more rapidly. However, Joe cannot simply exhale or urinate pure acid. Instead, his body

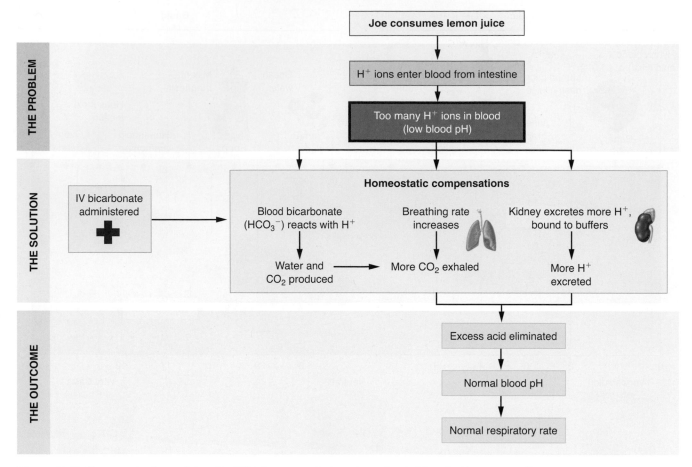

Figure 2.13. Homeostasis and Joe G. *Which organs help return Joe's blood pH to normal?*

must combine the acid with some other substance to convert it into a chemical that can be managed safely. Buffers to the rescue!

Buffers convert the hydrogen ions into compounds that can be safely excreted or exhaled. The excess H^+ ions in Joe's blood combine with blood bicarbonate (HCO_3^-), which is then converted into CO_2, which can be exhaled. The reaction is depicted below:

$$H^+ + HCO_3^- \rightarrow H_2CO_3 \rightarrow H_2O + CO_2$$

This reaction can proceed in either direction, but the high concentration of H^+ in Joe's blood drives the reaction to the right. In parallel, the hydrogen ions cause the brain to increase Joe's breathing rate so that the extra CO_2 can be exhaled rapidly. As Joe's lungs eliminate the excess acid as CO_2, Joe's blood pH rises and his breathing rate slows.

In the meantime, H^+ ions also combine with other buffers in the blood and kidneys ➡ (Chapter 16).

The Solution (Part 2): Medically Assisted Homeostasis. Unfortunately, Joe's symptoms suggest that his blood does not contain enough buffers to neutralize all of the hydrogen ions he has consumed, so the physician gave him extra bicarbonate buffer intravenously to assist in neutralizing and eliminating them. The success of the treatment and his homeostatic compensations is shown by Joe's reduced respiratory rate and normal blood biochemistry the following morning.

Case Notes

2.15. Bicarbonate decreases the concentration of hydrogen ions when it is dissolved in water. Is bicarbonate an acid or a base?

2.16. Look at the chemical reaction above. When Joe was given bicarbonate, did the reaction proceed to the left or to the right?

2.17. When Joe was given bicarbonate, what happened to his blood H⁺ concentration and his blood pH?

2.18. Which ion was responsible for Joe's fast breathing when he arrived at the hospital?

2.19. Would you consider rapid breathing to be a homeostatic mechanism in Joe's case? Explain.

Important Organic Molecules Are Carbohydrates, Lipids, Proteins, and Nucleotides

Organic molecules are created only by living things but can be found in nonliving things. Bread, for example, is made of the organic matter synthesized by wheat. In the main, organic molecules are formed from atoms of carbon (C), hydrogen (H), oxygen (O), and nitrogen (N).

There are four types of organic molecules: *carbohydrates, lipids, proteins,* and *nucleotides.* These molecules are generally quite large and complex and are frequently described as *macromolecules.*

Carbohydrates

Carbohydrates (Fig. 2.14) are organic molecules composed of atoms of carbon, hydrogen, and oxygen arranged in short or long chains or rings. Carbohydrates are an important source of energy.

The simplest carbohydrates contain a single unit consisting of three to seven carbon atoms. Not surprisingly, such sugars are called *simple sugars* (or *monosaccharides,* from Latin *saccharum,* "sugar"). For example, glucose (Fig. 2.14A), the main sugar in blood, most often occurs as a six-sided ring (hexamer). The formula is $C_6H_{12}O_6$. Glucose, like many organic molecules, is often illustrated in a simplified form, as in the left-hand depiction in the figure. Each line represents a covalent bond, with a carbon atom at either end of the bond. Hydrogen atoms are not shown.

Monosaccharides are monomers: more complex carbohydrates are polymers composed of monosaccharides strung together somewhat like the links in a metal chain. The bonds between the links are forged by dehydration synthesis reactions (Fig. 2.11).

The combination of two simple sugars is a *disaccharide* (Fig. 2.14B); for example, the combination of two

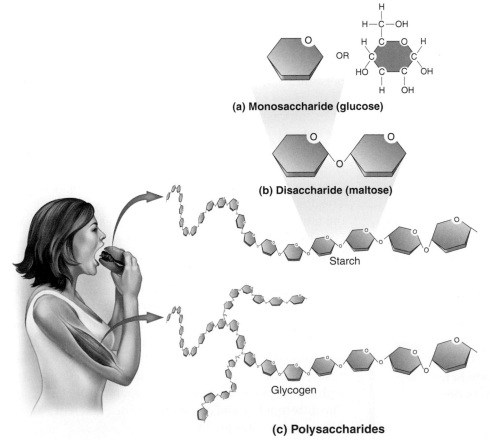

(a) Monosaccharide (glucose)

(b) Disaccharide (maltose)

Starch

Glycogen

(c) Polysaccharides

Figure 2.14. Carbohydrates.
A. Monosaccharides consist of a single sugar molecule. **B.** Two monosaccharides combine to form a disaccharide. **C.** Many monosaccharides combine to form polysaccharides such as glycogen or starch. Glycogen is used to store carbohydrate in muscle and liver tissue. *What is the difference between glycogen and starch?*

glucose molecules produces the disaccharide *maltose,* and glucose plus fructose (another simple sugar, the primary sugar in fruits and honey) yields a disaccharide called *sucrose* (table sugar). The other physiologically relevant mono- and disaccharides are summarized in Figure 14.1 in ➡ Chapter 14.

A polymer composed of more than two monosaccharides is called a *polysaccharide*. These long, branched chains of simple sugars serve as a store of energy in plants and animals. In plants, the energy-storing polysaccharide is called **starch;** it is formed of glucose molecules linked together in a long chain. Starch is abundant in wheat, beans, and potatoes. In animals, the energy-storing polysaccharide is **glycogen;** it is also formed of long chains of glucose but is more highly branched than starch. Glycogen is stored in muscle and liver tissues.

Carbohydrates provide a ready source of energy for all physiological activities. However, disaccharides and polysaccharides are too large to enter cells and cannot be burned for energy until they are broken down into monosaccharides by hydrolysis reactions. Enzymes in the small intestine's lumen and wall break larger carbohydrates into monosaccharides, which are small enough to be absorbed by intestinal epithelial cells. Monosaccharides (especially glucose) are used for energy, in a reaction where each molecule of glucose is combined with six oxygen molecules. This reaction produces carbon dioxide, water, and energy, as shown in the following formula:

$$C_6H_{12}O_6 + 6\ O_2 \rightarrow 6\ CO_2 + 6\ H_2O + energy$$

Excess glucose not required to fuel cell activities is assembled into *glycogen* for storage in liver or muscle, or it is converted into *fat*.

A very small amount of carbohydrate is used to build cell structures. For example, some sugar molecules are used to build nucleotides (discussed below), which in turn are used to construct the DNA of our genes. Other carbohydrate molecules are combined with protein or lipid molecules to make glycoproteins or glycolipids, respectively. These important molecules are discussed later in the text.

Case Note

2.20. The most abundant disaccharide in lemon juice is sucrose. How many sugar units does sucrose contain?

Lipids

Lipids are slick, greasy, nonpolar substances that are hydrophobic. Solid lipids are frequently described as *fats;* liquid lipids as *oils*. They are composed mainly of atoms of carbon and hydrogen, with only a few oxygen atoms. For example, one common lipid has the formula $C_{57}H_{110}O_6$. Unlike carbohydrates, all of which have a similar chemical structure, lipids vary greatly in their *chemical* structure and are alike mainly in their *physical* characteristics: they are oily and not soluble in water. Some lipids important to human physiology are identified in Table 2.2.

Common Lipids Contain Glycerol and Fatty Acids

Despite their dissimilarity, many lipids contain two common building blocks: a molecule called **glycerol** and one or more *fatty acids* (Fig. 2.15A). *Glycerol* is an alcohol (it contains an OH group). A **fatty acid** is a long chain of carbon atoms (and their attached hydrogens) with an acidic molecule (the carboxyl group, COOH) at one end.

If all of the carbon atoms in a fatty acid chain are linked by single bonds, each carbon atom will have two hydrogen atoms attached to it. Such a fatty acid (and the lipids and storage fat made from it) is said to be *saturated* (with hydrogen). These fatty acids are quite straight; thus, any fat containing mostly saturated fatty acids will contain tightly packed, parallel chains of fatty acids and will be rather dense and solid at room temperature. Butter, lard, and the visible fat in meats are examples of saturated fats.

In contrast, if *any* carbon atoms in a fatty acid chain are linked by double bonds, these carbons will have only one hydrogen atom attached instead of two. Such fatty acids are said to be *unsaturated;* that is, they have less than the maximum possible number of hydrogen atoms. The double bond forms a bend in the fatty acid chain, so the fatty acids cannot be packed into neat parallel rows. Unsaturated fats, such as olive and corn oils, are thus less dense and tend to be liquid instead of solid at room temperature. Monounsaturated fatty acids have one double bond in the chain, and polyunsaturated fatty acids have more than one double bond.

The distinction between saturated and unsaturated fats also has important health implications: diets high in saturated fats promote harmful deposits of fat in blood vessels (atherosclerosis), a condition associated with high blood pressure, heart disease, and other diseases. The different types of dietary fat are discussed in greater detail in ➡ Chapter 14.

Table 2.2   Examples of Common Lipids

Name	Structure	Location	Function(s)
Triglycerides	Glycerol + 3 fatty acids	Primary lipid in fat tissue	Store energy; cushion bones; keep organs warm and protected
Phospholipids	Glycerol + 2 fatty acids + phosphate	Cell membrane	Make up cell membrane
Sphingolipids	Sphingosine + fatty acid + phosphate + amino acid	Cell membrane	Cell recognition
Cholesterol	Steroid	Cell membrane, liver, blood	Contributes to cell membrane; precursor to steroid hormones; component of bile
Vitamin D	Steroid	Formed in skin	Necessary for bone health
Sex hormones	Steroid	Formed in testes, ovaries, adrenal glands	Necessary for normal sexual and reproductive function
Corticosteroids	Steroid	Formed in adrenal glands	Regulate aspects of growth and metabolism
Aldosterone	Steroid	Formed in adrenal glands	Regulates salt and water balance
Prostaglandins	Derived from arachidonic acid (a fatty acid)	Formed everywhere	Many diverse functions; involved in pain and inflammation
Leukotrienes	Derived from arachidonic acid (a fatty acid)	White blood cells	Assist in inflammatory response
Thromboxanes	Derived from arachidonic acid (a fatty acid)	Produced by platelets	Involved in blood clotting
Fat-soluble vitamins (A, E, K)	Derived from isoprene (a small molecule containing C and H)	Obtained in the diet	Necessary for vision (A), blood clotting (K); vitamin E is an antioxidant

Case Note

2.21. The most abundant fatty acid in lemon juice is called linoleic acid. It contains two double bonds. Is this fatty acid saturated or unsaturated?

Triglycerides Are Common Lipids in Foods and in the Body

Glycerol and fatty acids can be combined by dehydration synthesis (Fig. 2.11) to form common lipids. Depending on the number of fatty acids attached to the glycerol molecule, this reaction will result in a monoglyceride (one fatty acid), diglyceride (two fatty acids), or **triglyceride** (three fatty acids) (Fig. 2.15B). A triglyceride is a type of lipid that is stored efficiently by the body to be burned later for energy. Triglycerides are the most space-efficient energy-storage molecules because they pack the most potential energy into the smallest space. Triglycerides contain nine calories per gram; carbohydrates contain only four. A **fat** is a lipid that is solid at room temperature. Triglycerides are commonly described as fats, because they are stored in the body in their solid form. Fats do not include phospholipids or steroids, two other lipids discussed next.

Case Note

2.22. The fatty acid limonine is responsible for the odor of lemons and lemon juice. Would this compound be hydrophilic or hydrophobic?

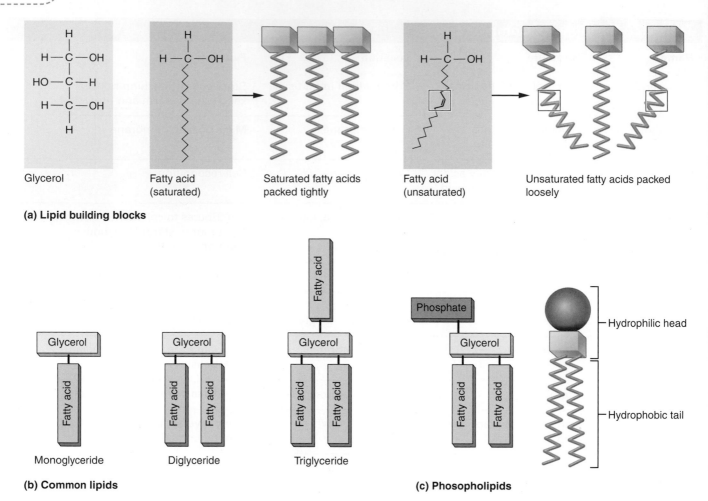

(a) Lipid building blocks

Glycerol   Fatty acid (saturated)   Saturated fatty acids packed tightly   Fatty acid (unsaturated)   Unsaturated fatty acids packed loosely

(b) Common lipids

Monoglyceride   Diglyceride   Triglyceride

(c) Phospholipids

Phosphate — Hydrophilic head — Hydrophobic tail

Figure 2.15. Common lipids. A. Common building blocks for lipids include glycerol (light blue squares) and fatty acids (orange lines). **B.** These components can be built into mono-, di-, or triglycerides. **C.** They can also combine with phosphate (blue circle) to form phospholipids. *How many fatty acids are found in triglycerides?*

Phospholipids Are Common Lipids That Contain Phosphate

Phospholipids (Fig. 2.15C) are similar to triglycerides. Each is formed of a molecule of glycerol to which there are attached three other molecules: two fatty acid molecules (not three, as in triglycerides) and a phosphate group. Phospholipids have a unique quality: one end of the molecule (the phosphate end) is polarized and is, therefore, hydrophilic (soluble in water). The other end, the end containing the fatty acids, is not polarized and is not soluble in water (hydrophobic), but it is soluble in lipid.

It is true that oil and water don't mix, but where oil and water meet, the hydrophobic end of a phospholipid molecule can mix with oil and the hydrophilic end can mix with water. This quality is very important in everyday life and in the body. Dish detergent, for example, consists of molecules similar to phospholipids. Since one end of the soap molecule is soluble in water and the other in grease, the addition of soap molecules to water allows us to wash a sink full of greasy dishes. In the body, a phospholipid-containing chemical called *bile* acts in much the same way to help break down the fats in our foods.

Steroids Are Ring-Shaped Lipids

Steroids (Fig. 2.16) are similar to triglycerides and phospholipids in that they, too, are formed mainly of carbon and hydrogen and are slick, oily substances. However, steroids are structurally very different: they consist of four interlocking rings of carbon and hydrogen (the steroid nucleus) to which are attached short chains of other molecules. It is these attached short chains that govern the specific function of each steroid.

Cholesterol is one of the most important steroids in the human body. It is synthesized by the liver and need not

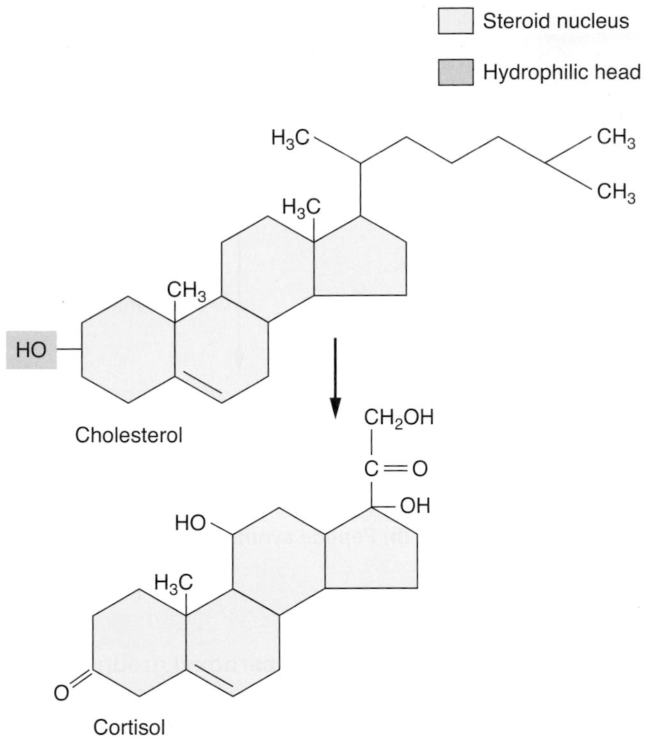

Figure 2.16. **Steroids are ring-shaped lipids.** Cholesterol contains the steroid nucleus of four carbon rings and a hydrophilic head group. Cholesterol can be converted into other steroids, such as cortisol, without modifying this steroid nucleus. *Which atom replaces the hydroxyl group as cholesterol is converted into cortisol?*

be consumed in the diet. However, it is present in foods of animal origin such as meat and cheese. Cholesterol is the basic building block for many other steroids called *hormones*. As discussed in ➡ Chapter 4, hormones act as signaling molecules that travel from one place to another and act as messengers in the regulation of body processes. For example, the steroid hormone *cortisol* helps regulate blood glucose levels.

Cholesterol, like phospholipids, contains hydrophilic and hydrophobic components. Cholesterol and phospholipids are the primary building blocks of the cell membrane, which is discussed in detail in ➡ (Chapter 3).

Prostaglandins Are Modified Fatty Acid Signaling Molecules

Prostaglandins are hydrophobic signaling molecules synthesized from fatty acids (Fig. 2.17). Their effects are both powerful and diverse. For instance, they cause blood vessels to dilate, are important in the functioning of the male and female genitalia, sensitize nerves to pain, and assist in regulating inflammation.

Nonsteroidal anti-inflammatory drugs (NSAIDs) such as ibuprofen treat pain by blocking the nerve-sensitizing effects of prostaglandins.

Proteins

Proteins comprise over half of the organic matter in the body. For example, skeletal muscle is mainly protein. Like carbohydrates and lipids, proteins are mainly composed of atoms of carbon, hydrogen, and oxygen; however, proteins also contain atoms of *nitrogen* (N) and sometimes *sulfur* (S). Proteins are much more diverse chemically and functionally than lipids and carbohydrates. They can be constructed of any combination of about 20 small building blocks called **amino acids** (Fig. 2.18A).

Amino acids get their name from the fact that they are always formed of an *amine* (or amino) group (HNH, or NH_2) and an acidic *carboxyl* group (COOH). These two functional groups are joined by a carbon atom. Amino acids differ from one another according to the type of molecule (called the *side chain* or *R group*) attached to the central carbon atom. Side chains can be hydrophilic or hydrophobic, they can be large or small, and they can be positively or negatively charged.

To form proteins, amino acids are joined into chains by dehydration synthesis (Fig. 2.18B). The bond between the carbon and nitrogen of adjacent amino acids is a *peptide bond*. A **peptide** is two or more amino acids linked by peptide bonds. The side chains project out from the amino acid chain. Chains shorter than about 100 amino acid molecules are called *polypeptides* (Fig. 2.18C). Chains of amino acids longer than about 100 molecules are called **proteins.**

Of all organic molecules, proteins have the most diverse roles because they can be made with so many different combinations of amino acids. Think of the 20 amino acids as an alphabet for building proteins. One sequence could be SOAP and another SOUP; a small change, just one letter, but a huge difference in function and taste. By varying the "spelling" of amino acids in

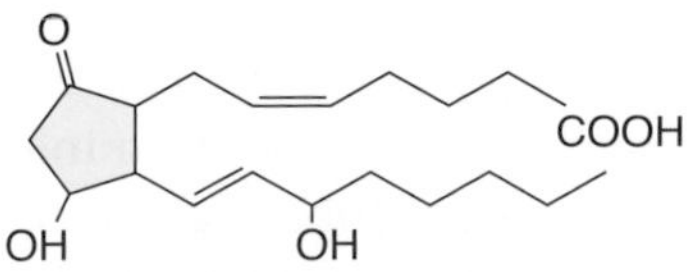

Prostaglandin (PGE₂)

Figure 2.17. **Prostaglandins.** Prostaglandin E2 (PGE2) is derived from a fatty acid. *Which functional group is COOH?*

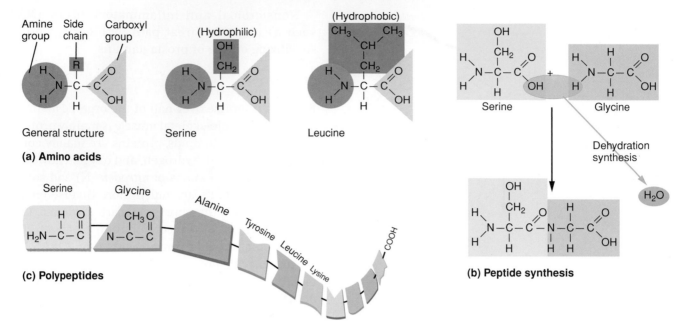

Figure 2.18. Proteins. A. Amino acids contain three functional groups: an amine group, a carboxyl group, and a variable side chain. **B.** Amino acids are monomers joined together into polymers by peptide bonds. **C.** A short polymer of amino acids is called a polypeptide; a longer polymer is called a protein. *How many amine groups are present in the assembled peptide in part B?*

proteins, the body can construct thousands of different proteins using the same amino acid ingredients.

The function of our body proteins is determined in part by their shape. The precise sequence of amino acids in a protein determines how it folds and twists into its characteristic three-dimensional shape, because the side chain of each amino acid will form certain types of bonds only with the side chains of certain other amino acids. The sulfur found in the amino acid *cysteine* (Fig. 2.18B), for instance, forms covalent bonds with other cysteine side chains. Hydrogen and ionic bonds also form between side chains and help fold proteins. This highly specific shaping gives proteins the capacity to perform a wide variety of highly specific tasks.

We can distinguish two main types of proteins according to their shape, structure, and general function. **Fibrous proteins (structural proteins)** are formed of long strands. They are tough, fixed in place, and used to build body structures. The most abundant fibrous protein in the body is *collagen*, which is found in skin, tendons, ligaments, and bones and is the main ingredient in scar tissue. Three interlocking fibrous proteins form collagen (Fig. 2.19A). On the other hand, **globular (functional) proteins** are more rounded molecules that perform specific tasks. **Hemoglobin,** for instance, is a complex globular protein that transports oxygen in the bloodstream (Fig. 2.19B). It contains four separate protein strands, each coiled together into a rounded shape.

Again, a critical aspect of all proteins is that their *form* is critical to their *function*. Just as the shape of a key must be exactly correct if it is to open a lock, a minor alteration in the shape of a protein can disable it from functioning. For instance, a single amino acid change in the hemoglobin gene can significantly alter the shape of the hemoglobin molecule. This abnormally shaped molecule results in the deformed (crescent- or sickle-shaped) red blood cells that are the result of a genetic disease known as sickle cell anemia.

Case Note

2.23. Notice in the case study figure that 1 cup of lemon juice contains 1 g of protein. What are the building blocks of proteins?

Nucleotides

Nucleotides, the fourth class of organic molecule, play important roles in cell processes such as information storage and the transfer of energy. Each nucleotide contains three building blocks: a simple sugar, a base (a circular molecule containing carbon and nitrogen), and one or more phosphate groups (Fig. 2.20A). The sugar is one of two related sugars, ribose or deoxyribose, that differ only by one hydroxyl group. Five bases

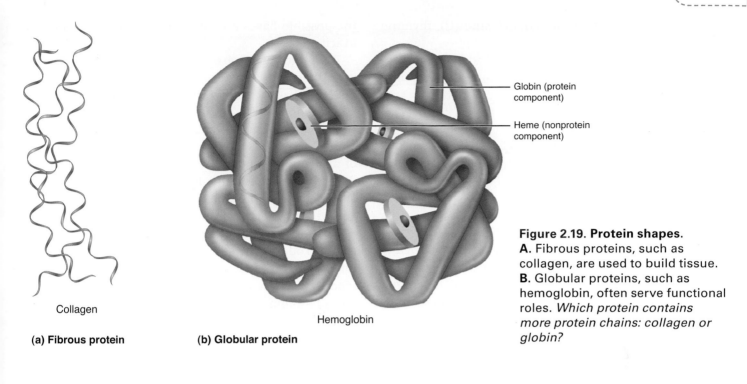

Figure 2.19. Protein shapes.
A. Fibrous proteins, such as collagen, are used to build tissue. **B.** Globular proteins, such as hemoglobin, often serve functional roles. *Which protein contains more protein chains: collagen or globin?*

Collagen

(a) Fibrous protein

Hemoglobin

(b) Globular protein

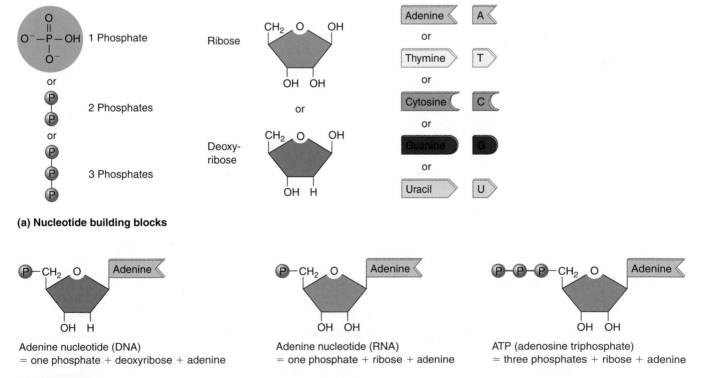

(a) Nucleotide building blocks

Adenine nucleotide (DNA)
= one phosphate + deoxyribose + adenine

Adenine nucleotide (RNA)
= one phosphate + ribose + adenine

ATP (adenosine triphosphate)
= three phosphates + ribose + adenine

(b) Nucleotides

Figure 2.20. Nucleotide structure. A. Nucleotides are built from a phosphate, a sugar, and a base. Two different depictions of each base are shown. **B.** These parts can be assembled into DNA nucleotides, RNA nucleotides, or adenosine triphosphate. *Which nucleotide contains more phosphate groups: the adenine nucleotide or ATP?*

exist: adenine (A), guanine (G), cytosine (C), thymine (T), or uracil (U).

Two types of nucleotides essential to human functioning are nucleic acids and adenosine triphosphate (Fig. 2.20B).

> *Remember This!* **Four classes of organic molecules are: carbohydrates, lipids, proteins, and nucleotides.**

Nucleic Acids Include DNA and RNA

Nucleotides are used to build **nucleic acids,** specifically deoxyribonucleic acid (**DNA**) and ribonucleic acid (**RNA**) (Fig. 2.19B). Notice that nucleic acids contain the three essential building blocks of all nucleotides: either deoxyribose (in deoxyribonucleic acid, DNA) or ribose (in ribonucleic acid, RNA), plus one of the five bases and a single phosphate group. In DNA, the possible bases are adenine (A), guanine (G), thymine (T), and cytosine (C). These are, so to speak, the alphabet of life. In RNA,

the possible bases are adenine, guanine, cytosine, and uracil (U).

DNA nucleotides are bound together in long strands, with alternating sugar and phosphate groups forming the DNA backbone and the bases projecting from the strand (Fig. 2.21A). These bases form hydrogen bonds with bases on another strand of DNA to form a "spiral ladder" shape called a double helix. DNA's two strands of nucleotides contain the blueprints to make all of the different body proteins. RNA can act as an agent for DNA, carrying DNA messages, or it can act on its own accord, performing many tasks in the cell.

The sequence of nucleotide molecules in DNA is the genetic code. In the same way that the sequence of letters in this sentence spells out the meaning of the sentence and the sequence of amino acids determines the type of protein, the sequence of nucleotides in DNA and RNA spell out the instructions that our cells follow to build proteins. The sequence is all-important—CTA codes for one amino acid, ACT for a completely different one. Long sequences of these three-letter bases spell out how amino acids are to be strung together to form

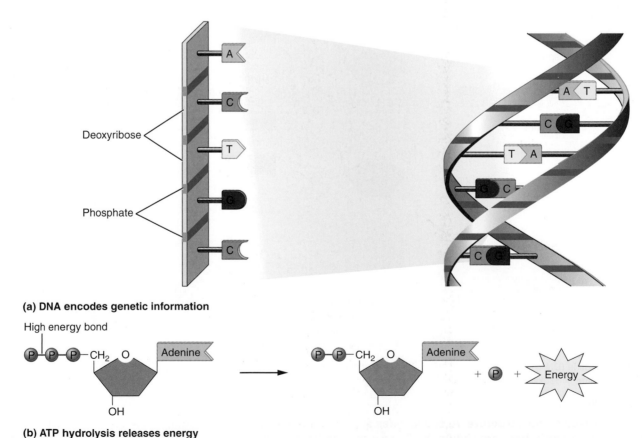

(a) DNA encodes genetic information

(b) ATP hydrolysis releases energy

Figure 2.21. Nucleotide function. A. Nucleotides found in DNA or RNA can encode information. **B.** Nucleotides such as ATP store energy in high-energy bonds. *Name the base found in ATP.*

proteins, and proteins, in turn, regulate every aspect of anatomy and physiology. Again, in DNA as in every other aspect of anatomy and physiology, form and function are inseparable.

Adenosine Triphosphate Is Involved in Energy Transfer

The nucleotide **adenosine triphosphate (ATP)** is involved in energy transfer (Fig. 2.20B). ATP is formed of the base *adenine* and the sugar *ribose*, to which are attached three phosphate groups. ATP fulfills a very different purpose than the nucleic acids: ATP is the energy currency of the body (Fig. 2.21B).

Anabolic chemical reactions and many transport mechanisms require energy, and this energy must be in the form of ATP. That is, the energy found in the chemical bonds of nutrients must first be converted into a specific bond in ATP. Most cells use glucose as a nutrient, but some prefer fatty acids or even amino acids. The energy in ATP is contained in a high-energy bonds between two phosphate (PO_4) groups. This bond takes a large amount of energy to create, so it holds a large amount of energy, which is liberated whenever the bond is broken.

2.18   If a substance has the chemical formula $C_{18}H_{36}O_2$, is it organic or inorganic?

2.19   Is dehydration synthesis an anabolic or a catabolic reaction?

2.20   How would you classify an amino acid: as a monomer or as a polymer?

2.21   True or false: The assembly of a disaccharide from two monosaccharides generates a water molecule.

2.22   If you dissolve an acidic substance in a solution, will the pH increase or decrease?

2.23   Name the polysaccharide used to store carbohydrate in human muscles.

2.24   List the two most common building blocks of lipids.

2.25   Name a lipid that contains two fatty acid chains.

2.26   Name the two functional groups that are identical in every amino acid.

2.27   Name the three building blocks used to construct nucleotides.

Word Parts

Latin/Greek Roots	English Equivalents	Examples
Hydr-/o-	Water	Hydrolysis: breakdown of water
Lip-/o-	Fat	Lipophilic: lipid-loving
-phobia	Fear	Lipophobic: fat-fearing
Phil-	Love	Hydrophilic: water-loving
-lysis	Breakdown	Lipolysis: breakdown of fat
Sacchar-/o-	Sugar	Monosaccharide: one sugar molecule
Mono-	One	Monosaccharide: one sugar molecule
Di-	Two	Disaccharide: sugar molecule containing two saccharide monomers
Tri-	Three	Triglyceride: fat containing three fatty acids and glycerol
Poly-	Many	Polypeptide: protein containing many amino acid subunits
Glyc-/o-	Sugar, glucose	Glycogen: storage form of glucose

Chapter Challenge

CHAPTER RECALL

1. A positively charged particle is called a(n)
   a. electron.
   b. proton.
   c. neutron.
   d. isotope.

2. Two radium atoms contain the same number of electrons and protons, but one atom contains two additional neutrons. These two atoms are
   a. isotopes.
   b. ions.
   c. polar.
   d. salts.

3. The hydrogen ion
   a. has two protons.
   b. is negatively charged.
   c. does not have any electrons.
   d. is an isotope.

4. Covalent bonds form when
   a. one atom donates an electron to another atom.
   b. a cation binds with an anion.
   c. polar molecules interact with water.
   d. two atoms share electrons.

5. Hydrophilic substances are
   a. frequently dipoles.
   b. often nonpolar.
   c. easily dissolved in fats.
   d. always ions.

6. Salts are formed by
   a. two atoms participating in a covalent bond.
   b. two anions.
   c. an anion and a cation.
   d. two cations.

7. Which of the following substances contains a sugar molecule?
   a. Triglyceride
   b. Iron
   c. ATP
   d. Carbon

8. A hydrolysis reaction
   a. results in the formation of a new water molecule.
   b. assembles monomers into polymers.
   c. liberates energy.
   d. none of the above.

9. The pH of a solution containing more hydrogen ions than hydroxyl ions might be
   a. 2.
   b. 7.
   c. 9.
   d. 13.

For each of the following macromolecules, indicate the building blocks that can be used to build them. Each letter may be used more than once.

Macromolecule	Building Blocks
10. Protein	a. Glucose
11. Triglyceride	b. Glycerol
12. Glycogen	c. Phosphate
13. Phospholipid	d. Fatty acid
14. ATP	e. Amino acid
15. Starch	f. Adenine (base)
16. Ribonucleic acid	g. Ribose

CONCEPTUAL UNDERSTANDING

Use the periodic table in Figure 2.2 to answer questions 17 through 20.

17. The atomic mass of phosphorus (P) is 31. How many neutrons are found in its nucleus?

18. A common ion in the body is K^+. How many electrons orbit the nucleus of this ion?

19. How many electrons will be found in the outer shell of lithium (Li)?

20. Name two macromolecules that contain a phosphate group and describe the purpose of the phosphate group in each macromolecule.

21. Discuss how function is determined by form at the molecular level, using fibrous and functional proteins as examples.

22. Compare and contrast the form and function of phospholipids and triglycerides.

APPLICATION

23. Consider the following reaction:

$$H^+ + HCO_3^- \rightleftharpoons H_2CO_3 \rightleftharpoons H_2O + CO_2$$

It is possible to reduce the concentration of CO_2 in the blood by hyperventilating (breathing rapidly). Will this action increase or decrease the pH of blood?

24. Draw the structure of a peptide formed with the sequence leucine–serine.

25. Calcium phosphate ($CaPO_4$) is formed by an ionic bond between a calcium cation and a phosphate (PO_4) anion. Phosphate is formed by covalent bonds between one phosphate atom and four oxygen atoms. State which of the following terms describe phosphate, defining each term in your answer: ion, molecule, compound.

You can find the answers to these questions on the student Web site at
http://thepoint.lww.com/McConnellandHull

3

Cells and Tissues

Major Themes

- The cell is the basic unit of every living thing.

- Cells vary in size, shape, and function.

- Cell activities are carried out by minute structures called organelles, the largest of which is the nucleus.

- The cell's interior is separated from the extracellular fluid by the cell membrane, which acts as a gate determining which substances can enter or leave the cell.

- The nucleus contains deoxyribonucleic acid (DNA), which codes for every possible protein the cell can synthesize.

- Proteins control all aspects of cell function.

- Cells reproduce by dividing into two.

Chapter Objectives

Case Study: "She's such a sickly child."

As you read through the following case study, assemble a list of the terms and concepts you must learn in order to understand Julia's condition.

Clinical History: Julia G., a 6-month-old girl, was brought to the emergency room by her parents because of recurrent episodes of wheezing and "chest colds."

"They come one after another and just drain the energy right out of her," her mother added. "We have three other children, but poor Julia doesn't have as much energy as the others. She's such a sickly child."

Physical Examination and Other Data: Julia weighed 14 lb, 10 oz, and was 25 in. long (normal, about 17 lb and 29 in.). Her body temperature was 101°F (normal, 98.6); heart rate 110 beats per minute (normal, 90 to 130); and respiratory rate 55 per minute (normal, 25 to 40). Listening to her lungs, the physician heard wheezes. Her skin was crusted with salt crystals. A chest x-ray revealed pneumonia.

Clinical Course: Cystic fibrosis was the preliminary diagnosis, pending completion of sweat chloride testing. Antibiotics and mucous-loosening inhalants were prescribed, and her parents were taught how to help clear her lungs of mucus by placing her in a head-down position and gently thumping on the back of her chest to shake loose the retained mucus.

Julia responded favorably—by late the next day her fever, rapid respirations, and heart rate returned to normal. On a follow-up chest x-ray a week later, the pneumonia was gone, but mild wheezing continued. Lab reports later showed sweat chloride to be markedly increased. The family was referred to a special cystic fibrosis clinic for follow-up care and genetic counseling.

After you have read Chapter 3, you will be able to explain Julia's symptoms and signs, understand the cellular basis of cystic fibrosis, and appreciate potential treatments.

Need to Know

It is important to understand the terms and concepts listed below before tackling the new information in this chapter.

■ Gradients, pressure ◀ (Chapter 1)

■ Enzymes, chemical reactions, proteins, carbohydrates, lipids, nucleotides, hydrophobic and hydrophilic substances, solutes and solvents ◀ (Chapter 2)

A **cell** is the basic structural and functional unit of an organism. In this chapter, we first discuss the many ways cells share certain *forms*—their shape, construction, and the minute structures they contain. We then discuss the many similar ways in which cells perform their various *functions*. The chapter closes with a discussion of the various types of **tissues**—groups of cells that work together to perform a discrete function.

As Robert Hooke discovered in 1665, living things are composed of cells (see the Web site for the whole story, described in the feature box *Animalcules and Cells*). Hooke, studying cork and seeing all cells alike, thought human cells would be also. Hooke had no way of knowing that his simple observation was the first step in a scientific journey that continues to this day. Nowadays we understand that some microscopic organisms, such as amoebae, consist of a single cell; but most living things are formed of many cells. Large animals, humans and whales alike, contain many trillions (10^{12}) of individual cells; in humans, there are roughly 200 different types of cells, including skin cells, heart cells, and brain cells.

Just as motorcycles differ from family vans and heavy trucks, according to their function, the forms of cells vary enormously, reflecting their varied functions. Spinal cord neurons (nerve cells), for instance, have a thin extension up to 3 ft long that carries messages from the spinal cord to the rest of the body; and the cells lining the small intestine have fingerlike projections to maximize the absorption of nutrients (Fig. 3.1).

Nevertheless, all cells are alike in some aspects of form and function. Despite the great variation among motor vehicles, they share some common features: all have wheels, axles, a frame, a steering mechanism, and a motor. Similarly, in spite of their great number and variety, cells share certain characteristics. Most are roughly the same size (microscopic), they require nutrients and waste disposal, and they reproduce by dividing into two (one cell becomes two). We begin with the common component of all cells—a membrane that separates the cell from its environment.

"... I could exceedingly plainly perceive it to be all perforated and porous, much like a Honey-comb, but that the pores of it were not regular ... these pores, or cells ... were indeed the first microscopical pores I ever saw, and perhaps, that were ever seen, for I had not met with any Writer or Person, that had made any mention of them before this...."

Robert Hooke (1635–1703), English microscopist, writing in 1665 in *Micrographia*, a collection of scientific observations. Hooke was reporting his observations of thin slices of cork (the bark of a cork tree) and was the first to perceive that living things are composed of small units, or cells, as he named them, because the hollow spaces occupied by cork cells reminded him of monks' cells in a monastery.

The Cell Membrane

The **cell membrane,** illustrated in Figure 3.2, forms the boundary of the cell and keeps its contents separated from the fluid outside of it, the *extracellular fluid.*

Phospholipids Form the Layers of the Cell Membrane

The membrane is a thin, pliable film composed of two layers of phospholipid molecules. As you learned in ◀ Chapter 2, the phosphate "head" of a phospholipid molecule is water-soluble (hydrophilic, or water-loving); the double-stranded lipid "tail" is not—it is hydrophobic and repels water.

In the cell membrane, phospholipid molecules are arranged in two layers oriented in opposite directions, like two rows of dolls laid foot to foot. The hydrophilic phosphate heads at either edge of the membrane face the watery cytoplasm internally and the watery extracellular fluid externally. At the same time, the oily hydrophobic tails (dolls' feet) form an internal oily layer that keeps the intracellular and extracellular fluids from mixing with each other. This design ensures that water-soluble substances, the most common type, cannot get into or out of the cell except through specific controlled gateways.

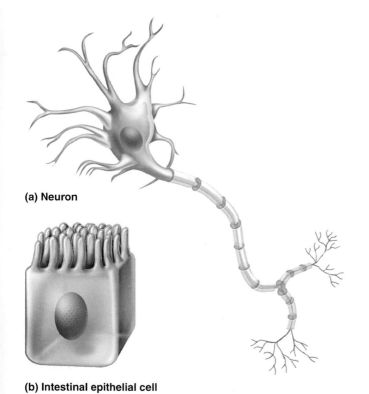

(a) Neuron

(b) Intestinal epithelial cell

Figure 3.1. Cell form and function. Form and function are linked even at the cellular level. **A.** Neurons have long, slender extensions that can carry signals over large distances. **B.** The squat shape and numerous short, fingerlike projections of a cell lining the small intestine increase its surface area and maximize absorption. *Which cell would be able to convey a signal over long distances?*

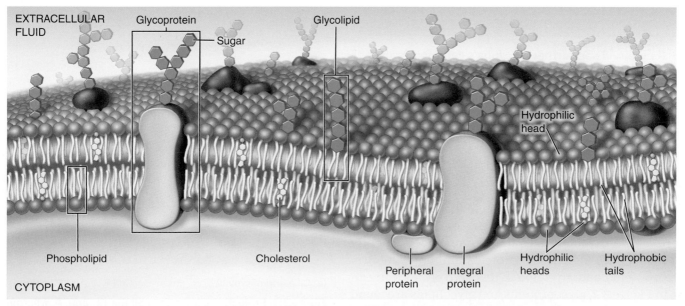

Figure 3.2. Cell membrane. The cell membrane separates the cytoplasm from the extracellular fluid. *Are sugar molecules found on the extracellular side or the intracellular side of the cell membrane?*

Cholesterol Adds Stability to the Cell Membrane

The cell membrane also contains a substantial amount of cholesterol, another lipid (◀ Chapter 2). Cholesterol molecules align themselves parallel to the phospholipids, with their hydrophilic heads next to the phosphate heads and their hydrophilic tails near the fatty acid tails. Cholesterol stabilizes the membrane and makes it more fluid.

> *Remember This!* **The lipid portion of the phospholipid bilayer prevents hydrophilic substances from freely crossing the cell membrane.**

Membrane Proteins Serve Structural and Functional Roles

The cell membrane also contains two types of proteins: integral and peripheral proteins.

Integral proteins extend through the full thickness of the membrane. Some enable water and water-soluble substances to cross the cell membrane (see Figs. 3.19 and 3.20 on pages 90 and 91). Others serve as receptors, binding specific substances that alter cell activity or that are required for cell use. For example, integral membrane receptors of liver cells latch onto passing cholesterol so that the liver cell can metabolize it.

Peripheral proteins are attached to the inner or outer membrane surface, often to integral proteins. They serve as anchors, attaching the cell to other cells or extracellular structures, or they may be involved in cell signaling.

> **Case Note**
>
> **3.1.** As we will see further on, Julia's cell membranes lack a protein that extends through the membrane, contacting both the extracellular fluid and the cytosol. Is this a peripheral or integral protein?

Carbohydrates Add "Fingerprint" Identity to Cells

Since lipids and proteins are found in the cell membrane, it is not surprising that carbohydrates are also found there. These carbohydrates don't occur independently but instead are attached to proteins and lipids to form glycoproteins and glycolipids, respectively. As you can see in Figure 3.2, glycoproteins and glycolipids are found exclusively in the outer layer of the cell membrane. The carbohydrate part of these molecules projects out of the membrane, forming a sugary coating called the *glycocalyx*

(Greek *kalux* = "husk"; thus a sugary, protective envelope). The glycocalyx serves two important functions:

- Because it is slippery, it protects the cell from mechanical damage.
- Because it differs from one type of cell to another and from one person to another, it acts as a molecular "fingerprint" to distinguish, say, liver cells from kidney cells or this person from that one. This last characteristic also enables the body to recognize what is "self" and what is "nonself," a matter of supreme importance: the entire task of the immune system rests on the ability of the body to recognize the difference between self and nonself ➡ (Chapter 12). For example, the immune system fights infectious agents because it recognizes that they are nonself. Except for identical twins, the same is true for organ transplants—the recipient must take drugs to dull the body's immune reaction.

> *Remember This!* **Projections from membrane glycoproteins and glycolipids form the glycocalyx, which protects the cell from damage and serves in identification.**

Cell Membranes May Have Surface Projections

The cell membrane of some cells may be formed into specialized projections (Fig. 3.3). **Microvilli** are short,

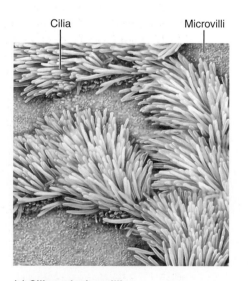

Cilia   Microvilli   Flagellum

(a) Cilia and microvilli   **(b) Flagellum**

Figure 3.3. Projections from the cell membrane. A. Microvilli are much smaller than cilia. **B.** Human sperm cells have a single cilium-like projection called the flagellum. *Which projection could create currents in the surrounding fluid?*

hairlike projections that facilitate absorption by increasing the cell membrane's surface area (Fig. 3.3A). For example, the margins of cells facing the lumen of the small intestine are covered with microvilli, which increases their absorptive surface area and enhances the ability of the intestine to absorb nutrients.

Cilia are much larger than microvilli and are long, hairlike projections (Fig. 3.3A). Some cells that line hollow body spaces, such as the airways, are ciliated. Cilia of neighboring cells move in a coordinated fashion, like waves of wheat shafts swaying in the wind. Their coordinated movement serves to brush material and fluid from one point to another. For example, particles entering the airways are trapped by mucus, which is swept upward by ciliary motion so that it can be coughed out or swallowed.

A **flagellum** is a large, tail-like membrane extension found on one type of human cell—the spermatozoon (sperm cell) (Fig. 3.3B). As the flagellum whips back and forth, it propels the sperm forward.

Case Note

3.2. Julia is having trouble clearing mucus from her lungs. Which cell membrane projections are involved in this process?

3.1   Identify the three major components of animal cell membranes.

3.2   Name two functions of integral proteins.

3.3   Which are shorter, microvilli or cilia?

Cell Organelles

The cell membrane surrounds the cell much like a wall surrounding a medieval city. And like the city, the cell proper is a hive of activity. Many cell functions are carried out by cell **organelles,** specialized cellular subunits that perform distinct functions. Some organelles are bounded by a membrane, which enables them to perform their chemical reactions without interference from the reactions of other organelles. These organelles include the nucleus, mitochondria, vesicles, endoplasmic reticulum, Golgi apparatus, lysosomes, and peroxisomes (Fig. 3.4). Three organelles—ribosomes, proteasomes,

and the cytoskeleton—are not surrounded by a membrane. With the exception of the nucleus, the contents of the cell are known as the **cytoplasm.**

All of the organelles are embedded in a viscous fluid called the **cytosol,** a water-based solution containing salts, bicarbonate, oxygen, CO_2, and minerals. This "inner sea" of our cells, like the oceans, is not simply an inert liquid: it teems with activity—enzymes are busy breaking down nutrients and synthesizing new molecules, and proteins, amino acids, and countless other molecules are swimming about. The cytosol of some cells also contains **inclusions,** which are globules of substances such as lipid, mucus, and pigments, many of which are formed as the result of disease.

The Nucleus Controls Cell Activity

The nucleus acts as a vast library of life, containing all of the information necessary for the body to function properly. This information exists in the form of *DNA,* a very long organic molecule that contains instructions to make proteins (see ◄ Chapter 2). DNA is visible in nondividing cells as **chromatin,** a dense tangle of DNA strands (Fig. 3.4). DNA is not scattered uniformly throughout the nucleus; instead, it is organized into dark-colored chromatin regions interspersed with regions of **nucleoplasm,** the semiliquid substance that fills the nucleus much as the cytoplasm fills the cell. A particularly dense chromatin tangle is found in the **nucleolus** ("little nucleus"), a small, darkly colored region of the nucleus where *ribosomes* (an organelle discussed below) are partially assembled.

Just as the cytoplasm is surrounded by the cell membrane, a two-layered membrane called the **nuclear envelope** surrounds the nucleoplasm. Like the cell membrane, the nuclear envelope acts as a gatekeeper: molecules pass into or out of the nucleus through multiple tiny holes in the membrane called **nuclear pores** or, in some cases, through the membrane itself.

Every type of human cell has a single nucleus during at least part of its life cycle. For example, liver cells and most other cells have a nucleus for their entire lives. In contrast, red blood cells lose their nucleus before entering the bloodstream, thereby gaining more space for carrying oxygen. Two of the three types of muscle cells are also unusual: they fuse into a single cell that contains multiple nuclei.

Nuclear DNA Is the Library of Life

Just as a library must be orderly to be useful, DNA is highly organized. The long threads of DNA are wound

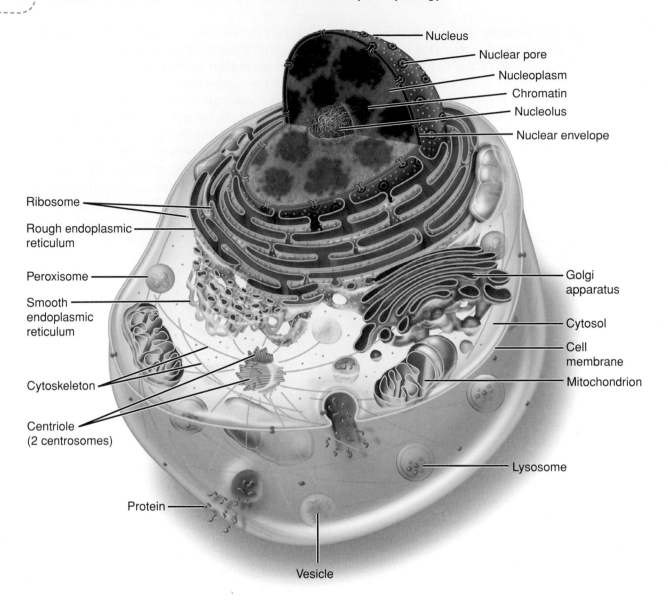

Figure 3.4. Cell organelles. Examples of each organelle are shown. Real body cells contain many more copies of each organelle (except the nucleus) than shown here. *How many mitochondria are shown in the figure? Which structure is within the nucleus—chromatin or centrioles?*

around nuclear organizing proteins, called **histones,** like thread around a spool (Fig. 3.5, bottom). DNA in the nucleus is called **chromosomal DNA,** because it and its accompanying histones are organized into 46 distinct units called **chromosomes.** The 46 chromosomes consist of 23 matched pairs, one of each set from the father, the other from the mother. DNA also exists in cellular organelles called mitochondria, discussed below, where it is referred to as **mitochondrial DNA.**

For most of the life span of a cell, the chromosomes lie together in such a dense tangle that the 46 individual chromosomes, though present, are not individually identifiable with even the most powerful microscopes

(Fig. 3.5, top right). If we think of each chromosome as a rope of DNA, we can see the piles of ropes (that is, the piles of chromatin) but cannot visually distinguish one rope from another.

In between cell divisions, DNA goes about its daily chores as messy piles of chromatin "ropes." However, when it's time for a cell to divide, a signal is tripped and DNA quickly aggregates itself into distinct forms, like soldiers organizing into various military battalions for deployment (Fig. 3.5, top left). Each chromosome duplicates itself, and the duplicates split up and march in opposite directions, eventually populating two daughter cells. We describe the details of cell division further on in this chapter.

All proteins in the body are formed under the direction of particular genes located at specific sites on individual chromosomes. However, only 5% of our DNA is functional as genes: 95% is noncoding DNA. In short, there are a lot of blank cards in the recipe box. The function, if any, of noncoding DNA is under investigation.

Genes merely carry information: they do not make proteins themselves but rather tell the remainder of the cell how to make them. Defective genes carry defective instructions and cause the cell to build useless or damaging proteins. Thus, to have a genetic defect is to fail to produce a particular protein or to produce a defective one.

> **Remember This!** DNA is important because it contains the templates for protein production. Proteins accomplish most cell functions.

Case Note

3.4. Julia has a defect in the DNA segment that encodes a particular protein, the CFTR protein. What is this segment called—a genome or a gene?

Vesicles Are Transport Containers

Like a city, a cell needs an efficient transport system to haul goods from one place to another. *Vesicles* (see Fig. 3.4) do the job of transporting substances into or out of the cell and between organelles.

Vesicles are watery bubbles with a phospholipid membrane wall. This membrane is derived from pinched-off bits of membranous organelles or the cell membrane; anything that was in the pinched-off section becomes part of the vesicle wall. After separating from its "parent" membrane with its goods for delivery, a vesicle usually travels through the cytosol on "tracks" made of *microtubules* (see the discussion of the *cytoskeleton*, below). It merges with another membranous organelle or the cell membrane and releases its contents (see the vesicle releasing a protein in Fig. 3.4). You can see this process in detail in Figure 3.7.

Ribosomes Synthesize All Cell Proteins

Ribosomes are tiny granules composed of protein and a special variety of RNA, *ribosomal RNA* (rRNA, Fig. 3.6). They are critically important because they assemble proteins.

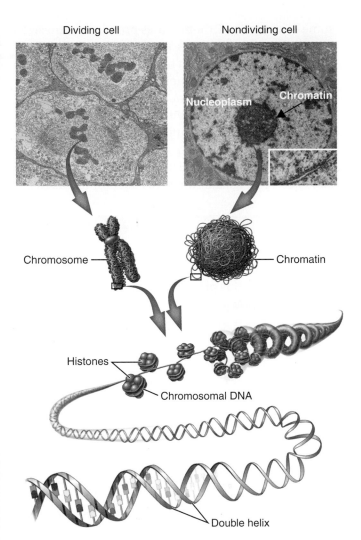

Figure 3.5. Organization of DNA. DNA is tightly coiled into chromosomes during cell division. In nondividing cells, it exists in a less compact form called chromatin, in which the individual chromosomes, though present, are not visible. Chromosomes and chromatin both consist of chromosomal DNA coiled around histones. *Are histones found in dividing cells, nondividing cells, or both?*

Case Note

3.3. Cystic fibrosis is a genetic disease. In Julia's case, it results from a minor defect on chromosome 7. Is it possible to distinguish chromosome 7 from the other chromosomes in nondividing cells?

Genes Are Specific Segments of DNA

Think of DNA as a recipe box and each *gene* as a recipe card in the box; this card carries the instructions for making a certain protein. A **gene** is a specific segment of coding DNA that encodes a particular protein. All of an individual's genes, together, constitute that person's **genome.**

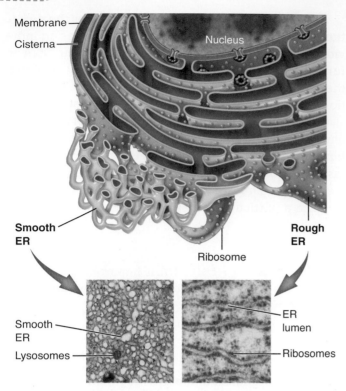

Figure 3.6. Endoplasmic reticulum and ribosomes. The *rough* endoplasmic reticulum consists of intricately folded membranes studded with ribosomes and the fluid-filled canals (cisternae) within the membranes. Ribosomes are composed of protein and rRNA organized together into two subunits. The *smooth* endoplasmic reticulum consists of a series of tubes and channels, which contain no ribosomes. *What is the term describing the canals of the endoplasmic reticulum?*

Recall that ribosomes are partially manufactured in the nucleolus. They exit through the nuclear pores into the cytoplasm, where they assemble themselves into functional units. As we will see, ribosomes remain in the cytoplasm or migrate over to the rough endoplasmic reticulum, depending on which type of protein they are assembling.

The Endoplasmic Reticulum Modifies Proteins and Other Molecules

The **endoplasmic reticulum (ER)** (Fig. 3.6) is an interconnected network of membranes involved in the synthesis of new molecules, modification of existing molecules, and removal of potentially harmful substances. The ER is folded into tubules and canals (*cisternae*) that connect to the nuclear envelope. There are two types of endoplasmic reticulum:

- *Rough endoplasmic reticulum* (RER) is a series of interconnected sacs (*cisternae*) coated with ribosomes. It owes its "rough" appearance to this ribo-

some coating. The RER modifies proteins synthesized by the attached ribosomes. For instance, it adds sugars to form glycoproteins and helps the proteins fold into the correct three-dimensional structure.
- *Smooth endoplasmic reticulum* (SER) is "smooth" because it is not coated with ribosomes. Its interconnected tubes synthesize lipids and in some cells break down nutrients and toxins. For example, liver cells have a large amount of SER because they break down and excrete drugs and waste products of metabolism.

Case Note

3.5. **The gene responsible for cystic fibrosis encodes a membrane protein. Which organelle would add sugar residues to this protein—the smooth or the rough endoplasmic reticulum?**

The Golgi Apparatus Modifies Proteins and Packages Them for Transport

The **Golgi apparatus** is the cell's distribution center; it further modifies proteins it receives from the rough endoplasmic reticulum and packages them into transport vesicles. Structurally, the Golgi apparatus is a hollow, membranous organelle somewhat like a stack of partially deflated balloons (Fig. 3.7). Unlike the interconnected cisternae of the ER, the Golgi cisternae are physically separate. Vesicles carry substances between different cisternae and transport modified proteins to other organelles or the cell membrane.

Lysosomes, Peroxisomes, and Proteasomes Degrade Toxins and Wastes

If vesicles are a cell's transport system, *lysosomes, peroxisomes,* and *proteasomes* are its sanitation department.

Lysosomes are membrane-bound packets of digestive enzymes that *lyse* (digest, break apart) other substances (Fig. 3.4). The enzymes are synthesized by the RER and packaged by the Golgi into a packet of digestive enzymes—a lysosome.

Among the many tasks of lysosomes, one is to rid the cell of worn-out parts, which, if allowed to accumulate, could impair cell function and cause disease. For example, *Tay–Sachs disease* is an inherited degenerative condition of the nervous system. The underlying problem is a defective waste-control mechanism: nerve cells are damaged by the accumulation of a fatty substance called *ganglioside.* A lysosomal enzyme digests worn-out

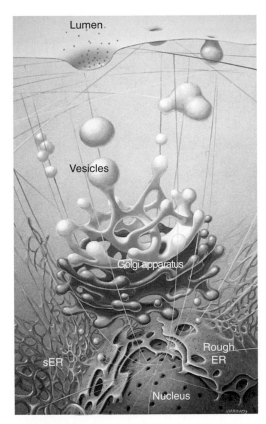

Figure 3.7. Golgi apparatus. Protein-containing vesicles originating from the rough endoplasmic reticulum merge with the nearest cisterna of the Golgi apparatus. Vesicles transport the proteins through successive cisternae, where they are modified and sorted. Finally, proteins are packaged into a transport vesicle and sent to other organelles or to the extracellular fluid. *How many cisternae are visible in this figure?*

gangliosides in normal cells, but Tay–Sachs patients don't make this enzyme.

Lysosomes also break apart and destroy harmful material that enters the cell from the extracellular fluid. For example, it is the job of certain white blood cells, called neutrophils, to "eat" infective agents such as bacteria. Neutrophils contain large numbers of lysosomes and are specifically designed to migrate to the site of an infection, ingest the infective agent, and kill it by digesting it.

Peroxisomes are another type of cytoplasmic enzyme sac (Fig. 3.4); however, their task is far different from that of lysosomes. The enzymes in peroxisomes detoxify alcohol and other cellular toxins and are especially abundant in liver cells. A common target of peroxisomes is free radicals, unstable molecules containing unpaired electrons (➡ Chapter 18). Free radicals are dangerous because they seek to steal electrons from the first available molecule, no matter if the molecule is damaged in the process. Free radicals are a product of normal cellular metabolism, but they must be kept in low concentration because they can damage any part of the cell.

Peroxisome enzymes convert free radicals into nontoxic molecules. A by-product of this reaction is hydrogen peroxide (H_2O_2), hence the name peroxisome.

Proteasomes are a third organelle of degradation, but they're very different structurally from lysosomes and peroxisomes. Rather than bubblelike membranous vesicles, proteasomes are tiny barrel-shaped structures composed of protein. Proteasomes capture and degrade proteins that are no longer required by the cell or proteins that have not been synthesized properly. (You can see proteasomes in Figure 3.17).

Case Note

3.6. Julia's cells synthesize a defective CFTR protein. Which organelle degrades defective proteins?

Mitochondria Produce Cell Energy

Mitochondria produce most of the energy required for all cell activities and are the powerhouses of body metabolism (Fig. 3.8). Very active cells, such as those in the liver and kidneys, are packed with mitochondria; relatively inactive cells, such as those in bones and tendons, have few. Under the microscope, mitochondria look remarkably like elongated bacteria; indeed, scientific evidence suggests that they may be descendants of primitive bacteria. This view is supported by the fact that they even have their own DNA, called mitochondrial DNA, which is not mixed with nuclear DNA and is passed almost exclusively from mother to daughter. This makes mitochondrial DNA useful in studying family trees and the migration of early hominids from our origin in east Africa. For more detail, see the nearby Basic Form, Basic Function box, titled *Mitochondria and the History of Humankind*. Mitochondria also have their own special ribosomes, which they use to synthesize some of their own proteins.

Mitochondria are composed of an outer, smooth membrane and an inner membrane with many folds (*cristae*) studded with enzymes. Inside the inner membrane is a cytosol-like fluid, called the *matrix*, which contains ribosomes, mitochondrial DNA, and additional enzymes. Mitochondrial enzymes generate energy (in the form of ATP) from nutrients and oxygen. In turn, ATP is used as fuel to power all of the activities of every cell.

The Cytoskeleton Is the Skeleton of the Cell

The **cytoskeleton** is a lacy network of protein fibers spread throughout the cell. It serves the cell in the

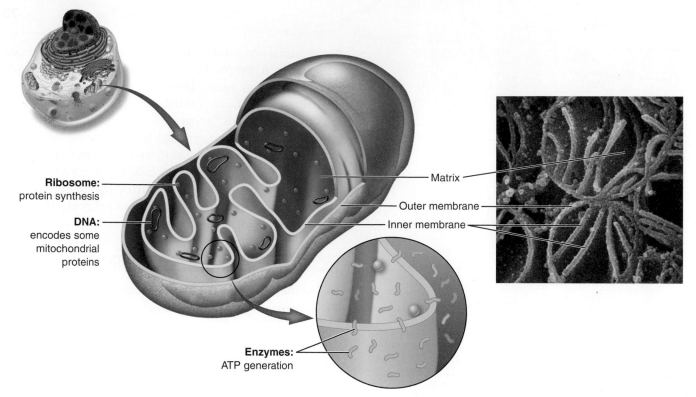

Ribosome:
protein synthesis

DNA:
encodes some
mitochondrial
proteins

Matrix

Outer membrane

Inner membrane

Enzymes:
ATP generation

Figure 3.8. Mitochondria. The enzymes used to generate ATP are found in the matrix and in the inner membrane. *Which other organelle is bound by two layers of membrane?*

BASIC FORM, BASIC FUNCTION

Mitochondria and the History of Humankind

Among the more interesting facts about mitochondria is that they have their own DNA, *mitochondrial DNA (mDNA)*, which is completely independent of *nuclear DNA*. Stranger still, this is inherited primarily from our mothers—human eggs are full of mitochondria, whereas sperm have only a few. That is, we got our mitochondrial DNA from our mothers, who got it from theirs, who got it from theirs—and so on back in time. This unique fact has helped answer one of the most fundamental questions all humans share: "Who am I?"

The immediate answer depends on knowing your ancestors, but with the passage of each generation, the trail becomes murkier and is soon lost. Analysis of nuclear and mitochondrial DNA of families and ethnic groups has been helpful in clarifying relationships and extending genealogical trees. This analysis depends on the regularity of innocent mutations (changes in DNA) of nuclear and of mitochondrial DNA that occur in every person, producing a unique DNA "fingerprint" that is passed along to subsequent generations.

But when our personal "trail of ancestors" dies out, we're all still left with the grandest question: "Who are we as a species?" It is into this question that analysis of

mitochondrial DNA offers its most dazzling insights. Mitochondrial DNA mutations occur at very regular time intervals, so that it is possible to calculate with considerable accuracy certain events in the history of humankind. This research has provided strong scientific evidence suggesting that *modern* humans (*Homo sapiens*) appeared first on the east African plain between 100,000 and 200,000 years ago. Premodern species appeared in Africa and spread to Europe and Asia much earlier: *Homo neanderthalensis* (Neanderthal man) spread from Africa to Europe about 350,000 years ago, and *Homo erectus* (upright man) spread from Africa to Asia about 1.8 million years ago.

Studies of these mutations in mitochondrial DNA show that all of us are sprung from a "mitochondrial Eve." This grandest of all grandmothers of every human lived about 140,000 years ago.

Incidentally, similar studies have been done of the *nuclear* (not mitochondrial) DNA on the male's Y chromosome, which is passed exclusively from father to son (females do not have a Y chromosome). These studies show that our "Y-chromosomal Adam" lived about 60,000 years ago.

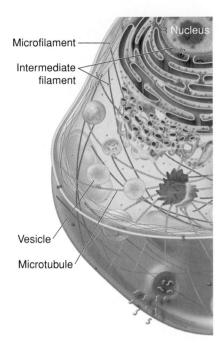

Microfilament

Intermediate
filament

Nucleus

Vesicle

Microtubule

Figure 3.9. Cytoskeleton. The cytoskeleton is composed of microfilaments, microtubules, and intermediate filaments. *Which type of filament lines the cell membrane?*

same way muscles, bones, and ligaments serve the body (Fig. 3.9).

Microfilaments are polymers of *actin*, a globular protein (◀ Chapter 2). The actin molecules are strung together into a long string, much like a beaded necklace.

Microfilaments are small but mighty—they form a latticework supporting the cell membrane, participate in the contraction of muscle cells, and aid the movement of certain mobile cells whose job it is to crawl about as they do their work.

Intermediate filaments consist of long protein fibers interwoven into strong ropes. For example, *keratin*, the main component of hair and nails, is an intermediate filament. These fibers stabilize the placement of organelles and strengthen the nuclear envelope. Both intermediate filaments and microfilaments also participate in the specialized *junctions* that hold neighboring cells together, as we discuss below.

Microtubules are polymers of *tubulin* molecules, much as microfilaments are polymers of actin molecules. However, unlike the other filaments, they are hollow. Microtubules attach to certain cell parts and act as pulleys to move parts around in the cell. For instance, they move vesicles between organelles and pull chromosomes apart during cell division.

Microtubules are built by the **centrioles,** each of which is a short bundle of tubules bound together like a bundle of sticks (Fig. 3.4). The two centrioles combine to form a small organelle called the **centrosome,** which organizes microtubules into an array for separating chromosomes during cell division. The components of the cytoskeleton are summarized in Table 3.1.

Table 3.1	Components of the Cytoskeleton		
	Form	**Function**	**Illustration**
Microfilaments	Two strands of actin molecules coiled together	Maintain cell shape; assist muscle contraction; form cilia/microvilli	Actin molecule — Microfilament (a)
Intermediate filaments	Cables composed of coiled fibrous proteins, such as keratin	Maintain cell shape; strengthen nuclear envelope	Fibrous protein molecule — Intermediate filament (b)
Microtubules	Hollow tubes composed of tubulin molecules	Organize organelles; move chromosomes and organelles (cell division); move vesicles; move cilia and flagella; assist muscle contraction	Tubulin molecule — Microtubule (c)

Case Note

3.7. In normal cells, microfilaments anchor the CFTR protein in the cell membrane. Do microfilaments contain tubulin, actin, or keratin?

3.4   Which type of nucleic acid—DNA or RNA—is found in chromatin?

3.5   The cytoplasm is to the nucleoplasm as the cell membrane is to what?

3.6   Why are mitochondria often compared with power plants?

3.7   How many phospholipid layers are found in the wall of a vesicle?

3.8   Name the type of endoplasmic reticulum that is associated with ribosomes.

3.9   You take a birthday gift for your aunt to a shipping facility in town. There, the gift is packaged in a shipping box and released to an overnight delivery service. How are the functions of the shipping facility similar to those of the Golgi apparatus?

3.10   Which organelle degrades bacteria: lysosome, peroxisome, or proteasome?

3.11   Which type of cell filament is hollow?

Cell Reproduction and Differentiation

Up to this point we have discussed the components of cells and the way they are organized (their form). Now we turn our attention to cell activities (their function). The activities of mature cells vary widely according to their type and location: stomach cells produce acid, kidney cells filter urine, and so on. But mature cells don't just spring into being. They have to start somewhere, and they start with reproduction.

Cell reproduction is necessary for growth from embryo to adult, but it is also important to maintain tissue health in adults, because cells age and die and must be replaced. After living out their natural lives of a few days, a few months, or many years, cells die by "natural suicide" in a carefully regulated, orderly process called **apoptosis.**

Cells can also die an early, unnatural death due to injury or disease, a process called **necrosis.** For example, the natural life span of most superficial skin cells is just a few days, after which they die naturally (apoptosis) and are shed imperceptibly every day. However, severe sunburn causes early death (necrosis) of skin cells, which are shed visibly as "peeling" skin.

The body replaces dead cells via reproduction. Cells reproduce by dividing into two. Those two become four, and so on, the process continuing until the dead or damaged cells have been replaced. Division of the cytoplasm is relatively simple: Half the organelles go to one new "daughter" cell and half to the other. But the nucleus is a special case: There is only one nucleus, and its DNA exists as a messy pile of chromatin that must be organized into two *exactly* matching sets so that the DNA of each offspring cell will *perfectly* match parent DNA. The process of organizing and dividing the nucleus into two daughter nuclei is called *mitosis,* which is discussed in detail further on.

The Cell Cycle Describes the Steps in Cell Division

The **cell cycle** (Fig. 3.10) is the orderly sequence of events by which one cell reproduces to form two. It can be thought of as a merry-go-round with a twist: Each round trip, one cell goes in, but two cells come out. A round trip in the cell cycle is divided into two major phases: the *interphase* and the *mitotic phase.*

The Interphase Is the Period between Cell Divisions

The **interphase** is the period of time between cell divisions, during which the cell prepares for its next division. It is by far the longest phase in the cell cycle, lasting from a few hours to many years. It begins after cell division is complete and ends when the cell begins a new division.

Each new offspring cell needs its own cellular equipment, so the cell must duplicate organelles and DNA and make extra cytosol during interphase. Among the things duplicated early in the interphase are the centrosomes, so that for much of the interphase the cell has two centrosomes. In addition to assembling material for division, cells are also busy carrying out their usual daily activities: making proteins, clearing waste, digesting nutrients, and so on.

The interphase proceeds in three subphases, which are:

● First growth phase (G1 phase): New daughter cells grow to adult size and duplicate their organelles.

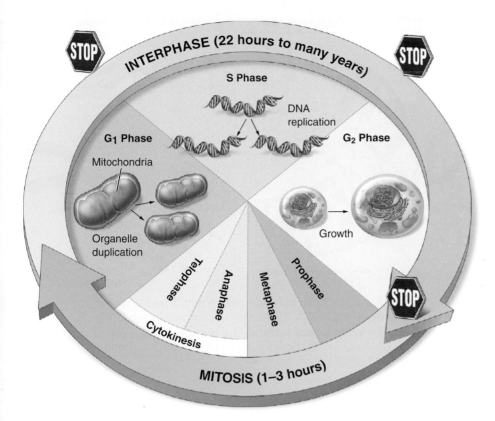

Figure 3.10. Cell cycle. Dividing cells proceed through the steps of the cell cycle in order, beginning with the G1 phase, the S phase, and the G2 phase (cumulatively described as the interphase) and terminating with mitosis. Interphase can last from 22 hours (h) to many years. Stop points between each of the phases ensure that the cell is healthy and prepared to enter the next phase. *Does the G1 phase occur just before or just after the mitosis phase?*

● *DNA synthesis phase* (S phase): As the time for cell division nears, the DNA of each of the 46 chromosomes duplicates itself precisely into two exactly matching strands—the original and a duplicate—by a process discussed in detail below. Each DNA strand coils tightly into a barlike **chromatid.** The two chromatids remain "stapled" together at a single point by a tiny structure— the **centromere.**

● *Second growth phase* (G2 phase): The cell continues to grow, adding more organelles and cytoplasm so that it will have enough bulk to divide into two new daughter cells.

> *Remember This!* **The centromere anchors two chromatids together. The centrosome organizes microtubules.**

Regulatory proteins tightly control each of these steps. Cells are, so to speak, being pushed to proliferate by one set of regulatory proteins and being held back by the opposing force of a second set of regulatory proteins. The balance of these forces controls the cell's growth rate. By way of contrast, cancer is an uncontrolled growth of cells. Some cancers are caused when growth-opposing proteins fail to do their job or growth-promoting proteins become overactive.

DNA Is Replicated in the S Phase of the Interphase

As mentioned above, before a cell can divide, it must prepare a complete and exact copy of its DNA for distribution to the two new daughter cells. Earlier, we said that in nondividing cells (that is, cells in the interphase), DNA exists in the nucleus as chromatin—a dense tangle of 46 separate strands, each one of which is a chromosome. During the S phase, the DNA in these strands duplicates itself (Fig. 3.11).

> *Remember This!* **DNA replication must occur before a cell can divide.**

So how does this duplication of chromosomes occur? Recall from ◀ Chapter 2 that DNA is a very long molecule composed of vast numbers of four different nucleotides and that the sequence of nucleotides is the genetic code. Nucleotides are named according to the *base* they contain: adenine (A), cytosine (C), guanine (G), and thymine (T). The cell makes nucleotides as they are needed from basic ingredients (sugars, nitrogen, and phosphate) in the cytosol. These nucleotides link to form two very long strands of DNA—for reasons to be discussed, these are called the *sense* strand and the *template* strand,

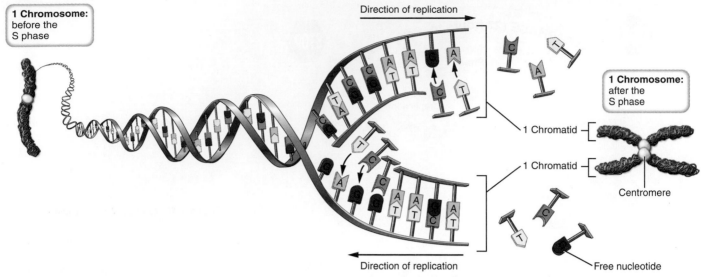

Figure 3.11. DNA replication. The DNA double helix is unwound, and base pairing rules (A with T and G with C) operate to assemble a new DNA strand on each original strand. After DNA replication is complete, each DNA molecule (chromatid) consists of one old strand and one newly synthesized strand. The two chromatids are joined together at the centromere. *How many strands of DNA are in the chromosome before DNA replication? How many strands of DNA are in the chromosome after DNA replication?*

which in turn are intertwined with each other like a twisted zipper into a double helix.

The "teeth" of each side of the zipper are the nucleotide bases in each strand of DNA. The teeth on the sense strand are linked loosely to the teeth of the template strand by weak hydrogen bonds. In forming this weak connection, each base links to its *complementary partner*: adenine (A) always links with thymine (T) (and vice versa) and cytosine (C) with guanine (G) (and vice versa). For example, the base sequence ATCG in the sense strand will be linked to the sequence TAGC in the template strand.

DNA replication in the S phase begins with the breaking of the weak hydrogen bonds that hold the sense and template strands together—an unzipping of the DNA. The unzipped nucleotide strands are forcibly held apart by an enzyme—**DNA polymerase** (not visible in Fig. 3.11). As a result, the nucleotides of each unzipped strand become exposed and slightly charged (because their bond has been broken), and rapidly attract their complementary nucleotides from the chemical soup of the cytosol.

That is, the sense strand attracts new nucleotides to form a new template strand, and the template strand attracts new nucleotides to form a new sense strand. DNA polymerase links the free nucleotides together and … Voila! Two identical sets of DNA, each a "sister" chromatid, exist where one used to be.

Case Note

3.8. **The gene responsible for cystic fibrosis, the CFTR gene, contains the sequence AAA in the DNA template strand. What is the sequence in the DNA sense strand?**

Mitosis Is the Period When the Nucleus Divides

After the DNA is duplicated, *mitosis* occurs. **Mitosis** is division of the original "mother" nucleus into two identical twin "daughter" nuclei, each of which has exactly the same DNA. Mitosis takes little time, usually a few hours, and consists of four distinct steps (Fig. 3.12):

● **Prophase.** The nuclear envelope disappears and individual DNA strands organize into visibly discrete pairs of chromatids joined by a centromere. Meanwhile in the cytoplasm, the two centrosomes separate, each moving to an opposite pole of the cell. Each centrosome sprouts an array of microtubules, which extend toward the center of the cell, seeking the centromeres that hold the chromatids together. By the end of prophase, each chromatid pair is attached at its centromere to microtubules from the opposing centrosomes, much like teams in a tug-of-war at opposite ends of the same rope.

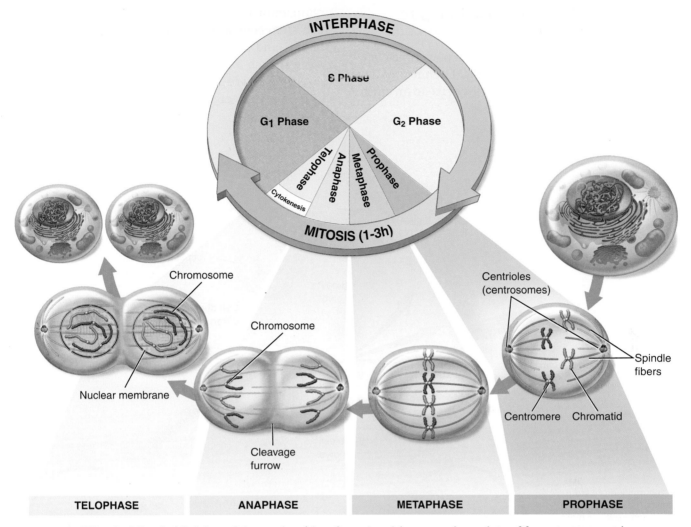

Figure 3.12. Mitosis. Mitosis (division of the nucleus) last from 1 to 3 hours and consists of four stages: prophase, metaphase, anaphase, and telophase. Cytokinesis (division of the entire cell) occurs toward the end of mitosis. *When are chromosomes lined up in the middle of the cell?*

- **Metaphase.** The microtubules pull the sister chromatid pairs into the equatorial plane of the cell. Each set of microtubules pulls equally on the chromosomes, and, as at the beginning of a tug-of-war, they are held tightly between the opposing centrosomes.
- **Anaphase.** The centromeres break apart, freeing the sister chromatids from one another. Each centrosome then reels its prize—one of the sister chromatids, now again called a chromosome—into its half of the cell.
- **Telophase.** Two separate nuclei form during telophase. The two separated chromosome sets uncoil into two chromatin piles, each surrounded by a nuclear envelope.

Cytokinesis Is the Division of the Cytoplasm

The cell cycle ends with **cytokinesis,** the division of the cytoplasm and the emergence of two new, inde-pendent daughter cells. A band of microfilaments forms about the equator of the cell and begins to tighten like a belt. Soon the cytoplasm is pinched into two parts, with one new nucleus resting in each. Two new daughter cells now exist where the single parent had been.

All Cells Arise from Stem Cells

Up to this point, we've discussed how cells divide. But there is more to the story, which brings us to stem cells. A **stem cell** is an unspecialized (undifferentiated) cell capable of reproducing itself plus a more specialized (differentiated) cell. The fertilized egg (zygote) is the granddaddy of all stem cells—and of all other cells. It evolves into an entire organism, including more than 200 types of cells in humans.

Stem Cells Are Classified According to Their Power to Form Other Types of Cells

Stem cells are classified according to their power (potency) to develop into more specialized cells. The most potent have the broadest powers: they can give rise to an entire organism or to any particular type of cell in the body. Such stem cells are *totipotent* stem cells (Latin *totus* = "whole," and *potens* = "to be able"). As just noted, the zygote is such a cell (Fig. 3.13). It divides into two cells, those two divide into four, and those four into eight. Up to the first eight cells, all are alike: They retain

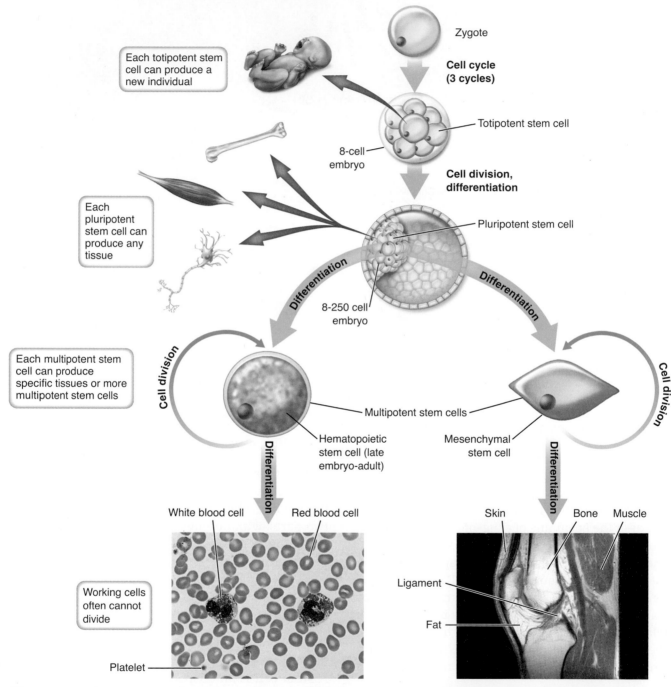

Figure 3.13. Cell differentiation and stem cells. The first three rounds of the cell cycle produce an eight-cell embryo containing totipotent stem cells. As the embryo grows by cell division, these stem cells differentiate into pluripotent and then multipotent stem cells. Two types of multipotent stem cells are illustrated—hematopoietic and mesenchymal stem cells. When a multipotent stem cell divides, the daughter cells can remain as stem cells or can differentiate into a working cell, such as a muscle cell, by turning some genes on and other genes off. *Which type of stem cell could be used to make an ear for a burn victim (which contains many different tissues)— pluripotent or multipotent?*

totipotency and each has the capacity, if separated from the others, to reproduce an entire human being.

But subsequent stem cell generations become more specialized and less potent. As the initial eight cells divide, each differentiates into a more specialized version—the *pluripotent* stem cell (Latin *plur* = "more, extra"). Each pluripotent cell can produce any type of tissue—heart, brain, liver, skin, you name it—but cannot form an entire new human being. These pluripotent stem cells are usually the type of stem cells in question when the term *embryonic stem cells* is used.

As the cell divisions continue, all of the pluripotent stem cells differentiate into less potent, more specialized *multipotent* stem cells. Each type of multipotent stem cell can produce a limited range of cell types. For example, *mesenchymal stem cells* can produce muscle, bone, ligament, or fat cells, while *hematopoietic stem cells* produce the many types of blood and bone marrow cells.

Multipotent stem cells frequently divide **by a symmetric division**—the two daughter cells are not alike. One daughter cell becomes a new stem cell to maintain the stem cell population and the other becomes a more specialized cell. For instance, the mesenchymal stem cell in Figure 3.13 can divide to produce one fibroblast and one new mesenchymal stem cell.

That being said, even stem cells cannot reproduce endlessly; that is, they cannot ride the cell cycle merry-go-round forever. Although science does not know exactly how many cell cycles each cell is allowed, it is a finite number that is tracked by a special "counter" (the *telomere*) at the ends of every chromosome; when a cell's "ticket to ride" expires, the cell dies and its tissue is less robustly functional.

Case Note

3.9. An experimental cystic fibrosis therapy uses stem cells to produce new airway cells. Airway cells are produced from the same stem cell population as fibroblasts. Would this therapy use hematopoietic or mesenchymal stem cells?

Some Stem Cells Persist into Adulthood

Small numbers of multipotent stem cells, often referred to as *adult* stem cells, persist in adults, acting as a ready reserve to replace dead or dysfunctional cells. Indeed, a critical factor in human health is the ability of injured and diseased tissues and organs to repair themselves by generating new cells from the population of adult stem cells.

Adult stem cells can be recruited into action if new cells are needed. Tissues containing short-lived cells—such as those in the skin, digestive tract, and airways—maintain large stem cell populations that are continually dividing to produce replacement cells. The short life and quick death and replacement of these cells is protective; the rapid shedding of dead cells refreshes tissue vigor and carries away hostile microbes and harmful agents. For example, ingested toxins frequently damage cells lining the digestive tract. Stem cells in the digestive tract's wall are always busy dividing asymmetrically, with some of the offspring cells remaining as stem cells and others differentiating into new functional digestive tract cells.

Muscles and nerves are at the other end of the spectrum: they are composed of cells designed to last a lifetime. Their ability to generate replacement cells is limited because they retain few stem cells into adulthood. Although these few stem cells are activated in response to tissue damage, their function is relatively limited and they cannot regenerate enough new cells to correct severe damage. The devastating effect of stroke, for instance, reflects the death of many neurons and the functions they control. Although limited functional recovery may occur as some dead neurons are replaced by new ones, the damage cannot be fully repaired in severe strokes because the regenerative ability of neuronal stem cells is limited. See ➡ Chapter 8 for a clinical snapshot on stem cells and regeneration in the central nervous system. In like manner, severe damage to skeletal or cardiac muscle is followed by incomplete repair because of the limited capacity of muscle stem cells.

Intermediate between these two extremes are other tissues, such as liver and kidney, whose cells have an intermediate life span of several months or years and are replaced by stem cells relatively infrequently. Their stem cells remain quiescent for a long time and spring into action in response to loss or damage. For instance, if half of the liver is taken from, say, a parent for transplant into a child, the entire missing half of the liver can regenerate in a few months.

Case Note

3.10. Julia's lung infection has damaged the cells lining her small airways. Will she be able to replace the damaged cells relatively quickly? Explain why or why not.

Stem Cells Can Be Used for Scientific and Therapeutic Purposes

Because of their powers of differentiation, stem cells hold great promise for scientific and therapeutic advancement. We do not have room here to debate the many ethical issues surrounding stem cell research or stem cell therapy. For instance, is it or is it not ethical to obtain stem cells from human embryos? And apart from the universal agreement that it would be immoral to generate a new human being from a totipotent stem cell, what limits should restrain the use of stem cells?

Nevertheless, it is clear that the powers possessed by stem cells can be directed toward therapeutic purposes. For example, in mice, it has been shown that bone marrow stem cells can convert into muscle stem cells that will partially correct a genetic muscle disease. In humans, bone marrow stem cell transplants have proven effective in the treatment of some types of leukemia. Other promising results have been recorded in the treatment of cancer, nervous system disease, and spinal cord injuries.

3.12   True or false? A zygote consists of two cells, an ovum and a sperm cell.

3.13   What is the name for natural cell death?

3.14   What is the difference between growth phases G1 and G2?

3.15   What is the complementary sequence for the nucleotide base sequence ACTG?

3.16   Which structure links sister chromatids together—the centrosome or the centromere?

3.17   When do sister chromatids separate—in prophase or anaphase?

3.18   What is the difference between multipotent and pluripotent stem cells?

3.19   Which organ retains more stem cells in adulthood—the intestine or the brain?

Cell Specialization

The many types of body cells all contain the same DNA. But cells differ greatly according to their task. The most basic task distinction is between *germ* cells and *somatic* cells. **Germ cells** are reproductive cells found only in the ovary or testis. They develop into ova (eggs) or sperm. All other cells are **somatic** cells (Greek *soma* = "body")—liver, muscle, brain, and every other type. Very

important is the fact that germ cells transmit the genetic code from one generation to another. Changes (mutations) in germ cell DNA are inheritable. Changes to somatic cell DNA (as in cancers) are not transmissible to the next generation.

If all cells contain exactly the same DNA, how do somatic cells differentiate into skin cells or intestinal cells or cells of the brain? The answer: they develop their different structures and functions because they synthesize different proteins. They accomplish this bit of magic by turning their genes on or off. That is, skin cells activate the genes that code for proteins that perform skin cell tasks, and avoid activating genes that code for proteins that perform intestinal cell tasks.

Genes Are Segments of DNA That Code for Particular Proteins

As explained earlier in this chapter, a gene is a relatively short segment of DNA (300 to 3,000 nucleotide pairs long, on average) that codes for one or more particular proteins; each protein, in turn, has a highly specific and limited role. For example, a gene in chromosome 17 encodes *growth hormone*, a protein necessary for normal growth of body parts. If this gene is absent or defective, growth hormone is not produced and growth is significantly impaired. To have a genetic defect is to have defective production or function of a particular protein.

Recall from ◀ Chapter 2 that proteins are composed of various combinations of amino acids linked into a chain that can vary in length from 100 to several thousand amino acid molecules. In contrast, a gene—a segment of DNA—consists of a specific sequence of the four different nucleotides, identified by abbreviations of their base names A, C, G, and T. But how do we get from nucleotides to amino acids? Before we answer that question, we must discuss one more player in protein synthesis—RNA.

RNA Carries Out DNA Instructions

DNA has been called the "master molecule"; however, it is a relatively inactive substance, somewhat like a library filled with books of instructions about how to perform tasks. The "other" nucleic acid, RNA (◀ Chapter 2), reads the DNA "book" and carries out the instructions with the help of cell proteins.

RNA differs structurally from DNA in three respects:

- RNA is a single strand of nucleotides; DNA is a double strand.

Table 3.2   Types of RNA

Type of RNA	Function
Messenger RNAs (mRNAs)	Each mRNA acts as a template for the synthesis of a specific protein.
Transfer RNAs (tRNAs)	Each tRNA transports a specific amino acid to the ribosome to be used in protein synthesis.
Ribosomal RNAs (rRNAs)	These fold together with proteins to produce ribosomes, the organelles responsible for protein synthesis.
Small nuclear RNAs (snRNAs)	These form part of enzymes that modify RNA.

- The sugar component of RNA is ribose; in DNA it is deoxyribose.
- In RNA, the nucleotide *uracil* replaces the *thymine* of DNA so that the "letters" of the RNA alphabet are A (adenine), C (cytosine), G (guanine), and U (uracil). Uracil (like thymine) forms bonds with adenine.

All RNAs are made by the same process—*transcription.* However, there are four different types of RNA, which serve very different roles. The structure and function of each RNA variety are summarized in Table 3.2.

Transcription Makes an RNA Copy of a Gene

To *transcribe* is to copy something from one medium to another, much as a music fan might write down the words to a song heard at a concert. In like manner, the process of transferring the DNA code to RNA is called **transcription.** It occurs in the nucleus by a process similar to DNA replication.

Transcription begins with the separation of the *template* strand of DNA from the *sense* strand. The exposed bases on the DNA template strand form hydrogen bonds with the complementary bases of free nucleotides (Fig. 3.14). Recalling that RNA contains U where DNA contains T, the complementary partners are G with C, and A with U or T. An enzyme (RNA polymerase, not shown) sews together the individual nucleotides into a long strand of RNA.

> *Remember This!* **RNA is assembled on the DNA template strand, so that its sequence resembles that of the sense DNA strand.**

For example, a DNA *sense* strand triplet GAG would have a complementary *template* strand sequence CTC. In RNA, transcription assembles the nucleotides complementary to the template strand—the RNA sequence is GAG. Thus, the DNA sequence GAG has

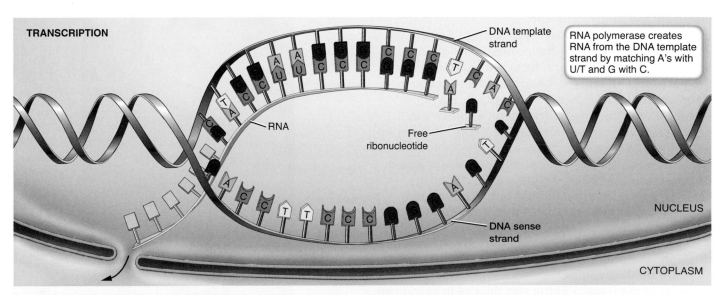

TRANSCRIPTION

DNA template strand

RNA polymerase creates RNA from the DNA template strand by matching A's with U/T and G with C.

RNA

Free ribonucleotide

NUCLEUS

DNA sense strand

CYTOPLASM

Figure 3.14. Transcription. The DNA helix unwinds, and a new mRNA strand is assembled on the template strand of DNA. The mRNA molecule contains the same base sequence as the DNA sense strand except that T bases are replaced with U bases. The mRNA leaves the nucleus through pores in the nuclear envelope. *Where is uracil found—in the mRNA or the DNA?*

been transcribed as GAG in the new RNA, bringing us back to where we started. Finally, we see the reason for calling the two DNA strands the template and the sense strands—the sense strand contains the permanent code that must be communicated as instructions to other cell parts. It "makes sense," so to speak. The template strand is a "negative mold" used to prepare a replica of the sense strand, much as the mold of a tooth is used to make a crown.

Case Note

3.11. The CFTR gene contains the sequence TAAAAT in the DNA sense strand. What will be the matching sequence in the DNA template strand and the mRNA strand produced?

Translation Assembles Proteins from Amino Acids

The word *translation* means the conversion of one language into another; for example, English into French. In biology, **translation** describes the process of translating the nucleotide sequence of mRNA into the amino acid sequence of a protein. For many years, scientists lacked the "dictionary" that linked a particular nucleotide "word" with an amino acid "word." Then they discovered that the code for amino acids was contained in *triplets*, three-letter sets of nucleotides that encode different amino acids. The corresponding nucleotide set in mRNA for each DNA triplet is called a **codon.** For instance, the DNA sense strand triplet ATG produces the mRNA sequence AUG, which encodes (is the codon for) the amino acid methionine. You can see other examples in

THE HISTORY OF SCIENCE

Decoding the Rosetta Stone

It may be helpful to think of DNA, RNA, and proteins as languages written in different alphabets: DNA is written with ATCG letters, RNA with AUCG letters, and proteins with amino acid letters. The relationship of these three molecules and the codes they carry is more than passingly similar to the famous Rosetta Stone, now on display in the British Museum.

One of the great mysteries of ancient Egypt was its writing, which used symbols called hieroglyphics. For more than 2000 years, from ancient Greece until the early 19th century, scholars tried and failed to translate (decode) Egyptian hieroglyphic writing. However, in 1799, one of Napoleon's soldiers in Egypt discovered a remarkable fragment of stone in the ruins of a monument in the Egyptian city of Rosetta (now Rashid). Etched into its surface was a decree written in three languages: Egyptian hieroglyphics, Demotic (an old but more modern Egyptian language than hieroglyphics), and Greek. Because Greek was and remains widely understood, scholars were able to work backward from Greek through Demotic to translate, for the first time, Egyptian hieroglyphics.

Because the Rosetta Stone was the key to solving an enduring mystery, the term *Rosetta Stone* has entered popular language as an expression for the key clue in solving any difficult problem.

The Rosetta Stone. The Rosetta Stone is inscribed with the same message in three languages and enabled linguists to translate from ancient Greek to Demotic to finally crack the code of ancient Egyptian hieroglyphics. In the same way, DNA triplets correspond to mRNA codons, which in turn are complementary to anticodons on tRNA molecules that transfer the correct amino acid to the growing protein.

Table 3.3 Some Amino Acids and Their Codons[a]

Amino Acid	DNA Triplet (sense strand)	mRNA Codon	tRNA Anticodon
Glycine	GGG	GGG	CCC
Isoleucine	ATT	AUU	UAA
Isoleucine	ATC	AUC	UAG
Leucine	CTC	CUC	GAG
Lysine	AAA	AAA	UUU
Phenylalanine	TTT	UUU	AAA
Proline	CCC	CCC	GGG
Valine	GTG	GUG	CAC

[a]It should be noted that some amino acids are encoded by more than one triplet.

Table 3.3. Note that some amino acids (such as leucine) are encoded by several different codons, just as the English word *snow* can be translated into many different words in the Inuit language.

If you find the relationship of the ATCG alphabet of DNA, the AUCG alphabet of mRNA, and the amino acid alphabet of proteins somewhat confusing, it may help to consider the Rosetta Stone, a stone fragment found in Egypt, that has on its surface the same passage written in three ancient languages. For more about this bit of history, see the accompanying History of Science box, titled *Decoding the Rosetta Stone*.

To recap, a sequence of three DNA nucleotides (the *triplet*) is transcribed into a sequence of mRNA nucleotides (the *codon*), which in turn translates the nucleotides of mRNA into an alphabet of amino acids to produce a protein. As shown by the discussion of Julia's case in the next section, changes in the DNA sequence can change the amino acid sequence, sometimes to devastating effect.

Case Note

3.12. In people with cystic fibrosis, the defect in the CFTR gene is the deletion of three nucleotides—CTT. How many amino acids will be deleted in the CFTR protein?

Case Discussion

The Genetic Code and Cystic Fibrosis: The Case of Julia G.

 Recall that Julia has cystic fibrosis, a disease caused by a genetic mutation that results in the accumulation of thick mucus in the lungs. Severe lung infections and difficulty breathing are the result.

A **mutation** is an error in the DNA code and is permanent if not repaired. Mutations happen frequently; however, they are usually detected and repaired by an inherent process before they cause damage. Mutations that are not repaired can cause cancers and other problems in ordinary body cells. However, if the mutation occurs in a germ cell (ovum or sperm), the defect is heritable and can be transmitted to offspring ➡ (Chapter 16). It is a germ cell mutation that is the cause of Julia's cystic fibrosis.

Part 1: Missing Nucleotides Result in a Missing Amino Acid

The gene involved is called the cystic fibrosis transmembrane conductance regulator, or CFTR gene (Fig. 3.15). This large gene, which contains a staggering 250,000 nucleotides, is found on chromosome 7.

Look first at the left-hand drawing, illustrating a small portion of the normal CFTR gene. The DNA sense strand reads ATC TTT GGT and the mRNA reads AUC UUU GGU. These three codons encode the amino acids isoleucine (Ile), phenylalanine (Phe), and glycine (Gly), respectively, so the amino acid sequence reads Ile-Phe-Gly. The next amino acid in the protein sequence is valine (Val).

The short DNA sequence C TT, shaded gray, is deleted in Julia's mutated CFTR gene, shown on the right. The mutated DNA sequence now reads ATT GGT GTT and is transcribed as AUU GGU GUU. The new amino acid sequence reads Ile-Gly-Val. Note that one amino acid—Phe—is deleted. This amino acid is the 508th amino acid in the chain, so Julia's mutation can be abbreviated as F508del.

This minute change—one amino acid out of 1,448—changes the shape of the CFTR protein ever so slightly. Actually, the change is so minute that the protein could still function. However, as we will see later in the text, this minute change is all it takes to cause serious disease.

Case Note

3.13. Name the codon encoding isoleucine in the normal CFTR protein and in the mutated CFTR protein.

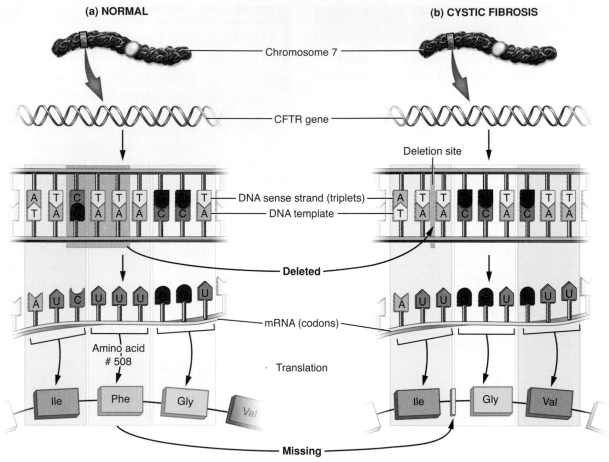

Figure 3.15. Cystic fibrosis and the genetic code. A. The CFTR gene is a segment of the DNA double helix of chromosome 7. The DNA is organized in nucleotide triplets. Each triplet corresponds to an mRNA codon, which corresponds to a single amino acid. **B.** Three nucleotides are deleted in the CFTR gene mutation F508del, resulting in the deletion of phenylalanine (F) at position 508. The sequence of amino acids determines how the protein will fold into a three-dimensional shape. *Which amino acid corresponds to the mRNA codon AUC?*

Ribosomes Synthesize Protein

There is more to this story than understanding the code. How is it, exactly, that proteins are made?

Translation occurs on ribosomes in the cytoplasm, but DNA is imprisoned in the nucleus. Messenger RNA earns its name by carrying the coded message for protein synthesis from the nucleus into the cytoplasm (Fig. 3.16). After leaving the nucleus through pores in the nuclear envelope, the mRNA strand attaches to a ribosome. It is then fed through the ribosome like a piece of coded tape. With the help of yet another type of RNA, *transfer RNA* (tRNA), the ribosome translates the mRNA codons one by one and adds the correct amino acid to the growing protein chain.

The raw material for protein synthesis is a pool of about 20 amino acids floating in the cytoplasm. For each amino acid there exists a transfer RNA designed

to carry it and it alone. Each tRNA has two ends: one end binds to its unique amino acid and the other end, the *anticodon,* consists of three nucleotides that bind to the mRNA codon for that particular amino acid. For instance, the mRNA codon GGG encodes glycine. The anticodon of a glycine tRNA consists of the complementary sequence (CCC) (Fig. 3.16, far right). The tRNA anticodons for different mRNA codons are shown in Table 3.2.

Figure 3.16 illustrates the steps in protein synthesis. The ribosome has already read a fair bit of the mRNA, and assembled a number of amino acids. It has just added proline to the growing protein chain and is preparing to add the next amino acid.

1. The ribosome reads the next codon—GGG—which encodes glycine (Gly). It brings in the glycine tRNA with its attached glycine.

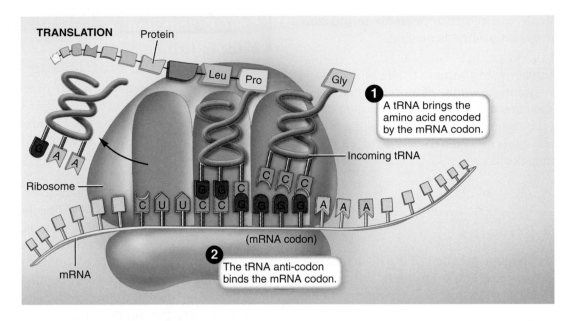

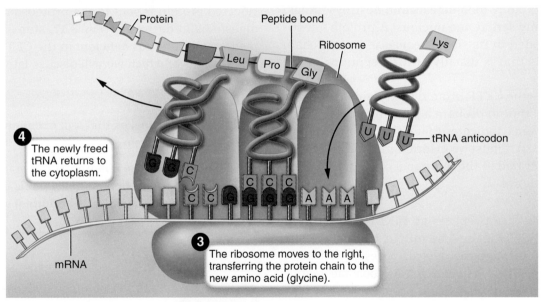

Figure 3.16. Translation. A ribosome reads the mRNA sequence. Each group of three bases (a codon) corresponds to an amino acid. Each codon bonds with a tRNA, with the complementary anticodon carrying the matching amino acid. The ribosome forms a peptide bond between the new amino acid (in this case, a proline) and the assembled peptide chain. Addition of the next amino acid to the chain (in this case a glycine) will simultaneously break the bond between the proline amino acid and the tRNA. The tRNA will then return to the cytosol to pick up another amino acid. *If the next mRNA codon reads AAA, what is the corresponding tRNA anticodon?*

2. The CCC of the glycine tRNA anticodon binds the GGG of the mRNA codon.

3. The ribosome joins glycine to the chain of previously assembled amino acids by forming a peptide bond. Glycine, now attached to the growing protein chain, remains attached to its tRNA, which anchors the protein to the ribosome.

4. Simultaneous with step 3, the ribosome moves to the right by one codon and releases the proline tRNA that carried the previous amino acid, proline.

5. The proline tRNA, now "emptied" of its amino acid, leaves the ribosome and floats back into the cytoplasm, where it picks up another proline. The proline

tRNA molecule can return repeatedly to the ribosome whenever its proline is needed.

The cycle repeats with the next tRNA, which in our example carries lysine. Assembly of a protein ends when the ribosome reaches a region of mRNA called a *stop codon*. This special codon signals that the process is complete. The ribosome releases both the strand of mRNA and the new protein molecule.

Organelles Work Together to Prepare Proteins

The ribosome is only one member of the team that prepares functional proteins and sends them to the correct location. This is a critical job, and many organelles work together to get it right: imagine the damage if a lysosomal enzyme (which digests and destroys things) were mistakenly sent to the nucleus! A protein's destination is encoded in its *signal sequence*, a short chain of amino acids that is usually at the beginning of the protein.

We return to the CFTR protein to explain how organelles work together to prepare a protein for its mission. Its signal sequence indicates that it is an integral membrane protein, so it must be inserted into the cell membrane. As illustrated in the left panel of Figure 3.17, the steps are as follows:

1. RNA polymerase prepares an mRNA transcript of the *normal* CFTR gene. The mRNA molecule leaves the nucleus through the nuclear pores.
2. Based on mRNA instructions, a ribosome begins assembly of the *normal* CFTR protein.
3. The ribosome attaches to rough endoplasmic reticulum (RER) where it completes assembling the protein. Parts of the protein are hydrophobic (soluble in fat but not in water). The hydrophobic portions remain in the RER membrane and do not feed into the RER lumen, embedding the CFTR protein in the RER membrane.
4. The RER helps the CFTR fold properly, adds sugar residues, and exports the folded CFTR protein embedded in the wall of a vesicle.
5. The CFTR-containing vesicle fuses with a Golgi cisterna. The Golgi apparatus further modifies the CFTR and packages it into a transport vesicle.
6. The transport vesicle with its CFTR protein fuses with the cell membrane. The CFTR protein is now integrated into the cell membrane, and becomes functional.

Exchange of Substances Across the Cell Membrane

All of the cell processes discussed thus far—reproduction, differentiation, and protein synthesis—require chemical signals and building block materials to be imported into the cell, and cells are constantly generating

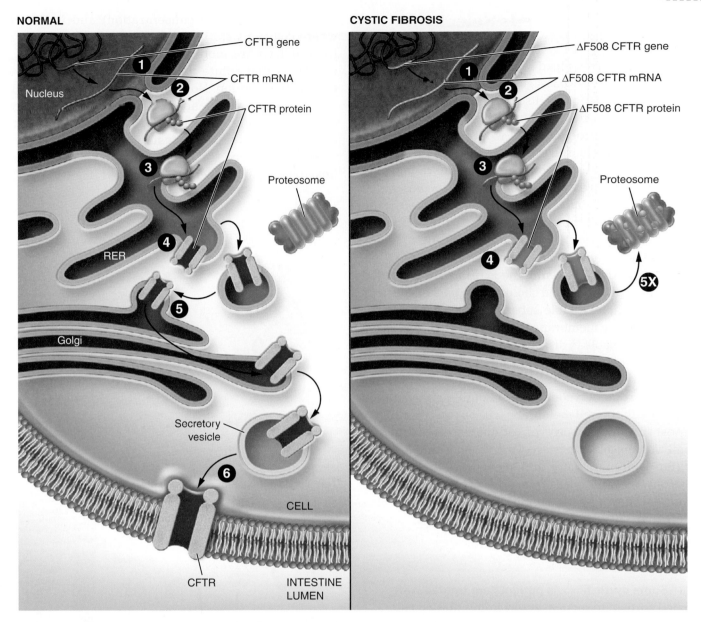

NORMAL

CFTR gene

CFTR mRNA

Nucleus

CFTR protein

Proteosome

RER

Golgi

Secretory vesicle

CELL

CFTR

INTESTINE LUMEN

CYSTIC FIBROSIS

ΔF508 CFTR gene

ΔF508 CFTR mRNA

ΔF508 CFTR protein

Proteosome

5X

(a) Normal CFTR protein

(b) Abnormal CFTR protein

Figure 3.17. Cystic fibrosis and protein production. A. The normal CFTR protein is assembled by ribosomes, modified by the RER, packaged by the Golgi apparatus, and sent in a secretory vesicle to be inserted into the cell membrane to assist in chloride transport. **B.** Upon assembly by ribosomes, the protein with the missing phenylalanine differs slightly in shape from the normal protein. This mutation is detected in the RER, and the protein is sent to a proteasome to be degraded. This impairs chloride transport between intestinal cells and the intestinal cavity (lumen) in patients with cystic fibrosis. *Where is the normal CFTR protein threaded into a membrane—the RER, the Golgi, or the cell membrane?*

waste products and secretions that must be exported from the cell. **Membrane transport** is the movement of substances across the cell membrane.

In this section, we discuss transport methods that move solutes across the membrane one molecule (or ion) at a time. The passive movement of substances down their concentration gradient from an area of high concentration to low requires no energy. This process is called **diffusion.** Conversely, energy is required to move substances up their concentration gradient from a region of low concentration into a region of higher concentration. This process is called **active transport.**

The Cell Membrane Is Selectively Permeable to Solutes and Solvents

Uncontrolled borders lead to trouble. Between nations they result in the easy passage of criminals and fugitives; the spread of contagious diseases of humans, animals, and plants; and a host of other ills. That's why border patrol agents control who (and what) gets to pass. The border between cells and the extracellular fluid—the cell membrane—is no exception. The cell membrane tightly controls the coming and going of many molecules, which must be "escorted" into and out of the cell by membrane proteins. Yet other molecules can cross the cell membrane freely. This variation in the membrane's "admittance policy" is called **selective permeability.**

Recall that the cell membrane is primarily composed of fats—phospholipids and cholesterol—packed together tightly. Thus, lipid-soluble substances can pass through easily; that is, the membrane is permeable to them. These substances are able to dissolve in the fatty interior of the cell membrane as easily as a drop of sunflower oil dissolves in olive oil. In addition to lipids, gases such as oxygen and carbon dioxide are lipid-soluble and can thus pass through the cell membrane without an escort.

Most substances, however, cannot dissolve in the cell membrane because they are *water-soluble* and *lipophobic* (fat-avoiding). Water and substances that can dissolve only in water need an escort to cross the cell membrane. As shown below, membrane proteins fulfill this escort role.

Case Note

3.16. Some cystic fibrosis patients are treated with inhaled cortisol to reduce airway inflammation and mucus production. Cortisol is hydrophobic; are airway cell membranes permeable to it?

Diffusion Moves Substances Down Concentration Gradients

The behavior of solutes in body fluids obeys the rule that substances naturally want to move down their concentration gradient: Substances in high concentration want to move into an area of lower concentration until the concentrations in the two areas become equal.

What prompts this movement of solutes? As with a bumper-car ride at an amusement park, the more particles, the more frequent the collisions. Over time, the net effect of these random collisions is that solute particles move away from areas where there are many collisions

(areas of high solute concentration) and into areas where collisions are less frequent (low solute concentration). They are, more or less, bounced out. The consequence is that solute particles distribute themselves evenly in a solution. No mixing is required; it happens naturally. This independent movement of substances down a concentration gradient from areas of high to low concentration is called **diffusion.**

We experience diffusion every day. For example, imagine a freshly baked tray of cookies (Fig. 3.18A). The

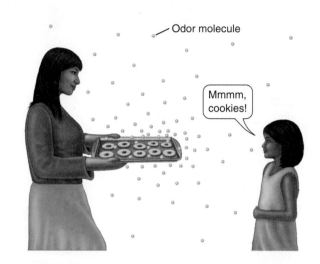

(a) Simple diffusion of odor molecules

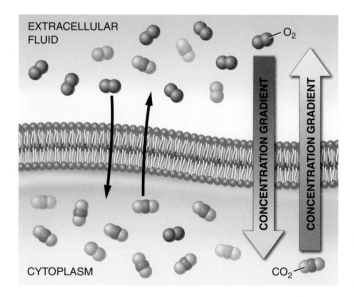

(b) Simple diffusion of gases

Figure 3.18. Simple diffusion. A. Cookies emit odor molecules, which move throughout the room by simple diffusion. **B.** Oxygen and carbon dioxide freely pass through the cell membrane. In working cells, oxygen enters cells and carbon dioxide exits cells down their respective concentration gradients. *Is the oxygen concentration higher inside or outside the cell?*

cookies are emitting odor molecules, which we interpret as smells. These odor molecules are initially concentrated next to the cookies, but in time they diffuse down a concentration gradient toward air that contains no odor molecules, and soon the room is filled with the scent of cookies. Diffusion is a passive process—in cells it occurs without expending energy from the cell's ATP reserves.

> **Remember This!** A solute diffuses down its concentration gradient.

Lipid-Soluble Substances Cross the Cell Membrane by Simple Diffusion

The diffusion of the odor molecules from the cookies occurs even if no one carries the cookies around. This form of diffusion, which does not involve membrane proteins, is called **simple diffusion.** Simple diffusion also occurs across the cell membrane, but only for lipid-soluble substances. For them, it is as if the cell membrane did not exist. The force of the concentration gradient is all it takes to propel their movement from one side of the membrane to the other, and escorts (e.g., membrane proteins) are not involved. In addition to the fact that it costs no energy, simple diffusion has no maximum speed—the steeper the concentration gradient, the faster the diffusion.

Oxygen and carbon dioxide are fat-soluble (and water-soluble, too) and move into and out of the cell with ease by simple diffusion (Fig. 3.18B). Oxygen is in high concentration in the extracellular fluid. Because cells consume oxygen during energy production, the intracellular oxygen concentration is lower than the extracellular concentration, and oxygen readily diffuses down its concentration gradient from the extracellular fluid into the cells. The opposite is true for carbon dioxide, which is in high concentration inside the cell because it is a waste product of energy production. It diffuses down a concentration gradient, first into the extracellular fluid and then into blood, where the concentration is lowest. Note that the movement of carbon dioxide *does not* affect the movement of oxygen, and vice versa. Each solute diffuses down its own concentration gradient, independent of the movements or gradients of other solutes.

Other than oxygen and carbon dioxide, only lipids such as cholesterol and lipid-soluble molecules such as certain vitamins can cross the membrane by simple diffusion.

Case Note

3.17. Julia's pneumonia interferes with oxygen diffusion from her lungs to her blood. Where is oxygen at a higher concentration—in the lungs or the blood?

Water-Soluble Substances Can Cross the Cell Membrane by Facilitated Diffusion

Gases and lipid-soluble molecules are, of course, not the only substances that cross the cell membrane. Untold numbers of ions, water, and nutrients must also pass through the cell membrane down their respective concentration gradients, but they cannot use simple diffusion. Instead, chemicals that are not lipid-soluble move down their concentration gradients by **facilitated diffusion**—their transport is facilitated (aided) by integral membrane proteins. Recall that integral proteins span the entire thickness of the membrane; their inner side is exposed to the cytosol and their outer side to the extracellular fluid. Proteins involved in facilitated diffusion can be divided into two classes: *channels* and *carrier proteins* (Fig. 3.19).

Channels are watery tunnels through the cell membrane. They have different shapes, which permit the passage of specific substances. Sodium and potassium ions, for example, cross the membrane through sodium and potassium channels (respectively). Water crosses using water channels called **aquaporins.** Channel-mediated diffusion can be regulated, because some channels (such as sodium channels) have gates that can be opened or closed (Fig. 3.19A).

Other substances, most importantly glucose, diffuse across the membrane through **carrier proteins.** Carrier proteins, like channels, only allow passage of specific solutes. However, unlike channel proteins, carriers must change shape in order to enable transport (Fig. 3.19B). In muscle cells, for example, glucose enters cells with the help of a carrier protein, which operates somewhat like a revolving door. As glucose binds to the extracellular end of the carrier protein, the carrier changes shape (the door revolves a half turn), and the binding site is now exposed to the intracellular compartment. Glucose dissociates from the binding site and continues its diffusion into the cytoplasm.

In contrast to simple diffusion, facilitated diffusion has a maximum rate of transport. Much like a hotel with only a few revolving doors, a cell has only has so many carrier proteins, and the carriers can admit only so many particles in a given period of time.

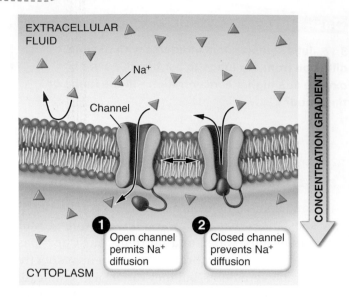

EXTRACELLULAR FLUID

Na+

Channel

1 Open channel permits Na+ diffusion

2 Closed channel prevents Na+ diffusion

CYTOPLASM

CONCENTRATION GRADIENT

(a) Channel-mediated facilitated diffusion

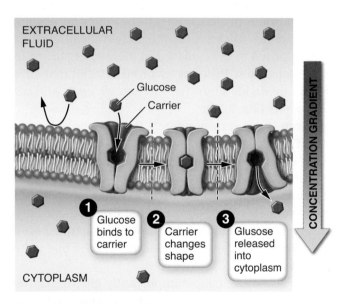

EXTRACELLULAR FLUID

Glucose

Carrier

1 Glucose binds to carrier

2 Carrier changes shape

3 Glusose released into cytoplasm

CYTOPLASM

CONCENTRATION GRADIENT

(b) Carrier-mediated facilitated diffusion

Figure 3.19. Facilitated diffusion. In facilitated diffusion, solutes move down their concentration gradients into or out of the cell with the help of integral proteins. **A.** Gated channels can open and close, controlling the movement of solute through the membrane. **B.** Carriers change shape in order to move substances through the membrane. *If this channel stays open, what will happen to the sodium gradient?*

Case Notes

3.18. The problems in Julia's airways reflect abnormal membrane transport of chloride ions. Given that chloride ions are hydrophilic, will they move by simple diffusion or facilitated diffusion?

3.19. The CFTR protein transports chloride ions across the cell membrane without changing shape. Is it a channel or a carrier protein?

3.20. The CFTR protein enables chloride ions to diffuse out of the cell. Which compartment must have the higher concentration of chloride ions—the cytosol or the extracellular fluid? Explain your reasoning.

Active Transport Requires Energy

Active transport augments or maintains concentration gradients by moving solutes "uphill" from an area of low concentration into an area of high concentration. Transport proteins called *pumps* force particles across a membrane from the low concentration side to the high concentration side. Unlike carriers involved in facilitated diffusion, these protein structures do not rely on the random movement of particles. Instead, pumps use energy obtained from ATP.

The most important pump is the **sodium–potassium pump,** abbreviated **Na+/K+-ATPase.** The abbreviation indicates that an ATP chemical bond is broken to generate energy for transport of sodium and potassium across the cell membrane (Fig. 3.20). In human cells, sodium is more concentrated outside the cell and potassium is more concentrated inside the cell. The Na+/K+-ATPase maintains this gradient against the steady "leak" by facilitated diffusion of these ions down their concentration gradients—potassium leaking out; sodium leaking in. It does this by forcing sodium ions out of the cell and potassium ions into the cell against their respective concentration gradients. We learn more about the importance of the Na+/K+-ATPase and the Na and K gradients in ➡ Chapter 4. By the way, substances that can pass through the cell membrane freely, like gases and other fat-soluble substances such as steroid hormones, cannot be actively transported—their transport is, in essence, uncontrollable.

> *Remember This!* Active transport does NOT depend on a concentration gradient.

Osmosis Is the Passive Movement of Water across a Semipermeable Membrane

Recall that the fatty, hydrophobic cell membrane is impermeable to hydrophilic solutes, which require specialized channels or carriers to cross the membrane.

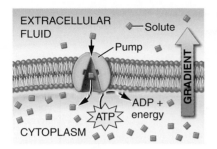

(a) Active transport

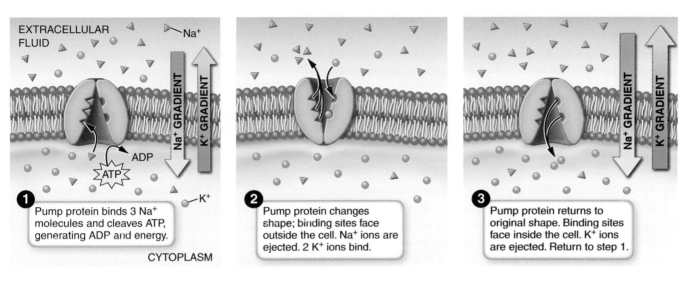

Figure 3.20. **Active transport. A.** Active transport moves solutes against their gradients, using the energy from ATP cleavage. **B.** The sodium concentration is higher outside the cell; potassium is more concentrated inside. The sodium–potassium pump maintains this gradient by exporting three sodium ions and importing two potassium ions during each transport cycle. *After step 3 has occurred, have the concentration gradients for sodium and potassium increased or decreased?*

① Pump protein binds 3 Na$^+$ molecules and cleaves ATP, generating ADP and energy.

② Pump protein changes shape; binding sites face outside the cell. Na$^+$ ions are ejected. 2 K$^+$ ions bind.

③ Pump protein returns to original shape. Binding sites face inside the cell. K$^+$ ions are ejected. Return to step 1.

(b) The sodium-potassium pump

No matter how strong the concentration gradient is between hydrophilic solutions across the cell membrane, solutes can diffuse to neutralize the gradient *only* if channels or carriers are present. Solutes that cannot cross the membrane to neutralize their gradient are called *nonpenetrating solutes.*

Any hydrophilic solute (such as NaCl) is considered to be nonpenetrating, because the cell permits only small amounts to enter or leave by regulated transport mechanisms. As in the adage, "If Mohammed cannot come to the mountain, then the mountain must come to Mohammed," if a solute cannot leave the more concentrated solution, water will move into it to dilute it. In essence, water is moving down the water gradient because less concentrated solutions have more water than more concentrated ones. Importantly, water's concentration depends on the total *number* of particles in the solution. The relative size of the particles does not matter: the movement of water does not discriminate between the largest protein particle and the smallest ion.

Solute concentration is expressed as **osmolarity:** the number of particles of solute per liter of solvent (water).

It is usually expressed as milliosmoles per liter (mOsm/L). An **osmotic gradient** exists between two solutions of different osmolarities. **Osmosis,** therefore, is the movement of water across a semipermeable membrane from the area of low solute osmolarity (higher water concentration) to an area of higher solute osmolarity (lower water concentration). Osmosis changes the volume of a solution, and in confined spaces (such as a cell) volume changes change the pressure (think of filling a water balloon until it explodes). The driving force of osmosis can thus be described as **osmotic pressure.**

To understand osmosis, imagine a semipermeable membrane separating two saltwater solutions of different concentrations (Fig. 3.21A). If the membrane does not permit salt to cross, water will move from the less salty side (which has the higher water concentration) to the saltier side (which has a lower water concentration) until enough water has moved to the salty side to dilute it to the point that the salt concentrations equalize (Fig. 3.21B).

It is possible to observe osmosis in action by placing red blood cells into solutions containing varying amounts of salt, glucose, or other nonpenetrating solutes (Fig. 3.22).

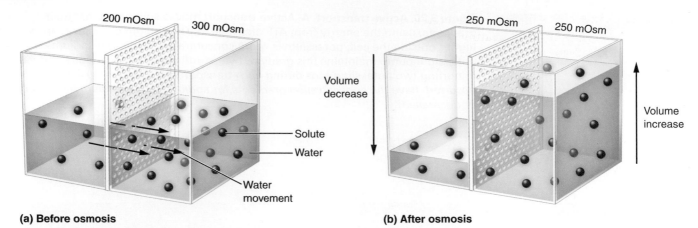

(a) Before osmosis

(b) After osmosis

Figure 3.21. Osmosis. A. Two compartments in a container are separated by a membrane that permits water but not solute to pass. The left-hand compartment has two-thirds the concentration of solute molecules (and correspondingly more water molecules) than the right-hand compartment. Before osmosis occurs, each compartment contains exactly 1 L of solution; that is, the volume of the two compartments is equal. **B.** After osmosis, water has moved from the left side to the right, equalizing the solute concentration but greatly changing the volume on either side. *If more solute were to be added to the right-hand compartment, what would happen to the volume of the left-hand compartment?*

Hypertonic solutions have a greater solute concentration than the intracellular fluid, which means that the water concentration is higher inside the cell than outside. Cells placed in a hypertonic solution will lose water and shrink, because water will exit the cell into the solution. **Hypotonic** solutions have lower solute concentration than the intracellular fluid, which means that water is in higher concentration outside the cell than inside. In this instance, water flows from the extracellular fluid into the cell; as a consequence the cell will swell and maybe even explode. **Isotonic** solutions are just right— the solute concentration is equivalent between the solution and the cell, and the cell volume will stay the same. Cells placed in isotonic solutions neither gain nor lose water by osmosis.

The osmolarity of most body cells and fluids is relatively fixed at about 300 mOsm, despite the fact that cells regularly import and export solute. In essence, the transport of each solute molecule is followed by the osmosis of one water molecule, so the osmolarity of the cell and the extracellular fluid remain unchanged. A good principle to remember is that *water follows solute*. For example, intestinal cells import nutrients, and, as shown later in the text, airway cells export chloride. Thus, as an intestinal cell absorbs glucose, it absorbs water as well, and as the airway cell secretes chloride it also secretes water.

Administration of intravenous fluids is a mainstay of medical treatment for any condition in which the body is short of water: excess loss by sweating, vomiting, or

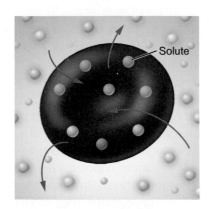

Isotonic: no volume change

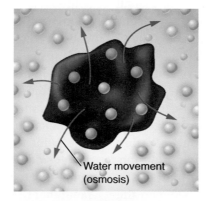

Hypertonic: cell shrinks

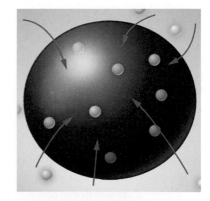

Hypotonic: cell swells

Figure 3.22. Effect of osmosis on volume of a red blood cell. Notice the changes in volume following immersion of a blood cell in isotonic, hypertonic, and hypotonic solutions. *Which solution causes cells to shrink?*

diarrhea, for example. The fluid administered is usually an isotonic solution of nonpenetrating solutes; that is, the solutes (usually a mixture of sodium, potassium, and bicarbonate, sometimes with glucose) have a combined osmolarity that matches body osmolarity. These substances can enter cells by regulated transport as required. The reason is twofold: (a) the patient has usually lost solute (salts) along with the water and (b) isotonic solutions of penetrating solutes do not cause shifts of water from one compartment to another. For example, the administration of pure water would cause a shift of water into cells and could swell them dangerously. Conversely, administration of a hypertonic solution would draw water out of cells with an equally dangerous effect.

> **Remember This!** Water moves from low osmolarity to high osmolarity down the *water gradient*.

Case Note

3.21. The ER nurse gives Julia an intravenous solution of glucose and electrolytes. The solute concentration of this solution was carefully determined so that it would not alter the volume of Julia's red blood cells. Was the solution hypotonic, isotonic, or hypertonic?

Case Discussion
Part 3: Julia's Mutation Affects Membrane Transport

Recall that to have a genetic defect is to have a defective (or absent) protein. The symptoms of cystic fibrosis that Julia exhibited—persistent cough and wheezing—can be traced back to a lack of CFTR proteins in the cell membrane. In normal airway cells, this channel enables chloride to exit the cell into the airway lumen (Fig. 3.23A). Recalling the rule of thumb that "water follows solute," water normally follows chloride ions into the airway lumen, diluting the respiratory secretions and keeping them sufficiently fluid.

Julia's airway cells do not have the CFTR protein in their membranes (Fig. 3.23B). Less chloride will move from the airway cell to the airway lumen, thus less water will enter the airway lumen to dilute the respiratory secretions. In other words, the chloride is trapped inside airway cells and becomes a nonpenetrating solute, so water remains within the airway cells to dilute the nonpenetrating solute. The lack of water renders respiratory secretions highly viscous (thick and gluelike). In the lungs, the thick mucus blocks airways and causes wheezing, coughing, and an increased risk of infection. Cystic fibrosis could, theoretically, be treated if the normal

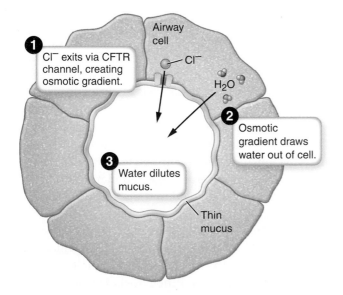

(a) Normal airway

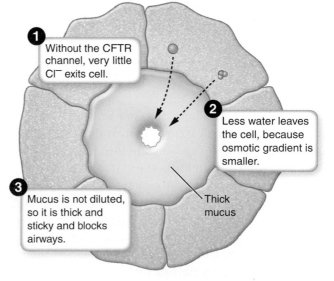

(b) Julia's airway

Figure 3.23. Cystic fibrosis and osmosis. Julia inherited a defective CFTR gene, which is missing three nucleotides. The CFTR protein synthesized from this gene is missing a phenylalanine and is degraded instead of being inserted into the cell membrane. The absence of the CFTR protein in the cell membrane impairs chloride transport, which impairs sodium diffusion and osmosis, resulting in airway blockages and salty sweat. *Which organelle degrades the abnormal CFTR protein?*

CFTR gene could be inserted into these troubled cells. See the Web site feature "Stem Cells and Cloning: A Cure for Cystic Fibrosis?" for more information.

In sweat gland cells (not shown), the chloride channel operates in the opposite direction: it is normally responsible for moving chloride from the sweat gland lumen into the cells. The chloride ions thus remain in the sweat. In cystic fibrosis patients, therefore, sweat has abnormally high chloride and sodium content. As you learned in ◀ Chapter 2, sodium chloride is ordinary table salt; thus, very salty sweat is a sign of cystic fibrosis. A laboratory test can detect abnormally high sodium and chloride in sweat when cystic fibrosis is a suspected diagnosis.

Incidentally, failure of chloride transport affects other organ systems as well. Unusually thick fluids tend to "gum up" pancreatic secretions in the intestine, which causes digestive problems. Similarly, thickened secretions in the male genital tract may be a cause of infertility.

Case Note

3.22. Why does a defect in the chloride channel reduce osmosis?

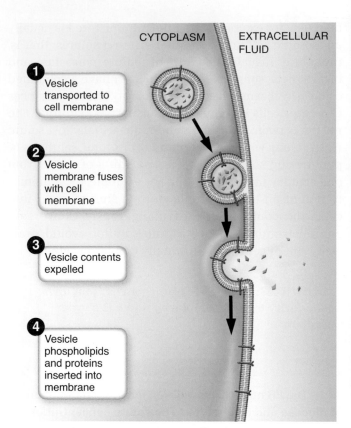

CYTOPLASM   EXTRACELLULAR FLUID

1 Vesicle transported to cell membrane

2 Vesicle membrane fuses with cell membrane

3 Vesicle contents expelled

4 Vesicle phospholipids and proteins inserted into membrane

Figure 3.24. Exocytosis. Exocytosis secretes substances from cells and inserts phospholipids and proteins into the cell membrane. *If exocytosis occurs in large amounts, what will happen to the cell membrane—will it grow or shrink?*

Cells Exchange Bulk Material with the Extracellular Fluid

Simple diffusion, osmosis, facilitated diffusion, and active transport move substances across the cell membrane particle by particle. *Exocytosis* and *endocytosis* are methods by which the cell takes in or expels substances in bulk.

Exocytosis—literally meaning "out of a cell"—is the method by which a cell expels material in bulk. The material inside the cell can be something the cell synthesized or it can be something the cell "ate" (ingested). In either case, the product to be expelled is first wrapped in a membrane to form a vesicle, which moves to and merges with the cell membrane, as illustrated in Figure 3.24. The vesicle then opens to the outside and ejects its contents into the extracellular space. The vesicle's membrane wall remains as a part of the cell membrane. By this device the cell membrane is modified, repaired, patched, or expanded. Also, as shown in Figure 3.17, exocytosis is the mechanism by which membrane proteins are inserted into the cell membrane.

Endocytosis—literally meaning "into a cell"—is the opposite of exocytosis. It, too, is an active process requiring

energy obtained from ATP. Endocytosis begins as the cell membrane bulges inward (Fig. 3.25). The membrane fuses and closes around a trapped substance, forming a vesicle in which the substance is enclosed in a sac of cell membrane. This vesicle then merges with a lysosome, whose enzymes digest the sac's contents. Some of the liberated building blocks—amino acids, for example—are recycled and used by the cell to synthesize new substances such as proteins. Digested material that cannot be recycled is repackaged and ejected from the cell by exocytosis.

Fluid-phase endocytosis (pinocytosis)—literally "cell drinking"—is a type of endocytosis in which the cell ingests small quantities of extracellular fluid and the substances it contains. Pinocytosis is nonspecific and does not operate to ingest any particular substance in the extracellular fluid, but it enables the cell to evaluate conditions in the extracellular fluid.

By contrast, **receptor-mediated endocytosis** is highly selective. In this process, a specific receptor on the cell membrane binds with a particular substance, usually a

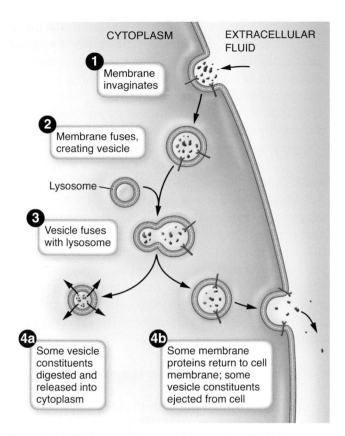

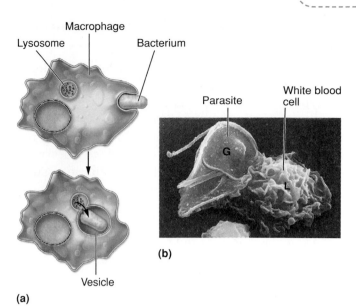

Figure 3.25. Endocytosis. In endocytosis, cells bring material from the extracellular fluid into the cytoplasm by packaging it in a vesicle made of cell membrane. *If endocytosis occurs in large amounts, what will happen to the cell membrane—will it grow or shrink?*

Figure 3.26. Phagocytosis. A. Some cells engulf microbes or worn-out cells using phagocytosis. The engulfed substance is enclosed in a vesicle, which follows the same path as the vesicle in Figure 3.25. **B.** A white blood cell engulfing a parasite. *Does phagocytosis resemble endocytosis or exocytosis?*

molecule too large to enter by other means. Once bound, the membrane folds inwardly and encapsulates the substance in a vesicle. Complexes made up of cholesterol and proteins (*lipoproteins*), for example, enter cells in this way.

Some cells combine endocytosis and exocytosis to move substances *across* the cell, a process sometimes described as *transcytosis*. Endocytosis brings the substance into the cell, which is subsequently transferred to another vesicle for exocytosis. Viruses such as HIV and herpes appear to pass through body linings via transcytosis.

Case Note

3.23. **Which process inserts the CFTR into the cell membrane—exocytosis or endocytosis?**

Phagocytosis—cell eating—like endocytosis, also brings substances, usually bacteria or other potentially harmful

substances, into the cell. Phagocytosis is restricted to mobile scavenger cells of the immune system—neutrophils and macrophages, for instance—called phagocytes. These cells ooze around a foreign particle, completely engulfing it in a large vesicle (Fig. 3.26). The vesicle fuses with a lysosome, which digests the vesicle's contents. Phagocytosis is used to destroy invading microbes and to clean up dead cell debris in diseased or injured tissue.

3.24 Which process transports a substance from a compartment where it is in short supply to an area where it is already highly concentrated?

3.25 What is the difference between hydrophilic and hydrophobic substances?

3.26 What is the difference between a carrier and a channel protein?

3.27 A vesicle membrane merges with the cell membrane at the end of what method of bulk transport?

3.28 Which mode of endocytosis is used to bring specific substances into the cell?

Tissue Types

We are built on bones, moved by muscles, controlled by nerves, and dressed in skin. And there you have it: the four types of tissue—bone is one form of *connective tissue*, muscles represent *muscle tissue*, nerves represent *nervous tissue*, and the top layer of skin, epidermis, represents *epithelial tissue*. In this chapter, we discuss epithelial tissue and connective tissue. Muscle tissue is covered in ➡ Chapter 6, and nervous tissue is taken up in ➡ Chapter 8.

Tissues are distinguished by (a) the cells of which they are formed and (b) the type and amount of extracellular material, called **extracellular matrix,** that surrounds and supports the cells.

Epithelial Tissue Covers the Body Surface, Lines Cavities, and Forms Glands

Epithelial tissue, also called **epithelium,** is a thin layer of cells lining the surface of the skin and hollow internal organs. Typically, epithelium is formed of cells that are packed closely together and function as a barrier (Fig. 3.27). There is minimal, if any, extracellular matrix separating the cells. Instead, epithelial cells rest on a thin sheet of specialized extracellular matrix, the **basement membrane,** composed mostly of protein. A sheet of epithelial cells together with its basement membrane is called an **epithelial membrane.**

Typically, epithelial membranes rest on a bed of connective tissue, which contains nerves, blood vessels, and supporting cells. (Note: In some circles, the bed of connective tissue is considered part of the epithelial membrane.) Notice that the epithelium itself does *not* contain blood vessels—oxygen and nutrients must diffuse across the basement membrane to reach the cells, a fact of extreme importance in cancer behavior. Epithelium also does not contain nerves. The most superficial layer of your skin is an epithelial membrane called the epidermis: If you've ever worked a tiny splinter out of your skin with a needle, you know that you can lift away the top layer without bleeding or pain.

Epithelial cells serve two major anatomic roles. First, they form membranes. Second, they form the secretory tissue of **glands,** special collections of cells that synthesize and secrete a product.

Epithelial cells have three functional roles:

- They serve as a *barrier* to separate tissues from the environment and protect the body. As epithelial membranes they cover and line; that is, they form the most superfi-

cial layer of the skin, for example, as well as the innermost lining of body cavities. No substance or microorganism can reach the internal environment of the body (blood and other fluids) without crossing a barrier formed by an epithelial membrane. At any given moment, thousands of bacteria are present on your skin and in your airways and digestive system, but infection will not occur unless they penetrate the epithelial layer.

- They *absorb* substances. For example, the epithelial membrane lining the intestines absorbs water and nutrients. What's more, epithelium acts as a gatekeeper of sorts, allowing beneficial substances to cross while rejecting others. For example, the epithelium that lines the intestines absorbs nutrients and water but keeps out bacteria and other particulate matter.

- They *secrete* substances. Secretion is a specialty of glands, groupings of epithelial cells that synthesize and secrete a product. For example, the salivary glands synthesize and secrete saliva.

Case Note

3.24. **Julia has a problem with the closely packed cells that line her airways. Are these cells part of epithelial tissue or connective tissue?**

Remember This! Epithelial membranes are composed of sheets of epithelial cells that rest on a basement membrane.

Epithelial Membranes Are Classified by Appearance

Epithelial membranes are classified according to the number of cell layers and the shape of the cells making up those layers. A *simple epithelial membrane* (or simple epithelium) is composed of a single layer of epithelial cells. Other membranes are composed of multiple layers of cells and are called *stratified epithelial membranes*. Each epithelial membrane is composed of different types and arrangements of cells according to its function.

Squamous cells are flat like a pancake. They occur in membranes either as a single layer, called *simple* squamous epithelium, or as multiple layers, called *stratified* squamous epithelium.

Simple squamous epithelium occurs where diffusion is important (Fig. 3.27A). For example, the walls of capillaries and the air sacs of the lungs are formed of a single layer of simple squamous epithelium to facilitate the passage of oxygen and carbon dioxide between blood and tissue on the one hand and between blood and air on the other.

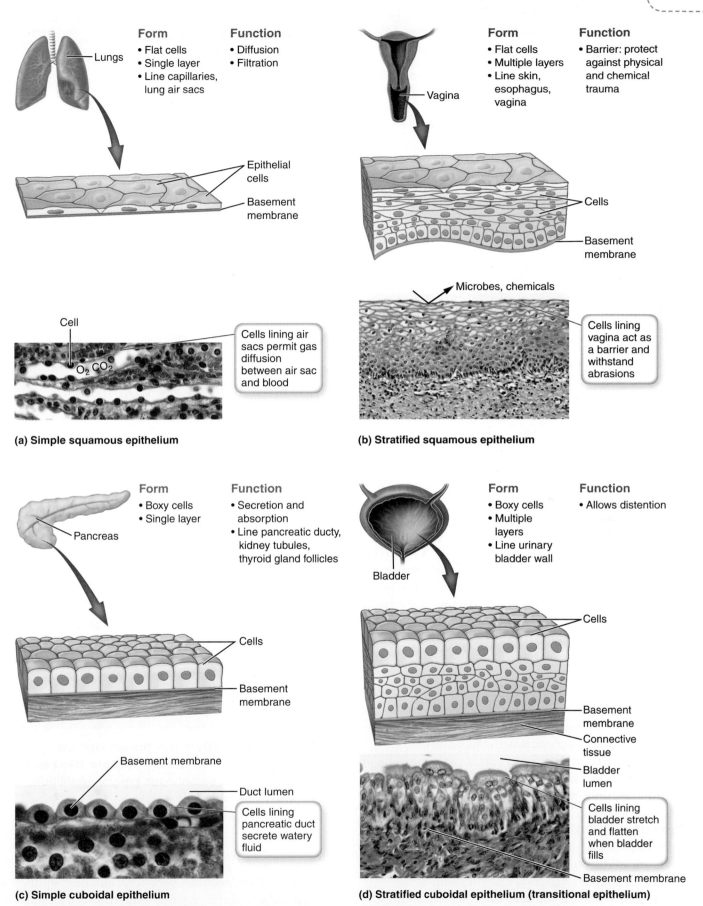

Form
- Flat cells
- Single layer
- Line capillaries, lung air sacs

Function
- Diffusion
- Filtration

Lungs

Epithelial cells

Basement membrane

Cell

O_2 CO_2

Cells lining air sacs permit gas diffusion between air sac and blood

(a) Simple squamous epithelium

Form
- Flat cells
- Multiple layers
- Line skin, esophagus, vagina

Function
- Barrier: protect against physical and chemical trauma

Vagina

Cells

Basement membrane

Microbes, chemicals

Cells lining vagina act as a barrier and withstand abrasions

(b) Stratified squamous epithelium

Form
- Boxy cells
- Single layer

Function
- Secretion and absorption
- Line pancreatic ducty, kidney tubules, thyroid gland follicles

Pancreas

Cells

Basement membrane

Basement membrane

Duct lumen

Cells lining pancreatic duct secrete watery fluid

(c) Simple cuboidal epithelium

Form
- Boxy cells
- Multiple layers
- Line urinary bladder wall

Function
- Allows distention

Bladder

Cells

Basement membrane

Connective tissue

Bladder lumen

Cells lining bladder stretch and flatten when bladder fills

Basement membrane

(d) Stratified cuboidal epithelium (transitional epithelium)

Figure 3.27. Epithelial tissue: Form and function. *Which term describes the number of layers—stratified or squamous? (continued)*

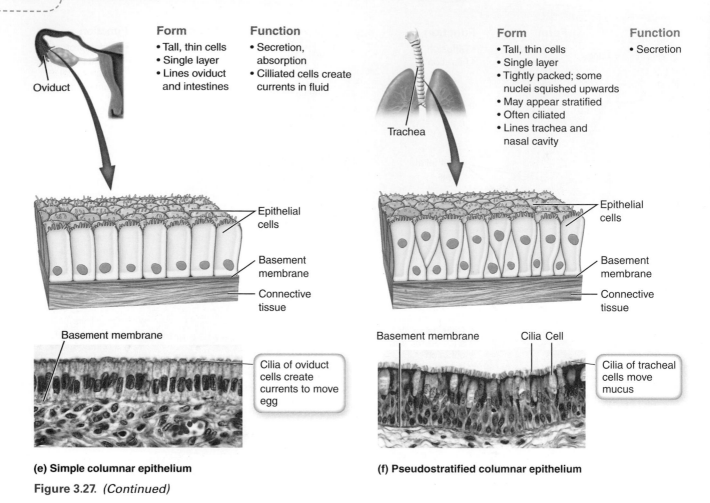

Form
- Tall, thin cells
- Single layer
- Lines oviduct and intestines

Function
- Secretion, absorption
- Cilliated cells create currents in fluid

Oviduct

Epithelial cells

Basement membrane

Connective tissue

Basement membrane

Cilia of oviduct cells create currents to move egg

(e) Simple columnar epithelium

Form
- Tall, thin cells
- Single layer
- Tightly packed; some nuclei squished upwards
- May appear stratified
- Often ciliated
- Lines trachea and nasal cavity

Function
- Secretion

Trachea

Epithelial cells

Basement membrane

Connective tissue

Basement membrane   Cilia Cell

Cilia of tracheal cells move mucus

(f) Pseudostratified columnar epithelium

Figure 3.27. *(Continued)*

Stratified squamous epithelium occurs in membranes where the barrier function is important (Fig. 3.27B). This tissue type is commonly subject to direct contact with the environment or indirect contact with environmental products, such as food (which may contain bacteria, irritants, or hard particles). For example, recall that the outermost layer of skin is an epithelial membrane, the **epidermis.** It is formed of many layers of flat squamous cells that, much like the shingles on a roof, resist the "weather" of bacteria, chemicals, rainwater, and other environmental insults. Also, a stratified squamous epithelium lines the esophagus (the food tube from mouth to stomach) to resist the irritations that sometimes accompany the passage of food.

Cuboidal cells are boxy and can occur in a single or multiple layers. A single layer of **simple cuboidal epithelium** often lines the ducts of glands; for example, the salivary glands (Fig. 3.27C).

Transitional epithelium lines the bladder and other organs capable of distention (Fig. 3.27D). Its name reflects its ability to transition to a less boxy, flatter shape as the bladder distends and fills with urine.

Columnar cells are tall and thin and stand upright, packed shoulder to shoulder. Columnar epithelium occurs most commonly as a single layer of cells, called **simple**

columnar epithelium (Fig. 3.27E). The intestines, for example, are lined by simple columnar epithelium.

In the lining of the respiratory tract, columnar cells are so thin and tightly packed that their nuclei, which are large, get shoved up or down in the cytoplasm in order to fit in the crowded space. Under the microscope, this arrangement gives a false (*pseudo*) appearance that multiple layers of cells are present. This type of epithelium is called **pseudostratified columnar epithelium** (Fig. 3.27F).

Case Note

3.25. Mucus frequently blocks the airways and pancreatic duct in cystic fibrosis patients. Use Figure 3.27 to identify and describe the type of epithelium lining each of these tubes.

Intercellular Junctions Join Cells Together

Two types of connections between cells help epithelial tissues form a waterproof barrier that is resistant to mechanical stress: tight junctions and desmosomes (Fig. 3.28). A **tight junction** is formed of a band of proteins

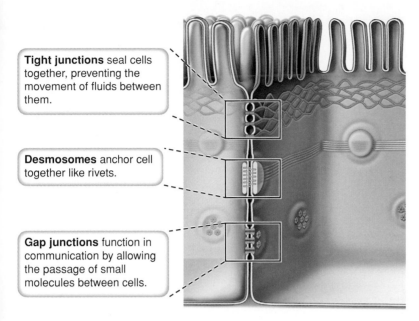

Tight junctions seal cells together, preventing the movement of fluids between them.

Desmosomes anchor cell together like rivets.

Gap junctions function in communication by allowing the passage of small molecules between cells.

Figure 3.28. Cell junctions. Tight junctions, desmosomes, and gap junctions structurally and functionally join together the cells of the intestinal wall. These junctions are also found individually or together in other tissues. *Which type of junction blocks the intercellular space the most?*

that fuse together the cell membranes of neighboring cells. Tight junctions create a waterproof seal between two cells, forming an impenetrable wall to prevent substances from moving between the cells. On the other hand, **desmosomes,** like buttons or rivets, hold cells tightly together at specific points but do not block the movement of substances between the cells. Intermediate filaments act as braces by joining desmosomes together.

A third type of junction—a **gap junction**—connects neighboring cells, not to seal or bind them but to serve as a means of communication. Gap junctions are special protein tunnels between neighboring cells. Small particles, such as ions, can pass from one cell to another through them to relay messages without entering the extracellular fluid. Intercellular junctions are found in nonepithelial tissues, too, particularly some types of muscle tissue.

As illustrated in Figure 3.28, tight junctions, desmosomes, and gap junctions structurally and functionally join together the cells of the intestinal wall. These junctions are also found individually or together in other tissues.

Case Note

3.26. Studies show that normal CFTR proteins are necessary for normal gap junction activity. Which of the following would be impaired in Julia's epithelial membranes: cell-to-cell communication or structural integrity?

Glands Are Made of Epithelial Cells

A gland is a mass of cells that synthesizes and secretes a product (the *secretion*) from raw materials obtained

from blood. There are two types of glands: those with ducts and those without ducts (Fig. 3.29). In **exocrine** glands, the epithelial cells are arranged into small sacs (*acini*) that secrete their product into ducts, which are lined by an epithelial membrane and convey the secretion to a body surface or a body cavity (Fig. 3.29A). For example, the breast contains mammary glands that, following childbirth, secrete milk into ducts, which deliver it to the nipple.

Endocrine glands are ductless, solid masses of epithelial cells (Fig. 3.29B). They secrete their products (chemicals called **hormones**) into tiny blood vessels; no duct is involved. Hormones circulate widely in the bloodstream and act as chemical messengers (➡ Chapter 4) that can affect any cell in the body. However, not all endocrine cells are found in endocrine glands. Some endocrine cells are scattered about in various organs, but in these tissues they are not gathered together to form a discrete gland. For example, certain cells in the stomach wall secrete hormones that stimulate other stomach cells to secrete acid into the lumen of the stomach.

Remember This! Exocrine glands secrete their product into ducts. Endocrine glands secrete their product directly into blood.

Case Note

3.27. Julia has trouble with her sweat glands. Are these glands exocrine or endocrine?

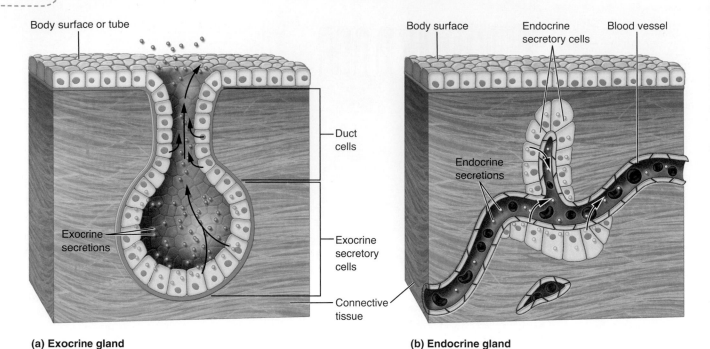

(a) Exocrine gland

(b) Endocrine gland

Figure 3.29. Glands. Glands consist of modified epithelial membranes. **A.** Exocrine glands. **B.** Endocrine glands. *Which type of gland secretes substances into the blood?*

Connective Tissue Connects Cells, Tissues, and Organs

The definition of **connective tissue** is tricky. It is a category of exclusion rather than one of inclusion; that is, connective tissues are those that are not muscle, nerve, or epithelium. Also, the word *connective* doesn't fit the entire category very well because blood, a connective tissue, does not connect cells, tissues, or organs except in the very broad sense of the word: blood facilitates communication between cells, and one definition of *to connect* is "to communicate." However, connective tissue has three other characteristics that make it distinctive:

- It is structural or supportive.
- It is derived from *mesoderm*, a primitive embryonic tissue (the other two are ectoderm and endoderm).
- It contains a large amount of extracellular substance, the extracellular matrix. Most connective tissue cells are not in direct contact with one another because they are separated by the matrix.

Remember This! A key property of connective tissue is its abundance of extracellular substance, the matrix, which separates the cells.

Connective Tissue Contains a Large Amount of Extracellular Matrix

The **extracellular matrix** of connective tissue consists of two components—an amorphous (structureless) *ground substance* and *fibers*. Connective tissue cells manufacture both components, which can be seen in Figure 3.30.

Ground substance is composed of water, minerals, and small glycoproteins. Both connective tissue cells and the fibers they synthesize are embedded within it. In fact, the ground substance often has a sticky character that holds the cells and fibers together. According to its content of certain proteins, carbohydrates, minerals, and water, the ground substance can be liquid, gelatinous, rubbery, or rock-hard.

Fibers are elongated protein filaments made by a special type of cell, the **fibroblast.** Fibroblasts make two types of protein: *collagen* and *elastin*.

Collagen protein is woven into two types of fibers, both of which are single, unbranched filaments— collagen fibers and reticular fibers. *Collagen fibers* consist of collagen protein woven into thick, parallel bundles designed to withstand the high tension of musculoskeletal movements. They are mainly found in tendons and ligaments. *Reticular fibers* consist of collagen protein woven into thin, delicate fibers that form the lacy supporting structural network of various organs.

Elastin is woven into *elastic fibers*, which are thin, branching filaments. Although they are weak, elastic fibers can stretch easily and recoil like a spring to resume their original shape and length. Elastic fibers give springy resiliency to some tissues, such as the vocal cords.

There Are Several Types of Connective Tissue

Connective tissue occurs in two broad forms: connective tissue proper and specialized connective tissue (Fig. 3.30).

Connective tissue proper consists of:

- Loose connective tissue
- Dense connective tissue

Specialized connective tissue consists of:

- Cartilage
- Bone
- Blood

Connective Tissue Proper

Loose connective tissue (or **areolar,** meaning "open," tissue) is lacy and filamentous like a spiderweb (Fig. 3.30A). It contains a few fibroblasts and lymphocytes, some fat cells (adipocytes, which are capable of storing fat), and a thin mixture of collagen and elastin fibers. It loosely binds skin to underlying organs, fills gaps between

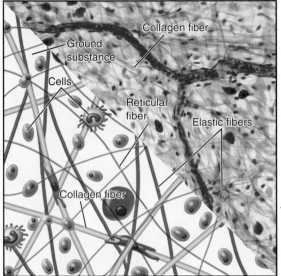
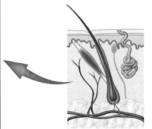

Form
- Cells: fibroblasts, adipocytes, lymphoid cells
- Extracellular matrix: collagenous, reticular, elastic fibers in disorganized network; semiliquid ground substance

Function
- Binds skin to underlying organs
- Forms middle wall of intestine
- Fills gaps between muscle fiber bundles
- Forms one layer of the meninges (protective brain cover)

(a) Loose (areolar) connective tissue

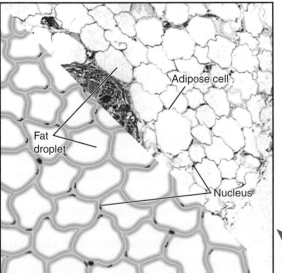
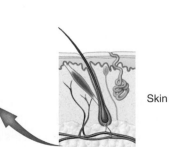

Form
- Modified loose connective tissue
- Cells: adipocytes
- Minimal extracellular matrix

Function
- Insulates and cushions internal organs, joints
- Stores energy in the form of triglycerides

(b) Adipose tissue

Figure 3.30. Connective tissue: Form and function. *(continued)*

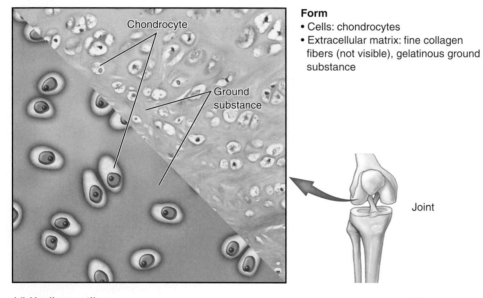

Form
- Cells: a few fibroblasts
- Extracellular matrix: collagenous fibers arranged in bundles (shown here) or irregular networks

Function
- Withstands strong forces
- Attaches structures together
- Forms scar tissue, ligaments, tendons, joint capsules, organ coverings (fascia)

Joint

(c) Dense connective tissue

Form
- Cells: chondrocytes
- Extracellular matrix: fine collagen fibers (not visible), gelatinous ground substance

Function
- Protects and cushions bone ends
- Provides structure to the voice box
- Binds rib ends to sternum

Joint

(d) Hyaline cartilage

Figure 3.30. *(Continued)*

muscle bundles and between certain organs, forms the important middle layer of the wall of the intestines (the submucosa, ➡ Chapter 14), and forms the outer sheath of large arteries.

Normally, the adipocytes in loose connective tissue store little fat. Sometimes, however, they accumulate so much intracellular fat that they crowd out other cells and the extracellular matrix. When this occurs, the tissue is described as **adipose tissue** (see Fig. 3.30B). Adipose tissue is still considered connective tissue despite the lack of extracellular matrix. It stores energy in the form of triglycerides.

Dense connective tissue (also called *fibrous* connective tissue) contains few fibroblasts but has a much higher fiber content than loose connective tissue (Fig. 3.30C). It is packed with collagen fibers woven into thick parallel bundles, is tough, and can withstand strong forces. It is the stuff of *scars* (which bind together the edges of a wound), *tendons* (which connect muscle to bone), *ligaments* (which usually connect bones across a joint), the *capsules* that encase many organs (the meninges that cover the brain and spinal cord, for example), and *fascia* (broad sheets of fibrous tissue that cover skeletal muscles and other tissues).

Specialized Connective Tissue

Cartilage is composed of *cartilage tissue*, which contains cartilage cells (*chondrocytes*), varying amounts of

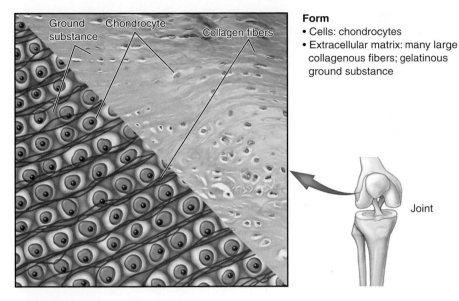

Form
- Cells: chondrocytes
- Extracellular matrix: many large collagenous fibers; gelatinous ground substance

Function
- Joins bones together
- Forms pads between vertebrae (discs)
- Joins pelvis bones to spine
- Joins bones to hyaline cartilage

Joint

(e) Fibrocartilage

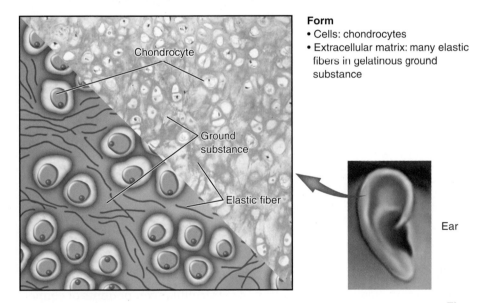

Form
- Cells: chondrocytes
- Extracellular matrix: many elastic fibers in gelatinous ground substance

Function
- Forms supple, flexible structures like the earlobe and upper part of voice box

Ear

(f) Elastic cartilage

Figure 3.30. *(Continued)*

collagen and elastin fibers, and a rubbery ground substance that imparts springy resiliency and keeps it lubricated. Cartilage is firm and somewhat flexible but not as hard as bone. It is designed to yield to force and return to its normal shape. It provides cushioning for joints (where bones meet) and shape to certain structures (the tip of the nose, for example). Types of cartilage vary according to the type and amount of fiber:

- **Hyaline** (meaning "glassy") cartilage contains fine collagen fibers and gelatinous ground substance and provides protective covering and cushioning for the ends of bones where they meet to form joints, such as the knee (Fig. 3.30D). It also provides structure to the

voice box (larynx), and binds the tips of the ribs to the breastbone (sternum).
- **Fibrocartilage** is very tough and contains more and coarser collagen fibers than hyaline cartilage (Fig. 3.30E). It joins pelvic bones to the spine and forms the cushioning pads (discs) that occur between the stack of bones (the vertebrae) that form the spine.
- **Elastic cartilage** is supple and flexible because of its high content of elastin fibers (Fig. 3.30F). It is found mainly in the external ear and parts of the larynx.

Bones are composed of *bone tissue*, which contains bone cells called *osteocytes* and a calcified extracellular

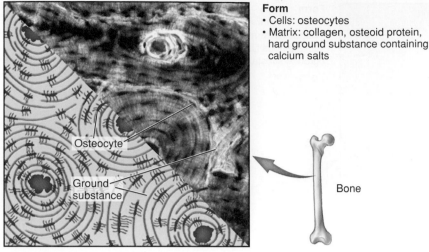

Form
- Cells: osteocytes
- Matrix: collagen, osteoid protein, hard ground substance containing calcium salts

Function
- Forms bones

(g) Bone

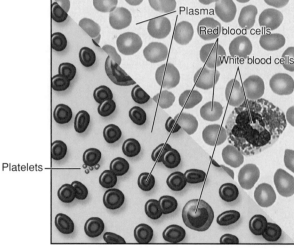

Osteocyte

Ground substance

Bone

Form
- Cells: red blood cells, white blood cells, platelets
- Matrix: liquid plasma

Function
- Forms blood

Plasma

Red blood cells

White blood cells

Platelets

Heart

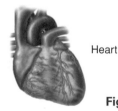

(h) Blood

Figure 3.30. *(Continued) Which type of connective tissue contains more collagen—elastic cartilage or fibrocartilage?*

matrix (Fig. 3.30G). The extracellular matrix is composed of collagen fibers and a specialized protein, *osteoid*, which becomes encrusted with calcium salts. Bone is rigid but not rocklike; it is very slightly flexible and designed to yield a bit under great force, a feature that avoids fragility and fractures.

Surprising to most, **blood** qualifies as connective tissue, but it is a special case. Blood "connects" only in the sense that it "communicates" by moving from one place to another to deliver water, cells, and dissolved molecules. However, like all connective tissues, it contains a large amount of extracellular material, the plasma, which is formed mainly of water and proteins and accounts for about 60% of blood volume (Fig. 3.30H). It also contains potential fibers: in response to trauma, a special plasma protein called *fibrinogen* links together to form a web of *fibrin* fibers as blood clots.

The cells and extracellular components of connective tissue are summarized in Table 3.4.

Pop Quiz

3.29 Name three functions of epithelial tissues.

3.30 Describe the cell shape and number of cell layers in simple cuboidal epithelium.

3.31 Which junction blocks water movement between cells?

3.32 Name three types of fibers found in connective tissue.

3.33 Name five types of connective tissue.

Table 3.4 Common Connective Tissue Components

Component	Form	Function
Cells		
Fibroblasts	Spindle-shaped cells	Synthesize components of extracellular matrix
Adipocytes	Cells containing large fat droplets	Cushion, insulate
Lymphoid cells	Mast cells, macrophages, plasma cells	Defend against foreign microbes
Extracellular Matrix		
Collagen fibers	Thick ropes of collagen	Help tissues resist physical deformation
Reticular fibers	Thin, delicate strings composed of collagen	Help support fragile tissues
Elastin fibers	Thin cables formed of elastin	Provide elastic recoil for tissues that become distended
Ground substance	Nonfibrous, amorphous substance that contains fibers and cells; consistency varies between tissue types	Provides support

Word Parts

Latin/Greek Roots	English Equivalents	Examples
cyt-/-o, -cyte	Cell	Endocytosis: bring something *into* the *cell*
End-/o-	Inside	
Ex-/o-	Outside	Exocrine: a *secretion outside* the body
-crine	Secretion	
Lys-/-o	Break down	Lysosome: a *body* that *breaks down* ingested substances
-some	Body	
Inter-	Between	Interphase: *between* cell divisions
Pro-	Before, forward, precursor	Prophase: the first stage, or *precursor,* of mitosis
Meta-	After, beside, with	Metaphase: chromosomes line up *beside* each other
Ana-	Upward, apart	Anaphase: chromosomes move *apart*
Telo-	End	Telophase: the *end* of mitosis
Iso-	Equal	Isotonic: a solution of *equal* osmotic pressure as the cell
Hypo-	Below normal, below	Hypotonic: a solution of *lower* osmotic pressure than the cell
Hyper-	Above normal, above	Hypertonic: a solution of *higher* osmotic pressure than the cell
Epi-	On	Epithelial: tissue *on* all body surfaces

(continued)

Word Parts (continued)

Latin/Greek Roots	English Equivalents	Examples
Fibr-/-o	Fiber	Fibroblast: cell that makes *fibers*
Chondr-/-o	Cartilage	Chondrocyte: *cartilage* cell
Pseudo-/-o	False	Pseudostratified: tissue that *falsely* appears to be stratified (in multiple layers)

Chapter Challenge

CHAPTER RECALL

Match up the organelles with their function:

1. **Ribosome**

2. **Lysosome**

3. **Proteosome**

4. **Smooth endoplasmic reticulum**

5. **Mitochondria**

6. **Rough endoplasmic reticulum**

a. Degradation of misfolded proteins

b. Synthesis of proteins

c. Synthesis of lipids

d. Protein modification

e. Degradation of ingested materials

f. Generation of ATP

7. **Tiny projections of the cell membrane that increase the absorptive surface of the cell are called**
   a. cilia.
   b. microvilli.
   c. flagella.
   d. centrioles.

8. **The cytoskeletal filaments made from intertwined fibrous proteins are called**
   a. intermediate filaments.
   b. microfilaments.
   c. macrofilaments.
   d. microtubules.

9. **Multipotent stem cells**
   a. are found in adults.
   b. can produce any tissue.
   c. can produce an entire organism.
   d. are present in the eight-cell embryo.

10. **Osmosis is**
    a. the diffusion of solute from the area of lower water concentration to the area of higher water concentration.
    b. a force that can change the volume of a cell.
    c. the movement of water from an area of high osmolarity to an area of low osmolarity.
    d. dependent upon ATP.

11. **The correct order of the stages of mitosis is**
    a. metaphase, anaphase, telophase, prophase.
    b. anaphase, metaphase, prophase, telophase.
    c. prophase, metaphase, anaphase, telophase.
    d. telophase, prophase, metaphase, anaphase.

12. **The nucleotide sequence AUGGAA**
    a. is synthesized by translation.
    b. contains codons.
    c. contains triplets.
    d. would be found only in the nucleus.

13. **The type of RNA that carries amino acids is**
    a. messenger RNA.
    b. transfer RNA.
    c. ribosomal RNA.
    d. small nuclear RNA.

14. **Which of the following processes occurs entirely in the cytoplasm?**
    a. Transcription
    b. Replication
    c. Translation
    d. All of the above

15. The sequence AAAGGGCCC would encode
    a. one amino acid.
    b. three amino acids.
    c. six amino acids.
    d. nine amino acids.

16. **Membrane proteins (such as the CFTR protein) move between the endoplasmic reticulum and the Golgi apparatus**
    a. by facilitated diffusion.
    b. using vesicles.
    c. by active transport.
    d. by exocytosis.

17. **The concentration gradient is irrelevant when substances move by**
    a. active transport.
    b. simple diffusion.
    c. osmosis.
    d. facilitated diffusion.

18. **Membrane proteins are used to transport substances by**
    a. facilitated diffusion.
    b. simple diffusion.
    c. osmosis.
    d. none of the above.

19. **Substances exit the cell by**
    a. pinocytosis.
    b. exocytosis.
    c. receptor-mediated endocytosis.
    d. phagocytosis.

20. **A membrane composed of multiple layers of flat cells is called**
    a. simple squamous epithelium.
    b. simple columnar epithelium.
    c. stratified squamous epithelium.
    d. stratified cuboidal epithelium.

21. **Desmosomes are used to**
    a. enable neighboring cells to communicate.
    b. block the space between adjacent cells.
    c. rivet epithelial cells together.
    d. make an epithelial membrane waterproof.

22. **Which of the following is a cell found in connective tissue?**
    a. Reticular fiber
    b. Chondrocyte
    c. Collagen
    d. Ground substance

23. **Adipose tissue is a modified form of**
    a. dense connective tissue.
    b. cartilage.
    c. loose connective tissue.
    d. epithelial tissue.

24. **Which of the following is NOT classified as connective tissue?**
    a. Cartilage.
    b. Bone.
    c. Blood.
    d. All of the above are connective tissue.

CONCEPTUAL UNDERSTANDING

25. **"Following DNA replication, one DNA molecule contains two new nucleotide strands and the other molecule contains the two original nucleotide strands." Is this statement true or false? Explain.**

26. **Explain why large steroids, such as cortisol, can pass through the cell membrane freely, whereas tiny hydrogen ions cannot.**

27. **Compare and contrast facilitated diffusion and active transport. List at least two similarities and two differences.**

28. **What is the difference between mitosis and cytokinesis? What would happen if one event happened without the other?**

29. **Compare and contrast blood and bone. List at least one similarity and two differences.**

APPLICATION

30. **Based on your understanding of stem cells, explain why skin injuries regenerate more easily than heart injuries.**

31. **Using Figure 3.12 as a guide, sketch the four stages of mitosis in a cell with a single chromosome. Use different colors for the two sister chromatids in each stage. Which stage of mitosis determines which daughter cell will receive which chromatid?**

32. **A solution has a lower osmolarity than the cytoplasm of a cell (assume that all solutes are nonpenetrating). State whether each statement is true or false and defend your answer.**
    a. The solution has a higher water concentration than the cell.
    b. The cell would swell up if placed in the solution.
    c. The solution is hypertonic, has a higher osmolarity, and shrinks cells.

You can find the answers to these questions on the student Web site at **http://thepoint.lww.com/McConnellandHull**

4

Communication: Chemical and Electrical Signaling

Major Themes

■ Cell-to-cell communication is critical for homeostasis and life.

■ Communication requires a *sender*, a *signal*, a *medium* to carry the signal, and a *receiver* to accept the signal.

■ Signals must be "translated" from a code into action.

■ The effect of the signal is determined by the receiver, not merely by the signal.

■ There are two kinds of physiological signals: chemical and electrical.

Chapter Objectives

8. Sketch a neuron, labeling the cell body, dendrites, axon, and myelin.

9. Compare and contrast action potentials and graded potentials.

10. Name the four phases of the action potential, and discuss the involvement of sodium and/or potassium channels in each phase.

11. Compare action potential propagation in myelinated and unmyelinated neurons.

12. Using the example of gamma aminobutyric acid (GABA), list all of the events that occur at a chemical synapse.

13. Explain why Andy's synaptic activity is reduced when he drinks decaffeinated coffee. In your explanation, use the following terms (not necessarily in order): *caffeine, adenosine, first* and *second messengers, endogenous ligands,* and *receptor antagonists;* also explain the importance of changes in receptor number and action potential thresholds.

14. Using examples from the case study, discuss the roles of senders, signals, mediums, and receivers in person-to-person and cell-to-cell communication.

Case Study: "I must be getting the flu."

As you read through the following case study, assemble a list of the terms and concepts you must learn in order to understand Andy's condition.

During the yearly Christmas visit with his in-laws, Andy M. began to feel bad.

"I must be getting the flu," he said to his wife. He hadn't felt right since the day after they'd arrived earlier in the week. "Max pounded me yesterday," he added, referring to a squash match with his brother-in-law. "I can usually whip him without much trouble, but yesterday I just couldn't get going. I'm tired and cranky, I can't maintain enough focus to read a paragraph, and to top it all, I have this splitting headache. I've been popping aspirin all day, but nothing helps!"

After dinner that night, his mother-in-law offered coffee. "It's decaffeinated," she said, "so it won't disturb your sleep."

Suddenly, Andy's muddled brain began to focus. *Decaffeinated.* "Barbara," he asked, "what kind of coffee have you been serving during the day?"

"Oh, it's decaf!" his mother-in-law answered cheerily. "After all, we're on holiday. Dick and I drink regular coffee during the work week and decaf on weekends. Too much caffeine isn't good for your health."

Eureka! Andy retreated to the kitchen, where he fished out some regular coffee and fired up the coffeemaker. As soon as the brew was ready, he guzzled two cups filled to the brim. Then he returned to the dining room, smiling and relieved. His "flu" had magically disappeared. "Sorry if I've been a bit testy," he said to his mother-in-law. "I feel better already. I confess—I'm a caffeine addict, and believe me, caffeine withdrawal is no picnic."

Andy's story illustrates the adverse effects of chronic overconsumption of caffeine, which is a type of drug called a *stimulant.* It all comes back to communication—drugs often interfere with signals passing from one cell to another. Caffeine, for instance, blocks sleep-inducing signals carried by a molecule called *adenosine.* Used moderately, caffeine enhances performance and improves mood and energy levels. But as with many drugs, a regular high intake can result in *addiction,* and caffeine withdrawal is an unpleasant experience. Read on to discover how caffeine exerts its effects by altering chemical and electrical communication.

111

Need to Know

It is important to understand the terms and concepts listed below before tackling the new information in this chapter.

- Homeostasis and negative feedback (← Chapter 1)

- Ions, hydrophobic, hydrophilic, proteins, steroids, and enzymes (← Chapter 2)

- Glands, cell membrane, concentration gradients, diffusion, active transport, and exocytosis (← Chapter 3)

Communication is so much a part of our daily life that we forget about it: we live in a hurricane of television images, music, text messages, conversations, phone calls, and e-mails . . . and we scarcely give a thought to anything but the message. We fail to appreciate the marvel of communication itself—its many parts or its varied means. And deep inside our bodies rages an even greater storm of communication: cells exchanging billions of messages every day. All communication is in code. The letters of this sentence, which your brain decodes and assigns a meaning, are really nothing more than curious shapes of ink on paper. And in our case, the spoken word *decaf* is nothing more than a sound that describes a type of coffee, which your brain decodes because you understand the sound codes called English. Your brain assigns meaning—"a drug that makes me alert"—to the sound code *caffeine* according to your stored knowledge of caffeinated beverages.

This chapter examines our internal communications—how signals originate and how they are transmitted, received, and interpreted. From the external environment we receive a variety of signals by means of our senses: eyes, light; ears, sound; nose, odors; tongue, taste; and skin, touch. In our internal environment, however, there are but two ways of signaling—chemical and electrical.

The medium is the message.

Marshall McLuhan (1911–1980), Canadian educator, scholar and philosopher, in his book *Understanding Media* (1964), arguing that the medium itself—television, for example—is an essential part of the message.

The Nature of Communication

Communication is the transmission and reception of information by a signal. It requires a *sender*, a *signal*, a *medium* (through which the signal is transmitted), and a *receiver*. Let's go back to the case, when Barbara tells Andy that his coffee was decaffeinated (Fig. 4.1).

1. Barbara is the *sender*. She wants to send Andy a message that the coffee is decaffeinated.
2. The coordinated contraction of Barbara's muscles expels air from her chest and manipulates the resulting sound into a *signal*—a specific series of sound waves that will be interpreted as "It's decaf."

3. The sound waves (the signal) travel through the *medium* of air to Andy.
4. Andy is the *receiver*. His ears receive the sound waves, and his brain decodes them with the meaning *I have been drinking decaffeinated coffee.*

Bodily Signals Are Chemical or Electrical

In everyday life we send many types of signals, from e-mails to smiles. Cell-to-cell signals take two forms, and these are the topic of the two main divisions of this chapter: *Chemical signals* are proteins, lipids, or even gases secreted by cells that prompt an effect in neighboring or distant cells. *Electrical signals* are changes in the overall balance of negative and positive ions inside

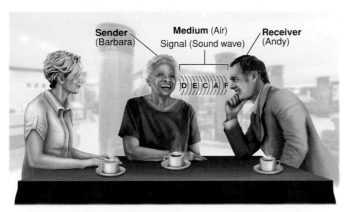

Figure 4.1. Communication. All communication requires a sender, a signal, a medium, and a receiver. *What is the medium in the example?*

and outside a cell that transmit signals along the cell membrane.

Messages can be carried by a series of electrical and chemical signals, much as a note is passed hand to hand across a classroom. For example, Barbara's brief words "It's decaf!" began as an electrical signal traveling from one end to the other of a brain cell (a *neuron*). Next, a chemical signal passed the signal to the next neuron in the series, which used an electrical signal to transmit the signal down its entire length. The signal continued to pass from neuron to neuron in this way until it reached Barbara's speech muscles, causing them to contract and produce and manipulate sound waves.

Communication Is Critical in Homeostasis

Recall that in ◀ Chapter 1 we introduced the term *homeostasis* and defined it as "The body's collective *communication* and control effort to maintain constant, healthy internal conditions." Homeostasis is the core goal of all physiological activity and depends on the ability of every cell to send and receive communications.

All cells participate in many homeostatic signal loops; that is, they both transmit and receive signals that help maintain the body in good health. Recall that all physiological conditions—blood pressure, for example—have a set point; that is, a value near which the condition must be maintained for optimal health. The cells that regulate each of the body's physiological conditions have sensors (receivers) that detect deviations from the set point. When a cell's sensor detects a deviation, it generates a signal asking for an opposing change. This activity, which

you will recognize as negative feedback, requires continual back-and-forth communication.

Problems with communication can cause disease. For instance, caffeine normally blocks the activity of a calming chemical signal called *adenosine*. Without his normal intake of caffeine, Andy was suffering from too many adenosine signals. The first gulp of caffeinated coffee stopped the adenosine signals like the cutting of a telephone line, and Andy suddenly felt much better.

Case Note

4.1. **In Andy's case, what disrupted his homeostasis?**

 4.1   Name the two types of homeostatic signals.

4.2   If a dolphin sends a sound wave signal to another dolphin, what is the medium?

Chemical Signaling

Chemical signals are molecules that serve bodily communication. Some chemical signals are small molecules, such as adenosine. Recall from Chapter 2 that adenosine is a building block of adenosine triphosphate (ATP), deoxyribonucleic acid (DNA), and ribonucleic acid (RNA). It is also a chemical signal that, as we will see throughout this chapter, plays a critical role in our case study. Other chemical signals include *amines*, which are modified amino acids; *steroids*, which are modified cholesterol molecules; and *proteins*. Even gases can act as chemical signals.

Regardless of their structure, all chemical signals share the same travel itinerary—they are released from a *secreting cell*, travel through a fluid to a *target cell*, and affect the activity of the target cell by binding a specific *receptor*. Because they bind to receptors, chemical signals are also referred to as **ligands** (Latin *ligare* = "to bind").

Chemical Signals Require Specific Receptors

Could you read a message in Braille? Probably not. If you're like most sighted people, you lack the "decoder" (information in your brain) necessary to interpret the raised dots of Braille code (you can read more about

Braille in the nearby History of Science box, titled *Braille: A Reading Code for the Blind*). In the same way, chemical signals come into contact with every cell in the body, but only those with the correct decoder can understand the message. Decoders of chemical signals are cell **receptors,** proteins that change the activity of the cell when bound by the chemical signal.

To put it another way, chemical signals exert an effect only on cells that contain the correct receptor; that is, cells to which they can bind. Cells without the correct receptor do not bind the signal molecule and do not respond. This principle is illustrated on the right side of Figure 4.2. One chemical signal (circles) will bind to the receptor on the upper cell, conveying the message that the cell should divide. A different signal (triangles), conversely, will bind to the receptor on the lower cell, conveying the message to undergo apoptosis and die. Neither cell receives the message intended for the other cell, because they do not have the other type of receptor.

> *Remember This!* **Ligands are chemical signals; receptors are receivers that bind the ligand.**

Chemical Signals Affect Neighboring and Distant Cells

Communication between different cells of the body works the same way as communication between people: there is a sender, a signal, a medium, and a receiver. Chemical signals can be classified according to the sender and the medium.

- **Hormones** are released by body cells and travel through the bloodstream (the medium) to act on *distant* cells (Fig. 4.2). They are also described as endocrine secretions (endo = internal), or endocrines for short. Although virtually all cells release hormones, the main purpose of specialized *endocrine glands* is hormone production (see Fig. 1.6). For example, the testes make *testosterone,* which travels through blood to muscle cells, where it binds to receptors and stimulates the growth of muscle cells. Some hormones are released by nerve cells, in which case they are sometimes called *neurohormones.*
- **Paracrine factors** (or paracrines) are chemical signals released by body cells that act on *nearby* cells. Paracrine signals reach their target by diffusion through the extracellular fluid. For example, the

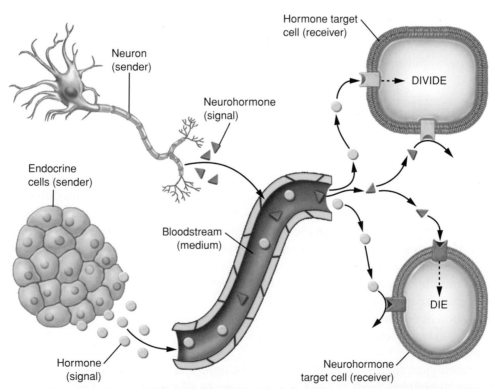

Figure 4.2. Chemical signaling. Endocrine glands or neurons send signals (hormones or neurohormones) through the medium of blood to target cells (receivers). Only cells with the correct receptor will receive the signal and be affected. *Which signal contains the message to divide—the hormone or the neurohormone?*

THE HISTORY OF SCIENCE

Braille: A Reading Code for the Blind

The Braille system, which enables blind people to read by deciphering (decoding) a code with their fingertips, is a product of the genius of Louis Braille (1809–1852), the son of a saddle and harness maker living in the French countryside near Paris. At age 3, while playing in his father's workshop, he accidentally punctured an eye with an awl, a sharp tool for making holes in leather. He quickly lost sight in the injured eye and, by the time he was 4, was completely blind in both eyes presumably because of an immune reaction (➡ Chapter 12) in his other eye.

Young Braille was very bright and kept up his studies orally in a standard classroom. At age 10, he was enrolled in the Royal Institution for Blind Youth in Paris, which also taught students orally; however, the library contained a few volumes of bulky books written with very large, raised conventional letters, which Braille used for a while but abandoned as much too slow for practical use. Louis performed well in his regular studies and displayed exceptional musical talent at the piano, which honed his tactile skills.

Then, one day in 1821, when Braille was only 12, in one of those casual moments that change world history, he met Charles Barbier, a former French soldier who was visiting the school. While in the army, Barbier had invented a system of coding military messages in raised dots so that they could be read (decoded) by touch at night, without having to wait for dawn or lighting a flame, which might reveal their position to an enemy. Barbier recognized that his system might help the blind and discussed it with Braille, who quickly realized its potential. However, Barbier's system was crude, cumbersome, and slow. Young Braille simplified and streamlined the code, and the rest is history.

Braille's system, now known simply as Braille, has been adopted worldwide and has evolved to enable the

Braille Alphabet

The six dots of the braille cell are arranged and numbered:	1 ● 4 2 ● 5 3 ● 6
The capital sign, dot 6, placed before a letter makes a capital letter.	1　4 2　5 3 ● 6
The number sign, dots 3, 4, 5, 6, placed before the characters a through j, makes the numbers 1 through 0. For example: a preceded by the number sign is 1, b is 2, etc.	1 ● 4 2 ● 5 3 ● 6

Decoder for Braille. This decoder enables users to understand the Braille signal, much as receptors enable body cells to decode chemical signals.

blind to write and read music and mathematical formulas as well as to use computers. A skilled Braille reader can read a Braille version of this sentence almost as fast as you can read it visually.

---

uncomfortable effects of hay fever reflect the release of *histamines* from white blood cells in response to pollens. Histamines signal nearby blood vessels to enlarge (red eyes) or nearby epithelial cells to secrete mucus (runny nose).

● **Neurotransmitters** are a special class of paracrine factors. They are released by neurons and travel a very short distance across a narrow intercellular space (a *synapse*, discussed further below) to another neuron or an effector cell to induce an electrical sig-nal. For example, the neurotransmitter *acetylcholine*

travels from a nerve to a muscle cell to induce an electrical signal that stimulates contraction.

Case Note

4.2. Recall that caffeine affects brain function by interfering with the action of adenosine. If adenosine is released by a neuron and travels a short distance to change the activity of a neighboring neuron, is it acting as a paracrine factor or a hormone?

Chemical Signals Vary According to Their Lipid Solubility and Binding Site

Recall that all chemical signals must bind to protein receptors in order to alter cell function. Some receptors are hidden deep within the cell, in the cytosol or even within the nucleus. And also recall from ⬅ (Chapter 2) that *hydrophilic* substances are soluble in water and *hydrophobic* substances are soluble in lipid but not in water. Only hydrophobic signals—lipid-soluble molecules that can cross the lipid layer of the cell membrane—can reach these intracellular receptors. Conversely, it is much more difficult for hydrophilic molecules to reach intracellular receptors because they are not lipid-soluble and cannot cross the cell membrane. Instead, hydrophilic signals usually bind membrane receptors—integral membrane proteins oriented to bind chemical signals in the watery extracellular fluid. As discussed below, intracellular and membrane receptors utilize very different strategies to alter target cell activity.

Hydrophobic Chemical Signals Bind to Intracellular Receptors

Hydrophobic signal molecules typically act by binding intracellular receptors and stimulating the synthesis of specific proteins. This is a relatively slow process that may take an hour or more to produce a response. The most common hydrophobic signal molecules are steroid hormones. The steroid hormone testosterone, for instance, stimulates the production of proteins that, in turn, stimulate sperm production and muscle development.

The signal transduction pathway for hydrophobic hormones and other hydrophobic ligands is essentially as follows (Fig. 4.3):

1. The hormone crosses the cell membrane by simple diffusion.
2. The hormone binds to a receptor in the cytosol or nucleus; the receptor changes shape.
3. If the receptor is in the cytoplasm, the hormone–receptor complex enters the nucleus through a nuclear pore.
4. The hormone–receptor complex binds to the regulatory region of a particular gene. Remember that genes are segments of DNA that code for a particular protein.
5. The gene is transcribed into mRNA strands.
6. Ribosomes translate the mRNA into a protein, which exerts its effect within the cell or by traveling to other cells.

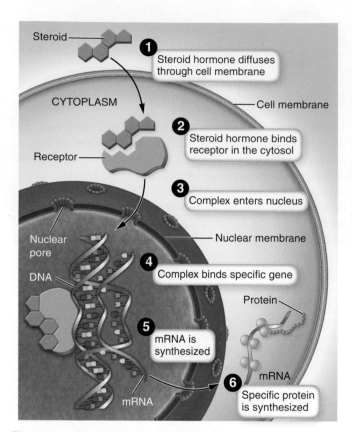

Figure 4.3. Signaling: intracellular receptors. *What binds to DNA—the ligand or the receptor?*

Hydrophilic Signal Molecules Bind to Receptors on the Cell Surface

We've said that hydrophilic signal molecules, which cannot cross the lipid layer of the cell membrane, must bind with protein receptors in the cell membrane. These receptors span the full thickness of the membrane: the ligand binding site is external and exposed to the extracellular fluid; the internal portion is exposed to the cytosol. There are three main categories of cell membrane receptors, each with a unique function (Fig. 4.4):

● Ligand-gated channel receptors
● Enzyme-linked receptors
● G-protein–linked receptors

Recall that hydrophobic signal pathways can require an hour or more to produce an effect. In contrast, the response to hydrophilic signal pathways can be very rapid, often a fraction of a second.

Case Note

4.3. Adenosine is water-soluble. Do you think it acts upon membrane receptors or intracellular receptors?

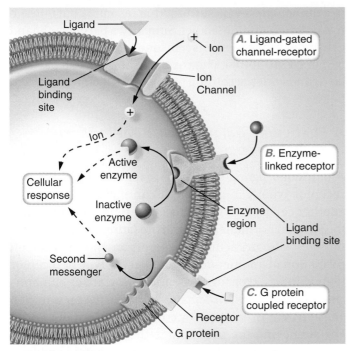

Figure 4.4. Signaling: cell membrane receptors. *Which ligand results in the production of a second messenger— the sphere, the triangle, or the square?*

Ligand-Gated Channel Receptors Modify Ion Flux

Ligand-gated channel receptors act as gates that can open to allow ions to cross the cell membrane (Fig. 4.4A). Binding of the ligand opens or closes the channel. These receptors are large proteins that contain two parts—one part binds the ligand and the other constitutes an ion channel that spans the cell membrane from outside to inside. Binding of the ligand can open or close the channel and modulate the flow (flux) of ions into or out of the cell. The ions subsequently cause a cellular response. As discussed later in the text, ligand-gated channels frequently convert a chemical signal (the ligand) into an electrical signal.

Enzyme-Linked Receptors Generate Active Enzymes

Enzyme-linked receptors also have two functional regions: the binding site is exposed to the extracellular fluid and the enzyme portion is exposed to the cytosol (Fig. 4.4B). Ligand binding to the extracellular side of the receptor activates the intracellular enzyme, which in turn activates another enzyme, which in turn activates a third enzyme, and so on. Eventually, the activated enzyme induces a functional change in the cell, such as the breakdown of glucose for energy.

G Protein Receptors Activate Intracellular Second Messengers

A ligand conveys a message, so it is a messenger. In the case of ligand gated channels, the ligand acts alone and the information it contains is sufficient to achieve the end result—opening or closing the gate. There are, however, certain reactions in which the ligand is merely a *first messenger,* which activates a **second messenger**— a small molecule that transmits a cell surface signal to sites of action in the cytoplasm or nucleus. You can think of second messengers as secret agents that carry coded messages from outsiders (i.e., hydrophilic ligands) who cannot penetrate the cell's border security.

G protein–coupled receptors (GPCRs) are a class of membrane receptors that utilize second messengers to propagate intracellularly an extracellular signal. GPCRs have two structural components: an extracellular portion that binds to the ligand and an intracellular part that interacts with a protein complex called a **G protein** (Fig. 4.4C). Note that the GPCR and the G protein itself are not the same. The ligand binds with the GPCR, which in turn activates the G protein. The G protein subsequently regulates the production of a specific second messenger.

Many GPCRs use *cAMP* (cyclic AMP, cyclic adenosine monophosphate) as their second messenger. For example, the hormone *glucagon* activates the cAMP pathway when the body needs more glucose to burn for energy. Figure 4.5 illustrates the steps in glucagon action.

1. Glucagon, the first messenger in this system, travels through the bloodstream to liver cells.
2. Glucagon binds to its receptor (a GPCR).
3. The bound GPCR activates a G protein.
4. The G protein, through a sequence of intervening steps, prompts production of the second messenger, cAMP.
5. Through a number of intervening steps, cAMP activates enzymes.
6. The enzymes increase glucose production by the liver.

This example also highlights one of the advantages of second messenger systems—at certain stages, succeeding products in the cascade are produced in greater quantity than the preceding products; that is, the signal is amplified. A single glucagon molecule can prompt the synthesis of many cAMP molecules, each of which stimulates ever-increasing activity from many enzymes. Hormones that act via second messenger systems are very effective in extremely small amounts.

It should be noted that other substances, even ions, can also function as second messengers. Calcium (Ca^{2+}) ions, for instance, that enter the cell through ion channels

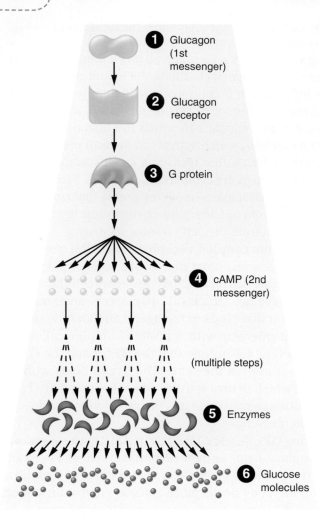

Figure 4.5. Second messengers and amplification. A single glucagon molecule can prompt the production of many glucose molecules. *How many G proteins are activated by one glucagon molecule?*

can tell an intestinal muscle cell to contract or a gland cell to secrete.

Case Note

4.4. The adenosine receptor uses cAMP as a second messenger. Is this receptor a G protein–coupled receptor or an enzyme-linked receptor?

Remember This! An enzyme is part of the receptor protein in enzyme-linked receptors but not in G protein–linked receptors.

Target Cell Response Is Determined by the Cell or the Signal

People respond differently to different signals. For example, the statement, "It's decaf!" might prompt relief

in a dinner guest who suffers from insomnia but outrage in Andy, our caffeine-addict. Similarly, a particular ligand can induce one response in one cell and a different response in another. The response of a cell to a particular ligand varies according to any of the factors discussed below.

A Particular Ligand Can Bind to More than One Type of Receptor

A particular ligand can incite reaction X by binding to receptor type X in one cell; it can incite reaction Y in another cell by binding to receptor type Y. For example, the adrenal hormone epinephrine (adrenalin) can cause muscle cell contraction in one organ and muscle cell relaxation in another depending on the type of receptor it encounters. (For further information about epinephrine, see *The Many Talents of Epinephrine* on our Web site at **http://thepoint.lww.com/McConnelland Hull.**) For adenosine the principle is the same—one ligand, two effects. Adenosine binds type II receptors on blood vessels but type I receptors on brain cells. The type II receptor (the A2R) in blood vessels increases cAMP production when adenosine binds to it, which widens the vessel, delivering more blood. In contrast, the type I receptor (the A1R) in brain cells decreases cAMP production when adenosine binds it, which reduces the electrical activity of the nerve cell.

Receptor Activity Can Be Modified by Agonists and Antagonists

The naturally occurring ligand for a particular receptor is called the *endogenous* ligand (Fig. 4.6A); that is, a ligand that originates within the body. By contrast an *exogenous* ligand is one originating outside the body. For example, epinephrine produced by the body is an endogenous ligand, while epinephrine administered as a drug is an exogenous ligand. Both bind to the same receptors (called *adrenergic receptors*), but their origin is not the same.

An **agonist** is a ligand that mimics the effect of an endogenous ligand. Some exogenous agonists bind with and activate a receptor in the absence of the endogenous ligand. For example, a common asthma medication (albuterol) is an adrenergic agonist. It binds to the same site on the adrenergic receptor as endogenous epinephrine and, like endogenous epinephrine, causes the airways to widen. Other agonists enhance the response to the endogenous ligand by binding to a different part of the receptor (Fig. 4.6B). The Clinical Snapshot later in this chapter, titled *Relaxing with GABA*, explains how alcohol acts as an agonist in this manner.

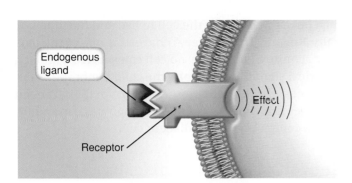

(a) The endogenous ligand

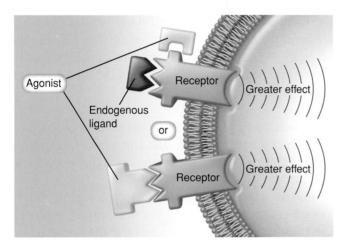

(b) Agonists

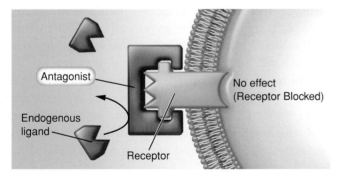

(c) Antagonists

Figure 4.6. Ligands, agonists, and antagonists. A. The endogenous ligand is produced by the body. It binds and activates the receptor, producing an effect. **B.** Some agonists bind to the receptor when the endogenous ligand is also bound, amplifying its effect. Other agonists bind to and activate the receptor, much like the endogenous ligand. **C.** Antagonists inhibit the ability of the endogenous ligand to activate the receptor, often by blocking the ligand binding site. *True or false: Both antagonists and agonists sometimes bind to the ligand-binding site of the receptor.*

Antagonists have an effect opposite to that of agonists. Some exogenous antagonists bind to a receptor and block the binding site, preventing the endogenous ligand from binding (Fig. 4.6C). Alternatively, antagonists can bind to a related receptor and *decrease* the effect of the endogenous ligand even if the endogenous ligand binds to its receptor. Either way, they don't turn off the receptor; they just prevent it from turning on or limit its effect. For example, certain breast cancer cells need estrogen to survive and have estrogen receptors to bind estrogen. For such breast cancers, the drug tamoxifen is useful treatment because it is an estrogen antagonist: it interferes with cancer cell survival by preventing estrogen action.

Case Note

4.5. Caffeine binds to the adenosine receptor and prevents adenosine from binding. Is caffeine an agonist or an antagonist for the adenosine receptor?

Cells Can Vary the Number and Sensitivity of Their Receptors

Cells can have thousands of receptors for one ligand and can increase or decrease this number in response to changing conditions. Cells with more receptors of a particular type are able to bind more ligand, which produces a more intense reaction to the ligand; the reverse is also true.

Commonly, if a ligand is present in excess for a long time, cells decrease their responsiveness by reducing the number of receptors. Alternatively, the cell may subtly alter the receptor's structure so that it responds less strongly to the ligand, much in the same way that our eyes respond to bright light by becoming less sensitive. The "high" from abused drugs typically lessens with time because cell receptors decrease or become less sensitive to the drug (ligand). People in this condition are said to be drug-tolerant, a condition that can have serious, even fatal consequences.

For example, heroin binds to brain cells and induces a dreamy euphoria, but at the same time it also binds to other cells that suppress breathing. Heroin addicts frequently become so drug-tolerant (partly because of the reduced receptor number) that to achieve the desired mental effect they must take doses that depress breathing. Addicts must walk a fine line between killing themselves and not getting the high they crave—a little too much drug and respirations become incapable of sustaining life. What's more, after withdrawal from heroin, the receptors return to normal (high) sensitivity. This, too, poses a danger: a recovered addict suffers a relapse and "shoots up" again, thinking that a big dose is needed. But the recovered receptors are numerous and sensitive, including the respiratory ones, and the addict sinks dreamily into respiratory paralysis and death.

Case Note

4.6. Remember that caffeine antagonizes adenosine action. How do you think Andy's body will compensate for the constant caffeine exposure? Will it increase or decrease the number of adenosine receptors? Explain.

4.3 What is the name of a chemical signal released from a gland into blood in order to influence cells in a distant part of the body?

4.4 True or false: If the concentration of a hormone is high enough, every body cell will respond to it.

4.5 Which type of ligand usually binds intracellular receptors—hydrophobic or hydrophilic?

4.6 Does the steroid receptor bind to DNA or mRNA?

4.7 Which portion of the receptor activates the G protein—the extracellular or intracellular portion?

4.8 Imagine that a young student is conducting research that she hopes will enable her to win a scholarship for graduate school. She is assisted in her studies by a professor who believes in her abilities, but she is blocked by an envious fellow student who steals her research data. Identify the agonist and the antagonist in our scenario.

Electrical Signaling

Chemical signals transmit most messages through blood or other fluid, but they are slow—too slow for some purposes, such as sending a message to the legs to run when danger appears. Such fast signals are transmitted electrically—somewhat like a lightning bolt, though not quite as fast. Electrical signals are the "spark of life" in every cell. As shown below, the inside and outside of the cell membrane have differing electrical charges like the positive and negative poles of a battery, a condition that

must be maintained for the cell to live and function normally. Keeping this "battery" charged is one of the most basic requirements of life: no cell can survive without it. Electrical signals travel along the cell membrane as disturbances of these charges. As they do, they transmit messages from one part of the cell membrane to another or, less commonly, to an adjacent cell.

Neurons Signal Electrically

Electrical signals occur to one degree or another in all cells but are most important in muscle and nerve cells. So to facilitate our discussion of electrical signaling, let's take a brief look at neuron structure (Fig. 4.7). As discussed in ➡ Chapter 8, there is considerable variation in the appearance of neurons, but all share the same basic components:

● The **cell body** comprises the main mass of the neuron and contains the nucleus and most of the organelles. The cell body integrates signals received by dendrites.

● **Dendrites** are branched, short cytoplasmic extensions of the neuron cell body that convey electrical signals from another cell *toward* the neuron cell body.

● The **axon** is a single, usually quite long, cytoplasmic extension from the neuron cell body, which conveys electrical signals *away* from the cell body and to other cells. Small axon branches, called *collaterals*, may branch off the main axon.

● Sections of most axons are covered with **myelin,** a fatty whitish substance that accelerates the transmission of electrical signals. As you can see in ➡ Figure 8.4, the myelin coating is actually the cytoplasmic extensions of specialized nervous system cells.

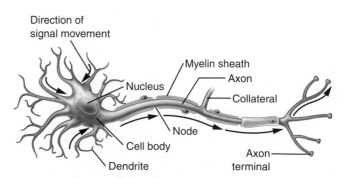

Figure 4.7. Structure of a neuron. This figure shows a myelinated neuron; unmyelinated neurons are similar but do not have a myelin sheath. *Does myelin wrap around the neuron's axon or its dendrites?*

Case Note

4.7. Adenosine receptors are present in the neuron component that commonly receives signals. Name this component.

Ions Carry Electrical Signals

Before we can understand how neurons and other cells send and receive electrical signals, we must review some electrical basics.

- *An excess of protons (+) or electrons (–) in an atom or molecule creates ions* that are positively or negatively charged. Positively charged ions such as sodium (Na^+) and potassium (K^+) are called **cations;** negatively charged ions such as chloride (Cl^-) or many proteins (abbreviated as Pr-) are called **anions.**
- *Similar charges repel one another*: positive repels positive, and negative repels negative.
- *Opposite charges attract one another*; that is, cations are attracted to anions. If a cation pairs up with an anion, the net charge is zero (0).
- *Energy is required to keep opposite charges apart.* A barrier that keeps positive and negative charges separated is an *insulator*. The cell membrane is an insulator, keeping opposite charges on opposite sides of the cell membrane.

Case Note

4.8. In some cells, adenosine increases the movement of potassium ions across the cell membrane. Is potassium an anion or a cation?

Electrical Signaling Relies on the Electrical Gradient

Recall from ◀ Chapter 1 that a gradient is a difference in the quantity of something between two areas. For example, a temperature gradient exists across a wall if it is hot on one side of the wall, cold on the other. An **electrical gradient,** reflecting different numbers of anions and cations on either side of the cell membrane, exists in every cell and is the basis of electrical communication. This electrical gradient is also known as the **membrane potential** because (a) it exists at the cell membrane and (b) like a battery, it is a source of *potential energy* (◀ Chapter 2).

There are millions of positively and negatively charged ions in the cytosol and extracellular fluid. Except under very unusual circumstances, the overall electrical balance of the body is neutral: for every negative charge there must be a positive one. Ions in the same fluid will pair up, each cation pairs with an anion neighbor, so the net charge is zero. However, the cell membrane prevents some ions from pairing up. Thus, if the cytosol accumulates an excess of anions (–), it follows that the extracellular fluid *must* have an excess of cations (+). As a result, the cytosol will be negatively charged (Fig. 4.8A). Conversely, a relative excess of cations in the cytosol (and anions in the extracellular fluid) would result in positively charged cytosol (Fig. 4.8B).

In both cases, an electrical gradient, the membrane potential, is created. This gradient exerts a force on all ions inside and outside the cell. If the cytosol is negatively charged, it will attract cations (+) and repel anions (–). Conversely, a positively charged cell interior will attract anions and repel cations. The insulating lipid of the cell membrane prevents these ions from rushing into or out of the cell to join one another. However, certain types of ion channels exist within the cell membrane to control ion passage into or out of the cell. These ion channels are discussed below.

The strength of the electrical charge, the membrane potential, is measured in millivolts (mV). By convention such measurements quantify the unpaired charges in the cytosol. That is, a negative membrane potential means that the cytosol contains unpaired anions. The value of the membrane potential in nerve or skeletal muscle cells "at rest" (that is, not transmitting electric signals), known as the **resting membrane potential,** is about –70 mV. This value informs us that the strength of the electrical gradient is 70 mV. The minus sign signifies that the cytosol contains excess anions—it is negatively charged. The membrane potential is absolutely necessary for electrical communication and thus for many body functions, from the heartbeat to respiration. Before discussing why this membrane potential exists and how the cell uses it for signaling, it is important to understand two other elements governing electric signaling: *concentration gradients* and *ion channels.*

Case Note

4.9. When Andy was drinking decaf, the resting membrane potential of some of his brain cells changed from –70 mV to –71 mV. Does the cytosol of his cells contain more or fewer anions than usual?

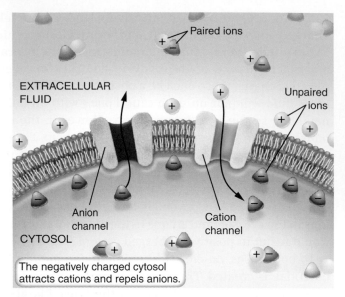 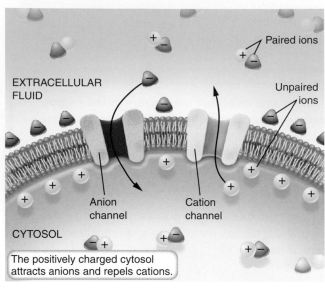

(a) Membrane potential = −70 mV (Resting membrane potential)   (b) Membrane potential = +30 mV

Figure 4.8. Membrane potential. An electrical gradient, or membrane potential, is created by an electrical imbalance between the extracellular fluid (ECF) and the cytosol. If ion channels are open (in this example, one ion channel for cations and one ion channel for anions), ions will cross the membrane. **A.** A negative membrane potential is created by a relative excess of negative ions in the cytosol. If open ion channels are present, anions will leave the cytosol and positive ions will enter the cytosol. **B.** A positive membrane potential is created by a relative excess of positive ions in the cytosol. If open ion channels are present, cations will leave the cytosol and anion ions will enter the cytosol. *Calcium is a cation. In cell B (assuming open calcium channels), will calcium enter or leave the cell?*

Sodium and Potassium Concentration Gradients Exist in Every Cell

Electrical signaling not only relies on the *electrical* gradient but also requires well-defined *concentration* gradients for numerous ions. The extracellular fluid bathing all cells, for instance, contains a high concentration of sodium (Na^+) and a low concentration of potassium (K^+) (Fig. 4.9). The intracellular fluid also contains sodium and potassium, but the concentrations are reversed: potassium is in high concentration and sodium low. Thus, if ion channels were to permit them to cross the cell membrane:

● Sodium would flow down its concentration gradient *into* the cell.
● Potassium would flow down its concentration gradient *out* of the cell.

The sodium and potassium concentration gradients are created and maintained by active transport (◄ Chapter 2), specifically the sodium–potassium ATP pump (abbreviated as the Na^+/K^+-ATPase). The Na^+/K^+-ATPase uses energy to force sodium ions out of the cell and potassium ions into the cell, each traveling "uphill" against their respective concentration gradients. Because of constant

activity of the Na^+/K^+-ATPase, these concentration gradients are fixed—they do not change in the living body.

Although sodium and potassium are the star players in electrical communication, negative ions play supporting roles. Intracellular protein anions help keep the cytosol negatively charged, and, as discussed further on, chloride anions in the extracellular fluid can participate in electrical signaling.

> **Remember This!** Sodium is always more concentrated outside the cell. Potassium is always more concentrated inside the cell.

Ions Move Down Gradients Using Ion Channels

Recall from ◄ Chapter 3 that ions cannot freely pass across the cell membrane; they require the assistance of membrane proteins such as ion channels. Many types of channels exist for each ion. Of these, three are particularly important in our discussion of electric signaling:

● *Leak channels:* Leak channels are always open, allowing ions to "leak" in or out of the cell down a concentration or electrical gradient. Most cells contain many

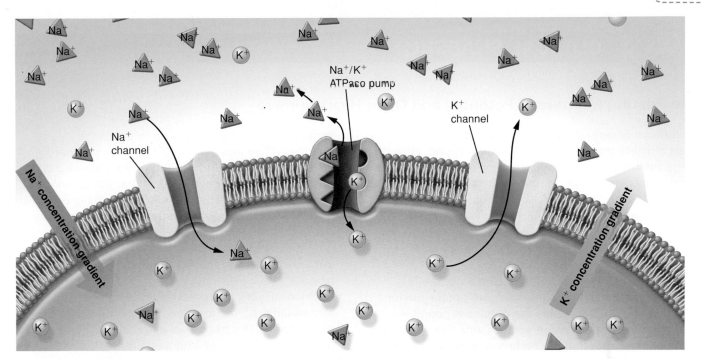

Figure 4.9. Sodium and potassium gradients. A. The sodium concentration gradient attempts to force sodium ions into the cell. **B.** The potassium concentration gradient attempts to force potassium out of the cell. These two gradients are opposed by the sodium–potassium ATPase pump. *True or false: The sodium–potassium ATPase pump actively transports sodium out of the cell.*

potassium leak channels to let K^+ out of the cell but very few sodium leak channels to let Na^+ in.

- *Ligand-gated channels:* These channels open in response to a specific chemical signal—a *ligand*. They play starring roles in converting a chemical signal (the ligand) into an electrical signal (a change in membrane potential).
- *Voltage-gated channels:* These channels open in response to a change in membrane potential and are responsible for long-distance electrical signals, discussed further on.

Because Na^+ and K^+ are ions and differ in their concentration inside and outside the cell, their movement through these channels will be affected by both concentration and electrical gradients. Consider, for instance, a negatively charged cytosol with open Na^+ and K^+ channels. Na^+ would enter the cell down both its concentration and electrical gradients. K^+, conversely, would be tugged in both directions: the negative interior pulling inwardly and the chemical gradient acting in the opposite direction to draw K^+ out of the cell. The net movement of K^+ would depend on whether the electrical or the chemical gradient were the stronger one.

The opposite would apply if the cytosol were positively charged. Assuming that Na^+ and K^+ channels are

open, this time potassium would be driven out of the cell by both electrical and concentration gradients. Sodium, on the other hand, would be the one tugged in both directions: the positive cytosol would be acting to force Na^+ outward and the chemical gradient would be acting to keep it inside. The net flow of Na^+ would depend on the balance of these two conflicting forces.

> ***Remember This!*** **Ion movement by facilitated diffusion requires (a) a gradient and (b) an open ion channel.**

The Resting Membrane Potential Is Determined by Potassium

The resting membrane potential, in which the cytosol is negatively charged in comparison to the extracellular fluid, is based on two simple elements:

- K^+ is more concentrated inside the cell than outside it.
- K^+ leak channels are always open, permitting K^+ to leak out of the cell.

The net result of potassium leakage is that the interior of the cell, the cytosol, becomes electrically negative.

CLINICAL SNAPSHOT

The Resting Membrane Potential and Open Heart Surgery

It's difficult to operate on the beating heart; so for most heart surgery the heartbeat must be stopped. How do surgeons do it, and even more important, how do they get it going again? It's actually quite simple: just infuse some potassium chloride (KCl; that is, K^+ and Cl^- in solution) into the coronary arteries (➡ Chapter 11) and the heart stops; wash it out and the heart begins beating again. To understand why, let's revisit the topic of resting membrane potential.

Recall that the interior of the cell membrane is negatively charged and contains more potassium ions (K^+) than the outside, while the outside is positively charged and contains more sodium ions (Na^+) than the inside. Recall, too, that the resting membrane potential is critical for cell function—in this case, for the contractions of the heart muscle that create the heartbeat.

Now, what happens down at the level of the cell membrane if we drastically increase the *extra*cellular K^+

concentration, as surgeons do to arrest the heartbeat? The injection of KCl into the coronary arteries increases *extra*cellular K^+ to a level equal to *intra*cellular K^+. The net result is that *the chemical concentration gradient that would normally urge the flow of K^+ out of cells is destroyed.* K^+ will not flow out of the cell; therefore a resting membrane potential cannot be maintained. The heart muscle immediately stops contracting and the surgeons go to work.

Once surgery is finished, surgeons wash out the extracellular K^+ by using an artificial fluid that is the rough equivalent of normal extracellular fluid, which is high in sodium, low in potassium. As *extra*cellular potassium falls, the chemical concentration gradient is reestablished, potassium begins to leak out of cells as described in this chapter, and the resting membrane potential is restored.

Without the potassium gradient, the membrane potential does not exist. See the nearby box, The Resting Membrane Potential and Open Heart Surgery, for the clinical implications of this fact. Below we discuss the many and varied changes in the electrical character of the cell membrane as a method of electrical signaling.

For a more detailed discussion of the resting membrane potential, read the text box *The Resting Membrane Potential: It's All About Potassium,* at <u>http://thepoint.lww.com/McConnellandHull</u>.

> *Remember This!* **The resting membrane potential is simply an electrical gradient, reflecting an excess of anions inside the cell and a corresponding excess of cations outside the cell.**

A Cell's Electrical Signal Is a Wave of Change in the Membrane Potential

Unlike concentration gradients, electrical gradients are exquisitely sensitive—the movement of just a few ions can change the membrane potential with lightning speed and without appreciably altering the concentration gradient.

What's more, since positive and negative ions are attracted to each other, any change in membrane potential

will spread to neighboring regions. As discussed in the next section, electrical signals in cells are transmitted as very short-lived changes in membrane potential, which spread from one region of the cell membrane to another. This wave of temporary change in membrane potential is what we have been referring to as the "electrical signal," and it is the means of electrical communication between cells.

It is important to realize that molecules racing along from one end of the cell to the other is not the way electrical waves travel. Rather, waves reflect temporary local activity—a disturbance—that moves from one place to another. It is similar to fans in a stadium participating in a wave that moves around a stadium: the fans stay in place and stand up briefly to continue the wave. They do not race from one section to the other. So it is with membrane potential: The electrical potential changes briefly, propagating the wave, and then returns to its previous status.

> *Remember This!* **In cell signaling, the sodium and potassium concentration gradients do not change significantly. However, the membrane potential (the electrical gradient) *does* change, and when it does, it travels as an electrical signal.**

Changes in Membrane Potential Are Depolarization, Hyperpolarization, and Repolarization

In many cells, the resting membrane potential never changes; the cells are able to do their work in ways that do not disturb their electrical balance. But in other cells, especially (but not exclusively) nerve and muscle cells, electrical activity is ceaseless, as changes in membrane potential convey signals during every millisecond of life.

Understanding the changes in membrane potential requires that you understand the concept of **electrical polarity.** In daily life, we use the adjective *polarized* to describe a situation in which groups, such as political parties, are on opposite sides of an issue. In reference to the cell membrane, a *polarized membrane* is one with an excess of negative charges on one side and an excess of positive charges on the other. The cell membrane of every cell in the body is polarized in this way.

Changes in membrane potential can be classified as one of three types, according to their effect on the strength of the electrical gradient (Fig. 4.10):

- *Depolarization.* A change that *reduces* the strength of the electrical gradient—that makes the cell's interior less negative—is called **depolarization.** For example, a change of the electrical charge from −70 mV to −40 mV is partial depolarization; and a change from −70 mV to 0 mV is complete depolarization. In the transmission of electrical signals across the cell membrane, the depolarization process actually "overshoots" a bit and the interior of the cell becomes positively charged for an instant. In this case the cell interior may go from −70 mV to +30 mV, but the process is still described as depolarization.
- *Hyperpolarization.* Conversely, a change that makes the cell's interior even more negative is called **hyperpolarization.** That is, it is a change that *increases* the strength of the electrical gradient. For example, a change from −70 mV to −80 mV is hyperpolarization.
- *Repolarization.* Finally, a change in the electrical gradient that returns a cell to its original resting membrane potential is described as **repolarization.** Again, a cell's resting state is *not* electrical neutrality (0 mV) but polarization (negative cell interior, −70 mV).

As shown in Figure 4.10A, we can use a graph to illustrate changes in membrane potential: notice that depolarization moves the voltage closer to neutral, whereas hyperpolarization moves the voltage away from neutral. After these changes, the membrane returns to the resting state by repolarization.

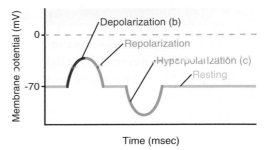

(a) Changes in membrane potential

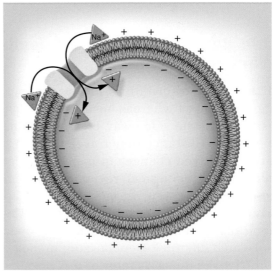

(b) Depolarization

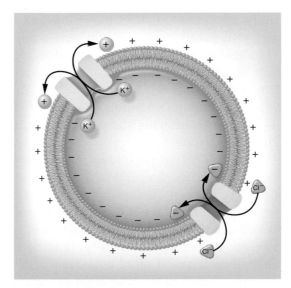

(c) Hyperpolarization

Figure 4.10. Membrane potential changes. A. Depolarization decreases the membrane potential, whereas hyperpolarization increases it. In the graph illustrated, the resting membrane potential is −70 mV. **B.** Sodium entry into the cell causes a depolarizing graded potential. **C.** Potassium exit or chloride entry causes a hyperpolarizing graded potential. *If the membrane potential moves from −70 mV to −65 mV, is the membrane hyperpolarized or depolarized?*

Case Notes

4.10. Adenosine acts on some neurons to increase the magnitude of the membrane potential. Does adenosine hyperpolarize or depolarize these neurons?

4.11. When Andy took his first gulp of caffeinated coffee, the membrane potential of certain nerve cells changed from −71 mV to −70 mV. Is this change described as hyperpolarization or depolarization?

Graded Potentials Remain Localized and Vary in Strength

Recall that neurotransmitters and paracrines work over short distances, whereas hormones are long-distance signals. In the same way, electrical signals are classified by the distance they travel. **Graded potentials** are short-lived changes in membrane potential that work locally over a small region of cell membrane or cytoplasm. They occur in most cells and are particularly important in neuronal dendrites. The most common cause of a graded potential is the opening of a ligand-gated channel. Depolarizing graded potentials, which make the cell's interior less negative, usually result from the activation of ligand-gated Na^+ channels that allow Na^+ into the cell (Fig. 4.10B). As Na^+ cations enter the cell, they match up with excess negative ions and reduce the polarity across the membrane. On the other hand, hyperpolarizing graded potentials, which make the cell's interior even more negative, result from K^+ leaving the cell through ligand-gated potassium channels (Fig. 4.10C) or from chloride (Cl^-) ions entering the cell through ligand-gated chloride channels.

The magnitude of an individual graded potential depends on how many channels in the cell membrane are open. If, for example, many sodium channels open to allow a large temporary influx of positive sodium ions, the cell will depolarize more than if fewer sodium ions entered. Conversely, a hyperpolarizing graded potential will be larger if more potassium or chloride channels are opened to let potassium out or chloride in. Graded potentials are important, for instance, in vision: the eye detects different light intensities because bright light stimulates a stronger hyperpolarizing graded potential in the retina than does dim light (➡ Chapter 9). Graded potentials, therefore, are one of the ways a cell responds to the strength of an incoming signal: the stronger the signal, the stronger the graded potential.

Many cells, neuronal dendrites in particular, experience multiple, simultaneous graded potentials. In the same way that a competitive diver's overall score reflects points gained for each successful dive and points lost for mistakes, the overall change in a cell's membrane potential depends on the sum of all individual graded potentials. For instance, let's say that a neuronal dendrite is exposed to two different ligands, one depolarizing the membrane by 10 mV and the other hyperpolarizing the membrane by 15 mV. The net result is a 5-mV hyperpolarization. The summation of graded potentials allows neurons to integrate information from many different sources.

Like any signal, a graded potential must travel to be effective. Graded potentials spread to neighboring regions of the cell membrane because negative and positive ions attract each other. The positive ions that enter the cell during an initial depolarization will attract nearby negative ions; as these negative ions leave their original position, they will leave some positive ions unpaired and thus depolarize the adjacent region. The graded potential spreads to this new region, which will depolarize adjacent new regions in the same way. That said, graded potentials travel only a very short distance. They lose strength as they travel and eventually fade to nothing, like ripples spreading out from a pebble dropped in a pond.

Case Note

4.12. Adenosine hyperpolarizes neurons. Does it increase or decrease the number of potassium ions leaving the cell?

Action Potentials Are Large Changes That Can Travel Long Distances

We've just discussed how graded potentials convey signals over very short distances—say, from the dendrite to the cell body of a neuron. In contrast, an **action potential** is a large change in membrane potential, which travels the length of the cell until it reaches the end, no matter the distance. For example, a neuron's axon can be several feet long, and action potentials travel the entire distance to communicate electrical signals. Action potentials occur only in neuronal axons, skeletal muscle cells, and a few other excitable cells.

Action Potentials Have Four Phases

As an action potential evolves at a given point on the cell membrane, we can graph the changes that occur, as shown

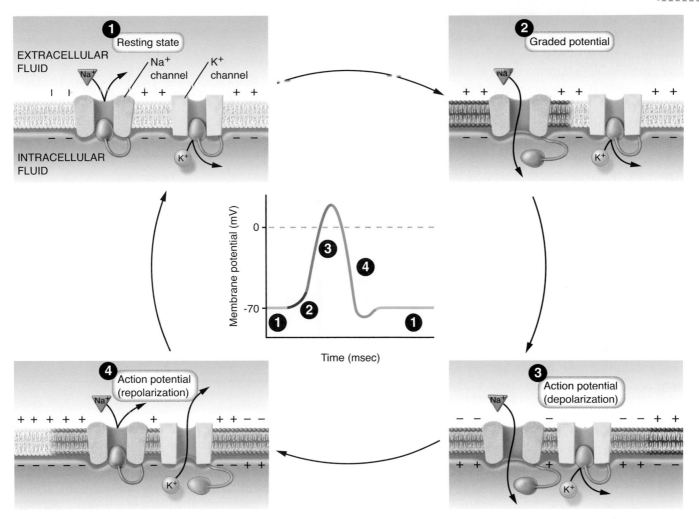

Figure 4.11. Action potential. The color of the membrane corresponds to the color on the action potential tracing. *Name the phase of the action potential when voltage-gated potassium channels are open.*

in Figure 4.11 (middle drawing). Time is plotted on the horizontal axis, the voltage of the membrane potential on the vertical axis. Notice that the events of the action potential are divided into four numbered phases, each with a corresponding color: (a) resting state; (b) graded potential; (c) depolarization; and (d) repolarization. If you study both parts of Figure 4.11 closely, you'll see that the colors for the phases on the graph in part A match the colors for the phase affecting the cell membrane in part B. For instance, red corresponds to the depolarizing phase of the action potential, so red membrane regions are depolarizing. Now let's take a closer look at the four phases of the action potential:

1. *Resting state.* The inside of the cell is negatively charged (yellow). Recall that this resting negative voltage is maintained because potassium *leak* channels are constantly open (these channels are not shown in the figure).

In contrast, also recall that voltage-gated channels are special membrane channels that open only in response to changes in membrane potential. During the resting state, voltage-gated channels for both Na^+ and K^+ are closed, as shown (Fig. 4.11, phase 1).

2. *Development of a graded potential.* Action potentials begin as graded potentials. Depolarization of a nearby membrane segment (say, a signal initiated by a touch) causes a few sodium channels to open. Sodium ions enter (down their concentration and electrical gradients), depolarizing the cell slightly (dark red, phase 2). Notice that, as the cell becomes a bit less negative, the tracing on the graph rises slightly toward neutrality (zero). Voltage-gated potassium channels are still closed.

3. *Depolarization.* As the graded potential depolarizes the membrane, more voltage-gated sodium channels open and more sodium rushes into the cell. Although

only one voltage-gated sodium channel is depicted in phase 3, the actual number that open depends on the magnitude of the depolarization—larger depolarizations open more channels. So many sodium ions enter that, at its peak, depolarization causes the inside of the cell temporarily to become positively charged (light red). Notice that the curve of the graph rises above zero. As depolarization peaks, voltage-gated potassium channels finally begin to open—they are much slower to respond to depolarization than voltage-gated Na^+ channels—and Na^+ channels close. The opening of K^+ channels and closure of Na^+ channels stops depolarization.

4. *Repolarization.* When the voltage-gated potassium channels open, positively charged potassium ions flood out of the cell and down their concentration and electrical gradients (blue); the cell membrane now returns to resting membrane potential (–70 mV; yellow). This lag time for potassium channels gives sodium a chance to flood in and completely depolarize the membrane before potassium floods out to repolarize the membrane back to the resting state.

> ***Remember This!*** **Sodium ions entering the cell cause depolarization; potassium ions exiting the cell cause repolarization.**

Action Potentials Are "All or Nothing"

Upon depressing the flush lever of a toilet, water begins to swirl around the bowl and, if the lever is depressed far enough, a full flush ensues. So it is with an action potential: it is all or nothing and requires a stimulus, in the form of a depolarizing graded potential, of sufficient strength to push it over the edge, so to speak. This critical electrical point that must be crossed is called the *threshold.* In most cells, the threshold is about 15 to 20 mV above resting membrane potential. If the graded potential remains subthreshold, it does not open enough voltage-gated sodium channels and an action potential will not develop (Fig. 4.12A). The graded potential will die out somewhat like a failed attempt to flush causes only a temporary swirl of water. The greater the difference between the resting membrane potential and the threshold, the more difficult it will be for the neuron to fire an action potential. It follows, therefore, that anything that hyperpolarizes the cell (i.e., makes the resting potential more negative) will make it more difficult to initiate an action potential.

Returning momentarily to toilet mechanics, pushing harder on the lever doesn't increase the magnitude of

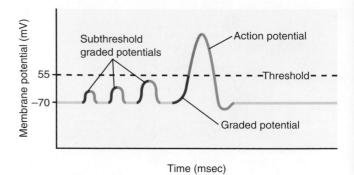

(a) Action potential thresholds

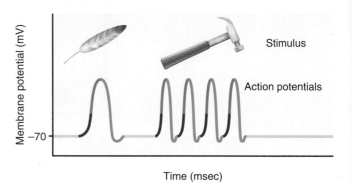

(b) Action potential frequency

Figure 4.12. Characteristics of action potentials. A. Graded potentials sufficiently large to reach the threshold will initiate an action potential. **B.** Stimulus intensity is encoded in the frequency of action potentials, not in the size of individual action potentials. The touch of a feather (*left*) might initiate a single action potential, whereas the whack of a hammer (*right*) would initiate many. *If the membrane is depolarized to –50 mV, will an action potential result?*

the flush—all successful flushes are of the same magnitude. Similarly, all action potentials are identical, regardless of the strength of the initiating stimulus. Action potentials are thus described as all or nothing. Because the strength of the signal—its *amplitude*—is uniform, the message it carries says, in effect, "Do it," for whatever "it" is. The signal cannot say, "Do a little" or "Do a lot" (that's only for graded potentials). For example, if "it" is activation of a motor neuron, the neuron fires or it doesn't; there is no partial firing. This does not mean, however, that the target cell's *response* is all or nothing. Although a single action potential cannot invoke a graded or variable response, a series of action potentials can. The more frequent the signal, the greater the response (Fig. 4.12B). For action potentials, intensity is encoded in their *frequency*—stronger sensory input (such as stronger touch or more intense smells) will be signaled by very frequent action potentials, while weaker stimuli will be signaled by infrequent action potentials.

Action Potentials Are Self-Regenerating

Action potentials feed upon themselves like a row of falling dominoes, so that they become self-regenerating and travel to the physical "end of the line," whatever that may be. Theoretically, a signal encoded by action potentials can be transmitted over an infinite distance without any loss in the strength or clarity of the signal, much the way that a very long line of dominoes will continue to fall after the first one is pushed over. This amazing feat is possible because action potentials are continually regenerated as they are transmitted.

Here's how. An action potential in one region of the membrane will depolarize neighboring sections, thereby opening voltage-gated sodium channels. Sodium entry through these channels will depolarize the region further (to the threshold) and propel the action potential, which depolarizes additional membrane, and on and on until the signal reaches the end of the axon or other cell region (Fig. 4.13A). Small neurons, muscle cells, and other excitable cells employ this method of transmission.

Action Potentials Travel Faster via Saltatory Conduction

Some neurons have an even faster method of transmitting an action potential—**saltatory conduction** (Latin *saltare* = "to jump from place to place"). As mentioned earlier, some axons (called *myelinated axons*) are covered by myelin sheaths. Myelinated axons look a bit like a row of sausages on a stick, the sausage representing the myelin cell's wrapping and the stick representing the axon. The minute collar of unwrapped space between adjacent cells (or sausages), where the axon is "naked," is called a **node of Ranvier** (Figs. 4.7 and 4.13).

The myelin sheath prevents sodium ions from crossing the membrane and causing action potentials. Thus,

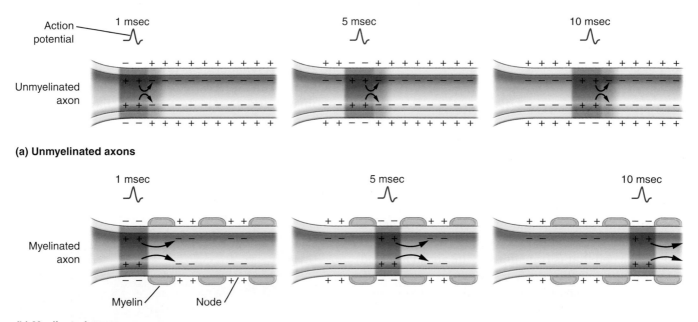

(a) Unmyelinated axons

(b) Myelinated axons

Figure 4.13. Action potential propagation. A. Action potential propagation for unmyelinated neurons. The dark red region of the membrane indicates where voltage-gated sodium channels are open and the action potential is occurring. At 1 ms, the action potential has started to move down the axon; at 5 ms, the action potential has moved a short distance; and at 10 ms, the action potential has moved slightly farther. **B.** Action potential propagation for myelinated neurons. The position of the action potential is indicated at 1 ms, 5 ms, and 10 ms. Note that the action potential has moved much farther in the myelinated neuron than in the unmyelinated neuron at 5 and 10 ms. In real neurons, the signal travels up to 200 times faster in a myelinated neuron than in an unmyelinated neuron. *Which channels are open in the depolarized (red) regions of the membrane?*

action potentials occur only in the exposed collars of membrane located at the nodes. The electrical disturbance caused by an action potential at one node jumps to the next node at the speed of light, as a spark of static electricity jumps between your finger and a metal object. At the node, it initiates a new action potential, which travels the very short distance down the exposed collar of axon before jumping again.

As shown in Figure 4.13B, this process is much faster than action potential conduction in unmyelinated nerve membranes. An action potential takes a certain amount of time—say, 5 milliseconds (ms)—to complete, whereas depolarization can spread from node to node virtually instantaneously. In the unmyelinated axon of Figure 4.13A, the action potential in the far left portion of the neuron is completed by 5 ms. It invokes a second action potential in the adjacent membrane section, consuming an additional 5 ms, which subsequently invokes a third action potential in the next membrane region. The action potential has traveled only a short distance down the axon during 15 ms. During this same time period, three action potentials have also occurred in the myelinated neuron (part B). However, since action potentials occur only at the nodes, the signal has spread much further down the axon.

Diseases in which the myelin sheath is damaged or missing, such as multiple sclerosis, result in abnormally slow impulse transmission in nerves supplying skeletal muscles. Muscle movements are correspondingly slow, or do not occur at all.

Synapses Transmit Electrical Signals between Cells

Remember that chemical signals use blood and extracellular fluid as their medium and travel through body fluids to affect other cells near and far. However, what happens to an electrical signal when it reaches the end of the cell? The answer is that it reaches a **synapse**, a site where an electrical signal passes from one cell to the next. The cell that is transmitting the signal is called the *presynaptic cell*; the receiver is the *postsynaptic cell*. Examples of synapses include the junction between a neuron and a skeletal muscle cell, between two cardiac muscle cells, or between two spinal cord neurons. In each case, the electrical signal must bridge the physical gap between the two cells.

Two options exist for a signal to traverse this physical barrier. *Electrical synapses* bridge the gap directly by *gap junctions*. *Chemical synapses,* conversely, use a chemical messenger—a *neurotransmitter*—to convey the message between the two cells.

Electrical Synapses Convey Signals through Gap Junctions

Gap junctions (◀ Chapter 3) are watery protein tunnels that permit ions and other small chemicals to pass from one cell to another. In an electrical synapse, ions pass from the presynaptic to the postsynaptic cell through the gap junction to propagate the action potential (Fig. 4.14). That is to say, the postsynaptic cell depolarizes when positive ions flow from the presynaptic cell to the postsynaptic cell, and negative ions flow in the opposite direction.

Electrical synapses are quick and efficient because the action potential spreads between neighboring cells as easily as from one membrane region to another. Electrical synapses link muscle cells in the heart, uterus, and intestines, so that all of the neighboring cells are activated and contract at essentially the same time. They also occur between some brain regions, so that all of the neurons fire at the same time.

Chemical Synapses Convert Electrical Signals into Chemical Ones

In contrast to cells linked by electrical synapses, cells participating in a chemical synapse are not physically linked; instead, they are separated by an exceedingly thin *synaptic cleft*. Electrical signals cannot directly cross this cleft; instead, neurotransmitters carry the signal across the synaptic cleft between the presynaptic and postsynaptic cells (Fig. 4.15). Familiar examples of neurotransmitters include acetylcholine, norepinephrine, and glutamate. See ➡ Chapter 8 for more information.

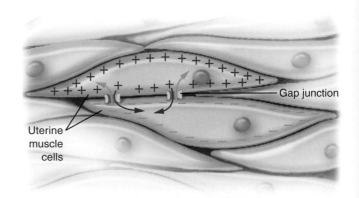

Figure 4.14. Electrical synapse. Gap junctions between two adjacent muscle cells in the uterus enable the action potential to pass directly from the upper cell into the lower cell. *Where is the action potential occurring—in the top cell or the bottom cell?*

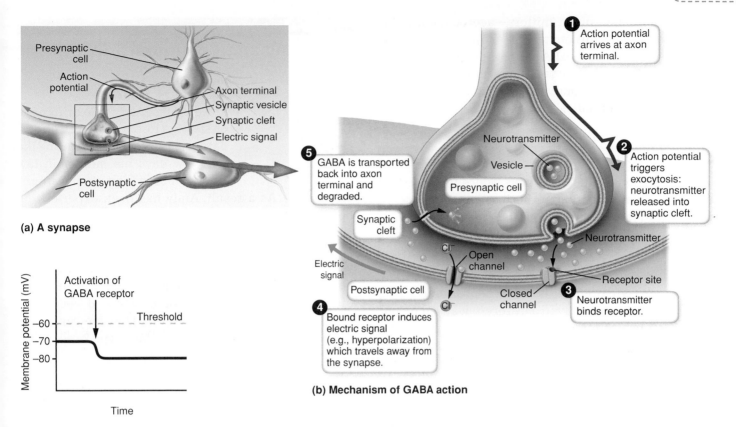

(a) A synapse

(b) Mechanism of GABA action

(c) GABA-induced hyperpolarization

Figure 4.15. Chemical synapses. A. The parts of a chemical synapse. **B.** The events at a chemical synapse. This example uses GABA, which is an inhibitory neurotransmitter that hyperpolarizes the postsynaptic cell. **C.** Changes in membrane potential resulting from the activation of a GABA receptor. *Is the fluid in the synaptic cleft extracellular or intracellular fluid?*

Chemical synapses carry signals between two neurons or between a neuron and a target cell—usually a gland cell or a muscle cell. The end of the presynaptic neuron, the **axon terminal,** contains many **synaptic vesicles,** which are packed with neurotransmitters. On the other side of the synaptic cleft, the cell membrane of the postsynaptic cell contains numerous protein receptors that bind the neurotransmitter. The signal is transmitted from the presynaptic to the postsynaptic cell as follows (Fig. 4.15B):

1. The action potential arrives at the axon terminal.
2. The action potential induces exocytosis of synaptic vesicles; that is, the synaptic vesicles fuse with the cell membrane and the neurotransmitter molecules are released into the synaptic cleft. The neurotransmitter diffuses across the synaptic cleft.
3. The neurotransmitter binds receptors on the postsynaptic cell membrane.
4. The bound receptors induce a change in the postsynaptic cell (usually a graded potential, either hyperpolarizing or depolarizing). The signal travels away from the synapse.

5. The neurotransmitter is rapidly removed from the synaptic cleft (into neurons or other nearby cells) and recycled or destroyed.

Upon binding with a receptor in the postsynaptic membrane, most neurotransmitters induce an electrical signal, either directly or via a second messenger (such as cAMP). For instance, Figure 4.15B illustrates the particular chloride channel receptor for the neurotransmitter GABA (a modified amino acid). The chloride channel is open only when GABA is bound to the channel receptor. Chloride is more concentrated outside the cell, so the open channels permit negative chloride ions to enter the cell, making the cell's interior even more negative than before. In this condition the cell membrane is hyperpolarized (Fig. 4.15C) and the resting membrane potential is farther away from the threshold required to produce an action potential. The cell is inhibited and it will take a greater depolarization to get this cell to fire. It's as if the GABA were saying, "Relax! Stop firing so many action potentials!" GABA acts at many points in the brain to slow down brain cell activity (see the Clinical Snapshot for more information).

Many other neurotransmitters—acetylcholine, norepinephrine, and glutamate, for example—exert the opposite effect from GABA on the postsynaptic cell. These so-called *excitatory* neurotransmitters depolarize the postsynaptic cell, increasing the chance that it will fire an action potential.

Case Note

4.14. One action of adenosine is to reduce the release of excitatory neurotransmitters such as glutamate and norepinephrine. Do you think it binds the presynaptic cell or the postsynaptic cell to exert this effect?

If the postsynaptic cell is not a neuron, the signal will alter the cell's activity in a different way. For example, if the postsynaptic cell is a gland cell, the signal could stimulate or inhibit hormone secretion from a gland. If it's a muscle cell, it could stimulate or inhibit contraction.

 4.9. Calcium ions are positively charged. Assuming that they can cross the membrane, will they be pulled into or pushed out of a resting cell?

4.10. If sodium moves down its concentration gradient, is it moving into the cell or out of the cell?

4.11. Name three types of ion channels.

4.12. Which sort of change—depolarization or hyperpolarization—is occurring if the membrane potential changes from –70 mV to –40 mV?

4.13. Does sodium entry into the cell result in depolarization or hyperpolarization?

4.14. In a single action potential, which event occurs first—depolarization or repolarization?

4.15. Which type of electrical signal can vary in amplitude: action potentials or graded potentials?

4.16. Which neuron would transmit a signal faster—a myelinated neuron or an unmyelinated neuron?

4.17. Which type of synapse relies on a physical bridge between the cells—chemical or electrical?

4.18. What is the space between the presynaptic and postsynaptic cells called?

Caffeine and Communication: The Case of Andy M.

Let's return to our case.

Recall that Andy, who drinks five or six big (16-oz) mugs of coffee a day, has just learned that during the past few days his mother-in-law has been brewing only decaffeinated coffee. As a result, Andy has been experiencing headaches, fatigue, and an inability to concentrate. So, what, exactly, does caffeine do that explains this case?

To understand the effect of caffeine we must first understand adenosine. As illustrated in Figure 4.16, adenosine is an endogenous chemical signal (ligand) that binds to a family of G-protein–coupled receptors. Depending on the receptor type, adenosine can either inhibit (type 1 receptors) or stimulate (type 2 receptors) cAMP production.

1. Adenosine modifies electrical signaling in the brain by acting synapses (Fig. 4.16, left). It binds to type 1 receptors on the presynaptic cell and reduces the release of neurotransmitters such as acetylcholine, dopamine, and glutamate. These are excitatory neurotransmitters that increase the chance that the postsynaptic cell will fire action potentials. It also binds to type I receptors on the postsynaptic cell to increase potassium influx, resulting in hyperpolarization. The resting membrane potential is now farther away from the threshold, so the neuron fires fewer action potentials. The net result of these two actions is less synaptic transmission, less communication, and decreased arousal. Moreover, dopamine normally stimulates a feeling of well-being and enhances locomotor abilities. Thus, *adenosine induces drowsiness, reduces the natural "high" associated with feelings of well-being, and interferes with locomotion.*
2. Adenosine binds type 2 receptors on blood vessel cells everywhere, including the brain (Fig. 4.16, right side). When adenosine binds to these receptors, the muscle cells controlling vessel diameter relax and the vessel widens. Thus, *adenosine induces blood vessels to swell and become congested with blood.*

In short, adenosine promotes rest and relaxation. Caffeine is an exogenous chemical antagonist that binds to adenosine receptors in brain neurons and blood vessels and blocks adenosine from exerting its normal

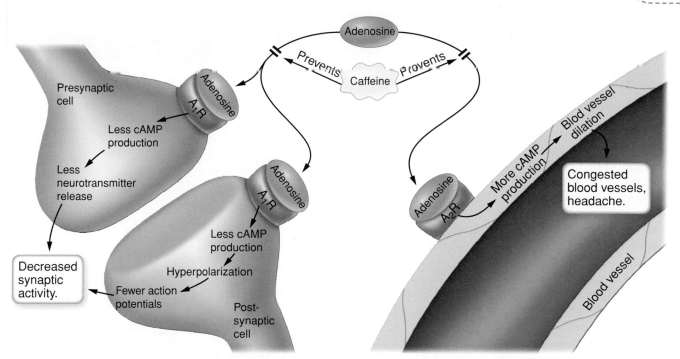

Figure 4.16. Caffeine withdrawal and Andy M. Adenosine binds to type 1 receptors (A1R) on neurons, resulting in decreased synaptic activity. Adenosine binds to type 2 receptors (A2R) on blood vessels, resulting in blood vessel widening. Caffeine antagonizes both effects. *How does adenosine decrease synaptic activity at the postsynaptic cell?*

functions. By interfering with adenosine binding, caffeine invokes the opposite of rest and relaxation. It enhances synaptic transmission and constricts blood vessels, resulting in increased alertness, increased locomotor ability, blood vessel constriction, and feelings of well-being.

Over time, Andy's body compensated for the presence of caffeine by increasing the number and sensitivity of adenosine receptors. This adaptation limits the effect of caffeine in people who drink a lot of it. Thus, in people like Andy, if caffeine is suddenly withdrawn, the brain goes into adenosine overload because all of those extra receptors are free to soak up adenosine.

So, when Andy's brain was suddenly deprived of its adenosine antagonist (caffeine) and went into adenosine overload, here is what he experienced:

- Brain alertness signals declined dramatically: Andy felt tired and was unable to concentrate normally.
- Blood vessels dilated excessively, increasing intracranial pressure: Andy got a pulsing headache.
- Dopamine synthesis fell sharply: Andy lost his sense of well-being, felt fatigued and clumsy, and couldn't compete well in an athletic event.

In the absence of caffeine, it takes about 5 to 7 days for adenosine receptor levels to return to normal and for caffeine withdrawal symptoms to cease. If you want to reduce your dependence on caffeine but avoid the unpleasant symptoms Andy experienced, reduce your intake gradually. And remember, coffee isn't the only thing containing caffeine—many sodas are loaded with it, and it is present in many other foods and drinks, including chocolate and tea.

Case Notes

4.15. By blocking adenosine's action, how does caffeine alter the resting membrane potential of the postsynaptic cell?

4.16. By blocking adenosine's action, how does caffeine alter the release of excitatory neurotransmitters?

4.17. By blocking adenosine's action, how does caffeine change the size of blood vessels and the amount of blood flow in the brain?

CLINICAL SNAPSHOT

Relaxing with GABA

Brain cell activity—that is, action potentials racing this way and that from cell to cell—is a good thing; it is the most important attribute of life. Epinephrine is an important stimulant of brain activity. It works by increasing the firing rate of action potentials in certain neurons. Excess brain cell activity is, however, not so desirable. For example, cocaine stimulates the activity of certain brain cells whose action potentials are normally associated with feelings of well-being and happiness, and overstimulation produces an unnatural euphoria. Overstimulation of other cells can provoke seizures; and stimulation of still other cells can induce panic or heightened anxiety.

Given that the whole of human physiology is a delicate balance of homeostatic forces, it should come as no surprise that, just as the body synthesizes neurotransmitters like epinephrine that *stimulate* action potentials, it also synthesizes neurotransmitters that *inhibit* action potentials. These inhibitory neurotransmitters include gamma aminobutyric acid, or **GABA**. A chemical ligand, GABA exerts its influence by binding to a certain type of GABA receptor (there are several) that controls flow of chloride (Cl^-) ions into brain neurons. When GABA binds to this particular type of GABA receptor, negatively charged ions flood into the cell and change the resting membrane potential from, say, –70 mV to –80 mV. This increased negativity makes the cell harder to stimulate. Thus, the ability of the cell to initiate action potentials is reduced and the cell generates fewer of them. So, if the cell sends signals that create anxiety (a normal emotion in certain situations, such as just before an important exam), GABA is going to suppress cell activity—and anxiety goes away.

The GABA receptor also has binding sites for other substances, including ethanol (found in alcoholic beverages), barbiturates, and benzodiazepines (members of a group of drugs commonly called tranquilizers). These substances cannot open the chloride channel themselves, but they augment the ability of GABA to keep the channel open; they are receptor agonists. Thus, alcohol, barbiturates, and benzodiazepines work (at least in part) by increasing the movement of chloride through GABA receptors and thereby reducing the activity (i.e., action potentials) in certain neurons. For example, if you are anxious about something, you can be sure that somewhere in your brain neurons are firing action potentials to cause the

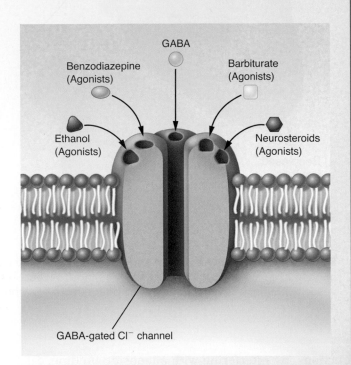

GABA-gated Cl^- channel

The GABA receptor. The chloride channel opens when GABA binds to it, resulting in chloride influx and hyperpolarization. Thus, GABA makes it harder for neurons to fire action potentials. GABA agonists (such as ethanol) increase this effect.

feeling. So how about a little extra intracellular chloride (courtesy of GABA) to make those bad feelings go away?

The existence of such binding sites raises an important question: Why did they evolve? Surely it was not for Neanderthals to stop at a Stone Age pharmacy for a quick pick-me-up to relieve the anxiety caused by saber-toothed tigers lurking in the bush. The most likely explanation is that they exist to bind to naturally occurring (endogenous) relaxers. The parts of the GABA receptor that bind pharmaceutical tranquilizers, for instance, also bind several substances synthesized in the brain—neurosteroids and cannabinoids. Cannabinoids, incidentally, are also the active ingredient in marijuana, which has a well-known relaxing effect. Naturally made relaxers like endocannabinoids and neurosteroids protect neurons from excessive stimulation.

Word Parts

Latin/Greek Word Parts	English Equivalents	Examples
Ant-/i-	Against	Antagonist: works against the agonist
-crine	Secretion	Endocrine: secreted within the body
De-	Remove	Depolarize: remove the polarization
Endo-	Within	Endogenous: generated inside the body
Exo-	Outside	Exogenous: generated outside the body
Neur-/o-	Neuron, nerve, nervous tissue	Neurocrine: secreted from a neuron
Para-	Beside	Paracrine: secreted to cells beside
Post-	After	Postsynaptic cell: cell after the synapse
Pre-	Before	Presynaptic cell: cell before the synapse
Re-	Again, back	Repolarize: polarize again

Chapter Challenge

CHAPTER RECALL

1. Chemical signals that travel through the bloodstream are described as
a. paracrine factors.
b. histamines.
c. neurotransmitters.
d. hormones.

2. The response of a cell to a particular hormone depends on
a. the quantity of ligand.
b. the quantity of receptors.
c. the type of receptors.
d. all of the above.

3. Hormone receptors in the nucleus
a. bind hydrophilic ligands.
b. interact directly with DNA.
c. activate second messengers.
d. usually bind protein hormones.

4. Ligand-gated ion channels
a. are found in the nucleus.
b. bind hydrophilic ligands.
c. interact directly with DNA.
d. usually bind steroid hormones.

5. An example of a second messenger is
a. a steroid hormone.
b. cAMP.
c. a graded potential.
d. a protein hormone.

6. G proteins are
a. components of enzyme-linked receptors.
b. components of ligand-gated ion channels.
c. linked to the intracellular portion of some receptors.
d. none of the above.

7. **Receptor agonists**
   a. can amplify the signal of the endogenous ligand.
   b. can activate receptors in the absence of the endogenous ligand.
   c. can bind a different part of the receptor than the endogenous ligand.
   d. all of the above.

8. **The part of the neuron that usually receives signals is the**
   a. cell body.
   b. axon.
   c. dendrite.
   d. myelin sheath.

9. **In living cells,**
   a. sodium is more concentrated inside the cell and potassium is more concentrated outside the cell.
   b. concentration gradients for sodium and potassium change significantly because of action potentials.
   c. potassium is more concentrated inside the cell, and sodium is more concentrated outside the cell.
   d. the sodium–potassium ATPase pump transports sodium into the cell and potassium out of the cell.

10. **If the inside of the cell contains an excess of negative charges,**
    a. the membrane potential will be positive.
    b. positive ions will be drawn into the cell (if they can cross the membrane).
    c. negative ions will be drawn into the cell (if they can cross the membrane).
    d. there is no electrical gradient.

11. **Graded potentials**
    a. do not change the membrane potential.
    b. do not appreciably change the concentration gradients for sodium and potassium.
    c. are used to transmit signals over large distances.
    d. none of the above.

12. **The depolarizing phase of the action potential results from**
    a. potassium ions entering the cell.
    b. potassium ions exiting the cell.
    c. sodium ions entering the cell.
    d. sodium ions exiting the cell.

13. **Subthreshold graded potentials**
    a. sometimes invoke an action potential.
    b. always invoke an action potential.
    c. never invoke an action potential.
    d. hyperpolarize the membrane.

14. *Saltatory conduction* **refers to**
    a. the transmission of graded potentials down a dendrite.
    b. the transmission of action potentials down myelinated axons.
    c. the transmission of action potentials down unmyelinated neurons.
    d. the "jumping" of an electrical signal from one cell to another.

15. **Which of the following components would not be found in an electrical synapse?**
    a. neurotransmitter
    b. presynaptic cell
    c. postsynaptic cell
    d. action potential

16. **Neurotransmitter receptors**
    a. are always ligand-gated ion channels.
    b. can induce depolarization or hyperpolarization of the postsynaptic cell.
    c. are floating in the synaptic cleft.
    d. are found in the nucleus of the postsynaptic cell.

CONCEPTUAL UNDERSTANDING

17. **List the steps involved when glucagon stimulates glucose production. Identify which steps amplify the signal.**

18. **Compare and contrast G protein–linked receptors and enzyme-linked receptors.**

19. **List the steps involved in an action potential. For each step, describe:**
    a. which channels are open and which are closed.
    b. which ions are moving and in which direction.

APPLICATION

20. **For each of the following examples, identify the type of signal (paracrine factor, hormone, or neurotransmitter).**
    a. A brain cell signaling a stomach cell to secrete acid.
    b. A stomach cell signaling the neighboring cell to secrete acid.

21. **In a cell with a membrane potential of +50 mV, which of the following statements would be true? Explain why or why not for each statement.**
    a. There would be a relative excess of negative charges outside the cell.
    b. Sodium ions would be drawn out of the cell down the electrical gradient.
    c. The sodium concentration gradient would be reversed, and sodium would be more abundant inside than outside the cell.

22. **During the action potential, sodium entry through voltage-gated sodium channels depolarizes the cell membrane, which opens more voltage-gated sodium channels. Is this an example of negative feedback or positive feedback? Explain why, using a diagram similar to Figure 1.7 (for negative feedback) or Figure 1.9 (for positive feedback).**

23. **A hormone called *growth hormone* stimulates growth. What are some of the different ways that the body can increase growth induced by this particular chemical signal?**

You can find the answers to these questions on the student Web site at **http://thepoint.lww.com/McConnellandHull**

5

Skin, Membranes, and Other Barriers to the Environment

Major Themes

- Skin is important in maintaining homeostasis.

- The external environment is a threat to the body's internal environment.

- Skin and other membranes separate the internal and external environments.

- Other physical and chemical mechanisms also help protect against the external environment.

Chapter Objectives

Case Study: "Blow on it."

As you read through the following case study, assemble a list of the terms and concepts you must learn in order to understand the case. Make a list of the patient's signs (such as low body temperature) and symptoms (such as pain), and (after studying the chapter) try to find reasons for them.

Clinical history: Sylvester P., a 19-year-old man, and a friend were working on the gas line on an antique car in the barn of a mountain home. As Sylvester, shirtless, worked on one end of the line and his friend on the other, he instructed his friend to "Blow on it" to see if the line was obstructed. As a result, he was sprayed with gasoline, which was ignited by the cigarette he was smoking. He sustained severe burns.

Because of the remote location of the home, it was nearly 4 hours from the time of the incident before he arrived at the burn unit of a large medical center.

Physical examination and other data: On admission, Sylvester's vital signs were as follows: blood pressure 90/60 (normal, 120/80), heart rate 130 (normal, 70), respiratory rate 26 (normal, 14), temperature 97.6°F (normal, 98.6). Partial- and full-thickness burns covered most of his head, neck, arms, and torso. Areas of his body that were not burned were pale. He was alert, in great pain, shivering with cold, and complaining of thirst.

Clinical course: In addition to pain medication, supplemental oxygen, and other supportive measures, an intravenous line was inserted and rapid infusion of a solution of water, glucose, minerals, and other nutrients was begun. In the first 24 hours, Sylvester received 15 L of fluid and his blood pressure was restored to normal. During the first week of hospitalization, he continued to require large volumes of intravenous fluid and nutrients to sustain normal blood pressure and provide fuel for the repair process.

By the end of the first week of hospitalization, Sylvester's burns had become infected and his wounds were oozing copious amounts of pus. By the end of the third week, laboratory analysis showed that bacteria had colonized his bloodstream. Despite intensive antibiotic therapy, blood transfusions, and intravenous feeding, he died of overwhelming infection on the 30th hospital day.

Other anatomy and physiology textbooks typically call this chapter "The Integumentary System." The word *integument* derives from the Latin *integumentum*, which means "a covering," and the integumentary system indeed includes our primary body covering, the skin, as well as its derivatives, the sweat and oil glands, nails, and hair. However, inasmuch as the principal role of the skin and its derivatives is to serve as a barrier to the environment, we have added to our chapter a discussion of nonintegumentary barriers, including membranes and chemical barriers. But why are barriers so important that we should devote an entire chapter to them?

Recall from ◄ Chapter 1 this major theme: *Life is sustained by specific things in the external environment.* These specific things provided by the environment include oxygen, water, food, pressure, and heat. However, the other side of the coin is this: The external environment is a daily threat to life. One defense we have against external threats is the way our body is organized. This recalls another concept from Chapter 1:

Organisms are organized; that is, every component has its place and must remain in place for the organism to thrive. Organization is maintained by barriers that create boundaries. Barriers keep the internal environment of living things organized and separated from the external environment in which they live.

In this chapter, we describe the form and function of skin and other body membranes. These tissues create physical barriers that prevent or limit exchange of things between the body's external and internal environment. We also discuss other mechanisms, both physical (cilia in the respiratory tract, for example) and chemical (gastric acid, for example), that aid in this task.

All the beauty of the world, 'tis but skin deep.

Ralph Venning (1620–1673), English clergyman, in his book of devotions, *Orthodox Paradoxes*. Venning's most famous quotation is "Better late than never."

The Functions of Skin

The most basic function of skin is to keep the inside in and the outside out. Like shrink-wrapping, skin retains body water and heat and keeps out environmental water. This homeostatic mechanism is so basic that most of us forget about it unless painfully reminded by something like a burn: when skin is burned, body fluids and heat escape rapidly. That's why Sylvester, our case study patient, complained "I'm thirsty" and "I'm cold": he was losing fluid and heat rapidly because his skin was burned away.

> *Remember This!* **The most basic function of skin is to create a boundary between the external environment and the body's internal environment.**

As just noted, skin passively retains heat. In addition, it cooperates with the nervous system in the regulation of body temperature. When the body core becomes either excessively warm or chilled, the brain's temperature regulation center transmits signals to blood vessels in the skin. In response, these blood vessels dilate to bring more blood to the body surface to lose heat, or they constrict to bring less blood to the body surface to preserve heat. Blood vessel dilation delivers warm blood to the surface for cooling, whereas constriction keeps warm blood circulating deep within the body core and away from skin, where heat can easily be lost (Fig. 5.1). The negative feedback loop regulating body temperature can be revisited in ⬅ Figure 1.7.

Figure 5.1 also illustrates the role of sweating in body cooling. In response to homeostatic signals from the nervous system, sweat is secreted onto the surface of skin, where it evaporates. This cools the skin and the blood flowing in the surface vessels.

The skin's rich supply of nerves and special sensors for heat, pain, and touch also play a protective role by enabling us to perceive danger. In ⬅ Chapter 4, you learned how action potentials generated by skin sensors are transmitted along nerve axons. These signals reach the brain, which interprets them and transmits signals coordinating a response to the threat.

Skin also provides protection against force and shields our inner workings from much of the brute force of the environment. As an example, a bullet penetrates skin with about as much difficulty as it penetrates bone. And

skin is aided in its protective task by the layer of fat upon which it rests. This fat layer provides cushioning over bony prominences and protection against blunt forces.

Skin is our primary defense against environmental contaminants. The thin outermost layer is epithelial tissue. It is virtually waterproof and serves as a barrier to threats such as bacteria, chemicals, fluids, pollens, and air pollution particles, keeping them from gaining access to the rich inner sea of body fluids. Moreover, the skin is populated with special immune cells ➡ (Chapter 12) that help protect the body from these environmental threats. The process of tattooing requires the tattoo artist to insert pigments beneath the top layer of skin, the epidermis. This breaks the skin barrier and can result in an infection or other serious problems. For more on tattooing, see the nearby Clinical Snapshot, titled *Tattoos*.

Case Note

5.1. Explain the following signs and symptoms from the case study based on the above discussion of skin function: low body temperature, thirst, infection.

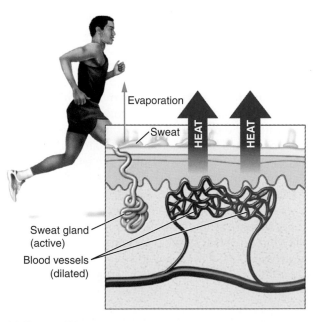
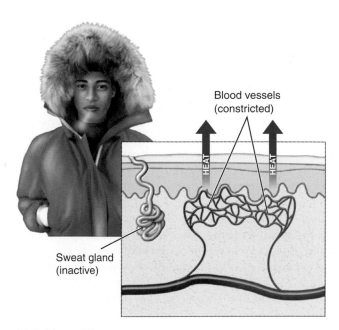

(a) Hot conditions   **(b) Cold conditions**

Figure 5.1. Skin and temperature regulation. A. Under hot conditions, the blood vessels of the skin dilate to promote flow of blood from the body core to the cooler body surface. At the same time, sweat glands produce watery sweat, which evaporates, carrying off more heat. **B.** Under cold conditions, the blood vessels of the skin constrict to reduce the flow of blood to the cooler body surface, thereby reducing heat loss. *In Chapter 1, we identified four functions of living organisms: organization, metabolism, adaptation, and reproduction. Which function does this figure represent?*

CLINICAL SNAPSHOT

Tattoos

The word *tattoo* has two meanings. It can signify the rapid tapping of a drum, as at a military funeral, or the process of marking the skin with pigment. Its use to describe skin art derives from the Polynesian Islands of the South Pacific. Captain James Cook (1728–1779), the legendary English explorer, first used the word in 1769 to describe the pigmented skin decorations he observed on the skin of the Polynesians. The term derives from the Tahitian and Samoan word *tatau*, meaning a puncture or mark on skin.

The practice of tattooing skin is at least 5,000 years old and is thought to have occurred in virtually every ancient culture, including Egypt, Greece, China, Japan, India, Africa, and South America as well as the Pacific islands. In 1991, the frozen corpse of a man was found high on a mountain along the Italian–Austrian border. "Otzi, the Iceman," as he came to be known, died about 5,000 years ago, and had 57 tattoo marks.

Tattooing basics have not changed much over the millennia: a nondissolvable colored material is inserted through the epidermis into the superficial dermis, where it remains visible permanently. This differs from temporary skin markings, both modern and ancient, in which the pigment is deposited in or on the epidermis and fades away in a few days as epidermal cells are shed.

The permanence of tattooing has traditionally been one of its most desirable features: in ancient days it served as a kind of "in group" code enabling members of a tribe or sect to identify one another, and in some cultures tattoos were applied for their presumed healing qualities. With the rise of scientific medicine, the decline of tribal societies, and the emergence of vast nation–states, the use of tattoos for healing and tribal identification has declined, although they are still used by some indigenous tribes and urban street gangs.

Nowadays, the permanence of tattoos has spawned a surge in the tattoo-removal business by plastic surgeons, as many people eventually decide they want their tattoo erased. Tattoos can be removed in one of two ways: surgical excision of the tattooed skin or bleaching of the

Tattoos. Tattoo inks can cause allergic reactions, as shown by the red raised bumps.

pigment with specialized lasers. Both techniques have drawbacks: surgical excision leaves a scar, and laser bleaching usually leaves the skin somewhat mottled.

To solve this problem, new tattoo inks have been developed. The pigment in one type of new ink is bonded to a plastic polymer, and the color will persist for a lifetime if left undisturbed; however, a laser beam easily breaks the bond, allowing the ink to be completely absorbed into the bloodstream and eliminated. Use of such inks has led to a heated controversy among tattoo artists and consumers: Some are so devoted to the idea of permanence that they decry the use of inks that can be removed.

Finally, a word about tattoo safety. All tattoos involve penetrating the protective barrier of the epidermis. Violation of this barrier carries risk, mainly of an allergic reaction or infection. In the United States, for example, a person who receives a tattoo will generally be prohibited from donating blood for 12 months in order to prevent the passage of an infection by transfused blood. Modern tattoo artists reduce infection risk by following universal sterile procedure and working with single-use tools. Almost any infective microbe can be transmitted, but by far the most dangerous are hepatitis viruses and the HIV virus of AIDS.

Finally, with the help of other body tissues and in the presence of adequate sunlight, skin synthesizes a hormone called cholecalciferol (vitamin D3), commonly known as *vitamin D* because it can also be obtained in the diet. (A *vitamin* is an organic substance that is present in foods and is necessary in minute amounts for life and health.) Vitamin D is critical for bone health because it enhances calcium absorption into the blood from the gastrointestinal tract. Calcium is a major mineral found in bone. Skin form and

Table 5.1   Skin Form and Function

Skin Layer	Important Elements	Form	Function
Epidermis		Surface portion of skin; no blood vessels	Protection from microbes, heat loss, mechanical trauma, dehydration, ultraviolet light
	Stratum corneum	Surface layer; keratin-filled dead cells	A part of the epidermal barrier
	Stratum basale	Deepest epidermal layer; columnar epithelium	Produces new epidermal cells
	Melanocyte	Cells found in stratum basale	Produces melanin
	Dendritic cells	Lymphoid cells found throughout epidermis	Immune defense
Dermis		Fibrous connective tissue; many blood vessels and nerves	Cushions, stretches, nourishes epidermis
Subcutaneous layer		Areolar connective tissue under skin; contains fat cells, vessels, nerves	Connects skin to surface muscle; cushions; insulates
Accessory structures	Sensory receptors	Free nerve endings or onionlike bulbs	Detect pressure, touch, temperature
	Sebaceous (oil) glands	Saclike; associated with hair follicles; produce sebum	Sebum lubricates skin; prevents dehydration
	Eccrine sweat glands	Secrete watery, salty sweat onto skin surface	Cooling
	Apocrine sweat glands	Coiled in armpits and groin; secretion contains cellular materials; released into hair follicle	Scent glands
	Hair	Composed of keratin; grows from cells in the hair follicle; arrector pili muscle elevates hair	Heat conservation, protection from ultraviolet light, lessens friction
	Nails	Composed of keratin synthesized by stratum corneum cells	Protect fingers and toes Facilitate grasping

function are summarized in Table 5.1. These features will be discussed in detail as we proceed through the chapter.

5.1 True or false? Blood vessel dilation delivers warm blood to the body surface for cooling.

5.2 Which hormone does skin make only in the presence of adequate sunlight?

Case Discussion

The Importance of Barriers: The Case of Sylvester P.

Sylvester suffered serious burns to more than 50% of his body surface, which means that he lost more than half of the protective covering of his body (Fig. 5.2). This loss of skin in turn prompted three further losses. First, water from his blood and tissues oozed out or evaporated into the

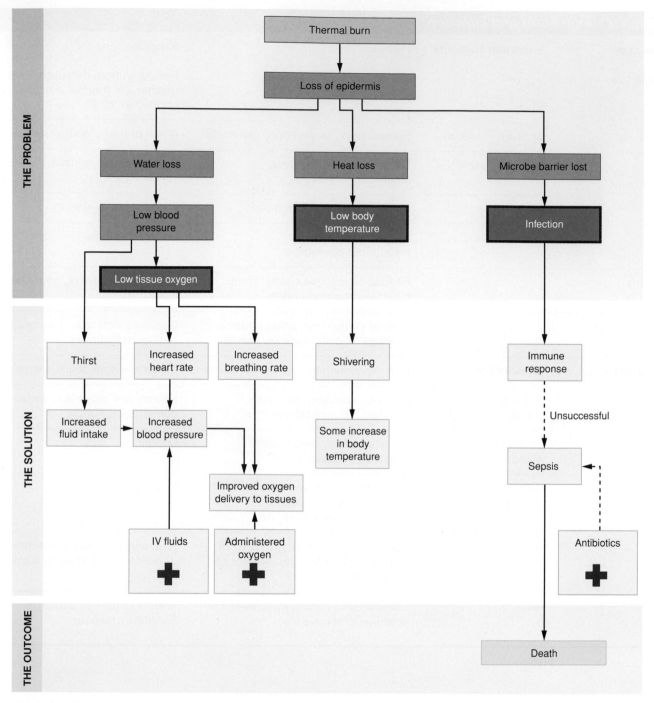

Figure 5.2. Homeostasis and Sylvester P. The burn destroyed Sylvester's epidermis, so water and heat could freely leave his body and microbes could enter. His body attempted to compensate for the resulting low blood pressure, low body temperature, and infections, and medical interventions also addressed these problems. However, the homeostatic and medical measures were not sufficient to restore homeostasis, and Sylvester died. *Which parameter is improved by intravenous fluids?*

air, which diminished his blood volume, lowered his blood pressure, and impaired oxygen delivery to his tissues. A second problem—low body temperature—reflected Sylvester's loss of body heat through water evaporation. Third, by being burned, Sylvester lost an important physical barrier to microbes. As a conse-

quence, microbes easily gained access to his tissues and blood.

Sylvester's body was able to compensate partially for his burns. Oxygen delivery to tissues was improved by increased heart rate (which increased his blood pressure) and respiratory rate. In addition, his thirst stimulated fluid intake (which also increased his blood pressure). Violent shivering (widespread, involuntary muscle contractions) generated the heat that raised his body temperature somewhat. Inflammation and other aspects of the immune system ➡ (Chapter 12) responded to cope with the invasion of infectious agents.

Medical interventions helped, too. To restore normal blood volume and pressure, Sylvester required 15 L (4 gallons) of intravenous fluid in the first 24 hours in the hospital. Supplemental oxygen increased the oxygen content of his blood and further improved oxygen flow to tissues. His energy needs were dramatically increased, in order to fuel the muscular contractions of shivering, fight infection, and heal his traumatized tissues. Intravenous administration of glucose and other nutrients helped address his negative nutrient balance. Sylvester did not experience shock (low blood pressure and vascular collapse) or hypothermia (low body temperature), because homeostatic and medical interventions were sufficient to meet these challenges.

However, loss of the epidermis over more than 50% of his body left Sylvester especially vulnerable to microbes, which gained access to and colonized his tissues and blood. The combined efforts of the body's immune response and administered antibiotics could not compensate for the loss of the defensive function of skin, and Sylvester died of a massive infection.

The Anatomy of Skin and Associated Structures

Skin is the largest human organ. Although literally as tough as leather, it is also soft and pliable and is custom-designed—very thin around the eyes and genitals, and unusually thick on the palms and soles and on the back. On the other hand, it is easily damaged because its outermost membrane, the *epidermis* (discussed below), is only about 0.1 mm thick and subjected regularly to scratches, bruises, cuts, scrapes, and other daily insults.

Embedded in its tissue layers, skin contains numerous other structures: It is richly supplied with blood vessels and nerves with special sense receptors attached to nerve endings. It also contains sweat glands, hair follicles, and oil glands (Fig. 5.3). Let's take a closer look at skin and its associated structures.

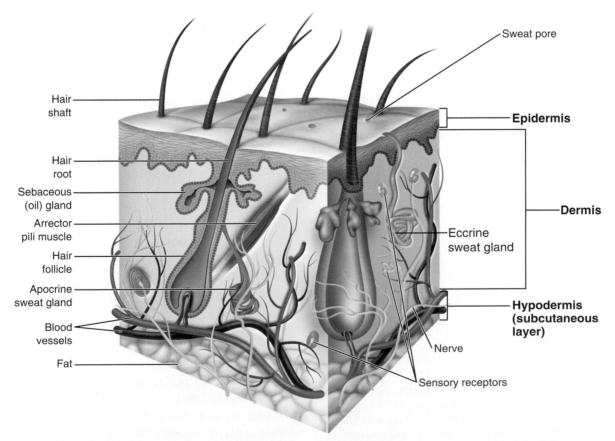

Figure 5.3. Anatomy of skin. *Which is larger—the sweat pore or the entrance to the hair follicle?*

Skin Is Composed of Two Layers of Tissue

Skin is composed of two layers: an outermost *epidermis* and a deeper *dermis* just beneath it. The epidermis is composed of epithelial cells and rests upon a *basement membrane* (Fig. 5.4), a thin protein membrane that separates the epidermis from the dermis. The epidermis does not contain blood vessels, nerves, glands or other structures, all of which are located deep to the epidermis in the dermis, and must get its nourishment and oxygen by diffusion across the basement membrane from blood vessels in the dermis.

> *Remember This!* **The epidermis is very thin. To appreciate exactly how thin, take careful notice of the next skin blister you see: the covering of the blister is the epidermis.**

The Epidermis Forms a Barrier

The **epidermis** is a surface sheet of epithelial cells arranged in five layers. Skin cells are constantly reproducing from a special pool of *stem cells* that divide continually to replace dead cells. From these stem cells, the five layers of the epidermis are produced.

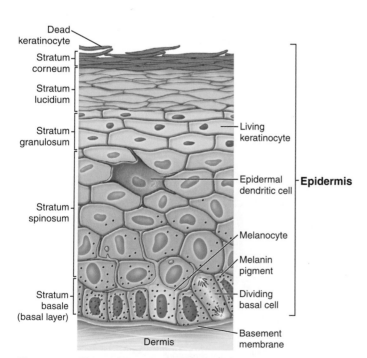

Figure 5.4. The epidermis. This figure illustrates a region where the epidermis is quite thick, like the palm of the hand. The stratum lucidum is absent in regions with thinner epidermis. *Which skin layer is deeper—the stratum corneum or the basal layer?*

Layers of the Epidermis

Epidermal stem cells are called **basal cells.** They rest on the basement membrane and form the **stratum basale (basal layer),** the deepest layer of the epidermis (Fig. 5.4). In healthy skin only basal cells divide; however, in skin cancers, all layers can participate in runaway cell growth (see the nearby Clinical Snapshot, titled *Skin Growths: What's Cancer and What's Not?*). The stratum basale also contains **melanocytes,** specialized cells that produce a pigment called *melanin*, discussed further on.

Each division of a basal cell produces a new stem cell, which remains in the stratum basale, and a second cell that is pushed away from the basement membrane to form the next layer of cells, the **stratum spinosum** (Latin *stratum* = "pavement," and *spinosum* = "spine-like"). In living skin, these cells are not spiny at all—they are relatively plump cuboidal cells joined together at many points by desmosomes. In microscopic preparations, conversely, these plump cells dry out and shrink, but the desmosomes hold tight and stretch, which gives the cells a spiny appearance. As these cells mature, they are pushed toward the skin surface by proliferating basal cells. Along the way, they accumulate increasing amounts of a tough, fibrous protein called *keratin*—the stuff of hair and nails—and are now called **keratinocytes.** Scattered among the keratinocytes are **dendritic cells** (also called Langerhans cells), octopuslike cells with tentacles (dendrites) that sit ready to intercept and phagocytize bacteria and other invading material. They are an important part of the immune system.

As cells of the stratum spinosum continue to mature, they become progressively drier because, as they are pushed toward the skin surface, they move further away from their source of fluid in the dermis. These flatter cells form the **stratum granulosum,** so named because they contain abundant protein-filled granules.

As keratinocytes continue to ascend, they die from apoptosis, and their nucleus and granules disappear. In the thick skin of the palms and soles, dying keratinocytes form a featureless sheet of cells called the **stratum lucidum** (Latin *lucidus* = "clear"), which is microscopically transparent. It is not present in the thinner skin of other areas.

Finally, at the skin surface, all that remains of the original cell is a dead, dry, acidic, flat packet of keratin protein. These dead cells are shed imperceptibly every day as they are replaced from below. This outermost layer of epidermis is called the **stratum corneum** (Latin *corneus* = "of horns") because in horned animals the stratum corneum condenses to form horns. The stratum corneum serves multiple purposes: (a) it is virtually

CLINICAL SNAPSHOT

Skin Growths: What's Cancer and What's Not?

Before discussing skin cancers, a few definitions are in order. A neoplasm (Greek *neo-* = "new" and *-plasm* = "matter") is an uncontrolled growth of new matter— that is, new cells. In everyday use, the word *tumor* (Latin for "swelling") means the same thing. A *malignant* neoplasm (tumor) is called **cancer;** it is a growth that is potentially fatal because of its ability to invade nearby tissues or spread widely by *metastasis*. A *benign* neoplasm is an uncontrolled growth of cells that is not usually potentially fatal, though an otherwise benign tumor, in the brain, for example, can be fatal by virtue of its critical location.

Despite the protection provided by the epidermis, skin is still exposed to many *carcinogens* (cancer-causing agents) such as ultraviolet light and environmental toxins.

Common benign skin tumors include lipoma, common wart, and common nevus. A *lipoma* is a benign tumor that arises from fat cells of the subcutis, which lies beneath the dermis. Strictly speaking it is a tumor of the hypodermis, but for practical purposes it is considered a tumor of skin. Most lipomas are about the size of a golf ball, but they may become even larger. They never become malignant and need not be removed except for cosmetic reasons or mechanical interference with body movement. A **wart** is a spiky overgrowth of epidermal cells caused by certain strains of human papillomavirus (HPV); other, high risk strains of HPV cause cancer of the female cervix. A *nevus* (commonly called a *mole*) is a benign tumor of melanocytes. Most moles are small, about the size of a pencil eraser, and are light brown, but some may contain little melanin pigment and appear pink.

Cancer does not leap into being in an instant; it must evolve through a precancerous stage from normal cells. A **keratosis** is a precancerous crusty overgrowth of keratinized epidermal cells that usually appears on sun-exposed skin. It is "cancer in the making"; that is, it evolves slowly from benign to malignant.

An important cause of skin cancer is sun exposure; it occurs most commonly in light-skinned people on the face, forehead, lower lip, back of the neck, and back of the hand. Skin cancer begins when chronic sun exposure alters the DNA in epidermal cells. Because melanin absorbs the sun's rays, people with more melanin are at a lower risk than people with fair skin, but dark-skinned people can develop skin cancer, too.

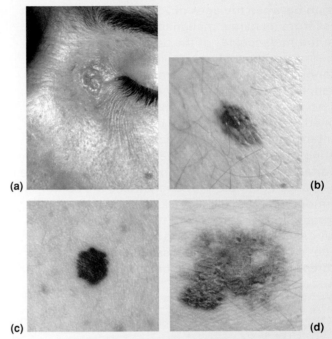

(a)   (b)

(c)   (d)

Skin neoplasms. A. Basal cell carcinoma. The pearly plaque has raised edges. **(B–D)** Melanoma. These malignant melanomas are characterized by irregular borders **(B–D)**, very dark pigmentation **(C)**, and/or irregular pigmentation **(B and D)**.

That's why the American Cancer Society recommends everyone use sunscreen on all exposed skin with a sun protection factor (SPF) of at least 15.

Any cancer that arises from an epithelium is called a **carcinoma,** so skin cancers are carcinomas. **Basal cell carcinoma** is an uncontrolled growth of cells from the stratum basale cells in the basal layer of the epidermis; **squamous cell carcinoma** is an uncontrolled growth of flat, keratinized, slightly more mature epidermal cells. Despite the fact that they qualify technically as cancers, basal and squamous cell cancers grow sluggishly and rarely spread to other sites. However, those that occur near critical structures, such as the eye, nose or ear, can be locally destructive. Death from basal or squamous carcinoma is an extreme rarity; so much so, that they are not counted in cancer statistics.

Malignant melanoma (often shortened to *melanoma*) is behaviorally the opposite of basal

(continued)

Skin Growths: What's Cancer and What's Not? *(Continued)*

and squamous carcinomas. Viciously malignant, it accounts for only about 5% of skin cancers but virtually 100% of skin cancer deaths. It is also the most common malignancy of any kind occurring in women between the ages of 25 and 29. As you might guess from its name, malignant melanoma arises from the *melanocytes* of the basal layer of the epidermis—almost all are darkly pigmented with melanin. Like basal and squamous carcinomas, melanomas are stimulated by exposure to sunlight and occur most often in sun-exposed areas of light-skinned people. About half of melanomas arise spontaneously, and another half arise from pre-existing common moles; indeed, one of the most important signs of early melanoma is a change in the appearance of a mole. However, moles are so common that there is little risk of an individual mole becoming malignant. Cure of malignant melanoma requires early recognition and surgical excision.

The "ABCD rule" encompasses the most important features to look for in a melanoma:

- *A* is for *asymmetry:* One half of a mole or birthmark does not match the other.
- *B* is for *border:* The edges are irregular, ragged, notched, or blurred.
- *C* is for *color:* The color is not the same all over and includes shades of brown or black, sometimes with patches of red, pink, white, or blue.
- *D* is for *diameter:* The area is larger than 6 mm (about ¼ in.) across, or the area has been growing.

Other important things to be aware of include:

- Rapid enlargement of an existing pigmented (brown-black) skin lesion (usually a mole)
- Itching or pain in a previously asymptomatic pigmented lesion
- Development of a new pigmented lesion in adult life
- Satellite areas of pigmentation around a pigmented lesion

---

waterproof, retaining body fluid and preventing absorption of environmental water; (b) it repels a variety of environmental agents, such as oils and other substances; (c) it has little nutrient content and thus doesn't readily sustain the growth of microbes; and (d) as its surface cells are shed, they carry away microbes and other hazards. Incidentally, it takes only a few weeks for the offspring of a basal cell to reach the skin surface, die, and fall away.

Most of the extra thickness in the skin of the palms and soles reflects the augmented stratum corneum and the presence of a stratum lucidum. However, the thickness of the skin of the back is due to increased thickness of the dermis.

Remember This! **Why does the skin of your fingers wrinkle after a long bath? Recall that the epidermis is virtually waterproof; however, its most superficial layer can hold water for a while. Thus the stratum corneum swells, increasing its surface area relative to underlying layers. The stratum corneum cannot detach from underlying skin layers, so it wrinkles instead. As soon as the water evaporates, the stratum corneum shrinks back to size and the wrinkles disappear.**

Case Note

5.2. The epidermis on Sylvester's palms was burned down to the stratum spinosum. Which epidermal layers have been lost?

Skin Color

Skin is the natural canvas of human life, revealing details of our personal histories and lifestyle choices. For example, years of smoking and excessive exposure to sunlight are reflected in marked premature wrinkling. And the appearance of our skin is of supreme social importance in our culture and many others. For example, Americans so value clear and youthful skin that we spend large sums of money to maintain and improve its appearance—we cleanse it, moisturize it, paint it with cosmetics, lighten or darken it with chemicals, decorate it with jewelry and tattoos, and reshape it with surgery.

Of all the attributes of skin, color has played the most important role historically. The most important determinant of skin color is the amount of melanin pigment it contains. Melanin is produced by **melanocytes,** specialized cells located in the stratum basale that arise from the embryonic nervous system (Figs. 5.4 and 5.5A). Genetic differences in skin color reflect differences in melanocyte

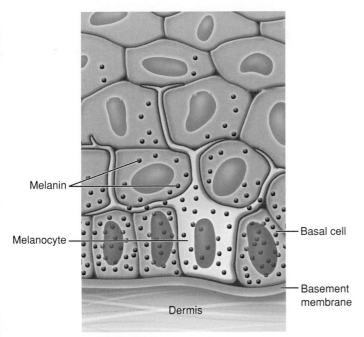

Melanin

Melanocyte

Basal cell

Basement membrane

Dermis

(a) Melanocyte

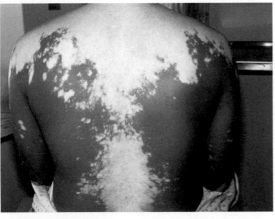

(b) Vitiligo

Figure 5.5. Melanocytes and pigmentation. A. Melanocytes provide pigment to neighboring cells, in which it shields the nucleus from damaging ultraviolet rays. **B.** A man with vitiligo. Melanocytes are not active in the unpigmented (white) regions. *Do melanocytes protect cells in the stratum corneum?*

activity, not in the number of melanocytes. Melanocytes produce varying amounts of dark brown melanin pigment and deliver it into nearby basal cells, where it resides and gives skin some of its color (blood provides most of the remaining color). Melanin protects the basal cells, the underlying cells of the dermis, and the melanocytes themselves by absorbing cancer-causing ultraviolet rays from the sun.

In healthy people the amount of melanin pigment varies greatly according to the person's genetic makeup

and the amount of sun exposure; sun exposure stimulates melanin production as a protective mechanism, and melanin absorbs the sun's rays, protecting tissue from radiation damage. Pregnancy and certain other conditions affect melanin production as well. A general darkening of skin is associated with some endocrine diseases, and numerous skin diseases can cause local darkening or lightening of skin due to increased or decreased local melanin production. For example, *vitiligo* is a patchy local blanching of skin to near whiteness owing to a local loss of melanin pigmentation (Fig. 5.5B).

> **Remember This!** **The varying shades of brownness of skin depend on the amount of melanin pigment produced by melanocytes in the epidermis.**

The general color of skin can also be influenced by blood; for example, the blush of embarrassment is due to a rush of blood into skin. In very dark-skinned people with a large amount of melanin, blood does not significantly affect skin color; however, in persons with light skin, blood is an important contributor. Oxygen is carried by a red pigment called **hemoglobin** ➡ (Chapter 10), which becomes redder according to its oxygen content—poorly oxygenated hemoglobin is dark red–blue, whereas fully oxygenated hemoglobin is bright red. When the blood's hemoglobin level is normal, it contributes a healthy appearance, no matter the individual's skin color. Light-skinned people who are anemic—who have low blood hemoglobin—appear pale, whereas anemia, especially mild anemia, in people with darkly pigmented skin may be more difficult to detect with the naked eye alone. In both light- and dark-skinned people poor oxygen content (from lung disease, for example) turns blood dark bluish-red, in contrast to the bright red of well-oxygenated blood. This condition imparts a bluish cast to skin called **cyanosis** (Greek *kauneos* = "dark blue").

Of the other conditions that affect general skin color, one of the most notable is an abnormal accumulation in blood of yellow **bilirubin** pigment. Called *jaundice*, this yellowish coloration occurs in association with some liver and blood diseases. Finally, skin normally contains a small amount of **carotene,** an orange–yellow pigment found in vegetables such as carrots. Excessive ingestion of foods high in beta-carotene will give the skin an orange coloration.

For more information about other aspects of skin color, you may wish to visit our Web site (**http://thepoint. lww.com/McConnellandHull**) to read "Skin Color."

Case Note

5.3. The unburned portions of Sylvester's skin were very pale, reflecting minimal blood circulation in his skin. Which blood pigment contributes to skin color?

The Dermis Supports the Epidermis

The **dermis** is the "tough as leather" part of skin; indeed, most of the thickness of leather is the preserved and treated dermis of animals. The dermis is a framework of fibrous and elastic tissue that lies deep to the epidermis and rests on a layer of subcutaneous fat (Fig. 5.6). The dermis is home to many other structures (nerves, specialized nerve endings, blood vessels, sweat and oil glands, and hair follicles), discussed below. Remember, too, that the epidermis has no blood vessels and gets all of its oxygen and nutrients by diffusion from the dermis.

The interface between dermis and epidermis is irregular, somewhat like the fit of a waffle iron and a waffle. The underside of the epidermis forms downward-projecting ridges that form a pattern somewhat like the ridges of a waffle. Projecting upward from the dermis are dermal papillae, which fit between the epidermal ridges like the teeth of a waffle iron. The interface between the epidermal ridges and the dermal papillae can be visualized in the thin skin of the fingertips, where it results in fingerprints.

The dermis is about 1 to 2 mm thick, which is about 10 to 20 times thicker than the epidermis. It is especially thick in the skin of the upper back and especially thin in the skin of the scrotum in males, the labia of females, and around the eyes. The superficial dermis, the dermis between the epidermal ridges, is called the **papillary dermis** because it extends from the tips of the dermal papillae to a short distance below the lower edges of the epidermal ridges. It is less densely fibrous and most often involved in skin disease. The deeper **reticular dermis** is more densely fibrous than the papillary dermis and less often involved in skin disease. It is home to the skin's main blood vessels, nerves, and skin appendages, such as sweat glands.

The primary cell type in the dermis is the **fibroblast,** which produces the collagen and elastin fibers that impart strength to fibrous tissue and account for skin toughness and elasticity. Also present in the dermis are macrophages and dendritic cells, which activate an immune response against any invaders that may penetrate the epidermal barrier; they also watch out for the development of skin tumors.

Skin Contains Blood Vessels, Nerves, Nails, Hair, and Glands

Remember that the skin is an organ, just like the liver or the stomach; like all organs, it contains blood vessels and nerves. The flow of blood into skin capillaries is controlled by *precapillary sphincters*. The sphincters clamp down to keep blood away from the skin when the body must retain heat, or they relax to allow greater blood flow if the body temperature is high (Fig. 5.1).

Skin also contains a rich supply of nerves; these have several types of endings, each of which senses the environment in a certain way (Fig. 5.3). We learn more about the senses in ➡ Chapter 9; for now, a quick preview will suffice. Some free nerve endings are adapted to sense changes in temperature and pain. Other free nerve endings are attached to hair follicles to sense hair movement—for example, the feel of a light breeze across an arm. Other nerve endings are surrounded by layers of epithelial tissue, much like the layers of an onion (they are not considered "free"). These structures detect sensations such as light touch, deep pressure, and vibration.

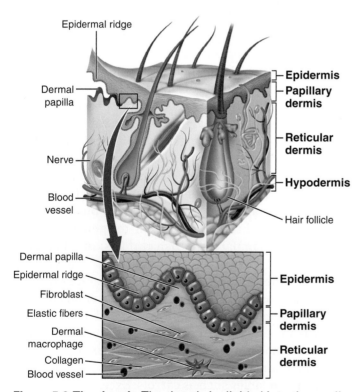

Figure 5.6. The dermis. The dermis is divided into the papillary dermis and the reticular dermis. Part of the epidermis has been peeled away to show the epidermal ridges and the dermal papillae. *Which layer is the most superficial— the papillary dermis or the reticular dermis?*

In addition to blood vessels and nerves, skin contains a set of *skin appendages*: These are nails, hair, and skin glands.

Nails Are Modified Stratum Corneum

Hair and nail salons might as well be called keratin spas—hair and nails are composed entirely of keratin, the same protein found in epidermal cells all over the body. The only difference is that the keratin of hair and nails is especially abundant and stiff.

In other animals, the structure that is our nail is a hoof or a claw. Nails protect sensitive fingers and toes from trauma, aid in the manipulation of small objects, and are useful for other tasks, like scratching an itch.

The visible part of each nail is the **nail plate,** which is actually nothing more than an especially thick, stiff layer of stratum corneum. The nail attaches at the **nail root,** which is covered with skin, and the **cuticle,** a thin layer of stratum corneum that stretches out from the skin around the edge of the nail (Fig. 5.7). The nail rests on a thin layer of epidermis, called the **nail bed,** beneath which a rich supply of dermal blood vessels is visible, giving nails a pink glow. The half-moon near the base of the nail, the *lunula,* appears white because the epidermal layer is thicker in this region, obscuring the dermal blood vessels.

The proximal portion of the nail bed is called the **nail matrix.** Cells in the matrix divide, and some of the new cells differentiate into nail cells. Nail matrix cells are stem cells similar to basal cells of the epidermis. As new nail cells are produced, old nail cells are pushed forward (about 1/10 mm per day). As they migrate outward from the matrix, they accumulate more keratin and die. Thus, the nail plate is composed only of keratin fibers and a tiny residue of dead cell material.

Remember This! When you groom your hands, your nail and cuticle clippings are bits of stratum corneum.

Hair

Hair is present in skin everywhere except the palms of the hands and the soles of the feet. Human hair has lost much

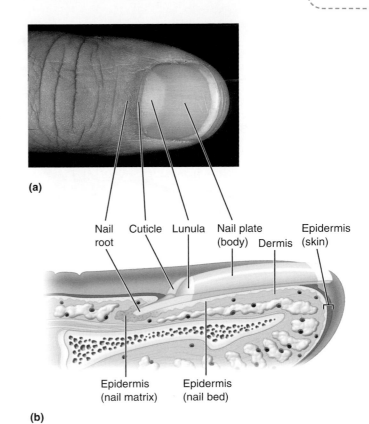

(a)

(b)

Figure 5.7. Nails. A. Fingernail, external view. In this photo, note that the nail root is hidden under skin and the cuticle. **B.** Finger, longitudinal section. Notice that the nail bed is an extension of the epidermis. *What part of the nail is whitish in color?*

of its insulating ability; evidence suggests that humans have been relatively hairless for perhaps 3 million years. However, as any bald male can attest, scalp hair has at least some insulating value in cold weather and serves as an effective sunscreen. Also, nose hairs trap particulate matter, eyebrows and eyelashes offer some help in keeping foreign matter out of the eyes, and hair in the armpits and around the anus and genitalia prevents chafing as skin rubs against skin.

Hair grows from hair follicles, which are embedded mainly in the dermis but may extend into the subcutaneous fat (Fig. 5.8A). Hair follicles are specialized structures that arise in the embryo as downward invaginations of epidermis, to which they remain attached. The hair follicle contains a reserve of skin stem cells in the aptly named **bulge** below the sebaceous gland. These stem cells can produce **matrix cells,** which migrate downward to the base of the hair follicle and differentiate into the different cell types that make up hair. In an emergency, they can also produce **basal cells,** which migrate upward to the basal layer of the epidermis and make keratinocytes. For instance, in severe burns, pressure sores, and

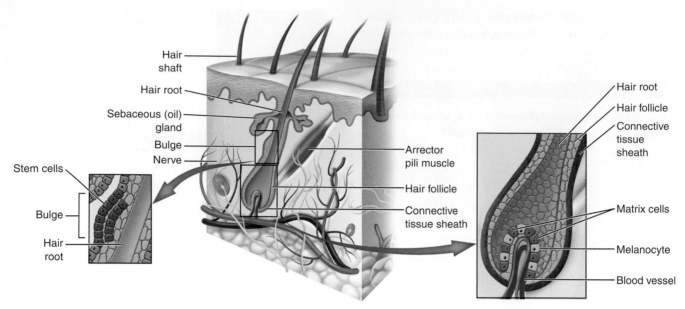

Figure 5.8. Hair and associated structures. This longitudinal section of a hair follicle reveals how deeply the invagination of the epidermis penetrates the dermis (*middle*). Hair cells arise from the division of matrix cells at the bottom of the hair follicle (*right*). Stem cells that produce matrix cells (or epidermal basal cells) are found in the bulge region (*left*). *Which tissue layer is furthest away from the hair—the hair follicle or the connective tissue sheath?*

other wounds in which the basal layer of the epidermis is destroyed, stem cells hidden within the bulge can migrate upward to form new epidermis.

Like nails, hair grows from a *matrix* of proliferating epidermal cells in the base of the follicle (Fig. 5.8B). And like all epidermal cells, hair cells mature and accumulate keratin in an increasing amount as they are pushed upward in the follicle to form first the **hair root** and finally the hair **shaft.** The hair shaft is the visible portion of the hair; it is composed entirely of keratin protein and does not contain any cells. Hair color is determined largely by the amount of melanin pigment added by melanocytes in the matrix. The openings of hair shafts, which are commonly called "pores," are visible to the naked eye, especially when inflamed by acne.

Attached to the hair follicle is a bundle of smooth muscle, the **arrector pili,** which is anchored in dermal fibrous tissue. Hair follicles and the projecting shafts are angled so that when the arrector pili muscle contracts, hair literally "stands on end" and goose bumps are visible on the skin. In humans this reaction is initiated by involuntary nerve signals and has no biological utility. It is an evolutionary remnant of the "fight or flight" reaction to threat.

Very fine hairs (*lanugo hairs*) cover the body at birth and are shed quickly and replaced by slightly darker fine hairs (*vellus hairs*), which cover the body from childhood onward. For example, the facial hairs of

women are vellus hairs. *Terminal hairs* are the thick hairs of the scalp and the male face, chest, and legs. They are thicker than vellus hairs and are usually deeply pigmented. Both vellus and terminal hairs go through a growth cycle, which varies from a few months (eyelashes) to several years (scalp). The natural life cycle of hair growth has three phases. The *growth phase* is longest. It is followed by a very short *transitional phase* before coming to the final *resting phase*, after which the hair is shed and the cycle repeats. In the normal scalp, about 90% of hair follicles are in the growth phase, about 1% to 2% are in the transitional phase, and about 10% are in the resting phase.

The number of hair follicles is fixed at birth, and no product, no matter how expensive, will cause hair to grow more densely on the scalp. That said, it is true that a few drugs will stimulate limited new growth from follicles that had previously stopped growing hair. Normal hair growth is determined by genetic factors and the influence of hormones from the endocrine system. Thus, hair loss (*alopecia*) can be a sign of certain endocrine-related diseases, although it can also occur in the absence of other disease. For example, male-pattern baldness is a natural consequence of aging attributable to the effect of male hormones, which on the one hand stimulate the *rate* of hair growth but on the other shorten the growth phase until, ultimately, the growth phase disappears and no hair grows. For additional reading about human hair, you may find interesting two of the stories on our Web site (**http://thepoint. lww.com/McConnellandHull**). One relates the tale of tycoon John D. Rockefeller, arguably the richest person in world history, who lost all of his body hair. Another tells a tale of lice and how their evolution provides clues about human evolution.

Case Note

5.5. In Sylvester's burned areas that have lost the epidermis but not the dermis, which cells can proliferate to produce a new epidermal layer?

Glands Produce Watery Sweat or Oily Sebum

Skin contains two types of glands: *sebaceous* glands and *sweat* glands.

Sebaceous Glands

Sebaceous glands are embedded in the dermis and attach to the upper part of hair follicles near the epidermis. They secrete an oily substance called **sebum,** which flows out of the follicle onto the skin surface (Fig. 5.9). Because oil is hydrophobic—it repels water—the oil in sebum contributes to the waterproof nature of the epidermis. It keeps skin soft and supple and improves the skin's resistance to chemicals, microbes, and other threats. Sebaceous glands are particularly large in skin of the forehead and nose, shoulders, upper chest and arms, and upper back. Testosterone and estrogen increase the activity of these glands. The steroid surge that occurs just before birth increases sebum production and may make the fetus more oily and slippery, facilitating childbirth. To the bane of many teenagers, sebaceous glands also ramp up their activity when sex steroid levels rise at adolescence, and the accumulated sebum blocks hair follicles. Acne is the result when blocked follicles get infected.

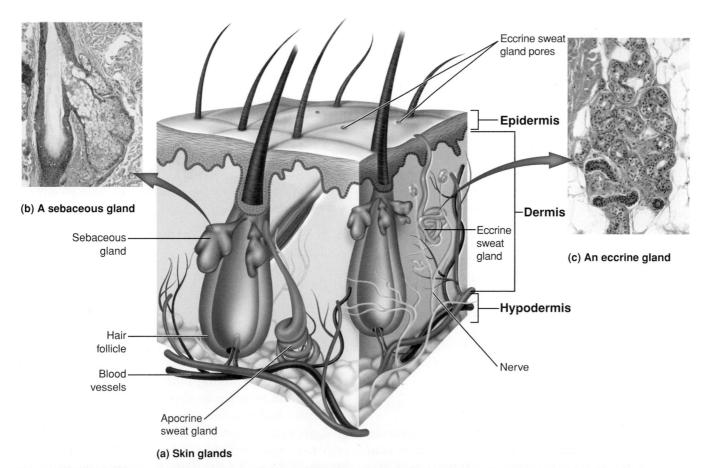

(b) A sebaceous gland

Sebaceous gland

Hair follicle

Blood vessels

Apocrine sweat gland

(a) Skin glands

Eccrine sweat gland pores

Epidermis

Dermis

Eccrine sweat gland

(c) An eccrine gland

Hypodermis

Nerve

Figure 5.9. Skin glands. A. Skin glands are found in the dermis. **B.** Sebaceous glands secrete an oily substance called sebum. **C.** Eccrine (shown) and apocrine (not shown) glands secrete sweat. *Which glands penetrate deeper than the dermis—apocrine or eccrine glands?*

Sweat Glands

Skin contains several million *sweat glands*, which are embedded in the dermis and open directly onto the skin surface. There are two types of sweat glands: *eccrine* and *apocrine*.

Eccrine sweat glands are by far the most numerous and occur all over the body. They are not associated with hair follicles, and they produce **sweat**, a salty fluid. Sweat flows from eccrine glands through ducts and empties onto the skin surface via sweat pores so small that they cannot be seen by the naked eye. Sweat is mildly acidic and antibacterial, but its main function is to contribute to homeostatic regulation of body temperature (Fig. 5.1A). Involuntary nerve signals stimulate the eccrine glands to secrete sweat when body temperature rises. Heat must be added to water to evaporate it, and as sweat evaporates it carries large amounts of body heat with it. However, sweating depletes body water and salt stores, which require replenishment—many liters of body water and grams of salt can be lost during hard work on a hot day. Failure to replenish fluids and salts can cause severe consequences, even death.

Apocrine sweat glands are far less numerous than their eccrine counterparts and are concentrated in the axillae (armpits) and anogenital regions. They are attached to and empty into hair follicles, along with sebaceous glands. The sweat they secrete contains water, salts, and a mixture of proteins and fats. Apocrine glands do not begin to function until axillary and anogenital hair growth occurs at puberty. In contrast to eccrine sweat glands, their main function is not to rid the body of heat. Rather, their job is to produce odor. In other words, they are scent glands. When skin bacteria metabolize the fats in apocrine secretions, we emit the characteristic odor of body sweat. Emotional stress increases the production of sweat from the apocrine glands. Like body hair, apocrine sweat glands are largely an evolutionary remnant. Other mammals use them to leave a distinctive scent to mark their presence.

Some other body structures are modified sweat glands. These include the milk (mammary) glands of the female breast and the wax (cerumen) glands of the external ear canal.

Subcutaneous Tissue Supports Skin

Below the dermis is a layer of fatty **subcutaneous tissue** (the hypodermis or *subcutis*), which is not part of skin but is intimately bonded to it (Fig. 5.3). It consists of a fat layer above a layer of dense *fascia* (a broad sheet of dense connective tissue) that sheaths the entire body.

Fibrous bands connect subcutaneous tissue to underlying structures and fascia.

The fatty portion of subcutaneous tissue serves as insulation, as a cushion against physical force, and as a store of fat for energy use. The thickness of the fat layer varies with age, sex, and nutrition. Very thin, underweight people have little subcutaneous fat; by contrast, the hypodermis can be 8 to 12 cm thick in very obese people. The body shape differences between men and women are due mainly to thicker subcutaneous fat around the hips of women. And as they age, men tend to develop thicker subcutaneous fat around the midbody and abdomen.

5.3   Which skin layer contains blood vessels—the epidermis or the dermis?

5.4   Which cell type evolves into keratinocytes—basal cells or dendritic cells?

5.5   Name the epidermal layer where melanocytes are found.

5.6   Which dermal cell type synthesizes collagen and elastin fibers?

5.7   It is a hot day. Will the skin's precapillary sphincters be open or closed?

5.8   Name the portion of the nail bed that gives rise to new nail cells.

5.9   Is the hair follicle formed of fibroblasts or epidermal cells?

5.10   Name the sweat glands that produce an oily or waxy secretion.

5.11   Name the sweat glands that are attached to hair follicles.

Healing of Skin Wounds

Healing is the natural repair of an injury. Whether that injury is a broken bone, influenza, or sunburn, healing reflects the body's collective efforts to restore normal structure and function.

Injury can result not only from physical trauma but also from radiation, chemicals, bacteria, pollen, and other agents. Figure 5.10 illustrates the events that occur after injury involving the skin. In this case, the injurious agent is thermal trauma (a serious burn); however, the body's

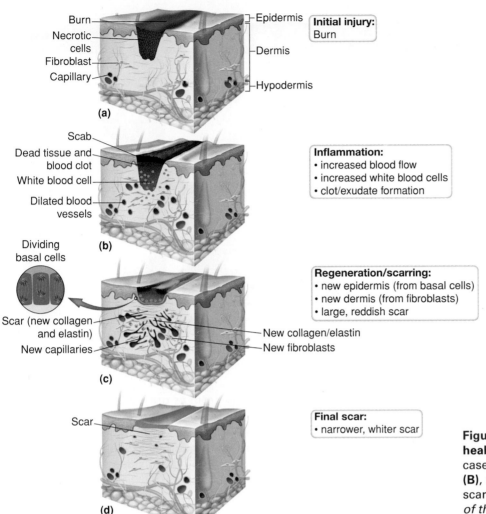

(a)

Burn
Necrotic cells
Fibroblast
Capillary

Epidermis
Dermis
Hypodermis

Initial injury:
Burn

(b)

Scab
Dead tissue and blood clot
White blood cell
Dilated blood vessels

Inflammation:
• increased blood flow
• increased white blood cells
• clot/exudate formation

(c)

Dividing basal cells
Scar (new collagen and elastin)
New capillaries
New collagen/elastin
New fibroblasts

Regeneration/scarring:
• new epidermis (from basal cells)
• new dermis (from fibroblasts)
• large, reddish scar

(d)

Scar

Final scar:
• narrower, whiter scar

Figure 5.10. Injury, inflammation, and healing. The response to an injury (in this case, a burn **[A]**) involves inflammation **(B)**, regeneration **(C)**, and in some cases scarring **(C** and **D)**. *What is the purpose of the white blood cells that have migrated to the site of injury?*

response to skin wounds involves the same general steps no matter what agent may be involved. The injured cells undergo **necrosis,** a form of premature cell death in which the cell dies and releases digestive enzymes (lysosomes) normally contained in cell organelles (Fig. 5.10A). These enzymes kill neighboring cells, so that damage can spread beyond the initial site of injury.

5.6. Sylvester's epidermal cells died from apoptosis. True or false?

Inflammation Is the Initial Response to Injury

The body's initial response to injury is **inflammation,** a coordinated series of responses involving white blood cells and blood vessels (Fig. 5.10B). Precapillary sphincters

open, bringing increased blood flow to the site of injury. White blood cells arrive in droves to destroy or neutralize any injurious agent (bacteria, for example) remaining at the site and to clean up the debris of injured and dead (*necrotic*) cells. If blood vessels have been broken, red blood cells and plasma are released into the site and form a clot (➡ see Fig. 10.10). In skin injuries, a protective crust (commonly called a *scab*) begins to form as water evaporates from oozing extracellular fluid, the blood clot, and inflammatory cells, collecting on the surface.

For more insight into inflammation and skin disease, refer to the nearby Clinical Snapshot, titled *Dermatitis: The Variations of Inflammation.*

5.7. Sylvester's burns were covered with pus, which contains large numbers of white blood cells. What were these white blood cells doing?

CLINICAL SNAPSHOT

Dermatitis: The Variations of Inflammation

In the catalog of human ills, skin has more entries than any other organ. This is explicable at least in part because it is so visible—it is, so to speak, an artist's palette upon which disease is colorfully painted. In addition, it comes into daily contact with countless environmental substances, microbes, and variations in heat and humidity. Skin is also secondarily affected by internal conditions such as nutritional status, pregnancy, disease, or stress. Considering all these factors, it's not surprising that about one-third of the people in the United States develop a skin condition each year, and skin complaints account for about 10% of annual physician visits.

What might be surprising is that most skin disorders are variations on a single theme: inflammation. Tumors and burns are exceptions and are discussed elsewhere in this chapter. Here, we focus on **dermatitis,** the clinical term for skin inflammation of any kind.

Dermatitis commonly arises from infection, whether by a fungus, bacterium, or virus. For example, *athlete's foot, "jock itch,"* and *"ringworm"* of the scalp are fungal infections of skin. *Impetigo* is a superficial infection of skin due to the *Streptococcus* bacterium, which causes a small cluster of blisters, often around a child's nose and mouth. A "boil" is an abscess; most are due to the *Staphylococcus* bacterium. Oral cold sores are caused by infection with the *herpes simplex virus*; herpes also causes infection of genital skin, where the condition is known as *herpes genitalis*.

Acne and seborrheic dermatitis are common forms of skin inflammation associated with sebaceous glands. *Acne* is an inflammation of hair follicles and sebaceous glands that is most common in adolescents with oily skin. It is characterized by hair follicles plugged with oily sebum, which appear as "whiteheads" or "blackheads." The black top of a blackhead is not due to dirt but to oxidation of the surface layer of sebum because the outlet of the follicle is open to the air; whiteheads are the same as blackheads but the outlet of the pore is closed and not exposed to oxygen from the air, so the sebum does not turn black. *Dandruff* is white flakes of dead superficial epidermal cells; when accompanied by dermatitis, the diagnosis is *seborrheic dermatitis*, a very common condition of oily, hairy skin that typically appears on the scalp, central face, upper

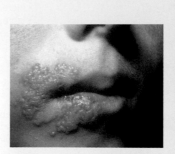
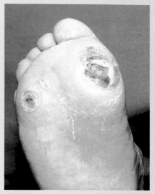

Common skin lesions. A. Cold sores, caused by the herpesvirus, are a striking visual manifestation of a viral infection. **B.** Decubitus ulcers are caused by inflammatory changes subsequent to pressure trauma.

anterior chest, and groin as greasy, scaly, red patches. The cause is unknown.

Dermatitis can also result from trauma. For example, a sunburn is inflamed skin due to thermal trauma. Chemical trauma can result in dermatitis when substances such as the oily secretions on the leaves of poison ivy, poison oak, or poison sumac contact the skin. Chronic pressure traumatizes the skin, too: Any amount of continuous pressure greater than blood pressure deprives skin of the blood flow necessary for the tissue to remain alive. This is often a problem for patients confined to a bed or wheelchair. For example, bedridden patients lying on their backs will develop "pressure sores" (*decubitus ulcers*) on their buttocks and heels if not turned regularly from side to side.

Eczema is a particular pattern of dermatitis characterized by weepy, blistering, crusted, itchy inflammation. It is caused by a great number of conditions, including allergies, caustic chemicals, and infection.

Allergic reactions are a very common cause of dermatitis. They result when substances that are not harmful to most people—such as dust mites, cat dander, or wool fibers—contact the skin and prompt a hypersensitivity reaction, which is characterized by hot, itchy welts called *hives*. Allergic reactions are discussed in more detail in ➡ Chapter 12.

Tissue Repair Is a Series of Orderly Steps

After inflammation subsides, tissue repair can begin. The body has two distinct processes for repairing tissues: *regeneration* and *scarring*.

Regeneration is the replacement of dead or injured cells with new functional cells. It can occur only in tissues that contain adequate reserves of stem cells. It begins when injured or dead cells send chemical signals to nearby stem cells. These stem cells respond by stepping up the pace of reproduction. The newly produced cells differentiate into new functional cells, which expand along the available scaffolding until the need for new cells is satisfied.

Scarring (fibrous repair) occurs when the injury is too severe or few stem cells are present. Dead or injured functional cells are replaced by scar tissue composed of fibroblasts and collagen fibers (Fig. 5.10C). Scarring develops as follows:

- Fibroblasts migrate into the damaged area (the *wound*), reproduce, and begin to synthesize new collagen or elastin fibers. These new fibers fill any void left by unreplaced functional cells and physically stabilize the wound by binding its edges together.
- New blood vessels sprout from existing ones nearby and grow into the area to nourish the healing process.
- A scar develops as fibroblasts synthesize additional collagen. At first the scar is bulky and red, reflecting the significant number of fibroblasts and blood vessels present. This accumulation of fibroblasts and blood vessels is called *granulation tissue* and marks a distinctive phase of wound repair, after which the scar begins to shrink and tighten into a smaller, whitish version. At first the scar doesn't "fit" and feels tight. Gradually, however, the scar reshapes itself to fit the site comfortably and matures to full strength (Fig. 5.10D).

Tissue healing has one of two results: complete regeneration or a mixture of regeneration and scarring. If the injury is mild and leaves the basic anatomic scaffolding (the *stroma*) of the tissue intact to support and direct regenerating cells, healing by regeneration may restore the tissue to complete anatomic and functional normalcy. In contrast, if the injury distorts or destroys some or all of the stroma upon which regenerating cells must align themselves, then the repair process will involve a mixture of regeneration and scarring.

The degree of regeneration also depends on the cells involved. Tissues with abundant stem cells, such as the epidermis, regenerate much more successfully than muscle tissue, which has fewer stem cells. Regeneration of damaged nerve cells (neurons) can be particularly problematic. If the sheath surrounding the neuron is undamaged, the axon will usually grow back and restore normal function. Damage to this sheath will impair and possibly prevent neuronal regeneration, and function might be permanently lost.

> *Remember This!* **Fibroblasts produce new connective tissue matrix. Stem cells in the basal epidermis or in the nail or hair matrix produce new epidermis.**

Case Notes

5.8. Sylvester's burns destroyed epidermis and muscle cells. Which of these tissues regenerates more easily, and why?

5.9. Some of Sylvester's burns destroyed the entire epidermis and dermis, including the epidermal stem cells. Is there any chance that the epidermis in this region will regenerate without medical intervention? Explain.

5.10. Some of Sylvester's less serious burns had begun to scar prior to his death. Which cell type do scars contain—basal cells or fibroblasts?

The Natural Repair of Burns Is a Good Example of Healing

Burns are among the most common and serious skin injuries. Severe burns are life-threatening because they eliminate the protective barrier of skin that separates the body's interior from the environment. Severely burned patients lose heat and fluid at a very rapid rate, sometimes losing so much fluid so quickly that blood pressure collapses and the patient dies of shock. For severely burned patients who survive the first week, infection is a major problem, because bacteria can reproduce rapidly in the fluids that ooze from the body surface where skin is absent. Moreover, since the skin barrier is gone, there is little to prevent the bacteria from invading tissue or blood.

The deeper the burn, the more severe the injury. Burns are classified according to the depth of the burn and the surface area of the burned skin (Table 5.2 and Fig. 5.11). To understand the classification system, it's worth recalling that epidermis consists of an epithelial tissue resting on a basement membrane, which in turn rests on the fibrous tissue of the dermis. Also recall that

Table 5.2   Classification of Burns

New Classification	Classic Classification	Example	Skin Regions Involved	Sensation	Appearance	Illustration
Superficial	First degree	Sunburn	Epidermis	Painful	Redness, swelling	(a)
Superficial partial thickness	Second degree	Scalding (a splash of boiling water)	Epidermis, papillary dermis	Painful to air and temperature	Blisters, clear fluid	(b)
Deep partial thickness	Second degree	Fire burn	Epidermis, reticular dermis	Only pressure perceived	Pale	(c)
Full thickness	Third degree	Electrical burn	Epidermis, dermis, underlying tissues	Only deep pressure perceived	Hard, black or purple	(d)

epithelial cells dip deep into the dermis to form hair follicles and glands. The following terms describe the depth of burns (Fig. 5.11A):

- *Superficial-thickness burns* (formerly known as *first-degree burns*) involve only the epidermis; most sunburns are good examples.
- *Partial-thickness* (or *second-degree*) burns involve injury to the dermis. These can be further subdivided into *superficial partial-thickness burns*, involving just the papillary dermis, and *deep partial-thickness burns*, also involving the reticular dermis.
- *Full-thickness* (or *third-degree)* burns involve the subcutaneous tissue and may also involve underlying bone and muscle.

Even superficial-thickness burns can be dangerous if they cover a significant portion of the body. A rough estimate of the percentage of a patient's surface area that is burned can be calculated using the *rule of nines*, which divides the body into regions based on multiples of nine (Fig. 5.11B). The entire head and neck (front and back), for instance, is equivalent to 9% of the entire surface area of the body.

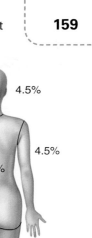

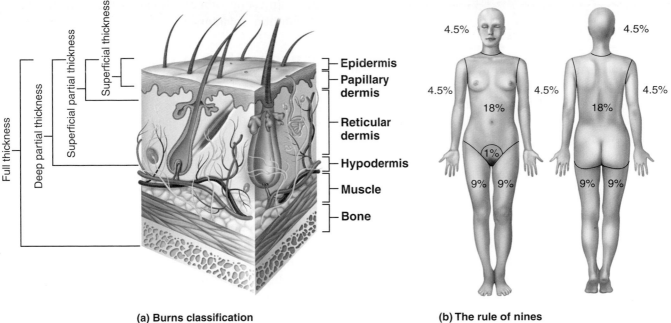

(a) Burns classification

(b) The rule of nines

Figure 5.11. Classification of burns. Burns are classified according to thickness **A.** and the surface area involved **B.** The numbers on the torso indicate the percentage of total body area. *Which type of burn penetrates the dermis—superficial or superficial partial thickness?*

Case Notes

5.11. Sylvester's burns covered the front and back of his upper limbs, head, and torso. What percentage of his body area is involved?

5.12. Sylvester suffered full-thickness burns to his torso. Which skin layers are involved?

Survival after burn injury depends on three critical variables: the patient's age (mortality rates rise with age), the percentage of surface area burned (burns involving more than 40% of surface area are usually fatal), and whether lung injury has also occurred (e.g., from inhalation of hot air or toxic fumes). The mortality rate can be roughly estimated by adding the age in years to the percent of surface area burned. For example, a 20-year-old person with a 30% burn has roughly a 50% chance of dying from the injury.

The cosmetic success of the healing process similarly depends on the burn depth. Immediately after superficial-thickness burns, blood vessels dilate and inflammatory cells appear. Dead or damaged epidermal cells slough away, and stem cells go into overdrive producing new cells in response to the injury. Within a week or two, the epidermis has been restored perfectly by regeneration and without scarring because the epidermis regrows smoothly across the intact basement membrane.

On the other hand, in partial-thickness burns, the injury destroys the epidermis, the basement membrane, and at least some of the dermis. The epidermis can still regenerate, thanks to the stem cells hidden in the bulge of hair follicles. However, without the scaffolding of an intact basement membrane, the new epidermal cells follow the uneven surface of the dermis, which is being scarred by fibrous repair. The repaired epidermis thus follows the uneven surface of dermal scars, resulting in the ridges and whorls of healed severe burns.

Skin grafting from unburned areas of skin is required to reestablish growth of epidermis in areas of full-thickness burn because most of the hair follicles (and the associated stem cell repositories) have been destroyed.

 5.12 Which protein is most abundant in a scar—keratin or collagen?

5.13 Which process occurs first—inflammation or scarring?

5.14 True or false: Injuries heal by either regeneration or scarring.

5.15 What is the difference between a superficial-thickness burn and a superficial partial-thickness burn?

Environmental Barriers Other than Skin

It is not immediately obvious that the external environment extends deep into the body, but it does: air from the environment extends into the deepest recesses of our lungs, and the water and food we ingest pass down the tube of our digestive tract. With every breath and bite we bring in bacteria and other microbes, dust particles, pollen, chemicals, and other potentially harmful environmental substances. Albeit to a lesser degree, the vagina is also exposed to irritants and microbes in the environment: for example, during intercourse, as a result of tampon use, and in childbirth, physical examination, or the insertion of therapeutic drugs. Finally, to at least some degree, the urinary tract is also open to the environment—bacteria and other infective agents can ascend the urethra ➡ (Chapter 16) to infect the bladder or climb even further up to infect the kidney. This is especially true for females, whose urethra is short and opens near the vaginal orifice, which normally harbors bacteria. As environmental substances travel within the airways, digestive tract, and vagina, they are never more than a few cells away from invading the internal environment. Fortunately, like skin, the membranes that line

these and some other body cavities are effective barriers to potential invaders.

Mucous Membranes Are Internal Barriers to the External Environment

Recall from ◀ Chapter 3 that epithelial membranes are sheets of closely packed epithelial cells that rest on a basement membrane. Epithelial membranes cover body surfaces and line body cavities. The epidermis is the body's main external epithelial membrane (the epithelium of the eyes covers a tiny percent). In contrast, internal epithelial membranes lining body cavities that are open to the external environment are called **mucous membranes,** or **mucosa,** because they are moist owing to mucous secretions from glands or mucous cells in the epithelium. Figure 5.12 illustrates epidermal and mucous membranes.

Like the epidermis, mucous membranes are composed of short-lived cells, which regenerate regularly because they have a short life span and are shed daily to be replaced with new progeny from stem cells. And just as the epidermis rests on the dermis—which contains blood vessels, nerves, glands, and accessory structures—mucous membranes rest on a bed of loose fibrous tissue (the *lamina propria*) that contains blood vessels, nerves,

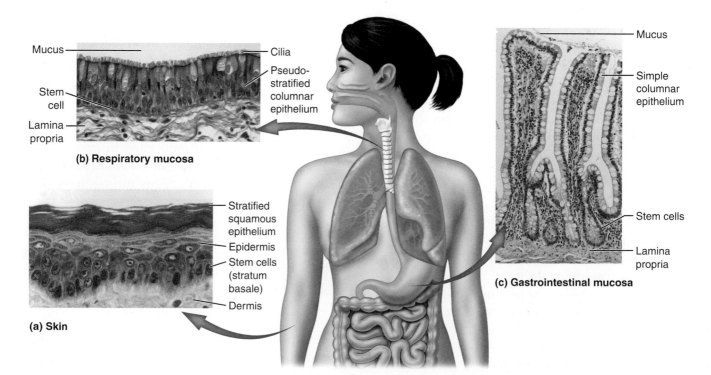

Figure 5.12. Epithelial membranes. The epidermis **(A)** (skin) shares many characteristics with mucous membranes in **(B),** the respiratory system, such as those lining the trachea and smaller airways, and **(C)** the gastrointestinal system, such as those lining the digestive tract. *What sort of epithelial membrane is found in the respiratory mucosa—columnar or squamous?*

glands, and other accessory structures. However, epidermis and mucous membranes differ as follows:

- Epidermis is dry and water-resistant; mucous membranes are wet.
- Epidermal cells contain much more keratin than the epithelial cells of mucous membranes.
- Epidermis protects against infection to a greater extent than do mucous membranes.
- Epidermis is composed of stratified squamous epithelium—multiple layers of cells ranging in shape from cuboidal at the deepest layers to very flat at the surface. The mucous membranes of the mouth, esophagus, and vagina are also stratified squamous epithelium. This friction-resistant design facilitates food passage, childbirth, and intercourse, respectively. However, other mucous membranes are usually composed of simple columnar epithelium.

Case Note

5.13. The lining of Sylvester's respiratory tract was relatively intact, because he didn't inhale much smoke (aside from cigarette smoke). Is this lining composed of mucosa or epidermis?

Remember This! **Epidermis and the lining epithelium of the intestines and bronchial airways share important characteristics: All are composed of short-lived cells that, as they die, are constantly being replaced by new ones.**

Hairs, Mucus, Cilia, and Acidic Fluids Also Act as Barriers

As mentioned earlier, human hair has lost much of its usefulness; however, hairs in the nostrils catch particulate matter and hairs in the external ear canal offer similar, limited protection to the ears.

Accessory glands in the airways secrete mucus; this covers the respiratory membrane with a thin film that traps microbes, pollen grains, and atmospheric dust and other particles. At this point cilia become important. Recall that cilia are microscopic, hairlike projections of cell membrane ← (Chapter 3) that wave to and fro in unison to move mucus superiorly into the upper airway and throat, where it can be swallowed, coughed out, or discharged from the nose.

Stomach acid has a low pH and is a chemical barrier capable of destroying microbes ingested in food or water. Although it is somewhat higher, the pH of vaginal mucus is also acidic and inhospitable to most microbes. A common cause of vaginal infection is change of the pH from acidic to basic.

5.16   Does the lamina propria underlie mucosa or epidermis?

5.17   Mucosal cells contain large amounts of keratin: true or false?

5.18   Name the cell membrane projection that moves mucus.

Word Parts

Latin/Greek Word Parts	English Equivalents	Examples
Adip-/o-	Fat	Adipocyte: fat cell
-crine	Secretion	Exocrine: secretion out of the body
-cyte	Cell	Adipocyte: fat cell
Derm-/o, dermat-/o, cutane-/o	Skin	Subcutaneous: under the skin
Ec-/exo-	Out; away from	Eccrine: secretion out of the body
Endo-	Within	Endocrine: secretion within the body
Epi-	Upon, over, outside	Epidermis: over the dermis

(continued)

Word Parts (continued)

Latin/Greek Word Parts	English Equivalents	Examples
Kerat-/-o	Keratin	Keratinocyte: keratin-filled cell
Melan-/-o	Black; melanin	Melanocyte: melanin-filled cell
-oma	Tumor	Melanoma: tumor of melanocytes
Seb-/-o	Sebum (oil)	Sebaceous: oil-producing
Squam-/-o	Scale	Squamous epithelium: scaly epithelium
Sub-	Beneath	Subcutaneous: under the skin

Chapter Challenge

CHAPTER RECALL

Match up each structure with its function:

1. **Eccrine sweat gland**
2. **Apocrine sweat gland**
3. **Sebaceous gland**
4. **Basal layer**
5. **Langerhans cell**
6. **Melanocyte**

a. Defends against invaders

b. Secretes watery, salty sweat onto the skin surface.

c. Synthesizes brown pigment.

d. Synthesizes keratinocytes.

e. Secretes sweat into hair follicles.

f. Secretes an oily substance.

7. **Which of the following is *not* a function of skin?**
   a. Protection against water loss
   b. Regulation of heat loss
   c. Defense against microbes
   d. Synthesis of red blood cells

8. **The epidermis and dermis are separated by the**
   a. stratum corneum.
   b. basement membrane.
   c. subcutaneous layer.
   d. matrix.

9. **The epidermis contains**
   a. blood vessels.
   b. collagen fibers.
   c. melanocytes.
   d. fat cells.

10. **Which of the following structures contain stem cells?**
    a. Nail matrix
    b. Basal layer
    c. Bulge of the hair follicle
    d. All of the above

11. **A beard is composed of**
    a. terminal hairs.
    b. lanugo hairs.
    c. matrix hairs.
    d. vellus hairs.

12. **Most scalp hairs are in the**
    a. apoptic phase.
    b. transitional phase.
    c. resting phase.
    d. growth phase.

13. **The visible pores of the skin are the openings of**
    a. eccrine sweat glands.
    b. sebaceous glands.
    c. hair follicles.
    d. all of the above.

14. **A superficial partial-thickness burn**
    a. involves only the epidermis.
    b. is not painful because pain receptors are burned away.
    c. is never life-threatening.
    d. may involve the upper portion of the dermis.

15. **Mucous membranes**
    a. are superior to the dermis.
    b. are composed of epithelial tissue.
    c. prevent water loss.
    d. are always composed of squamous cells.

CONCEPTUAL UNDERSTANDING

16. **Compare and contrast the structure and function of apocrine and eccrine sweat glands.**

17. **Describe the role of skin in maintaining a constant body temperature.**

18. **Compare and contrast the processes used to form new skin cells, nail cells, and hair cells.**

APPLICATION

19. **Full-thickness burns are not perceived by the patient. Explain why this is so.**

20. **All four tissue types discussed in Chapter 3 (muscle, nervous, epithelial, and connective) are found in skin. Give an example of each.**

You can find the answers to these questions on the student Web site at
http://thepoint.lww.com/McConnellandHull

6

Bones and Joints

Major Themes

- To remain healthy, bones require certain hormones, nutrients, and daily mechanical stress.
- Bones are actively involved in the regulation of blood calcium.
- Unhealthy bones grow abnormally, fracture easily, and heal poorly.
- Some joints allow movement; others do not.
- Freely movable joints are inherently unstable.

Chapter Objectives

Case Study: "You'd think I'd crashed a motorcycle!"

As you read through the following case study, assemble a list of the terms and concepts you must learn in order to understand Maggie's case. Make a list of her signs (such as swelling) and symptoms (such as pain) and (after studying the chapter), try to find reasons for them.

Clinical History: Maggie H., a 57-year-old grandmother, was training for her 10th marathon. She was finding training more difficult than in past years because of stiffness and pain in her knees and generalized muscle weakness, but she attributed these changes to aging. During a 10-mile training run, she made a misstep that caused her to catch herself awkwardly on the lateral edge of her left foot, with the sole turned inward. As she tumbled to the ground, she reached out to break her fall, and caught herself mainly on the heel of her left hand. Both her ankle and wrist were fractured in the fall. "I could hear my ankle snap," she said, "but my wrist was a surprise! I thought I'd just sprained it. I can't believe such a small thing could break so many bones. Look at me!" She managed a wry smile. "You'd think I'd crashed a motorcycle! All I was trying to do was get some exercise."

Physical examination and other data: Vital signs were unremarkable. The fracture site in her ankle was about 5 cm above the medial prominence of the left ankle (the "ankle bone"), where the foot and lower leg twisted and angled abnormally. Blood had pooled at the site and swollen it into a sizable, bluish bruise. Her left wrist was swollen and painful, particularly over the distal radius. She also had bruises to her left hip and the left side of her face. The radiologist's report on x-rays of her skull, spine, hip, ankle, and wrist described complete fracture of the tibia and fibula, which occurred through an area on the tibia where the bone appeared "hollowed out" into a cyst. The radius showed a hairline fracture extending completely across the head of the radius about 3 cm above the wrist joint. X-ray examination also showed the bones to be "low

Need to Know

It is important to understand the terms and concepts listed below before tackling the new information in this chapter.

- Negative feedback and anatomical terminology ← (Chapter 1)
- Connective tissue ← (Chapter 3)
- Hormone action ← (Chapter 4)

density and mottled, with other areas of cystic change." The radiologist also noted degenerative changes in the knee joint and generalized osteoporosis, which was most notable in the vertebrae, where an old fracture was noted. The final diagnosis was "(a) Complete fracture of tibia and fibula. (b) Hairline fracture of the distal left radius. (c) Extensive cystic change and bone thinning. Consider hyperparathyroidism. (d) Generalized osteoporosis with old fracture of the second thoracic vertebra."

Clinical course: The bones were realigned and the limbs were placed in casts. Laboratory tests revealed that blood calcium was 11.9 mg/dL (normal, 8.5 to 10.3), and blood parathyroid hormone was 101 pg/mL (normal, 11 to 54).

Imaging studies of the neck revealed a small mass near the left upper quadrant of the thyroid gland. A benign parathyroid tumor was surgically removed. The fractures healed slowly, but Maggie noticed that her muscle weakness disappeared. To prevent further fractures, Maggie was advised to take up speed-walking for exercise, and she was prescribed a treatment regimen for osteoporosis that included supplemental calcium, vitamin D, and other drugs.

The phrase "bred-in-the-bone" conveys the universal understanding that heredity and early childhood influences are so deeply ingrained in our being that they become part of our bones. The saying reflects the common knowledge that bones are the most enduring parts of our bodies. Indeed, most of what we know about the evolution of humankind is a tale written in the scattered bones of our apelike ancestors, bones that have endured hundreds of thousands of years. What is this imperishable "stuff" of bones? How do bones form and grow, and why did Maggie's bones fracture so easily? Finally, what are the bones and joints of the human skeleton? Ahead, we explore these questions and many more.

What is bred in the bone will never come out of the flesh.

From the *Panchatantra*, a collection of animal fables from the Indian subcontinent written about 200 BCE., and known in the West as the *Fables of Bidpai*, an Indian sage.

Bones and Bone Tissue

The word **bone** has two meanings: "a type of tissue (*osseous* tissue)" or "an anatomical structure." Anatomical structure is indicated in this sentence: "The tibia is a bone." In contrast, this sentence refers to tissue: "New bone grew into the fracture site." The term *cartilage* can also refer either to an anatomical structure or a tissue type. Indeed, you may recall from ← Chapter 3 that bone and cartilage are both types of specialized connective tissue. In a sense, most bone tissue "emerges" from cartilage: As you'll learn in this chapter, the initial limb "bones" of the developing fetus are formed of cartilage, and growth of new cartilage plays an important role in the healing of broken bones.

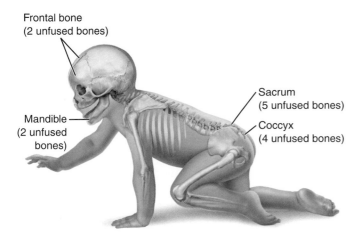

Figure 6.1. The infant skeleton. The skeleton in a newborn consists of over 300 bones. *As we grow, bones fuse in some areas. Name some of them.*

Frontal bone
(2 unfused bones)

Mandible
(2 unfused
bones)

Sacrum
(5 unfused bones)

Coccyx
(4 unfused bones)

Humans are born with over 300 bones (Fig. 6.1). As we grow, some bones fuse and the number falls to 206 in adults. For example, the frontal bones of the forehead and the jaw (mandible) are paired in the newborn, but in the adult, these bones have fused into one. The sacrum at the base of the spine consists of five bones at birth, and the coccyx (tailbone) consists of four, but each is a single bone in most adults.

Adult bones vary greatly in size: the tiny bones of the middle ear ➡ (Chapter 9), are only a few millimeters long and weigh a few milligrams; on the other hand, the large femur (the thigh bone) is about 45 cm long and weighs hundreds of grams.

The form of our skeleton (Greek *skeletos* = "dried up") is intimately wed to the most basic functions of life. It is no accident, for example, that merely folding an upper limb at the elbow brings the hand to the face, an act that is part of eating. If the elbow did not exist, no matter how long or short the distance from shoulder to fingertip, eating would require far more effort.

Bones are light, strong, and very slightly flexible. In our skeleton, they are vertically stacked and tied together by ligaments into an assembly that ought to be tottering, yet when coupled with the brain's precise control of skeletal muscle, this gives humans the advantage of upright posture, which enabled early humans to see over tall grasses in the African savannah.

Remember This! **By attempting to consume a bite of food with your arm outstretched, you can see for yourself the relevance of the elbow to easy eating. It's impossible! Even if your arm were very short, such an attempt would be highly frustrating.**

Bones and Bone Tissue Have Varied Functions

There is more to bones than you might imagine. They offer:

- *Support and stability.* As lumber is the frame of a home, so bones are the frame of the human body. Bones give us shape and maintain stability.
- *Controlled movement.* Without the stability and support of bones, our sophisticated activities would be impossible. We'd be something more akin to worms.
- *Protection.* The brain and spinal cord are vulnerable tissues, but they are protected from trauma by bone: The skull encases the brain and the spinal cord is threaded through a stack of thick vertebrae. Similarly, the ribs encircle the heart and lungs and offer some protection to the liver, spleen, and other contents of the upper abdomen.
- *Storage.* The central, hollow cavity present in the long bones of the limbs stores fat in the form of *yellow marrow.*
- *Blood cell production.* A honeycomb of minute spaces in many bones, including those of the limbs, holds the *red marrow,* which produces blood cells.

Apart from the roles outlined above, bone *tissue* has two additional functions:

- To maintain and repair bones.
- To act as a reserve pool of calcium, magnesium, and phosphorus. The concentration of these minerals in blood must be kept constant or else certain vital functions—such as muscle contraction, blood clotting, and the transmission of electrical signals—will be disrupted.

Case Note

6.1. Maggie H. fractured two bones in her lower leg. Which of the above functions do these bones normally perform?

Bones and Bone Tissue Are Classified According to Their Gross and Microscopic Structure

Healthy, mature bone tissue takes two forms. **Compact bone** is dense, with an orderly microscopic structure that has a smooth appearance and no grossly apparent spaces (however, it does have microscopic "pores").

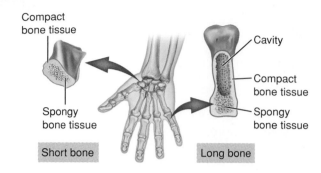

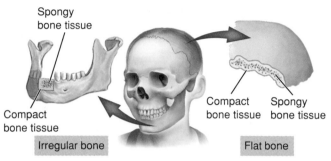

Figure 6.2. Bone shapes. All bones consist of an outer layer of compact bone tissue and an inner layer of spongy bone tissue. *Which type of bone contains a central cavity?*

Spongy bone, as its name implies, has many small open spaces that give it the appearance of a sponge. These spaces make the skeleton lighter than it could be if it were composed entirely of compact bone. Each of these types of bone tissue occurs in most bones: an outer thick layer of compact bone and an inner region of spongy bone.

Bones come in different shapes, according to their structural roles in the skeleton. There are five types of bones (four of which are shown in Fig. 6.2). Each contains different proportions of outer compact and inner spongy bone:

- **Long bones** are longer than they are wide. They consist of a long central shaft with a bulge at either end that forms a joint with another bone. All bones in the limbs are long bones except for the bones of the wrist and ankle. Their length makes these bones ideally suited to act as levers and to bear weight. Long bones are composed of an outer shell of compact bone. The ends contain an inner core of spongy bone, but the shaft contains a thin inner layer of spongy bone surrounding a large cavity.
- **Short bones** are boxy, lightweight, non–weight-bearing bones. Composed of spongy bone with a thin shell of compact bone, they make up the wrist and the ankle. Their small size and many flat surfaces make them useful in these complex joints—the many little

bones can slide easily upon each other, providing movement in many directions.
- **Flat bones** are thin, flat, and usually curved where they mold protectively around vulnerable anatomical structures. They are formed of two surface layers of compact bone and a thin central layer of spongy bone. The bones of the skull, mandible, ribs, and breastbone are flat bones.
- **Sesamoid bones** (not illustrated) are small bones that form within tendons or ligaments (strong tissues that bind other tissues together). The most important sesamoid bone is the patella (kneecap).
- **Irregular bones** are bones that do not fit into the above categories. Nevertheless, they, too, are composed of an outer region of compact bone surrounding a core of spongy bone. The vertebrae and the bones of the pelvis are irregular bones.

Case Note

6.2. Maggie's injured tibia extends from her knee to her ankle; her injured radius extends from her elbow to wrist. Classified by shape, what sort of bones did she injure?

Long Bones Contain a Diaphysis and Two Epiphyses

Again, long bones are found in our limbs. The humerus (the sole bone of the arm) is a good example. A long bone has a main shaft called the **diaphysis,** which widens into a funnel-shaped area called the **metaphysis** (Fig. 6.3A). At the end of a long bone is the **epiphysis,** the broadest part of the bone at each end. A tough, fibrous membrane, the **periosteum,** covers most of the bone. Where the bone meets another bone to form a joint, a cap of **articular cartilage** (joint-related cartilage) replaces the periosteum. Composed of hyaline cartilage, it provides a layer of lubricated, cushioning tissue to ensure smooth joint movement.

Underneath the periosteum lies a thick layer of compact bone surrounding a latticework of spongy bone (Fig. 6.3B). The spaces in the spongy bone latticework contain *red marrow,* which synthesizes blood cells. The metaphyses and epiphyses retain their spongy bone. In contrast, most of the spongy bone has been removed from the diaphysis, resulting in the inner **medullary cavity** (Fig. 6.3C). The medullary cavity of adult bones fills with *yellow marrow,* which is primarily fat. The **endosteum** is a layer of bone-forming and connective tissue cells that lines the medullary cavity.

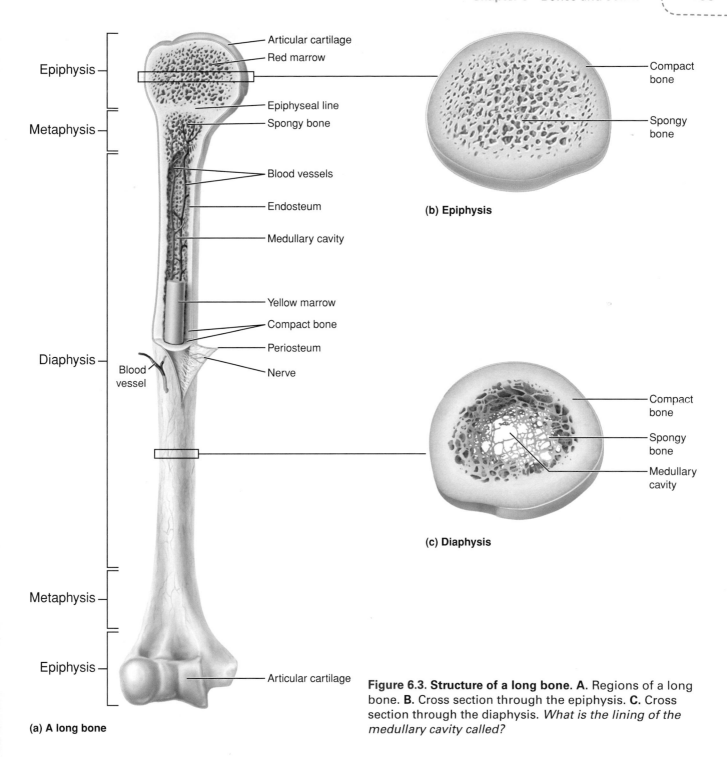

(a) A long bone

(b) Epiphysis

(c) Diaphysis

Articular cartilage
Red marrow
Epiphyseal line
Spongy bone
Blood vessels
Endosteum
Medullary cavity
Yellow marrow
Compact bone
Periosteum
Nerve
Blood vessel
Articular cartilage

Epiphysis
Metaphysis
Diaphysis
Metaphysis
Epiphysis

Compact bone
Spongy bone

Compact bone
Spongy bone
Medullary cavity

Figure 6.3. Structure of a long bone. A. Regions of a long bone. **B.** Cross section through the epiphysis. **C.** Cross section through the diaphysis. *What is the lining of the medullary cavity called?*

Looking more closely at the bone end in Figure 6.3A, a faint line (the *epiphyseal line)* can be perceived separating the epiphysis and metaphysis. This faint line is composed of compact bone and informs us that the bone is from an adult. In the growing bones of children and young adults, this structure is the *epiphyseal growth plate* (see Fig. 6.6 on page 172), a layer of cartilage and newly forming bone by which bones add length year by year until maturity.

Bones are metabolically active and must have ample access to blood, which is provided by a rich supply of blood vessels that penetrate the hard surface shell. And, as anyone who has broken a bone can attest, a broken bone is painful because bones are also richly supplied with nerves.

Bones have various anatomical features: projections, surfaces, ridges, grooves, holes, depressions, and other features with special names. See Table 6.1 for a review of these features.

Table 6.1   Bone Markings

Bone Marking	Description	Examples
Projections: Sites of Joint Formation or Muscle/Ligament Attachment		
Head	Rounded, knoblike end attached to narrow neck; forms joints	
Condyle	Rounded projection on the end of the bone; forms joints	
Facet	Smooth, flat surface; forms joints	
Epicondyle	Small projection above or on a condyle; attachment site	
Tuberosity	Large, rounded projection	
Spine	Sharp, slender projection	
Process	Prominent raised area of bone	
Crest	Distinct ridge along bone	
Line	Narrow ridge along bone; smaller than a crest	
Depressions and Holes		
Foramen (foramina; plural)	Opening permitting passage for nerve or vessel	
Fossa (fossae; plural)	Shallow depression; may form part of a joint	
Meatus	Opening to a short channel	
Sinus	Air-filled space in bone	

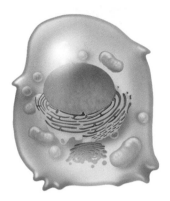

Bone Tissue Is Composed of Specialized Cells and Extracellular Matrix

Recall from Chapter 3 that bone is a connective tissue and, like all connective tissue, contains relatively few cells and a large amount of extracellular matrix. Three types of differentiated bone cells are present in bone tissue; you can remember them using the following mnemonic: osteo<u>b</u>lasts <u>b</u>uild bone; osteo<u>c</u>ytes <u>c</u>are for bone; and osteo<u>cl</u>asts <u>cl</u>eave (break down) and recy<u>cl</u>e bone (Fig. 6.4). These cells are present in all types of bone tissue.

Osteoblasts (Greek *osteo* = "bone" + *blast* = "new growth" or "sprout") are bone-building cells (Fig. 6.4A). They are derived from a fourth type of bone cell, the **osteogenic stem cell** (Greek *-genic* = "producing"). In adult bone, osteogenic stem cells are embedded in the periosteum and endosteum.

Osteoblasts produce **osteoid,** a special type of bone collagen. Recall from ◀ Chapter 3 that collagen fibers are able to withstand a high degree of tension. This important property gives bones a certain springy flexibility that protects them from injury as a person lands from a jump. Osteoid fibers are woven into a network somewhat like the beams that form the framework of a building. Osteoblasts lay down new bone by adding calcium and other mineral crystals to these osteoid beams.

As they form new bone, osteoblasts gradually become trapped in tiny, single-cell spaces called **lacunae,** where they mature into **osteocytes,** which nourish and care for bone (Fig. 6.4B). Osteocytes connect to one another by cytoplasmic extensions that travel through a maze of tiny tunnels called *canaliculi*, which radiate outward in all directions from the lacunae. Canaliculi and lacunae are also illustrated in Figure 6.5.

Osteoclasts (Greek *-clast* = "break") are very large cells with multiple nuclei (Fig. 6.4C). They are the offspring of *monocytes*, which also produce immune cells called *macrophages* ➡ (Chapter 10). Osteoclasts break down bone and release the calcium, phosphorus, and other components into the bloodstream for recycling. Concentrated mainly around the edges of spongy bone, osteoclasts work together with osteoblasts in a continual process of bone destruction and reconstruction that dissolves (resorbs), renews, remodels, and repairs bone.

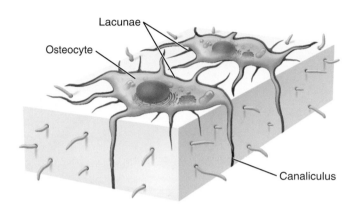

(a) Osteo<u>b</u>lasts <u>b</u>uild bone

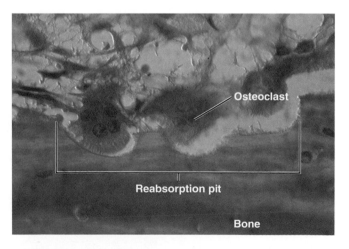

(b) Osteo<u>c</u>ytes <u>c</u>are for bone

(c) Osteo<u>cl</u>asts <u>cl</u>eave bone

Figure 6.4. Bone cells. Osteo<u>b</u>lasts **A.** which <u>b</u>uild bone, develop into osteo<u>c</u>ytes **B.** which <u>c</u>are for bone. Osteo<u>cl</u>asts **C.** <u>cl</u>eave, or break down, bone. *If osteoblasts are more active than osteoclasts, would a bone grow or shrink?*

Bone Tissue Is Precisely Structured

The actions of bone cells result in the intricate microscopic anatomy of a long bone, as illustrated in Figure 6.5. Recall that long bones contain a central medullary cavity lined by a layer of cells, the **endosteum** (not shown). Surrounding the medullary cavity is a thin layer of spongy bone and a thicker layer of compact bone.

Compact bone is built in concentric layers called **lamellae,** which, like tree rings, encircle the *central canal,* which runs lengthwise in the bone and carries nerves and blood vessels (Fig. 6.5, middle). Separating the lamellae are rings of osteocytes in their tiny lacunae,

which are linked with each other through the canaliculi (Fig. 6.5, bottom right). Note that the central canals are different from the medullary cavity: Each long bone has one large medullary cavity but many microscopic parallel central canals. Radiating outward at right angles from each central canal are small tunnels called **perforating canals,** which traverse from one central canal to another carrying smaller branches of nerves and blood vessels.

A central canal and the bone arrayed around it together form a pencil lead-thin cylindrical column of bone called an **osteon** (Fig. 6.5, middle). The compact bone tissue of long bones is composed of osteons

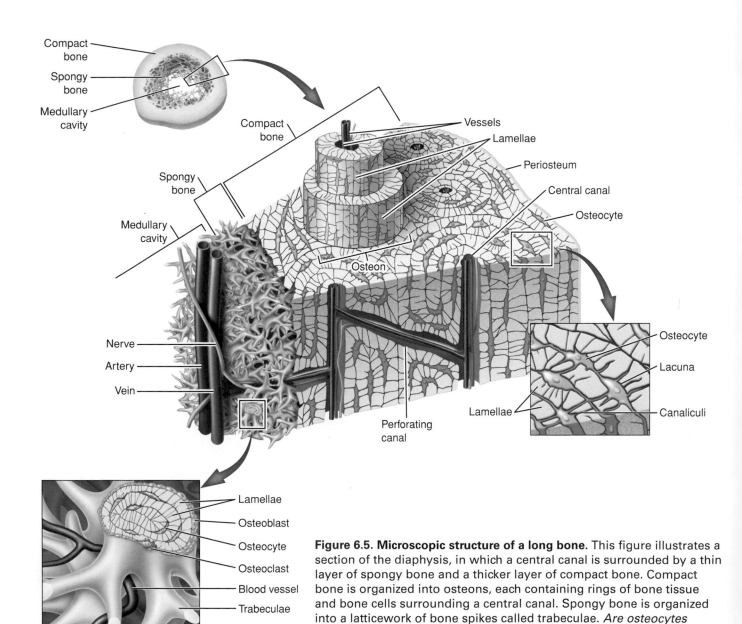

Figure 6.5. Microscopic structure of a long bone. This figure illustrates a section of the diaphysis, in which a central canal is surrounded by a thin layer of spongy bone and a thicker layer of compact bone. Compact bone is organized into osteons, each containing rings of bone tissue and bone cells surrounding a central canal. Spongy bone is organized into a latticework of bone spikes called trabeculae. *Are osteocytes found in lacunae or canaliculi?*

arranged in parallel bundles aligned with the long axis of the bone, which imparts strength and flexibility like a bundle of sticks.

However, in spongy bone, the microscopic pattern of lamellae is irregular and osteons are not present (Fig. 6.5, lower left). Although lamellae are still evident, there are no central canals because vessels can run freely through the open spaces of spongy bone. The bony lattices of spongy bone are called **trabeculae;** despite their apparent randomness, they are oriented to maximize the strength of the bone tissue. Osteoblasts and osteocytes line the trabeculae, where they are actively involved in bone remodeling.

> *Remember This!* **All (spongy and compact) bone tissue is organized into lamellae, but only compact bone tissue is organized into osteons.**

The net effect of the gross and microscopic organization of bone tissue is that bones are surprisingly light and strong. The many macroscopic and microscopic spaces reduce the weight of bone tissue. At the same time, the mineral crystals, though not as hard as tooth enamel, provide great durability and strength. And although bone is stiff, it is not brittle, because the elastic properties of its collagen fibers give it flexibility. Bones are difficult to break precisely because they are hard but will flex slightly. For example, *marble bone disease* is a congenital condition characterized by extremely dense, hard, inflexible bones . . . and by the ease with which they fracture.

Case Note

6.5. Maggie's wrist fracture extended across the distal diaphysis of the radius. Would the fracture line encounter a medullary canal, a central canal, or neither?

Ossification Is the Formation, Growth, Remodeling, and Repair of Bone

Bone formation is called **ossification** and occurs in four situations:

- the initial formation of bones in the embryo and fetus
- the growth of bones in children and adolescents
- the remodeling (reshaping) of bones in response to the stress of daily life or to unusual chronic stress.
- the repair of damaged bone, usually a fracture.

Fetal Bone Forms by Endochondral or Membranous Ossification

The fetal skeleton first appears in the 6-week-old embryo. Thin fibrous membranes appear where the skull and collarbones (clavicles) will form; in the remainder of the body bone-shaped islands of primitive cartilage appear. Bone that develops from cartilage is called **endochondral bone;** bone that develops from fibrous membranes is called **membranous bone.**

Endochondral Ossification

Most bones form by a process known as **endochondral ossification,** in which cartilage forms initially and is replaced by bone (Fig. 6.6). As with our earlier discussion of bone anatomy, long bones also offer a good example of bone growth by endochondral ossification. Bone synthesis begins in the middle of the shaft—the **primary ossification center.** The sequence of events is as follows:

1. *Chondroblasts prepare a cartilage model.* Like an artist preparing a plaster model before making a statue, most bones begin as a cartilage model. Cartilage, like plaster, is much more amenable to remodeling and growth than bone (or marble!). This early cartilage prototype is made by connective tissue cells called **chondroblasts** (*chondro* = "cartilage"), which secrete the fibrous extracellular matrix giving cartilage its strength and flexibility. A layer of cells condenses at the edge of the model to form a covering called the **perichondrium.** This cartilage, like all cartilage, even in adults, is *avascular*—it does not have its own blood supply and gets oxygen by diffusion from nearby tissues.

2. *A bony collar forms and the model enlarges.* Blood vessels invade the perichondrium midway down the model's shaft. This sudden influx of nutrients stimulates osteogenic stem cells to differentiate into bone-forming osteoblasts. Osteoblasts form a collar of spongy bone, which stabilizes the cartilage model until bone formation is well underway. The membrane covering the bony collar converts from perichondrium (which covers cartilage) to periosteum (which covers bone).

   Chondroblasts produce additional cartilage, which enlarges the cartilage model. Cells in the center of the model, now called *chondrocytes*, enlarge and stimulate the deposition of calcium salts in the nearby extracellular matrix. The chondrocytes become trapped in calcified lacunae, and, cut off from their nutrient supply, begin to die. The lacunae merge to form small hollow spaces.

3. *Ossification begins.* Blood vessels from the periosteum invade the primary ossification center. Accompanying

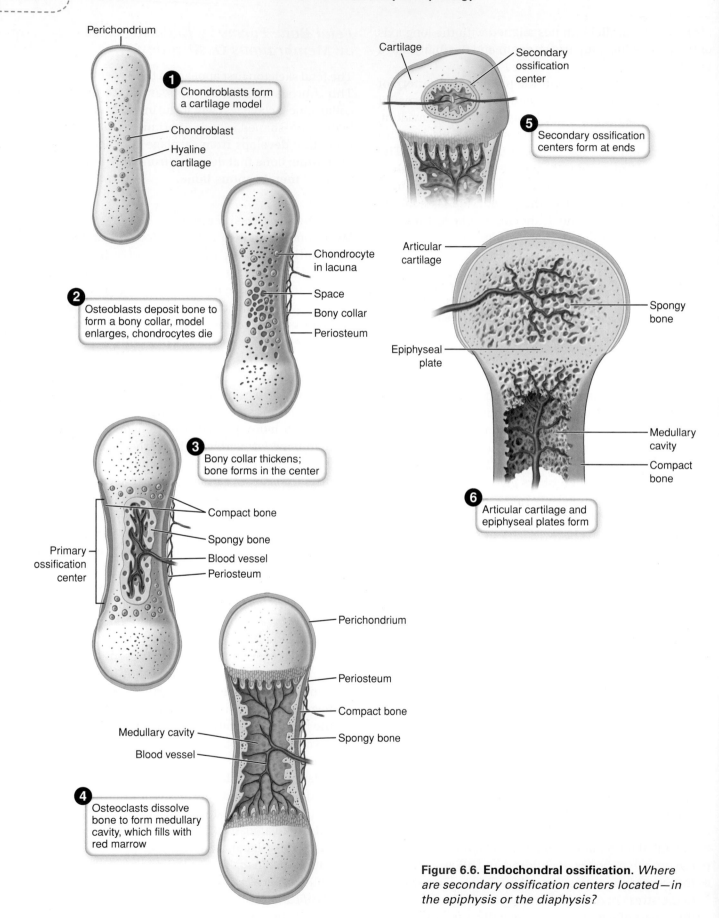

Perichondrium

1 Chondroblasts form a cartilage model

Chondroblast

Hyaline cartilage

2 Osteoblasts deposit bone to form a bony collar, model enlarges, chondrocytes die

Chondrocyte in lacuna

Space

Bony collar

Periosteum

3 Bony collar thickens; bone forms in the center

Compact bone

Spongy bone

Blood vessel

Periosteum

Primary ossification center

4 Osteoclasts dissolve bone to form medullary cavity, which fills with red marrow

Medullary cavity

Blood vessel

Perichondrium

Periosteum

Compact bone

Spongy bone

Cartilage

Secondary ossification center

5 Secondary ossification centers form at ends

Articular cartilage

Epiphyseal plate

Spongy bone

Medullary cavity

Compact bone

6 Articular cartilage and epiphyseal plates form

Figure 6.6. Endochondral ossification. *Where are secondary ossification centers located—in the epiphysis or the diaphysis?*

these blood vessels are both osteoclasts and osteoblasts. The osteoclasts degrade the calcified cartilage and the osteoblasts deposit new spongy bone tissue.

4. *The marrow space enlarges and bone marrow appears.* Osteoclasts continue their constructive destruction by breaking down some of the new spongy bone to create the medullary (marrow) cavity. Red marrow, which migrated into the ossification center along with the blood vessels and bone cells, fills the space. Osteoclasts and osteoblasts continue their creative remodeling by replacing much of the remaining spongy bone into compact bone.

5. *Secondary ossification centers appear (at each end of a long bone).* Shortly after birth, blood vessels invade the bone epiphyses and other (secondary) ossification centers appear. These growth regions are similar to the primary ossification center except that the newly formed spongy bone is not broken down to create a medullary space. Also, much of the newly formed spongy bone will remain as spongy and will not be converted into compact bone.

6. *Articular cartilage and an epiphyseal plate form.* At the ends of bones that will form joints with other bones, the perichondrium develops into articular cartilage. As discussed below, the primary and secondary ossification centers remain separated by cartilage until adulthood. This cartilage remnant is the only site where bone growth can occur and is called the *epiphyseal growth plate.* Bone growth at the epiphyseal plate follows the same principles of cartilage growth followed by replacement with bone, as explained further on.

Case Note

6.6. By what process did Maggie's tibia form?

Intramembranous Ossification

The flat bones of the skull, the clavicles, and the jawbone (mandible) are formed from fibrous tissue membranes by **intramembranous ossification.** Nevertheless, the final product is similar to bone formed by endochondral ossification: an outer shell of dense, compact bone and an inner core of spongy bone.

Initially, a bone-shaped fibrous membrane appears. Within this membrane fibroblasts and osteoblasts begin to deposit a haphazard weave of collagen fibers and bone tissue (respectively). This is a temporary type of bone called **woven bone,** which stands in stark contrast to the orderly microscopic structure of compact bone. As growth progresses, compact bone develops on the surfaces of the membrane and spongy bone develops in the center. The result is a layer of spongy bone sandwiched between outer layers of compact bone.

New bone formation by intramembranous ossification differs in two respects from endochondral ossification: (a) no intermediate cartilage stage develops and (b) intramembranous ossification passes through a stage characterized by woven bone. The fetal development of flat bones is the only *healthy* circumstance in which woven bone appears. In all other instances, woven bone appears as a reparative reaction to bone disease or injury.

Case Note

6.7. Which of Maggie's bones first developed as a fibrous membrane—her bruised facial bones or her broken radius?

Bones Grow until Early Adulthood

Under the influence of growth hormone, insulin-like growth factor, and other hormones (see Table 6.2), an infant's bones continue to grow throughout childhood and adolescence. Some bones, such as those of the pelvis, achieve full growth only in young adulthood.

Longitudinal Growth of Long Bones

From birth until adult-size bones are achieved, long bones grow in both length (longitudinal growth) and width (appositional growth) (Fig. 6.7).

Longitudinal growth occurs at the epiphyseal plates much as a crowd marching down an avenue swells as newcomers join. Continuing the process begun in the fetus, chondrocytes continue to multiply and add extra cartilage to the side of the epiphyseal plate closer to the end of the bone, adding to the overall bone length. Meanwhile, on the trailing, diaphyseal side of the plate, osteoblasts replace the new cartilage with bone, leaving some members of the "crowd" behind to be imprisoned by bone tissue and convert into osteocytes. Prior to puberty, the osteoblasts cannot catch up with the chondrocytes, so the bone continues to lengthen and the epiphyseal plate is pushed outward. However, osteoblast activity is stimulated by the hormones of puberty, so as puberty proceeds, the osteoblasts replace cartilage cells at a quicker pace.

Eventually, at about age 17 to 21 years (earlier for girls, later for boys), the osteoblasts effectively overtake the chondrocytes, replacing all of the cartilage with bone and bone growth stops. All that remains of the epiphyseal plate is the *epiphyseal line,* which appears as a thin layer of dense bone running across the spongy bone of the epiphysis.

Appositional Growth of Long Bones and Intramembranous Bones

Meanwhile, as the bone is lengthening by endochondral ossification, it is widening by *appositional growth* (Fig. 6.7). All along the length of the bone, osteoblasts beneath the periosteum add layers of new, compact bone somewhat as trees add new rings. All the while, at a slower rate, osteoclasts along the edge of the medullary space are dissolving and recycling bone minerals to expand the medullary cavity, making room for bone marrow. The net effect is that the bone grows wider, the medullary cavity grows larger, and the cortex becomes somewhat thicker.

Intramembranous bones also grow by appositional growth. Osteoblasts under the periosteum enlarge and thicken the bone by depositing new compact bone tissue, and osteoclasts enlarge the cavities in the spongy bone centers to keep the bone relatively light in weight.

Bone Remodeling Is the Continual Replacement of Old Bone by New

Remodeling is the normal "creative destruction" of bones as they are dissolved (resorbed) and remade ever so slightly each day. About 5% of bone mass is remodeled each year. In a delicate balancing act, osteoclasts are ever busy resorbing bone, which is replaced by osteoblasts, which are equally busy replacing it by synthesizing new osteoid and encrusting it with mineral crystals.

Bone remodeling is essential if bones are to maintain the flexibility and strength they need to respond to stress. Bones remold themselves according to the *physical stresses* (forces) placed upon them; if the stress is unusual, they will become deformed. For example, a boy with legs of different lengths may walk in a way

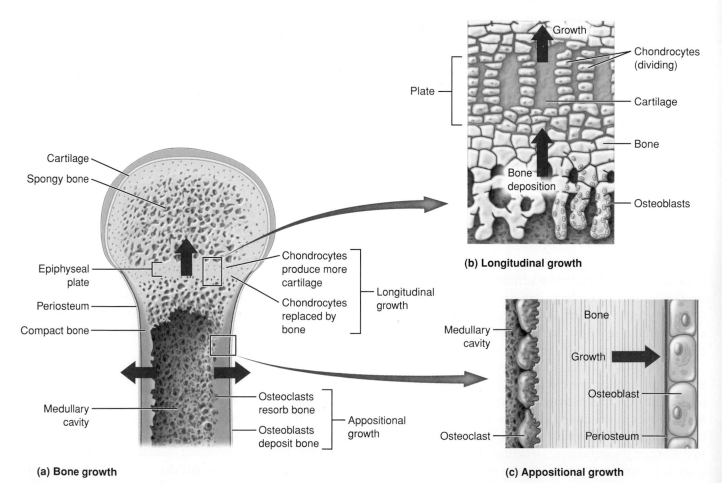

(b) Longitudinal growth

(a) Bone growth

(c) Appositional growth

Figure 6.7. Bone growth. A. Bones grow longitudinally at the epiphyseal plate and appositionally in the diaphysis. **B.** Longitudinal growth lengthens the diaphysis. **C.** Appositional growth thickens the compact bone layer and enlarges the medullary cavity. *Are osteoblasts below the periosteum involved in longitudinal or appositional growth?*

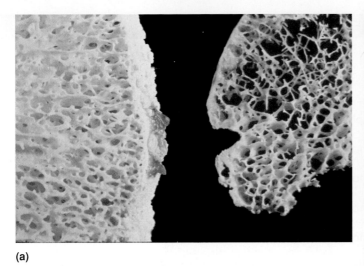

(a)

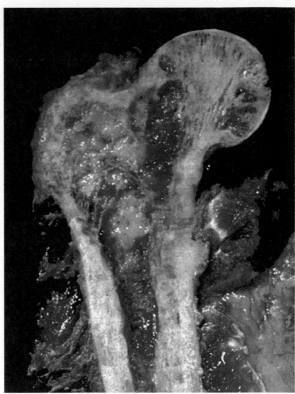

(b)

Figure 6.8. Bone disease. A. Bone sections from a normal (left) and osteoporotic (right) bone. The bone sections are the same thickness. **B.** Cross section of a femur (thigh bone) from a patient with Paget disease. Note the abnormal appearance and thickness of the compact bone of the diaphysis. *In osteoporosis, which process is overactive, bone resorption or formation?*

that places unusual stresses on his spine. Vertebral bones will remodel themselves to respond, and the result may be an abnormal sideways curvature of the spine (*scoliosis*).

Imbalance of osteoblastic creation and osteoclastic resorption of bone can result in bone disease (Fig. 6.8). In Maggie's case, bone resorption exceeds bone formation. Her bones have become weak and fragile, a condition called **osteoporosis** (Latin *poros* = "passageway"), which literally means porous (low-density) bone. In contrast, *Paget disease of bone* is an abnormal acceleration of remodeling in which there is overactivity of both bone formation and bone resorption. The result is dense, deformed bones that break easily.

Multiple Factors Affect Bone Health and Growth

Bone growth in children, bone remodeling in adults, and bone repair at any age depend on good bone health, which is influenced by multiple factors (Table 6.2). The most important are:

- *Physical stress.* An unstressed bone is an unhealthy bone, and an unhealthy bone is not durable and does not repair well. For example, patients rendered immobile by paralysis, dementia, lower limb disease, or

another problem often do not get enough physical exercise to maintain healthy bones.

- *An adequate supply of vitamins and minerals.* Vitamins A, C, and D are required for healthy bone, as is an adequate intake of calcium, phosphorus, and magnesium. For example, intestinal disease can prevent normal intestinal absorption of any of these essential nutrients even though dietary intake is normal, resulting in soft, deformed bones.
- *Hormones.* A surge of hormones—estrogen from the ovaries, testosterone from the testes—accounts for the growth spurt associated with puberty. Estrogen is especially important in the mineralization of new bone in adult females; testosterone is less of a bone health factor in adult males. Growth hormone and thyroid hormone also stimulate bone development and growth. These hormones are discussed in detail in ➡ Chapter 15.
- *Age.* From birth through the end of adolescence, our total bone mass increases because osteoblasts make more bone than osteoclasts degrade. During young adulthood, bone mass stabilizes; however, as sex hormones decline in middle age, especially in women after menopause, bone mass begins a long, slow decline as bones gradually demineralize. Osteoporosis—low bone density—is much more of a problem for women than men because estrogen is much more important for mineralization of new bone than testosterone, and

Table 6.2 Factors Involved in Bone Health

Component	Function
Dietary Components	
Calcium (Ca) and phosphorus (P)	Calcium phosphate crystals deposited on osteoid to make bone hard
Magnesium	Required for normal osteoclast activity
Vitamin D	Required for normal calcium absorption in the small intestine
Vitamin A	Required for normal osteoclast activity
Vitamin C	Required for normal collagen production
Vitamin K	Assists in the production of osteocalcin, a protein that helps form bone tissue
Fluoride	May contribute to bone mineralization and stimulate bone growth and repair
Lifestyle Components	
Exercise (weight bearing or strength training)	Bones subjected to stress become stronger; unstressed bones become weaker
Healthy weight maintenance	Low body weight results in less stress on bones; bones become weaker
Hormones	
Calcitriol (the active form of vitamin D in the body)	Required for normal calcium absorption in the small intestine
PTH	Stimulates osteoclast activity (bone resorption) and vitamin D activation
Calcitonin	Stimulates osteoblast activity and bone deposition
Insulinlike growth factor I (IGF-I)	Stimulates bone growth
Growth hormone	Stimulates bone growth, directly and by stimulating IGF-I production
Thyroid hormone	Required for normal bone growth
Insulin	Promotes normal bone growth
Estrogen, testosterone	Preserves bone mass by prolonging the life span of osteocytes and osteoblasts; inhibits osteoclast activity

women's bones are lighter than men's to begin with. In addition, during childhood and adolescence, when bone density is building, girls may be less likely than boys to participate in athletic activities that stress the bones. In females, bone mass begins its decline about age 30; in males it begins at about age 60.

That said, both males and females have an increased risk of fracture as they age. In part this is due to an age-related decline in growth hormone, which stimulates the production of bone collagen. As a result, less collagen is available for normal bone renewal, and bones become more brittle.

> *Remember This!* **An unstressed bone is an unhealthy bone.**

Local conditions that affect growth or repair are also important. For example, poor blood flow to a body part, such as a lower limb, can cause affected bone to grow or repair slowly. Diabetes ➡ (Chapter 15) and atherosclerosis ➡ (Chapter 11) are notable causes of low blood flow to the feet and lower limbs. Bone health can also be affected by other local conditions, such as infection or inflammation.

Case Notes

6.10. Maggie must repair her fragile bones. Remember that she is 57 years old. Which hormone could she take to improve her bone health—estrogen or testosterone?

6.11. Has Maggie's intensive exercise regime helped or harmed her bones?

A Healing Fracture Is a Model of Bone Repair

Healthy bones require the stress of daily life, which stimulates osteoblast and osteoclast activity to rebuild and refresh bone. Lack of stress, as with space flight or prolonged bed rest, causes bones to dissolve partially and leads to weak, brittle bones. However, force from an unusual direction or force that is unusually strong can stress a bone to the point of breaking. A **fracture** is a broken bone. A **pathological fracture** is a break that occurs when the bone is under normal, daily stress. Pathological fractures occur in bones with a condition that causes local weakening—a bone tumor, for example. Fracture prevention and classification are the subject of the accompanying clinical snapshot, titled *Fractures*.

Figure 6.9 illustrates the process required to heal Maggie's broken bones:

1. The break tears blood vessels and a blood clot (*hematoma*) accumulates at the site. Bone near the break dies, because its blood supply has been interrupted.
2. By the end of the first week, the clot is partially replaced with *granulation tissue* consisting of collagen fibers and other elements of extracellular matrix secreted by invading fibroblasts, new blood vessels, and white blood cells. Repair progresses with the appearance of *woven bone*, which ossifies by intramembranous ossification. This rich mixture of granulation tissue, cartilage, and woven bone is called a **soft callus,** which loosely unites the ends of the broken bone but cannot bear weight. Repair continues as osteoclasts reabsorb dead bone and smooth off the edges of the fracture.
3. After a couple of weeks, the soft callus matures into a **bony callus,** as more bone is deposited and the woven bone and cartilage are increasingly replaced by spongy bone. The bony callus binds the bone ends more securely and is capable of limited weight bearing.
4. Finally, excess spongy bone and callus are resorbed and replaced by dense, compact bone, which remodels into the previous anatomical outline under the influence of local mechanical forces.

Case Notes

6.12. Maggie's bone repair was very slow. Two weeks after her accident, a soft callus had formed around the broken bones. What did this soft callus contain?

6.13. Why did Maggie's physician consider her ankle fracture pathological?

Calcium Homeostasis Is Critical to Body Functioning

Recall from ⬅ Chapter 1 that homeostasis is "the body's collective communication and control effort to maintain constant, healthy internal conditions." The blood calcium concentration, like the concentration of many other ions, is regulated by a negative feedback loop—if it rises too high, the body reacts to lower it, and vice versa. Abnormally *high* blood calcium can cause confusion, kidney stones, muscle weakness, and other problems. Abnormally *low* blood calcium affects nerve impulse transmission and causes muscle spasms and tingling sensations. As you can see, a stable blood calcium concentration is essential for normal functioning, and the body uses a variety of mechanisms to maintain it.

Case Note

6.14. Refer back to the case introduction to check Maggie's blood calcium concentration. Is it abnormally high or low, and which of her symptoms are directly related to this abnormal change?

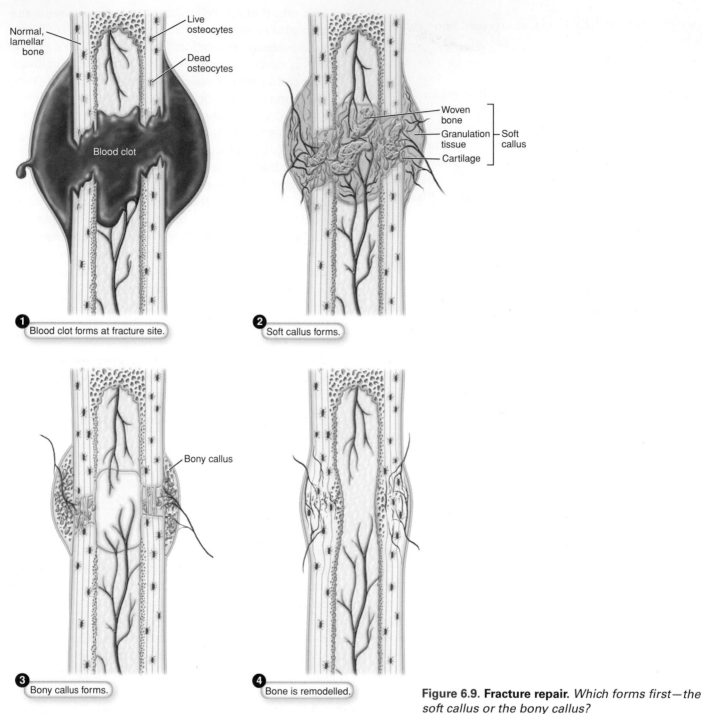

1 Blood clot forms at fracture site.

2 Soft callus forms.

3 Bony callus forms.

4 Bone is remodelled.

Figure 6.9. Fracture repair. *Which forms first—the soft callus or the bony callus?*

Bone Tissue Is Important in Calcium Homeostasis

Bone tissue, which stores 99% of the body's calcium, is intimately involved in calcium homeostasis. It acts as a "calcium bank" in which calcium can be stashed when blood levels are too high and from which calcium can be withdrawn if needed to replenish blood calcium. Bone tissue synthesis by osteoblasts lowers blood calcium levels, whereas bone tissue resorption by osteoclasts raises blood calcium concentrations (Fig. 6.10, left side). Osteoclasts will break down bone to maintain calcium homeostasis even if bone health suffers as a consequence.

Intestines and Kidneys Are Important in Calcium Homeostasis

The normal functioning of two other systems—the digestive system and the urinary system—is important

CLINICAL SNAPSHOT

Fractures

A fracture is a broken bone. Fractures occur when the stress on a bone exceeds the ability of the bone to resist that stress. Unhealthy bones, like those of Maggie H. in this chapter's case, are particularly susceptible to fracture; but even the healthiest bones, those in young adults, can fracture under sufficient stress, as, say, in a car accident or a fall.

Because bones demineralize and weaken with age, fractures are a major health problem for the elderly. The most important risk factors for fracture are listed below. Interestingly, many of these factors are present with increased frequency in aging populations.

- Age over 80 years
- Weight less than 130 lb
- Long-term use of benzodiazepines (widely used sedatives)
- No walking for exercise
- Poor vision
- Brain disease that affects physical stability or mental capacity

Fractures are classified according to whether or not bone has broken through skin and the pattern of the break—some examples are shown in the accompanying figure. A *closed* fracture is one in which bone has not broken through skin. If bone protrudes through skin, the fracture is *open* or *compound*. The pattern of the fracture is more variable:

- A single fracture line is a *simple* fracture.
- A simple fracture line extending all the way across the bone is a *complete* fracture; otherwise the fracture is *incomplete*.
- Multiple fractures in a single site form a *comminuted* fracture.
- Children's bones are more flexible than those of adults and tend to bend or break partially (incompletely), in a manner known as a *greenstick* fracture.
- Twisting force can cause a *spiral* fracture.
- Sudden end-to-end force that causes bone to collapse upon itself is an *impacted* fracture, also called a *compression* fracture (especially in vertebrae).

Most fractures occur suddenly, but stress fractures appear slowly as a result of repeated microfractures caused by significant repetitive stress; for example, in the foot of a long-distance runner.

A force powerful enough to break bone also can damage nearby tissue. Fractures are often accompanied by injury to muscles, blood vessels, nerves, and ligaments, so that the amount and degree of bleeding and injury is often greater than might be expected from a quick glance at an x-ray image.

(a) Closed, complete, spiral fracture

(b) Closed, incomplete, greenstick fracture

(c) Open, complete, simple fracture

(d) Closed, incomplete, impacted (compression) fracture

(e) Closed, complete, comminuted fracture

Fractures.

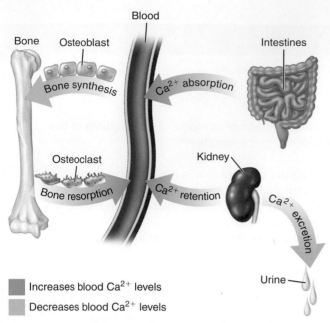

Figure 6.10. Calcium homeostasis. The body maintains a constant blood calcium concentration by modifying three processes: intestinal absorption of calcium; kidney excretion or retention of calcium; and bone synthesis or resorption. *Do osteoblasts take calcium away from the blood or deposit it in the blood?*

more commonly known as vitamin D, which is obtained from fortified dairy products and fish oils or synthesized by the skin in response to ultraviolet radiation in sunlight. Regardless of its source, the liver and kidneys convert cholecalciferol into its active form, known as **calcitriol** or **1,25-dihydroxyvitamin D3.** Calcitriol stimulates calcium uptake by intestinal cells. Vitamin D deficiency in children is a distinctive disease known as *rickets*, which is discussed in more detail on our Web site at **http://thePoint.lww.com/McConnellandHull.**

In addition, all blood (including the calcium dissolved in it) is filtered through the kidneys, paired organs of the urinary system. The kidneys "determine" how much calcium is sent back to the blood—that is, how much will be retained by the body—and how much is excreted in the urine.

Hormones Are Important in Calcium Homeostasis

We've just said that actions of the bones, intestines, and kidneys all influence blood calcium levels, but what controls their activities? If you recall our discussion of chemical signals in ◄— Chapter 4, you won't be surprised to learn that the most important minute-to-minute regulator of blood calcium is a hormone, **parathyroid hormone** (PTH), which is secreted by the parathyroid glands. These are four pea-sized glands posterior to the thyroid gland, a gland at the base of the throat (Fig. 6.11). The net effect of PTH's action is increased blood calcium concentrations. The opposite is also true: If blood calcium rises above its set point, the parathyroids secrete less

in calcium homeostasis. Calcium is an element, and of course the body cannot synthesize it. Instead, it must be consumed in the diet and absorbed through the intestine. Malabsorption disorders such as celiac disease can significantly decrease the intestinal absorption of calcium. In addition, adequate intestinal absorption of calcium is dependent upon the chemical signal *cholecalciferol*,

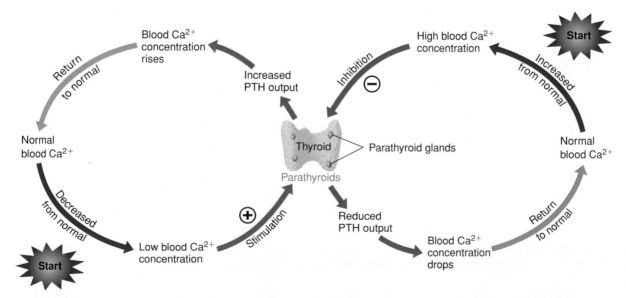

Figure 6.11. PTH regulates blood calcium levels. Red arrows illustrate a disruption to homeostasis, blue arrows illustrate the regulation and response of the parathyroid glands, and green arrows show the return to homeostasis. *Does high blood calcium inhibit or stimulate the parathyroid glands to secrete PTH?*

PTH and blood calcium settles back to its set point. To understand how PTH controls calcium homeostasis, we turn back to Maggie's case.

Case Discussion

Hyperparathyroidism: The Case of Maggie H.

Maggie was diagnosed with *hyperparathyroidism;* that is, her parathyroid glands were secreting abnormally high amounts of hormone. PTH acts at three different sites to increase blood calcium concentrations (Fig. 6.12):

1. PTH stimulates bone breakdown by increasing osteoclast activity.
2. PTH stimulates vitamin D activation by the kidneys. Vitamin D, in turn, increases intestinal calcium absorption.

3. PTH increases calcium retention by the kidney, reducing the amount of calcium normally lost in urine.

Maggie's osteoclasts were thus abnormally active, breaking down excessive amounts of bone tissue and producing weak, easily broken bones. The liberated calcium ions entered her bloodstream. To make matters worse, Maggie was absorbing too much dietary calcium and not excreting enough calcium in her urine. Together, these processes pathologically elevated blood calcium levels at the expense of bone tissue.

Why wasn't Maggie's body able to maintain homeostasis? Normally, Maggie's high blood calcium concentration would have inhibited PTH secretion by negative feedback. However, in Maggie's case, the PTH was secreted from a tumor, and tumors are notoriously insensitive to any form of negative feedback. By definition, a tumor is an uncontrolled growth of cells, which may be uncontrollable even by medical therapy. Once Maggie's tumor was surgically removed, PTH secretion and its action rapidly returned to normal. The bone damage, however, could not be cured. Maggie's bones will always be fragile, and her running days are over.

Case Note

6.15. Before her surgery, would taking vitamin D supplements be more likely to improve or worsen Maggie's condition? How did hyperparathyroidism weaken Maggie's bones?

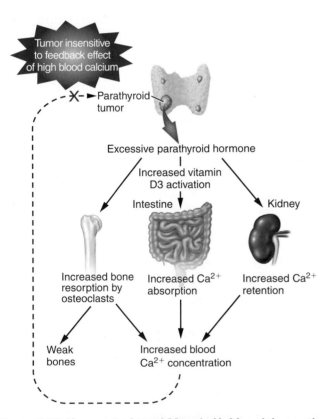

Figure 6.12. Homeostasis and Maggie H. Maggie's parathyroid tumor secreted large amounts of PTH, resulting in kidney stones, high blood calcium levels, and bone breakdown. Unlike normal parathyroid glands, the tumor did not respond to the high blood calcium concentration by secreting less PTH. *True or false. PTH stimulates calcium absorption by the intestines.*

6.1 Newborns have more than 300 bones, but adults have only 206. What happens to the missing bones?

6.2 What shape of bone is the frontal bone of the skull—flat or irregular?

6.3 Which bone feature projects outward from a bone—a foramen or a condyle?

6.4 What do bones have to do with blood cells?

6.5 Name the fibrous membrane that covers the shaft of bones.

6.6 If you wanted to extract some stem cells to make bone tissue, where would you look?

6.7 Name the bone cell that matures into an steocyte.

6.8 What happens to the calcium removed from bone tissue by osteoclasts?

6.9   Which type of canal travels longitudinally in a bone—perforating or central?

6.10   What is an osteon?

6.11   Are trabeculae found in spongy bone or compact bone?

6.12   Which type of bone develops from a cartilage model—membranous bone or endochondral bone?

6.13   Name the first site of bone deposition in endochondral ossification.

6.14   Does articular cartilage develop from perichondrium or periosteum?

6.15   Where does longitudinal bone growth occur—in the diaphysis or the epiphysis?

6.16   Would a medication that stimulates osteoblast activity be used to treat osteoporosis or Paget disease?

6.17   During fracture repair, what type of callus forms first?

6.18   List the components of granulation tissue.

6.19   Does PTH stimulate the activity of osteoblasts or osteoclasts?

6.20   Would an intravenous calcium injection increase or decrease PTH secretion?

Joints

With the exception of the hyoid bone in the neck and the sesamoid bones (such as the kneecap), every bone in the body articulates (meets) with another bone. The location at which bones meet is called a **joint,** an *articulation*, or an *arthrosis* (Greek *arthroun* = "to join or fasten"). Although all joints serve to hold bones together, not all joints allow movement. A few of the body's joints are entirely or nearly immovable. For example, the bones of the skull are fused together as if they were one—a vital attribute necessary to protect the brain.

In most cases, however, joints work together to allow movement of body parts. For example, the marvelous activities we perform with our hands reflect the coordinated movements of nearly every one of the joints in the upper limb. To get a feel for this, reach down and untie and retie your shoelaces while paying attention to the motion of the joints involved. So many movements occur simultaneously that you'll probably have to do it several

times to fully appreciate how the shoulder, elbow, wrist, and every joint in the hand work together to make it possible. Even one injured joint would make this simple activity quite difficult.

Joints Can Be Classified by Form or Function

Joints can be classified according to their form (structure) or their function (degree of movement allowed) (Fig. 6.13). For structural classification, two variables are important: (a) the presence or absence of a *space* between the bones and (b) if no space is present, the *type of connective tissue* that binds the bones together. According to their structure, joints are classified as:

- **Fibrous:** joints with no space; bones are joined by fibrous tissue. The joints between cranial bones are tight, immobile fibrous joints (called *sutures*).
- **Cartilaginous:** joints with no space; bones are joined by cartilage. The joints between vertebrae are cartilaginous joints.
- **Synovial:** joints with a space, the *synovial cavity*, separating the ends of the bones. Synovial joints have several other characteristics described below. All but a few joints in the limbs are synovial joints.

According to its degree of movement, a joint is classified as a:

- **Synarthrosis** (Greek *syn* = "united"): a fixed joint that allows no movement. The sutures between cranial bones are synarthroses.
- **Amphiarthrosis** (Greek *amphi* = "both sides"): a joint that allows slight movement. The joints between vertebrae of the spine and those between the wrist bones are amphiarthroses.
- **Diarthrosis** (Greek *di* = "between"): a freely movable joint. The shoulder is a freely movable joint, as are the joints of the fingers.

Not surprisingly, form and function go hand in hand, so that the function of joints varies according to their form. Accordingly, most fibrous joints are synarthroses, most cartilaginous joints are amphiarthroses, and most synovial joints are diarthroses. Let's look at each of the three types in a little more detail.

Case Note

6.16. **Maggie is experiencing pain in the freely movable knee joint. Is this joint a diarthrosis or a synarthrosis?**

FORM:
Structural classification

FUNCTION:
Functional classification

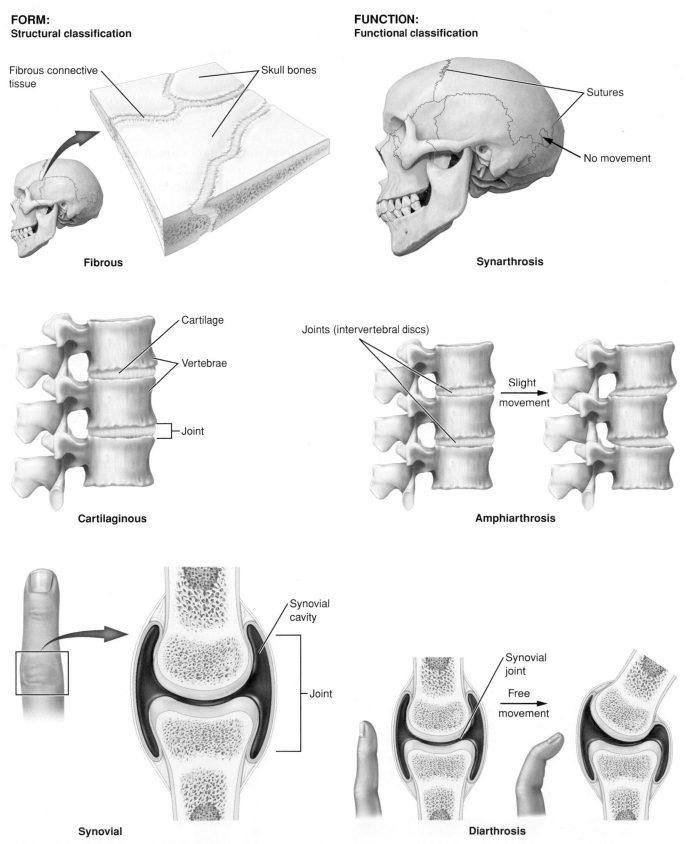

Figure 6.13. Joint classification. Joints can be classified by structure and function. *What is the structural and functional classification of the skull sutures?*

Fibrous Joints Are Sutures and Syndesmoses

Fibrous joints allow little or no motion. There are two types: a *suture* and a *syndesmosis*.

A **suture** is the tightest of all joints. Sutures join the bones of the cranium and consist of a thin layer of very dense fibrous tissue, which allows no motion (Fig. 6.13, top row). What's more, in a suture the bones interlock like the edges of a jigsaw puzzle, which brings added strength and stability to the union. Because no movement is allowed, all sutures are synarthroses.

A **syndesmosis** (Greek *syndesmos* = "bond") is a fibrous joint that allows very limited movement (an amphiarthrosis). The distance between the articulating bones is greater than that in a suture and the space is filled with a greater amount of fibrous tissue. The two bones of the forearm (the radius and the ulna) are joined along their length in a syndesmosis (see Fig. 6.37).

> *Remember This!* **Syndesmoses allow a little movement; sutures do not.**

Cartilaginous Joints Allow Almost No Movement

Cartilaginous joints usually allow limited or no movement. The bones are united by cartilage. Although not usually thought of as joints, the epiphyseal growth plates of growing long bones create a cartilaginous union between pieces of bone that technically fits the definition of a joint. They are synarthroses, which allow no movement and that, of course, disappear when bone growth ends.

However, for practical purposes, the only type of cartilaginous joint is a symphysis, in which a mixture of cartilage and fibrous tissue joins the bones. One such joint is the symphysis pubis, the anterior point of union between the pelvic bones (see Fig. 6.39). Ordinarily it allows virtually no movement, but during pregnancy the fibrous tissue relaxes allowing the joint to expand and the pelvis to enlarge to accommodate the uterus during pregnancy and the passage of the fetus during birth. The joints between the bodies of vertebrae are also symphyses, which are formed by *intervertebral discs* of fibrous tissue and cartilage (Fig. 6.13, middle row).

Case Note

6.17. Maggie's injured leg bones are joined along their length by a fibrous membrane, and each can move to a very small degree relative to the other. Use three terms to describe this joint.

Synovial Joints Allow a Wide Range of Motion

Synovial joints allow a wide variety and range of motion because the articulating bones are separated by a joint cavity (Fig. 6.14). They are characterized by the following features:

- A **joint (articular) capsule,** which surrounds and reinforces the joint. The outer *fibrous capsule* is composed of tough connective tissue that attaches to the surface of the bones on either side of the joint. Some of these connective tissue fibers are organized into thick bundles called *ligaments* (discussed next). The inner layer of the joint capsule is lined by a special membrane, the *synovial membrane,* discussed below.
- A **ligament,** which is a thick band of fibrous tissue extending from one bone to another at a joint. Some ligaments form part of the joint capsule that surrounds the joint. Other ligaments are outside the joint capsule, while still others are located within the joint space itself. The main ligaments usually bear distinctive names, such as the *cruciate ligaments* of the knee.

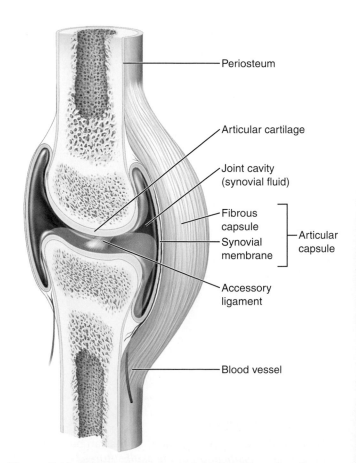

- Periosteum
- Articular cartilage
- Joint cavity (synovial fluid)
- Fibrous capsule — Articular capsule
- Synovial membrane
- Accessory ligament
- Blood vessel

Figure 6.14. Structure of a synovial joint. Note that ligaments help form the articular capsule and are also found within the joint cavity. *What covers the bone ends within the joint—synovial membrane or articular cartilage?*

CLINICAL SNAPSHOT

Arthritis

The joint lubrication was not what it was when I was competing, and I decided that not having arthritis or rheumatism for the rest of my life was a lot more important to me than returning to the track.

Edwin Moses (b. 1955), legendary American track-and-field athlete, explaining why he chose not to return to competition after retiring.

Strictly interpreted, the word *arthritis* means joint inflammation, but in everyday usage it means joint pain. And with good reason: Arthritis is always painful.

Arthritis is a very old disease among animals with joints: dinosaur bones and Egyptian mummies show unmistakable evidence of it. And it is common: arthritis is one of the most common ailments for which patients seek medical care.

Inflammation is the body's cellular reaction to injury, so it follows that an inflamed joint has been injured in some way. Although there are many ways a joint can be injured, the two most common are mechanical wear and autoimmune disease.

Osteoarthritis is a "wear and tear" condition that usually develops in response to chronic physical stress on joints. Although recreational exercise does not appear to increase the risk of osteoarthritis in general, elite track and field athletes like Edwin Moses are at increased risk of osteoarthritis in the knees because of their intensive training and performance. A much more common cause of repetitive joint stress in the American population is morbid obesity: the crushing weight of the body's bulk grinds relentlessly on the cushioning cartilages in the knees, hips, and ankles until they are destroyed. At this point, walking is possible only by grinding the periosteum of one bone on the other, a painful proposition, because periosteum is rich with nerve fibers. Osteoarthritis is also more common as we age, occurring to some degree in 80% of 65-year-olds. Treatment should focus on rehabilitation and physical therapy techniques aimed at preventing disability. Drug therapy for pain relief is a relatively minor aspect of treatment and typically includes *nonsteroidal anti-inflammatory drugs* (NSAIDs) or related compounds.

Another variety of arthritis is **rheumatoid arthritis.** The word *rheumatoid* derives from the Greek root *rheum* for "flowing" and the suffix *-oeides* for "like," a reference to the watery fluid that flows into the joints of patients with this form of arthritis. Rheumatoid arthritis

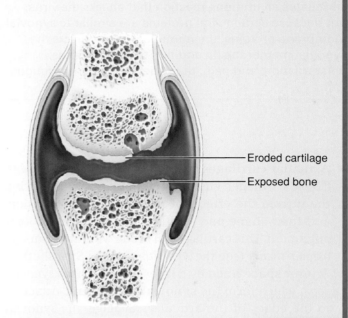

Eroded cartilage
Exposed bone

(a) Osteoarthritis

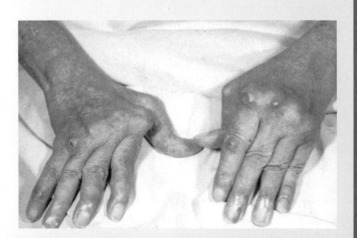

(b) Rheumatoid arthritis

A. A joint from a patient with osteoarthritis. Note the exposed bone surfaces. **B.** Rheumatoid arthritis results in significant joint swelling.

is an *autoimmune disease*, a condition in which the body's immune system attacks host tissues. As you'll learn in ➡ Chapter 12, normally the immune system attacks only foreign (*nonself*) proteins (from, say, viruses or bacteria); but in rheumatoid arthritis, it attacks *self*—the synovial membrane lining synovial joints. Why this occurs is not completely

Arthritis *(Continued)*

understood, but it appears to be a kind of molecular mistaken identity. It seems that an infection initially stimulates an immune reaction that attacks the virus, but because certain viral proteins are similar to synovial membrane proteins, the immune system attacks the synovial membrane. In addition to joint stiffness, inflammation, and pain, the patient experiences fatigue and weakness. The symptoms can remain mild or can progress to full disability. Treatment of rheumatoid arthritis includes many varieties of drugs—NSAIDs, steroids, gold, and immunosuppressive agents—most of which are aimed at dampening the autoimmune reaction; surgery may be required to rehabilitate severely inflamed joints.

- **Articular cartilages** that cover the ends of the bones. These cartilages cushion the bone ends, protecting them from the effects of friction and compression, and their shape and position help to maintain joint alignment. This cartilage can erode with time, causing *osteoarthritis* (see the *Arthritis* Clinical Snapshot).
- A **joint space** lined by a layer of specialized *synovial cells*, which form the **synovial membrane.** Attached to the edges of the articular cartilage, the synovial membrane is folded back on itself many times to increase its surface area. In the knee joint, for instance, the synovial membrane would cover 100 m² if it were completely unfolded. Synovial cells secrete a lubricating **synovial fluid,** which coats the interior of the joint. This thin film of very slippery fluid looks and feels somewhat like uncooked egg white.

Although the synovial membrane is well supplied with blood vessels, the articular cartilage, the joint space, and the associated ligaments are not. Instead, oxygen and nutrients are carried to cartilage and ligament cells by synovial fluid. However, in some types of arthritis, especially rheumatoid arthritis (see the *Arthritis* Clinical Snapshot), blood vessels grow into the joint in response to inflammation and bring with them destructive enzymes that erode joint cartilage.

Other structures are also at work around synovial joints (see Fig. 6.36). **Tendons** are dense bands of fibrous connective tissue, sometimes quite long, which attach the ends of muscles to bone and slide back and forth between tissues as movements occur. To accommodate the complex interaction of muscle, tendon, and nearby tissues during joint movement, special structures exist in the soft tissue surrounding the joint. **Bursae** (*bursa,* singular, from Latin *bursa* = "bag or purse") are small, thin-walled fibrous sacs, somewhat like collapsed balloons, that contain a small amount of synovial fluid. They are positioned between skin and bone and between other moving soft tissue parts around joints, where they roll back and forth with tissue movement associated with joint action. **Tendon sheaths** are elongated bursa-like sacs that wrap around tendons to smooth their back-and-forth movements.

Case Notes

6.18. Maggie's x-ray revealed that some of the cartilage coating her bone ends was worn away. What is this cartilage called?

6.19. Maggie's wrist injury also tore some of the connective tissue connecting her wrist bones. Which type of connective tissue connects bones?

Synovial Joints Allow Four Main Types of Movements

Specialized descriptive language is important for every field of knowledge, including joint movements, which are divided into four main categories: *gliding, rotation, angular,* and *special.*

Gliding is a movement of one flat surface over another. The motion can be in any direction and is usually over a short distance. Movements between the short bones of the wrist and ankle are gliding movements.

Rotation is the revolving of a bone about its long axis. For instance, shaking your head "no" rotates the uppermost vertebra on its axis.

Angular movement is an increase or decrease in the *angle* between the articulating bones. In describing angular movement, remember that all movements begin from standard anatomical position (Fig. 6.15A). The increase or decrease of an angle refers to the angle in reference to the front of the body. For example, the angle formed by the thigh and torso at the hip joint is 180 degrees. Bringing the thigh forward and upward toward

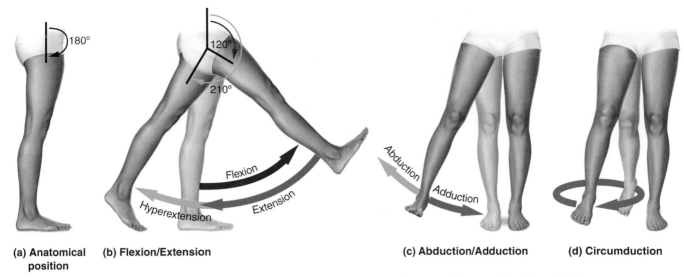

(a) **Anatomical position**   (b) **Flexion/Extension**   (c) **Abduction/Adduction**   (d) **Circumduction**

Figure 6.15. Angular movements at synovial joints. Movements are illustrated for the hip joint. *Which movement swings the leg to the side, away from the midline?*

the front of the body reduces the angle of the hip joint to 120 degrees (Fig. 6.15B). In contrast, sweeping the thigh backward increases the angle of the hip joint to more than 180 degrees.

There are five angular movements. These are described below, using the hip joint as an example.

- **Flexion** is an angular movement that *decreases* the joint angle, such as folding the thigh forward and upward. **Extension** is an angular movement that *increases* the joint angle. A body in the anatomical position is considered to be in *full extension,* so *extension* usually refers to the unfolding of a prior flexion back to the standard anatomical position. Lowering the raised thigh is an example of extension. *Hyperextension* is continuation of extension beyond the standard anatomical position. Moving the thigh toward the rear of the body hyperextends the hip joint. All three of these movements are shown in Figure 6.15B.
- **Abduction** is an angular movement of a bone away from the midline of the body; that is, it is "abducted" (kidnapped) away from the body. **Adduction** is angular movement of a bone toward the midline of the body (it may help to think that the bone is brought back and "added" to the body). Both are shown in Figure 6.15C. Lateral movement of the thigh, which brings it upward, is abduction of the hip joint. Returning it to the standard anatomical position is adduction.
- **Circumduction** is circular movement of the end of the bone (Fig. 6.15D). The hip joint circumducts slightly as a toe traces a circle on the ground.

> *Remember This!* **Movements at the shoulder joint can be difficult to describe. Remember that all body parts are in full extension in the anatomical position. Raising the arm to point at the sky is flexion; lowering the arm back to the anatomical position is extension.**

Special movements are movements other than those described above and are usually specific to one or a few joints (Fig. 6.16).

- **Protraction** and **retraction.** Protraction is a forward (anterior) thrust, such as jutting the jaw outward. Retraction is the opposite.
- **Depression** and **elevation.** Depression is downward (inferior) movement, such as lowering the jaw to open the mouth. Elevation is the opposite.
- *Dorsiflexion* and *plantar flexion* and *inversion* and *eversion* are movements of the foot. **Dorsiflexion** is bending the foot upward (superiorly), as when standing on the heel. **Plantar flexion** is the opposite, as when standing tiptoe—the foot is bent in the direction of the *plantar* surface (the underside of the foot). **Inversion** is turning the soles inward to face one another. **Eversion** is the opposite.
- **Supination** and **pronation** are rotational movements of the forearm that turn the hand. The hand is supine (palm forward) in the standard anatomical position (you SUPinate to carry your SUPper). Turning the palm posteriorly is pronation; turning it back is supination.

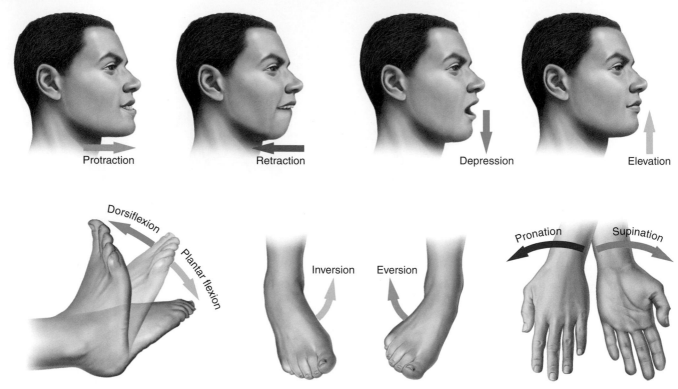

Figure 6.16. Special movements at synovial joints. *Which movement points the toes?*

These movements are illustrated again for different synovial joints in Displays 7.1 to 7.9, which identify the muscles that accomplish each movement.

Synovial Joints Are Constructed Mechanically in Six Ways

Once again, form and function go hand in hand: synovial joints are called upon to enable a variety of movements, and their distinctive construction accommodates a wide variety of movements. The six types of synovial joints are classified according to the type of movement they allow (Fig. 6.17):

- A **pivot joint** is a joint formed where the rounded or pointed end of a bone fits into a sleeve formed by bone and ligament (Fig. 6.17A). The joint between the two forearm bones (the radius and the ulna) at the elbow is an example; it allows rotation about the long axis of the radius as the hand supinates and pronates.
- A **gliding joint** is a flat joint that allows a sliding motion, usually in any direction; no rotation occurs (Fig. 6.17B). The joints between wrist bones are sliding joints, but actions at these joints are difficult to perceive because of the other movements that also occur at the wrist. The sliding motion associated with gliding joints can be perceived in some cartilaginous joints, such as those between the vertebrae (Fig. 6.13, middle row).
- A **condyloid joint** is one in which the egg-shaped end of one bone fits into a shallow depression of its partner (Fig. 6.17C). The joints between the fingers and the palm of the hand are condyloid joints: they allow flexion and extension, as in the clenching and unfolding of a fist, and side-to-side movement, as with adduction/abduction of the fingers.

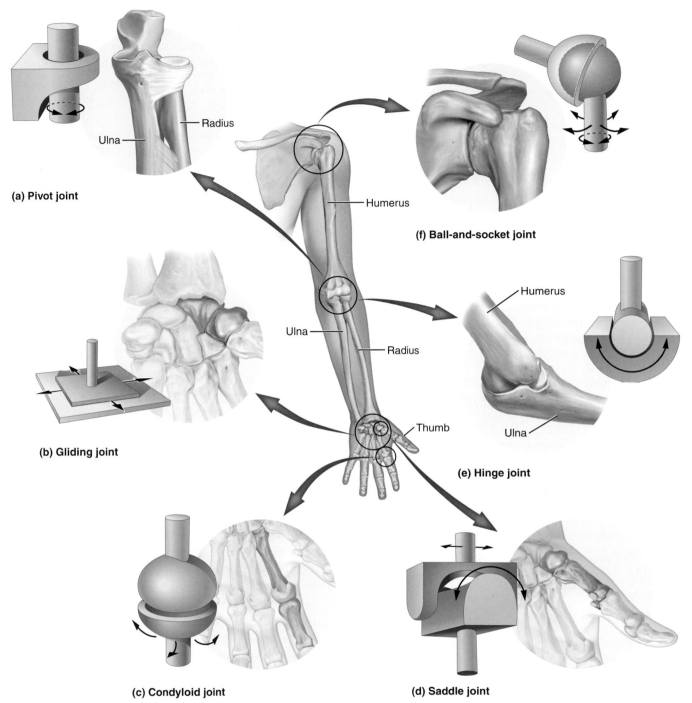

(a) Pivot joint

Radius

Ulna

(b) Gliding joint

(f) Ball-and-socket joint

Humerus

Humerus

Ulna

Radius

(e) Hinge joint

Ulna

Thumb

(c) Condyloid joint

(d) Saddle joint

Figure 6.17. Synovial joint classification. Synovial joints are classified by the types of movement they permit. *How would you classify the joint between the thumb and the palm of the hand?*

- A **saddle joint** is formed by a saddle-shaped depression in one bone that fits into the saddle-shaped depression of a second bone (Fig. 6.17D). The joint between the wrist and thumb is a saddle joint, which permits movement in two directions: the thumb can sweep across the palm to touch the tip of the little finger, or it can move up and down, as it does to press the space bar while typing.

- A **hinge joint** is formed where the cylindrical surface of one bone fits into a matching concavity in the opposite bone (Fig. 6.17E). The elbow and knee are hinge joints; each permits flexion and extension in one direction, like a door opening and closing.
- A **ball-and-socket joint** is formed where the spherical head of one bone fits into the spherical socket of the other bone (Fig. 6.17F). The shoulder and hip

joints are ball-and-socket joints: each can rotate about the long axis of the limb, and each can move in a multitude of folding and unfolding actions.

Case Notes

6.22. Maggie fractured her ankle and wrist. The small bones of the ankle and wrist slide over each other but do not permit rotation. Based on the movements they permit, what sort of joints are these?

6.23. Are Maggie's injured arm and leg bones part of the axial or the appendicular skeleton?

6.21 The pubic symphysis permits some movement during pregnancy. What is its structural and functional classification?

6.22 What is the difference between a suture and a syndesmosis?

6.23 Name the two parts of the joint capsule.

6.24 Name the small fluid-filled sacs found within and surrounding some joints.

6.25 A ballerina dancing on her tiptoes is performing which foot movement?

6.26 If you raise your arm vertically to forage in a high cupboard, are you flexing or extending your arm at the shoulder joint?

6.27 Pugs are small dogs with an underbite—their chin juts out so that their lower teeth sometimes extend over their upper lip. Is the pug's chin retracted or protracted?

6.28 Which joint type permits abduction—a hinge joint or a condyloid joint?

The Anatomy of Bones and Joints: The Axial Skeleton

The bones of the human skeleton are arranged into two primary groups (Fig. 6.18). The **axial skeleton** (shown in green) includes the skull, spine, sternum, and ribs; that is, the bones that constitute the axis of the body. The **appendicular skeleton** (Latin *appendere* = "to hang upon") comprises the bones of the shoulders, hips, and limbs, which hang from the axial skeleton. In this section, we locate and describe the 206 bones of the adult skeleton.

With the exception of the patella and other sesamoid bones and the hyoid bone, all bones articulate with other bones. These articulations are summarized in Table 6.3, and representative joints are discussed in greater detail in the narrative below.

Radiologists are physicians who use medical imaging technologies to diagnose and sometimes treat disease. They typically use x-ray technology to visualize bones and joints. The History of Science box, titled *Bones and the Discovery of X-rays* introduces the first radiologist and discusses the discovery of this revolutionary technical advance.

The Skull and Hyoid Bone

The **skull** is the apex of the axial skeleton and consists of 22 bones in two sets (see Fig. 6.18):

- 8 **cranial bones** form the *cranium*, which encloses the brain.
- 14 **facial bones,** including the mandible (lower jaw), support the face.

The vertebrae of the spine support the cranium, and the cranium, in turn, serves to support the facial bones. Collectively the cranium and facial bones support and protect our delicate special sense organs for sight (the eyes), for smell (the nose), for hearing and balance (the outer, middle, and inner ears), and taste (the tongue) as well as the nerves that serve them.

Bones that span the midline are single; those that do not are paired, one on each side, and are mirror images of each other. For example, the *mandible* and the *frontal bone* (of the forehead) span the midline and are single; but the *maxillary bones* (which form part of the cheek) occur as right and left specimens.

The **hyoid bone** is U-shaped and sits in the soft tissues of the anterior neck (see Fig. 6.18). It does not articulate with any other bones but is attached by a ligament to the base of the skull at the styloid process of the temporal bone (see below). The larynx (voice box) is immediately inferior to hyoid and attached to it by a membrane that runs from the inferior edge of the hyoid to the superior edge of the larynx.

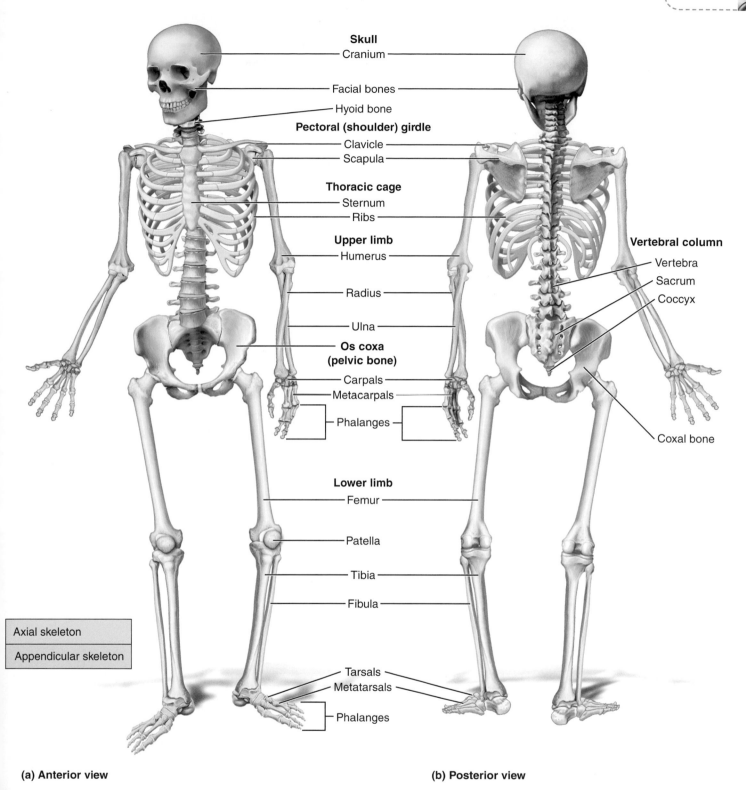

(a) Anterior view

(b) Posterior view

Figure 6.18. Axial and appendicular skeletons. The axial skeleton consists of the skull, the vertebral column, and the thoracic cage. The appendicular skeleton consists of the pectoral girdles, the pelvic girdle, the upper limbs, and the lower limbs. Cartilage is shown in light blue. **A.** Anterior view. **B.** Posterior view. *Are the metatarsals part of the upper limb or the lower limb?*

Table 6.3 Selected Joints

Joint Name	Articulating Bones	Structural Classification	Functional Classification	Movements Permitted
Axial Skeleton				
Skull sutures	Cranial bones	Fibrous	Synarthrosis	None
Temporomandibular joint	Mandible/temporal bone	Synovial	Diarthrosis (hinge)	Elevation, depression, retraction, protraction, lateral movement
Occipitovertebral joint	Occipital bone/atlas	Synovial	Diarthrosis (hinge)	Flexion, extension of head
Intervertebral joints	Vertebral bodies (not C1 or C2)	Cartilaginous (intervertebral discs)	Amphiarthrosis	Slight lateral movement
	Vertebral articular processes	Synovial	Diarthrosis (gliding)	Rotation, lateral movement
	Atlas/axis	Synovial	Diarthrosis (pivot)	Partial rotation
Costovertebral joints	Rib/thoracic vertebra (transverse processes)	Synovial	Diarthrosis (gliding)	Gliding
Sternocostal joints	First rib (costal cartilage)/ sternal manubrium	Cartilaginous	Synarthrosis	Joint fused in adults
	Second to seventh ribs (costal cartilage)/sternum	Synovial	Diarthrosis (gliding)	Gliding (during respiration)
Appendicular Skeleton				
Pectoral Girdle				
Sternoclavicular joint	Sternum (manubrium), clavicle	Synovial	Diarthrosis (gliding)	Gliding
Acromioclavicular joint	Clavicle/scapula (acromion)	Synovial	Diarthrosis (gliding)	Gliding
Upper Limb				
Shoulder	Humerus (head)/ scapula (glenoid cavity)	Synovial	Diarthrosis (ball and socket)	Flexion, extension, abduction, adduction, rotation, circumduction
Elbow	Humerus (trochlea)/ ulna (trochlear notch) and humerus (capitulum) and radius (head)	Synovial	Diarthrosis (hinge)	Flexion, extension
Proximal radioulnar joint	Radial head, radial notch of ulna	Synovial	Diarthrosis (pivot)	Supination and pronation (forearm)
Distal radioulnar joint	Ulnar head, ulnar notch of radius	Synovial	Diarthrosis (pivot)	Supination and pronation (forearm)
Fibrous radioulnar joint	Radius/ulna	Fibrous (syndesmosis)	Amphiarthrosis	None

Table 6.3 Selected Joints (continued)

Joint Name	Articulating Bones	Structural Classification	Functional Classification	Movements Permitted
Appendicular Skeleton (continued)				
Upper Limb				
Wrist joint	Radius, carpals	Synovial	Diarthrosis (condyloid)	Flexion, extension, abduction, adduction, circumduction
Carpal joints	Between carpals	Synovial	Diarthrosis (gliding)	Gliding
Carpometacarpal joint (thumb)	Carpals, first metacarpal	Synovial	Diarthrosis (saddle)	Flexion, extension, abduction, adduction, circumduction, opposition
Carpometacarpal joints (fingers)	Second to fifth metacarpals, carpals	Synovial	Diarthrosis (condyloid)	Flexion, extension, limited abduction and adduction
Metacarpophalangeal joints	Metacarpals/proximal phalanges	Synovial	Diarthrosis (condyloid)	Flexion, extension, abduction, adduction, circumduction
Interphalangeal joints	Between phalanges	Synovial	Diarthrosis (hinge)	Flexion, extension of fingers
Pelvic Girdle				
Sacroiliac joint	Sacrum/ilium	Synovial	Amphiarthrosis	Minimal movement
Pubic symphysis	Pubic bones	Cartilaginous	Amphiarthrosis	Slight movement (pregnancy)
Lower Limb				
Hip joint	Femur (head)/coxa (acetabulum)	Synovial	Diarthrosis (ball and socket)	Flexion, extension, abduction, adduction, rotation, circumduction
Knee joint	Femur, tibia	Synovial	Diarthrosis (hinge)	Flexion, extension, slight degree of rotation
Fibrous tibiofibular joint	Tibia/fibula	Fibrous (syndesmosis)	Amphiarthrosis	None
Proximal tibiofibular joint	Tibia/fibula	Synovial	Diarthrosis (gliding)	Gliding
Ankle	Tibia and fibula/talus	Synovial	Diarthrosis (hinge)	Dorsiflexion, plantar flexion
Tarsal joints	Between tarsals	Synovial	Diarthrosis (gliding)	Inversion, eversion
Tarsometatarsal joints	Tarsal/metatarsals	Synovial	Diarthrosis (gliding)	Gliding
Metatarsophalangeal joints	Metatarsals/phalanges	Synovial	Diarthrosis (hinge)	Flexion, extension, some abduction and adduction
Interphalangeal joints	Between phalanges	Synovial	Diarthrosis (hinge)	Flexion, extension

THE HISTORY OF SCIENCE

Bones and the Discovery of X-rays

The ancients knew a lot about human anatomy, especially about bones, which they understood well because of the endurance of bones after death. Three thousand years ago, Egyptians knew enough anatomy to ready bodies for mummification by removing the brain through the nose. And, as discussed in Chapter 1, anatomists have long used corpses to study the inner structures of the human body.

The main problem with acquiring knowledge in this way is that it requires death of the subject. Therefore, it is easy for us to imagine that, confronted with crisis deep in the body of a living patient, ancient physicians must have longed to see beneath the skin. Nowadays, however, we take it for granted that every cubic millimeter of our bones and internal organs can be examined without spilling a drop of blood, and x-rays were the first step.

For this remarkable fact we owe thanks to Wilhelm Roentgen, a Dutch physicist. Roentgen discovered x-rays as he experimented with electricity by passing an electrical current from one metal pole to another in a vacuum tube (much like a modern lightbulb). When the electrical current was at low power, the apparatus produced ordinary light plus some other mysterious rays. When Roentgen focused these rays on a panel coated with barium (a metal) that he had placed across the room, he found that they caused the barium to glow.

On a Friday afternoon, November 8, 1895, Roentgen conducted an experiment that initially required the vacuum tube to be covered completely with tinfoil and cardboard to keep all of the light and mysterious rays from escaping. To be sure that no rays were getting out, he darkened the room before turning up the current. Sure enough, the apparatus remained dark; no light was escaping. He was about to turn off the apparatus and turn on the lights when, far across the room, a glow caught his eye—the barium panel was glowing despite the fact that no light was escaping the apparatus. Roentgen instantly recognized that the mysterious rays—he later dubbed them x-rays—were passing from the tube, through the foil and cardboard, and striking the barium panel, making it glow.

On closer inspection of the glowing panel, Roentgen noticed a dull black line running across it. Looking carefully in the path of the mysterious rays, he

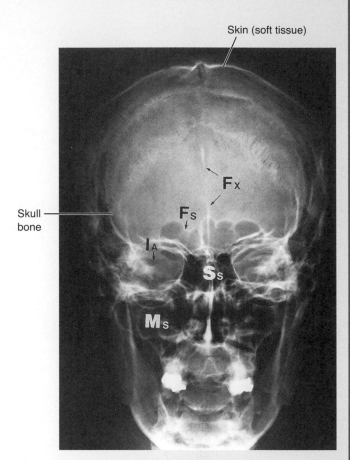

An x-ray, showing bone and soft tissue.

discovered a wire that was absorbing some of the rays. He wondered what else might block the rays, so he put a piece of paper in the beam's pathway. It had no effect; the panel still glowed. Next he tried a playing card, then a book, finding that the book dimmed the beam slightly. Finally, he held a small lead disc in the beam. Because it required him to place his hand in the beam, this simple act later earned him the first-ever Nobel Prize in physics. Lead stopped the strange rays completely, and—in a moment that literally changed the world—he noticed glowing on the panel the eerie but unmistakable image of his fingers, the bones clearly visible beneath the hazy outline of his flesh.

The hyoid serves to support the tongue above and anchor the larynx below. Numerous neck muscles also attach to the hyoid. It ensures smooth swallowing by serving as a fixed point for balancing the wave of muscular forces generated by the muscles of the neck, tongue, and larynx.

The hyoid is frequently the subject of intense scrutiny in medicolegal investigations when strangulation is the suspected cause of death. Finding a broken hyoid bone at autopsy is strong evidence of strangulation.

> *Remember This!* **The skull consists of the cranial bones and the facial bones.**

The Skull Contains Many Cavities

The cranium is the largest of the many skull cavities. The anterior surface of the skull contains the two **orbits,** which house the eyeballs, and the **nasal cavity,** which connects the nose to the back of the throat (Fig. 6.19; also see Fig. 6.24). Innumerable smaller openings (foramina) allow blood vessels and nerves to pass into or out of the bony cranium. The cranium also contains numerous **paranasal sinuses** (often called just *sinuses*) — air-filled spaces within bones that connect to the nasal cavity by foramina. Each sinus bears the name of the bone in which it is located; the details of each sinus is discussed in the context of the relevant bone.

Air moves between the nasal passages and paranasal sinuses in order to equalize pressure between the sinuses and the atmosphere. Because they are connected to the environment, sinuses are vulnerable to infection. Infection of a sinus is called *sinusitis*. The sinuses form part of the *upper respiratory system* ➡ (Chapter 13), which is vulnerable to viral infections ("the common cold"). When "a cold" occurs, the resulting inflammatory swelling and discharge easily block the tiny foramina, trapping air and fluid inside the sinus. Trapped fluid can become a breeding medium for bacteria, which can turn an otherwise minor viral infection into a much more severe bacterial one. What's more, blockage prevents equalization of air pressure between the sinuses and outside. If pressure cannot equalize and if environmental pressure changes (most often because of a change in altitude — up or down — in an elevator or airplane ride), the pressure difference presses on the sinus wall, creating a very uncomfortable or painful sensation. When this happens, it is sometimes helpful to hold the nostrils closed and blow gently to force air into the sinus foramen so as to unplug it and allow the pressure to equalize.

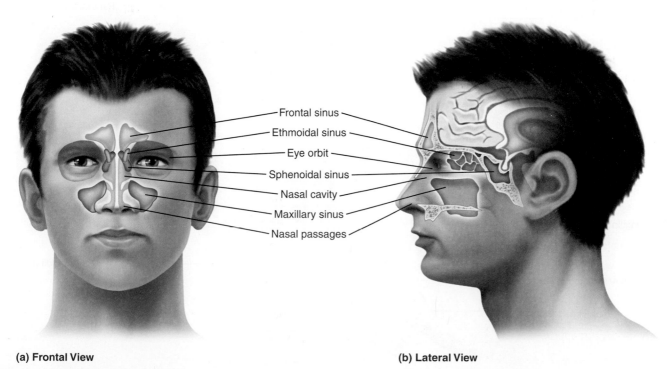

(a) Frontal View **(b) Lateral View**

— Frontal sinus —
— Ethmoidal sinus —
— Eye orbit —
— Sphenoidal sinus —
— Nasal cavity —
— Maxillary sinus —
— Nasal passages —

Figure 6.19. Sinuses. The sinuses are air-filled spaces in skull bones linked to the nasal passages. **A.** Frontal view. **B.** Lateral view. *Which sinuses are superior to the eyes?*

Finally, with the exception of the maxillary sinuses, which sit forward, every sinus is separated from the brain only by a relatively thin wall of bone. It is easy for bacterial sinus infections to penetrate bone and infect the brain.

> *Remember This!* **There are four sets of sinuses: sphenoidal, ethmoidal, frontal, and maxillary.**

Cranial Bones Protect the Brain

The cranium protects the brain. You can see different views of the cranium and facial bones in Figures 6.20 to 6.24. Although a bone or bone feature may be visible in multiple views, we refer to the view that shows a particular feature best.

The external surface of the cranium serves for attachment of muscles that move the head and provides the supporting framework for the facial bones. The eight bones of the cranium include the paired parietal bones, paired temporal bones, occipital bone, frontal bone, sphenoid bone, and ethmoid bone.

Parietal Bones

Attached to the rear of the frontal bone are the paired **parietal bones** (Fig. 6.20). These three bones connect to one another at the apex of the skull. From there, each parietal bone curves inferiorly, laterally, and posteriorly to form the largest portion of the roof and sides of the skull.

Temporal Bones

The inferior sides of the cranium and part of the floor are formed by the **temporal bones** (beneath the temples). The temporal bones contain several important anatomical features, all visible in Figure 6.21:

- The **zygomatic process,** a thin arch of the temporal bone that connects anteriorly with the **temporal process** of the **zygomatic bone** to form the **zygomatic arch.**
- The **external auditory meatus,** which leads inward (medially) from the external ear to the eardrum and middle ear.
- The **styloid process,** a narrow, cone-shaped downward projection to which neck muscles attach.

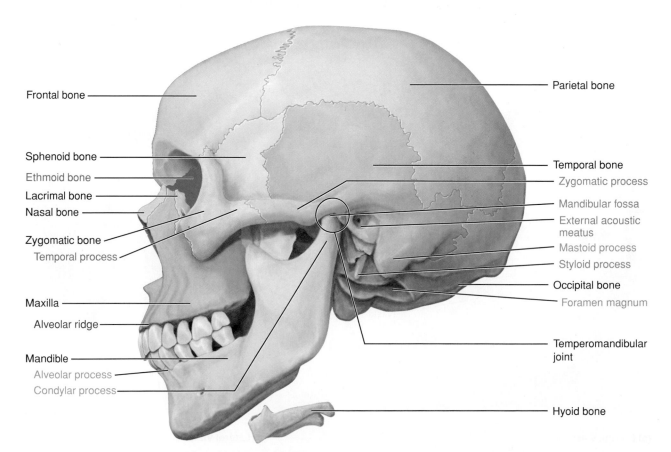

Figure 6.20. Skull and hyoid bone, lateral view. The names of the bones are in black type; the names of the bone features are in the same color as the bone under the relevant bone. *Which bone contains the condylar process?*

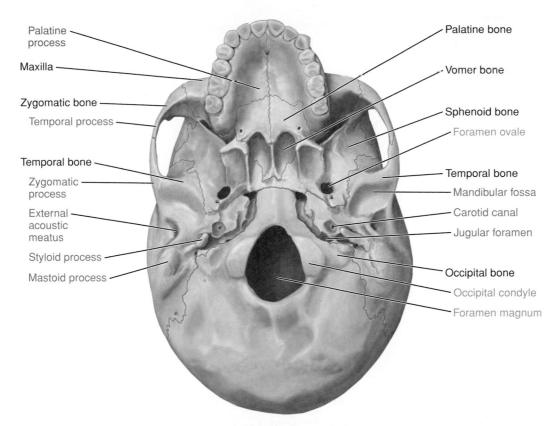

Figure 6.21. Skull: inferior view. *Which bone contains the external acoustic meatus?*

- The **mastoid process,** a thick bulge of bone behind the ear to which neck muscles attach. It sits posterior and inferior to the external auditory meatus and is honeycombed with small air spaces, the **mastoid air spaces.** Although these air cells are sometimes called the **mastoid sinuses,** they are not paranasal sinuses because they do not connect to the nasal cavity. Instead, they connect to the air in the middle ear ➡ (Chapter 9). Infections can spread from the middle ear to the mastoid air spaces, causing *mastoiditis,* and from the mastoid air cells to nearby brain tissue.
- The **mandibular fossa** is an indentation on the inferior side of the zygomatic process that articulates with the mandible to form the *temporomandibular joint* (discussed below).
- The **carotid canal** is an opening that allows passage of the internal carotid artery, which carries the majority of blood to the brain.
- The **jugular foramen** is situated in the suture between the temporal and occipital bones. It allows passage of the internal jugular vein, which drains blood away from the brain.

Occipital Bone

The **occipital bone** forms the posterior wall and posterior floor of the cranium and articulates with each parietal and temporal bone (Figs. 6.20 and 6.21). It contains the largest opening in the cranium, the **foramen magnum,** through which the spinal cord and brain connect. Lateral to the foramen magnum, two bean-shaped platforms, the **occipital condyles** (Greek *kondlylos* = "knuckle"), connect with the first cervical (neck) vertebra, and are the points upon which the skull rests on the neck.

Sphenoid Bone

The **sphenoid bone** (Greek *sphen* = "wedge") is a "keystone bone," so to speak. As you can see in Figure 6.22, it has a complex butterfly shape and sits at the base of the skull. From this location, it *connects to all other cranial bones* and also forms the majority of the orbit. The outstretched wings of the sphenoid reach out to connect laterally with the temporal and parietal bones and anteriorly to connect with the frontal and ethmoid bones. The tail end of the sphenoid connects posteriorly to the occipital bone. The butterfly's body, at the base of the

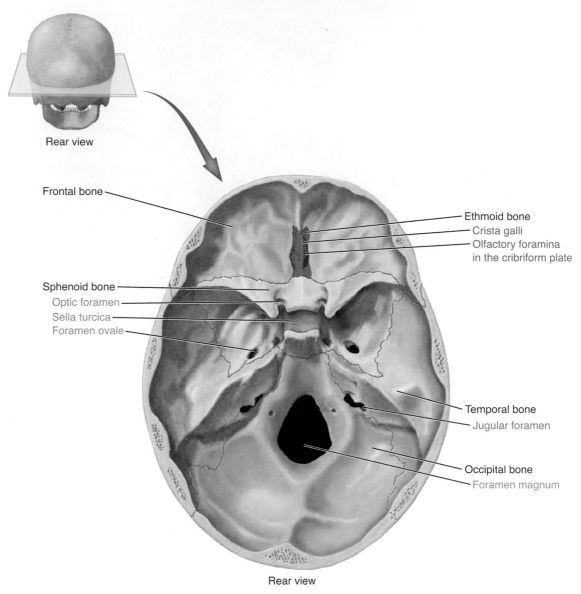

Rear view

Rear view

Figure 6.22. Skull: cranial floor. *Which bone contains the foramen ovale?*

skull, contains the large **sphenoid sinuses.** Riding the back of the butterfly's body is a cradle of bone, the **sella turcica,** which holds the pituitary gland (Fig. 6.22).

Two sets of nerves pass through the sphenoid. The mandibular nerve passes through the **foramen ovale** and the optic nerve through the **optic foramen** (Fig. 6.22).

Case Note

6.24. Maggie's skull x-ray revealed a bone cyst adjacent to the opening that permits the internal carotid artery to enter the cranial cavity. Name the bone and the feature.

Ethmoid Bone

The delicate **ethmoid bone** occupies the critical space between the nasal cavity and the orbits on either side.

Very little of the ethmoid bone is visible in the intact skull; its features are best visualized in a sagittal (Fig. 6.23) or frontal (Fig. 6.24) section. Its name comes from the Greek *ethmos,* meaning "sieve" or "strainer," and indeed the ethmoid bone is filled with small air cells. These air cells form the ethmoid sinuses, which sit in front of the sphenoid sinuses and high in the rear part of the nasal passage.

The superior edge of the ethmoid bone also has two narrow horizontal plates, one on each side of the perpendicular plate. These horizontal plates, called the **cribriform plates** (Latin *cribrum* = "sieve") are perforated with numerous tiny holes (olfactory foramina) that transmit fibers from the olfactory nerve into the nasal passages (Fig. 6.23). The **crista galli** (literally, "the cock's crest" or "rooster's comb") projects superiorly between

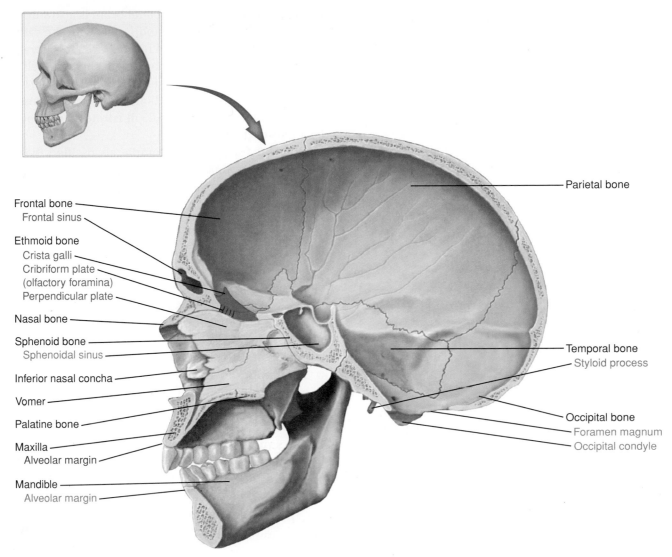

Figure 6.23. Skull: sagittal section. *Which bone is adjacent to the frontal bone—the occipital bone or the parietal bone?*

Frontal bone
Frontal sinus

Ethmoid bone
Crista galli
Cribriform plate
(olfactory foramina)
Perpendicular plate

Nasal bone

Sphenoid bone
Sphenoidal sinus

Inferior nasal concha

Vomer

Palatine bone

Maxilla
Alveolar margin

Mandible
Alveolar margin

Parietal bone

Temporal bone
Styloid process

Occipital bone
Foramen magnum
Occipital condyle

the cribiform plates. It serves as the site of attachment for the brain *meninges* (covering membranes). A large vertical plate of bone, the **perpendicular plate,** extends inferiorly into the midline of the nasal cavity and forms the superior portion of the *nasal septum,* which divides the interior of the nose into right and left nasal passages. (The inferior portion of the nasal septum is formed by the *vomer,* discussed below). In the nasal cavity, this plate bears two curled, fragile rolls of bone, the **superior nasal concha** and the **middle nasal concha** (Greek *konchos* = "shell"). The **inferior nasal concha** is a separate bone, discussed below. The conchae are covered with warm, moist mucous membrane: as we inhale, air is swirled, warmed, and humidified by its passage across the conchal membranes, and inhaled particles are trapped by contacting nasal mucus before they have a chance to enter the lungs.

Frontal Bone

The **frontal bone** forms the anterior cranium, including the forehead and the roofs of the orbits (eye sockets) (Figs. 6.23 and 6.24). The **frontal sinuses** sit behind each eyebrow (Fig. 6.23).

Cranial Bones Articulate to Form Fontanels or Sutures

A **fontanel** is an opening (a "soft spot") between the skull bones of an infant. Recall that bones of the cranium began their embryonic existence as membranes and at birth remain flexible and not fused to one another. As the bones harden during ossification, the edges of the membrane remain soft and span the distance between the growing bones of the fetal cranium. Fontanels allow the fetal cranium to be compressed and molded for passage

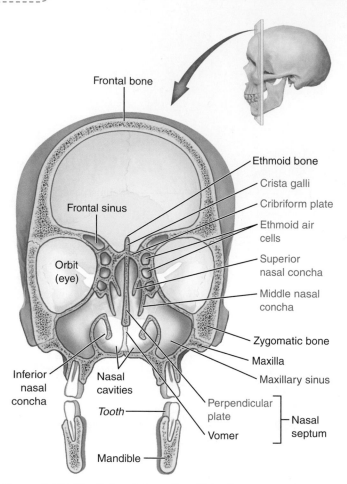

Figure 6. 24. Skull: frontal section. This diagrammatic representation of the skull illustrates how the ethmoid bone, vomer, and inferior nasal conchae articulate to form the bony structure of the nose. *Which two bones form the nasal septum?*

through the birth canal. The fetal brain is also soft, much softer than an adult brain, and molds with the cranium. Fontanels also provide room for the brain to grow during the first 2 years of life. At about age 2, the fontanels close as cranial bones fuse together. Afterward, the capacity of the cranium grows as skull bones grow to accommodate an enlarging brain. Figure 6.25 illustrates the names and locations of the fontanels.

Once the fontanels are completely ossified (typically by age 2), the cranial bones are fused together by immovable sutures. The cranium contains numerous sutures, but four deserve mention because they are anatomic landmarks (Fig. 6.26):

- The **sagittal suture** is formed where the two parietal bones meet. It runs in the midline of the cranium from the apex anteriorly to the occipital bone posteriorly.
- The **coronal suture** is formed where the two parietal bones meet the frontal bone. Somewhat like a girl's headband (*corona* = "crown"), it extends roughly across the top of the cranium from one ear to the other.
- The **lambdoid suture** is formed where the occipital bone meets the two parietal bones. It runs roughly horizontally from one side of the posterior cranium to the other.
- The **squamous suture** is formed where each temporal bone unites with the parietal and sphenoid bones. Roughly speaking, it traces an arc above each ear.

Remember This! **The fontanels in fetuses and infants ossify to form sutures.**

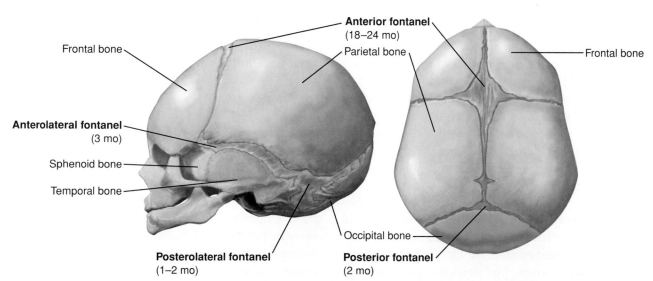

Figure 6.25. Fontanels. The infant skull at birth, showing the fontanels. The fontanels close at the age shown in the brackets. *Which fontanel closes last?*

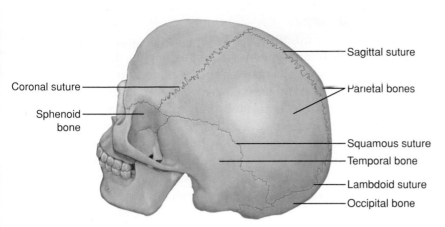

Figure 6.26. Cranial sutures. *Which suture outlines the temporal bone?*

Facial Bones Support the Face

The 14 facial bones support the soft tissues of the face and serve as attachment points for muscles that animate our expressions. All facial bones but the mandible are fused tightly to the cranium.

Maxillae

The keystone bones of the face are the paired **maxillae** (or maxillary bones), to which all facial bones except the mandible attach (Fig. 6.27). Where they fit together in the center of the face, the maxillary bones create the borders of the nasal cavity. The **maxillary sinuses** are the most anterior of the sinuses and sit on either side of the nose behind the cheeks.

The inferior **alveolar margin** contains **alveoli** (pockets) into which teeth insert. The *palatine process* of each maxilla is a horizontal extension that meets its partner in the midline to form the anterior part of the **hard palate** (the roof of the mouth) (Fig. 6.21). The superior part of each maxilla forms the inferomedial wall of each orbit (see Fig. 6.24).

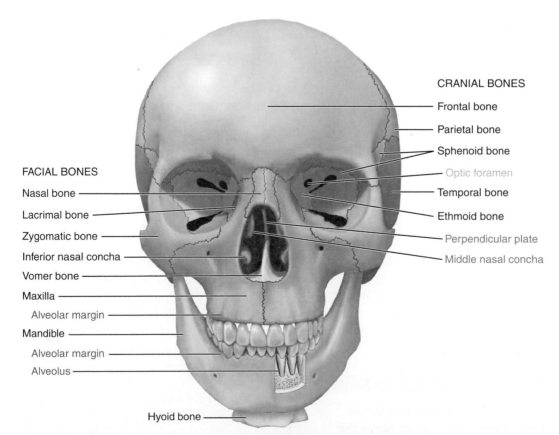

Figure 6.27. Skull, frontal view showing facial bones. *Which bone forms the roof of the eye orbit?*

Palatine Bones

The paired **palatine bones** attach to the posterior edge of the palatine processes of the maxillae (Fig. 6.21). Their name reflects the fact that they form the posterior aspect of the hard palate. Embryological failure of the palatine bones to develop fully and meet in the midline is the cause of cleft palate, which may or may not be associated with *cleft lip*.

Zygomatic Bones

The **zygomatic bones** sit at the lateral aspect of each maxilla and form the bony bulge of each cheek (the cheekbone) (Fig. 6.27). They also form the inferolateral floor of the orbit.

Lacrimal Bones

The paired **lacrimal bones** are fingernail-sized plates of bone wedged into the anteromedial aspect of the orbit (Fig. 6.27). The Latin word *lacrima* means "tear," and lacrimal bones have a tiny passageway for the **lacrimal duct** (tear duct), which carries tears from the eye into the nose.

Vomer Bone

The **vomer bone** (Latin *vomer* = "plowshare") is the only midline bone of the superior facial bones (Fig. 6.27). It is a roughly triangular vertical plate of bone that forms the inferior portion of the *bony* part of the **nasal septum** in connection with the perpendicular plate of the ethmoid bone, which forms the superior part (Fig. 6.24). The anterior end of the nasal septum is formed not by bone but by a plate of hyaline cartilage, the **nasal cartilage,** which extends to the tip of the nose (Fig. 6.23).

Nasal Bones

The **nasal bones** are a small pair of bones wedged between the superior edges of the maxillary bones (Fig. 6.27). They form the bony "bridge" of the nose. The remainder of the nose is composed of cartilage plates called **alae** (Greek *ala* = "wing") on each side and the nasal cartilage in the midline.

Inferior Nasal Conchae

Each **inferior nasal concha** is attached to the medial edge of the maxillary bone in the nasal cavity (see Fig. 6.24). Along with the superior and middle conchae of the ethmoid bone, the inferior nasal concha increase the surface area of the nasal mucosa.

Case Note

6.25. Maggie had a bruise over the two bones forming the lateral aspect of the orbit. Name the bones underlying the bruise.

Mandible

The **mandible** is U-shaped and hinged posteriorly to the temporal bones. Like the maxilla, the mandible contains an alveolar margin of bone into which the lower teeth are inserted. Each posterosuperior end of the mandible is formed into a capped post, the *condylar process*, also called the mandibular ramus (Fig. 6.20).

The **temporomandibular joint (TMJ),** commonly known as the jaw joint, consists of the condylar process of the mandible and the mandibular fossa of the temporal bone (Fig. 6.20). It can perform five movements:

- Elevation: close mouth
- Depression: open mouth
- Protraction: jutting the chin anteriorly
- Retraction: bringing the chin posteriorly
- Lateral movement: moving the chin from side to side

The TMJ, like any other joint, can suffer from a variety of problems: torn ligaments, internal derangement of joint parts, arthritis, or spasm of jaw muscles. Recurrent pain in the TMJ is known as *TMJ syndrome*, which can also be associated with a limited range of motion, a "popping" sound on opening the jaw, and headache. This complex problem can be due to overuse (excessive gum chewing, for example), trauma, grinding of the teeth, or dental problems.

The Vertebral Column

The *vertebral column*, also called the spine or backbone, is a stack of somewhat circular bones, the **vertebrae,** each of which has a central foramen through which the spinal cord and spinal nerves run. The vertebrae are held together by ligaments and articulate with one another in a way that gives the spine stability and strength.

During fetal development and childhood, the vertebral column usually consists of 33 bones, but fusion of bones 25 to 29 to form the sacral bone and fusion of bones 30 to 33 to form the coccyx usually reduces the adult count to 26 (24 vertebrae, 1 sacrum, and 1 coccyx). Sometimes the lowest four bones do not fuse completely, resulting in between 27 and 29 vertebrae. The spine supports the skull, the contents of the chest and abdomen, and the upper limbs, and is the axis upon which all body movement turns.

The Vertebral Column Includes Five Regions

From superior to inferior, the vertebral column is divided into five regions (Fig. 6.28):

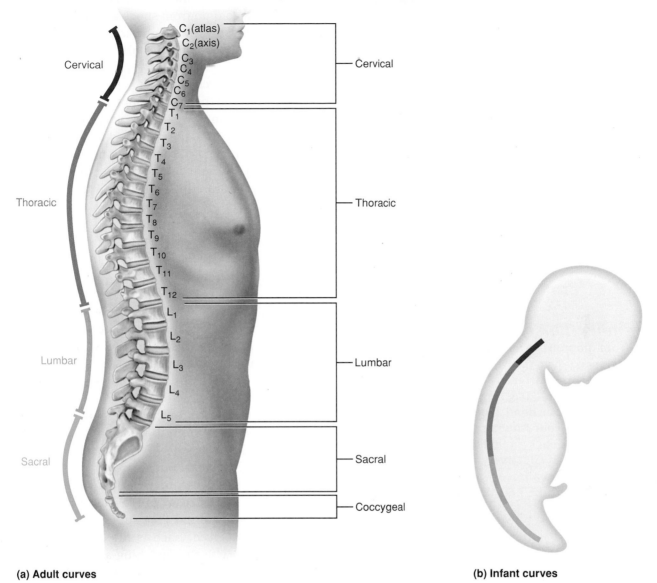

(a) Adult curves   **(b) Infant curves**

Figure 6.28. Vertebral column, lateral view. A. Adult. **B.** Infant. The same colors are used for the vertebral regions in the infant and the adult. *How many thoracic vertebrae are found in the adult spine?*

- The **cervical** (neck) region contains 7 vertebrae.
- The **thoracic** (chest) region contains 12 vertebrae.
- The **lumbar** (lower back) region contains 5 especially strong vertebrae.
- The **sacral** region contains 5 vertebrae in children, but a single *sacrum* in adults.
- The **coccygeal** region contains multiple vertebrae in children but usually a single *coccyx* in adults.

Note that each vertebra is named based on the region (e.g., "C" means cervical) and numbered from superior to inferior. Thus, the bottom cervical vertebra is called C7.

As you can see in Figure 6.28A, each region of the adult vertebral column is curved slightly in alternating fashion: the cervical spine is curved anteriorly; the thoracic spine, posteriorly; the lumbar spine, anteriorly; and the sacrum and coccyx, posteriorly.

In contrast, in the fetus and newborn, the spine is C-shaped (Fig. 6.28B). The cervical curve, which is a necessary component of the human development of upright posture to look across the landscape, develops when an infant begins to hold its head up (around 3 months). The lumbar curve is necessary for upright locomotion and develops as the infant learns to stand upright and walk (around 1 year). Because most of the weight of the chest and abdomen is anterior to the spine, development of the lumbar curve is necessary to shift the weight of the torso posteriorly, so that it remains in neutral vertical alignment over the feet.

Sometimes the spine may curve abnormally. *Scoliosis* is an abnormal side-to-side curvature that typically develops during childhood and adolescence. *Lordosis* is an exaggeration of the lumbar curve and is a temporary condition of pregnancy, when the weight of the fetus anterior to the spine shifts the woman's center of gravity. It is also common among people with excessive abdominal fat. *Kyphosis* is an excessive anterior curving of the spine. Kyphosis is especially common in older people with osteoporosis and is characterized by a bent-over posture.

> *Remember This!* **Your lumbar curve can adjust from moment to moment to keep your body weight balanced over your feet. Try carrying a heavy load in three different ways: in a backpack worn on the back, in a bag hugged to the chest, and finally on top of your head. Notice how placement of the load alters your lumbar curve. Carrying the load in front or in back exaggerates or diminishes the lumbar curve, respectively, as your skeleton adjusts to maintain balance. The head load should not alter the curvature, because the load is in line with the vertical axis.**

A Typical Vertebra Is a Circle of Bone

Each vertebra consists of a circle of bone surrounding the spinal cord. The vertebral circles form in the embryo when the edges of a trough of tissue grow up and over the spinal cord to meet in the middle. Failure to complete this process sometimes leaves the vertebral arch open dorsally, a condition called *spina bifida*. Minor degrees of this condition are fairly common, usually unknown to the patient, and not associated with neurological disease. However, severe degrees can cause paralysis and infection.

Almost every vertebra has the same basic features, as illustrated in Figure 6.29A and B:

● A thick, disc-shaped **body.** The bodies of all vertebrae except C1 and C2 are joined by the **intervertebral discs** (Fig. 6.29F): tough, flexible, round cushions that form cartilaginous joints. These have a rim of dense fibrous tissue surrounding a core of semisolid cartilaginous ground substance. They cushion the vertebral column from shock and allow limited rocking and twisting motion between the vertebrae. They are classified functionally as amphiarthroses (Fig. 6.13).

● A **vertebral arch** that arcs horizontally from one side of the body to the other, completing the circle of bone. It is divided into two stubby **pedicles** extending posterolaterally from the body and two **laminae** extending from the pedicles and meeting at the arch's apex.

● The central hole in the bone circle is called the **vertebral foramen.** The aligned foramina form the **spinal canal.**

Projecting from the vertebral arch are seven bony processes, which create four joints plus three sites for spinal muscle attachment:

● A single, midline, dorsal projection of bone that arises from the apex of the arch (where the laminae join), called the **spinous process.** As you run your finger down the center of your back, you are feeling the tips of spinous processes. The spinous processes serve as sites of muscle attachment.

● Two bony projections, the **transverse processes,** extend laterally from the sides of the vertebral arch and also serve as sites of muscle attachment.

● Four stubby plates of bone project from the vertebral arch and serve to connect to the vertebrae above or below. These **articular processes** occur in pairs, one right and one left, on the superior and inferior surfaces of the arch. Each is fitted with cartilage and articulates with the matching articular processes of the vertebrae above and below to form four synovial joints complete with articular cartilage and a small joint space (Fig. 6.29F). These joints are gliding joints, since they permit the spine to rotate and bend anteriorly or posteriorly, and right or left.

Finally, each vertebra has notches on the inferior surface of each pedicle, which interact with the vertebra below to create the **intervertebral foramen.** This small opening allows spinal nerves to pass from the spinal cord out into the body (Fig. 6.29F).

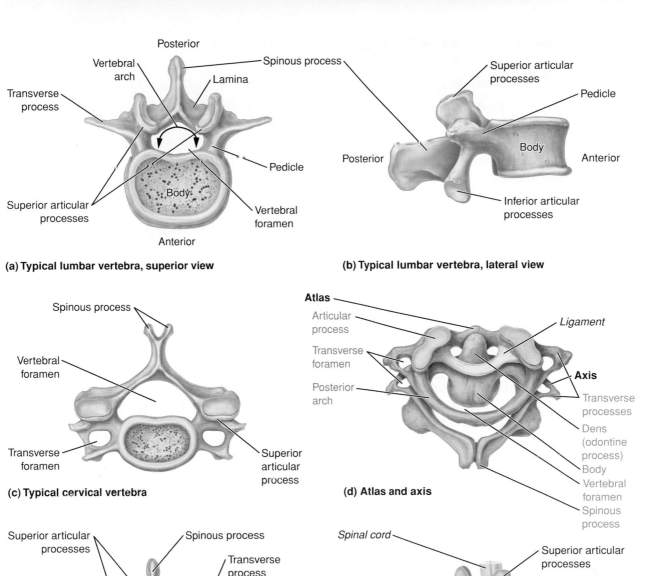

(a) Typical lumbar vertebra, superior view

(b) Typical lumbar vertebra, lateral view

(c) Typical cervical vertebra

(d) Atlas and axis

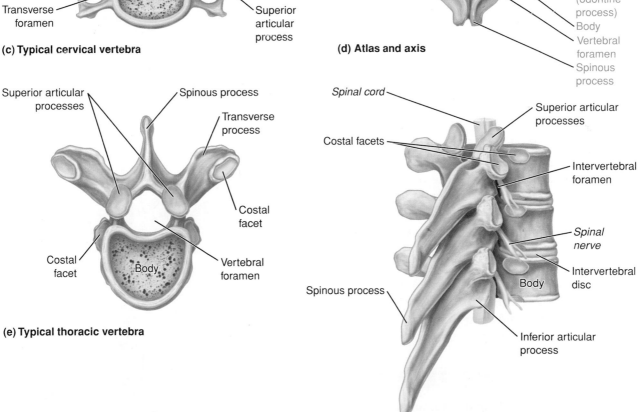

(e) Typical thoracic vertebra

(f) Articulated thoracic vertebrae

Figure 6.29. Vertebral structure. A. A lumbar vertebra is used to illustrate the basic features from a superior view and **B.** from a right lateral view. **C.** A typical cervical vertebra, as seen in C3 through C7. **D.** The atlas (C1) rotates around a post provided by the axis (C2). **E.** A typical thoracic vertebra. Note the modified transverse process, which articulates with a rib. **F.** Vertebrae stack on top of each other to form the vertebral column. In all figures, note that cartilage is shaded light blue. *Name the structure connecting the vertebral body to the transverse process.*

There Are Seven Cervical Vertebrae

In addition to the standard features described above, each cervical vertebra contains two extra foramina, one in each transverse process, through which blood vessels pass (Fig. 6.29C). The dorsal spinous processes of cervical vertebrae are short and often divided into two branches.

The first two cervical vertebrae differ anatomically from the others and perform different functions (Fig. 6.29D). C1 is highest and is called the **atlas,** after the Greek god of the same name who was thought to hold the world on his back and neck. On its superior surface are two articular facets that fit with the occipital condyles of the occipital bone. The resulting synovial (hinge) joint between the skull and vertebral column permits flexion, extension, and limited rotation and lateral movement.

The atlas has no body; instead, a thin *anterior arch* of bone is present, which loops around a vertical projection, the **dens** (or *odontoid process*), from the body of the C2 vertebra below. The C2 vertebra is called the **axis,** because the atlas rotates around the *axis* of the dens much as the earth rotates on its polar axis.

There Are 12 Thoracic and 5 Lumbar Vertebrae

There are 12 *thoracic vertebrae*, T1 to T12, which are larger and considerably stronger than their cervical counterparts above. Each has four **costal facets** that articulate with the ribs, one on the tip of each transverse process and one on either side of the body (Fig. 6.29E and F). The thoracic spine is not as flexible as the cervical spine above or the lumbar spine below because the ribs restrict its motion. Figure 6.29F shows articulated thoracic vertebrae.

There are five *lumbar vertebrae*, L1 to L5, which are the largest and heaviest of all (Fig. 6.29A and B). They bear much of the body's weight during activities such as bending, carrying loads, and even simply maintaining an erect seated or standing posture.

Case Note

6.28. Which processes on Maggie's fractured vertebra (T2) articulate with the ribs—the superior articular processes or the transverse processes?

Remember This! **Only the thoracic vertebrae articulate with the ribs.**

The Sacrum and Coccyx Are Fused

The **sacrum** is a single triangular bone formed from the gradual fusion of five fetal vertebrae, a process that begins in the teenage years and is complete by about age 30 (Fig. 6.30). It is a stout fixture to which the big muscles of the hips and thighs attach. Although it is a single bone, it has five regions, S1 to S5, one for each original sacral vertebra.

As with the other vertebrae, superior articular processes at the top of the sacrum articulate with the fifth lumbar vertebra. In its center is the hollow *sacral canal*, a continuation of the spinal canal. The inferior end of the sacrum is an opening, the *sacral hiatus*, which represents the end of the spinal canal. Spreading outward from its hollow center are two wings of bone, each of which is perforated by four foramina to allow passage of sacral spinal nerves S1 to S4 ➡ (Chapter 8). Sacral spinal nerve S5 exits through the sacral hiatus.

The **coccyx** (Greek *kokkux* = "cuckoo bird") is a small triangular bone shaped like a bird's beak (Fig. 6.30), which forms the inferior end of the axial skeleton. The original three to five fetal coccygeal bones may not fuse completely, so that the adult coccyx may occur as two to four separate bones. Apex downward, the coccyx is fused to the inferior end of the sacrum. It is, in effect, a human tailbone—an anatomical reminder of our common ancestry with tailed animals.

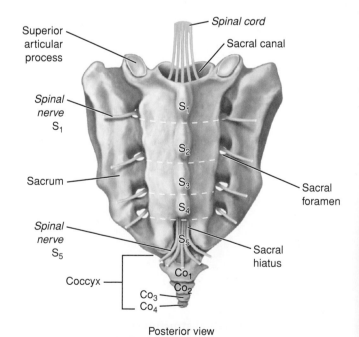

Figure 6.30. **Sacrum and coccyx.** The adult sacrum consists of five fused bones (S_1 to S_5); the coccyx is formed of four bones that may be only partially fused (Co_1 to Co_4). *Name the opening through which spinal nerves exit the sacrum.*

The Thorax

The word **thorax** (Greek *thoracos* = "breast plate") refers to the entire chest—the ribs and other chest bones plus the heart and lungs. The phrase **thoracic cage** (or bony thorax) is a part of the axial skeleton and refers to the ribs, sternum (breastbone), and thoracic vertebrae (Fig. 6.31).

Sternum

The **sternum** (Greek *sternon* = "breastbone") is a flat, slim, vertical bone situated in the center of the anterior thoracic wall. Each of its three parts was a separate bone in the fetus. From superior to inferior, they are the *manubrium,* the *body,* and the *xiphoid process* (Fig. 6.31). The **manubrium** is uppermost and is fused to the **body** at the **sternal angle,** an easily identifiable transverse ridge of bone that is very useful as an anatomical marker for medical examinations of the chest. The **xiphoid process** (Greek *xiphos* = "sword") is an inferior projection of the sternum.

The **clavicular notches** are hollows on either side of the upper end of the manubrium that articulate with the clavicle (collarbone) to form the *sternoclavicular joint* (not shown). Between the two clavicular notches is the **suprasternal (jugular) notch,** a V-shaped indentation in the upper end of the manubrium in the base of the anterior neck.

> *Remember This!* **The sternum is easily palpated and is an important medical landmark. You can feel the clavicular notches where the manubrium meets each of your clavicles and the V-shaped suprasternal notch above and between them. Superior to the suprasternal notch is the thyroid gland. About 4 cm inferior to this notch is the sternal angle, which you can feel as a small horizontal ridge. The second ribs joins the sternum at the sternal angle. This landmark represents the site beneath which the trachea splits into two bronchi.**

Because the sternum lies immediately beneath the skin and because it is a flat bone that usually contains red bone marrow, the sternum is a convenient site from which a bone marrow specimen can be obtained. The technique involves use of a short needle with a collar guard to prevent deep penetration into the heart or great blood vessels immediately beneath (Fig. 6.31B). Under local anesthesia, the needle is forced through the compact bone into the marrow cavity, and red marrow for study is aspirated in a syringe and spread onto a slide for microscopic study.

Finally, during cardiopulmonary resuscitation (CPR), a rescuer's hands must be kept on the inferior part of the sternal body and not on the xiphoid process. If the rescuer were to apply pressure over the xiphoid, it could fracture and be driven into the liver, which lies beneath it, possibly lacerating it (Fig. 6.31B).

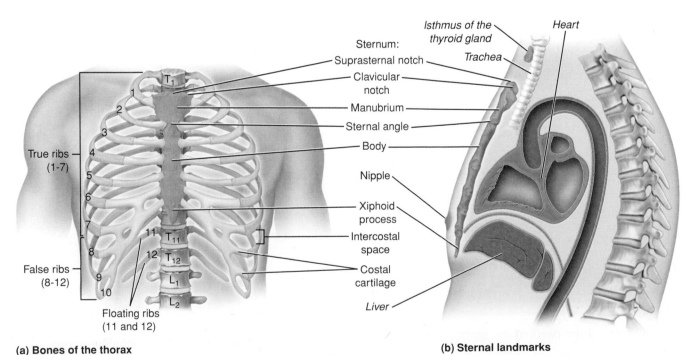

(a) Bones of the thorax

(b) Sternal landmarks

Figure 6.31. Thorax. A. Bones of the thorax. **B.** Sternal landmarks. *Name the most inferior part of the sternum.*

Ribs

Twelve pairs of **ribs** form the wall of the thoracic cage (see Fig. 6.31A). Contrary to Biblical myth, females do not have one fewer rib than males. Each rib attaches posteriorly to the transverse processes of the thoracic vertebrae. The resulting **costovertebral joints** (Greek *cost-* = "rib") are synovial gliding joints. Spaces between ribs are called **intercostal spaces.** By convention, the ribs are numbered from the most superior (rib 1) to the most inferior (rib 12). Rib 7 is the longest; others shorten progressively up and down the thoracic cage. They are described as follows:

- Ribs 1 through 7 are called the **true ribs** because they connect to the side of the sternum by individual cartilage rods, the **costal cartilages.**
- Ribs 8, 9, and 10 share a forked costal cartilage that merges with the costal cartilage of rib 7. They are sometimes called **false ribs** because they are not individually connected to the sternum.
- Ribs 11 and 12, called **floating ribs,** have no cartilage. They are also false ribs because they are not connected to the sternum.

 6.29 Is the mandible a cranial bone or a facial bone?

6.30 Which bone articulates with all the other cranial bones?

6.31 Name the suture formed where the two parietal bones meet the frontal bone.

6.32 Name the bones that contain teeth.

6.33 Name the four skull bones that contain sinuses.

6.34 Which vertebral curve is the last to develop after birth?

6.35 Name the three vertebral processes that attach to muscles.

6.36 Name the vertebra that articulates with the skull.

6.37 Which vertebrae are stronger—lumbar or cervical?

6.38 Name the opening at the inferior end of the sacrum.

6.39 Name the structure that separates the bodies of adjacent vertebrae.

6.40 Name the three parts of the sternum.

6.41 What connects true ribs to the sternum?

6.42 Name the joint connecting a rib to a vertebra.

The Anatomy of Bones and Joints: The Appendicular Skeleton

Having explored the axial skeleton—skull, vertebral column, and thoracic cage—it's time to turn to the appendicular skeleton.

The Pectoral (Shoulder) Girdle

Its most superior portions are the paired **pectoral (shoulder) girdles,** which attach the bones of the upper limbs to the axial skeleton. Each consists of two bones: the *clavicle* (collarbone) on the upper anterior chest, and the *scapula* (shoulder blade) on the upper posterior chest (Fig. 6.32A). Only the clavicle directly attaches,

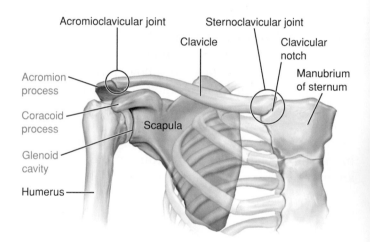

(a) Pectoral girdle, anterior view

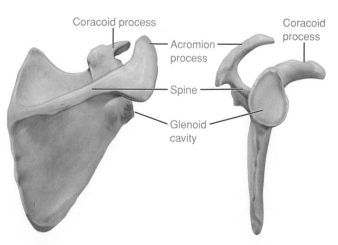

(b) The scapula, posterior view   **(c) The scapula, lateral view**

Figure 6.32. Bones of the pectoral (shoulder) girdle. A. An anterior view reveals that the pectoral girdle is created by the scapula and clavicle. **B.** Scapula, posterior view. **C.** Scapula, lateral view. *Which feature of the scapula articulates with the clavicle?*

bone-to-bone, to the thorax—it meets the sternum at its upper, lateral edge to form the sternoclavicular joint. The scapula directly connects to the thorax by muscles only.

The **clavicle** lies horizontally across the top of the chest and reaches from the clavicular notch of the sternum across the top of the shoulder, where it articulates with the superior aspect of the scapula (Fig. 6.32A). Heavy muscles anchor the pectoral girdle to the thorax, bracing it against outward-directed forces (such as someone yanking your arm) that might otherwise detach the pectoral girdle from the axial skeleton. The clavicle braces the shoulder anteriorly, preventing the strong chest muscles from pulling the shoulder towards the sternum. Many fractures of the clavicle result from a fall onto an outstretched arm, because the force of the fall is transmitted up the arm to a weak spot at the lateral end of the clavicle.

The **scapula** is a large triangular bone that lies flat along the superoposterior aspect of the thorax. It consists of a flat body and a dorsal ridge of bone, the **scapular spine** (Fig. 6.32B), which ends in a bony prominence, the **acromion process.** The acromion forms a bony bulge at the top of the shoulder, where it articulates with the clavicle to form the **acromioclavicular joint.** The nearby **coracoid process** (Greek *korax* = "crow," because it is shaped like a crow's beak) serves as an attachment point for certain chest and shoulder muscles. On the lateral face of the scapula is the **glenoid cavity** (or *glenoid fossa*), a shallow crater where the bone of the arm, the *humerus*, meets the scapula to form the shoulder joint (see below).

Case Note

6.29. Maggie braced herself with her hand when she fell. Which bone is frequently broken in this situation, because it conveys the force of the fall to the axial skeleton?

The Upper Limb

The **upper limb** consists of the **arm** (the limb above the elbow), the **forearm** (the limb between elbow and wrist), and the **hand,** which consists of the *wrist*, the *palm*, and the *fingers* (Fig. 6.33).

The Arm Bone Is the Humerus

The **humerus** is the longest and heaviest of the bones in the upper limb (Fig. 6.34). At the proximal end of the humerus is a bulbous *head*, which nestles into the

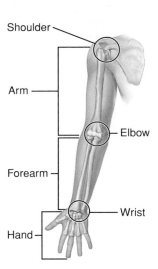

Figure 6.33. Upper limb. The upper limb extends from the shoulder to the fingertips. *What is the term used to describe the upper limb between the elbow and wrist?*

shallow cup of the glenoid cavity of the scapula. Circling the head is a shallow groove, the **anatomical neck,** which is the site of the former epiphyseal plate. Partway down the shaft is a small prominence, the **deltoid tuberosity,** where the large *deltoid muscle* overlying the shoulder joint attaches.

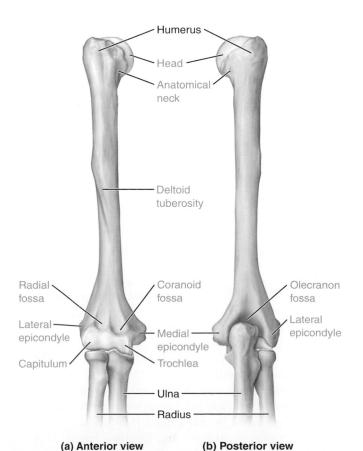

(a) Anterior view   **(b) Posterior view**

Figure 6.34. Humerus. A. Anterior view. **B.** Posterior view. *Name the ridge on the humeral diaphysis.*

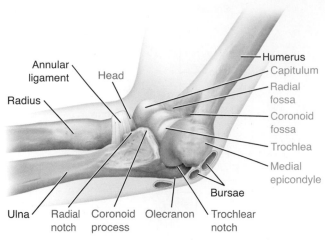

Figure 6.35. Elbow joint, medial view. The right elbow is illustrated in the flexed position. *Name the ligament that wraps around the radius.*

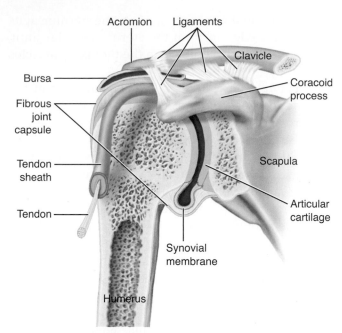

Figure 6.36. Right shoulder joint, cross section (anterior view). *Which structure is continuous with the synovial membrane—the tendon sheath or the ligaments?*

At the distal end of the humerus are two processes that articulate with the forearm bones to form the elbow joint. The lateral **capitulum** articulates with the lateral forearm bone (the *radius*); the medial **trochlea** articulates with the medial forearm bone (the *ulna*). Superior to the capitulum and trochlea are the lateral and medial **epicondyles,** respectively, where tendons for forearm muscles attach. The tendons of the posterior forearm muscles attach to the lateral epicondyle. Overuse of these muscles, such as hyperextension of the wrist as it readies the arm for a tennis stroke, can result in inflammation and pain in the tendons attached to the lateral epicondyle, a condition called "tennis elbow."

Immediately superior to the trochlea, on the anterior side, is a depression, the **radial fossa,** which makes room for a projection of the radius as the elbow joint is flexed (Fig. 6.35). A second depression superior to the capitulum, the **coracoid fossa,** receives the ulna when the elbow joint is flexed. Similarly, the posterior side contains a depression, the **olecranon fossa,** which makes room for a projection of the ulna as the elbow is extended.

> *Remember This!* **Find some of these bone features on your own body. Hold out your extended right forearm, palm upward. Place your left hand under your elbow and grip with your thumb and pointer finger. Now, flex the forearm. You should be able to feel the medial epicondyle of the humerus under your thumb, and the lateral epicondyle under your finger. The ulnar nerve passes over the medial epicondyle. Blows to this region produce an uncomfortable numbness and tingling; hence, it is called the "funny bone."**

The Humerus and Scapula Form the Shoulder Joint

The head of the humerus and the glenoid cavity of the scapula articulate to form the ball-and-socket shoulder joint (Fig. 6.36). The shallowness of the glenoid cavity contributes to the shoulder joint's wide range of motion, which is much greater than that of the hip joint, a much deeper ball-and-socket joint (discussed below). Its status as the most frequently dislocated joint highlights the trade-off between joint stability and range of motion.

A tough fibrous capsule envelops the entire joint, but the thick sheath of surrounding muscles plays the most important role in the stabilization of the shoulder. The fibrous capsule is continuous with some accessory ligaments (not shown in Fig. 6.36) and muscle tendons (only one is shown). The synovial membrane not only lines the joint cavity but also extends to form a sheath around the tendon that attaches the biceps muscle (remember that tendon sheaths are modified bursae). Other bursae reduce friction from the shoulder muscles, humerus, clavicle, and scapula rubbing against each other.

> *Remember This!* **Ligaments and tendons are often extensions of the joint fibrous capsule. Bursae and tendon sheaths are often extensions of the synovial membrane.**

The Forearm Contains the Ulna and Radius

The forearm contains two bones: the *ulna* and the *radius*. Recall that in standard anatomical position the palm of the hand faces anteriorly with the thumb lateral. In this position the *radius* is lateral (on the thumb side) and the *ulna* is medial (Fig. 6.37A). However, when the forearm is pronated so that the palm faces posteriorly and the thumb is medial, the radius crosses over the ulna to form an elongated X. The radius and ulna are connected and stabilized by a sheet of connective tissue, the *interosseous membrane*, forming a syndesmosis.

The **ulna** is longer than the radius and overlaps the humerus when the elbow is extended. The proximal end of the ulna is squared off posteriorly to form the *olecranon*, the bony prominence under the skin of the elbow, to which muscles of the posterior arm attach (Fig. 6.37C). The anterior arm muscles attach to a similar process on the other side of the ulna, the **coronoid process.** The C-shaped articular surface immediately distal to the

olecranon, the **trochlear notch,** articulates with the humerus to form the elbow joint (Figs. 6.37B and C). The **radial notch,** conversely, articulates with the radius to form the proximal **radioulnar joint** (Fig. 6.37A). The **annular ligament** extends from the ulna and wraps securely around the radial head, maintaining it snugly within the radial notch of the ulna and permitting stable rotation of the radius during pronation and supination of the forearm. At the distal end of the forearm, a second radioulnar joint exists, where the ulna and radius meet again. At the distal end of the ulna is the round, flat **head,** which articulates with wrist bones (Fig. 6.37B). On the medial side of the head is the **ulnar styloid process,** to which medial wrist ligaments attach.

The **radius** is oriented in the opposite direction to the ulna, with the **radial head** articulating with the humerus (Fig. 6.37B). The radial head also articulates with the radial notch of the ulna to form the proximal radioulnar joint. Beneath the head on the anterior surface is a roughened, slightly raised area, the **radial tuberosity,**

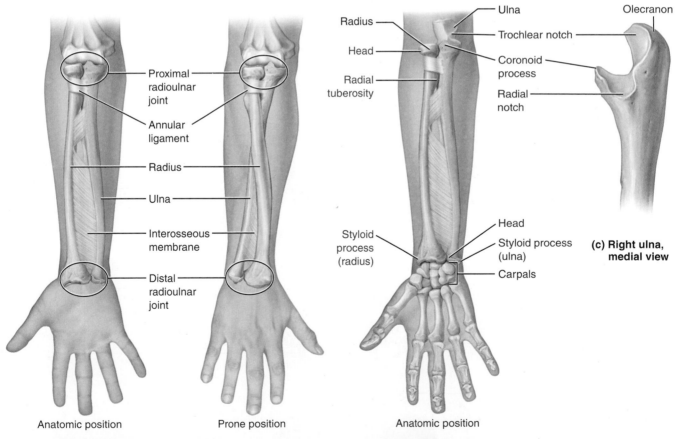

(a) Movements at the right radioulnar joints

(b) Right radius and ulna, anterior view

(c) Right ulna, medial view

Figure 6.37. Bones of the forearm. The forearm consists of the radius and ulna. **A.** The radius and ulna are uncrossed in anatomical position, but they cross over when the arm is pronated. **B.** Right radius and ulna, anterior view in the anatomical position. **C.** Right ulna, medial view. *Which bone has the coronoid process?*

which marks the site of attachment of the biceps muscle of the arm.

The distal end of the radius articulates with the small bones of the wrist. On the lateral side of the distal end is the **styloid process** of the radius, to which lateral wrist ligaments attach.

Case Note

6.30. Maggie's arm is in a sling, with the palm facing upward and the arm tight against her abdomen. Where is her injured radius—closest to her abdomen or on the opposite side of her arm?

The Humerus, Radius, and Ulna Form the Elbow Joint

Figure 6.35 illustrates how the elbow joint is formed by articulations between the trochlea of the humerus and trochlear notch of the ulna, as well as between the head of the radius and the capitulum of the humerus. The elbow is a hinge joint capable only of flexion and extension. Ligaments not shown in Figure 6.35 restrict all other movements.

Multiple bursae surround the elbow. The *olecranon bursa* lies over the tip of the olecranon and can become inflamed by direct trauma or by overuse of the joint, resulting in another form of *tennis elbow* (from repeated flexion and extension of the joint) or *student's elbow* (from prolonged leaning on the elbow while studying or writing).

The Hand Consists of the Wrist, Palm, and Fingers

The **hand** consists of the wrist, the palm, and the fingers (Fig. 6.38). The skeleton of the **wrist** is composed of two irregular rows of four bones each, the **carpal bones** (or *carpals*), which are bound firmly to one another by

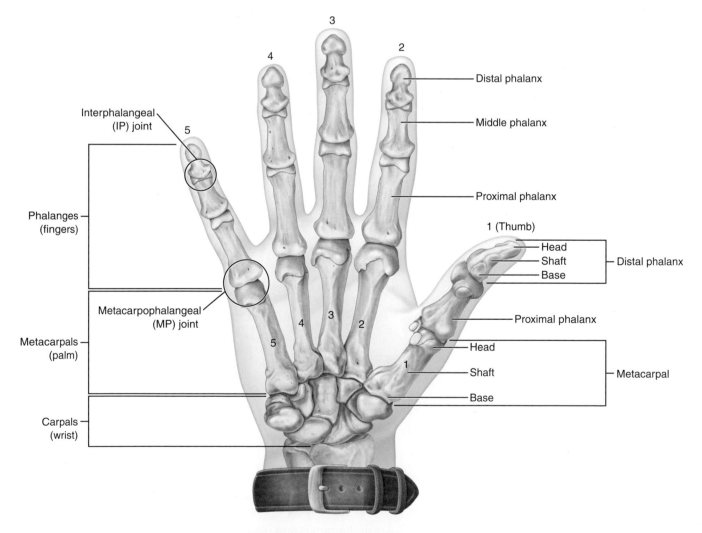

Right hand, palmar view

Figure 6.38. Hand. *How many phalanges are found in the thumb?*

ligaments that allow for slight shifting and twisting. The carpals are located in the heel of the hand—a "wrist"-watch is actually worn around the distal ends of the radius and ulna. Thus, Maggie's broken "wrist" reflected a fracture in the head of her radius, which articulates with the carpal bones to form the wrist joint. She did not actually fracture her wrist (carpal) bones. The bones of the palm are **metacarpals,** which are numbered 1 to 5, beginning with the metacarpal that links the wrist to the thumb. All metacarpals are similar: each is a long bone with central *shaft,* a proximal *base,* and a distal *head.* The metacarpal heads interact with the fingers at **metacarpophalangeal joints,** usually abbreviated to *MP joints.* The knuckles in a clinched fist are the metacarpal heads of these joints.

The fingers are numbered 1 to 5 beginning with the thumb. The bones in the fingers are **phalanges** (*phalanx* is singular). Each phalanx is a long bone with a proximal *base,* a distal *head,* and an intervening *shaft.* The thumb has two phalanges, proximal and distal, but all other fingers have three: proximal, middle, and distal. The joints between phalanges are the **interphalangeal (IP) joints.**

Case Note

6.31. Maggie landed on the heel of her hand. Which bones make up this region?

The Pelvic (Hip) Girdle

The **pelvic girdle,** to which the lower limbs attach, consists of a pair of **coxal bones** (*hip bones* or *coxae*). The coxal bones articulate anteriorly with each other and posteriorly with the sacrum to form the basinlike **bony pelvis** (Fig. 6.39).

The bony pelvis is the keystone structure of the skeleton—it provides a stable foundation for movements of both the upper body and the lower body. It also protects the pelvic organs.

Remember This! **Observe that the term** *pelvic girdle* **includes only the coxal bones, but that** *bony pelvis* **includes the sacrum.**

The Pelvic Girdle Includes the Ilium, the Ischium, and the Pubic Bone

At about age 16 to 18 years, each of the coxal bones is formed by the fusion of three bones: the *ilium,* the *ischium,* and the *pubic bone* (or *pubis*). Parts of each of the three original bones contribute to the **acetabulum,** a socket into which the bone of the thigh (the femur) fits to join the lower limb to the pelvis. A bony ridge, the **pelvic brim,** runs continuously across the sacrum, the two ilia, and the pubic bones (Fig. 6.39A). The significance of this important anatomical landmark is discussed below.

The **ilium** is largest and uppermost of the three. Each ilium is a wing-shaped bone that attaches firmly to the sacrum at a **sacroiliac joint** (Fig. 6.39A). The upper edge of each wing is the **iliac crest,** where your hands naturally rest from time to time. The iliac crest begins posteriorly as a bony prominence, the **posterosuperior iliac spine,** where the ilium and sacrum meet, and ends anteriorly at a second bony bulge, the **anterosuperior iliac spine** (Fig. 6.39B). Each serves as the attachment for muscle tendons and is an important anatomical landmark for various medical procedures.

The arc-shaped **ischium** forms the inferoposterior portion of each coxal bone. It has a large inferior bulge, the **ischial tuberosity,** to which large muscles and ligaments attach, and which bears our body weight when we are seated. As shown in Figure 6.39B, the anterior **pubic bone** is roughly the mirror image of the ischium. On each side, the ischium and pubis meet superiorly and inferiorly to create a large hole in each coxal bone, the **obturator foramen,** through which blood vessels and nerves pass to reach the anterior thigh. The two pubic bones meet anteriorly at the **symphysis pubis** to form a midline bony prominence above the genitalia. The usually immovable symphysis pubis is classified as an amphiarthrosis, because the connective tissue forming the joint relaxes in response to pregnancy hormones. As a result, the pubic bones can separate slightly to ease the baby's passage through the pelvis.

Viewed from above, the pelvic girdle has the appearance of a "turned up" shirt collar. The snug edge of the collar is the pelvic brim and the wings of the ilia form the turned up portion of the collar (Fig. 6.39A). The collar "buttons" at the symphysis pubis.

Case Note

6.32. The bruise on Maggie's left hip was just inferior to the waistband of her running shorts. Did she bruise her ilium, pubis, or ischium?

The True Pelvis Is Inferior to the Pelvic Brim

The pelvic brim is an important anatomical landmark. Recall from ◀ Chapter 1 (Fig. 1.13) that the upper edge of the pelvic girdle divides the abdominal and pelvic cavities. However, further anatomical differentiation

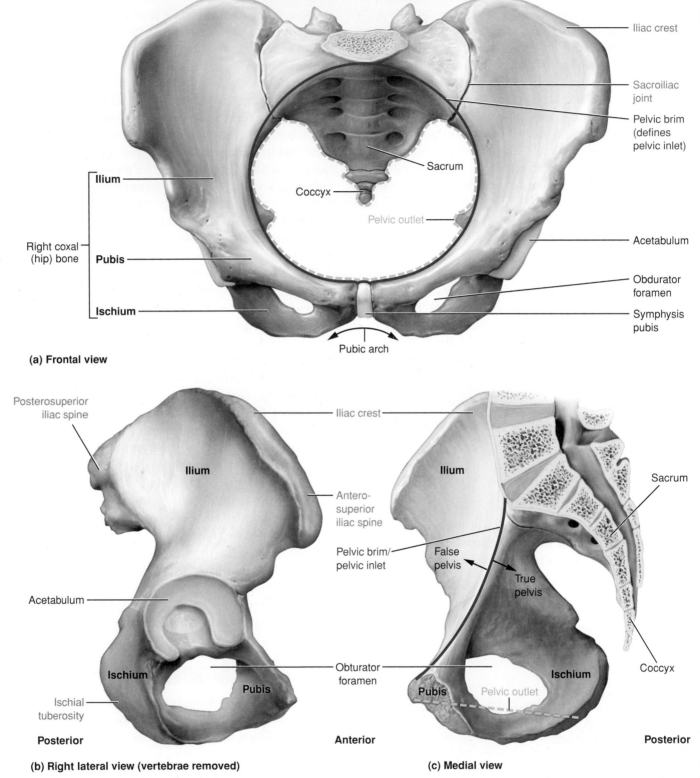

(a) Frontal view

Iliac crest

Sacroiliac joint

Pelvic brim (defines pelvic inlet)

Sacrum

Coccyx

Pelvic outlet

Acetabulum

Obdurator foramen

Symphysis pubis

Pubic arch

Ilium

Right coxal (hip) bone

Pubis

Ischium

(b) Right lateral view (vertebrae removed)

Posterosuperior iliac spine

Ilium

Acetabulum

Ischium

Ischial tuberosity

Posterior

Pubis

(c) Medial view

Iliac crest

Antero-superior iliac spine

Pelvic brim/ pelvic inlet

Obturator foramen

Anterior

Ilium

False pelvis

True pelvis

Sacrum

Coccyx

Ischium

Pubis

Pelvic outlet

Posterior

Figure 6.39. Bones of the pelvic girdle. The pelvic girdle consists of two coxae fused together. Each coxa consists of one ilium, one ischium, and one pubic bone fused together. **A.** Frontal view. **B.** Right lateral view. **C.** Medial view. *Which bone is not part of the pelvic brim?*

is important. The part of the pelvis above the pelvic brim but below the superior edges of the ilial wings is properly part of the abdomen, it is, therefore, called the **false pelvis** (Fig. 6.39C). The **true pelvis** lies below the pelvic brim and shelters the uterus and other internal reproductive organs and the urinary bladder.

The entry to the true pelvis is the **pelvic inlet,** the opening encircled by the pelvic brim (Fig. 6.39A and C). A companion opening, the **pelvic outlet,** is formed below by the inferior margins of the sacrum and coccyx and the coxal bones. The urinary and gastrointestinal tracts enter the true pelvis through the pelvic inlet and exit through the pelvic outlet to the genitalia and anus. The fetus, too, must pass through the pelvic inlet and outlet at birth. Note in Figure 6.39C that this pathway is not straight—the inlet is much closer to vertical than the outlet—so the fetus must bend in order to pass.

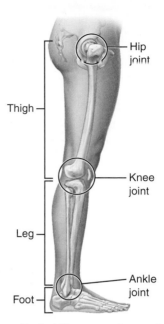

Figure 6.40. Lower limb. *What term is used to describe the region between the hip and the knee?*

> *Remember This!* The anterior iliac spine (the "hip bones") can be felt by placing your hands on your hips. The posterior iliac spine in many people is marked by skin dimples just above the buttocks. A line drawn between the uppermost points of the right and left iliac crests marks the intervertebral space between lumbar vertebrae 4 and 5; it is often used as a landmark for a procedure called *lumbar puncture*, in which a needle is inserted into the lower back between vertebrae and into the spinal canal to obtain spinal fluid for diagnostic testing.

The Lower Limb

The **lower limb** consists of the **thigh** (the limb above the knee), the **leg** (the limb between knee and ankle), and the **foot** (Fig. 6.40).

The lower limb bears full body weight and the considerable forces associated with standing, walking, running, and climbing, so it should be no surprise that the bones of the lower limb are heavier, thicker, and stronger than those of the upper limb.

The Thigh Bone Is the Femur

The bone of the thigh is the **femur,** the longest and heaviest bone in the body (Fig. 6.41). The proximal end of the femur is a stubby post, the **neck,** which is angled about 45 degrees and is capped by a spherical **head,** which articulates with the acetabulum of the coxal bones to form the hip joint (discussed below). The neck is the weakest part of the femur; it is a common site of fracture in elderly people, an injury frequently called a "broken hip."

The **femoral shaft** extends downward to the knee. It angles and bends inward slightly to bring the knee joints closer to the midline than the hip joints. In females, the angle is slightly greater because the female pelvis is slightly broader. Where the upper end of the shaft joins the neck are two projections, the lateral greater trochanter and the medial lesser trochanter, which form attachment points for muscles and ligaments.

The distal end of the femur expands into two knuckle-like projections, the **medial and lateral condyles** (Fig. 6.41B). The knuckled ends fit onto the slightly concave upper surface of the tibia to form the knee joint. Posteriorly between the condyles is a deep depression, the **intercondylar fossa.** Anteriorly is a slightly depressed smooth area, the patellar surface, which accommodates a bone embedded in tendon, the *patella* or kneecap (Fig. 6.41A).

> *Remember This!* To feel the greater trochanter of the femur, stand on one leg and allow the other leg to passively adduct. This bone feature can be felt in the lateral surface of the hip.

The Femur and Coxa Form the Hip Joint

The hip joint is formed by the head of the femur articulating with the acetabulum of the coxal bone (Fig. 6.42). It provides the best example of an articular capsule, because it completely envelops the hip joint and is very strong. The articular capsule is reinforced by accessory

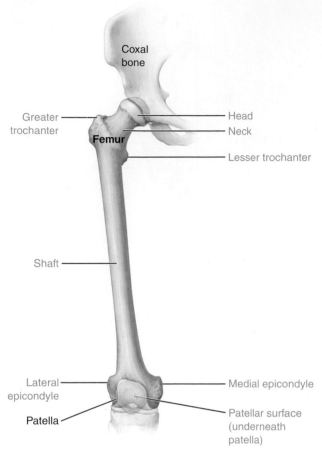

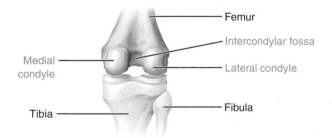

(a) Right femur, anterior view

(b) Right knee, posterior view

Figure 6.41. Femur. A. Right femur, anterior view. **B.** Distal portion of the right femur, posterior view. *Is the medial condyle of the femur found in the knee area or in the hip area?*

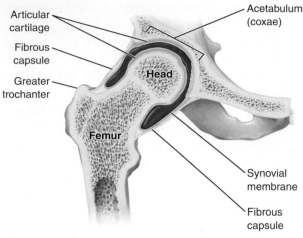

(a) Sectional view

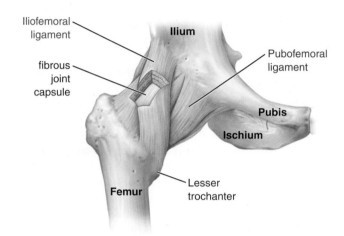

(b) Anterior view

Figure 6.42. Right hip joint. A. Sectional view. **B.** Anterior view. *What is the name of the socket into which the ball of the femoral head fits?*

The Leg Contains the Tibia and Fibula

The leg contains two bones, the medial *tibia* or shin bone, which is weight-bearing, and the lateral *fibula*, which stabilizes the ankle joint (Fig. 6.43). The two bones are connected along their length by a sheet of connective tissue, the *interosseous membrane*, which forms a syndesmosis.

The proximal end of the **tibia** expands to form two articular surfaces, the **medial** and **lateral condyles,** upon which the femoral condyles rest. Anteriorly, between the condyles is a nubbin of bone, the **tibial tuberosity,** which serves as a point of attachment for the patellar tendon (discussed below).

The distal end of the tibia articulates with the *talus*, the most superior of the ankle bones. Medially a bulging lip of tibia overlaps the ankle joint to form the medial

ligaments that originate from all three regions of the coxa (Fig. 6.42B). The joint is further stabilized by strong muscles that surround it.

The ball-and-socket construction of the hip joint grants it a wide variety of motions: flexion, extension, adduction, abduction, and rotation (see Fig. 6.16). Nevertheless, its full range of motion is narrower than that of the less stable shoulder joint.

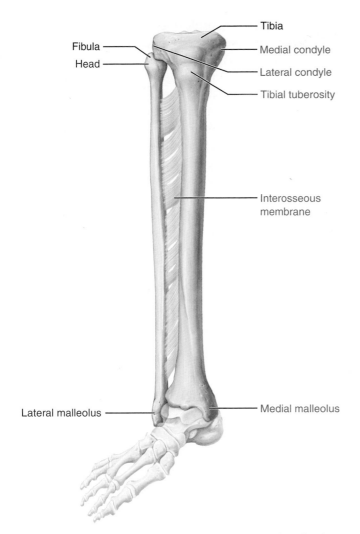

Figure 6.43. Bones of the leg. *Where is the lateral malleolus— on the fibula or on the tibia?*

prominence of the ankle, the **medial malleolus** (Latin *malleus* = "hammer").

The **fibula** is a slim bone, about as long as the tibia, but shifted somewhat inferiorly. It expands into an upper **head,** which articulates with the tibia below the lateral condyle and does not form part of the knee joint. Distally, however, the fibula extends below the inferior end of the tibia and expands to form an articular surface that connects to the tarsus. Laterally, a small bulge of fibula overlaps the ankle joint to form the lateral bony prominence of the ankle, the **lateral malleolus.**

Case Note

6.33. Maggie's injury was located just above the medial bony prominence of the ankle. Which bone forms this prominence, and what is its anatomical name?

The Patella Is the Kneecap

The **patella** (kneecap) is a sesamoid bone that develops within the **patellar tendon** connecting the quadriceps muscles to the tibia (Fig. 6.44A). In practice, the portion of this tendon between the patella and the tibia is called

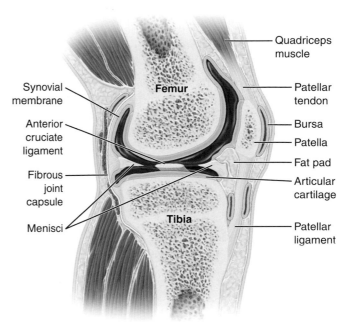

(a) Sagittal section

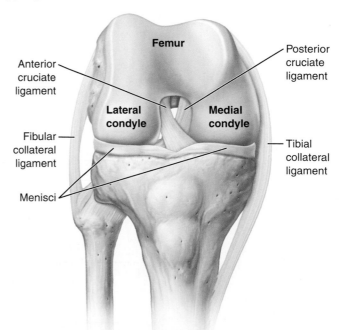

(b) Flexed right knee, anterior view

Figure 6.44. Knee joint. A. Sagittal section. **B.** Flexed right knee, anterior view. The patella and other anterior structures have been removed. *Which two ligaments are found within the synovial cavity?*

the *patellar ligament*, since tendons attach muscles to bones but ligaments attach bones to bones. As the knee joint is bent back and forth, the patellar tendon and patella glide up and down over the articulating surface of the tibia between the anterior ends of the condyles.

The Femur Articulates with the Tibia at the Knee Joint

The **knee joint** formed between the femur and tibia is subjected to great force—the whole of the standing body weight rests on it and large forces play upon it constantly during upright movements. Despite strong muscles above and below, the knee joint is relatively unprotected by muscles and is frequently injured. Important aspects of the joint are as follows:

- The fibrous joint capsule does not completely surround the joint, but it is supplemented by the thick tendons of heavy muscles (Fig. 6.44A). Among the most important of these is the tendon for the quadriceps muscle (which continues downward as the patellar ligament), which reinforces the anterior aspect of the joint.
- Fat pads and 13 bursae are scattered around the knee to reduce friction between the tissues that slide across one another during knee movements. Some bursae are continuations of the joint cavity.
- Six strong ligaments cross the joint, binding the femur to the tibia and fibula. The **cruciate ligaments** (cross-like, Latin *crux* = "cross") cross one another in the joint space and prevent the femoral and tibial condyles from sliding anteriorly or posteriorly across one another (Fig. 6.44B). The **tibial collateral** and **fibular collateral**

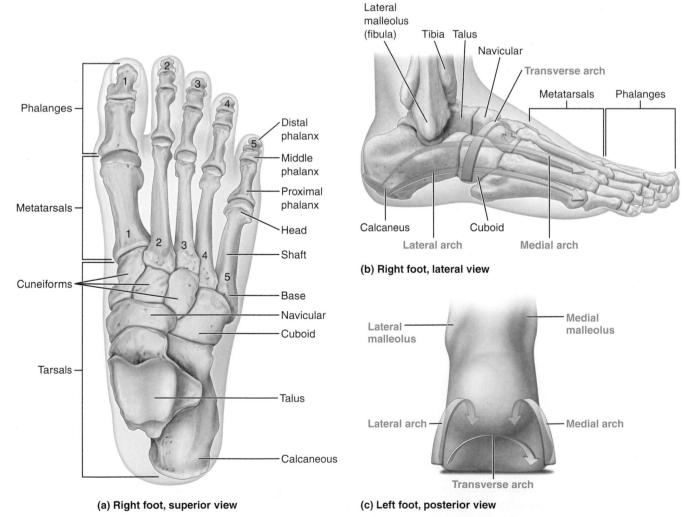

(a) Right foot, superior view

(b) Right foot, lateral view

(c) Left foot, posterior view

Figure 6.45. Foot. A. Right foot, superior view. **B.** Right foot, lateral view. **C.** Left foot, posterior view. Parts B and C illustrate the arches of the foot, which help to distribute body weight when standing upright. *Name the tarsal bone that articulates with the fifth metatarsal.*

ligaments also prevent a mediolateral rocking motion. Two **popliteal ligaments** (not shown) cross in the back of the knee to prevent mediolateral sliding.

● Positioned on each side of the tibial plateau is a **meniscus,** a pad of cartilage to cup the femoral condyles for smooth articulation.

The organization of the ligaments restricts the range of motion of the knee joint to flexion and extension. It is, therefore, a hinge joint.

The Foot Contains Tarsals, Metatarsals, and Phalanges

The bones in the **foot** consist of seven *tarsal bones* (*ankle bones*), five elongated *metatarsal bones,* connect the ankle bones (bones of the instep) to the toes, and the bones of the toes, the *phalanges* (Fig. 6.45).

The region of the lower limb commonly considered the "ankle" is actually the distal ends of the tibia and fibula, which bulge medially and laterally as the medial and lateral malleoli. The **tarsals** are bound to one another by ligaments to form an arch, a feature that allows the foot to bend somewhat to cushion the forces of walking, running, and jumping. The largest and strongest bone of the tarsals is the **calcaneus,** or heel bone. The uppermost tarsal bone is the **talus,** which articulates with the tibia and fibula to form the ankle joint. The remaining five bones are small and form a cluster that articulates with the metatarsals.

The **metatarsals** are numbered 1 to 5, beginning with the metacarpal that links with the first toe. All metatarsals are similar (except that the first metatarsal is substantially heavier than the others): each is a long bone with central *shaft*, a proximal *base,* and a distal *head*.

Like the metatarsals, the toes are numbered 1 to 5, beginning with the great toe. The bones in the toes are **phalanges** and resemble those in the hand. Each set consists of three bones except in the great toe, which has only two: they are larger and heavier than the others. As

Table 6.4 Differences Between the Male and Female Skeleton

Characteristic	Male	Female
Overall characteristics	Heavier, thicker bones	Lighter, thinner bones
	Larger muscle attachment sites	Smaller muscle attachment sites
	Larger joint surfaces	Smaller joint surfaces
Skull	Larger facial area	Smaller facial area
	More pronounced processes	More subtle processes
	Relatively larger maxillae and mandible	Relatively smaller maxillae and mandible
Pelvis		

Male pelvis — Narrower; Smaller, rounder pelvic inlet and outlet; Less moveable coccyx; Lateral facing acetabulum; Round obturator foramen; Narrow pubic arch

Female pelvis — Wider; Larger, wider pelvic inlet and outlet; More moveable coccyx; Anterior facing acetabulum; Oval obturator foramen; Wide pubic arch

| Coccyx | Less movable | More movable |

with the phalanges of the hand, each consists of a proximal *base*, a distal *head*, and an intervening *shaft*.

The skeleton of the foot is arranged into three arches: two that run lengthwise and one transverse (Fig. 6.45B and C). The medial lengthwise arch extends from the heel to the base of the great toe (the medial "ball" of the foot); the lateral one extends from the heel to the base of the fifth toe (the lateral ball of the foot). The transverse arch is composed of five small tarsals that fit against the proximal ends of the metatarsals.

Male and Female Skeletons Differ

In keeping with the (usually) heavier male body, the bones of males are generally larger and heavier than those of females, the articular ends are larger compared to the size of the shaft, and many of the markings—ridges, tuberosities, and lines—are more prominent because they are the attachment points for heavier muscles.

In keeping with their different reproductive roles, male and female pelvic structure is different. The female pelvis is broader and shallower than that of the male, and the female true pelvis is roomier: the pelvic inlet and outlet are larger to accommodate the passage of the head of the fetus during childbirth.

Table 6.4 identifies other points of comparison between the male and female skeleton. Overall, these differences enable archaeologists to identify quite readily the gender of skeletal remains.

6.43 Which part of the scapula can be felt as a bony bulge at the top of the shoulder?

6.44 Which bone contains the trochlea and capitulum?

6.45 Which feature of the scapula articulates with the humerus?

6.46 Where is the radial notch located—on the radius or on the ulna?

6.47 Which part of the humerus articulates with the ulna?

6.48 Which bones articulate to form the metacarpophalangeal joints?

6.49 Name the joint between the two pubic bones.

6.50 Name the features of the coxa and the femur that form the hip joint.

6.51 Complete this sentence: The humerus is to the upper limb as the _____ is to the lower limb.

6.52 In a patient with osteoarthritis in the knee joint, name the bones that are rubbing against each other to produce pain.

6.53 Which tarsal bone articulates with the tibia and fibula?

6.54 Whose pelvis would be shallower—Jack's or Maria's?

Word Parts

Latin/Greek Word Parts	English Equivalents	Examples
Ab-	Away from	Abduction: movement away from the midline
Ad-	Towards (adds to)	Adduction: movement towards the midline (adds the limb to the body)
Arthr-/o-, articul-/o-	Joint	Arthrosis: a joint
-blast	New growth	Osteoblast: cell that builds new bone
Carp-/o-	Wrist	Carpals: wrist bones
Chondr-/o-	Cartilage	Chondrocyte: cartilage cell
Circum-	Around	Circumduction: movement of limb around its origin
-clast	Break, fragment	Osteoclast: cell that breaks down bone

(continued)

Word Parts (continued)

Latin/Greek Word Parts	English Equivalents	Examples
Cost-/o-	Rib	Costal cartilage: rib cartilage
Epi-	Upon, over	Epicondyle: a projection over (above) a condyle
Os-/oss-	Bone	Ossification: process of becoming bone
Oste-/o-	Bone	Osteocyte: bone cell
Syn-	Together	Synarthrosis: joint close together, immovable joint
Synov-/i-	Synovial	Synovial joint: note that this term uses the root "synov/i", not the prefix "syn"
Tars-/o-	Instep	Tarsal bones: bones of the instep (foot arch)

Chapter Challenge

CHAPTER RECALL

1. Spongy bone is
 a. organized into osteons.
 b. organized into lamellae.
 c. found only in flat and short bones.
 d. all of the above.

2. The endosteum
 a. covers the outer surface of the bone.
 b. contains osteocytes.
 c. contains osteoblasts.
 d. is composed primarily of fat.

3. Place the events of endochondral ossification in the correct order.
  I. The epiphyseal line forms.
  II. Spongy bone forms.
  III. Cartilage forms.
  IV. Compact bone forms.

 a. I, II, IV, III
 b. III, II, IV, I
 c. III, IV, I, II
 d. II, IV, I, III

4. Intramembranous ossification
 a. can occur in adults.
 b. never happens in flat bones.
 c. never happens in long bones.
 d. involves the formation of a cartilage bone model.

5. Parathyroid hormone
 a. inhibits bone breakdown.
 b. is released by the thyroid gland.
 c. inhibits vitamin D activation.
 d. stimulates calcium retention by the kidneys.

6. A freely moveable joint is called a:
 a. diarthrosis
 b. synarthrosis
 c. amphiarthrosis
 d. syndesmosis

7. A joint that only permits flexion and extension is described as
 a. ball-and-socket.
 b. condyloid.
 c. hinge.
 d. pivot.

8. **The cheekbone is formed by the**
   a. frontal bone.
   b. maxillary bone.
   c. zygomatic bone.
   d. vomer bone.

9. **Which suture is formed by the line between the anterior and posterior fontanels?**
   a. Coronal suture
   b. Squamous suture
   c. Lambdoid suture
   d. Sagittal suture

10. **The knobby bone projections that can be felt in the middle of the back are called the**
    a. articular processes.
    b. spinous processes.
    c. pedicles.
    d. transverse processes.

11. **The intervertebral discs**
    a. are amphiarthroses.
    b. join together the first two cervical vertebrae.
    c. are hinge joints.
    d. are synovial joints.

12. **The most superior portion of the sternum is the**
    a. xiphoid process.
    b. sternal angle.
    c. body.
    d. manubrium.

13. **The pectoral girdle consists of the**
    a. sternum and clavicle.
    b. scapula and clavicle.
    c. humerus, scapula, and clavicle.
    d. humerus, sternum, and clavicle.

14. **Forearm tendons attach to a part of the humerus called the**
    a. trochlea.
    b. capitulum.
    c. epicondyles.
    d. coronoid fossa.

15. **The most important structure(s) that stabilize the shoulder joint is (are) the**
    a. articular capsules.
    b. tendon sheaths.
    c. bursae.
    d. sheaths of surrounding muscles.

16. **The proximal and distal radioulnar joints are**
    a. pivot joints.
    b. condyloid joints.
    c. cartilaginous joints.
    d. saddle joints.

17. **The true wrist is composed of the**
    a. tarsal bones.
    b. carpal bones.
    c. metacarpal bones.
    d. metatarsal bones.

18. **The obturator foramen of the coxal bones**
    a. forms a large socket for the femur.
    b. outlines the pelvic inlet.
    c. outlines the pelvic outlet.
    d. is a large hole formed by the ischial and pubic bones.

19. **The femur articulates with the tibia at the**
    a. greater trochanter.
    b. femoral head.
    c. lateral and medial condyles.
    d. trochlea.

20. **The lateral malleolus of the ankle is formed by the**
    a. calcaneus.
    b. fibula.
    c. tibia.
    d. talus.

21. **The cruciate ligaments**
    a. ensure that the knee functions as a saddle joint.
    b. are found within the articular capsule.
    c. help form the articular capsule.
    d. bind the patella to the femur, tibia, and fibula.

22. **Which of the following is a tarsal bone?**
    a. Calcaneus
    b. Hallux
    c. Phalanx
    d. None of the above

CONCEPTUAL UNDERSTANDING

23. **Compare and contrast with each other each of the bone markings in the following pairs:**
    a. Spine and line
    b. Condyle and head
    c. Foramen and sinus

24. **Compare the following terms that describe the pelvis: pelvic brim, pelvic inlet, pelvic outlet, true pelvis, false pelvis.**

APPLICATION

25. **List in order the movements (e.g., abduction) that must occur in order to accomplish the following actions:**
    a. A soccer player brings her leg far behind her and then kicks the ball, bringing her leg in front of her.
    b. A birdwatcher hears a bird sing and turns his head to the right in the direction of the sound.
    c. A gardener scoops up some loose soil in his hand, then pats it down next to a seedling.
    d. A child lying down in the snow raises her arms to make a "snow angel" and then brings them down to her sides again.

26. **Answer the following questions relating to the hip joint.**
    a. Classify the hip joint in terms of the degree of movement permitted.
    b. Classify the hip joint based on the types of movement permitted.
    c. Classify the hip joint in terms of the material between the adjoining bones.
    d. Identify the bones—and the specific features of these bones—that articulate within the capsule of the hip joint.
    e. List the types of movements that can occur at the hip joint.
    f. List the features of the hip joint that help to stabilize it.

27. **As a forensic scientist, you are evaluating a newly excavated skeleton from an archeological site. You note the following features of the bones: round obturator foramen, narrow pelvic inlet, fused coxal bones.**
    a. Is the skeleton male or female? Explain why.
    b. Is this the skeleton of an adult or a child? Explain why.

You can find the answers to these questions on the student Web site at
http://thepoint.lww.com/McConnellandHull

7

Muscles

Major Themes

- Muscle cells shorten on command; no other cells do.

- There are three types of muscle cells: skeletal, cardiac, and smooth.

- Skeletal muscle contracts voluntarily to produce body movements.

- Adenosine triphosphate (ATP), most of which is derived from glucose and fat metabolism, is the energy currency for muscle action.

- Smooth muscle contracts involuntarily to power many internal functions.

Chapter Objectives

Overview of Muscle   228

1. List five functions of muscle tissue.

2. Compare and contrast skeletal, smooth, and cardiac muscle.

Structure of Skeletal Muscle Tissue   231

3. Using a drawing, identify and describe the special features of a skeletal muscle cell, and explain how many such cells, along with connective tissue membranes, are built into a skeletal muscle.

Skeletal Muscle Contraction   233

4. Define *sarcomere* and explain how muscle contraction results from sarcomere shortening.

5. Describe the composition of the thin and thick filaments, and label the parts of the sarcomere.

6. List all the steps involved in muscle contraction, beginning with an action potential in a neuron and ending with the events of cross-bridge cycling.

7. List the steps involved in muscle relaxation.

Muscle Energy   243

8. Identify three uses for ATP in muscle contraction.

9. Explain the benefits and disadvantages of different energy sources (creatine phosphate, glycolysis, and mitochondrial respiration); compare anaerobic and aerobic metabolism.

10. Compare the structure and function of fast glycolytic fibers and slow oxidative fibers.

Case Study: Muscle Energy Metabolism: The Case of Hammid S.   246

11. List different causes of muscle fatigue, referring to the case study.

The Mechanics of Muscle Contraction   248

12. Explain how a stronger contraction results from modifying fiber length and/or recruiting additional motor units.

13. Provide examples of isometric, concentric isotonic, and eccentric isotonic contractions.

14. Discuss the effects of resistance training and endurance exercise on muscles.

Smooth Muscle   252

15. Describe the structural and functional differences between skeletal and smooth muscle.

16. List the steps involved in smooth muscle contraction, including the different types of stimuli that can induce contraction.

Skeletal Muscle Actions   255

17. Identify the prime mover, synergist, and/or antagonist for different body movements at each joint.

The Major Skeletal Muscles   256

18. For each body region (head and neck, upper limb, torso, and lower limb), label the major skeletal muscles on a diagram and indicate their insertion and origin.

"He's had these pains all of his life."

As you read through the following case study, assemble a list of the terms and concepts you must learn in order to understand Hammid's condition.

Clinical History: His mother brought Hammid S., a 10-year-old boy, to a pediatrician's office. The family had emigrated to the United States from Afghanistan 10 months earlier. With the aid of an interpreter she explained that Hammid was becoming increasingly upset by his inability to keep up with the other boys on his soccer team because of the painful muscle cramps that occurred in his legs with strenuous exercise. He had also been complaining that everyday activities requiring significant muscle effort, such as climbing stairs, caused him pain.

She explained further, "He's had these pains all of his life, but not so bad as now. The doctor in Herat told me he probably had liver disease because his urine is red or brown sometimes. But the dark urine always appears after the muscle cramps come. The cramps go away if he rests for a while."

Further questioning revealed that Hammid's older brother and younger sister were not affected by similar symptoms.

Physical Examination and Other Data: Hammid was of normal height and weight for his age, and his vital signs were unremarkable. Muscle size and tone were unremarkable. Mild proximal muscle weakness was present in all extremities and he had difficulty walking on his heels or toes more than 8 or 10 steps because cramps developed in his legs.

Laboratory evaluation revealed abnormally high levels of creatine kinase (a muscle enzyme) in the blood. A presumptive diagnosis of McArdle syndrome (type V glycogen storage disease, due to a genetic deficiency of muscle glycogen phosphorylase, another muscle enzyme) was made and an appointment with a specialist in muscular diseases was arranged.

Need to Know

It is important to understand the terms and concepts listed below before tackling the new information in this chapter.

- Nutrients and ATP ← (Chapter 2)
- Neuron structure, neurotransmitters, and chemical synapses ← (Chapter 4)
- Movements at synovial joints ← (Chapter 6)

Clinical Course: At the specialty clinic, a forearm ischemia test was performed, in which a blood pressure cuff was inflated to cut off blood flow and Hammid was asked to squeeze a rubber ball for a minute or until cramps appeared. Study of blood lactic acid in a forearm vein was abnormal: the normal increase of lactic acid did not occur during the test. A muscle biopsy was performed; it showed increased amounts of glycogen in the muscle fibers and a severe decrease in muscle content of glycogen phosphorylase.

Specialists at the clinic explained to Hammid's parents that his genetic defect was inherited and that no treatment was currently available. The parents were also reassured that although Hammid would have difficulty with strenuous exercise all of his life, he would be unlikely to suffer other problems.

Just as the word *bone* can refer either to an organ or to a tissue, so can the word *muscle;* that is, the biceps muscle (a muscle in the arm) is an organ primarily composed of muscle tissue. We derive the word *muscle* from the Latin *mus* (for "mouse"), a reference to the rippling motion of muscles, which was thought to resemble the movement of mice beneath the skin. In turn, *mus* was derived from earlier Greek, where *mys* (meaning both "mouse" and "muscle") gives us the prefixes *myo-* and *mys-*, which refer to muscle. A **myofilament,** for example, is a specialized cytoskeletal filament of muscle cells. Words referring to muscle tissue may also have the prefix *sarco-*, which is derived from Greek *sarx* (for "flesh"). For example, the cytoplasm of a muscle cell is called the **sarcoplasm.**

Courage is like a muscle strengthened by its use.

Ruth Gordon, American writer and actress (1896–1985)

Overview of Muscle

Muscle comprises about 40% to 50% of body weight. No other cell can do what muscle cells do: they contract (shorten) on conscious command. This ability makes muscle cells responsible for our movements, both visible and invisible: walking, talking, bowel movements, urination, breathing, heartbeats, the dilation and constriction of the pupils of our eyes, and many others. And when we are still—sitting or standing—muscle cells keep us erect.

Functions of Muscle

The core function of muscle is to convert chemical energy into mechanical force. Muscle acts to:

- *Move body parts.* Every movement of our body requires skeletal muscle action, from large movements like walking, to smaller movements like breathing or following a tennis match with our eyes.
- *Maintain body posture.* Although it is not immediately obvious, an uninterrupted sequence of tiny, silent contractions of postural skeletal muscle keeps us erect when we are standing or sitting and keeps our heads from slumping on our shoulders. A related activity is the *stabilization of joints*: In every activity, joints must be stabilized so that they do not swing out of control but operate in a smooth, steady fashion.
- *Adjust the volume of hollow structures.* By their response to unconscious autonomic commands, muscles in the

walls of hollow structures relax to increase volume and contract to decrease it. For example:

- Muscle in the bladder wall relaxes to allow the bladder to expand to accommodate more urine or contracts to expel it.
- Muscle in blood vessel walls relaxes to dilate blood vessels and allow more blood flow or contracts to reduce blood flow.

- *Move substances within the body.* The self-stimulated, automatic contractions of cardiac muscle pump blood through blood vessels; waves of smooth muscle contractions propel intestinal contents down the intestinal tract; and similar waves of smooth muscle contraction power male ejaculation and female orgasm.
- *Produce heat.* Whether it is the conversion of gasoline into vehicular motion or the conversion of glucose into muscle contraction, conversion of energy from one form to another always produces heat as a waste product. The body generates ATP to power muscle contraction. When it does so, about three-fourths of the nutrient energy consumed escapes as heat.

Because nearly half of body mass is skeletal muscle, most body heat comes from skeletal muscle contractions. And just as waste heat from an automobile engine is used to warm the car's interior on a frosty day, the heat from muscle contraction is the major source of heat to maintain body temperature. For example, when we shiver with cold, the shivers are involuntary skeletal muscle contractions that generate extra heat to raise body temperature.

Case Note

7.1. Based on his symptoms, which of the muscle functions just discussed is impaired in Hammid, our patient?

There Are Three Types of Muscle

There are three types of muscle: *skeletal, cardiac,* and *smooth* (Table 7.1). Their most important common characteristic

Table 7.1	Muscle Tissue		
Characteristic	**Skeletal**	**Cardiac**	**Smooth**
Location	Often attached to bones	Heart	Walls of blood vessels, visceral organs
Appearance	Long, cylindrical fibers	Branching cylindrical fibers	Small cells; sometimes branched
	Thin	Striated	Not striated
	Striated	Single nucleus	Single nucleus
	Multiple nuclei		

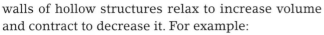

(a) Striation, Nucleus, Muscle fiber

(b) Muscle fiber, Striation

(c) Muscle cell, Nucleus

Control	Voluntary	Involuntary	Involuntary
Contraction	Rapid contraction and relaxation	Moderate contraction and relaxation	Slow contraction and relaxation; can maintain for extended periods
Fatigue?	Yes	No	No

is their ability to contract. Their most important differences relate to four qualities:

- *Location*
- *Microscopic appearance*
- Whether or not they are subject to *conscious control*
- The *type of contraction* they generate

Skeletal Muscle Moves the Skeleton

As the name suggests, most **skeletal muscle** is attached to bone and moves the skeleton (Fig. 7.1).

Skeletal muscles form the bulk of our muscle mass and add shape to the body. They make up the body wall, hence the alternate term *somatic muscle* (*soma* = "wall").

Microscopically, skeletal muscle is **striated muscle;** that is, it has cross-stripes (striations) on microscopic examination, an appearance that is intimately related to its function. Mature muscle cells are especially long and thin—up to a foot long—and are typically called **muscle fibers.** Keep in mind throughout this chapter that a muscle fiber is a single mature skeletal muscle cell.

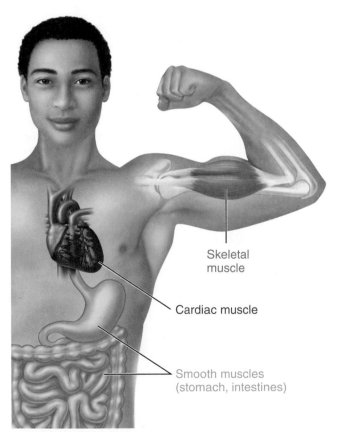

Skeletal muscle

Cardiac muscle

Smooth muscles
(stomach, intestines)

Figure 7.1. Muscles in action. Cardiac muscle (shown in dark red) keeps blood moving, smooth muscle (orange) enables food digestion and urine retention, and skeletal muscle (bright red) moves the body. *Which type of muscle makes up the stomach wall?*

What's more, skeletal muscle is **voluntary muscle;** that is, we can contract and relax it at will. Skeletal muscles can also function outside of our conscious control; for example, the diaphragm contracts and relaxes to keep us breathing while we sleep, and our neck and back muscles maintain our seated posture while our attention is devoted to our work.

Skeletal muscle fibers contract quickly and forcefully and then relax to become ready to contract again. They do not maintain contraction for an extended period of time, and they fatigue after repeated contraction. Further on, we discuss the precise nature of muscle fatigue, which is unique to skeletal muscle; cardiac and smooth muscle do not tire.

Cardiac Muscle Propels Blood through the Body

As its name suggests, **cardiac muscle** tissue is found only in the heart, and accounts for most of its mass (see Fig. 7.1). We perceive cardiac muscle contractions as our heartbeat, which propels blood through the blood vessels of the body.

Microscopically, cardiac muscle is striated, like skeletal muscle. Cardiac muscle cells are much shorter than skeletal muscle fibers, but they are branched and interconnected. The end of one branch is connected intimately to another, producing long cardiac muscle fibers. The result is that cardiac muscle is a network that in many ways behaves like a single huge muscle cell.

Of course, cardiac muscle is **involuntary muscle:** we cannot control its contractions by force of will. That said, some things we deliberately do—such as engaging in meditation—can slow our heartbeat, whereas other things—such as vigorous exercise—can increase it.

Like skeletal muscle fibers, cardiac muscle cells contract quickly and then relax. However, unlike skeletal fibers, they do not fatigue. We will return to cardiac muscle in ➡ Chapter 11 in our discussion of the heart.

Smooth Muscle Powers the Actions of Viscera

Smooth muscle tissue is found in thick layers in the walls of hollow organs such as blood vessels, the urinary bladder, the uterus, and the intestines (Fig. 7.1). The intestines and other abdominal organs are frequently described as the *viscera;* therefore, smooth muscle is also called *visceral muscle.* As noted earlier, it adjusts the volume of hollow structures and helps move substances—from food to blood—throughout the body.

Smooth muscle is named for its microscopic appearance: It is a **nonstriated muscle;** that is, it has a uniform, smooth appearance without cross-striations. As discussed further on in this chapter, its lack of striations is intimately related to its function.

Smooth muscle is involuntary: we do not command the wall of our stomach to relax to accommodate a large meal. It just happens as the stomach responds automatically to the mechanical stimulation of food bulk, which is but one of many stimuli that govern smooth muscle function. Usually, smooth muscle contracts slowly and can maintain the contraction over a long period. Generally, smooth muscle fibers do not fatigue.

Case Note

7.2. Which type of muscle is affected by Hammid's disease?

All Muscle Tissue Is Extensible

A final important characteristic of all three types of muscle tissue is its *extensibility*—its ability to stretch without tearing. Consider what happens when you open your mouth wide to bite an apple. The muscle that normally brings the jaws together must relax and lengthen in order to permit you to do it. If this muscle were not extensible, your simple action would cause the muscle to tear. The stretchable internal organs of the body—the heart, the bladder, the intestines, the uterus, and so on— are made of cardiac or smooth muscle, which also has this property. By contrast, if a surgeon were to inadvertently stretch the tissue of the brain, the liver, the spleen, or the kidney, it would tear.

7.1 What is the name of a mature skeletal muscle cell?

7.2 Name the two types of striated muscle.

7.3 Name two types of involuntary muscle.

7.4 Which type of muscle tissue experiences fatigue?

Structure of Skeletal Muscle Tissue

Since muscle cells are unique in their ability to contract and lengthen without tearing, it's not surprising that they have an unusual path of development and unique structural features. Again, the structure of cardiac mus-

cle is discussed in ➡ Chapter 11, and that of smooth muscle is discussed later in this chapter.

Myoblasts Fuse to Form Muscle Fibers

During embryonic development, stem cells produce immature muscle cells called **myoblasts** (*blast* = "precursor"). Several myoblasts fuse together to produce each skeletal muscle fiber, so each muscle fiber contains multiple nuclei. Some muscle stem cells persist into adulthood, hidden between the muscle cell membrane and the surrounding connective tissue. These adult muscle stem cells are called **satellite cells,** because of their location at the muscle fiber's periphery. Although mature skeletal muscle cells are fully differentiated and cannot divide, satellite cells can be activated by exercise, injury, or disease to produce new myoblasts that fuse to form new muscle fibers. However, satellite cell activity is not sufficient to repair major skeletal muscle injuries.

Remember This! **Adult muscle stem cells are called satellite cells; they produce myoblasts, which fuse to form skeletal muscle fibers.**

The Structure of a Muscle Cell Reflects Its Function

The following elements of a skeletal muscle cell (fiber) are essential to its function (Fig. 7.2A):

- The cell membrane is called the **sarcolemma.** Like the membrane of any body cell, it acts to contain the cell's contents and shield it from the extracellular environment. As discussed below, this function is especially important in muscle contraction.
- The sarcolemma in muscle cells not only surrounds the cytoplasm but also tunnels deep into the interior of the muscle fiber as a network of **T-tubules.** Action potentials travel down these T-tubules, which enable them to reach every part of the fiber virtually simultaneously to trigger a coordinated muscle contraction.
- Multiple cigar-shaped nuclei reside along the periphery of the cell, immediately beneath the sarcolemma. This location keeps them out of the way of muscle fiber contractions.
- The cytoplasm of the muscle cell, the **sarcoplasm,** is densely packed with the following structures, which are described in more detail further on:
  - **Myofibrils.** These slender, threadlike organelles accomplish the work of muscle contraction. Each

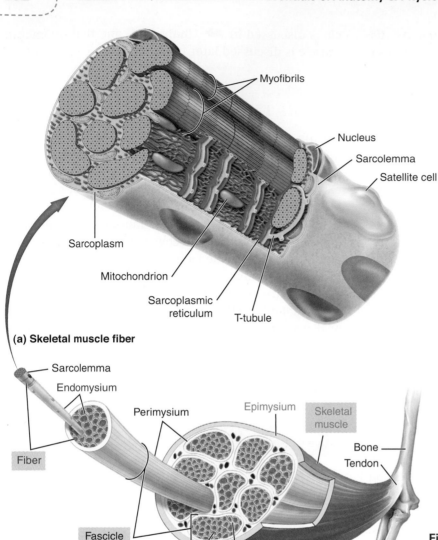

(a) Skeletal muscle fiber

(b) Skeletal muscle

Myofibrils

Nucleus

Sarcolemma

Satellite cell

Sarcoplasm

Mitochondrion

Sarcoplasmic reticulum

T-tubule

Sarcolemma

Endomysium

Perimysium

Epimysium

Skeletal muscle

Bone

Tendon

Fiber

Fascicle

Fiber   Endomysium

Figure 7.2. Skeletal muscle cells and muscles.
A. Muscle fibers are packed with myofibrils.
B. Numerous muscle fibers are packed together to form a fascicle, and many fascicles are packed together to form a muscle. *Name the connective tissue layer surrounding a fascicle.*

myofibril is a bundle of different proteins that runs the entire length of the muscle fiber. Each muscle fiber contains hundreds or thousands of myofibrils.

- **Sarcoplasmic reticulum** (SR). This organelle is a lacy network of fluid-filled tubules similar to the smooth endoplasmic reticulum in other body cells. It stores calcium ions necessary for muscle contraction. T-tubules are in close contact with the SR, separated by a small region of intracellular fluid.
- **Mitochondria.** These organelles generate the ATP that fuels muscle contraction.
- **Myoglobin** (not shown on Fig. 7.2), an iron-containing compound, stores oxygen used to generate energy for muscle contraction.

Case Notes

7.3. Hammid's blood chemistry showed elevated levels of creatine kinase, which is usually confined to the interior of the muscle cell. The cell membrane of his muscle cells must have ruptured, releasing the cell contents into his blood. What are the specific terms used to describe the muscle cell membrane and cytoplasm?

7.4. Hammid's red urine reflects the presence of the compound that stores oxygen inside skeletal muscle cells. Name this compound.

Connective Tissue Wraps Muscle Fibers, Fascicles, and Whole Muscles

Individual skeletal muscle fibers are delicate, and every one of them is wrapped by a sheath of connective tissue called the **endomysium,** which covers, insulates, supports, and protects them (Fig. 7.2B). Satellite cells reside between the endomysium and the sarcolemma.

Groups of about 100 muscle fibers are formed into structural and functional bundles called **fascicles.** These are wrapped with a thicker, tougher sheath of connective tissue called the **perimysium.**

In turn, groups of fascicles form muscles, which are wrapped by a tough and very substantial outer layer of connective tissue, the **epimysium.** As a muscle terminates near its attachment to bone, its epimysium binds together to form a tough and exceptionally strong collagenous tissue that attaches muscle to bone. When formed into a thick, tough cord for attachment at a single point, it is called a **tendon** (see Fig. 7.2B); you learned about tendons attaching to bone in ← Chapter 6. When formed into a sheet for broader, linear attachment, the epimysium is called an **aponeurosis.**

7.5 Are tendons examples of epithelial tissue or connective tissue?

7.6 What is the difference between a muscle fiber, a fascicle, and a myofibril?

7.7 What is the name of the membrane extensions that dip deep into the sarcoplasm?

7.8 What is the difference between the perimysium and the endomysium?

Skeletal Muscle Contraction

You're reading this chapter, and it's time to turn the page. As you lift your hand, you don't consciously direct your muscles to contract to produce your movements. It just happens. But how?

A Motor Unit Is a Motor Neuron and the Muscle Fibers It Controls

Contraction of a skeletal muscle requires communication. A **somatic motor neuron** carries a signal that stimulates contraction in skeletal muscle (a *visceral motor neuron* carries a similar signal to smooth muscle or glands). The cell bodies of motor neurons are located in the brain or spinal cord and send long cytoplasmic extensions called axons out to communicate with muscle fibers. As shown in Figure 7.3, the axon of a motor neuron branches toward its end to make contact with several muscle fibers. These branches are called *axon terminals*. A **motor unit** comprises a somatic motor neuron and the skeletal muscle fibers it controls.

Muscles that require small, highly precise movements (as in the muscles that control eye movement) may have as few as three muscle fibers per motor unit. Muscles responsible for large, powerful movements (in the thigh, for example) may have several thousand muscle fibers per motor unit.

Case Note

7.5. What type of neuron carries the signal to Hammid's muscles?

Motor Neurons Connect to Muscle Fibers at the Neuromuscular Junction

Near its tip, each axon terminal enlarges into a pancake-like swelling called a *synaptic bulb,* which lies flat on the surface of the muscle fiber. A single synaptic bulb meets a skeletal muscle fiber at a chemical synapse called the **neuromuscular junction** (Fig. 7.3B). The components of the neuromuscular junction are (Fig. 7.3C):

- The *synaptic bulb* of the neuron
- The *motor end plate* of the muscle fiber, which is that part of the fiber's sarcolemma across from the synaptic bulb
- The *synaptic cleft*, an exceedingly narrow space that separates the synaptic bulb from the motor end plate—the nerve and muscle fiber do not actually touch.

Recall from ← Chapter 4 that chemical synapses use neurotransmitters to transmit the signal between two adjacent cells—in this case, the motor neuron and the muscle fiber. In all synapses the basic process is the same: in response to an action potential in the presynaptic cell, neurotransmitter is released into the synaptic cleft; it then binds to specific receptors on the postsynaptic cell, altering its electrical activity. The neuromuscular junction is more specific—an action potential in the presynaptic cell *always* results in an action potential

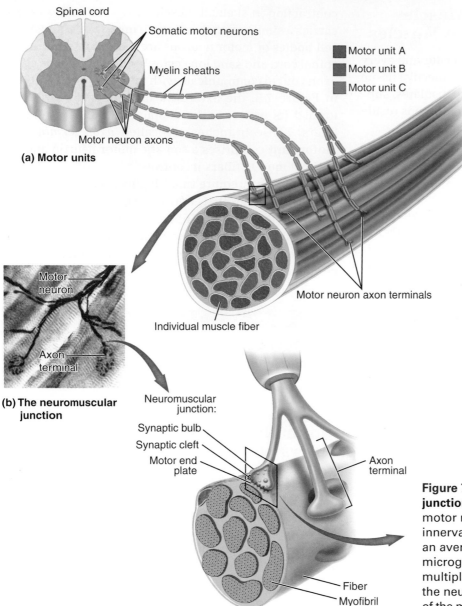

Figure 7.3. Motor units and the neuromuscular junction. A. A motor unit consists of a somatic motor neuron and the skeletal muscle fibers it innervates. This figure shows three motor units; an average muscle will have many more. **B.** This micrograph shows an axon branching to supply multiple muscle fibers. **C.** The synaptic bulb of the neuron synapses with the motor end plate of the muscle fiber. *Name the space that separates the neuron and the muscle cell.*

in the postsynaptic cell. Moreover, every skeletal neuromuscular junction uses the same neurotransmitter (*acetylcholine*) and the same neurotransmitter receptor—the *nicotinic cholinergic receptor* (Fig. 7.4)

This receptor is a ligand-gated ion channel ◀━ (Chapter 4), which opens to allow sodium (Na^+) ions to enter the cell when acetylcholine (the ligand) is bound to it.

The events are as follows:

1. The action potential arrives at the synaptic bulb of the somatic motor neuron (the presynaptic cell). The resulting depolarization triggers acetylcholine release into the synaptic cleft.

2. Acetylcholine (ACh) encounters one of two proteins. Some molecules meet with and are inactivated by *acetylcholinesterase*, an enzyme present in the synaptic cleft and embedded in the sarcolemma. This enzyme is always active, but it cannot keep up with ACh release from firing neurons, so ACh accumulates in the synaptic cleft. The neurotransmitter molecules that escape acetylcholinesterase's grasp meet and bind with a second protein, cholinergic nicotinic receptor in the motor end-plate membrane (the postsynaptic cell).

3. ACh binding opens the channel in the nicotinic receptor. Na^+ entry depolarizes the membrane enough to

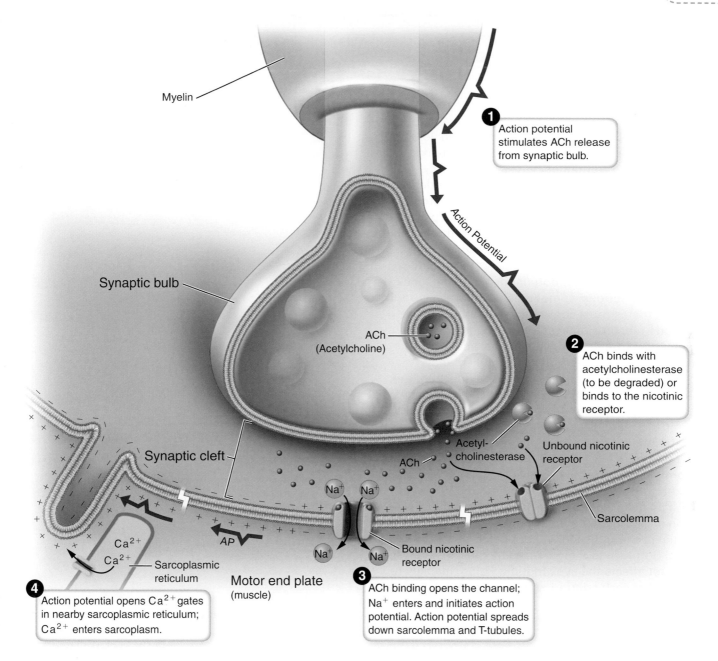

Myelin

① Action potential stimulates ACh release from synaptic bulb.

Action Potential

Synaptic bulb

ACh (Acetylcholine)

② ACh binds with acetylcholinesterase (to be degraded) or binds to the nicotinic receptor.

Synaptic cleft

Acetyl-cholinesterase

ACh

Unbound nicotinic receptor

Sarcolemma

Na⁺   Na⁺

Ca^{2+}
Ca^{2+} — Sarcoplasmic reticulum

AP

Na⁺   Na⁺

Bound nicotinic receptor

Motor end plate (muscle)

④ Action potential opens Ca^{2+} gates in nearby sarcoplasmic reticulum; Ca^{2+} enters sarcoplasm.

③ ACh binding opens the channel; Na⁺ enters and initiates action potential. Action potential spreads down sarcolemma and T-tubules.

Figure 7.4. Events at the neuromuscular junction. An electrical signal (an action potential) travels down the motor neuron. A chemical signal (ACh) carries the signal across the synaptic cleft and initiates an electrical signal (an action potential) in the muscle cell. *Name the enzyme that terminates ACh's action.*

cause an action potential. The action potential sweeps rapidly over the sarcolemma and races through the network of T-tubules deep within the cell.

4. The action potential triggers the opening of calcium gates in the membrane of the sarcoplasmic reticulum (SR). This releases calcium ions from the sarcoplasmic reticulum (SR) into the sarcoplasm. A specialized calcium (Ca^{2+}) transporter called the *calcium pump* actively transports Ca^{2+} back into the SR.

However, in a contracting fiber, the pump cannot keep up with Ca^{2+} release, so Ca^{2+} accumulates in the sarcoplasm. As shown later, it is these calcium ions that stimulate muscle contraction.

The function of chemical synapses can be affected by disease or manipulated or inactivated by drugs or poisons—see the nearby Clinical Snapshot, titled *Beauty and the Beasts,* for more information.

CLINICAL SNAPSHOT

Beauty and the Beasts: Attacking the Neuromuscular Junction

In 2006, untold numbers of women and men voluntarily poisoned the neuromuscular junction of certain facial muscles in order to rid themselves (temporarily) of frown lines. The poison? A toxin, marketed under the name Botox©, which is derived from the anaerobic bacterium *Clostridium botulinum*. Botox©, a protein, is one of the most potent toxins known, one that paralyzes muscles by preventing them from receiving nerve action potentials.

In clinical medicine *C. botulinum* poisoning, or botulism, is a serious, sometimes fatal paralytic condition that is most commonly encountered after ingestion of insufficiently sterilized (undercooked) home-canned meats, fish, vegetables, and fruits contaminated with *C. botulinum*. Botulism may also occur as a consequence of wound infection. The term *botulism* derives from Latin *botulus,* meaning "sausage": the name reflects the fact that the illness was initially recognized as resulting from consumption of contaminated sausage.

C. botulinum toxin acts at the *nerve* side of the neuromuscular junction to prevent synaptic vesicles in the axon from releasing their ACh into the synaptic cleft. If an action potential arrives at the synapse and no ACh is released into the synaptic cleft, the action potential is extinguished without being transferred to muscle. Botulism is characterized by muscle paralysis, which first affects the eyes (double vision, inability to focus) and speech (slurred words) and may cause fatal respiratory paralysis.

However, in small, local doses, the toxin causes limited muscle paralysis, which achieves a pleasing cosmetic effect by relaxing the facial muscles associated with facial wrinkles. For example, following an injection of Botox into the frontalis muscle of the forehead, frown lines disappear; they do not reappear until the effect of the toxin wears off in 4 to 6 months. Botox is also used therapeutically to prevent the muscle spasms that accompany migraine headaches, facial tics (involuntary or habitual contraction of facial muscles), and cervical dystonia (abnormal contractions of neck muscles that move the head).

Other animals exploit the fragility of the neuromuscular junction to paralyze their prey. For

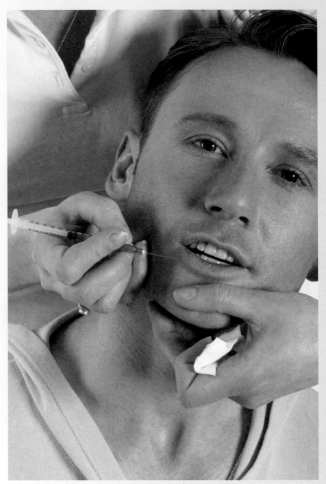

Neuromuscular junction toxins. Botulinum toxin is used to treat frown lines and other facial wrinkles.

example, Taiwanese cobra venom contains a toxin that binds tightly to the ACh receptor on the *muscle* side of the synapse and prevents ACh from binding, which interrupts propagation of the signal. Conversely, venom of the black widow spider causes motor axons to release all of their stored ACh, which overwhelms muscle receptors and interferes with controlled signal transfer across the synapse.

Recall from Chapter 4 that signals can be electrical or chemical. The sequence involved in stimulating muscle contraction is as follows:

1. An electrical signal in the somatic motor neuron
2. A chemical signal (ACh) in the synapse
3. An electrical signal in the sarcolemma
4. A chemical signal (calcium) in the sarcoplasm.

But how does a chemical signal—calcium—initiate force generation in the muscle fiber? In order to answer this question, we must delve deeper into the microscopic structure of the muscle fiber, paying particular attention to the myofilaments.

Case Note

7.6. What chemical is released by Hammid's somatic motor neurons to convey nerve signals to muscle cells?

Sarcomeres Are the Functional Units of Myofibrils

Recall that myofibrils are the organelles within the muscle fiber that accomplish the work of muscle contraction. To understand how they contract, we must examine their unusual structure. Each myofibril is, in essence, a bundle of two types of long **myofilaments:** *thick filaments* and *thin filaments.* You can visualize their precise arrangement, which is essential to their function, by imagining myofilaments as thick and thin pencils. Here's how:

● Imagine that the thick filaments (thick pencils) are sharpened on both ends and that the thin filaments (thin pencils) are sharpened on one end and with an eraser at the other (Fig. 7.5A).
● Next, imagine holding a bundle of thin pencils in each hand, with the erasers pointing outward and the sharpened tips pointed at one another.

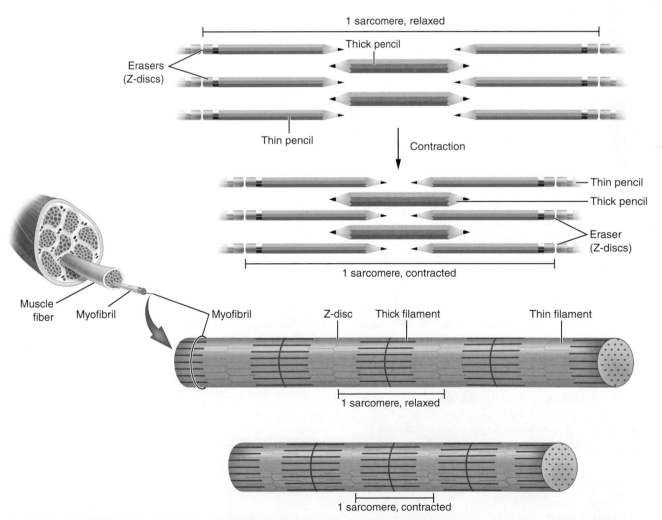

Figure 7.5. Myofibrils are composed of myofilaments. Pencils can be used to model a sarcomere. When the overlap between the pencils increases, the sarcomere shortens. A myofibril consists of many sarcomeres lined up end to end. When individual sarcomeres shorten, the entire muscle fiber (and thus the muscle) shortens. *Which structure is the same length as the muscle—the sarcomere or the myofibril?*

- Now, imagine placing a bundle of thick pencils (sharpened at both ends) between the two bundles of thin pencils. Notice that pointed ends of the thick pencils in this middle bundle face pointed ends of the thin pencils on either side.
- Finally, imagine pushing the thin pencil bundles into the thick pencil bundle in such a way that the thick and thin sharpened tips overlap slightly like interlocked fingertips.

And there you have it: a pencil replica of one **sarcomere,** which is the basic unit of skeletal muscle (Fig. 7.5B). A muscle fiber contains thousands of end-to-end sarcomeres, each a set of interdigitated bundles of thick and thin "pencils" joined at the "erasers" on each end. The joined eraser ends of the thin pencils are analogous to the **Z-discs** of a myofibril, which are found on either side of the sarcomere.

To imagine muscle contraction, imagine sliding the two sets of thin pencils toward each other over the center bundle of thick pencils. As the *overlap* of thick and thin pencils *increases,* the length of the whole sarcomere shortens. This is the essence of muscle contraction— the degree of *overlap* of thick and thin bundles increases as the sarcomere contracts, but the length of each thick and thin myofilament remains unchanged. This model of muscle contraction is called the *sliding filament mechanism,* since filaments are sliding over each other.

A single sarcomere is very small, only a few micrometers in length, but sarcomeres lined up end to end produce a myofibril that runs the entire length of a muscle fiber (Fig. 7.5B). As each sarcomere shortens, the entire myofibril shortens, the muscle fiber shortens, and thus the muscle shortens. Voila! Muscle contraction. Later we will see how the contraction of every sarcomere, myofibril, and muscle fiber of a motor unit occurs at the same time—a property that ensures smooth contraction.

The striated appearance of skeletal muscle examined under a light microscope is an orderly series of light and dark bands produced by the overlap of thick and thin filaments and the end-to-end junctions of sarcomeres. Details are presented in the accompanying Basic Form, Basic Function box, titled *How the Muscle Got Its Stripes.*

> ***Remember This!*** **During muscle contraction, sarcomeres and myofibrils shorten, but myofilaments do not change in length.**

Myofilaments Are Composed of Contractile Proteins

The molecular structure (form) of thick and thin filaments is essential to their contractile nature (function) (Fig. 7.6).

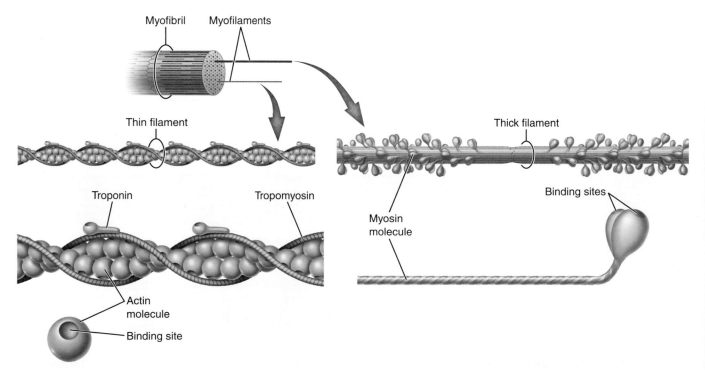

Figure 7.6. Thick and thin filaments. Thick filaments are composed of myosin molecules; thin filaments of actin, troponin, and tropomyosin. *Which protein covers the myosin binding site on the actin molecules?*

BASIC FORM, BASIC FUNCTION

How the Muscle Got Its Stripes

Recall that striated muscle can be identified by its stripes; it is crossed by alternating light and dark microscopic bands (stria) when examined under a light microscope. The stripes are also useful for another reason—they can help us visualize the minute movements of thick and thin filaments during muscle contraction.

The dark bands, called A bands, are dark because they contain the more opaque thick filaments. The light bands, called I bands, are light because they are composed exclusively of thin filaments. Recall, however, that thin and thick filaments overlap. The dark (A) bands are darkest on each end, where they overlap with thin filaments: this packing of thick and thin filaments blocks the most light. The H zone is the relatively paler region within the A band where only the thick filaments are present. In the center of the light (I) band is a zigzag line; which is the Z-disc, where the bundles of thin fibers meet and which marks the place where sarcomere units join together.

Note that a sarcomere is the space between Z-discs and is formed by half of a light (I) zone at each end and dark (A) band in the center; and that the light (I) zone is formed of the butting ends of two sarcomeres. Note further that when a sarcomere shortens, the Z-discs move closer together, I bands shorten, but the A band remains the same length. Why? Because the A band is a thick filament, which always stays the same length. The I bands, conversely, are thin filaments that do not overlap with thick filaments. As we increase the overlap between thick and thin filaments, more of the thin filament slides into the A band, where it is obscured by

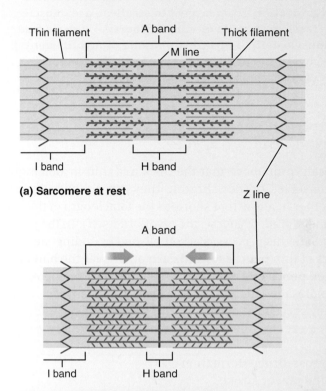

(a) Sarcomere at rest

(b) Contraction fiber and filament sliding

Muscle fiber zones and lines.

the thick filaments. Finally, what happens to H zone, representing thick filaments not overlapping with thin filaments? As with the I band, it shrinks as the sarcomere shortens.

Thick filaments are bundles of **myosin** protein. Each molecule of myosin is composed of a long shaft (the tail), one end of which terminates in two globular *heads*, somewhat like a two-headed golf club, one head up the shaft a bit from the other. Each myosin head has two important binding sites, one for ATP and one for thin filaments. When a myosin head is bound to the thin filaments it forms a *cross-bridge*. Many myosin molecules, with their heads pointing in opposite directions, are bundled together to form a thick filament. The molecules overlap like golf clubs taped together to form a chain, with the heads protruding over much of the length—at one end the shafts of the myosin molecules

are joined end to end to form the headless central segment of the thick filament.

Thin filaments are composed of three proteins—*actin, tropomyosin,* and *troponin*. The main constituent is **actin,** a small globular protein. Each thin filament contains two long strands of actin molecules that are twisted together, much like a necklace composed of two intertwined strings of pearls. Each actin molecule (that is, each "pearl") contains a binding site for a myosin head on a thick filament. In the resting state, however, this binding site is covered by **tropomyosin,** which prevents myosin binding until a signal for contraction arrives from the nerve that innervates the muscle. **Troponin,** the third constituent, controls the tropomyosin molecules, keeping them in place over the binding sites in relaxed muscle but moving them out of the way for contraction to occur.

> **Remember This!** Levels of skeletal muscle organization, from largest to smallest, are: muscle → fascicle (bundle of muscle fibers) → muscle fiber (muscle cell) → myofibril (bundle of myofilaments) → myofilament (strands of contractile proteins) → contractile protein.

Sarcomeres Shorten via the Cross-Bridge Cycle

Recall from above that the thick and thin myofilaments themselves do not shorten; they merely slide by one another in a way that shortens the total length of the sarcomere (and, of course, the myofibril itself). In the pencil analogy, this process is accomplished by sliding the bundles of thin pencils toward each other over the bundle of thick pencils. In the muscle cell, the task of sliding the bundles toward each other is the job of the myosin heads. They succeed in producing this movement via a series of three events collectively called the *cross-bridge cycle*:

- cross-bridge formation
- the power stroke
- cross-bridge detachment

The *power stroke* is the part of the cycle in which the thin filament actually moves. Of the body's many molecular movements, this is among the strangest and most effective. So let's take a close look at how the power stroke occurs before considering the cross-bridge cycle as a whole.

The key operators in the power stroke are the myosin heads. Each head serves as a claw that grabs a "pearl" of actin on a thin filament, anchors itself to it, and snaps backward, pulling the thin filament along the myosin tail a short distance. After this short pull, the myosin heads release, recock, and reattach to another actin pearl further along the thin filament, ready to snap backward again. In this way, the thick and thin filaments ratchet along one another, like someone (the thick filament) pulling up a rope (the thin filament) arm over arm.

Now let's review the full sequence of events that produce muscle contraction (Fig. 7.7). In a muscle fiber at rest, myosin-binding sites on actin molecules are covered by tropomyosin. In response to an action potential in the sarcolemma and T-tubules, Ca^{2+} is released from the sarcoplasmic reticulum (SR). Ca^{2+} binds to and activates troponin, which moves tropomyosin out of the way, exposing the myosin-binding site on each actin molecule (steps 1 to 3 on Fig. 7.7). Once these binding sites are exposed, the cross-bridge cycle (steps 4 to 7) can begin.

Cross-bridge formation occurs when "energized" myosin heads bind actin (step 4). Why do we characterize the myosin heads as energized? Recall from ◄ Chapter 2 that energy is released when ATP is cleaved into adenosine diphosphate (ADP) and phosphate. In a resting muscle fiber, the ATP has already been cleaved and the products, ADP and phosphate, are bound to the myosin heads. The energy released by ATP cleavage is stored in the "cocked" position of the myosin heads; that is, the myosin head is energized (step 4).

This stored energy is used in step 5, the power stroke, to pivot the myosin heads and move the thin filament. The ADP and phosphate molecules diffuse away immediately after the power stroke, but the cross-bridge remains in place.

The final step in the cross-bridge cycle, detachment, can occur only with the help of additional ATP. Only when a fresh ATP molecule binds to the myosin head (step 6) does the myosin head release from the actin, ready to begin another cross-bridge cycle (step 7).

Cross-bridge cycling occurs in waves, somewhat like a centipede's gait, so that the sliding motion is smooth, not jerky, as it would otherwise be if every myosin head pulled simultaneously like a rowing team. Such smooth waves of molecular increments, repeated quickly thousands of times, cause muscle fibers to shorten. Also, at any point in the contraction, some of the myosin heads are attached to the actin, so that the thin filaments cannot slide back to their original positions.

Case Note

7.7. Hammid's parents were told that his muscles could not get enough energy (i.e., ATP) for prolonged effort. Where does ATP bind in the myofilament?

It might help you to remember the events of the cross-bridge cycle if you understand that **rigor mortis,** the muscle stiffening that begins a few hours after death, is due to the lack of ATP. In death, the body can no longer generate ATP. Therefore the cross-bridge cycle can proceed up to step 5, where the myosin heads are firmly bound to the actin binding sites. And there things stop: relaxation cannot occur because, without a fresh ATP molecule, the myosin heads cannot detach from the

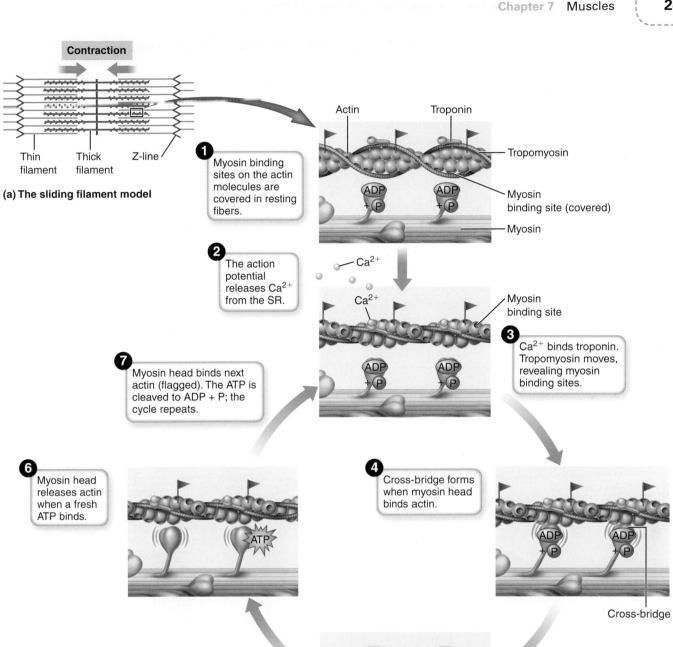

(a) The sliding filament model

(b) The cross-bridge cycle

Figure 7.7. Muscle contraction. A. Thick filaments pull thin filaments toward each other during muscle contraction. **B.** The steps in muscle contraction. Notice that the flagged actin molecule has moved (from step 4 to step 6) relative to the thick filament head. *Does ATP bind to actin or myosin?*

actin. Rigor mortis loosens its death grip on the skeleton after about 24 hours, as enzymes escape from lysosomes and digest myofibrils, allowing muscle to relax.

> **Remember This!** ATP binding causes the cross-bridge to release. The energy from ATP cleavage is necessary for the power stroke.

Muscle Relaxes When Cross-Bridge Cycling Ceases

We have now covered all of the elements of a successful muscle contraction, from the arrival of an action potential at the neuromuscular junction to the cross-bridge cycle. You can review these events in Figure 7.8A. Muscle relaxation, an equally important component of any muscle contraction, is essentially the reverse of these steps (Fig. 7.8B).

1. Without continued action potentials in the motor neuron, ACh release ceases. The constant efforts of acetylcholinesterase finally degrade all ACh molecules in the synaptic cleft.
2. Without ACh, the nicotinic receptor channels close, and action potentials in the sarcolemma cease.
3. The SR calcium channels close when the action potentials cease. The Ca^{2+} pump takes up remaining Ca^{2+} ions into the SR.
4. As the sarcoplasmic Ca^{2+} concentration drops, Ca^{2+} dissociates from the troponin. Tropomyosin resumes its previous position over the myosin binding sites.

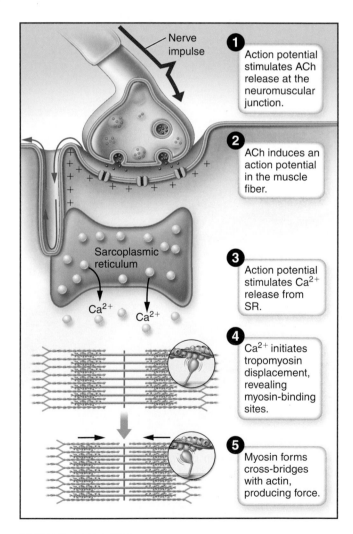

(a) Muscle contraction

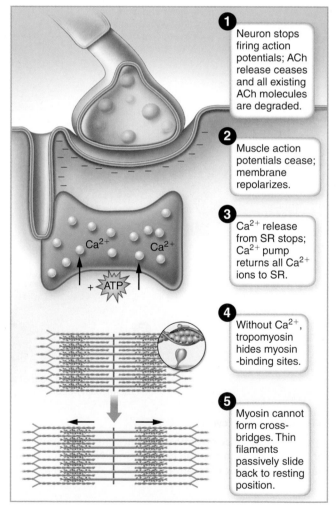

(b) Muscle relaxation

Figure 7.8. Muscle contraction and relaxation. A. Muscles contract by sarcomere shortening when calcium is present in the sarcoplasm. **B.** Muscles relax when calcium is pumped out of the sarcoplasm. *Which organelle stores calcium in muscle cells?*

5. Myosin can no longer bind actin—the thick filaments "lose their grip" on the thin filaments. Remember that muscle tissue is elastic, so the sarcomere rapidly returns to its resting length.

7.9   What is a motor unit?

7.10   How does the electrical signal in the neuron create an electrical signal in the muscle fiber?

7.11   Name three proteins found in thin filaments.

7.12   Does calcium bind thick filaments or thin filaments?

Muscle Energy

A steady supply of ATP is required to maintain every one of our cells, but muscle cells have particularly high energy needs. ATP fuels three important aspects of muscle activity:

- *Sarcolemma membrane potential:* Recall from ⬅ Chapter 4 that Na^+/K^+-ATPase is responsible for maintaining the Na^+ and K^+ gradients across the cell membrane, which are required for action potentials.
- *Cross-bridge cycling:* The myosin heads use the energy from ATP cleavage for the myosin head power stroke, and the cross-bridge breaks when a fresh ATP molecule binds.
- *Muscle relaxation:* The calcium pump uses ATP to actively transport calcium into the sarcoplasmic reticulum.

Recall that ATP stores energy in a chemical bond. The energy in this bond is released when a phosphate is removed from ATP, generating ADP, as shown in this reaction:

$$A\underline{T}P + H_2O \rightarrow A\underline{D}P + H_2O + PO4 + energy$$

Study of this reaction reveals that energy is required to force it in the opposite direction; that is, to convert energy-depleted A\underline{D}P back into A\underline{T}P. As shown next, we get most of this energy from the chemical bonds in nutrients.

Different Processes Can Generate ATP

Muscle cells are constantly generating ATP by a variety of processes. In general, processes that produce larger amounts of ATP involve more chemical reactions and thus require more time to complete. A contracting muscle fiber may use all of the processes to varying degrees, depending on the muscle type, the intensity of contraction, and the duration of the muscular activity.

ATP Stores and Creatine Phosphate Provide Immediate Energy

Muscles store a small amount of ATP (generated earlier by nutrient metabolism) to fuel the first few seconds of activity. However, since muscle fibers rupture if muscle ATP stores fall too low, various protective mechanisms usually prevent excessive depletion of ATP stocks. One of these mechanisms involves **creatine phosphate,** a molecule unique to muscle (Fig. 7.9A). It works by converting some of the energy-depleted ADP molecules back into ATP molecules by transferring its phosphate to ADP, a reaction that generates creatine plus ATP. Muscle cells contain only enough creatine phosphate to fuel about 10 seconds of activity. However, when the muscle fibers are at rest, they can regenerate their creatine phosphate stores by using ATP obtained from nutrients. A high-energy phosphate molecule is transferred to a creatine molecule, producing ADP plus a new molecule of creatine phosphate.

Glycolysis Produces Pyruvate and ATP

Glycolysis (*glyco-* = "sugar"; *-lysis* = "to break"), the breakdown of glucose into *pyruvate*, is the fastest method of generating ATP from nutrients (Fig. 7.9B). The initial source of glucose is *glycogen,* a glucose polymer stored within muscle fiber. Glycogen must be broken into individual glucose molecules (actually, glucose-6-phosphate), which are then used to generate ATP. This reaction is called **glycogenolysis,** and is catalyzed by an enzyme, *glycogen phosphorylase.* Blood glucose can also be used, but glycogen is more abundant and supplies glucose-6-phosphate at a faster rate.

Glycolysis occurs in the cytosol of muscle cells and is an *anaerobic* process; that is, it does not *require* oxygen, although it can also occur in the presence of oxygen. It generates three ATP molecules per glucose molecule derived from glycogen. When blood glucose is used, only two ATP molecules are generated per glucose molecule, because it costs one ATP molecule to convert blood glucose into glucose-6-phosphate.

Pyruvate, the end-product of glycolysis, can be a source of additional ATP. However, for reasons discussed in ➡ Chapter 15, pyruvate is frequently converted first into **lactic acid.** About half of this lactic acid will be converted back into pyruvate within the same muscle cell, during the

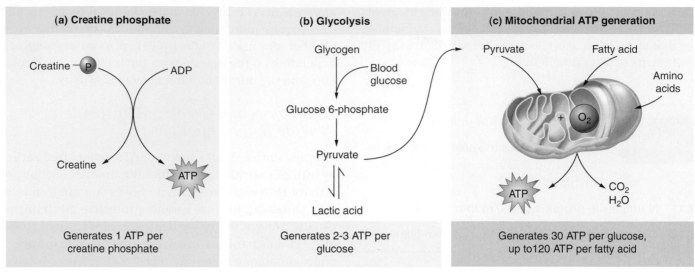

Figure 7.9. Muscle energy. A. Creatine phosphate transfers its phosphate group to ADP to generate ATP. When ATP is abundant, this reaction runs in reverse to regenerate creatine phosphate molecules at the expense of ATP. **B.** Glycolysis converts glucose produced by glycogen breakdown (or arriving in blood) into pyruvate. Pyruvate can be converted into lactic acid, and lactic acid can be converted back into pyruvate. **C.** Mitochondria generate large amounts of ATP from pyruvate, fatty acids, or amino acids. *Which substance can be used directly to generate ATP—lactic acid or pyruvate?*

infinitesimally brief rest between individual contractions (muscle fibers in a contracting muscle take turns producing force). Most of the remaining lactic acid will travel to nearby muscle cells where it, too, will be converted back into pyruvate. However, a very small amount of lactic acid travels to the liver and is converted into glucose.

Case Note

7.8. Is Hammid suffering from a shortage of ATP, creatine phosphate, or calcium?

Mitochondrial ATP Production Meets Long-Term Energy Needs

Mitochondria contain a host of enzymes that completely break down various nutrients and generate large amounts of ATP (Fig. 7.9C). The complex series of chemical reactions performed by these enzymes can be divided into two stages—the *citric acid cycle* and *mitochondrial respiration*—and is discussed in detail in ➡ Chapter 15.

Mitochondrial ATP generation is described as *aerobic* because, unlike glycolysis, it *requires* oxygen. Most of the required oxygen comes from oxygen bound to hemoglobin in blood, but some of it is obtained from oxygen bound to myoglobin in muscle. Although their oxygen need is absolute, mitochondria are not picky about their nutrient

source—they effectively metabolize pyruvate (generated by glycolysis) and fatty acids. The fatty acids can come from blood or from lipid droplets within the muscle fiber.

The reaction is as follows:

$$\text{Pyruvate or fatty acids} + O_2 \rightarrow CO_2 + H_2O + ATP$$

Mitochondria provide a slow and steady supply of ATP—they generate 30 ATPs per glucose molecule (recall that glycolysis also generates 2 to 3 ATPs per glucose molecule), or a staggering 120 ATPs per fatty acid molecule.

Remember This! Mitochondria do not directly break down glucose to generate ATP. Instead, they use pyruvate generated by glycolysis.

Especially in individuals consuming more protein than their body requires, blood amino acids are taken up by muscle fibers and used by mitochondria to generate ATP. However, body protein is not usually broken down to generate amino acids for energy. Most body organs are built on a framework of protein; hence, proteins are used for fuel only as a last resort—using amino acids to generate ATP is akin to burning the house down to keep warm. This is why, for example, people who are starving lose muscle mass—they are burning muscle protein to stay alive.

Case Notes

7.9. Our patient Hammid cannot convert glycogen into glucose. Name the enzyme that accomplishes this reaction.

7.10. Which process is defective in Hammid's muscle cells—glycogenolysis or glycolysis?

Remember This! The terms *anaerobic metabolism* and *glycolysis* are often used synonymously, but erroneously, since glycolysis is the necessary first step in both aerobic metabolism and anaerobic metabolism.

Muscle Cells Contract Aerobically or Anaerobically

Jogging and other endurance activities are often described as "aerobic exercise" because oxygen-dependent mitochondria generate most of the required ATP, from glycolysis-derived pyruvate, fatty acids, and perhaps amino acids. Muscle cells function aerobically if three conditions are met:

1. The muscle cell contains abundant mitochondria.
2. The muscle cell is supplied with adequate oxygen.
3. The ATP needs of the muscle cell are low or moderate.

Conversely, athletic activities requiring short-lived, powerful contractions are often described as "anaerobic exercises," because they meet their ATP needs using processes that do not require oxygen (stored ATP, creatine phosphate, and glycolysis). Anaerobic metabolism depends on muscle glycogen stores, since blood glucose delivery is too slow to keep up with demand. Most of the lactic acid produced as a glycolytic end product travels to nearby muscle cells for further metabolism. However, since lactic acid is generated faster than the noncontracting cells can convert it back into pyruvate, lactic acid often accumulates in blood. Most investigators do not believe that this lactic acid has any deleterious effects on muscle function. Nevertheless, for reasons to be discussed, muscle cells cannot generate ATP by anaerobic metabolism for long without tiring.

Anaerobic metabolism occurs in three circumstances. The first is a matter of imposed demand; that is, hard work. Anaerobic metabolism provides an extra energy kick when oxygen delivery to the muscle cell cannot keep up with the needs of mitochondrial respiration. The second is a matter of anatomy: anaerobic metabolism preferentially occurs in some muscle cells, called *glycolytic muscle fibers* (discussed later in the text). The third is a matter of timing: we use anaerobic metabolism when we begin to exercise, because the mitochondria take a few minutes to make enough ATP. It is important to note that the latter two circumstances do not reflect inadequate oxygen supply.

Skeletal Muscle Fibers Are Oxidative or Glycolytic

Muscle fibers can be classified according to their primary method of ATP generation. **Slow-twitch (oxidative, type I) fibers** are optimized for aerobic metabolism (Fig. 7.10).

They contain many mitochondria and an abundant supply of myoglobin, which stores oxygen. Slow-twitch fibers are packed with blood vessels that keep them supplied with glucose, oxygen, and fatty acids. Slow-twitch fibers are generally thin, slow to contract, and slow to fatigue. They are thus well suited to muscles that are continuously at work, such as the muscles that maintain posture. They also come into play during endurance exercise. Myoglobin is reddish, and slow-twitch fibers, reflecting their high myoglobin content, are dark reddish-brown.

On the other hand, **fast-twitch (glycolytic, type II) fibers** are optimized for anaerobic metabolism (Table 7.2). They need large supplies of creatine phosphate, glycolytic enzymes, and glycogen because the muscle fiber will generate only three ATPs per glucose molecule. They have less myoglobin, fewer mitochondria, and

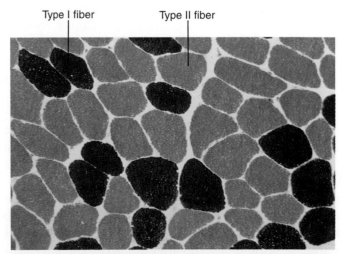

Type I fiber   Type II fiber

Figure 7.10. Muscle fiber types. The muscle fibers in this micrograph have been stained for the slow type of myosin found in slow-twitch (oxidative) muscle fibers. *Which fibers would contain fewer mitochondria—the darker cells or the lighter cells?*

Table 7.2   Muscle Fiber Types

Characteristic	Fast-Twitch Glycolytic	Slow-Twitch Oxidative
Appearance	White	Red
Primary ATP source	Anaerobic metabolism	Aerobic metabolism
Mitochondria/capillaries	Few	Many
Glycogen reserves	High	Low
Myoglobin content	Low	High
Rate of fatigue	Rapid	Slow
Fiber size	Large	Small
Contraction speed	Fast	Slow

fewer blood vessels than slow-twitch fibers. Thus, fast-twitch fibers are pale or whitish. Although they tire quickly, they are large and strong; thus they are well suited to explosive, large movements (such as lifting a heavy box or sprinting). Want to know more? Refer to the *Type IIa muscle fibers: The Best of Both Worlds* box on **http://thepoint.lww.com/McConnellandHull.com** for information about "superfibers" that combine the advantages of slow- and fast-twitch types.

To memorize these distinctions, it may help you to recall that chicken, turkey, and quail breast is "white meat" because it is composed mainly of fast-twitch fibers to power intense wing motion for short flights. By contrast, ducks and doves are migratory birds and their breast meat is reddish "dark meat" because it is composed of slow-twitch fibers to power sustained flights over hundreds of miles.

Most human skeletal muscles are a mixture of slow- and fast-twitch fibers; however, the fibers of any given motor unit are all of the same type. The percentage of fast and slow fibers in each muscle is genetically determined: some people have more fast-twitch fibers in certain muscles; others have more slow-twitch fibers in the same muscles. What's more, proportions vary according to muscle location and function. For example, the muscles of the upper limb and shoulder are predominantly fast-twitch fibers because they are used intermittently and briefly to produce large amounts of force for activities such as manipulating tools, lifting, or throwing. The muscles that power eye movements are composed entirely of

fast-twitch fibers. By contrast, the muscles of the spine and neck are predominantly slow-twitch fibers, because these muscles are in constant use maintaining posture.

Case Note

7.11. Recall that Hammid cannot walk on his toes without cramping. The muscle required for toe-walking, the *gastrocnemius,* has few mitochondria and large muscle fibers. Is it primarily composed of fast-twitch or slow-twitch fibers?

Case Discussion

Muscle Energy Metabolism: The Case of Hammid S.

Hammid suffers from a genetic defect in which he lacks an enzyme—glycogen phosphorylase—that is essential for the breakdown of glycogen. This reaction, called glycogenolysis, is necessary to provide the large amounts of glucose required for strenuous muscle activity.

Understanding how muscle gets its energy supplies is the key to understanding Hammid's signs and symptoms (Fig. 7.11). Recall that muscle obtains energy in three different ways:

1. ATP stores and creatine phosphate fuel the first few seconds of any contraction.
2. Glycogenolysis (glycogen breakdown), followed by glycolysis (pyruvate generation from glucose) can also generate energy relatively quickly at the beginning of the contraction. This process also provides an extra "kick" of energy when large amounts of ATP are needed in a short time period.
3. Aerobic metabolism, which requires oxygen to metabolize pyruvate (generated by glycolysis) or fatty acids, provides a steady supply of ATP over the long term. This process can use muscle stores of glycogen and fat or blood supplies of glucose and fatty acids.

Note that Hammid has no difficulty *initiating* muscle contractions, because his muscles have a normal, small store of ATP and creatine phosphate. This enables him to get under way. Nor is his long-term daily activity impaired—he is okay as long as demand is low. Using mitochondrial respiration, he can burn fatty acids from fat, or amino acids from protein, and he can even burn glucose obtained from his blood. But when demand is

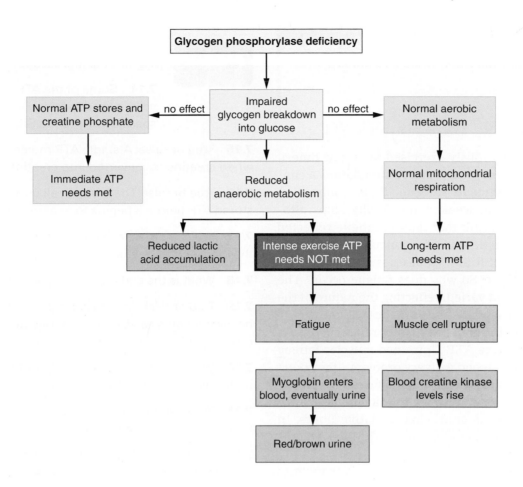

Figure 7.11. Muscle energy metabolism and Hammid S. *How do we know that some of Hammid's muscle cells have ruptured?*

high, his body cannot supply glucose by breaking down his abundant stores of glycogen because his defective gene cannot make the enzyme necessary to do the job. Hammid's problem therefore arises when he engages in sustained, vigorous activities that exhaust the available fuel. After a few minutes of strenuous effort, he consumes his entire supply of blood glucose; mitochondrial respiration is too slow to supply all of the demands for ATP, and his ability to obtain glucose from glycogen is defective.

Confirming the diagnosis is the important observation that Hammid's blood lactic acid did not rise as it normally should with strenuous activity. Why? Because during intense exercise, a normal person can use glycogenolysis to generate the large amount of glucose required for anaerobic metabolism. As glycolysis rapidly breaks down many glucose molecules into pyruvate, the pyruvate is converted into lactic acid. However, Hammid's metabolism is not normal—he can't break down glycogen to supply glucose. His glycolytic system must rely on blood glucose alone and quickly exhausts

the supply before excess pyruvate can accumulate and be converted into lactate.

When Hammid's muscles call for large amounts of fuel, the call goes unanswered, and the ATP levels in the muscle cell fall dangerously low. As a result, muscle cells rupture, releasing their contents (which include myoglobin and the creatine kinase enzyme) into the blood and eventually into the urine. Muscle cramps occur, creatine kinase levels are elevated in Hammid's blood, and myoglobin stains his urine brown.

Hammid's parents were advised to steer him away from vigorous activities like sprinting and soccer and to encourage moderate exercise such as jogging or hiking, which would increase the ability of his muscles to perform mitochondrial respiration. They were also instructed to make sure that he consumes a candy bar or a sugar-containing drink such as orange juice about 30 minutes before exercise in order to elevate his blood glucose. Finally, they were advised to insist that Hammid stop exercising if cramping occurred.

Skeletal Muscle Experiences Fatigue

When muscle is vigorously exercised for a long time, it loses the ability to respond to nerve stimulation, a condition known as **muscle fatigue.** Fiber contraction becomes weaker and weaker and finally stops altogether. We used to think that muscle fatigue reflected ATP depletion or lactic acid accumulation, but we now know that neither of these hypotheses can account for most cases of fatigue. So why does fatigue occur? The causes are many and varied, reflecting the nature of the exercise and the training state of the individual.

The major limit in submaximal endurance exercise is the ability to generate ATP. Untrained muscles fatigue because they have a blood delivery problem—they don't have enough capillaries perfusing their oxidative fibers. One of the benefits of endurance training is the growth of more blood vessels supplying oxidative fibers. In these trained individuals, glycogen stores then become the limiting factor.

Fatigue in maximal anaerobic exercise is thought to reflect phosphate accumulation. Recall that the energy is liberated from ATP by cleaving off one phosphate. Maximal exercise uses a lot of ATP in a short time, resulting in the accumulation of many phosphates. Phosphate interferes with contraction directly, by blocking cross-bridge formation, as well as indirectly, by reacting with calcium in the SR and reducing its release into the sarcoplasm.

However, we rarely see true muscle fatigue of the types described above, which are also known as *peripheral fatigue.* As accomplished athletes say, "The mind wears out before the muscle." Essentially, untrained athletes find the sensations created by exercise unpleasant; they, therefore, lessen their effort in order to gain relief. Also, many conditions (such as increased body temperature) lead the brain to send fewer signals to muscles. Thus, the most common cause of fatigue originates in the central nervous system and is thus called *central fatigue.*

The Mechanics of Muscle Contraction

The force of a muscle contraction is exquisitely controlled; we can use the same muscles to hold a delicate glass ornament and to wring water from a face towel. The force an individual muscle exerts depends on:

● The force exerted by each contracting fiber
● The number of motor units contracting

Individual Fibers Provide Force

Recall that muscle contraction is accomplished by cross-bridges formed between the myosin heads of the thick filaments and the actin binding sites of thin filaments.

Contraction strength depends, therefore, upon how many cross-bridges form. Cross-bridge number, in turn, depends on how many myosin heads can reach the thin filaments and how many binding sites on the thin filaments are available.

Contractile Power Depends on Muscle Fiber Length

Sarcomere length, and thus muscle length, is one determinant of the force developed by an individual muscle fiber. At the optimal sarcomere length, all of the myosin heads are positioned to be in contact with actin molecules and form cross-bridges, and the contraction will generate the maximum amount of tension possible (Fig. 7.12A, middle). This property of muscle is called the *length–tension relationship*. At very short sarcomere lengths, the thin filaments are pulled so close that they meet in the middle and overlap, which covers their binding sites and

interferes with the ability to form cross-bridges with thick filaments (Fig. 7.12A, left).

At very long sarcomere lengths, the exact opposite occurs—the thin filaments are so far apart that they lose most of their contact with the thick filaments (Fig. 7.12A, right). Thus, they contract poorly.

If we remember that sarcomeres are lined up end to end in a muscle, it is possible to extrapolate this length–tension relationship to the behavior of an entire muscle. Try performing a biceps curl. Holding a weight in your hand, start with the arm straight, elbow extended. In this position, the biceps muscle is relaxed and lengthened. Lift the weight, palm up, by flexing your elbow. As you do, the muscle shortens. Notice that the action is most difficult at the very beginning and very end of the curl because the sarcomeres are too long at the outset and too short at the end. Conversely, the middle portion of the curl is relatively easy, because the sarcomeres are at their optimal length and can generate the most force.

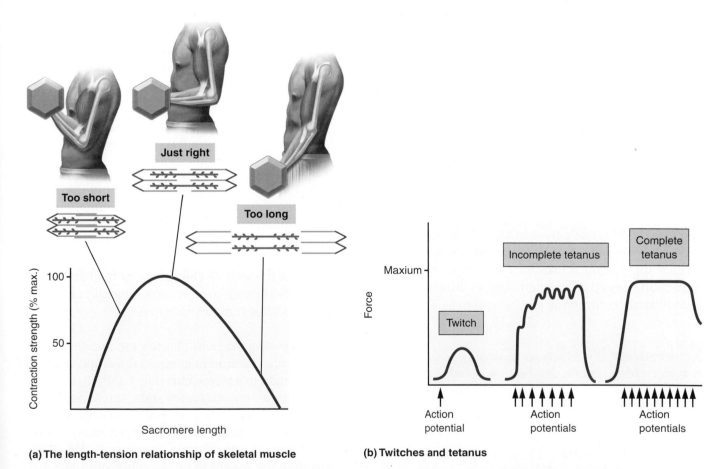

(a) The length-tension relationship of skeletal muscle

(b) Twitches and tetanus

Figure 7.12. Determinants of force. A. The force generated by individual fibers varies according to the muscle length, which determines the sarcomere length. At the optimum length, all myosin heads are able to form cross-bridges with actin molecules. **B.** The force generated by individual fibers depends upon the frequency of stimulation. Everyday productive muscle contractions usually involve incomplete tetanus. *Which type of contraction is invoked by a single action potential?*

Physiological Contractions Are Unfused Tetanus

A single action potential in a muscle fiber results in a weak, transient muscle contraction called a *twitch* (Fig. 7.12B, far left). A slightly stronger state of contraction results if a second action potential occurs before the twitch is finished; that is, the force of the two twitches is *summed* together. Subsequent action potentials result in progressively greater force, until a third state called *incomplete tetanus* is reached, in which the muscle fiber only relaxes slightly between subsequent contractions (Fig. 7.12B, right). Only in maximal contractions, such as lifting the heaviest weight possible for a single repetition, do we see the fourth state of contraction, *complete tetanus,* in which action potentials arrive so frequently that the fiber does not relax at all between contractions (Fig. 7.12B, right).

These responses to different action potential frequencies underline the importance of *calcium* in force generation—recall that calcium enables cross-bridge formation, and calcium reuptake into the sarcoplasmic reticulum results in relaxation. A single action potential does not release enough calcium to bind all of the troponin molecules, so not enough cross-bridges can form to generate maximum force. However, with repeated stimulation, the rate of Ca^{2+} release is greater than the rate of Ca^{2+} reuptake, so Ca^{2+} levels rise progressively higher with each successive action potential. The rate of calcium release is so high in complete tetanus that all binding sites are continually occupied, generating continuous, maximal force.

In everyday contractions, each skeletal muscle fiber receives action potentials at a high enough frequency to induce incomplete tetanus. In other words, contraction in an individual muscle fiber is all or none—*individual muscle fibers contract maximally or not at all.* We do not perceive the partial relaxations between subsequent contractions, because muscle fibers in different motor units alternate contracting and relaxing.

> **Remember This!** In an everyday contraction at a given fiber length, contraction of individual muscle fibers is all or none, as the fiber contracts in incomplete tetanus.

Contractile Power Depends on Number of Motor Units Involved

Recall that a motor unit is a group of muscle fibers innervated by a single motor neuron (Fig. 7.3). Motor units vary in size and in the force they can generate: slow-twitch muscle fibers are usually grouped into small motor units, whereas motor units containing fast-twitch muscle fibers are usually larger. Motor units, like individual muscle fibers, contract maximally or not at all. Thus, the amount of contractile power generated by an entire muscle depends on the number and type of motor units involved. The process of adding additional motor units to produce a graded increase of force is called *recruitment*.

As a skeletal muscle contracts, first only a few motor units are stimulated, and they are recruited in specific order. Slow-twitch fibers are recruited first; fast-twitch fibers are recruited if more force is necessary. Even at peak muscle force, not all motor units are active at the same time: they rotate in and out of service, some relaxing after using up their resources while others fill the need for contractile force until they, too, need a break.

Muscle fibers of various motor units are intermingled, so that two fibers of the same motor unit are not adjacent to each other—some will be on one side of the muscle or deep within, others on the other side or superficial. This means that even a weak contraction (which recruits only a few motor units) will recruit muscle fibers scattered throughout the muscle to ensure symmetrical contraction. Otherwise, a weak contraction would activate only one region of the muscle and the contraction would pull unevenly on the bone.

Muscle Fiber Contraction May or May Not Produce Movement

So far our assumption has been that contraction of a muscle fiber causes it to shorten. These dynamic or **isotonic contractions**—literally, "same tone" or "same force" contractions—are the stuff of everyday movement. Constant force is maintained over the course of a contraction, but the length changes. For example, lifting a weight in the gym or chewing your food are motions powered by isotonic contractions. Isotonic contractions can be classified into two subtypes:

- *Concentric contractions* shorten the muscle, bringing the muscle attachment closer to the origin, as in raising a weight in a biceps curl (Fig. 7.13B). In concentric contractions, myofilaments slide; sarcomeres, fibers, and muscles shorten; and movement occurs.
- *Eccentric contractions*, conversely, generate a restraining force as the muscle lengthens (Fig. 7.13C), enabling the weight to be smoothly and controllably lowered following a biceps curl. In eccentric contractions, the myosin heads grab onto the actin filaments and slow the rate of movement, somewhat like applying a brake. Contrary to intuition, eccentric contractions

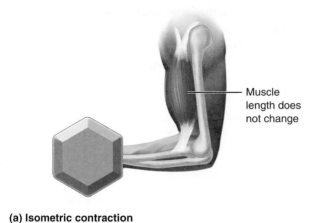

Muscle
length does
not change

(a) Isometric contraction

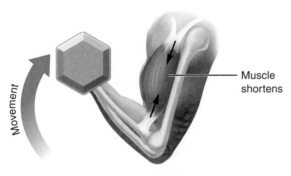

Movement

Muscle
shortens

(b) Concentric isotonic contraction

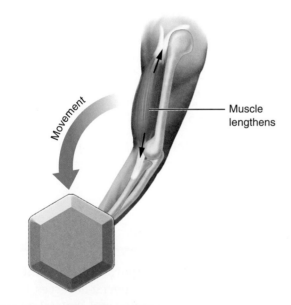

Movement

Muscle
lengthens

(c) Eccentric isotonic contraction

Figure 7.13. Isometric and concentric contractions. A. Isometric contractions, such as those keeping a heavy weight stationary, generate force but do not change muscle length. **B.** In a concentric contraction, the muscle shortens as it generates force to (in this example) raise a weight. **C.** In an eccentric contraction, the muscle lengthens as it generates force to (in this example) lower a weight. *During which type(s) of contraction does cross-bridge cycling occur?*

are actually more powerful than concentric contractions; that is, you use greater strength in lowering a heavy object than in lifting it.

However, the thing common to all muscle contraction is *force*, not movement. For example, if you try to lift a weight far beyond your strength, your muscles will contract but the weight won't budge: the fibers are generating force but not shortening because you are attempting to move an object that is—at least for you—immovable. Contractions that do not alter muscle length are called **isometric contractions**—literally, "same length" contractions (Fig. 7.13A). Force is generated and the muscle tenses; however, the myofibrils do not slide and the muscle does not change length. We perform isometric contractions all the time in order to oppose the downward force of gravity. For instance, the weight lifter in Figure 7.13C is exerting just enough upward force to offset the force of gravity pulling the weight downward. Similarly, isometric contractions maintain our upright *body posture.* Think about it: you do not have to concentrate on contracting your neck muscles in order to keep your head erect during the day, nor do you have to think about keeping your spine erect while sitting or standing. Subconscious, imperceptible isometric contractions do the job so you can focus on other matters.

Muscle tone is a state of subconscious isometric contraction that occurs even in voluntarily relaxed muscle. It maintains muscle in a healthy state, much the way that normal physical stress maintains healthy bone. If the nerve supply to a muscle is interrupted, perhaps because of an accident, the muscle loses its tone and becomes flaccid (soft, flabby). If the nerve connection is not reestablished, muscle fibers begin to shrink (*atrophy*). Complete lack of muscle tone is called *flaccid paralysis* and occurs when somatic motor nerves are unable to deliver action potentials to the muscle. For example, flaccid paralysis occurs with administration of Botox, which blocks the release of ACh from the somatic motor neuron at the neuromuscular synapse. The loss of facial wrinkles is due to induced flaccid paralysis of facial muscles that bunch skin into wrinkles. Flaccid paralysis also occurs with the severing of a peripheral nerve, or with severe spinal cord injury. In each of these examples the brain is not involved. By contrast, *spastic paralysis* is due to damage to the brain, which impairs the control of muscles. With brain lesions, voluntary control is lost, leaving the spinal cord to send uncontrolled action potentials to muscle, which causes uncontrolled muscle contraction. For example, the awkward, stiff gait of some patients with brain damage from stroke, cerebral palsy, or head injury is a manifestation of spastic paralysis.

> *Remember This!* The thing common to all muscle contraction is force, not movement.

Exercise Has a Positive Effect on Muscles

The saying "use it or lose it" applies to muscles just as it does to the practice of a skill. A worked muscle is a healthy muscle, and muscle improves its health according to the type of work it performs. Exercise improves the power and endurance of skeletal muscle. But the greatest benefit of exercise lies elsewhere: every system in the body is improved by physical exercise ➡ (see Chapter 18). Among nonsmokers, regular exercise is arguably the most important single activity for improving general health. Smokers benefit from exercise too, but the gain is small compared with the positive effect of quitting smoking.

Muscle *power* is improved by strength training regimes (also called *resistance training*), such as weight lifting, that increase muscle size. These exercises require repeated short bursts of powerful muscle action that overload and stress the muscle. We used to think that adult muscles grew only by enlarging existing muscle fibers with new myofibrils. Although this process does occur, it now seems certain that significant muscle growth reflects the participation of muscle stem cells, the satellite cells. Recall that the stem cells of adult muscle are located at the periphery of the muscle fiber. Exercise stimulates these stem cells to proliferate, producing new myoblasts that fuse with existing muscle fibers to make them larger. Myoblasts may also fuse with each other to produce entirely new muscle fibers.

Muscle power is critical in athletic endeavors requiring a large amount of force output, including the 100-yard dash, the pole vault, the high jump, and weight lifting. Note that these activities are often called *anaerobic* because they rely on anaerobic metabolism. Anaerobic exercises also enhance the ability of the larger, stronger muscle cells to produce ATP, using creatine phosphate and glycolysis.

Muscle *endurance* (resistance to fatigue) is improved by *aerobic* exercise that relies on mitochondrial ATP generation. These exercises require sustained low-level muscle action to improve muscle blood supply and increase the number of mitochondria. Endurance exercise also activates satellite cells, but muscles do not grow significantly bigger. Athletic performances that rely on aerobic conditioning include long-distance running, cross-country skiing, cycling, and long-distance swimming events. As discussed in later chapters, aerobic exercise also exerts beneficial effects on many other body systems, particularly the cardiovascular and respiratory systems.

 7.25 What is the difference between incomplete and complete tetanus, and which occurs more frequently?

7.26 True or false: Muscle contraction is always the strongest when the muscle is as long as possible.

7.27 Which type of motor unit is recruited first—that containing slow-twitch (type I) or fast-twitch (type II) fibers?

7.28 To generate a stronger contraction in skeletal muscle independent of muscle length, do we vary the force produced by each muscle fiber, alter the force produced by each motor unit, or vary the number of motor units recruited?

7.29 Give an example of an isometric and an isotonic muscle contraction.

7.30 Name an aerobic and an anaerobic exercise.

Smooth Muscle

Despite its functional importance, it is difficult for smooth muscle to get the respect it deserves. In the gym or on the athletic field, cardiac and skeletal muscles get all of the attention, as sweaty athletes admire their muscles and count their heart rates. Meanwhile, smooth muscle labors along, slow and reliable, tirelessly and quietly doing various jobs, such as massaging food through the gut to provide energy for the show, regulating blood flow by adjusting the diameter of blood vessels, and tightening sphincter muscles to hold urine and feces for release at another time.

Sheets of smooth muscle occur in the walls of all but the smallest blood vessels and in the walls of hollow organs: the intestines, the bronchial airway, the urinary and reproductive tracts, and others.

Smooth muscle takes about 25 times as long to contract as skeletal muscle and consumes only about 1% as much energy. Since smooth muscle contractions are relatively slow and do not generate the explosive force characteristic of skeletal muscle, aerobic metabolism, using nutrients from the blood, can easily meet smooth muscle's low energy needs—no need for anaerobic metabolism or stored glycogen here. Actin–myosin cross-bridges may latch semipermanently in a **latch state,** not unlike the rigor mortis that occurs after death, in which the cross-bridge cycle ceases while actin and myosin remain bound together. This latch state enables smooth muscle to maintain muscle tension without expending any energy at all, a state called *smooth muscle tone.* This low-level contraction is necessary for the proper function of blood vessels and other hollow structures that must maintain their size or shape against constant pressure.

The structure of smooth muscle cells and tissue is fundamentally different from that of skeletal and cardiac muscle (Fig. 7.14; Table 7.1). Not surprisingly, these structural differences account for the different contraction characteristics of smooth muscle: its slow, sustainable contraction; its tirelessness; its stretchiness; and its ability to propagate automatic waves of contraction.

Smooth Muscle Differs Structurally from Skeletal Muscle

Recall that in skeletal muscle the muscle cells are called fibers because they are very long and thin. In contrast, smooth muscle cells are short and plump. They have pointed ends and a bulge in the middle to accommodate a single nucleus, which lies squarely in the center of the cell, not to one side as in cardiac and skeletal muscle fibers. They are small for two reasons: their contractions are relatively weak, requiring fewer myofibrils, and they rely primarily on aerobic metabolism, which means they don't require large stores of glycogen.

Smooth muscle cells are formed upon a three-dimensional criss-cross structure of noncontractile **intermediate filaments** ◀ (Chapter 3), which are interconnected somewhat like a schoolyard jungle gym (intermediate filaments also strengthen skeletal muscle fibers but are organized differently). The filaments are interconnected by *dense bodies,* small dense protein discs scattered over the sarcolemma (muscle cell membrane). Dense bodies are the functional equivalent of the Z-disc in skeletal muscle; that is, they are anchor points for the filaments. Smooth muscle contraction, like that of skeletal muscle, is enabled by myofilaments—thick myosin filaments and thin actin filaments. These myofilaments are not arranged in perfectly ordered ranks, so that, unlike skeletal muscle, no dark-and-light pattern of striae (stripes) is created. Because of the

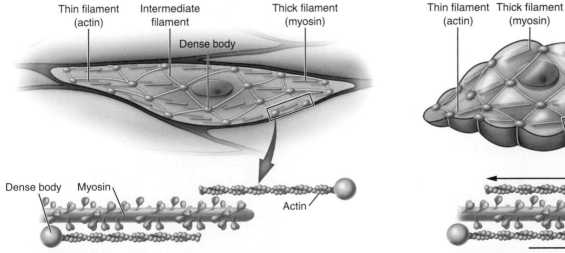

(a) Relaxed smooth muscle

(b) Contracted smooth muscle

Figure 7.14. Smooth muscle. A. A relaxed smooth muscle cell. Myosin molecules are interspersed between the actin molecules. **B.** A contracted cell. Myosin heads pull on thin filaments, increasing the overlap between the two filament types and shortening the cell. *How are actin molecules anchored—by Z-lines or dense bodies?*

arrangement of myofilaments and their association with the dense bodies, smooth muscle cells bulge out as they shorten (Fig. 7.14B). Even though smooth muscle cells are much shorter than skeletal muscle cells, the myofilaments inside of smooth muscle cells are longer. In addition, the thick (myosin) filaments in smooth muscle have protruding heads along their entire length, so there is no headless zone like the "golf club shaft" in skeletal myosin. As a result, the length–tension relationship illustrated in Figure 7.12 does not apply to smooth muscle. Even when smooth muscle cells are stretched greatly, at least some of the myosin heads can still contact actin, so the filaments can continue to claw out contractile force regardless of cell length.

The arrangement of smooth muscle cells into tissues also contributes to the muscle's stretchiness. Most smooth muscle cells are layered upon one another to form sheets of cells similar to multiple layers of shingles on a roof. This arrangement enables smooth muscle to be stretched in many directions without tearing as the cells slide across one another to accommodate the stretch.

> *Remember This!* **Intermediate filaments form the scaffolding of a smooth muscle cell, and myofilaments contract the cell.**

In Smooth Muscle, Calcium Acts on Myosin, Not Actin

To understand smooth muscle contraction and how it differs from skeletal muscle contraction, recall some of the details of the latter. Cross-bridge cycling requires that myosin heads in thick filaments bind to actin in thin filaments in order to claw out a contraction, but access to the thin-filament binding sites is controlled by troponin. A surge of Ca^{2+} ions stimulates troponin to expose the binding site. The myosin head then engages the actin binding site for the power stroke of contraction. In smooth muscle cells, the steps of the cross-bridge cycle detailed in Figure 7.7 are still relevant. However, smooth muscle differs in both the *source* and the *role* of the Ca^{2+} ions:

- *Source of Ca^{2+} ions.* Smooth muscle cells have very little SR. Instead, in smooth muscle, Ca^{2+} influx comes mainly through the cell membrane from extracellular fluid.
- *Role of Ca^{2+} ions.* Smooth muscle cells contain no troponin, so myosin binding sites on the thin filaments are always exposed. Instead of controlling the access to thin-filament binding sites, calcium in smooth muscle regulates the activity of the myosin heads on thick filaments. That is, only if calcium is present does the myosin cleave ATP and move through the cross-bridge cycle.

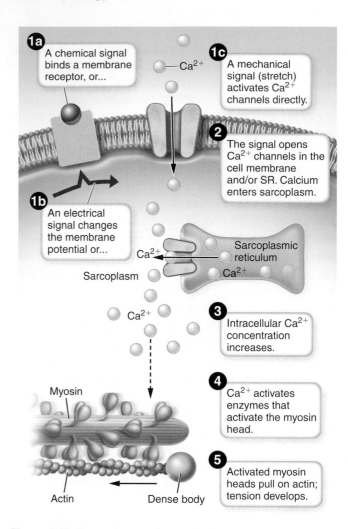

Figure 7.15. Smooth muscle regulation. Calcium enters the cytoplasm in response to a chemical, mechanical, or electrical signal and indirectly stimulates activity of the myosin head. The activated myosin molecules form cross-bridges with actin and contract the muscle. *True or false: Most of the calcium comes from the extracellular fluid.*

Because of these two important differences, the events of smooth muscle contraction differ from those in skeletal muscle (Fig. 7.15). Considerable variation exists in the mechanism of smooth muscle contraction, but a typical sequence is as follows:

1. An event—a chemical signal (e.g., neurotransmitter), an electrical signal (e.g., graded or action potential), or a mechanical signal (e.g., stretch)—activates calcium channels in the cell membrane and, in some cases, in the SR. Chemical signals must use a second-messenger system. These signals are discussed further below.
2. Calcium enters the cytoplasm from the extracellular fluid and possibly the limited amount of SR.
3. The intracellular Ca^{2+} concentration increases.
4. Through a number of intervening enzymatic steps, calcium activates myosin heads.

5. Activated myosin heads form cross-bridges with actin molecules, and the filaments slide upon one another, causing muscle contraction.

> **Remember This!** The myosin heads are regulated in smooth muscle; the binding sites on actin molecules are regulated in skeletal muscle.

As with skeletal muscle, relaxation of smooth muscle begins when calcium is actively removed from the cytoplasm. In the case of smooth muscle, it is accomplished primarily by membrane transport proteins. However, recall that the myosin heads were enzymatically activated to initiate muscle contraction. They must therefore be enzymatically deactivated in order to stop cross-bridge cycling and induce smooth muscle relaxation. The enzyme *myosin phosphatase* does the job.

Case Note

7.17. Based on the information provided here, will Hammid have trouble with smooth muscle function? Why or why not?

Smooth Muscle Contraction Is Involuntary

Smooth muscle movement is involuntary; that is, it is not subject to conscious control, like skeletal muscle. Some smooth muscle is innervated by the autonomic nervous system, an important division of the nervous system that itself is not subject to voluntary control (see ➡ Chapter 8).

However, autonomic nerves do not innervate all smooth muscles. Some smooth muscles are stimulated to contract by hormones or by local chemical signals such as prostaglandins, hydrogen ions, and gases (carbon dioxide, oxygen, and nitric oxide). Consider, for example, the smooth muscle lining blood vessels ➡ (Chapter 11). Smooth muscle cells in the walls of blood vessels contract or relax in response to locally produced paracrine factors secreted by neighboring cells that signal their need for more or less blood flow. Contraction of these muscle cells constricts the blood vessel, reducing blood flow, whereas relaxation expands the vessel, increasing blood flow.

Smooth muscle is also stimulated by mechanical signals. This homeostatic mechanism prevents overstretching of blood vessels and other tissues and thereby prevents injury. Consider, for instance, a stomach overstretched by a very large meal. The stomach muscle begins to contract as the stomach is filled to capacity, preventing tearing of the stomach muscle (and, incidentally, inducing discomfort that prevents further food consumption).

Finally, the cells in some smooth muscles have unstable membrane potentials, which generate self-stimulating action potentials called *pacemaker activity*. In the gastrointestinal tract, for example, pacemaker activity generates waves of smooth muscle contraction (*peristalsis*) that massage food from one end of the tract to the other ➡ (Chapter 14). As we will see in ➡ Chapter 11, cardiac muscle is also self-stimulating.

Smooth Muscle Contracts as a Single Unit

Groups of smooth muscle cells contract in unison because the cells are connected to one another by gap junctions (⬅ Chapter 4), tiny liquid tunnels from one cell to the next, which allow rapid spread of the signal through all cells. When an electrical or chemical signal stimulates one cell, the change sweeps through the entire network of muscle cells and they contract as a single unit. Thus, contraction strength in smooth muscle cannot be varied by changing the number of contracting cells, as in skeletal muscle, which contains muscle fibers that are electrically insulated from one another. Instead, the amount of tension generated by individual smooth muscle cells varies according to the amount of calcium allowed into the cell from the extracellular fluid, which in turn activates greater or fewer numbers of myosin heads.

 7.31 True or false: The calcium causing smooth muscle contraction usually comes from the extracellular fluid, but the calcium causing skeletal muscle contraction usually comes from the SR.

7.32 Would you find troponin in smooth muscle?

7.33 To generate a stronger contraction in smooth muscle, do we vary the force produced by each muscle fiber or vary the number of muscle cells contracting?

Skeletal Muscle Actions

Skeletal muscles move bones or stabilize them in certain positions, and (in the case of facial muscles) move skin and associated fascia. Most muscles cross a joint and act to move one bone in relation to the other. The end of the muscle that serves as an anchor for the movement is called the *origin*; the end that moves a body part is the *insertion*.

The contraction of a muscle pulls (never pushes!) the insertion toward the origin. Consider, for instance, the *masseter* muscle, with its origin on the zygomatic process

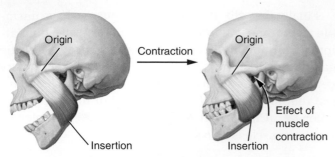

Figure 7.16. Origin and insertion. Most muscles span a joint and attach to two bones. The origin of the muscle attaches to the less movable bone; the insertion to the more movable bone. *In this illustration, does the muscle insert into the mandible or the temporal bone?*

of the temporal bone, and its insertion on the mandible (Fig. 7.16).

Contraction of this muscle closes the jaw, pulling the mandible (insertion) closer to the zygomatic process (origin). The words *origin* and *insertion* may not have literal meaning for the ends of certain muscles of the torso because the body part at both ends move. For example, some muscles attach to the spine at both ends and bend the spine, so it is arguable which end is the origin and which the insertion.

The action exerted by a particular muscle varies according to where it is attached and how the fibers are oriented. For example, a circular muscle surrounds the mouth. When it contracts, it purses the mouth, as in a kiss.

In producing movement, the actions of different muscles often complement or oppose each other. The role of a muscle in a particular movement can be described as follows:

- *Prime mover* (or *agonist*): the main muscle responsible for a given movement. The large quadriceps muscle on the anterior thigh is the prime mover that extends (straightens) the leg at the knee.
- *Antagonist:* a muscle that opposes the action of the prime mover. Antagonists must relax and lengthen to permit the movement, and they often exert the opposite action when they contract. The hamstrings muscles on the posterior leg must relax and lengthen when the quadriceps femoris straightens the leg.
- *Synergist:* a muscle that assists the action of the prime mover. Some synergists, called *fixators*, prevent the movement of a nearby joint. Remember that muscles shorten when they contract, bringing the insertion closer to the origin. Without fixators, the origin would also move toward the insertion. For instance, when we inhale deeply, several neck muscles stop the neck from flexing so that other muscles can elevate the rib cage.

To put these interactions together, let's consider how we raise the arm laterally at the shoulder (abduction). The deltoid muscle is the prime mover; the supraspinatus (a rotator cuff muscle deep to the deltoid) is a synergist important for the initiation of the movement. Gravity is the major antagonist, but muscular antagonists to the deltoid include the pectoralis major and the latissimus dorsi muscles (both muscles adduct the arm). Many muscles act as fixators by stabilizing the scapula, including the trapezius and pectoralis minor. All of these muscles can be visualized in Plate 7.5 at the end of the chapter.

Case Note

7.18. When Hammid walks on his heels, the gastrocnemius muscle contracts and the peroneus longus muscle relaxes. Which muscle is the prime mover and which is the antagonist?

7.34   When a muscle contracts, which part moves more—the origin or the insertion?

7.35   What is the name of a muscle that assists the action of a prime mover?

The Major Skeletal Muscles

The human body contains hundreds of muscles, ranging in size from the large, powerful thigh muscles to the tiny muscles that move our eyes. We cover a subset of these muscles, which we've chosen because they are important in body posture or movement or because they are important landmarks. As discussed in the History of Science box, titled *Medical Art and the History of Human Dissection,* artists have represented the human body for various purposes for millennia, but only began rendering its internal structures for scientific study a few hundred years ago. Use our illustrations and the accompanying tables to learn the location and shape of the major skeletal muscles. You can make the task of learning muscle anatomy easier by (a) learning the word parts used to name muscles and (b) performing the actions of each muscle as you read about it. Figures 7.17 and 7.18 provide an overview of the major superficial muscles. Plates 7.1 to 7. 9 provide more detailed views of the muscles in each region and summarize their important actions.

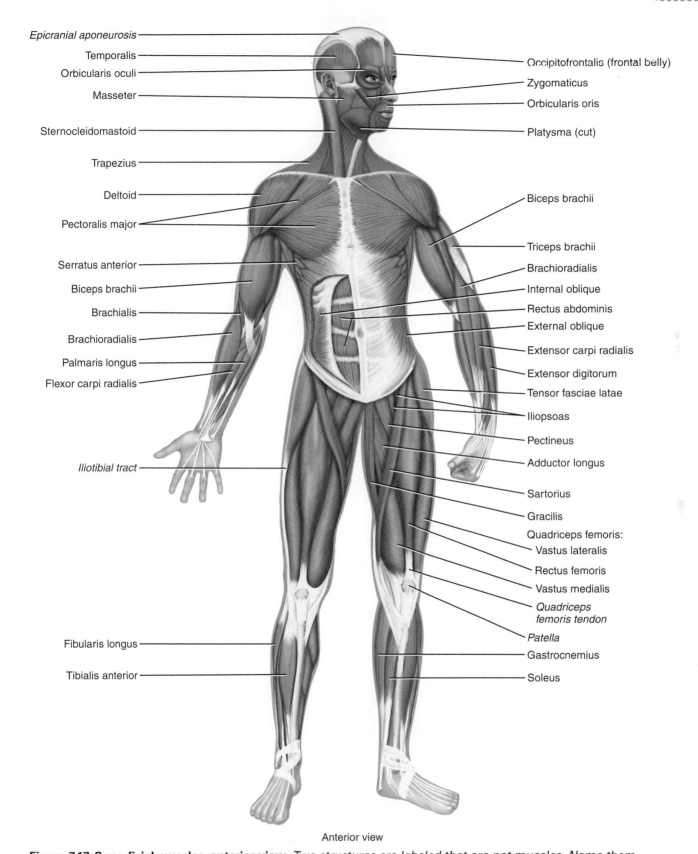

Epicranial aponeurosis
Temporalis
Orbicularis oculi
Masseter
Sternocleidomastoid
Trapezius
Deltoid
Pectoralis major
Serratus anterior
Biceps brachii
Brachialis
Brachioradialis
Palmaris longus
Flexor carpi radialis
Iliotibial tract
Fibularis longus
Tibialis anterior

Occipitofrontalis (frontal belly)
Zygomaticus
Orbicularis oris
Platysma (cut)
Biceps brachii
Triceps brachii
Brachioradialis
Internal oblique
Rectus abdominis
External oblique
Extensor carpi radialis
Extensor digitorum
Tensor fasciae latae
Iliopsoas
Pectineus
Adductor longus
Sartorius
Gracilis
Quadriceps femoris:
Vastus lateralis
Rectus femoris
Vastus medialis
Quadriceps femoris tendon
Patella
Gastrocnemius
Soleus

Anterior view

Figure 7.17. Superficial muscles, anterior view. *Two structures are labeled that are not muscles. Name them.*

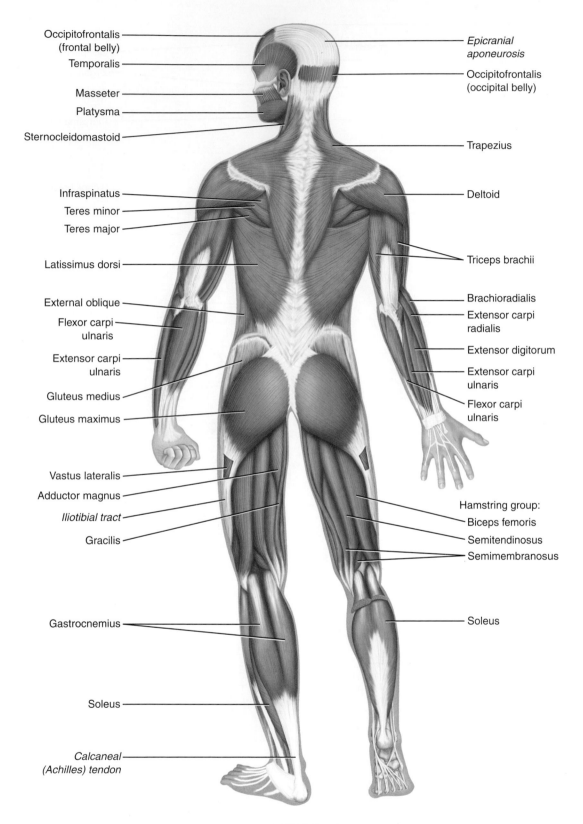

Occipitofrontalis (frontal belly)

Temporalis

Masseter

Platysma

Sternocleidomastoid

Infraspinatus

Teres minor

Teres major

Latissimus dorsi

External oblique

Flexor carpi ulnaris

Extensor carpi ulnaris

Gluteus medius

Gluteus maximus

Vastus lateralis

Adductor magnus

Iliotibial tract

Gracilis

Gastrocnemius

Soleus

Calcaneal (Achilles) tendon

Epicranial aponeurosis

Occipitofrontalis (occipital belly)

Trapezius

Deltoid

Triceps brachii

Brachioradialis

Extensor carpi radialis

Extensor digitorum

Extensor carpi ulnaris

Flexor carpi ulnaris

Hamstring group:
Biceps femoris
Semitendinosus
Semimembranosus

Soleus

Posterior view

Figure 7.18. Superficial muscles, posterior view. *Based on its name and your knowledge of movements at synovial joints, find a muscle that brings the lower limb closer to the midline.*

THE HISTORY OF SCIENCE

Medical Art and the History of Human Dissection

This book is filled with wonderful medical art, without which we would have an impoverished understanding of human form and function. These illustrations of muscles and other organs depict a reality documented by repeated dissections over many centuries. Can you imagine going through your daily life without knowing what your muscles look like? Or your heart? Or your brain? Until about 500 years ago, very few people knew such things.

The oldest depictions of the human form were not much more than stick figures rendered many thousands of years ago on the walls of caves (part A). They served an artistic purpose, perhaps for religious rites, and had no scientific intent. No early civilization attempted to depict the body's internal structure because every culture held that the sanctity of the human body forbade human dissection. However, there was deep interest in the human form as an object of art. In the last few centuries before the Common Era (BCE) the Greeks sculpted unparalleled masterpieces of the human form—strong, youthful figures predominated, their muscles clearly depicted beneath the surface, but the interest was artistic, not scientific (part B).

Then in the fourth century BCE, Herophilus of Chalcedon (350–280 BCE), a Greek, dissected human corpses. Herophilus described the brain, spinal cord, and nerves, speculating that they were of central importance to human function. The Egyptians soon followed when Alexander the Great (356–323 BCE) authorized dissections in Alexandria. But the descriptions produced by these ancient anatomists were largely narrative, not pictorial, and they were not informative by modern standards. The illustrations that were included were flat; they lacked perspective and mainly served to decorate the manuscript.

Although bodies continued to be dissected for the next thousand years, the knowledge that dissection could have provided was largely ignored as irrelevant by the physicians of the day. That's because they were steeped in the theories of Hippocrates (460–370 BCE), who defined good health as a proper balance among four supposed humors: phlegm (mucus), blood, black bile, and yellow bile. An excess of one or more of these humors, Hippocrates believed, caused illness. Thus, an understanding of anatomy was of no great use in this medical system.

With the coming of the Renaissance in western Europe in the 14th century, the modern scientific method was born and the facts revealed by human dissection began to be understood correctly for the first time. In the 16th century, Andreas Vesalius (1514–1564), a Dutchman, performed dissections, retained artists to depict the findings, and in 1543 published his momentous *De Humani Corporis Fabrica* (*On the Workings of the Human Body*), which for the first time depicted muscles, bones, and other body parts with remarkable clarity and artistic ingenuity (part C).

(a)

(b)

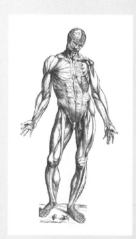

(c)

Muscle portrayals. A. Cave drawings. **B.** Greek sculpture. **C.** Vesalius's drawings.

Plate 7.1   Muscles of Facial Expression

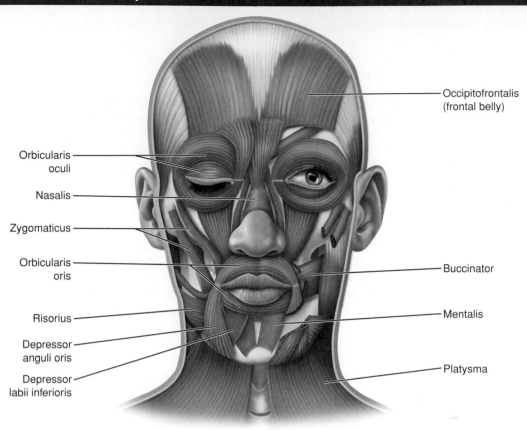

Occipitofrontalis
(frontal belly)

Orbicularis oculi

Nasalis

Zygomaticus

Orbicularis oris

Risorius

Depressor anguli oris

Depressor labii inferioris

Buccinator

Mentalis

Platysma

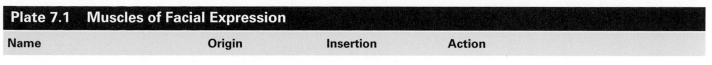

Plate 7.1   Muscles of Facial Expression

Name	Origin	Insertion	Action

Facial muscles attach to skin or other muscles rather than bones. Minute contractions of these muscles produce the subtle movements of skin and muscle we interpret as facial expressions.

Name	Origin	Insertion	Action
Occipitofrontalis, frontal belly *(occipit = base of skull; frontal = forward part)*	Epicranial aponeurosis (tendon)	Eyebrow, forehead skin	Raises eyebrows, wrinkles forehead
Occipitofrontalis, occipital belly *(occipit = base of skull; frontal = forward part)*	Occipital and temporal bones	Epicranial aponeurosis	Pulls scalp backward
Orbicularis oculi *(orb = circular; ocul = eye)*	Frontal bone, maxilla (eye orbit, medial wall)	Skin encircling eye	Closes eyelid
Nasalis *(nasal = nose)*	Maxilla	Bridge of nose (cartilage)	Brings sides of nose towards nasal septum
Zygomaticus *(zygoma = cheekbone)*	Zygomatic bone	Skin, muscle at lip corners	Raises corner of mouth, as in smiling
Orbicularis oris *(orb = circular; oris = mouth)*	Maxilla, deep surface of skin	Skin at mouth corners	Closes and protrudes lips (kissing, sucking), shapes lips (speech)
Depressor labii inferioris *(depressor = downward; labi = lip; infer = below)*	Mandible	Orbicularis oris	Depresses lower lip (when showing impatience)
Mentalis *(mentum = chin)*	Mandible	Chin skin	Elevates, protrudes lower lip (pouting)
Depressor anguli oris *(depressor = downward; anguli = corner; oris = mouth)*	Mandible	Mouth (angle)	Brings down mouth corners (frowning)
Buccinator *(bucia = cheek)*	Maxilla, mandible (alveolar processes)	Orbicularis oris	Flattens cheek (smiling, pushes food against molars, whistling, wind instruments)
Risorius *(risor = laugher)*	Platysma, masseter	Mouth angle	Draws mouth corner laterally (grinning)
Platysma *(platys = flat)*	Fascia covering deltoid, pectoralis major	Mandible	Tenses skin when teeth are clenched (resulting in skin ridges), depresses mandible, helps depressor anguli oris

Plate 7.2   Muscles Controlling the Jaw and Moving the Head

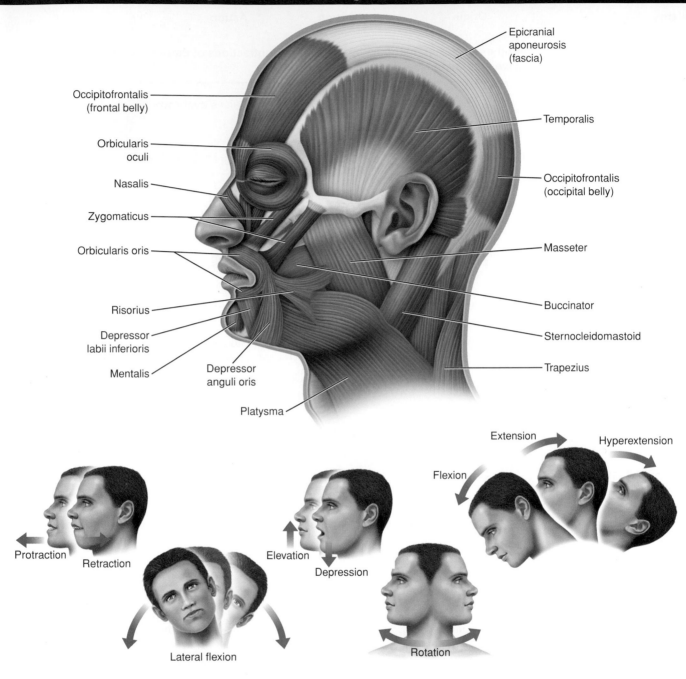

Epicranial aponeurosis (fascia)

Occipitofrontalis (frontal belly)

Orbicularis oculi

Nasalis

Zygomaticus

Orbicularis oris

Risorius

Depressor labii inferioris

Mentalis

Depressor anguli oris

Platysma

Temporalis

Occipitofrontalis (occipital belly)

Masseter

Buccinator

Sternocleidomastoid

Trapezius

Protraction   Retraction

Lateral flexion

Elevation

Depression

Rotation

Extension

Flexion

Hyperextension

Plate 7.2   Muscles Controlling the Jaw and Moving the Head

Name	Origin	Insertion	Action

The powerful muscles of the jaw move it upwards and sideways for speech and chewing (gravity is the prime mover for depressing the jaw). Yet other muscles move the entire head – flexing, extending, and rotating it on the cervical axis.

Name	Origin	Insertion	Action
Masseter *(maseter = chewer)*	Temporal bone (zygomatic process)	Mandible	Elevates jaw (biting, chewing)
Temporalis *(temporal = of the side of the head)*	Temporal bone	Mandible	Elevates jaw, retracts chin
Pterygoids *(deep muscles; not shown)*	Sphenoid bone	Mandible	Elevates jaw, protrudes chin
Sternocleidomastoid *(sternon = breastbone; cleido = clavicle; mastoid = mastoid process of temporal bone)*	Sternum, clavicle	Temporal bone (mastoid process), occipital bone	Together: flexes neck (brings chin to chest) Separately: laterally flexes, rotates neck (ear approaches shoulder on same side)
Trapezius (also see Plate 7.5) *(trapezoid = flat with four sides)*	Occipital bone, vertebrae (C7, thoracic)	Clavicle, scapula (acromion, spine)	Extends neck; also moves shoulder
Erector spinae (see Plate 7.3) *(erector = raise; spinae = of the spine)*	Ribs and vertebrae	Occipital bone, temporal bone, ribs, vertebrae	Extends neck (also moves vertebral column)

Plate 7.3 Muscles of the Thorax: Muscles that Move the Vertebral Column, Abdominal Muscles, and Respiratory Muscles

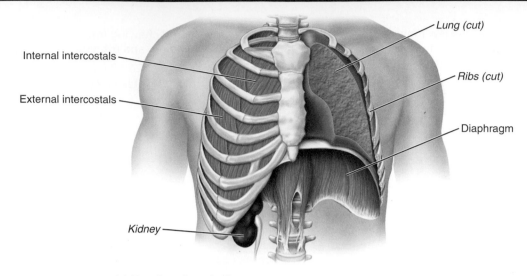

Internal intercostals

External intercostals

Lung (cut)

Ribs (cut)

Diaphragm

Kidney

(a) Muscles of respiration

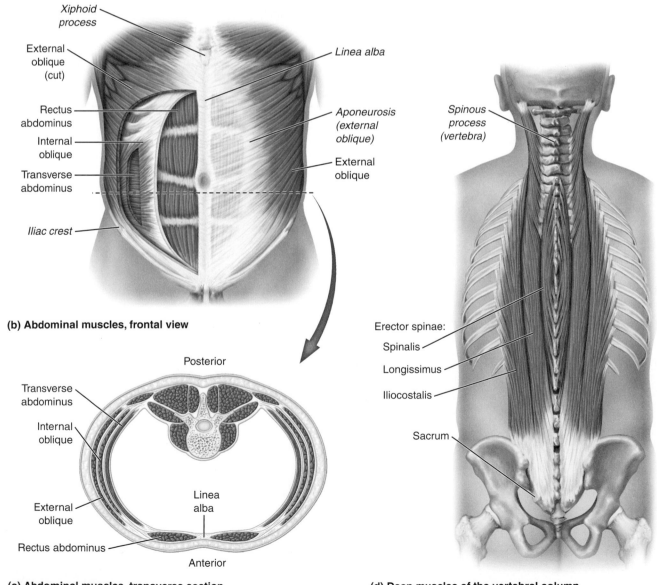

Xiphoid process

External oblique (cut)

Rectus abdominus

Internal oblique

Transverse abdominus

Iliac crest

Linea alba

Aponeurosis (external oblique)

External oblique

(b) Abdominal muscles, frontal view

Posterior

Transverse abdominus

Internal oblique

External oblique

Rectus abdominus

Linea alba

Anterior

(c) Abdominal muscles, transverse section

Spinous process (vertebra)

Erector spinae:

Spinalis

Longissimus

Iliocostalis

Sacrum

(d) Deep muscles of the vertebral column

Plate 7.3   Muscles of the Thorax: Muscles that Move the Vertebral Column, Abdominal Muscles, and Respiratory Muscles

Many muscles illustrated on this plate do not move bones. Instead, they are involved in the minute movements of respiration and in compressing the abdominal contents.

Name	Origin	Insertion	Action
Rectus abdominus *(rectus = straight; abdominus = abdomen)*	Pubis	Xiphoid process (sternum), ribs	Flexes spinal column, compresses abdomen
External oblique *(external = closer to the surface; oblique = slanting)*	5th–12th ribs	Ilium, pubis, linea alba	Both: flex spinal column, compress abdomen One: rotate, laterally flex vertebral column
Internal oblique *(internal = farther from the surface; oblique = slanting)*	Iliac crest	10th–12th ribs, linea alba	Same as external obliques
Transverse abdominis *(transverse = across; abdominis = abdomen)*	Iliac crest, intercostal cartilage of 7th–12th ribs	Xiphoid process, linea alba, pubis	Compresses abdomen
Erector spinae: spinalis, longissimus, and iliocostalis groups *(erector = raise; spinae = of the spine; longissimus = longest; iliocostal = related to the ribs)*	Tendon arising at ilium, sacrum, and lumbar vertebrae	Occipital bone, temporal bone, ribs, vertebrae	Both sides: extends vertebral column (also extends head) One side: laterally flexes vertebral column
Sternocleidomastoid (see Plate 7.2)	Sternum, clavicle	Temporal bone (mastoid process), occipital bone	Together: flexes cervical section of vertebral column (also moves head)
Diaphragm *(diaphragma = barrier or partition)*		Xiphoid process, costal cartilage of inferior ribs, lumbar vertebrae	Contracts to expand thorax, resulting in inhalation; relaxes to shrink thorax, resulting in exhalation
Internal intercostals *(internal = farther from the surface; intercostal = between the ribs)*	Superior border of ribs	Inferior border of rib above	Depress ribs; active exhalation
External intercostals *(external = closer to the surface; intercostal = between the ribs)*	Inferior border of ribs	Superior border of rib below	Elevate ribs during inhalation

Plate 7.4   Muscles of the Perineum

Name	Origin	Insertion	Action
Transverse perineus (*transverse = across; perineum = region between anus and genitals*)	Ischial tuberosity	Perineal body	Stabilizes perineum
Levator ani (*levator = raiser; ani = anus*)	Pubis, ischial spine	Coccyx, urethra, rectum, perineum	Aids defecation; stabilizes perineum
External anal sphincter (*external = closer to the surface; anal = anus; sphincter = tightener*)	Anococcygeal ligament, coccyx	Perineal body	Closes anus
Ischiocavernosus (*ischio = pelvis; cavernosus = hollow tissue of penis or clitoris*)	Ischial tuberosity, pubis	Clitoris (females), penis (males)	Maintains clitoral or penile erection by compressing veins
Bulbospongiosus (*bulbo = swollen; spongiosus = like a sponge*)	Penis (males) or Perineal fascia (females)	Perineal body, clitoris (females), penis (males)	Maintains clitoral or penile erection by compressing veins; aids in expelling last drops of urine or semen (males); constricts vagina (females)
Coccygeus (*coccyx = lower tip of spine*)	Ischium	Coccyx, lower sacrum	Stabilizes perineum; pulls coccyx forward during defecation, childbirth

The perineum lies below the pelvic outlet, and is the pelvic floor. Although they receive little attention, the muscles of the perineum support the weight of the abdominal organs, ensure urinary and rectal continence (that is, bladder and bowel control), and participate in reproductive behaviors. The perineal body is a small, but complex, structure composed of connective tissue and muscle; it is the origin for many perineal muscles; this structure is sometimes damaged during childbirth.

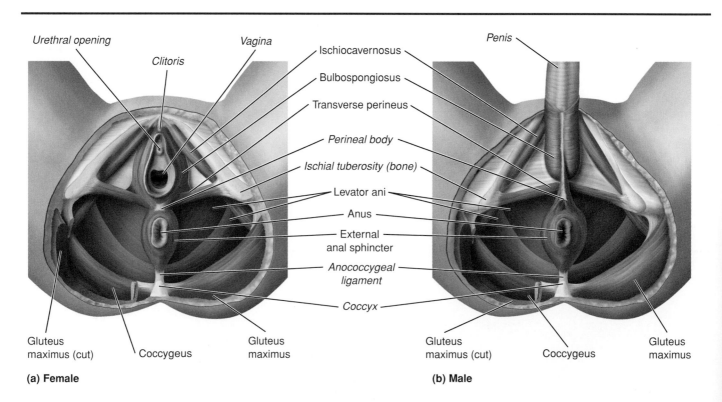

(a) Female

(b) Male

Plate 7.5   Muscles that Move and Stabilize the Pectoral Girdle

The muscles shown here anchor the upper limbs to the body by fixing the scapula in place. They also move the scapula, thereby moving the shoulder up or down, forward or backward. They also rotate the glenoid cavity, containing the head of the humerus, to enable lateral movements of the humerus.

Name	Origin	Insertion	Action
Levator scapulae *(levator = raiser; scapulae = scapula)*	Vertebrae C1–C4 (transverse processes)	Scapula (coracoid process)	Elevates and rotates the scapula inferiorly; fixes scapula (also flexes neck laterally)
Trapezius *(trapezi = shaped like a trapezoid)*	Occipital bone, vertebrae (C7, thoracic)	Clavicle, scapula (acromion, spine)	Superior part elevates scapula, inferior part depresses scapula; both parts together retract scapula
Pectoralis minor *(pector = chest; minor = lesser)*	2nd–5th ribs	Scapula (coracoid process)	Protracts scapula
Rhomboid major *(rhomboid = shaped like a rhombus; major = greater)*	Vertebrae T1–T4	Scapula	Retracts and rotates scapula inferiorly; used for forcible downward movements (like hammering)
Serratus anterior *(serratus = saw-toothed; anterior = before)*	Superior ribs	Scapula	Called the boxers muscle; important in punching and pushing because it protracts and stabilizes the scapula so that the shoulder moves down and forward; rotates scapula superiorly

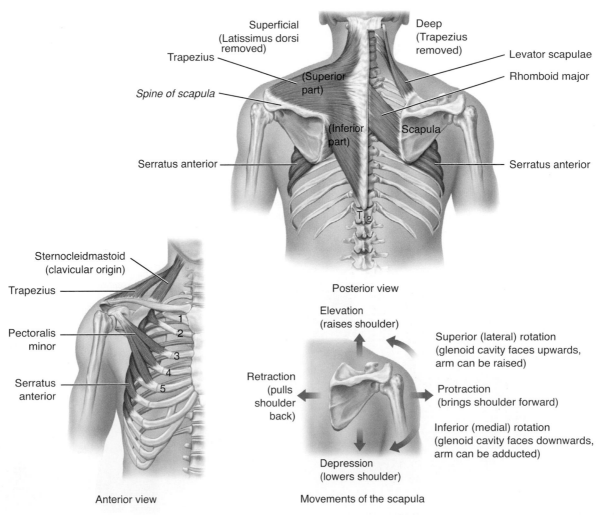

Posterior view

Anterior view

Movements of the scapula

Plate 7.6   Muscles that Move the Arm (Humerus) at the Shoulder Joint

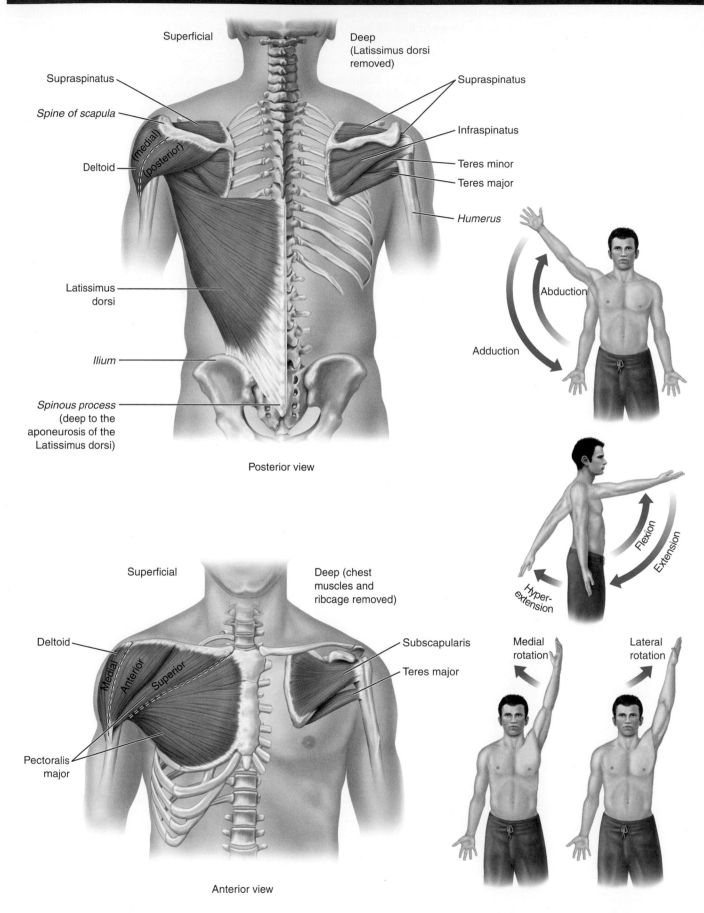

Superficial

Deep
(Latissimus dorsi
removed)

Supraspinatus

Spine of scapula

Deltoid

(medial)

(posterior)

Supraspinatus

Infraspinatus

Teres minor

Teres major

Humerus

Latissimus
dorsi

Ilium

Spinous process
(deep to the
aponeurosis of the
Latissimus dorsi)

Posterior view

Abduction

Adduction

Flexion

Extension

Hyper-
extension

Superficial

Deep (chest
muscles and
ribcage removed)

Deltoid

Medial

Anterior

Superior

Subscapularis

Teres major

Pectoralis
major

Anterior view

Medial
rotation

Lateral
rotation

Plate 7.6   Muscles that Move the Arm (Humerus) at the Shoulder Joint

Name	Origin	Insertion	Action

The shoulder joint is the most mobile joint in the body, moving the humerus in all possible dimensions. The shoulder joint is stabilized in part by the "rotator cuff" muscles, also known as the SITS muscles (supraspinatus, infraspinatus, teres minor, subscapularis). Their tendons form the rotator cuff by blending with the articular capsule to strengthen and stabilize the shoulder joint. Also, the muscles are tonically contracted to hold the humeral head in the shallow glenoid cavity.

Name	Origin	Insertion	Action
Latissimus dorsi *(latissimus = widest; dorsi = of the back)*	Vertebrae, sacrum, ilium	Humerus	"Climbing muscle"; extends and hyperextends humerus; adducts humerus behind back (i.e., to scratch an itch over the scapula); medially rotates humerus
Pectoralis major *(pector = chest; major = greater)*	Clavicle, sternum, cartilage of ribs	Humerus	Adducts and medially rotates humerus; superior portion flexes humerus
Teres major *(teres = long and round; major = greater)*	Scapula	Humerus	Adducts, medially rotates humerus; helps in extension from flexed position; helps stabilize shoulder joint when deltoid is active
Supraspinatus* *(supra = above; spina = spine of scapula)*	Scapula	Humerus	Assists deltoid to complete abduction
Infraspinatus* *(infra = below; spina = spine of scapula)*	Scapula	Humerus	Laterally rotates humerus
Teres minor* *(teres = long and round; minor = lesser)*	Scapula	Humerus	Laterally rotates humerus
Subscapularis* *(sub = beneath; scapularis = scapula)*	Subscapular fossa	Humerus	Medially rotates humerus
Deltoid *(deltoid = shaped like a triangle)*	Clavicle, scapula (spine and acromion)	Humerus	Forms rounded contour of shoulder; entire muscle abducts humerus; swings arms during walking (anterior part helps pectoralis major flex humerus; posterior part helps latissimus dorsi extend humerus)

*Part of the rotator cuff

Plate 7.7 Muscles that Move the Forearm, Hand, and Fingers

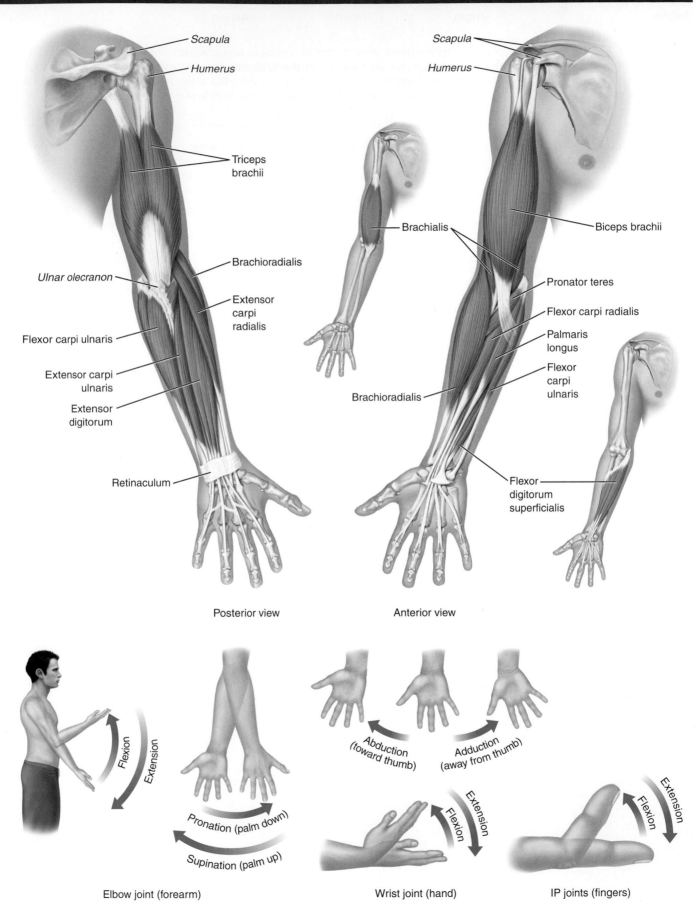

Posterior view

Anterior view

Elbow joint (forearm)

Wrist joint (hand)

IP joints (fingers)

Plate 7.7   Muscles that Move the Forearm, Hand, and Fingers

The muscles of the forearm can be separated into four groups: 1) those that flex and extend at the elbow joint; 2) those that flex or extend at the wrist joint; 3) those that flex or extend the fingers and thumb; and 4), those at that pronate or supinate the forearm.

Name	Origin	Insertion	Action
Brachialis (*brachi = arm*)	Humerus	Ulna	Flexes elbow (primary mover, all positions)
Brachioradialis (*brachi = arm; radi = radius*)	Humerus	Radius	Flexes elbow; assists brachialis when quick movements are required
Biceps brachii (*biceps = two heads; brachi = arm*)	Scapula (coracoid process and tubercle)	Radius, forearm fascia	Supinates elbow (primary mover), flexes elbow when forearm is supine (not when pronated)
Triceps brachii (*triceps = three heads; brachi = arm*)	Scapula, humerus	Ulnar olecranon	Extends elbow (primary mover)
Extensor carpi radialis (*extensor = increases joint angle; carpus = wrist; radi = radius*)	Humerus	2nd metacarpal	Extends, abducts wrist; necessary to clench fist
Pronator teres (*pronate = turn palm down; teres = long and round*)	Humerus, ulnar coronoid process	Radius	Pronates, flexes elbow
Flexor carpi radialis (*flex = decreases joint angle; carpus = wrist; radi = radius*)	Humerus	2nd and 3rd metacarpals	Flexes, abducts wrist (hand moves anterolaterally)
Palmaris longus (*palma = palm; longus = long*)	Humerus	Fascia	Weak wrist flexor
Flexor carpi ulnaris (*flex = decreases joint angle; carpus = wrist; ulnaris = ulna*)	Humerus, ulna	5th metacarpal	Flexes, adducts wrist
Extensor carpi ulnaris (*extensor = increases joint angle; carpus = wrist; ulnaris = ulna*)	Humerus, posterior ulna	5th metacarpal	Extends, adducts wrist; necessary to clench fist
Flexor digitorum superficialis (*flex = decreases joint angle; digit = finger or toe; superficial = near the surface*)	Humerus, ulna, radius	Middle phalanx, each finger	Flexes four fingers at proximal IP joint
Extensor digitorum (*extensor = increases joint angle; digit = finger or toe*)	Humerus	Distal and middle phalanges, each finger	Extends four fingers at all IP joints

Plate 7.8   Muscles that Move the Thigh and Leg

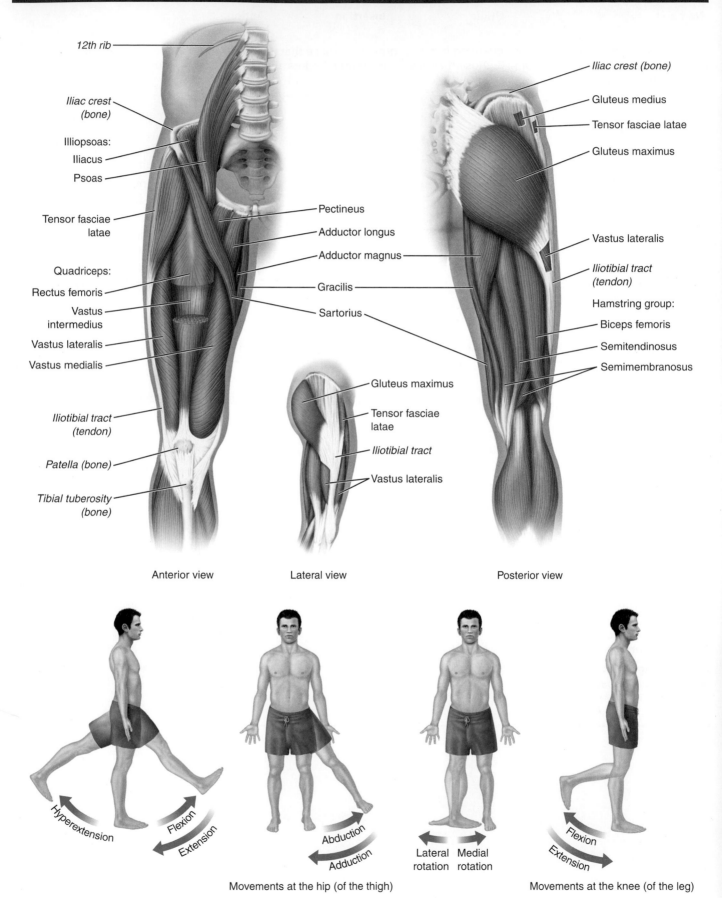

12th rib

Iliac crest (bone)

Illiopsoas:
Iliacus
Psoas

Tensor fasciae latae

Quadriceps:
Rectus femoris
Vastus intermedius
Vastus lateralis
Vastus medialis

Iliotibial tract (tendon)

Patella (bone)

Tibial tuberosity (bone)

Pectineus
Adductor longus
Adductor magnus
Gracilis
Sartorius

Iliac crest (bone)
Gluteus medius
Tensor fasciae latae
Gluteus maximus

Vastus lateralis
Iliotibial tract (tendon)

Hamstring group:
Biceps femoris
Semitendinosus
Semimembranosus

Gluteus maximus
Tensor fasciae latae
Iliotibial tract
Vastus lateralis

Anterior view

Lateral view

Posterior view

Hyperextension   Flexion
Extension

Abduction
Adduction

Lateral rotation   Medial rotation

Flexion
Extension

Movements at the hip (of the thigh)

Movements at the knee (of the leg)

Plate 7.8 Muscles that Move the Thigh and Leg

Name	Origin	Insertion	Action

Muscles of the thigh are large and powerful, befitting their role in maintaining an erect stance, in walking or running, or in lifting heavy loads. Some of these muscles act only to move the hip joint; others, the knee; and some move both. The anterior muscles generally act to flex the hip and extend the knee, as in the foreswing of walking. The posterior muscles generally act to extend the hip and flex the knee, as in the backswing of walking. A third large and powerful group, located medially, acts to adduct the thigh and has no effect on the leg. Abduction and rotation of the thigh is relatively weak and attained by small muscles or is a minor function of large muscles.

Name	Origin	Insertion	Action
Iliacus *(iliac = ilium)*	Ilium	Femur (lesser trochanter)	Flexes, laterally rotates hip; flexes vertebral column
Psoas *(psoas = muscle of the loin)*	Lumbar vertebrae	Joins iliacus to insert into femur (lesser trochanter)	Flexes, laterally rotates hip; flexes vertebral column
Sartorius *(sartor = tailor, referencing their traditional cross-legged position)*	Iliac spine	Tibia	Crosses the leg: flexes, abducts, and laterally rotates hip; flexes knee
Quadriceps Femoris: *(quadriceps = four heads; femoris = femur)*			
Rectus femoris *(rectus = straight; femoris = femur)*	Iliac spine	Four muscles join and insert into the patella, then the tibial tuberosity	Extends knee; flexes hip
Vastus lateralis *(vastus = large; lateralis = lateral)*	Femur (greater trochanter, linea aspera)		Extends knee
Vastus medialis *(vastus = large; medialis = medial)*	Femur (greater trochanter, linea aspera)		Extends knee
Vastus intermedius *(vastus = large; intermedius = middle)*	Femur		Extends knee
Gracilis *(gracile = slender)*	Pubis	Tibia	Adducts and medially rotates hip; flexes knee
Adductor longus *(adduct = move toward the centerline; longus = long)*	Pubic crest and symphysis	Femur (linea aspera)	Adducts, medially rotates, and flexes hip
Adductor magnus *(adduct = move toward the centerline; magnus = large)*	Pubis, ischium	Femur (linea aspera)	Adducts, medially rotates, and extends hip
Pectineus *(pectin = comb)*	Pubis	Femur	Adducts, flexes hip

Plate 7.8   Muscles that Move the Thigh and Leg (continued)

Name	Origin	Insertion	Action
Tensor fasciae latae *(tensor = tightener; fasciae = fascia; lat = wide)*	Ilium	Iliotibial tract, eventually tibia	Abducts, flexes hip
Gluteus medius *(glute = buttock; medius = middle)*	Ilium	Femur (greater trochanter)	Abducts, laterally rotates hip
Gluteus maximus *(glute = buttock; maximus = largest)*	Iliac crest, sacrum, coccyx	Iliotibial tract, femur (linea aspera)	Extends, laterally rotates hip
Hamstring group: *(referring to the tendons behind the knee)*			
Biceps femoris *(biceps = two heads; femoris = femur)*	Ischial tuberosity, linea aspera of femur	Fibula (head) and tibia (lateral condyle)	Flexes knee; extends hip
Semitendinosus *(semi = half; tendo = tendon)*	Ischial tuberosity	Proximal tibia	Flexes knee; extends hip
Semimembranosus *(semi = half; membran = membrane)*	Ischial tuberosity	Tibia (medial condyle)	Flexes knee; extends hip

Plate 7.9   Muscles that Move the Foot and Toes

Name	Origin	Insertion	Action

Muscles of the leg move the foot and toes and are divided into anterior, posterior and lateral groups according to their position in relation to the interosseous membrane that joins the tibia and fibula. Muscles in the anterior compartment extend the toes and dorsiflex the ankle, a weak movement but one critical to avoid dragging the toes while walking. Muscles in the posterior compartment flex the toes and plantarflex the foot, powerful movements when walking. Muscles in the lateral compartment evert the foot. The small muscles of the foot aid in all movements of the toes and help fine tune and stabilize body movements involving the foot.

Name	Origin	Insertion	Action
Tibialis anterior *(tibialis = tibia; anterior = front)*	Tibia: lateral condyle/body	1st cuneiform and metatarsal	Dorsiflexes, inverts ankle
Extensor digitorum longus *(extensor = increase joint angle; digitorum = finger or toe; longus = long)*	Tibia	Distal phalanges, 2nd to 5th toes	Extends 4 toes, dorsiflexes ankle
Extensor hallucis *(extensor = increase joint angle; hallux = great toe)*	Fibula	Phalanx of great toe	Extends great toe, dorsiflexes ankle
Fibularis longus *(fibularis = fibula; longus = long)*	Fibula, tibia (lateral condyle)	Medial cuneiform and first metatarsal of foot	Everts ankle; keeps leg steady when balancing on one foot
Gastrocnemius *(gastro = belly; cnem = leg)*	Femur: lateral, medial condyles	Calcaneus (via Achilles tendon)	Plantarflexes ankle; raises heel when walking; flexes knee; important in rapid movements (running, jumping)

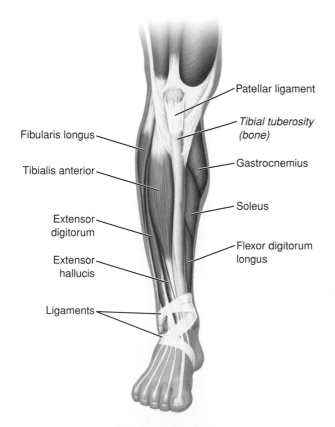

Fibularis longus
Tibialis anterior
Extensor digitorum
Extensor hallucis
Ligaments

Patellar ligament
Tibial tuberosity (bone)
Gastrocnemius
Soleus
Flexor digitorum longus

Anterior view, right leg

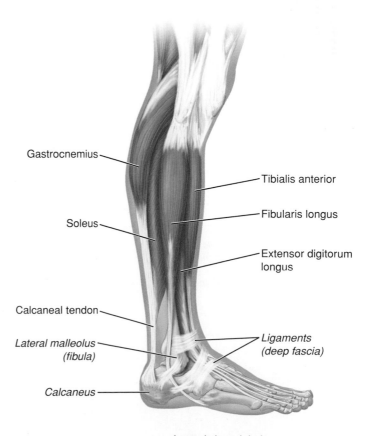

Gastrocnemius
Soleus
Calcaneal tendon
Lateral malleolus (fibula)
Calcaneus

Tibialis anterior
Fibularis longus
Extensor digitorum longus
Ligaments (deep fascia)

Lateral view, right leg

Plate 7.9   Muscles that Move the Foot and Toes (continued)

Name	Origin	Insertion	Action
Soleus (*soleus = a flat fish*)	Fibula (head) and proximal tibia	Calcaneus (via Achilles tendon)	Plantarflexes ankle (slow); contracts alternately with leg extensor muscles to maintain balance when walking
Tibialis posterior (*tibialis = tibia; posterior = rear*)	Tibia, fibula	Multiple tarsals and metatarsals	Plantarflexes and inverts ankle
Flexor digitorum longus (*flexor = decreases joint angle; digitorum = finger or toe; longus = long*)	Posterior tibia	Distal phalanges, 2nd to 5th toes	Flexes lateral 4 toes, plantarflexes ankle, supports longitudinal foot arches
Flexor hallucis (*flexor = decreases joint angle; hallux = great toe*)	Posterior fibula	Base of great toe	Flexes great toe; supports longitudinal foot arches; push-off muscle during running and jumping

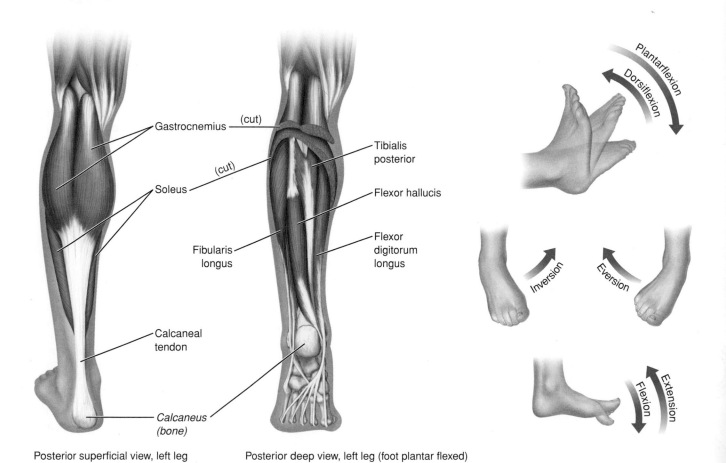

Posterior superficial view, left leg   Posterior deep view, left leg (foot plantar flexed)

Word Parts

Latin/Greek Word Parts	English Equivalents	Examples
my/o	Muscle	myoglobin; a globular protein found in muscle
sarco-	Flesh; muscle	sarcolemma: membrane (-lemma) of a muscle cell
tropo-	To turn	troponin: molecule (-in) that turns (moves tropomyosin out of the way) in order to produce muscle contraction
-metric	Length	isometric: contraction with no change (iso-) in length
-ton/o	Tension	isotonic: contraction with no change (iso-) in tension
con-	Together	concentric: muscle contraction that brings two bones together (a shortening contraction)
ec-	Away	eccentric: muscle contraction that moves two bones away from each other (a lengthening contraction)
syn-	Together	synergist: muscle working together with the prime mover
ant-	Against	antagonist: muscle working against the prime mover

Chapter Challenge

CHAPTER RECALL

1. **Which of the following characteristics apply to skeletal muscle (SK) and which to smooth muscle (SM)? Write all that apply.**
   a. Muscle is striped (striated) in appearance.
   b. Muscle contractions cannot be consciously controlled.
   c. Muscle tissue found in the stomach and intestinal wall.
   d. Muscle fibers that fatigue after repeated contractions.

2. **The outer membrane of a muscle cell is called the**
   a. sarcoplasm.
   b. sarcolemma.
   c. sarcoplasmic reticulum.
   d. endosomal membrane.

3. **Which of the following statements applies to the neuromuscular junction of skeletal muscles?**
   a. Uses norepinephrine as the neurotransmitter.
   b. Consists of multiple varicosities scattered over numerous muscle fibers.
   c. Its activation results in calcium entering the cell from the extracellular fluid.
   d. The neurotransmitter receptors are also sodium channels.

4. **Thin filaments**
   a. are anchored by dense bodies in smooth muscle.
   b. are composed only of actin.
   c. are found in skeletal muscle but not smooth muscle.
   d. are composed of myosin.

5. **A sarcomere**
   a. is a neuron and the muscle fiber it innervates.
   b. runs the entire length of a muscle.
   c. is joined to adjacent sarcomeres by the Z disc.
   d. is the functional unit of both skeletal and smooth muscle.

6. **During muscle contraction,**
   a. thick filaments shorten.
   b. thin filaments shorten.
   c. both thick and thin filaments shorten.
   d. neither thick nor thin filaments shorten.

7. **The role of calcium in smooth muscle contraction involves**
   a. activating enzymes that activate the myosin heads.
   b. providing energy for cross-bridge cycling.
   c. revealing myosin binding sites on actin molecules.
   d. initiating action potentials in the muscle cell.

8. **In skeletal muscle, fresh molecules of ATP are required for**
   a. detaching the myosin heads from the actin.
   b. maintaining the sodium concentration gradient.
   c. providing the energy for movement of the myosin heads.
   d. all of the above.

9. **Type I fibers**
   a. are slow to fatigue.
   b. contain large stores of glycogen.
   c. are the strongest type of fiber.
   d. contain few mitochondria.

10. **An example of an isometric contraction during a pushup would be**
    a. holding yourself immobile in the pushup position.
    b. contracting the triceps brachii as you lower your body to the floor.
    c. contracting the biceps brachii as you raise your body from the floor.
    d. b and c.

11. **The eyelid is closed by the actions of the**
    a. orbicularis oris.
    b. orbicularis oculi.
    c. mentalis.
    d. occipitofrontalis.

12. **Contraction of the sternocleidomastoid muscle on one side of the body only will**
    a. raise one shoulder.
    b. lower one shoulder.
    c. bring one ear closer to the shoulder on the same side.
    d. move one shoulder anteriorly.

13. **The most superficial abdominal muscle is the**
    a. transverse abdominis.
    b. internal oblique.
    c. external oblique.
    d. rectus abdominis.

14. **The muscle that retracts the scapula is the**
    a. serratus anterior.
    b. pectoralis minor.
    c. pectoralis major.
    d. rhomboid major.

15. **Which of the following muscles is *not* part of the rotator cuff?**
    a. Deltoid
    b. Supraspinatus
    c. Teres minor
    d. Subscapularis

16. **The primary mover for forearm flexion is the**
    a. biceps brachii.
    b. brachialis.
    c. brachioradialis.
    d. triceps brachii.

17. **The thigh muscle that originates on the ilium and inserts into the greater trochanter of the femur is the**
    a. sartorius.
    b. psoas.
    c. gluteus medius.
    d. tensor fascia lata.

18. **The muscle that abducts the thigh is the:**
    a. gacilis.
    b. pectineus.
    c. adductor longus.
    d. gluteus medius.

19. **The levator ani originates on the**
    a. ischium.
    b. pubis.
    c. perineal fascia.
    d. ilium.

20. **Contraction of the extensor hallucis would result in**
    a. inversion.
    b. eversion.
    c. plantarflexion.
    d. dorsiflexion.

21. **Which of the following muscles is part of the hamstring group?**
    a. Gracilis
    b. Vastus lateralis
    c. Semimembranosus
    d. Rectus femoris

22. Which of the following muscles extends the hand?
a. Palmaris longus
b. Extensor digitorum
c. Extensor carpi ulnaris
d. Pronator teres

23. Hiking to the top of a mountain, you encounter a view so beautiful that your jaw drops, opening your mouth. The prime mover for this action is
a. the temporalis muscle.
b. the masseter muscle.
c. the depressor labii inferioris muscle.
d. gravity.

24. You have your eye on a delicious chocolate chip cookie at a bake sale. But as your hand reaches out to get the cookie, someone else snatches it. You walk away, pouting. The prime mover in this action of the lips is the
a. mentalis.
b. buccinators.
c. zygomaticus.
d. risorius.

25. Which of the following muscles strongly flexes the spinal column?
a. rectus abdominis
b. erector spinae
c. transverse abdominis
d. internal intercostals

CONCEPTUAL UNDERSTANDING

26. List five functions of muscle tissue. Specify the type of muscle tissue involved—skeletal, cardiac, and/or smooth.

27. Compare and contrast anaerobic metabolism and aerobic metabolism under the following categories:
a. nutrient types used
b. approximate number of ATP molecules produced from one glucose molecule (and, if appropriate, fatty acid molecule)
c. requirement for oxygen

28. Returning for your second year of college, you notice that your friend has significantly "bulked up" and his muscles are visibly larger.
a. What sort of exercise results in significant muscle growth?
b. Discuss the role of satellite cells in muscle growth.

APPLICATION

29. A new drug has been developed that blocks acetylcholinesterase. You are looking for something to relax your muscles. Would this drug be appropriate to use? Explain why or why not.

30. While attending the ballet, you notice a dancer raising her heels to stand on her tiptoes.
a. Name this action using the movement terminology you learned in Chapter 6.
b. What is the prime mover for this action?
c. Name a synergistic muscle involved in this action.
d. Name an antagonistic muscle involved.

You can find the answers to these questions on the student Web site at
http://thepoint.lww.com/McConnellandHull

8

Nervous System

Major Themes

- The language of the central nervous system (CNS) consists of electrical signals.

- The brain and spinal cord receive, integrate, and react to information.

- The brain and spinal cord communicate with the body via the peripheral nervous system.

Chapter Objectives

12. List the two roots and the three branches of a spinal nerve, and describe the type of signal conveyed by each.

The Autonomic Nervous System   312

13. List similarities and differences between the autonomic and somatic divisions of the peripheral nervous system (PNS).

14. For each division of the autonomic nervous system, describe the location of the ganglia, the nerves conveying the signals, the neurotransmitter(s) used, and the effects on different target organs.

Pathways of Neural Function   320

15. Describe the types of signals carried by the anterior corticospinal tract, the lateral corticospinal tract, and the posterior column, and explain why sensory and motor signals from the right hand are primarily received by and sent from the left cerebral hemisphere.

16. Give an example of a spinal reflex, and name all of the components.

"The kid's bleeding and out cold."

As you read through the following case study, assemble a list of the terms and concepts you must learn in order to understand Larry's condition.

Clinical history: Larry H., age 17, an expert skateboarder, lost control in a downhill race against a friend and tumbled into a parked car, hitting his head against the bumper. He was not wearing a helmet.

A witness called 911 from his cell phone, shouting, "The kid's bleeding and out cold. Hurry!"

Physical examination and other data: On arrival at the hospital, Larry was found to have a moderately increased blood pressure, heart rate, and respiratory rate. He was unconscious and responded to pinprick stimulation by thrashing about. A deep gash was present in the left scalp. Imaging studies revealed a depressed fracture involving the left side of the frontal bone, the coronal suture, and the anterior aspect of the left parietal bone. In the same region, blood had accumulated in the cranial cavity immediately beneath the skull, and the underlying brain tissue was swollen.

Clinical course: Larry was immediately taken to surgery. A neurosurgeon temporarily removed a plate of bone and evacuated a large blood clot that had developed between the dura mater and the skull. The neurosurgeon also elevated the depressed bone and bored a hole into the nearby skull, leaving it open to relieve increased intracranial pressure. The neurosurgeon's notes also mentioned "lacerations and substantial brain swelling of the left frontal and temporal lobes that may have neurological implications."

Larry recovered slowly and was discharged to rehabilitation care with serious speech disorder called Broca's aphasia. Larry's speech was slow and labored and characterized by an inability to string together sets of words to make complete sentences. He typically uttered single nouns that were appropriate, but he had difficulty with verbs and grammar. For example, he might say "eat" when he actually wanted to drink. If asked to describe his accident, he might say only "skate," "car," and "head," and fill in the blanks with "uh" or other sounds. However, he could make all of the sounds necessary for speech, could understand what was spoken to him, and could read and

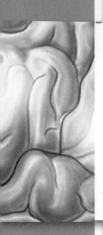

write normally. His eyesight and hearing were also normal. His intellect was not impaired, but he was emotionally less stable than before the accident. He graduated from high school with difficulty but was unable to stay enrolled in college or to hold a job for more than a few months. When lost to follow-up, he was living with his parents.

Communication is the key to our survival. External communication, which we call language, is a tool that enables us to cooperate with one another to fend off external threats and organize the environment for stability and mutual benefit. Civilization is the result. In like manner, internal communication enables our tissues and organs to respond cooperatively to external and internal environmental changes that threaten homeostasis.

Recall from Chapter 4 that the body contains two main communication systems. The *endocrine system*, discussed in ➡ Chapter 15, communicates by releasing chemicals into blood and other body fluids. The *nervous system*, the topic of this chapter, combines electrical and local chemical signals to communicate more urgent messages. In ← Chapter 4 we discussed the form and function of electrical signaling as the primary means of communication within *neurons*, the main functional cells of the nervous system. The cell bodies of these amazing cells receive and "make sense" of incoming sensory information and respond by issuing signals that allow us to think and act. Parts of neurons can extend several feet, transmitting electrical signals from one end of a cell to the other down long cytoplasmic extensions called *axons*. They employ local chemical communication to transfer electrical signals from one cell to the next—from neuron to neuron, from neuron to muscle cell, or from neuron to gland cell. In this chapter, we put neurons and their associated cells into context as part of the *nervous system*.

Brain, n. An apparatus with which we think that we think.

Ambrose Bierce (1842–1914?), American satirist, journalist, and short-story writer, in *The Devil's Dictionary*

Overview of the Nervous System

The **nervous system** is a unified collection of neurons and supporting cells that, with the endocrine system, regulate the body's response to change. Changes can be internal or external, conscious or unconscious—varying from a rush of conscious pleasure at the smell of freshly brewed coffee to an imperceptible change in blood pressure. The main structures of the nervous system are the *brain*, *spinal cord*, and *nerves*:

● Although the ancients used to believe that the heart was the "master controller" of the body, we now know that this honor goes to the **brain,** the organ housed in the skull.

● The **spinal cord** is an extension of nervous tissue that extends from the base of the brain through the center of the spine.

● **Nerves** are bundles of *axons* (cytoplasmic extensions) that extend from neurons in the brain or spinal cord to every area of the body, including the skin of the big toe and the innermost cells of the gastrointestinal tract.

We begin this chapter with an overview of the nervous system's structure and function, using broad brushstrokes. Then, we fill in the details by discussing each component of the nervous system in greater detail.

The Nervous System Senses, Integrates, and Responds to Information

The activities of the nervous system are as follows (Fig. 8.1):

- *Sensing.* Sensing is the detection of change by **sensory receptors,** specialized structures of the nervous system that are stimulated by changes within the body and outside of it. Two categories of receptors are those from our special senses (eyes, ears, etc.), of which we are aware, and those from our internal environment, of which we have no conscious perception. For example, sensory receptors in our ears detect the sound waves of a bear's roar. However, changes in many internal conditions—blood pressure and pH, for example—never rise to the level of consciousness; instead, they provide data to homeostatic systems, which silently monitor internal signals. Signals from both types of sensory receptors are relayed to the CNS by a particular class of neurons, known as **sensory (afferent) neurons.**
- *Integrating.* Integrating, which occurs primarily (but not exclusively) in the brain and spinal cord, is the combining and coordinating of sensory signals. Signals from our internal sensors are silently integrated into unconscious homeostatic responses. On the other hand, signals from our eyes, ears, and other special sensors are registered and blended by the brain to create a record of events. These records, called **percepts,** may be conscious or unconscious. A conscious percept may take the form of a thought ("There is a bear!") or an emotion (fear). Unconscious percepts influence behavior behind the scenes, so to speak. In any case, integration requires the help of a second class of neurons, known as **interneurons** (also called *association neurons*), which integrate sensory signals by relaying signals within the CNS.
- *Reacting.* Reaction is the generation of diverse outgoing signals by the brain and spinal cord in response to an incoming signal. For instance, on hearing the bear's roar, we may consciously decide to activate our skeletal muscles in order to climb a tree. Our blood pressure will rise as well, although we will not be conscious of it. Whether we are conscious of them or not, these signals are conveyed outward by **motor (efferent) neurons** to the heart, blood vessels, and skeletal muscles. Tissues and organs that respond to signals from motor neurons are referred to as **effectors** because they *effect* (cause) a change. Effectors are not part of the nervous system.

> *Remember This!* To help you distinguish between efferent and afferent neurons, you can remember that **a**fferent neurons **a**ccept stimuli from the environment and **e**fferent neurons **e**ffect change.

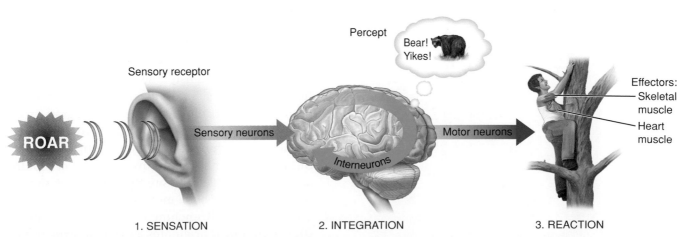

Percept — Bear! Yikes!

Sensory receptor

ROAR

Sensory neurons

Interneurons

Motor neurons

Effectors: Skeletal muscle / Heart muscle

1. SENSATION   2. INTEGRATION   3. REACTION

Figure 8.1. Functions of the nervous system. 1. Sensation. The ear senses the sound waves of the bear's roar and sends signals through sensory (afferent) neurons. 2. Integration. The brain and spinal cord integrate this sensory input with memories of what a bear sounds like to produce a percept (Yikes! A bear!). 3. Reaction. The brain and spinal cord send signals through motor (efferent) neurons to increase heart rate and climb a tree. *Is the heart muscle part of the nervous system?*

Our ability to react to change in our surroundings is dependent on all three steps. If disease or trauma blocks any one of these steps, reaction does not occur. If sensing is impaired, no signal is generated. If afferent nerves are damaged, the incoming sensing signal is blocked. If integration is impaired, no percept forms and no return signal can be generated to command a response. Finally, if efferent nerves are damaged, the outgoing command signal is blocked.

Case Note

8.1. Does Larry's speech problem appear to be mainly a matter of sensation, integration, or response?

The Central and Peripheral Nervous Systems Are the Primary Divisions

There are two main divisions of the nervous system (Fig. 8.2).

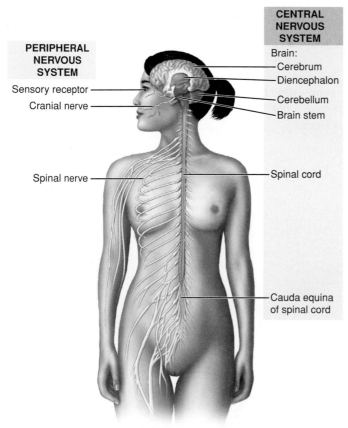

PERIPHERAL NERVOUS SYSTEM

Sensory receptor

Cranial nerve

Spinal nerve

CENTRAL NERVOUS SYSTEM

Brain:
Cerebrum
Diencephalon
Cerebellum
Brain stem

Spinal cord

Cauda equina of spinal cord

Figure 8.2. Divisions of the nervous system. Structurally, the nervous system is divided into the CNS (brain and spinal cord) and PNS (sensory receptors and nerves). *Are cranial nerves part of the PNS or the CNS?*

- The **central nervous system (CNS)** consists of the *brain* and *spinal cord.*
- The **peripheral nervous system (PNS)** consists of nerves and sensory receptors.

Let's look at each of these divisions in more detail.

The Central Nervous System Integrates Information

The CNS receives and integrates incoming (sensory) signals and responds with outgoing (motor) command signals. Anatomically, its components are vertically stacked according to the degree of signal integration they provide.

Lowest and providing the least integration is the **spinal cord,** a cylinder of nervous tissue about three feet long that is encased by vertebral bone and extends down the back from the base of the skull. The cord ends at about the first or second lumbar vertebra, breaking up into a fan of nerves called the *cauda equina* (Latin *cauda* = "tail" + *equina* = "horse": literally, tail of the horse). The spinal cord integrates and reacts automatically to some incoming sensory signals; it relays others upward to the brain for further integration and voluntary action. For other signals, it performs no integration and merely serves as a highway for nerve signals passing between the brain and the body.

Atop the spinal cord sits the **brain,** a 3-lb grayish organ that in the living has the consistency of a soft banana. It consists of about 100 billion *neurons* (the cells with which "we think that we think") and about 30 trillion other cells, the *glia,* which provide support to neurons. As we identify the major subdivisions of the brain, notice that they are stacked according to the degree of integration they provide:

- The most inferior part of the brain is the **brainstem,** a thumb-sized piece of tissue that merges with the spinal cord. The brainstem serves as a central clearinghouse for nerve signals, relaying them to other parts of the brain and spinal cord. It also integrates signals specific to several core physiological processes, including sleep, breathing, and blood pressure.
- Sitting posterior to the brainstem is the **cerebellum,** a fist-sized wedge of tissue that integrates signals in the regulation of body movement.
- Superior to and branching from the brainstem is the **diencephalon.** It originates as a single structure from the brainstem below but quickly divides into two pecan-sized masses, one on each side. The diencephalon contains the hypothalamus, which integrates and regulates additional core physiological processes

such as hunger, thirst, and body temperature, and the thalamus, which relays and integrates information from other brain areas.

- Sitting above the diencephalon is the **cerebrum** (Greek *cere-* = "soft like wax"), which consists of right and left **cerebral hemispheres,** each about the size and shape of a small boxer's glove. The cerebrum provides maximum information assimilation and is home to the highest degree of integration: *consciousness—* the queen of brain attributes. That there are two cerebral hemispheres has functional implications: by and large they are connected to opposite sides of the body. For more discussion of this topic, you may refer to the Clinical Snapshot later in this chapter, titled *Recovery from Brain Damage.*

We discuss each of these regions and their functions in more detail later in the rest of this chapter.

Case Note

8.2. Larry's injury occurred in the brain region providing the highest level of integration. Name this region.

The PNS Transmits Information

The PNS is a network of *sensory receptors* that detect internal and external environmental change, and *nerves,* that transmit information between the CNS and the periphery. The bundles of neurons and axons called **nerves** carry two-way traffic: some of the axons within a nerve convey incoming sensory signals to the CNS from sensory receptors while others convey outgoing motor command signals from the CNS to effectors, such as muscles and glands.

Anatomically, the PNS consists of *cranial nerves,* which originate from the brain, and *spinal nerves,* which originate from the spinal cord. Functionally, the PNS includes the *somatic* (voluntary, conscious) nervous system, which regulates the activity of skeletal muscles, and the *autonomic* (involuntary, unconscious) nervous system, which regulates the activity of the heart and viscera (internal organs). Figure 8.3A illustrates these divisions.

Case Note

8.3. Larry's speech and visual problems are due to injuries to which part of the nervous system?

The Somatic Division Is Voluntary

The **somatic nervous system** is voluntary; that is, it operates *consciously* (Fig. 8.3B, left side). It receives signals

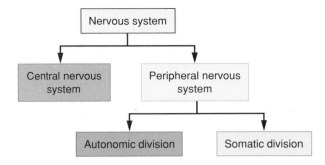

(a) Nervous system divisions

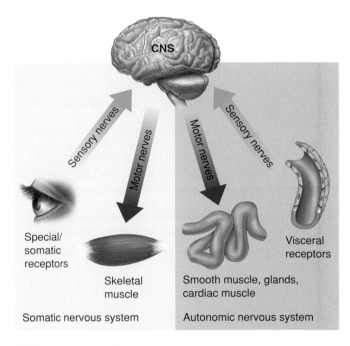

(b) Sensory and motor nerves

Figure 8.3. The nervous system. A. The main divisions of the nervous system are the central nervous system (CNS) and the peripheral nervous system (PNS). **B.** The autonomic division of the PNS receives information from visceral receptors and regulates the activity of smooth and cardiac muscle and glands. The somatic division of the PNS receives information from all other types of sensory receptors and regulates the activity of skeletal muscle. *Which neurons send signals to the brain—sensory neurons or motor neurons?*

from *special receptors* in the special sense organs (eyes, ears, etc.) and from *somatic receptors* in other body tissues (skin, muscles, etc.) that monitor sensations such as temperature and pain. Sensory receptors and sensory pathways are the subject of ➡ Chapter 9. Somatic sensory nerves convey sensory signals to the brain, where they are integrated into conscious experience. In response, voluntary motor signals are sent along somatic motor nerves to skeletal muscles, which contract on command.

The Autonomic Division Is Involuntary

The **autonomic nervous system** is involuntary; that is, it operates automatically and *unconsciously*. Autonomic sensory (afferent) nerves receive signals from visceral receptors that monitor internal conditions such as blood pressure and the stretching of walls of internal organs (Fig. 8.3B, right side). They then convey these data to the CNS. Autonomic motor (efferent) nerves convey signals from the CNS to effect changes in smooth muscle, cardiac muscle, and glands. Take special note of the following: autonomic motor signals may cause *either* increased activity (excitation) or decreased activity (inhibition) of the target tissue or organ.

8.1 Is the CNS involved in voluntary activities, involuntary activities, or both?

8.2 Name the two divisions of the PNS.

8.3 You have eaten a large extra-spicy pizza. Which type of nerve carries the signal from your painful intestines to your brain?

8.4 You decide to write a note reminding yourself never to eat such a pizza again. Which type of nerve carries the signal from your brain to your finger muscles?

8.5 Identify the main parts of the CNS, from inferior to superior.

Nervous System Cells and Tissues

Nervous tissue consists of two types of cells: **neurons,** the main functional cells of the nervous system, and **neuroglia,** the supporting cells. Neurons are the cells "with which we think that we think." They carry incoming sensory signals, integrate information, and carry outgoing command signals. Neuroglia are strictly supportive and do not carry nerve signals.

Neurons Carry Electrical Signals

Neurons specialize in electrical communication. Neurons have three parts: (a) a *cell body*, (b) *dendrites*, and (c) an *axon* (Fig. 8.4). Each neuron is a microcosm of the entire nervous system: dendrites generally *receive* information, the cell body *integrates* information from many

dendrites, and the axon *transmits* a signal to another neuron or an effector organ. Most axons are covered with **myelin,** a fatty whitish substance that is wrapped around axons to facilitate accelerated nerve impulse conduction. Before proceeding with this chapter, you should review the description of these parts in ◀ Chapter 4 as well as the discussion of electrical signaling.

Nervous tissue in the CNS is highly organized, with neuronal cell bodies clustered in some areas and neuronal axons in others. Where neuronal cell bodies predominate, the tissue is gray and is called **gray matter.** Gray matter occurs in three forms: as the surface layer (the *cortex*) of the cerebrum, as the central core of gray matter in the spinal cord, and as small nodules deep within the brain called **nuclei.** Nearly all CNS axons are sheathed by myelin, a fatty substance that gives them their distinctive white appearance. Thus, where axons are bundled together, the tissue is white and called **white matter.** What's more, axons in white matter are organized into special bundles that, as they travel from one region to another, are called **tracts.**

The nervous tissue of the PNS is organized in a similar fashion, but the descriptive terms differ. Small nodules of cell bodies (gray matter) in the PNS are called **ganglia.** In the PNS, bundles of axons are called **nerves.** Gray matter and ganglia are sites of signal integration. Tracts and nerves only *transmit* signals—no signal integration occurs along axons.

> *Remember This!* **Collections of neuron cell bodies in the CNS are called nuclei. Collections in the PNS are called ganglia. Neuronal axon bundles are called tracts in the CNS and nerves in the PNS.**

Neuroglial Cells Support Neurons

Neuroglia (Greek *glia* = "glue") play a subordinate role: they maintain the integrity of the brain and spinal cord by gluing neurons together, and they sustain neurons by maintaining homeostasis of the extracellular fluid that bathes them. Commonly referred to as *glial cells,* they are much more numerous than neurons, and they are not directly involved in electrical signaling.

The CNS contains several types of glial cells: *astrocytes, microglia, oligodendrocytes,* and *ependymal cells* (Fig. 8.4).

The most abundant glial cells are **astrocytes,** so named because of their starlike shape (Greek *astron* = "star"). From a central cell body they send out numerous projections that grip and stabilize nearby neurons and capillaries (the smallest of blood vessels), anchoring them in place. Astrocytes also control the composition of

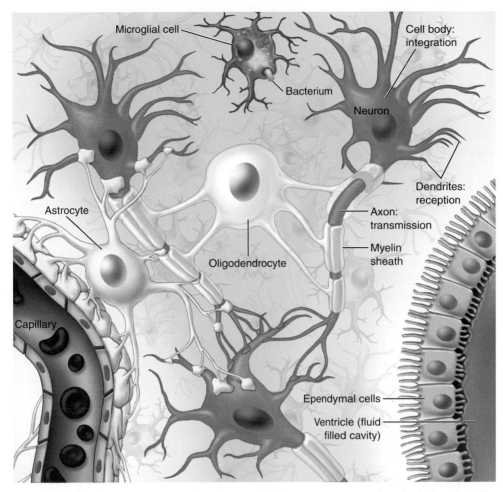

Figure 8.4. Neurons and neuroglia of the CNS. Neurons convey electrical signals; glia serve supporting roles. *Which type of cell forms the myelin sheath of brain neurons?*

the extracellular fluid by absorbing and degrading neurotransmitters and excess ions that would otherwise interfere with healthy functioning. Finally, astrocytes serve as the neuronal stem cells.

Microglia are the roaming scavenger cells—the phagocytes—of the CNS. They engulf, digest, and destroy invading organisms and other foreign matter and clean up the debris of dead cells that remains after injury.

Oligodendrocytes (Greek *oligo* = "few"), literally cells with few dendrites, synthesize the myelin sheath of CNS axons. Each oligodendrocyte covers (myelinates) short segments of multiple axons.

Ependymal cells line brain cavities called *ventricles*. As discussed below, these cells produce *cerebrospinal fluid,* a plasma-like liquid that cushions and protects brain tissue (see Fig. 8.7.), and their cilia facilitate the movement of this fluid. Ependymal cells may also act as neuronal stem cells.

The PNS is mainly composed of axons and thus does not require a multicellular supporting staff of glial cells,

like the CNS. The major PNS glial cell is the **Schwann cell,** which forms the myelin sheath of the PNS axon. As we discussed in ◀ Chapter 4, each Schwann cell myelinates a single axon.

Case Note

8.4. **Would Larry's brain injury involve damage to Schwann cells or to oligodendrocytes?**

 8.6 Which of the following is white matter: CNS axons, CNS cell bodies, or PNS axons?

8.7 True or false? Glial cells transmit nerve impulses.

8.8 Which glial cell(s) can produce new neurons?

Protection of the Nervous System

Because most neurons are almost irreplaceable, and because their tasks are demanding and precise, extraordinary protection is provided for the CNS—protection against physical trauma and disruptions of normal homeostasis by disease or other conditions.

Bone and Connective Tissue Membranes Offer Physical Protection

On a larger scale, it is astonishing that neurons survive the traumas of modern life: skateboard crashes, athletic collisions, bicycle accidents, falls at home—you name it. That they do so speaks for the body's remarkably effective system of physical protection. As we mentioned in ◀ Chapter 6, the skull encases the brain and the vertebrae shield the spinal cord. Occasionally, however, the very bones that protect become a mechanism of injury. For example, our patient, Larry H., suffered a depressed skull fracture: the depressed, sharp edges of fractured skull cut into his brain, crushing neuronal cell bodies in the gray matter and slicing through tracts, which accounts for the symptoms of brain injury that plagued him after the accident. Even more devastating injury can occur when vertebrae are fractured. In 1995, Christopher Reeve, the late actor who played Superman, fell from a horse and landed on the crown of his head with such force that it fractured the first and second cervical vertebrae. No longer locked in a stable configuration by solid bone, the vertebrae shifted upon one another, crushing the spinal cord and leaving him paralyzed from the neck down.

PNS axons have no comparable bony armor. They are, however, flexible and cushioned by the soft tissues in which they are located. PNS axons are bundled into nerves covered by three layers of protective connective tissue. Each axon (either myelinated or unmyelinated) in a nerve is sheathed in a thin fibrous membrane, the **endoneurium.** Bundles of axons are gathered together into *fascicles*, which in turn are sheathed by a fibrous **perineurium.** Multiple fascicles are gathered to form a nerve, which is sheathed by a fibrous **epineurium** (Fig. 8.5).

> *Remember This!* **Compare Figure 8.5 and Figure 7.2 (depicting a muscle cell). Note the similarities.**

Meninges Sheath the Brain and Spinal Cord

Connective tissue membranes also surround the CNS. Inside the skull and vertebrae, three membranes, collectively known as the **meninges,** sheath the brain and spinal cord (Fig. 8.6). We discuss the spinal cord meninges in more detail later in the chapter.

The outermost membrane is the **dura mater** (Latin *dura* = "tough" + *mater* = "mother," literally "tough mother"), a strong, thick, fibrous membrane that extends from the brain to the most inferior region of the vertebral column. The dura mater consists of outer and inner layers. The outer layer is tightly applied to the interior wall of cranial bones. The blood vessels running between the two dural layers can be ruptured by traumatic brain injuries. Bleeding from Larry's torn arteries had forced its way into the potential space between the skull and dura to form the blood clot mentioned in the case study.

The inner layer of the dura has multiple folds that follow the outline of the brain, forming a rough mold around it (Fig. 8.6A):

- The **falx cerebri** is a fold of the inner layer of dura that runs in the sagittal plane down the center of the skull, separating the two cerebral hemispheres. Within the falx cerebri are two large veins, the *superior* and *inferior sagittal sinuses*.
- The **tentorium cerebelli** is a fold of inner dura that extends in a horizontal plane across the back of the skull and separates the cerebral hemispheres above from the cerebellum below.

The middle meningeal membrane, known as the **arachnoid mater** (commonly shortened to *arachnoid*), very closely lines the inner side of the dura mater (Fig. 8.6B). It sends projections called *arachnoid villi* into the superior sagittal sinus. It is considerably thicker than the innermost meningeal layer, the *pia mater*, but is less tough than the dura mater. The **pia mater** (Latin *pia* = "tender," literally "tender mother") is an exceedingly thin membrane only a few cells thick. It is tightly fitted over every prominence and depression in the irregular surface of the brain and spinal cord.

A space, the **subarachnoid space,** exists between the arachnoid and the pia. Extending across the subarachnoid space is a thicket of tens of thousands of tiny fibers; these connect the arachnoid mater to the pia mater adhering to the brain and spinal cord. The result is a cobweb-like network from which the arachnoid gets its name (Greek *arakhnoeides* = "cobweb-like"), and which

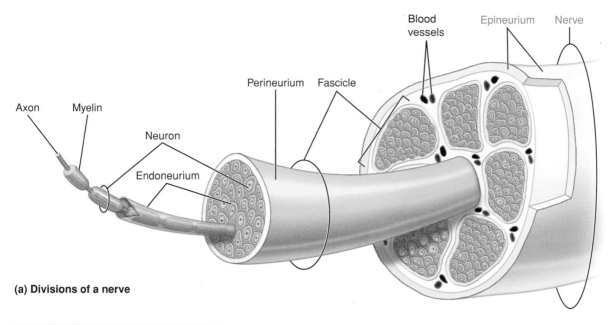

(a) Divisions of a nerve

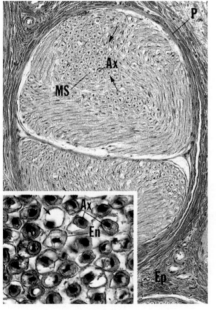

(b) Micrograph of a nerve

Figure 8.5. Connective tissue coverings of a nerve. A. Nerves contain three connective tissue layers. **B.** This micrograph shows two fascicles, each surrounded by perineurium (P). You can also see the myelin sheath (MS) surrounding an axon (Ax) and the perimysium (P) surrounding the entire nerve. The inset shows the endoneurium (En) around a single neuron. *Which connective tissue layer touches myelin?*

serves to stabilize the brain and spinal cord within their bony casing. The subarachnoid space contains cerebrospinal fluid (see below). Most of the large blood vessels that supply the brain travel along the surface of the brain in the subarachnoid space.

Case Note

8.5. Blood leaked out of broken vessels in Larry's brain and accumulated between the dura and the skull. Is there normally a space between the dura and the skull?

Cerebrospinal Fluid Bathes the Brain and Spinal Cord

The brain and spinal cord are filled with and surrounded by a watery fluid, the **cerebrospinal fluid** (commonly "spinal fluid" or CSF), which cushions the brain and spinal cord and mediates exchange of substances with the blood. The CSF is mainly water and contains only a very small amount of protein and a very few white blood cells; red blood cells are not present. Otherwise, its chemical composition is close to that of blood plasma; for example, the concentration of CSF glucose is normally close to that of blood glucose.

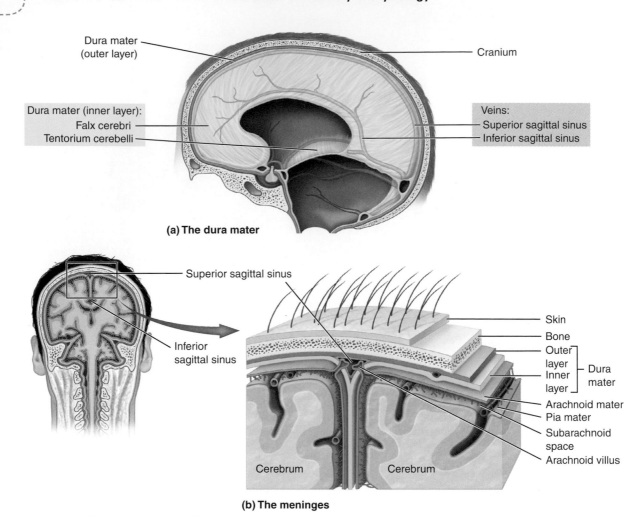

(a) The dura mater

(b) The meninges

Figure 8.6. The meninges. A. The innermost layer of the dura mater folds to form the falx cerebri and tentorium cerebelli. Large veins, the venous sinuses, run within these folds. **B.** Deep to the dura mater are two more meninges: the arachnoid mater and the pia mater. *Which membrane is attached to the cranium?*

CSF circulates around the brain and spinal cord through the subarachnoid space; it can exchange fluid and other substances with brain tissue through the somewhat permeable pia mater. Within the brain, it circulates through the **ventricular system,** a series of interconnected chambers (ventricles) that, during embryonic development, were in the center of the neural tube.

The **lateral ventricles** (also known as the first and second ventricles) form hollow spaces in the cerebral hemispheres (Fig. 8.7). Each ventricle forms an arc that is shaped something like a ram's horn: it extends posteriorly from the frontal lobe to curl inferiorly and then anteriorly to enter the temporal lobe (Fig. 8.7B). The lateral ventricles connect in the center to a thin vertical space, the **third ventricle,** which separates the two parts of the diencephalon. The **cerebral aqueduct** joins the third ventricle to the **fourth ventricle,** which sits between the cerebellum and the brainstem. The **central canal** extends inferiorly from the fourth ventricle, down

the center of the spinal cord. The relationship between the cerebral ventricles and brain tissue becomes clearer if we examine brain development in the embryo. Refer to the nearby Basic Form, Basic Function box, titled *The Developing Brain,* for more information.

Projecting into each ventricle are small, delicate, cauliflower-like collections of blood vessels and ependymal cells that together form the **choroid plexus.** The ependymal cells make CSF from blood plasma and secrete it into the ventricles.

The production, flow, and absorption of CSF are illustrated in Figure 8.7C.

1. CSF formed by the choroid plexus of the lateral ventricles flows into the third ventricle, where it is joined by CSF produced in this ventricle.
2. CSF subsequently flows through the cerebral aqueduct into the fourth ventricle, where it is joined by more newly synthesized CSF.

BASIC FORM, BASIC FUNCTION

The Developing Brain

The relationship between the cerebral ventricles and the different brain regions is much easier to understand if we consider how the brain develops embryonically. The upper figure resembles the human brain during embryonic development as well as the brain of other quadripedal mammals.

The embryonic CNS consists of a long hollow tube of tissue. The hollow interior of the tube forms the CSF-containing ventricular system. The most *rostral* (= "toward the nose") portion of the tube is enlarged into two blobs, the two cerebral hemispheres, each containing a fluid-filled lateral ventricle. If we could float on a raft through the ventricular system, we would proceed from the lateral ventricles into the third ventricle, which is surrounded by the diencephalon. Next would be the cerebral aqueduct, surrounded by the midbrain. Continuing in a *caudal* (= "toward the tail") direction through the fourth ventricle, we would find the cerebellum located dorsally and the upper brainstem ventrally. Finally, our raft would exit the brain and enter the central canal of the spinal cord.

As the human brain develops, an amazing event occurs—the long neural tube gets a kink in it. The cerebrum grows and essentially flops forward. The skull bones restrict the growth of the cerebral hemispheres toward the crown of the head, so they curve outward (towards the ear) and inferiorly. The result? The brain we see in part B of this figure, and the ventricles that we see in Figure 8.7.

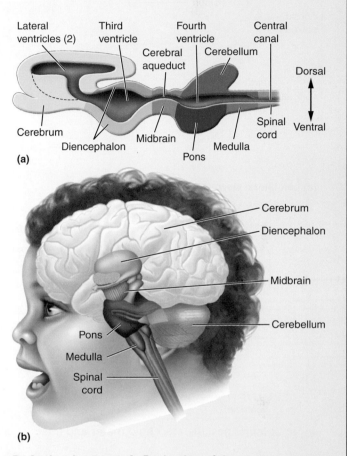

Brain development. A. Basic plan of the nervous system, as in an early embryo. The ventricular system is shown in dark blue. **B.** The human brain.

3. A small amount of CSF flows into the central canal of the spinal cord. The rest exits the ventricular system through small *apertures* (holes) in the medial and lateral walls of the fourth ventricle and enters the subarachnoid space. It circulates freely in this space around the brain and spinal cord.

4. CSF is absorbed from the subarachnoid space, through arachnoid granulations, into the blood of the superior sagittal sinus.

Normal adults have about 150 mL of CSF volume but at any one time produce about 500 mL per day, a daily turnover of three to four volumes.

The rate of CSF synthesis is regulated in order to maintain constant CSF pressure. Because of the rapid rate of CSF production and reabsorption, obstruction of CSF flow or reabsorption causes increased intracranial pressure. Increased CSF pressure is a serious matter—it injures or kills neurons and can cause severe and permanent brain damage. In young children whose skull bones have not fused, increased CSF pressure pushes unfused bones apart and causes the skull to enlarge and fill with CSF. This condition, called *hydrocephalus* (Greek *hydro* = "water" + *kephale* = "head"), is usually associated with brain atrophy and severe mental impairment. Conversely, in adults, who have rigid skulls, increased

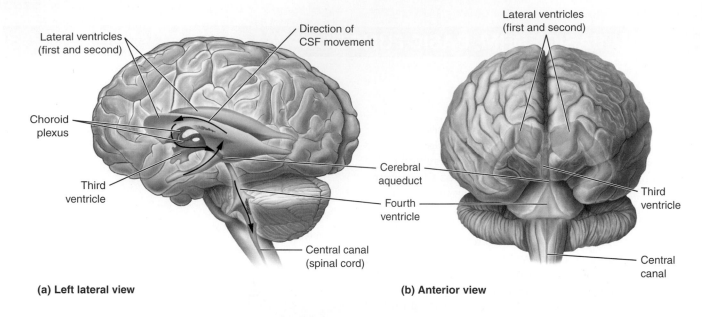

(a) Left lateral view

(b) Anterior view

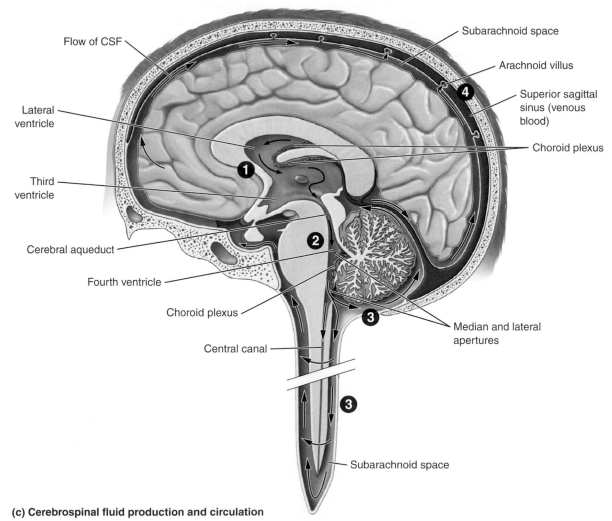

(c) Cerebrospinal fluid production and circulation

Figure 8.7. The ventricular system and CSF. A. Left lateral view. Each ventricle contains a choroid plexus, which synthesizes CSF. **B.** Anterior view. **C.** CSF circulates from the ventricular system into the subarachnoid space, and subsequently enters the bloodstream by diffusing from arachnoid granulations into the superior sagittal sinus. The numbers correspond to the narrative on pages 290 and 291. *Name the connecting vessel between the third and fourth ventricles.*

CSF pressure acts to force brain tissue, most of which contains blood vessels, out of the skull through any available opening—a condition called a brain *herniation*. When brain tissue herniates, the protruding tissue may press on blood vessels and block blood flow, or the herniated brain may be damaged directly, sometimes with devastating results.

Case Note

8.6. The superior left portion of Larry's brain was swollen. Would this swelling affect the lateral ventricle or the fourth ventricle?

The Blood–Brain Barrier Offers Molecular Protection

We've talked about structures and fluids that protect the nervous system from physical trauma, but the body provides protection against molecular disruption as well. For example, blood chemistry is constantly changing in response to eating, medication, or exercise. These variations in blood chemistry, which are well tolerated by other organs, can interfere with the functioning of the brain; for example, by causing "unauthorized" neural signaling.

Fortunately, capillaries in the brain have a unique structure that acts like a firewall to prevent most substances from entering vulnerable brain tissue. The cells that form the walls of capillaries in the brain are cemented together so tightly that many blood-borne substances simply cannot pass. This extremely low permeability of brain capillaries is known as the **blood–brain barrier.** Adding to this molecular barricade is the effect of the pia mater, which plays a role in keeping some molecules from diffusing into the brain from the CSF.

Only gases and other fat-soluble substances can move across these barriers without assistance. Since alcohol, caffeine, and nicotine are fat-soluble, they easily gain access to the brain. Certain small, water-soluble substances, such as glucose and amino acids, are allowed to pass only by regulated membrane transport processes (facilitated diffusion and active transport; see ◀ Chapter 3). Large molecules, proteins in particular, and many drugs, cannot enter brain tissue and remain in the bloodstream. The blood–brain barrier also prevents most bacteria from infecting delicate brain tissue. But some, such as those that cause bacterial meningitis, are able to slip through by disguising themselves as "self" proteins and poking holes in the barrier.

 8.9 Name in sequence the three layers of meninges starting with the outermost.

8.10 Which ventricle drains into the central canal?

8.11 Where is CSF produced? Where is it absorbed?

8.12 If someone drinks a rum and coke (containing sugar, caffeine, and alcohol), which chemicals can pass through the blood–brain barrier freely, without using specific transport mechanisms?

The Brain and Cranial Nerves

The brain appears deceptively bland—grayish and off-white, with a soft and gelatin-like consistency. It has no moving parts, and is without easily discernible distinctive features, such as the valves inside the heart or the respiratory branches within the lungs. Indeed, in preserving a corpse for the afterlife, ancient Egyptians thought so little of the brain that they drilled holes in the roof of the nose, inserted straws, sucked out the brain, and discarded it. We now know that our brain sets us apart from every other living thing, primarily because it grants us consciousness, the very essence of what it means to be human. In this section, we'll identify the regions of the brain and the functions they allow, including consciousness.

Cerebral Gray Matter Includes the Cerebral Cortex and Basal Nuclei

Recall that the cerebrum consists of two large masses of tissue, the right and left cerebral hemispheres (Fig. 8.8). Mirror images of each other, they resemble two boxer's gloves, thumbs inferolateral, flattened against each other where they meet in the midline. Although they are mirror-image "twins" anatomically, the two cerebral hemispheres differ somewhat functionally. You can read about the special talents of each hemisphere in the student resource section of the web site **https://thepoint./lww. com/McConnellandHull.**

The bulk of cerebral mass is a central core of *white matter* composed of tracts (Fig. 8.8B). These tracts transmit nerve impulses within one cerebral hemisphere, between cerebral hemispheres, and between the cerebrum and the cerebellum and brainstem.

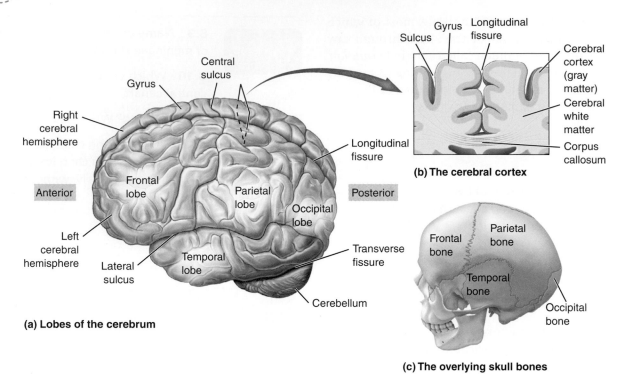

(b) The cerebral cortex

(c) The overlying skull bones

(a) Lobes of the cerebrum

Figure 8.8. The cerebrum. A. Fissures and sulci divide the cerebrum into four lobes. **B.** The highly folded cerebral cortex overlies the cerebral white matter. **C.** The lobes are named after the overlying skull bones. *Which lobe is more anterior—the temporal lobe or the occipital lobe?*

The remainder of the cerebral mass is *gray matter*, which occurs in two main forms: The first is a thin layer of neuron cell bodies that covers the core mass of white matter, like the rind of an orange. This layer is called, appropriately enough, the **cerebral cortex** (*cortex* = "rind"). Other gray matter resides as nodules, called **nuclei,** at various sites in the brain. The **basal nuclei** (see Fig. 8.11 on page 300) are a collection of nuclei deep within the base of the cerebrum. We begin our tour of the cerebral gray matter by exploring the cerebral cortex, home to the highest brain functions, such as intellect, reason, language, and the initiation of voluntary actions. Further in the text, we turn our attention to the basal nuclei.

Grooves Divide the Cerebrum into Lobes

The surface of each cerebral hemisphere is folded into a series of interconnected ridges and grooves, much like mountain ridges separated by valleys (Fig. 8.8A). The ridges are called **gyri** (plural of gyrus) and the grooves are **sulci** (plural of sulcus). Especially deep grooves are called **fissures** (Fig. 8.8B).

The **transverse fissure** separates the cerebellum from the cerebrum and the **longitudinal fissure** separates the right and left cerebral hemispheres. Deep in the longitudinal fissure, the cerebral hemispheres are joined by a thick bridge of tracts, the **corpus callosum,** which conduct signals between the two (Fig. 8.8B). From the longitudinal fissure, the **central sulcus** extends laterally and inferiorly from the apex of the cerebral hemisphere; it divides the cerebral cortex into anterior and posterior halves. The **lateral sulcus** separates the "thumb" of the "boxer's glove" from the remainder of the cerebral hemisphere.

These fissures and sulci divide each cerebral hemisphere into four *lobes* (Fig. 8.8A), which are named after the skull bones that cover them (Fig. 8.8C):

- The **frontal lobe** is the anterior part of each hemisphere and extends anteriorly from the central sulcus. It is separated from the temporal lobe below by the lateral sulcus.
- The **parietal lobe** is the middle part of each hemisphere and extends posteriorly from the central sulcus. There is no distinct boundary between the parietal lobe and the *occipital lobe*, which forms the posterior part of each hemisphere.
- The **temporal lobe** is the most lateral and inferior lobe (the thumb of the boxer's glove). It is anterior to the occipital lobe and is separated from the frontal and parietal lobes above it by the lateral sulcus. The **insula** of the temporal lobe is located deep in the fold

of the lateral sulcus, hidden from easy view, because it is covered by parts of the frontal, parietal, and temporal lobes. It is, so to speak, in the web of the thumb of the boxer's glove. The insula is illustrated in Figure 8.9B. Note that some anatomists consider the insula to be a separate lobe.

● The **occipital lobe** is the posterior part of each cerebral hemisphere and is separated from the cerebellum below by the transverse fissure. There is no anatomical feature delineating where the occipital lobe ends and the parietal and temporal lobes begin.

Case Note

8.7. Larry's brain injury lies anterior to the central sulcus, on either side of the lateral sulcus. Which two brain lobes were injured?

All Areas of the Cerebral Cortex Participate in Consciousness

What most sets humans apart from other living things? Famously difficult to define, **consciousness** is perhaps best described by what it does—it integrates the senses into a composite picture of reality and one's place in it. To *integrate* is to join two or more things together to make them part of a larger whole. The cerebral cortex receives sight, sound, smell, taste, touch, and other signals and binds them into an internal picture of a "self" apart from the environment ("nonself"). This sense of self is made up of sensations, memories of the past, and expectations of the future and has a concept of "time" based on immediate sensation and stored memory. Consciousness grants us an internal dialogue with ourselves ("thinking") that is uniquely human and endows us with the ability *to decide*. The lion sees the zebra but doesn't *decide* to attack in the sense that humans decide. The lion is on autopilot: the zebra will be a meal if the lion *feels* hungry; otherwise not. This is not to say that all animals operate by reflex only. Some animals—dogs, for example—can be trained to disobey their instinctual (reflexive) urges: instead of chasing a squirrel, a dog can be taught to stop with a "stay" command. But did the dog *decide* to stop, or was that, too, a reflex? So, are dogs conscious? Maybe in a certain way, but not in the way humans experience consciousness.

Ambrose Bierce's quotation at the opening of this chapter is clever and funny, but it is more profound than you might imagine, as it slyly raises one of the most vexing and difficult problems in all of science and philosophy: Do human beings really make choices, or are we merely reacting to the environment in a chain of causality (automaticity), like a row of dominoes falling in response to the fall of the first one? Are we on a very sophisticated autopilot, like the lion seeing the zebra? Or do we really decide? Common sense argues that of course we decide: I will have a hamburger—no pickles, please. However, that is this time; next time maybe I will have pickles. But scientific discoveries—from Copernicus's heliocentric universe to Einstein's theory of relativity—have often made history by rejecting common sense. When Copernicus did it, we realized that the earth orbits the sun. With Einstein, we discovered that space is bent and time may be an illusion.

Think about these facts: the stars and planets wheel in their majestic orbits unaffected by our intervention, reacting to cataclysmic events of the inconceivably remote past. Meanwhile, within our bodies, molecules race this way and that without our intervention, reacting in a chain of homeostatic activity, one action leading automatically and inevitably to the next. On a larger scale, our heart, lungs, intestines, liver, kidneys, and other organs quietly go about their jobs without our help—they, too, are on autopilot. So where, *exactly*, between Mars and mitochondria does automaticity end? Or does it end? Was I destined to order my hamburger without pickles (this time)? Was it really a choice? The nearby Clinical Snapshot, titled *The Power of the Subconscious Mind,* describes some intriguing studies into this question.

The truth is that we don't have an answer. However, in this book, we take the familiar position that we do, in fact, make decisions. Precisely how we do so is equally uncertain. But as we explore further, we know for a fact that certain regions of the cerebral cortex contribute to specific aspects of our conscious experience.

Discrete Regions of the Cerebral Cortex Govern Specific Functions

Cells in the right lobe of the liver perform exactly the same tasks as those in the left lobe. That is to say, the function of liver cells is not site-specific. This is not true for the brain as a whole, nor is it true for the cerebral cortex: signals related to specific functions—speech, vision, hearing, motor control, touch—are integrated at specific sites. Many brain functions are governed by a series of interconnected sites, but even so, each subsite makes a unique contribution to function that is not duplicated anywhere else. For example, optical signals from the eyes are integrated into visual images by certain parts of the brain. Destroy those parts of the brain and vision is gone forever, because no other part of the brain can perform that function, even if the eyes and other parts of the brain perform normally.

CLINICAL SNAPSHOT

The Power of the Subconscious Mind: Overrated or Understated?

The idea of subconscious influences on human behavior is recorded in writings thousands of years old. That we have a subconscious mind is common knowledge, yet precisely because we are unaware of it, most of us do not believe we are much influenced by it. The facts, however, suggest otherwise.

Consider the following experiment at a leading university. Experimenters arranged to have a lab assistant appear to casually encounter a series of students in a hallway going to class. The assistant, burdened with books, papers, and a drink in a cup, asked a series of students to help by holding the drink for a few seconds, which on some occasions was hot coffee, on others iced coffee.

A short time later, in the classroom, these same students read a story about a fictitious person and were asked to rate the person as warm or cold, sociable or distant, and generous or selfish. Students who held the hot coffee were much more likely to rate the person as warm, sociable, and generous. Students who held the iced coffee tended to rate the person as cold, distant, and selfish.

 OR  ?

Consciousness. Hot or cold drink? Do we really decide?

Or consider the following experiment. Students played a one-on-one investment game across a long table. Sitting at the far end of the table—and never mentioned—was either a briefcase and black leather portfolio or a casual backpack. Students who played the game in the presence of the briefcase and portfolio were much more businesslike, aggressive, and stingy with their money than those who played the game in the presence of the backpack.

What conclusion can we draw from such studies? The unconscious brain is far more purposeful, involved, and effective than we realize.

However, as shown below, to a certain limited degree some nearby portions of the brain can be retrained or "remapped" to perform the duties of dead or damaged neurons, but only if the nearby neurons normally perform a closely related function. For example, on some occasions, damage to neurons controlling muscle movement in one part of the body can be effectively replaced by nearby neurons controlling other muscle movements. But visual neurons in the occipital lobe cannot be retrained to substitute for neurons in some other part of the brain that are important in the sense of hearing.

Brain physiologists distinguish three primary types of functions, each accomplished by discrete regions of the cerebral cortex:

- *Sensory areas* receive sensory signals.
- *Motor areas* initiate voluntary motor movements.
- *Association areas* integrate information from other brain regions.

As illustrated in Figure 8.9, these broad categories can be further broken down into smaller, anatomically specific areas related to the function they perform. The primary visual cortex (also known as the primary visual area), for instance, is a sensory area responsible for processing visual input.

Before we discuss these areas, we should point out that the cerebral cortex is fed information by many other sites in the cerebrum, cerebellum, diencephalon, and brainstem. The cortex provides the last and highest order of integration that gives humans their unique mental characteristics—consciousness, emotions, reasoning. But it does not provide these attributes alone; it relies on help from other parts of the brain.

Case Note

8.8. What type of functional area of Larry's cerebral cortex appears the most damaged?

Sensory Areas

Sensory signals are interpreted in several areas of cerebral cortex. Because many sensory fibers cross from one

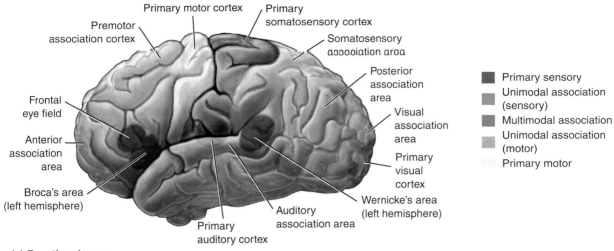

(a) Functional areas

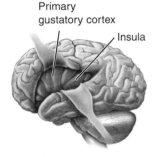

(b) The gustatory cortex and insula

Figure 8.9. Functional areas of the cerebrum. A. Primary sensory areas of the cerebrum receive sensory information. Unimodal association areas integrate sensory information from a single sense. Multimodal association areas integrate signals from many sources and make decisions. The motor association cortex organizes movements, and the primary motor cortex sends commands to individual muscles. **B.** The gustatory cortex and insula can be visualized by pulling back parts of the parietal and temporal lobes. *Which area receives input directly from the primary visual cortex?*

side to the other in the spinal cord or brainstem, the right cerebral cortex receives most of its signals from the left side of the body, and vice versa (this is not entirely true for the eyes, as discussed in ➡ Chapter 9). You can see a depiction of sensory fibers crossing over in Figure 8.22 on page 319.

Sensory input from skin receptors (touch, pain, pressure, and temperature) is interpreted in the **primary somatosensory cortex,** which runs along the posterior edge of the central sulcus in the most anterior part of the parietal cortex (Fig. 8.9). Notice that the primary somatosensory cortex is located directly across the central sulcus from the primary motor cortex. The primary somatosensory cortex is divided into regions according to the body area served. As you might expect, areas that are highly sensitive to touch, such as the lips and index finger, take up much more cortex than less sensitive regions (such as the back). This concept is illustrated in Figure 8.23 on page 320.

Sensory signals for vision, hearing, taste, and smell are interpreted in other parts of the cortex. The **primary visual cortex,** located in the most posterior cortex of each occipital lobe, weaves sensory signals from the eyes

into visual images. The **primary auditory cortex** is located on the superior surface of each temporal lobe, where the central sulcus meets the lateral sulcus. The **primary gustatory areas** are located low in the postcentral gyrus deep in the lateral sulcus of each hemisphere; they interpret sensory signals from taste buds on the tongue and can be seen in Figure 8.9B.

Each primary sensory area projects to a nearby, more integrative region of sensory cortex called a **unimodal association area.** *Unimodal* means "one mode," and these areas integrate signals for one discrete sense; that is, one mode of sensing. For instance, the visual cortex is immediately adjacent to the *visual association area*, which integrates visual information coming from different sites in the primary visual cortex about the color, shape, and movement of an object. Each unimodal association area thus provides an integrated perception of the world based on a single sense—for example, all the complexities of sound without visual or touch input, or all of the complexities of a handled object without the distractions of sight or sound. Selective damage to a unimodal association area results in a specific type of *agnosia* (= "not knowing"). A person with damage to the visual

Finally, Larry's emotional instability suggests the possibility of some damage to the limbic system, which we have learned adds emotional content and control to perceptions.

Incidentally, we used to think that people who have sustained a brain injury, like Larry, could not make new neurons. Now we know that neuron growth does occur in some brain regions. As discussed in the accompanying Clinical Snapshot, titled *Recovery from Brain Damage,* researchers are working to harness this neuron production to other areas of the brain so as to improve recovery following injury.

8.13   Name two types of gray matter in the brain.

8.14   With which lobe of the cerebral cortex do we see (perceive visual signals)?

8.15   Which association area is more involved in planning—the posterior association area or the anterior association area?

8.16   What is the difference between the premotor association cortex and the primary motor cortex?

8.17   True or false: Movement of the right hand is controlled by the left primary motor cortex.

8.18   Name four anatomical components of the limbic system.

8.19   Which of the following structures is part of the basal nuclei: mamillary body, globus pallidus, cingulate gyrus?

8.20   Name the brain lobe and brain area where Broca's area is found.

The Diencephalon Includes the Hypothalamus, Thalamus, and Pineal Gland

The **diencephalon,** is almost completely encased by the cerebral hemispheres, which fit over it like a thick helmet over a very small head. The diencephalon consists of several small structures, the three most important of which are the thalamus, hypothalamus, and pineal gland (Fig. 8.13).

- The **thalamus** consists of two pecan-sized masses sitting atop of the brainstem. It is an important relay station for sensory signals (pain, for example) and motor signals passing to and from the cerebral cortex, cerebellum, and brainstem. Thalamic nuclei integrate this information into a preliminary, crude perception of the world; they also play secondary roles in higher brain functions such as cognition, learning, memory, and consciousness.
- The **hypothalamus** is inferior and anterior to the thalamus. It contains 12 nuclei that regulate core vegetative functions such as body temperature, water balance, hunger, and sexual drive. It also governs the autonomic nervous system and **diurnal rhythms**; that is, the daily rhythms of life such as waking and sleeping and the regular rise and fall of blood hormones (cortisol, for example) throughout the day (compare with the pineal gland, below). Some classification schemes group the hypothalamus with the limbic system, since it responds to input from the amygdala related to negative emotions such as anger or fear. The hypothalamus also serves an endocrine role, secreting hormones that regulate other endocrine glands and exert effects in their own right (see ➡ Chapter 15).
- The **pineal gland** is about the size of a coffee bean and sits in the midline atop the midbrain. It is largest during childhood; it then shrinks and often calcifies during adulthood. It is important in the establishment of diurnal rhythms. In the absence of light, either at night or in other dark circumstances, the pineal gland secretes the hormone *melatonin*, which promotes sleepiness and is important in maintaining the daily sleep cycle. The out-of-kilter sleep patterns (jet lag) we suffer when flying across several time zones is caused by a temporary mismatch of melatonin secretion and daylight hours. A regular sleep pattern returns when pineal melatonin secretion adjusts to the new "clock."

Case Note

8.11. Larry's heart rate was elevated, a function controlled by the autonomic nervous system. Which part of the diencephalon was involved in his increased heart rate?

The Brainstem Contains Tracts and Nuclei

The **brainstem** is the lowest and most primitive part of the brain. It is about the size of your thumb and connects the spinal cord below to the diencephalon above

Recovery from Brain Damage: Is Regeneration Possible?

Not too long ago the brain was considered immutable: what you were born with is what you had forever, and if some of it was destroyed, it could not be replaced. Now we know that this is not completely true: although lost brain tissue is largely irreplaceable, this is not necessarily true for lost brain function. The brain is constantly being "rewired" as it adapts to experience or injury—thinking, learning, and doing actually alter the microanatomy of the brain. It turns out that the brain, like muscles, withers or grows according to use—the phrase "use it or lose it" is quite literally true.

For example, some patients who have lost a limb will report feeling that the limb is still there, a condition known as *phantom limb syndrome*. What's more, the cerebral cortex that once received sensory signals from the missing limb will be put to other use, a fact proved by the following: in a patient who has lost a limb, touching a certain point on the face produces a sensation that some part of the missing limb is being touched. Indeed, careful touching experiments on such patients reveal a map of the arm on the face—touch here on the cheek, and it feels as if the missing thumb were being touched; touch a few millimeters away, and it feels as if the wrist were being touched—a clear indication that the arm cortex has been reallocated (remapped) to the face.

But perhaps the ultimate question is this: Does *thinking* remap the brain? Formerly the answer was a resounding "no," but now it seems to be "maybe." For example, consider the following experiment involving Buddhist monks who had meditated on love and compassion for many years. The electrical activity of their brains was compared with that of volunteers who meditated on love and compassion for a week. The monks' brains showed a particular type of meditative electrical activity that was permanent and remained active even when they were not meditating. By comparison, similar electrical activity in meditating volunteers disappeared quickly. It seems as though long-term meditation permanently altered the monks' brains.

On a more concrete level, when someone sustains a head injury or suffers a stroke, the devastating effects of the resulting brain damage often persist; our patient, Larry H., is an example. Recall from ← Chapter 3 that neurons are long-lasting cells, and tissue regeneration following injury is difficult because few stem cells are present. Recent evidence suggests, however, that limited neuron regeneration does occur and could be harnessed to improve recovery following brain injury.

Recall also that stem cells must meet two criteria—they must give rise to at least two different cell types and be

Brain plasticity. Meditation, as practiced by these Buddhist monks, may remodel the brain.

capable of self-regeneration. Cells meeting these criteria have been found in two locations in the adult brain: (a) adjacent to each lateral ventricle near the olfactory bulb, a structure important to the sense of smell, and (b) the hippocampus. These stem cells are astrocytes that are able to give rise to more astrocyte stem cells, oligodendrocytes, or neurons. Under normal conditions, these stem cells cycle slowly, replacing neurons in the hippocampus and olfactory bulb that (unlike most neurons) are not permanent but die off regularly. These new neurons form connections with the other neurons and are involved in the sense of olfaction (smell) and in memory. When a brain injury occurs and neurons die, the stem cells step up production, and, amazingly, some of the new neurons migrate to the damaged area and attempt to establish new connections. Some of the slow, partial recovery following brain injury is due to regrowth of these new neurons.

It would be of enormous benefit if this process could be enhanced—more neurons, more migration, and more connections could result in faster and more complete recovery. The good news is that neuron proliferation is improved by positive life experiences, such as physical exercise, mental stimulation, and healthy eating. Studies in rodents and in vitro also suggest that certain growth factors, which could be given as medicine, stimulate neuronal proliferation and migration. It is also possible that stem cells could be introduced directly into the damaged area to enhance recovery.

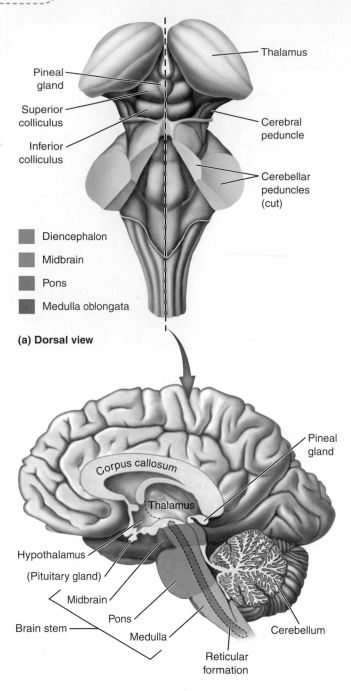

(a) Dorsal view

(b) Mid-sagittal section

Figure 8.13. Diencephalon and brainstem. A. Dorsal view. The cerebrum and cerebellum have been removed. **B.** Mid-sagittal view. *Name the three primary components of the diencephalon.*

(Fig. 8.13). The brainstem conveys signals from the spinal cord to the brain and also contains the nuclei for some cranial nerves (see Table 8.1). It is divided into the *midbrain,* the *pons,* and the *medulla oblongata.*

The **midbrain** is the most superior portion of the brainstem and connects directly to the diencephalon above. The anterior part of the midbrain consists of two huge bundles

of tracts, the **cerebral peduncles,** which transmit sensory signals traveling from below and motor signals traveling from above. Anatomically, the cerebral hemispheres seem to sprout from the cerebral peduncles the way a head of broccoli sprouts from its thick stem. The midbrain also contains multiple nuclei, which are organized in pairs on either side of the midline. Many nuclei are concentrated in the superior and inferior **colliculi** (Latin = "little hills"), which appear as upper and lower pairs of small mounds on the posterior surface of the midbrain (Fig. 8.13A). Nuclei in the superior colliculi coordinate tracking and scanning movements of the eyes and govern the movement of the head in response to visual signals. Nuclei in the inferior colliculi are part of the hearing apparatus and relay signals for hearing from the ears to the thalamus. They also relay signals for the startle reflex, which account for the sudden movements of body parts in response to loud, unexpected noise. The *substantia nigra* is a dark nucleus located near the basal nuclei (see Fig. 8.11).

The **pons** (Greek = "bridge") is a bulge that forms the middle and largest part of the brainstem. Like the medulla, it consists of bundles of tracts and several nuclei. The tracts connect above to the diencephalon, below to the spinal cord, and posteriorly to the cerebellum (the *cerebellar peduncles*). Its nuclei help the medulla regulate breathing.

The **medulla oblongata** (or just *medulla*) is the lowest portion of the brainstem and connects directly to the spinal cord. Most of its mass is composed of thick bundles of tracts that convey signals between the spinal cord and the higher regions of the brain. Recall that cerebral tracts cross to the opposite side of the body: most do so in the medulla. Also, the medulla contains several small but important nuclei:

- The *cardiovascular center* regulates the rate and force of heartbeats and the diameter of blood vessels (which in turn directly influences blood pressure).
- The *respiratory rhythmicity center* maintains the breathing rate.
- Other nuclei are important in (a) sensations of touch and vibration and (b) control of reflexes for swallowing, coughing, vomiting, sneezing, and hiccupping.

Finally, in addition to the structures described above, a somewhat sausage-shaped network of neuronal cell bodies criss-crossed by myelinated axons extends vertically throughout the entire brainstem, running immediately ventral to the fourth ventricle. These cell bodies are not condensed enough to form an anatomic nucleus, but they serve as one. This network, the **reticular formation,** contains both sensory and motor neurons that connect widely to the cerebral cortex (Fig. 8.13B). The descending motor neurons regulate muscle tone, the low-level contraction that occurs in resting muscles. Input passing through the sensory neurons is filtered so that background stimuli (such as the continual buzz of an electric fan) are ignored but a novel stimulus (such as the ringing of an alarm clock) is not. A subset of these sensory neurons forms the **reticular activating system.** Signals from the reticular activating system are important in producing wakefulness and general awareness—activity keeps us awake; inactivity induces sleep.

Case Note

8.12. Larry's heart rate and breathing rate were both increased when he arrived at the hospital. Name the brainstem nucleus that regulates each function.

The Cerebellum Coordinates Motor Activities

The **cerebellum** (literally "little cerebrum") sits posterior to the brainstem and immediately under the occipital poles of the cerebrum (Fig. 8.13B). It is attached to the brainstem by two thick bundles of nerve tracts, the right and left **cerebellar peduncles** (Fig. 8.13A). It is divided into two hemispheres, each of which is composed of central white matter and a deeply convoluted cortex of gray matter, which on cross section imparts a cauliflower-like appearance. The fourth ventricle, filled with CSF, separates the dorsal side of the brainstem and the ventral side of the cerebellum.

The cerebellum has no *direct* role in motor activities: it does not conceive or issue commands for body parts to do this or that. Rather, it is like the overseer of a factory who *modifies* the workers' efforts in response to executive commands to maximize efficiency. The cerebellum, working closely with the basal nuclei, monitors and makes corrective adjustments of musculoskeletal activities while they are under way. To accomplish this task, the cerebellum must sense the precise position of body parts. This position sense is called *proprioception* ➡ (Chapter 9), which is the product of sensory signals from muscles, tendons, and ligaments transmitted to the cerebellum via the somatosensory cortex. The cerebellum compares this input with signals from the cerebral cortex about the desired position of body parts and sends corrective motor commands to the brainstem, which relays them on to muscles. The result is smooth, coordinated musculoskeletal motion.

The ingestion of alcohol impairs the cerebellar proprioceptive sense; this is the basis for a standard law enforcement test for sobriety. The subject is asked to close his or her eyes and touch the tip of the nose with a finger. People significantly impaired by alcohol miss the mark and have to feel their way across the face to find the tip of the nose. Permanent impairment of the proprioceptive sense due to infection or injury has been reported. Such patients are severely handicapped and have difficulty with speech, which requires very precise manipulation of lips and tongue, which in turn requires proprioception. Such people also have difficulty grasping objects because they have an impaired sense of whether their fingers are properly located to ensure a grip.

Case Note

8.13. Before Larry's accident, he was capable of performing flips on his skateboard. Do you think the cerebellum was involved in this maneuver?

Cranial Nerves Arise from the Cerebrum and Brainstem

Twelve pairs of **cranial nerves** exit the cranium. Like all nerves, they are part of the PNS, conveying sensory and motor signals between the brain and the periphery. They are numbered by Roman numerals (I through XII) according to where they attach to the brain: the most superior is cranial nerve I, the most inferior is cranial nerve XII (Fig. 8.14). Cranial nerves I (*olfactory*) and II (*optic*) are attached to the cerebrum, and cranial nerve XI (*accessory*) is attached to the cervical spinal cord. The remainder are attached at the brainstem. The status of the accessory nerve is currently under investigation because anatomists now realize that, contrary to popular belief, it does not attach to the brainstem and perhaps should not be classified as a cranial nerve.

Remember This! The brain is part of the CNS; the cranial nerves are part of the PNS.

Some of the names of cranial nerves reveal the most important structures they control. For example, cranial nerve II is the optic nerve, which carries sensory signals from the eyes, and cranial nerve VII is the facial nerve, which carries sensory and motor signals from and to the face. All but one of the cranial nerves serve the head and neck. The exception is X, the *vagus* nerve, which, like the *vagabond* its name suggests, wanders away from the head and neck to serve diverse sites in the chest and abdomen.

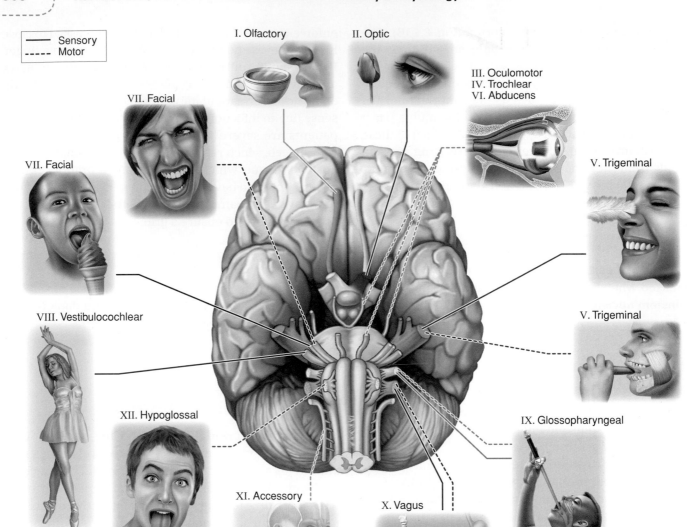

Figure 8.14. Cranial nerves. The 12 cranial nerves transmit sensory signals from sensory organs (solid lines) and motor signals to effector organs (dotted lines). *As you sniff a stinky cheese and make a face, which **sensory** nerve is active?*

Cranial nerves, like all nerves, can be classified as *sensory, motor,* or *mixed* according to the signals they carry (Table 8.1). Most nerves are mixed, but nerves I (olfactory), II (optic), and VIII (vestibulocochlear) are purely sensory. None are purely motor, although several are almost entirely motor: III (oculomotor), IV (trochlear), VI (abducens), XI (accessory), and XII (hypoglossal). Having significant sensory and motor function are V (trigeminal), VII (facial), and X (vagus).

The cranial nerves pass over uneven surfaces within the skull and exit the skull through openings (foramina). At first glance the olfactory nerve (I) does not appear to leave the skull—it seems to terminate on the floor of the

skull above the nose. However, it sends extensions through the tiny foramina of the cribriform plate to serve the olfactory epithelium in the roof of the nasal cavity.

At the point of exit, cranial nerves are vulnerable to skull fractures or increased intracranial pressure resulting from brain swelling or abnormal brain growths. As intracranial pressure increases, the cranial nerves can be compressed against uneven surfaces or the edges of the foramen through which they pass as they exit the skull. Such pressure can cause temporary or permanent cranial nerve damage. For example, one cause of anosmia, loss of smell, is a fracture of the cribriform plate that shears off the extensions of the olfactory nerve into the nose.

Table 8.1   Cranial Nerves

Nerve	Mnemonic*	Name	Sensory	Motor	Clinical Examination
I	Oh	**Olfactory**	Transmits olfactory (smell) impulses	None	Subject identifies various odors using each nostril
II	Oh	**Optic**	Transmits visual impulses	None	Test vision using a vision chart and a visual field test
III	Oh	**Oculomotor**	None	Controls voluntary eye and eyelid movements; controls pupil size	Test papillary reflex (pupils constrict in bright light); ask subject to blink and to follow a moving object medially and vertically
IV	To	**Trochlear**	None	Controls voluntary eye movements (superior oblique muscle)	Ask subject to follow a moving object inferiorly and medially
V	Touch	**Trigeminal**	Transmits impulses for touch, pressure, pain, and position from the face	Controls chewing muscles	Test corneal reflex by touching eye with cotton swab; test sensation by touching face with cotton swab, heat, cold, and a pinprick; test motor function by asking subject to move jaw against resistance
VI	And	**Abducens**	None	Controls voluntary eye movements (lateral rectus muscle)	Ask subject to follow a moving object laterally
VII	Feel	**Facial**	Transmits taste impulses from the anterior 2/3 of the tongue	Controls muscles of facial expression	Test ability to taste sweet, sour, salty, and bitter; test motor function by asking subject to smile, frown, whistle, etc.
VIII	Very	**Vestibulo-cochlear**	Transmits impulses from the ear (equilibrium, sound)	None	Test hearing by audiography; test balance by rotating patient
IX	Good	**Glosso-pharyngeal**	Transmits taste impulses and sensations from back of tongue Transmits blood pressure changes from the carotid sinus	Controls muscles involved in swallowing; controls some salivary glands	Test gag reflex by touching throat; test ability to taste
X	Velvet	**Vagus**	Transmits sensations from the throat, thoracic, and abdominal organs	Controls muscles involved in swallowing, coughing, speech; controls smooth and cardiac muscle and gastrointestinal glands	Test gag reflex by touching palate

(continued)

Table 8.1. Cranial Nerves (continued)

Nerve	Mne-monic*	Name	Sensory	Motor	Clinical Examination
XI	A	**Accessory**	None	Controls muscles that move the head and neck	Ask subject to shrug shoulders, rotate head against resistance
XII	H!	**Hypoglossal**	None	Controls tongue movements for swallowing and speech	Check for abnormalities of tongue appearance; ask subject to stick out tongue

*Or try *On One's Own Trying To Acquire Foreign Vocabulary Gives Very Agonizing Headaches.* Daniel Casse, University of Toronto.

8.21 Which part of the diencephalon controls appetite, water balance, body temperature, and sexual drive?

8.22 Name the most superior part of the brainstem.

8.23 Name two structures apart from the cerebral cortex that are important in motor control.

8.24 Which brain region controls alertness: the reticular formation or the thalamus?

8.25 Give the name and number of the cranial nerve that supplies the chest and abdomen.

The Spinal Cord and Spinal Nerves

Immediately inferior to the brain is the spinal cord, which connects to the medulla through the foramen magnum (see Fig. 8.2). The two primary functions of the spinal cord are illustrated by the following example: You are walking barefoot on the beach with a friend, when you step on a sharp stone. You jerk your foot away, say "Ouch! I stepped on a stone!" and then resume your walk and conversation. These simple acts involve the spinal cord, which sometimes acts merely as a pathway for nerve signals and other times actively integrates signals and directs responses. Let's look more closely at these two roles:

1. *The spinal cord can, without assistance from the brain, receive and <u>integrate</u> some incoming sensory signals; it can then generate involuntary motor signals to tissues or organs to effect a response.* When you stepped on the sharp stone, you didn't have to think "I should jerk my foot away" — the action had already occurred by the time the pain signal reached your cerebral cortex and you perceived pain. This rapid, involuntary response is a **reflex,** discussed later in this chapter.

2. *The spinal cord acts as a <u>pathway</u>, carrying sensory signals from the PNS to the brain, and conscious (voluntary) motor signals from the brain to the PNS.* The sharp stone stimulated sensory signals from your foot that traveled upward through the spinal cord to your cerebral cortex, where they were integrated into consciousness. You perceived an unpleasant sensation. You conceived an idea to tell your friend what happened, and you did so. Then you conceived the idea to continue your walk, and you did so. Motor signals from your cerebral cortex traveled through your spinal cord to various muscles involved in speech and walking. In the perception of pain and the decision to talk and walk, the spinal cord acted merely as a pathway. The brain did all of the integrating.

Case Note

8.14. Is there any evidence that Larry suffered damage to his spinal cord? Why or why not?

Vertebrae and Meninges Protect the Spinal Cord

Recall that the brain is protected by the bony cranium and by a series of tough connective tissue membranes called the *meninges.* The spinal cord is similarly protected by the bony armor of the vertebrae and the three meninges (Fig. 8.15). The pia mater and arachnoid mater of the spinal cord are continuous with those of the brain. However, only the outer layer of dura mater continues

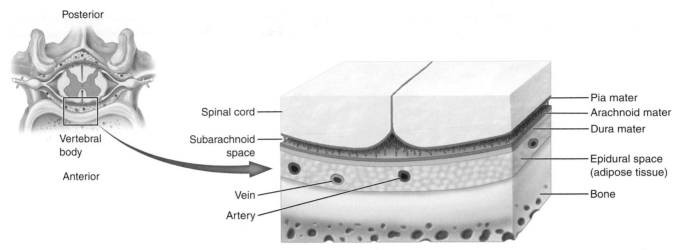

Figure 8.15. The spinal cord, vertebrae, and meninges. *How many layers of dura mater surround the spinal cord?*

into the vertebral canal as a tube that surrounds the spinal cord. In contrast to the region of dura lining the skull, vertebral dura is not physically attached to vertebral bone (Fig. 8.6C). Instead, the dura and bone are separated by a thin space containing fat—the **epidural space**—which is often used in obstetrics as a site where the physician injects anesthetic drugs that blunt the pain of childbirth. As an injection site, the epidural space is preferable to the subarachnoid space, because drugs injected into the epidural space migrate very little after injection. Conversely, drugs injected into CSF in the subarachnoid space could be washed up the spine or even into the brain, with potentially serious consequences, such as respiratory paralysis.

Although the spinal cord ends at about the first or second lumbar vertebra, the dural sheath continues inferiorly as a tube to about the level of the second sacral vertebra, forming a sac that covers the nerves of the cauda equina. This extension of the dura does not contain spinal cord, which makes it an ideal site for a *spinal tap*: a needle can be safely inserted between vertebrae and through the meninges into the subarachnoid space to measure spinal fluid pressure, introduce drugs or diagnostic fluids, or obtain CSF for analysis, without worry about puncturing the spinal cord.

The Spinal Cord Contains Gray and White Matter

A cross section of the spinal cord reveals a central core of gray matter arranged roughly in the shape of a butterfly with wings outspread (Fig. 8.16A). In the center of the gray matter is a tiny canal, the **central canal,** which contains CSF. Each wing is called a *horn*. The **ventral horns** contain the cell bodies of motor neurons. The **dorsal horns** contain the axons of unmyeli-

nated sensory neurons; their cell bodies are located outside the spinal cord, in the *dorsal root ganglion*. (Recall that a ganglion is a collection of cell bodies outside the CNS.) The ventral horn can also contain unmyelinated interneurons, which relay signals between motor and sensory neurons (not shown).

Fitted between and around the horns are thick bundles of myelinated nerve axons called **white columns** (Fig. 8.16A). These tracts convey sensory and motor signals up and down the cord (see Fig. 8.23 on page 320). Two grooves plunge deeply into the white matter, separating the cord into right and left halves. Dorsally, the **posterior median sulcus** makes a shallow indentation. Ventrally, the **anterior median fissure** extends deeply into the cord, nearly to the central canal.

Spinal Nerves Contain Roots, Trunks, and Branches

Spinal nerves have three portions, much like the roots, trunk, and branches of a tree (Fig. 8.16B). Each nerve has two *roots*— a *ventral root* through which motor axons exit the cord and a *dorsal root* through which sensory axons enter the cord. Just after exiting the cord, dorsal and ventral roots join to form a short, thick combined nerve *trunk*, which exits the vertebral column as a spinal nerve and then divides into multiple *branches* (or *rami*) that spread outward like tree branches to connect with body structures.

Ventral and Dorsal Roots Emerge from Their Corresponding Horns

The **ventral root** contains axons of motor neurons emerging from the ventral horn of spinal cord gray matter. These axons conduct motor (efferent) signals to

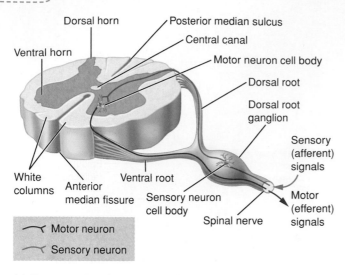

(a) Cross-sectional anatomy of the spinal cord

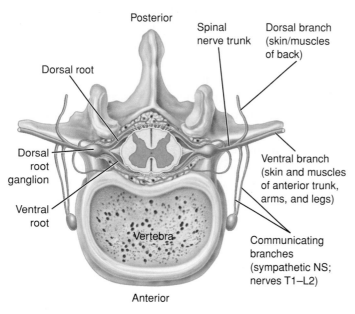

(b) Spinal nerve branches

Figure 8.16. The spinal cord, cross section. A. Sensory nerves enter the spinal cord through the dorsal horn; motor nerves leave the spinal cord through the ventral horn. **B.** Spinal nerves divide into numerous branches. *Which branch carries signals for the autonomic nervous system?*

skeletal muscle. Many ventral roots also contain axons of autonomic motor neurons that conduct motor signals to the heart and other viscera and glands.

The **dorsal root** conveys incoming sensory (afferent) signals from skin, muscle, and body organs. Contained within the dorsal root is a swelling, the **dorsal root ganglion,** which is composed of the cell bodies of autonomic and somatosensory neurons bringing signals to the cord (Fig. 8.16A). Notice that the cell bodies of sensory neu-

rons are in the dorsal root ganglion, whereas the cell bodies of motor neurons are in the ventral horn of the cord.

Ventral and Dorsal Roots Join to Form a Nerve Trunk

After the dorsal and ventral roots join to form a spinal nerve, the nerve splits into multiple *branches*, each serving a different purpose (Fig. 8.16B). The **ventral branch** travels to the skin and muscles of the ventral surface of the torso and the upper and lower limbs. In contrast, the **dorsal branch** travels to the skin and muscles of the dorsal surface of the torso but not the dorsal surfaces of the limbs. The third branch, the **communicating branch,** is found only in spinal nerves T1 through L2. This branch contains nerves involved in the sympathetic nervous system, discussed below.

> **Case Note**
>
> **8.15. In the ER, Larry responded to pinpricks on his chest by thrashing about, even when he was unconscious. Name the branch and root that carried the sensory signal from the pricked skin into the spinal cord.**

Thirty-One Pairs of Spinal Nerves Emerge From the Spinal Cord

Thirty-one pairs of spinal nerves arise from the left and right sides of the spinal cord (Fig. 8.17). With the exception of the cervical nerves, they are named according to the vertebra *superior* to their point of exit from the vertebral canal. For instance, the first lumbar nerve (L1) exits inferior to the first lumbar vertebra. However, for cervical nerves the opposite applies. Nerves 1 through 8 are named for the vertebra *inferior* to their point of exit because the first spinal nerve exits inferior the cranium and superior to the first cervical vertebra (the atlas). Notice in Figure 8.17 that, from superior to inferior, the spinal nerves have progressively more downward slope as they travel from the cord to their exit point from the vertebral column. This observation reflects the fact that, as we grow from childhood to adulthood, the vertebral column grows more rapidly than the spinal cord. Another result of this uneven growth rate is that the spinal cord ends at about the first or second lumbar vertebra, below which is a fan of spinal nerve roots, which we identified earlier as the *cauda equina.*

Each spinal nerve serves a particular anatomical site. For example, sensory signals from the thumb travel up

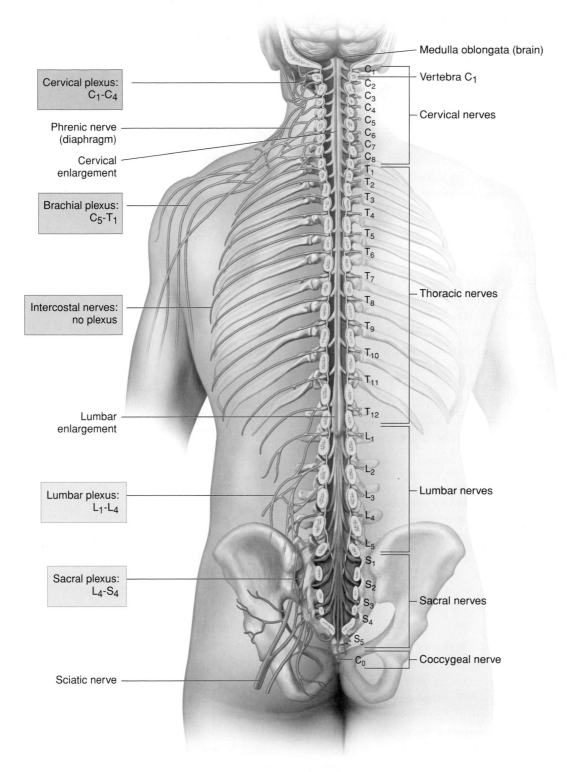

Medulla oblongata (brain)

Cervical plexus:
C_1-C_4

Vertebra C_1

C_1
C_2
C_3
C_4
C_5
C_6
C_7
C_8

Cervical nerves

Phrenic nerve
(diaphragm)

Cervical
enlargement

Brachial plexus:
C_5-T_1

T_1
T_2
T_3
T_4
T_5
T_6
T_7
T_8
T_9
T_{10}
T_{11}
T_{12}

Thoracic nerves

Intercostal nerves:
no plexus

Lumbar
enlargement

L_1
L_2
L_3
L_4
L_5

Lumbar nerves

Lumbar plexus:
L_1-L_4

Sacral plexus:
L_4-S_4

S_1
S_2
S_3
S_4
S_5
C_0

Sacral nerves

Coccygeal nerve

Sciatic nerve

Posterior view

Figure 8.17. The spinal nerve plexuses. The spinal nerves are numbered on the right. The spinal plexuses and selected spinal nerves formed by the ventral branches are shown on the left. *Which thoracic nerve participates in a plexus?*

a nerve branch in the arm and enter the spinal cord via the dorsal (sensory) root of the sixth cervical nerve (C6). Outbound C6 motor signals to muscles and other tissues in the arm travel through the ventral motor root. They work together: burn your thumb and the muscles of your arm jerk it away from the heat.

The spinal nerves supplying the upper and lower limbs contain especially large numbers of neurons. To accommodate these, the regions of the spinal cord supplying the upper and lower limbs are thickened into the *cervical enlargement* and the *lumbar enlargement,* respectively.

Some Ventral Branches Intermingle to Form Plexuses

The ventral branches of thoracic nerves 2 through 12 extend directly to the structures they supply. However, the ventral branches of cervical, lumbar, sacral, and coccygeal spinal nerves intermingle to form **plexuses** (Latin = "braids"), or webs of interlaced axons that recombine to form new nerves that carry axons from multiple spinal cord segments (Fig. 8.17). For example, branches of some lumbar nerves and most sacral nerves unite from the *sacral plexus.* The axons of many of these branches recombine to form the *sciatic nerve,* the largest nerve in the body.

> *Remember This!* **Branches of nerves from several spinal cord levels intertwine to produce a plexus, from which fibers recombine to form peripheral nerves.**

The four major plexuses of the somatic nervous system are:

- The **cervical plexus,** which serves the skin and muscles of the posterior head, the neck, the upper part of the shoulder, and—important to note—the diaphragm (via the phrenic nerve). This plexus is composed mainly of axons from cervical nerves C1 to C4, with a few axons from C5. Damage to the spinal cord above the origin of the phrenic nerve (which originates from nerves C3, C4, and C5) can paralyze the diaphragm and cause respiratory failure. Such an injury occurred to the late Christopher Reeve.
- The **brachial plexus,** formed by nerves C5 to C8 and T1, serves the upper limb and some neck and shoulder muscles.
- The **lumbar plexus,** formed by nerves L1 to L4, serves the abdominal wall, parts of the lower limb, and the external genitalia.

- The **sacral plexus,** formed by nerves L4 to S4, serves the buttocks, perineum, and lower limb. Note that nerve L4 contributes axons to both the lumbar and sacral plexuses.

8.26 What is the name of the event when the spinal cord integrates a sensory signal into an immediate motor action?

8.27 In which horn of spinal cord gray matter are the cell bodies of motor neurons located?

8.28 What separates the dura from the vertebral bone?

8.29 Which groove extends more deeply into the spinal cord—the posterior median sulcus or the anterior median fissure?

8.30 Is the dorsal root ganglion part of the PNS or the CNS?

8.31 True or false: The communicating branch of nerves T1 through L2 transmits signals for the autonomic nervous system.

8.32 Nerves to the lower limb pass through which two plexuses?

The Autonomic Nervous System

We learned at the beginning of this chapter that the somatic division of the PNS carries sensory signals that the cerebral cortex *consciously* perceives and motor signals that the cerebral cortex *voluntarily* issues. The sole effectors of the somatic division are skeletal muscles. By contrast, the autonomic division, called the **autonomic nervous system (ANS),** carries sensory and motor signals to and from the hypothalamus that operate *automatically,* without conscious awareness and beyond voluntary control. The ANS uses a wide variety of effectors, including cardiac muscle, smooth muscle, and glands.

The Autonomic and Somatic Divisions Differ Anatomically

The somatic and autonomic nervous systems differ anatomically and functionally, as summarized in Table 8.2 later in the chapter and illustrated in Figure 8.18. For

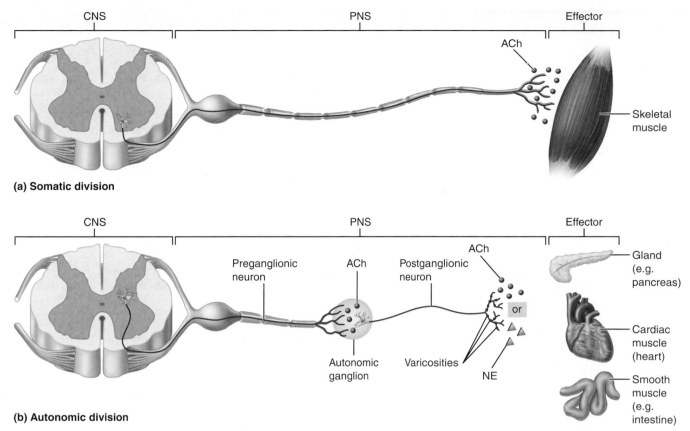

CNS PNS Effector

ACh

Skeletal muscle

(a) Somatic division

CNS PNS Effector

Preganglionic neuron

ACh

Postganglionic neuron

ACh

or

Gland (e.g. pancreas)

Cardiac muscle (heart)

Smooth muscle (e.g. intestine)

Autonomic ganglion

Varicosities

NE

(b) Autonomic division

Figure 8.18. Somatic and autonomic motor pathways. A. Somatic motor pathways use a single cholinergic neuron to convey signals between the spinal cord and skeletal muscle. **B.** Autonomic motor pathways use a cholinergic neuron and a second neuron that is either cholinergic or adrenergic to convey signals between the spinal cord and smooth muscle, cardiac muscle, or glands. Cholinergic neurons use acetylcholine (ACh) and adrenergic neurons use norepinephrine (NE). *Which type of neuron is always cholinergic—preganglionic or postganglionic?*

instance, they differ in the number of neurons connecting the CNS to the target organ:

- In the *somatic* division a *single* neuron connects the CNS to the target organ. For most cranial nerves, the cell body of the somatic motor neuron lies in the brain; for spinal nerves (and cranial nerve XI), the cell body lies in the spinal cord. However, for both, the axon extends all the way to the skeletal muscle fiber it stimulates.
- In the *autonomic* division, *two* neurons are required to connect the CNS to the target tissue. These two neurons synapse in an **autonomic ganglion,** a nodule of PNS gray matter. The cell body of the first neuron, the **preganglionic neuron,** lies in brain or spinal cord gray matter. This first, myelinated neuron synapses with a *second* neuron, the unmyelinated **postganglionic neuron.** The cell body of this second neuron lies in an autonomic ganglion.

The somatic and autonomic systems also use different *neurotransmitters*. Recall from Chapter 4 that neu-

rotransmitters are chemicals that relay, amplify, or alter electrical signals between a neuron and another cell.

- *Somatic* action potentials are relayed across synapses from neuron to skeletal muscle solely by *acetylcholine (ACh)*.
- *Autonomic* action potentials are relayed across synapses from neuron to neuron by acetylcholine, but signals from neuron to effector tissue (cardiac muscle, smooth muscle, or gland) are relayed by either *ACh* or *norepinephrine (NE)*.

Remember This! **To reach target tissue from the CNS, the somatic (voluntary) nervous system needs but a single neuron; however, the autonomic (involuntary) system requires two.**

What's more, the way that ANS motor neurons synapse with their target organs differs. Recall that somatic motor neurons divide into multiple tiny branches, or

Table 8.2 The Peripheral Nervous System (Motor Pathways): Characteristics of Form

Characteristic	Somatic Motor NS	Sympathetic NS	Parasympathetic NS
Number of neurons	One	Two	Two
Preganglionic cell body location	Ventral horn (all spinal nerves), brainstem/cervical spinal cord	Ventral horn of T1-L2	Ventral horn of S2–S4, brainstem (cranial nerves III, VII, IX, X)
Spinal nerve branch	Dorsal branch supplies dorsal body surface, ventral branch supplies rest of body	Communicating branch	Pelvic nerves sprout off from the ventral branch
Preganglionic neuron length	N/A	Short	Long
Preganglionic neuron neurotransmitter	N/A	Acetylcholine	Acetylcholine
Name of ganglia	N/A	Sympathetic chain, collateral ganglia	Terminal ganglia
Location of ganglia	N/A	Near vertebrae	Near target organ
Postganglionic neuron	N/A	Long	Short
Postganglionic neuron neurotransmitter	N/A	Usually norepinephrine	Acetylcholine
Plexuses	Ventral branches form the cervical, brachial, lumbar, and sacral plexuses	Postganglionic neurons join autonomic plexuses	Some preganglionic neurons join autonomic plexuses

axon terminals (see Fig. 7.3B on page 234), which release neurotransmitter into a small, discrete site, represented in muscle by the neuromuscular junction. However, postganglionic autonomic motor neurons do not have axon terminals. Instead, neurotransmitter is released from swollen axon portions called *varicosities*, which serve a broad area by allowing the neurotransmitter to diffuse more widely than in a motor synapse. This enables a relatively small number of ANS motor fibers to control large expanses of smooth muscle or glandular tissue.

Parasympathetic and Sympathetic Divisions Induce Opposing Effects

Like the somatic nervous system, the ANS includes both motor and sensory nerves. We discuss autonomic sensory nerves further on in this section.

Autonomic motor nerves are grouped into two divisions, based on the effects they induce and the neurotransmitter they use. The **parasympathetic division** stimulates a "rest and digest" reaction using acetylcholine as a neurotransmitter; the **sympathetic** division stimulates a fight-or-flight reaction using norepinephrine (see Table 8.2). Norepinephrine is very similar to *epinephrine,* a hormone secreted by the adrenal gland that is discussed further below.

Case Note

8.16. Larry's heart rate was elevated when he arrived at the hospital. Which branch of the autonomic nervous system induced this change, and which neurotransmitter(s) were involved in carrying the signal from his CNS to his heart?

Parasympathetic Signals Support the Rest-and-Digest State

Parasympathetic activity is evident when we are calm and relaxed (Fig. 8.19A). Imagine yourself lying in a field after a picnic. Your heartbeat and respiration rate settle into a slow, regular rhythm. Digestive juices flow,

(a) The rest-and-digest response

(b) The fight-or-flight response

Figure 8.19. Effects of the autonomic nervous system.
A. The parasympathetic division activates the rest-reproduce-digest response. **B.** The sympathetic division activates the flight-or-fight response. *Which division increases the heart rate?*

intestinal activity increases, and the internal anal sphincter relaxes for possible elimination. Blood pressure falls and lung airways constrict, because you are inactive and don't need much oxygen or blood flow. Pupils narrow and lenses thicken for clear close vision.

All of these diverse actions are induced by the same neurotransmitter—ACh—released from parasympathetic postganglionic neurons onto the different target organs. Hence, parasympathetic effects are sometimes described as *cholinergic.*

Remember This! **Both preganglionic and postganglionic neurons of the parasympathetic nervous system use ACh.**

Sympathetic Signals Stimulate the Fight-or-Flight Response

Sympathetic activity is evident when we are excited or stressed (Fig. 8.19B). Imagine a car careening toward you. Your sympathetic system kicks in immediately—your heart races, you breathe rapidly and deeply, your hair literally "stands on end," you break out in a sweat, and your pupils dilate. These perceptible changes are accompanied by equally dramatic imperceptible changes: your blood pressure rises, skin and intestinal blood vessels constrict (shunting blood to brain and muscle for more effective use), blood glucose rises (to fuel the escape), lung airways dilate (to gather more air for the escape), and the lenses of your eyes flatten for better distant vision.

Sympathetic postganglionic neurons induce these widespread effects by releasing the neurotransmitter norepinephrine. Sympathetic effects are often described as *adrenergic,* based on the alternate name for norepinephrine, noradrenaline.

But it's not just the fear of an impending accident that will stimulate sympathetic activity. Exercise, trauma (surgery or accident), excitement at a spectator sport, or emotional upset may also trigger an outpouring of sympathetic activity. Recent evidence suggests that sympathetic activity may cement memories more firmly in the cerebral cortex, making them more or less unforgettable.

Case Note

8.17. Recall that Larry's blood pressure was low after his accident. What would correct his problem—an injection of epinephrine or of ACh?

The Parasympathetic and Sympathetic Divisions Differ Anatomically

The sympathetic and parasympathetic divisions differ not only in the effects they produce and the postganglionic neurotransmitter they use but also anatomically. For instance, sympathetic ganglia lie close to the spine, whereas parasympathetic ganglia lie near or in the target tissue. The result is that in the sympathetic system the preganglionic axon is short, because the ganglia lie close to the spine, and the postganglionic axon is long. The reverse is true in the parasympathetic system, where the preganglionic axon is long, because the ganglia lie in or near the target tissue, and the postganglionic axon is short. You can see these differences in Figures 8.20 and 8.21.

Second, parasympathetic and sympathetic nerves leave the spinal cord at different segments—parasympathetic nerves leave from the brain and the sacral region of the spinal cord; sympathetic nerves leave from the thoracic and lumbar regions of the spinal cord.

Parasympathetic Signals Arise from Cranial and Sacral Regions

The parasympathetic division of the autonomic nervous system is also called the *craniosacral division,* because parasympathetic motor signals originate from cranial nerve nuclei and sacral segments of the spinal cord (Fig. 8.20).

The preganglionic neurons of the cranial portion of the craniosacral division originate in the brain nuclei of cranial nerves III, VII, IX, and X. Those of the sacral portion arise from gray matter in the S2 through S4 region of the spinal cord and travel through corresponding spinal nerves. These parasympathetic preganglionic neurons synapse with a second (postganglionic) neuron in a **terminal ganglion,** which lies very close to or actually within the target tissue.

The cranial nerves carry outgoing parasympathetic signals destined for the head, face, thorax, and abdomen. About 80% of total parasympathetic motor signal outflow is carried by the cranial nerve X, the vagus nerve,

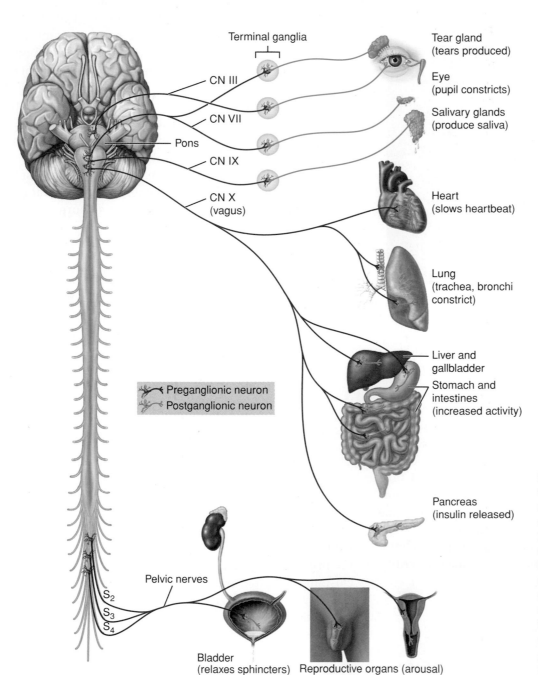

Figure 8.20. The parasympathetic division. Preganglionic neurons, shown in red, originate in the brainstem (cranial nerves III, VII, IX, and X) and spinal cord segments S2-S4 (pelvic nerves) and terminate at a synapse in the terminal ganglia. The short postsynaptic neuron (blue) terminates on a target organ. Only one side is shown. *What is the effect of parasympathetic activation on the trachea and bronchi?*

which is connected to multiple sites in the chest (heart, larynx, lung airways) and abdomen (intestines, pancreas, liver, gallbladder, and pancreas). Other cranial and sacral nerves carry the remaining 20%.

The parasympathetic neurons originating in sacral spinal cord segments S2, S3, and S4 extend for a short distance down the ventral branch of their spinal nerve; then they branch off to form the exclusively parasympathetic *pelvic nerves* (which have no somatic or sympathetic components). These nerves carry parasympathetic motor signals to parts of the large intestine and the pelvic organs.

The large intestine, which is responsible for holding and eliminating feces, contains parasympathetic ganglia in its muscular wall. Congenital absence of these ganglia in a segment of colon prevents normal waves of contraction (peristalsis) from passing down the wall to push fecal matter through the intestine. In such cases, the large intestine is effectively obstructed—although the lumen is open— and feces cannot pass through the affected segment. Upstream, the unaffected portion of the colon dilates markedly as it fills with fecal matter, a condition called *congenital megacolon*. Without surgical intervention to remove the defective segment of colon, rupture, infection and death can occur.

Sympathetic Signals Arise from Thoracic and Lumbar Regions

The sympathetic division of the autonomic nervous system is also called the *thoracolumbar division,* because sympathetic signals originate from thoracic and the first two lumbar segments of the spinal cord (Fig. 8.21).

The first sympathetic motor neuron (the preganglionic neuron) originates in the gray matter of the spinal cord. Its axon extends through the ventral root of a spinal nerve and down the communicating branch of a thoracic or lumbar nerve to the **sympathetic chain,** a series of paired ganglia that form a beaded chain along either side of the entire length of the vertebral column. Even though the sympathetic chain receives input only from spinal nerves T1 through L2 (inclusive), there is a ganglion in the chain at the level of each vertebra in the thoracic, lumbar, sacral, and coccygeal regions and three ganglia in the cervical region.

> **Remember This!** Communicating branches connect the ventral branch of thoracic and lumbar spinal nerves to the sympathetic chain ganglia.

The axons of some sympathetic motor neurons synapse with their second neuron (the postganglionic neuron) within the sympathetic chain. However, the axons of other neurons do not synapse in the sympathetic chain. Rather, they pass through the chain to synapse with their second neuron in one of several **collateral ganglia,** which occur as unpaired nodules located near the anterior edge of the vertebral column. The three largest collateral ganglia—the *celiac, superior mesenteric,* and *inferior mesenteric*—are located near the origins of large arteries branching from the aorta and are named after them. The nerves connecting the sympathetic chain and collateral ganglia are called **splanchnic nerves.**

After the synapse in the sympathetic chain or collateral ganglia, the postganglionic axon extends to the target organ (the intestines or bladder, for example). There is one notable exception: the adrenal gland. Neurons originating in the spinal cord pass through a collateral ganglion and terminate directly in the inner portion (medulla) of the adrenal gland ➡ (Chapter 15). Upon stimulation by sympathetic nerve impulses, the adrenal gland secretes epinephrine, which adds to the effect of other sympathetic neural signals to invoke the fight-or-flight sympathetic response discussed above. After the danger is over, sympathetic nerve activity ceases quickly; but because adrenal hormones take a while to be metabolized by the liver, it may take a while to wind down from their effect.

Although not illustrated in Figure 8.21 because it would be cluttered to show, every sympathetic chain ganglion sends fibers to a particular skin region to innervate sweat glands, muscles, and adipose tissue. By contrast, skin does not receive parasympathetic innervation. Selected sympathetic chain ganglia also project nerves to specific visceral organs. The cervical ganglia serve the sympathetic needs of the eyes, salivary glands, heart, and lungs. Signals destined for the gastrointestinal tract, reproductive organs, kidneys, and bladder originate from thoracic, lumbar, and sacral ganglia.

Case Note

8.18. Did the signal to increase Larry's heart rate pass through a sympathetic chain ganglion, a collateral ganglion, or both?

Parasympathetic and Sympathetic Neurons Mingle in Plexuses

Recall that the ventral branches of spinal nerves interlace to form the plexuses of the somatic nervous system. Autonomic nerves of the torso also form plexuses (Fig. 8.22). Because sympathetic ganglia are located close to the spinal cord but parasympathetic ganglia are located near the target organ, an autonomic plexus contains

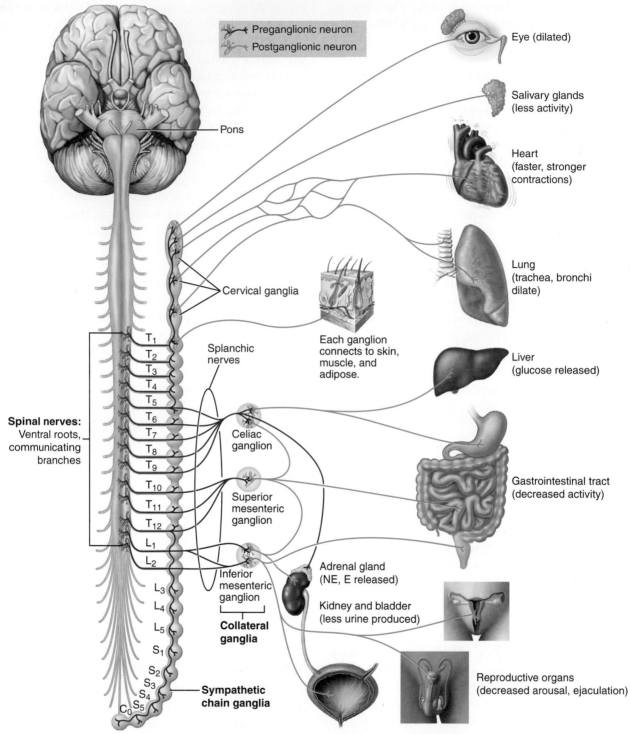

Figure 8.21. The sympathetic division. Preganglionic neurons, shown in red, originate in spinal cord segments T1–L2 and terminate in one or more ganglia. Postganglionic neurons, shown in blue, originate in the sympathetic chain or collateral ganglia and terminate on target organs. Note that the adrenal gland is directly innervated by a preganglionic neuron. Only one side is shown. *Which organs are innervated by the cervical ganglia?*

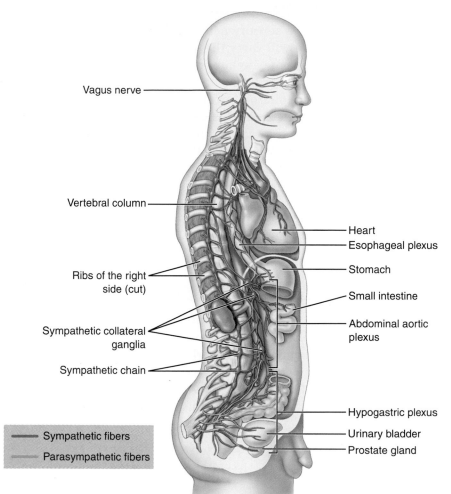

Vagus nerve

Vertebral column

Ribs of the right
side (cut)

Sympathetic collateral
ganglia

Sympathetic chain

Heart
Esophageal plexus
Stomach
Small intestine
Abdominal aortic
plexus

Hypogastric plexus
Urinary bladder
Prostate gland

— Sympathetic fibers
— Parasympathetic fibers

Figure 8.22. Autonomic ganglia and nerve plexuses. The sympathetic chain and sympathetic collateral ganglia contain cell bodies; the nerve plexuses contain axons. *Does the hypogastric plexus serve the sympathetic division, parasympathetic division, or both?*

mingled *postganglionic* sympathetic and *preganglionic* parasympathetic fibers, which create delicate networks as they intertwine on their way to target tissue. Two such plexuses—the *abdominal aortic plexus* and the *hypogastric plexus* supplying the pelvic organs—are illustrated on Figure 8.22. The esophageal plexus, however, contains only parasympathetic fibers, those of the vagus nerve.

Autonomic Sensory Nerves Are Diffusely Distributed

Thus far, our discussion has focused primarily on autonomic motor nerves. However, autonomic sensory nerves are also part of the ANS; they transmit visceral sensory information from organs innervated by the ANS. Some of these signals (such as bladder distention) enter into consciousness, whereas others (such as blood vessel distention by high blood pressure) do not. The sensory neurons conveying these signals follow many routes to arrive at the CNS; they may travel alongside ANS motor neurons

innervating the same region or even alongside somatic sensory neurons innervating nearby skin regions.

Pop Quiz

8.33 How many neurons are necessary in the somatic and autonomic divisions of the PNS to connect the CNS to target tissue?

8.34 True or false? Dilated pupils indicate a rest-and-digest state consistent with parasympathetic activity.

8.35 Which of the following nerves carry only ANS signals: cranial nerves or pelvic nerves?

8.36 The ganglia located close to the vertebral column belong to which division of the autonomic nervous system?

8.37 Name the three collateral ganglia.

Pathways of Neural Function

For the sake of simplicity and clarity, the preceding discussion considered brain regions, the spinal cord, and the nerves as independent anatomical entities. However, the axon of a single neuron can run through multiple brain regions and several feet of spinal cord, or through part of the spinal cord and the entire length of a spinal nerve. This section discusses some of the pathways that electrical signals follow between the periphery and the brain.

Sensory and Motor Pathways Form Spinal Cord White Matter

Recall from Figure 8.16 that the spinal cord white matter consists of myelinated axons running longitudinally through the cord. These axons are organized into tracts, each of which carries a particular signal type—sensory or motor. Three important tracts—the *anterior corticospinal tract*, the *lateral corticospinal tract*, and the *posterior column*—are shown in Figure 8.23A, but there are many more.

The *anterior* and *lateral corticospinal tracts* make up the corticospinal (motor) pathway (Fig. 8.23B). Consider the simple act of extending your index finger to touch a cat. A signal initiated in the primary motor cortex of the *left* cerebral hemisphere, in the region that controls the right hand, will travel to the medulla oblongata via *upper motor neurons*. Most of the axons of the upper motor neurons will cross over (*decussate*, from Latin *decussis* = for the symbol X) within the medulla to run down the *right* lateral corticospinal tract to the spinal cord where spinal nerve C6 originates. A few (less than 20%) will continue down the left side of the spinal cord, through the left anterior corticospinal tract, and will not cross over to the right side until they reach the spinal cord segment supplying nerve C6. Regardless of the pathway they followed to arrive at C6, all neurons synapse with a second motor neuron, the *lower motor neuron,* within the ventral horn of spinal cord gray matter. These neurons leave the cord through the ventral root to form part of spinal nerve C6. Other motor signals destined for other skeletal muscles exit from the brainstem (in the case of cranial nerves) or from different spinal cord segments.

The *posterior column* (Fig. 8.23C) is a tract that carries sensory signals for fine touch, pressure, vibration, and proprioception. Returning to our previous example, as you reach out your index finger, the soft cat fur will activate sensory receptors in your finger that are part of a sensory neuron. The process of this sensory neuron will carry the signal past the cell body in the dorsal root ganglion and through the dorsal root into the spinal cord and up through the posterior column to the medulla oblongata, where it will synapse with a second neuron. This second neuron will cross over to the left side within the medulla and proceed to the thalamus, where it will synapse with a third neuron. This third neuron takes the signal to the primary sensory cortex region responsible for the index finger. If, say, the cat were to respond to your attentions with a scratch, pain receptors in your finger would take a different pathway.

Case Note

8.19. **Physicians evaluated Larry's neurological state while he was unconscious. One test involved firmly stroking the plantar surface of his left foot from heel to toe with the blunt end of a ballpoint pen (Babinski test). Describe the pathway this sensory stimulus would take to arrive at his sensory cortex.**

Reflexes Are Involuntary Responses to Sensory Signals

Reflexes are fast, automatic motor reactions that are not subject to conscious control; that is, the action does not require a conscious decision. However, after the reflex has occurred, the *result* can enter consciousness. Some reflexes are inborn, such as jerking your foot away from a sharp stone, mentioned earlier in this chapter. Other reflexes are learned, such as the instantaneous move of your foot to the brake pedal if something unexpected happens as you are driving. The result of each of these examples comes into consciousness after the fact, but some visceral reflexes occur without our awareness—for example, the reflexive rise of blood pressure associated with excitement or fear.

Somatic reflexes are those in which skeletal muscle is stimulated to respond. **Autonomic reflexes** are those in which the ANS effects the response. The knee-jerk reaction in your doctor's office, in which the leg swings out involuntarily after a sharp tap on the patellar tendon, is a somatic reflex. On the other hand, salivation at the sight of tasty food and the rise of blood pressure and heart rate that occurs with excitement are autonomic reflexes.

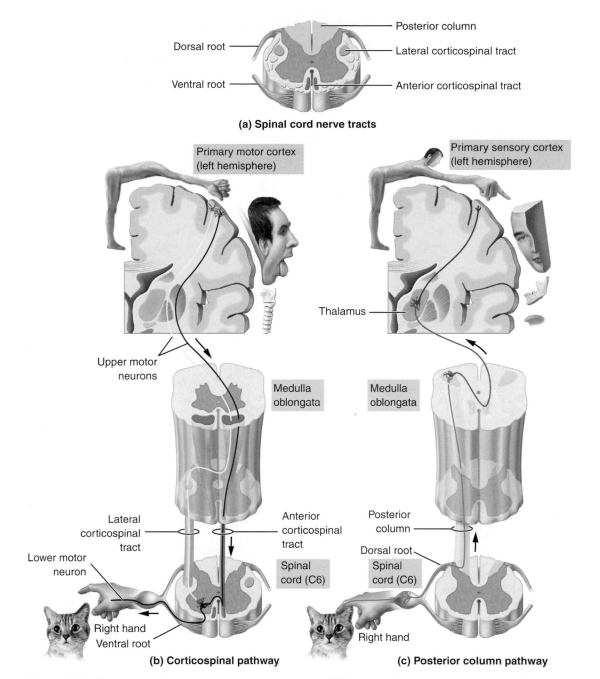

(a) Spinal cord nerve tracts

(b) Corticospinal pathway

(c) Posterior column pathway

Figure 8.23. Sensory and motor pathways. A. Spinal cord tracts. Three of the many tracts are illustrated on this cross section of the spinal cord. **B. Corticospinal (motor) pathway.** Upper and lower motor neurons carry the signal from the primary motor cortex to the finger muscles. Signals passing through the lateral corticospinal tract cross over at the medulla; those passing through the anterior corticospinal tract cross over within the spinal cord. **C. Posterior column pathway.** Three neurons carry the signal from touch receptors in the finger to the primary sensory cortex. This pathway crosses over at the medulla. *Which pathway is formed by neurons that cross over in the medulla oblongata—the anterior corticospinal tract or the lateral corticospinal tract?*

If the incoming sensory and outgoing motor signal for a reflex is integrated purely by the spinal cord, the reflex is called a **spinal reflex**; however, if the spinal cord transmits the signal upward and reflex integration occurs in the brain, it is a **cranial reflex.** Spinal reflexes are familiar to most who have had a physical examination: for example, the knee-jerk reaction just mentioned results from stimulation of a *spinal reflex*. On the other hand, it is a *cranial reflex* that is responsible for the lightning fast movement of your foot to the brake pedal that occurs in trying to avoid an automobile accident.

Some autonomic reflexes do not involve either the brain or spinal cord. Instead, the integration occurs within a ganglion, such as a sympathetic chain ganglion or a terminal ganglion. For example, ganglionic autonomic reflexes govern intestinal function in food digestion, which is discussed in ➡ Chapter 14.

The pathway followed by signals in a reflex is called a **reflex arc** (Fig. 8.24). In the discussion below, we use the knee-jerk reflex as an example. Regardless of the particulars of our example, the components of a reflex arc are:

1. *Sensory receptor*. A sensory receptor initiates the reflex. The doctor's tap on the patient's patellar tendon slightly stretches the quadriceps femoris muscle of the anterior thigh, and in doing so activates a sensory receptor, in this case a *stretch receptor* present in all skeletal muscles. The stretch receptor fires an action potential.
2. *Sensory neuron*. A sensory neuron transmits the signal to spinal cord or brain. In the example, the action

potential speeds along the axon of a sensory neuron to the spinal cord. There, one branch of the axon sends the signal to a motor neuron in spinal cord gray matter and another branch sends the signal upward to gray matter in the brain.

3. *Integrating center*. The sensory signal is integrated by gray matter in the spinal cord or brain. In the knee-jerk reflex, the sensory signal directly stimulates a motor neuron; meanwhile, the signal sent upward to brain gray matter is integrated into conscious awareness of the reflex.
4. *Motor neuron*. A motor neuron transmits the signal to that part of the body which will respond (skeletal muscle, heart, or intestine, for example). In the knee jerk example, the signal passes out of the spinal cord along the axon of a somatic motor neuron and is carried to the quadriceps femoris for action.
5. *Effector*. The motor signal activates a target organ or tissue to induce a response. In our example, the effector is skeletal muscle, but it could be virtually any tissue in the body depending upon the particular reflex.

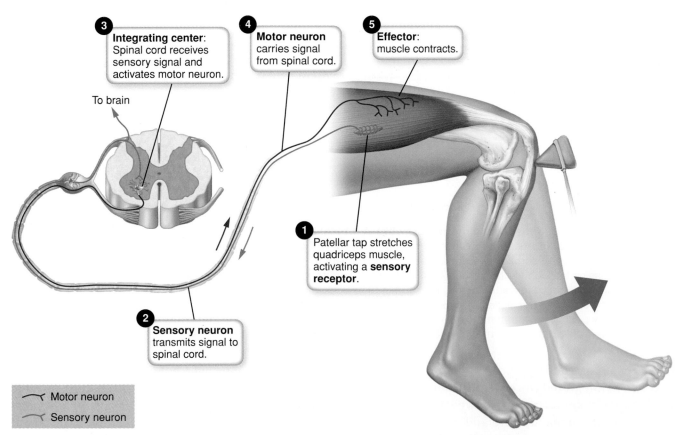

3 **Integrating center**: Spinal cord receives sensory signal and activates motor neuron.

To brain

4 **Motor neuron** carries signal from spinal cord.

5 **Effector**: muscle contracts.

1 Patellar tap stretches quadriceps muscle, activating a **sensory receptor**.

2 **Sensory neuron** transmits signal to spinal cord.

Motor neuron
Sensory neuron

Figure 8.24. Simple reflex arc. The response to the doctor's tap on the patellar tendon is an example of a simple reflex arc. *How many neurons are involved in this reflex?*

Neural Form and Function Are Tightly Integrated

Of necessity, we have analyzed the nervous system by breaking it down into smaller parts. However, it is likely that most of the brain is active at any given moment. Consider the act of hitting a baseball. As the player watches the ball approach (Fig. 8.25A):

- Visual information travels through the optic nerve to the visual cortices, which are used to identify the ball and determine its speed, position, and direction in order to predict where it will cross the plate.
- The somatosensory cortices determine the position of the arms, legs, and trunk. This information comes from sensory receptors in skin, muscle, and joints; passes down spinal nerves to the dorsal root ganglia; and enters the spinal cord via the dorsal horns. These sensory signals ascend to the brain through the posterior column.
- Other sensory areas provide less important sensory input about tastes, sounds, and smells.
- The posterior association area receives all of the sensory information, determines what is relevant to the situation, and sends an integrated view of body position, the moving ball, and the body surroundings to the primary association area.
- The primary association area produces a motor program to hit the ball, in association with the premotor cortex.
- The amygdala signals the hypothalamus to activate the sympathetic nervous system, and signals pass out of the spinal cord to the sympathetic root ganglia and subsequently to skin and visceral organs. Brainstem nuclei increase heart rate and respiration (for example) to improve the chance that the behavior will succeed. The amygdala also motivates the player to try his best.
- The thalamus processes and relays information between many different brain regions.

Once the player begins to swing, all of the above areas remain active. In addition, the execution of the behavior requires the activity of additional brain centers (Fig. 8.25B):

- The primary motor cortex carries out commands from higher motor areas, causing skeletal muscles to contract in a precise pattern that will (the batter hopes!) produce a result that moves the bat precisely to the

(a) Preparation

(b) Action

Figure 8.25. Neural form and function. A. Sensory and planning areas are involved as the baseball approaches. **B.** Planning and motor areas come into play as the player swings the bat. *Are the basal nuclei involved in sensory perception or motor activities?*

spot predicted where the ball will cross the plate. Signals pass down the anterior and lateral corticospinal pathways of the spinal cord to the ventral horns of particular spinal cord regions. Nerve impulses travel through the ventral branches of selected spinal nerves to stimulate skeletal muscle contraction.

- The cerebellum monitors outgoing and ingoing signals to coordinate and fine-tune the movements.
- The basal nuclei help automate the movement, so that each swing is roughly the same.
- The hippocampus records everything that happens, so that the player can relive the disappointment of a miss or the satisfaction of a hit.

8.38 Some motor impulses originating in the right brain cross over to the left side in the medulla. As they travel down the spinal cord, which tract are they using?

8.39 Does the posterior column carry motor or sensory signals?

8.40 True or false? Autonomic reflexes are those in which skeletal muscles respond.

8.41 In a spinal reflex, which neuron is activated first—the sensory neuron or the motor neuron?

Word Parts

Latin/Greek Word Parts	English Equivalents	Examples
neur/o, neur/i	Nervous system, nerve	Neuron: a nerve cell
auto-	Self	Autonomic nervous system: the division of the PNS carrying signals between the CNS and organs that function by themselves (without conscious control)
syn-	Together	Synapse: junction bringing two cells together
-glia	Glue	Neuroglia: cells that glue nerve cells together
mening/i/o	Membrane	Meninges: membranes covering the brain
cortic/o	Outer layer	Cerebral cortex: outer layer of the cerebrum
cephal/o; encephal/o	Brain	Hydrocephalus: excess fluid (water; hydro-) in the brain
cerebr/o	Cerebrum or brain	Cerebrospinal fluid: fluid in the spinal cord and brain

Chapter Challenge

CHAPTER RECALL

1. **Afferent neurons**
   a. convey information away from the CNS.
   b. can carry either motor or sensory signals.
   c. always carry sensory signals.
   d. are found only in the CNS.

2. **Gray matter**
   a. contains myelin.
   b. is frequently a site of signal integration.
   c. is found in tracts and nerves.
   d. forms the outer portion of the spinal cord.

3. **The connective tissue coverings of a nerve, from the most superficial to the deep, are**
   a. perineurium, epineurium, endoneurium.
   b. epineurium, perineurium, endoneurium.
   c. periosteum, epiosteum, endosteum.
   d. epimysium, perimysium, endomysium.

4. **Astrocytes**
   a. produce new neurons following brain injury.
   b. digest and destroy invading organisms.
   c. synthesize the myelin sheath of CNS neurons.
   d. synthesize the myelin sheath of PNS neurons.

5. **The fold of dural membrane that separates the two cerebral hemispheres is the**
   a. arachnoid villus.
   b. pia mater.
   c. arachnoid mater.
   d. falx cerebri.

6. **The third and fourth ventricles are connected by**
   a. arachnoid granulations.
   b. the central aqueduct.
   c. foramina.
   d. the lateral ventricles.

7. **CSF is made by the**
   a. subarachnoid space.
   b. choroid plexus.
   c. pia mater.
   d. dural sinus.

8. **The basal nuclei are found**
   a. within the white matter of the cerebrum.
   b. in the cerebral cortex.
   c. in the cerebellum.
   d. in the brainstem.

9. **The central sulcus**
   a. is a ridge of brain tissue next to the longitudinal fissure.
   b. is a groove separating the cerebral cortex into anterior and posterior halves.
   c. separates the cerebrum from the cerebellum.
   d. separates the temporal lobe from the occipital lobe.

10. **The primary auditory area is found in the**
    a. frontal lobe.
    b. occipital lobe.
    c. insula.
    d. temporal lobe.

11. **The primary somatosensory cortex of the left cerebral hemisphere receives sensory information from**
    a. skin on the right side of the body.
    b. skin on the left side of the body.
    c. the right inner ear.
    d. the left inner ear.

12. **Which of the following structures is *not* part of the limbic system?**
    a. Amygdala
    b. Thalamus
    c. Mammary body
    d. Cingulate gyrus

13. **The part of the brain directly controlling the autonomic nervous system is the**
    a. hypothalamus.
    b. cingulate gyrus.
    c. basal nuclei.
    d. thalamus.

14. **Moving from inferior to superior, the regions of the brainstem are the**
    a. pons, midbrain, medulla oblongata.
    b. midbrain, pons, medulla oblongata.
    c. medulla oblongata, pons, midbrain.
    d. midbrain, medulla oblongata, pons.

15. **The cranial nerve controlling the chewing muscles is numbered**
    a. I.
    b. III.
    c. V.
    d. VII.

16. **Sensory neurons enter the spinal cord via the**
    a. dorsal horn.
    b. central canal.
    c. ventral horn.
    d. anterior white column.

17. **The ventral branch of nerve S2**
    a. transmits signals for the sympathetic nervous system.
    b. carries signals for the autonomic and somatic nervous systems.
    c. innervates the skin of the back.
    d. contributes neurons to the lumbar plexus.

18. **Activation of the parasympathetic nervous system results in**
    a. slowing of the heart rate.
    b. flattening of the lens.
    c. dilation of the pupils.
    d. all of the above.

19. **Acetylcholine is used at synapses formed by**
    a. a motor neuron and a muscle cell.
    b. the preganglionic and postganglionic sympathetic neurons.
    c. the postganglionic parasympathetic neuron and its target cell.
    d. all of the above.

20. **Parasympathetic nerve signals would most likely travel through**
    a. a pelvic nerve.
    b. a collateral ganglion.
    c. the communicating branch of a spinal nerve.
    d. one of the lumbar spinal nerves.

21. **The left anterior corticospinal tract contains the axons of**
    a. motor neurons that originated in the left primary motor cortex.
    b. sensory neurons carrying signals from the left side of the body.
    c. motor neurons that originated in the right primary motor cortex.
    d. sensory neurons carrying signals from the right side of the body.

22. **A somatic, spinal reflex**
    a. may involve a gland.
    b. will not be consciously perceived.
    c. always involves skeletal muscle.
    d. is controlled by the autonomic nervous system.

CONCEPTUAL UNDERSTANDING

23. **Describe the journey of the CSF, beginning with its synthesis and ending with its entry into the circulatory system.**

24. **List the name, number, and sensory information conveyed for each of the purely sensory cranial nerves.**

25. **Compare and contrast a nerve plexus and a ganglion. Name one similarity and one difference, and give an example of each.**

APPLICATION

26. **You are waiting in a cafe for a long-lost friend. The sound of her voice from behind you fills you with joy, and you shout, "You're late!" Discuss the neural pathways involved in this scenario, beginning with activation of sound receptors in the ear and finishing with the contraction of tongue and throat muscles to produce speech. Also, mention the specific cranial nerves involved (list at least three nerves involved in speech).**

27. In babies, urination is controlled by a reflex. Bladder distention caused by urine accumulation activates a stretch receptor. A sensory neuron conveys the signal down the pelvic nerve to the sacral region of the spinal cord, where the sensory neuron complexes with an interneuron. The interneuron activates a motor neuron, which travels down the pelvic nerve to synapse with a postganglionic neuron on the bladder. Activation of this final neuron stimulates contraction of the bladder smooth muscle. Based on this description, answer the following questions:

a. Is it a cranial or a spinal reflex?

b. Is it an autonomic or a somatic reflex?

c. Identify each of the five components in this reflex.

You can find the answers to these questions on the student Web site at **http://thepoint.lww.com/McConnellandHull**

9

Sensation: The Somatic and Special Senses

Major Themes

- Sensing and sensation are different.

- The body has sensing mechanisms that detect change in the environment.

- Sense mechanisms generate nerve signals that are sent to the brain.

- The brain integrates sensing signals into conscious sensation.

- Adaptation, the lessening of sensing signals and perceived sensation in a steady-state environment, is an important aspect of sensing and sensation.

Chapter Objectives

"I observe, I feel, I think, I imagine."

As you read through the following case study, assemble a list of the terms and concepts you need to learn in order to understand Helen Keller's case.

Clinical history: Helen Keller was born in Alabama in 1880. She was a happy, healthy, intelligent child who was already saying a few words when, at the age of 19 months, she was stricken by a sudden feverish illness that left her deaf and blind.

Physical examination and other data: Lack of detailed medical records limits our understanding of Helen's precise defects. The little information we have about her original illness and subsequent medical evaluations suggests that the structure of her eyes and ears was unaffected and that the damage lay in her central nervous system (CNS), either in cranial nerves or in the brain itself.

Clinical course: After being struck deaf and blind, Helen became frustrated and impatient. Although she used certain symbols, such as a sawing motion with her hands when she wanted a piece of bread, her behavior was often more animal than human: raging, unpredictable, wild. But at other times she was placid and showed signs of uncanny intelligence: she would mimic things she could neither see nor hear, donning a hat before a mirror or perching glasses on her nose while holding a newspaper in front of her face.

When Helen was 6 years old her parents hired as her teacher Anne Sullivan, a gifted woman in her own right and herself nearly blind. Their relationship, memorialized in the 1962 movie *The*

Need to Know

It is important to understand the terms and concepts listed below before tackling the new information in this chapter.

- ■ Membrane potential, action potential, threshold, neurotransmitters, G protein–coupled receptors ◀ (Chapter 4)

- ■ Functional areas of the cerebral cortex, motor, and sensory areas ◀ (Chapter 8)

Miracle Worker, was famously effective and lasted nearly 50 years. Annie communicated with Keller by touch: she pressed sign-language finger signs into Helen's palm. A major breakthrough occurred when she signed "water" into Helen's palm while holding Helen's hand in a stream of water from a well. As she experienced the sensation of the cold water flowing over her hand and felt the strange pattern repeatedly striking her palm, she recognized for the first time that a certain finger sign represented a precise "something" in the world. In a moment that changed her world forever, Helen experienced a snap of insight, relating the "water" sign to the object causing this cold, wet, sensation. "Somehow," Helen recollected years later, "the mystery of language was revealed to me."

Aided by what ultimately proved to be exceptional intelligence and an astonishingly retentive memory, Helen grew, in the span of a single month, from an uncontrollable brat into a calm, affectionate child with an intense desire to learn. Within 12 months, news of her remarkable transformation made Helen world famous. Deprived of what most would say are the two most important senses, Helen became arguably the most famous and accomplished person ever to be both deaf and blind. When she graduated with high honors from Radcliffe College, she became the first deaf-blind person ever to earn a bachelor of arts degree. In the course of her long life, she was an outspoken and early advocate for women's voting rights, for disability and worker's rights, and for birth control.

Every living thing senses the things in its environment that assist its survival in its particular environmental niche. For example, migratory birds sense the earth's magnetic field to aid long-range navigation, sharks sense electrical currents in the sea to locate prey, and humans sense patterns of light, sound, textures, and other stimuli discussed in this chapter.

The brain, our organ of consciousness, sensation, and thought, is protected and isolated from the world and even from the rest of the body by its encasement in the skull. Like a king captured in a tower in a strange land, it relies on distant informants to gather news about the world outside and transmit it in an understandable language, in this case the language of electrical signals. The brain's informants are **sensory receptors,** specialized cells (in many cases modified neurons) or parts of cells that react to a specific type of stimulus, such as light or sound waves, and convert it into an electrical signal that the brain can understand.

> ***Remember This!*** Remember the *receptor proteins* discussed in ◀ Chapter 4 that are involved in chemical signaling? In contrast, sensory receptors are *entire cells* or *parts of cells* that detect changes in the environment.

It is a terrible thing to see and have no vision.

Helen Keller (1880–1968), American author, political activist, and lecturer; the first deaf-blind person to earn a bachelor of arts degree

Sensing and Sensation

Sensing and sensation are two separate affairs. **Sensing** is merely the detection of a stimulus by a sensory receptor. It does not require conscious perception. For example, sensory receptors in our bodies continually sense things of which we have no conscious awareness—blood pressure and blood pH, for example—because their signals are not relayed upward to the cerebral cortex for integration into consciousness.

In other circumstances, sensing leads to **sensation;** that is, to conscious perception. Sensory receptors—such as those that sense light, pain, or a bitter taste—transmit electrical signals to the cerebral cortex via nerves that travel to the cerebral cortex. Within microseconds, the cerebral cortex integrates the signals into a sensation. We perceive the position of our bodies in space, for example, as well as the sounds, smells, and sights of the world around us. This sensory information shapes our view of ourselves and our environment, which we may interpret as agreeable or disagreeable.

Case Note

9.1. **Did Helen Keller lose her ability to sense light and sound or her sensation of them?**

We Have More Than Five Senses

The notion of a "sixth sense," an intuition, a hunch, springs from the classic notion that there are five human senses: vision, hearing, taste, smell, and touch. But the truth of the matter is that we have more than five senses. Exactly how many is a matter of definition, but it is certainly more than five.

Consider the matter of the relative position of body parts. The athlete in Figure 9.1A knows precisely the location of her feet without seeing them, and she modifies their position in order to maintain balance. This position sense is called **proprioception**—our perception of the position of body parts relative to one another.

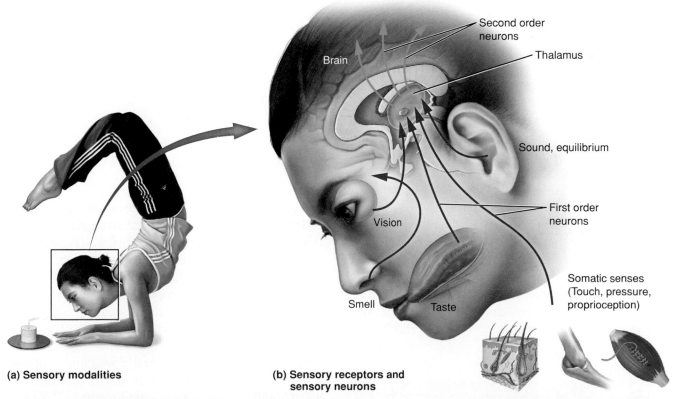

(a) Sensory modalities

(b) Sensory receptors and sensory neurons

Figure 9.1. Senses and sensation. A. The senses of equilibrium and proprioception help the athlete to maintain the pose. **B.** Sensory receptors (light blue) gather information about the external and internal environment. First-order (red) and second-order (green) sensory neurons convey this information to the brain for integration. Smell is processed seperately. *Which sense informs the athlete that her legs are bent—proprioception or equilibrium?*

Also consider the related matter of **equilibrium,** or our sense of balance. If we're off balance, we have mechanisms, both conscious and reflexive, to bring us back into balance. Being in equilibrium does not mean being still; it means being in control. The flips and turns of gymnastic routines are marvelous acts of controlled equilibrium. Failed equilibrium leads to a stumble or fall.

If you add proprioception and equilibrium, then there are seven senses. But wait: There is still the tricky matter of touch. We all agree that touch is a sense. What about pain, or the sensation of heat and cold? These, too, are separate senses.

It's easier to account for all of the body's senses if we classify them into three distinct groups:

- The **visceral senses** are structures and mechanisms that detect changes in our internal environment, such as an increase in blood pressure, a decrease in blood glucose, and so on. They operate entirely below the level of consciousness to maintain homeostasis; thus, they do not produce *sensation*. Given that visceral sensing is not relayed to the cerebral cortex for integration into consciousness, it is easy to understand that most of the back-and-forth signaling of homeostasis (Chapter 1) is from visceral senses.
- The **somatic senses** (Greek *soma* = "body") are simple structures with uncomplicated detection mechanisms. They are very widely distributed in skin, muscle, bones, joints, and other tissues and are not clustered into a special anatomical site or organ. The signals they generate are integrated into consciousness. Touch, for example, is a somatic sense.
- The **special senses** are complex systems with sophisticated detection apparatus. They occupy a particular anatomic site and their signals are integrated into consciousness. Vision, smell, taste, and hearing are special senses; equilibrium is another.

Although we treat these senses as discrete entities, in reality we integrate sensory information from separate modalities. Returning to Figure 9.1A, the athlete uses visual input and her somatic senses to keep her equilibrium. If she eats a restorative apple after her yoga practice, taste, smell, vision, and sound may all contribute to her enjoyment.

Case Note

9.2. Which senses did Helen's illness destroy? Are these visceral, somatic, or special senses?

All Sensations Result from the Same Sequence of Events

Although we exist in a swarm of somatic and special sensations, each results from the same basic sequence of events (Fig. 9.1B):

1. A stimulus, a *change* in the environment, occurs.
2. A particular structure on a sensory receptor, such as a receptor protein or hairlike projection from the cell membrane, detects the stimulus, and the cell converts it into an electrical signal. For instance, the odor molecules from a burning candle are sensed by sensory receptors in the nose. Sensing *change* is the key: most sensory receptors do not effectively detect features of the environment that do not *change*, such as constant pressure or constant speed (see *adaptation*, below). The different sensory receptors are shaded blue in Figure 9.1B.
3. One or more neurons carry the signal to the brain. The first neuron to carry the signal from the sensory receptor is referred to as the *first-order neuron* (represented by the red arrows); most signals are relayed through additional *second-order* (green arrows) and possibly *third-order* neurons before they reach the brain.
4. The brain integrates the signal into a conscious perception.

Case Note

9.3. Is it likely that Helen's eyes continued to react to light and her ears to sound?

Sensory Adaptation Is a Decrease of Sensory Signals

Important in the function of sensory receptors is **adaptation**—a decrease in the signal strength generated by the receptor during prolonged steady-state stimulation. Signal strength fades under constant stimulation, so that the sensory apparatus can be ready to signal something new—a change. For example, a man who carries a fat wallet in one hip pocket soon becomes unaware of it because the pressure signal fades. But shift it to a new pocket and it becomes very noticeable for a while. Touch and smell sensors adapt rapidly; pain and proprioceptive sensors adapt slowly. Later in this chapter you will learn about the eye's adaptation to light.

Adaptation has evolutionary survival value because, by reducing the competition from continuing stimuli, it heightens our ability to detect *change*, and it is change that our sensing systems must detect if we are to survive. Threats to our survival are announced by change: the characteristic smell of an approaching thunderstorm is a change, a new odor, that signals a need to seek shelter—it is the change that counts, not the odor itself, which fades quickly as nasal sensory receptors adapt, readying themselves for another new odor.

9.1   Identify a key difference between the visceral senses and other senses.

9.2   Is equilibrium a special sense or a somatic sense?

9.3   What is the name of our sense of the relative position of body parts?

9.4   Define *first-order neuron.*

9.5   What is meant by sensory adaptation?

Somatic Senses

As noted earlier, the somatic senses detect changes in skin, muscle, bones, joints, and other tissues and send signals to the cerebral cortex for integration into consciousness. Touch, temperature, pain, and relative position of body parts are somatic sensations.

Tactile Receptors Detect Touch and Pressure

Tactile receptors detect touch and pressure. They are among the most numerous sensory receptors; they are found in skin, muscle, joints, and internal organs. Anatomically, they consist of the tips of neuron dendrites that are sometimes wrapped, onionlike, in multiple layers of protein. (Review Fig. 4.11 on ◀ page 125 for the structure of dendrites.) They are classified as **mechanoreceptors,** because they are activated when a mechanical force pushes on or stretches the cell. They transmit signals that the brain interprets as touch, pressure, texture, stretching, or vibration. They are especially numerous and varied in skin. Of the several types of tactile skin receptors, the five most important are the following (Fig. 9.2):

- **Free sensory nerve endings** detect touch and pressure. Some endings are wrapped around hair roots and detect movements caused by the wind or the light touch of an insect crawling on the skin.

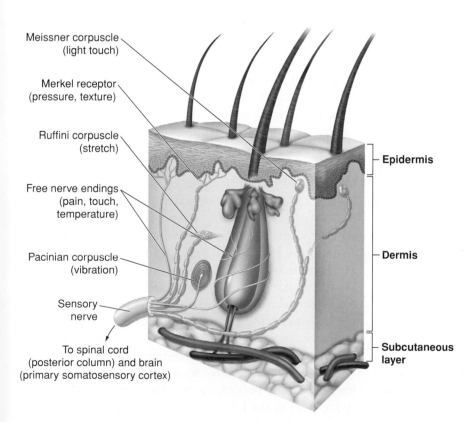

Figure 9.2. Skin receptors. Free nerve endings, modified nerve endings, and specialized receptor cells detect different tactile stimuli. *Which receptor type is found in more superficial skin regions—Meissner corpuscles or Ruffini corpuscles?*

- **Merkel receptors** are enlarged nerve endings associated with modified epithelial cells that detect pressure and texture; they adapt slowly.
- **Meissner corpuscles** are small oval bodies at the tips of certain sensory neurons. They are concentrated in hairless skin, especially the lips, fingertips, palms, soles, nipples, and external genitalia. They sense light touch, such as fluttering or stroking, and adapt rapidly.
- **Pacinian corpuscles** are larger than Meissner corpuscles and are located deep in skin and subcutaneous tissue as well as in tendons and ligaments. They sense vibration or sudden pressure but adapt quickly and do not detect steady pressure.
- **Ruffini corpuscles** are located in the dermis and sense stretching. They are activated by minute displacements of the fingernails or other small movements; thus, they are important in identifying grasped objects.

Touch, seemingly so simple, is actually a complex sensation: a single "touch" activates one or more of the receptors listed above. As we discussed in ◀ Chapter 8, signals from somatic receptors travel through the posterior column to the primary somatosensory cortex.

Case Note

9.4. Helen's tutor touch-signed into her hand. Which receptors are concentrated in this region and detected the gentle stroking of the tutor's fingers?

Thermal Receptors Are Free Nerve Endings

Temperature receptors are free nerve endings located mainly in skin (Fig. 9.2), although some are present in the upper digestive tract, the cornea of the eye, and the urinary bladder. The sensation of temperature depends on *thermoreceptors*, which respond to alterations in skin temperature.

At normal skin temperature (about 34°C), thermoreceptor neurons send action potentials to the cerebral cortex at an intermediate, continuous rate. Their firing rate changes as skin temperature moves away from this value. Warmer stimulates more rapid firing, cooler less rapid firing; the cerebral cortex interprets these changes in firing rate as warmer or cooler.

Thermal sensors adapt very rapidly. For example, your bath water initially may feel particularly hot, but the sensation soon fades to comfortable warmth.

Case Note

9.5. In the incident at the water well, Helen described the sensation as cool, not painful. Would the firing rate of her thermoreceptor neurons have increased or decreased when the cool water was applied?

Chemical and Physical Trauma Causes Pain

Pain is difficult to define, study, and describe because it is so subjective. A painful condition that will completely disable one person may have much less effect on another. What's more, there is no standard language familiar to patients to characterize what they are feeling. Indeed, laboratory pain experiments are limited by the fact that animals lack language to describe their pain.

Pain is sensed by free sensory neuron endings called **nociceptors** (Latin *noci-* = "harmful"), which exist in every body tissue but the CNS. Pain is induced by tissue damage and is unpleasant, stimulating efforts to remove the offending stimulus. Pain receptors adapt slowly if at all: there is no survival advantage to dulling the reaction to tissue damage and the pain it produces. Were this the case, we might grow accustomed to the pain produced by such damage and even more damage might occur.

Nociceptors can be activated by physical deformation—such as that resulting from a cut or puncture wound, temperature extremes, oxygen deprivation, or by chemicals (*nociceptive chemicals*, such as bradykinin or histamine) released in response to tissue damage. This activation takes the form of an action potential in the sensory neuron's ending (the first-order neuron), which travels up the neuron to the spinal cord. A second-order neuron carries the signal to the brain.

We experience two types of pain, fast and slow, because we have two types of first-order nociceptive neurons. Consider the sensations when you step on a nail or a tack (Fig. 9.3). **Fast pain** is perceived very quickly, usually within 0.1 second of the stimulus, and rises almost instantly to peak intensity. Fast pain signals are carried by myelinated neurons that rapidly convey an acute, sharp, stabbing pain (recall that myelin accelerates the transmission of action potentials). **Slow pain,** on the other hand, begins more than 1 second after the stimulus and rises to a peak over several seconds or minutes. Slow pain signals are carried by unmyelinated neurons and are the throbbing, aching, sometimes burning sensations that typically arise from an injured site (your foot) as the initial sharp pain subsides. Slow pain

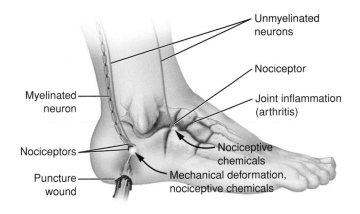

(a) **Mechanical and chemical stimuli activate nociceptors**

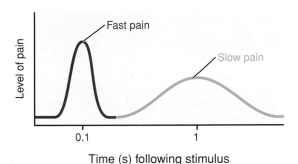

(b) **The brain receives fast pain signals before slow pain signals**

Figure 9.3. Fast and slow pain. A. Stepping on a nail activates nociceptors on myelinated neurons, which rapidly convey a stabbing sensation, and on unmyelinated neurons, which convey an aching sensation. In contrast, the chronic inflammation of arthritis activates only unmyelinated fibers. **B.** From the moment that the nociceptors are activated, fast pain is detected by the brain much more quickly than slow pain. *What types of stimuli activate slow-pain nociceptors?*

signals also characteristically arise from inflamed tissue, as with infection or arthritis. Slow pain signals are also generated by the rapid, marked stretching of visceral organs, such as the stretching of the cervix in childbirth or the stretching of a ureter with the passage a kidney stone.

When pain from tissue damage at a particular anatomical area appears to be coming from somewhere else, the pain is said to be **referred pain.** If you cut your finger, it's the finger that hurts; but if you are having a heart attack, the pain may appear to come from—be referred to—your left jaw, neck, or left arm. Referred pain occurs when visceral sensory fibers from an affected organ enter the spinal cord along with sensory fibers from another part of the body and both fiber types converge within a single ascending tract. For example, sensory fibers from the heart enter the spinal cord with sensory fibers from the left side of the jaw and neck and the left arm.

Pain relief is **analgesia** (Greek *an-* = "without" ι *algos* = "pain"), and it is a complex topic. The best way to relieve pain is to remove the offending stimulus: if you have a thorn in your skin, you will pluck it out. But often this is not possible. Some cancers, for example, are painful and cannot be removed. Neither can inflammation, whether from a chronic condition such as arthritis or an acute injury such as a sprained ankle. That's where pharmacological pain relievers come in. Anti-inflammatory drugs like aspirin, for example, block formation of local inflammatory chemicals (such as prostaglandins) that increase the sensitivity of pain receptors. Inflammation is discussed in more detail in ➡ Chapter 10. Other drugs, especially opiates such as morphine, work by altering the quality of pain perception in the brain: the pain doesn't go away, but the patient doesn't care about it as much. Alternatively, nerves can be deadened by administration of local **anesthesia** (Greek *aisthesis* = "feeling"). In this procedure, nerves are injected with a chemical that prevents action potentials from being transmitted; thus, pain signals cannot reach the CNS. On a grander scale, prior to surgery, the patient can be rendered temporarily comatose via general anesthesia, so that the pain signals are not perceived.

Pain signals may continue to arise from nociceptors even though the original offending stimulus has been removed. This reaction may reflect ongoing inflammation, since inflammation releases chemicals that sensitize nociceptors so that they respond even to light touch with a pain signal. Sunburned skin is an example. In some cases, patients who have injured a body part may continue to have pain from the site even though the damage is healed completely. Some chronic pain syndromes are thought to develop in this way. On the other hand, some people are born with a genetic defect that renders them unable to sense pain. They will not withdraw either consciously or reflexively from noxious stimuli. As you can imagine, injuries, usually self-inflicted, are a major problem in such cases. For example, a patient may bite off part of his tongue while chewing food or break a bone and yet be unaware of the injury.

Case Note

9.6. During her febrile illness, Helen experienced significant pain, likely resulting from inflammation. Which sort of receptors are activated by chemicals released during inflammation?

Proprioception Is the Body's Position Sense

As stated earlier, *proprioception* is the sensing and perception of the position of body parts *relative to one another*. Proprioception tells you, for example, whether your left leg is bent or straight, and whether your facial muscles are contracted into a smile or slack with boredom. The word derives from Latin *proprius* = "one's own" + *recipere* = "to receive"—therefore to receive information about oneself. Proprioceptive sensory receptors, called **proprioceptors,** sense the degree of muscle stretch or contraction and the angle of body joints—data that the brain integrates and interprets as the position of body parts relative to one another.

Proprioception is critically important in every task of life, from walking to throwing a baseball, so it's not surprising that proprioceptors adapt very slowly or not at all. You can test this easily: Close your eyes for a moment. Even if you remain very still for a long time, you can know with precision the exact position of every part of your left upper limb. Although we consciously attend to visual and auditory input to a much greater extent than we do by proprioception, we would be helpless without it. For example, when your arm "goes to sleep" because it has been compressed for an extended period, you have no idea whether your hand is palm up or palm down, because the proprioceptive signals are interrupted. You must look at it to know.

Proprioception is extremely sensitive, too. The most sensitive proprioceptive sense is in the mouth: you know that a piece of food lodged between two teeth feels much larger than its actual size because the slightest movement of a tooth feels more like a mile than a micron.

Proprioceptors are located in muscles and connective tissue. Skeletal muscle proprioceptors are called **muscle spindles;** they contain modified muscle fibers (*intrafusal* muscle fibers) that detect skeletal muscle length, because they are oriented in parallel with the contracting muscle fibers (Fig. 9.4). Intrafusal fibers stretch as the muscle elongates, resulting in an increased action potential firing rate in the associated neurons. Muscle spindles are responsible for the spinal reflex illustrated in ◄ Figure 8.24; tapping the patellar tendon stretches the quadriceps muscle and thus the muscle spindles within. The stretched muscle contracts in response. Muscle spindles thus protect against excessive muscle stretch; they activate a reflex that causes the muscle to contract and thereby reduce the amount of stretch.

Golgi tendon organs are collagen strands within tendons, near the point where they merge with muscle fibers (Fig. 9.4). Muscle tension activates sensory nerve

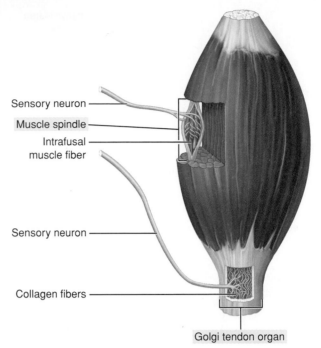

Figure 9.4. Proprioceptors. Muscle spindles detect muscle length, Golgi tendon organs detect muscle tension. *Which proprioceptor type is found within the muscle tissue?*

Labels: Sensory neuron · Muscle spindle · Intrafusal muscle fiber · Sensory neuron · Collagen fibers · Golgi tendon organ

endings that spiral around the collagen fibers. Golgi tendon organs protect against potentially damaging muscle tension that might tear muscle or tendon by use of excess force; their activation causes the muscle to relax, decreasing the tension.

Other proprioceptors located in joint capsules and ligaments relay information about joint positions and angles.

Case Note

9.7. Helen loved climbing trees, despite her lack of vision. As she hung from a branch, muscle tension, but not muscle length, would have changed dramatically. Which proprioceptors would have been activated?

9.6 What is the name of our sense of the relative position of body parts?

9.7 If you push your thumb and fingertip together for a prolonged period of time, which touch receptors will continuously respond to the pressure—Meissner corpuscles or Merkel receptors?

9.8   As you step into your bath, what happens to the firing rate of the thermoreceptors in the skin of your foot?

9.9   What kind of pain neuron will carry the fast pain signal from a pinprick—myelinated or unmyelinated?

9.10   Which type of proprioceptor activates a reflex that contracts the associated muscle—muscle spindles or Golgi tendon organs?

Taste

Taste (Latin *taxare* = "to judge") or **gustation** (Latin *gustare* = "to taste") is the detection of flavors by the mouth. Taste, one of the special senses, brings pleasure but also enhances survival, since it stimulates us to eat nutrient-rich foods and helps us avoid poisonous, rotten, and other dangerous foods. The elements of gustation are:

- Receptor: gustatory receptor cells (chemoreceptors)
- Stimulus: chemicals called *tastants*
- Pathway: gustatory pathway

Each of these elements is discussed in further detail below.

Taste Receptors Are Located in Papillae in the Mouth and Throat

The gustatory receptor cells (or taste cells) are **chemoreceptors**; that is, they respond to chemical ligands that bind to them. They are clustered in **taste buds,** which are largely confined to small projections on the tongue called *papillae* (Fig. 9.5A and B). A few taste buds can also be found in the roof of the mouth and on the epiglottis at the back of the mouth.

Of the three types of papillae on the tongue, two contain taste buds and one does not:

- **Vallate papillae** are the largest and least numerous (8 to 12). They contain taste buds and are arrayed in a V-shaped row at the back of the tongue. Vallate (Latin *vallum* = "wall") papillae are so named because they sit in the center of a circular pit, which has the form of a walled crevice surrounding the papilla. The crevice can be likened to a moat around a castle: it accumulates fluids for reaction with taste buds in the papillae.

- **Fungiform papillae** are much more numerous, have a mushroom-like shape, and are scattered over the entire surface of the tongue. Those on the back half of the tongue contain taste buds.
- **Filiform papillae** (Latin *filum* = "thread") are by far the most numerous; they are narrow and pointed and cover the entire surface of the tongue. They contain touch receptors but no taste buds. They impart a rough texture to the tongue mucosa, which is useful in manipulating food in the mouth.

As just noted, the primary functional cell within a taste bud is the **gustatory receptor cell,** an elongated epithelial cell that has a single microvillus, the *gustatory hair*, extending from its tip (Fig. 9.5C). Each taste bud consists of several dozen such cells nestled shoulder to shoulder beneath the epithelium of the tongue or other site. At the tip of each, the gustatory hair projects through an opening, the *taste pore*, into the oral space.

Gustatory receptor cells survive for about 10 days before dying naturally by apoptosis. Stem cells within the taste bud, called **basal cells,** are thus quite active in producing new gustatory cells.

> *Remember This!* **Gustatory receptors, unlike most sensory receptors, are epithelial cells, not neurons.**

Taste Stimuli Are Five Basic Chemical Tastants

Scientists have succeeded in identifying five basic tastes: *sour, salty, sweet, bitter,* and *umami.* Some also postulate a sixth taste—fatty acids—responsible for the oily mouth feel of French fries or butter. Chemicals that stimulate gustatory receptor cells are called **tastants,** and we have specific gustatory receptor cells for each of the five tastes.

There are five primary tastes, each reflecting a separate tastant acting on a distinct gustatory receptor cell:

- *Sour* taste is caused by hydrogen (H^+) ions found in acidic foods and is proportional to pH. The more acidic (the lower the pH), the more sour the taste. Sour gustatory cells are activated, for example, by lemon juice and vinegar, which are very acidic.
- *Salty* taste is elicited by ionized salts, mainly sodium ions (Na^+) from sodium chloride (NaCl, ordinary table salt).

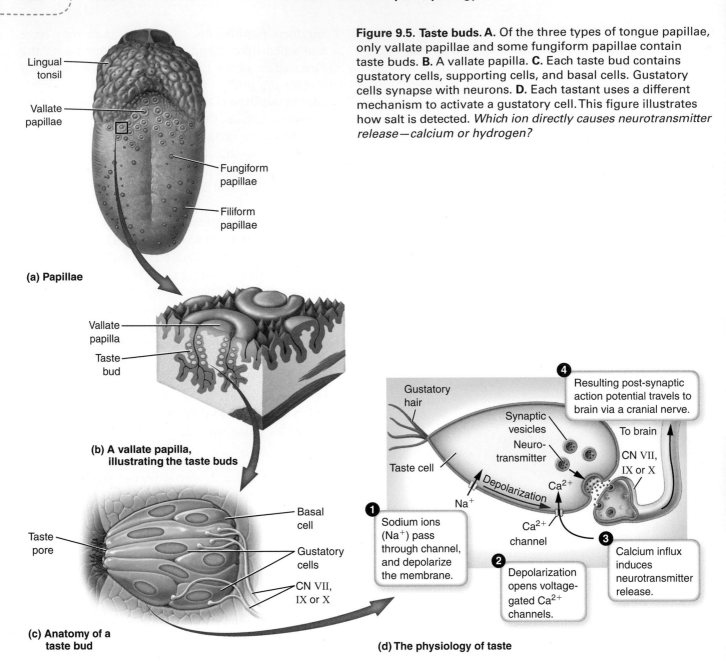

Figure 9.5. Taste buds. A. Of the three types of tongue papillae, only vallate papillae and some fungiform papillae contain taste buds. **B.** A vallate papilla. **C.** Each taste bud contains gustatory cells, supporting cells, and basal cells. Gustatory cells synapse with neurons. **D.** Each tastant uses a different mechanism to activate a gustatory cell. This figure illustrates how salt is detected. *Which ion directly causes neurotransmitter release—calcium or hydrogen?*

Lingual tonsil

Vallate papillae

Fungiform papillae

Filiform papillae

(a) Papillae

Vallate papilla

Taste bud

(b) A vallate papilla, illustrating the taste buds

Basal cell

Gustatory cells

CN VII, IX or X

Taste pore

(c) Anatomy of a taste bud

Gustatory hair

Synaptic vesicles

Neuro-transmitter

Taste cell

Ca^{2+}

Na^+

Depolarization

Ca^{2+} channel

To brain

CN VII, IX or X

❹ Resulting post-synaptic action potential travels to brain via a cranial nerve.

❶ Sodium ions (Na^+) pass through channel, and depolarize the membrane.

❷ Depolarization opens voltage-gated Ca^{2+} channels.

❸ Calcium influx induces neurotransmitter release.

(d) The physiology of taste

- *Sweet* taste is caused by a great variety of substances, most notably sugars. Other organic compounds, such as alcohols and amino acids, can also activate the "sweet" receptor cells and thus taste sweet.
- *Bitter* taste, like sweet, is elicited by many different compounds, most of them also organic chemicals. Substances found in chocolate or coffee, for instance, activate the bitter gustatory cells.
- *Umami* is a Japanese word meaning "delicious," "meaty," or "savory." It is a pleasant sensation attributed to the detection of specific amino acids (glutamate and aspartate), which are common in meat,

cheese, mushrooms, and other protein-rich foods. Stimulation of umami receptor cells explains why foods with added monosodium glutamate (MSG) have a more substantive, full sensation in the mouth.

Each tastant uses a slightly different mechanism to activate its receptor cell. Figure 9.5D illustrates how salt is detected. The Na^+ ions of the salt enter and depolarize the receptor cell, which opens voltage-gated calcium channels. Calcium entering the cell stimulates neurotransmitter release, and the neurotransmitter invokes an action potential in a sensory neuron (discussed further below). Other tastants depolarize the cell in different ways. Hydrogen ions, for instance, acidify the cytosol, which opens cation channels that permit depolarizing cations (such as Na^+) to enter the cell. In contrast, the organic compounds responsible for sweet, bitter, and umami tastes bind specific *G protein–coupled receptors*. As we discussed in ◀ Chapter 4, these specialized receptors activate second messenger systems. In gustatory cells, the second messengers open cation channels that depolarize the cell.

Adaptation of taste receptors is rapid. On first contact, the tastant instantly stimulates an intense burst of signals, but the frequency of signals begins to decline within seconds to a steady level as long as the tastant is present.

Tastant signals have survival value. Sweet, salt, and umami tastant signals encourage intake of carbohydrates, minerals, and proteins—all necessary for survival. Sour and bitter in small amounts can add to the pleasure of food and drink (think of lemonade); but in larger amounts they are so distasteful that they are rejected. It is no surprise, therefore, that many toxic plants and medicines have a sour or bitter taste.

Note that pure water does not stimulate taste receptors. Water is almost universally present in food and drink; if it stimulated taste receptors, we could not taste much of anything but water. Any time water has taste, it is due to some tastant in water, not to water itself.

> *Remember This!* **Each of the five tastants activates a different type of gustatory receptor cell.**

Flavors Reflect Tastant Combinations and Other Sensations

Every flavor—vanilla, coffee, garlic, you name it—reflects varying combinations of each of the five basic tastes. However, temperature, smell, texture, and even pain also play a very important role in how something tastes. Consider, for instance, the cooling effect of a peppermint or the pain associated with spicy peppers—it's a little dose of pain that makes spicy foods so tasty to some people.

Odors from food activate odor receptors, which are thousands of times more numerous and more sensitive than taste receptors. Most things that contain molecules stimulating taste also contain molecules stimulating smell, which leads to the saying that taste is "nine parts smell and one part taste." As a consequence, odor is far more important in the "taste" of food than taste itself. Try it for yourself: take a bite or sip of something while holding your nose. You will see how much "taste" is suddenly missing.

Case Note

9.8. Helen enjoyed eating ice cream. In addition to gustatory receptors, which other types of receptors would have been activated?

Taste Signals Follow a Special Nerve Pathway

Gustatory receptor cells synapse with a first-order neuron of a cranial nerve—the facial nerve (VII) and glossopharyngeal nerve (IX) serve taste buds on the tongue, and the vagus nerve (X) serves taste buds in the throat and epiglottis. Each of these cranial nerves synapses in the medulla oblongata with second-order neurons that carry the signal to the thalamus. Third-order neurons relay the signal to the **primary gustatory area** located low in the postcentral gyrus of the parietal lobe of the cerebral cortex, deep in the lateral sulcus (see Fig. 8.8). There, depending on which gustatory receptor cells were activated and to what extent, the brain interprets the sensory signal as a particular taste. Taste information is also relayed to the limbic system, which vests the signals with emotional implications, and to the hypothalamus, which controls autonomic functions (such as swallowing and gagging) in response to pleasing or toxic tastes.

9.11 Name three cranial nerves that convey taste signals.

9.12 Which type of papilla does not contain taste buds?

9.13 Name three brain structures that encounter gustatory signals.

9.14 Which tastant depolarizes the membrane by activating a G protein–coupled receptor—sugar or salt?

9.15 True or false. Every flavor is due to stimulation of a gustatory receptor cell unique to that flavor.

Smell

Olfaction (Latin *olfacere* = "to smell") is the detection of volatile chemicals called **odorants** in air by olfactory chemoreceptors in the nasal cavity. It is a *special sense*.

Not all molecules stimulate odor receptors. Notably the gases that make up air—nitrogen and oxygen—are odorless. If it were otherwise, we could not smell much of anything but air. Nor are all volatile substances odorants. Water, for example, has no odor. The odor of an approaching thunderstorm is not due to water but to ozone (O_3) generated by the interaction of lightning and oxygen in the air. And one of the features of carbon monoxide that makes it such a deadly poisonous gas is that it is odorless; since it stimulates no odor signals, we fail to take action to protect ourselves from it.

Important though it is, the sense of smell in humans is poor compared with that in many other animals. For example, the olfactory epithelium of dogs contains 40 times more sensory cells than that of humans. Not only can dogs' more sensitive noses detect odors in tiny amounts, they can distinguish odors with much greater specificity than humans, even to the smell of a particular person.

Olfactory Receptors Are Located in the Olfactory Epithelium of the Nose

Detection of odorants depends on receptors in the **olfactory epithelium,** which is located on the inferior surface of the ethmoid bone high in the nasal cavity (Fig. 9.6A). Human olfactory epithelium contains three types of cells: *basal cells, supporting cells,* and *olfactory receptor cells* (Fig. 9.6B).

The basal cells are stem cells, which divide to produce both supporting cells and olfactory receptor cells. The supporting cells serve similar roles to glia in the CNS—they physically support and cushion the receptor cells, modulate the ion content of the surrounding fluid, and phagocytize cellular debris.

Olfactory receptor cells are modified first-order sensory neurons whose cell bodies lie in the olfactory epithelium. The olfactory receptor cell sends an axon upward through the tiny holes of the cribriform plate to synapse with a second-order neuron in the **olfactory bulb,** a swelling at the tip of the olfactory nerve (cranial nerve I) inside the skull. Olfactory receptor cells also send dendrites downward into the nasal cavity, where the tips of the dendrites protrude into the nasal cavity as cilia.

The ciliated nasal surface of the olfactory epithelium is coated by mucus secreted by nearby mucous glands, which forms a protective barrier that also traps odor molecules for detection. Olfactory receptor cells live about a month before dying naturally by apoptosis; they are then replaced by new cells generated from basal cells in the olfactory epithelium.

9.9. Helen's favorite tree was a mimosa, which she identified by smell. If her illness had destroyed the basal cells in her olfactory epithelium, would she have retained the ability to identify a mimosa? Explain.

Olfactory Receptor Cells Detect Odorant Molecules

The average human nose can distinguish among more than 10,000 different odors. Although most odors result from a specific combination of multiple odorant molecules, sometimes a single molecule causes an odor. For example, the stench of a skunk's musk is created by one small odorant molecule, as shown in Figure 9.6C. Odorant molecules bind to specific G protein–coupled receptors (GPCRs) in the cilia of olfactory receptor cells. Activation of the GPCR depolarizes the olfactory receptor cell, increasing neurotransmitter release.

Because taste and smell are related senses, physiologists long believed that they operated on similar principles. Because they proved long ago that taste is due to signals from a combination of only five or six primary taste receptor types, scientists assumed that we also have only a small number of odorant receptor types, each of which reacts with different odorant chemicals in various combinations to account for a particular smell. We now know that this is false: there is no five-letter "alphabet" of basic odor receptor types, each one supposedly responsible for signaling detection of one basic odor.

Instead, humans have hundreds of types of odorant receptor proteins, which are all G protein–coupled receptors. Each odorant receptor cell makes one type of receptor protein, which binds (reacts to) only a few different odorant molecules. Some odorants are a perfect fit with the receptor and stimulate an intense response; others don't fit as well and have a lesser response, which the cortex perceives as less odorous. We identify a smell based on which olfactory cells are activated, and to what extent. Thus, the odorant molecules that convey the smell of a rose will react with odorant receptors on a certain set of cells, but the odorant molecules that convey the smell of sour milk will react with a different set of receptors on a different set of cells.

Olfactory cells adapt very quickly, losing about 50% of their sensitivity in the first second or so. Although they adapt very little beyond that, we all know from personal experience that smell sensations fade almost to the point

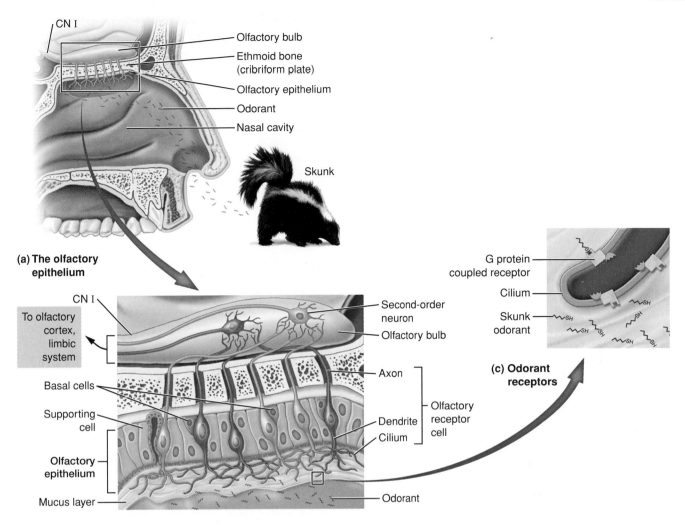

(a) The olfactory epithelium

(b) Olfactory neurons

(c) Odorant receptors

Figure 9.6. Smell. A. Smells are sensed by the olfactory epithelium, located in the superior portion of the nasal cavity. Neurons from the olfactory epithelium project into the olfactory bulb and through cranial nerve I (the olfactory nerve). **B.** Odorant molecules are detected by olfactory cells, modified first-order sensory neurons. Olfactory cells recognizing a particular odorant converge to synapse with the same second-order neuron, which conveys the signal through cranial nerve I to the brain. The skunk musk odorant activates the red subset of olfactory cells, but not the green. **C.** Odorant receptors reacting with a particular odorant (in this case, skunk musk) are found in the cell membrane of olfactory cell cilia. *Which cell type produces new olfactory cells—basal cells or supporting cells?*

of extinction after a few minutes. Though not completely understood, it is clear that this further smell adaptation reflects adaptation in the brain, which becomes insensitive to the continuing signals sent by the olfactory epithelium. For example, cigarette smokers' olfactory cells and cerebral cortex have so adapted to the odor of smoking that they scarcely notice the smell of the tobacco they are smoking. Not so for nonsmokers, who can detect burning tobacco a block away.

Olfactory Signals Follow a Special Nerve Pathway

The detection of olfactory signals is similar to the detection of gustatory signals—odorant molecules activate an odorant-specific palette of odorant receptor proteins, resulting in the depolarization of particular receptor cells. The axons of cells with the same odorant receptor protein converge like roots gathering into a tree trunk, synapsing with a single second-order neuron in the olfactory bulb. In our hypothetical example in Figure 9.6B, the skunk musk odorant molecule binds to receptors on the red cells but not the green cells, so the red second-order neuron (but not the green one) is activated.

Unlike other sensory signals, olfactory signals do not have to pass through the thalamus before reaching the primary sensory cortex. Smells thus invoke a perception much more quickly than other types of sensory input.

Recall from ⬅ Chapter 8 that the primary olfactory cortex consists of several regions of the temporal and frontal lobes. Each region receives input from specific second-order neurons. So, depending on which brain regions are activated and to what degree, the brain identifies a distinctive smell—the "odorant pattern." Other axons from the olfactory bulb connect to the limbic system and to the hypothalamus, which vest odors with emotional content. For example, the odor of diesel fuel exhaust may take you back to an outdoor roadwork job experience or, if enchiladas once made you violently ill, the smell of enchiladas with green chili sauce may strike you as nauseating for years thereafter.

Case Note

9.10. After her illness, Helen could find her house by smelling the roses growing on the front porch. Name the receptor cell and neural pathway used to transmit this information to her brain.

9.16 Which olfactory epithelial cells are capable of cell division: olfactory receptor cells or basal cells?

9.17 About how many different types of odorant receptors are there—five, several hundred, or ten thousand?

9.18 Where do odorant molecules bind—to the dendrite or the axon terminus?

9.19 True or false: Odorants activate receptor cells by binding to receptor channels.

9.20 Which type of signal—olfactory or gustatory— passes through the thalamus?

The Ear and Hearing

Deafness: a malady affecting dogs when their person calls them and they want to stay out.

Source: The first and most common result found in an Internet search in 2005 for "definition: deafness."

Hearing another of the special senses, is the conscious perception of sound waves; **deafness** is loss of this sense. Hearing depends upon the integrated function of the three parts of the ear: the *external, middle,* and *inner,* each of which is discussed in detail further on.

However, the inner ear has a completely separate role in our sense of **equilibrium**—our sense of balance. We discuss hearing and equilibrium separately because they involve such completely different sensations. But although the sensations are very different, the sensors share similar properties. Both are functions of the ear, and both depend upon mechanical actions to stimulate mechanoreceptors: membranes and bones move, fluid sloshes this way and that, and tiny hairs sway back and forth. Both communicate with the brain via cranial nerve VIII, the vestibulocochlear nerve. Finally, some ear diseases can affect both hearing and equilibrium. However, the sensations themselves are distinctly separate— hearing, or the lack of it, has no impact on balance, and balance does not influence hearing.

The Ear Consists of Three Interconnected Organs

The ear consists of three interconnected organs (Fig. 9.7):

● The *external ear*, which collects and focuses sound waves and channels them toward the middle ear
● The *middle ear*, which converts sound waves into movements of tiny bones and membranes, which, in turn, convert these movements into waves in the fluid that fills the inner ear
● The *inner ear*, which senses fluid movement created by sound waves (the *cochlea*) or by head motion (the *vestibular apparatus*). The *vestibulocochlear nerve* transmits impulses from the inner ear to the brain.

The External Ear

The **external ear** begins laterally with the **pinna** (also called the *auricle*), the fleshy part of the ear attached to the side of the head (Fig. 9.7). The pinna is covered by skin and composed of elastic cartilage, fat, and fibrous tissue. Its funnel-like form reflects its function, which is to collect sound waves and channel them into the external auditory canal.

About 2.5 cm long, the **external auditory canal** leads from the pinna to the tympanic membrane (eardrum) of the middle ear. The lateral half of the external auditory canal is surrounded by soft tissue, the medial half is encased in the temporal bone of the skull, and the entire canal is lined by skin. **Cerumen** (ear wax) is a pasty secretion produced by sebaceous and apocrine glands in the outer third of the external auditory canal. Cerumen lubricates the ear canal and protects it by virtue of its antimicrobial properties, but accumulation of excessive cerumen can impair hearing by blocking the canal.

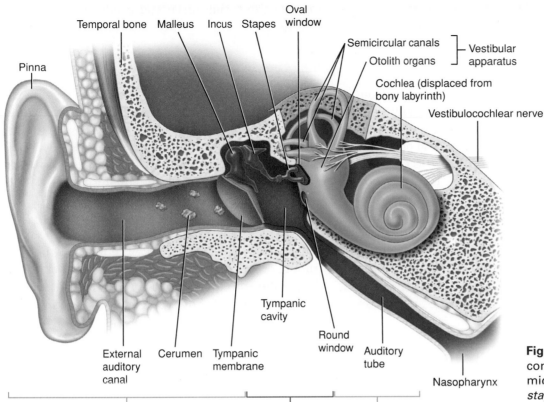

Figure 9.7. The ear. The ear consists of the external, middle, and inner ear. *Is the stapes found in the middle or inner ear?*

The Middle Ear

Hollowed out of the solid mass of the temporal bone is the air-filled chamber of the **middle ear** (see Fig. 9.7). Its lateral wall is the **tympanic membrane** (eardrum), which separates it from the external auditory canal.

The tympanic membrane vibrates when struck by sound waves, just as a drumhead vibrates when struck with drumstick. These vibrations mark the initial step in the detection of sound waves by our hearing apparatus.

On the medial side of the tympanic membrane is the **tympanic cavity,** an air-filled space in the temporal bone that is lined by simple cuboidal epithelium.

Spanning the tympanic cavity are three tiny bones (**ossicles**), which transfer sound waves from the tympanic membrane to the inner ear. The *malleus* (Latin for "hammer") attaches to the tympanic membrane. It has a rounded head like a mallet, which fits into a depression on the next tiny bone, the *incus* (Latin for "anvil"). The incus, in turn, connects to the smallest bone in the body, the *stapes* (Latin for "stirrup"), so named because it is shaped very much like a stirrup. The stapes has a narrow "strap" of bone that connects it to the incus, and a broad "footplate" that inserts into the **oval window,** an opening between the middle ear and the inner ear. The footplate of the stapes is attached to the edges of the

oval window by a delicate rim of fibrous tissue that allows small back-and-forth movements of the footplate in the window. Further on, we discuss how the vibrations of the stapes transmit sound waves to the inner ear. Near to the oval window is a second window between the middle and inner ears, the **round window,** which is also discussed in more detail below.

The **auditory** (or *eustachian*) **tube** runs medially and inferiorly from the tympanic cavity of the middle ear to the nasopharynx. This connection enables air to enter or leave the middle ear in order to normalize air pressure within the tympanic cavity. As discussed in ➡ Chapter 13, gases at lower pressure have a larger volume (that is, they take up more space), and vice versa. So, as we ascend in a fast elevator or an airplane and atmospheric pressure drops, the air within the tympanic cavity expands. Normally, some of this expanded volume of air will leave the tympanic cavity via the auditory tube in order to prevent pressure buildup from the expanding air. However, a stuffy nose can block the nasal end of the auditory tube and prevent air from escaping. The resulting pressure gradient puts tension on the tympanic membrane, so that it can no longer vibrate easily, and a somewhat uncomfortable pressure develops that is accompanied by a temporary hearing loss. Moving the jaw back and forth or yawning can help open the auditory tube to relieve the pressure. However, in severe cases, if the person were ascending, for instance, to thousands of feet from low altitude in an unpressurized airplane, the air trapped in the middle ear can cause very painful stretching or rupture of the tympanic membrane.

The auditory tube is also a conduit for microbes from the nose and throat, which can infect the middle ear, causing the painful ear infections (*otitis media*) that commonly afflict young children. Otitis media often follows an upper respiratory infection like the common cold. The excess mucus produced by the infection obstructs the auditory tube and serves as a breeding ground for bacteria. Children are more often affected than adults because their auditory tubes are smaller and more susceptible to obstruction. Moreover, the *facial nerve* (cranial nerve VII) passes through the temporal bone very near the tympanic cavity and is separated from it by only a very thin layer of bone. Otitis media can spread to the facial nerve and produce facial muscle paralysis.

The Inner Ear

Deepest in the temporal bone are the interconnected tubes and chambers of the **inner ear,** collectively called the *labyrinth* (Fig. 9.8A), a term from Greek mythology that describes an impossibly complex maze. Functionally, the labyrinth can be divided into the *cochlea* for hearing and the *vestibular apparatus* for equilibrium, or balance:

- The **cochlea** (Latin for "snail") is the hearing part of the inner ear and is a snail-like hollow tube coiled upon itself for 2.5 turns. Like the turns in the brass tubing of a trumpet, the 2.5 turns of the cochlea fit a long tube into a small space. The cochlea captures sound waves and converts them into nerve impulses.

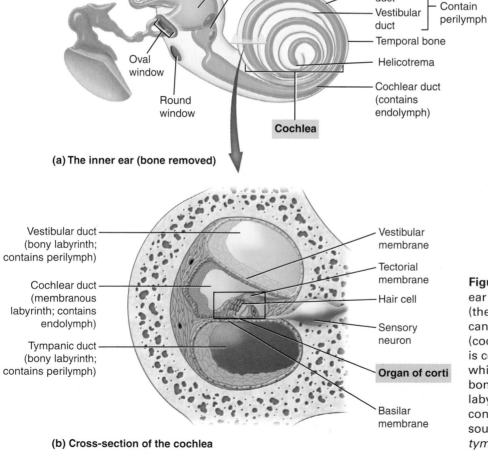

(a) The inner ear (bone removed)

(b) Cross-section of the cochlea

Figure 9.8. The inner ear. A. The inner ear consists of the organs of equilibrium (the otolith organs and semicircular canals) and the organ of hearing (cochlea). The membranous labyrinth is contained within the bony labyrinth, which is carved out of the temporal bone. **B.** In the cochlea, the membranous labyrinth is called the cochlear duct; it contains the hair cells that detect sounds. *What fluid is contained in the tympanic duct?*

- The **vestibular apparatus** is the equilibrium part of the inner ear and consists of five fluid-filled structures—the three *semicircular canals* and the two *otolith organs* (the *utricle* and *saccule*). Below, we discuss how these structures sense both motion and the pull of gravity, literally telling us which way is up and where we are going. Together they allow us to remain balanced as we move about—whether standing up from a chair, walking, or turning a somersault.

Structurally, the cochlea, the semicircular canals, and the otolith organs are actually formed of *two* labyrinths, one inside the other, hand in glove. Both are filled with fluid. The outer labyrinth, called the **bony labyrinth,** is a space hollowed out of temporal bone and filled with **perilymph** (peri = "around"). Perilymph is actually cerebrospinal fluid that fills the labyrinth via a connection to the subarachnoid space ⬅ (Chapter 8). Floating in the perilymph and fitted perfectly inside the bony labyrinth is the **membranous labyrinth,** a soft-walled, membranous set of chambers and tubes. The membranous labyrinth is filled with a second fluid, **endolymph** (endo = "inside"), made by the membranous labyrinth itself. Endolymph and perilymph flow freely through the cochlea, semicircular canals, and otolith organs, but they are always separated from each other by the membrane—endolymph inside the membranous labyrinth, perilymph outside of it.

Case Note

9.11. Helen was able to climb trees, indicating that she could perceive head motion and gravity, although she could not hear. Was her brain able to process information from the cochlea, the vestibular apparatus, or both?

Structure of the Cochlea

In the cochlea, the membranous labyrinth is called the **cochlear duct.** The membrane of the cochlear duct attaches to either side of the bony labyrinth like an inflated balloon stretched between two sides of a hollow space. Notice in Figure 9.8B that the cross-sectional shape of the inner labyrinth is somewhat wedge-shaped rather than round. This arrangement creates three tubular spaces in the cochlea that coil side by side around the cochlear turns. The center tube is the cochlear duct filled with endolymph. The second and third tubes are created outside the cochlear duct between the walls of the cochlear duct and the wall of the bony labyrinth on each side; they are filled with perilymph. These three

spaces follow the cochlear spiral inward to the center point, where the cochlear duct ends in a closed tip. However, the two outer spaces join at an opening, the *helicotrema* (see Fig. 9.8A), so that the perilymph filling them is a continuous body of fluid.

Each outer tube connects to one of the two windows between the middle ear and the inner ear. The outer tube originating at the *oval* window is called the **vestibular duct.** The other outer tube originating at the *round* window is called the **tympanic duct.** The common wall between the cochlear duct and the vestibular duct is called the **vestibular membrane,** whereas the common wall between the cochlear duct and the tympanic duct is called the **basilar membrane** (Fig. 9.8B).

Running the length of the basilar membrane inside the cochlear duct is the **organ of Corti,** a strip of tall receptor cells with cilia (hair cells) that detect sound waves. A third membrane, the **tectorial membrane,** lies entirely within the cochlear duct. It originates from one side of the cochlear duct and spreads outward and lies upon the hairs of the receptor cells like a wet blanket. As discussed below, this contact between the tectorial membrane and the hair-cell cilia is important in changing sound waves into nerve signals.

> *Remember This!* **The outer and middle ears are involved in sound sensing only. The inner ear is involved in sound sensing and equilibrium.**

Hearing Is the Detection and Conscious Integration of Sound Waves

Hearing is the detection of sound waves and integration into consciousness of the sensory signals produced by the ear. Our sense of hearing is exquisitely sensitive—the ear reacts far more quickly to sound than the eye does to light.

What Are Sound Waves?

Sound waves are alternating layers of *compressed air*, in which the gas molecules are closer together, and *decompressed air*, in which the gas molecules are spread apart (Fig. 9.9A). For example, a blast from a speaker is nothing more than a wave of compressed air created by the explosion that moves through air much like a "wave" moves through spectators at a stadium. Continuous sounds—a musical note, or the howl of a coyote, for example—are a succession of very closely spaced sound waves created by rapidly vibrating structures, such as the strings of a violin, the speaker cones in an audio system, or the vocal cords

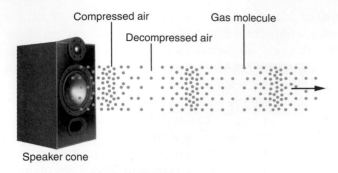

Compressed air   Gas molecule

Decompressed air

Speaker cone

(a) The nature of sound waves

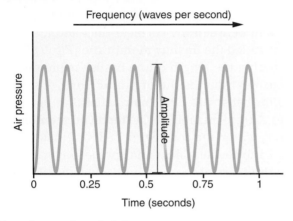

Frequency (waves per second)

Air pressure

Amplitude

0   0.25   0.5   0.75   1

Time (seconds)

(b) Sound wave characteristics

Figure 9.9. Sound waves. A. Sound waves are pressure waves caused by a vibrating structure, in this case a speaker cone. **B.** A graph representing changes in air pressure over time. The amplitude of the pressure change corresponds to the loudness of the sound. The number of waves passing by a particular site in one second is the frequency, or pitch of the sound. *If the speaker emitted a higher-pitched sound, would we see more or fewer waves on the graph?*

of a coyote. The strings, the speaker cones, and the vocal cords compress and decompress air as they rapidly move back and forth, creating alternating waves of compressed and rarified air that our ears convert into nerve signals that our brain interprets as sound.

Sound waves (and other waves, like radio and TV signals) are typically displayed in graphic form (Fig. 9.9B). The vertical axis is signal strength or intensity—the *amplitude*, which is displayed by the vertical distance from wave peak to trough. The horizontal axis is time. The number of waves per unit of time is the *frequency*.

The marvelous variety of sound depends on these two key characteristics: the number of waves per second (the frequency or pitch, measured in *hertz*), and the degree of air compression (the amplitude, intensity or loudness). For high-frequency waves (10,000 waves per second, or 10,000 Hz), such as the whistle of a piccolo, the distance between successive wave peaks is short. For low-frequency

waves (20 Hz), such as the rumble of a bass drum, the distance between wave peaks is long. Waves with high amplitude (highly compressed air) are loud; low-amplitude sounds have less air compression and are softer. The detection range of human hearing is astounding—the loudest sounds we can hear without ear damage are *1 trillion* (10^{12}) times more intense than the quietest sounds we can hear. But even so, our range is less astounding than that of many of our animal brethren.

Case Note

9.12. Helen was able to "hear" the low-pitched sound of large branches hitting the ground during a thunderstorm. Is she detecting a low-frequency sound or a high-frequency sound?

The Organ of Corti Detects Sound Waves

We are able to detect sounds and determine their pitch and loudness because of the phenomenon of *resonance*. Resonance is the sympathetic vibration of one part in concert (in tune) with another. For example, windows rattle in response to the pounding of a jackhammer. Likewise, we can feel the rumble of a jackhammer in our bones because our bones also vibrate sympathetically.

In the same way, sound waves cause vibrations in a series of anatomical structures in the ear, vibration in one structure causing sympathetic vibrations in the next. Eventually, the amplitude and frequency of the vibration is detected by mechanoreceptor cells of the organ of Corti, which relay the signal to the brain. Importantly, the amplitude and frequency of the vibrations in ear structures are directly proportional to the amplitude and frequency of sound waves.

From sound waves in air to nerve signals racing to the brain along the vestibulocochlear nerve, the sequence of events involved in hearing is as follows (Fig. 9.10):

1. The external ear channels sound waves to the tympanic membrane.
2. The tympanic membrane vibrates "in tune," setting up corresponding vibrations in the ossicles of the middle ear. The intensity and frequency of the vibrations of the malleus are transmitted to the incus and subsequently to the stapes. Loud sounds cause large vibrations, soft sounds cause small vibrations, high-pitched sounds cause rapid vibrations, and low-pitched sounds cause slow vibrations.
3. The footplate of the stapes vibrates in the oval window. Movement in the oval window creates waves in

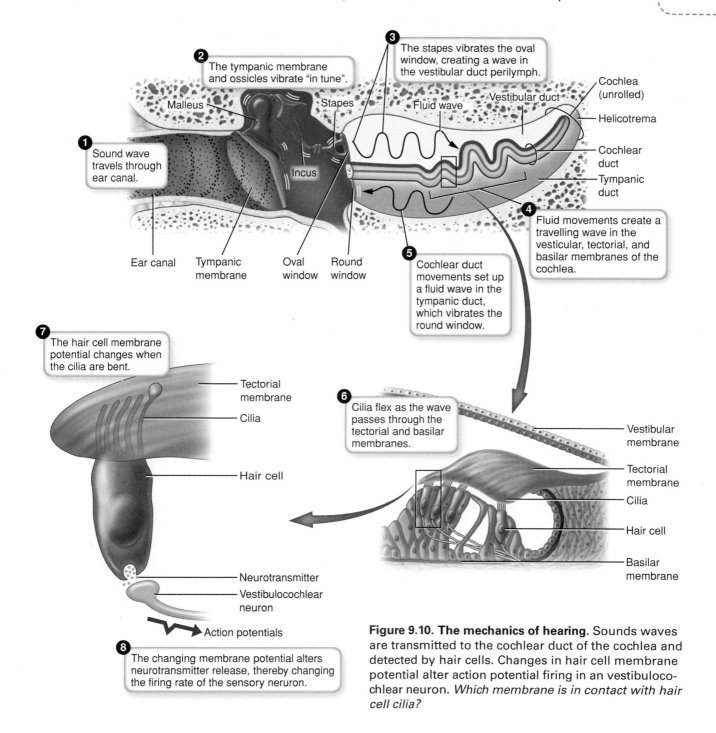

1 Sound wave travels through ear canal.

2 The tympanic membrane and ossicles vibrate "in tune".

3 The stapes vibrates the oval window, creating a wave in the vestibular duct perilymph.

4 Fluid movements create a travelling wave in the vesticular, tectorial, and basilar membranes of the cochlea.

5 Cochlear duct movements set up a fluid wave in the tympanic duct, which vibrates the round window.

6 Cilia flex as the wave passes through the tectorial and basilar membranes.

7 The hair cell membrane potential changes when the cilia are bent.

8 The changing membrane potential alters neurotransmitter release, thereby changing the firing rate of the sensory neruon.

Malleus · Stapes · Incus · Ear canal · Tympanic membrane · Oval window · Round window

Fluid wave · Vestibular duct · Cochlea (unrolled) · Helicotrema · Cochlear duct · Tympanic duct

Tectorial membrane · Cilia · Hair cell · Neurotransmitter · Vestibulocochlear neuron · Action potentials

Vestibular membrane · Tectorial membrane · Cilia · Hair cell · Basilar membrane

Figure 9.10. The mechanics of hearing. Sounds waves are transmitted to the cochlear duct of the cochlea and detected by hair cells. Changes in hair cell membrane potential alter action potential firing in an vestibulocochlear neuron. *Which membrane is in contact with hair cell cilia?*

the perilymph of the vestibular duct that are proportional to the amplitude and frequency of the original sound wave.

4. The fluid wave in the vestibular duct induces a *traveling wave* in the membranes of the cochlear duct, much like snapping a rope produces a traveling wave in the rope.

5. The cochlear movements initiate a second fluid wave in the tympanic duct, which is dissipated when it hits the round window. If not for movement of the round window—if only solid bone were there—the waves would bounce back, making a return trip, activating the organ of Corti again to create an internal "echo." If this were the case, clear hearing would be impossible.

6. Recall that hair cells perch on the basilar membrane, and the hair-cell tips are embedded in the tectorial membrane. These two membranes differ somewhat in their flexibility, so they move a bit in relation to each other as the traveling wave passes through

them. As a result, the hair-cell tips are flexed back and forth with each wave, as the basilar membrane moves in relation to the tectorial membrane.

7. Bending of the cilia changes the membrane potential of hair cells: bending in one direction hyperpolarizes the hair cell, whereas bending in the other direction depolarizes the hair cell. These membrane potential changes cause cyclical changes in neurotransmitter release from hair cells into their synapse with a first-order neuron.

8. Neurotransmitter molecules released into the synapse change the membrane potential in a neuron of the vestibulocochlear nerve, altering the frequency of action potential firing. Action potentials travel through the vestibulocochlear nerve to the brain, as described below.

Case Note

9.13. If Helen were in a noisy room, which membrane would vibrate first—the cochlear membrane, vestibular membrane, basilar membrane, or tympanic membrane?

We Detect Three Qualities of Sound Waves

Before we can understand the audible world, our brain must assess the pitch, volume, and direction of the sound waves.

Pitch

Recall that the pitch of a sound reflects its frequency. In stringed instruments such as guitars and violins, short, stiff (tight) strings vibrate very rapidly to produce a high-pitched sound, and long, limber (loose) strings vibrate slowly to produce a low-pitched sound. So it is with the basilar membrane: at its base (near the oval window) it is narrow and stiff, and it vibrates sympathetically with high-pitched sounds; at its tip it is wide and limber, and it vibrates sympathetically with low-pitched sounds. Thus, different membrane regions and groups of receptor hair cells vibrate according to the pitch of the sound being sensed, stimulating action potentials in different sets of neurons, which in turn are interpreted by the brain as high- or low-pitched sound.

Volume

The loudness of a sound reflects the amplitude of sound waves. Stronger vibrations of the eardrum in response to loud sounds result in larger vibrations of the basilar membrane, causing more hair-cell bending and hence more action potentials. Softer sounds, with their minute vibrations, cause less displacement and fewer action potentials. This variation in the number of action potentials sent to the brain is interpreted as loud or soft sound.

Remember This! **Pitch is communicated by the location of activated hair cells; the frequency of action potential firing conveys loudness.**

Direction

Determination of sound *direction* is dependent on the gross anatomy of the external ears. Except for sound coming from directly in front or behind or above or below, sound reaches one ear a fraction of a second before the other and is louder in the near ear than in the far one. These differences are detected by the brain and interpreted directionally. Discrimination of sound coming from directly ahead or behind, above or below, depends on the shape and orientation of the pinna. It faces forward, and its cupped shape slightly alters the pitch and loudness of sound, which is interpreted accordingly by the brain.

Signals for Hearing Follow the Auditory Pathway

Signals travel from the cochlea through the cochlear branch of the vestibulocochlear nerve to the *primary auditory cortex* on the superior surface of the temporal lobe and then on to the *auditory association area* nearby. Signals take multiple pathways from one ear to the cortex. Most (but not all) signals cross over to the opposite cortex. The net result is that each hemisphere of the cerebral cortex receives auditory input from both ears. This special feature is a biological "insurance policy" for hearing: destruction of the auditory cortex in one cerebral hemisphere does not completely destroy hearing in one ear. One of the most common pathways is shown in Figure 9.11.

Case Note

9.14. If Helen was deaf in only one ear, would that indicate a defect in the auditory cortex on the affected side or the ear itself? Explain.

The primary auditory cortex is divided into *pitch maps*—distinct regions that are sensitive to a particular frequency range—bass rumbles here, piercing whistles

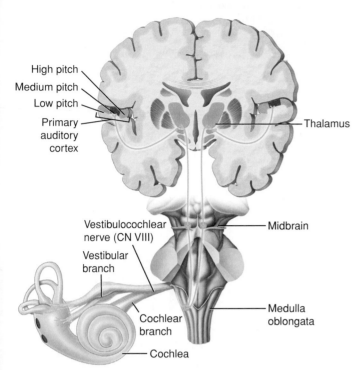

Figure 9.11. Auditory pathway. Signals from the left cochlea travel to the primary auditory cortex in both hemispheres. *How many neurons are used to convey an auditory signal from the cochlea to the right auditory cortex?*

there. Signals from hair cells in a particular area of the basilar membrane (which respond to a particular frequency range) are projected to a related pitch region of the auditory cortex. The brain, therefore, perceives pitch based on the particular area of cortex activated. The primary auditory cortex may also be divided into *volume maps,* with some neurons responding best to the frequent action potentials induced by loud sounds, while others are most sensitive to the less frequent action potentials of quiet sounds.

After our auditory apparatus has detected the various qualities of sound and relayed them to the primary auditory cortex, the brain must still "make sense" of what would otherwise be meaningless noise. This is the job of the nearby auditory association cortex. This part of the cortex discriminates *patterns* of sound according to the complex variations of pitch, volume, and direction and ascribes meaning to them. Our ability to recognize instantly the distinctive quality of a person's speech is evidence of the remarkable ability of the auditory cortex to assign meaning to sound. The auditory association cortex also receives input from other brain regions, such as the visual and somatosensory cortices, to help in this task. For instance, it is easier to understand sounds if we see the person speaking.

From this discussion it should be apparent that hearing is a very complicated matter. For more about impaired hearing (deafness), see the nearby Clinical Snapshot titled *Hearing Abnormalities*.

9.21 Which type of receptor is involved in hearing and balance—nociceptor, mechanoreceptor, or chemoreceptor?

9.22 Other than holding up sunglasses, what is the function of the pinna?

9.23 Which ossicle is attached to the tympanic membrane?

9.24 The cochlear duct contains perilymph—true or false?

9.25 As we get further away from the sound source, the loudness decreases but the pitch stays the same. What is changing—frequency or amplitude?

9.26 Which membrane is in contact with hair-cell cilia—the basilar or the tectorial?

9.27 You hear two sounds of equal volume—one high-pitched and one low-pitched. What varies in the basilar membrane between these two sounds—the region of the basilar membrane that is vibrating or the degree of vibration?

The Inner Ear and Equilibrium

Equilibrium, our sense of balance, informs us of our orientation in space relative to gravity and whether or not we are in control of our movements. If we're off balance, we have mechanisms, both conscious and reflexive, to bring us back into equilibrium. Being in equilibrium does not mean being still; it means being in control. The graceful turns and leaps of a ballerina are acts of superbly controlled equilibrium. In this section we discuss the contributions of the inner ear, which detects changes in speed and the direction of gravity, to equilibrium. Other sensory input—visual cues, proprioceptive input, and even tactile input such as the pressure of our legs on a seat—is also key. You can see the importance of visual input for yourself, by standing on one foot with your eyes open (relatively easy) and with your eyes closed (much harder).

CLINICAL SNAPSHOT

Hearing Abnormalities

Deafness

Deafness is an inability to hear normally. Hearing loss attributable to interference with the conduction of sound waves in the ear canal or middle ear is **conductive deafness**. Hearing loss due to cochlear disease is **sensory deafness**. Hearing loss due to disease of the vestibulocochlear nerve, CNS nerve tracts, or auditory cortex is **neural deafness**.

Conductive deafness may be due to cerumen (ear wax) obstructing the ear canal or impacted on the ear drum or to disease of the middle ear. Stiffening of the ossicles (*otosclerosis*) so that they do not vibrate properly is a hereditary condition in some adults, but it can be caused by chronic middle ear infection, especially in children. Often conductive deafness can be corrected surgically.

Sensory deafness is a result of damage to the organ of Corti in the cochlea. The most common identifiable cause is chronic exposure to very loud sounds. For example, exposure to a loud car horn for 1 minute or chronic exposure to anything as loud as a lawnmower can cause permanent sensory deafness. But by far the most common cause is idiopathic (not identifiable); it commonly occurs in an elderly person with no known chronic exposure. Most hearing aids sold to adults are for sensory hearing loss. Hearing aids typically include a microphone that picks up sounds, an amplifier that makes them louder, and a speaker that sends them toward the inner ear. For profoundly deaf patients who do not benefit from the usual hearing aid, sensory deafness can be treated by inserting an electronic probe into the cochlea (*cochlear implant*) to directly stimulate the organ of Corti.

Neural deafness is the least common type of hearing loss and is due to lesions in the vestibulocochlear nerve or, less commonly, in the CNS. Hearing aids or cochlear implants may offer some help.

Tinnitus

Tinnitus (Latin *tinnire* = "to ring") is an unwelcome ringing or buzzing in the ears. *Objective tinnitus* is an unwelcome sound that arises outside of the auditory

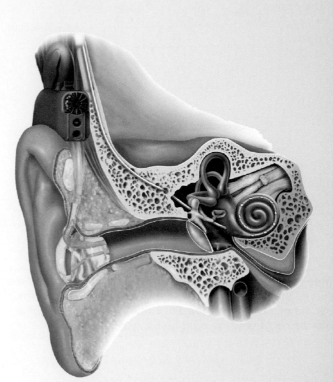

A cochlear implant.

system, often within the body itself; it can frequently can be heard by an examiner. For example, some patients with vascular abnormalities can hear blood rushing through the vessels in their neck. Heart patients with artificial valves may have a similar problem.

But far more common and far more unwelcome is *subjective tinnitus*, in which the sound arises from somewhere within the auditory system. The sounds are described most frequently as a metallic ringing or a buzzing or blowing. Subjective tinnitus is very common; indeed, almost anyone who pays strict attention in a very quiet place can hear a faint, high-pitched, metallic ring. Sometimes tinnitus is associated with the development of deafness. Whatever the cause, it can be very distracting in some people. Severe tinnitus can even be disabling.

The Vestibular Apparatus Senses Acceleration

The vestibular apparatus maintains equilibrium in part by detecting **acceleration,** the *rate of change* in speed with time. We experience several types of acceleration:

- *Positive acceleration* is speeding up.
- *Negative acceleration* is slowing down, or deceleration.
- *Linear acceleration* occurs when we change speed in a straight line, as in an accelerating elevator or in a car as it slows to a stop.
- *Angular acceleration* represents a change in speed while you rotate around a fixed point, such as turning your head or spinning in a chair.

Acceleration is always associated with force—consider how an accelerating car forces you back in your seat or your tendency to slide forward with a quick stop.

Note that acceleration reflects *changing* speed. At *constant* speed, the vestibular apparatus is inactive. For example, in a jet airplane, we may know intellectually that we are moving at 500 mph, but the constant speed does not activate our equilibrium sensors. That said, most human activities involve the starting and stopping of motion, prompting the speed changes that do activate our equilibrium sensors. Equilibrium also depends on the ability of the inner ear to detect changes in speed during continuous movement. Detection of the direction of gravity, however, operates even when we are motionless.

9.15. How can we be reasonably certain that Helen's inner ear was not affected by the illness that made her deaf?

9.16. When Helen turned her head side to side to indicate "no," was her head experiencing linear acceleration or angular acceleration?

The Vestibular Apparatus Consists of the Otolith Organs and the Semicircular Canals

Like the cochlea, the vestibular apparatus is hollowed out of bone, filled with perilymph, and fitted internally with a membranous labyrinth filled with endolymph (Fig. 9.8). The vestibular apparatus has two main subdivisions: the otolith organs and the semicircular canals (Fig. 9.12). The **otolith organs** are two interconnected balloonlike chambers of the membranous labyrinth, the **saccule** and the **utricle,** each containing a sensory receptor called a *macula.* Branching out from the utricle are three **semicircular canals:** two vertical, one each roughly in the coronal and sagittal planes, and one roughly in the horizontal plane, all of which fit together at right angles like the corners of a box. A sensory receptor called a *crista ampullaris* resides within the opening of each semicircular canal. Signals from all parts of the vestibular apparatus travel to the brain via the vestibular branch of the vestibulocochlear nerve.

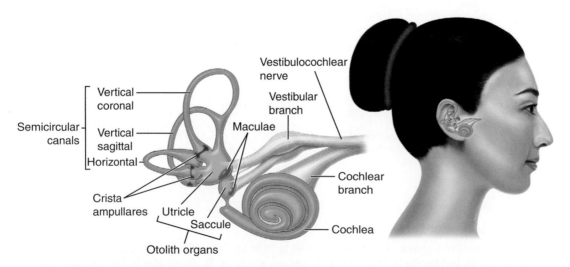

Figure 9.12. The vestibular apparatus. The vestibular apparatus consists of the semicircular canals and the otolith organs. *Where would a crista ampullaris be found—in the utricle or in a semicircular canal?*

The Otolith Organs Sense Gravity and Linear Acceleration

The utricle and saccule are structurally very similar. Within each chamber is a flat patch of tall sensory hair cells called a **macula.** The cilia of the hair cells protrude into a gelatinous membrane, the otolithic membrane, which is weighted down by tiny crystals called **otoliths** (literally "ear rocks") (Fig. 9.13). Notice that the maculae in the utricle and saccule are oriented perpendicularly to each other—the macula in the utricle is mainly horizontal; the macula in the saccule is mainly vertical.

The otolith organs answer two sorts of questions:

- How is my head oriented relative to gravity? That is, am I standing up, lying down, or somewhere in between? In effect, which way is up?
- Is my speed changing? That is, am I accelerating horizontally or vertically?

Sensing Gravity: Which Way Is Up?

Even with our eyes closed, we can distinguish up from down because of the effect of gravity on the maculae. When the head is upright, the utricle's macule is horizontal. The

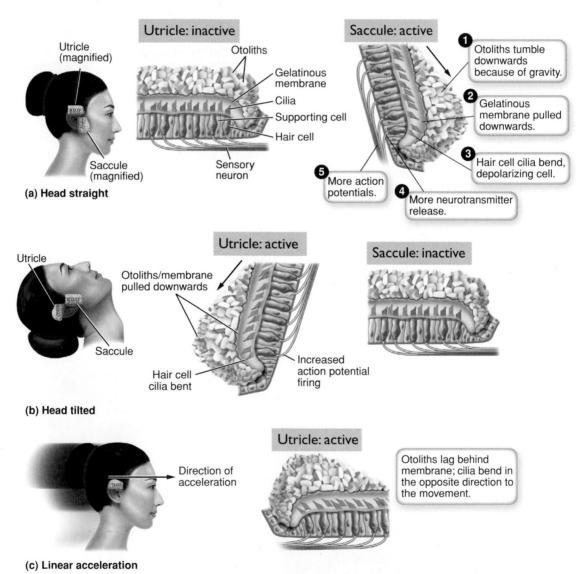

Utricle (magnified)

Utricle: inactive

Otoliths

Gelatinous membrane

Cilia

Supporting cell

Hair cell

Saccule (magnified)

Sensory neuron

(a) Head straight

Saccule: active

1 Otoliths tumble downwards because of gravity.

2 Gelatinous membrane pulled downwards.

3 Hair cell cilia bend, depolarizing cell.

4 More neurotransmitter release.

5 More action potentials.

Utricle

Utricle: active

Otoliths/membrane pulled downwards

Saccule

Hair cell cilia bent

Increased action potential firing

Saccule: inactive

(b) Head tilted

Direction of acceleration

Utricle: active

Otoliths lag behind membrane; cilia bend in the opposite direction to the movement.

(c) Linear acceleration

Figure 9.13. The otolith organs. A. When the head is held upright, the hair cell cilia in the utricle's macule are upright, but those in the saccule's macule are bent. **B.** When the head is horizontal, the hair cells in the saccule are straight and those in the utricle are bent. **C.** During linear acceleration (as at the beginning of a sprint), the heavy gelatinous membrane lags behind the hair cells and bends the cilia. *If you lie down, are the utricle hair cell cilia bent or straight? How about the saccule hair cell cilia?*

hair cell cilia are straight (Fig. 9.13A). The saccule's macule, conversely, is vertical. Gravity pulls the otolith-studded membrane downward, bending the hair-cell cilia. Just as in cochlear hair cells, the bending of hair-cell cilia—depending on which way they're bent—depolarizes or hyperpolarizes the cell and alters action potential firing in the sensory nerve. The brain perceives this situation—saccule hair cells bent, and utricle hair cells straight—as indicating that our head is "up."

When the head is horizontal, either when we're lying down or standing and looking up at the sky, the utricle is nearly vertical and the saccule nearly horizontal (Fig. 9.13B). The saccule hair cells are straight and the utricle hair cells bent, resulting in a characteristic pattern of action potential firing that indicates the head is "sideways." Interestingly, in contrast to smell and taste and some other sensory receptors, maculae do not adapt—if you've been lying down for an hour, you still receive the sensation that you are flat.

Sensing Linear Acceleration

The maculae also detect linear (straight line) acceleration or deceleration—the forces that push you back into your seat in an accelerating car, or fling you forward with a quick stop, or—to a lesser degree—make you feel heavier or lighter at the beginning and end of a trip in an elevator.

To understand how the otolith organs sense the force of linear acceleration requires an understanding of *inertia*, or resistance to movement. Anything at rest—such as a rock—remains at rest unless an outside force acts on it. The gelatinous otolithic membrane and its load of "ear rocks" has a lot of inertia—it resists both positive and negative acceleration. This means that, during acceleration, the otolithic membrane lags behind the hair cells embedded in it, causing the hair cells to bend backward. During deceleration, the otolithic membrane slides forward, bending the hair cells forward. Let's look closely at this in the utricle and saccule.

Consider first the utricle, whose macule is oriented horizontally when the head is erect (Fig. 9.13C). As a gymnast begins her forward movement and accelerates into a run, the inertia of the heavy, otolith-studded gelatinous membrane lags behind and pulls backward on the hair-cell cilia. The cilia thus bend opposite to the direction of acceleration. If she settles into a run at constant speed, the gelatinous membrane will return to its normal position and the cilia will straighten up again. Then, as she decelerates at the end of her run, the otolithic membrane shifts forward, bending the hair-cell cilia forward. The macule of the saccule, conversely, detects ver-

tical accelerations as in an elevator ride. As the elevator accelerates upward, hair-cell cilia are bent downward toward the earth and we sense upward acceleration; as it decelerates, the cilia are bent away from the earth and we sense deceleration. If the elevator is ascending or descending at a constant rate, the hair-cell cilia are straight and we don't sense any movement.

> ***Remember This!*** **Hair cells in the cochlea detect sound waves; hair cells in the vestibule detect gravity and linear acceleration.**

Semicircular Canals Detect Angular Acceleration

The organs of equilibrium that respond to angular (rotational) acceleration are the three *semicircular canals*. They are oriented at right angles to one another like the sides of a corner of a box so that, no matter the plane of rotation, the sensors in at least one canal will be activated (Fig. 9.14A). For instance, nodding "yes" activates the vertical canals, nodding "no" activates the horizontal canal. The membranous labyrinth of the canals is filled with endolymph and contains a nodule of hair cells, the **crista ampullaris,** which projects into the lumen of the canal. Atop each crista is a tall cap of gelatinous material, the *cupula* (Fig. 9.14B). In turn, the cupula extends into the ampulla, more or less "waving in the wind" of the endolymph filling the membranous labyrinth.

When your head turns, the cristae and cupulae move at the same speed; however, the inertia of the endolymph causes it to lag behind. As a consequence there is fluid drag on the cupula, which bends the microvilli of the hair cells and alters the membrane potential. As we saw with the otolithic membrane, the cupula bends opposite to the direction of the movement. Bending in one direction (counterclockwise) hyperpolarizes the cell, whereas bending in the other direction (clockwise) depolarizes the cell. The resulting change in action potential frequency in the sensory neurons of the three canals, each with its unique orientation, tells the brain that the head is moving, and in which direction (Fig. 9.14C).

The semicircular canals perform a simple but critical task: they detect the *beginning* and the *end* of head rotation, when angular acceleration occurs. You can maintain your equilibrium without them as long as you are not making any sudden or intricate rotational movements, because the vestibule calculates the position of your head in space. However, daily human activity is a jumble of complex head movements, many of which involve rotation. It is the job of the semicircular canals

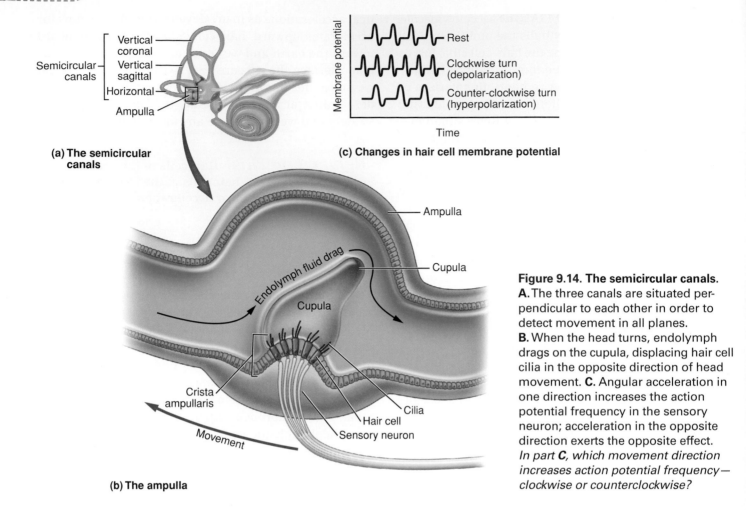

(a) **The semicircular canals**

(c) **Changes in hair cell membrane potential**

(b) **The ampulla**

Figure 9.14. The semicircular canals. **A.** The three canals are situated perpendicular to each other in order to detect movement in all planes. **B.** When the head turns, endolymph drags on the cupula, displacing hair cell cilia in the opposite direction of head movement. **C.** Angular acceleration in one direction increases the action potential frequency in the sensory neuron; acceleration in the opposite direction exerts the opposite effect. *In part **C**, which movement direction increases action potential frequency— clockwise or counterclockwise?*

to detect the rotation the *instant* it occurs, so that, in effect, it *anticipates* how turning motions will disrupt your balance so that adjusting movements can be made by other body parts. Without this "early warning" mechanism, we would stumble about in our daily tasks even though the utricle and saccule were perfectly normal.

So, add it all up and what do you have? A marvelously integrated system that keeps daily life from being a series of pratfalls—a system that allows for the beauty of human movement. Consider the feats of a gymnast. While waiting for the signal to begin her routine, she stands still. The signal sounds. She begins running until she reaches full speed, then does a forward flip, and sticks a perfect landing into a stock-still finishing position. She has accelerated, reached a constant speed, rotated her head and body 360 degrees, and decelerated to a full stop in an instant, thanks to the remarkable ability of the vestibular apparatus of her inner ear to maintain her equilibrium.

On a lighter note, it is the continued movement of endolymph within the semicircular canals that explains the brief dizziness that many children seem to enjoy

after whirling their bodies around or spinning in a chair. When the movement stops, endolymph temporarily continues to flow, dragging on the crista ampullaris and imparting a sensation of movement that our eyes tell us is not occurring.

Case Note

9.17. In Helen Keller's autobiography, she mentions the crude signs she invented shortly after her illness. For instance, she would shake her head side to side to indicate "no." Would this action preferentially activate maculae or hair cells in the semicircular canals?

Signals for Equilibrium Follow Specific Pathways

Bending of the sensory hair cells sends action potentials to first-order sensory neurons in the vestibular branch of the vestibulocochlear nerve (VIII), which relays signals to the medulla oblongata and cerebellum. From the medulla, some impulses are relayed to nuclei associated with cranial nerves controlling eye and neck movements. The cerebellum accepts signals from the utricle, saccule, and semicircular canals as well as from the visual system and proprioceptive sensors. It integrates these signals and relays them to the cortex for integration into consciousness. Voluntary motor commands can then be added to ensure ongoing control of body position and movement.

9.28 Name the two chambers of the vestibule.

9.29 Which part of the inner ear enables a spinning ice skater to maintain the control necessary to earn a good score from the judges?

9.30 Name the gelatinous structure covering the hair-cell cilia in the semicircular canals.

9.31 Are otiliths involved in the detection of linear acceleration or angular acceleration?

Vision

Vision is the detection of light by the eye *and* integration into consciousness of the sensory signals produced. It is a special sense, and the one many people value most highly.

The **eye** is the organ of vision. In its entirety, it is referred to as the **eyeball,** a sphere about 24 to 25 mm (1 inch) in diameter. The specific light-sensing cells in the eye are **photoreceptors,** and the pathways between photoreceptors and the brain are the *visual pathways.* By sensing minute variations of color, light, and dark, we perceive one aspect of the world around us—the world revealed by light. Unlike all other sensory organs, the eyes can be manipulated by the body: they can be aimed in a certain direction and focused near or far—a clue to their survival value.

Accessory Structures Protect and Move the Eyeball

The eyeball and related structures occupy a bony socket in the skull, the **orbit.** Fatty connective tissue cushions the eyeball within the orbit (Fig. 9.15B). For obvious

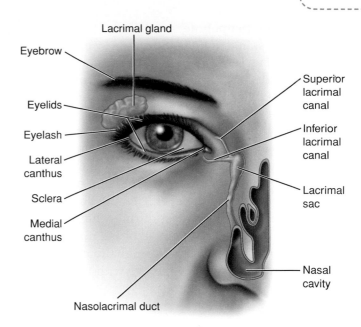

(a) The eyelid and lacrimal apparatus

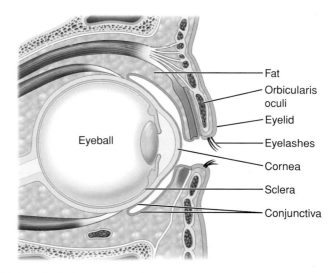

(b) A sagittal view of the orbit

Figure 9.15. Accessory structures of the eye. A. External accessory structures and the nasolacrimal apparatus. **B.** Sagittal view. The conjunctiva lines the eyelids and covers the anterior surface of the eyeball. *What is the white part of the eye called?*

reasons, the anterior surface of the eyeball lacks bony protection and must rely on the *eyebrows* and *eyelids* to protect it from potentially injurious agents.

The Eyebrows and Eyelids Protect the Anterior Eyeball

The **eyebrows** are bushy lines of coarse hair that sit on the supraorbital ridges of the frontal bones and keep sweat, scalp hair, and debris from falling into the eyes (Fig. 9.15A).

The **eyelids** (palpebrae) are flaps of tissue covered externally by skin and lined internally by a mucosa, the *conjunctiva* (Fig. 9.15B). The eyelids meet at the medial and lateral corners of the eye, called the **medial canthus** and **lateral canthus.** The **orbicularis oculi** muscle moves the eyelids to blink or squint (Fig. 9.15A). Attached to the free edges of the eyelids are the **eyelashes.** The eyelash follicles are richly supplied with sensory nerves; the slightest puff of air moves the eyelashes and invokes a reflexive contraction of the orbicularis oculi that closes the eyes. The function of eyelids is protective: they cover most of the eyeball, allowing only a small portion of it to be exposed directly to the environment. Their blinking continually cleanses and lubricates the front part of the eye with tears and the oily secretions of eyelid glands.

The Nasolacrimal Apparatus Produces Tears

The **nasolacrimal apparatus** consists of the almond-shaped **lacrimal glands,** which produce tears, and the associated tear drainage ducts (Fig. 9.15A). The lacrimal glands are located inside each orbit above the lateral canthus. Tears (lacrimal fluid) wash across the front of the eye and drain into the nasal cavity through a series of drainage ducts located in the eyelid near the medial canthus. Tears pass first through tiny openings on the margin of each eyelid, superior and inferior **lacrimal canals.** Tears then flow into an elongated, vertical sac. The large upper end of the sac, the **lacrimal sac,** collects tears, which subsequently pass through the small, lower end of the sac, the **nasolacrimal duct,** into the nasal cavity. Lacrimal fluid is slightly salty and does more than wash away debris and maintain lubrication. It contains two types of antimicrobial agents: antibodies and lysozyme. Antibodies are immune system proteins that attach to viral and bacterial proteins to mark them for destruction. Lysozyme, on the other hand, is a digestive enzyme that dissolves bacterial cell walls.

The Conjunctiva Coats the Eyeball and Eyelids

The **conjunctiva,** a delicate membrane composed of cuboidal epithelial cells, lines the eyelids and much of the anterior part of the eyeball (Fig. 9.15B). It is continuous with the transparent epithelial membrane covering the **cornea,** the anterior, domed, clear part of the eyeball. The conjunctiva extends from the undersurface of the eyelids into deep pockets around the periph-

ery of the eyeball. In the deepest part of the pocket, it turns anteriorly to provide a transparent cover for the white outer layer of the eyeball (the *sclera*). A highly vascular structure, the conjunctiva also contains mucus-producing cells that keep the eyeball moist and lubricated.

The Extraocular Muscles Move the Eyeball

The external muscles of the eye (*extraocular muscles*) are also found within the orbit. These six small skeletal (voluntary) muscles attach to the eyeball and move the eye in all directions (Fig. 9.16 and Table 9.1). They work together to keep both eyes pointed at an object. The superior, inferior, lateral, and medial rectus muscles all originate in a tendon ring posterior to the eye. One muscle, the superior oblique, traverses a small fibrocartilage "sling" before inserting into the eyeball.

The name of each muscle describes its location and provides clues as to its action. The superior rectus, for instance, elevates the eye to view the sky, while the medial rectus turns the eyeball inward toward on the nose. However, the superior and inferior rectus muscles cannot turn the eyeball upward or downward (respectively) without also rotating it medially. The superior and inferior obliques provide an opposing lateral pull that lets you focus your gaze directly upward or downward.

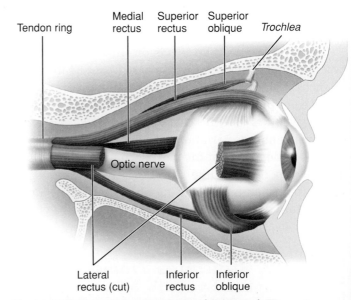

Figure 9.16. External eye muscles (right eye). The extraocular eye muscles originate on the bones (not shown) or the tendon ring of the eye orbit and insert into the sclera. *Which muscle passes through the trochlea?*

Table 9.1   The Eye Muscles

Name	Origin	Insertion	Action	Nerve
Superior rectus	Tendon ring in eye orbit	Sclera, just superior to the cornea	Elevation	Oculomotor (III)
Inferior rectus	Tendon ring in eye orbit	Sclera, just inferior to the cornea	Depression	Oculomotor (III)
Lateral rectus	Tendon ring in eye orbit	Sclera, just lateral to the cornea	Lateral rotation	Abducens (VI)
Medial rectus	Tendon ring in eye orbit	Sclera, just medial to cornea	Medial rotation	Oculomotor (III)
Superior oblique	Sphenoid bone in eye orbit	Tendon passes through trochlea (a ligament "sling"), inserts in superior, lateral portion of sclera	Lateral rotation and depression	Trochlear (IV)
Inferior oblique	Maxilla in eye orbit	Inferior, lateral portion of sclera	Lateral rotation and elevation	Oculomotor (III)

Case Note

9.18. Before her teacher came, Helen would often cry tears of frustration. Starting with the lacrimal gland and ending with the nasal cavity, list the structures through which tears pass.

The Eye Is Formed of Three Tissue Layers Plus the Lens

Most of the diverse eye structures are formed from and are part of one of three tissue layers, or *tunics*. From the outermost to inner they are the *fibrous tunic*, the *vascular tunic*, and the *neural tunic* (or *retina*) (Fig. 9.17A). The *lens* is not part of the three tunics.

The Lens

The **lens** is a flattened spheroid composed of many layers of stretchy elastic fibers; it is suspended by minute *suspensory ligaments* in the anterior portion of the eyeball. As we will see further on, the lens bends (refracts) light to focus it on the retina. It is crystal-clear and free of blood vessels and lymphatics in order to maximize light transmission. A **cataract** is an opacified lens through which light cannot pass normally. The cause of most cataracts is unknown, but age is an important factor. Many diseases, most notably diabetes, are associated with cataract formation. Cataractous

lenses can be removed and replaced easily by artificial ones.

The Fibrous Tunic: Sclera and Cornea

The **fibrous tunic** is the tough, outer coat of the eyeball, which gives it shape and protects the delicate structures inside (Fig. 9.17A). In most of the eyeball, it is called the **sclera**, which is visible as the "white of the eye." At the very front of the eyeball the fibrous tunic forms the transparent *cornea*.

As noted earlier, the **cornea** is a dome of crystal-clear tissue at the anterior center of the eye. Like the lens, it contains no blood vessels or lymphatics that would otherwise obstruct light transmission. Along with the lens, the cornea bends light to focus it on the retina. Worldwide, a scarred or damaged cornea is among the most common causes of blindness. Corneal transplants are the most successful tissue transplants because the need is great, the operation simple, and there is little risk of immune rejection. Rejection is rare because immune cells travel in blood, and there are no blood vessels in the cornea.

Case Note

9.19. Helen's eyes were examined by numerous specialists, who would shine a light through the transparent part of the fibrous tunic. Name this part.

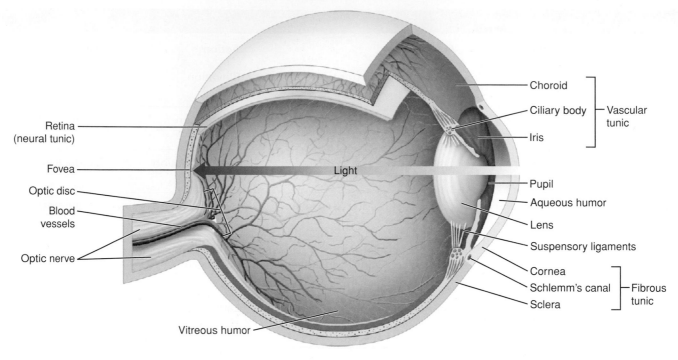

(a) The eyeball

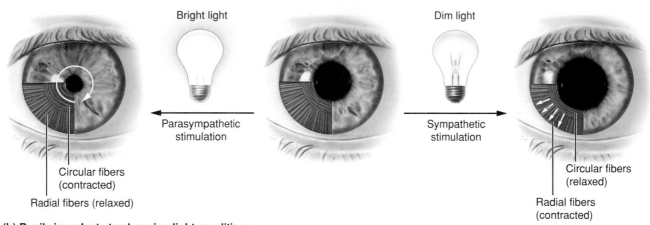

(b) Pipil size adapts to changing light conditions

Figure 9.17. Eye anatomy. A. The eye is composed of three layers, or tunics, plus the lens, and contains two types of fluid. **B.** One iris muscle contracts to dilate the pupil; the other contracts to constrict the pupil. *In dim light, which muscle fibers contract to enlarge the pupil—the radial fibers or the circular fibers?*

The Vascular Tunic: Choroid, Ciliary Body, and Iris

The **vascular tunic** (also called the *uvea*) is the middle layer and has three distinct parts: the *choroid*, the *ciliary body*, and the *iris* (Fig. 9.17A). The **choroid** is a highly vascular membrane that constitutes the posterior three-fourths of the vascular tunic; it provides oxygen and nutrients to the retina and sclera. The dark brown color of this layer is due to melanin production by choroidal melanocytes. Melanin absorbs light to avoid reflection and glare back into the retina. The degree of choroidal pigmentation is proportional to the degree of pigmentation of the skin; that is, people with dark skin have dark choroids, and their eyes are less sensitive to light than those of people with fair skin.

In the front part of the eye the vascular tunic forms the *ciliary body* and *iris*. The **ciliary body** is a doughnut-shaped ring of tissue that encircles the lens. It is composed of a rich network of blood vessels mixed with lesser amounts of fibrous tissue and smooth muscle fibers. Miniscule **suspensory ligaments** extend between the lens and the ciliary body. The ciliary body has two functions, both discussed in more detail further on—it produces some of the eye fluids and adjusts the shape of the lens to focus images on the retina.

Arising from the anterior edge of the ciliary body is the **iris** (the colored part of the eye), a thin, pigmented disc composed of fibrous tissue and smooth muscle fibers, which sits between the cornea and the lens. Eye color is largely determined by the amount of brown pigment in the iris. People with a large amount are brown-eyed; those with little are blue-eyed. Those with middling amounts have other colors, according to other features of the iris that disperse light.

In the center of the iris is the **pupil,** an opening that enlarges (dilates) or shrinks (constricts) to vary the amount of light that enters the eye. Two sets of smooth muscles control pupil size. Immediately encircling the pupil is a circular set innervated by parasympathetic nerve fibers that, as they contract, constrict the pupil (Fig. 9.17B). Outward is a radial set innervated by sympathetic nerve fibers that, as they contract, dilate the pupil. Light receptors in the iris gauge the amount of light striking the iris and constrict the pupil in bright light or enlarge it in dim light. However, other factors affect the pupil. For example, sympathetic stimulation enlarges the pupil as a part of the "fight or flight" reaction discussed in ← Chapter 8.

Case Note

9.20. Would Helen's pupils have reacted to bright light? If so, which set of muscle fibers would have contracted, and how would her pupil size have changed?

The Neural Tunic: The Retina

Innermost is the **neural tunic,** or **retina,** the image-sensing layer of the eyeball, which is the origin of the visual pathway (see Fig. 9.17A). The retina lines the inside of the eyeball and extends as far anteriorly as the edge of the ciliary body. The retina is called the neural tunic because it contains three layers of neurons. Some of these neurons are *photoreceptors,* specialized neurons that detect light; the others process and transmit signals originating in the photoreceptor layer. We investigate the properties of these neurons in greater detail further on.

The axons of the superficial neuron layer of the retina gather to form the **optic nerve (cranial nerve II),** which

transmits visual signals to the brain (see below). Where the optic nerve leaves the eye it forms the **optic disc,** a coin-shaped white spot on the retina. Light rays falling on the optic disc are not detected; this fact accounts for the phenomenon of the *blind spot* (see Remember This!, below). Blood vessels also connect with the eyeball at the optic disc and radiate outward to serve the retina. These blood vessels block light where they cross the retina, but the brain "accounts" for their presence and we do not normally perceive them.

The sharpest vision occurs in the center of the retina. Dead center, straight back from the cornea, is the **macula lutea** (Latin *macula* = "spot" + *lutea* = "yellow"), a small and highly sensitive part of the retina responsible for central vision. In the center of the macula is the **fovea centralis** (Latin *fovea* = "small pit"). It is about 1 mm in diameter, densely packed with the most sensitive photoreceptor cells, and is responsible for the sharpest vision. Retinal blood vessels approach the fovea from every direction, but they end at its edge, so that blood vessels do not cross the field of our sharpest vision. One of the most severe problems of diabetes is retinal disease caused by the growth of new retinal blood vessels, which prevent light from reaching the retina. Head or eye trauma or the vagaries of aging can cause the retina to detach from the vascular tunic, resulting in quick and permanent loss of vision unless immediately corrected by surgery.

Remember This! On a piece of paper, draw a cross (on the left) and a dot (on the right) about 7 cm (2.5 in.) apart. Holding the paper at a comfortable reading distance, close your left eye, stare at the cross, and simply notice the dot. Continue staring, and move further away from your picture. Eventually the dot will disappear, because the light rays from this area are falling on your optic disc.

The Eye Is Filled with Fluid

The eye is filled with two low-pressure fluids, which are separated by the lens (see Fig. 9.17A). Anterior to the lens is the **aqueous humor,** a clear, watery fluid secreted by the ciliary body. Aqueous humor flows from behind the lens, through the pupil, and into the space beneath the cornea; it provides oxygen and nutrients to the lens and cornea. The pressure created by its production and flow maintains *intraocular pressure,* normally 11 to 21 mm Hg, which helps the eyeball to remain "inflated" and retain its spherical shape. Aqueous humor drains out of the eye through tiny pores into **Schlemm's canal,** a duct that encircles the eye at the base of the ciliary body.

From Schlemm's canal, aqueous humor is absorbed into blood.

A variety of conditions can obstruct the flow or absorption of aqueous humor. The resulting increase in intraocular pressure can cause degeneration of the optic nerve and progressive blindness, a condition known as **glaucoma.** However, not all patients with increased intraocular pressure develop glaucoma, and some patients with normal intraocular pressure develop glaucoma apparently because their eyes are unusually sensitive to normal levels of intraocular pressure.

Posterior to the lens is a large chamber filled with **vitreous humor,** a transparent gel that is 99% water and 1% collagen fibers. It serves to transmit light, keep the eyeball inflated and stable, keep the retina firmly pressed against the choroid, and support the posterior surface of the lens.

Light Behaves as a Wave and as a Particle

By definition, **light** is a form of electromagnetic radiation detectable by the human eye. Other forms of radiation—the microwaves from your kitchen microwave oven, the x-rays at your doctor's office, the radio waves picked up by your automobile antenna—are also electromagnetic radiation, but they are not detectable by the human eye. Electromagnetic radiation (including light) has a peculiar property: it behaves both as a wave and as a particle. It helps to think of the particle as carrying or transmitting the wave, just as air carries sound waves.

In its behavior as a *particle,* light comprises packets of energy called a *photons*. Photons can be emitted from any molecule if energy is added. For example, if heat is added to metal, it first glows red as it emits low-frequency, low-energy photons. As more heat is added, it glows white (a mix of colors) and finally blue as it emits high-energy photons. Also, like other particles, photons can interact with other substances: the perception of light by the retina depends on the interaction between photon particles and molecules in the retina, which we discuss below.

In its behavior as *waves,* light has various frequencies, just like sound waves. For some waves, the distance between the peaks is short because there are many waves in a given length. These are high-frequency waves (also called short waves). They carry the most energy, and the eye interprets them as blue. For other waves, the distance between peaks is long; that is, there are fewer waves in a given length. These are low-frequency waves (long waves), which carry less energy. The eye interprets low-frequency waves as red. Wavelengths in between these convey other colors.

Sunlight and most artificial light is a mix of all wavelengths, from red to blue. That is to say, white light is a mix of all light wavelengths. The perceived color of cool objects depends on the color of light reflected from them. A red shirt is red because it absorbs all but low-frequency red wavelengths, which it reflects and we see as red. However, some objects can be heated to a point that they radiate light, not just reflect it. The blue of the cooking flame in your kitchen is due to the radiation of light from methane (natural gas) heated to a very high temperature.

Light Is Bent to Focus Images on the Retina

One feature of the wave behavior of light is that it bends as it passes from a substance of one density, such as air, to another at a different density, such as water. Bending, known as **refraction,** occurs at the interface between two materials. Curved surfaces can refract light rays toward or away from one another; the lens and the cornea bend light rays so that they converge. (A Clinical Snapshot, further on, shows how this property of light is exploited to correct visual defects.) Most of the light-bending power of the eye is in the cornea, but its shape is fixed; that is, it cannot be adjusted for near or distant visual fields. Further bending of light for perfect focus is the task of the lens (discussed next), which can change shape. The **focal point** is where rays meet that have been bent toward one another.

Remember This! **Refraction can be demonstrated easily by extending your index finger and placing it part way into a basin of water. If you look from the correct angle, your finger will appear to bend at the point where air meets water.**

Accommodation Is a Focusing Adjustment by the Lens

Accommodation is the ability of the lens to vary its thickness in order to precisely focus light rays from objects far or near on the retina. Below the level of consciousness, the brain perceives blurry images and modifies lens thickness to improve focus. The lens has far less focusing power than the cornea; it serves to "fine tune" focusing for the sharpest possible vision.

Recall that suspensory ligaments extend between the outer rim of the lens and the inner edge of the ciliary body, somewhat like the springs that hold a trampoline tight within its encircling frame. In the absence of external forces, the lens tends to assume a plump, ovoid shape. However, the suspensory ligaments maintain a constant outward tension that keeps it somewhat flattened.

The tension of the suspensory ligaments is controlled by the **ciliary smooth muscle** within the ciliary body. As the muscle relaxes, the ciliary body thins out into a larger circle. Tension in the ligaments increases and the lens flattens (Fig. 9.18A, right side). Contraction flattens the muscle and narrows the diameter of the circle, releasing tension on the ligaments and allowing the lens to assume a thicker, more spherical shape (Fig. 9.18B, right side).

Light rays arriving from objects more than 20 ft away are very nearly parallel to one another, so minimal refraction is required for distant vision. In this case, the ciliary muscle relaxes, the suspensory ligaments tighten, and the lens flattens (Fig. 9.18A, left side). However, as near objects become increasingly close, the rays from them become increasingly divergent and need to be refracted (bent) to a much greater extent (Fig. 9.18B, left side). In this case, the ciliary muscle contracts, the sus-

pensory ligaments go slack, and the lens rounds. Thus, the lens accommodates for precise focus by becoming thicker or thinner as necessary.

> *Remember This!* **Relaxation of the ciliary muscle increases tension in the suspensory ligaments, flattening the lens.**

If accommodation fails, the lens cannot refract light rays to converge on the retina, and the image will be blurry. The causes and treatments for these so-called *errors of refraction* are discussed in the nearby Clinical Snapshot titled *Disorders of Refraction*.

Also note in Figure 9.18 that images arrive at the retina upside down and with left-to-right reversal. The visual cortex inverts and reverses images so that we perceive the world correctly.

To summarize, in its trip through the eye, light first strikes the cornea. The curved shape of the cornea bends rays toward the center of the eye. Light then passes

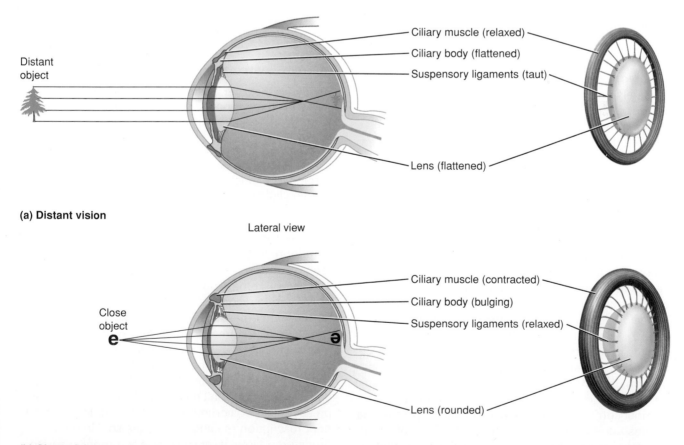

(a) Distant vision

Lateral view

(b) Close vision

Figure 9.18. Accommodation. A. The lens is flattened to lessen refraction when viewing distant objects. **B.** The lens is rounded to maximize refraction when viewing close objects (such as letters). *When the lens is rounded, is the ciliary muscle contracted or relaxed?*

CLINICAL SNAPSHOT

Disorders of Refraction

In the normal eye, objects beyond 20 ft are focused sharply on the retina when the ciliary muscle is relaxed. In this state there is tension on the suspensory ligaments that hold the lens in a slightly flattened shape. As objects move closer, the ciliary muscle contracts, the diameter of the ciliary body narrows, tension on the suspensory ligaments lessens, and the lens naturally thickens, becoming a bit more spherical to bend light rays to a greater degree and focus the image correctly on the retina. The point at which light rays converge is called the *focal point;* in the healthy eye, the focal point is always on the retina. However, this presumes that the eyeball is normally shaped and that the lens can easily change shape, which is not always the case.

If the eyeball is too long, the person cannot focus sharply on distant objects. Light rays meet in front of the retina and have diverged again (gone out of focus) by the time they reach the retina. However, the abnormally elongated eyeball has little effect on the ability of the lens to focus on near objects, so close vision is normal. This common condition is called **myopia,** or *nearsightedness.* Genes influence eyeball shape—myopic parents often have myopic children. Myopia is also more common in people who do a lot of near work, such as reading, writing, or drawing.

Just the opposite is true for people with a foreshortened eyeball. They cannot focus sharply on near objects because light rays fail to converge by the time they reach the retina. Instead, they come into focus behind the retina. However, the abnormally short eyeball has little effect on the ability of the lens to focus on distant objects, so the person's distant vision is normal. This condition is called **hyperopia,** or *farsightedness.*

Sometimes, however, misfocused images are due to irregularities in the curvature of the cornea. The normal cornea has the same curvature in every imaginary cross section. For example, visualize a clock face imposed on the cornea: the curvature across the cornea in the 1 o'clock to 7 o'clock line should be the same for the line through the cornea from 4 o'clock to 10 o'clock, and so on for every other line across the cornea. If the curvature is not uniform, the image is properly focused in some areas but blurred in others, a condition called **astigmatism.**

The most common of all visual problems is **presbyopia** (Greek **presbus** = "old man" + *opia* = "eye"), a condition that commonly develops as we age. Presbyopia is due to increasing inflexibility of the lens, which eventually leaves it "frozen" for distant vision and unable to focus on near objects. Recall that light rays from near objects are

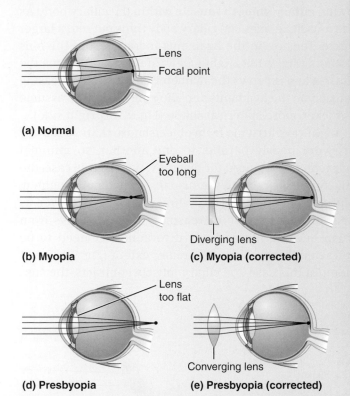

(a) Normal

(b) Myopia

(c) Myopia (corrected)

(d) Presbyopia

(e) Presbyopia (corrected)

Refractive disorders. A. In the normal eye, light rays are refracted to focus on the retina. **B.** The myopic eye is too long, so the focal point is before the retina. **C.** Myopia can be corrected with a lens that causes light rays to diverge. **D.** In presbyopia, the lens cannot bend light effectively, so the focal point is behind the eye. **E.** A converging lens functionally replaces the presbyopic lens, bending light to focus on the retina.

diverging as they reach the cornea and thus need to be bent more toward the center in order to meet at the retina. If the lens can no longer bend them, near objects will remain out of focus.

Refraction disorders can be treated with prescription lenses (either glasses or contact lenses) that bend light. For myopia, the prescription is lenses that diverge light, that is, bend rays away from center, so they won't meet in front of the retina. The opposite is true for hyperopia or presbyopia—the prescription is lenses that converge light so the rays won't meet behind the retina. Because presbyopia is so common and quite simple to correct, nonprescription reading glasses are sold in most drugstores. Now, laser surgery is commonly used to precisely reshape the cornea to bend rays more toward or away from the center in order to optimize light refraction and focus rays exactly on the retina.

through the aqueous humor and falls on the iris, which gauges the intensity of light and dilates or constricts the pupil to allow more or less light to pass. Rays then strike the lens, which further bends them toward the center. Rays then travel through the vitreous humor and come together at a focal point, which is ideally the surface of the retina.

Rods and Cones of the Retina Detect Light

Recall that the retina is composed of three layers of neurons (Fig. 9.19). Separating the neurons from the choroid is a layer of epithelial cells packed with melanin, the *pigmented epithelium* (or *pigmented layer of the retina*). The pigmented epithelium, like the choroid, absorbs light to prevent reflection back into the retina.

Somewhat surprisingly, the photoreceptors (light-sensing neurons) are not the first retinal cells to encounter light but lie beneath two other neuron layers against the pigmented epithelium. The two types of photoreceptors are called *rods* and *cones*. About 100 million rods are present, versus about 6 million cones. Rods and cones are not distributed uniformly throughout the retina—rods are concentrated mainly in the periphery of the retina, whereas cones are found in the center.

The neuronal layer superficial to the photoreceptors is composed of **bipolar cells,** neurons with two extensions. One extension (the dendrite) receives signals from the deepest neural layer, the rods and cones; the other extension (the axon) transmits signals to **ganglion cells,** the neurons on the retinal surface. The short dendrites of ganglion cells receive signals from the bipolar

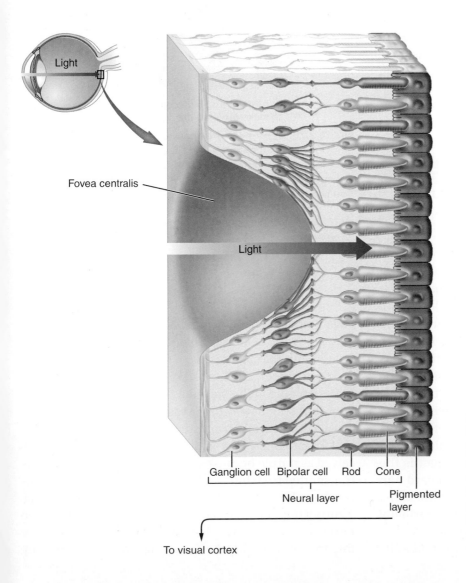

Figure 9.19. Photoreceptors of the retina. Except at the fovea centralis, light must pass through the ganglion cells and the bipolar cells to reach the photoreceptor cells—rods and cones. Rods and cones send their signals through bipolar cells to ganglion cells, whose axons run along the retinal surface and form the optic nerve. *Does a single bipolar cell synapse with multiple rods or multiple cones?*

cells; their long axons run across the surface of the retina and gather at the optic disc to form the optic nerve, which transmits visual signals to the brain. All in all, light must first traverse two nerve cell layers to reach light-sensing cells. However, bipolar and ganglion neurons and their axons are unmyelinated and sufficiently transparent that we do not perceive them.

Rods Discriminate Only Light or Dark

Rods are exquisitely sensitive to light: a single photon can stimulate a rod. They do not sense color but are mainly involved in separating shades of gray in dim light. They are most concentrated at the periphery of the retina and are responsible for night vision, motion detection, and peripheral vision, features that have high survival value in dangerous settings.

Although they are sensitive to the smallest amount of light, rods are not very discriminating; that is, they do not differentiate detail well. This poor discrimination occurs because *many* rods convey their signals through a *few* bipolar cells to a *single* ganglion cell—an arrangement that concentrates light from a large patch of retina to a single point in the brain. Thus peripheral vision is very sensitive to light, but at low resolution. If we had only rods, our view of the world would be grainy and bereft of color, like an old, low-resolution black-and-white movie.

Case Note

9.22. Is it likely that Helen's rods continued to react to light after she was blinded?

Cones Detect Color

Cones sense color but require bright light to function properly. Thus, they are responsible for detailed vision in brightly lit conditions. Cones are concentrated in the center of the retina, in the fovea centralis (see Fig. 9.19). Recall from our discussion above that multiple rods connect to a single ganglion cell. In contrast, in the fovea centralis, a *single* cone passes its signal through a *single* bipolar cell to a *single* ganglion cell, which conveys the information from that single cone to the brain. This process allows for the finest possible detail and sharpest image resolution. Adding to the effect is the fact that the axons of ganglion cells radiate outward—an arrangement that gets them out of the way of incoming rays and leaves the center of the fovea with a virtually unobstructed view. What's more, retinal blood vessels approach the fovea from every direction but end at its

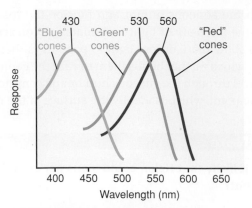

(a) Different light wavelengths activate different cones

(b) Red, green and blue combine to make other colors

Figure 9.20. Color perception. A. Color depends on wavelength; short-wavelength light is blue, long-wavelength light is red. Different wavelengths activate the three different cone types to different degrees, resulting in the perception of blue, green, and red. **B.** The brain combines the input from blue, green, and red cones to perceive a specific color. *Which "color" activates red, green, and blue cones equally?*

edge, so that nutrients and oxygen reach the most critical area by diffusion and blood vessels do not cross the field of our sharpest vision.

There are three types of cones, called *blue, green,* and *red,* according to the wavelength (and thus the color) of light that stimulates them most vigorously (Fig. 9.20A). Color vision results from stimulation of various combinations of cones. For example, a purple object is perceived as purple because the brain combines the signals transmitted from blue and red cones, which were stimulated by blue- and red-frequency light waves (Fig. 9.20B). Thus, the brain behaves much like an artist, who mixes basic colors to achieve more complex colors. Prove this to yourself with the following simple experiment:

shine red light into one eye and green into the other and you will see yellow as a result of integration of the signals by the brain. Incidentally, when all three types of cones are equally activated, you see white.

> *Remember This!* **Cones provide higher image resolution than rods because one cone communicates with one ganglion cell, but many rods communicate with one ganglion cell.**

Photons Activate Rods and Cones

Before we consider how rods and cones detect light, a brief discussion is in order about the electrical properties of these unique cells. You may want to review the discussion of electrical signaling in ← Chapter 4 before proceeding. Recall that an action potential is an all-or-none phenomenon, whereas graded potentials are of variable strength. Recall also that the interior of all body cells is negatively charged, and that the interior can become less negative (depolarization) or more negative (hyperpolarization).

Rods, cones, and bipolar cells, unlike ganglion cells, do not fire action potentials because they lack voltage-gated sodium channels. Instead, they have *graded potentials*—local changes in membrane potential that can be depolarizing or hyperpolarizing, small or large. The amount of neurotransmitter released by rods, cones, and bipolar cells varies in proportion to graded changes in the membrane potential—depolarization increases neurotransmitter release; hyperpolarization decreases it.

In their resting (dark) state, the membrane potential of rods and cones is about -30 mV (compared with about -65 mV in most other body cells) (Fig. 9.21A). This partially depolarized state is maintained because sodium channels are open, permitting sodium (a positive ion) to "leak" continuously into the negatively charged interior of the cell. This depolarizing flow of sodium is called the *dark current* because it occurs in the absence of light striking the receptor cell. The partially depolarized state of resting rod and cone cells stimulates continual release into their synapse with bipolar cells of an *inhibitory* neurotransmitter called *glutamate*. Glutamate inhibits bipolar cells by hyperpolarizing them. Bipolar cells cannot stimulate ganglion cells until they are released from this restraint. In the meantime, ganglion cells wait passively for stimulation by bipolar cells and send no signals to the brain. Thus, in the total absence of light, the brain perceives nothing but darkness. This state continues until a ray of light releases bipolar cells from the paralyzing effect of photoreceptor cell inhibition created by the sodium "leak."

Phototransduction is the process by which photons initiate events in a photoreceptor cell that culminate in the firing of an action potential by a ganglion cell. Here's how it happens. Each rod and cone contains a certain type of *photoreceptor protein*, a molecule that changes shape and breaks down when it absorbs a photon. **Rhodopsin,** the photoreceptor protein found in rods, is activated by light of any wavelength (color). Cones, on the other hand, contain one of three types of **photopsins,** which react to red, green, or blue light.

The reaction of rhodopsin or photopsin with a photon of light initiates a cascade of chemical reactions that closes the sodium channels of rods and cones and hyperpolarizes them. Notice that the reaction is *hyperpolarization* (a *more negative* cell interior) of the rod or cone, not *depolarization* (Fig. 9.21B). This hyperpolarization reduces glutamate release, allowing the bipolar cell to depolarize and stimulate a ganglion cell, which in turn sends an action potential racing to the brain with the message "light."

The intensity of the light (that is, the number of photons) determines the degree of photoreceptor cell hyperpolarization: more photons stimulate the breakdown of more rhodopsin or photopsin molecules, thus closing more sodium channels and hyperpolarizing the cell to a greater extent. The corresponding reduction in glutamate release depolarizes the bipolar cell more, causing ganglion cells to fire more action potentials, which is interpreted as more intense light by the brain.

The entire pathway rapidly resets itself, so that rhodopsin is regenerated to react again. That's why we can perceive a rapidly moving vehicle as moving: we perceive the continual arrival of new photons reflected from the vehicle as it moves from one position to another. The "resetting" of rhodopsin depends on vitamin A. Vitamin A deficiency can result in *night blindness,* (impaired vision in low light) because it interferes with the ability of each rod to perceive sequential photons of light.

Our eyes also adapt—at varying speeds—to darkness and bright light, as discussed in the nearby Basic Form, Basic Function box titled *Light Adaptation.* It is quite amazing, really: the simple and transient hyperpolarization of a photoreceptor cell allows us to perceive the minute gradations of shading that endow sight with its rich color and vivid motion.

> *Remember This!* **In the presence of light, rod cells hyperpolarize, bipolar cells depolarize, and ganglion cells fire action potentials.**

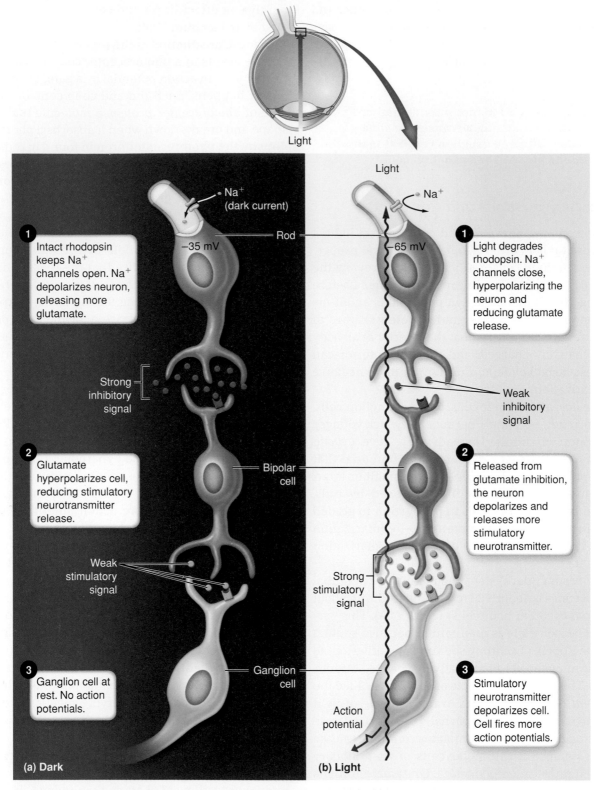

Figure 9.21. Photoreception. This figure illustrates photoreception in rods. Photoreception in cones occurs by a similar mechanism. **A.** The dark state. **B.** The light response. *If a photoreceptor cell is depolarized, will the action potential frequency in the ganglion cell be increased or decreased?*

BASIC FORM, BASIC FUNCTION

Light Adaptation

A daytime trip to a movie theater provides a lesson in visual adaptation to bright and dark environments. Walk into the theater from the sunlit lobby and it is difficult to see people and seats; walk out into the sunlight after the feature and the glare is almost blinding. But in each case the pupils dilate or constrict as the eyes adapt (accommodate) quickly to the new (changed) lighting condition. How so? Pupil reaction is just the beginning. The real story involves changes in the retinal photoreceptors themselves.

In bright light, rods and cones become less sensitive to light, an adaptation that has survival value—glare from bright light impairs clear vision. They decrease intracellular calcium levels, which interferes with the visual signaling pathway. They sequester or inactivate some of rhodopsin and other visual pigments where light cannot access them. Light rays find fewer photopigment molecules, and those they do find need more photons to activate phototransduction. As we walk from daylight into a darkened theater, the light level plummets, but it's not totally dark. Nevertheless, we lack the light sensitivity to pick up the few light rays that are there. As a result, we're almost blind for a while. Dark adaptation is relatively slow; our "night" vision improves over the course of 30 minutes, as calcium levels slowly rise and photopigments accumulate. This increased sensitivity to

Visual Adaptation. Over time, we can see in a dim theater because of visual adaptation.

every stray light ray enables us to see our popcorn and our neighbors.

In contrast, as we walk out into the sunlight after the feature, our rods are loaded with rhodopsin, our cells are exquisitely sensitive to light, and thus we are almost blinded by the glare. Fortunately, light adaptation occurs relatively quickly as the calcium concentration and photopigment availability fall to the reduced levels suited for bright light.

Case Note

9.23. Assuming that Helen's photoreceptor cells were normal, would her bipolar cells have been depolarized or hyperpolarized in the dark?

Depth Perception Is a Product of Binocular Sight

Sight from a single eye conveys no perception of depth. For depth perception, sight from both eyes—*binocular vision*—is required. Depth perception is due to the fact that the eyes each see an object from a slightly different perspective: they are separated by about 65 mm (2.5 in.).

Binocular vision produces two images, one from the right eye and one from the left. If the brain does not automatically adjust the eyes so that the images overlie

exactly the same area of each retina, "double vision" occurs. As we move closer to an object that we're viewing, our eyes automatically rotate medially toward the nose. Called **convergence,** this automatic rotation assures that the images strike each retina in exactly the same manner. This, in turn, assures a single image. When objects are far away, the lines of sight are parallel and no convergence is required.

Convergence plays an essential role in our perception of depth. The brain calculates the distance of an object by triangulation based on the degree of inward rotation of the two eyes: the more inwardly rotated, the closer the object is perceived to be. The base of the triangle is a line between the eyeballs, and the sides of the triangle are imaginary lines drawn from the lens of each eye to the object being viewed. The angle at each end of the base line is the degree of convergence. It's really just high school trigonometry to calculate the distance.

The Visual Pathway Ends in the Occipital Cortex

One of the better examples of the relationship of form and function is reflected in the anatomy of the optic nerve (cranial nerve II) and the processing of visual sensory signals. The optic nerve from each eye leaves the back of each eyeball and converge at the base of the brain to form a large X (the **optic chiasm**, Greek *khiazein* = "to mark with an X") before joining the brain, an unusual arrangement not found with any other set of cranial nerves (Fig. 9.22).

To understand how this anatomical layout promotes binocular vision, imagine this scene in front of your eyes: a tree to your left, a car to your right, with each at the edge of your peripheral vision. Now trace the paths of the light rays as they pass through your eyes to your retina and you will discover that the rays from the tree on your *left* strike the *right* side of *both* retinas, and the rays from the car on your *right* strike the *left* side of *both* retinas.

It is apparent, then, that for these images to be brought together for integration, signals from the *left* side of both retinas must go to *one place* and the images from the *right side* of both retinas must go to *another*. This means that nerve tracts from the optic nerves must separate and go to two different locations. And, indeed, that is just what happens—axons from each eye separate at the optic chiasm. Signals from the left side of both retinas (signals from the car on your right) are routed to the left occipital cortex, and signals from the right side of both retinas (signals from the tree on your left) are routed to the right occipital cortex. Each axon of the optic nerve synapses in the thalamus with secondary neurons called *projection fibers*, which carry the visual signals to the primary visual cortex in the occipital lobes of the brain.

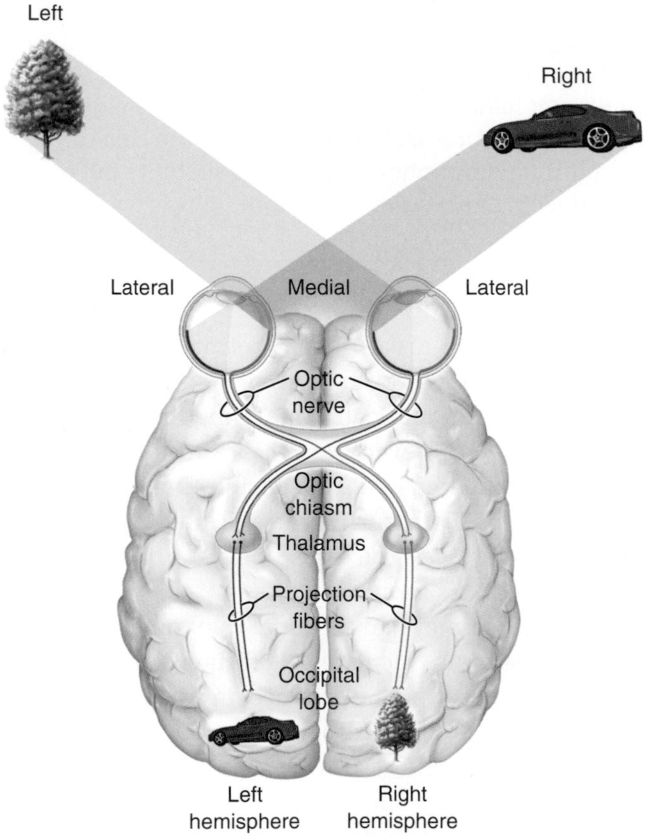

> **Remember This!** Neurons from the medial side of the eye (closest to the nose) cross over in the optic chiasm; neurons from the lateral side do not.

Case Note

9.24. Is it reasonable to speculate that Helen's blindness was due to damage to her occipital cortex?

9.32 Are tears produced by the lacrimal sac, lacrimal gland, or lacrimal canal?

9.33 Rewrite this statement to make it true by replacing the *underlined* terms. The sclera is part of the *vascular* tunic; the iris is part of the *neural* tunic.

9.34 Which layer of the neural tunic covers the posterior iris?

9.35 Name the ligaments attached to the lens.

9.36 Where would you find the most photoreceptors—in the macula lutea or the optic disc?

9.37 Which task requires more refraction—reading or watching a hawk in the sky?

9.38 As you read these words, are your ciliary muscles contracted or relaxed?

9.39 Which neuron type is farthest away from the pigmented retinal layer—ganglion cells or bipolar cells?

9.40 Name the retinal region in which light contacts photoreceptor cells without passing through other cells.

9.41 When does the greatest convergence occur—when the lens is rounded or when it is flattened?

Figure 9.22. Visual pathway. Regardless of the eye involved, information from the left visual field is transmitted to the right cerebral hemisphere and vice versa. *In the left eye, which neurons will cross over at the optic chiasm—neurons from the left half or the right half?*

Loss of Sensory Function: The Case of Helen Keller

In her time Helen Keller was the most famous and accomplished deaf-blind person in world history. Before reading this discussion of her case, you may want to revisit the case of Larry H. in ◄ Chapter 8 for comparison. Recall that Larry's brain damage affected the brain's ability to integrate sensory signals, not its ability to receive them.

The perception of any external change (a *stimulus*) requires normal functioning of the sensory organ, the sensory receptor, the nerve pathway, and the relevant brain region (Fig. 9.23). Helen was able to receive and interpret signals from numerous sensory modalities. She could smell mimosa flowers, feel cool water on her palm, and climb trees without concern for her equilibrium. Helen's olfactory receptors, tactile receptors, and

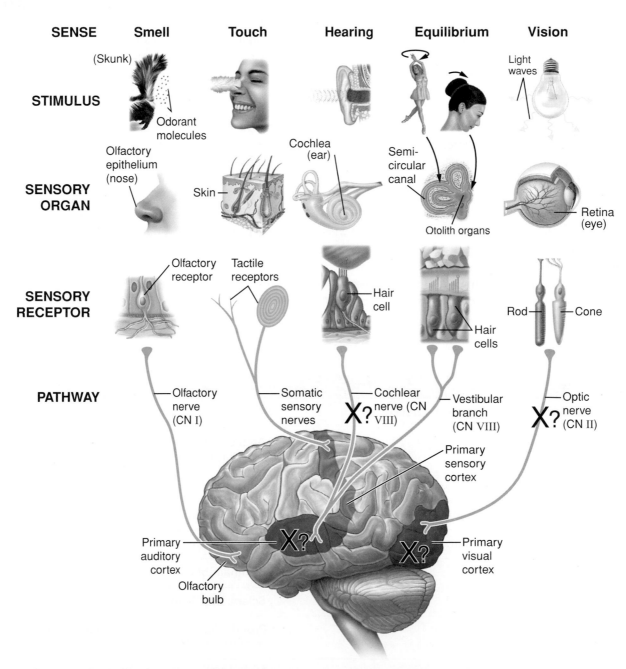

Figure 9.23. Sensation and Helen Keller. Helen's sensory organs and sensory receptors were, in all likelihood, intact. However, signals originating in the hair cells of the cochlea and the photoreceptors of the retina failed to reach the brain. Taste, which was not affected by Helen's disorder, is not illustrated. Note that the pathways are simplified. *Name three stimuli that activate hair cells.*

hair cells of her otolith organs were thus intact, as were the pathways and brain regions transmitting and interpreting these signals. Only two senses—hearing and vision—were lost. Although we cannot be certain, the evidence is overwhelming that her sensory organs (the cochlea and retina) were unaffected. The acute feverish illness that struck her at age 2 must have somehow damaged her cranial nerves or the brain itself. Meningitis (infection of the meninges) or encephalitis (infection of the brain itself) could have caused inflammatory swelling of the brain, damaging both her visual and auditory cortices. Alternatively, the increased intracranial pressure associated with inflammation may have "crushed" the optic and vestibulocochlear nerves.

Despite profound disability, Helen Keller became famous for her intellect, writing skill, and humanitarian activities. To study her life is to confront what it means to be human; that is, what it means to have a mind. Recall from our discussion in Chapter 8 that consciousness is difficult to define but clearly depends on the brain integrating signals from sensory receptors into a composite "picture" of the environment and our place in it. If an individual's two most sophisticated senses are disabled, yet the individual retains superior cognition, a rich imagination, and a remarkable ability to make sense of the world in the same way that hearing and sighted people do, then upon what, precisely, does consciousness depend?

For much of her life, Helen was tethered to her tutor, Annie Sullivan, with such closeness that some people accused Helen of being an imposter and Annie of being a ventriloquist. The accusation prompted a variety of questions that reflected an ongoing debate about the nature of consciousness: Was Helen Annie's stage dummy, a counterfeit copy of Annie's mind, or did she have a mind of her own? Whose mental "picture" of reality was it: Helen's or Annie's?

The matter was gradually settled as Helen learned to read Braille ← (see the History of Science box in Chapter 4 titled *Braille: A Reading Code for the Blind*) and to speak. These accomplishments enabled Helen to become gradually less dependent on Annie. As a mature adult, Helen outlived Annie and flourished in her absence, once again proving that she had a mind of her own. Helen knew of the controversy and provided the best answer to her detractors, saying, "Blindness has no limiting effect on mental vision." Nevertheless, the ancient debate goes on about the nature of consciousness: Is it merely sensation and perception, or is there something more?

What do you think—are we more than the sum of the signals we receive from our sensing receptors? Is our conscious "picture" of reality merely a paint-by-the-numbers replica of the outside world, or does thought—that internal dialogue with ourselves—create something more? What is the role of memory? Of imagination? In short, what is the hidden element that the mind adds to sensation? Keller's life seems proof that the mind, too, has an eye.

Word Parts

Latin/Greek Word Parts	English Equivalents	Examples
aque/o	Water	Aqueous humor: watery humor filling the eyeball anterior to the iris
-esthesia	Sensation	Anaesthesia: lack of (an-) sensation
-algesia	Pain	Analgesic: drug that inhibits pain
lacrim/o	Tear, lacrimal apparatus	Lacrimal gland: produces tears
ophthalm/o	Eye	Ophthalmologist: eye doctor
scler/o	Hard	Sclera: the "hard" (fibrous) layer of the eye
proprio-	Self	Proprioception: sense (-ception) of self (the position of one's own body parts)
noci-	Pain	Nociception: sensation of pain

Word Parts (continued)		
Latin/Greek Word Parts	**English Equivalents**	**Examples**
naso-	Nose	Nasolacrimal duct: drains tears (lacrim) into the nose
-ception	Receive, sense	Nociception: sensation of pain
conjunctiv/o	Conjunctiva; to join together	Conjunctiva; membrane that joins together the eyelid and eyeball
photo-	Light	Photoreceptor: detects light
vitreo-	Glass	Vitreous humor: viscous (glassy) humor filling the eyeball posterior to the iris

Chapter Challenge

CHAPTER RECALL

1. **Prolonged pressure would be most accurately sensed by the**
   a. Merkel receptors.
   b. Meissner corpuscles.
   c. Ruffini corpuscles.
   d. Pacinian corpuscles.

2. **Nociceptors detect**
   a. mechanical pressure.
   b. pain.
   c. chemicals.
   d. light.

3. **Proprioceptors**
   a. include the semicircular canals.
   b. sense head position relative to gravity.
   c. sense the relative position of body parts.
   d. are responsible for the sense of hearing.

4. **Muscle spindles are**
   a. collagen strands interlaced with neuron terminals.
   b. the sensory organs of equilibrium.
   c. the sensory receptors that measure muscle stretch.
   d. specialized muscle fibers that produce force.

5. **Fungiform papillae**
   a. are found everywhere on the tongue.
   b. always contain taste buds.
   c. are restricted to the back of the tongue.
   d. are involved in olfaction.

6. **A gustatory receptor cell**
   a. usually hyperpolarizes when stimulated by a particular tastant.
   b. detects one of five primary tastes.
   c. detects one of 350 different tastants.
   d. can undergo cell division to generate new gustatory receptor cells.

7. **Olfactory receptor cells**
   a. are found entirely within the skull.
   b. have cell bodies within the skull but axons that project outside of the skull.
   c. have cell bodies outside the skull but send axons inside the skull.
   d. are found entirely outside the skull.

8. **An odorant receptor is a(n)**
   a. modified first-order sensory neuron.
   b. modified second-order neuron.
   c. membrane protein.
   d. none of the above.

9. **Which of the following sensory signals does *not* pass through the thalamus?**
   a. Vision
   b. Olfaction
   c. Hearing
   d. Gustation

10. **The stapes connects the**
    a. tympanic membrane to the malleus.
    b. malleus to the incus.
    c. incus to the oval window.
    d. oval window to the round window.

11. **The wall of the cochlear duct is part of the**
    a. vestibular apparatus.
    b. membranous labyrinth.
    c. bony labyrinth.
    d. organ of Corti.

12. **Alternating layers of compressed and decompressed air are detected as**
    a. light.
    b. smells.
    c. tastes.
    d. sound.

13. **Hair-cell cilia involved in hearing are embedded in the**
    a. basilar membrane.
    b. tectorial membrane.
    c. tympanic membrane.
    d. vestibular membrane.

14. **The utricle**
    a. is filled with perilymph.
    b. detects head rotation.
    c. is oriented perpendicularly to the force of gravity.
    d. contains the crista ampullaris.

15. **The cupula is**
    a. a cluster of hair cells in the semicircular canals.
    b. a gelatinous membrane containing otoliths.
    c. a gelatinous cap within an ampulla.
    d. an enlarged area of the membranous labyrinth.

16. **Hair cells are found in the**
    a. saccule.
    b. nasal epithelium.
    c. retina.
    d. all of the above.

17. **The conjunctiva covers**
    a. the inner surface of the eyelids.
    b. the cornea.
    c. the posterior surface of the eyeball.
    d. all of the above.

18. **Peering upward to watch an airplane is accomplished by the**
    a. superior oblique.
    b. superior rectus.
    c. orbicularis oculi.
    d. lateral rectus.

19. **Tears drain from the lacrimal sac into the**
    a. lacrimal gland.
    b. nasolacrimal duct.
    c. lateral canthus.
    d. lacrimal canal.

20. **The vascular tunic forms the**
    a. choroid and iris.
    b. ciliary body and cornea.
    c. iris and retina.
    d. sclera and ciliary body.

21. **Vitreous humor**
    a. fills the space between the lens and the iris.
    b. is only found posterior to the lens.
    c. is drained via Schlemm's canal.
    d. fills the space between the lens and the cornea.

22. **Contraction of the ciliary muscle**
    a. increases the tension on the suspensory ligaments.
    b. constricts the pupil.
    c. facilitates close vision.
    d. stimulates aqueous humor production by the ciliary body.

23. **The optic disc**
    a. contains many cones.
    b. appears as a white spot on the retina.
    c. contains the fovea centralis.
    d. A and C.

24. **In a lit room,**
    a. rods are hyperpolarized, bipolar cells are depolarized, and ganglion cells fire action potentials.
    b. rods are depolarized, bipolar cells are hyperpolarized, and ganglion cells do not fire action potentials.
    c. rods are depolarized, bipolar cells are depolarized, and ganglion cells fire action potentials.
    d. rods are hyperpolarized, bipolar cells are hyperpolarized, and ganglion cells do not fire action potentials.

CONCEPTUAL UNDERSTANDING

25. **Explain the difference between a sensory receptor and a hormone receptor.**

26. **Explain why we can detect 10,000 different odors using only 350 different olfactory receptor cells.**

APPLICATION

27. **You are babysitting a 2-year-old who suddenly emits a loud, high-pitched, piercing wail of alarm.**
    a. Describe the amplitude of the resulting sound wave—will it be small or large?
    b. Describe the frequency of the resulting sound wave—will it be high or low?
    c. Which portion of the basilar membrane will resonate most strongly—near the oval window or near the apex of the cochlea?

28. **You are scuba diving in the Caribbean. Name the sensory receptor cell(s) responsible for each of the following sensations.**
    a. Spotting a seahorse
    b. Knowing how quickly you are descending to the ocean floor
    c. Identifying the position of your left flipper
    d. Perceiving the sound of the dive boat speeding away

You can find the answers to these questions on the student Web site at **http://thepoint.lww.com/McConnellandHull**

10

Blood

Major Themes

- The cellular elements of blood have a short life span and require continuous replacement.

- White blood cells play a critical role in defending the body against microbes and other foreign substances.

- Red blood cells transport oxygen from the lungs to tissues and return carbon dioxide.

- Platelets are cell fragments critical in the prevention and control of hemorrhage.

- Certain plasma proteins are important in body defenses, in the transport of essential substances, in maintaining blood volume, and in blood clotting.

- Numerous nutrient and waste substances are dissolved in plasma.

Chapter Objectives

7. Describe the life cycle of a red blood cell.

8. Discuss the diagnosis and possible causes of anemia.

9. Discuss the pathway that results in increased red cell production in individuals living at high altitudes.

Platelets 392

10. Describe the regulation and mechanism of platelet synthesis and the problems associated with thrombocytopenia.

Hemostasis 393

11. Use diagrams to explain the three steps of hemostasis.

12. Use flowcharts to illustrate the tissue factor pathway, contact activation pathway, and common pathway of blood clotting, and explain how two different anticoagulants work.

13. List three differences between coagulation and thrombosis.

Blood Groups and Transfusion 398

14. Determine which blood group can be transfused into patients with A, B, O, or AB blood, and explain why agglutination reactions occur.

Bone Marrow Failure: The Case of Eleanor B. 403

15. Use the case study to explain the central role of bone marrow in blood cell production and to discuss the roles of erythrocytes, platelets, and leukocytes.

"My life is running out of my nose!"

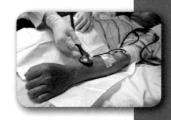

As you read through the following case study, assemble a list of the terms and concepts you must learn in order to understand Eleanor's case.

Clinical History: Eleanor B., a 52-year-old professor of anthropology, presented to the emergency room of a large metropolitan hospital with a severe nosebleed. "I never have nosebleeds. This takes the cake; it just won't stop. My life is running out of my nose!"

Questioning by hospital staff revealed that 8 years earlier she had surgery for breast cancer. Physicians had followed her closely until 3 years earlier, when she divorced and moved to her current job in a new city. "I'm embarrassed to admit that I haven't seen a doctor in three years," she said. "I've just been too busy to have my regular checkups." Further questioning revealed nothing medically unusual. She mentioned, however, having felt unusually tired in the last few months. "I seem to wear out at even the smallest tasks. Last week I stopped for a rest on a park bench on my way home. That's never been necessary before."

Physical examination and other data: Eleanor was pale, and her skin contained numerous pinpoint hemorrhages. Otherwise her physical examination was unremarkable. Blood analysis revealed a marked deficiency in red blood cells, white blood cells, and platelets. Her blood type was determined to be O positive.

Clinical course: In the emergency room her nose was packed with cotton strips and she was transfused with platelets, which stopped the nosebleed. She was admitted to the hospital and transfused with red blood cells. Knowing that breast cancer has a tendency to spread to bones, the examining physician suspected that cancer cells had taken over her red bone marrow, the site of blood cell production. This theory was confirmed by a bone marrow biopsy, which showed nearly complete replacement of normal bone marrow by cancer cells.

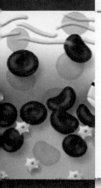

She was treated with additional chemotherapy but continued to need red blood cell transfusions to maintain hemoglobin near the normal level, and she required antibiotic treatment on several occasions for bacterial infections—pneumonia, skin abscesses, and recurrent diarrhea. Nine months later she was brought to the emergency room by ambulance for severe bloody vomiting. She was pale and confused, with blood pressure 60/20 mm Hg (normal: 120/80) and heart rate 140 beats per minute (normal: 70). Despite heroic efforts to save her, her heart stopped and could not be restarted.

Lab studies from blood collected before her death showed the counts of red and white blood cells and platelets to be very low. *Staphylococcus aureus*, a bacterium, was cultured from her blood. At autopsy, the bone cavities normally containing red marrow were filled with tumor; little normal marrow remained. She was also found to have severe bacterial pneumonia and an extensive fungal infection in her esophagus. The latter had produced a large esophageal ulcer, which was the source of her fatal hemorrhage.

Our earliest ancestors recognized that blood was vital to life: whether gushing from man or mastodon, it was a warm, red, sticky fluid that carried away with it the victim's life force. Honoring its importance, ancient peoples used blood in sacrificial rites, and the color of blood has come to symbolize valor and vitality. Eastern European folklore describes vampires—creatures who, by drinking fresh blood, could stave off death. It's not such a far-fetched idea—even today we take blood from one person to give to another for that very purpose. So what, exactly, is this life-sustaining fluid that we hold so dear?

Don't worry; the bleeding always stops.

Advice from older surgeons to worried younger ones.

Overview of Blood

Blood is the fluid circulated by the heart through the blood vessels. Blood cells are made by the bone marrow and released into blood.

Recall from ◀ Chapter 3 that we defined blood as liquid connective tissue. It fits that classification for two reasons: (1) functionally, it connects various parts of the body by carrying chemical signals, fluids, and nutrients from one place to another; (2) structurally, it contains cells and a large amount of extracellular matrix (plasma),

which in the case of blood, is liquid rather than solid. In addition, like cells in solid connective tissues, blood cells require nutrients, produce wastes, die, and are replaced by new cells. But although classified as connective tissue, blood has certain distinctive properties that set it apart from other tissues:

● Blood cells are in continuous motion.
● The life span of blood cells is unusually short, varying from a few hours to a few months.
● Blood is red—bright red if well oxygenated (on the way to tissues from the lungs), dark reddish-blue if

oxygen-depleted (on the way back to the lungs from tissues).

- Although it is a fluid, blood is thicker (more viscous) than water because it contains proteins and cells; it feels slightly oily and sticky for the same reason.
- Blood is slightly alkaline: pH averages about 7.40 (water is 7.0).
- Blood accounts for about 7% of body weight—about 5 kilograms (5 liters or 6 quarts of volume) in a 70-kg (155 lb) person.

Before we explore the structure of blood in detail, let's take a look at its functions.

Blood Has Five Main Functions

In the broadest sense, blood is a transportation system, truly an "inner river" that carries the essentials of life to tissues, returns metabolic wastes for elimination, and conveys chemical messages (Fig. 10.1). In particular, the functions of blood are:

- *Transport.* Blood transports gases (such as oxygen), nutrients, waste products, and chemical messages between organs, tissues, and cells.
- *Hydraulic force.* This is the pressure created by fluid flowing in a closed space. Blood propelled out of the heart into blood vessels creates a form of hydraulic pressure called *blood pressure*. Without adequate blood pressure, human life is impossible—the vital substances blood transports will not reach distant tissues. Moreover, blood pressure enables the kidney to make urine and provides the hydraulic force for male and female erection.
- *Defense.* Blood cells and other blood-borne substances defend against threats originating both externally (microbes) and internally (cancer).
- *Heat transfer.* The circulation of blood through the skin exposes blood to skin temperature, which is normally much cooler than the core temperature of organs, where the metabolism of nutrients generates heat. Warm blood passes through skin, heat is lost, and cooler blood returns to the body core.
- *Prevention of blood loss.* Blood cells and elements form short-lived blockages in damaged vessels in order to prevent blood loss following tissue injury.

Until about 150 years ago, few physicians had more than a vague understanding of these functions and their relationship to health. Indeed, their fanciful theories of illness sometimes led them to bleed patients of large quantities of blood in a misguided effort to promote

Figure 10.1. Blood functions. The many functions of blood, as illustrated by a wounded hiker. *The hiker's face is flushed, reflecting increased blood flow. How does this increased superficial blood flow help him?*

healing. Sometimes this worsened the patient's illness or led to the loss of the patient's life, as we discuss in the accompanying History of Science box, titled *The History of Blood Transfusion.*

Blood Is Composed of Plasma and Cellular Elements

As we just noted, blood is composed of two main compartments: a fluid extracellular matrix called *plasma* and the solid *cellular elements* suspended within it. The cellular elements are whole blood cells and fragments of cells called *platelets*. Clinicians refer to plasma and its cellular components together as *whole blood*. Because the cellular elements are heavier (denser) than plasma, we can separate the two by *centrifugation* in a tube—the heavy cellular elements settle to the bottom of the tube

THE HISTORY OF SCIENCE

The History of Blood Transfusion

The circulation of blood in blood vessels through the body had been worked out correctly in the 13th century by an Arab physician named Ibn Al-Nafis. He rejected the prevailing view, originating in the second century with the Greek physician Galen, that bright red arterial blood originated in the heart and was consumed by the tissues, and that dark venous blood originated in the liver and served another purpose. Ibn Al-Nafis's findings were not known in Europe for centuries. Thus, the idea of blood flowing in blood vessels was not understood in the West until English physician William Harvey discovered the facts in 1628.

Even before physicians understood the anatomy and physiology of the circulatory system, however, they made crude attempts to infuse blood into the sick. The first recorded attempt occurred in 1492, as Columbus was discovering the Americas. Pope Innocent VIII lay comatose in Rome and physicians attempted to rescue him by collecting blood from three 10-year-old boys and infusing it into the pope. Since the concept of blood circulating in veins had not yet been established, the blood was infused into his mouth. It was not a success—the Pope and all the boys died.

In 1665, British physician Richard Lower devised instruments to shunt blood between surgically joined dogs. He observed that the ill effect of blood loss on one dog could be reversed by shunting the blood of a second dog into the first.

A few years later, in 1668, Dr. Jean-Baptiste Denys, personal physician to King Louis XIV of France, transfused a small amount of sheep's blood into a 15-year-old boy. The boy survived and the experiment was counted a success, but modern physicians recognize that his survival was likely due to the small amount of blood transfused. One of Dr. Denys's less fortunate patients survived two such transfusions but died after the third, amid great controversy.

The first human-to-human intravascular blood transfusion would not occur for about another 150 years. In 1818, Dr. James Blundell, a British obstetrician caring for a woman who was hemorrhaging after

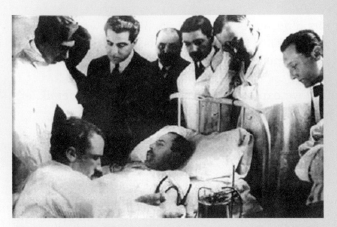

Blood transfusions. Direct person-to-person transfusions.

childbirth, recruited the patient's husband as a donor and extracted four ounces (about 120 mL) of blood from the man's arm to transfuse into his wife. In the ensuing years, Dr. Blundell performed 10 additional transfusions, 5 of which were judged beneficial.

For another 100 years, all attempts at blood transfusion were direct: from the body of one person directly into another. Two advances—blood banking and compatibility testing—were required before indirect blood transfusions could be widely used. In about 1918, perhaps stimulated by the awful carnage of World War I (1914–1918), it was discovered that blood could be anticoagulated, refrigerated, and stored ("banked") for a few days before transfusion. But blood banking was of limited use until the concept of donor and recipient blood compatibility came to be understood. This leap forward can be attributed to the work of the Austrian scientist Karl Landsteiner. He discovered ABO blood groups in 1900, and, with his colleague Alexander Wiener, discovered the Rh blood group in 1937. With this discovery, all of the pieces were at last in place and modern transfusion practice spread widely.

beneath the lighter liquid at the top (Fig. 10.2A). When we do this, here's what we see:

- At the top of our tube is **plasma,** the liquid part of blood, which accounts for about 55% of blood volume. It is a golden, syrupy mix composed mainly of water in which are dissolved proteins, nutrients, minerals, and other life essentials.
- At the bottom of the tube are *red blood cells* (clinically referred to as *erythrocytes, erythro* = "red"), the heaviest and most numerous cells. They normally account for about 45% of blood volume, a percentage called the **hematocrit** (Greek *haima* = "blood"; *krites* = "judge", as in someone who *separates* things, such as right from wrong).
- At the interface between plasma and red cells is a thin, tan layer, the **buffy coat.** This layer contains cells that are not as heavy as red cells—*white blood cells* (clinically referred to as *leukocytes, leuko* = "white"): these include monocytes, lymphocytes, neutrophils, eosinophils, and basophils. The buffy coat also contains *platelets*, fragments of a bone marrow cell called a *megakaryocyte.* The buffy coat normally accounts for less than 1% of blood volume.

> *Remember This!* **The two major compartments of blood are fluid (plasma) and solid matter (cellular elements).**

The cellular elements of blood are easily visualized by microscopic examination of a *blood smear*. In this procedure, the lab technician spreads a drop of blood thinly over a microscope slide (Fig. 10.2B) and then stains the smear with Wright's stain, a mixture of blue and red dyes that impart colors to the various elements of the cells. It includes *hematoxylin*, a dark blue, alkaline dye, which stains the nucleus blue, and *eosin*, a bright red, acidic dye, which stains the cytoplasm pale red or pink. Cell organelles may stain a mixture of red, blue, or neutral tan. Based on the size and shape of the cells, the presence or absence of a nucleus, and the color of cytoplasmic granules (if any), the various cellular elements can easily be identified (see Table 10.1 on page 383).

In addition to centrifugation to determine the hematocrit, other laboratory procedures can be employed to count the number of red blood cells, white blood cells, and platelets; measure the average size of the red blood cells; determine the amount of oxygen-carrying hemoglobin in the red blood cells; and determine the percentage of the various subtypes of white blood cells. Typically these measurements are done as a group by automated machines, and the procedure is typically referred to as a *complete blood count* (CBC).

Case Notes

10.1. **In Eleanor's case, the initial blood analysis found that cellular elements accounted for about 27% of her blood volume. What percent of her blood volume was composed of plasma?**

10.2. **Based on the information in the previous case note, what was Eleanor's approximate hematocrit?**

Plasma Contains Water and Solutes

Plasma is the extracellular fluid of blood. It is nearly identical to the interstitial fluid of solid tissues ← (Chapter 5) except for the large amount of blood proteins it contains. Transparent and straw-colored, plasma is about 90% water; about 9% is composed of specialized proteins.

There are three major types of **plasma proteins:**

- *Albumin.* The most abundant (about 55%) of all plasma proteins is **albumin.** Albumin accounts for most of the plasma *osmotic pressure* ← (Chapter 3), the force that tends to hold water in blood and draw water across the blood vessel wall from tissues into the bloodstream. This ability to keep water inside blood vessels is very important in maintaining blood volume. Albumin also acts as a *binding protein* that transports fatty acids, steroids, and other substances in blood.
- *Fibrinogen.* Somewhat over 5% of plasma protein is **fibrinogen,** a small protein involved in blood clotting, as explained later in the text.
- *Globulins.* Most of the other blood proteins are *globulins,* a catchall category that includes specialized binding (transport) proteins, enzymes, protein hormones, and clotting factors. For example, *transferrin* is a special globulin that transports iron, an important function discussed further on. Of particular interest is a subgroup of globulins, the *gamma globulins*, also called *antibodies*, made by specialized white blood cells to attack harmful microbes.

About 1% of plasma is a rich mixture of other solutes. These include:

- Glucose
- Cholesterol and other lipids
- Vitamins and other essential compounds
- Calcium, iron, sodium, potassium, and other minerals
- Metabolic wastes
- Dissolved gases such as oxygen, nitrogen, and carbon dioxide.

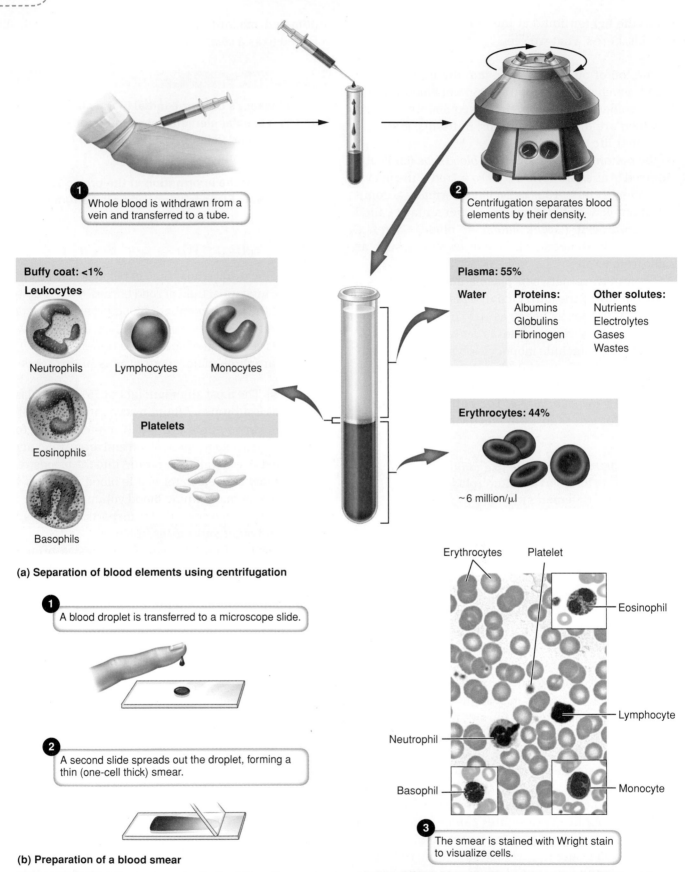

1 Whole blood is withdrawn from a vein and transferred to a tube.

2 Centrifugation separates blood elements by their density.

Buffy coat: <1%

Leukocytes

Neutrophils   Lymphocytes   Monocytes

Eosinophils

Platelets

Basophils

Plasma: 55%

Water	Proteins:	Other solutes:
	Albumins	Nutrients
	Globulins	Electrolytes
	Fibrinogen	Gases
		Wastes

Erythrocytes: 44%

~6 million/µl

(a) Separation of blood elements using centrifugation

1 A blood droplet is transferred to a microscope slide.

2 A second slide spreads out the droplet, forming a thin (one-cell thick) smear.

(b) Preparation of a blood smear

Erythrocytes   Platelet

Eosinophil

Lymphocyte

Neutrophil

Monocyte

Basophil

3 The smear is stained with Wright stain to visualize cells.

Figure 10.2. Blood components. A. Centrifugation separates out the different blood elements to permit blood analysis. In a tube of centrifuged blood, cellular elements (erythrocytes, platelets, and white blood cells) constitute the bottom 45% (or so) and plasma constitutes the upper 55%. **B.** Blood smears are used to visualize the different cellular elements of a patient's blood. The small insets show how monocytes, basophils, and eosinophils (which are few) would appear in a blood smear. *Name the only cellular element that is not found in the buffy coat.*

> **Remember This!** Plasma is 90% water, 9% protein, 1% nonprotein solutes.

Cellular Elements Are Produced in the Bone Marrow

Like all cells, blood cells have a life cycle—they are produced, they work, and then they die by apoptosis. Compared with cells in most other tissues, the life span of all blood cells is short. Among blood cells, erythrocytes have by far the longest life span, about 120 days; leukocytes have life spans varying from a few hours to 2 weeks, and platelets live a week or two. Old blood cells and platelets die a natural death from apoptosis and are removed from circulation by the spleen. It is important to remember that the number of a particular type of cell in a blood sample depends upon its life span as well as its rate of production—cells with longer life spans and greater production will be present in greater numbers.

The generation of blood cells is called **hematopoiesis** (Greek *haima* = "blood"; *poiesis* = "formation"). All blood cells arise from a common ancestor, a *pluripotent hematopoietic stem cell*, literally a blood cell with "many powers" (Fig. 10.3). It resides in the bone marrow. This master cell, in turn, gives rise to two types of specialized stem cells with narrower powers:

- *Lymphoid* stem cells give rise to lymphocytes, a type of leukocyte.
- *Myeloid* stem cells give rise to all other blood cells—the four other types of leukocytes as well as erythrocytes. They also produce megakaryocytes, which produce platelets.

The harvesting of pluripotent stem cells holds great therapeutic promise if they can be collected in abundance and coaxed into forming new blood cells or perhaps other cells, such as heart or liver cells. It turns out that blood collected from the umbilical cord of a newborn infant is rich in pluripotent stem cells and can be frozen for future use or cultivated immediately for benefit to the infant itself or to someone else.

Recall from our discussion of bones in ◀ Chapter 6 that bone marrow is either yellow (fatty) or red (hematopoietic). Red marrow actively produces blood cells; yellow marrow normally does not. In adults, most red marrow lies in the marrow cavity of bones of the spine, pelvis, ribs, cranium, and the proximal ends of long bones.

Despite the fact that RBCs are 1,000 times more abundant than WBCs in blood, only about 25% of red marrow is composed of developing red cells; the remaining 75% consists of developing white cells. This owes to the fact that white cells have much shorter life spans than red blood cells, so they must be replaced much more often.

The production of cellular blood elements in the fetus differs materially from that in an adult. In the embryo and developing fetus, cellular elements of blood are produced primarily in the liver and spleen; but by the time of birth, production has gradually shifted to the red marrow. However, the liver and spleen retain their ability to produce cellular elements throughout life and will do so, even in older adults, under some circumstances. Blood cell production by the liver and spleen in an adult is called *extramedullary hematopoiesis* (hematopoiesis away from the medullary cavity of bones). It can occur in certain conditions that wipe out the bone marrow. For example, in Eleanor's case, metastatic breast cancer in her bones replaced most of her red marrow, forcing the liver and spleen to take over hematopoiesis.

Case Note

10.3. At autopsy, Eleanor was found to have blood cell production in her liver and spleen. What is the name of this condition and what does it mean?

10.1 True or false: Blood, like cartilage, is classified as connective tissue.

10.2 List four substances transported by blood.

10.3 Explain why the hydraulic force of blood is important.

10.4 Name two types of threats that blood defends against.

10.5 True or false: Increasing the circulation of blood through skin warms the body.

10.6 Name the two main compartments of blood.

10.7 If you leave a tube of blood at room temperature, the red blood cells will settle to the bottom of the tube. Why?

10.8 Is a test result of "47%" from a hematocrit or a blood smear?

10.9 Change the underlined term to make this statement true: <u>Albumin</u> is a specialized protein that transports iron.

10.10 If we wanted to make neutrophils in the test tube, which type of stem cell would we use—lymphoid or myeloid?

10.11 Which blood element lives the longest—red blood cells, white blood cells, or platelets?

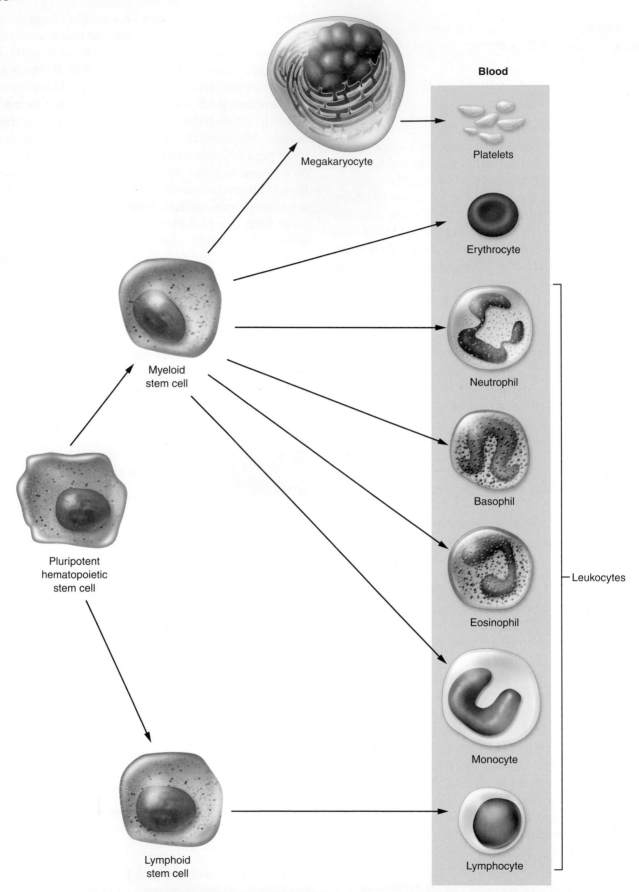

Figure 10.3. Hematopoiesis. All blood cells are the offspring of pluripotent hematopoietic stem cells that arise in the bone marrow. Numerous intermediate cell types between the stem cells and the mature blood cells are not illustrated. *Which blood cell is not produced from a myeloid stem cell?*

Leukocytes, Inflammation, and Immunity

Leukocytes (*white blood cells, or WBCs*) defend the body against infections, airborne particulate matter, and newly hatched tumor cells; when tissue is damaged, WBCs clean up the debris and assist with repair. Such functions would not be possible for cells confined within blood vessels, but leukocytes are motile. In fact, their unique form of movement has its own name, **diapedesis** (Greek *dia* = "through" + *pedan* = "leap"), which nicely captures their amebalike ability to crawl between adjacent cells in the walls of blood vessels and squirm out into the tissues. There, they roam at large, performing their tasks. Normal blood contains about 5,000 to 10,000 leukocytes per microliter (one thousandth of a milliliter, also defined as a cubic millimeter or mm^3).

Leukocytes are the only formed elements of blood that are complete cells—each one has a nucleus and a complete set of organelles. Their appearance varies considerably: some are large, others small; some have a prominent cytoplasm with visible organelles, others have minimal cytoplasm with few organelles; and some have a simple, round nucleus, whereas others have a multilobed nucleus (Table 10.1).

Table 10.1 Cellular Elements of Blood

Cell Type	Appearance (Wright's Stain)	Function	Illustration
Erythrocytes (RBCs)	No nucleus or granules Cytoplasm: red	Gas transport, acid–base balance	
Leukocytes (white blood cells)			
Granulocytes			
Neutrophils (54% to 62% of WBCs)	Nucleus: lobed, dark purple Granules: fine, light tan or not visible Cytoplasm: pale pink	Phagocytosis of bacteria, other invaders	
Eosinophils (1% to 3% of WBCs)	Nucleus: lobed, purple Granules: large, red Cytoplasm: pale pink	Allergic inflammatory reactions; defense against parasites	
Basophils (<1% of WBCs)	Nucleus: lobed, dark blue, often obscured by granules Granules: large, dark blue Cytoplasm: pink	Allergic inflammatory reactions	
Cells Without Granules			
Lymphocytes (25% to 38% of WBCs)	Nucleus: unlobed, deep purple Granules: few if any Cytoplasm: scant, blue	Involved in specific immune responses	
Monocytes (3% to 7% of WBCs)	Nucleus: purple—large and irregular Granules: few if any Cytoplasm: light blue	Precursor to macrophages	monocyte → In tissues → macrophage
Platelets	Granules: purple or none Cytoplasm: light blue	Blood clotting	

Just as leukocytes are more complex than erythrocytes or platelets, their production (**leukopoiesis**) is more complex and depends heavily on cell-to-cell signaling by *cytokines*, large protein or glycoprotein molecules released by cells to influence the behavior of other cells locally or far away. Leukocytes release cytokines that stimulate the production and activity of other leukocytes. These WBC-to-WBC cytokines are called *interleukins* (literally "between leukocytes"). Dozens have been identified and named, such as IL-1, IL-2, and so on. Interleukins are discussed in more detail in ➡ Chapter 12.

Leukocytes can be classified into three main groupings: granulocytes, lymphocytes, and monocytes (see Table 10.1).

Granulocytes Contain Large Cytoplasmic Granules

The majority of leukocytes, about 65%, are **granulocytes,** large WBCs with an irregularly shaped nucleus and abundant cytoplasm containing large cytoplasmic granules (*lysosomes*, ⬅ Chapter 3). Granulocytes are actively mobile cells that defend the body against bacteria and other threats, such as inhaled particulate matter, and assist in cleaning up and repairing damaged tissue.

Granulocytes are further classified according to the color of their cytoplasmic granules when stained with a standard red/blue dye mixture:

- Most granulocytes—about 60% of all WBCs—are **neutrophils,** so called because their cytoplasmic granules stain a nearly colorless (that is, *neutral*) bluish-tan. Neutrophils have an elongated, twisted, dark purple nucleus of five or six lobes shaped like a chain of linked sausages. Since their cytoplasmic granules contain packets of digestive enzymes, neutrophils are able to defend the body by phagocytizing and digesting bacteria, particulate matter, and other foreign threats. And in damaged tissue, they ingest and clear away cell debris to pave the way for tissue repair and healing. Their life span is the shortest of all blood cells—only a few hours.
- About 3% of all WBCs are **eosinophils,** granulocytes that have "eosin-loving" cytoplasmic granules (eosin is a red dye), which stain bright red. These granules release packets of enzymes that defend against two types of threats: (a) they destroy parasites (multicellular microbes, such as intestinal worms), and (b) they neutralize offending agents in allergic reactions. The eosinophil nucleus stains dark purple and has two lobes somewhat like two sausage links.

- Less than 1% of all WBCs are **basophils,** granulocytes with alkaline (basic) cytoplasmic granules that stain dark blue-black because of their affinity for hematoxylin (a blue dye). Basophils are important in allergic reactions: they release *histamine*, which dilates blood vessels, and *heparin*, a natural anticlotting (anticoagulant) compound that also has wide use as a drug, discussed further on. Other chemicals act as signals to recruit eosinophils and other cells to the site. The nuclei of basophils stain dark purple and are shaped somewhat like a long sausage bent into a C-shape.

Case Note

10.4. Eleanor's bone marrow is not making adequate numbers of neutrophils, among other cells. Which process best describes neutrophil synthesis—leukopoiesis or diapedesis?

Lymphocytes Are Cells of the Immune System

About 30% of WBCs are **lymphocytes,** cells of the *immune system* (➡ Chapter 12) that circulate in blood. However, they are also normally found in other tissues, especially lymph nodes, the spleen, and bone marrow. Lymphocytes are smaller and more uniform than granulocytes. They have round nuclei that stain dark blue and a small amount of light blue cytoplasm without granules (see Table 10.1).

Lymphocytes defend against external threats—bacteria, viruses, and fungi—and the internal threat of newly forming tumor cells. However, unlike granulocytes, they are only sluggishly mobile, are not phagocytes, and do not release packets of destructive chemicals. Instead, they defend the body by a group of highly specialized molecular mechanisms collectively called *immunity*—the subject of ➡ Chapter 12.

Monocytes Develop into Macrophages

About 5% of WBCs are **monocytes,** which, like lymphocytes, do not contain granules. Monocytes are somewhat larger than granulocytes; they have large, darkly staining U-shaped nuclei and abundant light blue cytoplasm (see Table 10.1). After crawling out of the bloodstream, monocytes become highly mobile *phagocytic* cells called **macrophages,** with a gargantuan appetite for ingesting and destroying certain microbes and foreign material. Additionally, macrophages play a key role in helping lymphocytes do their job in the immune system.

Leukocytes Are Involved in Inflammation

In the broadest sense, some type of injury—physical trauma, molecular toxin, lack of oxygen, radiation, or infection—causes every disease. **Inflammation** is the body's collective cellular and vascular response to injury, and leukocytes play an important role in it (Table 10.2). Inflammation precedes and is intimately linked to the process of tissue repair, which was discussed in ◀ Chapter 5 (see Fig. 5.9). Repair begins as the immediate effects of injury begin to fade.

Acute Inflammation

Acute inflammation is the result of short-term, intense injury and persists for a few hours or days. In acute injury, inflammation fades gradually as the damaged tissue is regenerated from stem cells ◀ (Chapter 3) or, if regeneration is not possible, replaced by scar tissue.

Injured cells or invading microbes release substances that cause blood vessels to dilate and become "leaky." Plasma oozes from blood into the injured tissue, bringing nutrients and other molecules to assist with defense and repair. At the same time, leukocytes crawl out of blood vessels into the damaged tissue to attack and remove the offending agent, to clean up cell debris, and to initiate the repair process. The leukocytes most commonly involved in acute inflammation are neutrophils, which are quite abundant in blood and can respond quickly. Moreover, neutrophils are attracted by—and effectively defend against—the bacteria that frequently enter the body through an injured barrier such as skin. In severe acute inflammation, neutrophils accumulate to form *pus*, the creamy white fluid for which leukocytes were originally given the name *white blood cells*. Depending on the degree and duration of injury, the tissue becomes distended, red, warm, and painful as it becomes engorged with blood and leaked plasma and as nerve endings become irritated by the process ◀ (Chapter 9).

Chronic Inflammation

Chronic inflammation is the result of longer-term, less intense injury, and can persist for weeks or years. In the case of certain chronic diseases, the injury itself is ongoing. In fact, injury, inflammation, and repair often coexist, sometimes indefinitely, as is the case with joints inflamed by chronic arthritis.

The leukocytes most commonly involved in chronic inflammation are lymphocytes and macrophages, because chronic damage is often caused by slow-acting immune reactions (▶ Chapter 12) or by viruses, fungi, irritants, or other agents that do not attract neutrophils. Neutrophils, therefore, are rarely present.

Although chronic inflammation is certainly painful and can be debilitating, it is usually not as intensely hot, swollen, red, or tender as acute inflammation. For example, *rheumatoid arthritis* is a chronic inflammatory condition of joints caused by faulty immune reactions. Smoking is another example: it damages and causes chronic inflammation in bronchial airways, which results in coughing, wheezing, and shortness of breath.

Table 10.2   Leukocytes and Inflammation

Type of Inflammation	Cause	End Result	Primary Leukocytes Involved
Acute	Short-term, intense injury	Repair or scarring	Neutrophils
Chronic	Long-term injury or chronic inflammatory disease	Persistent inflammation and repair	Lymphocytes, macrophages
Allergic	Allergies (immune system hypersensitivity)	Varies from mild, local inflammation to systemic shock and death	Eosinophils, some basophils
Parasitic	Parasites	Parasite may be destroyed, or inflammation may harm the host	Eosinophils

Intermediate Types of Inflammation

Falling somewhere between acute and chronic injury and inflammation are tissue injuries caused by parasites and allergic reactions, each of which attracts *eosinophils* and *basophils*. Hay fever is a common allergic condition; microscopic study of nasal mucus will reveal an abundance of eosinophils and some basophils. Or, if a complete blood count is done in a child with chronic abdominal pain, it may reveal large numbers of eosinophils; in that case intestinal parasites would rank high on the list of diagnostic considerations.

Case Note

10.5. We know that Eleanor, between hospital admissions, suffered from repeated bacterial infections. Lack of which blood cell is likely to blame?

Infection Is a Risk with Low Numbers of Leukocytes

Recall that granulocytes attack invading bacteria and help to clean up damaged tissue; also that lymphocytes resist viral, fungal, and other nonbacterial infections. Given these roles, it's not surprising that infection is a major risk for patients with low numbers of leukocytes or dysfunctional leukocytes. A low blood leukocyte count is called **leukopenia** (Greek *penia* = "poverty"). Patients can be deficient in any particular type of WBC. **Neutropenia**—a deficiency in neutrophils—also impairs wound healing, because neutrophils facilitate the healing process by removing debris. Eleanor's body was unable to repair her esophageal ulcer because of neutropenia.

The cardinal example of a disease marked by low lymphocyte counts is acquired immunodeficiency syndrome (AIDS), a condition in which the human immunodeficiency virus (HIV) virus kills lymphocytes, leaving the body vulnerable to recurrent infections and the development of malignant tumors. AIDS is discussed in greater detail in ➡ Chapter 12.

Case Note

10.6. Eleanor's blood count revealed a low number of lymphocytes. Which term do you think describes this condition—lymphocytopenia or eosinopenia?

Leukemia and Lymphoma Are Malignant Diseases of Leukocytes

Recall from our discussion of the cell's life cycle and cell growth in ◀ Chapter 3 that a malignancy is an uncontrolled growth of cells. **Leukemia** is malignant growth of any type of leukocyte, including lymphocytes, in which *malignant leukocytes are present in blood.* The total number of leukocytes in blood is increased and microscopic examination reveals increased numbers of abnormal-appearing malignant leukocytes (Fig. 10.4).

Lymphoma is a malignant growth of lymphocytes; however, *no malignant cells are not detectable in blood.* Instead, masses of malignant lymphocytes occur in lymph nodes, bone marrow, or other organs.

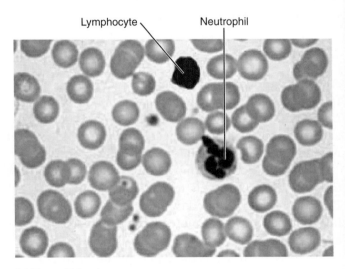

(a) Normal blood

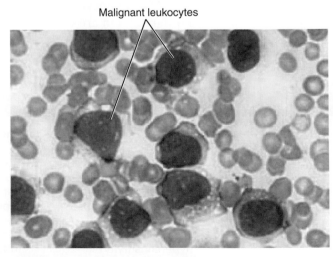

(b) Blood from a patient with leukemia

Figure 10.4. Leukemia. A. Normal blood, containing a normal neutrophil and lymphocyte. **B.** Blood from a patient with leukemia, showing numerous abnormal (cancerous) leukocytes. *Which blood cell in part A is a granulocyte?*

10.12 Rank the following leukocytes from the most abundant to the least abundant: basophils, eosinophils, lymphocytes, neutrophils, monocytes.

10.13 Which of the above leukocytes are granulocytes?

10.14 What is the clinical term for a low blood leukocyte count?

10.15 Are neutrophils the major players in acute inflammation or chronic inflammation?

10.16 An adult patient has large numbers of malignant lymphocytes in her blood. What would you call the patient's disease—lymphoma or leukemia?

Erythrocytes and Oxygen Transport

Erythrocytes (*red blood cells, or RBCs*) are workhorse transportation cells—their major task is to transport oxygen from the lungs to tissues and carbon dioxide waste from tissues to the lungs. They are nonmotile cells that normally remain entirely within blood vessels; escape of erythrocytes from a blood vessel (hemorrhage) is abnormal. Erythrocytes are the most abundant cellular element in blood. A microliter (1 mm^3) of blood contains about 5 million RBCs (a measurement called the **RBC count**). Recall that only about 5,000 WBCs are present in the same volume.

Erythrocyte Form Contributes to Function

The physical characteristics of RBCs are perfectly suited to their function (Fig. 10.5). First, circulating RBCs have no nuclei—a nucleus would just occupy otherwise useful space for transporting oxygen. The RBC spits out its nucleus just before leaving the bone marrow. The net result is that a circulating RBC is essentially a sack of *hemoglobin*—a molecule to be discussed in more detail below—that temporarily binds either oxygen or CO_2 for transportation. RBCs also have few cytoplasmic organelles and a very low rate of metabolism. Without mitochondria, they must rely on anaerobic respiration, and without endoplasmic reticulum, they cannot synthesize most proteins. This almost lifeless state reflects their passive role—their main job is to transport oxygen and carbon dioxide, not use it or generate it. Any oxygen used by a RBC is oxygen that doesn't reach tissues.

Second, each RBC is shaped into a flattened, biconcave disc (Fig. 10.5A). This design is efficient—a lot of membrane surface area and relatively little volume; it keeps the hemoglobin and its precious load of oxygen near the cell membrane for easy diffusion into tissues. Were RBCs spherical, the gases bound in the center of the cell would have a long way to diffuse before reaching the cell membrane.

Third, RBCs are flexible and can safely travel through the tiniest capillaries. This ensures that, in a healthy person, even the most peripheral body tissues are perfused with blood.

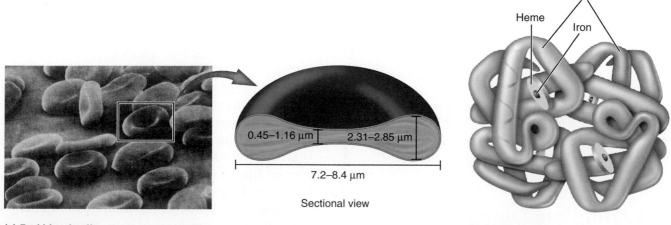

(a) Red blood cells

0.45–1.16 μm   2.31–2.85 μm

7.2–8.4 μm

Sectional view

Globin
Heme
Iron

(b) Hemoglobin

Figure 10.5. Physical characteristics of RBCs. A. RBCs are biconcave discs without nuclei. **B.** Hemoglobin consists of four subunits, each containing a long protein chain (globin) and a complex organic molecule containing iron (heme). *Why are RBCs biconcave instead of spherical?*

Oxygen Is Bound to Hemoglobin

The effectiveness of oxygen transport by blood is directly proportional to the number of erythrocytes and the amount of *hemoglobin* they contain. **Hemoglobin** is a large iron-containing molecule that binds oxygen and carbon dioxide and completely fills the RBC's cytoplasm. Each molecule of hemoglobin is composed of four folded chains of *globin*, a large protein molecule, and four molecules of *heme*, a red pigment, each of which contains an iron ion (Fig. 10.5B). Each of these four ions of iron combines with one molecule of oxygen and each RBC contains about 250 million hemoglobin molecules; therefore, each RBC carries about a *billion* oxygen molecules. As discussed in ➡ Chapter 13, hemoglobin can also carry carbon dioxide and hydrogen ions in blood. Hemoglobin deficiency impairs oxygen transport to tissues, a condition called *anemia,* discussed in detail further on.

Hemoglobin does not function well outside of RBCs. When released into plasma from damaged erythrocytes, it quickly diffuses out of the vascular system and is unavailable for O_2/CO_2 transport.

Hemoglobin in the fetus (*hemoglobin F*) differs molecularly from adult hemoglobin (*hemoglobin A*): it has a greater affinity for oxygen than hemoglobin A—a fact of great physiological importance. The fetus gets its oxygen from its mother via the placenta, where a membrane separates fetal and maternal blood. Oxygen crosses this membrane by diffusion. The higher-affinity hemoglobin F lures the oxygen away from the maternal, lower-affinity hemoglobin A, just as a stronger magnet will lure a metal ball away from a weaker magnet. After birth—that is, by the end of the third month—hemoglobin F is usually completely replaced by hemoglobin A.

The Erythrocyte Life Cycle Spans 120 Days

Erythrocytes are produced by the bone marrow by a process called **erythropoiesis;** they circulate for about 120 days before dying a natural death by apoptosis. The stages in an erythrocyte's life cycle are diagrammed in Figure 10.6.

1. Marrow production of erythrocytes (or any other cell) requires raw material—amino acids, lipids, sugars, vitamins, and minerals. Synthesis of hemoglobin is critical to erythrocyte production, and hemoglobin cannot be synthesized without iron, zinc, copper, vitamin B_{12}, vitamin B_6, and folic acid. Iron is particularly important, because iron is a vital part of hemoglobin and virtually all of the body's stores of iron are used to make the heme of hemoglobin. The body stores iron in the bone marrow, liver, and spleen, bound to a specialized iron storage protein called **ferritin.**

2. As RBCs near maturity in the bone marrow, they eject their nuclei and become **reticulocytes,** ready to enter the bloodstream. This name derives from the fact that reticulocytes retain a spidery (reticular) network of ribosomes. Once in circulation, reticulocytes mature in a day or two into adult erythrocytes and lose their distinctive appearance. The level of reticulocytes in normal blood is usually less than 1%. A reticulocyte percentage greater than about 1% indicates that the bone marrow is making new erythrocytes at a greater than normal pace because the oxygen-carrying capacity of the blood is too low.

3. Erythrocytes survive for about 120 days. Recall that they do not have any ribosomes, so they cannot make new proteins to repair themselves. Old (senescent) erythrocytes have been battered by their high-risk voyage through the tumult of the heart and the twisting, narrow passages of the capillaries; macrophages seek out these decrepit cells and destroy them.

4. Macrophages in the bone marrow, spleen, and liver phagocytose aged erythrocytes.

5. The iron atoms from metabolized heme (as well as dietary iron) are transported in blood by a specialized plasma protein, called **transferrin,** to the bone marrow, liver, and spleen for storage.

6. The noniron components of the heme molecule are metabolized into *bilirubin*, a bright yellow pigment that is removed from blood by the liver and excreted into the intestines. When present in excess, bilirubin accumulates in blood and imparts a yellow discoloration to skin, a condition called *jaundice.*

7. The amino acids in the globin part of hemoglobin are recycled to make other proteins.

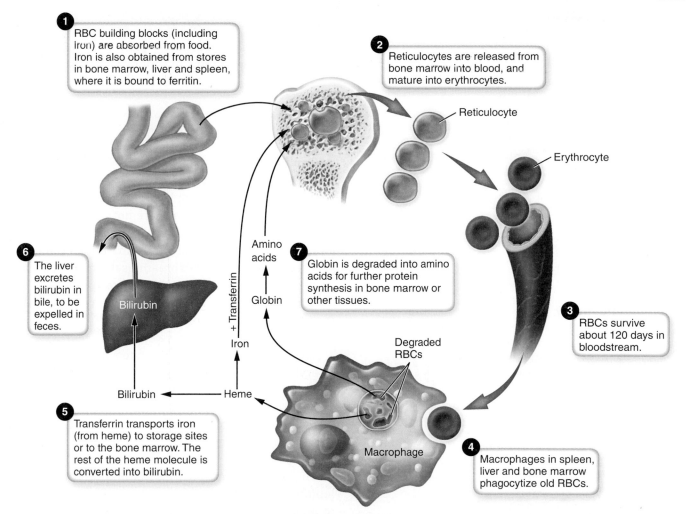

1 RBC building blocks (including Iron) are absorbed from food. Iron is also obtained from stores in bone marrow, liver and spleen, where it is bound to ferritin.

2 Reticulocytes are released from bone marrow into blood, and mature into erythrocytes.

Reticulocyte

Erythrocyte

Amino acids

7 Globin is degraded into amino acids for further protein synthesis in bone marrow or other tissues.

Globin

6 The liver excretes bilirubin in bile, to be expelled in feces.

Bilirubin

+ Transferrin

Iron

Bilirubin ← Heme

5 Transferrin transports iron (from heme) to storage sites or to the bone marrow. The rest of the heme molecule is converted into bilirubin.

3 RBCs survive about 120 days in bloodstream.

Degraded RBCs

Macrophage

4 Macrophages in spleen, liver and bone marrow phagocytize old RBCs.

Figure 10.6. Erythrocyte life cycle. Some components required for erythrocyte synthesis are recycled from old erythrocytes; others must be obtained from the diet. *Name the binding protein that transports iron in blood.*

Case Note

10.9. Recall that Eleanor's bone marrow is not producing normal numbers of erythrocytes. In the absence of other problems, do you think her bilirubin level would be elevated or decreased?

Erythropoietin Stimulates Erythropoiesis

Erythropoiesis is stimulated by **erythropoietin** (*EPO*), a protein hormone produced by the kidneys. EPO is not stored; its production is regulated in a classic negative feedback loop based on the availability of oxygen—production increases in response to *hypoxia* (low blood oxygen) and decreases as oxygen content rises (Fig. 10.7).

People who live at high altitude have higher RBC counts (and higher hemoglobin and hematocrit, too) than

those who live at lower altitudes. This is because they are mildly hypoxic from the thin mountain air. Likewise, in anemic patients, the low level of transported oxygen stimulates the homeostatic production of EPO, rousing the bone marrow to pour out new erythrocytes to correct the anemia. Synthetic EPO is now available for the treatment of anemia and is especially useful in patients with chronic kidney disease, who cannot make enough natural EPO to avoid becoming anemic. Blood loss, be it the result of donating blood or hemorrhage, similarly activates EPO production because not enough erythrocytes are present to carry the expected amount of oxygen.

The use of synthetic EPO injections to boost RBC production and oxygen-carrying capacity was especially prevalent in international cycling events in the 1980s—that is, until laboratory methods were devised to detect it. Some athletes then turned to "blood doping," a practice in which blood is withdrawn and stored in advance of competition with the expectation of

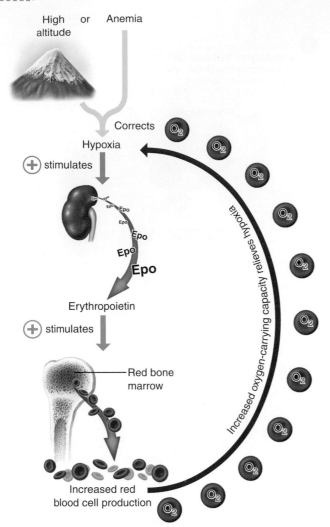

Figure 10.7. Erythropoietin stimulates erythropoiesis.
Hypoxia stimulates EPO production by the kidneys, which increases the oxygen carrying capacity of blood by stimulating the production of RBCs. Completing the negative feedback loop, increased oxygen carrying capacity then decreases EPO production. *What type of marrow is the target tissue for EPO—yellow or red?*

reinfusion immediately before competition. The depletion of RBCs renders the athlete temporarily anemic; however, this condition stimulates natural RBC production, which returns total circulating erythrocytes toward normal. Later, the stored blood is reinfused, giving the athlete an extra charge of oxygen-carrying capacity and the expectation of improved performance. Blood doping is more difficult for athletic organizations to detect than EPO injections, but a hematocrit over 50% or hemoglobin in excess of 17 g/100 mL may be grounds for disqualification. Blood doping is dangerous, too: unusually high numbers of erythrocytes can thicken blood to the point that it does not flow smoothly, which may cause blood vessel obstruction and a stroke or heart attack.

Case Note

10.10. If we measured Eleanor's plasma erythropoietin, would we expect it to be low, normal, or high?

Anemia Is Too Little Hemoglobin in Blood

Anemia is too little hemoglobin in blood, which means that less oxygen is delivered to tissues. As a result, patients with anemia are pale from lack of red hemoglobin coloration. And because their cells are not provided with sufficient oxygen for normal function, they are at an increased risk for infection, tire easily, and can't think clearly. They also become easily short of breath, as their heart and lungs work overtime trying to make up the oxygen deficit.

Anemia may be a manifestation of any of the following three basic conditions:

- There are too few erythrocytes in the blood (the erythrocyte *count* is low).
- The *size* of the average erythrocyte is too small.
- The *concentration of hemoglobin* in the average erythrocyte (the amount of hemoglobin per unit of RBC volume) is too low.

These measurements are the stuff of routine medical laboratory practice and offer valuable clues to the cause of anemia in any given patient. More about RBC measurements can be found on our website at **https://lww.thepoint/McConnellandHull.com.**

When there are too few erythrocytes in blood or their size is too small, the erythrocyte volume absent from blood is replaced by plasma, usually within a few hours. This response expands blood volume back to normal, but of course it also dilutes the erythrocyte volume even further. This is why the most common initial clinical indication of anemia is a reduced hematocrit (Fig. 10.8A).

Anemia is not a disease; rather, it is merely a *sign*, a clue, that something more fundamental is amiss. These more fundamental conditions include the following:

- Hemolytic anemia: Abnormally rapid destruction of RBCs (life span of less than 120 days)
- Hemorrhagic anemia: Bleeding
- Production failure anemia: Impaired production of new RBCs or failure to produce enough hemoglobin in them.

Remember This! Anemia is a sign of disease, not a disease itself. In all anemias, the missing RBC volume is replaced by plasma.

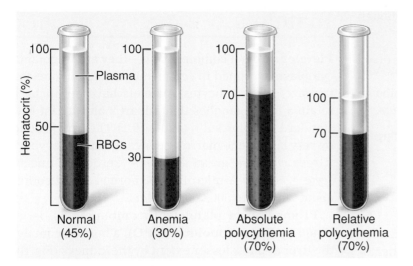

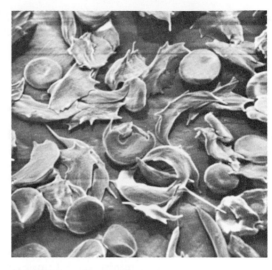

(a) Hematocrit disorders

(b) Sickle-cell anemia

Figure 10.8. Anemia and polycythemia. A. The hematocrit can diagnose anemia or polycythemia. **B.** Sickle cell anemia. *Which tube represents the hematocrit of a dehydrated individual?*

Hemolytic Anemia

Hemolytic anemia is due to the abnormal destruction of erythrocytes (before they die a natural death by apoptosis)—a process known as **hemolysis (**Greek *lysis* = "loosening"). For example, malaria is a parasitic infection of erythrocytes that destroys them as they circulate.

Hemolysis is the cause of anemia in *sickle cell anemia.* The cells contain enough hemoglobin, but it is flawed because of a genetic defect of hemoglobin synthesis. The hemoglobin molecules tend to crystallize into long, thin rods that deform erythrocytes into elongated, pointy, curved, stiff cells that look like an ancient sickle (scythe) for harvesting wheat (Fig. 10.8 B). The life span of these inflexible cells is as short as 4 days, since macrophages detect their abnormal shape and destroy them. Sickle cells also tend to jam together and block blood flow, like sticks in a beaver dam. Such a blockage is painful and starves tissues of oxygen; it can lead to cell death.

Hemorrhagic Anemia

Hemorrhagic anemia is anemia due to the loss of RBCs by bleeding. It can take weeks or months for the bone marrow to replace all of the lost erythrocytes. For serious hemorrhage, transfusion or replacement erythrocytes is necessary to preserve life, as we saw with our case study of President Reagan in Chapter 1.

Chronic blood loss, such as undetected intestinal bleeding, is a different matter. If bleeding is very slow, hemorrhagic anemia will not develop because newly produced erythrocytes can keep up with the loss. However, anemia *will* develop if bleeding exceeds the body's capacity to produce new erythrocytes. In these cases,

anemia develops slowly and patients may become severely anemic without realizing it. The only symptom they may notice is that they tire or become short of breath easily.

Because RBCs contain iron and iron is necessary for hemoglobin production, chronic bleeding may deplete the bone marrow of iron and impair hemoglobin production (discussed next).

Production Failure Anemia

Normally, all of the cellular elements of blood are made in the bone marrow and released into blood. **Production failure anemia** is due to failure of the bone marrow to produce normal erythrocytes in sufficient numbers to replace those dying a natural death (by apoptosis).

Aplastic anemia (Greek *a-* = "without" + *plasis* = "formation") results from destruction of the marrow by drugs, toxins, radiation, or invasion by metastatic cancer.

In **iron deficiency anemia,** the bone marrow is normal, but it lacks enough iron to synthesize a normal amount of heme for hemoglobin. There are two main causes of iron deficiency anemia—lack of dietary iron and iron loss due to chronic bleeding. Because they lose blood (and iron with it) normally with every menstrual period, women of childbearing age are especially vulnerable to iron deficiency anemia. Normal menstrual bleeding is usually not severe enough to produce anemia because dietary iron is usually adequate to keep up with the demand for replacement hemoglobin. But for women with sustained excessive menstrual bleeding, even oral iron supplementation may not suffice to prevent the development of iron deficiency and the anemia that accompanies it. After body iron stores are exhausted, the marrow cannot make normal RBCs, nor can it pack them with enough hemoglobin; iron deficiency anemia is the result.

Case Notes

10.11. Which two of the following terms best describe Eleanor's anemia: *aplastic, hemolytic, production failure.*

10.12. Are Eleanor's frequent nosebleeds and bleeding ulcer contributing to her anemia? If so, how?

Polycythemia Is Too Many RBCs

The opposite of anemia is **polycythemia** (poly- = "many"; -cyte = "cell"; Greek *haima* = "blood")—too much hemoglobin in blood. As you might expect, measurements of hemoglobin, hematocrit, and RBC count are high (Fig. 10.8A). The most common cause of polycythemia is dehydration, which reduces plasma volume—erythrocytes are concentrated in less than a normal amount of plasma, a situation described as *relative polycythemia.* However, some people have increased RBC counts because there is an *actual* increase in the number of erythrocytes in the body. They are said to have *absolute* polycythemia. In many such people, this is a normal adaptation to life at a high altitude caused by increased erythropoietin production. A few patients with absolute polycythemia have a bone marrow malignancy called **polycythemia vera** (Greek *vera* = "true"), an uncontrolled, malignant proliferation of primitive erythrocyte precursor cells in the bone marrow.

10.17 What is the difference between a reticulocyte and an erythrocyte?

10.18 Fill in the blanks: Iron is transported by _____ in blood and is stored attached to _____ in tissues.

10.19 Identify three structural characteristics of RBCs that directly support their function.

10.20 True or false? A single molecule of hemoglobin can bind to about a billion molecules of oxygen.

10.21 Name the three categories of anemia.

10.22 If an anemic patient drinks a large volume of fluid and increases his plasma volume, will his hematocrit increase or decrease?

10.23 How can lung disease lead to polycythemia?

10.24 Is EPO used to treat polycythemia or anemia?

Platelets

Platelets are not complete cells—they are fragments of cytoplasm wrapped in cell membrane. About one-tenth the size of erythrocytes, platelets have no nuclei but retain a few organelles that impart a bluish cast with a standard red/blue stain (Table 10.1). They are produced by very large bone marrow cells called **megakaryocytes** (Greek *megas* = "great"; *karuon* = "kernel" or "nucleus"; *-cyte* = "cell"). A microliter of normal blood contains about 200,000 to 500,000 platelets.

Production of platelets, **thrombopoiesis,** is governed by **thrombopoietin (TPO),** a hormone made by the liver and to a lesser extent by the kidneys (Fig.10.9). Thrombopoietin stimulates megakaryocyte production by the bone marrow and megakaryocytes release platelets into blood. Platelets have a life span of about a week and they, too, are phagocytized by macrophages, mainly those in the spleen. Thrombopoietin levels are inversely related to the platelet count in a classic

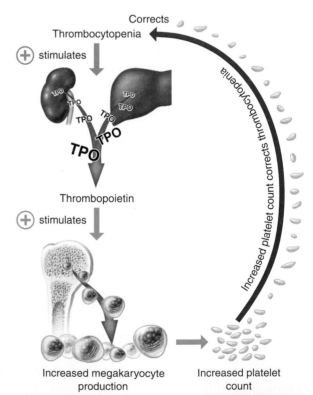

Figure 10.9. Thrombopoietin stimulates thrombopoiesis. The platelet count is maintained at a relatively constant level by a negative feedback loop involving the hormone thrombopoietin. *If platelet count were artificially increased by a platelet transfusion, what would happen to megakaryocyte production?*

negative homeostatic feedback mechanism: as the number of platelets declines, thrombopoietin levels rise and thrombopoiesis is stimulated.

Platelets are very important in hemostasis: they help prevent hemorrhage and help stop it when it occurs. The role of platelets in hemostasis is discussed further on. An abnormally low number of platelets is called **thrombocytopenia.** Spontaneous hemorrhage becomes a danger when the platelet count falls below 50,000/μL.

Case Note

10.13. What is the cause of Eleanor's measles-like spots and nosebleed?

 10.25 Correct the italicized terms in the following statement to make it true: *Erythropoiesis* is the synthesis of platelets; *leukopoiesis* is the synthesis of RBCs.

10.26 Name the two cellular components of blood that do not have DNA.

10.27 True or false: A patient with thrombocytopenia is at increased risk of hemorrhage.

Hemostasis

Hemostasis is the collective name for a group of activities that together prevent or stop bleeding. *Blood clotting* is the best known of these activities, but the formation of a blood clot is only one of many factors in the prevention and control of hemorrhage.

Blood vessels play an important role in hemostasis. The diameter of small vessels, called *arterioles,* determines blood flow to a particular region (such as a small patch of skin). Contraction of smooth muscle in the vessel wall is *vasoconstriction,* which narrows the vessel diameter and reduces blood flow to a particular tissue or organ. Smooth muscle relaxation is *vasodilation,* which increases vessel diameter and blood flow.

All blood vessels and the heart are lined by a smooth endothelial membrane, which prevents blood from touching any tissue or fluid outside the confines of the vascular system. Normally, therefore, blood is confined to the interior of the heart and blood vessels (together known as the *vascular space*) and flows smoothly and clot-free. Disruption of this barrier allows blood to contact extravascular tissues, initiating the events of hemostasis.

Remember This! Hemostasis, the control of hemorrhage, is different from homeostasis, the maintenance of a stable internal environment.

Blood Loss Threatens Life

Sudden blood loss, as with a serious wound, is an immediate threat to health and life. It has two ill effects. The loss of RBCs results in *reduced oxygen transport capacity.* At the same time, the loss of blood volume decreases blood pressure and results in *low blood flow to tissues.* Lack of oxygen transport and low blood flow combine to make severe hemorrhage a serious problem. Sudden loss of 10% of blood volume (about a pint, or 500 mL) will cause a drop of blood pressure, weakness or fainting, and fatigue with normal effort. Sudden loss of about 30% (1½ quarts, or 1,500 mL) will dramatically reduce blood pressure and may be fatal if the lost blood volume is not replaced quickly by transfusion of either plasma or whole blood (discussed below).

Case Note

10.14. If Eleanor receives a plasma transfusion, which blood function will be improved: oxygen carrying capacity or blood pressure?

Gradual loss of blood, as with a slowly bleeding intestinal tumor, has a less dramatic effect because the lost *volume* of whole blood is continually replaced in less than a day by an equal volume of *cell-free* fluid (plasma). Although the number of RBCs is diluted and the patient's hemoglobin, hematocrit, and RBC count fall, the patient's blood pressure may remain normal because of the increased production of plasma. But without adequate RBCs, the oxygen-transport function of blood remains impaired. Patients may fatigue easily but be able to go about their daily affairs. If blood loss stops and no erythrocytes are infused (transfused), it may take months for the bone marrow to produce enough RBCs for blood hematocrit, hemoglobin, and RBC count to return to normal. Oxygen-transport capacity can be improved by new RBCs, whether by natural production or transfusion.

Remember This! **Significant blood loss is associated with low oxygen transport capacity and low blood flow. Slow blood loss reduces oxygen-transport capacity, but blood flow and pressure may remain stable.**

Neither a stabilized sudden blood loss nor a gradual blood loss significantly reduces the other functions of blood—defense, heat transfer, and clotting. The leukocytes and other elements responsible for defense against microscopic invaders are quickly replaced. Heat transfer mainly depends on the volume of blood flowing to skin, and, as we have seen, after blood loss, the missing volume is usually restored within a day by plasma. Finally, clotting factors are so abundant in plasma that the amount lost in nonfatal blood loss is insignificant.

Vasoconstriction Is the First Action in Hemostasis

Hemostasis has three components (Fig. 10.10):

1. Vasoconstriction
2. Formation of a platelet plug
3. Coagulation (blood clotting)

When a blood vessel is broken, damaged endothelial cells release local chemical signals (*paracrine factors*, ◀ Chapter 4). These chemical signals, along with reflex autonomic nerve signals, stimulate smooth muscle contraction in the vessel wall, which reduces the diameter of the lumen (Fig. 10.10, step 1). Vasoconstriction limits but does not stop blood loss from the ruptured vessel. The reaction is instantaneous but short-lived; nevertheless, it buys precious time for platelets and coagulation to effect a lasting solution.

Platelets Slow Hemorrhage by Forming a Temporary Plug

When a blood vessel breaks, blood cells and plasma leave the vascular space and come into contact with collagen in the connective tissue surrounding the blood vessel (Fig.10.10, step 2). Platelets stick to the collagen and aggregate to form a *platelet plug*, which physically fills the break in the blood vessel wall. The plug may completely fill a small, capillary-size break and completely stop the bleeding, or it may limit the rate of hemorrhage from larger vessels. Chemicals released from damaged endothelial cells make platelets stickier, encouraging further platelet aggregation. Platelets in

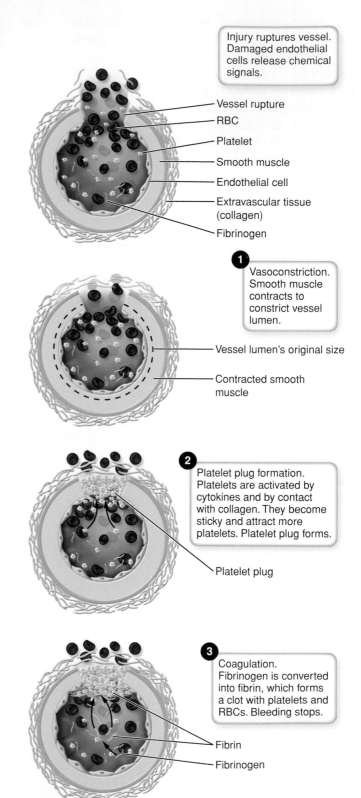

Injury ruptures vessel. Damaged endothelial cells release chemical signals.

Vessel rupture
RBC
Platelet
Smooth muscle
Endothelial cell
Extravascular tissue (collagen)
Fibrinogen

1 Vasoconstriction. Smooth muscle contracts to constrict vessel lumen.

Vessel lumen's original size

Contracted smooth muscle

2 Platelet plug formation. Platelets are activated by cytokines and by contact with collagen. They become sticky and attract more platelets. Platelet plug forms.

Platelet plug

3 Coagulation. Fibrinogen is converted into fibrin, which forms a clot with platelets and RBCs. Bleeding stops.

Fibrin
Fibrinogen

Figure 10.10. The steps of hemostasis. Bleeding is stopped by sequentially vasoconstricting the vessel, forming a platelet plug and forming a clot. *Which protein forms the clot—fibrinogen or fibrin?*

the plug, in turn, release chemical signals that (a) stimulate further blood vessel contraction, (b) attract additional platelets, and (c) accelerate coagulation.

In a patient with a bleeding problem, lab technicians can count the number of platelets and test for adequate platelet function. If the number of platelets is too low, a platelet plug cannot form. What's more, low numbers of platelets cannot accelerate coagulation as they do when present in normal numbers. The *bleeding time* test examines the ability of platelets to form a plug and release their coagulation factors; it does so by determining the time it takes a tiny standardized pinprick wound to stop bleeding.

Case Note

10.15. Because of her thrombocytopenia, would Eleanor have a normal result in a bleeding time test?

Coagulation Results in a Fibrin Clot

Coagulation is a chain of events that produces a **clot,** a gel-like, semisolid web of *fibrin* filaments, platelets, and trapped blood cells that obstructs further escape of blood from an injured vessel (Fig.10.10, step 3). **Fibrin** is a long filamentous protein created by polymerization of *fibrinogen*, one of the three major types of plasma protein mentioned earlier.

Coagulation Factors Control Clot Formation

Coagulation is the result of interactions among about 50 blood **coagulation factors,** most of which circulate in plasma, although some are present in platelets, endothelial cells, and tissues. Most coagulation factors are proteins made by the liver; they are denoted by names and Roman numerals: factor I, factor II, and so on. These factors interact with one another in a cascade of steps: one reaction prompts another until the process is forced to its end point—a clot formed from a tangle of very long, thin strands of fibrin, which entrap blood cells and platelets and obstruct blood flow through the break. Absence of one factor can interrupt the entire process.

For example, factor VIII is also called *hemophilia factor* because its absence is the cause of *hemophilia*, a sex-linked genetic condition found almost exclusively in males. The absence of factor VIII interrupts the coagulation cascade and prevents the production of fibrin. Platelets function normally, but platelets alone cannot stop hemorrhage, and blood loss continues. Hemophilia is associated with easy bruising or severe spontaneous hemorrhage, especially into joints, from otherwise minor trauma.

Coagulation Factors Interact in Pathways

The coagulation cascade can be divided into two initial pathways. The *tissue factor pathway* is most important. The *contact activation pathway* serves a secondary role (Fig. 10.11).

These two initial pathways can be likened to two rivers joining to make a larger one—they merge into a single, final pathway called the *common pathway*. Each pathway involves a distinct set of coagulation factors.

The **tissue factor pathway** is initiated when extravascular tissue or fluid, which ordinarily do not come into contact with blood, are exposed to plasma. The cell membrane of extravascular cells contains a protein called **tissue factor** (or *thromboplastin*), which initiates clotting when it comes into contact with plasma. When blood vessel damage exposes plasma to tissue factor, tissue factor activates a plasma clotting factor, *factor VII*. Factor VII then activates the common pathway, discussed below. The tissue factor pathway is fast, producing a clot in a few seconds.

The **contact activation pathway** is so named because it is most often initiated by contact with foreign material, especially when blood is removed from the body (in vitro), as in laboratory testing. This pathway is initiated when plasma factor XII comes into contact with foreign substances, such as glass, plastic, or metal. Some of the chemical reactions of the contact activation pathway require calcium, which acts as a clotting factor itself (factor IV). Indeed, in laboratory practice, one way to prevent blood from clotting (*anticoagulation*, discussed further below) is to add a chemical that binds calcium and prevents its participation in the clotting process. The contact activation pathway is slow and takes a few minutes to produce a clot. This pathway can also be activated in vivo, but it is of secondary importance to the tissue factor pathway because individuals lacking one or more factors in this pathway do not necessarily have bleeding disorders.

The **common pathway** begins with the activation of *thrombokinase* (factor X), an enzyme that converts another clotting factor, **prothrombin** (factor II), into **thrombin.** In the final step of coagulation, thrombin acts on fibrinogen (factor I) to convert it to the weave of fibrin that forms the clot.

Clotting is a *rapid* homeostatic response that normally occurs only *outside* of the vascular space and is the consequence of normal physiological systems. However, in some pathological circumstances, blood clots *inside* blood

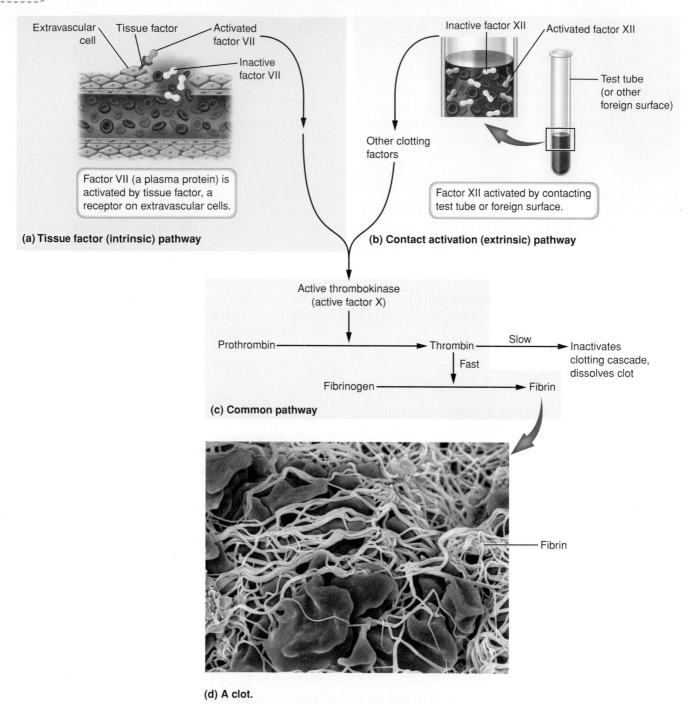

(a) Tissue factor (intrinsic) pathway

Extravascular cell — Tissue factor — Activated factor VII — Inactive factor VII

Factor VII (a plasma protein) is activated by tissue factor, a receptor on extravascular cells.

(b) Contact activation (extrinsic) pathway

Inactive factor XII — Activated factor XII — Test tube (or other foreign surface)

Other clotting factors

Factor XII activated by contacting test tube or foreign surface.

(c) Common pathway

Active thrombokinase (active factor X)

Prothrombin → Thrombin — Slow → Inactivates clotting cascade, dissolves clot

Fast

Fibrinogen → Fibrin

(d) A clot.

Fibrin

Figure 10.11. The coagulation pathway. A. Exposure of clotting factor VII to tissue factor on extravascular cells activates the tissue factor pathway. **B.** The contact activation pathway is activated when clotting factor XII contacts collagen or a foreign surface such as glass. Activated factor XII initiates a series of enzymatic reactions. **C.** Activated factor VII or factors activated by the contact activation pathway activate the common pathway, which terminates in the formation of a web of fibrin fibers forming a clot. **D.** A photomicrograph of a blood clot. *Which enzyme converts fibrinogen into fibrin?*

vessels (*intravascular coagulation*) even if no hemorrhage is occurring. Intravascular coagulation is always secondary to some other serious disease and is often fatal.

In a patient with a bleeding problem that might be due to faulty hemostasis, lab technicians can test the quality of coagulation by artificially activating each of the clotting pathways in a blood specimen and measuring the time required to form a clot. A deficiency of any of the plasma factors will extend the time it takes to form a clot. By knowing which factors are involved in each pathway, the deficiency can be identified or narrowed to a few possibilities.

Plasmin Dissolves Clots

After it forms, a clot slowly contracts, pulling together the wounded tissue and making the clot denser and more secure. Later, as the body's repair process regenerates new tissue or scar tissue to provide permanent closure, the clot slowly dissolves. Dissolution is due to the action of another blood protein, *plasmin*, which digests the fibrin holding the clot together. Thrombin, the protein that stimulates clot formation, also promotes clot dissolution by stimulating plasmin production.

As a clot contracts, it expresses a fluid called **serum,** which is identical to plasma except that it contains no fibrinogen or other clotting factors. Serum is the fluid that seeps, for example, from a fresh skin abrasion. Serum is also harvested from whole blood for laboratory analysis because it contains most of the constituents of plasma but does not form clots that could interfere with laboratory tests. For more detail about blood versus plasma versus serum, see the nearby Clinical Snapshot titled *Serum? Plasma? Blood?*

Anticoagulants Prevent Coagulation

Whether or not blood coagulates depends upon the balance of forces promoting or opposing coagulation.

Normally, blood is maintained in a smooth, clot-free state because circulating anticoagulants prevent clot formation. Clotting occurs when procoagulation forces prevail. Given that clotting is a chain reaction, it is reasonable to ask: Once it starts, why doesn't it proceed to coagulate the entire blood supply? The answer is that natural anticoagulants are released at the site of the clot and restrict clotting to the locality of the injury and hemorrhage. For example, basophils and other defensive cells release heparin, a polysaccharide that acts as a quick, short-lived anticoagulant by interfering with the action of prothrombin (factor II).

Pharmaceutical anticoagulants are frequently administered as therapeutic drugs. For example, patients undergoing kidney dialysis are treated with a form of heparin to keep blood circulating through tubes that would otherwise cause activate the contact activation pathway. *Warfarin* (Coumadin™) is a widely used anticoagulant that interferes with the action of vitamin K, which is required by the liver for the production of several coagulation factors, most notably prothrombin (II). Because it takes a few days for existing coagulation factors to disappear, warfarin anticoagulation takes a few days to develop.

Finally, *aspirin* is a weak anticoagulant—it interferes with the ability of platelets to become "sticky" and form

CLINICAL SNAPSHOT

Serum? Plasma? Blood?

Serum is the fluid remaining after blood has clotted. Serum and plasma are identical in every respect except that plasma contains fibrinogen and serum contains none—all of the fibrinogen is consumed in forming fibrin to make the clot. For example, the bloody scab on a skin scrape is a dried clot, and the yellowish fluid that sometimes seeps from around the edges of the scab is serum.

Plasma is tricky to use for laboratory analyses for two reasons:

● Anticlotting chemicals must be added to blood in order to obtain unclotted plasma. These chemicals alter plasma such that it no longer is an accurate representation of plasma as it was in the body. For example, most laboratory anticoagulants bind calcium to prevent clotting, and this prevents analysis of the plasma's calcium content.

● Anticlotting chemicals do not always work perfectly; small fibrin clots can develop and "gum up" delicate laboratory instruments.

Because serum contains no fibrinogen and cannot clot again, it is much more practical to use it for laboratory analyses of the constituents of plasma. Here's how it's done. The lab technician collects blood into a glass or plastic tube, which activates the contact activation clotting pathway and allows the blood to clot. Just as a clot shrinks after formation in the body, it shrinks in a specimen tube—after an hour, the clot is about half its original size. Serum is the remaining fluid, which can easily be harvested by centrifugation of the clot to the bottom of the tube.

Because serum and plasma are identical for most purposes and laboratory analysis of serum so accurately reflects values in whole blood, the words "serum," "plasma," and "blood" are often used interchangeably. For example, when people speak of "blood glucose" or "plasma glucose," the actual measurement was almost always done on serum.

a platelet plug. As we have seen, aggregation of sticky platelets is a critical step in hemostasis. As we discuss next, platelet aggregation is also one of the initial steps in forming the cellular layers of a *thrombus*—a risk factor in cardiovascular disease. For this reason a daily aspirin is effective in preventing some heart attacks and strokes caused by thrombus formation.

Anticoagulants are also used for laboratory tests that require unclotted blood. More information can be found about this topic in the nearby Clinical Snapshot titled *Serum? Plasma? Blood?*

Case Note

10.16. The emergency room physicians advised Eleanor to stay away from aspirin. Why?

Coagulation and Thrombosis Differ

A **thrombus** is an abnormal, localized *intravascular* collection of platelets and blood cells, but it is not a clot—coagulation factors are not necessary for its formation. And whereas clotting is a normal process that occurs outside of blood vessels, a thrombus is always abnormal and occurs only within the lumen of a blood vessel.

In the formation of a clot, blood cells are trapped in a rapidly developed, gelatinous weave of fibrin at a site of hemorrhage. By contrast, a thrombus forms inside a blood vessel at a point where the vessel lining is injured or blood flow as normal—no vascular tear or hemorrhage is required. The injured lining of the blood vessel becomes "sticky" and platelets begin to accumulate slowly. Later, WBCs begin to accumulate and finally some RBCs become trapped in the growing mass. As cells accumulate, they do so in layers, creating a *visible internal architecture* that is distinctly different from a blood clot, which is a featureless red gel.

Thrombi tend to occur in arteries damaged by *atherosclerosis* (➡ Chapter 11), an accumulation of cholesterol, other lipids, and scar tissue that occurs in damaged arteries. And because thrombi form inside blood vessels, they are capable of obstructing blood flow and are often a key element in a chain of events leading to heart attack or stroke.

The danger of intravascular thrombus formation during long flights prompts airlines to urge passengers to keep their seatbelts loose and to move about the plane periodically. Tight seatbelts and lack of activity cause blood flow to stagnate in the legs, thus favoring thrombus formation. These can break away and be car-

ried to the lungs. If such a thrombus is large enough, it can completely and fatally block blood flow to the lungs.

Remember This! **A clot and a thrombus are distinctly different.**

10.28 Which term describes the processes that stop bleeding—*hematopoiesis, hemostasis, or homeostasis?*

10.29 Identify in order the three steps involved in hemostasis.

10.30 Which cellular elements of blood form a plug in broken blood vessels?

10.31 Which plasma protein is involved in blood clotting—fibrinogen, albumin, or gamma globulin?

10.32 Which element is a clotting factor—calcium or iron?

10.33 True or false: Tissue factor is a soluble blood protein.

10.34 Fill in the blank: A deficiency in the _____ pathway does not necessarily result in a bleeding disorder.

10.35 Do the following terms apply to thrombosis or clotting? *Intravascular, organized structure,* and *pathological.*

10.36 Do heparin and warfarin block the tissue factor pathway, contact activation pathway, or common pathway? Explain.

Blood Groups and Transfusion

We noted at the beginning of this chapter that humankind has long associated loss of blood with illness or death and understood that blood was somehow vital to life. So it is no surprise that history is littered with attempts to restore health by transfusing blood. The nearby History of Science box, titled *The Ancient Practice*

THE HISTORY OF SCIENCE

The Ancient Practice of Bloodletting

Skin and blood were the first body tissues to be recognized by the ancients: skin was constantly visible, whereas blood became visible following trauma. It was easy enough to see that injury produces bleeding, severe injury produces severe bleeding, and severe bleeding causes death. The conclusion necessarily followed that the "life spirit" must be in blood, and blood, therefore, must be important in maintaining life.

But as civilizations evolved, ancient eastern and western traditions came to differ in their views about blood. Asian physicians recognized that the flow of blood transported "energy" to body parts. However, beginning with the Greeks, western physicians came to see blood as the carrier of harmful substances and thus practiced *bloodletting*, a therapy in which blood is drained from the body. Bloodletting was a recorded practice from at least the time of the Sumerians (1800 BCE) until the 19th century.

For more than 1,000 years, Western physicians believed that the excess accumulation of certain "humours"—black bile, yellow bile, phlegm, and blood itself—was the cause of all illness. These physicians advocated and perpetuated the practice of bloodletting as a mainstay of treatment to rid the body of imagined excesses of humours. Two methods were common: bloodsucking leeches were applied to skin, or veins were sliced open. For example, in 1799, on what proved to be his deathbed, George Washington was repeatedly bled of perhaps a liter or two of blood by a series of physicians. His ailment: a severe throat infection, which leaves us to speculate that he might have survived but for the ministrations of his doctors.

Bloodletting persisted well into the 19th century until it was largely displaced by treatments backed by empirical evidence of success. Nevertheless, it survives in a very narrow niche in modern medicine. Patients with a surprisingly common inherited condition called *hemochromatosis* (literally "blood colored") accumulate

Bloodletting in the 19th century.

too much iron in their blood and other tissues, which can be toxic to the heart and other organs. RBCs contain large amounts of iron, and there is no good way to remove iron from the body except by draining away blood from time to time. In such cases, bloodletting is a very effective practice that allows patients to lead otherwise normal lives.

Bloodletting is also used to treat a rare type of malignancy that produces an excess of certain blood proteins, which causes blood to become so thick that it cannot flow properly. However, in this instance, after patients are bled, the proteins are removed from the blood and the "thinned blood" is returned to the body.

of Bloodletting, offers an insight into the evolution of modern transfusion medicine.

If you've ever donated blood, you know that it's collected by venipuncture into sterile bags; but you might not be aware of the other steps that go on behind the scenes. First, donated blood is mixed with anticoagulant to prevent clotting; then it is refrigerated for up to 35 days until used. Before it's transfused, blood is tested to make sure that it's free from any infectious microbes and to determine its *blood group*. Without these steps, transfusion would be a highly risky business because (a) any bacteria, viruses, or other microbes in the donor blood could be transferred to the recipient and (b), the body of the recipient will reject the transfused (donor) cells unless they contain certain special proteins that match the recipient's own. We discuss the issues and challenges of safe transfusion in this section.

Incidentally, whole blood and RBCs are not the only transfusable substances. Modern technique enables transfusion of selected components of plasma (e.g., coagulation factor VIII into hemophiliacs) or cellular elements (e.g., platelets).

Erythrocyte Antigens and Plasma Antibodies Determine Transfusion Compatibility

For the purposes of transfusion, each person's blood is distinctive from the blood of many but not all other people.

Transfusion traits are due to *antigens* and *antibodies*, which are discussed in detail in ➡ Chapter 12. For now we will define an **antigen** as a molecule capable of provoking a defensive reaction by the immune system. An **antibody** is a specialized blood protein generated specifically to mount an attack against an antigen. The type of antigen important to our story of blood transfusion is a glycoprotein on the surface of red blood cell membranes. Persons possessing a certain RBC antigen are said to belong to a certain **blood group** (or *blood type*). For practical purposes, when the term *blood group* is commonly used, it refers to *RBC blood groups*. WBCs and platelets also have surface antigens that give them distinctive characteristics, but their numbers are so few that they are of little importance in most blood transfusions.

> **Remember This!** Blood group is determined by RBC antigens.

There are two main sets of RBC blood group antigens: the *ABO group* and the *Rh group,* both of which are inheritable. The **ABO group** is based on the presence or absence of two antigens, A and B (Fig. 10.12). Persons with neither A nor B antigen on their RBCs form blood group O (or blood *type* O). Persons with both A and B form blood group AB. Persons possessing only A antigen form group A; patients with only B form blood group B. Among Americans, the prevalence of the major RBC ABO blood groups, from most to least common, are blood types O, A, B, and AB.

The **Rh group,** which is so named because the antigen group was originally discovered in *rh*esus monkeys, is determined by the presence or absence of a different antigen. There are eight antigens in the Rh group, but only one (antigen D) is important. About 80% of Americans have the Rh D antigen on their RBCs and are said to be *Rh-positive;* those without the Rh D antigen are said to be *Rh-negative.*

The complete description of a person's blood type requires both ABO and Rh type. Thus a person may be O-positive (meaning type O, Rh-positive), AB-negative (meaning type AB, Rh-negative), and so on.

By virtue of the presence of A and B antigens in bacteria, food, and other components of the environment, every person early in life develops antibodies against the A or B antigens *not* present in their own blood. Therefore:

- The plasma of type A blood contains anti-B antibodies.
- The plasma of type B blood contains anti-A antibodies.
- The plasma of type O blood has both anti-A and anti-B antibodies (recall that type O blood cells do not contain A or B antigens).
- The plasma of type AB blood has neither anti-A nor anti-B antibodies.

The situation is quite different for the Rh D antigen. The environment does not contain Rh D antigen. Therefore, antibodies to the Rh D antigen develop *only* in an Rh-negative person who is exposed to large amounts of the Rh D antigen—say, by transfusion with Rh-positive blood.

The result is that anti-A and/or anti-B antibodies are present in almost everyone (except those with the rare AB blood type), and very, very few persons have anti Rh D antibody—only those exposed in some way to RBCs containing Rh D antigen.

Agglutination Reveals Blood Types

In the presence of the particular type of antigen they are designed to attack, antibodies attach to the antigens on the surface of RBCs and bind the RBCs together. The result is that interlocking chains of RBCs form in blood, a process called **agglutination** (Fig. 10.13). For example, agglutination occurs if type A blood is transfused into a recipient with type O blood because the type O patient has anti-A antibodies in plasma. Such a reaction is extremely serious.

Agglutination of RBCs is visible to the naked eye and is used to choose correctly matched blood groups for safe transfusion. To determine a patient's blood type, a procedure called **blood typing** is performed. Anticoagulated patient blood is placed on each of two slides. Anti-A

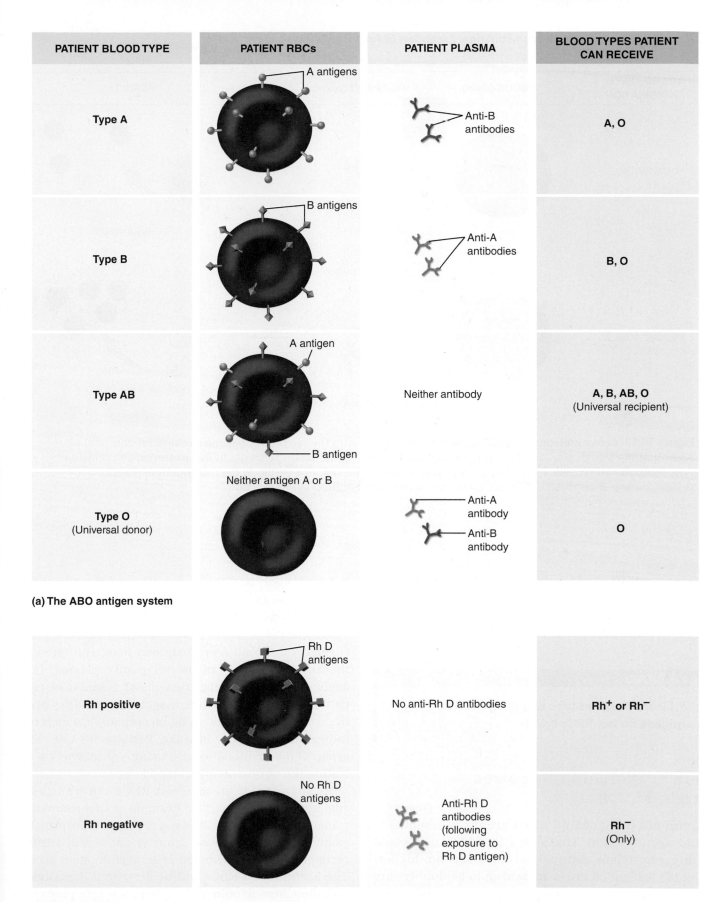

PATIENT BLOOD TYPE	PATIENT RBCs	PATIENT PLASMA	BLOOD TYPES PATIENT CAN RECEIVE
Type A	A antigens	Anti-B antibodies	**A, O**
Type B	B antigens	Anti-A antibodies	**B, O**
Type AB	A antigen / B antigen	Neither antibody	**A, B, AB, O** (Universal recipient)
Type O (Universal donor)	Neither antigen A or B	Anti-A antibody / Anti-B antibody	**O**

(a) The ABO antigen system

Rh positive	Rh D antigens	No anti-Rh D antibodies	**Rh⁺ or Rh⁻**
Rh negative	No Rh D antigens	Anti-Rh D antibodies (following exposure to Rh D antigen)	**Rh⁻** (Only)

(b) The Rh antigen

Figure 10.12. Common blood groups. A. ABO blood groups. An individual's plasma contains antibodies against A or B antigens not present on his or her erythrocytes. **B.** Rh blood groups. Antibodies against antigen Rh D are present only in Rh-negative people who have been exposed to Rh-positive blood (perhaps from a transfusion). *If you wanted plasma with both anti-A and anti-B antibodies in it, which blood group would you use?*

401

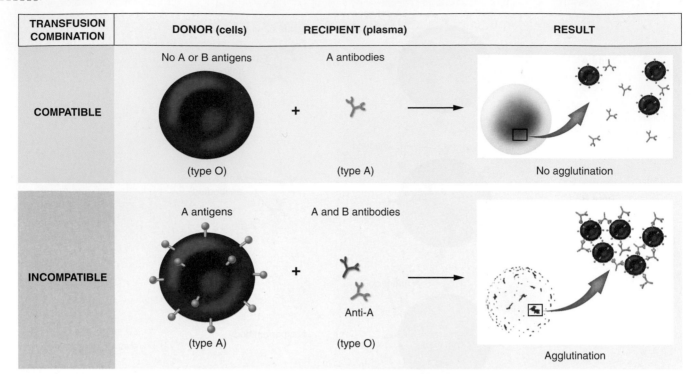

Figure 10.13. Cross-matching. Donor blood cells are mixed with the recipient's plasma to check for compatibility. A compatible transfusion combination will not result in an agglutination reaction. In this example, someone with type A blood can receive a transfusion with type O blood, but not vice versa. *In the incompatible transfusion, would the anti-B antibody of the recipient participate in the agglutination reaction?*

plasma is added to one and anti-B to the other. If RBCs in both agglutinate, the patient belongs to blood group AB; if there is no agglutination on either slide, the patient is type O, and so on.

Case Note

10.17. Eleanor's blood type is O-positive. What AB or Rh antigens do her RBCs have?

Safe Transfusion Requires Cross-Matching

Careful matching of RBC blood groups is essential to avoid agglutination. That's why, in addition to blood typing, physicians routinely order a second, higher-level lab test called **cross-matching** to be doubly sure that no agglutination will occur. The most important step in a cross-match involves mixing a sample of the proposed donor's RBCs with a sample of the proposed recipient's plasma to ensure that the two are actually, not just theoretically, compatible (see Fig. 10.13). For example, we might propose to transfuse RBCs from a group O donor into a patient with group A RBCs. We expect the cross-match to be compatible because group O cells contain no A or B antigen to agglutinate with the anti-B antibody in the recipient's plasma. If no agglutination occurs, the cross-match is said to be compatible. On the other hand, if agglutination does occur, the cross-match is said to be incompatible, which may indicate a laboratory mistake. Perhaps it's not really group O donor blood we are using—or maybe we are testing the wrong recipient blood.

In any case, if mismatched RBCs are transfused, agglutination will occur. An example of an incompatible transfusion combination is the reverse of our previous example; if we transfuse group A donor RBCs into a recipient with type O blood cells, agglutination occurs. The agglutinated RBCs will be destroyed (hemolysis), and their hemoglobin will be released into plasma in what is called an *acute hemolytic reaction*. This free hemoglobin must be cleansed (cleared) from plasma by the kidneys, but kidney capacity is limited and can easily be overwhelmed, resulting in kidney failure. In addition, even a small amount of agglutinated RBCs can clog blood vessels, thus severely reducing blood flow to tissues. The result is often fatal.

Case Note

10.18. Based on the information in the previous case note, could Eleanor be safely transfused with type A blood?

10.37 Which blood group does not have any of the A, B, or Rh antigens?

10.38 If someone has anti-A antibodies and Rh antibodies, what is his or her blood group?

10.39 If someone has anti-A antibodies but no Rh antibodies, can we be certain of his or her blood group? Explain.

10.40 Explain how an incompatible transfusion could cause the death of the patient.

Case Discussion

Bone Marrow Failure: The Case of Eleanor B.

Eleanor died of bone marrow failure—breast cancer spread to her bones and replaced most of the bone marrow. At the time of her death, every one of her major cell lines was affected: platelet and leukocyte cell counts were very low, and she was severely anemic even before her final, fatal hemorrhage. Let's discuss each of these elements in turn (Fig. 10.14).

- *Leukocytes* When she first appeared in the emergency room, Eleanor's total white cell count was very low. From the time of her initial emergency room visit until the time of her death, she required antibiotic treatment for pneumonia, skin infections, and recurrent diarrhea, suggesting that the leukopenia impaired her resistance to disease.

The medical record does not indicate the counts of specific WBCs, but it is certain that the numbers of both neutrophils and lymphocytes (the most abundant blood leukocytes) were markedly decreased. Autopsy findings are confirmatory: bacterial infection of blood and lungs, which can be attributed to low neutrophil and lymphocyte counts, and fungal infection of her esophagus, which is sure to be a reflection of a low lymphocyte count. Moreover, healing of the ulcer, which had been caused by the fungal infection, was also impaired by neutropenia.

- *Platelets* Note that on Eleanor's initial hospitalization, her platelet count was very low; she also had a severe nosebleed and tiny skin hemorrhages (*petechiae*, which are characteristic of platelet-induced bleeding). Physicians transfused her with platelets, which had the desired effect of stopping the bleeding. On her final admission with a massive gastrointestinal hemorrhage, the platelet count was again very low. The thrombocytopenia and bleeding esophageal ulcer combined to produce uncontrollable acute esophageal hemorrhage.
- *Erythrocytes* When she initially appeared with a nosebleed, Eleanor was significantly anemic, likely reflecting the failure of marrow production even before the nosebleed occurred. The nosebleed (and the later esophageal hemorrhage) introduced to the existing production-failure anemia a new problem of hemorrhagic anemia.

Although Eleanor's final blood hemoglobin level was extremely low, some patients with such low hemoglobin levels can survive if their blood volume is adequate—that is, if their bleeding is slow enough (or therapy quick enough) for their plasma volume to expand to make up for the lost blood volume. But Eleanor's blood volume was not adequate—she bled so rapidly that both blood volume and blood pressure fell to undetectable levels and she expired from lack of oxygen (hypoxia).

Case Notes

10.19. On her final emergency room visit, why is Eleanor's blood pressure low?

10.20. Eleanor is confused because her brain is not getting enough oxygen. In addition to her low blood pressure, can you think of a second reason?

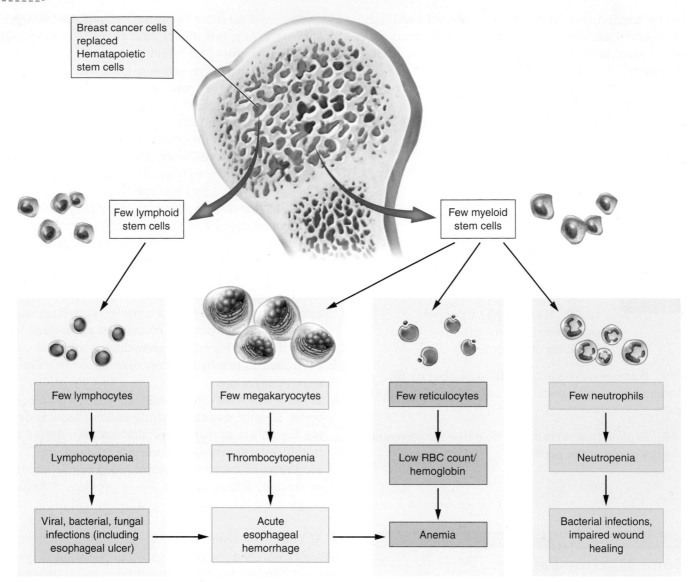

Figure 10.14. Bone marrow failure and Eleanor H. Eleanor's bone marrow can no longer produce enough blood cells to fight infection, prevent hemorrhage, and prevent anemia. *The absence of which cell type increases Eleanor's susceptibility to fungal infections?*

Word Parts		
Latin/Greek Word Parts	**English Equivalents**	**Examples**
-cyte	Cell	Leukocyte: white blood cell
-emia	Blood	Anemia: lack (an-) of red blood cells
erythr/o	Red, red blood cell	Erythropoietin: hormone stimulating red blood cell synthesis
hemat/o, hem/o	Blood	Hemostasis: maintenance of a constant Blood volume
leuk/o, leukocyt/o	White blood cell	Leukocytopenia: too few white blood cells

Word Parts (continued)

Latin/Greek Word Parts	English Equivalents	Examples
lymph/o	Lymphocyte	Lymphopenia: lymphocyte deficiency
macro-	Large	Macrophage: large cell
mono-	One	Monocyte: contains a single (unlobed) nucleus
myel/o	Bone marrow	Myeloid stem cell: stem cell originating in bone marrow
-oxia	Oxygen	Hypoxia: low oxygen concentration (in blood)
-penia	Deficiency	Thrombocytopenia: platelet deficiency
-poiesis	Synthesis	Erythropoiesis: erythrocyte synthesis
pro-	Before, in front of	Prothrombin: a precursor to thrombin
thromb/o, thrombocyto	Blood clot, platelet	Thrombopoietin: hormone stimulating platelet synthesis

Chapter Challenge

CHAPTER RECALL

1. **If you have 5 L of blood and cellular elements account for 40% of your blood volume, what is your total plasma volume?**
   a. 5 L
   b. 4 L
   c. 3 L
   d. 2 L

2. **Which of the following is *not* a role of albumin?**
   a. Blood clotting
   b. Transportation of fatty acids
   c. Maintaining the osmotic pressure of blood
   d. None of the above

3. **Serum contains**
   a. RBCs.
   b. clotting factors.
   c. leukocytes.
   d. none of the above.

4. **Which of the following cellular elements contains a nucleus?**
   a. Erythrocyte
   b. Megakaryocyte
   c. Platelet
   d. All of the above

5. **Hemoglobin**
   a. is found in leukocytes.
   b. contains iron.
   c. transports fatty acids and other substances in blood.
   d. is an inorganic molecule.

6. **Monocytes are**
   a. granulocytes.
   b. the most abundant WBCs.
   c. the precursor to a type of phagocyte.
   d. the least abundant WBCs.

7. **The generation of blood cells is called**
   a. hemostasis.
   b. homeostasis.
   c. hematopoiesis.
   d. hemolysis.

8. **Erythropoietin is**
   a. a clotting factor made by the liver.
   b. produced in response to hypoxia.
   c. produced by erythrocytes.
   d. the hormone that stimulates platelet synthesis.

9. **Lymphoid stem cells produce**
   a. all WBCs.
   b. only lymphocytes.
   c. only granulocytes.
   d. all blood cells.

10. **Bilirubin is**
    a. produced from globin degradation.
    b. recycled in the bone marrow into new hemoglobin molecules.
    c. excreted into the intestine by the liver.
    d. the protein that stores iron in tissues.

11. **Which of the following cells is the most important player in acute inflammation and wound healing?**
    a. Eosinophils
    b. Basophils
    c. Lymphocytes
    d. Neutrophils

12. **Reticulocytes mature into**
    a. basophils.
    b. macrophages.
    c. erythrocytes.
    d. platelets.

13. **The increased hematocrit resulting from dehydration is called**
    a. relative anemia.
    b. absolute anemia.
    c. relative polycythemia.
    d. absolute polycythemia.

14. **You have just cut your finger. The body's first physiological response to this trauma is the**
    a. formation of a platelet plug.
    b. activation of clotting factors.
    c. formation of a thrombus.
    d. vasoconstriction.

15. **The tissue factor pathway is**
    a. initiated when cells that do not usually encounter blood proteins are exposed to plasma.
    b. stimulated by interactions between clotting factors and foreign substances.
    c. of secondary importance to the contact activation pathway in responding to tissue injury.
    d. the only clotting pathway that involves thrombin.

16. **Thrombin**
    a. converts fibrinogen into fibrin.
    b. binds tissue factor.
    c. forms a meshwork of fibers within the blood clot.
    d. is involved in the formation of thrombi but not of blood clots.

17. **Transfusion reactions largely reflect the presence of antigens on**
    a. RBCs.
    b. WBCs.
    c. platelets.
    d. all blood cells.

18. **An individual with type A blood will have antibodies against**
    a. B antigen.
    b. A antigen.
    c. both A and B antigens.
    d. neither A nor B antigens.

19. **An individual with type O blood can safely receive a transfusion with**
    a. type A blood.
    b. type B blood.
    c. type AB blood.
    d. none of the above.

CONCEPTUAL UNDERSTANDING

20. **Name three structural characteristics of RBCs and explain the functional implications of each.**

APPLICATION

21. In a person who is experiencing chronic slow bleeding in the gastrointestinal tract, which of the following functions would be adversely affected? Defend your answers.
    a. Fighting off a cold
    b. Flushing red on a hot day
    c. Hemostasis of a paper cut
    d. oxygen transport

22. Your friend has recently learned that that 75% of her bone marrow is involved in leukocyte production, and she is worried that she might have leukemia. Is she right to worry? Explain.

23. The most common blood type in Norway is A+. In an A+ individual,
    a. which antigens are present on RBCs?
    b. which antibodies are present in plasma?
    c. would a transfusion with A- blood induce agglutination? Explain why or why not.

You can find the answers to these questions on the student Web site at
http://thepoint.lww.com/McConnellandHull

11

The Cardiovascular System

Major Themes

- Blood flows in a closed circuit powered by muscular contractions of the heart.

- One-way valves in the heart and veins act as gates to ensure unidirectional blood flow.

- The heart has a specialized electrical system to stimulate and coordinate contractions.

- Blood flows under high pressure away from the heart through arteries; blood flows back to the heart under low pressure through veins.

- Arteries and veins are connected by a vast network of tiny capillaries.

- Blood pressure is determined by volume of blood in the arteries and the stretchiness of the arterial walls. In practice, it changes according to the amount of blood pumped by the heart and the resistance to its flow through blood vessels.

Chapter Objectives

The Organization of the Cardiovascular System 411

1. Describe the pathway a blood cell takes from the lungs to the tissues and back.

Structure and Function of the Heart 413

2. Accurately describe the anatomical location of the heart, and describe the layers of the heart wall and the pericardium.

3. Compare skeletal and cardiac muscle in terms of their microscopic appearance, the role of calcium in muscle contraction, and the method of generating stronger or weaker contractions.

4. Explain the importance of the longer duration of the action potential in cardiac muscle than in skeletal muscle.

5. Explain the importance of pressure gradients and valves in governing blood flow through the heart and great vessels.

6. Label the coronary arteries and the coronary sinus on anterior and posterior views of the heart.

The Heartbeat 422

7. For each stage of the cardiac cycle, know which valves are open, which chambers are contracting, and the direction of blood flow (if any).

8. List the structures that make up the cardiac conducting system, and explain why the sinoatrial (SA) node is usually the pacemaker.

9. Link the events of an electrocardiogram (ECG) to the cardiac cycle.

Cardiac Output 427

10. Discuss the effects of the autonomic nervous system on heart rate, stroke volume, and cardiac output.

11. Explain the relationship between preload, cardiac muscle fiber length, and stroke volume. Discuss the relevance of this relationship to cardiovascular function.

Structure and Function of Blood Vessels 431

12. Describe the different tunics present in the five types of vessels.

13. Compare the function of elastic arteries, muscular arteries, arterioles, and precapillary sphincters.

14. List four methods of capillary exchange.

15. Predict the effects of changes in hydrostatic and osmotic pressure on bulk flow and fluid balance.

16. List three mechanisms that assist venous return to the heart.

Blood Flow and Blood Pressure 438

17. Compare diastolic pressure, systolic pressure, mean arterial pressure, and pulse pressure.

18. Explain how arterial blood volume and vessel compliance are the major determinants of blood pressure.

19. Explain how overall blood volume, cardiac output, and peripheral resistance all modify blood pressure by altering arterial blood volume.

20. Describe how the sympathetic nervous system acts to increase blood pressure.

Myocardial Infarction: The Case of Bob W. 443

21. Use the case study to illustrate the relevance of heart rate and the strength of myocardial contraction to maintaining adequate blood pressure.

The Major Blood Vessels 444

22. Discuss the advantages of portal circulations and anastomoses, giving an example of each.

23. Label diagrams of the major veins and arteries of the head, upper limb, torso, and lower limb.

24. Trace the pathway that a blood cell would take in leaving from and returning to different body regions, such as the big toe, large intestine, and left cerebral hemisphere.

"He's been having a lot of chest pain."

As you read through the following case study, assemble a list of the terms and concepts you must learn in order to understand it.

Clinical History: Bob W., a 55-year-old male, was wheeled into to the emergency room by his wife. Semiconscious and unable to speak for himself, he was moaning with apparent pain. His wife reported that he was having severe deep anterior chest pain that spread to his left neck, left jaw, and left elbow. The pain had wakened him from sleep about an hour earlier.

Physical Examination and Other Data: Bob was severely obese was very short of breath. Vital signs were as follows: blood pressure 75/45 (normal 120/80), heart rate 30 beats per minute (normal 72), and respirations 30 per minute (normal 14). An ECG showed evidence of myocardial infarction.

Clinical Course: An intravenous drip of norepinephrine was begun in an effort to raise his blood pressure, which rose to 105/70 but began to fall again after 30 minutes. An external pacemaker was applied to his chest in an effort to stimulate his heart rate, which rose above 60 for a short while but fell back despite continued stimulation. A catheter was inserted in his right femoral artery and threaded into the root of his aorta. Dye injected for x-ray visualization of the coronary arteries revealed complete obstruction of the right coronary artery and severe narrowing of the left coronary. It was also noted that the posterior wall of his left ventricle was flaccid and not contracting.

Bob's blood pressure fell further. Within the hour it became undetectable and he suffered cardiac arrest. Resuscitation was not successful. His wife granted permission for an autopsy, which revealed that the right main coronary artery was severely narrowed by atherosclerosis and completely occluded by a fresh thrombus ⬅ (Chapter 10). The remainder of his coronary arterial tree and his systemic arterial tree were severely atherosclerotic.

The final autopsy report concluded that Bob had suffered an acute coronary occlusion with myocardial infarction.

The **cardiovascular system** consists of the *heart* and *blood vessels* and the fluid they contain—*blood*. Since blood is discussed in Chapter 10, this chapter focuses on the heart and blood vessels.

Among the ancients, the heart was widely conceived of as the seat of emotion—after all, the heart races with excitement, fear, anger, and other strong emotions. Some felt it was the seat of the soul: ancient Egyptian tomb paintings show the deceased's heart being weighed against a large ostrich feather. If the heart was lighter, it was presumed to be free of weight of the impurities of sin and the person as well as the heart would go on to the afterlife, but if it was heavier, it would be devoured by a demon and the person would be eternally damned. In southern Mexico and Central America, ruins of the Mayan culture include magnificent stucco reliefs that depict religious rites in which the still-beating heart is ripped from the chest of a sacrificial victim.

That the heart was important did not mean, however, that the ancients understood the workings of the cardiovascular system. Even the simple fact that blood flowed within vessels was not entirely clear—bleeding was envisioned like water seeping from a leaky bucket. A few early thinkers suspected that blood flowed internally, but they thought of it as being manufactured by the heart or liver to take a one-way trip to tissues, where it was consumed. For a quick history of scientific theories about the cardiovascular system, see the nearby history of science box titled *Does Blood Flow in and out Like the Tide?*

[The heart] is the household divinity which, discharging its function, nourishes, cherishes, quickens the whole body, and is indeed the foundation of life, the source of all action.

William Harvey (1578–1657), English physician, in *Exercitatio Anatomica de Motu Cordis et Sanguinis in Animalibus* (*An Anatomical Exercise on the Motion of the Heart and Blood in Living Beings*), 1628

The Organization of the Cardiovascular System

Recall from Chapter 10 that blood is a "river of life," carrying nutrients and other essentials to tissues and returning metabolic waste for elimination. The **heart** is a muscular pump in the center of the chest, which supplies the force for blood distribution. It is composed of two upper chambers—the **atria** (Latin = "entrance halls"), which receive blood as it enters the heart, and two lower chambers, the **ventricles** (Latin *venter* = "belly"), which provide the main force for expelling blood from the heart (Fig. 11.1).

Blood vessels are tubes through which blood flows from the heart through the lungs and all other tissues and back to the heart in a continuous loop. There are three primary types: **arteries,** which carry blood away from the heart; **veins,** which carry blood to the heart; and minute **capillaries,** which join arteries and veins.

The Systemic and Pulmonary Circulations Are Separate

The heart and blood vessels are divided into two transport systems, each beginning and ending in the heart:

● The **pulmonary circulation** is in charge of waste drop-off and supply pickup—it carries blood to and from the lungs in order to drop off carbon dioxide waste and pick up vital oxygen supplies. The right side of the heart receives CO_2-rich, oxygen-poor

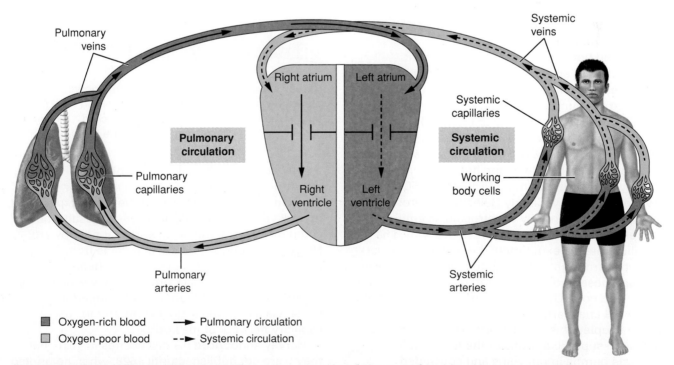

Figure 11.1. **The cardiovascular system.** The pulmonary circulation transfers blood between the heart and the lungs; the systemic circulation transfers blood between the heart and the other tissues of the body. Arteries carry blood from the heart, and veins deliver it back to the heart. *Which veins contain oxygen-rich blood—the pulmonary veins or the systemic veins?*

blood from every part of the body and propels it through the lungs and into the left side of the heart.

- The **systemic circulation** is in charge of supply delivery and waste pickup—it carries blood to and from every part of the body in order to deliver oxygen and other supplies to working tissues and to pick up waste products such as carbon dioxide. The left side of the heart receives oxygen-rich, CO_2-poor blood from the lungs and propels it through the furthest reaches of the body and back to the right side of the heart.

Also note in Figure 11.1 that, in the systemic circulation, arteries carry oxygen-rich blood and veins carry oxygen-poor blood. The opposite is true in the pulmonary circulation: the arterial system carries oxygen-poor blood and veins carry oxygen-rich blood.

> *Remember This!* **The heart is two pumps, side by side: one pumps oxygen-poor blood to the lungs, the other pumps oxygen-rich blood to the rest of the body.**

THE HISTORY OF SCIENCE

Does Blood Flow In and Out Like the Tide?

Nowadays students learn in elementary school that blood flows in an endless loop—from the heart to tissues and back. The ancients, however, had a different view, proposed by Galen, a Greek physician, in the second century A.D. Galen correctly observed that arterial and venous blood are different. He proposed that venous blood originates in the liver and arterial blood in the heart, and that blood moves outward to other organs—which consume it—from these manufacturing sites. This misunderstanding held sway for nearly 1500 years until English physician William Harvey (1578–1657) reasoned otherwise.

Harvey used both mathematics and scientific experimentation to show that blood could not be manufactured and consumed as Galen postulated. First, he estimated the number of heartbeats in a day to be 48,000, although it's actually closer to 100,000, and that the volume moved with each beat was 5 mL, though it's actually closer to 80 mL. Even using these low estimates, Harvey proved that if the liver and heart were producing blood, they would have to manufacture over 500 pounds of new blood each day. Clearly Galen had it wrong.

Harvey also did carefully constructed scientific experiments. As you can see in the figure, which he used to illustrate his experiment, he tied a tourniquet on the upper arm tightly enough to stop blood flow into the arm and observed that the limb below the stricture grew pale and cool. Upon release of the tourniquet, the arm grew red and warm.

Harvey coupled this observation with what he saw happening in superficial veins of the forearm. He noticed little bumps in the veins and concluded correctly that they were the valves that Fabricius, his teacher in Italy, had discovered earlier. Harvey tried to massage blood toward the hand by running his finger

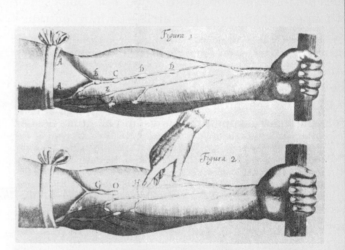

Harvey's experiment. This experiment showed that blood flows only toward the heart. If blood flow from the extremity (toward the heart) is blocked, the vein cannot refill by blood flowing from the heart.

downward along the vein, but to no avail. However, the same technique applied upward readily emptied the vein of its contents. He did the same experiment in neck veins, but found the opposite result—he could massage blood downward but not upward. Harvey concluded correctly that veins from every part of the body moved blood to the heart and that the heart pumped it back out. But he was never able to understand how blood passed from arteries to veins, via a system we now know as the vast capillary network of the body.

Harvey announced his discoveries in 1616, but they were not published until 1628, when he printed *Exercitatio Anatomica de Motu Cordis et Sanguinis in Animalibus* (*An Anatomical Exercise on the Motion of the Heart and Blood in Animals*).

Pressure Gradients Govern Cardiac Blood Flow

This chapter is concerned with the movement of fluid—the flow of blood through the heart and blood vessels in an endless loop. But before we get into the specifics, we must deal with an important topic: *how* does blood flow? The short answer is that blood flows down a *pressure gradient* ◄ (Chapter 1), a difference in pressure between two areas. Be it from one heart chamber to the next or from near to far along the length of an artery, blood is always moving down a pressure gradient from a region of higher pressure to a region of lower pressure.

Two mechanisms create pressure gradients within the cardiovascular system:

● Additional blood is forced into a compartment.
● Muscle surrounding the compartment contracts.

Consider a balloon filled with water. Add more water, and pressure within the balloon increases. Eventually the pressure will be greater than the balloon wall can withstand. It will burst and water will flow where it is allowed to go—from inside to outside. Alternatively, we can compress the wall with our hands enough to produce the same result—increasing internal pressure and rupturing the wall. Similarly, the fluid pressure in a heart compartment (say, the left atrium) can be increased by filling it with more blood or by contracting the wall of the atrium to squeeze the compartment. When the pressure in the atrium is greater than the pressure in the adjacent ventricle, blood flows from the atrium into the ventricle.

Case Notes

11.1. Did Bob W. have trouble with his heart, his blood vessels, or both?

11.2. Part of Bob W.'s left ventricle could not contract. What effect does this have on the pressure gradient between the left ventricle and the artery leaving the heart?

11.1 True or false: Any blood vessel that carries blood away from the heart is an artery.

11.2 Which circulation carries oxygen to the bones of your hand—pulmonary or systemic?

11.3 Name the type of gradient that governs blood flow through vessels and heart chambers.

Structure and Function of the Heart

The heart lies between the lungs, anterior to the vertebral column, superior to the diaphragm, and inferior to the superior edge of the sternum (Fig. 11.2A). It is about the size and shape of a clenched fist, two thirds of which lies to the left of the midline. The narrow *apex* points inferiorly and laterally toward the left hip. The wider *base* is situated opposite to the apex, where the great vessels attach to the heart.

The heart lies in a supple, closed sac, the **pericardium** (Greek *peri* = "around"), which is about 1 to 2 mm thick. It is composed of two layers (Fig.11.2B and C). The thick outer *fibrous pericardium* is anchored inferiorly to the diaphragm and superiorly to the large blood vessels where they attach to the base of the heart. The inner *serous pericardium* folds back on itself to form a double-layered membrane that lines the inner surface of the sac as well as the surface of the heart itself. The layer of the serous pericardium attached to the fibrous pericardium is the *parietal layer;* the layer covering the heart is the *visceral layer,* also known as the *epicardium.* The potential space between these two layers of the serous pericardium is the *pericardial cavity.* You can see how this space was formed during fetal development in Figure 1.14. Cells of the serous pericardium secrete a minute amount of *pericardial fluid,* a lubricant that permits the two serous layers to slide over each other as the heart moves with each beat.

Under some circumstances the potential space of the pericardial cavity becomes a real space filled with fluid— for example, when the pericardium is inflamed (pericarditis) and inflammatory fluid accumulates. In some instances the accumulated fluid compresses the heart and prevents it from beating properly.

The Heart Wall Has Three Layers

Virtually the entire thickness of the heart wall—and therefore the entire mass of the heart—is composed of cardiac muscle, or **myocardium** (Fig. 11.2C), which is consider in detail below.

The outermost layer of the heart wall is the **epicardium,** which is actually the same thing as the visceral layer of the serous pericardium discussed above. The inner layer of the heart wall, the **endocardium,** is also only a few cells thick and represents a continuation of the similar cells that line all blood vessels. The flat epithelial cells of the endocardium are the only cells of the heart wall that come into direct contact with blood. As they do in blood vessels, these cells form a smooth,

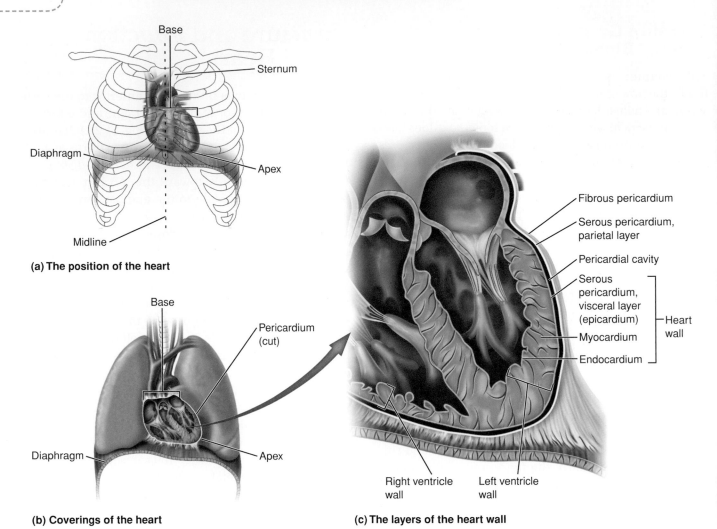

(a) The position of the heart

(b) Coverings of the heart

(c) The layers of the heart wall

Figure 11.2. The heart and its coverings. A. The heart's apex points toward the left hip, and its base points toward the posterior right shoulder. **B.** The heart is firmly attached to the diaphragm and the major blood vessels by the pericardium. **C.** The heart wall contains three layers and is surrounded by the pericardium. *Which layer of the pericardium is part of the heart wall?*

friction-free surface to ensure a smooth flow and prevent the formation of blood clots (⬅ Chapter 10).

Where it forms the atria, the wall is thin, about 2 to 3 mm. The wall of the right ventricle is about 4 to 5 mm thick. In contrast, the wall of the left ventricle is about 11 to 13 mm thick, reflecting its capacity to generate enough force to propel blood from the heart to the toes.

Case Note

11.3. Bob's cardiac muscle was damaged. Which heart layer was affected?

A Fibrous Skeleton Provides Insulation and Support

An important feature of the heart wall is its *fibrous skeleton*, which is composed of dense connective tissue. In

the myocardium it is a delicate lacework that serves to bind cardiac muscle fibers together. However, at the junction of the atria and ventricles, it condenses into a plane of fibrous tissue that cuts completely across the heart and separates the atria above from the ventricles below (see Fig. 11.5A). With a single notable exception, this layer of fibrous tissue provides complete electrical separation of the atria from the ventricles. The exception is a bundle of specialized cardiac muscle, part of the *cardiac conduction system,* discussed further on, which pierces the fibrous tissue to carry electrical signals from the atria to the ventricles.

Cardiac Muscle Has Unique Properties

Cardiac myocytes (muscle cells) are organized in concentric layers that wrap around the heart (Fig. 11.3,

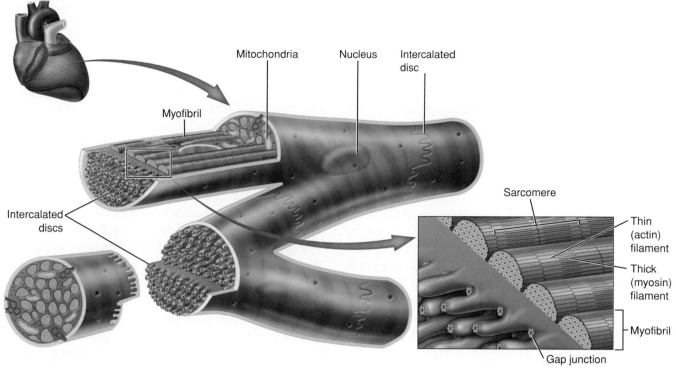

Figure 11.3. Cardiac muscle. Cardiac muscle cells are joined end to end with intercalated discs. *What is the purpose of gap junctions—to hold cells together physically or to couple cells together electrically?*

upper left). This arrangement enables them to squeeze blood out of the heart like toothpaste out of a tube. Cardiac muscle shares some characteristics with skeletal smooth muscle. Like skeletal muscle, cardiac muscle is *striated*, reflecting the fact that its myofilaments are organized into sarcomeres, each containing an orderly arrangement of thin and thick filaments (Fig. 11.3). Cardiac muscle is also regulated similarly to skeletal muscle—calcium binds to troponin, which frees myosin binding sites on the actin molecules of the thin filaments. However, cardiac muscle and skeletal muscle differ structurally and functionally in important ways.

Heart Cells Are Electrically Coupled

Skeletal muscle fibers are electrically insulated from one another; one fiber can contract while its neighbor relaxes. *Cardiac* muscle fibers, conversely, are interconnected and electric signals pass easily between them. Cardiac cells have multiple blunt-ended branches that link one cell to another at flat plates called *intercalated discs* (Fig. 11.3, right). Each intercalated disc contains gap junctions—tiny fluid-filled tunnels—that permit ions to carry electrical signals directly from cell to cell. As a result, the heart behaves like smooth muscle: a wave of contraction flows quickly through every cardiac muscle cell. However, it is worth repeating that the wave

does not spread directly from atrial muscle cells to ventricular muscle cells. Instead, the signal is channeled through fibers of the cardiac conduction system, which penetrate the electrical insulation of the fibrous skeleton that separates the atria from the ventricles.

Cardiac Contraction Strength Depends on Cross-Bridge Number

Certain circumstances, such as exercise, require stronger cardiac contractions in order to propel more blood. However, the mechanism for stronger contraction in cardiac muscle differs from that in skeletal muscle. To illustrate the difference, consider a team of workers moving a shipment of boxes. In skeletal muscle, each worker always exerts maximal effort, but the number of workers varies. A light box could be lifted by one worker (i.e., one muscle fiber), whereas a heavier box would have to be lifted by many workers (i.e., many muscle fibers). In cardiac muscle, conversely, all of the lifters work together. Each worker exerts a bit of effort to lift a light box or a greater effort to lift a heavy box.

Here's how it works. Recall that skeletal muscle fibers are not electrically connected and each contraction is "all or none"—a small force is generated by the contraction of a few fibers. More force requires more contracting fibers, which must be recruited by conscious effort ("lift harder") from the pool of noncontracting fibers. This method does not apply in cardiac muscle because all cells are electrically linked and contract together, so there is no pool of noncontracting fibers from which to recruit more fibers to work. Also recall from Chapter 7 that the force generated by a muscle fiber depends on how many cross bridges (attachments) form between thick (myosin) and thin (actin) filaments. During a skeletal muscle contraction, each contracting fiber forms the maximum number of cross bridges—each fiber is exerting maximum effort. Instead, cardiac muscle generates more force when more cross bridges form in each cell.

A resting person's heart does not contract very strongly because there is a limited amount of calcium in the sarcoplasm. As a result, many troponin molecules remain unbound and many binding sites on the actin molecules remain covered and unattached; that is, few cross-bridge attachments have formed. However, exercise (or other stress) increases the amount of calcium in the sarcoplasm, which frees up more binding sites to form cross bridges.

Calcium Sustains Myocardial Action Potentials

Myocardial action potentials differ from skeletal muscle action potentials in two respects. As you can see by comparing parts A and B of Figure 11.4, the myocardial action potential lasts much longer, and calcium plays an important role. The initial depolarization of cardiac muscle, like that in skeletal muscle, results from Na⁺ entry through voltage-gated channels. The final, repolarizing phase of the action potential is also familiar and is due to K⁺ exit

from the cell. However, notice the sustained depolarization between the initial depolarization and repolarization. This phase of the action potential, called the *plateau*, is sustained by the entry of Ca^{2+} ions into the cell through voltage-gated calcium channels.

These Ca^{2+} ions not only keep the cell depolarized but also open Ca^{2+} channels in the sarcoplasmic reticulum, releasing a flood of calcium into the cytoplasm. As in the familiar adage "It takes money to make money," the initial dose of calcium releases more calcium in a positive feedback loop that sustains cardiac muscle in a contracted state. In skeletal muscle cells, calcium is *not* involved in the action potential, and calcium channels in the sarcoplasmic reticulum are opened by membrane depolarization.

The distinctive nature of the cardiac action potential also affects the interval between successive contractions. Action potentials cannot be initiated in a depolarized cell; that is, one action potential must end before a second action potential can begin. In *skeletal* muscle, the action potential ends very quickly, during the initial phase of muscle contraction (Fig. 11.4A). As a result, the cell is ready for another contraction before the current contraction is complete—an arrangement that facilitates smooth contractions called *tetanus* (see Fig. 7.12). In contrast, *cardiac* action potentials last as long as the cardiac muscle is contracted (Fig. 11.4B). Recall that in skeletal muscle an individual cycle of contraction and relaxation is called a *muscle twitch*. The long cardiac action potential means that a muscle twitch is the only form of cardiac muscle contraction; tetanus, a prolonged state of contraction, never occurs. Each heart chamber fully relaxes between twitches, allowing enough time for filling to occur.

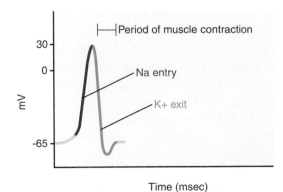

(a) Skeletal muscle action potential

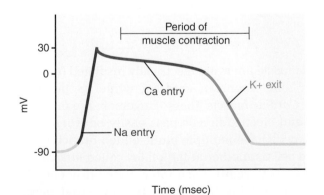

(b) Cardiac muscle action potential

Figure 11.4. Cardiac muscle action potentials. A. The action potential in skeletal muscle cells is of short duration. **B.** The action potential in cardiac muscle cells has an extended plateau period. *Which ion participates in the action potential in cardiac muscle but not in skeletal muscle?*

11.4. Bob was given epinephrine in an attempt to increase the strength of cardiac muscle contraction. Based on the information in this section, do you think epinephrine recruited more cardiac muscle fibers or increased the number of cross bridges forming in each contracting fiber?

Some Myocardial Cells Are Autorhythmic

Unlike skeletal muscle, some cardiac muscle is autorhythmic; that is, it is capable of initiating contraction without an outside stimulus. Regions of myocardium that possess this ability are called *autorhythmic myocardium*. Here the cells are smaller than other myocardial cells (those of the *contractile myocardium*, just discussed). Autorhythmic myocardial cells contract quite weakly, but they excel at transmitting electrical signals. The specialized action potentials of autorhythmic myocardial cells are discussed below, in the section on the cardiac conduction system.

Cardiac Muscle Generates Energy Aerobically

Another unique characteristic of cardiac muscle is that it relies exclusively upon aerobic metabolism and is therefore packed with mitochondria. It will burn almost anything for energy—glucose, amino acids, lactic acid, or any other sort of nutrient available, so that oxygen availability (not nutrient availability) is the limiting factor in cardiac energy metabolism. It does not fatigue under normal conditions but can do so as a consequence of disease.

Blood Flows through Cardiac Chambers and Great Vessels

Recall that the heart is actually two side-by-side pumps, the right heart and the left heart, each consisting of an upper atrium and a lower ventricle. The atria are divided by the **interatrial septum** (Fig. 11.5), a thin wall that contains an oval depression, the *fossa ovalis*, which is a remnant of the *foramen ovale*, an opening between the atria in the fetal heart that allows blood to bypass the lungs in fetal life. The ventricles are separated by a thick muscular septum, the **interventricular septum**, which narrows at its upper end to a thin fibrous membrane where it joins the fibrous skeleton.

You can follow the pathway that blood follows through the heart and great vessels in Figure 11.5. The right atrium is connected to three large veins bringing oxygen-poor blood from the body (1). The **superior vena cava** (Latin *vena* = "vein" + *cava* = "hollow") brings blood from the head, neck, and upper limbs; the **inferior vena cava** brings blood from the viscera, trunk, and lower limbs; and the **coronary sinus** drains blood from the heart muscle. The right atrium (2) delivers blood into the right ventricle (3), which pumps it into the **pulmonary trunk (4).** The pulmonary trunk divides into the right and left pulmonary arteries (5), which deliver blood to the lungs.

Four **pulmonary veins** (6) deliver oxygen-rich blood to the left atrium (7). The left ventricle (8) pumps the blood it receives from the left atrium into the *ascending* portion of the **aorta** (9) for distribution to the systemic capillaries of the entire body (10). The pulmonary trunk and the aorta are referred to as the *great vessels* of the heart, a tribute to their large size.

Heart Valves Ensure One-Way Flow

The heart contains four valves that act as one-way gates controlling the direction of blood flow (See Fig. 11.5A). Cardiac valves are flaps (*leaflets*) of thin, strong, flexible fibrous tissue. The base of each leaflet is supported by the fibrous skeleton (Fig. 11.6A); the other edge can move back and forth like a flag in the wind. The free edges of the valve leaflets are pushed out of the way by the flow of blood in the desired direction, but they are pushed back together to form a leakproof seal when blood attempts to flow in the opposite direction.

The **atrioventricular valves** sit between the atria and ventricles. The right atrioventricular valve is known as the **tricuspid valve** (Latin *tri* = "three" + *cusp* = "point") and is a three-leaflet valve separating the right atrium and ventricle. The left atrioventricular valve is also known as the **mitral valve,** because its two leaflets resemble a bishop's hat (a mitre). Each AV valve opens when atrial pressure exceeds the pressure in the ventricle below. This pressure gradient pushes the valves apart, which allows blood flow from the atrium to the ventricle (Fig. 11.6B). The AV valves close when the opposite occurs: when the ventricular pressure suddenly becomes greater than pressure in the atrium above, which pushes upward on the valves, slamming them closed to prevent backflow (Fig. 11.6C).

Considering the high pressure within the ventricle when it's filled with blood, you might wonder what keeps the valve leaflets from inverting like an inside-out umbrella in a strong wind. The answer is that the free edge of each leaflet is anchored by thin, tendonlike cords of fibrous tissue (**chordae tendineae**) to small mounds

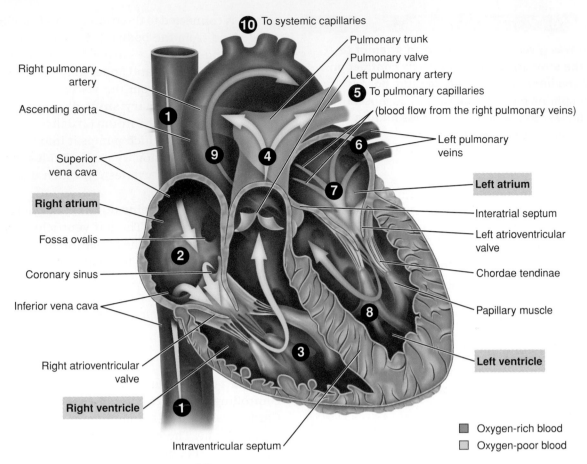

(a) Anatomical view of cardiac blood flow

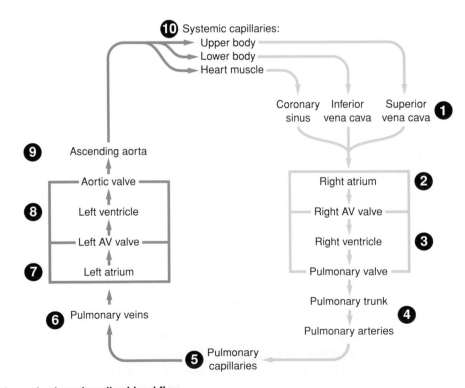

(b) Schematic view of cardiac blood flow

Figure 11.5. Blood flow through the heart. A. Unidirectional flow through the chambers of the heart and out through the aorta or pulmonary trunk is ensured by a series of valves. **B.** A schematic diagram of the direction of blood flow. *Which chamber receives blood from the heart muscle itself?*

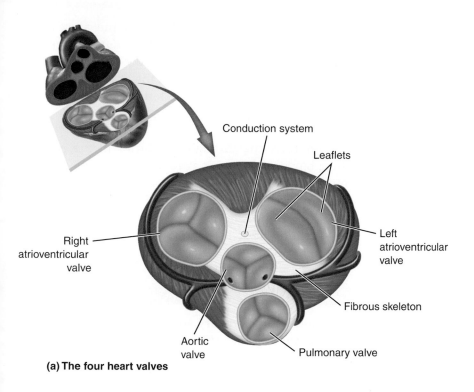

Conduction system

Leaflets

Right
atrioventricular
valve

Left
atrioventricular
valve

Fibrous skeleton

Aortic
valve

Pulmonary valve

(a) The four heart valves

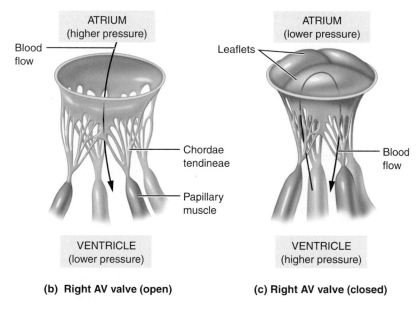

ATRIUM
(higher pressure)

Blood
flow

Chordae
tendineae

Papillary
muscle

VENTRICLE
(lower pressure)

(b) Right AV valve (open)

ATRIUM
(lower pressure)

Leaflets

Blood
flow

VENTRICLE
(higher pressure)

(c) Right AV valve (closed)

**Figure 11.6. Valves and the fibrous skeleton.
A.** The fibrous skeleton separates the atria
from the ventricles and supports the heart
valves. **B.** In response to higher pressure in
the atrium than in the ventricle, the right
atrioventricular (AV) valve opens, enabling
blood to flow from the atrium to the ventricle. **C.** The right AV valve closes, preventing
blood flow from the ventricle to the atrium.
Which valve has only two flaps?

of cardiac muscle tissue (**papillary muscles,** from Latin
papula = "small protuberance" or "nipple") (Fig. 11.5A;
also see Fig. 11.6B).

The *semilunar valves* (Latin *semi* = "partial" + *lunar* =
"moon") sit in the base of the aorta and pulmonary trunk,
where they connect to the ventricles. No chordae tendineae
or papillary muscles exist on these valves. The **pulmonary valve** is composed of three cup-like leaflets with
the open end of the cup facing into the lumen of the
main pulmonary artery (Fig. 11.5A). When closed, the

edges of the valve cusps meet to form three back-to-
back half-moon arcs. The leaflets are easily pushed
aside when ventricular pressure exceeds that in the pulmonary artery, permitting blood to flow out of the right
ventricle. As soon as the ventricle ends its part of the
beat and relaxes, pressure in the pulmonary artery
exceeds that in the ventricle and the valve slams shut,
preventing recently ejected blood from reentering the
ventricle. Between the left ventricle and the aorta is a
similarly constructed, three-leaflet valve, the **aortic valve**

(Fig. 11.5A). Valve function is discussed further below, along with the cardiac cycle.

> *Remember This!* Note that there are no valves between the atria and the veins that deliver blood to them.

The Coronary Circulation Supplies Blood to the Heart

Although heart muscle comprises only 1/200 of body mass, it consumes 1/20 of the blood supply. That's not surprising when you consider how tirelessly the cells work. But where does the heart get *its* blood? It cannot meet its needs from the blood that fills the heart chambers, because the muscle is thick and oxygen and nutrients can diffuse only a very short distance into the cardiac wall. Instead, heart muscle is supplied by its own blood supply, the **coronary circulation.** The name derives from Latin *corona*, for "wreath" or "crown," a reference to the fact that coronary vessels encircle the heart's surface as a crown encircles the head.

Before reading on, take a moment to orient yourself with the location of the atria and ventricles shown in Figure 11.7.

Indicative of the importance of the *coronary arteries* is that they are the first vessels to branch from the aorta, each arising within the first centimeter. There are two main **coronary arteries;** they arise on either side of the base of the aorta and loop around the heart in the *coronary sulcus*, a groove that circles the heart where the atria join the ventricles. They and their main branches run on the surface of the heart and send penetrating arteries downward into the myocardium as they go. Though patterns vary, the most common anatomy is as follows:

- The **left coronary artery** extends to the left of the aorta and within a centimeter or so branches into the *anterior interventricular (descending) artery* and the *circumflex artery*. The anterior interventricular artery runs down the anterior surface of the heart to the apex of the left ventricle and supplies the anterior part of both ventricles and the interventricular septum. The circumflex artery supplies the left atrium and the lateral wall of the left ventricle.
- The **right coronary artery** extends to the right of the aorta for 2 or 3 cm before branching into two. The *right marginal artery* is a small branch that extends down the right side of the heart to supply the lateral aspect of the right ventricle. The right coronary artery continues its loop around the heart in the coronary sulcus until it reaches the posterior side,

where it turns downward toward the apex of the heart as the *posterior interventricular (descending) artery* (Fig. 11.7B). The right coronary artery supplies the right atrium, most of the right ventricle, and the posterior aspect of the left ventricle and interventricular septum.

After traveling through capillaries in the myocardium, blood recollects in coronary veins, whose return route is roughly parallel to that of the companion artery. Coronary veins join together to form the **coronary sinus,** which runs in the posterior coronary sulcus and empties into the right atrium near its junction with the superior and inferior venae cavae.

> ### Case Note
>
> **11.5.** Bob's right coronary artery was blocked. Which of the branch arteries still received blood—the right marginal artery or the circumflex artery?

At many points near the end of their travel, small branches of the right and left coronary arteries join to create an *anastomosis*, a union of two or more blood vessels by a vessel larger than a capillary. These coronary anastomoses offer an alternative (collateral) route for blood delivery if one small branch is occluded.

The branching pattern of the coronary vessels can vary significantly between individuals. For example, in some people the left coronary artery continues in the coronary sulcus around to the back of the heart and gives rise to the posterior interventricular coronary artery. Such anatomic variation may require imaging of the coronary arteries to remove uncertainty before undertaking some medical or surgical procedures.

Interruption of the blood supply to any tissue usually causes tissue dysfunction or death. Blockage of coronary blood flow is serious and often fatal. Partial blockage of coronary blood flow causes *angina pectoris* (Greek *ankhone* = "strangling" + Latin *pectoris* = "chest"), an intermittent and very distinctive pain beneath the sternum that has a crushing, choking, and aching character. Much more serious is complete blockage of blood flow, which deprives cardiac muscle of required oxygen. The result is usually death (necrosis) of the myocardium supplied by the artery. A localized area of dead tissue resulting from failure to receive blood supply is an **infarct**—in this instance a myocardial infarct, more commonly called a *heart attack*. The pain of a heart attack is similar to that of angina but much longer lasting. Because dead muscle cells cannot contract, the pumping action of the heart is directly impaired, often fatally.

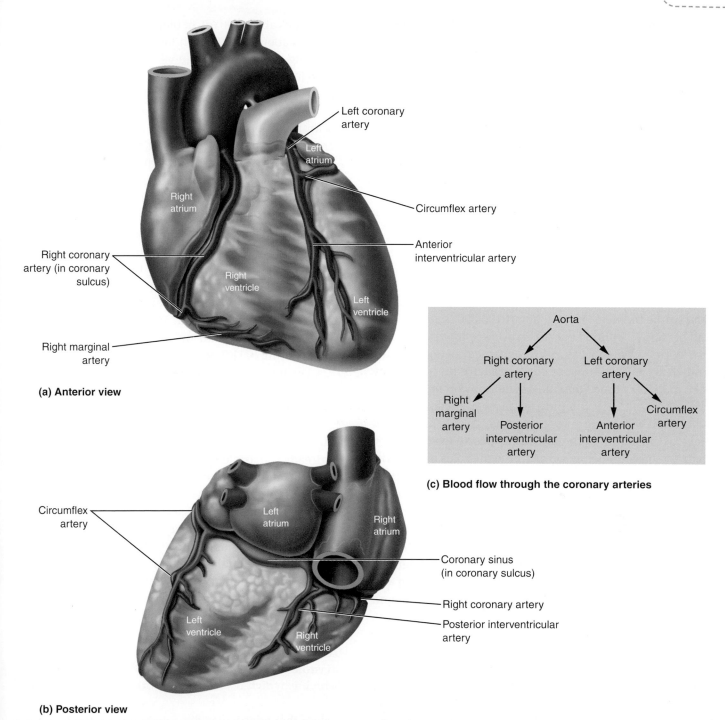

(a) Anterior view

(b) Posterior view

(c) Blood flow through the coronary arteries

Figure 11.7. External heart features and the coronary circulation. The coronary circulation supplies most of the heart muscle. Only the major vessels are illustrated here; there are many more. **A.** Anterior view. **B.** Posterior view. *Name the artery that descends along the posterior surface of the left ventricle.*

11.4 True or false: The base of the heart lies on the diaphragm.

11.5 Pericardial fluid is secreted by the _____ layer of the pericardium.

11.6 Which heart layer is part of the pericardium: the endocardium or the epicardium?

11.7 The dense connective tissue separating the atria from the ventricles is called the_____.

11.8 True or false: Cardiac muscle contracts when calcium binds the myosin heads.

11.9 Fill in the blanks: In cardiac muscle cells, the initial depolarization results from _____ ion influx and the prolonged depolarization results from _____ ion influx.

11.10   Are normal cardiac muscle contractions in the form of twitches or tetanus?

11.11   Where is the fossa ovalis—in the interventricular septum or the interatrial septum?

11.12   Which vessel or vessels receive blood from the right ventricle?

11.13   Name the two structures that strengthen the atrioventricular valves.

11.14   Is the coronary sinus an artery or a vein?

The Heartbeat

Every time the heart beats—about 100,000 times in a day and perhaps 3 billion beats in an average lifetime—a precise series of electrical and mechanical events occurs. And these events must occur in exactly the same way each and every time. Imagine a symphony orchestra playing a short but very complicated piece of music time and time again without fail. To the listener it's just a beautiful piece of music, but to the musicians it is an exacting task. In the discussion below, we examine the roles of the individual "musicians" that "perform" the human heartbeat.

The Cardiac Cycle Describes One Heartbeat

The **cardiac cycle** is the sequence of events between the beginning of one heartbeat and the beginning of the next. These events are coordinated to optimize blood flow from the atria into the ventricles, from the ventricles to the great arteries (the aorta and pulmonary arteries), and from the great veins (the venae cavae and pulmonary veins) back into the atria. Each cycle includes two contraction events: atrial contraction followed by ventricular contraction. That is, the left and right atria contract to send blood from the atria to the ventricles, and then the left and right ventricles contract to send blood into the great arteries. The contraction period for each pair of heart chambers is called **systole** (Greek *sustellein* = "contract"); the relaxation stage is called **diastole** (Greek *diastellein* = "expand"). The events of the cardiac cycle can be divided into the following steps, as illustrated in Figure 11.8:

1. *Atrial systole*. The atria contract while the ventricles are relaxed (ventricular diastole) and ready to receive blood. At the beginning of atrial systole, each ventricle contains about 100 mL of blood. Atrial contraction then forces about 20% (20 mL) more blood (and even more during exercise) into the ventricles. No valves guard the junction of the atria and vessels; therefore, as the atria contract, some blood is pushed back into the pulmonary veins and venae cavae. This backflow wave can be observed as a pulsing of the neck veins in a recumbent person. In contrast, the aortic and pulmonary valves are closed during atrial systole, preventing backflow from the great vessels into the ventricles.

2. *Ventricular systole*. The ventricles begin to contract after the atria are fully relaxed (atrial diastole). Pressure builds quickly as the ventricles contract, which forces closed the atrioventricular valves and forces open the semilunar valves of the aorta and pulmonary artery. Blood then gushes from the right ventricle into the pulmonary artery and from the left ventricle into the aorta. While the ventricles are ejecting blood, the atrioventricular valves are closed, preventing backflow.

    During ventricular systole, the atria are relaxed and are passively filling with blood from the pulmonary veins and the venae cavae.

3. *Complete Diastole*. After ventricular systole, the ventricles relax into ventricular diastole. The atria are already in diastole, so that in this phase the entire heart is relaxed and refilling before the cycle begins again with another atrial systole (step 1). As the ventricle relaxes, ventricular pressure falls below the pressure in the aortic and pulmonary arteries, so the semilunar valves close. Although the atria remain relaxed in this stage, the accumulated blood entering from the vena cava and pulmonary veins increases atrial pressure above ventricular pressure. The atrioventricular valves open, and blood flows from atria to ventricles.

> *Remember This!* **Systole means contraction; diastole means relaxation. It will help you remember which is which if you think of the "systolic squeeze."**

Note that ventricular diastole begins at the beginning of stage 3 and continues until the beginning of stage 2 of the subsequent cardiac cycle. Since the ventricles continue to receive blood while they are relaxed, they contain the most blood just before they contract. This maximum load—about 120 mL in an average, resting heart—is called the **preload,** because it is the amount of blood loaded into the ventricle and ready for ejection. Since this volume occurs at the end of ventricular diastole, it is also known as the **end-diastolic volume (EDV).** Preload (EDV) is a very important figure in cardiac physiology. It is discussed again further on.

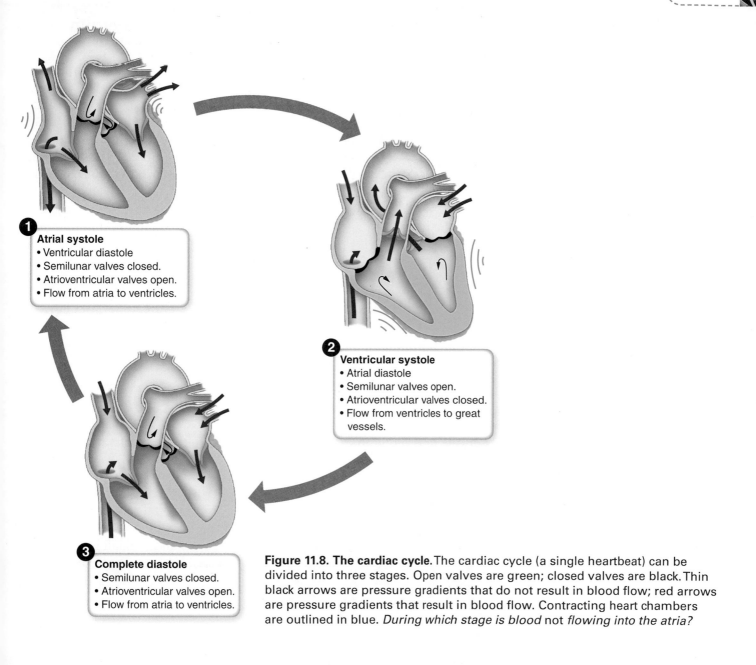

1 Atrial systole
- Ventricular diastole
- Semilunar valves closed.
- Atrioventricular valves open.
- Flow from atria to ventricles.

2 Ventricular systole
- Atrial diastole
- Semilunar valves open.
- Atrioventricular valves closed.
- Flow from ventricles to great vessels.

3 Complete diastole
- Semilunar valves closed.
- Atrioventricular valves open.
- Flow from atria to ventricles.

Figure 11.8. The cardiac cycle. The cardiac cycle (a single heartbeat) can be divided into three stages. Open valves are green; closed valves are black. Thin black arrows are pressure gradients that do not result in blood flow; red arrows are pressure gradients that result in blood flow. Contracting heart chambers are outlined in blue. *During which stage is blood not flowing into the atria?*

The volume of blood ejected from the left ventricle during ventricular systole, normally about 70 mL, is the **stroke volume** (and it must be the same for the right ventricle, of course). In each ventricle, some blood (about 50 mL) remains unejected at the end of systole—an amount called the **end-systolic volume.** The percent of total ventricular volume ejected in one contraction is the **ejection fraction** (stroke volume divided by end-diastolic volume). For example, if 70 mL is ejected and 50 mL remains, the total is 120 mL and the ejection fraction is 70/120 = 58%. The resting ejection fraction is about 60%, but it can increase substantially during exercise.

In healthy hearts, the flow of blood from one chamber to another during the cardiac cycle is smooth and silent. The only sound a clinician normally hears when listening to (*auscultating*) the heart is that of the valves slapping together. The atrioventricular valves make the **first heart sound (S1)** as they snap closed at the beginning of ventricular contraction; the aortic and pulmonary valves make the **second heart sound (S2)** as they snap closed when ventricular relaxation begins. The sounds are soft—*lub dub, lub dub*. Flow makes no audible sound. But in some hearts in people with valve disease, flow is turbulent and noisy and can be heard as a **murmur**—a soft, swishing sound—*shh, shh*—as blood passes through the valve. For example, if inflammation has welded the aortic cusps together and narrowed the aortic opening, the turbulence of blood passing through creates a murmur. The heart sounds in this instance would be *lub-shhh-dub, lub-shhh-dub*.

Case Note

11.6. What happened to Bob's stroke volume and ejection fraction as a result of his heart attack?

The Cardiac Conduction System Initiates Each Heartbeat

The **cardiac conduction system** (CCS) is a branching network of specialized myocardial cells that functions as an "express lane" for electrical signals that control cardiac contraction (Fig. 11.9). Recall that an action potential in a somatic motor neuron induces an action potential in a skeletal muscle fiber. The CCS is very different—each cell can fire a constant pattern of action

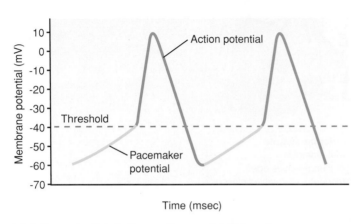

(a) Action potentials in the cardiac conducting system

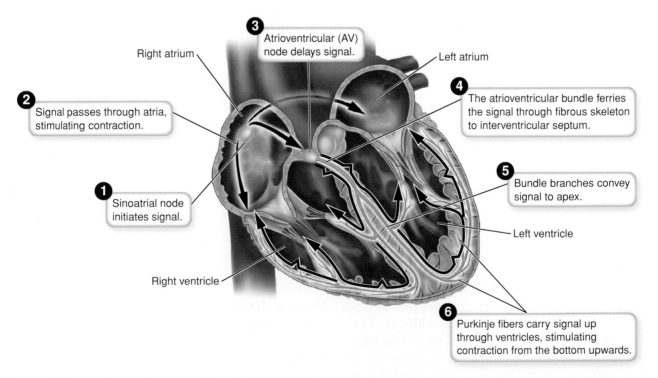

(b) The cardiac conducting system

Figure 11.9. The cardiac conduction system. A. Pacemaker potentials in autorhythmic cells. **B.** The anatomical components of the cardiac conduction system and the steps involved in one heartbeat. The black arrows indicate the spread of electrical impulses through the conducting system. *Which structure sends electrical impulses directly to ventricular cells—the bundle branches or the Purkinje fibers?*

potentials without any external stimulation. This property is called *autorhythmicity*.

Cells in the conduction system are autorhythmic because of an unusual feature of their cell membrane potential (Fig.11.9A). Like all cells that fire action potentials, they will fire an action potential when they are depolarized to a particular voltage, the *threshold*. However, contrary to the stable resting membrane potential of most cells, conduction cells have an *unstable* membrane potential. Between action potentials, a steady influx of positive ions "leak" into the cells and causes a gradual rise in the membrane potential toward the threshold. This *gradual* change in membrane voltage is called the **pacemaker potential.** When the pacemaker potential reaches the threshold, an action potential fires. The membrane rapidly repolarizes, positive ions begin "leaking" in again, and the pacemaker potential begins once more.

Conducting cells, like all cardiac muscle cells, are linked by gap junctions, so that action potentials pass from one conducting cell to the next as positive ions flow through gap junctions and depolarize the next cell to threshold. Thus, cells of the conducting system can fire action potentials because of their own innate autorhythmicity or in response to an action potential in a neighboring cell.

> **Remember This!** Although all conducting cells are autorhythmic, most fire action potentials as a result of action potentials in a neighboring cell.

Now let's look at the steps involved in a single heartbeat (Fig. 11.9B).

1. The cells of the **sinoatrial (SA) node** fire an action potential first. The SA node is located in the right atrium near the opening of the superior vena cava.
2. A wave of action potentials passes from the SA node through the contractile cells of the atria, stimulating contraction as it goes.
3. Some action potentials take a "shortcut" through scattered conducting cells within the atria to arrive at a second cluster of conducting cells, the **atrioventricular (AV) node,** located in the interatrial septum. The AV node is narrow and does not contain as many gap junctions as the rest of the conducting pathway, so the electrical signal pauses here for a critical fraction of a second, known as the *AV delay*. This allows the atria to complete their contraction before ventricular contraction begins.
4. Extending inferiorly from the AV node is the **atrioventricular bundle** (also called the **bundle of His)**—a short bundle of conduction fibers that is the *only* electrical connection between the atria and the

ventricles. It pierces the fibrous skeleton that insulates the atria from the ventricles (Fig. 11.6A), carrying the signal into the interventricular septum.
5. As it continues downward, the AV bundle splits into right and left **bundle branches,** which extend down the interventricular septum to the apex of the heart. Note that the atrioventricular bundle and the bundle branches do not stimulate contraction of cells of the ventricle because they are not connected by gap junctions to neighboring contractile cells.
6. At the apex of the ventricle, the bundle branches divide into smaller **Purkinje fibers,** which turn upward and carry the signal to the ventricular contractile cells. As a result, the wave of contraction spreads upward from the ventricular apex, squeezing blood from the ventricles like toothpaste from a tube.

An important feature of the CCS is that each component has its own intrinsic autorhythmic firing rate. If we were to isolate the different parts of the CCS from each other, the SA node would fire at a rate of 100 beats per minute (nearly twice per second), the AV node at 50 beats per minute (about once per second), and the AV bundle at about 30 beats per minute (about once every 2 seconds). The normal resting heart rate is less than 100 beats per minute because of the slowing influence of the autonomic nervous system, a point discussed further on.

This arrangement—lower is slower—ensures electrical orderliness: in healthy people the SA node always sets the pace of the heartbeat; the lower areas don't get a chance to depolarize at their slower rate because they are depolarized first by a signal from above. Every time the SA node fires an action potential, all downstream conducting cells must also fire an action potential. If the SA node fires nearly twice per second, the AV node never has time to initiate its own action potential because it is always depolarized first in response to the SA node action potential. If, perchance, the SA node is inactivated owing to disease or injury, the AV node sets the pace and the intrinsic rate becomes slower. Delay or block of signal generation or transmission through the AV node is called **heart block.**

Case Note

11.7. Why was Bob's heart rate so slow?

The Electrocardiogram Is an Electric Record of the Heartbeat

Recall from ◀ Chapter 4 that voltage is the difference in electrical potential between two points, and that membrane potential is the difference (expressed in millivolts) between the cytosol and extracellular fluid. An **electrocardiogram** (**ECG,** or EKG from the original German) is a graphical tracing of the voltage *changes*—greater or lesser than the moment before—caused by each heartbeat. It depicts not just a single action potential but the grand sum of all of the electrical activity of the many cardiac muscle fibers as they polarize and depolarize in the beating heart. If the voltage is not *changing*, even if the heart is depolarized, the ECG shows a flat line.

There are four major voltage changes with each heartbeat: (a) atrial depolarization, (b) atrial repolarization, (c) ventricular depolarization, and (d) ventricular repolarization. However, only three corresponding waves are detectable in a normal ECG, because the small wave of atrial repolarization (b) occurs at the same time and is masked by the large wave of ventricular depolarization (c). The three waves are referred to as follows (Fig. 11.10):

- The **P wave** is first. It is small and represents atrial depolarization, thus the beginning of the atrial action potential.
- The **QRS complex** is next. It is a rapid series of three waves that represent ventricular depolarization and the beginning of the ventricular action potential. The repolarization of the atria is hidden in this large wave, so the QRS complex also marks the end of the atrial action potential.
- Finally comes the **T wave,** a midsize wave that represents ventricular repolarization and the end of the ventricular action potential.

The two time periods separating these three waves are called *intervals* or *segments*.

- First is the **P-Q segment,** from the beginning of the P wave to the beginning of the downward deflection of the Q wave. The P wave indicates atrial depolarization and the QRS complex indicates atrial repolarization, so the P-Q segment represents the duration of the atrial action potential. Recalling that the QRS complex also indicates ventricular depolarization, this segment also represents the time it takes for the electrical signal to spread through from the atria to the ventricles. Although the ECG measures only electrical events, we know that contraction follows depolarization. Atrial systole begins approximately at the peak of the P wave and ends at about the Q wave, so the duration of atrial contraction is a bit shorter than the P-Q segment.
- Next is the **Q-T segment,** from the beginning of the QRS complex to the end of the T wave. This segment indicates the duration of the ventricular action potential. Ventricular systole begins at the R wave and ends at the midpoint of the T wave.

The time period between the beginnings of successive P waves is one complete cardiac cycle. This period can be used to calculate the heart rate. For instance, a P-P interval of 1 second indicates a heart rate of 60 beats per minute.

ECGs are invaluable aids in the diagnosis of heart conditions. For example, an enlarged P wave suggests enlarged atria, because there are more cells contributing to the voltage changes. For the same reason, an enlarged T wave suggests an enlarged left ventricle. A prolonged

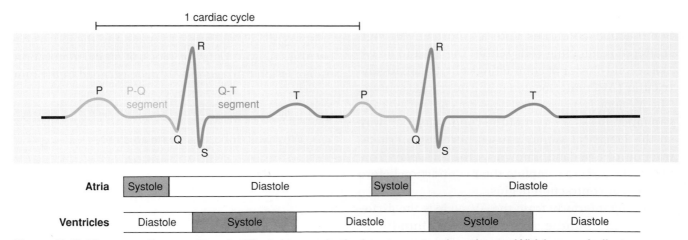

Figure 11.10. Electrocardiogram. The electrical changes in the heart over two heartbeats. *Which wave indicates the beginning of ventricular diastole?*

P-R segment indicates some degree of pathological delay of the signal (heart block) near the AV node.

 11.15 When do the ventricles contain the end-diastolic volume—at the end of atrial diastole or at the end of ventricular diastole?

11.16 Which of the following could be induced by potassium exit from the cardiac cell—the pacemaker potential or the repolarization phase of the action potential?

11.17 In a normal heartbeat, would the AV bundle or the Purkinje fibers be activated first?

11.18 Which segment represents atrial systole—the P-Q segment or the Q-T segment?

Cardiac Output

Cardiac output (CO) is the *volume* of blood ejected *per minute* by the left ventricle into the aorta. It is a measure of the total amount of blood *flow* in the body. Keep in mind that the right ventricle operates in the same way, but the left ventricle is the benchmark for this measurement.

CO is the product of the heart rate (HR, beats per minute) and the volume ejected per beat (milliliters per beat), called the *stroke volume* (SV). In a healthy resting person, CO is about 5 L/min:

$$CO = HR \times SV$$
$$= 70 \text{ beats per minute} \times 70 \text{ mL per beat}$$
$$= 4,900 \text{ mL/min, or approximately 5 L/min}$$

Inasmuch as normal blood volume is about 5 L, this means that our entire blood volume circulates through the body about once per minute. CO can increase up to fivefold with exercise—up to about 25 L/min or even more in exceptional athletes.

The Autonomic Nervous System Regulates Cardiac Output

Cardiac output changes minute to minute in order to satisfy demand and regulate blood pressure (this point is discussed more fully further on). These short-term changes in cardiac output are initiated by the autonomic nervous system, which alters cardiac function by modulating the activity of the *vasomotor center* (VMC) of the medulla oblongata (Fig. 11.11).

The Sympathetic Nervous System Increases Heart Rate and Stroke Volume

Stress activates the "fight or flight" aspect of the sympathetic nervous system, which increases cardiac output to provide more blood to working tissues. Signals are transmitted by sympathetic nerve fibers that descend from the vasomotor center downward through the spinal cord and sympathetic chain to join a network of sympathetic (and parasympathetic) fibers around the heart called the *cardiac plexus*. Some of the sympathetic fibers innervate the SA node, which, you will recall, is the supreme governor of heart rate. Therefore, as more sympathetic signals arrive at the SA node, the heart rate increases.

Other sympathetic fibers directly innervate ventricular muscle. The effect of these sympathetic signals is to increase the contraction strength of ventricular muscle cells, which increases the amount of blood expelled per beat (the stroke volume). This increase is independent of any change in *initial* fiber length. The contraction strength at a given *initial* fiber length is known as **contractility** and is an important factor in cardiac performance, a topic discussed in detail below. Sympathetic signals reaching cardiac cells exert their effect on contractility by influencing cellular calcium. Recall from our discussion above, about cross-bridge formation, that the intracellular calcium concentration alters the force generated by cardiac muscle cells. Sympathetic stimulation increases calcium influx, so more cross bridges form and more force develops. To continue our earlier example, each worker is lifting harder on the box. Epinephrine released from the adrenal gland under sympathetic stimulation has the same effect on heart rate and stroke volume as sympathetic nerve signals acting directly on the heart.

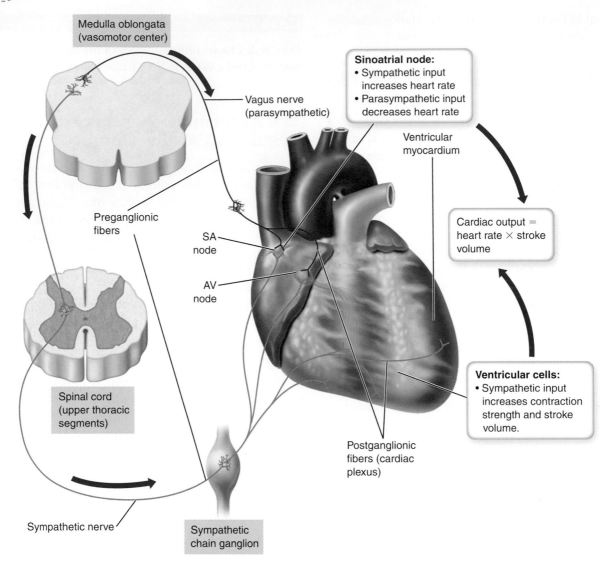

Figure 11.11. Autonomic innervation of the heart. The vasomotor center controls the heart rate and the force of cardiac muscle contraction. Sympathetic fibers are shown in green; parasympathetic fibers in red. *Which branch of the autonomic nervous system does not pass through the spinal cord?*

Calcium flow (and thus contraction strength) can also be medically manipulated. For example, administration of the drug *digitalis* increases contraction strength by increasing Ca^{2+} influx. First derived from leaves of the foxglove plant (*Digitalis purpurea*) over 200 years ago to treat edema, digitalis is now widely used to improve the power of failing hearts. Conversely, *calcium channel blockers*, as their name suggests, decrease calcium influx and are widely used to treat high blood pressure by decreasing contraction strength.

The Parasympathetic Nervous System Reduces Heart Rate

In the absence of a stressor, tissues need less blood. Cardiac output can be reduced by activating the parasympathetic nervous system, which releases acetylcholine to slow the heart rate. Parasympathetic nerve signals reach the heart via the vagus nerve (cranial nerve X), which arises in the medulla oblongata and passes inferiorly into the neck to join the cardiac plexus to innervate the heart. Fibers from the vagus nerve synapse with short postganglionic neurons on the heart surface to carry the signal the brief remaining distance to the SA and AV nodes. The parasympathetic nervous system has little effect on contraction strength, because the parasympathetic vagal nerve sends few fibers to innervate cardiac muscle.

It may surprise you to learn that, in the resting state, both stimulating sympathetic and suppressing parasympathetic signals are sent to the SA node. *The dominant influence is parasympathetic* and is referred to as **vagal tone.** It has the effect of suppressing the inherent rapid autorhythmic pace of the SA node. Recall that free of outside influence, the SA node would discharge about

Table 11.1 Factors Regulating Heart Rate

Factor	Effect on Heart Rate	Relevance
Sympathetic nervous system (or circulating epinephrine)	Increase	Increases cardiac output to deal with stress.
Parasympathetic nervous system	Decrease	Decreases cardiac output under resting conditions.
Increased body temperature	Increase	Fever increases heart rate; hypothermia decreases it.
Increased age	Decrease	The newborn heart rate is 120 per minute; in the elderly adult, it is less than 70 per minute.
Increased fitness	Decrease (resting heart rate)	Trained athletes have low heart rates (less than 50 per minute).
Thyroid hormones	Increase	Excessive thyroid hormones increase heart rate.
Altered Na^+, K^+, or Ca^{2+} concentrations in blood	Variable	Electrolyte imbalances disrupt heart rate.

100 times per minut in an adult; however, in healthy adults the rate is suppressed to about 75 by vagal tone. Moreover, as listed in Table 11.1, factors other than the autonomic system affect heart rate.

Preload Also Influences Stroke Volume

Recall from ◀ Chapter 7 that the length of a skeletal muscle fiber influences contraction strength. This relationship is even more important for cardiac muscle fibers: in the *healthy* heart, greater fiber stretch results in a stronger contraction.

The volume of blood in the ventricle determines the length of the muscle fibers that make up its wall—fibers stretch further to surround a larger volume of blood. The relevant heart volume is the preload (end-diastolic volume), since the ventricles are loaded with (contain) this volume when they begin to contract. This relationship

between preload and contraction strength is known as the **Frank–Starling law** of the heart.

Preload, in turn, is partially determined by *venous return*, the volume of blood flowing into the right atrium. It is also influenced by the *filling time,* the period between ventricular contractions: less filling time, less preload. In exercise, for instance, studies show that preload *does not change* despite increased venous return. Why? Because exercise increases heart rate, and the heart has less time to fill between beats. In other words, the exercising heart pumps the blood *out* so rapidly that no blood accumulates and preload doesn't change. In fact, when the heart rate is above about 150 beats per minute, further increases in heart rate do not increase cardiac output.

The length–tension relationship of cardiac muscle ensures that, under normal circumstances, each beat of the ventricle pumps out the same volume of blood it receives. Keep in mind, however, that this does not mean that the ventricle empties itself fully, because some blood (the end-systolic volume) remains in the ventricle at the end of each beat. This mechanism ensures that, under normal circumstances, blood does not pool in the heart chambers or in the pulmonary circulation. If the amount of blood flowing into the right atrium (venous return) is suddenly increased (say, by a blood transfusion), end-diastolic volume increases, fibers stretch, and the ventricle contracts more strongly. Stroke volume increases correspondingly to eject the increased inflow, and end-systolic volume does not change.

Case Note

11.10. **Given Bob's slow heart rate, was his preload likely to be higher or lower than normal? Why?**

Heart failure is a condition in which the heart is unable to eject the blood delivered to it and blood accumulates in the heart or lungs. It is usually due to a diseased heart muscle. This accumulation of blood stretches cardiac fibers, which *increases* contractile strength by maximizing the number of cross bridges that can form. However, as blood continues to accumulate, it stretches cardiac fibers so far that some of the myosin heads can no longer reach the actin filaments to form cross bridges. Thus, excessive stretching due to abnormally large end-diastolic volume *decreases* contractile power instead of increasing it. A vicious cycle ensues: further stretching weakens the muscle, less blood is ejected, more blood accumulates, the ventricle stretches more, and so on. Heart failure is discussed in more detail in the nearby clinical snapshot titled *Congestive Heart Failure: Bob's "Big Heart."*

CLINICAL SNAPSHOT

Congestive Heart Failure: Bob's "Big Heart"

At the memorial service for Bob, many people recalled his generosity, his kindness, his "big heart." Although they were speaking metaphorically, it's also true that Bob's heart was abnormally enlarged at the time of his death. Why?

Recall that the normal heart pumps all of the blood delivered to it, thanks to the relationship between preload and stroke volume. This relationship, known as the Frank–Starling curve, relies on healthy heart muscle. But what happens when, as with Bob, coronary artery disease interferes with heart muscle's blood supply and damages cardiac muscle? The result can be, as it was with Bob, congestive heart failure (CHF), a condition in which *the heart is unable to eject the volume of blood delivered to it* and becomes engorged with blood—the ventricle dilates and cardiac fibers are stretched. In such cases, the engorgement can cause the heart to enlarge so significantly that the abnormal size is detectable by chest x-ray.

Understanding heart failure requires understanding the Frank–Starling curve. As the heart is unable to eject the entire new load of blood delivered to it, preload increases, and the stretched myocardial muscle fibers compensate by contracting more forcefully, just as a stretched spring pulls ever harder as it lengthens. The resulting increase in contraction strength (and thus stroke volume) *compensates* for the impaired functioning of the heart and maintains normal cardiac output. Therefore this initial stage is called **compensated heart failure;** it usually produces no symptoms, because cardiac output is maintained.

However, sometimes the compensation is not enough and blood continues to accumulate. Eventually, the muscle fibers become stretched too far, and further stretching results in *weaker,* not stronger, contractions—just as an overstretched spring loses its power. The force of contraction weakens, stroke volume and cardiac output fall, and the heart enters a vicious cycle: with each beat, more blood accumulates, preload continues to increase, contractile power decreases, stroke volume decreases, cardiac output falls, and even more blood accumulates. This stage, called **uncompensated heart failure,** usually produces symptoms.

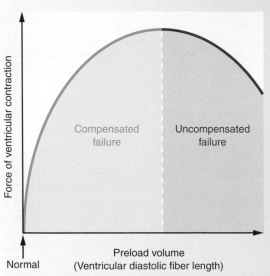

Heart failure. Early (compensated) heart failure increases the force of contraction; late (uncompensated) failure reduces it.

Bob's heart failure was due to two mechanisms. Initially it was muscle damage—there were fewer muscle fibers to eject blood. Therefore more blood accumulated with each beat. Second, as blood accumulated and preload increased, the remaining healthy ventricular fibers were stretched. We know from the Frank–Starling curve that the first bit of stretching helped; but as more blood accumulated, it stretched the fibers too far, which had a deleterious effect on the pumping power that remained. With each heartbeat, the volume of blood in the ventricle increased bit by bit. And, as in the traffic jam caused by one stalled car, the congestion "backed up" in Bob's vascular system, first in the left atrium and then in the pulmonary veins and capillaries. This increased pressure caused fluid to ooze from lung capillaries into his lungs, impairing gas exchange. His cardiac output continued to fall, which further starved his tissues for oxygen. Ultimately, Bob's "big heart" caused his demise.

11.19 True or false: If the cardiac output is 5 L, then the lungs receive 5 L of blood per minute.

11.20 What is the name of the brain center that regulates the heart rate?

11.21 Is the cardiac plexus part of the sympathetic or parasympathetic nervous system?

11.22 If vagal tone is increased, will heart rate increase or decrease?

11.23 Which branch of the autonomic nervous system uses blood chemicals to alter heart rate?

11.24 Which determinant of contraction strength is *directly* regulated by the sympathetic nervous system— myocyte calcium concentration or fiber length?

Structure and Function of Blood Vessels

We turn now from the *cardiac* part of the cardiovascular system to the *vascular* part—the network of blood vessels that carry blood to and from the tissues. **Blood vessels** are flexible tubes made up of several layers of tissue. Blood flows through their central space, which is called the **lumen.**

Recall that blood flow through the heart relies upon pressure gradients—blood flows from areas of high pressure to areas of lower pressure. Similarly, blood flow in blood vessels relies upon a pressure gradient. No pressure, no flow, no life.

Pressure Gradients and Resistance Govern Blood Flow

The flow of blood through the cardiovascular system is subject to the same laws of nature that propel water through a house supplied by a well—a network of pipes and a pump to supply pressure. Simple enough; but in the human body, several factors combine to make the physiology of blood flow rather more complex. But first, let's define some terms.

- **Blood flow** is the *volume* of blood flow *per unit of time* (usually liters per minute) through the entire circulatory system or a particular organ or vessel. For the entire circulatory system, it is the same as cardiac output (CO).

- **Resistance** is the opposition to flow created by the friction encountered by a fluid as it passes through a tube.

Resistance to flow, in water pipes or blood vessels, is determined by three factors:

- The *viscosity* of the fluid (its thickness or "gooeyness")
- The *length of the tube* through which the fluid flows
- The diameter of the tube

Normally blood viscosity and the aggregate length of blood vessels are unchanging, so by far the most important factor in determining resistance is blood vessel diameter. Most of the resistance is created by the *decreasing diameter* of vessels as they branch into smaller and smaller vessels in the peripheral circulation. This is referred to as **peripheral resistance.** Even a tiny narrowing, a single millimeter in a small vessel, can cause a dramatic increase of resistance to flow.

It makes sense that flow will increase if (a) the pressure gradient increases and/or (b) the resistance decreases. We can quantify this relationship by the simple equation

$$Flow = pressure/resistance$$

> *Remember This!* You can experience the determinants of vessel resistance by drinking fluids through a straw. A milkshake provides more resistance to flows than soda pop, and a long narrow straw creates more resistance than a short wide straw.

Blood Vessels Can Be Classified by Function and Tissue Properties

Recall that blood vessels carry blood in a full circle away from and back to the heart. These vessels can be divided into five types:

- *Arteries* carry blood away from the heart at relatively high pressure—from the right ventricle to the lungs; from the left ventricle to the body.
- Arteries branch successively into smaller arteries and then into *arterioles*, the smallest arteries.
- In turn, arterioles branch into the smallest blood vessels, *capillaries*, where all fluid, gas, and other molecular exchange occurs between blood and other tissues.
- Capillaries merge to form the smallest veins, called *venules*, which carry blood back toward the heart.

● Venules merge to form ever larger *veins*, which carry blood at relatively low pressure back to the heart

Now let's look at two very important tissue properties of blood vessels that are important in their function. **Compliance** is the ease with which blood vessels (or lungs or other tissues) stretch in response to increasing pressure. **Elastance** is the tendency of blood vessels (or lungs or other tissues) to recoil toward their original dimensions as blood pressure falls. These properties differ in arteries and veins as follows:

● Arteries and arterioles have low compliance and high elastance. It takes a lot of pressure to distend them, and when pressure falls, they readily snap back to their original dimensions.

● Veins have the opposite characteristics. They are highly compliant—they stretch easily with small increases of pressure; and they have low elastance—they do not spring back to their original shape as readily as arteries.

Socks provide an apt example. When new, socks are like arteries. They have low compliance and high elastance, so they resist stretching and fit snugly. When you take off a new pair of socks, they return readily to their original narrow shape. But when they get old and worn, socks are more like veins. They have high compliance and low elastance, so they stretch very easily and don't fit snugly, and when you take them off, they don't readily return to their original shape.

As well as accounting for the functional differences between arteries and veins, vessel compliance is altered in some medical conditions. The clinical snapshot near the end of the chapter titled *Atherosclerosis: We're Eating Ourselves to Death,* discusses *atherosclerosis,* a vascular disorder characterized by decreased compliance.

Blood Vessel Layers Are Called Tunics

In our discussion of blood vessel structure, we'll consider a large artery, which contains all of the possible components of a blood vessel, as illustrated in Figure 11.12. The arterial wall can be divided into three layers, or *tunics*. Beginning from the lumen of the vessel, these tunics are as follows:

● The *tunica interna* (or *intima*): The innermost layer composed of thin, flat cells, the **endothelium,** and a

supporting basement membrane. Its smooth surface facilitates blood flow through the vessel and prevents blood from coming into contact with tissue outside the blood vessel lumen, which can cause blood to clot (◀ Chapter 10).

● The *tunica media* (or *media*): The middle layer is composed of smooth muscle mixed with elastic fibers. As explained below, this layer determines vessel diameter.

● The *tunica externa* (or the *adventitia*): The outermost layer consists of collagen and elastic fibers. This layer is often continuous with the connective tissue of the surrounding organ.

The tunica media is separated from the tunica externa and tunica interna by external and internal elastic layers, respectively, which contribute to the compliance and elastance of the vessel. As large arteries branch into smaller ones and then into arterioles, the tunicas media and externa become thinner and elastic fibers begin to disappear. Eventually, in the smallest arterioles, the tunica externa fades away and the vessels are simply encapsulated in the connective tissue of the surrounding organ. Once the arterioles branch into capillaries, even the tunica media disappears, so that a capillary consists only of endothelial cells supported by a basement membrane.

Consider now how these tunics are modified in the venous system as blood returns to the heart. First, as capillaries merge into venules, the tunicas media and externa reappear. Venules are structurally similar to arterioles except that their walls are thinner and they have fewer smooth muscle cells. However, as veins become larger, they begin to differ substantially from arteries. They have thinner walls because they do not contain as much elastic tissue as arteries, they have less smooth muscle in the tunica media, and they have a thinner tunica interna. However, the tunica externa of veins is thicker.

Arteries are much thicker-walled and more rubbery than veins, which are thin-walled and flabby. Again, structure serves function: veins accommodate slow, low-pressure flow, whereas arteries must accommodate rapid, high-pressure flow. In addition, large veins have valves that, like the heart valves, ensure one-way blood flow by preventing backflow. Conversely, no valves are present in the arterial system because pressure is much too high to allow backflow. Finally, both venules and veins are larger than their companion arterioles and arteries at every level in order to accommodate more volume: *at any given time, about 65% of total blood volume is present in the venous system.*

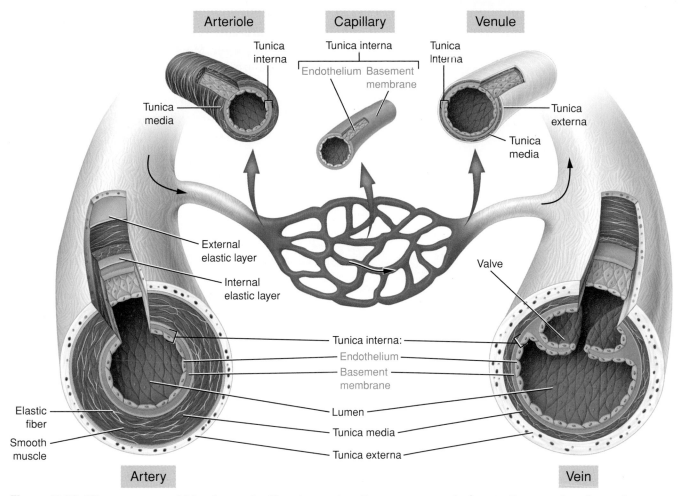

Figure 11.12. Microanatomy of blood vessels. Blood vessel walls are composed of up to three tunics; the tunica externa, tunica media, and tunica interna. *Which tunic is found in all types of vessels?*

Case Note

11.11. Bob's atherosclerosis began with damage to the inner lining of his blood vessels. Name this layer.

The Tunica Media Determines Resistance and Blood Flow

As you learned in ◀ Chapter 10, the average adult body contains 4.5 to 5 L of blood. That sounds like a lot, but it's not enough blood to completely fill all of our blood vessels all the time. Consider the advice not to swim after a meal. The logic behind it is that we do not have enough blood to simultaneously increase intestinal blood flow to digest a large meal as well as to increase flow to muscles to power swimming. It's important, then, that the vascular system have the ability to shift the flow of blood from one part of the body to another. It does this by constricting some blood vessels and relaxing others.

It is the tunica media of arteries, arterioles, veins, and venules that contracts or relaxes to narrow or enlarge the vessel lumen (Fig. 11.13). **Vasoconstriction** and **vasodilation** alter the resistance of the blood vessel. Blood will flow less freely in narrow, constricted vessels and more freely in dilated ones—the body reduces blood flow to a region by constricting the supplying arteries and arterioles; it increases blood flow to another by dilating the supplying vessels.

Remember This! Vasoconstriction increases resistance, and blood follows the path of least resistance.

The tunica media is innervated by sympathetic nerves, which maintain a low level of constant stimulation and smooth muscle contraction called *sympathetic tone*. Additional sympathetic stimulation, either neural or endocrine (from adrenal epinephrine), causes contraction of

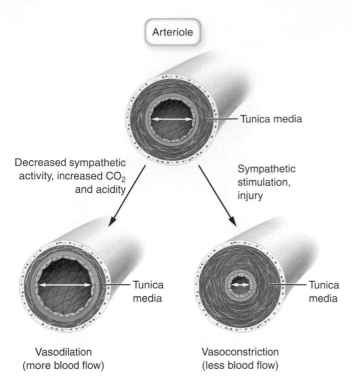

Figure 11.13. Vasoconstriction and vasodilation. The smooth muscle of the tunica media contracts or relaxes to alter the lumen's diameter (*white arrow*) and thereby modulate blood flow. *Nitric oxide relaxes vascular smooth muscle. Would nitric oxide increase or decrease blood flow?*

the fibers and vasoconstriction. Moreover, as part of hemostasis, vessels constrict reflexively when injured. In contrast, inhibition or interruption of sympathetic tone causes smooth muscle relaxation and vasodilation. Similarly, substances that accumulate as a result of intense metabolic activity, such as carbon dioxide and acid, induce vasodilation and increase blood flow. This topic is considered further in the discussion of arterioles.

Arteries Maintain Blood Pressure and Distribute Blood to Organs

Arteries are large blood vessels that carry blood under high pressure from the heart outward. The largest arteries are called **elastic arteries** because of their relatively high content of elastic tissue. Elastic arteries include the aorta and pulmonary arteries and their major branches. Elastic arteries play an important role in the maintenance of blood pressure. As the ventricles eject blood, the elastic arteries expand to accommodate it; then, as the ventricles relax during diastole, the arterial elastic tissue recoils, squeezing blood forward (it can't go backward because the pulmonary and aortic valves are closed), which maintains blood pressure until the ventricle pumps again.

Medium-size arteries, on the other hand, are called **muscular arteries** because of the relative thickness of the muscular layer. Most arteries supplying the limbs and viscera are muscular arteries and are capable of a much greater degree of vasoconstriction and vasodilation than the elastic arteries. The primary effect of these changes in diameter is altered flow to different body regions, as discussed below.

Arterioles Regulate Blood Pressure and Distribute Blood to Capillaries

Arterioles are small arteries that distribute blood to individual capillary beds. They are the major site of resistance in the vascular system. The body can vary arteriolar resistance either globally, to modify blood pressure, as discussed later, or selectively, to modify blood distribution. For example, local blood flow increases at the site of inflammation. Local chemical signals released by injured tissue and white blood cells induce vasodilation and increase blood flow. The throbbing redness of an infected toe is due to the effects of increased blood flow.

The amount of blood flowing through particular muscular arteries and arterioles is also governed by the metabolic needs of the tissue it supplies. For instance, when you start swimming your muscle fibers suddenly ramp up their metabolic activity. They require more oxygen and release more metabolic waste products, such as CO_2 and H^+ ions. These conditions (low O_2, high CO_2, and high H^+) relax vascular smooth muscle, dilating the arteries and arterioles and increasing blood flow. Eventually the additional blood flow will match the needs of the tissue's metabolic activity, and blood flow will stabilize. When you stop swimming, the increase in O_2 and decrease in metabolic wastes induces the opposite effect—the muscular arteries and arterioles constrict and blood flow lessens.

Capillaries Are the Smallest Blood Vessels

Capillaries are microscopic blood vessels that link the arterial and venous systems. Almost every body tissue contains capillaries. The exceptions are epithelial membranes, joint cartilage, the cornea and lens, and cardiac valves, each of which is nourished by diffusion from nearby vessels. The number of capillaries in a tissue depends upon its metabolic rate. Tissues with high metabolic rates—kidneys, muscle, liver, and brain—are packed with capillaries. On the other hand, tissues with low metabolic rates—ligaments and tendons especially—have relatively few capillaries.

Capillary Beds

Capillaries themselves branch to form interconnecting networks called **capillary beds,** which can be likened to three-dimensional clusters of neighborhood streets (capillaries) around a central thoroughfare (a *metarteriole*) (Fig. 11.14). Blood flows from an arteriole into a capillary bed, then out into a venule on the other side. In the resting state, most traffic (blood) stays on the main road (the metarteriole), kept there by bands of smooth muscle (*precapillary sphincters*), that constrict to keep traffic out of the neighborhood (Fig. 11.14A). However, when oxygen levels in the tissue are low or metabolic wastes accumulate, the sphincters relax and shunt blood into the neighborhood (Fig. 11.14B).

Precapillary sphincters are formed of smooth muscle, and regulated by the same factors that regulate the arterioles supplying the capillary beds. Thus, low oxygen will increase blood flow in a capillary bed by three mechanisms—it will relax the precapillary sphincters and dilate the muscular artery and arteriole that supply it.

Capillary Variants

The endothelial cells of capillaries join to one another like tiles in a floor. They are fitted together in a way that allows some fluid to "leak" through the seams; some capillaries leak more, others less. Some seams are very close-fitting, providing little opportunity for seepage of fluid or solutes. At the other extreme, large gaps, or *pores*, separate some endothelial cells, allowing especially free fluid exchange between blood and tissue. A **sinusoid** is a rather large, very leaky type of capillary that occurs in the liver, bone marrow, and lymphoid tissues. Sinusoids have large lumens and especially leaky endothelial walls. Their endothelial cells are not tightly joined and their basement membrane is incomplete; an arrangement that creates wide slits, allowing exceptionally free passage of fluid and solutes. They can even allow free passage of whole cells; for example, new blood cells enter the blood from bone marrow through the wide spaces in the walls of bone marrow sinusoids.

Substances Move between Capillary Blood and Tissues

Capillaries have the thinnest walls, so it makes sense that they are the sites of exchange between blood and tissues. To maximize exchange, capillary blood flow is particularly slow—think of the speedy flow in a large river (the aorta), compared with that in the delta, where the river fans out into hundreds of small channels filled with lazily moving water (the capillaries).

Substances can exchange between the blood and the interstitial fluid via one of four methods (Fig. 11.15A):

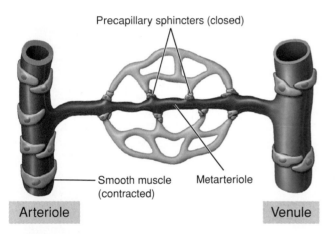

(a) Capillary bed in a resting muscle

Precapillary sphincters (closed)

Smooth muscle (contracted)   Metarteriole

Arteriole

Venule

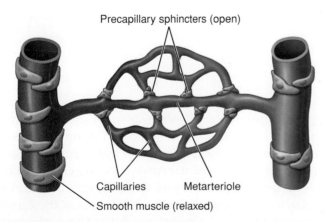

(b) Capillary bed in a working muscle

Precapillary sphincters (open)

Capillaries   Metarteriole

Smooth muscle (relaxed)

Figure 11.14. Capillary beds. The precapillary sphincters open to permit flow through the capillary bed **(A)** or close to restrict flow to the metarteriole **(B).** *Which figure shows vasoconstriction of the arteriole—(A) or (B)?*

1. *Directly through endothelial cells by simple diffusion or by use of transporter proteins.* That is, molecules cross the endothelial cell membrane into the cytoplasm and out the other side. Oxygen, CO_2, and lipids move down their concentration gradients by simple diffusion; ions, sugars, and amino acids are carried by facilitated diffusion using transporters. Some substances may also move into or out of endothelial cells by active transport, against their concentration gradients.
2. *Directly through endothelial cells by vesicular transport.* Small quantities of larger molecules, such as proteins, may cross the capillary wall packaged in vesicles ◀ (Chapter 3).
3. *Between endothelial cells, by simple diffusion.* Some substances, such as glucose or ions, may diffuse through the gaps between capillary cells because the gaps are permeable to all solutes.

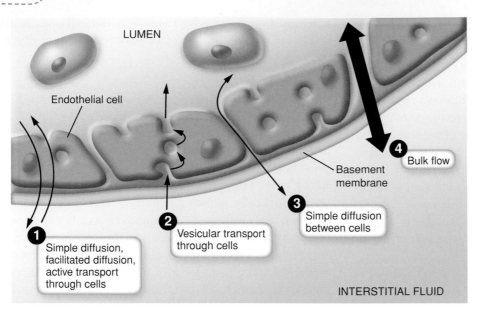

(a) Transport through the capillary wall

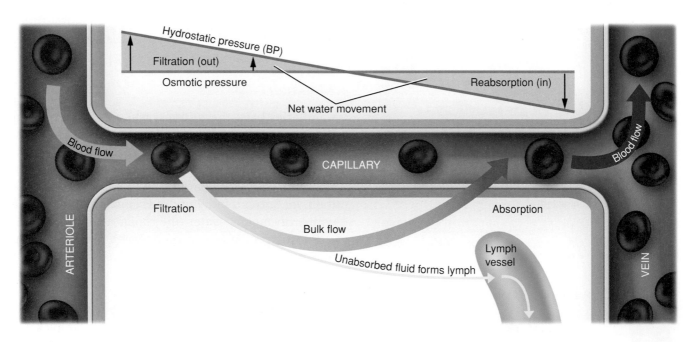

(b) Bulk flow

Figure 11.15. **Exchange across capillary walls. A.** Substances can move between the blood and the tissues by four different mechanisms. **B.** Bulk flow is determined by the balance between outward-directed hydrostatic pressure and inward-directed osmotic pressure. *If someone has high blood pressure, will he or she have greater filtration or greater reabsorption?*

4. *Bulk flow.* Water and dissolved substances can pass through gaps between capillary or sinusoid cells by a different passive mechanism, called *bulk flow.*

The first three mechanisms are discussed in ⬅ Chapter 3. **Bulk flow,** however, is a new topic—the movement of large volumes of fluid, including any substance dissolved in the fluid. Though not critical for nutrient uptake or waste disposal, bulk flow plays an important role in water balance between the different body compartments. It represents the net effect of two opposing pressure gradients: *hydrostatic* pressure, which tends to force water out of blood and into the intercellular space, and *osmotic* pressure, which tends to draw water from the intercellular space into blood (Fig. 11.15B).

Hydrostatic pressure is the physical pressure of fluid in a closed space. It propels water out of the garden hose and toothpaste out of the tube. The hydrostatic pressure within vessels is blood pressure. The increased hydrostatic pressure of blood compared with the interstitial fluid forces fluid out of the capillaries. Recall from Chapter 3 that osmotic pressure is the driving force of water moving from an area of high water concentration (low solute concentration) to an area of low water concentration (high solute concentration). Osmosis occurs only across semipermeable membranes; that is, membranes the solute cannot penetrate but which water penetrates easily. Since blood contains a much greater concentration of proteins (such as albumin) compared with the interstitial fluid, the osmotic gradient tends to draw water into the capillary.

The balance between hydrostatic and osmotic pressures determines the magnitude and the direction of bulk flow. **Filtration** is the net movement of fluid from blood across a capillary wall into a tissue or space. It occurs when hydrostatic pressure exceeds osmotic pressure. **Absorption** is the opposite: the movement of fluid from a tissue or space into blood, which occurs when osmotic pressure exceeds hydrostatic pressure. Intestinal capillaries can absorb water by bulk flow throughout the entire capillary length, because the osmotic pressure of blood consistently exceeds hydrostatic pressure. Conversely, in the kidney, the specialized capillaries of the glomerulus ➡ (Chapter 16) exclusively *filter* water out of blood into the lumen of renal tubules, a process necessary for urine production. Both filtration *and* absorption occur in other capillaries (Fig.11.5B). Filtration occurs at the arteriolar end, because hydrostatic pressure exceeds osmotic pressure. However, hydrostatic pressure falls as blood passes through the capillary, because filtration reduces capillary blood volume and blood pressure decreases with increasing distance from the heart. Eventually, hydrostatic pressure equals osmotic pressure (which remains relatively constant) and bulk flow stops. Hydrostatic pressure continues to fall as blood flows farther into the venous side; osmotic pressure now becomes high enough to overcome it and absorption results.

Each day about 20 L of water is filtered out of capillaries, of which 17 L is absorbed. The remaining 3 L finds its way into the lymphatic system ➡ (Chapter 12) and later drains back into blood.

Case Note

11.12. Bob's blood pressure was dangerously low, but the osmotic pressure of his blood was unchanged. Was filtration increased or decreased?

Venules and Veins Return Blood to the Heart

In the average person at any given moment, about 7% of blood is in the heart and 8% is in the pulmonary vascular system. The rest is in the systemic vascular system: 13% is in arteries and arterioles, 7% is in capillaries, and *65% is in veins and venules*. The most important reservoirs are the veins in abdominal organs, especially the liver and spleen, and in the skin. However, this relative distribution changes in response to exercise or other stressors. The tunica media of veins, like that of arteries, can contract to constrict the vessels. Sympathetic activation constricts veins, which reduces blood volume in the veins and increases it in arteries. As discussed further on, arterial blood volume is a key determinant of blood pressure.

The pressure gradients in veins are so low that they alone cannot propel blood back to the heart from the distant reaches of the legs. The *skeletal muscle pump* and the *respiratory pump* add extra force:

- The **skeletal muscle pump** is the massaging action of contracting muscles, which squeeze large veins in the limbs. Venous valves open to enable blood flow toward the heart but close to prevent blood flow away from the heart. For instance, when the leg is relaxed (Fig. 11.16, left side), the valves prevent blood from falling to the feet owing to gravity. Leg muscle contraction acts like squeezing a tube of toothpaste — blood squirts in both directions away from the squeeze (Fig. 11.16, right side). The distal valve closes, preventing blood from moving back into the foot, but the proximal valve opens and blood moves upward toward the heart. Venous return is thus enhanced.
- By contrast, the **respiratory pump** boosts venous return by creating a pressure gradient between the abdomen and the thorax. During inhalation, the diaphragm moves downward, reducing the abdominal volume and increasing abdominal pressure. This downward movement of the diaphragm, accompanied by an outward movement of the ribs, also expands the thoracic volume, reducing thoracic pressure. Blood flows down this pressure gradient from the increased pressure in abdominal veins into decreased pressure in the inferior vena cava in the chest.

The importance of venous valves is illustrated by **varicose veins** (Latin *varix* = "swollen vein"), a condition in which incompetent venous valves impair venous blood flow. The increased blood volume remaining in the distal veins causes venous distension, and also reduces blood leaving the capillaries. The augmented capillary blood volume increases bulk flow into the nearby tissue. The result is a familiar one: a person with visibly dilated veins in her legs and feet, whose ankles are swollen with fluid.

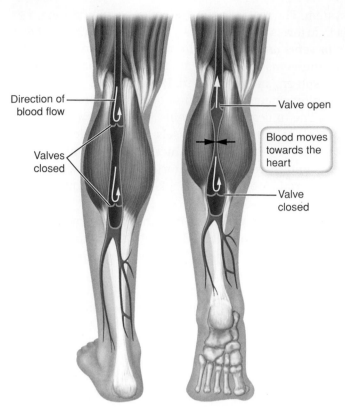

Figure 11.16. The skeletal muscle pump. Muscle contraction forces blood to flow through veins; valves ensure that blood flows only toward the heart. *In the right-hand diagram, what would happen if the lower valve did not close?*

11.25 Which vessels are formed when capillaries merge?

11.26 Name the vascular tunic composed of smooth muscle.

11.27 If a blood vessel contains only a tunica interna, what sort of vessel is it?

11.28 Which vessels determine blood distribution to individual capillary beds—muscular arteries or arterioles?

11.29 Which vessels play a critical role in blood pressure regulation—muscular arteries or arterioles?

11.30 Heat is produced by working muscles. Do you think that heat causes vasodilation or vasoconstriction?

11.31 What is the difference between a sinus and a sinusoid?

11.32 Does blood move faster in a capillary or an artery?

11.33 Which transport mechanism is used by oxygen in leaving a capillary—simple diffusion, vesicular transport, or facilitated diffusion?

11.34 Which type of gradient decreases between the arteriolar and venous ends of a capillary?

11.35 Which blood constituent is the major determinant of blood osmotic pressure?

11.36 Will an increase in blood osmotic pressure increase or decrease fluid entering the tissues?

11.37 Name two types of pumps that help veins return blood to the heart.

Blood Flow and Blood Pressure

We learned in ◀ Chapter 1 that pressure is necessary for life. **Blood pressure** is the force exerted by blood on the wall of the vessel that contains it. It is the force that propels flow as blood moves down the pressure gradient from areas of higher blood pressure to lower. Blood pressure is also the force behind filtration of substances from inside the vascular space into tissues or spaces outside.

Blood pressure is usually stated in millimeters of mercury (mm Hg). For example, blood pressure of 100 mm Hg is equivalent to the pressure beneath a 100-mm column of mercury. Unless stated otherwise, the general term "blood pressure" refers to pressure in large arteries near the heart.

Left Ventricular Contraction Is the Main Force Creating Blood Pressure

Life depends on the flow of blood propelled by the contractions of the left ventricle. As shown in Figure 11.17, normal left ventricular pressure changes from nearly zero to 120 mm Hg during every beat as the ventricles contract. Aortic pressure rises correspondingly, because increasing aortic volume increases pressure. The peak pressure developed in the aorta and large arteries is the **systolic pressure,** normally about 120 mm Hg. As the ventricle relaxes in diastole, ventricular pressure falls almost to zero, allowing low-pressure blood from the atria to fill the ventricles. However, the aorta and arteries are elastic—they stretch slightly to accommodate blood ejected from the ventricle, but they rapidly return back to their original size after ejection. Because the aortic valve has closed, this elastic recoil squeezes blood forward and maintains aortic pressure even though the ventricle has completely relaxed. This propels blood forward. Thus aortic pressure and arterial pressure never fall to zero. The lowest degree of pressure within the arteries, called the **diastolic pressure,** is normally about 80 mm Hg. The difference between systolic and diastolic blood pressure is called **pulse pressure.** Average (mean) blood pressure in the arteries (**mean arterial pressure,**

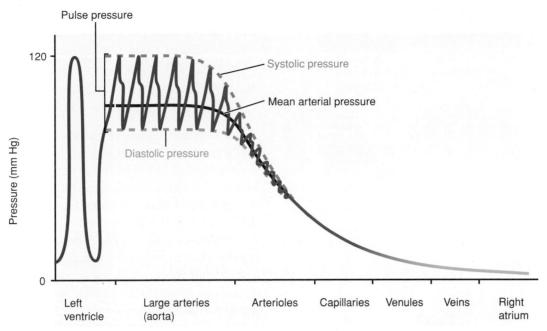

Figure 11.17. Pressures in the systemic circulation. Pressure decreases as the blood passes through the systemic circulation. *Is it possible to detect pulse pressure in the veins?*

or **MAP**) is closer to diastolic than systolic pressure because diastole lasts much longer than systole.

The pulse pressure represents a pressure wave created by ventricular contraction and arterial recoil. This wave does not represent a bolus of blood passing a point; rather, it is a disturbance, similar to an ocean wave, transmitted in fluid, and it travels about 10 times faster than blood flows. You can feel the arteries bulging as the pressure wave passes at *pulse points*—body regions where arteries pass close to the skin (See Fig. 11.22). Since each pulse represents one cardiac cycle, the number of pressure waves at the wrist or the neck indicate the heart rate. A wide pulse pressure—high systolic with low diastolic—is a sign of the stiffening (loss of compliance) associated with atherosclerosis.

Recall that the small diameter of arterioles dramatically increases resistance, which sustains higher pressure upstream but reduces it downstream. With increasing distance from the heart, both systolic and diastolic pressures fall greatly, and the difference between them narrows until pulse pressure disappears by the time blood enters capillaries. Mean pressure falls from 90 mm Hg in the aorta and large arteries to about 25 mm Hg in capillaries, 15 mm Hg in venules, and near zero in the venae cavae.

If we were to show a figure similar to Figure 11.17 for the pulmonary circulation, we would see a similar tracing, but all of the pressure values would be much lower. In the pulmonary artery, normal pressure is about 24 mm Hg systolic and 8 mm Hg diastolic. Pressure in pulmonary capillaries is about 7 mm Hg and falls to about 3 mm Hg in the pulmonary vein.

Arterial Compliance and Arterial Blood Volume Control Blood Pressure

Blood pressure must be tightly regulated—too low and tissues will die from the lack of blood, too high and blood vessels will be damaged. The two main physical factors controlling blood pressure are vessel compliance and arterial blood volume (Fig. 11.18A).

Earlier in this chapter, we talked about compliance. A very stretchy, compliant blood vessel will expand easily to accommodate a large increase in volume with minimal increase of pressure. In contrast, a noncompliant, stiff vessel wall will expand very little and pressure will rise quickly as more blood is pumped into the vessel.

Arterial blood pressure is directly related to arterial blood volume. If you think about adding more and more water to a balloon until the pressure bursts the balloon, it will make sense that low arterial blood volume is associated with decreased blood pressure and high arterial blood volume is associated with increased blood pressure.

Arterial blood volume, and thus arterial pressure, is also related to *total* blood volume. The relationship is easy enough to understand: arterial blood is about 13% of total blood volume in a resting person; therefore, if total blood volume increases, arterial blood volume increases with it. For instance, a high-salt diet increases plasma osmolarity, which in turn increases plasma water retention, and with it total blood volume. Arterial volume and blood pressure increase accordingly. This relationship is well known in medicine: people who consume a high-salt diet tend to

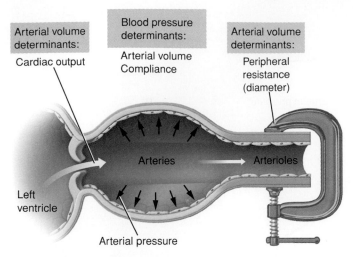

Arterial volume determinants:
Cardiac output

Blood pressure determinants:
Arterial volume
Compliance

Arterial volume determinants:
Peripheral resistance (diameter)

Arteries → Arterioles

Left ventricle

Arterial pressure

(a) Lower blood pressure

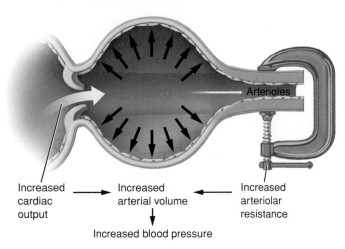

Arterioles

Increased cardiac output → Increased arterial volume ← Increased arteriolar resistance

↓

Increased blood pressure

(b) Higher blood pressure

Figure 11.18. Determinants of blood pressure. A. Arterial blood pressure depends on two primary determinants: arterial blood volume and vessel compliance. Arterial blood volume, in turn, can be altered minute by minute by changing cardiac output and arteriolar resistance. Note that overall blood volume also alters arterial blood volume. **B.** Increasing cardiac output and/or arteriolar resistance increases blood pressure by augmenting arterial blood volume. *If arteriolar radius increases, will blood pressure rise or fall?*

have high blood pressure, and patients with high blood pressure of any cause are encouraged to eat a low-salt diet. In like manner, hemorrhage reduces blood pressure by reducing total, and thus arterial, blood volume.

The body regulates overall blood volume by modifying water loss in the urine. We investigate the pathways that alter water retention (and thus blood volume and blood pressure) in ➡ Chapter 13. However, as discussed below, the body can regulate arterial blood volume minute to minute by altering the percentage of total blood volume in the arteries. A greater arterial blood

volume (and a correspondingly lower venous blood volume) increases blood pressure, and vice versa.

Cardiac Output and Peripheral Resistance Determine Arterial Blood Volume

Blood vessel compliance changes only in disease. Normally, the body regulates blood pressure by controlling the volume of blood in the arterial tree. Arterial blood volume, in turn, varies depending upon the balance between the amount of blood *entering* the arteries from the heart and the amount *leaving* the arteries through the arterioles (Fig. 11.18A). The volume of blood entering the arterial tree is the cardiac output. Increased cardiac output increases blood pressure (Fig. 11.18B). The resistance of arterioles (which varies according to their diameter) determines how readily blood will leave the arteries. As arteriolar resistance increases, blood "dams up" in the arterial tree and blood pressure rises.

Consider the crowded entranceway to a popular sports event or concert—there is the *flow* of people trying to enter, the *resistance* of the security guards checking bags, and the *pressure* of the crowd in the entryway. The crowd pressure will be increased if the flow of people trying to enter is increased or if the security guards increase their resistance by checking more thoroughly.

The minute-to-minute regulation of blood pressure thus comes down to these points:

1. Blood pressure is largely determined by arterial blood volume.
2. Changing cardiac output and/or arteriolar resistance alters arterial blood volume.
3. The arterioles are the sites of greatest resistance in the vascular system.

We can quantify the relationship between blood pressure, cardiac output, and peripheral resistance as follows:

Blood pressure (BP) = flow (cardiac output, CO)
$\times$ peripheral resistance (PR)

$$BP = CO \times PR$$

For instance, at a specific flow rate (that is, a given cardiac output), as peripheral resistance rises, so does blood pressure, and vice versa. Similarly, at a given peripheral resistance, any change in blood pressure will change flow rate—higher pressure will produce greater flow and lower pressure, lower flow.

These mechanisms are the backbone of short-term blood pressure homeostasis, which is regulated by autonomic nervous system controls that act to maintain systemic arterial blood pressure at a level appropriate for healthy gas and fluid exchange. Long-term blood pressure homeostasis is governed by renal ➡ (Chapter 16) mechanisms.

(Chapter 16)

> ### Case Note
>
> **11.13.** Bob's blood pressure was low. What was the main reason—low cardiac output or low peripheral resistance?

The Autonomic Nervous System Controls Cardiac Output and Peripheral Resistance

Blood pressure is monitored by **baroreceptors**, special pressure receptors located in the walls of the aorta and the carotid arteries in the neck. Baroreceptors are stretch receptors: increasing pressure stretches the vascular wall, which causes more baroreceptor signals to be sent the vasomotor center. Likewise, if blood pressure falls, arterial walls are not stretched as much and fewer signals are generated. Baroreceptors connect by sensory autonomic nerves to the vasomotor center in the brainstem, which in turn connects to arterioles and the heart by motor sympathetic and parasympathetic autonomic nerves. When the vasomotor center receives *fewer* baroreceptor signals, indicating lower pressure, it *increases* its output of sympathetic signals, and vice versa.

Blood pressure is maintained at a relatively constant level by a classical negative feedback loop in which baroreceptors *sense* a perturbation, the vasomotor center *integrates* information from the baroreceptors, and the heart and blood vessels *effect* changes to restore blood pressure to normal. Figure 11.19 illustrates how this feedback loop functions when a serious bleed reduces blood pressure.

1. The loss of blood reduces arterial blood volume, which reduces blood pressure.
2. The walls of the aorta (and carotid artery, not shown) are stretched less, so the baroreceptor firing rate decreases. The sensory nerve sends fewer action potentials to the vasomotor center.
3. The vasomotor center responds to this reduced input by activating the sympathetic nervous system and, though not illustrated in Figure 11.19, inhibiting the parasympathetic nervous system. Sympathetic nerves innervating the arterioles and heart increase their firing rate, while parasympathetic nerves innervating the heart reduce their firing rate. Sympathetic activation also stimulates the release of epinephrine from the adrenal medulla.

4a. The arterioles respond by contracting their smooth muscle, constricting their lumen, and increasing peripheral resistance.
4b. Sympathetic activation (and parasympathetic inactivation) increases heart rate by accelerating the depolarization of the SA node; it also increases stroke volume by increasing the contraction strength of ventricular cells. The result is increased cardiac output.
5. Blood pressure rises because both output and resistance have increased.

However, as we learned in President Reagan's case in ⬅ Chapter 1, this homeostatic reflex loop cannot completely correct for an especially large loss of blood—in our Chapter 1 case, President Reagan's blood pressure fell dangerously low despite homeostatic compensations. In such cases replacement of lost volume is absolutely necessary by infusion of blood, plasma, or salt solutions.

It should be noted that *not all* arterioles constrict in response to sympathetic activation. Only arterioles in nonvital tissues such as skin and muscle respond. This diminishes flow to those parts and reallocates it to more vital structures, mainly the heart and brain.

The baroreceptor reflex loop also guards against **hypertension** (sustained high blood pressure), which can damage blood vessels. A rise in blood pressure increases arterial stretch and thus the baroreceptor firing rate. This increased input alters the activity of the vasomotor center, shifting the balance toward increased parasympathetic activation and decreased sympathetic activation. The heart reduces cardiac output, the arterioles dilate, and blood pressure drops.

> ### Case Note
>
> **11.14.** Bob was given epinephrine to increase his blood pressure. What was the mechanism of action of this drug?

Arterial Blood Pressure Is Measured by Sphygmomanometry

Blood pressure is most commonly measured by use of a **sphygmomanometer,** a jawbreaker name derived from the Greek *sphugmos,* "pulse" (Fig. 11.20). Added to this root is the Greek *manos,* "thin," and *metre,* "measuring"—a reference to the ancient practice of inserting tall thin tubes into arteries of experimental animals to see how high the blood would climb.

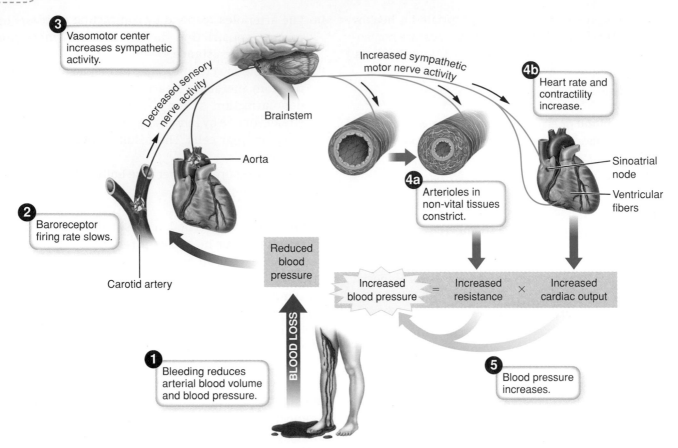

Figure 11.19. The baroreceptor reflex. Changes in blood pressure are detected by baroreceptors, which activate the vasomotor center. The vasomotor center attempts to reverse the change by modulating arteriole diameter and cardiac output. *If the baroreceptor firing rate increases, will the vasomotor center induce vasoconstriction or vasodilation?*

Conventional sphygmomanometers rely on a pressure gauge attached to an inflatable cuff, which is placed around the upper arm (at heart level). The cuff is inflated to a pressure certain to cut off all blood flow by squeezing shut the main (brachial) artery. The cuff is then deflated gradually as the examiner listens with a stethoscope placed over the artery at the lower end of the cuff just above the crook of the elbow. As cuff pressure falls below systolic blood pressure, blood begins to squirt through and makes a sound (the *Korotkoff sound*). The Korotkoff sound continues with each pulse wave until it is silenced as cuff pressure falls below diastolic blood pressure. The range between the pressure at the time of first appearance of the sound and the pressure at its disappearance is the blood pressure. High-tech devices that do not require a listener are also available.

Despite its simplicity, *accurate* high blood pressure measurements are difficult to obtain because they are influenced by posture, breathing, girth of the upper arm, clothing, anxiety, drugs, whether measured before or after a meal or exercise, and many other variables. Accurate readings are important because hypertension is a very common condition that causes no symptoms of its own but promotes heart attack, stroke, and other serious disease. It is generally

agreed that optimal blood pressure is less than 120/80 and that persistent readings greater than 140/90 are high enough to warrant a diagnosis of hypertension, which requires treatment. In the middle ground between 120/80 and 140/90 patients are said to be prehypertensive. A growing body of experts recommend treatment of prehypertension.

 11.38 Which parameter is equal to blood flow—stroke volume or cardiac output?

11.39 Which blood pressure determinant changes when cardiac output changes—arterial compliance or arterial blood volume?

11.40 In the absence of other changes, how will increased vessel radius affect blood pressure?

11.41 Which pressure value is lower—systolic or diastolic?

11.42 If blood pressure rises, what happens to the baroreceptor firing rate?

11.43 If the sympathetic nervous system is activated, which vessels will constrict—brain arterioles or skin arterioles?

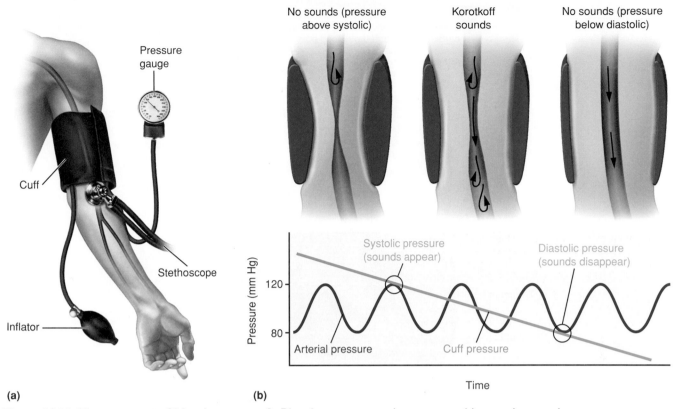

No sounds (pressure above systolic)

Korotkoff sounds

No sounds (pressure below diastolic)

Pressure gauge

Cuff

Stethoscope

Inflator

Systolic pressure (sounds appear)

Diastolic pressure (sounds disappear)

Pressure (mm Hg)

120

80

Arterial pressure

Cuff pressure

Time

(a)   (b)

Figure 11.20. Measurement of blood pressure. A. Blood pressure can be measured by a using a sphygmomanometer. **B.** As the clinician lowers the cuff pressure, Korotkoff sounds begin at the systolic pressure and disappear at the diastolic pressure. *What is the systolic pressure in (B)?*

11.44 If overall blood volume increases (in the absence of other changes), what will happen to blood pressure?

11.45 During a blood pressure measurement, which pressure is indicated by the disappearance of the Korotkoff sound?

Case Discussion

Myocardial Infarction: The Case of Bob W.

Let's return to our case. Bob suffered from hypertension, which damaged the endothelial lining of his arteries and resulted in *atherosclerosis*. This arterial disease is characterized by inflammation, scarring, and fat deposits in arterial walls The nearby Clinical Snapshot, titled *Atherosclerosis: We're Eating Ourselves to Death,* explains atherosclerosis in more detail.

The atherosclerosis in Bob's coronary arteries roughened the arterial walls and narrowed the vessel lumens. The abnormal arterial lining resulted in the formation of a thrombus ◀ (Chapter 10) in his right coronary artery, which supplied blood to his posterior left ventricle and the posterior part of the interventricular sys-

tem. This muscle tissue would still have received some blood from anastomoses, but that small amount of blood was insufficient and the posterior wall tissue died and was no longer contracting (Fig. 11.21).

That is, Bob had a heart attack. The pain in Bob's left neck, jaw, and elbow was *referred pain*—neurons conveying painful stimuli from the dying heart muscle converge with neurons from the jaw, neck, and arm, so the brain interprets the pain signals as originating from these peripheral sites.

A heart attack is more accurately described as a *myocardial infarction*—the death of heart muscle tissue caused by oxygen starvation. As a result of the loss of muscle tissue, the contractile force generated by the left ventricle was severely reduced. Stroke volume fell, which impaired cardiac output and caused a fall in blood pressure.

Recall also that the pacemaker potential depolarizes more slowly at successively lower parts of the cardiac conduction system. Bob's heart rate was 30 beats per minute—a rate so low that it indicates that the Purkinje fibers in the ventricle had taken up the autorhythmic task of stimulating each beat. Such a slow heart rate added to the difficulty in maintaining normal blood pressure, because an effective cardiac output could not be maintained.

Assuming that Bob's homeostatic mechanisms controlling blood pressure were intact, the fall in blood pressure

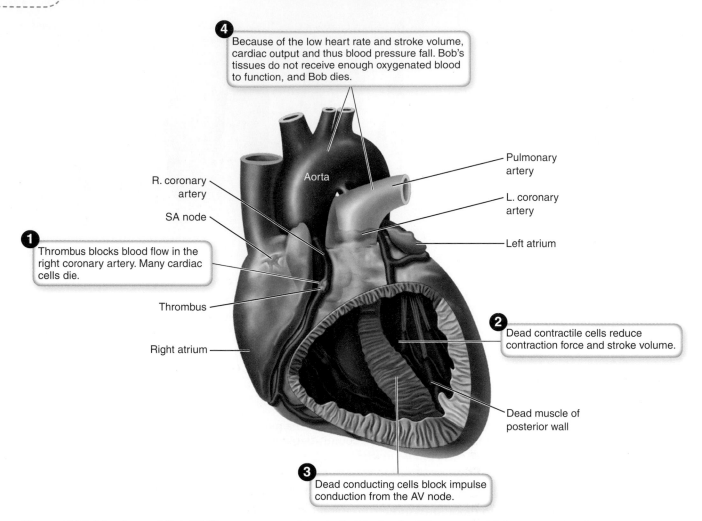

Figure 11.21. The Case of Bob W. The events resulting in Bob's death. *Why was Bob's heart rate low?*

would have decreased the baroreceptor firing rate. His vasomotor center would have sent sympathetic signals to increase vasoconstriction and thus increase peripheral resistance. Similar signals must have been sent to increase heart rate and cause more forceful myocardial contraction. However, recall that Bob's ECG showed a complete blockage of signal transmission from atria to ventricles, indicating that some of the dead tissue included fibers of the cardiac conduction system. No matter how many sympathetic signals arrived at the SA node, they could not increase his heart rate because they could not be relayed to the ventricles.

Bob's blood pressure of 75/45 was insufficient to propel blood through the systemic circulation, so his tissues did not receive enough oxygenated blood to function properly. The medically administered norepinephrine was able to stimulate the living cardiac cells, but they were not able to sustain blood pressure for long. Eventually, the lack of blood supply to Bob's brain and other vital organs proved fatal.

Case Note

11.15. Why was Bob's stroke volume reduced?

The Major Blood Vessels

We now turn to the names and features of the main blood vessels. The locations of some important arteries are illustrated in Figure 11.22 and the veins in Figure 11.23. The plates that follow illustrate these vessels (and a few more) in greater detail. As you go through the plates, notice the flowcharts accompanying many illustrations. You may find it helpful use the flowchart like a road map as you trace your path through the vessels in the corresponding illustration. As you study these discussions and figures, be aware that normal anatomical variants are common and do not cause problems. However, some deviations from normal can cause malfunction (disease).

CLINICAL SNAPSHOT

Atherosclerosis: We're Eating Ourselves to Death

Cardiovascular disease (CVD) is disease of the heart and blood vessels and is the leading cause of death (35% of deaths) among Americans. By comparison, all types of cancer account for about 23% of U.S. deaths. It is a diverse group of conditions, the most common of which are heart disease (29% of deaths) and strokes (6%). Most cases of cardiovascular disease can be traced back to a vascular disorder called *atherosclerosis.*

Atherosclerosis is a disease in which arterial walls accumulate deposits of lipids, inflammatory cells, and scar tissue, which impairs blood flow. Veins are not affected. It is a complex process that begins with injury to the endothelial cells that line all arteries. The physical pounding of high blood pressure and the ill effect of high blood glucose in diabetes are two well-known risk factors. Other factors, such as advanced age and family history, are not within our control. However, most risk factors are modifiable; that is, they are at influenced by lifestyle choices. These include:

- *Smoking.*
- *Physical inactivity.*
- *A diet high in saturated and "trans" fats.*
- *Obesity.*
- *High blood cholesterol,* especially low-density lipoprotein (LDL) cholesterol.
- *High blood pressure.*
- *Type II diabetes mellitus,* which is closely correlated with obesity. Type I diabetes is equally risky but not related to obesity.

Recall that endothelial cells function to keep blood and tissue safely separated, because blood will clot if it comes into contact with tissue. Also recall that inflammation is the body's response to injury. The inflammatory reaction incited by endothelial cell injury spreads deep into the wall of the artery, weakening and scarring (*sclerosing*) it. What's more, damaged endothelial cells allow lipids, mainly cholesterol, to seep through and into the wall of the artery, where they are trapped as pools of soft, mealy material. These deposits are called **atheromas,** from Greek *athere,* for a thick porridge or gruel.

In most instances the accumulation of fat and scar tissue causes narrowing of the blood vessel, which

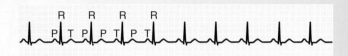

(a) Normal ECG

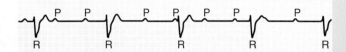

(b) Third-degree heart block

Atherosclerosis. An atherosclerotic aorta. Note the fatty deposits (atheromas).

slowly starves downstream tissue of blood. With slow starvation, the downstream tissue may die or wither gradually, which leads to chronic heart disease as the heart muscle loses its ability to function. Alternatively, the blockage may be sudden. Sudden blockages are often due to the bursting or cracking of an atheroma, which allows blood to come into contact with tissue on the other side of the endothelium. When this happens, platelets stick to the ruptured or cracked area and accumulate to form a thrombus
◄ (Chapter 10), which obstructs the artery. This is precisely what happened to our patient, Bob: an atheroma in his right coronary artery suddenly developed an occlusive thrombus, which caused the death of the heart muscle supplied by the artery.

What's more, an important consequence of atherosclerosis is loss of vascular compliance—the walls of arteries become stiff with scar tissue and lose their ability to stretch with each beat. They become less like a soft rubber hose and more like a metal pipe. A notable effect of lost compliance is systolic hypertension. This occurs because more ventricular power and pressure are required to force blood into a stiff vascular tree. Atherosclerosis also weakens the walls of blood vessels, allowing them to balloon outward permanently from the pressure inside. Such a weak, ballooned area is called an **aneurysm.** Sometimes the weakening is so great that the aneurysm bursts, causing a fatal hemorrhage.

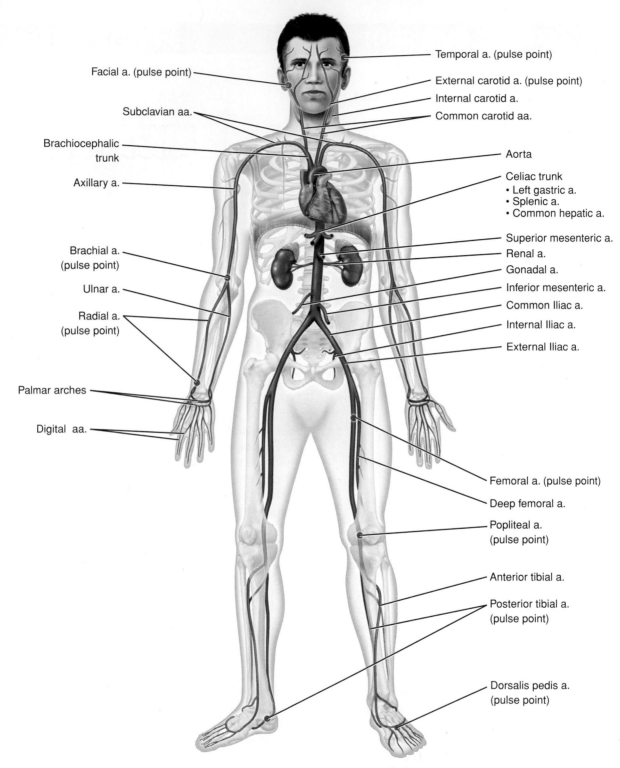

Temporal a. (pulse point)

Facial a. (pulse point)

External carotid a. (pulse point)

Internal carotid a.

Subclavian aa.

Common carotid aa.

Brachiocephalic trunk

Aorta

Axillary a.

Celiac trunk
• Left gastric a.
• Splenic a.
• Common hepatic a.

Superior mesenteric a.

Brachial a. (pulse point)

Renal a.

Gonadal a.

Ulnar a.

Inferior mesenteric a.

Radial a. (pulse point)

Common Iliac a.

Internal Iliac a.

External Iliac a.

Palmar arches

Digital aa.

Femoral a. (pulse point)

Deep femoral a.

Popliteal a. (pulse point)

Anterior tibial a.

Posterior tibial a. (pulse point)

Dorsalis pedis a. (pulse point)

Figure 11.22. Major systemic arteries. The circles indicate pulse points, where the heart rate can be measured. *Which artery passes behind the knee—the popliteal artery or the axillary artery?*

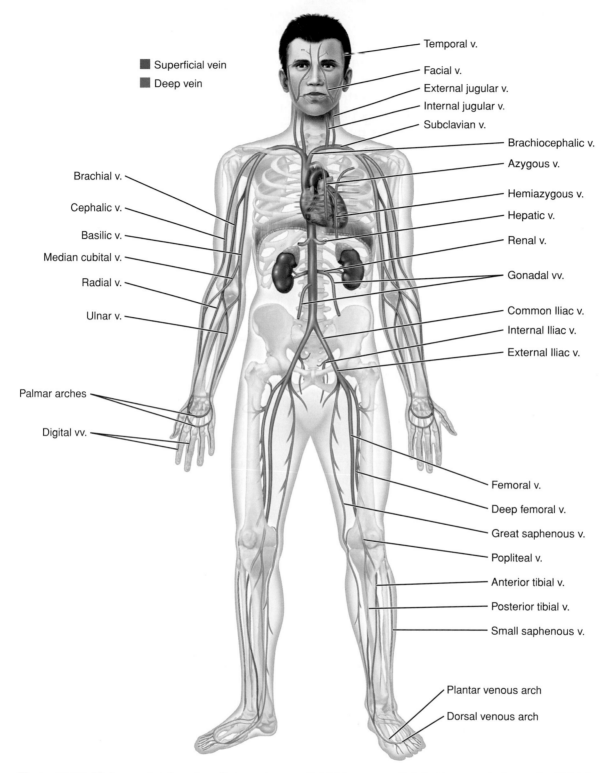

Figure 11.23. Major systemic veins. Deep veins are light blue; superficial veins are dark purple. *Which vein is superficial—the brachial vein or the basilic vein?*

Arteries of the Torso

All blood leaving the left ventricle enters the **aorta,** the largest blood vessel in the body. It extends upward from the heart and then arches to the left and downward, traveling along the anterior edge of the vertebral column. It ends at the rear brim of the pelvis (about the level of the umbilicus), where it divides into the right and left common **iliac** arteries, which supply the pelvis and lower limbs. For convenience, it is divided into four sections: the **ascending aorta** (the initial, vertical section), the **aortic arch** (the second section), the **thoracic aorta,** which extends to the diaphragm, and the **abdominal aorta,** which extends from the diaphragm.

The aorta averages about 1 in. (2.5 cm) in diameter— about the size of a garden hose. The first branches are the right and left coronary arteries, which originate within the first few millimeters of the ascending aorta and nourish the heart muscle. The next three branches arise from the aortic arch. First is the **brachiocephalic** artery, a short trunk that branches into the right **subclavian** artery, which supplies the right neck and right upper limb, and the right common **carotid,** which supplies the right side of the neck, head, and brain. There is no equivalent of the brachiocephalic artery for the left side of the body. Instead, the areas supplied on the right by the two branches of the brachiocephalic are supplied on the left by two arteries that branch directly from the aorta. The left common carotid artery is the second branch off the aortic arch and

supplies the left side of the neck, head, and brain. The third branch of the aortic arch is the left subclavian artery, which supplies the left neck and left upper limb.

Beneath the arch, most branches are generally paired right and left sets. For instance, right and left branches serve the chest cage and body wall, the bronchi and lungs, the adrenals, kidneys, and gonads. In contrast, single branches serve the liver, spleen, and intestines.

Regions Supplied by the Aortic Branches

Artery	Region
Common hepatic	Liver
Left gastric	Stomach, esophagus
Splenic	Spleen, stomach, pancreas
Suprarenal	Adrenal gland
Renal	Kidney
Superior mesenteric	Small intestine, large intestine, pancreas
Gonadal	Ovaries or testes
Inferior mesenteric	Descending large intestine, rectum

Plate 11.1 Arteries of the Torso

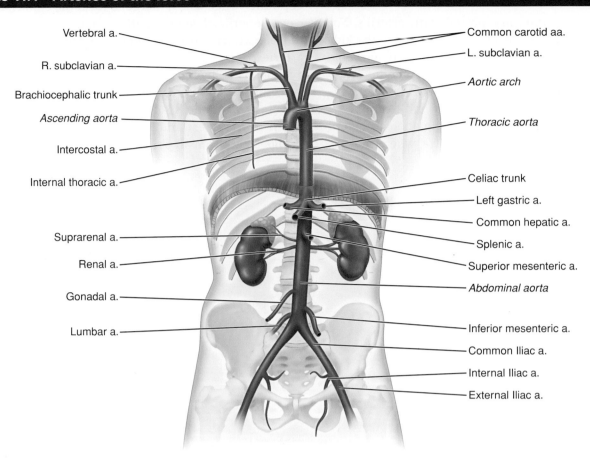

Vertebral a.
R. subclavian a.
Brachiocephalic trunk
Ascending aorta
Intercostal a.
Internal thoracic a.
Suprarenal a.
Renal a.
Gonadal a.
Lumbar a.

Common carotid aa.
L. subclavian a.
Aortic arch
Thoracic aorta
Celiac trunk
Left gastric a.
Common hepatic a.
Splenic a.
Superior mesenteric a.
Abdominal aorta
Inferior mesenteric a.
Common Iliac a.
Internal Iliac a.
External Iliac a.

(a) Anatomical view

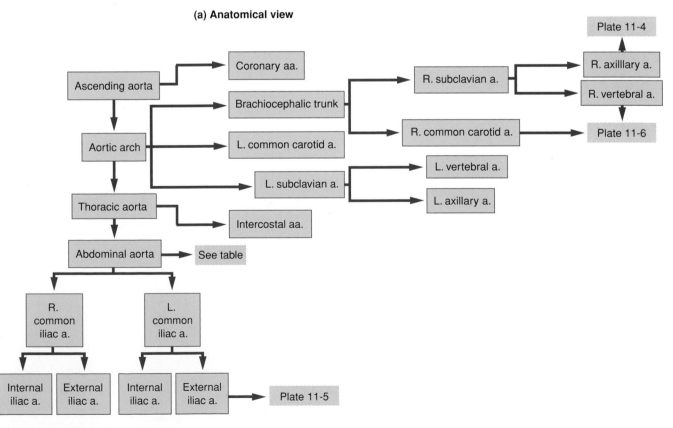

(b) Schematic view

Veins of the Torso

In general, deep veins are paired with arteries. For example, the right renal artery carries blood to the kidney and the right renal vein brings it back. A remarkable exception to this rule is venous return from the intestines, which is discussed in detail below. A lesser asymmetry involves the unpaired **azygos** and hemiazygos veins, which drain various parts of the abdominal and chest walls. And in contrast to the aorta, where there is only one brachiocephalic artery, there are two **brachiocephalic veins,** which merge to form the superior vena cava. Superficial veins usually have names unrelated to those of arteries, because there are very few superficial arteries.

Blood is delivered into the right atrium by the superior **vena cava,** which brings blood from all points above the diaphragm: the head, neck, brain, upper limbs, and much of the chest and lungs. The inferior vena cava brings blood from all points below the diaphragm.

Plate 11.2   Veins of the Torso

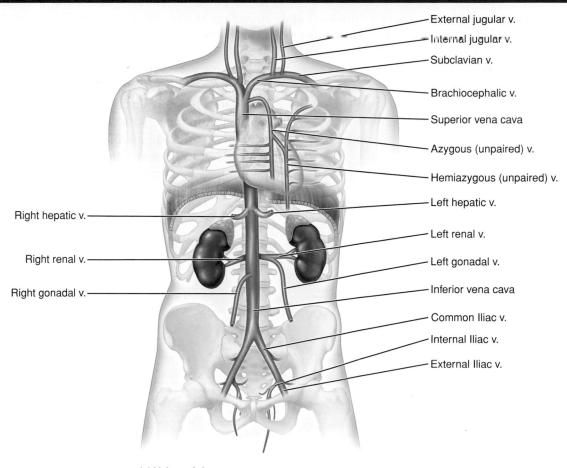

External jugular v.
Internal jugular v.
Subclavian v.
Brachiocephalic v.
Superior vena cava
Azygous (unpaired) v.
Hemiazygous (unpaired) v.
Left hepatic v.
Left renal v.
Left gonadal v.
Inferior vena cava
Common Iliac v.
Internal Iliac v.
External Iliac v.

Right hepatic v.
Right renal v.
Right gonadal v.

(a) Veins of the torso

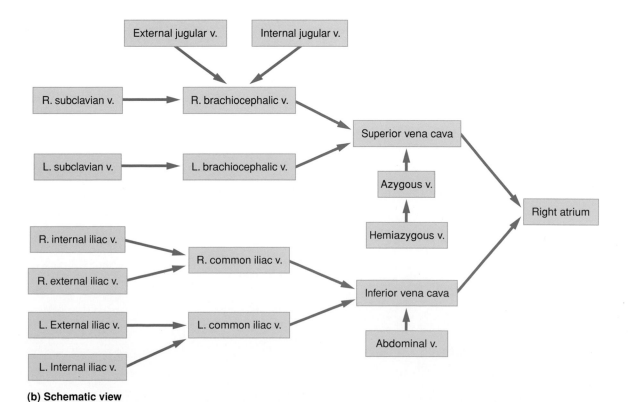

External jugular v.   Internal jugular v.

R. subclavian v. → R. brachiocephalic v.

L. subclavian v. → L. brachiocephalic v.

Superior vena cava

Azygous v.

Hemiazygous v.

Right atrium

R. internal iliac v.
R. external iliac v. → R. common iliac v.

L. External iliac v. → L. common iliac v.
L. Internal iliac v.

Inferior vena cava

Abdominal v.

(b) Schematic view

The Portal Circulation

A **portal circulation** is any circulatory pathway that includes two capillary beds: one that drains into another. Venous return from the intestines is such a portal system: veins returning from the digestive tract, pancreas, and spleen merge to form the **hepatic portal vein,** which does *not* join the inferior vena cava. Instead, it redivides and flows into a network of sinusoids that bring blood into intimate contact with hepatic cells, which metabolize nutrients from the intestine. This collection of *two* capillary beds (the intestinal capillaries and the hepatic sinusoids) and the hepatic portal vein connecting them is called the *hepatic portal circulation,* which is revisited in Chapter 14. Another portal circulation is found in the endocrine system (Chapter 15).

Not only venous blood flows into and out of the liver—arterial blood also flows into the liver from the hepatic artery to supply hepatic cells with oxygen. Venous blood derived from the hepatic artery flow merges with portal venous blood and is gathered together in small veins that merge to form the **hepatic** vein, which flows into the inferior vena cava.

Plate 11.3 The Portal Circulation

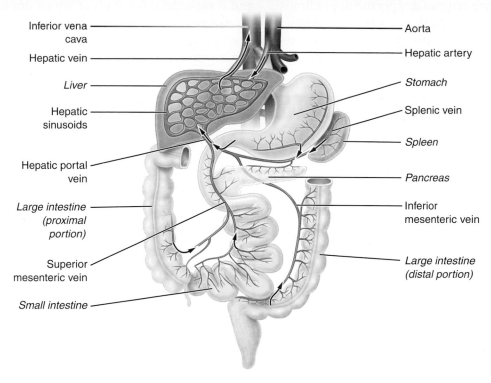

Inferior vena cava
Hepatic vein
Liver
Hepatic sinusoids
Hepatic portal vein
Large intestine (proximal portion)
Superior mesenteric vein
Small intestine

Aorta
Hepatic artery
Stomach
Splenic vein
Spleen
Pancreas
Inferior mesenteric vein
Large intestine (distal portion)

(a) Anatomical view

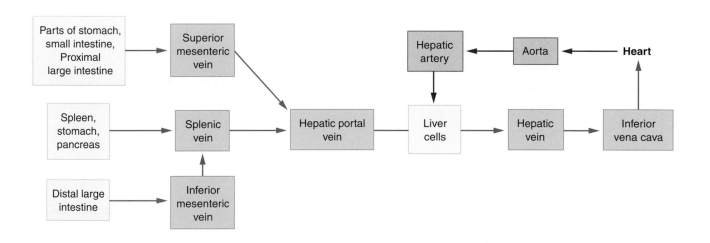

Parts of stomach, small intestine, Proximal large intestine → Superior mesenteric vein

Spleen, stomach, pancreas → Splenic vein

Distal large intestine → Inferior mesenteric vein

Hepatic artery ← Aorta ← **Heart**

Hepatic portal vein → Liver cells → Hepatic vein → Inferior vena cava

(b) Schematic view

Arteries and Veins of the Upper Limb

Branches of the axillary arteries supply the upper limbs. Note that the subclavian artery becomes the axillary artery at the lateral edge of the first rib, and subsequently the brachial artery at the inferior border of the teres major muscle. The veins of the upper limb are more complex than the arteries, consisting of a deep set and a superficial set, both of which drain into the subclavian vein. Many of the arm veins are paired, but only one vein is illustrated for clarity. For instance, there are two brachial veins that run parallel to each other.

Plate 11.4   Arteries and Veins of the Upper Limb

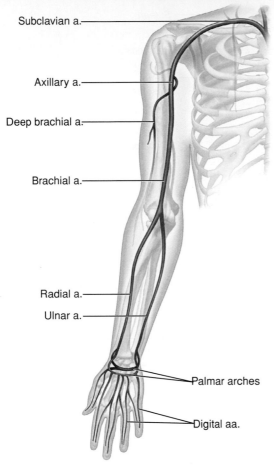

Subclavian a.
Axillary a.
Deep brachial a.
Brachial a.
Radial a.
Ulnar a.
Palmar arches
Digital aa.

(a) Arteries of the right upper limb

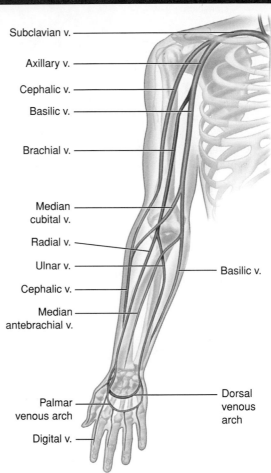

Subclavian v.
Axillary v.
Cephalic v.
Basilic v.
Brachial v.
Median cubital v.
Radial v.
Ulnar v.
Cephalic v.
Median antebrachial v.
Basilic v.
Palmar venous arch
Digital v.
Dorsal venous arch

(b) Veins of the right upper limb

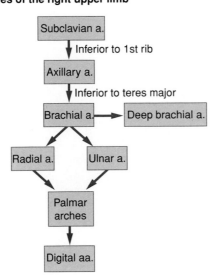

Subclavian a.
↓ Inferior to 1st rib
Axillary a.
↓ Inferior to teres major
Brachial a. → Deep brachial a.
Radial a.   Ulnar a.
Palmar arches
Digital aa.

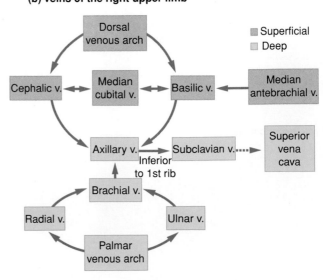

■ Superficial
□ Deep

Dorsal venous arch
Cephalic v. ↔ Median cubital v. ↔ Basilic v. ← Median antebrachial v.
Axillary v. → Subclavian v. ┈┈▶ Superior vena cava
Inferior to 1st rib
Brachial v.
Radial v.   Ulnar v.
Palmar venous arch

Arteries and Veins of the Lower Limb

The arteries and veins of the lower limb are organized similarly to those of the upper limb. Note that the arteries are illustrated for the left leg and the veins for the right leg. The saphenous vein is the longest vein in the body; parts of it can be surgically removed to replace a damaged vessel elsewhere. The plantar arches, like the palmar arches, are anastomoses.

Plate 11.5   Arteries and Veins of the Lower Limb

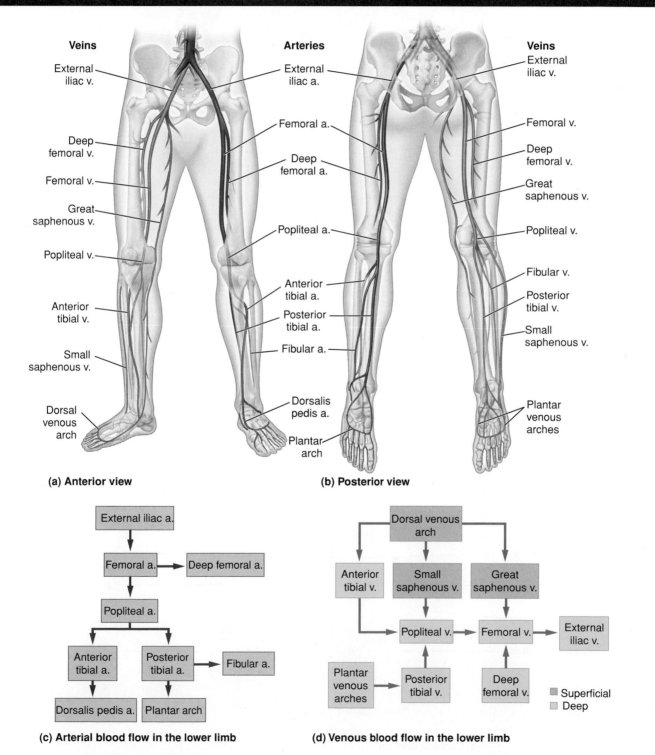

(a) Anterior view

(b) Posterior view

(c) Arterial blood flow in the lower limb

(d) Venous blood flow in the lower limb

Plate 11.6   Arteries of the Head and Neck

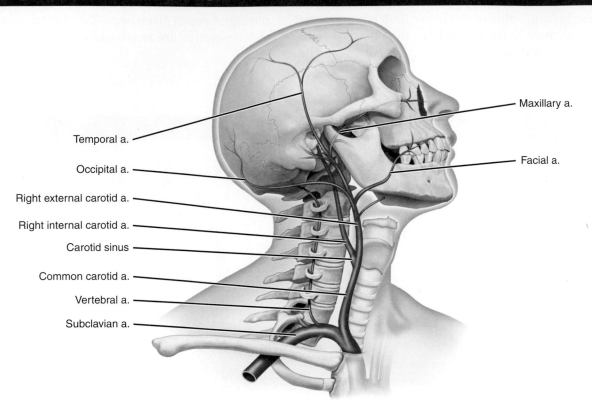

Temporal a.

Occipital a.

Right external carotid a.

Right internal carotid a.

Carotid sinus

Common carotid a.

Vertebral a.

Subclavian a.

Maxillary a.

Facial a.

(a) Superficial arteries of the head and neck

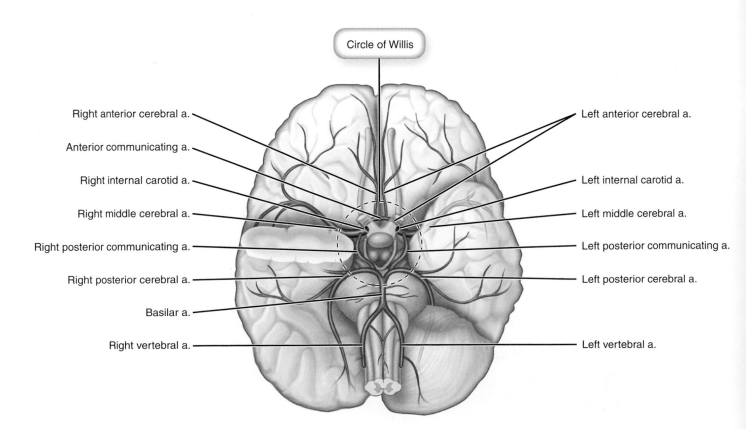

Circle of Willis

Right anterior cerebral a.

Anterior communicating a.

Right internal carotid a.

Right middle cerebral a.

Right posterior communicating a.

Right posterior cerebral a.

Basilar a.

Right vertebral a.

Left anterior cerebral a.

Left internal carotid a.

Left middle cerebral a.

Left posterior communicating a.

Left posterior cerebral a.

Left vertebral a.

(b) Cerebral arteries

Plate 11.6 (continued)

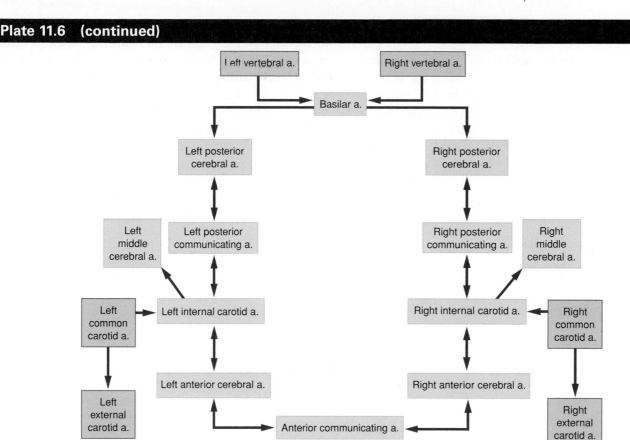

(c)

Arteries of the Head and Neck

The arteries of the head and neck are unusual. First, there are dual supply routes on each side—an arrangement ensuring that if one of the two is blocked, the other can still supply blood. Second, the arteries connect in a circle, the **circle of Willis,** on the floor of the brain, which interconnects the right and left circulations, so that blood from one side can supply the whole brain if the other side is obstructed. With the exception of small connecting (*communicating*) branches, the circle of Willis consists of short segments of arteries that form it as they pass through on the way to more distant sites.

One set of arteries supplying the brain arises from the right and left internal carotid arteries. Each internal carotid artery enters the skull and branches to form middle and anterior cerebral arteries, which supply the anterior and medial portions of the brain. The proximal regions of the left and right anterior cerebral arteries are joined by the anterior communicating artery to form the anterior aspect of the cerebral arterial circle.

A second paired set is the vertebral arteries, which arise from the subclavian arteries and travel a posterior route upward through foramina in the lateral processes of cervical vertebrae. They pass through the foramen magnum and join to form the basilar artery, which gives off small branches to supply the brainstem and continues upward on the anterior aspect of the brainstem. The basilar artery branches into the right and left posterior cerebral arteries, which supply the posterior portion of the brain. The initial segments of each combine to form the posterior part of the circle of Willis.

Veins of the Head and Neck

The scalp and face are drained by paired veins that empty into the internal or external jugular veins. The superior spinal cord and vertebrae are similarly drained by the paired vertebral veins. However, blood draining from the brain flows not into normal veins but into dural sinuses—large, elongated chambers formed between layers of the dura mater (Fig. 8.6). The **superior sagittal sinus** runs from anterior to posterior along the superior edge of the falx cerebri. The smaller **inferior sagittal sinus** runs along the inferior border of the falx and empties into the **straight sinus.** The straight sinus and the superior sagittal sinus meet at the aptly named **confluence of sinuses,** which separates into left and right S-shaped **sigmoid sinuses.** The sigmoid sinuses, in turn, drain into the right and left internal jugular veins, which travel with their companion carotid arteries down the neck.

Plate 11.7   Veins of the Head and Neck

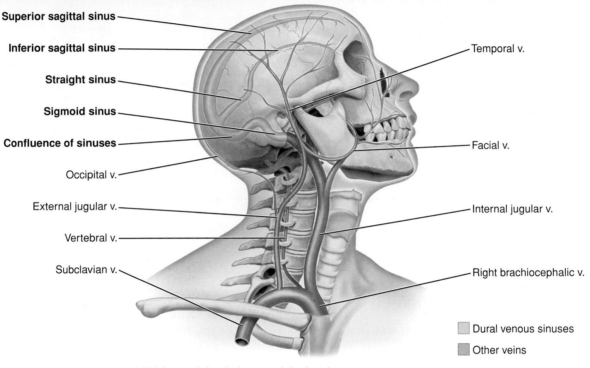

Superior sagittal sinus

Inferior sagittal sinus

Straight sinus

Sigmoid sinus

Confluence of sinuses

Occipital v.

External jugular v.

Vertebral v.

Subclavian v.

Temporal v.

Facial v.

Internal jugular v.

Right brachiocephalic v.

Dural venous sinuses

Other veins

(a) Veins and dural sinuses of the head

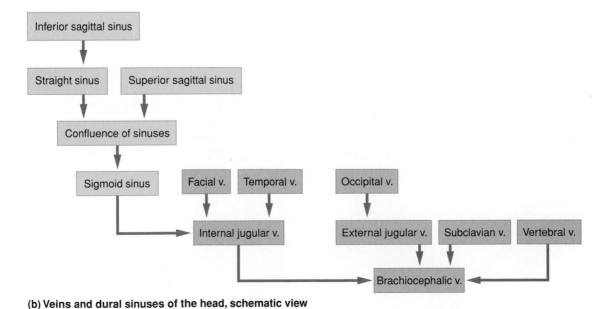

(b) Veins and dural sinuses of the head, schematic view

Word Parts

Latin/Greek Word Parts	English Equivalents	Examples
Cardi/o	Heart	Electrocardiogram: measurement of the electric activity of the heart
Atri/o	Atrium	Interatrial septum: wall dividing the atria
Vas/o	Vessel	Vasoconstriction: narrowing of a blood vessel
Coron/o	Crown	Coronary circulation: vessels that circle the heart like a crown
My/o	Muscle	Myocardium: muscle layer of the heart
Baro-	Pressure	Baroreceptor: a receptor that detects changes in pressure, such as blood pressure
Epi-	Upon	Epicardium: connective tissue layer directly upon the heart (cardium)
Endo-	Inner	Endocardium: inside layer of the heart wall
Peri-	Around	Pericardium: connective tissue membrane around the heart

Chapter Challenge

CHAPTER RECALL

1. **The systemic circulation carries**
   a. oxygen-rich blood in veins.
   b. oxygen-rich blood from the lungs to the heart.
   c. oxygen-poor blood from the body to the heart.
   d. blood from the capillaries to the arteries.

2. **The pressure in a closed fluid-filled compartment will decrease if**
   a. additional fluid is added to the compartment.
   b. the muscle surrounding it contracts.
   c. the volume of the compartment decreases.
   d. fluid is drained out of the compartment.

3. **The flexible sac surrounding the heart is called the**
   a. mediastinum.
   b. pericardium.
   c. pleura.
   d. myocardium.

4. **The action potential in a cardiac muscle cell**
   a. is about the same length as an action potential in a skeletal muscle cell.
   b. includes a sustained depolarization induced by sodium entry.
   c. includes a repolarization phase resulting from calcium leaving the cell.
   d. involves three ions: sodium, calcium, and potassium.

5. **The calcium that enters the cell during a cardiac action potential**
   a. depolarizes the cell.
   b. opens calcium channels in the sarcoplasmic reticulum.
   c. binds to troponin and initiates cross-bridge cycling.
   d. all of the above.

6. A valve guards the entrance to the
   a. pulmonary trunk.
   b. right atrium.
   c. superior vena cava.
   d. left atrium.

7. The right marginal artery is a branch of the
   a. coronary sinus.
   b. left coronary artery.
   c. posterior interventricular artery.
   d. right coronary artery.

8. The atrioventricular valves are closed during the entire period of
   a. ventricular systole.
   b. atrial systole.
   c. the cardiac cycle.
   d. atrial diastole.

9. The order in which impulses travel through the heart is
   a. atrioventricular bundle, Purkinje fibers, sinoatrial node, AV node.
   b. Purkinje fibers, atrioventricular bundle, AV node, sinoatrial node.
   c. sinoatrial node, atrioventricular node, atrioventricular bundle, Purkinje fibers.
   d. atrioventricular node, atrioventricular bundle, sinoatrial node, Purkinje fibers.

10. Ventricular filling occurs during the
    a. QRS complex.
    b. P-Q segment.
    c. Q-T segment.
    d. time between the T wave and the QRS wave.

11. Which determinant has the greatest potential to change cardiac output?
    a. Heart rate over 150 beats per minute
    b. Calcium concentration within muscle fibers
    c. Length of muscle fibers
    d. Blood pressure in the aorta

12. The parasympathetic nervous system
    a. controls the strength of ventricular contractions.
    b. decreases the slope of the pacemaker potential.
    c. uses norepinephrine to decrease heart rate.
    d. includes axons of the cardiac plexus.

13. You are viewing a blood vessel under the microscope. You see three distinct layers, one of which is a thick muscular layer. The layers are separated by extensive elastic tissue. This vessel is most likely
    a. a capillary.
    b. an arteriole.
    c. a vein.
    d. an artery.

14. Which of the following would stimulate vasodilation?
    a. High pH
    b. Low CO_2 concentration
    c. Low oxygen concentration
    d. Epinephrine

15. Assuming that all other factors remain the same, increased blood pressure would most likely result in
    a. increased interstitial fluid.
    b. decreased interstitial fluid.
    c. increased blood osmotic pressure.
    d. decreased blood osmotic pressure.

16. Which of the following does *not* help to move blood back to the heart?
    a. Vein valves
    b. Skeletal muscle contraction
    c. Chest expansion
    d. Vein dilation

17. The most important determinant of peripheral resistance is
    a. blood viscosity.
    b. vessel length.
    c. vessel diameter.
    d. heart rate.

18. Which of the following changes would decrease blood pressure?
    a. A steeper slope of the pacemaker potential in the sinoatrial node
    b. Increased calcium entry into muscle cells of the left ventricle
    c. Widespread vasodilation
    d. Increased blood volume

19. The Korotkoff sound can be detected when
    a. no blood is moving through the artery.
    b. blood flow through the artery is completely unobstructed by the blood pressure cuff.
    c. the pressure of the cuff is above systolic pressure.
    d. the pressure of the cuff is between systolic and diastolic pressure.

20. A system in which blood circulates through two capillary beds before returning to the heart is called
    a. a circulating system
    b. a portal system
    c. a sinus
    d. an anastomosis

21. **A blood cell is journeying from the left thumb to the heart. It would probably *not* pass through the**
    a. left brachiocephalic vein.
    b. superior vena cava.
    c. left jugular vein.
    d. left subclavian vein.

22. **The brachiocephalic vein is formed by the union of the**
    a. femoral and iliac veins.
    b. subclavian and jugular veins.
    c. gastric and splenic veins.
    d. carotid and subclavian veins.

23. **The _____ artery becomes the femoral artery.**
    a. external iliac
    c. superior mesenteric
    b. internal iliac
    d. deep femoral

24. **Which of the following arteries is *not* found in the circle of Willis?**
    a. Anterior cerebral
    b. Posterior communicating
    c. Vertebral
    d. Middle cerebral

25. **The inferior vena cava receives blood from the**
    a. hepatic portal vein.
    c. azygos vein.
    b. brachiocephalic vein.
    d. hemiazygos vein.

CONCEPTUAL UNDERSTANDING

26. **Is it possible for a cardiac muscle cell to contract independently of its neighbors? Explain why or why not.**

27. **List the route a blood cell would take to travel from the left ventricle to the right kidney.**

APPLICATION

28. **An artery branches into two arterioles of equal resting diameter. Arteriole A then receives sympathetic input; arteriole B does not. All other parameters are identical.**
    a. Which vessel will have the smaller diameter?
    b. Which vessel will have the higher resistance?
    c. Which vessel will have the higher blood flow?

29. **When she exercises at intermediate intensity, Ms. M., a trained triathlete, has a cardiac output of 15 L/min and a heart rate of 150 beats per minute.**
    a How much blood does her right ventricle pump per minute?
    b. What is her stroke volume?

30. **Mr. L. suffered a myocardial infarction that incapacitated his sinoatrial node. His atrioventricular node is now acting as the pacemaker. How will his heartbeat differ from normal? What is the usual treatment for his condition?**

31. **Mr. Q., age 72, has a blood pressure reading of 90/60 mm Hg and complains of dizziness, cold extremities, and episodes of fainting.**
    a. What is his systolic pressure?
    b. What is his diastolic pressure?
    c. Which two cardiovascular factors could be modified to increase Mr. Q.'s blood pressure?

You can find the answers to these questions on the student Web site at http://thepoint.lww.com/McConnellandHull

12

The Immune and Lymphatic Systems

Major Themes

- Every person has a set of proteins that define "self" and are unique to that individual. With the exception of identical siblings, self-proteins are "nonself" (alien, foreign) to every other person.

- Cells of the immune system distinguish between self and nonself and neutralize or destroy nonself substances.

- The lymphatic system houses immune cells but is also involved in fluid balance and nutrient transport.

Chapter Objectives

"She is still using."

As you read through the following case study, assemble a list of the terms and concepts you must learn in order to understand Miriam's case.

Clinical History: Miriam K., a 28-year-old single woman with no children, had immigrated to the United States from Botswana 3 years before her initial appearance at the emergency room with a friend. Her primary complaints included fever and diarrhea. Questioning revealed that over the previous 6 months she had lost 25 pounds and experienced poor appetite, recurrent night sweats, intermittent fever, and bouts of severe diarrhea. She reported that she was well before leaving Botswana—an African nation in which the most reliable statistics (as of 2010, collected in 2008) estimate that 20% to 25% of adults are infected with human immunodeficiency virus (HIV). She admitted to having had sex for money and that she had used intravenous drugs but claimed that she had quit 6 months earlier. However, her friend said, "She is still using."

Physical examination and other data: Physical exam revealed a lethargic, thin woman complaining of thirst and abdominal cramps. Her abdomen was tender but contained no masses. Enlarged lymph nodes were present in the cervical, axillary, and inguinal regions. Marks on her forearms suggested the possibility of self-injection. Lab studies in the emergency room showed her to have low blood hemoglobin and a low lymphocyte count. Blood tests for HIV and helper T-cell lymphocyte count were also ordered.

Clinical course: Miriam was admitted to the hospital for evaluation. Blood tests revealed that she was HIV-positive and her helper T-cell count was 375/mm^3 (normal: >500). Her diagnosis was

463

Need to Know

It is important to understand the terms and concepts listed below before tackling the new information in this chapter.

■ Proteins, carbohydrates, nucleic acids ← (Chapter 3)

■ White blood cells, inflammation, antibodies, antigens ← (Chapter 10)

■ Anatomy of blood vessels, basic plan of the circulatory system ← (Chapter 11)

"HIV infection; not yet AIDS," and she was treated with antibiotics and intravenous fluids. She recovered well. She was discharged with a starter pack of anti-HIV drugs, a prescription for more, and an appointment at a public health clinic, but she failed to fill the prescription and to keep the clinic appointment.

Fourteen months later, Miriam appeared in the emergency room again complaining of shortness of breath and severe diarrhea. Chest x-rays showed bilateral pulmonary shadows suggestive of pneumonia. Her helper T-cell count was 186/mm^3. She was given antibiotics and a clinic appointment but again failed to appear for follow-up care.

Her final appearance in the emergency room occurred 10 months later, when she was brought in by staff from a homeless shelter. She was semicomatose and caregivers reported that she suffered from severe memory problems, persistent coughing, diarrhea, painful urination, and weight loss. Clinical investigation revealed extensive bilateral pneumonia and a severe urinary tract infection, but she died before a full investigation could be completed.

An autopsy was performed. The body was very thin. The lungs were meaty, nearly airless, and severely congested. Microscopic study revealed that the air sacs were filled with a type of fungus. Given the degree of infection, there was relatively little inflammatory reaction. Abdominal lymph nodes were enlarged. The spleen was also enlarged and congested with blood. Both kidneys were riddled with small abscesses. Urinary cultures collected before death but reported postmortem showed overwhelming numbers of streptococcal bacteria. The mucosa of the small and large intestines were extensively ulcerated. Microscopic study revealed evidence of viral infection in intestinal epithelial cells. In the left frontal lobe of the brain was a 4-cm tumor, shown by subsequent study to be composed of B cells. The brain was mildly atrophied, and microscopic study revealed widespread, mild chronic inflammation and collections of immune cells suggestive of HIV infection of the brain.

The final autopsy diagnoses were AIDS with fungal pneumonia, HIV infection of the brain, B-cell lymphoma of the brain, bacterial infection of the urinary tract, and viral infection of the small and large intestines.

Throughout this chapter a single concept is of overriding importance: "It's us against them." "Us" is each person's normal cells and tissues, which we refer to as **self.** "Them" is an untold number and variety of external threats that attempt to penetrate our body's natural barriers and invade our internal environment. For example, daily we are swarmed by microbes. Many are harmless, but others, called **pathogens,** are capable of causing disease. We are also beset from within: cancer cells can spring anew from any tissue, then escape from their origin and spread to disrupt the functioning of other tissues. Pathogens, cancer cells, and other alien cells and substances are referred to as **nonself.**

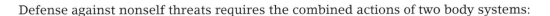

Defense against nonself threats requires the combined actions of two body systems:

- The **immune system** is the body's cellular defense system. Back in ◀ Chapter 10, we encountered its key players: the white blood cells, or leukocytes.
- The **lymphatic system** consists of a branching network of *lymphoid vessels* and *lymphoid organs* that house the immune cells.

These are the most widely dispersed of all organ systems, and portions of both are found in virtually every part of the body. Their smallest unit is an individual immune cell; their largest is an organ called the spleen. Collectively their mass is about equal to that of the brain.

The joy I felt at the prospect before me of being the instrument destined to take away from the world one of its greatest calamities (smallpox) was so excessive that I found myself in a kind of reverie.

Edward Jenner (1749–1823), English surgeon, in 1796, upon realizing that cowpox vaccination could prevent smallpox, one of the most deadly contagious diseases ever known. Note: Though not conclusively documented, evidence suggests that centuries earlier the Chinese practiced somewhat similar methods.

Functions of the Immune and Lymphatic Systems

The primary function of both the immune and lymphatic systems is defense. To understand when they come into play, it may help to imagine your normal cells—your *self*—as the inhabitants of a castle (Fig. 12.1). Any external threat must penetrate multiple lines of defense before it can cause disease.

Physical and Chemical Barriers Prevent Pathogen Entry

Just as the first lines of defense for our castle are the exterior walls, our body's initial defenses are the large-scale physical barriers discussed in ◀ Chapter 5: skin and mucous membranes, assisted by hairs, mucus, and cilia, which trap particulate matter. We also have chemical barriers, such as saliva and stomach acid, which digest or neutralize threats entering through the gastrointestinal tract.

These defenses are quite effective, allowing no admittance to the thousands of microbes we encounter each day. In contrast, think of how quickly pathogens can cause infection when the skin barrier is violated by an open wound. For example, our patient in ◀ Chapter 5 died because burns left him susceptible to overwhelming bacterial infection. Similarly, a genetic defect in the movement of cilia ("immotile cilia") predisposes affected

individuals to frequent lung infections, and inadequate saliva production results in increased tooth decay from excess bacterial activity.

Still, even castles have points of contact with the outside world and are, therefore, susceptible to sneaky invaders. The human body has similar weaknesses—pathogens frequently gain access to our internal tissues via our eyes, nose, gastrointestinal tract, or urinary tract.

Figure 12.1. Barriers. These walls protect the occupants from external threats, much as our physical and chemical barriers protect our body cells from pathogens. *Which body system physically blocks pathogens from penetrating our internal environment?*

The Immune System Has Three Tasks

Enemies that succeed in breaking through a castle's wall will encounter guards in the castle's interior. Similarly, pathogens that succeed in penetrating the body's physical and chemical barriers will face the cellular defenses of the immune system. In addition, the immune system combats the emergence of cancer cells, which arise inside our barriers like traitors among the inhabitants of a castle. And the inhabitants must also have a means of dealing with the dying and the dead. In short, the cells of the immune system have three major tasks:

1. To defend against pathogens and other external nonself threats
2. To identify and destroy any abnormal cells arising from within (cancer cells, for example)
3. To destroy damaged or aged cells and free their components for reuse or elimination

Performing these roles are white blood cells, or leukocytes, discussed in ◀ Chapter 10. Some leukocytes remain in blood for their entire life span, but the majority migrate out of blood and into other body tissues, where most of their work is done.

The Lymphatic System Has Three Functions

If the immune system provides the soldiers to defend the castle's inhabitants, the lymphatic system provides the infrastructure: sentry posts, training facilities, and tunnels linking all these locations. The lymphoid organs act as sites of lymphocyte development and action, and immune cells travel through the lymphoid vessels (as well as through blood vessels). Thus, the lymphatic system helps the immune system to defend against pathogens and cancer as well as to remove dying cells. The lymphatic vessels also serve two additional functions unrelated to defense: they transport fluids and specific nutrients. In short, the lymphatic system has three major tasks:

1. To house and support immune cells
2. To filter and return some interstitial fluid to blood, as discussed in ◀ Chapter 11
3. To absorb fats and fat-soluble vitamins from the intestines and deliver them to blood, which as discussed in ➡ Chapter 14.

Before looking at how the immune and lymphatic systems work together in the body's defense, we must identify the many structures involved. We begin with the cells of the immune system and then explore the organs and vessels of the lymphatic system.

Case Note

12.1. Some of Miriam's tissues are swollen, reflecting the accumulation of interstitial fluid. Which system returns interstitial fluid to blood—the immune system or the lymphatic system?

 12.1 For each of the following, state whether they are a function of the immune system, the lymphatic system, or both: fat absorption, destruction of old cells, defense against pathogens.

Cells of the Immune System

Leukocytes are the cells of the immune system. We identified in ◀ Chapter 10 three classes of leukocytes: granulocytes, lymphocytes, and monocytes. These leukocytes circulate in blood and migrate into tissues, especially lymphoid tissues, where they may reside permanently or may travel on demand to fight infection, for example. Recall also from Chapter 10 that inflammation "is the body's collective cellular and vascular response to injury," and leukocytes, especially granulocytes, are also the cells of inflammation. Inflammation not only acts as part of the immune response to defeat the invader, but it is also essential to the cleanup and repair of tissues that occurs in the healing process. Inflammation is an *immune* activity when it attacks and neutralizes or destroys alien substances.

Granulocytes Are Primarily Inflammatory Cells

Granulocytes are the most abundant leukocyte in blood; in healthy people, about 65% of blood leukocytes are granulocytes. However, their actions are more relevant to inflammation than immunity. For example, in tissue damaged or destroyed by blunt trauma, radiation, toxic chemicals, or poor blood supply, leukocytes perform healing tasks not related to immunity.

Granulocytes, especially neutrophils, are phagocytes that ingest and digest foreign material and cell debris as a part of the inflammatory and healing process. However, their large cytoplasmic granules contain a variety of chemicals that are important in killing and digesting foreign substances as part of the immune response.

Lymphocytes Are Immune Cells and Congregate in Blood and Lymphoid Organs

Although some lymphocytes are present in blood, the great majority (95%) are present in lymphoid organs and other tissues, where they defend the body against pathogens and other threatening substances.

Recall from ◀ Chapter 10 that in the fetus, all lymphoid cells originate initially in the spleen and liver; but by birth, production has shifted to the bone marrow. Also recall that in adults, all blood cells, including lymphocytes, are produced in the red bone marrow, most of which is found in the spongy bone of the spine, pelvis, ribs, cranium, and the proximal ends of long bones.

Lymphocytes complete their maturation in various lymphoid organs (Fig. 12.2):

● **T lymphocytes (T cells)** mature in a lymphoid organ in the chest called the *thymus*.

● **B lymphocytes (B cells)** complete their maturation with the red bone marrow.
● **Natural killer (NK) cells** mature in either the bone marrow or in lymphoid organs called *lymph nodes*. Note that NK cells are the only lymphocytes without the word *lymphocyte* in their name.

The immature lymphocytes released from the bone marrow retain the ability to divide, so one immature lymphocyte arriving at the thymus can produce many mature T lymphocytes. Once they mature, lymphocytes (including NK cells) circulate throughout the lymphatic and circulatory systems and congregate in lymphoid organs, such as the lymph nodes. The maturation and roles of the different lymphocyte classes are discussed in greater detail further on.

Case Note

12.2. **Miriam's helper T lymphocyte count was low. Did these cells mature in the bone marrow, the thymus, or both?**

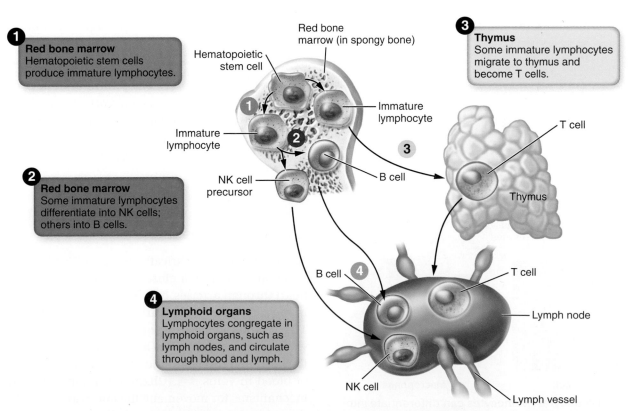

1 **Red bone marrow**
Hematopoietic stem cells produce immature lymphocytes.

Red bone marrow (in spongy bone)

Hematopoietic stem cell

Immature lymphocyte

Immature lymphocyte

2 **Red bone marrow**
Some immature lymphocytes differentiate into NK cells; others into B cells.

NK cell precursor

B cell

3 **Thymus**
Some immature lymphocytes migrate to thymus and become T cells.

T cell

Thymus

4 **Lymphoid organs**
Lymphocytes congregate in lymphoid organs, such as lymph nodes, and circulate through blood and lymph.

B cell

T cell

Lymph node

NK cell

Lymph vessel

Figure 12.2. Lymphocytes. The steps in the synthesis of three types of lymphocytes: B cells, T cells, and natural killer (NK) cells. *Do NK cells differentiate in the thymus or the bone marrow?*

Monocytes Develop into Macrophages and Dendritic Cells

Like all other blood cells, monocytes are produced by the red bone marrow; and like all leukocytes, monocytes can migrate from blood vessels into body tissues. For monocytes, however, it's a one-way trip—they differentiate into specialized cells, *macrophages* or *dendritic cells*, which remain within tissue (Fig. 12.3).

When monocytes differentiate into **macrophages,** they acquire certain new powers. The most significant of these is an enhanced capability for *phagocytosis*—they develop a colossal appetite for ingesting and destroying pathogens and other foreign substances. Macrophages can be found in virtually every body tissue, but they have special names and characteristics in certain tissues. In the brain, they develop into spidery cells called *microglia;* in the liver, they are called *Kupffer cells.* In lymphoid organs, however, they are simply called macrophages.

Dendritic cells constitute a second population of important tissue leukocytes. Like macrophages, they can develop from monocytes. However, unlike macrophages, they can also develop from lymphocyte precursors. Dendritic cells are concentrated in lymphoid organs and in body regions in direct contact with the environment: the epidermis and the lining of the respiratory and gastrointestinal tracts.

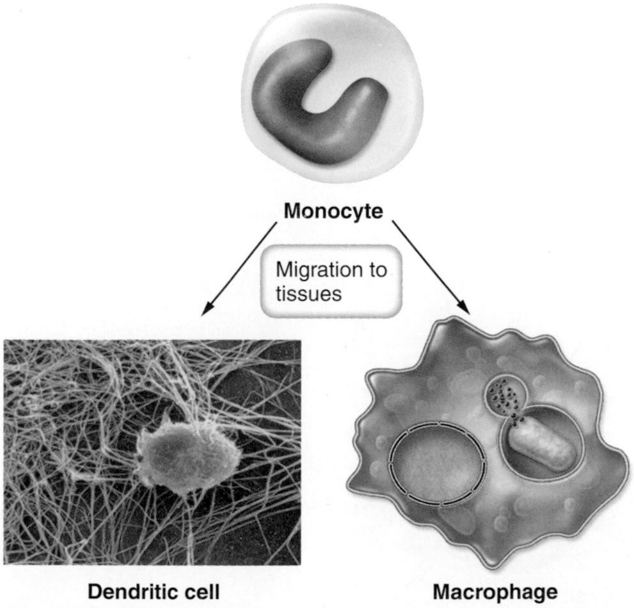

Both dendritic cells and macrophages are involved in *antigen presentation,* a process that helps mobilize the immune system to attack a particular invader (recall from ◀ Chapter 10 that an antigen is any substance capable of inciting an immune reaction). Antigen presentation is discussed in greater detail further on in this chapter.

12.2 Where are most lymphocytes found—in the blood or in the lymphoid organs?

12.3 You sprain your knee playing basketball (the skin is not broken). Does the swelling and pain that follow represent actions of the immune system and/or the inflammatory response?

12.4 True or false: T cells mature in the bone marrow.

12.5 Are NK cells granulocytes, monocytes, or lymphocytes?

12.6 Name two types of cells that can develop from monocytes.

Monocyte

| Migration to tissues |

Dendritic cell   **Macrophage**

Figure 12.3. Monocytes. Monocytes can differentiate into macrophages or dendritic cells. *Which monocyte descendant has large numbers of thin cytoplasmic extensions?*

Vessels and Organs of the Lymphatic System

As mentioned above, the **lymphatic system** consists of two main parts. First is a vast network of *lymphoid vessels,* which run through all tissues. Second are the *lymphoid organs,* which are largely composed of lymphocytes, macrophages, and related cells supported by epithelial or connective tissue.

Lymph Fluid Circulates through Lymphoid Vessels

As illustrated schematically in Figure 12.4A, the cardiovascular system is a closed network: blood leaves the heart through arteries, exchanges materials with tissues in the capillaries, and returns to the heart through veins. The lymphatic system, conversely, is an open or one-way system that does not have a dedicated pump. Fluid flow through the lymphatic system is similar to the flow of blood in veins ◀ (Chapter 11) and relies on three mechanisms for movement: the one-way assist of valves; the massaging action of the skeletal muscle pump; and the sucking action of the respiratory pump.

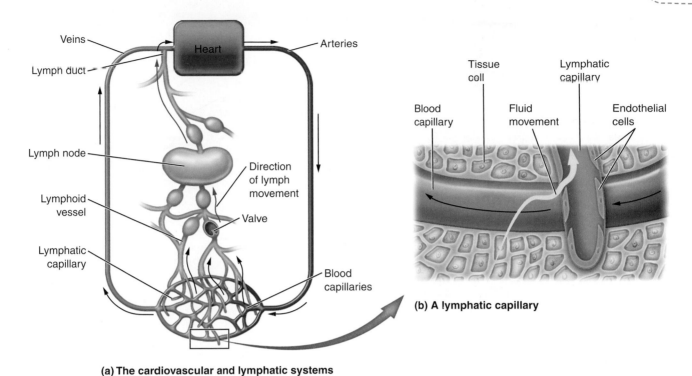

(a) The cardiovascular and lymphatic systems

(b) A lymphatic capillary

Figure 12.4. The lymphatic system. A. The vessels of the lymphatic system pick up excess tissue fluid and return it to the venous circulation. **B.** Tissue fluid enters lymphatic capillaries through small gaps between the cells in their walls. *Which type of lymphoid vessel is physically connected to cardiovascular vessels?*

The lymphatic system begins with the minute **lymphatic capillaries,** which pick up some of the interstitial fluid between cells. Lymphatic capillaries merge to form **lymphoid vessels,** which eventually drain into one of two large **lymph ducts,** which empty into the large veins near the heart. As lymph flows from the periphery to the torso, it percolates through multiple small lymphoid organs called *lymph nodes,* which house many lymphocytes. With this overview in mind, let us examine lymphatic circulation in more detail.

Lymphatic Capillaries Pick Up Tissue Fluid

We've just seen that the lymphatic network begins with lymphatic capillaries, which are closed-end vessels composed of endothelial cells (Fig. 12.4B). Recall from Chapter 10 that some fluid moves out of blood capillaries into tissues as blood passes from the arterial to the venous side. The resulting increase in interstitial fluid pressure creates a pressure gradient, which forces fluid into these lymphatic capillaries through the spaces between their endothelial cells. This fluid, now called *lymph,* moves up the lymphatic capillary and into larger lymphoid vessels.

Newly formed lymph is thus identical to the interstitial fluid, containing water, electrolytes, and a little protein.

However, lymph acquires other constituents on its journey to join blood. For example, as it passes through lymph nodes, lymph acquires leukocytes. Lymph draining the intestines also contains fats and fat-soluble vitamins. Lymph may also contain microbes or migrating (metastasizing) cancer cells.

Lymphoid Vessels Form a Complex Network

Figure 12.5 illustrates the complex web of lymphoid vessels and lymph nodes that permeates every body tissue except bone and nervous tissue. The overall arrangement of the vessels of the lymphatic system is much less orderly than that of the vessels of the venous system, in which venules merge into progressively larger veins. Lymphatic vessels also merge with other, similarly sized lymphatic vessels, but may subsequently branch again into smaller vessels. Eventually, however, all lymph drains into one of two lymphatic ducts.

The smaller of the two main ducts, the **right lymphatic duct,** drains the right arm, right side of the chest, and right side of the head and neck (Fig. 12.5C). The right lymphatic duct empties into the right subclavian vein just distal to the point where it merges with the internal jugular vein to form the right brachiocephalic vein (Fig. 12.5B). The larger of the two, the **thoracic duct,** serves the remainder of the

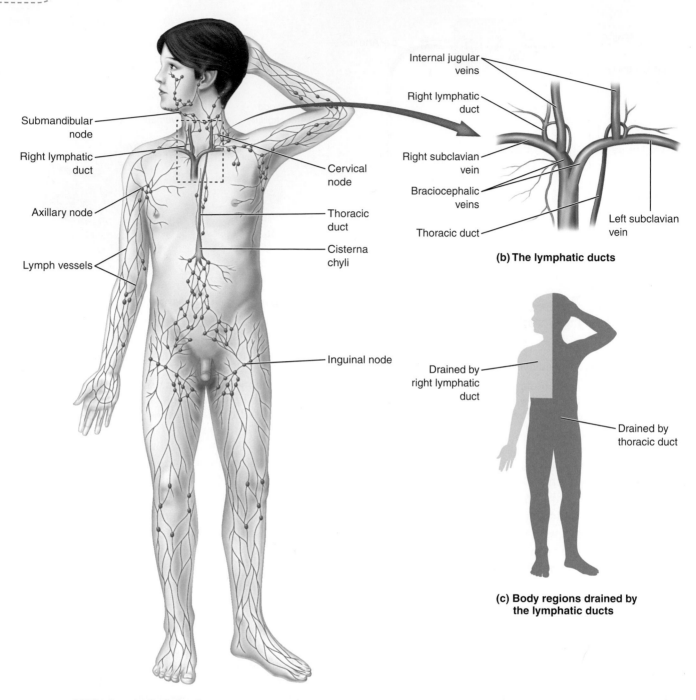

(a) The lymphatic network

(b) The lymphatic ducts

(c) Body regions drained by the lymphatic ducts

Figure 12.5. Lymphoid vessels. A. Lymphoid vessels connect lymph nodes. **B.** The thoracic and right lymphatic ducts drain into the subclavian veins. **C.** Body regions drained by the two lymphatic ducts. *How are lymph nodes in the armpit region named?*

body, including the intestines and other viscera. The thoracic duct forms a large pouch in the lumbar area, the *cisterna chyli,* which is large enough to be visible with the naked eye (Fig. 12.5A). It subsequently ascends the left side of the vertebral column, becoming about 5 mm in diameter as it empties into the left subclavian vein.

> *Remember This!* **The main lymphatic duct is the thoracic duct, which drains lymph fluid from most of the body. It travels up the left side of the vertebral column and empties into the left subclavian vein.**

Lymph Nodes Are Small Lymphoid Organs

About 500 lymph nodes are interspersed in fat and other soft tissue along lymph channels in every part of the body (See Fig. 12.5A). Prominent groups of lymph nodes occur immediately beneath the skin in the groin (inguinal nodes), the armpit (axillary nodes), and along the lateral neck beneath the ear and jaw (cervical and submandibular nodes). Other large clusters are present in the retroperitoneum, the mesenteries of the abdomen, and in the mediastinum.

Lymph Nodes Contain Many Lymphocytes

Each lymph node is an ovoid structure a few millimeters wide at its largest point, which is enclosed in a thin capsule of fibrous tissue (Fig. 12.6). Beneath the capsule is an arc of lymphoid tissue called the *cortex*, in which lymphocytes are arranged into small clusters called lymphoid *follicles* (or nodules). Most follicles contain a *germinal center* where B-lymphocytes multiply and are surrounded by dendritic cells. T lymphocytes and NK cells are concentrated deeper in the cortex. The innermost *medulla* contains parallel cords of a variety of lymphocytes.

Sinuses throughout the lymph node accommodate the flow of lymph. The sinuses contain a network of fine reticular fibers as well as a population of resident macrophages. Some pathogens and other particulate matter become trapped in the reticular fibers, where they become easy prey for phagocytosis by macrophages. A small amount of lymph also filters into the follicles and medullary cords, exposing the pathogens to the deadly attention of lymphocytes. Lymph from incoming (afferent) vessels enters the node through the cortex, filters through the sinuses, and exits from the medulla into an outgoing (efferent) lymphoid vessel.

Lymph Nodes Defend against Pathogens and Cancer Cells

Lymph nodes are the only lymphoid organs in direct contact with lymph. They are, therefore, ideally situated to serve as the "sentry posts" of the lymphatic system, monitoring lymph fluid for pathogens and trapping them for destruction by immune cells. Lymph nodes also trap cancer cells that spread from one tissue to another through lymphatic vessels. Cancer cells temporarily trapped in a lymph node continue to divide and reproduce and may spread from one lymph node to another up the chain of nodes, eventually entering the bloodstream. However, if discovered early, removal or radiation of affected lymph nodes and others nearby may prevent further spread of the cancerous cells into blood. For example, most breast lymphatics drain through a collection of lymph nodes in the axilla. If a breast cancer

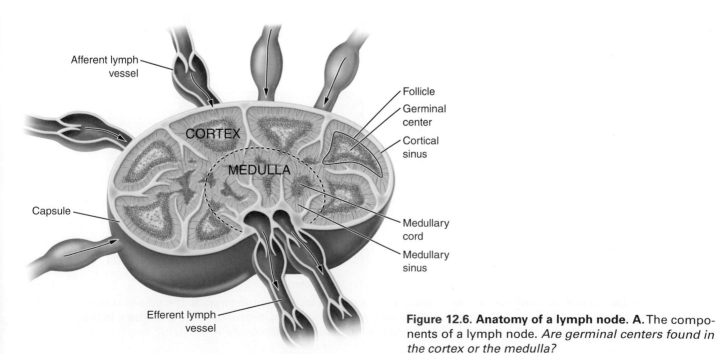

Figure 12.6. Anatomy of a lymph node. A. The components of a lymph node. *Are germinal centers found in the cortex or the medulla?*

spreads, it usually first spreads to axillary lymph nodes, where it may be confined for a while before spreading into the bloodstream and distant organs. In patients with breast cancer, it is common practice to biopsy the lowest axillary lymph node, the sentinel ("lookout") node to check for cancer cells. The outlook for recovery is better for patients with cancer-free lymph nodes. However, some patients survive to live a normal life span after removal of cancerous nodes, which is proof that the nodes have done their job of temporarily halting the spread of the cancer.

As mentioned above, lymph nodes also serve as "training camps" where some NK cells mature. Moreover, they act as lymphocyte breeding grounds, where lymphocyte precursors can congregate and produce new lymphocytes. Lymph nodes accelerate lymphocyte production when they encounter pathogens or antigens. Nodes enlarge as they fill with all of these new cells; they may, therefore, become easily visible or palpable, a condition called *lymphadenopathy*.

Case Note

12.3. Enlarged lymph nodes were detected in Miriam's neck, armpit, and groin. Use the correct anatomical terms to describe these three regions.

The Spleen Serves Immune and Nonimmune Purposes

The **spleen** sits in the upper left abdomen just deep to the posterolateral rib cage and inferior to the diaphragm (Fig. 12.7A). It has a thin, dense, fibrous capsule enclosing two types of tissue: red pulp and white pulp (Fig. 12.7B). The *red pulp* consists of broad venous sinuses filled with slow-moving blood as well as cords of splenic tissue composed of red blood cells, lymphocytes, and many macrophages. The *white pulp* consists of nodules of lymphocytes and macrophages that cuff splenic arteries.

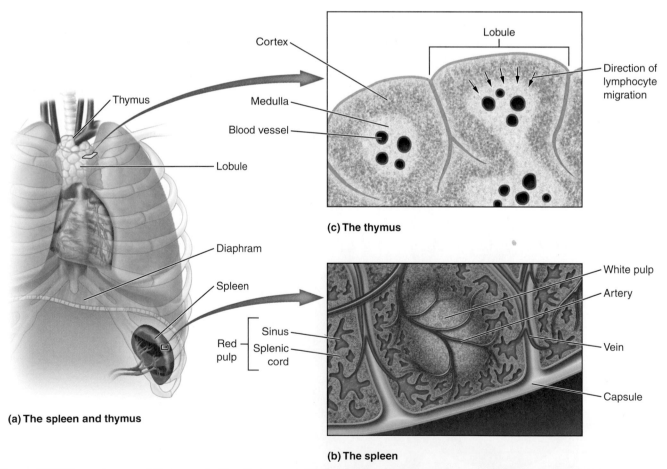

(c) The thymus

(a) The spleen and thymus

(b) The spleen

Figure 12.7. The spleen and thymus. A. The thymus and spleen are located in the torso. **B.** The spleen is composed of red pulp and white pulp. **C.** Lymphocytes make up most of the thymic tissue; they proliferate in the outer cortex and migrate into the medulla. *Are sinuses found in the red pulp or the white pulp?*

The leukocytes of the red and white pulp screen blood in much the same way that lymph node screens lymph; that is, macrophages engulf pathogens and lymphocytes perform their immune duties. Splenic macrophages also perform a housekeeping role, breaking down old red blood cells and capturing their iron for recycling. Recent studies show that splenic macrophages serve as a "standing army" of immune cells capable of flooding the body on demand—when there is infection, trauma, heart attack, or any other problem where many fresh immune cells are needed to combat infection or to enhance the inflammatory response.

The spleen also performs two nonimmune roles relevant to the cardiovascular system: (a) In fetuses or severely anemic adults, it is an important site of red blood cell production, and (b) it stores platelets to be released when needed to replace those used in stopping hemorrhage.

The Thymus Processes T Lymphocytes

Recall that the **thymus** is a lymphoid organ where immature lymphocytes mature into *T lymphocytes* before migrating to other lymphoid organs. It lies in the anterior chest immediately posterior to the center of the sternum (see Fig. 12.7A). It achieves peak size, about 30 g, at puberty and gradually shrinks to near nothing in elderly people; nevertheless, it remains active enough for normal purposes. The thymus is divided into lobules, each containing an outer cortex and an inner medulla (Fig. 12.7C). Lymphocytes proliferate in the cortex and migrate into the medulla as they mature. They leave the thymus via blood vessels in the medulla.

Neither the thymus nor the bone marrow is directly linked to the lymphatic network, and they do not participate *directly* in body defense. Instead, they are training centers from which newly matured "fighting cells" are formed before being released to migrate to other lymphoid organs.

Case Note

12.4. **Untreated AIDS patients like Miriam often have a significant reduction in white pulp. Is white pulp found in the thymus or the spleen?**

Mucosa-Associated Lymphoid Tissue (MALT) Monitors Pathogen Entry Points

Just as sentry posts at the castle gates stand on the lookout for invaders, collections of lymphoid tissue are found at the sites where pathogens frequently enter the body—the mouth, nose, and gastrointestinal tract (Fig. 12.8). These nodules of lymphoid tissue are described as **mucosa-associated lymphoid tissue (MALT)** because they are associated with the mucosal layer of the respiratory and gastrointestinal systems.

The **tonsils** are simple but rather large masses of lymphoid tissue that form a ring around the opening of the throat at the back of the mouth. Largest and easiest to see are the *palatine tonsils*, commonly just called the "tonsils," which sit on either side of the posterior end of the oral cavity. Out of sight, high in the posterior throat, is the *pharyngeal tonsil* (called the *adenoid* when enlarged). Several other small tonsils complete the ring. Nodules of MALT are also abundant in the small intestine, where they are called *Peyer's patches,* and in the *appendix,* a small outcropping of the large intestine.

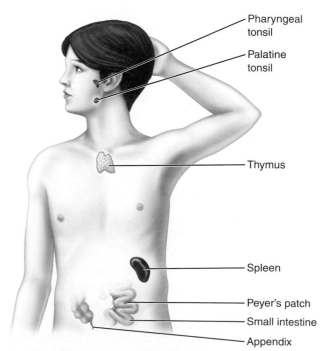

Figure 12.8. Mucosa-associated lymphoid tissue. Mucosa-associated lymphoid tissue guards common pathogen entry points. *Name the lymphoid tissue clusters found in the small intestine.*

12.7 Which component is not commonly found in lymph? Erythrocytes, leukocytes, or electrolytes?

12.8 What is the name of the enlarged portion of the thoracic duct?

12.9 Which lymphatic duct drains the right arm?

12.10 Which lymphatic duct drains the right leg?

12.11 Which organ shrinks after puberty? The spleen or the thymus?

12.12 Which lymphoid organ(s) contain both a cortex and a medulla? Lymph node, spleen, and/or thymus?

12.13 Name the lymphoid organ involved in T-lymphocyte maturation.

12.14 What is MALT?

Overview of the Immune Response

The term *immune* indicates exempt status; that is, to be "free of" or "protected against." In modern physiology and medicine, **immunity** refers to the actions of leukocytes that are able to distinguish between self and nonself substances and neutralize those that are nonself. This definition raises a crucial question: How do immune cells distinguish self from nonself?

Self and Nonself Are Defined by Proteins

The molecular definition of self lies in DNA. With the exception of identical siblings, every person's DNA is unique. And, because DNA codes for proteins, each person is composed of a unique set of proteins that define self and are unique to that *one* individual—they are nonself (alien) to *every other person*. This means that tissue transplanted from another person is nonself. And because *abnormal* DNA is the cause of all tumors, benign or malignant, every tumor contains alien proteins and can be recognized as nonself. Moreover, the proteins of other organisms—from bacteria to baboons—are even more distinctively nonself to humans. What about proteins we breathe in, such as those in dust mites or mold

spores? You guessed it—these are also nonself. Proteins we consume as food are also nonself, but the process of digestion breaks them down into their component amino acids, which are not recognized as nonself.

The first job of the lymphatic and immune systems is to distinguish between self and nonself proteins. When nonself proteins are recognized, these systems mobilize to neutralize or destroy the organisms or tissues that contain them. For example, they battle the nonself proteins in transplanted hearts, livers, and other transplants, which makes it critical to find donors whose proteins are as similar as possible to the recipient's.

The ability of the lymphatic and immune systems to recognize nonself protein rests on the ability of proteins to act as *antigens*. An **antigen** is a molecule that is capable of stimulating a defensive response from the immune system. Virtually any *large* organic molecule can stimulate an immune reaction, including glycoproteins, lipoproteins, and glycolipids. Glycoproteins, for example, are the antigens responsible for the distinctive characteristics of the ABO red cell antigen system discussed in ◀ Chapter 10. In our further discussions of immunity, we will refer to proteins with the understanding that what we say also refers to some other large organic molecules.

The term *antigen* derives from "*anti*body-*gen*erating," a reference to an antigen's ability to stimulate the immune system to produce **antibodies,** specialized proteins that help immune cells destroy antigens. Antibodies are discussed in more detail further on.

> ***Remember This!*** Nonself antigens come from two sources: from "outside," such as pathogens or transplants, and from "inside," such as the antigens of cancer cells.

Pathogens Contain Non-Self Antigens

Any organism is capable of causing injury and disease if it penetrates skin or mucous membranes and gains access to extracellular fluid or the interior of cells. However, only a small fraction of organisms are considered to be primarily pathogenic, and even among these, only a few varieties cause most human disease. Indeed, we live in a sea of potentially pathogenic organisms, most of which do us no harm most of the time. For example, brushing swollen gums releases a shower of oral bacteria into the bloodstream, and in the vast majority of instances nothing ill comes of it. On the other hand, the human immunodeficiency virus (HIV), the cause of AIDS, is potentially fatal every time it invades someone's body.

Only a few pathogens (mainly intestinal parasites) are large enough to be seen by the naked eye. Most are microscopic; indeed, some cannot be seen even with the most powerful light microscope because they are smaller than the shortest visible wavelengths of light. Viruses are an example. The development of the electron microscope in the 1930s enabled scientists to view many pathogens for the first time. Keep in mind, though, that no matter how small they are, *all pathogens stimulate an immune response because they contain nonself antigens.*

Pathogens fall into two broad classes: viruses, and nonviral pathogens—bacteria, fungi, amebae, worms, and other organisms. Viruses have been called "organisms at the edge of life" because are distinctive from other life forms in two ways:

- Outside of a host cell, they cannot reproduce: they cannot metabolize nutrients, replicate, or otherwise participate in the processes of life. They require the host cell to do everything for them, and outside of cells they are in a state of suspended animation.
- They have no cell membrane or organelles. Instead, they have an outer protein coat that surrounds an intricate arrangement of nucleic acids (DNA or RNA).

Because, like bacteria and other pathogens, viruses have DNA or RNA, they are able to replicate. However, unlike bacteria, viruses invade cells and hijack the cell's metabolic processes when they multiply, diverting the cell's efforts to their own selfish ends.

On the other hand, *nonviral pathogens* are organized as cells. They can live independently outside of other cells—on water fountains, skin, toilets, and in lakes and streams. They have a cell membrane and cytoplasm, and they have DNA and RNA, although some do not have a nucleus. The characteristics of some common pathogens are described in Table 12.1.

Viruses and nonviral pathogens injure tissues differently. Viruses invade cells and insert their genetic material into the cell nucleus, using it to make new copies of themselves. Eventually the cell dies and releases its viruses to infect other cells. Some viruses are more subtle: they stimulate the infected cell to proliferate rapidly. The result is the development of a tumor.

Nonviral pathogens do their damage in a variety of ways:

- Some secrete toxins that undermine vital cell processes nearby or invade the bloodstream to do damage at a distance.
- Some deplete nutrients needed by the cell or the host. For example, some intestinal parasites have such an appetite that the person infected can die of starvation.
- Upon their death, some pathogens release breakdown products that are more harmful than the living organisms.
- Finally, some incite a reaction by the immune system—inflammation—that can be as harmful as the pathogen itself.

Each pathogen class requires a different treatment. An **antibiotic** is a small molecule capable of killing or stopping the reproduction of *bacteria.* Antibiotics usually interfere with bacterial metabolism; for example, by interrupting the synthesis or activity of critical proteins. Antibiotics do *not* affect nonbacterial pathogens such as viruses. However, researchers have developed a few antiviral compounds capable of slowing viral replication by blocking the viruses' entry into host cells and otherwise shortening viral illnesses. For example, some can lessen the severity and shorten the length of an episode of influenza. Fungi, amebae, and worms require treatment by other specialized types of drugs.

Case Notes

12.5. Miriam is given antibiotics at her first appointment. Will these drugs attack her HIV? Why or why not?

12.6. Upon autopsy, microorganisms were found in Miriam's lungs and were deemed responsible for her pneumonia. The pathologist saw minute single-celled microorganisms with a cell wall and nucleus. Could this organism have been a virus?

MHC Glycoproteins Display Other Antigens to the Immune System

Among our most important self antigens are specific glycoproteins known as the **major histocompatibility complex (MHC).** These glycoproteins are found on the cell membranes of every *nucleated* cell. *Histo-* indicates body tissues, so the name *major histocompatibility complex* reflects the importance of these glycoproteins in determining compatible tissue matches among patients being considered for organ transplant. This importance derives from two factors:

- MHC antigens themselves are self-specific like all antigens. That is, except in of identical siblings, they are alien to every other person and can incite an

Table 12.1   Common Pathogens

Pathogen	Structural Characteristics	Functional Characteristics	Treatment	Common Diseases
Viruses	DNA/RNA and protein coat	Cannot reproduce outside of cells	Antivirals, which slow viral replication	Influenza, the common cold, measles, HIV/AIDS
Bacteria	Microscopic cell without nucleus	Common on keyboards, water fountains, toilets . . .	Antibiotics, which slow bacterial reproduction	Strep throat, some sinus and lung infections, some food poisoning
Fungi	Microscopic, unicellular (yeasts) or multicellular (molds)	Usually infect body surfaces and openings	Antifungals, which destroy the cell walls	Athlete's foot, yeast infections
Protozoa	Microscopic, unicellular	Common in water supplies of developing countries	Antiprotozoan drugs, which interfere with protozoan metabolism	Malaria, sleeping sickness
Worms	Multicellular	Prefer to live within body spaces and cells	Antihelminthics, which interfere with the worm's metabolism	Roundworms, tapeworms (helminths)

immune attack if tissue containing them is transplanted into another person. Incompatible MHC antigens are responsible for most tissue and organ rejections after transplant.

- It is the task of MHC antigens to bind *other* self and nonself antigens and display them on the cell surface somewhat like a billboard for the immune system to read and act upon.

Two Classes of MHC Glycoproteins Display Antigens

There are two types of MHC glycoproteins—class I and class II. Class I MHC glycoproteins (MHC I) are found in the membrane of *every* body cell except red blood cells (which do not have nuclei). Class II MHC glycoproteins (MHC II) are found only on certain *immune* cells that are capable of phagocytosis—for example, dendritic cells.

We have said that MHC glycoproteins can act like billboards, exhibiting antigens to the immune system. MHC I glycoproteins display antigens synthesized inside the cell—that is, the displayed antigens are obtained from the cell's cytosol. In healthy cells MHC I glycoproteins display normal self-antigens, sending the message: "This cell and all other cells like me are self; leave us alone" (Fig. 12.9A). In contrast, recall that cancerous or virus-infected cells synthesize *nonself* proteins. In these unhealthy cells, MHC I glycoproteins display these non-self (but internally produced) antigens. This sends a different message: "Alien antigen present; kill me" (Fig. 12.9B).

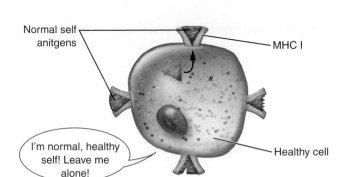

Normal self anitgens

MHC I

I'm normal, healthy self! Leave me alone!

Healthy cell

(a) Healthy cells

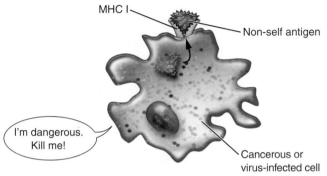

MHC I

Non-self antigen

I'm dangerous. Kill me!

Cancerous or virus-infected cell

(b) Infected or cancerous cells

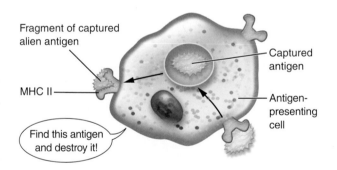

Fragment of captured alien antigen

Captured antigen

MHC II

Antigen-presenting cell

Find this antigen and destroy it!

(c) Antigen-presenting immune cells

Figure 12.9. MHC glycoproteins. MHC glycoproteins display antigens on the cell surface that convey messages. **A.** Healthy cells produce MHC I glycoproteins bound to normal cell proteins. **B.** Cancerous or virus-infected cells express MHC I glycoproteins bound to abnormal cell proteins. **C.** Some immune cells express MHC II glycoproteins linked to proteins from invading pathogens. *Which type of antigen is linked to MHC II glycoproteins—one produced by one's own cell or one produced by a foreign cell?*

The number of MHC I glycoproteins on a cell membrane is a message in itself. MHC I glycoproteins are abundant in the membranes of healthy cells but relatively lacking in the membranes of cancerous or virus-infected cells. Cells lacking a normal amount of MHC I glycoprotein are subject to attack by certain immune cells, yet to be discussed.

The process works in much the same way for MCH II glycoproteins on the surface of immune system cells, but the source of the antigen is different. MHC II glycoproteins do not display internally produced self-antigens. Nor is the immune cell invaded by alien antigen. Rather, the immune cell captures the antigen, from, say, an invading virus or bacterium. The immune cell then internalizes it into a vacuole and displays a fragment of the antigen linked to its MHC II glycoprotein (Fig. 12.9C). Cells that perform this function are described as *antigen-presenting cells*. This display serves a function similar to that of a picture of the criminal on a "Wanted" poster, stimulating other immune cells to track down and destroy the foreign cells that produced the antigen.

> **Remember This!** **MHC I glycoproteins display internally produced antigens; MHC II glycoproteins display antigens from external sources.**

MHC Glycoproteins Define Appropriate Transplant Recipients

Because they play such an important role in presenting *other* antigens to the immune system, it is easy to forget that MHC glycoproteins *are antigens themselves* to the immune systems of other people. As such, they can be powerful adversaries during organ or tissue transplantation.

Successful transplantation usually requires that donor and recipient be as much alike antigenically as possible in order to avoid immune rejection of the donated organ. An exception is corneal transplantation. Because the cornea has no blood supply (◀ Chapter 9), it is not exposed to the immune system and incites no immune reaction when it is transplanted.

As we noted at the beginning of this chapter, identical siblings, who share identical self-antigens, are the closest possible matches for each other. For the rest of us, although careful matching of MHC antigens can limit the severity of an immune reaction, every transplant causes some degree of immune rejection. The closest donor matches of MHC antigens lead to the fewest rejections and are most likely to be found in close relatives of the recipient, including siblings, parents, and children.

Evidence of rejection depends on the organ involved and varies from mild to severe. Mild rejection may be detected only by laboratory tests, but severe rejection can occur quickly and may lead to organ failure. For example, mild rejection of a transplanted kidney may appear only as abnormal urine tests, but severe rejection is manifest by kidney failure.

Drug therapy to suppress the recipient's immune system can limit the severity of the rejection reaction by inhibiting the production and action of leukocytes. But as you might suspect, immunosuppressive therapy has its problems: recipients may become more susceptible to infections and the development of tumors.

Immune Reactions Are Innate or Adaptive

Nonself antigens, either pathogens from the external environment that breach skin or mucosal barriers or tumor cells arising from within, encounter the immune system, which responds in four ways:

- It detects the nonself antigen.
- It communicates the discovery to other cells by chemical signals (*cytokines*).
- It stimulates other cells to proliferate and assist in defense.
- It destroys or neutralizes the invader or cancer cell.

These four steps of the immune response involve the two overlapping types of immunity (Fig. 12.10):

- **Innate immunity** (Latin *innatus* = "inborn"), which is present from birth and requires no programming. The reaction begins within minutes and draws an inflammatory response by recruiting macrophages, neutrophils, and NK cells to attack the invader. Innate immunity can be likened to an unaimed, quick shotgun blast of pellets not targeted at a *particular* pathogen—*any* pathogen will elicit a response and suffice as a target. Phagocytosis by neutrophils is an example.

- **Adaptive immunity** (*acquired* or *programmed* immunity), which is a *specific* immune response aimed at a *particular* nonself antigen. It can be likened to the carefully aimed rifle shot. Adaptive immunity must learn what to do; it is educated by an initial encounter with its particular target and goes after that target only. Just as any educational activity requires time, adaptive immunity is slower than innate immunity in its response upon encountering the threat for the *first* time. Leukocytes must first interact with the invader, take measure of its unique characteristics, and then program themselves to generate a specific response.

At one time the medical definition of the term *immunity* was confined to adaptive immunity. The modern view, however, is that immunity encompasses all the body's cellular and chemical activities that fight nonself invaders, both innate and adaptive. In the next two sections, we'll take a closer look at these two types of immunity.

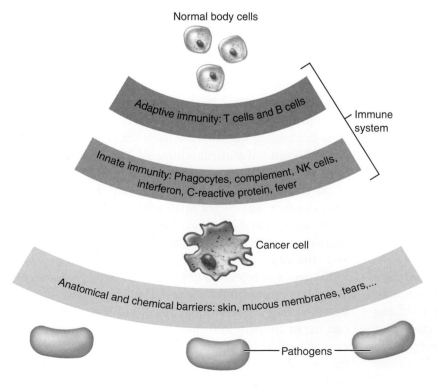

Normal body cells

Adaptive immunity: T cells and B cells

Immune system

Innate immunity: Phagocytes, complement, NK cells, interferon, C-reactive protein, fever

Cancer cell

Anatomical and chemical barriers: skin, mucous membranes, tears,...

Pathogens

Figure 12.10. Lines of defense. Physical barriers are the first line of defense against pathogens; the innate and adaptive arms of the immune system are the second and third lines. *Are T and B lymphocytes part of adaptive immunity or innate immunity?*

Case Note

12.7. When the fungus first invaded Miriam's lungs, which aspect of the immune system was the first to react: adaptive or innate?

12.15 Why do tumors contain nonself antigens?

12.16 Of the following—bacteria, viruses, fungi, worms—which are cellular organisms?

12.17 What is the principal function of MHC glycoproteins?

12.18 Which type of MHC glycoprotein displays antigens from invading pathogens?

12.19 Which type of MHC glycoprotein is expressed by all nucleated cells?

12.20 Why would transplantation of a kidney from a mother to her child elicit an immune response?

Innate Immunity

Innate immunity requires no "education" or "programming" in response to specific threats and is thus more rapid than adaptive immunity. When skin and mucosal barriers are breached by an invading pathogen, the cellular and molecular activities of the innate immune system try to hold off the invaders until the adaptive immune system can be programmed to respond. The innate immune response begins within minutes and prompts the broad range of actions, discussed here.

Inflammation Serves a Role in Innate Immunity

Recall from ⬅ Chapter 10 and our discussion earlier in this chapter that inflammation is "the body's collective cellular and vascular response to injury." Much of the inflammatory response has to do with cleaning up damage and fostering the healing process that follows. However, inflammation becomes an *immune* activity when it attacks and neutralizes or destroys alien antigens.

The inflammatory process relies on the actions of blood and tissue leukocytes and involves the following elements:

- Increased blood flow, which brings fresh immune cells (lymphocytes and other white blood cells), nutrients, oxygen, and other essentials to the injured region
- Leaky blood vessels, which permit blood phagocytes and plasma proteins to enter injured tissues
- Chemical signals released by the injury or the pathogen, which draw leukocytes to the injured site or pathogen

Leukocytes Play a Role in Innate Immunity

Some leukocytes have an inborn, innate ability to recognize nonself—they do not have to be "taught" what is nonself by meeting it; they know in advance. They were born and raised, so to speak, among a tribe of self cells and operate on this principle: "If you're not one of us, you're going to be dead."

Neutrophils and macrophages are phagocytic cells that have this capacity. Recall that some monocytes enter specific tissues and differentiate into macrophages without exposure to injury or infection. For example, liver macrophages (Kupffer cells) break down old blood cells, and lung macrophages (*dust cells*) destroy inhaled pathogens and particles. These resident macrophage populations can live a long time—14 months for a Kupffer cell—and may never encounter a pathogen.

Phagocytic cells work by wrapping themselves around their target—particulate matter, pathogen, or cell debris—and internalizing it into a vacuole (see Fig. 3.26). A lysosome merges with the vacuole and digests its contents. Later, the digested remains are expelled from the cell.

Another group of leukocytes important in innate immunity are natural killer (NK) cells, one of the three types of lymphocytes introduced earlier in this chapter. Like all lymphocytes, NK cells patrol blood and lymph, congregate in lymphoid organs, and make periodic forays into nonlymphoid tissues. Only about 5% of all lymphocytes are NK cells, but what they lack in numbers they make up for in speed and aggression. NK cells have an inborn capacity to attack and kill virus-infected cells and cancer cells. Unlike other lymphocytes, they do not require time to adapt to a specific enemy. Instead, their response is immediate because they easily recognize the nonself character of abnormal cells. Although the mechanism behind this recognition ability is not completely understood, the quantity of MHC I glycoproteins in the cell membrane appears to play a key role. MHC I glycoproteins inhibit NK-cell activity; therefore, normal cells are protected from NK-cell attack. However, virus-infected and cancer cells have abnormally low numbers of MHC I glycoproteins in their membranes; therefore, they are not protected from NK-cell attack. Chemical secretions from other immune cells and aspects of acquired immunity (such as antibodies) are also important regulators of NK-cell activity.

NK cells, like other lymphocytes, cannot perform phagocytosis. They kill by directly contacting the target cell and inducing it to undergo cell death.

Chemical Signals Are Important Regulators of Innate Immunity

Cytokines are small proteins released by immune cells (usually) that act as messenger molecules to regulate immune function. The cytokine family is a large and unruly one—each cytokine has multiple roles, including nonimmune ones, and multiple cytokines regulate the same immune function. Cytokines mediate their effect by binding to receptors on target cells. Cytokines frequently (but not always) act as *paracrine factors*—they travel only a short distance from the secreting cell to act on a neighboring cell. This characteristic is important for keeping some immune reactions, such as inflammation, confined to a small area.

The actions of cytokines in *innate* immunity are to:

● Stimulate the maturation and activity of macrophages, neutrophils, and NK cells
● Induce vasodilation
● Attract other leukocytes (chemotaxis)
● Stimulate the production of noncytokine inflammatory molecules, such as C-reactive protein (discussed below)
● Induce fever

As discussed below, cytokines also play important roles in the specific immune reactions of acquired immunity.

Fever is abnormally high body temperature, which is induced by specific cytokines called *pyrogens* that reset the brain's hypothalamic thermostat. Any type of tissue injury, not just infection, can cause fever—a burn or a heart attack, for example. By accelerating cellular reactions, fever enhances the positive effects of inflammation, speeds repair, and inhibits the growth of some pathogens.

Of the many types of cytokines, some of the most important are the following (Table 12.2):

● **Interleukins (ILs)** are so-named because of their signaling activity between (*inter-*) leukocytes. The interleukins are numbered (e.g., IL-1) based on the order of their discovery.
● **Interferons** are named for their ability to interfere with the spread of viruses to neighboring cells. Some interferons have broader roles, generally stimulating immune activity.
● **Tumor necrosis factors** are another group of cytokines with wide-ranging actions.

Other, inflammatory chemicals induce similar effects (see Table 12.2):

● The **kinins** are a group of inflammatory blood proteins activated by the same pathways that initiate clotting. Kinins such as *bradykinin* act locally at the site of injury or infection to increase vascular permeability. Kinins also activate nociceptors (pain receptors).
● *Prostaglandins* and *histamine* increase local blood flow, attract neutrophils, and activate nociceptors.

Case Note

12.8. What is the mechanism by which Miriam developed fever with her infection?

Complement and Other Chemical Effectors Attack Pathogens

Some of the chemicals involved in the inflammatory process have direct antimicrobial properties. For instance:

● **Lysozymes** are enzymes contained in the granules of granulocytes. They digest bacterial cell walls.
● **C-reactive protein (CRP)** is synthesized by the liver in response to cytokines released by macrophages in injured tissue (see the nearby Basic Form, Basic Function box titled *C-Reactive Protein: A New Role for an Old Test*). CRP attaches to pathogens and marks them for phagocytosis by neutrophils and macrophages. This ability, called **opsonization,** is common to several immune effector chemicals.

Table 12.2   The Chemicals of Innate Immunity*

Chemical	Description	Producing Cells	Action
Cytokines			
Interleukins	Signaling proteins	Usually leukocytes	Different interleukins activate different aspects of inflammation
Interferons	Glycoproteins	Released by virus-infected cells or leukocytes	Block viruses from infecting other cells; activate natural killer cells
Tumor necrosis factors	Glycoproteins	Wide range of leukocytes, infected or damaged cells	Damage vessels supplying tumors; stimulate death of infected cells; promote inflammation and phagocytosis
Noncytokines			
Kinins	Proteins	Synthesized in liver; found in blood in their inactive forms	Stimulate inflammation, pain
Lysozymes	Enzymes	Produced within granulocyte granules	Digest internalized pathogens
C-reactive protein	Protein	Liver	Marks pathogens and injured cells for phagocytosis; activates complement
Complement	Set of 20 blood proteins	Synthesized in liver; found in blood in their inactive forms	Enhances inflammation and phagocytosis, lyses bacteria

*Note that many of these chemicals are also active in acquired immunity.

Any molecule that acts as a binding enhancer for the process of phagocytosis is an **opsonin.** Opsonization comes up again in our discussion of the adaptive immune system. CRP is also one of multiple activators of the *complement system.*

The **complement system** (often referred to simply as **complement**) is yet another aspect of innate immunity. It consists of about 20 small proteins, most produced by the liver, that normally circulate in their inactive form in the bloodstream. When stimulated by the presence of a pathogen, they interact in a cascade that produces many active complement complexes. Complement does the following:

● It magnifies the inflammatory response by attracting immune cells, stimulating the proliferation and activity of immune cells and increasing vascular permeability.

● It forms the *membrane attack complex,* a multiprotein apparatus that drills a hole in the cell membrane of the offending organism or cell. Solutes and fluids pass through the hole into the cell, resulting in lysis (rupture, disintegration) and death of the pathogen or cell (Fig. 12.11A).

● It acts as an *opsonin* by attaching to the cell membrane of the offending cell or pathogen, marking it for destruction by macrophages and other phagocytes (Fig. 12.11B). As we see further on, antibodies also act as opsonins.

As mentioned previously, the chemical signal CRP is released during inflammation and can activate complement. The complement system can also be activated directly by nonself cells. Antibodies (which are key players in adaptive immunity) can also activate complement. Thus, complement is involved in both innate and adaptive immunity.

BASIC FORM, BASIC FUNCTION

C-Reactive Protein: A New Role for an Old Test

C-reactive protein (CRP) plays an important role in inflammation and immunity, particularly in the activation of complement. It was first discovered in the 1930s as a mysterious blood "factor" in patients suffering from different illnesses, including cancer and inflammatory diseases. Further study revealed that the liver made it in response to inflammation, and over the years it played a somewhat minor role in medicine as a marker of inflammation. For example, a painful knee can be due to inflammation (arthritis), in which case blood CRP would be abnormally high.

Nowadays, however, CRP is becoming familiar to many health-conscious people because it serves as a clear marker for atherosclerosis, which begins as an inflammatory reaction in blood vessels (see ◀ Chapter 11). Atherosclerosis is the underlying cause of most heart attacks and strokes. Public attention to atherosclerosis has focused mainly on high blood cholesterol; however, a about half of people having heart attacks have normal blood cholesterol, a fact that makes CRP especially useful as an additional marker for atherosclerosis risk. CRP is produced before this

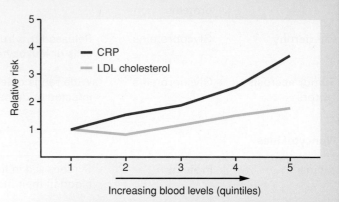

C-reactive protein.

inflammation provokes any symptoms, so elevated CRP levels can warn of heart disease in time to initiate treatment and avoid a fatal heart attack. However, since CRP is produced in response to all types of inflammation, other known causes of inflammation (such as arthritis) must be ruled out before attributing increased CRP to atherosclerosis.

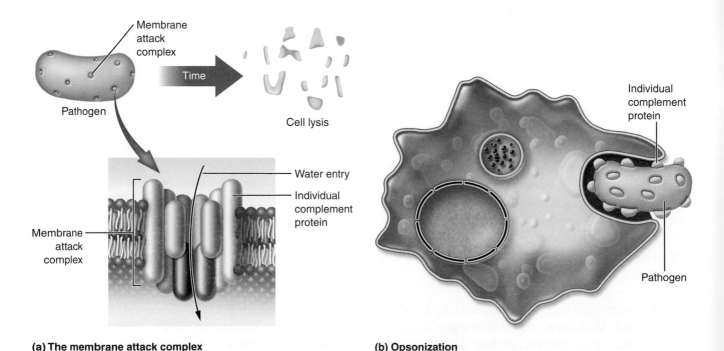

(a) The membrane attack complex

(b) Opsonization

Figure 12.11. Complement. A. Numerous complement proteins interact to form a membrane attack complex. **B.** Complement proteins coat pathogens to render them easy prey for macrophages. *How does the membrane attack complex kill the pathogen?*

12.21 Why does innate immunity respond more quickly than adaptive immunity?

12.22 Name three types of cells important in innate immunity.

12.23 Which chemical would not be found in blood— CRP, complement, or lysozyme?

12.24 Which chemical—complement or interferon— can break open the cell walls of some bacteria, causing their destruction?

12.25 Which types of immunity—innate, adaptive, or both—would respond if you developed an infection after puncturing your foot on a rusty nail?

Adaptive Immunity

Up to this point we have been discussing innate immunity: the nonspecific, inherent cellular defense mechanisms that attack pathogens and other nonself substances without previous training. Now we take up the topic of **adaptive immunity,** the body's targeted, programmed response to the presence of nonself substances. The cells active in adaptive immunity are B and T lymphocytes as well as antigen-presenting cells, such as macrophages and dendritic cells.

Each Lymphocyte Develops to Recognize a Single Antigen

During its "basic training" in the bone marrow or thymus, each B and T lymphocyte "learns" to respond to a particular antigen and no other. Each lymphocyte inserts into its outer membrane specialized receptors—*B-cell antigen receptors (BCRs)* or *T-cell antigen receptors (TCRs)*—capable of binding its specific antigen. Within our population of B and T lymphocytes, there is, for example, a lymphocyte with antigen receptors capable of recognizing only a certain antigen from a certain species of streptococcal bacterium, another lymphocyte with antigen receptors that can recognize only particular antigen from a particular influenza virus, and so on. Incidentally, we all have lymphocytes with receptors for these particular pathogen antigens—whether or not we have ever encountered these pathogens before. That's because our genes direct the production of over a billion different antigen receptors, which can respond to virtually *any antigen we will ever encounter*.

Another critical task of a lymphocyte's "basic training" is to learn to recognize *self* antigens and to avoid attacking them. If a lymphocyte does not master this skill of **self tolerance,** it will be destroyed. For instance, in its passage through the thymus, any T cell that cannot recognize self or that reacts strongly with self antigens is destroyed by apoptosis. As a result, the only T lymphocytes that remain alive are those that know what self looks like and leave it alone. Loss of self tolerance by one means or another is the cause of **autoimmune disease** (Greek *auto* = "self"), which is discussed in more detail below.

Antigens Activate Lymphocytes

Even after completing their "basic training," lymphocytes are not fully mature; that is, they are not yet capable of engaging a pathogen or other enemy. To reach their full powers, they must encounter their designated antigen in the extracellular space and bind it to their antigen receptor—an interaction called an *antigen challenge*. Upon challenge, two supremely important events occur: *activation* and *clonal expansion*.

Activation is the stimulation of a lymphocyte by an antigen. It is an enabling action without which the lymphocyte remains "asleep," incapable of immune activity. During activation, the lymphocyte binds for the first time to its target antigen—an action equivalent to waking a sleeping beast. Upon binding, the lymphocyte quickly multiplies to produce a collection of identical descendants of the original lymphocyte. This action, called **clonal expansion,** produces an army of identical lymphocytes to fight the invader. Figure 12.12 (step 2) shows an example of B lymphocyte activation and clonal expansion. B lymphocytes can be activated directly by raw antigen—that is, by antigen that is still part of the original pathogen. In contrast, T lymphocytes cannot react directly with raw antigen: the antigen must first be prepared and then presented to the T cell by other cells in the immune system. Antigen preparation and presentation are discussed in greater detail later in this chapter.

B and T Lymphocytes Provide Two Types of Adaptive Immunity

B and T cells provide alternative but complementary ways for adaptive immune defense: *antibody-mediated* and *cell-mediated immunity*.

In **antibody-mediated immunity,** B lymphocytes synthesize and secrete antibodies. As we saw in Chapter 10,

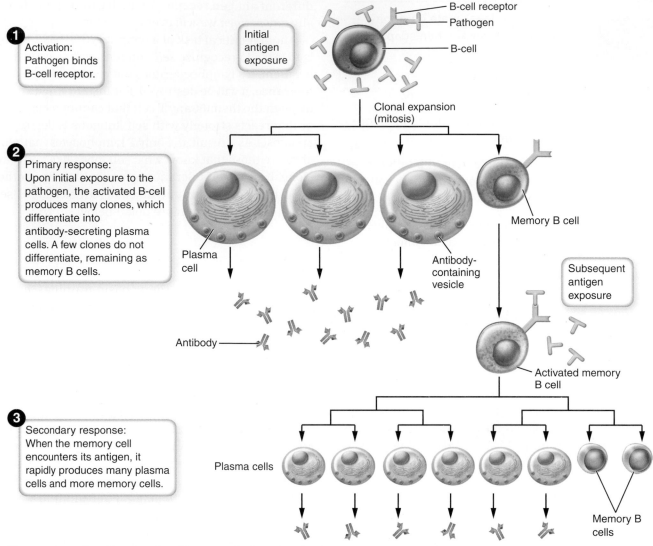

Figure 12.12. B-cell activation. The steps in B-cell activation and action. *What is the term that describes the rapid proliferation of identical lymphocytes?*

antibodies are highly specific blood proteins that bind to an invader, preparing it for destruction by other cells of the immune system.

In **cell-mediated immunity,** a certain type of T lymphocyte *directly* attacks and kills the invader. It is assisted in the task by the secretions and actions of other T cells, but antibodies are not involved.

We are now ready to explore each of these two types of adaptive immunity.

Case Note

12.9. **An HIV infection can be diagnosed by the presence of particular antibodies in blood. Are antibodies produced by humoral or cell-mediated immunity?**

Antibody-Mediated Immunity Is a Function of B Cells

Recall that B cells are activated when they encounter a specific antigen (Fig. 12.12, step 1). The activated B lymphocyte undergoes multiple rounds of cell division to produce many *clones* (identical cells), all of which have the same capacity to identify, bind to, and react with the particular antigen that started the chain reaction (step 2). As discussed further on, some of these B cells can activate other arms of the immune system by acting as antigen-presenting cells. However, most of these B lymphocytes evolve into **plasma cells,** which are B cells in the process of actively secreting antibodies directed against the antigen that originated the reaction. Every one of the secreted antibodies from a particular clone of B cells has precisely the same antigen-binding properties as the receptor molecules on the surface of the parent B lymphocyte.

A few members of the new army of cloned B cells do not evolve into plasma cells. Instead, they become very long lived **memory B cells,** which lurk throughout the lymphatic system. They proliferate rapidly upon another encounter with the same antigen.

The **primary immune response**—the *initial* B-cell activation, clonal expansion, and binding of antigen—takes about a week because the immune system has not previously encountered the particular antigen. In contrast, the response on *subsequent* exposure (step 3), called the **secondary immune response,** is much quicker and more intense because the memory cells are already "primed" to initiate the response. Consider what happens when a child first catches a cold from a particular type of virus. It takes about a week for the symptoms to disappear because the primary immune response takes that long to generate enough antibodies to eliminate the virus. However, if the child is exposed to the same cold virus again a month, a year, or 5 years later, the child will not develop a cold. That is, no fever or other symptoms will occur because the secondary immune response eliminates the virus before it can have any effect.

Antibodies Are Large Proteins with a Complex Structure

The backbone of each antibody is composed of two long molecules, called the *heavy (H) chains,* which join to form a Y-shaped molecule. The stem of the Y is composed of the side-by-side portion of the molecules. Above the stem, the heavy chains bend away from one another to form the arms of the Y. (Fig. 12.13). Two shorter molecules, the *light (L) chains,* are attached alongside the bent ends.

There are several classes of antibodies (discussed below). The stem and a small portion of the proximal branch of each arm of the Y is the same in every antibody of the same class, regardless of the antigen to which it is reacting. It follows, then, that these portions of the antibody molecule are called the *constant region* (or *constant fragment,* abbreviated *Fc*) because they are constructed identically. However, the tips of the Y are designated the *variable region* (*variable fragment, Fv*) because they are distinctive according to the target antigen. In an immune reaction, antigens bind to the Fv region; macrophages and complement can bind to the Fc region.

Antibody Classes
Antibodies are also known as **immunoglobulins** or *gamma globulins.* There are five classes (types) of immu-

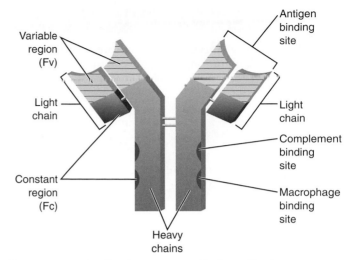

Figure 12.13. Antibody structure. Each antibody monomer consists of two heavy chains and two light chains. Parts of the light and heavy chain combine to form the variable region and the constant region. *Where is the complement binding site—in the Fc region or the Fv region?*

noglobulin: G, A, M, D, and E (abbreviated IgG, IgA, and so on). Each is defined by its heavy chain: IgG has a heavy chain type called *gamma,* IgA a heavy chain called *alpha,* and so on. Each class has a unique molecular structure and a distinct role in immune reactions. Some immunoglobulins are inserted into leukocyte membranes; others circulate in body fluids. One class (IgM) is produced quickly, leading the attack against a new invader, but it disappears after a few weeks. Another class (IgG) appears more slowly and continues to be produced almost indefinitely, providing permanent immunity. Still another class (IgA) is especially abundant in the respiratory and intestinal tract secretions. The structures, locations, and roles of the different antibody classes are summarized in Table 12.3.

Antibody Actions
Irrespective of type, an antibody alone cannot destroy an antigen. Instead, it neutralizes it or targets it for destruction by other means. An antibody bound to its target antigen is called an **immune complex** and is capable of stimulating a wide range of cellular and molecular activity against the antigen.

Neutralization is the most basic mechanism of antibody action—the antibody blocks the effect of the antigen. For example, some viruses and bacteria produce *exotoxins,* toxic products, which are capable of inflicting serious cell damage. Most exotoxins are antigens themselves and stimulate an antibody response. When bound by an antibody, the exotoxin typically loses its power and phagocytes later destroy the immune complex.

Table 12.3   Antibody Classes

Antibody Class and Illustration	Structure and Location	Importance	Major Actions
IgM	Monomer (B-cell membranes) or pentamer (blood)	First antibody released in immune reactions; replaced by IgG after a few weeks	B-cells: antigen receptor for activated B cells. Blood: complement activation, agglutination, precipitation
(a)			
IgG	Monomer (blood)	Most abundant antibody type; Conveys passive immunity to fetus; provides most of permanent immunity in later life	Complement activation
(b)			
IgA	Monomer (blood); dimer (breast milk, mucus, saliva, tears)	Forms protective barrier on mucous membranes; conveys passive immunity to breast-feeding infants	Prevents pathogen attachment to mucous membranes
(c)			
IgD	Monomer (B-cell membranes)	Activates B cells	Antigen receptor for inactive B cells
(d)			
IgE	Monomer: skin, mucous membranes	Inserted into mast cell membranes, enabling mast cell activity in allergic reactions	Stimulates histamine release by mast cells, inflammation
(e)			

An antibody coating surrounding cancerous or infected cells also activates the innate immune system.

Because antibodies have more than one antigen-binding site, they are capable of attaching to more than one antigen and can bind antigens together to form chains or clumps. For instance, when antibodies bind soluble antigens (such as toxins), the antibody–antigen complexes fall out of solution—a process called *precipitation*. Precipitated antigens lose their ability to affect cell processes and are easy targets for phagocytes. Alternatively, antibodies can bind antigens on the surface of groups of cells to form cell clumps—a process called *agglutination* (Fig. 12.14). Agglutinated cells, such as bacteria, for example, are immobilized and rendered more susceptible to phagocytosis than otherwise. IgM molecules in blood are particularly

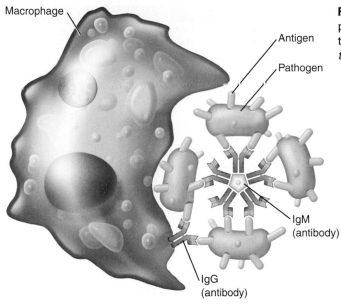

Figure 12.14. Antibody actions. A. Antibodies agglutinate pathogenic cells, rendering them more susceptible to phagocytosis. **B.** Antibodies activate the complement pathway. *Which two classes of antibody do you see in (A)?*

(a) Agglutination

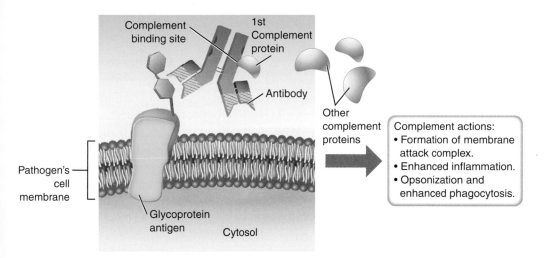

(b) Antibodies activate complement

adept at precipitation and agglutination, because each IgM antibody has 10 antigen-binding sites.

Another activity of IgG and IgM antibodies is *complement activation* (Fig. 12.14B). When they bind to antigens, antibodies change shape in a way that exposes complement-binding sites in the Fc stem. Binding of a complement protein to the antibody activates complement. Recall from our earlier discussion that complement acts as an opsonin, coating cells with complement proteins, which marks pathogens or cells for phagocytosis. Antibodies also act as opsonins when they attach to a target antigen, so that pathogens frequently "wear" a coat of both antibodies and activated complement proteins and are ripe targets for phagocytosis.

Remember This! The antigen–antibody complex leads to pathogen destruction by phagocytosis or cell lysis.

Case Note

12.10. Miriam's ability to make antibodies was impaired by HIV infection. But to the extent that she made antibodies against the lung fungus infection, which class of antibodies did she produce initially to fight the infection: IgG, IgA, or IgM?

Cellular Immunity Is a Function of T Cells

Like B cells, T cells have receptors capable of recognizing one and only one specific antigen.

That said, T cells greatly differ from B cells in their function. Recall that to be activated and multiply, B cells interact directly with antigen in the extracellular space. In addition, B cells secrete antibodies into the bloodstream. These antibodies circulate widely and freely to attach to pathogens, toxins, and other antigens, which are relatively easy prey in the extracellular space. However, if a pathogen—a virus, for example—is hiding inside a cell, it is safe from B-cell attack. That is where T cells come in. Their job is to destroy enemies that have gained access to host cells.

Because T cells make no antibodies to do the work for them, T cells themselves must attack the antigen directly. For this reason T-cell immunity is often called **cellular immunity.** What is more, as discussed below, T-cell activation is more complex than B-cell activation and takes considerably longer. For this reason, T-cell immunity is sometimes called **delayed immunity.**

T Cells Are Classified by Their Immune Roles

There are several classes of T cells:

- **Cytotoxic T cells (T$_C$ cells)** target and destroy any other cell in the body that the immune system has identified as containing alien antigen. This could be a cancer cell but most often is a cell infected by a virus.
- **Helper T cells (T$_H$ cells)** are so called because they facilitate the immune activities of both B cells and other T cells.
- **Regulatory T cells (T$_R$ cells** or *suppressor cells*) suppress the immune system. They act to suppress activation of the immune system and thereby maintain immune system homeostasis and self tolerance. They shut down the immune response after successful defense against an invading antigen, and in doing so they help prevent autoimmune disease.
- **Memory T cells (T$_M$ cells),** like memory B cells, these enable the cellular immune system to mount a rapid attack against previously encountered foreign antigens.

The first two classes are the most relevant to our discussion, and are discussed further below.

Case Note

12.11. Miriam's helper T-cell count fell steadily over time. Would this have had any effect on her humoral immune system?

Cytotoxic T Cells Attack Infected or Cancer Cells

We have said that cytotoxic T cells have a special capacity to attack intracellular antigens such as viruses or the abnormal proteins produced by cancer cells. But unlike B cells, T cells cannot recognize or react with freely circulating, unprocessed antigens. They must rely on **antigen-presenting cells (APCs)**—macrophages, dendritic cells, or B cells—to capture and process their specific antigen and present it to them. Dendritic cells are especially effective antigen presenters: they have large tentacles and sheets of cytoplasm that spread a wide net to catch antigens (see Fig. 12.3 on page 468). So let us look at antigen presentation using the example of a dendritic cell.

1. A dendritic cell captures and ingests a fragment of an infected cell or a cancer cell, breaks it into smaller antigens (Fig. 12.15, top), and combines an antigen with an MHC I glycoprotein.
2. The MHC I–antigen complex is transferred to the surface of the dendritic cell membrane. Then a cytotoxic T cell with the corresponding receptor (T-cell receptor, or TCR) binds the complex. This interaction partially activates the cytotoxic T cell, but complete activation also requires the efforts of helper T cells, which are discussed further on.
3. The fully activated cytotoxic T cell then divides into an army of cloned, identical cells to attack offending cells containing virus or tumor proteins.
4. The new T cell clones attach themselves to the MHC I of infected or cancer cells like an avenging spacecraft docking upon an alien invader. They then deploy any of a number of weapons to destroy the cellular integrity of the targeted cell.
5. Powerful weapons are the release of the proteins *perforin* and *granzyme* from the T cell into the extracellular fluid. *Perforin* opens the target cell membrane, enabling *granzymes* to enter the cytoplasm, where they attack key components of the cell and kill it.

Remember This! **Receptors on cytotoxic T cells bind only to MHC 1 proteins, which are present on all cells except red blood cells.**

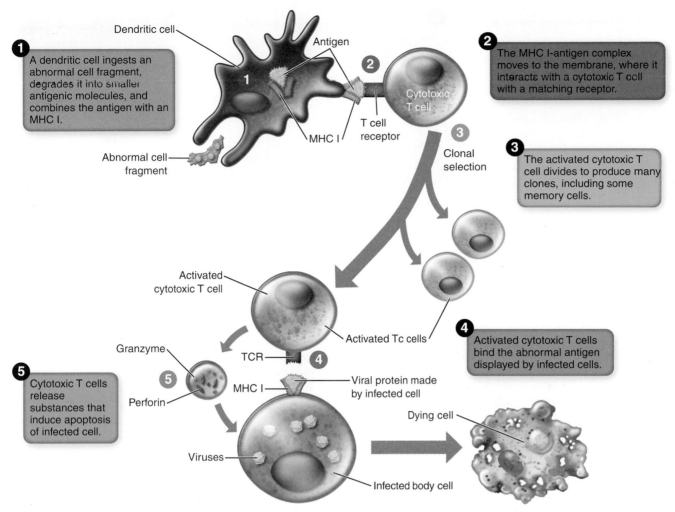

1 A dendritic cell ingests an abnormal cell fragment, degrades it into smaller antigenic molecules, and combines the antigen with an MHC I.

2 The MHC I-antigen complex moves to the membrane, where it interacts with a cytotoxic T cell with a matching receptor.

3 The activated cytotoxic T cell divides to produce many clones, including some memory cells.

4 Activated cytotoxic T cells bind the abnormal antigen displayed by infected cells.

5 Cytotoxic T cells release substances that induce apoptosis of infected cell.

Dendritic cell · Antigen · Cytotoxic T cell · T cell receptor · MHC I · Abnormal cell fragment · Clonal selection · Activated cytotoxic T cell · Granzyme · TCR · MHC I · Perforin · Viruses · Activated Tc cells · Viral protein made by infected cell · Dying cell · Infected body cell

Figure 12.15. Cytotoxic T cells. The steps involved in cytotoxic T cell (Tc cell) activation and action. *Which MHC glycoproteins are bound by cytotoxic T cells: MHC I or MHC II?*

After killing the cell, the cytotoxic T cell detaches and moves on to another target cell; but ultimately most of the new cytotoxic cells "wear out" and die by apoptosis. However, a small number go into hibernation as memory T cells, which can be awakened by reappearance of the offending antigen and can swing into action without time for "education" because they "remember" what to do.

Recall from our earlier discussion about innate immunity that NK cells have an inborn ability to recognize nonself antigen in the form of tumorous DNA and kill cells that possess it. In a process called **immune surveillance,** cytotoxic T cells and NK cells roam the body together looking for nonself cell markers to attack. Proof positive of the effectiveness of immune surveillance is that patients who have immunodeficiency diseases, such as HIV–AIDS, have an increased incidence of tumors because their immune surveillance function is faulty.

Case Note

12.12. **What aspect of Miriam's illness played a role in the development of her brain tumor?**

Helper T Cells Assist Other Immune Cells

Helper T-cell activation, like cytotoxic T-cell activation, relies on APCs: dendritic cells, B cells, or macrophages. However, unlike cytotoxic T cells, helper T cells recognize foreign antigens that exist outside of cells—usually pathogens or toxins that have invaded the body. As you might expect, APCs are abundant at sites where foreign antigen is most likely to be found: lymph nodes and other lymphoid organs, skin, and the lining mucosa of

Human Form, Human Function: Essentials of Anatomy & Physiology

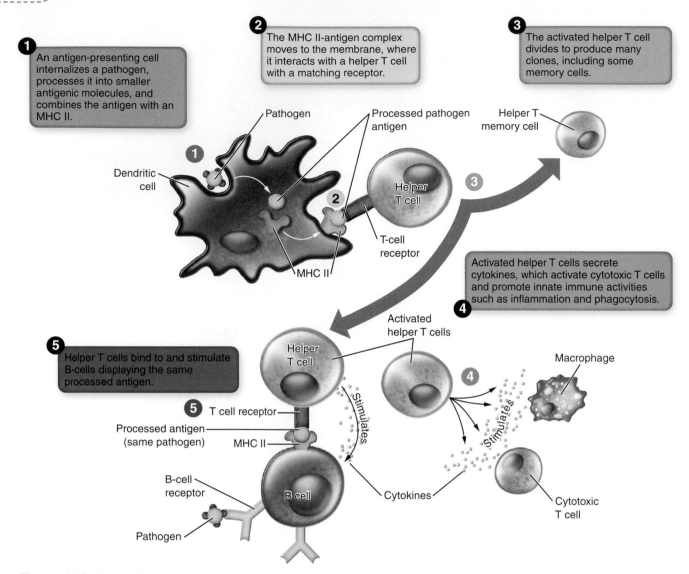

1 An antigen-presenting cell internalizes a pathogen, processes it into smaller antigenic molecules, and combines the antigen with an MHC II.

2 The MHC II-antigen complex moves to the membrane, where it interacts with a helper T cell with a matching receptor.

3 The activated helper T cell divides to produce many clones, including some memory cells.

4 Activated helper T cells secrete cytokines, which activate cytotoxic T cells and promote innate immune activities such as inflammation and phagocytosis.

5 Helper T cells bind to and stimulate B-cells displaying the same processed antigen.

Pathogen

Processed pathogen antigen

Helper T memory cell

Dendritic cell

Helper T cell

T-cell receptor

MHC II

Activated helper T cells

Helper T cell

Macrophage

Stimulates

Stimulates

T cell receptor

Processed antigen (same pathogen)

MHC II

B-cell receptor

B cell

Cytokines

Cytotoxic T cell

Pathogen

Figure 12.16. Helper T cells. The steps involved in helper T-cell activation and action. Any antigen-presenting cell can activate a helper T cell if it expresses the relevant antigen, attached to an MHC II, in its outer membrane. Helper T cells induce an immune response by secreting cytokines and physically interacting with B cells. *Do helper T cells interact with antigen-presenting cells expressing MHC I or MHC II glycoproteins?*

the respiratory, intestinal, urinary, and reproductive tracts. If an antigen breaks through skin or mucosa, it is likely to come into contact with and be captured by an APC. Even if it escapes, it is likely to be picked up by lymph fluid and delivered via lymphoid vessels to a lymph node, where additional APCs await.

Recall from our earlier discussion of major histocompatibility complexes that MHC II glycoproteins are expressed only on antigen-presenting cells. An APC ingests the pathogen, attaches it to an MHC II glycoprotein, and expresses the complex on its outer membrane

(Fig. 12.16, step 1). The MHC II–antigen combination activates helper T cells (step 2), thus stimulating the production of clones of helper T cells (step 3).

Helper T cells do not directly destroy pathogens. Instead, they coordinate and inspire other cells to attack pathogens, much the same way as military commanders coordinate and inspire soldiers to attack the opposing army. For instance, activated helper T cells secrete cytokines, which promote inflammation by attracting macrophages and other leukocytes to the infected area and stimulating phagocytosis (step 4). Helper T cell–released cytokines are also necessary to completely activate cytotoxic T cells. Finally, helper T cells directly interact with the B cells programmed to attack the same antigen (step 5). This interaction stimulates the B cell; it will produce more plasma cells, and the resulting plasma cells will produce more antibodies. That is why the B-cell response in the absence of helper T cells is relatively weak and does not last long. Indeed, it is diminished helper T-cell activity that accounts for the poor antibody-mediated immunity in HIV/AIDS patients.

As with cytotoxic T cells, the helper T cell response to an antigen peaks about a week after antigen exposure and subsequently declines as most of the new T cells die by apoptosis. However, some cells of the clone persist as helper T memory cells, enabling rapid initiation of cellular immunity upon subsequent antigen encounters. The continued presence of memory helper T cells means that the pathogen will be unable to establish an infection on subsequent visits because the person has become immune to the pathogen.

Case Notes

12.13. Which arm of the immune system would most effectively attack the HIV viruses hidden within Miriam's body cells?

12.14. Miriam's B-cells are normal, yet she does not produce a normal amount of antibodies in response to antigenic challenge. What is the basis of her impaired antibody-mediated immunity?

Lymphocytes Interact to Attack Pathogens and Abnormal Cells

Figure 12.17 summarizes the development and actions of the three types of lymphocytes—NK cells, B lymphocytes, and T lymphocytes. Note that NK cells, unlike B and T lymphocytes, are not directed against a specific target and are thus classified as players in innate immunity. Nevertheless, helper T-cell secretions greatly enhance NK cell activity. Although not shown on this figure, we have mentioned many other interactions between the innate and acquired immune systems. For instance:

- Antibodies activate complement, and complement and antibodies work together to opsonize pathogens.
- Lymphocyte-secreted cytokines attract and activate neutrophils and macrophages.

- Lymphocyte-secreted cytokines induce the vascular changes of inflammation, such as increased blood flow and vascular permeability.

Further on in this chapter we discuss how different aspects of immunity coordinate their attacks against bacteria and virally infected cells.

Long-Term Immunity Results from Memory Cells and Antibodies

Recall that *immunity* is the ability to resist infection by a particular pathogen. **Active immunity** is immunity provided by the actions of the one's own immune system as it produces antibodies and memory cells in response to antigen exposure (Fig. 12.18, left side). We develop **natural active immunity** to pathogens that have infected us—that we are exposed to in the ordinary course of life. That is, once infected with a particular type of cold virus, for example, you will not be reinfected by the same type. You may be *exposed* to it, but you will not be *infected* by it. However, there are many other viruses capable of causing an infection, and the natural immunity from infection by one virus does not extend to the others. Active immunity tends to last a very long time, but usually not a lifetime. So an infectious disease you fought off to as a child can sometimes come back to cause symptoms—though usually milder—in adulthood.

T and B memory cells are also the basis of **artificial active immunity,** immunity induced by purposeful exposure to antigen, commonly known as **vaccination** or **immunization**. Unlike natural active immunity, vaccination enables immunity to develop without causing an illness. Many vaccines contain a fragment of the pathogen that, in itself, cannot cause illness. Others contain an *attenuated pathogen,* one that has been altered so that it can stimulate an immune response but not cause illness. In any case, the administered antigen stimulates an immune reaction and the production of memory cells in the same way as natural infection. Upon subsequent exposure, the pathogen is killed or neutralized immediately by the quick reaction of an immune system that has been prepared ahead of time.

Most antigens are administered by injection, so that the gastrointestinal system does not destroy the antigen. However, vaccines with attenuated (not capable of causing infection) viruses can be administered orally. Since active immunity fades, the immune system can be reinvigorated against most pathogens by a second vaccination, called a "booster shot," many years after the initial immunization. For more about the remarkable history

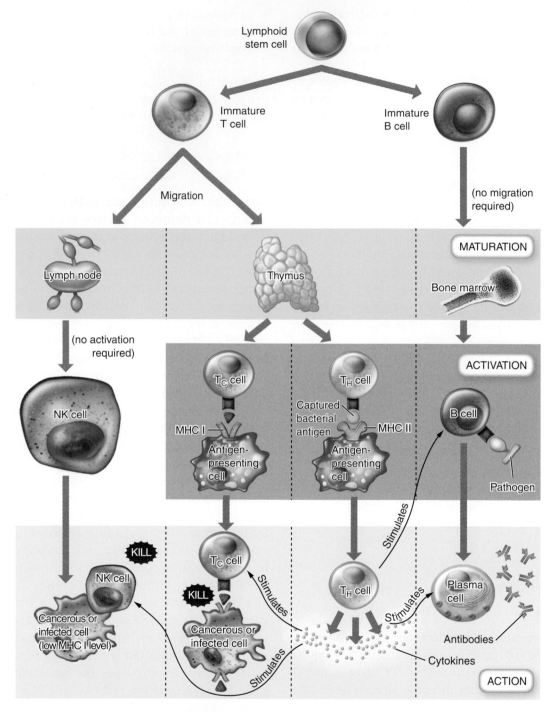

Figure 12.17. Lymphocyte production and action. A summary of the production, maturation, activation, and action of B and T lymphocytes and natural killer cells. Notice that helper T cells stimulate the activity of all other lymphocytes. *Which type of lymphocyte attacks cells with low MHC I levels?*

of vaccination, see the History of Science box titled *Edward Jenner's Joy* on page 498.

Memory cells are not involved in **passive immunity,** a type of immunity that can be transferred from one organism to another by transfer of premade antibodies into a nonimmune recipient (Fig. 12.18, right side). For

example antibodies can be transferred from a mother to her nursing child through the placenta or in breast milk (natural passive immunity) or collected from a patient already immune to a disease and injected into another patient (artificial passive immunity). Sometimes specific antibodies are prepared in large amounts for public

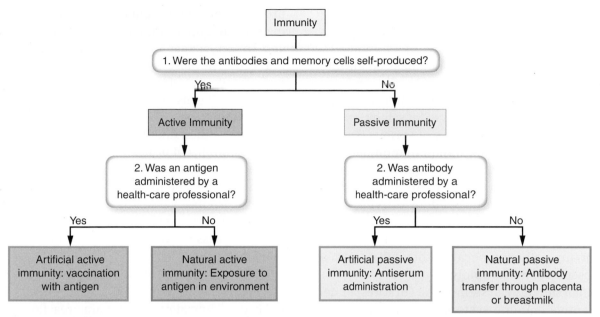

Figure 12.18. Types of acquired immunity. The classification of acquired immunity depends on (a) whether the antigen/antibody was administered and (b) whether or not the body actively synthesizes antibodies. *The injection of antiserum provides which type of immunity?*

inoculation programs by injecting animals with a particular pathogen (such as the toxin of the tetanus bacterium, for example) and collecting the resulting antibodies for injection. Whether prepared from animals or humans, the resulting antibody-rich solution is often called an *antiserum.* Passive immunity lasts only as long as the antibodies circulate—usually a few weeks—after which they disappear, having been metabolized by the host. For example, hepatitis A outbreaks can often be traced to contaminated food or water or an infected food worker, say a cafeteria worker in a school. In such situations, it is common practice to administer to students and other cafeteria workers injections that contain high concentrations of antihepatitis antibodies, which will confer passive immunity until the threat subsides.

12.30   Name three different proteins or protein complexes that participate in opsonization.

12.31   Name the T cell type activated by antigens attached to MHC I molecules.

12.32   Do helper T cells promote cell-mediated immunity, antibody-mediated immunity, or both?

12.33   If someone were deficient in memory cells, which would be more affected—the primary immune response or the secondary response?

12.34   Which of the following provokes active immunity—receiving antibodies through breast milk or receiving a vaccination against the chickenpox virus?

12.26   How many protein chains are in a single immunoglobulin monomer?

12.27   Which antibody action would be most effective against a toxin—complement activation or neutralization?

12.28   True or false: T cell activation always requires an antigen-presenting cell.

12.29   What is the name of the T cell membrane protein that interacts with a specific antigen?

An Integrated View of Body Defenses

So far we have learned how immunity relates to the body's natural barriers and to the inflammatory process, and we have learned the workings the two major branches of immunity: innate and adaptive. In this section we retrace out steps and take a look at how the immune system functions as a whole, as it does in the body day by day.

Innate and Adaptive Immunity Are Interdependent

Throughout this chapter we have emphasized the interdependent nature of innate and adaptive immunity. Figure 12.19 illustrates the point. To further your understanding of this interdependence, let's look at some examples of the reactions to injury of various types.

Recall from our earlier discussion that inflammation has roles in both immunity and repair. For example, while slicing an apple, you slash your thumb, and staphylococcal bacteria normally resident on your hands invade your tissues and proliferate (Fig. 12.19A). Bacteria elicit the quickest and most intense inflammatory reaction of any pathogen because they kill or damage the most tissue the most rapidly.

Your innate immune system is the "first responder": As the adaptive immune system is sampling the antigens of the bacteria, the innate immune system sends hordes of neutrophils, macrophages, protective proteins, and cytokines to the site of injury to attract more defensive cells and to neutralize or kill the invader.

Meanwhile, the B and T cells of the adaptive immune system are undergoing programming to kill the invading bacteria. The interrelatedness of the B- and T-cell systems is illustrated by the activity of helper T cells, which are programmed by their interaction with antigen-presenting cells that present them with captured bacterial antigen. In turn, helper T cells directly stimulate the activity of B cells. They also secrete cytokines that activate cytotoxic T cells and plasma cells and attract phagocytes to the site. As B cells interact with bacterial antigens, some evolve into plasma cells, which secrete antibodies into blood; these then find their way back to the invader and either neutralize it or make it susceptible to destruction by macrophages or other white blood cells.

Finally, after the invader is vanquished, the inflammatory response works to clean up the dead cells and bacteria by phagocytosis and paves the way for the repair process to knit the edges of your cut into a scar that closes the defect.

Other workings of the immune system are illustrated by viral infections (Fig. 12.19B). Viruses elicit a much less intense inflammatory and immune reaction because the damage they inflict is less acute and is spread out over a longer period of time. Let us say that someone with a cold sneezed into the air near you. Some of the viruses entering your airways would be ingested and killed immediately by phagocytes, and interferon would attempt to limit the spread of the virus. Meanwhile, dendritic cells and macrophages would ingest other copies of the virus and present viral antigens to T cells of the adaptive immune system.

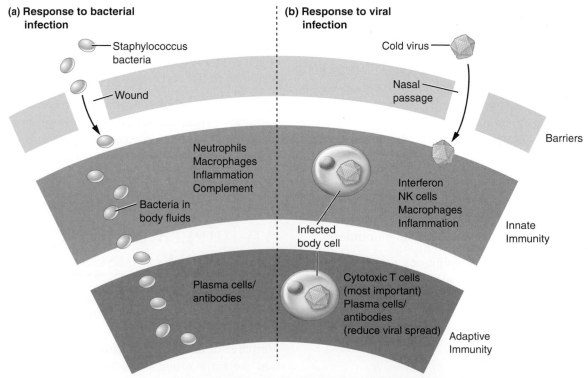

Figure 12.19. The integrated immune response to infection. A. The response to a bacterial infection. **B.** The response to a viral infection. *Is complement effective against bacteria or viruses?*

Antigen presentation in combination with cytokines produced by helper T cells would prompt the activation and proliferation of antiviral cytotoxic T cells, which would migrate into the damaged tissues over a period of a few days to kill infected cells, which would then be cleaned up by macrophages. B cells would also be activated and evolve quickly into antibody-secreting plasma cells. The antibodies cannot reach viruses inside cells, but they can prevent viruses released from infected cells from infecting other cells. Notice that, whereas neutrophils are key in the inflammatory response to bacterial invasion, few granulocytes are involved in the response to viral infection.

Case Note

12.15. Studies show that HIV infection impairs interferon production. Would this affect Miriam's susceptibility to infection by bacteria, viruses, or both?

The Immune Response Is Not Always Helpful

A healthy immune system is one of the cornerstones of good health—it wards off alien organisms and substances and purges the body of certain defective cells before they grow into tumors. But, as with other aspects of life, the sword—it cuts two ways. Recall that the immune system can create havoc by rejecting transplanted organs. Here, we discuss two other situations in which our immune system can work against us.

In Autoimmune Disease the Immune System Attacks Self Antigens

We have met the enemy and he is us.

Walt Kelly, American cartoonist (1913–1973), for his main character, Pogo, an opossum

In the quotation above, cartoonist Walt Kelly was alluding to environmental pollution; but the sentiment he expressed applies equally well to autoimmune disease. A somewhat similar but much more common problem than transplant rejection, **autoimmune disease** is a condition in which our own tissues become the enemy. That is, self tissues are regarded as alien and are attacked by the very system that we look to for protection. The resulting inflammation can lead to chronic inflammation, pain, organ dysfunction, and even death.

What is it that goes so terribly wrong? In most instances, it is not self that changes but the immune system. The topic is very complex and somewhat unsettled in some of the particulars, but there can be no doubt of its scientific validity. A single example will be instructive.

It is well known that the antigens in some infectious agents share common features with certain self antigens, so much so that the antibodies made against the invader *also* attack the self-antigens with similar characteristics. It's a simple case of mistaken identity. One of the best-known examples is the heart disease (rheumatic fever) that can follow some infections by the *Streptococcus* bacterium, which has antigens (proteins) similar to those in the human heart. As a result, in a very few streptococcal infections, antistreptococcal antibodies attack the heart, damaging the heart muscle and valves. However, the initiating cause of many autoimmune diseases (such as rheumatoid arthritis or type 1 diabetes) remains speculative. Increasing evidence suggests that invading microbes trigger the process, which then becomes self-perpetuating.

Allergies Are Exaggerated Immune Reactions

An **allergy** is an exaggerated (hypersensitivity) immune reactivity to certain substances (**allergens**) such as pollens, food ingredients, metals, or other substances in the environment that are ordinarily harmless. That is, it is the immune reaction, not the pathogen itself, that is harmful.

The hypersensitivity is preceded by an initial contact or "sensitizing dose." For people allergic to bee stings, for example, it is their first bee sting. The initial episode "acquaints" the immune system with the antigen: helper T cells stimulate B-cell production of IgE antibodies, which attach to mast cells and basophils in tissues. This response produces no immune symptoms. Upon the second and subsequent exposures, the antigen becomes bound to the IgE antibodies in the membranes of mast cells and basophils that have become primed and ready for action by virtue of the initial exposure. Antigen binding triggers the release of inflammatory chemicals such as histamine, which produce inflammation and vascular effects familiar as redness, heat, itching, and swelling.

For example, *hay fever* (seasonal allergic rhinitis) is a common allergy triggered by exposure to wind-borne pollens from trees, grasses, and weeds. It is characterized by sneezing, runny nose (rhinorrhea), and nasal congestion, all triggered by the outflow of inflammatory chemicals from mast cells. Although hay fever is not much more than a nuisance, allergies to specific foods, insect stings, and some medications are very dangerous. In such cases an acute inflammatory response called *anaphylaxis* causes dilation of blood vessels throughout the body. This, in turn, can prompt a dramatic fall in blood pressure or swelling in the throat or airway, either of which can prove fatal.

12.35 Which arm of the immune system would most effectively attack influenza viruses hidden within body cells?

12.36 What is the name of a disorder in which the immune system reacts against self?

12.37 What is the name of the condition in which a person has an immune reaction to an environmental substance that ordinarily does not cause problems for most people?

12.38 True or false? Inflammation is a critical first response to invasion by pathogenic bacteria.

12.39 Which event triggers the release of inflammatory chemicals from mast cells; the insert of IgE molecules into their membranes or the binding of an allergen to the IgE molecules?

Case Discussion

AIDS: The Case of Miriam K.

Miriam K. was *immunodeficient*—her immune system was unable to cope with pathogens. Immunodeficiencies can result from inherited genetic defects, but Miriam's immunodeficiency was *acquired* because of an infection. She suffered from the most common of all immunodeficiency diseases, **AIDS (acquired immune deficiency syndrome).** AIDS, which ranks among the most devastating epidemics in world history, is caused by infection with **human immunodeficiency virus (HIV).** Like all viruses, HIV contains a nucleic acid (in this case, RNA, not DNA), surrounded by a protein coat.

Although it can also infect macrophages and dendritic cells, HIV preferentially infects helper T cells (**T$_H$ cells**): it contains a protein that fits perfectly into a receptor on T-cell membranes (Fig. 12.20, step 1). Once internalized into the cell, the virus makes a DNA copy of its own RNA genome and incorporates the DNA into the infected cell's DNA. The new, corrupted DNA takes control, forcing the T cell to produce all of the necessary materials to make new HIV viruses, which are released

into the extracellular space to attach to more helper T cells (step 2). In the process, helper T cells die and B cells and cytotoxic T cells are deprived of the help they need to function effectively (step 3).

HIV infection is not necessarily the same thing as AIDS. Not everyone infected with HIV develops AIDS. AIDS is a *syndrome*—a group of signs and symptoms—characterized by low counts of helper T cells, brain disease, abnormal susceptibility to infection, and/or tumors. AIDS appears only at an advanced stage of HIV infection. Most patients with HIV infection do not develop AIDS for many years, especially if the infection is detected early and treated with anti-HIV drugs. Miriam, however, was noncompliant—she did not take the anti-HIV drugs prescribed for her and thus developed full-blown AIDS rather soon after the presumptive time of her initial infection (as best we can tell from her history).

Miriam died of overwhelming infections attributed to the destruction of her helper T-cell population by HIV, which in turn led to the broad failure of her innate, antibody-mediated, and cellular immunity (step 4). Such infections are called **opportunistic infections** because they are caused by organisms that do not ordinarily cause infection in patients with healthy immune systems.

We often focus on the negative aspects of inflammation—pain, redness, swelling, fever—but its saving grace is that it fights infectious pathogens. Despite her many infections, Miriam's tissues were only mildly inflamed. That is because, without the facilitating actions of cytokines produced by helper T cells, Miriam's inflammatory response was compromised, allowing pathogens easy access deep into her body.

Patients with deficient B-cell function (antibody-mediated immunity) do not produce an effective antibody response. Thus they often suffer from infections with ordinary pus-forming (*pyogenic*) bacteria such as *Streptococcus* or *Staphylococcus*. The effectiveness of the antibiotics administered at her first visit attest to the fact that Miriam was suffering from bacterial infections, and at death she was suffering from a severe bacterial infection of her urinary tract.

Patients with defective T cell function (cell-mediated immunity) are prone to infections by viruses and fungi and to the development of tumors as a result of failed immune surveillance. At the time of Miriam's death, she had extensive pneumonia due to overwhelming lung infection by *Pneumocystis jiroveci*, a fungus normally found in the throats of some healthy people. She also suffered from a severe intestinal infection by *cytomegalovirus*, a virus widely found in nature but rarely a cause of human infection. Autopsy also revealed a B-lymphocyte tumor in her brain, which can be attributed to failed T-cell immune surveillance.

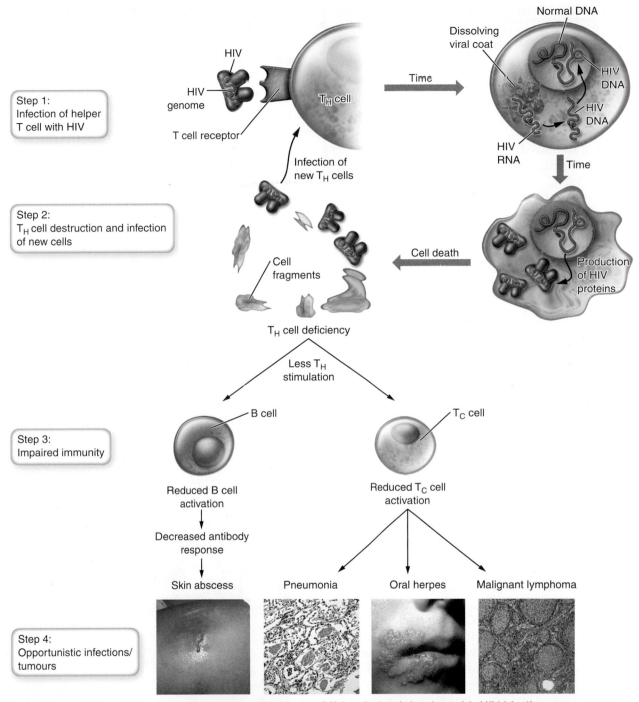

Figure 12.20. Miriam K. The cascade of events leading to Miriam's death begins with HIV binding to a receptor on a helper T cell. The virus invades the T cell and uses its genetic machinery to produce copies of itself. These copies flow out of the dying T cell and infect more host cells. As the cycle repeats, immunity is significantly impaired. *How does HIV bind to the helper T cell?*

Finally, Miriam's brain was also infected by HIV, which accounts for the dementia and coma present on her final admission.

HIV is transmitted in two primary ways: by intimate person-to-person contact and by needle injections, usually among users of illicit drugs who are sharing dirty needles. Transmission by blood transfusion is now very rare because of donor screening. *Casual contact will not*

transmit HIV, so although there is much about Miriam's life we do not know, we do know that she was not infected by shaking hands or sharing food with an HIV-positive acquaintance. It seems likely she contracted HIV from sharing needles with other drug users. Alternatively, it is possible that Miriam contracted HIV sexually: 25% of the adult population of Zimbabwe are infected by HIV, the great majority due to heterosexual transmission. The most reliable statistics available in 2010 (collected in 2008) conclude that one in four adults in Botswana is HIV positive. Throughout developing nations, most sexual transmission of HIV occurs during heterosexual contact. This is in part because people with a preexisting sexually transmitted disease such as syphilis or gonorrhea are more vulnerable to HIV infection, and there is a high prevalence of other sexually transmitted diseases in developing nations. However, in developed nations, most sexual transmission is among men having sex with men (54% in 2008).

Case Notes

12.15. Miriam suffered from both severe bacterial and viral infections, indicating failure of both B- and T-cell defenses. What is the common link between the B- and T-cell systems that is failing?

12.16. How do HIV infection and AIDS differ?

12.17. Miriam died from infections with ordinarily innocuous microbes. What is the name of these infections?

THE HISTORY OF SCIENCE

Edward Jenner's Joy

A clue to the importance of vaccinations in human history is suggested by the quotation that opens this chapter, in which Edward Jenner expresses the joy he felt when he realized that he had found a way to prevent smallpox, one of the most deadly of all contagious diseases.

Smallpox is an infectious viral disease known since antiquity—the mummified remains of Egyptian pharaoh Ramses IV (d. 1156 B.C.) bear evidence of it. As recently as 1967, smallpox caused epidemics resulting in 2 million deaths; however, a massive vaccination effort by the World Health Organization led to eventual elimination of the disease: the last recorded case was in Somalia in 1977. The only remaining smallpox viruses are believed to be in two tightly guarded laboratories, one in Russia and the other in the United States.

The tale of Jenner's discovery is well known, although there is some evidence that the Chinese may have engaged in similar practices hundreds of years earlier. It had long been observed that cowpox and smallpox were similar: both caused acute illnesses characterized by skin and mucous membrane eruptions, both were highly contagious and could be spread by skin-to-skin contact. What's more, it was known that dairymaids who had had cowpox did not catch smallpox.

In May 1796, Jenner found Sarah Nelmes, a young dairymaid, with fresh cowpox sores. He collected some of the fluid from her lesions and on May 14 inoculated James Phipps, an 8-year-old boy. The youngster became mildly ill for a few days but quickly recovered. On July 1, Jenner inoculated the boy with smallpox. Today we shudder at the danger to which the boy was exposed, but medical ethics was different in the 18th century, and Jenner's daring move may well have saved the boy's life. No disease developed. Over the next 2 years,

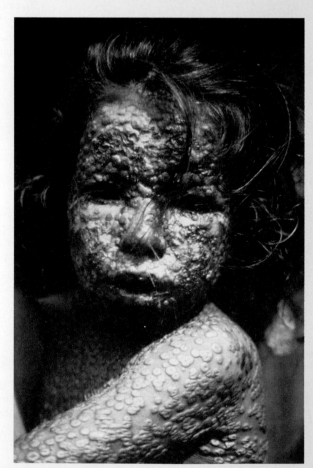

A child with smallpox.

Jenner repeated his experiment on more patients and was always successful. He published his results in 1798, the practice spread around the world, and smallpox infections and deaths plunged.

Word Parts

Latin/Greek Word Parts	English Equivalents	Examples
anti-	Antibody; also means against, opposing, curative	Antigen: something that produces (-gen) antibodies
-gen	Causing, generating	See path/o, anti-
immun/o	Immunity, immune system	Immunoglobulin: globular blood protein (globulin) involved in immunity
lingu/o	Tongue	Lingual tonsils: tonsils at the back of the tongue
lymph/o	Lymphatic system	Lymphoid: resembling (-oid) lymph or lymphatic tissue
path/o	Suffering, disease	Pathogen: organism causing (-gen) disease

Chapter Challenge

CHAPTER RECALL

1. **Which cells mature in the thymus?**
   a. B lymphocytes
   b. T lymphocytes
   c. NK cells
   d. All of the above

2. **Which of the following is *not* a function of the lymphatic system?**
   a. Red blood cell transport
   b. Lymphocyte transport
   c. Fat transport
   d. Water transport

3. **Which of the following lymphoid vessels drains into the right lymphatic duct?**
   a. Right mammary vessels
   b. Right tibial vessels
   c. Right iliac vessels
   d. Right femoral vessels

4. **Lymph flows through the**
   a. Thymus.
   b. Spleen.
   c. lymph nodes.
   d. all of the above.

5. **B cells multiply in the lymph node region called the**
   a. medullary cord.
   b. germinal center.
   c. medullary sinus.
   d. afferent vessel.

6. **Broad venous sinuses are found in the**
   a. white pulp of the spleen.
   b. red pulp of the spleen.
   c. cortex of a lymph node.
   d. medulla of a lymph node.

7. Antibiotics effectively treat diseases resulting from
   a. viruses.
   b. bacteria.
   c. worms.
   d. parasites.

8. Which of the following cells or substances participates in innate immunity?
   a. NK cells
   b. Antibodies
   c. Cytotoxic T cells
   d. None of the above

9. Macrophages participate in
   a. phagocytosis.
   b. antigen presentation.
   c. immunity.
   d. all of the above.

10. MHC I proteins are found in the membrane of
    a. most body cells except antigen-presenting cells.
    b. antigen-presenting cells only.
    c all body cells except red blood cells.
    d. only cancer or virus-infected cells.

11. Which of the following cells is *not* capable of phagocytosis?
    a. Macrophage
    b. Dendritic cell
    c. NK cell
    d. Neutrophil

12. C-reactive protein
    a. is an interleukin.
    b. stimulates fever.
    c. is an enzyme found in the granules of granulocytes.
    d. can activate the complement system.

13. Antibody-mediated immunity involves
    a. NK cells.
    b. interferon.
    c. cytotoxic T cells.
    d. plasma cells.

14. Clonal expansion describes the
    a. stimulation of a lymphocyte by an antigen.
    b. many cell divisions of an activated lymphocyte.
    c. activation of macrophages by cytokines.
    d. inactivation of T cells that recognize self proteins.

15. Passive immunity results from
    a. vaccination.
    b. pathogen exposure.
    c. memory cell formation.
    d. antibodies transferred between individuals.

16. On an antibody molecule, antigens bind to the
    a. Fv (variable) region.
    b. Fc (constant) region.
    c. heavy chain only.
    d. light chain only.

17. IgM antibodies would most likely be found in the
    a. blood of a newly infected individual.
    b. saliva of a newly infected individual.
    c. blood of an individual with a long-term infection.
    d. saliva of an individual with a long-term infection.

18. Antibodies
    a. form holes in the pathogen's cell membrane to induce cell lysis.
    b. bind complement proteins in the Fv region.
    c. coat pathogens to render them more sensitive to phagocytosis.
    d. all of the above.

19. Helper T cells are activated when they
    a. encounter a virus-infected cell.
    b. bind a surface antigen on a bacteria.
    c. interact with a macrophage expressing an antigen–MHC II complex.
    d. interact with a dendritic cell expressing an antigen–MHC I complex.

20. Unlike antigen-presenting cells, helper T cells display antigens bound to a
    a. T-cell antigen receptor.
    b. MHC I glycoprotein.
    c. MHC II glycoprotein.
    d. cytokine.

21. Perforin molecules are released from
    a. helper T cells.
    b. cytotoxic T cells.
    c. dendritic cells.
    d. macrophages.

22. The cells that release histamine during allergic reactions are called
    a. T lymphocytes.
    b. B lymphocytes.
    c. mast cells.
    d. neutrophils.

23. **Which of the following is more effective against bacteria than against viruses?**
    a. interferon.
    b. cytotoxic T cells.
    c. plasma cells.
    d. all of the above.

CONCEPTUAL UNDERSTANDING

24. **Compare and contrast veins and lymphoid vessels.**

25. **Explain how the following cells or chemicals play roles in both innate and adaptive immunity:**
    a. Macrophages
    b. Complement
    c. NK cells.

APPLICATION

26. **While playing for the Colorado Avalanche hockey team, Peter Forsberg had his spleen removed subsequent to an injury during a hockey game. A journalist stated that "You can function perfectly without a spleen. I don't think there should be any restrictions." Discuss this statement in light of your knowledge of the functions of the spleen.**

You can find the answers to these questions on the student Web site at
http://thepoint.lww.com/McConnellandHull

13

The Respiratory System

Major Themes

- The oxygen necessary for life is obtained from air.

- Oxygen is absorbed from air into blood in lung capillaries for transport to tissues.

- Carbon dioxide is absorbed from tissues into blood and transported to the lungs.

- The muscles of the diaphragm and chest wall contract to alter the volume of the chest, which moves air into and out of the lungs.

- Respiratory rhythm is controlled by the brain.

- Effective ventilation depends on compliant lungs and an unobstructed airway.

Chapter Objectives

Overview of Respiration   505

1. List the five stages of respiration.

The Anatomy of the Air Pathway   505

2. Describe the anatomy and function of the nose, pharynx, larynx, and trachea.

3. Explain how different lung subdivisions (such as a lobe) are supplied by different subdivisions of the bronchial tree (such as a lobar bronchi).

Pulmonary Ventilation   515

4. Explain how changes in thoracic volume alter the pressure in the lungs, referring to Boyle's law and the pleurae in your answer.

5. Explain how changes in thoracic pressure result in airflow.

6. Name the muscles involved in shallow and deep inspirations and active expirations.

7. Explain why resting exhalation does not require muscular effort.

8. Calculate alveolar ventilation from the minute ventilation, respiratory rate, and dead space.

"Cigarette asthma"

As you read through the following case study, assemble a list of the terms and concepts you must learn in order to understand it.

Clinical History: Luther M., a 61-year-old man, was accompanied to the emergency room by his son, who had been taking care of him for several years because of "lung troubles." Gasping for breath, Luther identified his problem as "cigarette asthma." His son explained that many years ago a physician had told Luther he was developing cigarette asthma and had advised him to quit smoking. His son said, "He tried, but it was just too hard. He would give up cigarettes for a few days or a few weeks, but he always went back."

In the last few years he had been visiting the hospital's emphysema clinic with increasing frequency.

Physical Examination and Other Data: Luther's temperature was 100.5°F (normal: 98.2°F), respirations 28 (normal: 14), blood pressure 115/75 (normal: 120/80), and heart rate 105 (normal: 72). He appeared to be of average stature but was very thin, almost skeletal. His upper airway appeared to be clear. His color was good but he struggled for each breath. He sat on the edge of the exam table, hunched forward, hands on knees and elbows spread outward like a bird. Despite his spindly arms and legs, his chest was large and barrel-shaped. His ribs showed through the skin and the muscles of his anterior neck bulged with every inspiratory effort as he struggled to breathe. Exhalation was prolonged and accompanied by a loud wheeze.

His emphysema clinic chart included the results of several spirograms. They revealed the typical emphysema pattern—very small inspiratory reserve volume (IRV) and very low forced expiratory volume in 1 second (FEV_1), which indicated severe expiratory obstruction.

A chest x-ray revealed loss of lung tissue consistent with severe emphysema. Also present were shadows consistent with acute pneumonia. Blood tests revealed low blood pH (acidosis), high arterial partial pressure of carbon dioxide, and low partial pressure of oxygen. Total white blood cell count was moderately elevated with an increased percentage of neutrophils. Hemoglobin and hematocrit were abnormally and unusually high.

Clinical Course: Luther was admitted to the hospital for ventilatory assistance, oxygen supplementation, and treatment of pneumonia. Within a few hours he was breathing easier and complaining about not being able to have a cigarette because smoking was prohibited in the hospital.

On the second day, his fever spiked and his respiratory distress returned. He became cyanotic. A chest x-ray revealed that the pneumonia had spread. Despite continuing treatment, he became difficult to arouse and died in a coma on the fifth hospital day.

At autopsy major abnormalities were confined to the lungs and heart. The right ventricle was moderately enlarged and the wall was abnormally thickened (hypertrophic). His lungs remained inflated and did not collapse as the chest was opened. They were flimsy and lacked substance, especially in the upper lobes, where some airspaces were several centimeters in diameter. The lower lobes were wet and boggy and contained many areas of yellowish pus mixed with bloody fluid. Microscopic study showed precancerous changes in the respiratory epithelium of the right main bronchus.

Final autopsy diagnosis was emphysema due to cigarette smoking, precancerous changes in respiratory mucosa, and severe acute pneumonia.

The respiratory system functions to supply cells with oxygen, which is necessary for all metabolic processes, and to rid the body of carbon dioxide, the most abundant of the wastes generated by metabolism. Exhalation of carbon dioxide is critical in maintaining proper blood pH. The respiratory system also plays an important role in the sense of smell by bringing odorants into the nose for detection. Finally, the movement of air in and out of the respiratory system vibrates the vocal cords, enabling us to produce the sounds of the human voice. Functionally, the respiratory system is divided into two main compartments: the *airways*, which conduct air, and the *lungs*, where oxygen is absorbed from air into blood and carbon dioxide passes from blood into air.

The ancients knew almost nothing of these particulars. Hippocrates, Aristotle, and other authorities of the ancient world did anatomical studies on humans but developed no useful understanding of lung function. Among their many mistaken suppositions was the idea that the esophagus conducted air directly to the heart. It was not until Dutch anatomist Andreas Vesalius published his detailed anatomical studies in 1543 that we began to develop a clearer understanding of pulmonary function.

"Having emphysema . . . (gasp) . . . is like drowning . . . (gasp) . . . only worse because . . . (gasp) . . . it is taking me . . . (gasp) . . . so long to die."

Emphysema patient, a 55-year-old man who had smoked two packs of cigarettes a day for 40 years, 1960, Parkland Hospital, Dallas, the day before his death

Overview of Respiration

At the heart of the respiratory process is the production of adenosine triphosphate (ATP), which uses oxygen and produces carbon dioxide as a byproduct. In its broadest meaning, **respiration** also includes a sequence of interrelated processes that bring oxygen from the atmosphere into body cells and move carbon dioxide in the opposite direction. These processes can be broken down into five distinct activities (Fig. 13.1):

1. **Pulmonary ventilation** (breathing) moves air into and out of the lungs.
2. **External gas exchange** (also called *external respiration*) is the absorption of oxygen from lung air into blood and the movement of carbon dioxide from blood into lung air.
3. **Gas transport** is the transport of oxygen from lungs to tissues via blood and the transport of carbon dioxide from tissues to lungs, also via blood.
4. **Internal gas exchange** (also called *internal respiration*) is the transfer of oxygen from blood to body cells and the transfer of carbon dioxide from cells to blood.
5. **Cellular respiration** is the utilization of oxygen and production of carbon dioxide by body cells to generate ATP ← (Chapter 2), which cells use for energy.

13.1 What is the physiological term for breathing: pulmonary ventilation or external gas exchange?

13.2 Name the five distinct activities involved in respiration.

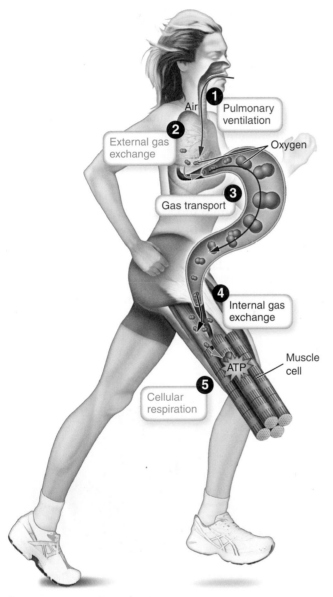

Figure 13.1. Respiration. The five steps of respiration bring oxygen from the atmosphere into body cells (muscle cells are highlighted in this illustration), which use it to generate ATP. Although not illustrated here, carbon dioxide follows the opposite path, from the body cells to the atmosphere. *Which process exchanges gases between the blood and tissue cells?*

The Anatomy of the Air Pathway

The respiratory system can be conveniently divided into upper and lower parts (Fig. 13.2). The *upper respiratory system* consists of structures in the head: the nose, nasal cavity, paranasal sinuses, and pharynx. The *lower respiratory system* consists of structures in the neck and chest: the *larynx* (voice box), *trachea* (windpipe), *bronchi* (a branching network of smaller air tunnels), and *lungs*.

However, it is actually more instructive to learn the anatomy by tracing the route air takes as it travels through two "zones." The *conducting zone* is made up of the airways—that is, the passages through which air travels on its way toward the tiny air sacs of the lungs. The air sacs themselves make up the *respiratory zone*, the zone in which gas exchange occurs.

Case Note

13.1. Clinical studies showed that Luther's external gas exchange was significantly impaired. Does external gas exchange occur in the conducting zone or the respiratory zone?

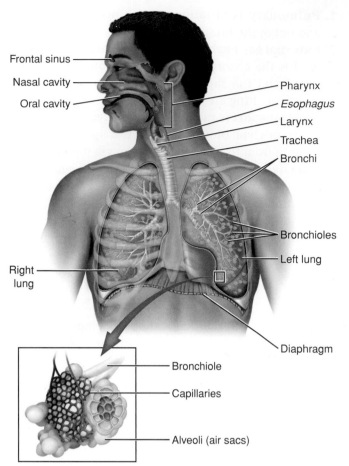

Frontal sinus

Nasal cavity

Oral cavity

Pharynx

Esophagus

Larynx

Trachea

Bronchi

Bronchioles

Left lung

Right lung

Diaphragm

Bronchiole

Capillaries

Alveoli (air sacs)

Figure 13.2. Structures of the respiratory tract. The respiratory tract consists of the conducting and respiratory zones. *Which zone includes the trachea?*

The Conducting Zone Carries Air into the Lungs

The airways of the **conducting zone** are shown in Figure 13.2. Notice that air sequentially passes through the nose, nasal cavity, nasopharynx, oropharynx, laryngopharynx, larynx, trachea, and bronchi before it passes into the respiratory zone. Each of these structures is discussed below.

The airways of the conducting zone are lined almost entirely by pseudostratified ciliated, columnar epithelium (*respiratory epithelium*) (Fig. 13.3). Between the ciliated epithelial cells are mucus-producing *goblet cells.* This epithelial membrane warms and humidifies air, and the sticky mucus traps smaller particulate matter that otherwise might be inhaled deep into the lungs. Rhythmic actions of the carpet of cilia move the particulate-laden mucus to the pharynx, where it can be swallowed or spat out. This combined function, which is sometimes referred to as the *ciliary mucus elevator*, is part of the nonspecific immune defense system ⬅ (Chapter 12).

The Nose Filters and Moistens Air

The nose consists of an external, visible part and an internal cavity within the skull.

The *external nose* is fitted to an opening in the anterior skull (see Fig. 6.27). Beneath the skin, it is composed of cartilage, bone, and dense fibrous tissue lined by respiratory epithelium (Fig. 13.4A). At its superior end, it is formed of the two nasal bones ⬅ (Chapter 6), which form the

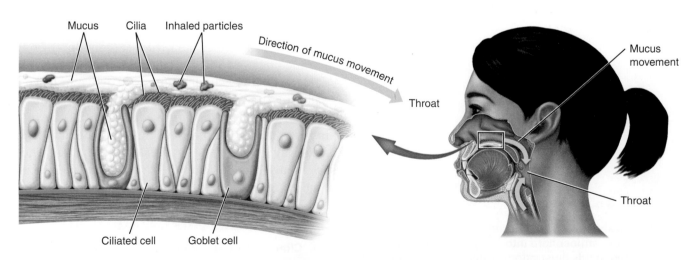

Mucus

Cilia

Inhaled particles

Direction of mucus movement

Throat

Ciliated cell

Goblet cell

Mucus movement

Throat

Figure 13.3. The lining of the conducting zone. The conducting zone is lined by mucus-producing pseudostratified columnar ciliated epithelium, which moves mucus and particulate matter containing inhaled pathogens toward the throat. *What are the mucus-producing cells called?*

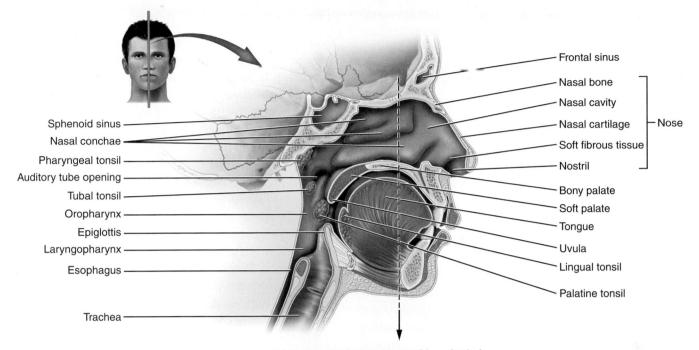

Frontal sinus

Nasal bone
Nasal cavity — Nose
Nasal cartilage
Soft fibrous tissue
Nostril

Sphenoid sinus
Nasal conchae
Pharyngeal tonsil
Auditory tube opening
Tubal tonsil
Oropharynx
Epiglottis
Laryngopharynx
Esophagus

Trachea

Bony palate
Soft palate
Tongue
Uvula
Lingual tonsil
Palatine tonsil

(a) Upper respiratory tract, mid-sagittal view

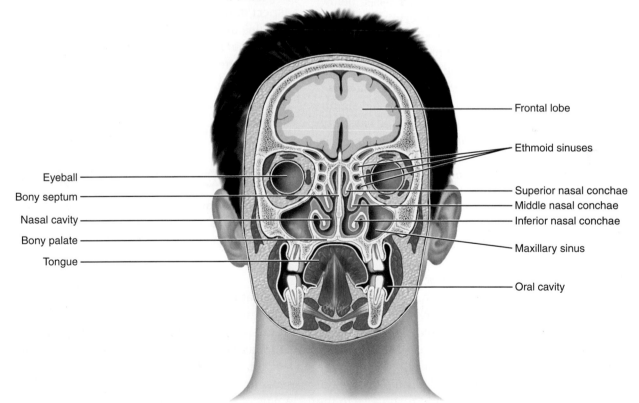

Frontal lobe

Ethmoid sinuses

Eyeball
Bony septum
Nasal cavity
Bony palate
Tongue

Superior nasal conchae
Middle nasal conchae
Inferior nasal conchae

Maxillary sinus

Oral cavity

(b) The nasal cavity, frontal section

Figure 13.4. The nose and throat. A. Structures of the nose and throat form the upper part of the conducting zone. **B.** The nasal cavity is continuous with the sinuses. *Name the most inferior portion of the pharynx.*

"bridge" of the nose between the eyes. Most of the remainder of the external nose is composed of plates of cartilage. It is divided vertically by a midline **nasal septum,** also composed of cartilage (Fig. 13.4B). The lower lateral bulbous edges of the nose are composed of soft fibrous tissue. The horizontal openings on either side are called *nostrils.* Numerous fine hairs project from the mucosa immediately inside the nostrils and are designed to trap particles too large to be handled by the ciliary mucus elevator.

The *internal nose* is a cavity in the skull, the **nasal cavity.** It is divided into right and left halves by the bony septum formed from the vomer and the perpendicular plate of the ethmoid (Fig. 13.4B). This bony septum connects anteriorly with the cartilaginous septum of the external nose. The roof of the nasal cavity is formed by the base of the skull. The anterior part of the floor is formed by the bone of the **hard palate** ← (Chapter 6). The posterior part of the floor is formed by the **soft palate,** a fleshy shelf that extends posteriorly from the bony palate and, during swallowing, swings upward to seal off the nasal cavity and the space behind it (the *nasopharynx,* discussed below) from food and water. The soft palate terminates in the **uvula,** the small fleshy pendulum visible at the back of the throat. Parts of several facial bones form the lateral wall of each nasal cavity. Projecting from the lateral walls are three delicate scrolls of bone, the **nasal conchae** (*concha,* singular), which are covered by mucosa and significantly increase the mucosal surface area of the nasal cavity; they also create turbulence in inhaled air to increase moistening, warming, and cleansing.

Nasal structure determines airflow. The nostrils are horizontal, which directs inhaled air upward to the roof of the nasal cavity, where the olfactory apparatus is located ← (Chapter 9). On the other hand, the posterior opening of the internal nose into the throat is vertical, a shape that directs exhaled air through the lower regions.

The nasal cavity is ringed by hollow, air-filled **paranasal sinuses** ← (Chapter 6), which are located in and named for the skull bones that contain them. The frontal and sphenoid sinuses are visible in Figure 13.4A; the ethmoid and maxillary sinuses in Figure 13.4B. The sinuses are lined by respiratory epithelium and connect to the nasal cavity by tiny openings (ostia). They make the skull lighter than it would be otherwise, add resonance to the voice, and add warmth and moisture to inhaled air.

Case Note

13.2. Like many emphysematous patients, Luther inhaled through his mouth, not his nose. Will this affect the quality of the inhaled air?

The Pharynx Is the Throat

Commonly called the *throat,* the **pharynx** is a hollow, tubular space lined by mucosa and skeletal muscles important in swallowing (Fig. 13.4A). It extends inferiorly from the base of the skull to the larynx (voice box) and is open to the nasal cavity anteriorly and superiorly. It is divided into an upper *nasopharynx,* a middle *oropharynx,* and a lower *laryngopharynx.*

The **nasopharynx** is lined by respiratory epithelium and lies directly posterior to the nasal cavity and anterior to the cervical spine. Entering its upper, lateral edge on each side are the auditory (eustachian) tubes ← (Chapter 9), which connect to the middle ears in order to equalize atmospheric pressure on each side of the tympanic membranes. Two collections of lymphoid tissue guard the nasopharynx: the *pharyngeal tonsil* (the adenoid), which sits high on the posterior wall, and the *tubal tonsils,* which guard the entrance to each auditory tube.

The **oropharynx** is directly posterior to the mouth. It extends from the soft palate down along the posterior aspect of the tongue. It is guarded on each side by large nodules of lymphoid tissue, the *palatine tonsils,* and by the *lingual tonsil,* a thick patch of lymphoid tissue on the base of the tongue. The **laryngopharynx** is a space immediately superior and posterior to the larynx and is the narrowest part of the pharynx. To accommodate the friction and food passage, the oro- and laryngopharynx are lined by stratified squamous epithelium.

Case Note

13.3. Luther breathed through his mouth. Which portion of his pharynx did not encounter freshly inhaled air?

Together, the oro- and laryngopharynx form the critical intersection where air and food passageways meet. That is because both the larynx and the upper end of the esophagus (the tube that conveys food to the stomach, → Chapter 14) connect here. Just as the soft palate seals the nasopharynx during swallowing, the *epiglottis,* a part of the larynx discussed below, folds downward to seal the entrance to the larynx.

The Larynx Produces Sound

The **larynx** (voice box) is a 5-cm-long complex tubular assembly of cartilage, skeletal muscle, and ligaments that connects the laryngopharynx above with the trachea below. It has three functions:

- Air conduction
- Diversion of food into the esophagus and air into the trachea
- Phonation (speech)

As shown in Figure 13.5, the larynx is suspended by ligaments from the *hyoid*, a horseshoe of bone in the anterior neck that arcs around the base of the tongue (see ◄ Chapter 6). It contains three cartilaginous structures:

- The most superior structure is the **epiglottis,** a leaf of flexible elastic cartilage that extends upward from the larynx and the base of the tongue and projects into the oropharynx. During swallowing, the larynx is pulled upward and the epiglottis tips downward to cover the laryngeal inlet. This action directs food into

the esophagus. This mechanism can be circumvented when we inhale or laugh while eating, permitting water or food to enter the larynx. Anything other than air entering the larynx stimulates the cough reflex, to expel the material from the airway.

- Forming the anterior part of the larynx is a large shield of cartilage called the *thyroid cartilage*. It forms a prominence in the anterior neck often called the "Adam's apple," a reflection of its prominence in males. Owing to the effect of testosterone, males have a larger larynx with longer and thicker vocal cords, which accounts for the deeper pitch of male voices.
- The most inferior part of the larynx is a ring of *cricoid cartilage*, which attaches by ligaments to the thyroid cartilage above and the first ring of tracheal cartilage below.

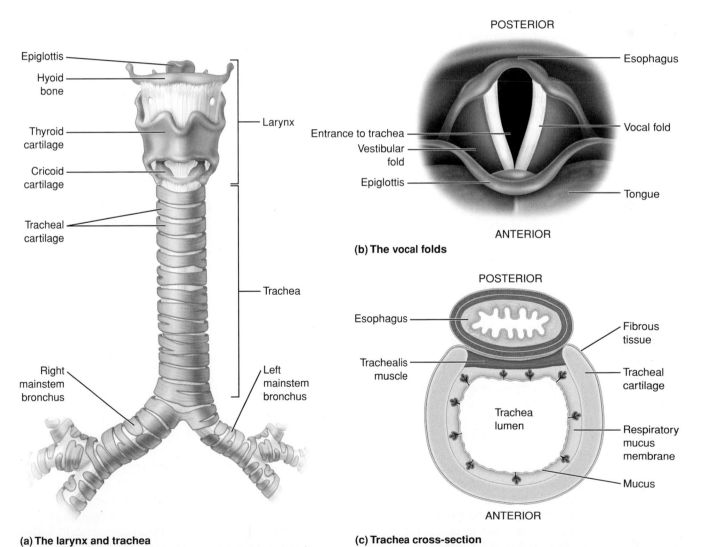

(a) The larynx and trachea

(b) The vocal folds

(c) Trachea cross-section

Figure 13.5. The larynx and trachea. A. The larynx and trachea are strengthened by cartilage rings. **B.** The vocal cords are supported by the vestibular folds. **C.** The trachea in relation to the esophagus. *Which is more superior— the thyroid cartilage or the cricoid cartilage?*

All together, the laryngopharynx, the epiglottis, and larynx are an anatomically complex and critical part of the conducting zone, one that under certain circumstances can interfere with breathing. For an interesting insight into this typically unheralded piece of anatomy, see the nearby History of Science box titled *The Strangulation of George Washington*.

The mucosa of the larynx forms two folds on each side of the airway: an upper set of **vestibular folds** (false vocal cords) and a lower set of **vocal folds (true vocal cords)** immediately below them (Fig. 13.5B).

The vocal folds produce the sounds of voice by vibrating as air is expelled past them. They contain tense ligaments strung between cartilaginous attachments, and they vibrate like the strings of a musical instrument. Voluntary skeletal muscles vary the tension on the cords by contracting or relaxing, thereby changing the pitch of the human voice. Contraction increases the

THE HISTORY OF SCIENCE

The Strangulation of George Washington

In 1796, after his second term as President of the United States, George Washington retired to his Mount Vernon estate. On December 12, 1799, still physically robust at age 68, he rode his horse most of the day in a heavy snowfall and near freezing temperatures. The following day he complained of a sore throat and hoarseness but again rode most of the day in the snow and cold. In the early hours of the next day, December 14, he woke with difficulty breathing. His assistant was summoned and prepared a medicinal mixture, which the general tried to drink but caused a convulsion of coughing and suffocation. Washington was a strong believer in bloodletting as a cure-all, so his estate manager was summoned and drained 350 to 450 mL of blood from his forearm.

Later physicians arrived and more blood was drained on several occasions and he was given a medicinal enema. A junior physician argued for cutting into the trachea (tracheostomy) to allow air to enter through the neck, a procedure known for centuries but never performed by any of those in attendance. The senior physician ruled against it.

Finally, late in the afternoon, the greatest Founding Father realized that he was dying and called his assistant. "I feel myself going. I thank you for your attentions but I pray you take no more troubles about me. Let me go off quietly. I cannot last long." At 10:10 P.M. after raising his arm to check his pulse, he died. Several weeks later his physician published an account of Washington's death, citing a fatal inflammation of the epiglottis, larynx, and upper trachea.

The exact cause of Washington's death has been the subject of much debate. Some argue it was diphtheria, a bacterial disease in which dead cells, clotted plasma, and other debris can form a membrane thick enough to

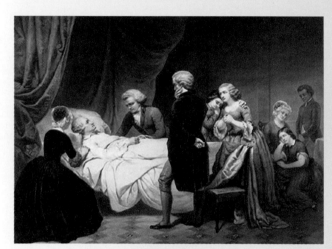

The death of George Washington.

occlude the airway. Others suggest it was an infection of the epiglottis, which we now know can be caused by certain bacteria. Still others suggest that his illness began as a simple bacterial sore throat, perhaps a "strep throat," that invaded the soft tissue of the back of the pharynx and caused so much swelling that the narrow passage closed.

The quantity of blood removed from Washington has been estimated by authorities to be 5 to 7 pints. Washington was a physically impressive man, 6 ft 3 in. and weighing about 230 lb, suggesting that his blood volume was about 14 pints. That he was drained of nearly half his blood volume surely played some role in his demise. The fact that Washington stopped struggling and appeared calm shortly before his death may have been due to weakness from excessive bloodletting.

tension, raising the pitch; relaxation reduces the tension, lowering the pitch. The true cords can also be moved closer to or farther away from one another. When the cords are close together, forming a narrow slit, they produce a higher pitch, and vice versa. The pressure of air forced over the cords controls the volume of sound: greater pressure produces louder sound.

The vestibular folds have no role in speech; however, they can be voluntarily closed to seal the airway in order to increase intrathoracic and intra-abdominal pressure. Closing the vestibular folds and trying to breathe out, a technique called the *Valsalva maneuver*, increases pressure in the trunk, stiffening it for heavy lifting or for expelling stool at defecation.

The Trachea Conducts Air to the Bronchi

The **trachea** (windpipe) is a flexible tube about 4 in. (11 cm) long and 2 cm in diameter that extends downward from the larynx (Fig. 13.5A). It passes through the anterior neck, behind the sternum, and into the mediastinum (the space between the lungs), where it divides into right and left mainstem *bronchi*.

The trachea is composed of three layers. The inner lining is respiratory mucous membrane consisting of epithelium and submucosa. A middle layer contains about 16 to 20 C-shaped rings of *tracheal cartilage,* with the open end of the C facing posteriorly. The thin outer layer is composed of fibrous tissue (Fig. 13.5C).

As just noted, the incomplete tracheal rings do not provide support to the posterior part of the trachea. However, the esophagus fits neatly into the open end of the C of each tracheal cartilage, which braces the trachea posteriorly, aided by a thin layer of smooth muscle (the trachealis muscle). This arrangement allows the esophagus to expand anteriorly if needed to accommodate a bolus of swallowed food. The trachealis muscle contracts slightly with each expiration, narrowing the diameter of the trachea with a gentle squeeze, which gives added velocity to air being expelled. The trachealis contracts more forcefully with the forceful expiration of a cough.

As with all respiratory epithelium, cilia in the trachea move in coordinated fashion to sweep away mucus and particles. In the trachea and bronchi, the direction of motion is upward, so that mucus can be coughed out (expectorated) or swallowed. Conversely, in the nasopharynx, the waves of ciliary motion propel mucus downward for swallowing. Smoking damages cilia, impairing or even paralyzing their motion and destroying them in the long run. Chronic smokers, therefore,

tend to retain inhaled contaminants and bacteria and cannot effectively expel accumulating mucus, which accounts for at least some of their increased risk for lung infections. Alcohol also impairs ciliary action and, in like manner, accounts for some of the increased risk of lung infections in chronic alcoholics.

Case Note

13.4. Could smoking have been partially responsible for Luther's pneumonia? Explain.

The Bronchial Tree Conducts Air into the Respiratory Zone

Because of its branched structure, the network of bronchi is called the *bronchial tree* (Fig. 13.6A). Branching from the trachea are the right and left **mainstem (primary) bronchi,** which angle laterally for several centimeters before branching into **lobar (secondary) bronchi** and then **segmental (tertiary)** bronchi. Segmental bronchi branch into successively smaller ranks of bronchi. Branches smaller than 1 mm are called **bronchioles.** These travel a short distance before branching into even smaller *alveolar ducts* in the respiratory zone (Fig. 13.6B).

The left mainstem bronchus is slightly smaller than the right because it carries air to the smaller left lung, which makes space for the heart (see Fig. 13.7). It is also somewhat more horizontally oriented because it is tilted up a bit to pass above the heart. By comparison, the right mainstem bronchus is wider and steeper. This seemingly minor anatomical fact is clinically important: an aspirated foreign object—a peanut, for example—is more likely to enter the wider, more vertical right side of the bronchial tree than the left.

In their larger branches, bronchi have the same structure as the trachea: a lining of respiratory epithelium, rings of cartilages, and a fibrous outer layer. However, important changes occur as branches become successively smaller:

● C-shaped rings of cartilage fade into small curved plates, which in turn disappear entirely in bronchioles.
● Epithelium changes from ciliated, tall, columnar epithelium into shorter and boxier cuboidal epithelium, which contains few cilia or goblet cells.
● The amount of smooth muscle increases, so that bronchioles are completely encircled by bands of smooth muscle fibers, which contract or relax to change the diameter of the bronchiole and with it the resistance

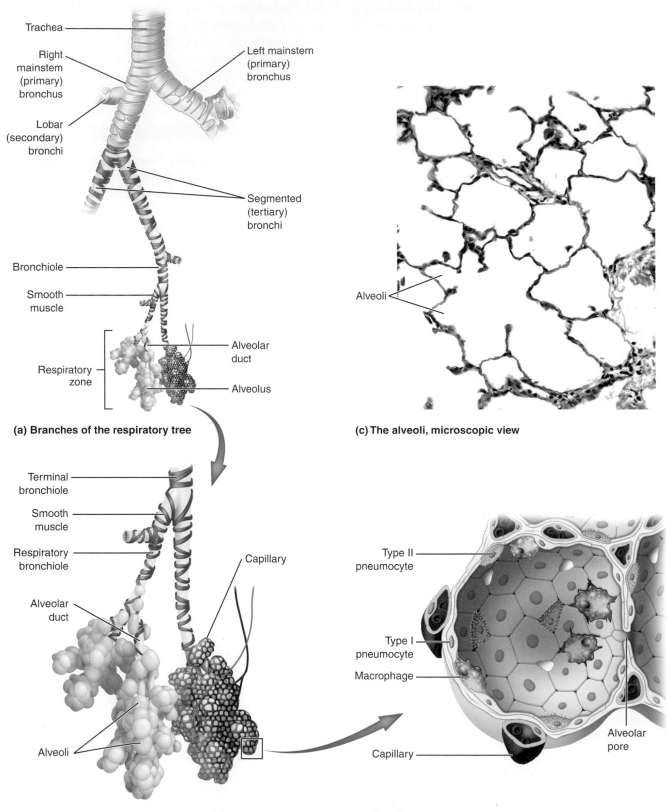

(a) Branches of the respiratory tree

(b) Structures of the respiratiory zone

(c) The alveoli, microscopic view

(d) Parts of an alveolus

Figure 13.6. The bronchial tree and alveoli. A. The mainstem bronchi branch into successively smaller branches, which terminate in alveolar clusters. **B.** This microscopic view of lung tissue illustrates how alveoli intersect to form a latticework **C.** Gas exchange occurs in the alveolar ducts and alveoli of the respiratory zone. **D.** Each alveolus contains cells for gas exchange and protection. *Does smooth muscle surround bronchioles or segmental bronchi?*

to airflow. For example, during exercise, this smooth muscle relaxes to facilitate greater airflow.

The Alveoli Constitute the Respiratory Zone

Gas exchange occurs in the **respiratory zone,** which is composed of tiny air sacs (**alveoli,** Greek *alveol* = "small hollow space"). Alveoli are arranged like nodules around a tiny central vestibule, the **alveolar duct** (Fig. 13.6B). Alveoli from adjacent clusters are tightly packed together to form a latticework of interspersed alveoli and capillaries (Fig. 13.6C). Air can flow freely between neighboring alveoli through tiny **alveolar pores** (Fig. 13.6D).

Alveoli are composed mainly of **type I pneumocytes**— exceedingly thin, flat (squamous) cells fitted together like tiles. These cells form the innermost layer of the alveolar wall. They rest on a basement membrane, which also supports the type I cells of adjacent capillaries. Interspersed among type I cells are scattered **type II pneumocytes**—plump cells that secrete a slick, soapy substance called *surfactant,* which keeps alveoli open (a topic explored below). Type II pneumocytes also secrete antimicrobial proteins that act as part of the innate immune system ◀ (Chapter 12) to protect against inhaled microbes. Between the pneumocytes that form each alveolus there is a very thin space, the **pulmonary interstitium,** which contains capillaries, lymphatic channels, a few fibroblasts, and collagen and elastin fibers. The fibroblasts make the collagen and elastic fibers, which provide structural support and give the lungs their elastic recoil.

Remember This! **Gas exchange occurs in the respiratory zone—the alveolar ducts and alveoli.**

The exposed surface of the alveolar wall is called the **respiratory membrane.** If laid out flat, an adult's respiratory membrane would cover the space of a four-car garage—about 750 sq ft. However, it is so thin (about 0.5 microns) that all 750 sq ft of it can be stuffed into the human chest, and oxygen and carbon dioxide can diffuse easily between blood and alveolar air.

By the time air reaches the respiratory zone, it has been warmed and humidified and purged of most of its particulate impurities by contact with the respiratory mucosa that lines the conducting zone. Any particulate matter—dust, bacteria, and the like—that survives the trip is usually gobbled up by **alveolar macrophages,** which roam freely about the alveoli doing their usual housekeeping chores. Exhausted macrophages and their

ingested debris are swept upward by the ciliary motion of bronchial epithelium and ultimately swallowed or coughed out. However, in some circumstances, the load of particulate matter overwhelms the capacity of alveolar macrophages to clean up debris; in that case, particles are deposited in the interstitial space. The "black lung" disease of coal miners is an example.

The Gross Anatomy of the Lungs

Each lung is somewhat cone-shaped, with a concave *base,* which sits on the diaphragm at the floor of the thoracic cavity, and a rounded *apex,* which rises high in the chest to the base of the neck behind the clavicle (Fig. 13.7A). The right lung is somewhat larger than the left, which must make room for the heart.

Each lung is divided into smaller divisions, called **lobes.** A single oblique fissure divides the left lung into halves: a slightly smaller superior lobe and an inferior lobe. Two fissures, one horizontal, the other oblique, divide the larger right lung into three lobes: a slightly smaller superior lobe, a large inferior lobe, and a small wedge-shaped middle lobe that sits anteriorly between the two. Each lobe is supplied with air by a lobar bronchus branching from the main bronchus.

Thin planes of dense fibrous tissue, in turn, divide each lobe into two to five *segments* according to lobar size. Each segment is supplied with air by a segmental bronchus. Finally, each segment is organized into smaller *lobules,* each about the size of a fingertip and served by a bronchiole. Segmentation has survival value: it often confines disease—pneumonia, for example—to only one or two segments.

Case Note

13.5. Luther's pneumonia began in the middle lobe; therefore, which lung was infected first?

Each Lung Is Covered by Pleura

A very thin serous membrane, the **pleura,** covers each lung (Fig.13.7B). The pleura is a single membrane with two faces:

- The **visceral pleura** covers the lungs, closely following its contours, including the fissures between lobes.
- The **parietal pleura** lines the inside of the chest wall and the superior surface of the diaphragm.

The two layers produce a slick, thin fluid, the **pleural fluid,** which enables them to slip smoothly across each

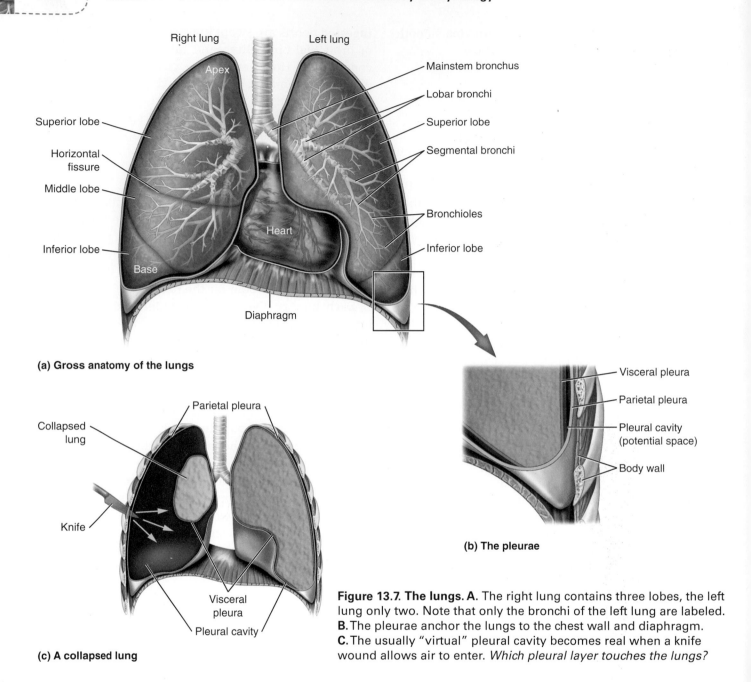

(a) Gross anatomy of the lungs

(c) A collapsed lung

(b) The pleurae

Figure 13.7. The lungs. A. The right lung contains three lobes, the left lung only two. Note that only the bronchi of the left lung are labeled. **B.** The pleurae anchor the lungs to the chest wall and diaphragm. **C.** The usually "virtual" pleural cavity becomes real when a knife wound allows air to enter. *Which pleural layer touches the lungs?*

other. Pleural fluid occupies the **pleural cavity** between the two membranes. Like the pericardial cavity discussed in ← Chapter 11, the pleural cavity is only a *potential space* a few microns wide. In a healthy individual, the pleural fluid glues the two membranes together like two plates of glass held together by a thin film of water. Like glass plates, the membranes cannot be easily separated, but they do slide across each other with ease. Their adhesion effectively "glues" the lungs to the chest wall, stretching them so that they fill the larger chest cavity.

As discussed further below, this adhesiveness is necessary for breathing. Outward movements of the chest wall stretch the lungs; all the while, the elastic recoil of the expanded lungs tugs the chest wall inward. If air gets into the pleural space, it breaks the seal and the visceral pleura covering the lungs pull away from the parietal pleura covering the chest wall. This condition, called *pneumothorax*, can occur, for instance, as a result of a penetrating chest wound. Once the lungs lose contact with the chest wall, they tend to collapse to a greater or lesser degree, depending on the volume of air that has entered the pleural space (Fig. 13.7C). Because each lung has its own pleural cavity unconnected to the other, it is possible to collapse just one lung. However, if both lungs are affected and if the volume of air is large, breathing becomes impossible.

Each Lung Is Supplied by Blood Vessels and Nerves

All of the blood arriving at the lungs through the pulmonary artery from the right ventricle is oxygen-poor. The lungs take this "used" blood and make it "new" again—they remove some carbon dioxide and add additional oxygen. However, lung cells, like all body cells, need oxygen themselves in order to function properly. So, each lung has two blood supplies: the *pulmonary arteries* carry oxygen-poor blood for processing (Fig. 11.5); and the *bronchial arteries* (which branch from the aorta) supply oxygen-rich blood for lung cell use.

The lungs are supplied by fibers from the sympathetic and parasympathetic nervous systems, both of which send nerve branches along the bronchial tree. Signals from sympathetic fibers relax bronchial smooth muscle, enlarging the airway for the fight-or-flight reaction. Parasympathetic signals cause bronchial smooth muscle to contract. A steady stream of background parasympathetic signals from the vagus nerve (vagal tone) maintains bronchioles in a slightly constricted state.

13.3 Which cells produce mucus: goblet cells or ciliated epithelial cells?

13.4 What forms the anterior part of the palate: the bony palate or the soft palate?

13.5 True or false: The nose does not contain any bones.

13.6 Which structure seals off the nasopharynx during swallowing: the epiglottis or the soft palate?

13.7 Name the cartilage that forms the Adam's apple.

13.8 What is the role of the vestibular folds?

13.9 Place the following structures in the order they would encounter air inhaled through the nose: mainstem bronchi, secondary bronchi, trachea, larynx, bronchioles, pharynx.

13.10 Which alveolar cells participate in gas exchange—alveolar macrophages, type I pneumocytes, or type II pneumocytes?

13.11 Each lobule of the lung is supplied by a separate lobar bronchus: true or false?

13.12 Which pleural membrane contacts the lungs: the parietal layer or the visceral layer?

13.13 Which blood vessels supply lung cells with oxygen: the pulmonary vessels or the bronchial vessels?

Pulmonary Ventilation

In the long chain of events that takes oxygen from air and into the interior of cells, pulmonary ventilation, or breathing, is the first step. Conversely, it is the last step in ridding the body of much of the carbon dioxide generated by metabolic processes of life. **Inspiration** (inhalation) is the drawing in of air; **expiration** (exhalation) is the expulsion of air. In both cases, air is moving. And, just like the movement of blood through vessels or ions through cell membranes, air moves down *gradients*. In particular, air moves down *pressure gradients,* from a region of higher pressure to a region of lower pressure.

Changes in Chest Volume Create Pressure Gradients

Understanding the movement of air into and out of the lungs requires mention of **Boyle's law,** which states that the pressure and volume of a gas are inversely proportional; that is, in a closed space, if volume decreases, pressure increases proportionally, and vice versa. The law can also be stated in a different manner: the product of pressure and volume (pV) is constant for a given number of gas molecules in an enclosed space (k):

$$pV = k$$

Consider a syringe as an example—a sealed, air-filled cylinder with a plunger at one end (Fig. 13.8A, part i). Pushing in the plunger decreases the volume of the cylinder (V) and thereby increases the pressure (p)—it becomes increasingly difficult to push in the plunger (Fig. 13.8A, part ii). If the cap is removed from the syringe, air flows from the region of high pressure inside the syringe to the region of low pressure outside the syringe until the pressures are equal (Fig. 13.8A, part iii).

Ventilation occurs when atmospheric pressure differs from **intrapulmonary pressure**—the pressure of air in the lungs. Air flows into the lungs when atmospheric pressure exceeds intrapulmonary pressure, and vice versa. We have no control over atmospheric pressure, so we move air into and out of the lungs by altering intrapulmonary pressure. We modify intrapulmonary pressure by altering chest volume, which determines lung volume (V). Decreasing lung volume increases intrapulmonary pressure (p), and vice versa, so that, in accord with Boyle's law, air flows out of or into the lungs

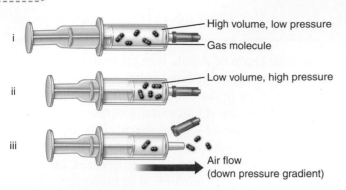

(a) Boyle's Law and air flow

Figure 13.8. Volume changes create pressure gradients, which cause airflow. A. Pressure is increased if the same number of gas molecules is confined into a smaller space. **B.** Air does not flow at rest, because there is no pressure gradient between the lungs and the atmosphere. During inspiration, intrapulmonary pressure decreases as the thorax and lungs expand, so air flows into the lungs. Lung size decreases during expiration, increasing the intrapulmonary pressure and forcing air out of the lungs. *When is intrapulmonary pressure greater than atmospheric pressure— during inhalation or exhalation?*

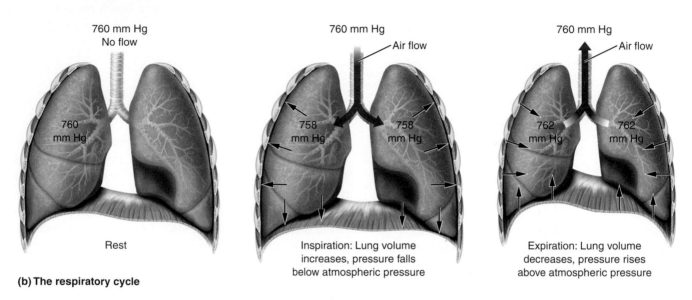

(b) The respiratory cycle

because the product of pressure (p) and volume (V) is constant (k).

The sequence of pressure and volume changes during a single breath is called the **respiratory cycle.** At rest between breaths, pressure in the airway and alveoli (intrapulmonary pressure) is 760 mm Hg (at sea level), in equilibrium with atmospheric pressure (Fig. 13.8B, left panel). During inspiration, we reduce the intrapulmonary pressure *below* atmospheric pressure by *increasing* lung volume (middle panel). As a consequence, air flows down the pressure gradient from the atmosphere into the lungs until enough volume of air has entered to raise the pressure back to equilibrium with atmospheric pressure. During expiration, the opposite occurs: we increase the intrapulmonary pressure *above* atmospheric pressure by *reducing* lung volume, so air flows from the lungs to the atmosphere. Air leaving the lungs reduces intrapulmonary pressure until it is once again equal with atmospheric pressure

at 760 mm Hg and airflow ceases. Intrapulmonary pressure *always* reaches equilibrium with atmospheric pressure twice in the cycle: at the end of inspiration and the end of expiration.

Remember This! **Compressing a gas into a smaller volume increases its pressure, and vice versa.**

Muscles Change Thoracic Volume

The skeletal muscles of the diaphragm, chest wall, neck, and/or abdominal wall contract to enlarge or reduce the volume of the chest cavity and, with it, the volume of the lungs. Since the lungs are attached to the thoracic cavity by the thin film of pleural fluid that holds the pleurae together, changes in thoracic volume also change lung volume.

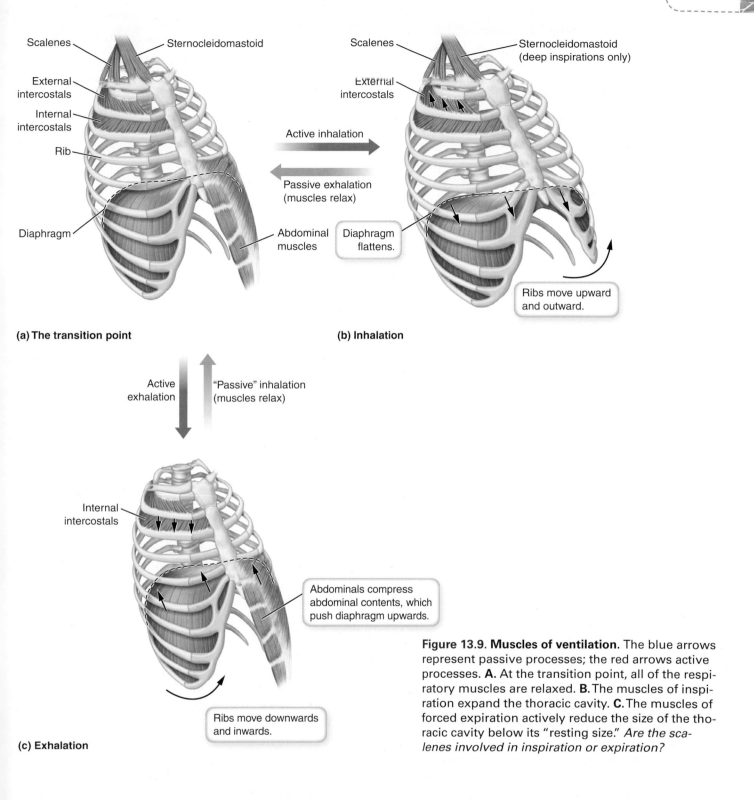

(a) The transition point

(b) Inhalation

(c) Exhalation

Figure 13.9. Muscles of ventilation. The blue arrows represent passive processes; the red arrows active processes. **A.** At the transition point, all of the respiratory muscles are relaxed. **B.** The muscles of inspiration expand the thoracic cavity. **C.** The muscles of forced expiration actively reduce the size of the thoracic cavity below its "resting size." *Are the scalenes involved in inspiration or expiration?*

The anatomy of the ribs and associated muscles is important (Fig. 13.9A). The scalene and sternocleidomastoid muscles of the neck suspend the sternum and first ribs from the skull. They elevate the rib cage when they contract. The other ribs are suspended from those above by two sets of **intercostal muscles,** the *external* and *internal* intercostals, whose fibers are oriented at roughly 90 degrees to one another. The *external* intercostal muscles raise the ribs and expand the chest when they contract. Conversely, *internal* intercostals pull the ribs downward and inward as they contract.

Focus on your breathing for a moment, noticing the pause *between* exhalation and inhalation. At this moment, called the **transition point,** the inward recoil of the lungs is perfectly balanced by the outward forces exerted by the chest wall, and all of the respiratory muscles are relaxed. **Inhalation** expands chest volume away from the transition point, which requires the activity of three muscle groups (Fig. 13.9B).

First is the diaphragm, the most important inspiratory muscle, especially when an individual is lying down. As it contracts it *flattens,* which pushes abdominal contents downward and outward. Second, the scalenes contract to elevate the sternum and the upper ribs. The actions of diaphragm and scalenes expand the thoracic cavity vertically. Third, the external intercostals contract and pull upward on the ribs, rotating them upward and outward to enlarge the chest cavity like the ballooning of the front of a skirt in a strong wind. And as the chest wall expands both horizontally and vertically, so do the lungs. The larger inhalations associated with strenuous activity require greater effort from the respiratory muscles but also bring additional *accessory* muscles into play. The sternocleidomastoid muscles, for instance, perform the same action as the scalenes to further increase the vertical dimensions of the thoracic cavity.

In quiet breathing, or **eupnea** (from Greek *eu-* = "good" and *pnein* = "breathe"), exhalation is passive. The inspiratory muscles relax, and the elastic recoil of the stretched lungs and thoracic wall tugs the thorax back to its original dimension at the transition point. The abdominal contents recoil, pushing the diaphragm upward. The resulting decrease in thoracic and lung volume pushes air out.

However, if we continue to exhale past the transition point, muscular effort is required to reduce the thoracic volume by force (Fig. 13.9C). Active or *forced* expirations are used during heavy exertion or when we laugh, cough, or sneeze. The muscles of the anterior abdominal wall contract, squeezing the abdominal contents, which increases intra-abdominal pressure and pushes up the diaphragm. The internal intercostals pull the rib cage downward and inward. Since eupneic breathing does not involve the muscles of forced expiration, the internal intercostals and abdominals are considered accessory muscles of respiration.

The accessory muscles of inspiration and expiration are used not only for heavy tasks but also to compensate for disease. *Chronic obstructive pulmonary disease* is an umbrella term for a small family of diseases, including asthma and emphysema, in which forced expiration is the norm.

Case Note

13.6. That Luther's lungs lost their elastic recoil was proven at autopsy—they did not recoil as the chest was opened. Throughout the course of his illness, what did Luther have to do in order to exhale properly?

Rate and Depth of Breathing Determine the Effectiveness of Ventilation

At rest (eupneic breathing) the average (155 lb) normal person breathes about 12 times per minute, each breath moving about 500 mL of air into and out of the conducting zone and respiratory zone. This 500 mL of air is referred to as the **tidal volume.** The total amount of air moved into and out of the lungs in 1 minute is called the **minute ventilation,** which at rest is about 500 mL × 12 breaths per minute, or 6,000 mL (or 6 L). However, only about 70% (350 mL of a 500-mL tidal volume) reaches the respiratory zone and participates in gas exchange. About 30% (150 mL) remains trapped in the conducting zone, which constitutes what is called **anatomical dead space** because no gas exchange occurs there. Alveoli that are not adequately ventilated or are not supplied with blood contribute to the dead space. For example, our patient Luther's dead space was increased because many of his small airways were collapsed or filled with inflammatory fluid from his pneumonia and could not be ventilated. Because of the effect of dead space, minute ventilation is only a rough gauge of breathing effectiveness. A much better measure is the **alveolar ventilation rate (AVR).** AVR is calculated like minute ventilation (rate × volume) but uses fresh air in the calculation. That is: *AVR = (tidal volume – dead space) × respiratory rate.* Thus, using the numbers above:

$$AVR = (500 \text{ mL} - 150 \text{ mL}) \times 12 = 4,200 \text{ mL } (4.2 \text{ L})$$

A moment's thought will reveal that the *depth* of breathing is the critical determinant of alveolar ventilation. Consider that if tidal volume is equal to dead space, no fresh air will reach the alveoli regardless of the number of breaths taken per minute. For this reason, shallow breathing is not very effective—much of the fresh air gets only as far as the dead space and little reaches the alveoli. Deep breathing is much more effective because dead space is constant and all of the extra fresh air enters the respiratory zone. Rapid, deep breathing is most effective of all.

One way to appreciate the importance of dead space is to breathe under water through a long tube (a garden hose, for example), which has the effect of markedly increasing dead space—say, from 150 to 3,000 mL. It is impossible to get fresh air into the respiratory zone in this way because no exhaled air escapes the tube and no fresh air reaches the lungs; every breath inhales nothing but the previous exhaled air. For a more detailed explanation, see the childhood experiment described in the Basic Form, Basic Function box, titled *Doctor McConnell's Breathing Lessons,* near the end of the chapter.

> ### Case Note
>
> **13.7. A lab technician recorded the following values for Luther's AVR and minute ventilation: 7 L and 3.5 L. However, he forgot to record which value is the AVR and which is the minute ventilation. Is his minute ventilation 7 L or 3.5 L?**

Physical Factors Affect Ventilation

Breathing is so natural and easy when we are healthy. However, it only takes a bad chest cold to appreciate that physical factors can interfere with easy filling and emptying. Three factors are most important:

- the ability of the diaphragm and the muscles of the torso to change the volume of the chest cavity
- the ability of the lungs to respond to musculoskeletal forces
- the ability of the airways to accommodate airflow

Muscle Paralysis Affects Ventilation

It seems almost too obvious to mention, but without muscular power, natural ventilation cannot occur. As discussed above, contraction of the diaphragm and muscles of the neck, chest wall, back, and abdomen all play a role in the mechanics of ventilation. Paralysis of any of these muscle groups can impair ventilation. For example, in the first half of the 20th century, entire hospital wings were occupied by polio patients who required mechanical ventilatory assistance because the poliovirus had destroyed somatic motor neurons ← (Chapter 8) and thus paralyzed the muscles of ventilation. Spinal cord trauma can have a similar effect.

Compliance and Elastance Affect Ventilation

Even if the muscles work normally, ventilation can be impaired by the lungs' inability to expand or to return to their normal size. Recall that in ← Chapter 11, we discussed certain qualities of blood vessels. We described *compliance* as the ability of blood vessels to stretch, and *elastance* is their ability to return to their original dimensions. These terms also apply to other flexible structures, including the lungs:

- **Compliance** is the ease with which the lungs can be distended to accommodate increased volume. Normally the lungs are very compliant and can be stretched easily. Of secondary importance to lung compliance is the compliance of the chest wall. In severe obesity, the weight of chest wall fat reduces chest wall compliance, making breathing more difficult. Or some arthritic conditions of the spine make it difficult for the ribs to move easily to expand chest volume.
- **Elastance** is the ability of the lungs to return to their original dimension at the transition point. Lung elastance is due mainly to the quantity of elastic fibers in the pulmonary interstitium. These fibers are stretched during inspiration and passively recoil to reduce lung volume during passive expiration.

Clinically, compliance is often inversely related to elastance—more compliant lungs are often less elastic (like our previous example of old stretched-out socks) and vice versa. Conversely, scarred lungs are more elastic but less compliant because the basic purpose of scar tissue is to knit tissue together with a firm, inelastic bond. Scarred lungs are difficult to stretch and quickly return to their original dimension.

Lung compliance reflects two distinct forces at work: (a) the compliance of pulmonary connective tissue and (b) the *surface tension* of the film of liquid that coats alveolar walls. The thin film of fluid that lines each alveolus is, in effect, a bubble of water that wants to contract into a spherical, airless water droplet. The force of surface tension tugs inward on the alveolar wall and resists ventilatory inflation. The net result is that alveoli ought to be extraordinarily difficult to inflate. But of course, they are not. *Surfactant* to the rescue!

Recall from our earlier discussion that about 10% of alveolar cells are type II pneumocytes, which secrete surfactant. Surfactant molecules intersperse themselves between water molecules, decreasing surface tension and counteracting the tendency of water to contract into a tiny droplet. This effect is easy to demonstrate. Everyone knows that soap is necessary to create bubbles. Why? Because soap is a surfactant. Without surfactant, water molecules are so strongly attracted to one another—their surface tension is so high—that trying to blow bubbles would create only a spray of water droplets. But add soap, and voila! A light puff of air is enough to cause the water molecules to expand into bubbles! Surfactant similarly increases alveolar compliance by making the alveoli easier to inflate.

The effect of decreased surfactant can be seen in some prematurely born infants suffering from *respiratory distress syndrome of the newborn*. In these infants the lungs have not matured enough to produce a normal amount of surfactant, and ventilation is difficult because alveolar surface tension is very high. They struggle to inhale because they do not have enough surfactant to "make bubbles." The struggle is so severe that their ribs show through the skin with each gasp of inhalation. Without mechanical ventilation, they may die of suffocation because their inspiratory muscles fatigue to the point that they cannot inhale enough air to sustain life. Women in premature labor are frequently given corticosteroids, which accelerate fetal lung development and surfactant production. Several treatment strategies can be effective after birth, including surfactant inhalation.

Case Note

13.8. Luther's lungs were abnormally compliant. Did this make it easier or harder for Luther to inhale? Explain.

Airway Resistance Affects Ventilation

Like blood flowing through vessels, air flows into and out of the lungs in response to changing pressure gradients through the tracheobronchial tree. Normally, air flows freely because bronchial tubes are extraordinarily large and air is much, much less viscous than blood. Resistance to airflow is primarily determined by the diameter of the tube through which it flows. Thus, if the airways narrow, airflow is reduced. In certain disease states, resistance to airflow can be severe, even fatal. In **asthma,** for instance, bronchiolar muscle contraction (*bronchospasm*) and excessive mucus production narrow and obstruct bronchioles, increasing airway resistance. An asthma "attack" can be brought on by inhaled irritants, allergens, cold air, an underlying viral infection, anxiety, or even exercise. Severe, untreated acute bronchospasm can be fatal.

The main adjustment point for homeostatic control of airway resistance lies in the bronchioles, which have no cartilage to stiffen their walls and are endowed with abundant smooth muscle fibers. Parasympathetic signals from the vagus nerve (vagal tone) maintain bronchioles in a slightly constricted state. The sympathetic nervous system, activated by stressors such as exercise, dilates bronchioles to improve ventilation. Recall that the sympathetic nervous system uses norepinephrine (from nerves) and epinephrine (from the adrenal medulla) to mediate its effects. It stands to reason, then, that asthma sufferers use inhaled epinephrine (or epinephrine agonists) to treat asthma attacks—it relaxes bronchiolar smooth muscle, which dilates bronchioles and allows increased airflow.

Case Note

13.9. Luther's bronchioles had the nasty habit of compressing when he breathed, because smoking destroyed much of the supporting structure of the pulmonary interstitium that held them open. Do you think Luther's resistance to airflow was increased or decreased?

Pulmonary Ventilation Is Quantified by Spirometry

The diagnosis of ventilatory disease relies on precise measurements of changes in lung volume during breathing. A simple instrument called a **spirometer** can quantify the *volume* and *rate of airflow* into and out of the lungs. The technique is simple: having been given a mouthpiece and tube connected to a measuring device, the patient is asked to perform different breathing actions. Figure 13.10A shows an example of the recorded result, known as a **spirogram.** Note that some measures are called *volumes* and others are called *capacities*. A capacity is a combination of two or more volumes.

We begin our spirometry tracing at the transition point between breaths. At that point the volume of air remaining in your lungs is called the **functional residual capacity.** The functional residual capacity of the average young adult is about 2,400 mL. This is the volume that fills the airways of the conducting zones and partially inflates the alveoli.

First, from the transition point, breathe quietly for four breaths. We noted above that the small volume of air moving in and out of your lungs in quiet breathing is the **tidal volume,** so named for its resemblance to the ebb and flow of the ocean's tide.

Next, from rest at the transition point, exhale as forcefully as you can. The volume of air exhaled from the

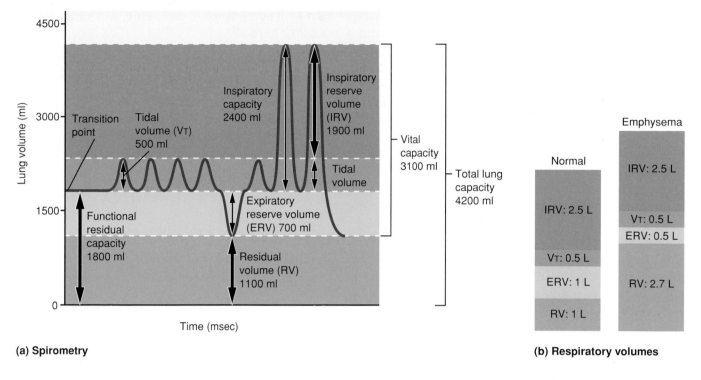

(a) Spirometry

(b) Respiratory volumes

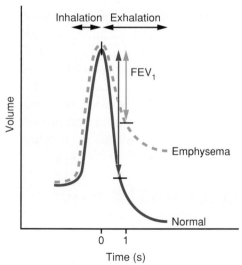

(c) Forced expiratory volume

Figure 13.10. Lung volumes and capacities. A. Lung volumes and capacities of an average adult female, as measured by spirometry. Upward deflections are inhalations, downward deflections are exhalations. Volumes and capacities indicated by thin arrows are directly measured in this tracing; thick arrows must be calculated. **B.** Respiratory volumes in a normal and an emphysemic person. See part A for the meaning of the abbreviations. **C.** The forced expiratory volume in 1 second (the FEV1) is used to diagnose many respiratory illnesses. *Which lung capacities include the residual volume?*

transition point is the **expiratory reserve volume,** which is a reserve because it is not called upon in quiet breathing. The accessory expiratory muscles, such as the abdominals, are called into play when you expel this reserve volume. However, try as you might, there is some air in your lungs that cannot be expelled even at the point of maximal exhalation. This volume, the **residual volume,** must be measured by specialized techniques. The residual volume and the expiratory reserve volume added together make up the functional residual capacity.

Returning to rest at the transition point, inhale as deeply as possible. The volume of air inhaled from the transition point is the **inspiratory capacity.** The *extra* amount of air inhaled above the tidal volume is called the **inspiratory reserve volume.**

The total amount of air that can be moved in one breath with maximum inhalation and maximum exhalation is the **vital capacity.** This capacity includes the *tidal volume* plus the *inspiratory reserve volume* and the *expiratory reserve volume.* You can also determine the vital capacity directly, by inhaling maximally and then exhaling maximally. But we still have not determined the **total lung capacity**—that is, the entire volume of air that the lungs can hold. To do that, we must add the *residual volume* to the vital capacity. With maximum expansion, average adult lungs can hold about 6 L of air for males and just under 5 L for females. That is, the total lung capacity in healthy adults is about 5 to 6 L.

The total lung capacity, vital capacity, and residual volume are fixed in healthy people because they are determined by individual anatomy. They typically change only in response to growth or disease. Emphysema, for instance, makes it more difficult for patients to exhale because small airways, lacking support, collapse. The resulting change in residual volume is shown in the right side of Figure 13.10B. Emphysema and other diseases also modify the lung volumes and capacities, as does activity level. Exercising increases the tidal volume severalfold; during maximal effort, the tidal volume reaches vital capacity. Since vital capacity remains the same, inspiratory and expiratory reserve volumes must decrease as tidal volume increases.

A patient's airflow *rate* is also an important diagnostic tool (Fig. 13.10C). The patient is asked to take the deepest possible breath and exhale as rapidly as possible until no additional air can be exhaled. The amount of air exhaled in the first second is the **forced expiratory volume in 1 second (FEV$_1$).** The FEV$_1$ is normally about 80% of the vital capacity, but it can be abnormally low in obstructive airway disorders. The green tracing in Figure 13.10C, for instance, shows what Luther's flow rate would have been.

Case Notes

13.10. Luther's tidal volume was normal but his inspiratory capacity was reduced. How could this be?

13.11. Luther's FEV$_1$ was reduced. Why?

13.14   If you want to double the gas pressure, how should you change its volume?

13.15   If the intrapulmonary pressure is 750 mm Hg and the atmospheric pressure is 752 mm Hg, will air enter or leave the lungs?

13.16   Which, if any, of the following muscles contracts during expiration of the tidal volume—scalenes, abdominals, diaphragm?

13.17   If compliance is decreased, will inhalation be easier or harder?

13.18   Which zone is equivalent to the anatomical dead space—the conducting zone or the respiratory zone?

13.19   The maximum volume of air that can be *inhaled* from the transition point is the inspiratory capacity. True or false?

13.20   The functional residual capacity is the only *capacity* that cannot be measured by spirometry. Explain why.

13.21   Which measure determines the airflow rate—the total lung capacity or the FEV$_1$?

Gas Exchange and Transport

Recall from the opening of this chapter that respiration is a multistep process in which pulmonary ventilation is but the first step. Now we discuss the next three steps—*external* gas exchange (between blood and alveolar air), *internal* gas exchange (between blood and body cells), and gas transport by blood between lungs and tissues. How cells use up oxygen and produce carbon dioxide (cellular respiration) is discussed in ◀ Chapter 15.

Gas Exchange Involves Partial Pressure Gradients

Air is a mixture of gases—nitrogen, oxygen, water vapor, and carbon dioxide—each of which contributes to total atmospheric pressure in direct proportion to their relative concentration in air. Pulmonary ventilation involves the bulk movement of air: the various air molecules move together in and out of the lungs, down the same pressure gradient. When it comes to internal and external gas exchange, however, each gas acts independently, moving down its *own* pressure gradient. That is, oxygen movement is determined by the oxygen pressure gradient between lungs and blood or between blood and tissues; it is unaffected by movements and gradients for carbon dioxide.

The pressure for a specific gas is called its **partial pressure.** Each partial pressure is determined by multiplying the atmospheric pressure by the percentage of the specific gas in the atmospheric air. For example, the atmospheric pressure at sea level is 760 mm Hg. Air is 20.9% oxygen, so the partial pressure of oxygen (the Po_2) is 760 mm Hg $\times$ 0.209 = 159 mm Hg.

To understand external gas exchange, it is not enough to know the partial pressure of oxygen (or a different gas) in the alveolar air. We also need to know the partial pressure of the same gas in the alveolar capillary blood. The partial pressure of a dissolved gas depends on two elements: its *concentration* and its *solubility* (ability to dissolve in blood). Obviously, the greater the concentration of a gas, the greater its partial pressure. Solubility, however, is another matter. Think of it this way: the more soluble a gas is in fluid, the less it wants to "escape" and, hence, the lower the partial pressure it creates at a given concentration in the fluid. For instance, carbon dioxide is much more soluble in water (or blood) than is oxygen. Therefore, it takes a much greater concentration of carbon dioxide in blood to create the same partial pressure exerted by a lower concentration of oxygen (Fig. 13.11). Thus, in a blood sample in which the partial pressure of both gases is 100 mm Hg, the concentration of dissolved oxygen will be much lower than that of the carbon dioxide because of their differing solubilities.

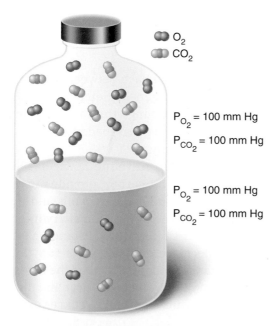

Figure 13.11. Gas partial pressures in liquids. It takes fewer gas molecules in liquid than in air to establish the same partial gas pressure gradient. Highly water-soluble gases (like carbon dioxide) require more gas molecules to establish a given partial pressure than do less water-soluble gases (like oxygen). *There are more oxygen molecules in the air than in the liquid. Will oxygen diffuse into the liquid?*

O$_2$
CO$_2$

P_{O_2} = 100 mm Hg

P_{CO_2} = 100 mm Hg

P_{O_2} = 100 mm Hg

P_{CO_2} = 100 mm Hg

Case Note

13.12. During his hospitalization, Luther was breathing air enriched with oxygen. The atmospheric pressure was 760 mm Hg and the oxygen concentration was 50%. What was the oxygen partial pressure in the inspired air?

External Gas Exchange Loads Oxygen and Unloads Carbon Dioxide

External gas exchange—the transfer of oxygen from alveoli to blood and of carbon dioxide from blood to alveoli—occurs entirely by diffusion. Gas diffusion across the pulmonary membrane depends on two main factors:

- The partial pressure gradients between alveolar air and blood
- The health of lung tissue

Larger Pressure Gradients Increase Gas Exchange

The most important determinant of external gas exchange is the partial pressure gradients between the alveolus and blood. The partial pressure of oxygen (Po_2) is higher in the alveolus (104 mm Hg) than in oxygen-depleted blood arriving at the lungs (Po_2 = 40 mm Hg), so oxygen diffuses down its partial pressure gradient from the alveolus to the blood (Fig. 13.12). The situation is reversed for carbon dioxide—it diffuses from carbon dioxide–rich blood (Pco_2 = 45 mm Hg) into the alveoli (Pco_2 = 40 mm Hg). In both instances equilibrium is quickly established between blood and air so that the Po_2 and Pco_2 in blood leaving the lungs is the same as in alveolar air.

Alveolar air contains many times more carbon dioxide and substantially less oxygen than atmospheric air, because alveoli contain a mixture of fresh air and "old" air that has already participated in gas exchange. However, as the alveolar ventilation rate increases and fresh air washes out more of the old, alveolar gas concentrations move toward those of atmospheric air; the oxygen concentration and partial pressure increase, and the carbon dioxide concentration and partial pressure decrease. This, of course, increases the pressure gradients, which causes more rapid diffusion of oxygen and carbon dioxide across the pulmonary membrane.

Conversely, anything that impairs pulmonary ventilation will reduce partial pressure gradients as well as gas exchange (Fig. 13.12, factor a). For instance, atmospheric pressure decreases at high altitude. At 10,000 ft above

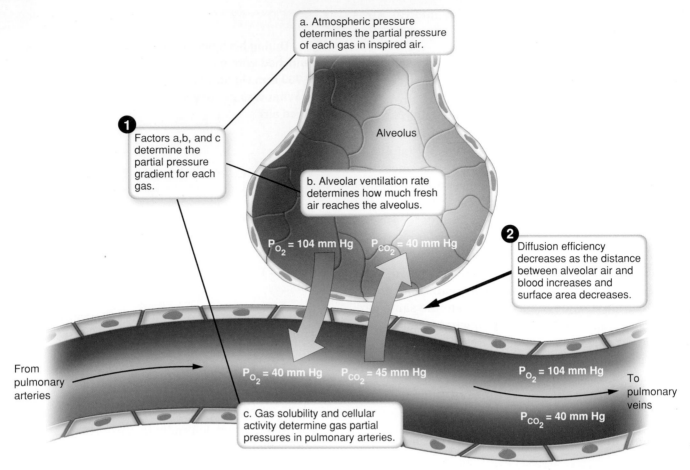

① Factors a,b, and c determine the partial pressure gradient for each gas.

a. Atmospheric pressure determines the partial pressure of each gas in inspired air.

Alveolus

b. Alveolar ventilation rate determines how much fresh air reaches the alveolus.

P_{O_2} = 104 mm Hg P_{CO_2} = 40 mm Hg

② Diffusion efficiency decreases as the distance between alveolar air and blood increases and surface area decreases.

From pulmonary arteries

P_{O_2} = 40 mm Hg P_{CO_2} = 45 mm Hg

P_{O_2} = 104 mm Hg

To pulmonary veins

P_{CO_2} = 40 mm Hg

c. Gas solubility and cellular activity determine gas partial pressures in pulmonary arteries.

Figure 13.12. External gas exchange. Note that the partial pressures in pulmonary veins and alveolar air are the same. *How will oxygen diffusion change if the alveolar oxygen level decreases (say, due to breath-holding)?*

sea level, the *percentage* of oxygen in air does not change (it remains 20.9%), but the atmospheric pressure falls from 760 mm Hg to 523 mm Hg. As a result, the P_{O_2} falls to 109 mm Hg (523 × 0.209). This, in turn, lowers the difference (the gradient) between the partial pressure of oxygen in the lungs and that in the blood, which makes it more difficult for blood to load oxygen. No wonder we get short of breath at high altitude.

Emphysema impairs gas exchange by altering how much fresh air reaches the alveoli (Fig. 13.12, factor b). For instance, recall from above that patients with emphysema have difficulty exhaling "old" (high carbon dioxide, low oxygen) air, which leaves less room for fresh (high oxygen, low carbon dioxide) air. The resulting decrease in alveolar oxygen and increase in alveolar carbon dioxide reduces gas exchange. As a result patients with emphysema have high blood carbon dioxide and low oxygen.

Altering partial gas pressures in blood can also increase the pressure gradient (Fig. 13.12, factor c). For instance, during intense exercise, body cells use up more oxygen and produce more carbon dioxide, increasing

the pressure gradient for each between blood and alveolar air. When this oxygen-depleted, carbon dioxide–rich blood reaches the lungs the rate of external gas exchange increases—oxygen is loaded and carbon dioxide unloaded more easily.

Finally, recall that in traveling between blood and alveolar air, oxygen and carbon dioxide must diffuse through the alveolar and capillary walls and the intervening connective tissue of the pulmonary interstitium. Any disease that increases the distance gases must diffuse or will decrease diffusion efficiency (Fig. 13.12, step 2). For instance, pneumonia coats the alveolar wall with inflammatory fluid, and toxins scar and thicken the pulmonary interstitium—and each reduces external gas exchange.

Efficient diffusion also depends on a large surface area. Emphysema, for instance, is frequently associated with alveolar destruction as well as impaired exhalation. The loss of alveoli reduces the overall surface area of the respiratory membrane available for diffusion: less membrane, less gas diffusion, less external gas exchange.

Oxygen Transport and Internal Gas Exchange

Oxygen diffuses from oxygen-rich alveolar air (P_{O_2} = 104 mm Hg) into oxygen-depleted pulmonary arterial blood (P_{O_2} = 40 mm Hg) (Fig. 13.13A). However, the watery part of blood (plasma) is a rather inhospitable medium for oxygen. Oxygen is not very soluble in water, so only about 1% of blood oxygen can dissolve in plasma. Fortunately, red blood cells and their load of hemoglobin are available to capture the remaining 99%.

Hemoglobin with oxygen molecules attached, known as **oxyhemoglobin,** is scarlet-red, the color of arterial blood. After losing its oxygen, hemoglobin is called *reduced hemoglobin* (or *deoxyhemoglobin*); it is a darker shade of red, the color of venous blood.

Each hemoglobin molecule contains four atoms of iron, each of which can combine with one molecule of oxygen. Each red blood cell contains about 250 million hemoglobin molecules; therefore, each can carry up to about a *billion* oxygen molecules. However, oxygen rarely fills all of the hemoglobin capacity available. The degree (percent) to which oxygen occupies available hemoglobin binding sites is called the **hemoglobin saturation**.

The degree of hemoglobin saturation, and thus the amount of oxyhemoglobin formed, depends on the partial pressure of oxygen. Because P_{O_2} is high in the lungs, oxygen readily binds hemoglobin in blood passing through the lungs (Fig. 13.13A). In tissues, as oxygen diffuses from capillaries into body cells, the partial pressure of oxygen in plasma falls, which stimulates

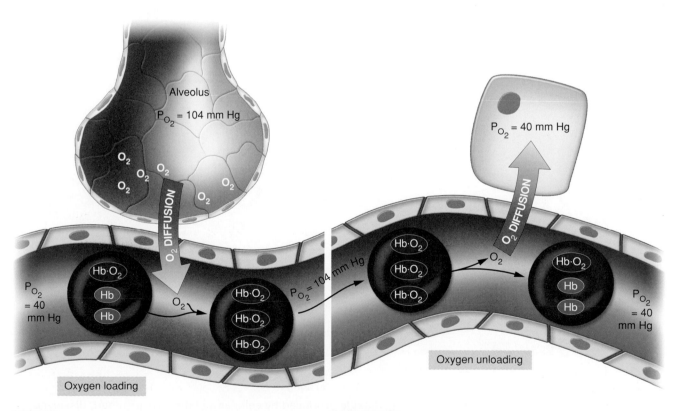

(a) Lungs   **(b) Body cells**

Figure 13.13. Oxygen transport and external gas exchange. Most oxygen is carried in blood bound to hemoglobin, but a small amount of oxygen (not shown) is carried dissolved in plasma. Note that each red blood cell contains many hemoglobin molecules; only three are shown. **A.** Hemoglobin binds oxygen in the lung capillaries, where the oxygen partial pressure is high. **B.** Hemoglobin releases oxygen at the tissue capillaries, where the oxygen partial pressure is low. *Where would an oxygen molecule have the best chance of getting picked up by hemoglobin – in blood arriving at the lungs or in blood arriving at the tissues?*

hemoglobin to release its oxygen load (Fig. 13.13B). These newly released molecules then diffuse into a body cell for use in metabolic reactions.

Hemoglobin saturation varies from a high of about 98% in pulmonary capillaries and falls to a low of about 45% in peripheral tissue capillaries. Figure 13.13 illustrates that not all hemoglobin-bound oxygen is unloaded in the tissues. Certain physical factors increase oxygen *unloading*—increased P_{CO_2}, increased hydrogen ions (increased acidity), and increased temperature. This should be no surprise: each of these variables is associated with increased tissue workload, increased metabolism, and increased need for oxygen.

As you would expect, the reverse is true, too. That is, oxygen *loading* is favored by decreased P_{O_2}, decreased hydrogen (less acidity), and lower temperature. And where might that be? In the lungs.

Carbon Dioxide Transport and Internal Gas Exchange

Carbon dioxide moves in the opposite direction to oxygen—carbon dioxide enters blood in tissues and leaves blood in the lungs. Within tissues, carbon dioxide passes into plasma by flowing down the carbon dioxide partial pressure gradient: tissue P_{CO_2} is about 45 mm Hg, and the P_{CO_2} of blood entering systemic capillaries is about 40 mm Hg. Carbon dioxide, like oxygen, can be carried dissolved in plasma and attached to hemoglobin. However, the proportions are very different (Fig. 13.14A,

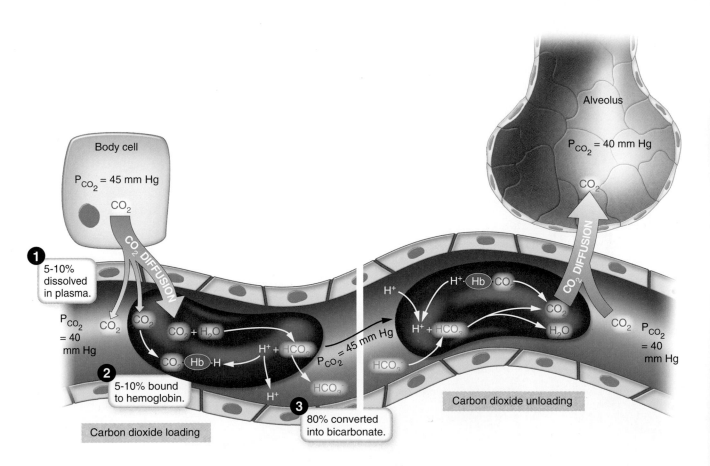

(a) Tissues

(b) Lungs

Figure 13.14. Carbon dioxide transport. A. Carbon dioxide produced by cellular metabolism is, in part, dissolved in plasma or attached to hemoglobin for transport. Most carbon dioxide is converted into bicarbonate ions in a reaction that produces hydrogen ions. Some of the hydrogen ions bind hemoglobin, but others diffuse into the blood and increase blood acidity. **B.** The carbon dioxide partial pressure gradient between the pulmonary blood and the alveoli drives dissolved carbon dioxide and hemoglobin-bound carbon dioxide out of the blood. Blood bicarbonate combines with hydrogen ions from the blood and/or from hemoglobin to produce additional carbon dioxide that diffuses into alveoli. *Which change would have a greater effect on the ability of blood to carry carbon dioxide: blocking bicarbonate production or destroying hemoglobin?*

step 1). About 5% to 10% of blood carbon dioxide is dissolved in blood, a figure much higher than the scant 1% of oxygen carried this way. Because carbon dioxide is much more soluble in water than oxygen. Another 5% to 10% of carbon dioxide is carried bound to hemoglobin (step 2).

The remaining 80% to 90% of carbon dioxide is carried not as a gas but as a highly soluble molecule called bicarbonate (step 3). Recall from ◀━ Chapter 2 that the enzyme *carbonic anhydrase* catalyzes the formation of bicarbonate from carbon dioxide and water:

$$CO_2 + H_2O \Leftrightarrow H_2CO_3 \Leftrightarrow H^+ + HCO_3^-$$

Note that the arrows point in both directions, indicating that the reaction can proceed in either direction: if carbon dioxide is high or hydrogen low, the reaction moves to the right. As a result, carbon dioxide levels drop and hydrogen levels rise. Conversely, if carbon dioxide is low or hydrogen high, the reaction proceeds to the left, producing carbon dioxide and eliminating hydrogen.

Since carbon dioxide partial pressure is high in tissue capillaries, carbon dioxide is converted into bicarbonate and hydrogen ions (Fig. 13.14A). Hemoglobin acts as a buffer ◀━ (Chapter 2) that dampens swings of blood pH and keeps it in a safe range. It does so by binding hydrogen ions, keeping them sequestered in red blood cells and limiting the number of hydrogen ions that enter plasma to alter pH. Nevertheless, because hemoglobin can bind only so many hydrogen ions, *large changes in blood carbon dioxide will change blood pH.* The bottom line is that carbon dioxide retention (from breath-holding or lung disease) increases blood acidity and likewise hyperventilation (excessive breathing) lowers blood carbon dioxide and makes blood more alkaline.

Once blood reaches the lungs, carbon dioxide again flows down a partial pressure gradient from blood into the alveoli: blood P_{CO_2} is about 45 mm Hg, and alveolar P_{CO_2} is about 40 mm Hg (Fig. 13.14B). The above chemical reactions reverse—plasma carbon dioxide diffuses into the alveoli; carbon dioxide detaches from hemoglobin and diffuses out as well. Bicarbonate in plasma reenters red blood cells and combines with hydrogen, resulting in the production of carbon dioxide and water. Carbon dioxide then diffuses down its pressure gradient out of the red cell into plasma and then into the alveoli.

Case Note

13.14. Why do you think Luther's blood was acidic?

13.22 If the atmospheric pressure is 700 mm Hg and the oxygen percentage is 20%, what is the partial pressure of oxygen?

13.23 A solution contains 100 mm Hg of both oxygen and nitrogen. Given that nitrogen is much less soluble than oxygen, will the solution contain more oxygen molecules or more nitrogen molecules?

13.24 If the partial pressure of oxygen in pulmonary capillaries decreased, would the oxygen pressure gradient between the alveoli and pulmonary capillary blood increase or decrease?

13.25 Which parameter would you increase in order to enhance gas diffusion: the alveolar wall thickness or the overall surface area available for gas diffusion?

13.26 Which change would most enhance the amount of oxygen carried in the blood: increasing the amount of oxygen dissolved in the plasma or increasing the hemoglobin concentration? Explain.

13.27 If the hydrogen ion concentration in blood drops, what will happen to the production of carbon dioxide from bicarbonate? Will it increase or decrease?

The Control of Respiration

"As natural as breathing," is a phrase most of us have heard, which is a testament to the effortlessness of normal breathing. It is effortless because it usually occurs automatically, responding to the needs of the moment without our awareness. Yet we can choose to make ourselves aware and assume a limited degree of conscious control. We can hold our breath in order to be in water without drowning or to temporarily avoid noxious fumes—or we can breathe deeply and rapidly whenever we so desire: to blow out candles on a birthday cake, for example.

Conscious control requires cerebral cortex command of voluntary muscles, which stands in contrast to the lack of conscious control in the regulation of heart rate and other functions over seen by the autonomic nervous system. But the cerebral cortex can exert only limited influence; the basic rhythm and rate of respiration lies deeper in the brain in more primitive regions over which we have no conscious control.

Respiratory Rhythm Is Controlled by the Brainstem

During quiet, normal respiration, we breathe at about 12 breaths per minute. Each cycle is about 5 seconds: 2 seconds of inspiration and 3 seconds of expiration.

The **respiratory center** is a collection of neurons in the brainstem that initiates the cycle and modulates it in response to chemical or physical factors (Fig. 13.15). Neurons in the respiratory center are clustered into the *ventral respiratory group*, the *dorsal respiratory group*, and the *pontine respiratory group*. While the dorsal and pontine respiratory groups help fine-tune the respiratory rhythm, it is the **ventral respiratory group (VRG)** that sets the basic rhythm. Despite the obvious importance of this, the mechanisms involved remain elusive. Current thinking is that, in adults, neuronal populations within the VRG stimulate each other in a positive feedback manner, which peaks with a burst of firing that activates motor nerves innervating the diaphragm and external intercostals. This burst of firing invokes negative feedback that shuts down the system, reducing motor nerve input to the inspiratory muscles. Muscles relax and expiration occurs as air is forced out by the elastic recoil of the lungs. If active expiration is required, additional neuron populations in the VRG recruit the expiratory muscles.

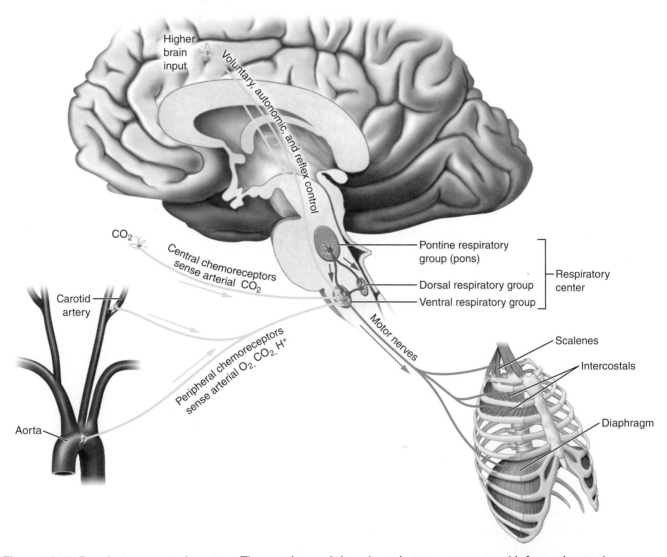

Figure 13.15. Respiratory control centers. The pontine and dorsal respiratory groups send information to the ventral respiratory group (brown neurons), which sets the basic respiratory rhythm by altering motor nerve activity supplying inspiratory muscles (blue neurons). The respiratory center, in turn, receives inputs from peripheral and central chemoreceptors and higher brain regions (yellow neurons). *Where are the peripheral chemoreceptors located?*

The basic respiratory pattern can be modified by conscious thought or by unconscious mechanisms. For instance, sneezing, coughing, laughing, singing, and speaking all require coordinated actions of the respiratory muscles. Moreover, respiration ceases when we swallow liquids and the respiratory cycle usually resumes with an exhalation, a bit of precise coordination that prevents choking. And there are many other factors, chief among them the arterial concentration of carbon dioxide, that also modulate breathing by altering the activity of the respiratory center.

Several Factors Influence the Rate and Depth of Respiration

Multiple factors influence breathing rate and depth according to body demands. Apart from voluntary cortical signals, the most important are chemical—arterial blood P_{CO_2}, pH, and, to a lesser extent, P_{O_2}. Respiratory center neurons fire more rapidly when the body's metabolic needs are high. The increased number of nerve action potentials reaching respiratory muscles recruit more muscle fibers to contract, increasing the development of force.

Elevated Blood Carbon Dioxide Stimulates Breathing

The central chemoreceptors in the medulla and, to a lesser extent, the peripheral chemoreceptors in the aortic arch and carotid arteries (in the neck) regulate unconscious ventilation (see Fig. 13.15). It is commonly thought that low arterial blood P_{O_2} is the primary stimulus for breathing, but this is not the case. Although low oxygen plays a small role, *the major stimulus to breathe is high arterial carbon dioxide*. Peripheral chemoreceptors respond directly to increased carbon dioxide, but central chemoreceptors respond only to the change in pH resulting from high blood carbon dioxide.

Recall that arterial P_{CO_2} directly and immediately influences pH via its reaction with water to form carbonic acid (H_2CO_3), which dissociates into hydrogen (H^+) and bicarbonate (HCO_3^-)

$$CO_2 + H_2O \Leftrightarrow H_2CO_3 \Leftrightarrow H^+ + HCO_3^-$$

Carbon dioxide (but not hydrogen ions) diffuses with great ease through fluid and cell membranes, so the cerebrospinal fluid (CSF) P_{CO_2} exactly matches arterial P_{CO_2} (Fig. 13.16, step 1). If arterial (and CSF) carbon

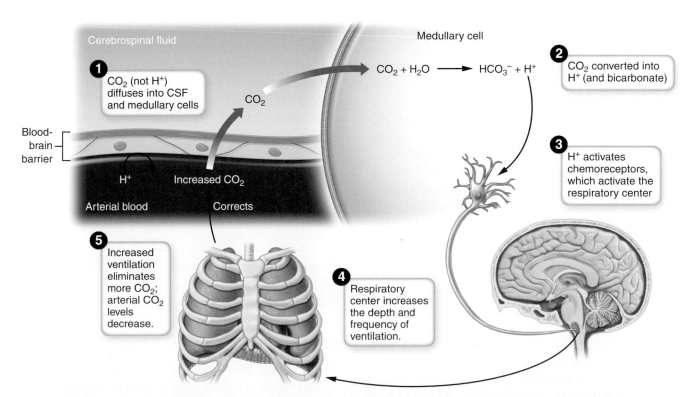

Figure 13.16. The central chemoreceptors. These stimulate respiration when carbon dioxide in arterial blood is high. Note that the conversion of carbon dioxide into bicarbonate and hydrogen ions can occur within the CSF as well as in medullary cells. *Would the central chemoreceptors respond to a change in peripheral arterial pH, CSF pH, or both?*

dioxide rises, the chemical reaction above is driven to the right and more hydrogen ions are produced (step 2). Medullary chemoreceptors sense the increase of hydrogen ions (step 3) and stimulate the respiratory center to drive more rapid and deep breathing (step 4). The increased ventilation reduces blood carbon dioxide in a classic homeostatic negative feedback loop (step 5). The reverse is also true: if arterial blood (and thus CSF) P_{CO_2} falls, the chemical reaction is driven to the left, hydrogen and bicarbonate are combined into carbon dioxide and water, and fewer hydrogen ions are available to stimulate the respiratory center. Respiration slows, and carbon dioxide accumulates.

As you can see, the central chemoreceptors are the primary regulators of the respiratory drive. Indeed, they account for up to 80% of inputs into the respiratory center.

Case Note

13.15. Luther's blood carbon dioxide concentration was abnormally high. Describe the mechanism by which increased carbon dioxide altered his breathing rate.

Very Low Blood Oxygen Also Stimulates Breathing

Surprisingly, arterial P_{O_2} can drop significantly without affecting the respiratory rate. Only major reductions in blood P_{O_2}, such as those observed in serious lung disease or high altitude, activate peripheral chemoreceptors and stimulate breathing. The lack of responsiveness to low arterial P_{O_2}, is rarely an issue, because the blood partial pressures of oxygen and carbon dioxide generally move in a see-saw fashion: one is up and the other is down. For example, reduced ventilation results in **hypoxia** (low arterial P_{O_2}) as well as **hypercapnia** (high arterial P_{CO_2}). Hypercapnia increases ventilation, which restores both carbon dioxide and oxygen arterial partial pressures to normal.

However, a technique used by some competitive swimmers—hyperventilation (breathing at a rate and depth greater than physiological conditions demand) followed by breath-holding—can dissociate P_{CO_2} and P_{O_2} levels. As discussed in the nearby Basic Form, Basic Function box titled *Doctor McConnell's Breathing Lessons,* because blood oxygen content is largely determined by the hemoglobin concentration, hyperventilation decreases arterial P_{CO_2} but does not significantly increase arterial P_{O_2}. As a result of the low initial arterial P_{CO_2}, swimmers can hold their breath much longer before P_{CO_2} rises high enough to stimulate breathing. But before this threshold is reached, arterial P_{O_2} (which was not significantly increased) can drop to dangerously low levels, and unconsciousness and drowning may ensue.

On the other hand, some chronic respiratory diseases—emphysema, for example—are associated with long-term carbon dioxide retention. In such circumstances peripheral P_{CO_2} chemoreceptors adapt to the hypercapnia, so that it no longer stimulates respiratory drive. In these circumstances P_{O_2} becomes the principal driver of respiration.

Other Factors Also Regulate Breathing

Arterial blood pH stimulates peripheral chemoreceptors independently of carbon dioxide stimulation. Although it is greatly influenced by carbon dioxide, blood pH can be affected by other variables, such as our lemon juice case in Chapter 2, intense exercise or the metabolic abnormalities associated with untreated, poorly regulated diabetes mellitus ➡ (Chapter 15). Regardless of the cause, low (acidic) arterial blood pH stimulates respiratory drive, which eliminates carbon dioxide and raises pH by driving the equation to left.

Factors other than pH, P_{CO_2}, and P_{O_2} that also modulate breathing include the following:

- *The limbic system.* Fear and other emotions modulated by the limbic system can stimulate rapid deep breathing.
- *Proprioceptive receptors.* As soon as exercise begins, proprioceptive receptors in our limbs sense the change and stimulate breathing before P_{CO_2} and other chemical signals exert their effects.
- *Body temperature.* Fever stimulates respiration; hypothermia suppresses it.
- *Pain.* Prolonged somatic pain stimulates respiration; visceral pain suppresses it.
- *Irritants.* Inhaled irritants stimulate coughing.
- *The inflation reflex.* Stretch receptors in the walls of bronchi sense inflation of the lungs and signal the respiratory center to stop inhalation. The dorsal respiratory group has been implicated in this reflex. Conversely, pneumothorax with lung collapse stimulates an attempt to breathe deeply.

13.28 Which brain region is primarily responsible for the basic respiratory rhythm—the brainstem, the cerebrum, or the cerebellum?

13.29 What does the pontine respiratory group do?

13.30 Which chemical *directly* activates chemoreceptors of the respiratory center—hydrogen or carbon dioxide?

13.31 True or false: When you hold your breath, your brain senses that blood oxygen levels are dropping and stimulates the urge to breathe.

BASIC FORM, BASIC FUNCTION

Doctor McConnell's Breathing Lessons

One of life's first lessons is that you can't breathe water. Everyone knows the truth of this lesson, but no one knows how or when he or she learned it. The consequences of getting even a small amount of water past the epiglottis and into the larynx is so alarming that the lesson it teaches is indelible. Of course, I don't recall when I learned that you can drown in water, but I do recall exactly how I learned about respiratory dead space, though at the time I didn't know it had a name. Even so, I realized its significance.

In our little town in the 1940s, only one home had a swimming pool. By modern standards it was very primitive: the owner filled it with a garden hose, chucked in enough chlorine to last for a while, and when the water turned green it was pumped out and the pool refilled. While swimming there one day I took interest in the garden hose and concluded that if I disconnected it from the faucet and drained it of water, I could use it as an underwater air supply. At first I was elated: I could draw a breath under water. But very quickly I learned that it didn't provide an advantage, because I grew short of breath just as quickly as before. After thinking on the matter for a while, it dawned on me: I wasn't getting fresh air because the hose was far too long, and I was reinhaling the air I'd just exhaled because it was trapped in the hose. On another visit I brought a much shorter length of garden hose cut from

an old hose a home. It worked, but not nearly as well as getting a breath of fresh air.

Later, as a teenager, I had another breathing lesson at the municipal pool. Macho water contests were a staple of our almost daily visits. Among them was a duel to see who could swim the most pool lengths under water. All of us knew that rapid, deep breathing enabled us to hold our breaths longer. We thought, erroneously as we now know, that we were "storing" oxygen. Little did we know that the key was what deep breathing did to our carbon dioxide, not our oxygen—it was not low oxygen that made us want to take a breath, but high carbon dioxide. If carbon dioxide is blown off to very low levels, so low that a person can go unusually long without breathing, it is possible for blood oxygen to fall so low that unconsciousness occurs before carbon dioxide rises high enough to make taking a breath irresistible. Luckily, before I drowned in one of these contests, I learned how dangerous it could be. During one episode I swam under water for an unusually long distance. As I surfaced at the pool's edge after an extraordinary effort, I felt faint and held on the pool's edge fearing that I would pass out. It was clear to me that I'd "run out of oxygen." It scared me enough not to try so hard again.

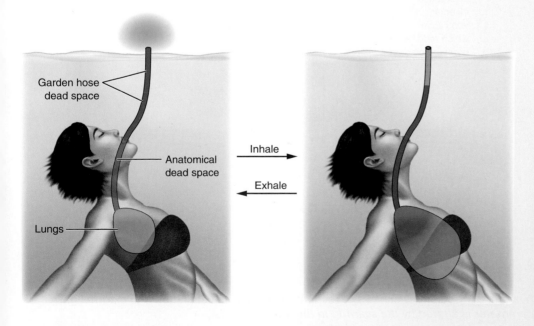

Swimming lessons. Increasing the anatomical dead space (by breathing through a garden hose) dramatically affects alveolar ventilation. Fresh air from the environment (orange) never reaches the lungs and exhaled air from the lungs (red) never reaches the atmosphere.

Case Discussion
Emphysema: The Case of Luther M.

Luther M. suffered from **emphysema**, a form of chronic obstructive pulmonary disease (COPD) most commonly caused by cigarette smoking. Other causes, such as inhaling polluted air or workplace pollutants, are much rarer.

Cigarette smoking irritates the bronchi and alveoli, inciting an inflammatory reaction (Fig. 13.17, middle column). White blood cells arrive but find no pathogens to fight. Nevertheless, they release their enzymes, which digest and destroy alveolar walls, much like popping the bubbles in bubble wrap used to protect goods in the mail. What's more, cigarette smoke incites scarring in bronchiolar walls and in the delicate interstitial tissue between alveoli. The reduced surface area resulting from

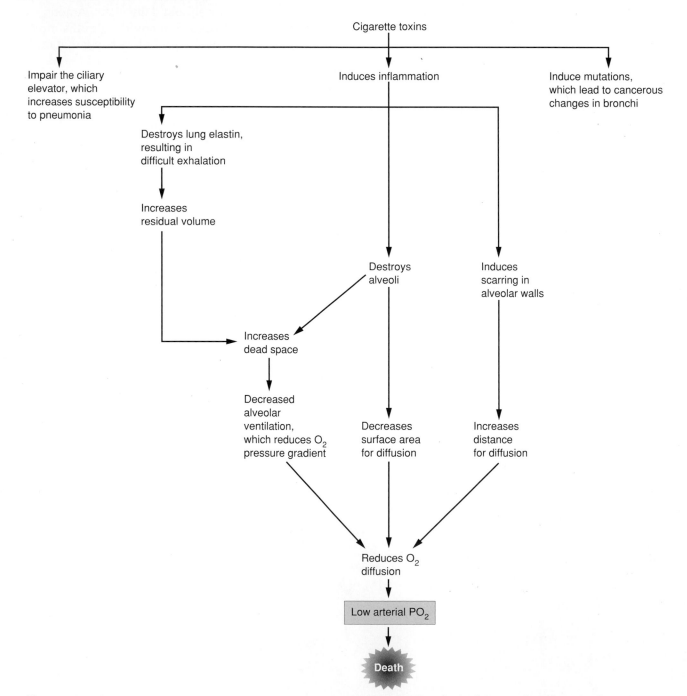

Figure 13.17. Luther B. Emphysema disrupts many aspects of respiration, including ventilation and external gas exchange. *Which diffusion parameter is altered by the scarring in the alveolar walls?*

alveolar destruction and the increased diffusion distance caused by alveolar scarring impair gas diffusion, so external gas exchange is abnormally inefficient.

External gas exchange is reduced even further by the effect of emphysema on alveolar ventilation rate. Remember that the alveolar ventilation rate determines how much fresh air reaches the alveoli, which in turn determines the partial pressure gradients between blood and alveoli, which affects gas diffusion. Alveolar ventilation rate is reduced in emphysemic patients because of their large dead space. Their dead space, in turn, reflects not only the destruction of alveoli but also their impaired exhalation. Since they cannot exhale all of their inhaled air, they keep more "used up" air in their lungs and cannot bring in fresh air on the next inhalation—essentially, dead space is increased.

Exhalation is difficult for three reasons. First, the loss of interstitial elastin reduces lung recoil, which normally accounts for exhalation. Moreover, small airways tend to collapse when thoracic pressure rises to expel air, trapping air within alveoli. Finally, the chronic smoke irritation of bronchioles makes them tend to contract with expiration in an asthmalike way, increasing the resistance to airflow.

As if this were not enough, cigarette smoke does other types of damage. It inhibits the rhythmic sweeping action of bronchial epithelial cell cilia (Fig. 13.17, left column). As a result, smokers cannot easily clear their lungs of mucus and bacteria. No wonder Luther developed pneumonia. In addition, cigarette smoke contains chemicals that damage the DNA of bronchial epithelial cells (Fig. 13.17, right column). The body has mechanisms to repair damaged DNA, but smoking for many years finally overwhelms the mechanism and precancerous changes in DNA begin to appear. Sure enough, Luther had precancerous changes in his bronchial epithelium. It seems reasonable to suppose that if he had lived another few years, he would have developed lung cancer.

In the end, Luther M. died of hypoxia. His P_{O_2}, low to begin with, fell further as pneumonia filled his emphysematous lungs with inflammatory fluid. Hemoglobin saturation fell to a point that his blood could not deliver enough oxygen to maintain consciousness or to supply his heart with enough oxygen for it to continue to beat.

Word Parts

Latin/Greek Word Parts	English Equivalents	Examples
oro-	Mouth	Oropharynx: the part of the pharynx behind the mouth
-capno, -capnea	Carbon dioxide	Hypercapnea: excessive (hyper-) carbon dioxide
naso-	Nose	Nasopharynx: the part of the pharynx behind the nose
pneumo-	Lungs, respiration	Pneumocyte: lung cell (-cyte)
pleur-, pleuro-	Ribs, side, pleura	Pleural fluid: fluid between pleura (and near ribs)
-pnea	Breath, respiration	Hypernea: excessive (hyper-) breathing
eu-	Good	Eupnea: "good" (quiet, normal) breathing
oxy-, oxo-	Oxygen	Hypoxia: low (hypo-) oxygen

Chapter Challenge

CHAPTER RECALL

1. **The transfer of oxygen from blood to cells occurs during**
   a. pulmonary ventilation.
   b. internal gas exchange.
   c. external gas exchange.
   d. cellular respiration.

2. **Which of the following is part of both the upper respiratory system and the conducting zone?**
   a. Trachea
   b. Bronchi
   c. Nasopharynx
   d. Alveoli

3. **Which of the following is *not* one of the paranasal sinuses?**
   a. Ethmoid sinus
   b. Frontal sinus
   c. Maxillary sinus
   d. Occipital sinus

4. **Which tonsils guard the oropharynx?**
   a. Tubal tonsils
   b. Pharyngeal tonsils
   c. Adenoids
   d. Palatine tonsils

5. **Structures forming the larynx include the**
   a. Epiglottis
   b. Thyroid cartilage
   c. Vestibular fold
   d. All of the above

6. **The vocal folds increase the pitch of a sound by**
   a. relaxing the associated skeletal muscles.
   b. moving the cords closer to one another.
   c. decreasing the tension of the ligaments.
   d. none of the above.

7. **The posterior surface of the trachea is**
   a. braced by the esophagus.
   b. supported by cartilage rings.
   c. adjacent to the stomach.
   d. rendered more flexible by the absence of smooth muscle.

8. **Surfactant is secreted by**
   a. type I pneumocytes.
   b. type II pneumocytes.
   c. alveolar macrophages.
   d. ciliated epithelial cells.

9. **The parietal pleura**
   a. lines the chest wall.
   b. produces surfactant.
   c. covers the lung surface.
   d. is separated by the visceral pleura by an air-filled space.

10. **The lungs receive oxygen-rich blood from the**
    a. bronchial circulation.
    b. pulmonary circulation.
    c. bronchi.
    d. none of the above.

11. **The vital capacity is equal to**
    a. the air volume in the lungs at maximum inhalation.
    b. the residual volume plus the total lung capacity.
    c. the inspiratory capacity plus the expiratory capacity.
    d. the sum of the tidal volume, expiratory reserve volume, and inspiratory reserve volume.

12. **Given the same breathing rate and tidal volume, an increased dead space will**
    a. reduce the minute ventilation.
    b. increase the minute ventilation.
    c. reduce the alveolar ventilation rate.
    d. increase the alveolar ventilation rate.

13. **What happens when the diaphragm contracts?**
    a. The lung volume decreases.
    b. The abdominal contents are pushed downward.
    c. Expiration.
    d. The lungs are compressed.

14. **Which of the following changes would make inhalation more difficult?**
    a. Increased surfactant
    b. Increased elastance
    c. Increased compliance
    d. Reduced elastic fibers

15. **Oxygen diffusion into blood from the lungs would likely increase if**
    a. airway resistance decreased.
    b. the alveolar membrane were thickened by scarring.
    c. the pulmonary blood oxygen partial pressure rose.
    d. the atmospheric pressure dropped.

16. **Which of the following changes would increase oxygen unloading at the tissues?**
    a. Chilling
    b. Decreasing the oxygen partial pressure in the cells
    c. Increasing the pH
    d. Decreasing the carbon dioxide concentration

17. **Hemoglobin has binding sites for**
    a. oxygen.
    b. carbon dioxide.
    c. hydrogen ions.
    d. all of the above.

18. **Most blood carbon dioxide is carried**
    a. attached to hemoglobin.
    b. dissolved in plasma.
    c. as bicarbonate.
    d. as carbonic acid (H_2CO_3).

19. **Which statement about the ventral respiratory group is false?**
    a. It sets the basic respiratory rhythm.
    b. It alters the respiratory rhythm when arterial carbon dioxide levels change.
    c. Its neurons are active only during inspiration.
    d. Its activity can be altered by the cerebral cortex.

20. **Which of the following concentration changes is *directly* sensed by the central chemoreceptors and communicated to the respiratory center?**
    a. Increased hydrogen ions
    b. Increased carbon dioxide
    c. Increased bicarbonate
    d. Increased oxygen

CONCEPTUAL UNDERSTANDING

21. **What is segmentation? Explain its relevance to Luther's pneumonia.**

22. **Beginning at the transition point, the maximal amount of inspired air is a *capacity*; but the maximum amount of expired air is a *volume*. Why?**

APPLICATION

23. **James, a 16-year-old boy, is having an attack of exercise-induced asthma. His bronchi are considerably constricted.**
    a. What effect will this bronchoconstriction have on his dead space? Explain.
    b. How will the change in dead space alter the ratio between his alveolar ventilation rate and minute ventilation?

24. **You are performing lung function tests on a wheezy 5-year-old child. The data are as follows:**

    Breathing rate: 20/min
    Tidal volume: 210 mL
    Dead space (determined by a different test): 50 mL
    Functional residual capacity: 900 mL
    Total lung capacity: 1.8 L

    Based on these values, calculate the following measures:
    a. Alveolar ventilation rate
    b. Inspiratory reserve volume

25. **Use the equation below to answer the following questions.**

$$CO_2 + H_2O \Leftrightarrow H_2CO_3 \text{ (carbonic acid)}$$
$$\Leftrightarrow H+ + HCO_3^- \text{ (bicarbonate)}$$

    a. Young Evan is holding his breath because his mother won't let him play with the can opener. What will happen to the pH of his blood? Explain.
    b. Maria, aged 19, has a nasty stomach flu and has vomited nine times in 12 hours. As a result of losing so much stomach acid, her blood pH is increased. How will this change affect the carbon dioxide concentration in her blood, and how will the change in carbon dioxide affect her breathing rate?

You can find the answers to these questions on the student Web site at **http://thepoint.lww.com/McConnellandHull**

14

The Digestive System

Major Themes

- Nutrients are carbohydrates, lipids, proteins, vitamins, minerals, and water.

- Energy is extracted from carbohydrates, lipids, and proteins. Vitamins, minerals, and water facilitate energy-producing reactions.

- The purpose of digestion is to break large, unabsorbable nutrient molecules into smaller, absorbable ones.

- Digestion is a multistep, sequential process involving both mechanical and chemical actions.

- Digestion occurs within the gastrointestinal tract.

- The liver and pancreas assist digestion by adding secretions to the intestinal tract, but they also have nondigestive functions.

Chapter Objectives

"I can't take these pills anymore."

As you read through the following case study, assemble a list of the terms and concepts you must learn in order to understand it.

Clinical History: Margot C. was a 19-year-old college sophomore with a history of mood swings and a medical diagnosis of bipolar disorder (also known as manic-depressive disorder). During the semester, she ran out of her medication and visited the campus clinic. There, the physician was able to convince Margot to take a substitute medicine because her regular prescription was not on hand.

Margot returned to the clinic the next day complaining that "the new pills upset my stomach." The physician urged her to continue the medication, telling her "your system will get used to it." She returned again 2 days later, saying, "I can't take these pills anymore. They give me diarrhea and gas." On further questioning she said, "This reminds me of the trouble I used to have with milk and other dairy products. They told my mom that I was lactose intolerant, so I grew up drinking soy milk. When I was a teenager I'd sneak pizza or an ice cream cone, but it always had the same result—diarrhea and gas—so now I avoid dairy products altogether."

Physical Examination and Other Data: Vital signs and laboratory data were unremarkable. On physical examination, Margot's abdomen was soft and nontender, but she asked, "Please don't press too hard! I'm full of gas and I need to go to the bathroom." No other noteworthy findings were recorded.

Clinical Course: The physician wrote a prescription for her regular medicine and instructed her to come back in a week. When she returned, she said the intestinal symptoms had disappeared completely. Intrigued with the clear association of intestinal symptoms with the substitute

medicine, the physician did some Internet research and could not find intestinal symptoms among the recognized adverse effects. Just as she was about to log off she happened to notice the "Each capsule contains" section of the drug maker's web information page. Below the listing of the active drug was a listing of "Inactive ingredients," which carried this entry: "Lactose filler."

Certain that she'd found the problem, the physician called Margot and told her that she could not tolerate lactose, the carbohydrate found in dairy products. In addition, she told her of a research study on lactose intolerance being conducted by the university's medical school and said she could be paid a stipend for participating. Margot agreed to participate and to take one of the troublesome pills as part of the study. First, she was given a standardized drink of milk, after which her blood sugar and the amount of hydrogen gas in her exhaled breath were measured. The test produced the expected symptoms: Margot's blood glucose failed to rise significantly, and hydrogen gas appeared in her exhaled breath. On the following visit, she was given one of the capsules, which produced a similar result.

We are awash in a sea of food and food information. Supermarkets, restaurants, cafes, and convenience stores abound. Even gasoline stations offer more fuel for humans than for automobiles: to go inside to pay your bill is to be confronted by racks of colorful packages containing foods of amazing variety. What's more, we're constantly bombarded with information about food: in news stories, on TV, at our health clubs, and so on. Even some churches have recently begun to offer programs on healthy eating!

And yet, despite all we know about how food tastes and how to prepare it, most people know very little of what happens to it from the time it goes in one end and the waste comes out the other. That sequence of events, which enables us to harvest the energy and nutrients in food, is most commonly referred to as **digestion.** However, in physiology, the term has a narrower meaning: digestion is the mechanical and chemical action that breaks down food into its component molecules. It is followed by two related processes: absorption of nutrients into the blood and elimination of wastes.

We begin this chapter by describing the nutrients themselves. We subsequently discuss digestion, absorption, and elimination. In the course of our study, we also explore the interrelationships of form and function and the overlapping, integrated activities of the neural and endocrine control of digestive processes.

"Tell me what you eat, and I will tell you what you are."
Anthelme Brillat-Savarin (1755–1826), *The Physiology of Taste,* 1825

Nutrients

A **nutrient** is any chemical in food or drink that an organism needs to live or grow. This certainly includes *water*—which is essential for the chemical reactions that sustain life, and it is part of our daily diet. Three organic molecules—*carbohydrates, lipids,* and *proteins*—are also classified as nutrients. Called **macronutrients** because we need to consume them in large amounts, these provide the energy and the raw material for the construction of cells, tissues, and organs. In addition to these four are two final classes of nutrients: *vitamins* and *minerals,* which are referred to as **micronutrients** because we consume them in minute quantities. We discuss each of the six classes of nutrient separately below.

We are what we eat: apart from the small number of molecules remaining from our newborn body, we are constructed from ingested nutrients. For example, remember from ← Chapter 3 that our cell membranes contain elements of each of the three major nutrient groups: phospholipids and cholesterol are lipids; ion channels and receptors are formed of protein; and carbohydrates are a key element of membrane glycoproteins. Bones are formed of collagen (a protein) and calcium (a mineral), and the liver and muscles store large amounts of glycogen (a carbohydrate). But it is important to understand that *the form of the ingested nutrient does not dictate the form of the final product.* That is, ingested carbohydrates can be burned for immediate energy or broken down into their components and reassembled as parts of new carbohydrates or even lipids.

The energy we derive from nutrients is measured in *calories,* a term with several versions. One version, the *kilocalorie* or "large" calorie, is 1,000 times the value of the "small" calorie. A large calorie (often written with a capital C) is the amount of heat required to raise the temperature of 1 kg of water by 1°C. In the United States and many other nations, food labels express caloric content in large calories. For this reason it is often called the *food calorie.* In this textbook, when the word *calorie* is used, it refers to the large calorie and is not capitalized.

Plants Supply Most Dietary Carbohydrates

Carbohydrates provide short-term energy for most body cells and are also added to proteins and lipids. Recall from ← Chapter 2 that the word root *saccharide* describes carbohydrates, with the prefix (mono-, di-, or poly-) specifying the number of sugars making up the molecule (Fig. 14.1, top). The ultimate goal of carbohydrate digestion is to break down larger saccharides into their component monosaccharides. Natural carbohydrate sources include the following:

- Grains, legumes (peas, beans, and lentils), and many vegetables, which contain starch (a polysaccharide).
- Fruits, which contain fructose (a monosaccharide). Fructose is also found in commercial syrups, such as high-fructose corn syrup, which are used to sweeten soft drinks and many other products.
- Table sugar, which contains sucrose (a disaccharide of glucose and fructose) and is extracted from sugar beets or sugar cane.
- Dairy products, which contain lactose (a disaccharide of glucose and galactose).

Many plant products also contain **dietary fiber** (nondigestible polysaccharides such as *cellulose*). Fiber is nondigestible in humans because we lack the necessary enzymes to break cellulose into its glucose pieces, which can be absorbed. Instead, fiber remains as a polysaccharide in the digestive tract and adds bulk to feces by retaining water. Fiber also binds intestinal fats, preventing them from being absorbed. The bulk created by fiber also creates a sensation of fullness, which reduces hunger.

Case Note

14.1. Margot has problems with lactose. How many sugars make up one lactose molecule?

Proteins Provide Amino Acids

The digestive tract breaks dietary proteins down into their constituent *amino acids,* which are used by body cells to build the proteins they need (Fig. 14.1, middle). Proteins provide structural components for muscle, tendon, bone, and skin; they also serve functional roles as enzymes, receptors, transport agents, and hormones. Amino acids can also be used for energy, but the amino group must first be removed and converted into a nitrogenous compound, such as urea, which can be excreted in the urine. Only when starved of carbohydrates and fats will the body break down tissue proteins to liberate amino acids. The western diet contains ample amino acids to build all necessary proteins and still provide about 10% to 15% of the body's energy needs.

Human proteins are constructed from specific combinations of 20 different amino acids. Of these, nine,

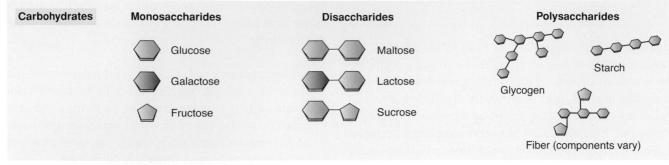

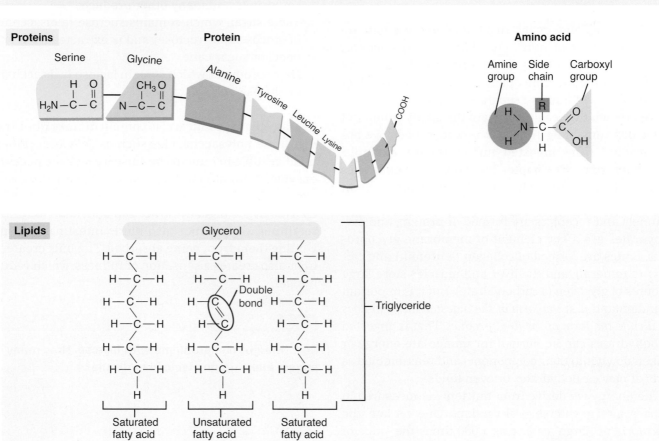

Figure 14.1. The macronutrients. Carbohydrates, proteins, and lipds are the three classes of nutrients that provide energy. Triglycerides are the lipids most abundant in our diet. *Name the two monosaccharide components of sucrose.*

called **essential amino acids** (EAA), must be obtained from our diet because the liver cannot synthesize them. The protein in meat, fish, milk, and eggs is called *complete protein* because it contains all of the essential amino acids (EAAs). With the exception of soybeans and some whole grains (such as quinoa), plant protein sources are not complete, but many plant foods complement each other in the amino acids they provide. For instance, rice and beans, a traditional vegetarian meal, provide a full complement of amino acids, as does a peanut butter sandwich. Thus, well-planned vegetarian diets contain enough variety to provide all of the EAAs.

Deficiency of a single EAA can be induced in experimental animals, but it is not clear if humans ever suffer from disease on account of EAA deficiency. However, the effects of protein deficiency are clear. If the diet contains some calories from carbohydrate but remains very low in protein—as in desperately poor areas where meat, beans, and other high-quality proteins are not available— the person develops a type of starvation known *kwashiorkor* (from the Niger–Congo language of Ghana, in West Africa). Lacking protein, the intestinal lining atrophies, which impairs nutrient absorption. In addition, the liver cannot synthesize albumin and other blood proteins. Recall from ◄ Chapter 11 that albumin provides most of the blood's osmotic pressure, which acts to retain water in blood capillaries and prevent its movement into the interstitial space (see Fig. 11.15B). Albumin deficiency thus increases the amount of interstitial fluid, giving the patient a swollen belly and overall puffy appearance that can mask underlying emaciation. Lack of protein also leaves the person, usually a child, highly vulnerable to infectious disease.

When diets deficient in protein are also deficient in calories, the patient develops *marasmus*. This is a more severe form of starvation characterized by wasting, skeletal thinness as fat stores and muscle tissues waste away; and stunted growth, since all energy is used solely to keep the body's cells alive. Children with marasmus also have an aged, wrinkled appearance because their bodies shrivel to the point that their skin no longer "fits."

Lipids Are Important Nutrients

Recall from ◄ Chapter 2 that there are three main types of lipids: triglycerides (commonly called fats), phospholipids, and steroids (cholesterol and others). Triglycerides are an important source of energy, and phospholipids and steroids act as structural elements in cell membranes. Some steroids act as hormones (cortisol and testosterone, for example), and some triglyceride components act in signal transduction pathways. Dietary fat also carries fat-soluble vitamins A, D, E, and K (discussed below) into the body. Finally, body lipid stores (fatty tissue) serve to cushion the body from trauma and provide insulation against heat loss.

Unsaturated Fats Are Healthiest

The great majority of dietary lipids are triglycerides (Fig. 14.1, bottom). Recall that triglyceride molecules consist of a glycerol base to which three fatty acids are attached. Each fatty acid is a chain of carbon atoms bonded to hydrogen. The fatty acids in a **saturated fat** contain carbons joined by single bonds, like this: $-CH_2-CH_2-CH_2-$. Each carbon atom can form only four bonds, and in this instance two are used in connecting to the carbons on either side, which leaves room for only two hydrogens. Therefore each carbon is *saturated* with hydrogen atoms, and no more can be added. Saturated fatty acids are straight chains that can be packed tightly against one another, which makes them dense, and they are solid at room temperature. Most animal fats, including butter, are primarily saturated fat, as are coconut and palm oil. A diet high in saturated fats has long been associated with an increased risk of cardiovascular disease.

In some fatty acid molecules, a few of the hydrogen atoms are absent; some of the carbon atoms have only one hydrogen atom and are connected to the next carbon by a double bond, like this: $-CH_2-CH=CH-CH_2-$. Fats containing these fatty acids are called **unsaturated fats** because some of the carbon atoms can accept more hydrogen atoms. *Monounsaturated fats* have one double bond in their fatty acid chains; *polyunsaturated fats* have more than one double bond. At the point along the chain where these double bonds occur, the fatty acid molecule is kinked. As a consequence, unsaturated fat molecules cannot be tightly packed; they are less dense and therefore liquid at room temperature. Canola oil, most nut and seed oils, and olive oil are composed primarily of monounsaturated fatty acids. Sunflower, corn, soybean, and fish oils are primarily polyunsaturated. Diets in which the predominant fats are unsaturated are associated with a decreased risk of cardiovascular disease.

A group of polyunsaturated fatty acids that have been receiving a great deal of attention recently are the *essential fatty acids*. These are also referred to as *omega fatty acids* because their first double carbon bond is located toward the end of the chain (*omega* is the last letter in the Greek alphabet). The body can synthesize most fatty acids from components of other molecules, but this is not true for the essential fatty acids: their special molecular structure requires that they be obtained from the diet. Fish, shellfish, and certain plant, seed, and nut oils are good dietary sources. People who consume a diet high in essential fatty acids have a lower risk of cardiovascular disease than people who do not. The body uses omega fatty acids to synthesize a number of essential biological compounds that regulate inflammation, blood clotting, and other important functions.

Unsaturated fats are cheap and relatively healthy, but they spoil quickly and are difficult to transport because they are liquid. As a consequence the food industry has for decades converted unsaturated fats to saturated fats by adding hydrogen, which keeps the fat from going rancid. Such fats are called *hydrogenated fats*, or *trans fats* (think of them as *trans*formed fat). Although this has enabled the food industry to provide food that is appealing and cheap, trans fats pose serious health risks. They impair blood glucose control and induce inflammation. For this reason, in 2006, the U.S. Food and Drug Administration (FDA) began requiring that all food labels identify the food's content of trans fatty acid on the Nutrition Facts panel. Shortly after this requirement went into effect, many food manufacturers replaced the trans fatty acids with more healthful fats.

Apart from the type of dietary fat, the amount of dietary fat is important because fat contains nine calories per gram, whereas protein and carbohydrate contain only four. It is easier, therefore, to ingest excess calories in the form of fat.

Fats Circulate in Blood as Lipoproteins

Unlike proteins and carbohydrates, fats are not soluble in blood. Triglycerides and cholesterol circulate in blood in large protein–lipid complexes called **lipoproteins.** Each lipoprotein consists of an inner core, composed of triglyceride and/or cholesterol, and an outer shell, composed of phospholipids, cholesterol, and specialized proteins called **apoproteins.** All three components of the outer shell are partially hydrophilic and partially hydrophobic. The hydrophobic portions of these molecules face the lipoprotein core, and the hydrophilic portions face outward to interact with watery blood. Thus the phospholipids and proteins render the lipoprotein soluble in blood.

Protein is denser than lipid, so a lipoprotein containing proportionally more protein is denser than one containing more lipid. Based on molecular density, lipoproteins are classified as follows:

- **Very low density lipoprotein (VLDL)** is mostly lipid and has very little protein. Most of the lipid is triglyceride.
- **Low-density lipoprotein (LDL)** is somewhat less lipid and has more protein. Most of the lipid is cholesterol.
- **High-density lipoprotein (HDL)** has little lipid and is mostly protein. Most of the lipid is cholesterol.

VLDL is the main transport form for triglycerides from the liver to the rest of the body (Fig. 14.2). Target tissues liberate free fatty acids from VLDL and use them for energy or store them for later use.

LDL is the major transport form for cholesterol. Cholesterol found within LDL is called *LDL cholesterol.* Cells needing cholesterol grab the LDL packets using LDL receptors located on their cell membrane. Unfortunately, LDLs also lodge in the blood vessel walls of some people and accumulate there as part of the cardiovascular disease process. Because it contributes to cardiovascular disease, LDL cholesterol is often referred to as the "bad" cholesterol.

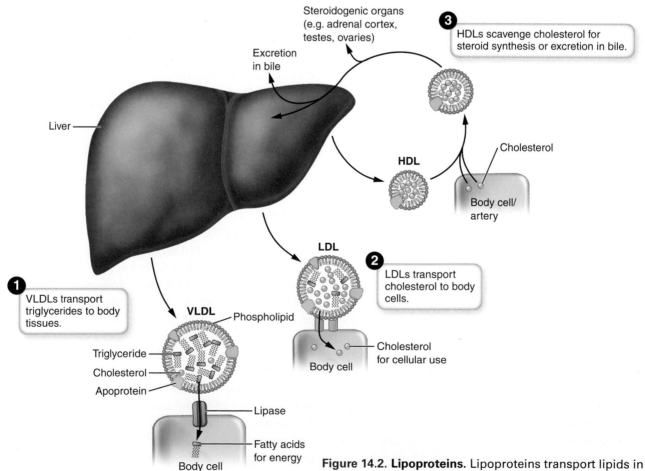

Figure 14.2. Lipoproteins. Lipoproteins transport lipids in blood. *Which type of lipoprotein picks up cholesterol from cells?*

CLINICAL SNAPSHOT

Desirable Plasma Lipid Concentrations

In regard to plasma lipid concentrations, the concept of *normal* ◄— (Chapter 1) is misleading because it is relies on calculations to determine the *average* cholesterol of presumably healthy people. The problem is this: the *average* cholesterol of many presumably healthy people is unhealthily high, especially in developed Western nations, because of diets rich in saturated and trans fats. *Desirable* is a better concept than *normal*.

In the United States prior to about 1970, what is now considered very high cholesterol was considered normal. However, we now know that total cholesterol should be kept below 200 mg/dL for good health. When it comes to total plasma cholesterol, lower is better—lowering total cholesterol has benefits at almost any initial cholesterol level (any starting point). To be specific, *decreasing total cholesterol by 40 mg/dL cuts the prevalence of atherosclerosis complications in half.* And lowering it another 40 mg/dL cuts the risk by half again. Evidence suggests that atherosclerosis may not develop at total cholesterol levels below 150 mg/dL, a level unreachable in most cultures without a strict low-fat vegetarian diet. However, useful though it is, the measurement of total cholesterol is less useful in predicting cardiovascular risk than determining the fractions that compose it: low, high, and very low density lipoprotein cholesterol.

Desirable *LDL cholesterol* (LDL-C) is considered to be <100 mg/dL. As with total cholesterol, lower is better.

Desirable *HDL cholesterol* (HDL-C) appears to be above 60 mg/dL—a level that is not associated with cardiovascular risk. On the other hand, values less than 40 mg/dL are associated with a very high risk of atherosclerosis. Low HDL-C is strongly influenced by genetics and is more difficult to change than total or LDL-C, which are more influenced by diet. However, low

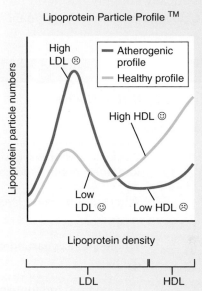

Lipid profiles, which separate lipid particles by their density.

HDL-C can be treated and the effort is worthwhile—increasing HDL-C lowers coronary risk. Quitting smoking is especially effective in raising HDL-C: this is one of the reasons why smoking cessation is a very effective strategy to reduce cardiovascular risk. Exercise and the consumption of moderate amounts of alcohol are also effective in raising HDL-C. (Moderate alcohol consumption is defined as no more than one drink per day for women and one to two drinks per day for men.)

Very little cholesterol is present in very low density lipoprotein (VLDL); therefore VLDL levels bear only a limited relationship to atherosclerotic cardiovascular risk.

HDL also transports cholesterol. HDL travels through the blood, picking up stray cholesterol from arterial linings or cells and delivering its cholesterol cargo to certain glands, which use it to make steroid hormones, or to the liver for secretion into bile. HDL improves arterial health by removing cholesterol from tissues, so HDL cholesterol is often called "good" cholesterol, despite the fact that the cholesterol molecule within HDLs and

LDLs is identical. That is, the *destination* of cholesterol, not its carrier, dictates its role in health or disease—LDL cholesterol is cholesterol that is staying in the body for various purposes; however, HDL cholesterol is "waste" cholesterol headed out of the body by secretion into bile. See the nearby clinical snapshot titled *Desirable Plasma Lipid Concentrations* for more information on blood cholesterol levels.

Vitamins Are Required for Good Health

Vitamins are just as important as carbohydrates, fats, and proteins for good health. Thirteen have been named: A, eight B vitamins, C, D, E, and K. They were originally named alphabetically, roughly based on their order of discovery, but gaps now exist in the alphabetical chain as knowledge has accumulated and substances were reassigned to other physiological categories.

A **vitamin** is a powerful small *organic* molecule (Chapter 2) required for a particular metabolic reaction and must be obtained in the diet because the body cannot synthesize it (Table 14.1). Exceptions to this rule are vitamin D, which can be made in skin upon exposure to sunlight, and vitamin K, which can be made by intestinal

Table 14.1   Vitamins

Name	Dietary Source	Function
Water-soluble Vitamins		
Vitamin C (ascorbic acid)	Citrus, strawberries, kiwi fruit, melons, raw tomatoes, green vegetables	Promotes protein synthesis, promotes wound healing, acts as an antioxidant
Vitamin B1 (thiamine)	Whole grains, yeast, liver, eggs, pork, nuts	Coenzyme in carbohydrate breakdown
Vitamin B2 (riboflavin)	Whole grains, yeast, meat, eggs, beets, peas, peanuts	Coenzyme component
Niacin	Derived from tryptophan an essential amino acid found in whole grains, yeast, meat, fish, peas, beans, nuts	Component of NAD and a coenzyme in lipid metabolism
Vitamin B6 (pyridoxine)	Whole grains, yeast, salmon, yoghurt, tomatoes, yellow corn, spinach	Coenzyme in amino acid metabolism
Vitamin B12 (cyanobalamin)	Liver, kidney, milk, eggs, cheese, meat; intestinal absorption requires *intrinsic factor,* a protein secreted by the stomach	Coenzyme in red blood cell formation and amino acid metabolism
Pantothenic acid	Liver, kidneys, yeast, green vegetables	Component of coenzyme A
Folic acid	Green leafy vegetables, broccoli, asparagus, beans, citrus	Coenzyme in RNA and DNA synthesis and blood cell formation
Biotin	Yeast, liver, egg yolk, kidney	Coenzyme in pyruvate metabolism and fatty acid synthesis
Fat-soluble Vitamins		
Vitamin A (retinol)	Produced from carotene, a provitamin found in orange, yellow, or dark green vegetables	Epithelial cell function
Vitamin D	Fish oils, egg yolk, fortified dairy products	Necessary for intestinal calcium absorption
Vitamin E (tocopherols)	Fresh nuts, wheat germ, seed oils, green leafy vegetables	Involved in nucleic acid and red blood cell synthesis; necessary for nervous system formation and maintenance; antioxidant
Vitamin K	Liver, spinach, cauliflower, cabbage	Coenzyme in clotting factor synthesis

bacteria, although some is present in the diet. What's more, some vitamins are ingested as inactive precursors, or *provitamins*, which are converted by the body into the active vitamin. For example, vitamin A (also known as retinol) is synthesized from the provitamin *beta carotene*.

The physiological roles of vitamins are varied and numerous. For example, vitamin C is known primarily for its role in protecting the body's connective tissues, but it also functions as an antioxidant and promotes iron absorption. The B vitamins enable us to utilize the carbohydrate, protein, and fat in our diet. They do this by participating in metabolic reactions as *coenzymes*—substances that promote the functions of enzymes. This function of B vitamins is discussed in greater detail in ➡ Chapter 15. Other vitamins act as hormones, assist in DNA synthesis, or have other functions (Table 14.1).

Some vitamins are soluble in water, others in fat. Water-soluble vitamins (the B vitamins and vitamin C) cannot be stored by the body. Deficiencies of water-soluble vitamins develop quickly because any daily excess is excreted in urine. In contrast, deficiencies of fat-soluble vitamins (A, D, E, and K) develop more slowly, because they can be stored in body fat. On the other hand, since they can be absorbed into the body only dissolved in dietary fat, anything that impairs fat digestion can cause deficiencies of the fat-soluble vitamins. For instance, some pancreatic diseases impair the production of pancreatic enzymes, which digest dietary fat. If these enzymes are lacking, the undigested fat passes into feces and carries the fat-soluble vitamins with it.

Because the body stores fat-soluble vitamins, prolonged excessive consumption can lead to dangerous accumulations. For example, too much vitamin D can increase blood calcium concentrations. The excess calcium passes into urine, where it may form calcium stones in the urinary tract.

The *dietary reference intake (DRI)* of each vitamin varies according to age and gender. Fruits and vegetables are good sources of many vitamins, and a few are found exclusively in animal products or grains (Table 14.1).

Case Note

14.2. The soy milk Margot consumes is supplemented with vitamin D. Is vitamin D fat-soluble or water-soluble, and what does it do?

Dietary Minerals Are a Necessity

A short, strict definition of **mineral** is difficult, but for our purposes it is considered a *nonorganic* element that can form a crystalline solid. Compounds composed of

carbon, hydrogen, oxygen, and nitrogen are not considered minerals because by definition they are organic.

About 4% of body weight is composed of seven **major minerals** calcium, phosphorus, potassium, sulfur, sodium, chloride, and magnesium. About three-fourths of this is accounted for by the calcium and phosphorus in bone. Sodium, potassium, and chloride are critical in the composition of body fluids; calcium and magnesium are important in muscle contraction; and sulfur occurs in a variety of physiological compounds (Table 14.2).

Other minerals are present in very small amounts, and account for less than 0.01% of body weight. These **trace minerals** include chromium, copper, fluoride, iodine, iron, manganese, selenium, and zinc. Some have highly specific and critical roles; for example, iron is an essential component of hemoglobin, which transports of oxygen in blood; fluoride helps maintain healthy teeth and bones; iodine is required to produce thyroid hormone; and chromium, manganese, selenium, iron, copper, and zinc are required in a variety of metabolic tasks.

Minerals are found in most foods with the exception of grains. Calcium and phosphorus, for instance, are abundant in dairy products, whereas meat is rich in iron and zinc (Table 14.2).

Certain medical conditions can cause a mineral deficiency. For example, large amounts of sodium can be lost by people with chronic diarrhea, which lowers blood sodium concentration. Low blood sodium results in low plasma volume and low blood pressure. Low blood pressure, in turn, causes fatigue, dizziness, and fainting.

Sometimes, however, the deficiency is due to a diet deficient in a particular mineral. Iron is a good example. Because iron is necessary for hemoglobin synthesis (⬅ Chapter 10), a diet deficient in iron results in hemoglobin deficiency (anemia). Women in their reproductive years are particularly vulnerable because they lose iron regularly by menstruation.

Vegan diets—which exclude meat, poultry, fish, dairy products, and eggs—are increasingly common and have well-documented health benefits, including a reduced risk of cardiovascular disease and type 2 diabetes. However, since animal products are excellent sources of calcium, zinc, and iron, vegans must plan their diet with care. They must also monitor their diet for adequate content of vitamins D and B12, protein, and the omega fatty acids.

Case Note

14.3. Natural soy milk contains iron, calcium, zinc, riboflavin, and sodium. Which of these substances are minerals, and which are vitamins?

Table 14.2 Major and Trace Minerals

Mineral	Dietary Source	Function
Major Minerals		
Calcium	Dairy products, leafy greens, shellfish, egg yolk	Calcium salts form bone and teeth; calcium ions are involved in heart action potentials, neurotransmitter release, muscle contraction, and signal transduction.
Phosphorus	All animal products, nuts, legumes, grains	Phosphate salts help form bone and teeth; component of many large biological molecules (e.g., ATP, proteins); acid–base balance.
Sodium	Table salt, salt-supplemented foods	Helps to maintain body water balance and blood pressure; necessary for action potentials, muscle contraction, and the active transport of other ions and nutrients.
Potassium	Many foods, especially avocados, dried apricots, and animal products	Electric signaling, protein synthesis, maintenance of intracellular osmotic pressure.
Chlorine	Table salt	Helps to maintain body water balance (along with sodium); acid-base balance.
Sulfur	All animal products, legumes	Component of vitamins, proteins, connective tissue.
Magnesium	Animal products, whole grains, nuts, legumes, leafy greens	Component of coenzymes; necessary for muscle contraction.
Trace Minerals		
Chromium	Animal products, whole grains, wine	Required for glucose metabolism.
Copper	Shellfish, whole grains, legumes, meat	Required for the synthesis of hemoglobin, melanin, myelin, and some enzymes.
Fluoride	Fluorinated water	May prevent dental cavities and osteoporosis.
Iodine	Shellfish, iodine-supplemented salt	A component of thyroid hormone.
Iron	Liver, meat, egg yolk, legumes	Part of hemoglobin and mitochondrial enzymes.
Manganese	Nuts, legumes, whole grains, leafy greens	Involved in many metabolic reactions.
Selenium	Seafood, meat	An antioxidant.
Zinc	Seafood, meat, legumes	A component of enzymes and structural proteins.

Water Is the Most Important Nutrient

If there is magic on this planet, it is contained in water.

Loren Eiseley (1907–1977), American anthropologist, educator, and natural science writer; in *The Immense Journey*, 1957

Water is the matter and the matrix of life: it is the most common molecule in every life form, and all of life's essential processes unfold within it. About 60% of an adult's body weight is water: it is the major constituent of every cell and of the fluid between cells; it forms most of our blood, lymph, cerebrospinal fluid, urine, and much of our feces; it contributes to cartilage and synovial joints; and we even see the world through eyes filled with watery fluid. We can live for weeks without food but only a few days without water.

Recall from Chapter 2 that water is a solvent, a lubricant, a cushion for cells and tissues, and a heat sink capable of holding large amounts of heat without much rise in temperature. And do not forget that water is critical in metabolism—water and carbon dioxide are the end products of the body's energy-producing reactions.

How much water do we need? Most experts agree that adults need between 8 and 12 cups per day (about 2.5 L), but requirements vary according to gender, age, level of physical activity, environmental temperature, and other factors.

A Healthy Diet Requires Attention

Dietary habits vary—from one culture to another and from one time to another. For example, the Vietnamese diet is low in fat and high in fiber, with plenty of whole grains and fresh vegetables. In contrast, the American diet is high in fat and low in fiber, with refined grains and few fresh vegetables. Diet varies from one era to another, too: in the mid-nineteenth century, the high-fiber, meatless Graham diet was all the rage across the United States, whereas the late twentieth century found us caught up in a meat-rich, "low carb" craze.

Evidence from scientific studies comparing diet and health in many cultures, as summarized by the Healthy Eating Pyramid from Harvard School of Public Health, suggests the following guidelines for good health (Fig. 14.3):

1. *Control your weight.* Matching calorie input with energy output should be the basis of every diet. Obesity is a toxic condition that increases the risk of diabetes, cancer, cardiovascular disease, and high blood pressure. Apart from stopping smoking, the single most effective way to improve long-term health is to maintain a healthful weight.

2. *Exercise regularly.* Regular exercise, even just a daily short walk, helps avoid excess weight by burning calories; but it also pays other big dividends—people who exercise regularly live longer, healthier, and happier lives.

3. *Rely primarily on whole grains for carbohydrates.* Whole grains—whole wheat bread, brown rice, and oatmeal, for example—have not been refined to strip them of their fiber and are therefore digested and absorbed more slowly than refined grains such as white flour. As a result, they do not raise blood glucose levels as quickly or as significantly as refined carbohydrates. What's more, their fiber content provides bulk that is more effective in satisfying hunger than the "empty calories" of refined carbohydrates.

4. *For fats, rely on oils from plants, nuts, and fish.* Most people think fat is bad, but it is an essential part of our daily diet. The problem is not fat itself, but saturated and trans fats. A healthy diet replaces these harmful fats with oils from plants, nuts, and fish. The result is lower LDL cholesterol, higher HDL cholesterol, and less risk of cardiovascular disease.

5. *Consume a "rainbow" of vegetables and fruits each day.* A diet high in vegetables and fruits, which provide vitamins, minerals, fiber, and beneficial plant chemicals called *phytochemicals*, decreases the risk of cardiovascular disease, some types of cancer, and age-related loss of vision.

6. *For protein, choose fish, poultry, eggs, nuts, seeds, beans, and tofu.* These foods are good sources of protein because they provide healthy fats, vitamins, and minerals. Some offer fiber, too. Tofu is made from soybeans and is a good source of high-quality protein—it is also low in saturated fat. Eggs have gotten a bad reputation. Although they do contain some cholesterol, dietary cholesterol is a minor factor in blood cholesterol. So in moderation, eggs are a good source of protein.

7. *Consume adequate calcium and vitamin D.* Calcium and vitamin D are important for bone health. As you learned in Chapter 5, people with dark skin and those who do not get much sun exposure may not be able to produce enough vitamin D naturally to support good bone health. For them, a vitamin D supplement may be necessary. Fortified milk is a good source of vitamin D and calcium, but regular versions contain saturated fat, so it is wise to choose low-fat or skim milk. Other good sources of calcium include cheese and yogurt, certain nuts, beans, and dark-green leafy vegetables.

THE HEALTHY EATING PYRAMID

©Department of Nutrition, Harvard School of Public Health

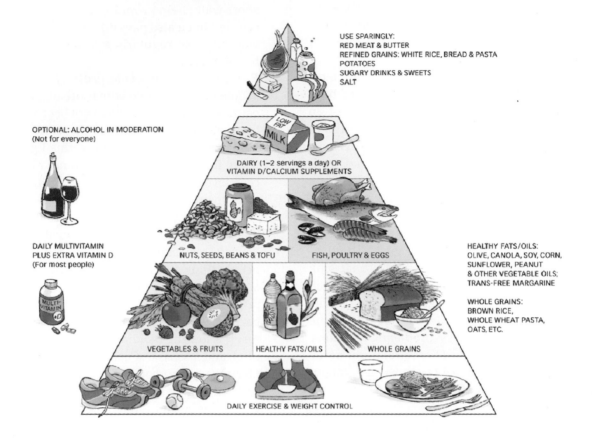

Figure 14.3. The Harvard Healthy Eating Pyramid. *Which oils should be consumed in larger amounts—butter or plant oils?*

8. *Consume red meat, butter, cheese, and whole milk sparingly because of their high content of saturated fat.* Red meat is often touted as a valuable source of iron—a trace mineral required for hemoglobin production ◀ (Chapter 10). However, poultry, fish, many legumes, and iron-fortified breads and cereals are rich in iron.

9. *Certain other foods should also be consumed sparingly.* The list includes refined grains (white bread, white rice, cakes, cookies), white potatoes (sweet potatoes are much healthier), candy and sugary drinks, and table sugar. It is wise, too, to add as little salt (sodium chloride) to food as possible. High-sodium diets increase the risk of high blood pressure and the cardiovascular disease associated with it.

10. *Multivitamin–mineral supplements* can fill in the occasional gaps in any diet. As mentioned above, vitamin D supplements are beneficial for some people. An iron supplement may be wise for women in their reproductive years because they lose iron with each menstrual period. Folic acid supplements are

recommended for all women of childbearing age because it helps prevent a type of birth defect that develops before the woman even knows she is pregnant. Otherwise, research evidence shows no benefit to taking supplements of individual vitamins or minerals unless medically prescribed.

11. *Alcohol in moderation* can contribute to a healthy diet, especially in middle-aged people. The data are clear: Americans who consume alcohol in moderation have less cardiovascular disease than those who do not. Moderation means no more than one drink (one beer, one glass of wine, one cocktail) daily for nonpregnant women and one to two drinks daily for men. Pregnant women should not consume any alcohol because of the risk of alcohol-related fetal abnormalities.

14.1   Which of the following are classified as nutrients? Water, amino acids, oxygen, lipids, vitamins, carbohydrates.

14.2   Can humans digest fiber, starch, or both?

14.3   True or false: The fatty acids in saturated fats contain one or more double bonds between the carbon atoms.

14.4   Which type of lipoprotein returns cholesterol to the liver?

14.5   Can humans make essential amino acids?

14.6   Which of the following is a *provitamin*: iron, beta carotene, retinol.

14.7   Which mineral is a critical structural component of hemoglobin?

14.8   Which lipids are healthier—meat fats or plant oils?

Overview of the Digestive System

Now that we have learned about what we eat, let us learn what happens when we do. *Digestion* breaks food into its molecular constituents, which are in turn metabolized for energy or used to build new molecules, organelles, cells, tissues, and organs. The *digestive system* consists of the *gastrointestinal tract* plus accessory organs.

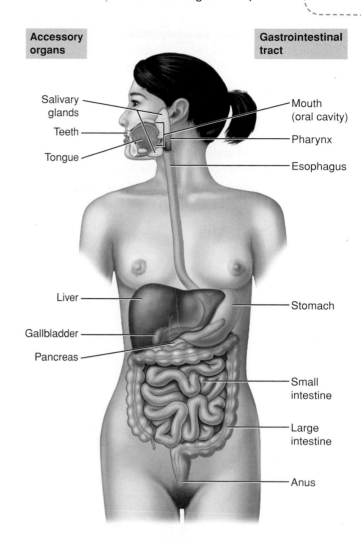

Figure 14.4. The gastrointestinal system—consisting of the gastrointestinal tract and the accessory organs. *Which of the following is an accessory organ: the gallbladder or the pharynx?*

The Gastrointestinal Tract and Accessory Organs Constitute the Digestive System

The **gastrointestinal (GI) tract** (also called the *digestive tract*) is a continuous sequence of muscular, tubular organs that extend from the lips to the anus. These are the mouth, pharynx, esophagus, stomach, small intestine, and large intestine (Fig. 14.4). In addition, digestion requires the assistance of the *accessory digestive organs*, which do not necessarily encounter food but provide mechanical or chemical tools. Accessory digestive organs include the teeth, tongue, salivary glands, liver, gallbladder, and pancreas.

The liver and pancreas deserve special mention because each has a dual nature—a digestive function and nondigestive activities that are varied and affect the entire body. Both secrete digestive juices into the GI tract. However, the liver also plays essential metabolic roles, and the pancreas has an endocrine gland function— it synthesizes and secretes hormones that regulate blood sugar. The nondigestive functions of the liver and pancreas are discussed in ➡ Chapter 15.

Case Note

14.4. When Margot claimed that "the pills upset her stomach," she was referring to pain in the umbilical region of the abdomen. Which part of the digestive tract is deep to the umbilicus (Fig. 14.4)?

The Peritoneal Cavity, Peritoneum, and Mesenteries Contain the Abdominopelvic Viscera

The GI tract bears an important relationship to the **peritoneum,** a translucent membrane composed of squamous epithelial cells. The portions of the peritoneum that cover the abdominal wall are described as the **parietal peritoneum.** Other portions, described as the **visceral peritoneum,** fold over all or part of the abdominal organs (Fig. 14.5). Between the parietal and visceral peritoneum lies a space, the **peritoneal cavity,** that is

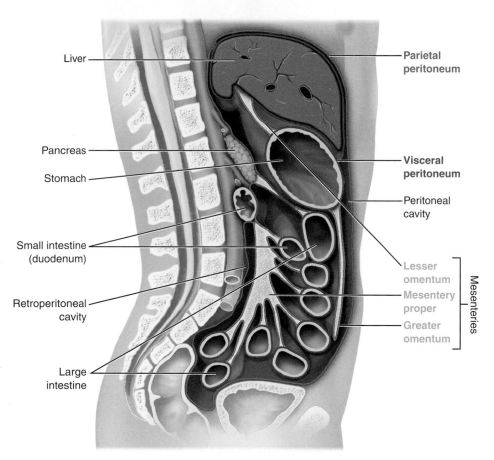

Liver

Pancreas

Stomach

Small intestine (duodenum)

Retroperitoneal cavity

Large intestine

Parietal peritoneum

Visceral peritoneum

Peritoneal cavity

Lesser omentum

Mesentery proper

Greater omentum

Mesenteries

(a) The peritoneum

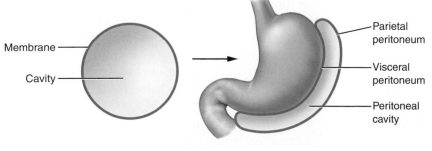

Membrane

Cavity

Parietal peritoneum

Visceral peritoneum

Peritoneal cavity

(b) Formation of the peritoneum

Figure 14.5. The peritoneum. A. The peritoneal membranes line the abdominopelvic cavity and cover most abdominal organs. Mesenteries are folded-over layers of peritoneum containing nerves and vessels. **B.** The peritoneal membranes form in the embryo when the abdominal organs push their way into the peritoneal cavity. *Which peritoneal layer covers the stomach?*

filled with slippery **peritoneal fluid.** This fluid not only enables the membranes to slide over each other but also allows the abdominal viscera to slide around in the abdomen relative to the chest wall and other visceral organs.

The embryonic formation of the peritoneum is very similar to the formation of the pericardium surrounding the heart (See ◀ Figure 1.14). The peritoneal cavity begins as a balloonlike sphere attached to the anterior body wall, with the peritoneum forming the wall of the balloon. All of the abdominal organs lie posterior and inferior to it. As these organs develop, the abdominal organs push their way anteriorly and the pelvic organs push superiorly into the peritoneal cavity like fists pushing into the wall of a balloon. The portion of the balloon wall directly touching the organ becomes the visceral peritoneum; the other side of the balloon wall remains anchored to the body wall as the visceral peritoneum. The space inside the balloon remains as the peritoneal cavity, which exists only as a thin potential space separating the two layers.

As the fetus grows and the organs advance into the peritoneal cavity, they carry their blood and lymphatic vessels and nerves, which lengthen but remain attached to the aorta, vena cava, and lymph ducts. Peritoneal membrane envelops these bundles of vessels and nerves to form a **mesentery,** a double fold of peritoneum that anchors and stabilizes them.

The major mesenteries are:

- The *greater omentum,* which hangs from the large intestine and stomach like an apron that covers the small intestines. Fat stored between the layers of the greater omentum accounts for much of the abdominal girth in overweight people.
- The *mesentery proper,* which arises in a vertical strip anterior to the lumbar vertebrae and attaches to the small intestine.
- The *lesser omentum,* which suspends the stomach from the undersurface of the liver.

The pancreas, the proximal part of the small intestine (called the *duodenum*), and parts of the large intestine do not advance but remain near the spine in the **retroperitoneal cavity** (*retro* = "behind").

Remember This! **A mesentery consists of two membrane layers, which fold over an organ like a towel folded over an arm—the towel is the peritoneal membrane and the arm is an organ.**

The Gastrointestinal Tract Has Four Tissue Layers

From the esophagus to the anus, the GI tract has the same basic structure: four layers of tissue, which vary somewhat in microscopic detail according to location and the task at hand. The layers, from the lumen outward, are mucosa, submucosa, muscularis, and serosa (Fig. 14.6).

- **Mucosa.** The mucosa lines the lumen of the canal and consists of three layers:
  1. The *epithelium* is an innermost layer of epithelial cells in direct contact with the luminal contents. Although it is composed of columnar epithelium throughout most of the digestive tract, the epithelium is made up of stratified squamous epithelial cells in the esophagus and anus. Modified epithelial cells called *goblet cells* secrete mucus, which lubricates food and provides a protective film over the epithelium. Other functions of the epithelium are to form a physical barrier that separates the body's internal environment from bacteria and other intestinal contents, to secrete digestive enzymes and hormones, and to absorb the end products of digestion.
  2. The *lamina propria* is a layer of loose areolar connective tissue. It contains blood and lymphatic vessels and collections of lymphocytes (*mucosa-associated lymphoid tissue* [*MALT*], discussed in ◀ Chapter 12).
  3. The *muscularis mucosae* is a wispy layer of smooth muscle. Contraction of these muscle fibers wrinkles the mucosa into folds.
- **Submucosa.** The submucosa is a broad layer of moderately dense connective tissue. It contains blood and lymphatic vessels as well as glands composed of deep infoldings of mucosal epithelium. It also contains a nerve network, the **submucosal nerve plexus,** which helps regulate mucosal function.
- **Muscularis externa.** The muscularis externa is the muscular workhorse of the digestive tract. The proximal GI tract (mouth, pharynx, and proximal esophagus) and the external anal sphincter contain skeletal muscle. The muscularis externa in the intervening regions (stomach, small intestine, large intestine, rectum) is composed of two layers of smooth muscle, an inner *circular* layer and an outer *longitudinal* layer. Between the two muscle layers is the **myenteric nerve plexus,** a second nerve network that regulates the muscularis externa.

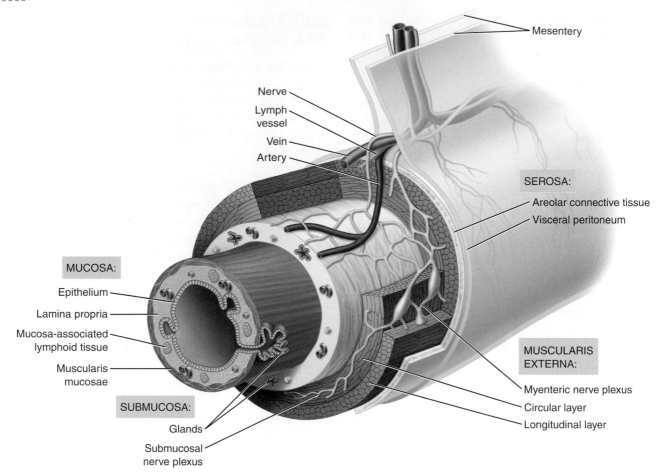

Figure 14.6. Layers of the gastrointestinal tract wall. The four major layers are the mucosa, submucosa, muscularis externa, and serosa. Each layer also contains blood and lymphatic vessels and nerves. *Name the two major gastrointestinal nerve networks.*

- **Serosa.** The outermost serosa is composed of the visceral peritoneum and a thin layer of areolar connective tissue between the peritoneum and the muscularis externa.

Together, the submucosal and myenteric nerve plexuses form the **enteric nervous system** (ENS), or "gut brain." This name is more apt than you might think: the ENS contains more neurons than the spinal cord. As discussed further on, the ENS interacts with the autonomic nervous system to regulate digestive function.

Case Note

14.5. Margot's digestive upsets sometimes involved inflammation in the cells lining the intestinal lumen. What type of cells are they, and which of the four GI layers is affected?

The Digestive System Performs Six Important Functions

The digestive system functions like a recycling plant, where objects moving along a conveyor belt are broken down into simple components that can be reused for other functions and the unusable leftover material is discarded. In like manner, the GI tract is divided into a series of distinct regions, each with a particular function. Muscular contractions of the tract wall are the equivalent of a conveyor belt, and the digestive tract and accessory organs provide the tools and procedures to accomplish digestion.

The Stages of Digestion

Collectively, the digestive system processes food through six basic steps (Fig. 14.7):

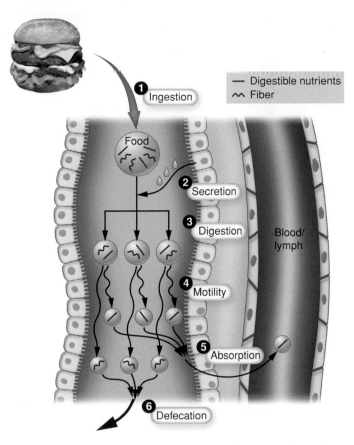

Figure 14.7. Gastrointestinal processes. The gastrointestinal tract extracts usable substances from food by performing six processes. *Motility only occurs after digestion is complete: true or false?*

1. **Ingestion** (eating) brings food into the oral cavity of the digestive tract.
2. **Secretion** is the release of fluids—mainly water with acid, buffers, and/or enzymes—into the lumen by epithelial cells and accessory digestive organs. Recall from ◀ Chapter 2 that enzymes are molecules, usually proteins, most with names ending in *ase*, that are designed to accelerate particular chemical reactions. For example, a protease accelerates the breakup of proteins. These exocrine secretions provide the chemical tools for digestion and help protect the digestive tract's wall.
3. **Digestion.** *Mechanical digestion* is the tearing and cutting of food into small pieces by the teeth and by the churning action of the stomach. *Chemical digestion* is the cleaving of large foodstuff molecules into smaller ones by the secretions mentioned above.
4. **Motility.** Muscular contractions of the GI tract provide motility—they physically move foodstuffs from place to place and mix them with digestive secretions.

5. **Absorption** is the uptake of small molecules from the GI tract into the blood or lymph.
6. **Defecation** is the passage of feces—compacted indigestible food material, bacteria, and shed epithelial cells—through the anus.

> ***Remember This!*** **Remember that "digestion" includes mechanical activities, such as chewing and churning, as well as chemical digestion.**

Segmentation and Peristalsis Are Types of Motility

Two types of muscular activity occur as food moves through the GI tract:

- **Peristalsis** is a *wave* of circular muscle contraction that passes down the GI tract, propelling a food bolus ahead of it (Fig. 14.8A). Peristalsis occurs throughout the GI tract, but to a lesser extent in the small intestine than in other regions.
- **Segmentation** is not a wave but a *stationary* constriction that features regular contractions and relaxations of circular muscle (Fig. 14.8B). In most instances segmentation serves to mix contents rather than propel them along—some contents are pushed down the bowel, others up, sloshing them back and forth. Segmentation occurs mainly in the small intestine.

Case Note

14.6. **When Margot consumes foods containing lactose, her intestinal contents are propelled abnormally quickly through her GI tract. Which of the six digestive processes moves food through the GI tract?**

Chemical Digestion Breaks Food into Its Building Blocks

During the process of chemical digestion, nutrients are broken down into their building blocks as follows (Table 14.3 on page 545 identifies the chemicals involved.):

- Carbohydrates are broken down into monosaccharides (single sugar molecules).
- Triglycerides, the most common form of ingested lipid, are each broken down into a monoglyceride (a glycerol with one fatty acid attached) and two *free fatty acids*.

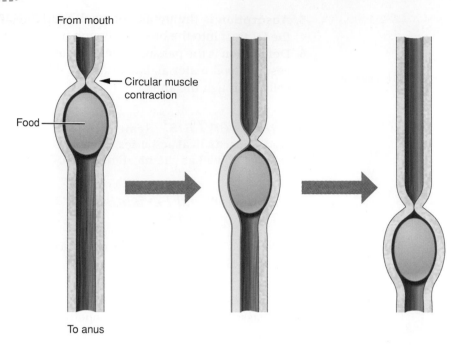

From mouth

Circular muscle contraction

Food

To anus

(a) Peristalsis

Circular muscle contraction

(b) Segmentation

Figure 14.8. Gastrointestinal motility. A. Peristalsis moves the luminal contents ahead of the contraction wave. **B.** Segmentation squishes the contents back and forth to mix them. *Which type of motility would be most useful in the esophagus, where we want the contents to move very quickly?*

● The proteins we consume, whether in plant or animal foods, are broken down into smaller chains called *polypeptides* or into individual amino acids.

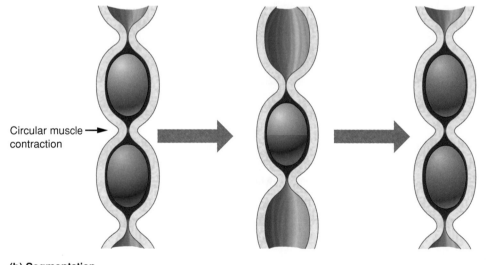

Pop Quiz

14.9 Which of the following is an accessory digestive organ: the esophagus or the pancreas?

14.10 Which of the following are classified as mesenteries: greater omentum, lesser omentum, mesentery proper?

14.11 Name the nerve plexus that regulates the function of the mucosa.

14.12 Are substances secreted into the digestive tract lumen endocrine or exocrine?

14.13 Which type of motility is best at propelling substances through the digestive tract—peristalsis or segmentation?

14.14 Is glucose a monosaccharide or a disaccharide?

Table 14.3   Digestive Enzymes

Name	Source	Function
Salivary amylase	Salivary glands	Digests starch into maltose
Pepsin	Gastric glands	Digests proteins into shorter polypeptides
Pancreatic amylase	Pancreas	Digests starches into shorter sugar chains
Pancreatic proteases (trypsin, chymotrypsin)	Pancreas	Digest polypeptides into shorter polypeptides
Pancreatic lipase	Pancreas	Digests triglycerides into monoglycerides and fatty acids
Nucleases	Pancreas	Digest nucleic acids (RNA, DNA)
Exopeptidases (carboxypeptidases, aminopeptidases)	Pancreas, intestinal brush border	Digest polypeptides to individual amino acids
Disaccharidases (lactase, maltase, sucrase)	Intestinal brush border	Digest disaccharides to monosaccharides

The Mouth and Associated Structures

The oral cavity (mouth) is framed by the cheeks, palate, tongue, lips, and pharynx (Fig. 14.9A). Every surface is lined by stratified squamous mucosa, which, like epidermis, is designed to withstand friction and take rough treatment because it sheds cells daily. For example, you scald your mouth with a hot pizza one day and it's fine the next because the injured cells have been replaced.

The lips are fleshy folds of tissue that form the anterior border and frame the opening of the mouth. On the outside they are lined by skin and on the inside by squamous mucosa. On the inside of each lip is a small fold of tissue, a labial *frenulum* (Latin *frenum* = "bridle"), which serves to reinforce the connection of the upper lip to the maxilla and the lower lip to the mandible. The lips seal the mouth during swallowing and chewing, manipulate food, and are important in speech. They merge with the cheeks to form the lateral border of the mouth. Both lips and cheeks contain strips of skeletal muscle that allow voluntary movement.

The palate forms the roof of the mouth (Figure 14.9B). The anterior palate, or hard palate, is roofed by bony plates of the maxillary and palatine bones and serves as a hard surface against which the tongue can manipulate food. The posterior palate, or soft palate, is a mobile shelf of tissue formed mainly of skeletal muscle that extends posteriorly from the hard palate. Hanging from the center of the palate is a finger of soft tissue, the uvula. With swallowing, the uvula and palate retract upward to prevent food from entering the back of the nose (the nasopharynx). Laterally, the soft palate attaches to folds of vertical tissue that contain the palatine tonsils (Chapter 12) and extend downward to attach to the base of the tongue.

The tongue occupies the floor of the mouth. Most of its bulk is formed of skeletal muscle, which enables chewing, swallowing, and speech. The superior aspect of the tongue contains taste buds (Chapter 9). The tongue tapers anteriorly and inferiorly into the labial frenulum, which anchors the tongue to the floor of the mouth and prevents it from moving too far posteriorly during swallowing or sleep.

Teeth Initiate Mechanical Digestion

The teeth are specialized, hard, bonelike organs (Fig. 14.9A). Humans have two sets of teeth. *Deciduous teeth* (Latin *decidere* = "to fall off") or *milk teeth* first appear at about age 6 months and regularly thereafter until all 20 are present by age 2 or 3 years. They are followed by *permanent teeth*, which begin to erupt from the gums about age 6. As they grow, they push upward, loosening the deciduous teeth, which fall out. Humans normally

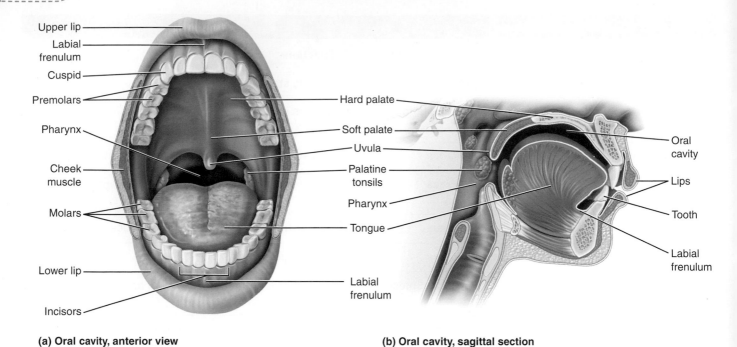

(a) Oral cavity, anterior view

(b) Oral cavity, sagittal section

Figure 14.9. The oral cavity. A. The oral cavity (mouth) is bordered by the palate, tongue, cheeks, and pharynx. **B.** This sagittal section highlights the structure of the tongue and palate. *Which structure is directly connected to the soft palate—the uvula or the frenulum?*

have 32 permanent teeth, which have differing shapes and functions according to location:

- Four *incisors* are present, top and bottom, nearest the midline in the anterior mouth. They have a single, sharp edge for cutting.
- Next laterally are the *cuspids* (canines), one on each side, top and bottom. These "vampire teeth" have a single, sharp point for piercing and tearing.
- Moving further away laterally and posteriorly are the two *premolars* (bicuspids), which have two points for tearing and crushing.
- At the far back on each side are three *molars* (Latin *molaris* = "millstone"), each of which has four or five points for grinding and crushing.

The teeth are set into sockets (*alveoli*) of the mandible and maxillae and rimmed by a collar of soft tissue, the *gums* or **gingiva** (Fig. 14.10). A tooth has two major parts, the crown and root. The **crown** is the visible part projecting out of the gums; the **root** is embedded in gingiva and alveolar bone. In the center of each tooth is the **pulp cavity,** which contains nerves as well as and blood and lymphatic vessels. The pulp cavity extends from the crown to the root tip, where nerves and vessels exit to make their external connections. The extension of the pulp into the root forms the **root canal.** Most of a tooth is composed of **dentin,** a bonelike specialized connective tissue that provides rigidity and surrounds the pulp cavity. The dentin of each root is covered by **cementum,** a second bonelike layer that

attaches to the **periodontal ligament** (Greek *odont* = "tooth"), a tough, fibrous sheath that anchors the tooth into alveolar bone. The dentin of each crown is covered by **enamel,** which owes its amazing hardness to its high calcium content and the orientation of calcium crystals.

Case Note

14.7. Margot was administered a very fast-acting medication at the clinic, which she placed adjacent to the frenulum of her tongue. In laypersons' language, where did she place the tablet?

Salivary Glands Secrete Saliva

Three pairs of *salivary glands* lie beyond the boundaries of the mouth and empty their secretions into the oral cavity through ducts (Fig. 14.11):

- The **parotid glands** are somewhat smaller than a deck of playing cards. These largest salivary glands are located anteriorly and inferiorly to the ear, over the masseter muscle. The parotid duct opens on the interior of the cheek at about where the molars meet.
- The **submandibular glands** are about one-third the size of the parotids and lie alongside the mandible in the floor of the mouth on either side of the base of the tongue. Each submandibular duct opens in the floor of the mouth beside the frenulum.

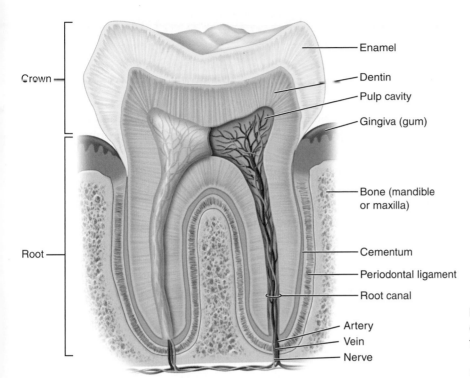

Figure 14.10. Anatomy of a tooth. The covering of the root canal and pulp cavity on the right side has been cut away. *What "cements" the periodontal ligament tot the tooth?*

- The smaller **sublingual glands** are anterior to the submandibular glands in the floor of the mouth. They connect to the frenulum by multiple small ducts.

Saliva is a watery fluid secreted into the mouth by the salivary glands. We secrete 1.5 L of saliva daily on average, containing:

- *Water,* which dilutes food, rinses the mouth, and dissolves food molecules so they can be detected by the taste buds.

- *IgA antibodies* and *lysozyme,* which keep the oral bacterial population under control.
- *Mucin,* which absorbs water to form mucus. Mucus lubricates food for easy swallowing.
- *Amylase,* the enzyme which initiates carbohydrate digestion. Amylase cleaves starches into the disaccharide maltose.

Case Note

14.8. Margot's tablets also include starch, which helps to bring water into the tablet to dissolve the active ingredient. If Margot holds the tablet in her mouth for a while, what metabolic process will happen to the starch?

Secretion, Motility, and Digestion Begin in the Mouth

Food is in the mouth for only a short time, but a number of digestive processes occur:

- *Secretion:* The salivary glands secrete saliva into the oral cavity.
- *Motility:* The tongue manipulates food for mastication and swallowing.
- *Digestion:* The teeth, tongue, and cheeks work together to initiate mechanical digestion, breaking up food into smaller pieces (*maceration*) and mixing it with saliva.

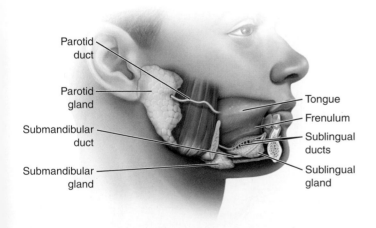

Figure 14.11. Salivary glands. The salivary glands secrete saliva into the oral cavity through small ducts. *Name the salivary gland directly inferior to the tongue.*

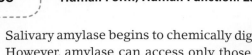

Salivary amylase begins to chemically digest starches. However, amylase can access only those starches on the outside of each relatively large food piece.

● *Absorption*: Food molecules are not absorbed in the mouth. However, some fast-acting medications can be absorbed through the mucous membranes. Nitroglycerin tablets, for instance, are used to treat heart disease by dilating the coronary vessels. They can be placed under the tongue at the first hint of trouble, and the medication rapidly enters the bloodstream.

14.15　Which part of the palate is more anterior—the soft palate or the hard palate?

14.16　How many molars are found in the adult mouth?

14.17　Does cementum cover the root or the crown?

14.18　Which salivary glands are the largest?

The Pharynx and Esophagus

When food is swallowed, it passes first from the mouth into the **oropharynx,** a funnel-shaped passage that sits directly posterior to the mouth (Fig. 14.12). Its wall is composed of *skeletal* muscle and it is lined by stratified squamous epithelium. Superior to the oropharynx is the *nasopharynx*. Immediately inferior to the oropharynx is the **laryngopharynx,** also composed of skeletal muscle and lined by squamous epithelium. It connects inferiorly to the esophagus for passage of food toward the stomach and anteriorly to the larynx for passage of air into the respiratory tree.

The **esophagus** is a thick-walled tube of smooth muscle, about 10 in. (25 cm) long, which begins at the lower end of the laryngopharynx, extends downward in front of the vertebral column, passes through an opening in the diaphragm (the *hiatus*), and ends at its attachment to the stomach. It is stabilized above by its connection to the laryngopharynx and below by the encircling diaphragm.

The esophagus is lined by stratified squamous epithelium, which facilitates the smooth downward slide of

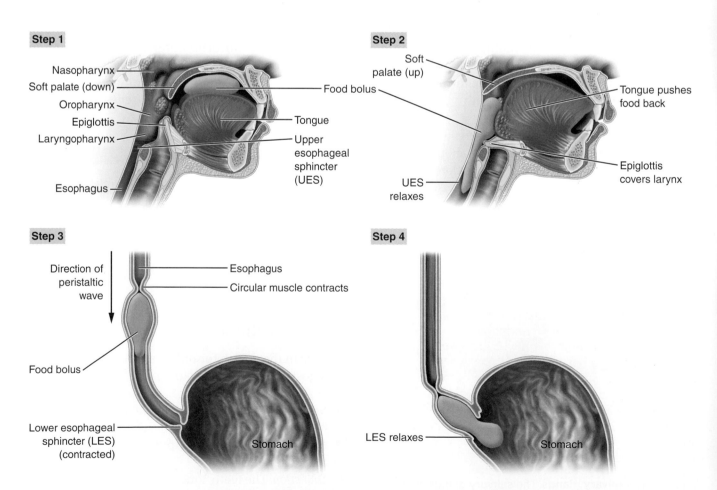

Figure 14.12. The pharynx and swallowing. Swallowing involves the coordinated efforts of skeletal muscles in the mouth, pharynx, and esophagus. *Does the epiglottis cover the larynx or the esophagus during swallowing?*

food. At the lower end of the esophagus, it joins the stomach at a region called the *gastroesophageal junction*; at this point the squamous epithelium abruptly changes to the acid-resistant, mucus-secreting columnar epithelium that lines the stomach.

In swallowing, the tongue pushes a bolus of food posteriorly toward the oropharynx (Fig. 14.12, step 1). Once the food reaches the oropharynx, it initiates a reflex contraction of the pharyngeal muscles that propels the bolus down into the laryngopharynx. The soft palate and uvula rise to seal the entrance to the nasopharynx, and the epiglottis folds neatly over the opening of the larynx to seal it and prevent food from entering the airway. The wave of electrical activity that contracts the pharynx wall relaxes the *upper esophageal sphincter (UES)*, a circular band of skeletal muscle. Relaxation of the UES allows the bolus to enter the esophagus (step 2). The electrical wave continues down the esophagus, triggering a peristaltic contraction that pushes the bolus ahead of it (step 3). Finally, the bolus reaches the *lower esophageal sphincter (LES)*, a circular band of smooth muscle. Relaxation of the LES allows the bolus to enter the stomach (step 4).

Immediately after the bolus has passed, the sphincter contracts again to prevent stomach contents from refluxing upward. This quick closure protects the squamous epithelium of the esophagus, which would otherwise easily be damaged by stomach acid. Failure of this mechanism and resultant gastric acid reflux is the cause of gastroesophageal reflux disease, which is discussed in the nearby clinical snapshot titled *Gastroesophageal Reflux Disease (GERD)*.

 ## CLINICAL SNAPSHOT

Gastroesophageal Reflux Disease (GERD)

A leaky lower esophageal sphincter can allow the acidic gastric contents to reflux back into the esophagus, particularly when a person is lying down. The highly corrosive gastric acid damages the lower esophagus, causing inflammation and pain, a condition called *gastroesophageal reflux disease (GERD)* or *reflux esophagitis.* The regurgitated gastric acid can even project all the way to the trachea and enter the lungs, scarring lung tissue and increasing the chances of lung cancer.

In developed countries, about 10% of adults, especially those over 40, have some degree of GERD. Overweight and obesity are risk factors, as is cigarette smoking. A hiatal hernia, in which the superior portion of the stomach protrudes through the hiatus, also increases the risk. The condition also afflicts up to 35% of babies, leading to frequent milk spit-ups and increased susceptibility to respiratory illnesses. Pregnant women also transiently suffer from GERD, since estrogen (which is produced in large amounts during pregnancy) relaxes the lower esophageal sphincter.

The dominant symptom is pain, which is often difficult to distinguish from heart pain because the heart is adjacent to the lower esophagus. Indeed, the pain from GERD is commonly called **heartburn**. Some cases of GERD are, however, asymptomatic. Complications include bleeding and fibrous scarring (stricture). In especially severe cases (about 10%), the squamous epithelium is so damaged that it changes into tall, columnar, acid-resistant epithelium like that found in the stomach, a process called *metaplasia* (Greek *metaplassein* = "to mold into a new

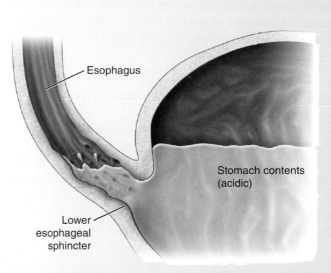

Gastroesophageal reflux disease (GERD). The lower esophageal sphincter fails to keep stomach contents away from vulnerable esophageal tissues.

form"). This change is associated with an approximate 40-fold increased risk for esophageal cancer.

GERD can often be successfully treated by lifestyle modifications, such as elevating the head of the bed and avoiding reflux triggers such as alcohol, caffeine, and spicy foods before bed. Losing weight and quitting smoking are often helpful. Common medications such as antacids can reduce the severity of GERD symptoms by reducing stomach acidity; more severe cases are treated with drugs that inhibit the secretion of gastric juice.

The Stomach

Most meals are eaten much more quickly than the body can digest them. Thus, the GI tract needs a "holding tank" from which food can be released in batches over several hours as intestinal capacity permits. The stomach serves this purpose and two more: while food is "waiting its turn," the stomach mixes and churns it and begins the process of protein digestion. Very little material is absorbed by the stomach; mainly some water, alcohol, minerals, and drugs.

The Stomach Has Four Regions

The **stomach** is a J-shaped funnel that meets the *esophagus* above and empties into the *duodenum* of the small intestine below. It lies in the upper left quadrant of the abdomen, tucked beneath the lower edge of the left rib cage. It is relatively immobile at each end but has considerable room for movement in between.

The stomach has four main regions (Fig. 14.13). In descending order they are as follows:

- The *cardia* is the ring of stomach that adjoins the esophagus. It is so named because the heart is immediately superior to this region, on the other side of the diaphragm.
- The *fundus* is the bulbous, superior part to the left of the cardia.
- Below the fundus is the *body*, the largest part of the stomach.
- Below the body is the pylorus (Greek *pulouros* = gatekeeper), a funnel-shaped region that narrows as it connects to the duodenum.

A thick ring of smooth muscle, the *pyloric sphincter*, circles the junction of the pylorus and duodenum.

The left (lateral) rim of the stomach is called the *greater curvature*. The right (medial) rim is the *lesser curvature*. Like other parts of the GI tract, the stomach is encased within folds of peritoneum, which anchor it to the posterior wall of the abdominal cavity.

The stomach is innervated by the autonomic nervous system. Sympathetic fibers reach the stomach from the celiac ganglion; parasympathetic fibers are supplied by the vagus nerve (cranial nerve X) (see ← Figures 8.21 and 8.22).

Arterial blood reaches the stomach from left gastric artery, a branch of the celiac trunk (see Plate ← 11.2). Venous return from the stomach travels through the left and right gastric veins to the portal vein. Portal blood passes through the liver for processing, as discussed below (see Plate 11.3).

The Stomach Wall Is Composed of Four Layers

The stomach is composed of the same four layers as other parts of the GI tract: mucosa, submucosa, muscularis, and serosa (Figs. 14.13 and 14.14). When the stomach is empty, tonic contractions of the muscularis "wrinkle" the mucosa and submucosa into large folds, called *rugae* (Latin *rugosus* = "wrinkle"). The muscularis externa relaxes when arriving food stretches the stomach, and the rugae flatten. The muscularis externa has the usual inner circular and outer longitudinal layers of smooth muscle, which combine to propel food into the intestine. However, the stomach has a third, innermost *oblique layer* that crosses the other two layers at about 45 degrees and enables it to churn food in a way unlike other parts of the digestive tract. The outermost layer of the stomach is the serosa, which is formed of visceral peritoneum and a thin layer of fibrous tissue between the peritoneum and the muscularis externa.

The Mucosa Secretes Mucus, Gastric Juice, Hormones, and Enzymes

The stomach's mucosa contains numerous tiny openings, or *gastric pits*, which are tubular invaginations of surface mucosa (Fig. 14.14). *Surface mucous cells* line the surface mucosa and the gastric pits. At the bottom, each gastric pit opens onto *gastric glands*. Each gastric gland contains *mucous neck cells*, which secrete mucus that protects the stomach lining. Depending on their location, gastric glands may contain additional cell types that empty other secretions into the gastric pit. For example, gastric glands in the body of the stomach secrete large amounts (about 1.5 L/day) of acidic *gastric*

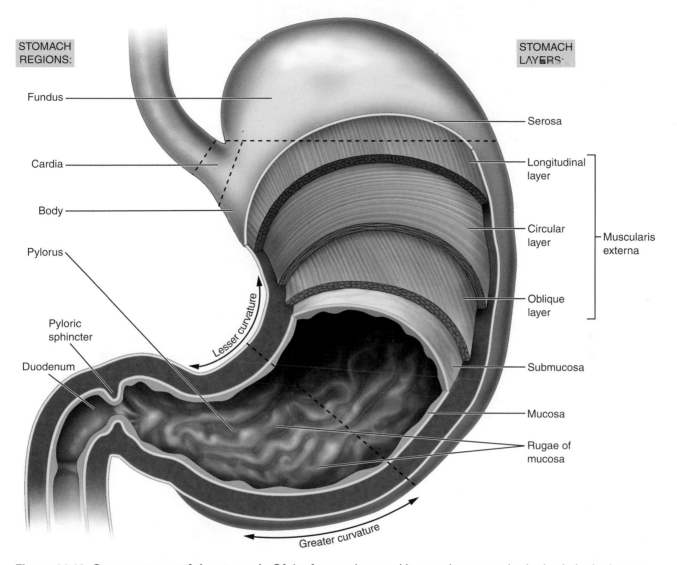

STOMACH REGIONS:

- Fundus
- Cardia
- Body
- Pylorus
- Pyloric sphincter
- Duodenum

Lesser curvature

Greater curvature

STOMACH LAYERS:

- Serosa
- Longitudinal layer
- Circular layer — Muscularis externa
- Oblique layer
- Submucosa
- Mucosa
- Rugae of mucosa

Figure 14.13. Gross anatomy of the stomach. Of the four regions making up the stomach, the body is the largest. Gases are stored in the most superior stomach region. *What is this region called?*

juice. Gastric juice contains secretions from two specialized cell types; *parietal cells* and *chief cells.*

Parietal cells simultaneously secrete *hydrochloric acid* and *intrinsic factor* into the gastric lumen. Hydrochloric acid (HCl) is a very strong acid (the pH of gastric juice ranges from 1 to 3.5). The acidity of gastric juice is important in the digestion of protein (discussed below) and kills most pathogens ingested in food or drink. *Intrinsic factor* is a glycoprotein that binds with vitamin B12 so that it can be absorbed by the small intestine. Since vitamin B12 is necessary for DNA synthesis, red blood cell synthesis, and neuron health, a deficiency of intrinsic factor can be devastating.

Chief cells secrete *pepsinogen* into the gastric lumen. It is an inactive precursor to **pepsin** (Greek *pepsis* = digestion), a protein-digesting enzyme discussed further on.

Gastric juice is so acidic that it can literally dissolve nails, so why does it not dissolve the gastric mucosa? First, the glands that secrete gastric acid are acid-resistant and very tightly joined together to prevent gastric juice from leaking between cells into the submucosa, where acid could do great damage. Second, gastric mucus forms an alkaline, viscous coating that physically prevents acid from reaching the epithelial cells and chemically neutralizes any stray acid molecules that get

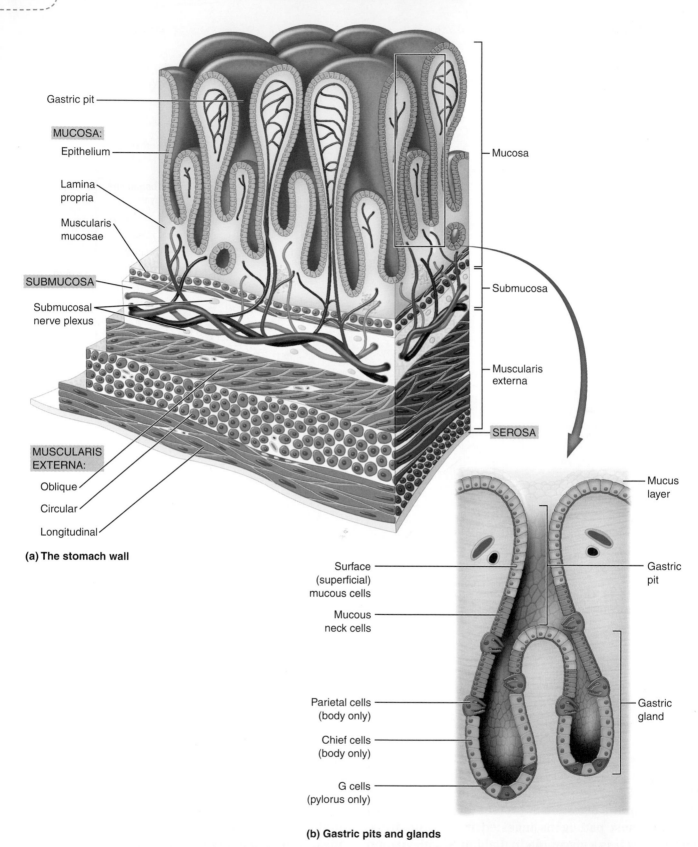

Gastric pit

MUCOSA:

Epithelium

Lamina propria

Muscularis mucosae

SUBMUCOSA

Submucosal nerve plexus

MUSCULARIS EXTERNA:

Oblique

Circular

Longitudinal

Mucosa

Submucosa

Muscularis externa

SEROSA

(a) The stomach wall

Mucus layer

Surface (superficial) mucous cells

Gastric pit

Mucous neck cells

Parietal cells (body only)

Gastric gland

Chief cells (body only)

G cells (pylorus only)

(b) Gastric pits and glands

Figure 14.14. The stomach layers and gastric pits. A. This cross-section of the stomach wall highlights the arrangements of the three layers of the muscularis externa. **B.** Gastric glands in the stomach body secrete hydrochloric acid and pepsinogen into the stomach lumen. Gastric glands in the pylorus secrete hormones into the blood and interstitial fluid. Mucous cells are found throughout the stomach. *Gastric glands are entirely exocrine glands: true or false?*

too close. Third, surface epithelial cells do not live long enough to be damaged by acid that circumvents the previous two mechanisms, since they are shed after a few days and replaced by new cells from a reserve of stem cells.

However, if gastric acid penetrates the epithelium, it can do severe damage. The persistent presence of acid beneath the mucosa eats away at underlying tissues, creating a craterlike defect in a mucosal surface called an *ulcer.* Surprisingly, ulcers are mainly due to infection by *Helicobacter pylori,* a bacterium that arrives in contaminated food. It evades the destructive gastric juice by burrowing deep into the protective mucous layer and even secretes bicarbonate to buffer its surroundings. Other bacterial secretions inflame and damage the mucosa, allowing gastric acid access to unprotected submucosal tissue. But other factors are also important. Peptic ulcers occur more frequently in men than in women, in those who abuse alcohol, in smokers, and in those taking long-term, high doses of nonsteroidal anti-inflammatory drugs (for example, aspirin or ibuprofen) for chronic pain. Treatment includes antibiotics to eliminate the *H. pylori* and other medications to reduce gastric acid secretion.

Secretions from chief cells and parietal cells work together to initiate protein digestion (Fig. 14.15A). Recall that proteins are long chains of amino acids coiled and folded into three-dimensional shapes. First, hydrochloric acid straightens out (denatures) proteins so that enzymes can access the bonds between the amino acids (step 1). HCl also converts inactive pepsinogen into active pepsin (step 2). Active pepsin can cleave and activate other pepsinogen molecules into pepsin (step 3). Finally, pepsin cleaves the peptide bonds between certain amino acids, producing shorter peptides (step 4).

> *Remember This!* **Chief cells are located deep in gastric glands. The pepsinogen they secrete remains inactive until it nears the lumen of the stomach, where gastric acid converts it to active pepsin. This activation mechanism is protective: were pepsin secreted in active form deep in the gastric wall, it would digest the stomach itself.**

The gastric glands in the pylorus do not secrete much gastric juice because they do not have many parietal or chief cells. Instead, they contain several types of specialized endocrine cells that secrete hormones into the interstitial fluid and bloodstream. Among them are **G cells,** which secrete the hormone *gastrin* into blood.

We discuss gastrin's role in regulating stomach function at the end of the chapter.

Case Note

14.9. Margot's previous medication was surrounded by an enteric coating, that it would dissolve only in an alkaline environment. Will this coating dissolve in the stomach? Explain.

The Muscularis Externa Digests Food Mechanically

The muscularis externa participates in mechanical digestion (maceration) by generating strong peristaltic waves (Fig. 14.15B), which are self-sustaining, somewhat like those of the heart's sinoatrial node. They move through the gastric wall every few minutes, beginning quite weakly in the fundus and gathering strength as they move toward the pylorus. Like a clothes dryer, the waves tumble food, thoroughly exposing its proteins to acid and pepsin for chemical digestion. With each wave, the pyloric sphincter relaxes slightly and temporarily to allow a small volume of stomach contents to enter the duodenum. In the stomach, food is macerated into ever smaller pieces until it enters the duodenum as a creamy pulp called chyme (Greek *khumos* = "juice").

Most of the stomach contents, however, are retained for tumbling and digestion. This is important: chyme leaving the stomach is packed with dissolved foodstuff; it is therefore hypertonic and highly acidic. If it were released too rapidly, the acidity could destroy the duodenal lining, resulting in a duodenal ulcer. In addition, the osmotic pull of the glucose and other solutes in a large amount of chyme would attract a large volume of water, more than the intestine could handle. The result would be diarrhea—and poor digestion and absorption. This has clinical implications. For example, some surgical procedures, especially for cancer, require removal of the lower stomach and pyloric sphincter and connection of the small intestine directly to the greater curvature of the stomach. This type of "gastric bypass" eliminates the slow, measured release of chyme through the pyloric sphincter into the small intestine. As a result, these patients can suffer from "dumping syndrome," in which the large volume of chyme entering the intestine attracts water and causes osmotic diarrhea within a short time after a meal. In other words, the patient "dumps" his or her meal shortly after eating.

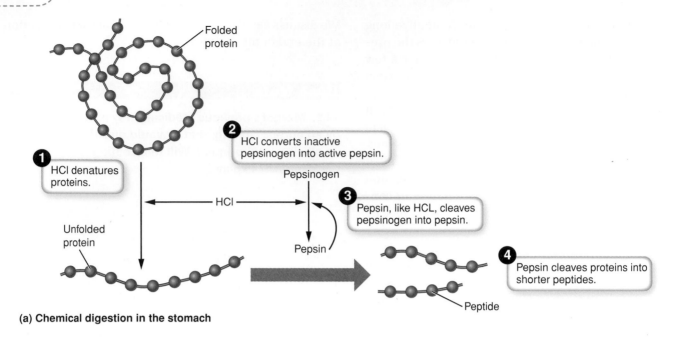

(a) Chemical digestion in the stomach

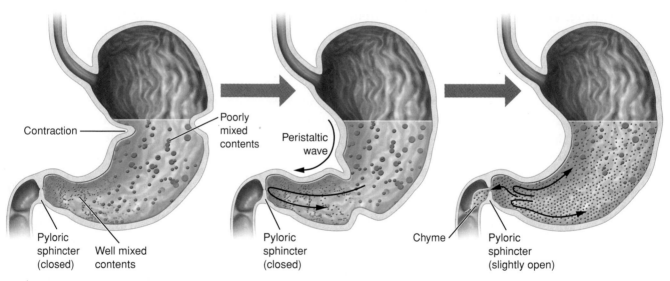

(b) Mechanical digestion in the stomach

Figure 14.15. Stomach function. A. Secretions from parietal and chief cells initiate protein digestion. **B.** The three layers of the muscularis externa contract to mix stomach contents and to eject small volumes into the duodenum. *What substance denatures proteins—hydrochloric acid or pepsin?*

14.22 Which sphincter is the most superior—the pyloric sphincter or the lower esophageal sphincter?

14.23 Name the muscle layer in the stomach that is closest to the mucosa.

14.24 Do chief cells secrete pepsinogen or pepsin?

14.25 Are G cells endocrine cells or exocrine cells? Explain.

14.26 Which form of motility occurs in the stomach—segmentation or peristalsis?

14.27 Name two substances capable of converting pepsinogen into pepsin.

The Small Intestine, Liver, and Pancreas

Upon exiting the stomach, chyme enters the lumen of the small intestine, where chemical digestion occurs with the aid of secretions from the *pancreas* and *liver*. These accessory organs are complex organs with non-digestive functions as well; these are discussed in ➡ Chapter 15. Here only their digestive functions are considered.

The Anatomy of the Small Intestine

The **small intestine** is a convoluted tube about 1 in. (2.5 cm) in diameter. It measures about 10 ft long in a living person (in vivo), but the smooth muscle relaxation that occurs at death lengthens the tube to close to 20 ft in cadavers (in our further discussions, we refer to in vivo measurements). It begins at the distal end of the stomach at the pyloric sphincter and ends where it joins the colon at the *ileocecal valve*.

The Small Intestine Consists of the Duodenum, Jejunum, and Ileum

The small intestine has three major parts, which from proximal to distal are the *duodenum*, the *jejunum*, and the *ileum* (Fig. 14.16).

The **duodenum** (Latin *duodeni* = "in twelves"—so named because it is about 12 fingerwidths or 10 in. long) is the initial, shortest segment. It extends in a C-shaped arc from the pyloric sphincter around the head of the pancreas to join with the jejunum. Just distal to the pyloric sphincter, the duodenum flexes posteriorly, exiting the peritoneal cavity and entering the retroperitoneum, which serves to anchor it in place. The duodenum specializes in chemical digestion and absorption.

The **jejunum** is about 3 ft long (Latin *jejunus* = "fasting," because the ancients usually found the jejunum empty at death). Where the duodenum ends and the jejunum begins is generally considered to be the point at which the intestine exits the retroperitoneal space and reenters the peritoneal cavity. At this point, the intestine becomes less restrained by mesentery and

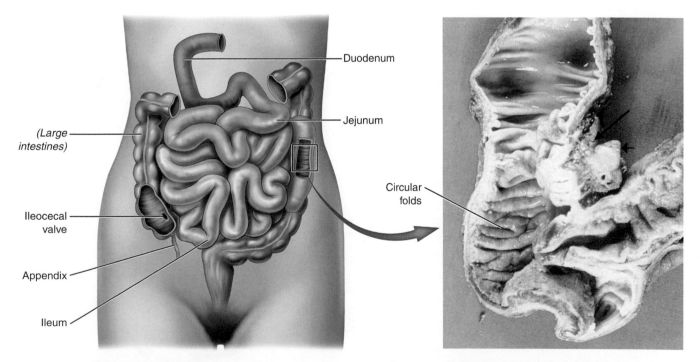

(a) Segments of the small intestine

(b) Circular folds of the small intestine

Figure 14.16. Gross anatomy of the small intestine. A. The small intestine consists of three segments. **B.** The circular folds are clearly visible in this cadaver photograph. *When chyme passes from the small intestine to the large intestine, which valve does it pass through?*

much more mobile. The jejunum specializes in nutrient absorption.

Even less precise is the transition from the jejunum to the final portion of the small intestine, the **ileum** (Latin *ilia* = "entrails"). The two regions join to form sausagelike coils that writhe in the abdominal cavity like snakes in a pit. The ileum is about 6 ft long and empties into the colon through a one-way valve, the *ileocecal valve*. The ileum absorbs any nutrients remaining in the chyme. It also absorbs *bile salts,* important factors in fat digestion, discussed further on.

The Wall of the Small Intestine

From inside to out, the layers of the small intestine are similar to other parts of the intestinal tract: a mucosa lined by epithelial cells, a submucosa containing blood vessels, lymphatics and lymphoid patches, a smooth muscle wall, and, outermost, a serosa of visceral peritoneum (Fig. 14.17). However, the anatomy of the small intestine's mucosa is distinctive so as to maximize the absorptive surface.

Although chemical and enzymatic digestion occur in the lumen of the small intestine, nutrient absorption

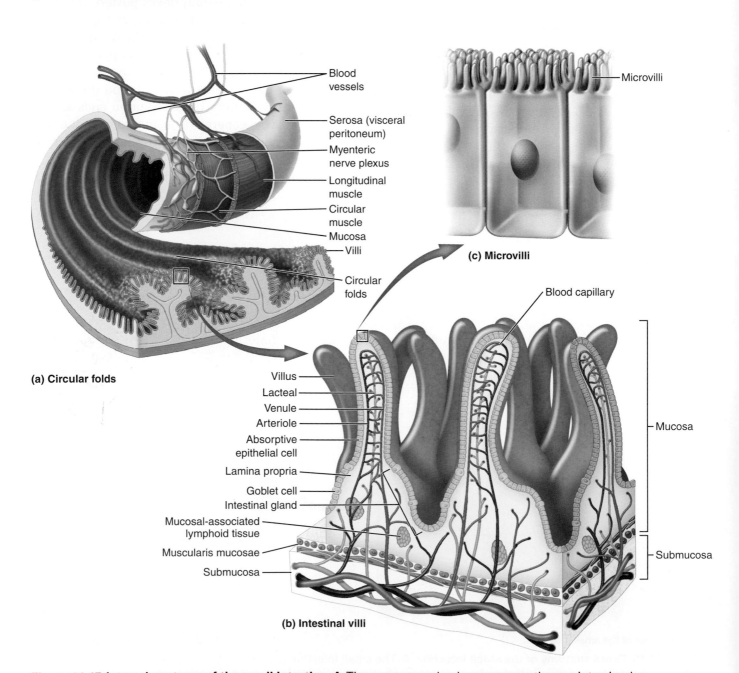

Figure 14.17. Internal anatomy of the small intestine. A. The mucosa and submucosa are thrown into circular folds. **B.** Each fold contains many villi. **C.** The plasma membrane of the absorptive cells in each villus is formed into microvilli. *Which vessel in **(B)** is part of the lymphatic system?*

occurs through the mucosa. Just as the honeycombed alveolar structure of the lungs crams enough respiratory membrane into the chest to cover the floor of a four-car garage, the mucosa of the small intestine provides several times that much surface for nutrient absorption in its 10-ft length.

First, the mucosa and submucosa are thrown into **circular folds,** like socks bunched at the ankle (Fig. 14.17A). These folds spiral down the length of the small intestine, and as chyme washes over them, the intestine twists around, delaying its transit. This increases the time and exposure for absorption.

Second, the lining epithelium of each fold is formed into innumerable tiny hairlike **villi,** which impart a velvety texture that greatly multiplies the surface area (Fig. 14.17B). Internally, each villus has a central core that contains a vascular loop: an incoming arteriole and an outgoing venule, bridged by capillaries. Nonlipid nutrients pass into blood through these capillaries. The core of each villus also contains a **lacteal** (Latin *lacteus* = "milk"), a blunt-ended lymph capillary devoted to the absorption of lipids.

Each villus is coated with a single layer of tall columnar cells, packed shoulder to shoulder. Most of these are *absorptive epithelial cells* that, as their name suggests, absorb nutrients. The surface area of each absorptive cell is magnified still further by the hundreds of microscopic **microvilli** that project from the cell membrane facing the lumen (Fig. 14.17C). When viewed microscopically, the absorptive surface resembles the bristles of a brush; it is therefore called the *brush border* and special digestive enzymes in the microvilli are called **brush-border enzymes.** The cell membranes of absorptive epithelial cells also contain many transporter proteins that participate in nutrient uptake. Interspersed among the absorptive cells are goblet cells, which provide mucus to lubricate and dilute chyme.

Between villi, the epithelium dips downward into the submucosa to form tortuous *intestinal glands*, which secrete a thin mixture of water and mucus (*intestinal juice*) that mixes with chyme and increases its fluidity. Intestinal glands in the duodenum (also called *duodenal* or *Brunner glands*) are particularly important, since their mucus-laden secretions are alkaline and help neutralize gastric acid. Note that intestinal glands, unlike gastric glands, do not secrete digestive enzymes.

In addition to absorptive and goblet cells, the intestinal glands also contain endocrine cells. These cells secrete *secretin* and *cholecystokinin* (discussed further on in this chapter) and other hormones that have an effect on the digestive process.

Case Note

14.10. As part of the research study, a surgeon takes a biopsy of Margot's small intestine and isolates the layer containing brush-border enzymes. Are these enzymes produced by the intestinal glands or the villi?

The Pancreas Secretes Digestive Juices

The pancreas is a fleshy, tan organ about 13 to 14 cm long that lies in the retroperitoneum, behind the stomach and in front of the aorta and inferior vena cava. It is tadpole-shaped, its head resting in the C-shaped curve of the duodenum (Fig. 14.18A). Its tail lies anterior to the left kidney and medial to the spleen. The pancreas contains two types of glands (◀ Chapter 3): endocrine glands, which secrete pancreatic hormones into the bloodstream; and exocrine glands, which secrete pancreatic juice into ducts draining into the duodenum.

The endocrine tissue of the pancreas constitutes less than 1% of the pancreatic mass. It is scattered into thousands of tiny islets of tissue called *pancreatic islets* (islets of Langerhans). Pancreatic islet hormones regulate blood glucose, among other things. Insulin, for instance, is released by pancreatic islets as blood glucose rises. The hormonal functions of the pancreas are discussed further in ◀ Chapter 15.

The great majority of pancreatic tissue consists of *exocrine glands* called *acini* (singular, *acinus*) and associated ducts (Fig. 14.18B).

The exocrine pancreas produces over 1 L of *pancreatic juice* daily—a clear, thin fluid composed mainly of water, bicarbonate, and digestive enzymes. The bicarbonate, secreted by epithelial cells lining small pancreatic ducts, renders pancreatic juice alkaline to help neutralize the acidic pH of chyme leaving the stomach. In addition, it creates a favorable environment for pancreatic enzymes, which work best at the alkaline pH (8.0 to 8.3) of pancreatic juice.

The acinar cells secrete a cocktail of digestive enzymes (see Table 14.3) to break down ingested nutrients. Some of these enzymes exist as inactive enzymes (zymogens), which become activated in the small intestine after secretion. This is particularly important for protein-digesting enzymes; if not secreted in inactive form, they would digest the pancreatic duct and the pancreas itself before reaching the intestine. Sometimes, however, things go awry. For example, if a tumor obstructs the pancreatic duct, pressure rises in the duct and zymogens leak into

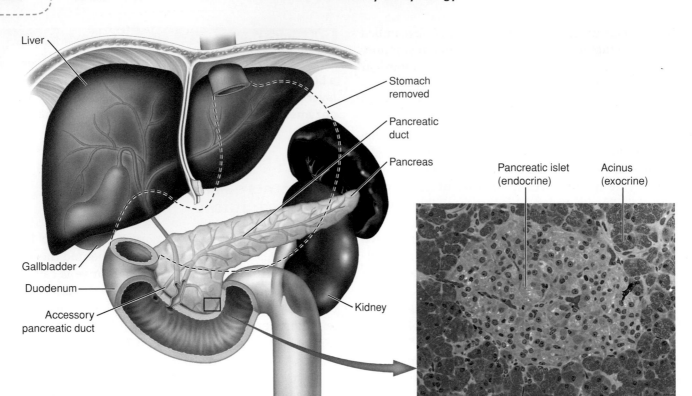

(a) Location of the pancreas   (b) Pancreatic glands

Figure 14.18. The pancreas. A. The pancreas is situated inferior to the stomach. **B.** The exocrine pancreas consists of acini of secreting cells that drain into ducts. *What are the exocrine portions of the pancreas called?*

the spaces between glands. This converts the zymogens into active enzymes, which begin to digest the pancreas itself. This, in turn, causes more zymogen leakage. A vicious cycle ensues and the result is pancreatitis, a serious, sometimes fatal condition.

Pancreatic juice travels through a branching network of ducts that join to form a central duct, the **pancreatic duct,** which empties into the duodenum (Fig.4.18A). A smaller duct, the **accessory pancreatic duct,** also channels pancreatic juice into the duodenum.

Case Note

14.11. The production of soy milk requires extensive boiling to destroy protease inhibitors present in soybeans. As a health-conscious teenager, Margot was consuming large amounts of unprocessed soy milk. Study Table 14.3. What are some enzymes that would be inhibited, and where are they produced?

The Liver Secretes Bile

The **liver** is far more than an accessory to the digestive system; it is arguably the queen of metabolic organs. It processes all of the nutrients absorbed from the intestine—a topic discussed in ➡ Chapter 15. Here we discuss only its digestive role, which is to secrete **bile,** a fluid that that contains *bile salts*. **Bile salts** are cholesterol-related chemicals that *emulsify* fats—they break large globules of fat into tiny droplets, which are more accessible to digestive enzymes.

Remember This! You can see the difference between large lipid droplets and an emulsion in a bottle of oil-and-vinegar salad dressing. An unshaken bottle contains a very large lipid droplet overlying an aqueous vinegar layer. Strong shaking, especially in the presence of an *emulsifier* like mustard, breaks the large droplet into many smaller droplets.

The liver is a highly vascular, meaty, brownish-red organ, second only to skin in size and weighing, on average, about 3 lb (1.4 kg) (Fig. 14.19). It sits under the right half of the diaphragm, tucked safely under the right rib cage, but a small part extends to the left of the midline.

Viewed from the exterior, the liver has four lobes. However, internally, the liver is uniform and without lobar boundaries. By far the largest is the *right lobe*, which is to the right of the midline. The *left lobe* is about half the size of the right and extends across the midline to form the narrow left side (Fig. 14.19A). Two small lobes are present on the inferior surface near the midline. The *caudate lobe* is posterior and the *quadrate lobe* is anterior and sits medial to the gallbladder (Figure 14.19B).

The liver is suspended from the diaphragm by the *falciform ligament*, which lies in the midline (Fig. 14.19A). This ligament divides the right and left lobes and arcs like a curved blade (Latin *falx* = "sickle") over the anterosuperior surface of the liver. With the exception of a round patch in the top center where the apex of the liver touches the diaphragm, the entire liver is covered by peritoneum.

Recall from ◀ Plate 11.3 that the liver and digestive tract participate in the *hepatic portal circulation*—which involves two capillary beds. Veins returning from the digestive tract, pancreas, and spleen merge to form the **hepatic portal vein,** which does *not* join the inferior vena cava. Instead, it redivides into a network of leaky capillaries in the liver, which enable liver cells to

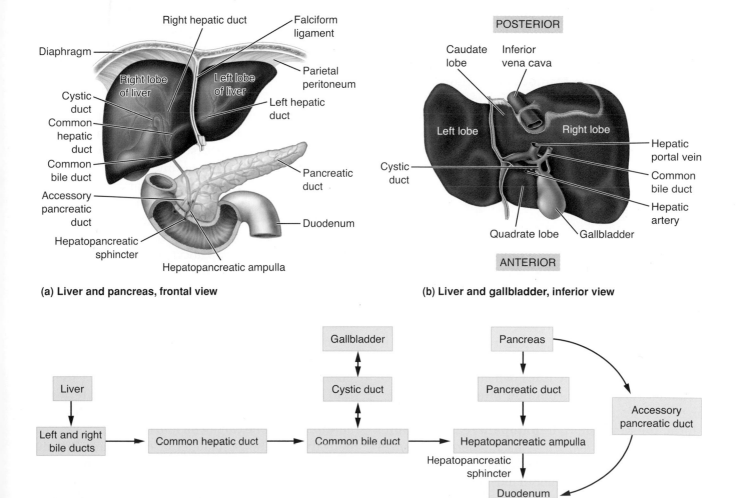

(a) **Liver and pancreas, frontal view**

(b) **Liver and gallbladder, inferior view**

(c) **The pancreatic and hepatic ducts**

Figure 14.19. The liver, gallbladder, and pancreas. A. The anatomical relationships between the liver, gallbladder and pancreas. **B.** This inferior view reveals two additional, smaller hepatic lobes and the many vessels that attach to the liver's underside. **C.** The flow of bile and pancreatic juices. Note that bile can flow in either direction in the cystic duct. *Name the ligament separating the right and left lobes of the liver.*

process the newly absorbed nutrients. The liver thus has a dual blood supply (Fig. 14.19B). The **hepatic artery** brings freshly oxygenated blood from the celiac branch of the aorta. The **hepatic portal vein** (or, commonly, the *portal vein*) brings nutrient-laden venous blood from the intestinal capillaries. The **hepatic veins** return all hepatic blood to the inferior vena cava.

The Gallbladder Stores Bile

The **gallbladder** is a "holding tank" for bile. It is a thin-walled, teardrop-shaped sac suspended from the underside of the liver (Fig. 14.19B). It is a bit larger than a golf ball and can store about 50 mL of bile until it is needed for the emulsification of fats. After doing their digestive work, the bile salts in bile are absorbed by the ileum and immediately returned through the portal vein to the liver, where they are secreted once again into bile.

Case Note

14.12. Like many drugs, Margot's medication is processed by the liver after its absorption from the intestine into the bloodstream. Which vessel initially carries the drug to the liver—the hepatic artery or the portal vein?

Hepatic and Pancreatic Secretions Drain into the Duodenum

An intricate system of ducts shuttles bile between the liver, the gallbladder, and the duodenum; it also shunts pancreatic juice from the pancreas into the duodenum (Fig. 14.19C). The liver continuously produces bile, which drains from the right and left lobes into the right and left hepatic ducts, respectively. These two ducts merge to form the **common hepatic duct.** The common hepatic duct joins with the **cystic duct** (which connects to the gallbladder) to form the **common bile duct.** The common hepatic duct subsequently joins the pancreatic duct to form the very short **hepatopancreatic ampulla** (diminutive of Latin *ampora* = "flask"; therefore, a small flask—also called the *ampulla of Vater*).

An encircling band of smooth muscle, the *hepatopancreatic sphincter*, controls the entry of bile and pancreatic juices from the ampulla into the duodenum. This sphincter closes when bile is not needed, so newly made bile backs up the cystic duct to be stored in the gallbladder. Signals to be discussed further on cause the gallbladder to contract and the sphincter to relax, releasing bile and pancreatic juice into the duodenum. A secondary

duct, the accessory pancreatic duct, bypasses the ampulla to drain pancreatic juices into the duodenum.

Microscopic Anatomy of the Liver

The main functional cell of the liver is the **hepatocyte.** The basic functional unit of the liver is the **hepatic lobule,** a cylindrical tube of hepatocytes surrounding a vein. Each is a few millimeters long and about 1 mm in diameter (Fig. 14.20A). An average liver contains about 50,000 to 100,000 lobules. The edges of each hepatic lobule are marked by **portal triads,** which consist of:

- A tiny bile duct of the liver duct system
- A venule that is a branch of the portal vein bringing blood from the intestine
- An arteriole that is a branch of the hepatic artery bringing blood from the aorta

Each portal triad serves more than one adjacent lobule.

Each lobule is constructed of plates of hepatocytes about two cells thick, which radiate outward from a **central vein** like spokes in a wheel (Fig. 14.20A and B). Between individual hepatocytes are **bile canaliculi,** tiny tubes into which bile is first secreted. Between plates of hepatocytes are the **hepatic sinusoids,** large, leaky capillaries that carry a mixture of portal venous blood and hepatic arterial blood. The portal blood contains nutrients and toxins from the intestines that must be processed and/or stored by the liver. The hepatic arterial blood supplies each hepatocyte with oxygen and fats for processing. Resident hepatic macrophages known as **Kupffer cells** populate hepatic sinusoids. They cleanse portal blood of intestinal bacteria and other foreign matter.

Arterial and portal blood merge and flow *inward* to the center of the lobule through the sinusoids before emptying into central veins. Central veins then join into hepatic veins, which empty into the inferior vena cava. Bile, however, flows in the opposite direction— *outward* from bile canaliculi into bile ducts for eventual delivery into the gallbladder or duodenum. Bile ducts join and exit the undersurface of the liver as the common hepatic duct.

The endothelial lining of hepatic sinusoids has very large pores between adjacent endothelial cells, which permit proteins and other large molecules to pass freely between the blood and the hepatocyte (Fig. 14.20C). The sinusoids are separated from the hepatocytes by a thin **perisinusoidal space,** which connects to hepatic lymphatic vessels.

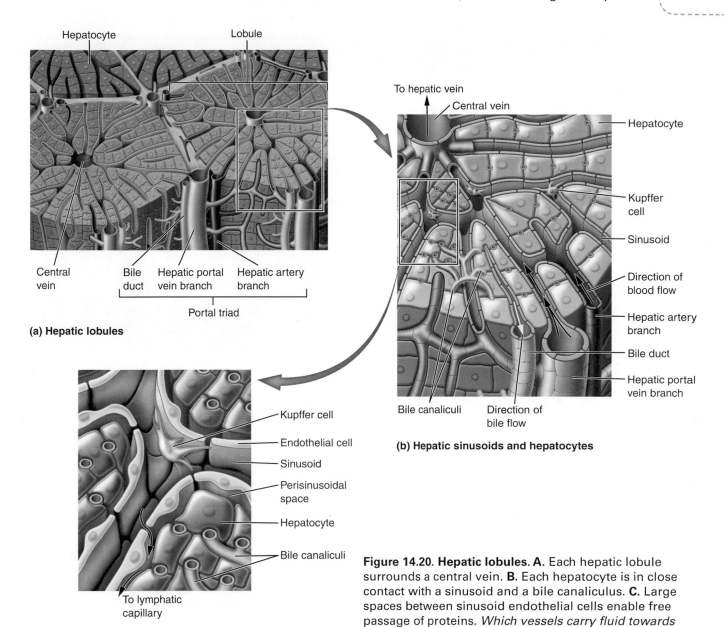

(a) Hepatic lobules

(b) Hepatic sinusoids and hepatocytes

(c) Lymphatic drainage of the liver

Figure 14.20. Hepatic lobules. A. Each hepatic lobule surrounds a central vein. **B.** Each hepatocyte is in close contact with a sinusoid and a bile canaliculus. **C.** Large spaces between sinusoid endothelial cells enable free passage of proteins. *Which vessels carry fluid towards the portal triad—bile canaliculi or sinusoids?*

> **Remember This!** Hepatocytes absorb nutrients from and secrete proteins into sinusoids, and secrete bile into the bile canaliculi.

The Hepatic Vascular and Lymph Systems Are Unique

Hepatic lymph and blood flow are unique on two counts.

First, about half of all the body's lymph is formed in the liver. The large pores between sinusoidal endothelial cells allow blood proteins and other large molecules to enter the perisinusoidal space, and water follows. This mix of protein, water and other substances becomes lymph as it empties into hepatic lymphatic vessels.

Second, at any given moment, the liver holds a relatively large volume of blood, about one-tenth of the total blood volume. Furthermore, hepatic sinusoids are distensible and contractible and can store or release blood as necessary. For example, in cases of severe hemorrhage, sinusoids constrict to shift blood from the liver into the general circulation.

The importance of portal venous blood flow and lymph formation is highlighted when blood and lymph flow is impaired by severe liver scarring, as it is in cirrhosis. See the nearby clinical snapshot titled *Cirrhosis of the Liver* for more information.

CLINICAL SNAPSHOT

Cirrhosis of the Liver

A large amount of blood flows through the liver from the intestines, and a large volume of lymph fluid is formed in the liver. Any factor that increases resistance to hepatic blood or lymph flow thus has a severe effect on the entire cardiovascular system. The most common cause is cirrhosis of the liver, a broad, patterned scarring of liver tissue (including blood and lymph vessels) usually caused by chronic alcoholism or chronic hepatitis. (The word *cirrhosis* derives from Greek *kirrhos* = "tawny," or "yellowish," because many affected livers are so colored, as in the accompanying illustration.) Scarred tissue does not function well, especially in scarring as broad and severe as that of cirrhosis. This scarring makes it harder for the blood arriving at the liver to leave the liver through the hepatic vein.

Because blood arriving at the liver must return to the vena cava by one route or another, the high vascular resistance of cirrhosis diverts blood through other venous channels—via the lesser omentum into veins in the stomach and esophagus and via veins in the lower rectum and anus. The high flow in these veins causes them to enlarge into varices (singular = varix; Latin = "dilated vein"). In patients with cirrhosis, varices are commonly found in the esophagus, abdominal wall, and anus, where they appear as hemorrhoids. Bleeding from esophageal varices is often the final, fatal complication in cirrhosis.

A further complication of cirrhosis is that the high portal venous pressure forces more fluid into the persinusoidal space; this, in turn, increases lymph flow and lymph pressure. Some of this high-pressure lymph oozes from the surface of the liver and into the abdominal cavity, where it accumulates as fluid (**ascites**, from Greek *askos* = "wine skin," because the bloated belly of patients with ascites resembles a wineskin filled with wine.).

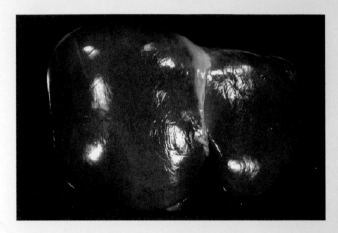

(a) Normal liver

(b) Cirrhotic liver

Cirrhosis. A. Normal liver. **B.** Cirrhosis. Note the small size of the cirrhotic liver.

Most Chemical Digestion and Absorption Occurs in the Small Intestine

By the time food leaves the stomach and enters the small intestine, it is a semifluid mix of food particles, acid, and digestive enzymes (chyme). Although a limited amount of chemical digestion occurs in the mouth with the addition of salivary amylase and in the stomach with the addition of hydrochloric acid and pepsin, most chemical digestion is accomplished in the small intestine.

Digestion and Absorption in the Small Intestine

There are two types of digestive enzymes in the intestine: *pancreatic enzymes* and *brush-border enzymes*. Pancreatic enzymes are secreted into the intestine and do their work in the intestinal lumen, as they mix with chyme. However, brush-border enzymes are embedded in the absorptive cell membranes and do their digestive work there, not in the lumen. Intestinal juice adds very little to the chemical processes but ensures the fluidity and alkalinity of chyme, without which chemical digestion would be much less effective.

Recall from ◄ Chapter 3 that substances move across cell membranes by simple diffusion, facilitated diffusion, osmosis, and active transport. All are involved in intestinal absorption. All nonlipid absorbed nutrients enter the portal blood system and percolate through the liver, where they may be utilized for certain purposes, or pass on into the general circulation. Lipids, however, are absorbed into the intestinal lymphatic system and delivered into blood.

Digestion and Absorption of Carbohydrates

Recall from Figure 14.1 that carbohydrates are constructed from three different monosaccharide building blocks. Intestinal cells can absorb only monosaccharides; thus poly- and disaccharides must be broken down. Figure 14.21A illustrates carbohydrate digestion and absorption in the small intestine.

Salivary amylase begins carbohydrate digestion in the mouth. Pancreatic amylase continues the job, breaking all starch molecules into *maltose*, a disaccharide consisting of two linked glucose molecules.

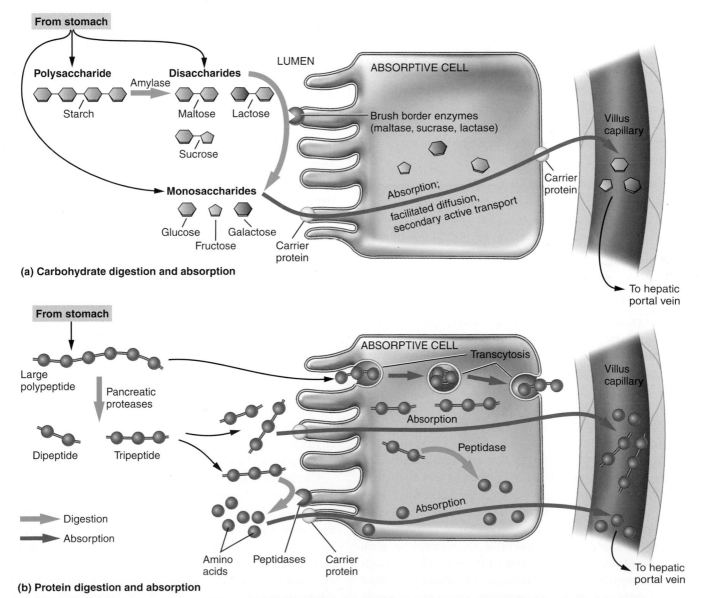

(a) Carbohydrate digestion and absorption

(b) Protein digestion and absorption

Figure 14.21. Digestion and absorption in the small intestine. A. Carbohydrates are digested into monosaccharides before absorption. **B.** Proteins are usually digested into small peptides or individual amino acids before absorption, but polypeptides can sometimes be absorbed. *Which form of protein is absorbed by transcytosis— amino acids or polypeptides?*

Brush-border enzymes digest disaccharides into monosaccharides. For instance, lactase digests lactose (milk sugar) into galactose and glucose. Sucrase digests sucrose (table sugar) into fructose and glucose, and maltase digests maltose into individual glucose molecules. Ingested monosaccharides, such as fructose in fruit and glucose in certain processed foods, need no further digestion.

All carbohydrates pass out of absorptive epithelial cells and into blood as monosaccharides. These can be taken up by body cells.

Case Notes

14.13. Margot cannot digest lactose. Which enzyme is Margot lacking, and where is the enzyme normally produced?

14.14. As part of the study, Margot's blood glucose levels were measured 30 minutes after she drank two glasses of milk. The blood glucose levels will be elevated in a lactose-tolerant person. Do you think that Margot's blood glucose levels will rise or stay roughly the same? Why or why not?

14.15. If Margot's blood glucose levels were measured after she drank soy milk, which contains sucrose instead of glucose, what result would you expect?

Digestion and Absorption of Proteins

Recall that the digestion of proteins begins in the stomach, where hydrochloric acid denatures proteins and pepsin breaks them down into small polypeptides. Pancreatic proteases (for example, trypsin and chymotrypsin) break these polypeptides into even shorter chains, such as dipeptides, tripeptides, and short polypeptides (Fig. 14.21B). Other enzymes, primarily in the brush border, can then cleave individual amino acids off of either end of the chains. The result is that digested protein is usually absorbed in the form of amino acids and short-chain peptides. Short peptides can be further digested into amino acids within epithelial cells, or they may pass intact into the portal bloodstream.

Infrequently, larger peptides can be absorbed intact into the bloodstream, especially in newborns with immature GI systems. They pass through cells by *transcytosis,* a vesicle-mediated transport mechanism. These larger peptides can act as *antigens,* inducing immune reactions that can lead to food allergies. One such example is large polypeptides derived from wheat protein (gluten), which can be absorbed intact into an infant's bloodstream. Wheat allergies can be a result. To avoid this situation, parents are advised to avoid feeding babies wheat prior to 7 to 12 months of age.

Digestion and Absorption of Lipids

Lipids pose a particular problem for the digestive system because of their hydrophobic nature. Lipid molecules huddle together into large *lipid droplets,* which have comparatively less surface area than many small droplets. These large droplets reduce the available area for surface contact with aqueous intestinal contents, thereby restricting enzyme access to only the outermost molecules. Before effective digestion can occur, bile salts must emulsify these large droplets into many smaller ones, which exposes all lipid molecules to enzymatic attack (Fig. 14.22; step 1).

The particular target of lipase is triglycerides, the most common dietary fat. Pancreatic lipase breaks off two of the fatty acids and leaves the third attached to glycerol (a monoglyceride) (step 2). Most of the other types of fat (such as short-chain fatty acids or cholesterol) can be used for metabolic purposes without further digestion. Also note that there is no brush-border enzyme activity involved in the digestion of fat.

The products of fat digestion are as hydrophobic as the original triglyceride, so they still require the protection of bile salts in the intestine. Monoglycerides, fatty acids, and some cholesterol aggregate into tiny disc-shaped **micelles** coated with bile salts (step 3). Fatty acids and monoglycerides diffuse out of the micelle across the absorptive cell membrane (step 4). Most cholesterol is not absorbed in this fashion; it is imported into cells by a specialized carrier protein.

Once inside the cell, the smooth endoplasmic reticulum reassembles monoglycerides and free fatty acids into triglycerides (step 5). Next, triglycerides, proteins, and cholesterol are assembled into delivery packets called **chylomicrons** (step 6), which, like other lipoproteins, are lipid droplets wrapped in a coat of protein. The chylomicrons are then expelled into the extracellular fluid by exocytosis.

Chylomicrons are too large to pass between the endothelial cells of capillaries, but they can easily slip into leaky lymphatic lacteal vessels (step 7). From there, they travel through the lymphatic system and enter the left subclavian vein via the thoracic duct. As blood-borne chylomicrons pass through adipose or liver tissue, they are captured and stored until they are need for fuel or other purposes.

Short-chain fatty acids escape packaging into chylomicrons and diffuse into intestinal blood, which subsequently enters the portal circulation. All other lipids,

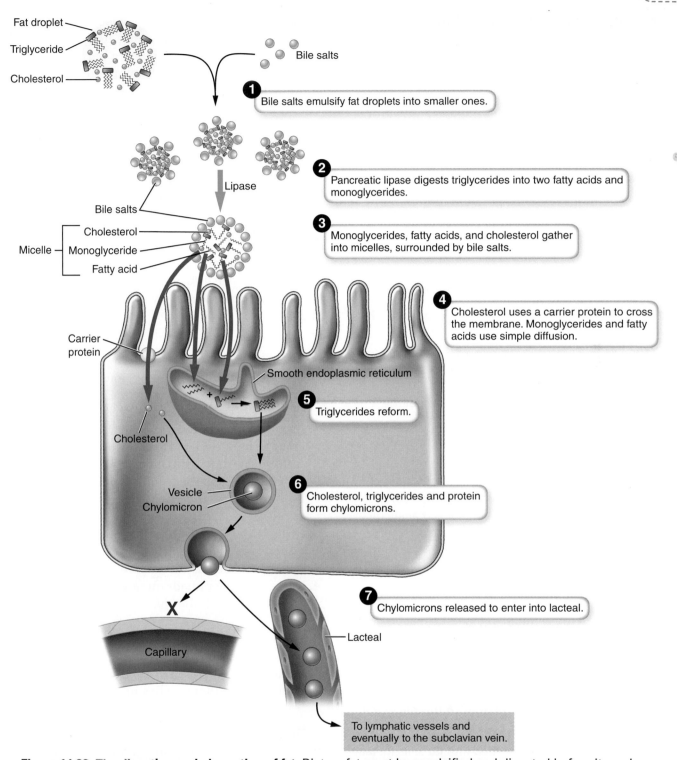

Figure 14.22. The digestion and absorption of fat. Dietary fat must be emulsified and digested before it can be absorbed into intestinal absorptive cells. *Micelles form within absorptive cells—true or false?*

which entered the lymph, bypass the portal blood and liver and directly enter the general circulation.

The majority of bile salts are actively reabsorbed into the portal circulation from the ileum. The liver "recycles" them into the next batch of bile and sends them back to the intestine.

Absorption of Other Substances

The absorption of vitamins varies. For example, vitamin B12 is absorbed mainly in the ileum and can be absorbed only attached to intrinsic factor, as discussed earlier. Fat-soluble vitamins such as A, D, E, and K are absorbed with lipids into the lymphatic system.

Water-soluble vitamins are absorbed by diffusion into the blood.

The small intestine directly absorbs minerals. Water moves from the intestinal lumen into absorptive cells and then the blood.

14.28   What is the shortest segment of the small intestine?

14.29   Does the small intestine contain rugae or circular folds?

14.30   What is part of the cell membrane of an individual cell—villi or microvilli?

14.31   Which solution would contain a lower acid concentration—intestinal chyme or gastric chyme?

14.32   Which organ produces digestive enzymes—the liver or the pancreas?

14.33   True or false: Bile release is regulated, but bile synthesis is not.

14.34   Which duct is formed by the junction of the common bile duct with the pancreatic duct?

14.35   Which liver lobe is the largest?

14.36   How are hepatocytes supplied with oxygen?

14.37   Name the fluids that flow through each of the following structures of the liver: sinusoids, canaliculi, and perisinusoidal space.

14.38   Which nutrient form can be absorbed, especially in newborns, when it is only partially digested—proteins or carbohydrates?

14.39   Which enzyme is found in the intestinal brush border—amylase or sucrase?

14.40   Which nutrient form can be absorbed by intestinal cells—dipeptides or disaccharides?

14.41   What is the difference between a micelle and a lipid droplet?

14.42   Why are most fats absorbed into lacteals instead of blood capillaries?

The Large Intestine

The small intestine joins the colon via the **ileocecal valve,** a one-way gate that is closed most of the time by the *ileocecal sphincter*, a ring of smooth muscle encircling the valve. Shaped somewhat like a lazy question mark, the large intestine runs in a partial loop around the edges of the abdomen, framing the small intestine in the center.

The Large Intestine Consists of the Colon, Appendix, Rectum, and Anus

Compared with the small intestine, the large intestine is shorter, has a larger diameter, and is much less convoluted. In life it is about 1 m long and about 6 cm in diameter.

The first part of the large intestine is the **colon,** which accounts for about 90% of its length. In turn, from proximal to distal, the colon is divisible into five regions (Fig. 14.23A).

1. The **cecum** lies in the right lower quadrant of the abdomen; it is a slightly bulbous area of the colon that is the first to receive chyme. Extending outward from the cecum near the ileocecal valve is a worm-like blind pouch, the *appendix*. Usually about 5 to 6 cm long and 1 cm in diameter, the appendix is fully formed intestine with the usual layers.
2. From the cecum, the colon extends vertically up the right side of the abdomen as the *ascending colon*, also called the *right colon*.
3. In the right upper quadrant beneath the liver the colon turns (the *hepatic flexure*) and runs across the upper abdomen as the *transverse colon*.
4. In the left upper quadrant near the spleen, the colon turns downward (the *splenic flexure*) as the *descending colon*, also called the *left colon*.
5. In the left lower quadrant, the descending colon gives way to the *sigmoid colon* (Greek *sigma* = "S"; *sigmoid* = "S-like"), a 1-ft-long, slightly tortuous segment of colon that empties into the *rectum*.

The final parts of the large intestine are the *rectum* and *anal canal*. The **rectum** (Latin = "straight") is a short, vertical, relatively straight section of intestine in the centerline of the body. It serves as a final holding point for intestinal contents before discharge and is partially crossed by three folds of mucosa (*rectal valves*) that act to restrain the passage of solid matter but allow gas to escape. The final and shortest (about 2 cm) part of the large intestine is the *anal canal*, which ends with an opening, the **anus** (Latin = "ring").

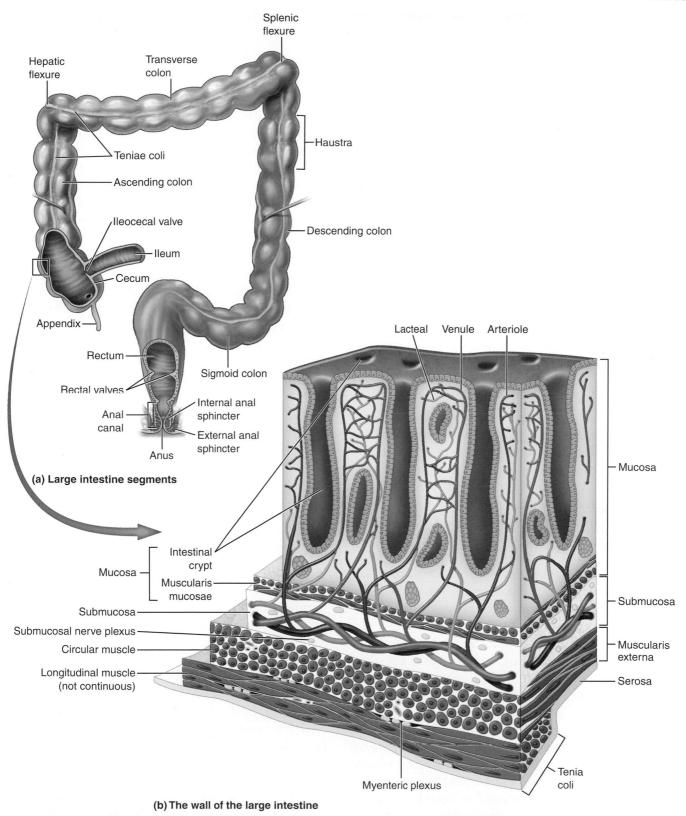

(a) Large intestine segments

(b) The wall of the large intestine

Figure 14.23. The large intestine. A. The large intestine consists of five segments. **B.** The wall of the large intestine lacks villi and a continuous longitudinal muscle layer. *What are the strips of longitudinal muscle called?*

The Muscularis and Mucosal Layers Are Modified in the Large Intestine

The layers of the wall of the large intestine are similar to the other gastrointestinal segments—mucosa, submucosa, muscularis and serosa—but they differ in some particulars.

The mucosal surface is flat and lacks the elaborate folds and villi of the small intestine; however, it is punctuated by deep, tubular *crypts*, the colonic version of intestinal glands (Fig. 14.23B). The lining epithelium is composed of columnar absorptive cells mixed with an abundance of goblet cells that secrete mucus to ensure the easy passage of compacted feces. The anus, however, is lined by stratified squamous epithelium (as in the mouth, throat, and esophagus) in order to withstand the abrasion of stool passage. Colonic absorptive epithelium contains virtually no enzyme- or hormone-secreting cells because its main function is simply to absorb water.

The submucosa of the large intestine is similar to that of the small intestine. It contains lymph and blood vessels and patches of lymphoid tissue.

The muscularis is formed of an inner circular layer and outer longitudinal layer, but with a distinctive difference: the outer, longitudinal layer is arranged into three parallel strips, the **teniae coli** (Greek *tainia* = "ribbons") (Fig. 14.23A). These strips shorten the colon, causing it to pucker laterally into broad shallow sacs or **haustra** (singular: haustrum), which impart a knobby appearance.

Teniae and haustra are absent in the rectum, where full, thick layers of circular and longitudinal muscle are present. The anus is ringed by two set of muscles, an internal involuntary *smooth* muscle ring (the *internal anal sphincter*), and an external ring of *skeletal* muscle (the *external anal sphincter*, Fig. 14.23A). These sphincters contract to close the anus and, as described further below, relax to enable defecation (the expulsion of feces).

The serosa of the large intestine is similar to the serosa of other layers of the intestine—a thin layer of visceral peritoneal cells.

The Main Functions of the Large Intestine Are Storage and Propulsion

By the time chyme reaches the ileocecal valve, almost all nutrients have been digested and absorbed. The pressure of chyme opens the valve, permitting it entry into the large intestine. Upon entry into the colon, chyme becomes *feces*, which consists of water, indigestible food matter (mainly fiber), salts, shed intestinal epithelial cells, and hordes of bacteria and their by-products. Very little usable nutrient remains. Feces remain in the colon for 12 to 24 hours. In the ascending colon, fecal matter is semisolid; but as it moves toward the rectum, water is absorbed by colonic epithelium and peristalsis compacts it into a pasty, formed mass.

Often lost in the headlines about smoking and lung cancer, about prostate cancer in men and breast cancer in women, is the fact that colon cancer is the third most common cause of cancer death in humans. It is also a fact that the colon is host to more tumors—most benign, but many malignant—than any other internal organ. The high death rate from colon cancer is especially tragic because colon cancer is more readily detected in its early stages than almost any other malignancy.

Colonic Bacteria

As chyme enters the colon, very few ingested bacteria remain—gastric acid, intestinal enzymes, or the immune activities of mucosa-associated lymphoid tissue have killed most of them. However, several hundred varieties of live bacteria inhabit the colon, living in a natural and mutually beneficial relationship with the body. In fact, by the time an infant is only a few weeks old, bacteria populate its intestinal tract. These bacteria are constantly reproducing and dying—dead bacteria make up about half of the dry mass of feces.

Bacteria play a central role in colonic physiology. They metabolize bilirubin, derived from the breakdown of hemoglobin, into a brown pigment that gives feces its characteristic color. They also produce significant amounts of vitamin K. Moreover, by sheer number, they are an important guard against infection: they are so numerous and claim so much of the limited supply of available nutrients and other environmental assets that only a large dose of infective bacteria could establish a foothold. Ingestion of a few pathogenic bacteria would not be likely to produce infection.

Bacteria also digest mucin and the small amount of unabsorbed carbohydrate and protein remaining in the colonic lumen, producing hydrogen ions and gases as by products. Some of the hydrogen ions enter the bloodstream and can be exhaled in the breath. The gases are expelled from the anus as *flatus*—perhaps 500 mL of flatus is produced in a given day. A large percentage—how large is debatable—of flatus comes from swallowed air. Some people swallow a lot; others only a little.

Case Note

14.17. Margot's lactose intolerance was detected by the hydrogen breath test. Why were her hydrogen levels elevated?

Large Intestinal Motility and Defecation

Motility in the large intestine occurs by *mass movements,* large peristaltic sweeps that propel the contents several feet at a time. Mass movements are stimulated by two factors:

- Arrival of food in the stomach. This reflex action is revealed in daily life by the frequent necessity for a toilet visit shortly after a meal.
- Filling of the large intestine.

Eventually feces reach the rectum, where they accumulate. Increased pressure from accumulating feces distends the rectum, activating the *defecation reflex. Defecation* is passage of feces through the anus. The external longitudinal smooth muscles in the rectal wall contract, shortening the rectum and increasing pressure even more in a positive feedback loop. Another loop of the same reflex causes relaxation of the internal anal sphincter. Defecation occurs when the cerebral cortex relaxes the external anal sphincter.

Normal bowel habits vary greatly, from two to three bowel movements per day to two to three per week. *Constipation* is difficult or infrequent evacuation, hard stool, or a feeling of incomplete emptying. *Diarrhea,* however, is famously difficult to define. It is always associated with excess water in stool, and among the lay public is equated with even a single loose watery stool. A strict medical definition, much less the causes, is beyond the scope of this textbook, but there is a brief discussion of the topic in our case review below.

14.43 Which portion of the colon empties into the rectum?

14.44 Which sphincter is under voluntary control—the external or internal anal sphincter?

14.45 The colonic epithelium does not produce enzymes, and accessory organs do not secrete into the colon. Does any chemical digestion occur in the colon? If so, how?

Regulation of Gastrointestinal Function

The GI tract is virtually 100% effective at its job—it digests and absorbs nearly all of the usable nutrients we consume, so that almost no nutrients are present in feces. These processes are optimized by the tight regulation of the other major GI processes—*motility* and *secretion*.

The Endocrine and Nervous Systems Regulate Motility and Secretion

Motility and secretion are regulated by the overlapping activities of the endocrine system and the nervous system.

Endocrine Feedback Pathways Involve Gastrointestinal Hormones

The GI system produces many different hormones—it has been called the most interesting endocrine gland in the body. Unlike the cells of traditional endocrine glands, such as the anterior pituitary gland, the hormone-producing cells of the GI system are not grouped together. Instead, endocrine cells are scattered throughout the intestinal mucosa. Their secretions may travel through the extracellular fluid to affect nearby cells, or they may enter the blood to act at a greater distance.

In general, the actions of each hormone negatively feed back to reduce the stimulus for its secretion. For instance, intestinal endocrine cells release *secretin* when the duodenal pH becomes too low. Secretin stimulates the pancreas to produce a more alkaline juice to neutralize the acidity. You can see these negative feedback loops in Figure 14.27.

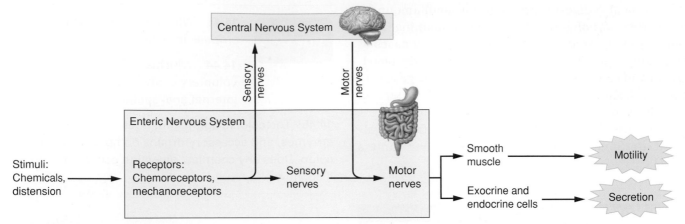

Figure 14.24. The enteric nervous system, which detects changes in the digestive tract's lumen and induces changes in intestinal motility and secretion in response. The enteric nervous system also exchanges sensory signals and motor signals with the central nervous system. *Which effector regulates motility—smooth muscle or secretory cells?*

Neural Reflexes Involve the Enteric Nervous System

Recall that the *myenteric* and *submucosal nerve plexuses* form the *enteric nervous system* (the "gut brain"). The enteric nervous system includes both sensory and motor nerves (Fig. 14.24). Sensory nerves receive information from chemoreceptors about the composition of the digestive tract contents and from mechanoreceptors regarding the degree of distention. Without leaving the enteric nervous system, these sensory nerves can activate motor nerves. Motor nerves control *motility* by regulating smooth muscle contraction; they control *secretion* by regulating exocrine and endocrine cell activity.

The enteric nervous system is interlaced with other branches of the nervous system. It sends sensory signals to the brain, so we perceive intestinal distention, for instance, as "feeling full." It also receives motor signals from the brain. Generally speaking, sympathetic motor nerves decrease digestive function and the parasympathetic motor nerve (the vagal nerve) stimulates digestive function. You can see an example of central nervous system involvement in digestive function in Figure 14.25.

Case Note

14.18. The lactose contained in Margot's pill passed through her small intestine undigested and unabsorbed. What type of receptor will lactose activate—a chemoreceptor or a mechanoreceptor?

Gastrointestinal Regulation Involves Three Phases

To the uninitiated, food goes in one end and waste comes out the other. Normal digestion is among the most silent of body processes, so it is sometimes difficult to imagine the elegance with which it is controlled. The regulation of the digestive process can be conveniently broken down into three phases: *cephalic*, *gastric*, and *intestinal*.

The Cephalic Phase

Digestion begins with the brain's generation of an impulse to eat. The **cephalic phase** of digestion is the body's response to the thought, smell, or taste of food (Fig. 14.25). The brain sends signals down the vagus nerve, which stimulates the activity of stomach smooth muscle to begin contracting in anticipation of a meal. Second, it stimulates the secretion of hydrogen ions and pepsinogen from the gastric mucosa, which initiate protein digestion.

Case Note

14.19. Margot is tempted by a luscious (but forbidden) piece of chocolate ice-cream cake. Will the activity of her parietal cells increase or decrease, and which nerve carries the signal from her brain to her stomach?

The Gastric Phase

The **gastric phase** begins when food enters the stomach. Chemoreceptors sense the arrival of peptides and

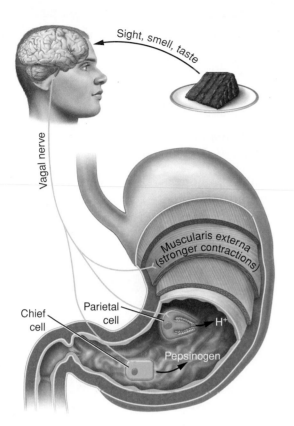

Figure 14.25. The cephalic phase, which prepares the stomach for food to come. *Which nerve carries signals to the stomach from the brain?*

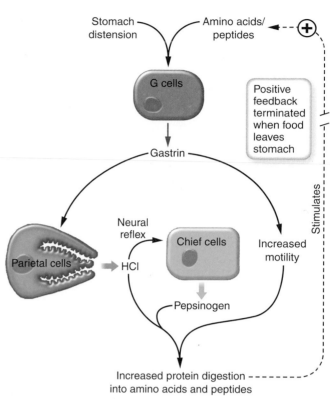

Figure 14.26. The gastric phase, which initiates protein digestion in response to food in the stomach. *Which hormone stimulates the events of the gastric phase?*

amino acids, and mechanoreceptors sense the distention of the stomach wall (Fig. 14.26). These stimuli induce gastrin production from G cells in the pylorus. Gastrin travels through interstitial fluid and blood to intensify the contraction of smooth muscle, which promotes the mechanical breakdown and mixing of the stomach contents. Gastrin increases the secretion of hydrochloric acid from parietal cells. Hydrochloric acid, in turn, stimulates pepsinogen secretion from chief cells via a neural reflex. The net result of gastrin activity is the production of *more* amino acids and peptides—which was the original *stimulus* for secretion. This positive feedback loop ends when food leaves the stomach to enter the duodenum.

Case Note

14.20. Margot eats a piece of ice cream cake in a moment of weakness. How does her stomach know that the cake has arrived?

The Intestinal Phase

The **intestinal phase** is stimulated by entry of food into the intestine and is a complex set of reflexes involving the stomach, small intestine, liver, and pancreas. The goals and activities of the intestinal phase are to (Fig. 14.27):

1. *Ensure the presence of adequate digestive enzymes and bile.* Fat and protein in chyme stimulate the release of the hormone **cholecystokinin (CCK)** from intestinal endocrine cells into the bloodstream. CCK subsequently stimulates pancreatic enzyme release. CCK also stimulates contraction of the gallbladder and discharge of its load of bile. Finally, CCK relaxes the hepatopancreatic sphincter, permitting the bile and pancreatic juice to enter the duodenum. Chemical digestion of fat and protein (as well as carbohydrate) ensues, removing the initial stimulus for CCK secretion.

2. *Protect the fragile intestinal mucosa from excessive amounts of stomach acid.* Acid in chyme stimulates

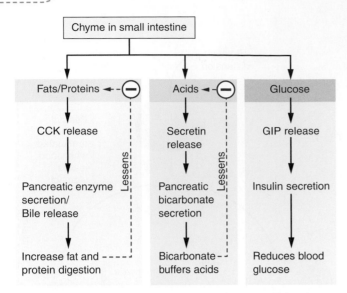

Figure 14.27. The intestinal phase, which promotes digestion and protects the intestine from excessive gastric emptying. Note that all intestinal hormones also decrease gastric motility. *Which hormone increases the pH of the intestinal contents—secretin or cholescystokinin?*

the release of the hormone **secretin** from duodenal epithelial cells. Secretin travels through blood to the pancreas, where it stimulates the production of bicarbonate from the cells lining the pancreatic ducts. The increased bicarbonate production in the resulting pancreatic juice neutralizes the acidic chyme, removing the stimulus for secretin secretion.

3. *Prepare the liver and other metabolic organs for the nutrients soon entering the blood.* A third hormone, called **gastric inhibitory peptide (GIP),** helps protect against abnormally high blood glucose following a sugary meal. Glucose in chyme stimulates the release of GIP from duodenal and jejunal mucosa. GIP then travels to the pancreas to stimulate *insulin* release. Insulin, as we learn in the next chapter, lowers blood glucose by stimulating glucose uptake into cells. Thus GIP prevents a surge in blood glucose by acting *before* the glucose is even absorbed.

4. *Provide adequate time for intestinal digestion and absorption.* Digestion and absorption are slow processes and cannot be rushed by the arrival of large amounts of new chyme from the stomach. The three GI hormones mentioned above inhibit stomach motility and release of chyme into the duodenum.

Case Note

14.21. Will GIP secretion increase when Margot consumes milk? Why or why not?

14.46 Does the cephalic phase of digestion primarily affect the stomach or the intestine?

14.47 Which hormone increases the activity of the gastric muscularis externa—secretin or gastrin?

14.48 Which hormone stimulates pancreatic enzyme secretion—cholecystokinin (CCK) or secretin?

14.49 Which hormone is released in response to decreased intestinal pH—gastrin or secretin?

Case Discussion

Lactose Intolerance: The Case of Margot C.

Let's return to our case.

Recall that Margot was diagnosed with lactose intolerance as a child—dairy products produced bloating, gas, and intestinal cramps—but she had not had definitive laboratory testing. These same symptoms occurred when she took the pills a physician gave her.

The key to understanding this case is to recall that dietary carbohydrates must be broken down into monosaccharides in order to be absorbed. Lactose is a disaccharide that is normally broken down by intestinal lactase into its components: glucose and galactose. The resulting glucose molecules stimulate intestinal GIP release, which stimulates insulin secretion. In lactose-tolerant people, glucose and galactose are absorbed completely and their blood concentration rises; none remains in the intestine for bacterial digestion. The rise in blood glucose further stimulates insulin secretion.

However, Margot's intestinal epithelium contains no lactase, and ingested lactose remains in the chyme (Fig. 14.28). No glucose will be produced from milk or other dairy products, so GIP secretion will not occur and the blood glucose level will not rise. Lactose remains in chyme as it passes into the colon, raising the osmolarity of the intestinal contents. The highly osmotic feces draws fluid from the colonic mucosal cells, resulting in watery feces (diarrhea). What's more, colonic bacteria metabolize lactose into short-chain fatty acids, which attract even more water. Bacterial activity also produces large amounts of hydrogen gas, which in turn stretches the intestine and causes cramping. Hydrogen gas enters blood and diffuses into alveolar air with carbon dioxide

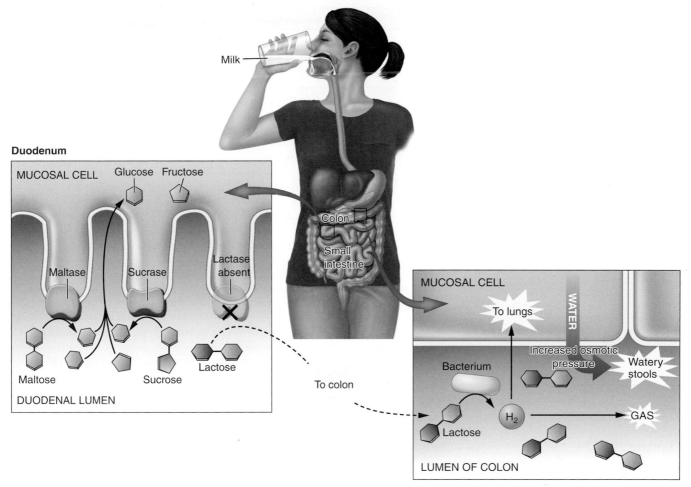

Figure 14.28. The case of Margot C. Margot's problems can be traced to the lack of lactase enzymes in her duodenal cells. As a result, lactose continues on to the colon, where it osmotically attracts water (causing diarrhea) and is metabolized by bacteria, producing acidity and gas. *Can Margot's intestine generate any glucose from ingested disaccharides?*

and is detectable in exhaled breath by a special laboratory instrument.

Thus, Margot suffered from *diarrhea*, a condition everyone recognizes when they have it but which proves surprisingly difficult to define with precision. A commonly used definition is more than one daily bowel movement of liquid stools.

Excess stool water is characteristic of all diarrheas. Excess water accumulates in stool for four reasons:

- *Increased osmotic load* (Margot's problem).
- *Increased secretion of water and electrolytes* by the intestinal mucosa, which can be caused by a variety of conditions. For example, certain bacteria release chemicals that have this effect.

- *Inflammation.* Damaged intestinal epithelium loses its integrity and weeps interstitial fluid.
- *Decreased absorption.* Decreased intestinal length does not allow for the complete absorption of chyme or fecal water. For example, surgical resection of a substantial portion of either the small intestine or colon can produce absorptive diarrhea.

Case Note

14.22. Why do Margot's feces contain short-chain fatty acids after she consumes ice cream?

Word Parts

Latin/Greek Word Parts	English Equivalents	Examples
-ase	Enzyme	Lactase: enzyme that digests lactose
cephal-	Head	Cephalic phase: phase of digestion involving the head
chol-, chole-	Bile	Cholecystokinin: hormone that stimulates bile release
entero-	Intestine	Myenteric plexus: nerve network in the intestine
gastr-	Stomach	Gastrin: hormone produced by the stomach
hepato-	Liver	Hepatocyte: liver cell
lingu-	Tongue	Sublingual salivary gland: gland under (sub-) the tongue
lip-/lipo-	Fat	Lipase: enzyme that digests fat
naso-	Nose	Nasopharynx: pharynx portion adjacent to the nose
oro-	Mouth	Oropharynx: pharynx portion adjacent to the mouth
sacchar-	Sugar	Disaccharide: molecule consisting of two (di-) sugars

Chapter Challenge

CHAPTER RECALL

1. Trans fatty acids
   a. contain double bonds between some of the carbons.
   b. are also called unsaturated fatty acids.
   c. are abundant in natural sources of fat.
   d. contain only single bonds between the carbons.

2. The essential amino acids
   a. can be synthesized by the liver from other amino acids.
   b. are found only in animal proteins.
   c. are found in both animal and plant proteins.
   d. number 20 in total.

3. Low-density lipoproteins
   a. contain cholesterol and triglycerides.
   b. are synthesized in the liver.
   c. contribute to arterial disease.
   d. all of the above.

4. Which of the following is *not* an accessory organ?
   a. Gallbladder
   b. Esophagus
   c. Liver
   d. Pancreas

5. **The greater omentum**
   a. covers the liver.
   b. is ventral to the mesentery proper.
   c. envelops the loops of the small intestine.
   d. connects the transverse colon to the posterior abdominal wall.

6. **The myenteric plexus is found**
   a. between the mucosa and submucosa.
   b. between the circular and longitudinal muscle layers.
   c. underneath the serosa.
   d. within the lamina propria.

7. **The tongue is anchored to the mouth floor by a(n)**
   a. uvula.
   b. palatine tonsil.
   c. soft palate.
   d. frenulum.

8. **The bony tooth sockets are called the**
   a. alveoli.
   b. gingiva.
   c. periodontal ligaments.
   d. root canals.

9. **Which salivary glands drain near the molars?**
   a. Submandibular
   b. Sublingual
   c. Palatine
   d. Parotid

10. **From superior to inferior, the parts of the pharynx are the**
    a. oropharynx, laryngopharynx, nasopharynx.
    b. laryngopharynx, nasopharynx, oropharynx.
    c. laryngopharynx, oropharynx, nasopharynx.
    d. nasopharynx, oropharynx, laryngopharynx.

11. **Rugae are found in the**
    a. stomach.
    b. small intestine.
    c. large intestine.
    d. all of the above.

12. **Hydrochloric acid is secreted from**
    a. neck cells.
    b. G cells.
    c. parietal cells.
    d. chief cells.

13. **The _____ controls the entry of chyme into the small intestine and the _____ controls the exit of the residual contents from the small intestine.**
    a. ileocecal valve, pyloric sphincter
    b. lower esophageal sphincter, pyloric sphincter
    c. pyloric sphincter, ileocecal valve
    d. lower esophageal sphincter, ileocecal valve

14. **From largest to smallest, the parts of the small intestine that expand its surface area are the**
    a. villi, microvilli, folds
    b. folds, villi, microvilli
    c. microvilli, villi, folds
    d. villi, folds, microvilli

15. **The enzymes found in the small intestine's lumen are produced by the**
    a. pancreas.
    b. intestinal glands.
    c. goblet cells.
    d. liver.

16. **The enzyme that breaks down sucrose is produced by the**
    a. pancreas.
    b. liver.
    c. salivary glands.
    d. intestinal cells.

17. **The hepatopancreatic ampulla receives secretions from the**
    a. liver.
    b. pancreas.
    c. gallbladder.
    d. all of the above.

18. **The cells that produce bile are called**
    a. Kupffer cells.
    b. endothelial cells.
    c. hepatocytes.
    d. ascites.

19. **Which of the following macromolecules contains glucose?**
    a. Lactose
    b. Glycogen
    c. Sucrose
    d. All of the above

20. Large proteins can enter the bloodstream of a newborn intact via the process of
    a. facilitated diffusion.
    b. transcytosis.
    c. active transport.
    d. none of the above.

21. **Which of the following structures is assembled *within* intestinal cells?**
    a. Micelles
    b. Lipid droplets
    c. Chylomicrons
    d. Emulsifications

22. **Bile**
    a. emulsifies large fat droplets into smaller droplets.
    b. digests triglycerides into their component parts.
    c. transports the products of fat digestion into intestinal cells.
    d. all of the above.

23. **The bend between the ascending colon and the transverse colon is called the**
    a. hepatic flexure.
    b. splenic flexure.
    c. sigmoid colon.
    d. rectum.

24. **The strips of longitudinal muscle in the long intestine**
    a. contract rhythmically to enable peristalsis.
    b. contract arhythmically to enable segmentation.
    c. are called the teniae coli.
    d. are located within the mucosal layer.

25. **Secretin secretion is highest**
    a. during the cephalic phase.
    b. during the intestinal phase.
    c. during the gastric phase.
    d. none of the above.

26. **Thinking about a delicious cookie stimulates the motility of the**
    a. small intestine.
    b. esophagus.
    c. stomach.
    d. large intestine.

27. **Cholecystokinin (CCK) stimulates**
    a. pancreatic enzyme secretion.
    b. pancreatic bicarbonate secretion.
    c. bile synthesis by the liver.
    d. acid production by parietal cells.

28. **Food in the intestine inhibits gastric activity by stimulating the release of**
    a. secretin.
    b. cholecystokinin.
    c. gastric inhibitory peptide.
    d. all of the above.

CONCEPTUAL UNDERSTANDING

29. **For each of the following dietary recommendations, explain the physiological rationale behind it.**
    a. Carbohydrates should be whole-grain.
    b. A diet including significant amounts of plant oils is healthier than a no-fat diet.

30. **List three differences between protein digestion and triglyceride digestion.**

31. **The blood in hepatic sinusoids is lower in oxygen content than blood in most capillaries. Why? Discuss the blood supply to the liver in your answer.**

32. **Do parietal cells produce any enzymes? Define *enzyme* in your answer.**

APPLICATION

33. **Mr. M is taking a drug that blocks the action of cholecystokinin (CCK). Discuss the impact of this drug on digestion.**

34. **You have just returned from India. Unfortunately, a nasty strain of *Escherichia coli* bacteria has made the trip with you, causing severe diarrhea. You take a large dose of broad-spectrum antibiotics and notice that your feces have become pale. Why?**

35. **Galactose is a monosaccharide common in dairy products, vegetables, and fruits. Galactose is also a product of the digestion of lactose, a disaccharide of galactose and glucose in high concentration in human breast milk and dairy products. However, regardless of its origin, galactose cannot be burned for energy unless it is first converted to glucose. Some infants have a genetic defect that prevents them from converting galactose into glucose, which results in a harmful accumulation of galactose in blood (galactosemia). These infants should not be breastfed because of the high content of lactose in breast milk. What would be a better infant formula for an infant with galactosemia: one based on cow's milk or one based on soy milk?**

36. **UV light stimulates vitamin D synthesis by the skin, so individuals exposed to adequate sunlight do not require dietary vitamin D. Is vitamin D really a vitamin for wintering Canadians, living in low-sunlight conditions? How about Australians, living in a much sunnier clime?**

To find out what you've mastered and where you need more study, take the following self-test, then check your answers on our website at **http://thepoint.lww.com/McConnellandHull**

15

Metabolism and Endocrine Control

Major Themes

■ Glucose is the principal source of the body's daily energy supply.

■ Insulin is necessary for glucose to enter cells.

■ Energy can become mass; mass can become energy.

■ Food can only do work and generate heat or become tissue.

■ Enzymes facilitate every pathway of energy and heat generation from food mass.

■ Hormones regulate tissue and gland activity.

Chapter Objectives

"He's having one of those acid attacks."

As you read through the following case study, assemble a list of the terms and concepts you must learn in order to understand it.

Clinical History: Santo G., a 47-year-old Hispanic man with a long history of obesity and type 2 diabetes of recent onset, required daily insulin injections. His wife brought him to the emergency room early in the morning because he was confused, breathing rapidly, and having to urinate far more often than normal. "He's having one of those acid attacks," she said, "when everything gets out of balance." She further revealed that he had not been careful about his diet—the night before he had gorged on pizza and ice cream and had consumed 8 to 10 bottles of beer, becoming so inebriated that he went to sleep without taking his insulin.

Physical Examination and Other Data: Vital signs included the following: temperature 39°C (102.2°F), heart rate 96 beats per minute (normal 72), and blood pressure 90/60 (normal 120/80). Respiratory rate was not recorded, but Santo's breath had an unusual "apple juice" odor. He was 5 ft 9 in. tall and weighed 285 pounds. He was drowsy and confused about time and place.

Lab tests revealed abnormally high blood glucose and low (acidic) blood pH. His urine was also unusually acidic and contained large amounts of glucose and ketone bodies, neither of which should normally be present.

Clinical Course: The emergency room physician made a diagnosis of diabetic ketoacidosis and admitted Santo to the hospital. He was given insulin injections and large volumes of intravenous electrolyte solution containing bicarbonate. Further studies failed to find infection or another underlying cause for his fever.

Within 24 hours, Santo's blood pressure had returned to near normal and he became fully awake and coherent. His blood glucose had decreased but was not yet in the normal range. Blood pH, bicarbonate and P_{CO_2} also had returned to near normal. Urinalysis revealed a stronger yellow color, less acidic pH, slightly positive glucose, and negative ketone bodies.

He was discharged on the third hospital day.

In ⬅ Chapter 14, we introduced the nutrients in food as a prelude to our discussion of the digestive system and its operations. In this chapter, we explore what happens to nutrients after they have been absorbed into the body. As everyone experiences but few of us understand, the nutrients in food have only two potential fates: (a) They can be converted into energy, which also produces heat, or (b) they can become mass—that is, components of new molecules, cells, or tissues.

To gain a better appreciation of these two fates, imagine the following scenario. Thirteen-year-old Lizzie travels with her parents from Seattle to Chile one July to hike in the Andes Mountains. It is winter in South America—therefore cold—and Lizzie shivers each morning until exertion warms her body and the sunlight warms the trail. Lizzie is in the middle of her adolescent growth spurt and, away from her favorite foods, is taking in fewer calories per day. After 6 weeks, when Lizzie and her family return to Seattle. Now, Lizzie's growth spurt seems to have come to a standstill. She has lost 5 pounds and has not gained a millimeter in height. What happened? Let us take apart this situation and find out (Fig. 15.1A).

- Lizzie's physical activity (or locomotion) uses energy.
- The cold environmental temperature causes her body to use more energy to generate heat.
- Her reduced calorie intake leaves her with inadequate energy.

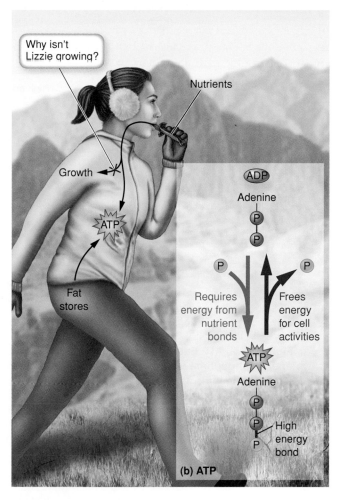

(a) Energy balance

Why isn't Lizzie growing?

Nutrients

Growth

ATP

Fat stores

ADP
Adenine
P
P

P   P

Requires energy from nutrient bonds

Frees energy for cell activities

ATP
Adenine
P
P
P — High energy bond

(b) ATP

Figure 15.1. The fate of nutrients. A. Lizzie's high energy needs divert all of her nutrient energy from growth to ATP production. **B.** ATP synthesis uses energy from nutrient bonds; ATP breakdown liberates energy for cellular use. *If Lizzie stops eating entirely, will she stop producing ATP?*

● As her body uses the nutrients she consumes to meet Lizzie's increased energy needs, it is deprived of the nutrients needed for the production of new molecules, cells, and tissues, so her growth stops. In fact, her body breaks down (catabolizes) her fat stores and she loses weight.

Lizzie's story is a simplified example of the interrelationship of metabolism and endocrine control. The story is actually a lot more complex, as you'll learn in this chapter.

Personally, I stay away from natural foods. At my age I need all the preservatives I can get.

George Burns (1896–1996), cigar-chomping, dead-pan American comedian, writer and actor first famous from 1920–1950. His career was reborn at age 80 and he was still making people laugh when he died at age 100.

The Generation of Energy

In our example immediately above, remember that most of Lizzie's nutrients are directed toward energy production instead of growth. All of her activities require energy: muscle contraction, maintenance of cell membrane polarity, transmission of nerve signals, fluid secretion by glands,

the synthesis of hormones . . . in short, life itself requires energy. We derive this energy from the macronutrients we ingest; that is, from carbohydrates, fats, and proteins. You may want to revisit the different types of energy described in ◀ Chapter 2, Box 2.1, before you begin this section.

Nutrient energy is contained in the *bonds* binding atoms together, specifically in the electrons that form the bond. However, just like a store that accepts only a particular currency, cells accept only a particular kind of energy. The energy currency of cells is **adenosine triphosphate,** or ATP. ATP is created from adenosine diphosphate (ADP) by adding a third phosphate, a reaction that uses the energy from nutrient bonds (Fig. 15.1B, down arrow). Energy to power cell activities is freed from ATP when an enzyme cleaves the bond connecting the third phosphate group to the rest of the molecule (Fig. 15.1B, up arrow). The energy contained in one of these ATP bonds, for instance, powers a single cross-bridge cycle in a muscle filament during muscle contraction ← (Chapter 7), or one cycle of active transport of the Na^+/K^+-ATPase ← (Chapter 3).

However, the generation of ATP is but the final step in a complex process that converts energy stored in nutrient bonds into the universal energy currency of ATP. During this process, nutrients are converted into a succession of other molecules called **metabolic intermediates.** A particular chain of reactions, from nutrient molecule to waste, constitutes a **metabolic pathway.** These pathways begin at different nutrient points—carbohydrate, fat. or protein—and branch and merge in a complex network like the routes in a highway map. Some energy flows down each, but all end at the same destination: ATP.

ATP Is the Molecular Fuel of Life

Nutrient energy is harnessed from a metabolic intermediate into an ATP molecule by two different methods. The first is quite direct—some metabolic intermediates have an available phosphate that can be used to make ATP without further steps. That is, the energy contained in a phosphate bond is used to directly convert ADP into ATP. Figure 15.2A illustrates the reaction.

The second method accounts for most ATP generation but requires more steps. It uses the energy of nutrient bonds to add a hydrogen atom to a specialized molecule called an **electron carrier.** The name *electron carrier* reflects the fact that the energy of the hydrogen atom's *electron* will eventually be used by mitochondria to convert ADP to ATP. Here we discuss two electron carriers, both derived from B vitamins: *NAD* (or NAD^+; *nicotinamide adenine dinucleotide*), a form of niacin, also called vitamin B3; and *FAD* (or FAD^+; *flavine adenine dinucleotide*), a form of riboflavin, also called vitamin B2. Figure 15.2B shows one such example, in which a hydrogen atom is transferred from a metabolic intermediate to an NAD^+ molecule to make NADH.

A grand summary of this second method of energy metabolism is this: Hydrogen atoms with their electrons

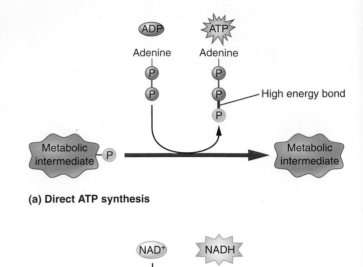

(a) Direct ATP synthesis

(b) Energy carriers

Figure 15.2. ATP Energy generation pathways. A. High-energy phosphate groups can be directly transferred from a metabolic intermediate to an ADP molecule, generating ATP. **B.** High-energy electrons (part of a hydrogen atom) can be transferred from a metabolic intermediate to an NAD^+ molecule, generating NADH. Mitochondria then use NADH to generate ATP. *Which molecule has more energy— NAD^+ or NADH?*

are stripped from nutrients, held for a while by an electron carrier, and then used by mitochondria to convert ADP into ATP. In the end the hydrogen protons and their electrons are accepted by oxygen to create H_2O.

If hydrogen is the star player in generating the energy we need to survive, then why is it that we can live for only a few minutes without oxygen? The answer is that if there is no oxygen to accept the hydrogen atoms and their electrons (to make H_2O), the flow of electrons stops, as do the chemical reactions necessary for maintaining life.

> *Remember This!* **Nutrients provide the energy to convert ADP into ATP. Usually this energy is transmitted through hydrogen atoms and their electrons.**

Case Note

15.1. Santo consumed a large meal of pizza and ice cream. Does his body use this food energy to add a hydrogen atom to NAD^+ or to remove a hydrogen atom from NADH?

ATP Generation Requires Up to Four Steps

Each nutrient can be broken down by any number of metabolic pathways. In ← Chapter 7, we discussed these energy-generating pathways as they provide energy to muscle cells; in fact, however, they fuel all cells.

ATP generation requires up to four sequential stages:

- *Stage 1: Glycolysis (carbohydrates only)*
- *Stage 2: Acetyl coenzyme A synthesis (most nutrients)*
- *Stage 3: Citric acid cycle (all nutrients)*
- *Stage 4: Oxidative phosphorylation (all nutrients)*

The first stage takes place in the cell's cytosol, but the remaining three stages occur in the mitochondria. We discuss each of these stages in turn, beginning with glycolysis.

Case Note

15.2. **Santo's blood contains large amounts of glucose, from the ice cream, beer, and pizza he consumed last night. However, as we will see, his muscle cells cannot use carbohydrate for energy. Which step of ATP generation will *not* occur in these cells?**

Glycolysis Is the First Stage in Carbohydrate Energy Metabolism

Carbohydrates (but not other nutrients) begin their metabolic journey with **glycolysis,** a series of reactions in the cell cytosol that breaks the six-carbon glucose molecule into two three-carbon molecules called *pyruvate* (Fig. 15.3 step 1). Pyruvate then enters the second stage of energy generation, which occurs in the mitochondria (step 2).

The conversion of glucose into pyruvate breaks certain chemical bonds, and the energy in these bonds results in the production of two new ATP molecules. In addition, two hydrogen atoms are transferred to NAD^+ to create two NADH, where the hydrogens and their prized electrons are temporarily "stored" for further use in energy production. As we will see, mitochondria convert these NADH molecules back to NAD^+ with the help of oxygen, generating several ATP molecules in the process.

As we discussed in ← Chapter 7, sometimes the mitochondria simply cannot keep up with the demand for energy—the pace of glycolysis exceeds the rate at which mitochondria can accept pyruvate. In such conditions pyruvate accumulates as mitochondria fall behind in converting NADH back to NAD^+. Without enough

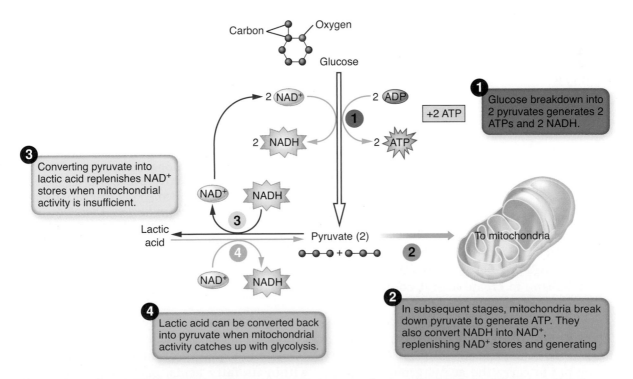

Figure 15.3. Glycolysis. This process generates two pyruvates from each glucose molecule, generating ATP and NADH as it does so. Pyruvate can be further metabolized by the mitochondria or temporarily converted into lactic acid. Each gray bead represents a carbon atom. *What happens to the NAD^+ molecule generated when pyruvate is converted into lactic acid?*

NAD^+ molecules, glycolysis could theoretically grind to a halt. Thankfully, pyruvate that cannot enter the mitochondria is converted into lactic acid, which takes the hydrogen from NADH and regenerates NAD^+ (step 3). The restored NAD^+ can then participate in another round of glycolysis. When the pace of glycolysis slows or mitochondrial activity accelerates, the reaction can flow again to the right and lactic acid can be converted back into pyruvate. Alternatively, lactic acid can be exported to other muscle cells with greater mitochondrial capabilities or (less commonly) to the liver for conversion into glucose.

Case Note

15.3. Santo's brain cells can use glucose for energy. If a brain cell converts one glucose molecule into two pyruvate molecules, how many net ATP molecules are generated directly?

Most Nutrients Can Generate Acetyl-CoA

The setting now turns from the cytosol to the cell's *mitochondria* ← (Chapter 3), complex organelles formed by two sets of membranes. The inner mitochondrial membrane faces inward and encloses a viscous soup of enzymes, DNA, and ribosomes called the *mitochondrial matrix*. The outer mitochondrial membrane faces the cytosol. The space between the two membranes is the *intermembrane space*.

The pyruvate ($CH_3COCOOH$) generated from glucose enters the mitochondrial matrix. There, one carbon atom and two oxygen atoms of pyruvate are lost as CO_2, which diffuses into the bloodstream and is exhaled from the lungs. In the same process a hydrogen atom with its electron is stripped from the remains of the pyruvate molecule and added to an NAD^+ molecule, generating NADH. This leaves three hydrogen, two carbons and an oxygen to form an *acetyl group* ($CH3CO^+$). These reactions are summarized in Figure 15.4. The acetyl group joins with *coenzyme A,* a product of vitamin B5 (pantothenic acid) metabolism, to create **acetyl coenzyme A (acetyl-CoA).** Acetyl-CoA then passes into the citric acid cycle, where the biggest energy payoff occurs.

In the following discussions, it is important to understand that *coenzyme A is a reusable transport molecule* that serves simply to carry the acetyl groups. Acetyl groups store energy in hydrogen atoms and chemical bonds. After delivering one acetyl group, coenzyme A picks up another and the cycle begins again.

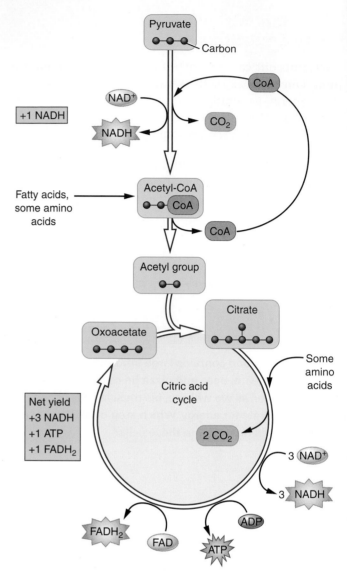

Figure 15.4. Acetyl-CoA synthesis and the citric acid cycle. Pyruvate, fatty acids, or some amino acids are converted into acetyl-CoA. The acetyl group enters the citric acid cycle by fusing with oxoacetate to form citrate. Numerous enzyme-catalyzed reactions process citrate into other molecules, generating ATP, NADH, FADH2, and carbon dioxide as they do so. *Remember that each glucose molecule generates two pyruvates. Therefore, how many NADH, FADH2, and ATP molecules are generated from a single glucose molecule in the citric acid cycle alone?*

Carbohydrates (glucose) are not the only source of acetyl groups to form acetyl-CoA; fat and protein can contribute too. Most of the energy in fat is contained within its fatty acids; each fatty acid can produce multiple acetyl CoA molecules. Some amino acids can also be used to generate acetyl-CoA, as discussed further on.

15.4. Remember that Santo's muscle cells have difficulty using glucose for energy. Will his muscle cells be able to generate acetyl CoA? If so, how?

Acetyl-CoA Enters the Citric Acid Cycle

Acetyl-CoA molecules enter the **citric acid cycle,** a circular series of chemical reactions that, like a merry-go-round, always ends where it began (Fig. 15.4). The reaction begins with a pool of *oxoacetate,* a four-carbon molecule that is consumed, regenerated, and reused in an endless cycle. Oxoacetate combines with the two carbons of acetyl CoA to form *citrate,* a six-carbon molecule. Seven more reactions (mercifully not illustrated in Fig. 15.4) convert citrate *back* to oxoacetate, during which the two carbons derived from the acetyl-CoA are lost as carbon dioxide waste. In the process, these reactions generate a wealth of energy intermediates—one ATP, three NADH, and one FADH2 for every acetyl-CoA molecule that enters the cycle (and recall that each glucose generates two acetyl-CoA molecules, and fatty acids generate many). The NADH and FADH2 temporarily hold energy (in the form of hydrogen atoms), which will be exploited in the next stage. Some amino acids can enter the citric acid cycle at different points, generating variable amounts of ATP, NADH, and FADH2.

> *Remember This!* **The conversion of pyruvate into acetyl CoA and the reactions of citric acid cycle generate the carbon dioxide that the respiratory system works so hard to eliminate.**

However, it is worth noting that the citric acid cycle is completely dependent on the supply of oxoacetate; the merry-go-round stops without it. And where do we get oxoacetate? From glucose. So although we can generate a lot of energy from fat and protein by giving them a ride on the citric acid cycle, nothing happens without glucose to provide the initial pool of oxoacetate. For this reason it can be fairly said that "fat and protein burn in a carbohydrate flame."

15.5. If Santo's muscle cells generate eight acetyl-CoA molecules from a fatty acid molecule, how many NADH molecules will be generated from these eight acetyl-CoA molecules?

Oxidative Phosphorylation Generates the Most Energy

In keeping account of the energy produced so far from one glucose molecule, we have tallied only four ATPs: the conversion of glucose into two pyruvates generates two ATPs, and further metabolism of the two pyruvates generates two more ATPs. The bulk of the ATP generation occurs by *oxidative phosphorylation* across the inner mitochondrial membrane.

Oxidative phosphorylation produces ATP using energy from H^+ ions of the NADH and FADH2 generated during glycolysis, acetyl-CoA production, and the citric acid cycle. The "oxidative" part of the name derives from the fact that the H^+ ions eventually react with oxygen to produce water. The "phosphorylation" part of the name refers to the fact that the energy released by the NADH/FADH2 reactions is used to add a phosphate to ADP, converting it to ATP.

In the first step, specialized **respiratory enzyme complexes** in the inner mitochondrial membrane use NADH and FADH2 to generate a hydrogen gradient. Remember that FAD and NAD^+ were "energized" by the addition of a hydrogen atom to create FADH2 and NADH, and that each hydrogen atom consists of a proton, which you can think of as a hydrogen ion (H^+), and an electron. One of the respiratory enzyme complexes splits each hydrogen atom into its proton and its electron (Fig. 15.5). The electrons stripped from the hydrogen atoms pass from one respiratory enzyme complex to the next, dropping off energy as they go. The energy released by the electrons is used to pump the protons (H^+) out of the mitochondrial matrix into the intermembrane space, thereby creating an H^+ gradient between the inside and outside of the inner membrane (step 1). *The net result of these reactions is that the chemical energy of FADH2 and NADH has been traded for the energy of an H^+ gradient across the inner mitochondrial membrane.*

The second step uses this gradient to generate ATP. A specialized inner membrane protein, *ATP synthase,* contains a hydrogen channel that permits H^+ ions to travel down their gradient back into the mitochondrial matrix. As hydrogen ions diffuse through the channel and down their gradient, ATP synthase uses the energy of their passage to synthesize ATP from ADP.

The third step disposes of the players in oxidative phosphorylation—the energy-depleted electrons and the well-traveled hydrogen ions. These combine with gaseous oxygen to form water. This final step is critical. If oxygen is not present, the system backs up and oxidative phosphorylation cannot occur.

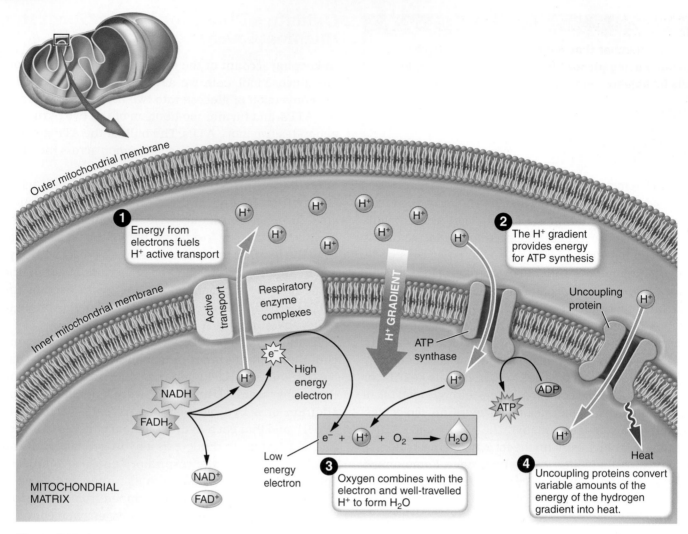

Figure 15.5. Oxidative phosphorylation. NADH and FADH2 provide electrons. Respiratory enzyme complexes use the energy from these electrons to drive active transport of hydrogen ions. ATP synthase uses the hydrogen gradient to provide energy for ATP synthesis. *Which protein transports hydrogen ions by facilitated diffusion—ATP synthase or respiratory enzyme complexes?*

The hydrogen gradient contributes to the production not only of ATP but also of heat. As any sweaty exerciser knows, increasing ATP production unavoidably increases heat production. Specialized channels in the mitochondrial membrane, called *uncoupling proteins,* let hydrogen leak down its gradient without the concomitant production of an ATP molecule (step 4). The energy of the gradient is, instead, "lost" as heat. Increasing the number of uncoupling proteins in the membrane increases heat production but also decreases the efficiency of ATP production. That is, the metabolism of one glucose molecule would produce fewer ATPs but more heat.

> *Remember This!* Although it is convenient to say that cellular metabolism uses oxygen and produces carbon dioxide, it actually happens in the opposite order—CO_2 is produced prior to and during the citric acid cycle, and O_2 is not consumed until later, at the end of oxidative phosphorylation in the mitochondria.

It is difficult to predict just how many ATPs will be generated from each energized FADH2 and NADH molecule because mitochondrial efficiency varies and there are some alternate pathways that electrons can

Table 15.1

	Glucose* (produces 2 acetyl-CoA)		Fatty Acid (palmitic acid; produces 8 acetyl-CoA)	
Glycolysis	2 NADH →	2 ATP 5 ATP	Not applicable	
Acetyl CoA synthesis	2 NADH →	5 ATP	7 FADH2 → 7 NADH →	10.5 ATP 17.5 ATP
Citric acid cycle	2 FADH2 → 6 NADH →	2 ATP 3 ATP 15 ATP	8 FADH2 → 24 NADH →	8 ATP 12 ATP 60 ATP
Total ATP generation per molecule		32 ATP		108 ATP

*Recall that glycolysis produces two pyruvate molecules, each of which generates an acetyl CoA molecule. All of the NADH and FADH2 are used for ATP production by oxidative phosphorylation.

15.1 Correct the underlined term to make this statement true: NAD^+, picks up a phosphate from metabolic intermediates and brings it to the mitochondria.

15.2 Which of the following nutrient types participate in glycolysis? Carbohydrate, protein, fat.

15.3 Can lactic acid be used in the muscle cell that produced it? If so, how?

15.4 Does the coenzyme A portion of an acetyl-CoA molecule enter the citric acid cycle?

15.5 Which process *generates* the largest number of NADH and FADH2 molecules—glycolysis, the citric acid cycle, or oxidative phosphorylation?

15.6 Why is oxygen necessary for oxidative phosphorylation?

15.7 True or false: All of the ATP molecules generated by glucose breakdown are generated during oxidative phosphorylation.

take. The generally used formula is that each FADH2 molecule generates 1.5 ATPs, whereas each NADH generates about 2.5 ATPs. The nearby Table 15.1 quantifies ATP generation from different energy pathways.

Carbohydrates, Fatty Acids, and Amino Acids Can Generate Energy

Figure 15.6 summarizes the different pathways that generate ATP—glycolysis, which occurs in the cytoplasm; the citric acid cycle, which occurs in the mitochondrial matrix; and oxidative phosphorylation, which involves the inner mitochondrial membrane. Note that:

1. Only carbohydrates participate in glycolysis; fatty acids and most amino acids enter later, usually just before the citric acid cycle.
2. Electron carrier molecules (NADH and FADH2) generated by glycolysis and the citric acid cycle are used to generate ATP during oxidative phosphorylation.
3. CO_2 is produced just prior to and during the citric acid cycle, but oxygen is not consumed until the final step of oxidative phosphorylation.

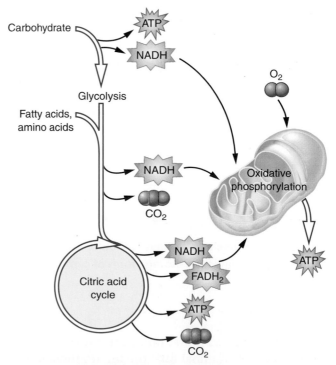

Figure 15.6. A Summary of ATP generation. Glucose, fatty acids, and amino acids feed into the ATP-generating pathways at different points. *True or false—the citric acid cycle directly generates one ATP per cycle.*

The Role of the Liver in Metabolism

The liver has rightly been called the queen of metabolism because of the variety of its tasks and the volume of product it processes. It breaks down some molecules (catabolism), builds new ones (anabolism), and converts one product into another. But before getting into the metabolic detail, let us review a bit of hepatic anatomy and physiology.

All venous blood leaving the intestine is gathered into a network of veins called the *hepatic portal system*, which rebranches into a second capillary network that percolates blood through the liver, bringing it into contact with hepatocytes. The above anatomy ensures that the liver gets the first pass at all absorbed carbohydrate and protein. Intestinal fat takes a longer path, since it is absorbed into intestinal lymph vessels that bypass the liver and empty in the systemic circulation. Fat is thus diluted in systemic blood before it arrives at the liver via the hepatic artery. If necessary, the liver breaks down these nutrient molecules and then uses the pieces to assemble new products.

In thinking about the complexities of hepatic nutrient metabolism, it is important to realize that each nutrient has both a *storage form* and a *usable form*. For example, liver and muscle store carbohydrate as glycogen, but the usable form of carbohydrate is glucose. For fat metabolism, the storage form is triglyceride deposited in adipose tissue or organs, but the usable form is mainly fatty acids and glycerol from triglyceride, with lesser amounts of cholesterol. When it comes to proteins, the storage form is whole protein (muscle fibers, plasma albumin, enzymes, and so on), and the usable form is the amino acid components of protein.

Finally, to a limited extent, the liver can convert one nutrient into another. For instance, it can convert amino acids into glucose and glucose into fatty acids.

Case Note

15.6. Santo's muscle cells have difficulty storing carbohydrate. What is the storage form of carbohydrate?

The Liver Metabolizes Fats

Recall from Chapter 2 that lipids are a diverse group of macromolecules. Animal fat—the fat around a juicy cut of beef, for example—is triglyceride (three fatty acids attached to a glycerol molecule), so it should be no surprise that the fat of human adipose tissue is also triglyceride. Most cells in the body can metabolize triglyceride,

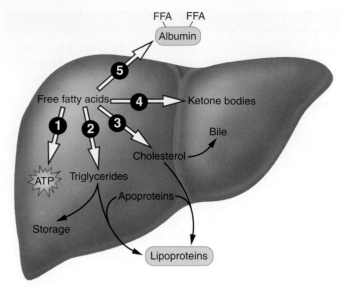

Figure 15.7. Role of the liver in lipid metabolism. The many fates of free fatty acids (FFAs). The numbers refer to the text. Glycerol (not shown) can be used to generate glucose or triglycerides. *How do free fatty acids travel in blood?*

usually for energy, and adipose tissue stores and processes triglycerides. However, the main site of fat metabolism is the liver.

Hepatocytes break triglycerides into glycerol and free fatty acids (FFAs). The fate of these molecules depends on body needs. Glycerol is usually reincorporated into triglycerides or converted to glucose. In contrast, FFAs can have any of five fates (the numbers below correspond to those on Fig. 15.7). FFAs can be:

1. Metabolized completely to provide hepatocytes with ATP.
2. Incorporated into triglyceride, for storage within the liver or export to adipose or other tissues. The synthesis of triglyceride from glycerol and free fatty acids is called **lipogenesis.** As we discussed in ◄ Chapter 14 , lipids circulate in blood within large protein–lipid complexes called *lipoproteins*. The specialized transport proteins incorporated into lipoproteins are called *apoproteins*.
3. Synthesized into cholesterol, which can be excreted in bile or exported to other tissues.
4. Converted into ketone bodies. **Ketone bodies** are partially metabolized fatty acids, which can be used by the brain and heart for energy when glucose stores are insufficient.
5. Exported to other tissues as fatty acids bound to albumin.

Just as most cells in the body can metabolize triglyceride, most can also use FFAs for energy. For example, cardiac and skeletal muscle can use FFAs to generate ATP. The brain, conversely, *cannot* use FFAs for energy—it depends on glucose, or on ketone bodies if glucose is not available.

Case Note

15.7. **Recall that Santo's blood contained ketone bodies. Are ketone bodies generated from carbohydrates or from fats?**

The Liver Metabolizes Carbohydrates

Dietary carbohydrates play the central role in satisfying the body's energy needs—about 90% of dietary carbohydrate is used for energy production. The intestine converts all carbohydrate into glucose, fructose, and galactose, which enter the hepatic portal vein and flow into the liver, where fructose and galactose are converted into glucose (Fig. 15.8, step 1).

One of the liver's main roles in energy metabolism is to act as a glucose buffer. As blood glucose rises, the liver takes glucose out of circulation. As blood glucose falls, the liver puts it back. The liver lowers blood glucose by converting it to a storage form, *glycogen*, by the process of **glycogenesis** (step 2a, green arrow). Other tissues also lower blood glucose by using it for energy, and, in the case of skeletal muscle, making and storing glycogen (Step 2b). Muscle, however, stores glycogen for its own energy purposes, not for homeostatic control of blood glucose. There is a limit to how much glycogen the liver and muscles can store or the body can use, and the excess is stored as fat—the acetyl-CoA produced from glucose breakdown can be diverted to synthesize fatty acids (step 3). See Figure 15.7 for the fate of these fatty acids.

Conversely, a fall of blood glucose stimulates the opposite reaction—hepatocytes break down stored glycogen into glucose, a process called **glycogenolysis** (step 2a, red arrow). The newly formed glucose molecules are released into blood in order to restore blood glucose homeostasis (step 4).

> *Remember This!* **Be careful not to confuse** *glycogenolysis*, **the breakdown of glycogen into glucose, with** *glycolysis*, **the breakdown of glucose.**

People with impaired liver function have poor blood glucose control. For example, after a carbohydrate-rich meal, blood glucose may rise to diabetic levels because the liver is unable to convert glucose to glycogen and

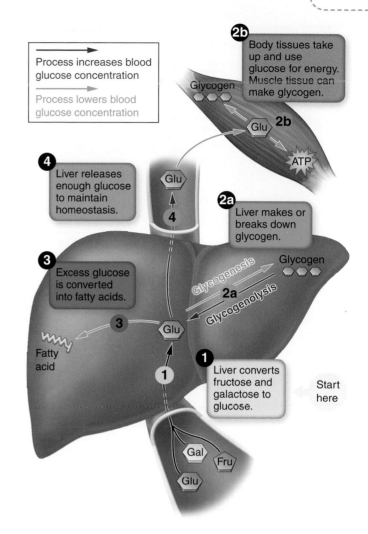

Figure 15.8. Role of the liver in carbohydrate metabolism. The liver stores excess glucose as glycogen or fatty acids, and releases it back into the bloodstream as glucose when the blood glucose concentration decreases. *Does glycogenolysis increase or decrease blood glucose concentrations?*

store it. Conversely, after a prolonged fast, blood glucose may fall to very low levels because the liver cannot convert glycogen to glucose quickly enough.

Case Note

15.8. **Diabetes mellitus is associated with increased glycogen breakdown and reduced glycogen synthesis. What are the scientific terms used to identify these two processes?**

The Liver Metabolizes Proteins

Recall that proteins are especially large polypeptides—long chains of amino acids linked to one another by peptide bonds. Almost all dietary protein is digested into

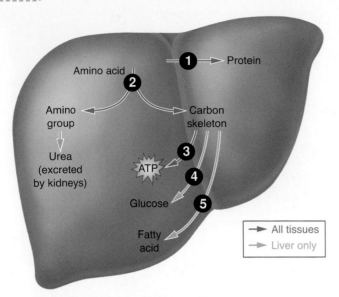

Figure 15.9. Role of the liver in protein metabolism. Amino acids from proteins can be used to generate new proteins. Alternatively, they are converted to other nutrient types for storage or energy, or used directly to generate ATP. *What is the fate of the amine group of amino acids?*

amino acids in the intestine before entering portal blood; however, a few polypeptides remain, and the liver finishes the job by breaking them into amino acids. Tissues use amino acids in two different ways: in anabolic reactions that build proteins, and in catabolic reactions that generate ATP or other nutrients (Fig. 15.9).

Amino Acid Anabolism Generates Proteins

All tissues use amino acids to build proteins (Fig. 15.9, ❶). Consider, for instance, the large amounts of protein required to build muscle tissue. The liver builds proteins collectively referred to as **hepatic proteins.** These include:

- Albumin
- Transport proteins such as transferrin, which transports iron
- Apoproteins, which transport cholesterol, triglyceride, and phospholipid in lipoprotein packets
- Clotting factors, such as fibrinogen

The liver must synthesize a large amount of protein each day because the half-life of plasma proteins varies—fibrinogen and other clotting factors a few days; albumin about a week. Plasma protein levels fall rapidly if liver function is significantly impaired by liver disease such as cirrhosis ◀ (Chapter 14). For example, albumin accounts for a majority of the osmotic pressure of plasma, and low plasma albumin allows fluid to escape from blood vessels into tissues (causing edema) or body spaces (causing ascites). In addition, low levels of clotting factors are associated with bleeding (bruising or intestinal bleeding, for example).

Amino Acid Catabolism Generates ATP and Other Nutrients

Body tissues other than the brain and heart can metabolize amino acids for energy; amino acids supply about 10% to 15% of our daily energy needs. The first step in amino acid breakdown (catabolism) is *deamination*: removal of the –NH2 amino group, leaving a carbon skeleton behind (Fig. 15.9, ❷). Hepatocytes metabolize the amino group into urea for excretion by the kidneys. The carbon skeleton can be used to generate ATP by entering the citric acid cycle (❸).

The liver can also convert the carbon skeleton into other nutrients. Of particular importance is the conversion of carbon skeletons into glucose molecules (❹). This process is called **gluconeogenesis,** because it creates (-genesis) a new (neo-) glucose molecule. The glucose molecules can be used to bolster blood glucose levels in times of starvation or converted into glycogen and stored in the liver. Finally, the liver can also convert amino acids into fatty acids, which can be stored as fat, whether in the liver or in adipose tissue (❺).

Importantly, excess amino acids cannot be stored as protein. That's why consuming more protein than you need will not help you build muscle: protein synthesis in muscle is controlled by the strain placed on the muscle. That being said, our requirement for essential amino acids must be provided in the diet because the body cannot make them. A dietary deficiency in any essential amino acid inhibits protein synthesis. Leucine, for instance, is deficient in many diets, and consuming leucine has been shown to increase muscle protein synthesis.

Case Note

15.9. Gluconeogenesis is increased by Santo's disease. Which part of the amino acid is used to synthesize new glucose molecules—the carbon skeleton or the amino group?

The Liver Has Other Metabolic Roles

Beyond its critical and diverse roles in carbohydrate, fat, and protein metabolism, the liver performs additional tasks.

First, it is a storehouse for a variety of substances. It stores excess carbohydrate and lipid beyond its own need for energy; it stores vitamins, most notably vitamins

A, D, E, K, and B12; it also stores iron as ferritin ⬅ (Chapter 10). The majority of body iron storage reserves are in the liver, which adds iron to the circulation or removes it for storage according to the bone marrow's need to synthesize hemoglobin.

Second, the liver clears the body of a long list of bodily wastes, drugs, and toxins by adding them to bile. Recall that bile is critical for normal fat digestion, but it also serves as a vehicle for excreting liver-processed substances into the feces. For instance, the liver excretes blood bilirubin (the waste product of the hemoglobin from dead red blood cells), steroid hormones, and even calcium into bile. What is more, the liver removes cholesterol from blood and uses it for synthesis of bile acids. Although most of these bile acids are reabsorbed by the intestine, some pass into feces and carry cholesterol out of the body.

In summary, the liver has three primary roles: (a) it is a clearinghouse that processes nutrients and delivers appropriate amounts of each to other tissues via the bloodstream; (b) it is a manufacturing facility that synthesizes important substances and spreads them via the bloodstream; and (c) it is a recycling center that gathers waste and disposes of it.

The Liver Directs Metabolism of the Fed and Fasting States

It is easy to think that each nutrient type does a certain job: that proteins build muscle, that fat gets stored as adipose tissue, or that glucose is the only fuel the body burns. Yet if the discussion above proves anything, it is that nutrition is a very integrated metabolic process. The flow of nutrients into or out of storage, and which nutrients are being actively metabolized for energy, depends completely on one thing: whether or not you have just eaten.

If you have eaten recently, you are said to be in the **fed state,** in which metabolic processes work together to lower the surge of blood glucose and to build tissue. After the consumption of a normal meal, nutrient molecules—amino acids, glucose, and fatty acids—flood into the bloodstream. In the fed state, cells use whichever nutrients they need and store the excess. Tissue is built (anabolism): cell membranes and muscle fibers are formed and other nonenergy uses are satisfied. Beyond these necessities, the remainder is stored as glycogen or triglyceride. See the left column of Table 15.2 for details.

> *Remember This!* **Protein synthesis to build big muscles needs not only the nutrients of the fed state but also exercise! All excess nutrients eventually find their way into fat.**

Table 15.2 The Fed and Fasted States

Fed State	Fasted State
All body cells take up and use glucose for energy	The brain and heart take up and use glucose for energy; other tissues (such as muscle) use fatty acids (or amino acids) instead
All tissues use amino acids to synthesize required proteins	In prolonged fasting, tissues break down proteins to generate amino acids
The liver converts excess amino acids into glucose destined for storage	The liver converts amino acids into glucose destined for tissue use
Liver and muscle convert glucose into glycogen until glycogen stores are full	Liver and muscle break down glycogen into glucose
The liver converts excess glucose into fatty acids	Adipose and liver convert triglycerides into fatty acids and glycerol
Liver and adipose assemble fatty acids (and glycerol) into triglycerides	The liver converts fatty acids into ketones for use by the brain and heart

Abstain from food and you are in the **fasting state,** in which the opposite is true: the liver (especially) and other tissues work together to raise blood glucose. Between meals nutrients cease to enter the blood from the digestive system and there is a fall in the blood concentration of fatty acids, amino acids, and glucose. The fall in fatty acids and amino acids is not especially problematic. Glucose, however, is another matter: the body satisfies its *immediate* energy needs from glucose. The brain is especially vulnerable because it cannot burn fatty acids or amino acids. When blood glucose is insufficient, the liver accelerates the conversion of fatty acids into ketone bodies, which the brain can use for energy. The brain can meet two thirds of its energy needs from ketone bodies, but, as our discussion of diabetes shows, ketone bodies acidify the blood and can cause problems. Therefore, mainly in defense of the brain, all of the metabolic changes of the fasting state work together to raise blood glucose. If fasting becomes starvation, the body will figuratively burn down the house (protein) to provide enough amino acids for glucose synthesis.

Case Note

15.10. Santo's liver is trying to raise blood glucose concentrations. Does this indicate that his liver in the fasting state or the fed state?

15.8 Name the primary storage form of fats.

15.9 Which blood vessel delivers blood rich in absorbed nutrients directly to the liver?

15.10 Which nutrient is directly converted into cholesterol—glucose or fatty acids?

15.11 What is the difference between glycolysis and glycogenolysis?

15.12 True or false: Excess carbohydrate or protein can be converted into fatty acids and eventually stored as fat.

15.13 Where is albumin synthesized?

15.14 Which process would be favored in someone who just ate a large meal: glucose synthesis or fatty acid synthesis?

15.15 In a fasting person, which nutrient source will muscle cells prefer to use for energy: glucose or fatty acids?

15.16 Name the process that synthesizes new glucose molecules from amino acids.

Energy Balance

Human physiology has not changed significantly for at least 100,000 years. Back then, agriculture and domesticated animals were unknown and humans were on their feet most of the day, walking to rivers and streams to get water, gathering edible plants, and stalking prey. Meat consumption was rare and meals might be days apart. So, it should be no surprise that the human body evolved the capacity to use energy efficiently and to hoard any excess nutrients ingested.

And today? In developed nations obesity has become an epidemic—food is cheap, plentiful, and tasty; and technology has taken most of the physical effort out of work. Instead of worrying about where to find the next fistful of berries, the average person is worried about excess weight.

You might not think that Einstein would crop up in this discussion, but his insight about matter and energy is relevant. Recall that Einstein proved that matter and energy are merely different expressions of the same thing: energy can be converted into mass (which happens when we gain weight) and mass can be converted into energy (which is why exercise 'burns" away weight). And when energy is spent, it produces only two outcomes: work and heat. The following equation expresses the idea:

$$\text{Nutrient input (mass convertible to energy)} = \text{energy output (work + heat) + new mass (carbohydrate, fat and protein of cells and tissues)}$$

The energy output (work and heat) keeps us alive, and the excess is metabolized into new mass: cells and tissues of every kind. If we take in more energy than we expend in work and heat, we gain mass (weight). If we expend more energy in work and heat than we take in, then, like Lizzie in our scenario at the beginning of this chapter, we lose mass (weight). In short, maintaining energy balance is all about matching input with output.

Recall that that oxidative phosphorylation produces ATP *and* heat. The balance between ATP and heat production is not constant. For instance, certain tissues can alter mitochondrial activity to produce *less* ATP and *more* heat per nutrient. This change is described as *decreasing* the efficiency of ATP production, because more of the energy contained in the original nutrient chemical bond is "lost" as heat. However, as we will see later, this waste heat plays a critical role in maintaining a constant body temperature.

Case Note

15.11. Recall that Santo is very obese and continues to gain weight. Which is greater—his energy input or his energy output?

Metabolism, Digestion, and Locomotion Use Energy

Energy output includes the *base amount* of energy the body expends just to stay alive, the energy consumed by *digestion* (yes, digestion is work and consumes energy), and the amount spent in *locomotion* (movement).

Basal Metabolic Rate Measures Resting Energy Needs

Our rate of energy consumption is lowest during rest, when there is little activity other than breathing. This rate is thus known as the **basal metabolic rate (BMR)** and is measured in an awake, fasting, supine individual in a warm (but not hot) room. Several factors affect a person's BMR, including:

- *Lean body mass:* Resting muscle burns more energy than resting fat. Even when inactive, people with more muscle and less fat have a higher BMR.
- *Gender:* BMR is generally higher in males (because of their higher lean body mass) than in females.
- *Age:* Metabolic rates decline with age, mostly because of reduced lean muscle mass.
- *Genetics:* Specific combinations of genes control the "efficiency" of ATP production. Leaner people often have a less efficient metabolism, producing proportionally more heat and less ATP per nutrient molecule than fatter people.
- *Hormones:* As explained further on, thyroid hormones modify BMR by accelerating ATP utilization but also decreasing the efficiency of ATP synthesis.
- *Dieting.* Dieting is slow starvation. The body recognizes it as such and lowers the BMR in order to preserve energy stores for future use. Which means that as soon as you start dieting, you begin to burn fewer calories in the basal state.

Case Note

15.12. Studies reveal that Santo's basal metabolic rate is extremely low for his weight and age. Considering only the fact that Santo's body contains proportionally more fat than that of a leaner person, would his obesity play a role in his low BMR?

Digestion Uses Energy

Recall from ← Chapter 14 the many processes that are required to digest food and absorb nutrients. These processes use significant amounts of energy, generally about 10% as much as an individual's basal metabolic rate.

Locomotion Can Be Controlled Voluntarily

Locomotion, body movement, is the only way we can consciously influence energy balance. Any activity—be it walking, swimming, driving a car, or vacuuming the floor—will increase the metabolic rate and burn more nutrients. Locomotion also includes relatively involuntary movements, such as fidgeting or pacing, that also burn calories.

Many Factors Influence Energy Input

There is survival value in the hunger we feel after existing without food for a while, but hunger does not justify all our decisions to eat. Of the factors governing food intake, many are sociocultural or psychological rather than physiological. Consider the overeating that routinely occurs during a sociable Thanksgiving dinner, or the "comfort" foods we eat during periods of stress. Here, we limit our discussion to a few of the physiological signals governing energy intake, focusing on those that physiologists have learned much more about in recent years. These are: *ghrelin, leptin, cholecystokinin,* and the *PYY family* of peptides (Fig. 15.10).

Why do we start eating? The most important signal appears to be **ghrelin,** a peptide hormone produced mainly by specialized cells in the mucosa of the fundus of the stomach. The empty stomach produces a large amount of ghrelin, which circulates in blood and stimulates the hypothalamus, which in turn signals the cerebral cortex with a hunger sensation (Fig. 15.10A). Ghrelin, therefore, is an eating initiator.

Why do we stop eating? One factor is gastric distention, which relays autonomic sensory signals to the brain as a sensation of fullness. Another factor is cholecystokinin (CCK), discussed in ← Chapter 14. The entry of food into the duodenum stimulates the release of CCK from intestinal epithelium, which enters the bloodstream and among other things stimulates the pancreas to secrete its digestive enzymes. CCK also acts on the hypothalamus to decrease hunger signals to the cortex (Fig. 15.10B). A third factor involved when we push aside our plates is a hormone called *peptide YY* (PYY). Food in the intestine stimulates intestinal mucosa to release PYY, which exerts a number of effects on the intestine, and also inhibits hypothalamic hunger signals. Other gut and pancreatic peptides also induce satiety in response to a full digestive tract.

What about long-term regulation of appetite, especially regarding obesity? The underlying mechanisms are not fully understood, but it is helpful to think of the hypothalamus as having a "fat thermostat" set to maintain a certain quantity of energy stored as fat. If the amount of stored fat changes, it can modify hunger and energy expenditure. One of the most important indicators of fat stores is circulating levels of **leptin,** a hormone produced by adipose tissue. Leptin acts on the

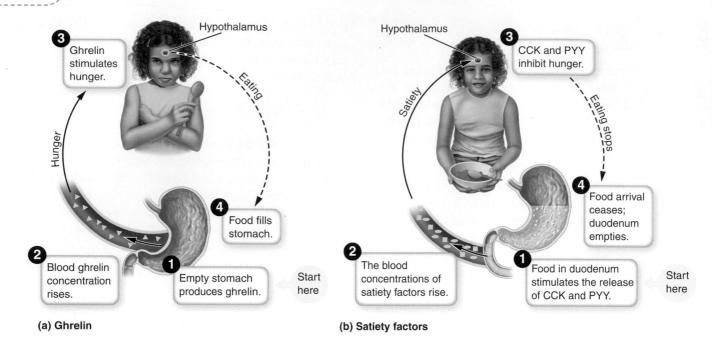

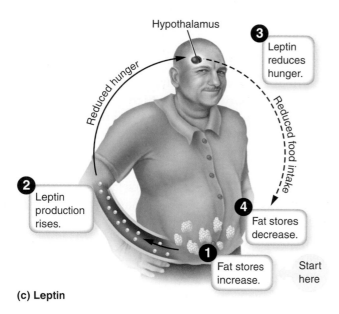

(c) Leptin

Figure 15.10. Regulation of food intake. A. Ghrelin initiates hunger. **B.** CCK and PYY inhibit hunger. **C.** Leptin governs appetite over the long term. *Which of the following would be produced more when we are hungry—ghrelin or CCK?*

hypothalamus to decrease hunger signals (Fig. 15.10C). Theoretically, the greater the fat stores, the higher the blood leptin concentration will be, and the less hungry we'll feel. Suppose, for example, you were a person without a weight problem and you went on a cruise, ate without caution, engaged in very little physical activity, and gained 5 pounds, most of it fat. The increased leptin secreted by the new fat would decrease hypothalamic hunger signals to the cortex and you would feel less inclined to eat until your fat stores and blood leptin had returned to their prior state. Paradoxically, many obese people have high blood leptin concentrations but keep on eating, which suggests that they are somehow resistant to the effect of leptin.

Case Note

15.13. Which of the following would theoretically help Santo's obesity: a leptin blocker or a hormone that increases the cellular responses to leptin?

The pathways governing appetite and body weight are complex and redundant; we have touched only lightly on the topic. This complexity renders the treatment of obesity very difficult. A proven but somewhat risky, method of quickly and permanently losing large amounts of body fat is *bariatric surgery*, which is discussed in the nearby clinical snapshot titled *Weight-Loss Surgery*.

CLINICAL SNAPSHOT

Weight-Loss Surgery

Surgery to lose weight (*bariatric surgery*, from Greek *baros* = "weight") is very serious business because the stakes are so high: obesity is a significant risk factor for premature death, yet surgery to relieve it is fraught with hazards, too. Nevertheless, bariatric surgery is increasingly common as a treatment for severe obesity for two reasons: first, the benefits usually outweigh the risks; and second, obesity is epidemic in the United States and most other industrialized societies—roughly 25% of Americans are frankly obese, and another 33% are overweight. There are two basic ways surgery can help:

1. *By diverting food around part of the small intestine* (part a). In this procedure, first the proximal stomach is closed into a small pouch; then the small bowel is transected and the distal part is sewn into the side of the newly created stomach pouch. This has the effect of bypassing the duodenum, which decreases the opportunity for food absorption and allows calories to pass into stool.
2. *By making the stomach smaller* (part b). This can be done in any number of ways, from cutting out part of the stomach to putting a kind of drawstring (a band) around the upper part and cinching it down so that the part of the stomach above the constriction is quite small. The result is that it takes a much smaller amount of food to distend the stomach and send an "I am full" signal to the cerebral cortex.

Perhaps the most important benefit of the surgery is that it reduces appetite. Recall that gut peptides such as ghrelin induce hunger. Bariatric surgery reduces ghrelin production, for reasons that are not yet well understood.

Recent advances in technology allow these procedures, especially the banding procedure, to be done by "keyhole surgery" through long thin tubes inserted through small cuts in the abdominal wall. The technical term for this approach is *laparoscopy*, which derives from Greek *lapara* = "flank" or "side" and Greek *skopein* = "to look at." Hence the term *lap band* has entered our language.

Successful bariatric surgery does not guarantee good health or even permanent weight loss. Postoperative complications are common, ranging from infections to nutritional problems related to dumping syndrome (see ◀ Chapter 14). What's more, some patients fail to restrict their food intake appropriately. Some eat numerous small meals—so many that they may not lose weight or may regain weight lost initially. Others

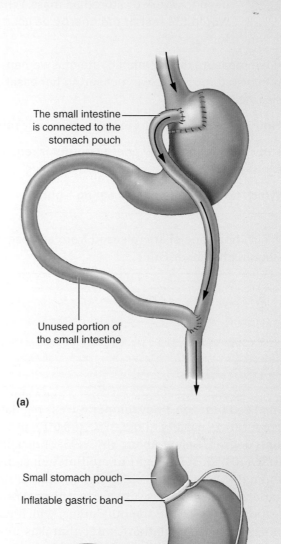

The small intestine is connected to the stomach pouch

Unused portion of the small intestine

(a)

Small stomach pouch

Inflatable gastric band

Larger stomach pouch

Port (for inflating band)

(b)

Bariatric surgery. Bariatric surgery is an accepted way of treating morbid obesity.

may stuff their small pouch to such an extent that it stretches, allowing them to overeat with increasingly greater comfort. Eventually many such patients regain the weight they lost initially.

15.17 If you just ate an orange, some of the energy contained in it would be used for work or stored as mass. What would the rest of the energy be used for?

15.18 If we measure the metabolic rate of a woman reclining in a cold room, are we measuring her basal metabolic rate?

15.19 How does dieting alter the basal metabolic rate?

15.20 How much of our daily energy output is required to digest our food?

15.21 Which hormone stimulates eating—ghrelin or leptin?

15.22 Which hormone(s) are released from adipose tissue—leptin, ghrelin, or PYY?

Regulation of Body Temperature

For optimal cell function, body temperature is regulated in a narrow range, usually about 37°C (98.6°F), plus or minus one degree. Below about 35°C (95°F) or above about 40°C (104°F), cells begin to malfunction significantly enough to cause illness or death. Maintaining body temperature within this very narrow normal range is the responsibility of the hypothalamus, which contains a temperature-regulation center that acts much like a room thermostat with a particular *set point*. Local hypothalamic temperature sensors detect variations from this set point in blood temperature; they also receive nerve signals from temperature sensors in skin and deeper structures of the body. They then relay homeostatic signals to muscle and other effectors that increase or decrease heat generation or heat loss.

This negative-feedback mechanism enables the hypothalamus to maintain normal set point temperature in very warm or very cold circumstances. In fact, despite the narrow limits for cellular function, the body as a whole can survive quite comfortably in ambient temperatures ranging from −35°C (−31°F) to +44°C (113°F) or even more extreme levels provided that appropriate clothing is being worn and precautions have been taken. Before we discuss temperature homeostasis, we must discuss how heat moves between the body and the environment.

Heat Transfers by Four Processes

Heat is a form of energy attributable to the motion of all of the molecules in a substance. The molecules in hot water or hot air are racing around much more rapidly than those in cold. Like everything else, heat moves down a gradient from warm (higher concentration of heat) to cold (lower concentration). There are four ways heat can be transferred from a warm area to a cooler one (Fig. 15.11).

First is heat transfer by **radiation**—that is, by electromagnetic waves such as light. Walk outside on a cold, sunny day and turn your face to the sun. The warmth you feel arises when light rays from the sun strike your skin, transferring their thermal energy to it. The body also loses heat by radiation, but the electromagnetic waves are in the infrared spectrum, so they are not visible to the unaided eye. However, don a pair of infrared-sensitive night-vision goggles and glowing humans leap out of the dark. Transfer of heat by radiation does not require contact between the source and the receiver.

Second is heat transfer by **conduction,** which is accomplished when high-energy, active molecules of a warm object bump into slower molecules in a cooler object and speed them up. Lying on a sun-heated rock will warm your skin by conduction.

Third is heat transfer by **convection,** a modified form of conduction. In convection, a solid structure loses heat to a gas or liquid, which then moves away, carrying the heat with it. For instance, muscle heat transfers to the blood by convection, which flows on and is replaced by new, cooler blood. Skin warms adjacent air; then a breeze moves the warm air away and replaces it with cooler air.

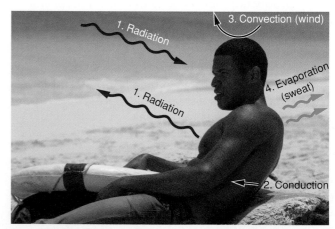

Figure 15.11. Heat transfer. Heat transfers between the environment and the body by four different mechanisms. *In this figure, which heat transfer mechanisms are adding heat to the body?*

Finally, heat transfer occurs by **evaporation,** a process by which liquid water molecules move into air to become vapor. An example is sweat: just as it takes energy to boil water, it takes energy to evaporate sweat, and the heat carried away in the evaporating water makes the skin cooler. Heat loss by sweating is a very important part of body temperature control. The potential heat loss from evaporation is considerable. But evaporation involves water gradients as well as heat gradients—the hotter and drier the air, the greater the evaporation. When the environmental air is humid (saturated with water vapor), our bodies cannot cool as efficiently, regardless of the volume of sweat we produce, because only *evaporating* sweat cools the body, and the more humid the air the less rapidly sweat evaporates. Incidentally, sweat is just one mechanism of evaporation—even in the coldest conditions, we lose about 500 mL of water daily from the skin via insensible evaporation of perspiration and the evaporation of water vapor in our breath.

Heat transfers from the body core to the environment by a combination of these processes. For example, consider the heat produced by a hard-working muscle such as the heart. This heat transfers to blood by conduction, moves away and into skin by convection, and can be lost into the environment by radiation, conduction, convection, or evaporation.

(a) Cold conditions

Case Note

15.14. Like most obese people, Santo sweats a lot, and he is always wiping his bald head with a towel. Does wiped-away sweat cool the body? Explain.

Cold Conditions Stimulate Heat Retention

Cold conditions prompt behavioral and physiological responses to limit heat loss (Fig. 15.12A). Covering body surfaces with insulating clothing limits radiant, conductive, and convective heat loss. Down or poly-fill insulating clothing or dressing in multiple layers provides the best insulation, because air is trapped between the feathers or clothing layers. Air conducts heat very poorly, so the heat never makes it out to the environment.

Physiological responses limit heat loss and, to a certain extent, stimulate heat production. In the cold, the hypothalamus signals skin blood vessels to constrict, which limits skin blood flow and the chance for blood to lose heat through the skin. But skin is designed to func-

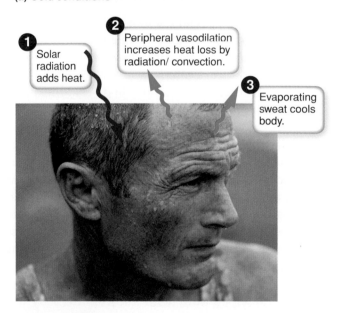

(b) Hot conditions

Figure 15.12. Thermoregulation. A. Under cold conditions, physiological and behavioral responses limit heat loss and generate additional metabolic heat. **B.** Under hot conditions, sweating and peripheral vasodilation help the body shed metabolic heat and heat absorbed from solar radiation. *How does sweating cool the body—by conduction or by evaporation?*

tion at low temperature and suffers no ill effect until it is exposed for a while to air temperatures near freezing. Tissue freezing, or *frostbite,* results when blood flow cannot keep tissue temperatures above 32°F. Cell cytoplasm freezes, ice crystals rupture cell membranes, and cell death occurs. High wind, which maximizes convection, and wet clothing, which maximizes evaporation, increase the risk of frostbite.

If behavioral and vascular responses cannot maintain body temperature, the hypothalamus stimulates other homeostatic responses to attempt to maximize heat generation (*thermogenesis*). *Shivering* is involuntary muscle contractions that are nondirectional and therefore mechanically inefficient; they generate heat from muscle activity. Thyroid hormones and signaling by the sympathetic nervous system can increase thermogenesis without shivering, particularly in newborns. *Nonshivering thermogenesis* relies on mitochondrial respiration to generate heat but does not involve muscle contractions. Instead, the electron transport chain in mitochondrial oxidative phosphorylation becomes modified slightly to produce less ATP and more heat from each high-energy electron transfer.

Hot Conditions Stimulate Heat Loss

Hot weather poses special challenges for heat regulation. The body must disperse heat gained from solar radiation and metabolic reactions. However, the increased ambient temperature minimizes or even reverses the temperature gradient between the body and the atmosphere, reducing the effectiveness of all modes of heat loss (Fig. 15.12B). So how do we stay cool?

You can remember the major mechanisms for heat loss by imagining people exercising in the heat—they will be red and sweaty. The redness reflects heat loss. As blood (body) temperature rises, the hypothalamus dispatches autonomic signals to dilate skin blood vessels. More blood flows to skin, where its heat can be lost to the environment by radiation and convection. Sweating enables evaporative heat loss. If sweat does not evaporate because it is too humid outside, or if it drips from the body or is wiped away with a towel, the body will lose water but stay hot.

Body temperature will rise if the body does not lose heat as fast as it is generated. *Heat exhaustion* is a body temperature over 39.5°C (103°F) in the absence of other illness or infection. Afflicted persons sweat profusely, but at a cost—lost fluid causes low blood volume and low blood pressure. *Heat stroke* is body temperature over 41°C (106°F) in the absence of an underlying illness. Dehydration becomes so severe that sweating slows down or stops, evaporative heat loss ceases, the hypothalamus ceases to operate properly, and body temperature soars out of control. Untreated heat stroke is invariably fatal. Even when it is treated properly, the fatality rate is near 50%.

Proper clothing choices can help the body's cooling efforts. Skin absorbs large amounts of radiant heat from the sun; therefore white, reflective clothing considerably reduces heat gain through radiation. Loose clothing enables air currents to come in contact with skin surfaces, enabling convective and evaporative heat loss. The relative frequency of death from heatstroke among football players in summertime scrimmages highlights the importance of proper clothing when exercising in the heat. Football uniforms are bulky, tight-fitting, and made of nonbreathable polyester. This combination retards air movement and traps heat and moisture, creating a layer of warm, humid air next to the skin. Heat loss by radiation, evaporation, and convection is reduced. Under such circumstances, in an environmental temperature of 32°C (89.6°F), the core temperature of an exercising football player in full uniform can rise from normal to 39°C (102.2°F) in 30 minutes.

Hyperthermia

Hyperthermia is a body temperature increased above the normal range, but "normal" varies more than you might think. Because they are deep and do a lot of work, core body organs such as the liver and muscles maintain a temperature that is slightly higher than the temperature in more superficial organs—skin, for example. What is more, temperature in the rectum, where infant temperature is often measured, is about 1°C higher than oral temperature. Finally, normal body temperature varies from one person to another. Nevertheless, a useful definition of **hyperthermia** is a body temperature *above* 38°C or 100°F.

There are but two fundamental reasons for hyperthermia:

1. *The body is unable to shed heat as fast as it is generated*. In this instance the usual situation is hot, humid weather, dehydration, and hard work.
2. *The hypothalamic set point moves to a higher point*. In this instance, an underlying condition causes the hypothalamic thermostat to reset to an abnormally high temperature, a condition known as **fever**. The usual cause is tissue injury, inflammation, or both, which cause the release of signal molecules from injured or dead cells that reset the hypothalamic set point. However, some pathogens, certain bacteria in particular, contain similar molecules and release them in the course of infection. Finally, toxic conditions of the brain may cause hypothalamic dysfunction.

15.23 Which form of heat transfer occurs between a solid and a liquid or a solid and a gas?

15.24 The elderly do not sweat as effectively as younger people. Which mode of heat transfer will be impaired?

15.25 Does thermogenesis warm or cool the body?

15.26 A hot person is red and sweaty. Why?

15.27 Which heat disorder is more serious—heat stroke or heat exhaustion?

The Endocrine Pancreas

The remainder of this chapter focuses on the **endocrine glands:** discrete organs that secrete hormones into blood. These are illustrated in Figure 15.13.

The first endocrine gland to be considered is the endocrine portion of the pancreas. Recall from ← Chapter 14 that the exocrine pancreas—about 99% of the mass of the pancreas—secretes digestive juices into ducts that empty into the intestine. The remaining 1% of pancreatic mass consists of scattered, small aggregations of endocrine tissue called *pancreatic islets* (*islets of Langerhans*). The pancreatic islets synthesize hormones and secrete them directly into blood.

The endocrine pancreas regulates many of the metabolic pathways discussed in this chapter. Pancreatic islets are richly supplied with blood, which brings with it information about blood glucose levels (Fig. 15.14A). Additionally, neurons from both the sympathetic and parasympathetic nervous system terminate on islet cells, permitting an additional level of neural control.

The endocrine pancreatic islets may be tiny, but they are powerful and necessary for life. Two types of pancreatic islet cells are relevant to our discussion (Fig. 15.14B):

● *Beta cells,* which release *insulin,* a hormone that predominates in the fed state
● *Alpha cells,* which release *glucagon,* a hormone that predominates in the fasting state

These two hormones interact antagonistically to keep blood glucose levels nearly constant (70 to 110 mg/dL blood). Importantly, the pancreas continually produces both glucagon and insulin. It is the ratio of the production of these two hormones that regulates nutrient metabolism.

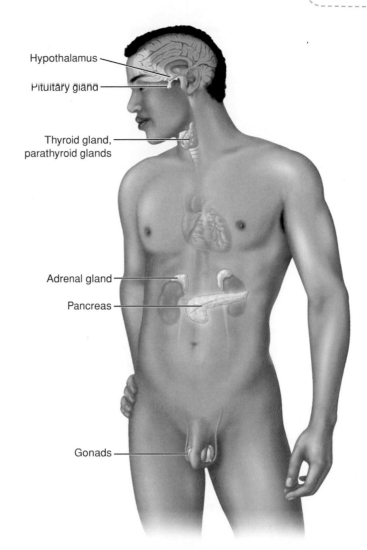

Figure 15.13. **The endocrine system.** Organs of the endocrine system are glands that release their secretions directly into the bloodstream. *What are the secretions of endocrine glands called?*

Insulin Predominates in the Fed State

Insulin is the hormone of glucose abundance—in effect, its message to body cells is "Use glucose for energy and store everything else."

Nutrient Abundance Stimulates Insulin Secretion

Insulin production and release increases in response to signals of nutrient abundance. These signals include the following (Fig. 15.15, left side):

● *Gut hormones* such as gastric inhibitory peptide (GIP). These hormones are secreted when food is in the intestine, even before nutrient absorption occurs.

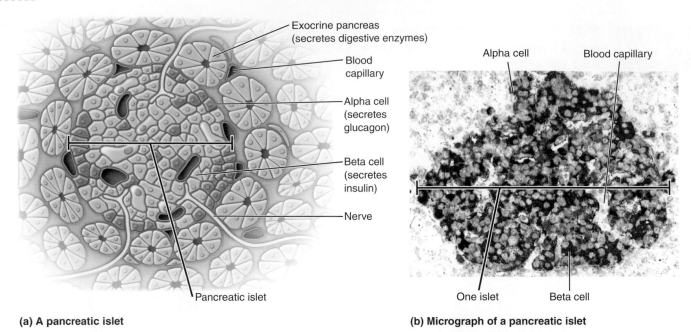

(a) A pancreatic islet

(b) Micrograph of a pancreatic islet

Figure 15.14. The endocrine pancreas. A. Each pancreatic islet contains alpha and beta cells. **B.** This micrograph of a pancreatic islet shows pinkish-purple alpha cells and reddish-brown beta cells. *Which hormone do beta cells secrete?*

These gut hormones are responsible for about 50% of postmeal insulin production.

- *High blood glucose concentration* resulting from absorbed nutrients. In the following discussion, note that insulin's actions reduce blood glucose concentrations, which inhibit insulin secretion in a classic negative feedback loop.

Insulin Promotes Nutrient Use and Storage

Insulin, in turn, acts on many body organs to stimulate nutrient uptake, use, and storage. These actions include (Fig. 15.15, left side):

1. *Insulin stimulates glucose uptake by fat cells and muscle cells*. In the absence of insulin, the cells in these tissues cannot extract glucose from blood because their glucose transporters are locked away in the cell interior. Insulin stimulates the movement of these transporters to the cell membrane, where they permit glucose to enter the cell. Glucose uptake into *exercising muscle* is not insulin-dependent—changes in exercising muscle cells favor movement of glucose transporters from the interior of the cell to the cell membrane. The liver, brain, and kidney are similarly adept at importing glucose regardless of insulin status.

2. *Insulin stimulates glucose utilization*. Insulin activates enzymes involved in glucose breakdown for energy

(glycolysis) and in glucose storage as glycogen (glycogenesis). It inhibits enzymes that increase glucose production from amino acids (gluconeogenesis) and or from glycogen (glycogenolysis).

3. *Insulin stimulates fat synthesis and inhibits fat breakdown*. Insulin "assumes" that the body has absorbed enough glucose to meet energy needs, so it promotes fat storage by inhibiting the breakdown of triglyceride into free fatty acids and the use of fatty acids for energy. Similarly, it promotes the conversion of glucose and amino acids into fatty acids and the synthesis of triglycerides from fatty acids and glycerol.

4. *Insulin promotes protein synthesis*. The presence of insulin encourages cells, especially liver and muscle, to make any protein they need from the pool of amino acids. Excess amino acids are converted into fatty acids and incorporated into fat.

Actions 1 and 2 promote the use of glucose and glycogen for energy, thus sparing tissues from being burned. Actions 3 and 4 promote tissue building. The net effect is therefore *anabolic*. As shown below, if insulin is absent or ineffective, blood glucose rises abnormally high.

Glucagon Predominates in the Fasting State

Glucagon is insulin's opposite—it is released by pancreatic alpha cells between meals and promotes cellular

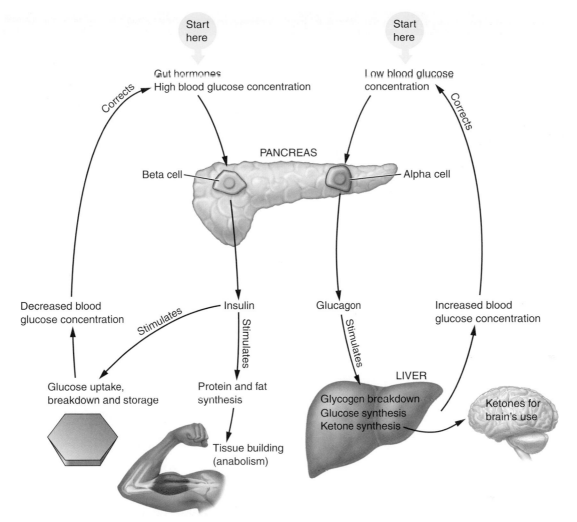

Figure 15.15. Insulin and glucagon. Insulin and glucagon regulate glucose homeostasis. Insulin also promotes anabolism (tissue building), but glucagon does not promote catabolism (tissue breakdown) to the same extent. *Which hormone stimulates glucose synthesis— glucagon or insulin?*

activities that increase blood glucose concentrations (Fig. 15.15, right side). However, the actions of glucagon are much more limited than those of insulin—glucagon does not promote catabolism as effectively insulin promotes anabolism. Many of the characteristics of the fasting state, such as reduced protein synthesis, reflect the absence of insulin rather than the presence of glucagon. In effect, glucagon's message to body cells is "Save nutrients for the brain if you can, but it's not an emergency."

Fasting Stimulates Glucagon Production

Glucagon secretion increases in response to nutrient deficiency. The most important trigger for the increased secretion of glucagon in alpha cells is a drop in blood glucose concentration below 100 mL/dL.

Glucagon Promotes the Synthesis of Glucose and Ketones

In contrast to insulin's widespread effects, glucagon's actions are largely confined to the liver. Within the liver, glucagon stimulates:

1. *The breakdown of glycogen to glucose* (glycogenolysis) from hepatic glycogen stores.
2. *The creation of glucose from amino acids* (gluconeogenesis).

Steps 1 and 2 raise blood glucose concentration.

3. *Production of ketone bodies from fatty acids* (ketogenesis). Ketogenesis begins only after about 8 hours of fasting. The brain uses the resulting ketone bodies for energy.

Table 15.3   Comparison of Type 1 and Type 2 Diabetes

Criterion	Type 1	Type 2
Age at onset	Usually teenage years	Usually adults, but increasingly prevalent in children
Rapidity and intensity at onset	Sudden, severe	Slow, subtle
Body weight	Normal to underweight	Usually overweight or obese
Genetic influence	Less than 20% of patients	More than 60% of patients
Underlying cause	Insulin deficiency; autoimmune destruction of pancreatic cells	Tissue resistance to insulin effect; linked to increased adipose stores
Blood insulin level	Marked decrease	Initially or increased; decreased in late-stage type II diabetes
Episodes of ketoacidosis	Periodic	Rare
Treatment	Insulin, diet, exercise	Initially, weight loss (diet and exercise) Secondary: oral hypoglycemic drugs Final: insulin
Prevalence	10% of all diabetics	90% of all diabetics

During the fasting state, insulin secretion drops significantly. Without insulin stimulating anabolic processes, the availability of amino acids and fatty acids increases. Glucagon promotes the conversion of available amino acids and fatty acids into glucose and ketone bodies, respectively. However, glucagon does not directly alter the synthesis or breakdown of fats or proteins.

Diabetes Mellitus Results from Insulin Insufficiency

Diabetes mellitus is the result of a deficiency of insulin *production* or *action*. As a consequence, blood glucose rises abnormally high.

The name *diabetes mellitus* derives from one of its most prominent signs: the production of large volumes of sweet urine. Diabetes derives from the Greek word *diabainein,* for "go through," and *mellitus,* the Latin word for "sweet"; thus "the sweetness that goes through"—a remarkably apt name. Diabetic urine is sweet because blood glucose rises to a very high level, higher than the kidney can manage, and some of it spills into urine.

Autoimmune destruction ◄ (Chapter 12) of pancreatic beta cells accounts for 10% of cases and is called

type 1 diabetes. It usually develops in childhood or early adulthood. The much more prevalent (90%) **type 2 diabetes** reflects a resistance of target cells to the effect of insulin. That is, target cells do not respond to insulin because of a problem with insulin receptors or signaling pathways. Obesity is a cause of type II diabetes but not type I. See Table 15.2 for a comparison of the two types.

Case Discussion

Diabetic Ketoacidosis: The Case of Santo G.

Santo has type 2 diabetes mellitus, a result of an unfortunate genetic susceptibility coupled with excess body fat. Santo's cells are not very responsive to insulin, so it takes an abnormally high insulin dose to make his cells respond. If Santo does not take supplemental insulin injections, many of his body cells (especially muscle cells) cannot take up and burn glucose (Fig. 15.16, ❶). His brain and heart can still use glucose and will not starve, but most

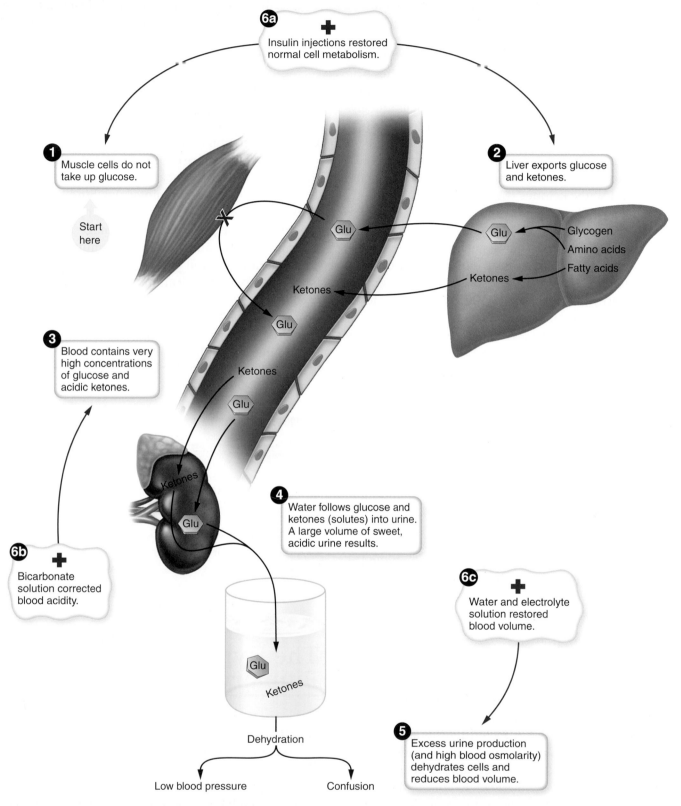

6a **+** Insulin injections restored normal cell metabolism.

1 Muscle cells do not take up glucose.

Start here

2 Liver exports glucose and ketones.

Glu — Glycogen — Amino acids — Fatty acids

Ketones

3 Blood contains very high concentrations of glucose and acidic ketones.

6b **+** Bicarbonate solution corrected blood acidity.

4 Water follows glucose and ketones (solutes) into urine. A large volume of sweet, acidic urine results.

6c **+** Water and electrolyte solution restored blood volume.

Dehydration

Low blood pressure

Confusion

5 Excess urine production (and high blood osmolarity) dehydrates cells and reduces blood volume.

Figure 15.16. The case of Santo S. Steps 1 to 5 illustrate the problems that result from inadequate insulin action. Steps 6a, 6b, and 6c illustrate how Santo's medical treatment corrected these problems. *What raw materials does the liver use to make glucose?*

of the glucose from his pizza and ice cream remains in his blood, causing high blood glucose, or *hyperglycemia*.

What is more, the body relies on insulin signaling to tell it when to switch from the fasted state into the fed state. Without insulin signals, Santo's liver switches over to fasted state metabolism (❷), which causes the production of even more glucose from amino acids and glycogen, and the synthesis of ketones from fatty acids (❷). As a result of his impaired glucose uptake and his increased hepatic production of glucose and ketones, Santo's blood contains abnormally high amounts of both glucose and acidic ketones (❸).

The high blood glucose and ketone concentrations also impact kidney function. As discussed in ➡ Chapter 16, the blood glucose concentration is normally low enough that glucose does not "spill" into urine. But as glucose rises to about 180 mg/dL and higher, glucose begins to appear in urine in ever greater amounts. Add ketone bodies to the mix, and urine contains much more solute than normal. This increased solute load "traps" water in urine—water that the kidney would otherwise reclaim and return to blood. The result is a flood of urine that causes Santo to become dehydrated (❹,❺).

Now, with these facts in mind, let us review Santo G's case on a point-by-point basis. Notice that many of the ill effects of diabetes mellitus are due not to the absence of insulin per se but to the high blood glucose concentration resulting from it.

- Santo was confused. His mental state can be explained by acidic blood and dehydration.
- His breath had an unusual apple juice odor. Some ketone bodies are volatile and can be expelled in exhaled air. They account for the peculiar odor of his breath.
- He was urinating far more often than normal. High blood glucose causes glucose to spill into the urine. Water is retained in urine by the osmotic power of glucose, which results in a large urine volume.
- Santo's body temperature was 39°C (102.2°F). In the absence of evidence of any underlying tissue damage or infection, we must assume that the thermostat in his hypothalamus was affected by the acidosis and tissue dehydration and reset to a higher level.
- His heart rate was 96 beats per minute (high) and his blood pressure was 90/60 (low). Failure to drink fluid in his stuporous condition and loss of fluid in urine dehydrated him and lowered his blood volume. The heart increased its rate to increase cardiac output and blood pressure, but it was not enough—his blood pressure remained low.

Therapy was straightforward. Injected insulin allowed his cells to metabolize glucose (❻ⓐ). This lowered his blood glucose concentration and reduced the burning of triglycerides. A bicarbonate solution corrected his blood acidity (❻ⓑ). Administration of water and electrolytes rehydrated his tissues and expanded his blood volume (❻ⓒ). Once these fundamentals were in place, everything else came back into line, including his brain function and body temperature.

If Santo is not more careful, he may begin to suffer the *chronic* diseases that afflict people with diabetes. The extra blood glucose molecules bind to proteins of every sort to create abnormal glycoproteins. Over the years, these glycoproteins damage blood vessel walls and nerves and account for a multitude of vascular, kidney, eye, and heart diseases. The good news is that lifestyle modifications reduce the severity of type 2 diabetes. If Santo manages to reduce his adipose tissue mass, the insulin responsiveness of all his cells will improve and he could, perhaps, stop taking supplemental insulin.

15.28 Which cell type secretes insulin?

15.29 Name the two major stimuli for insulin secretion.

15.30 Does insulin stimulate the breakdown of any nutrients?

15.31 List two ways by which glucagon increases blood glucose concentrations.

The Hypothalamus and Pituitary Gland

So far, we have learned much about the hypothalamus, how it controls body temperature and the autonomic nervous system. However, it also produces hormones and is thus an endocrine organ. The hypothalamus is anatomically linked with the **pituitary gland** (*hypophyis*), which hangs below it somewhat like a bean from the *pituitary stalk*. The gland itself nestles in a depression of the sphenoid bone called the *sella turcica* (Fig. 15.17). It is composed of two distinct parts—a smaller *posterior pituitary* and a larger *anterior pituitary*.

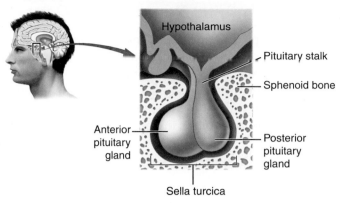

Figure 15.17. The pituitary gland. The posterior pituitary gland is neural tissue; the anterior pituitary gland is endocrine epithelial tissue. *Name the structure connecting the hypothalamus with the pituitary gland.*

The Posterior Pituitary Gland Secretes Hypothalamic Hormones

The **posterior pituitary gland** is actually an extension of the hypothalamus: it is brain tissue that consists of the axon terminals of hypothalamic neurons. These neurons span three structures—their cell bodies lie in the hypothalamus, their axons descend through the pituitary stalk, and their axon terminals form the posterior pituitary gland (Fig. 15.18). Hypothalamic hormones are synthesized and packaged into secretory vesicles within the cell body in the hypothalamus, travel down the pituitary stalk through axons, and are stored in axon terminals in the posterior pituitary gland, where they await release into blood.

Posterior pituitary hormones include *antidiuretic hormone (ADH)* and *oxytocin*, each of which is synthesized by a separate population of hypothalamic neurons under

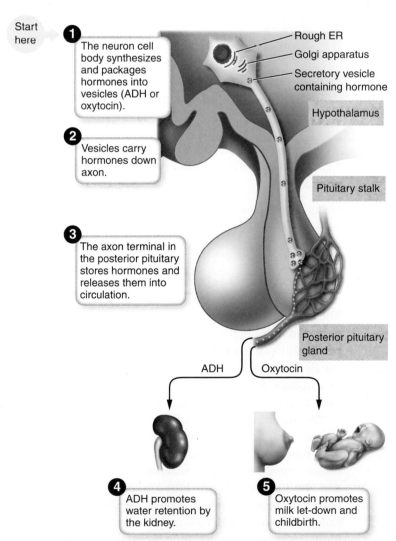

Figure 15.18. The posterior pituitary gland. This gland stores and releases hormones made in the hypothalamus. *Based on the organelles involved, are posterior pituitary hormones proteins or steroids?*

the influence of neural or blood-borne signals. ADH promotes water retention by the kidney; it is discussed in ➡ Chapter 16. Oxytocin facilitates childbirth and breast milk let-down; it is discussed in ➡ Chapter 17.

Case Note

15.15. Remember that Santo is dehydrated. Given the action of ADH on the kidney, do you think that dehydration would stimulate or inhibit ADH secretion?

The Anterior Pituitary Gland Is Regulated by Hypothalamic Hormones

In contrast to the hypothalamus and posterior pituitary, the **anterior pituitary gland** is not neural tissue—it is an epithelial gland like the pancreas. It is regulated by **releasing hormones** secreted by hypothalamic neurons. Just as the liver is fed by a *portal venous system,* so is the anterior pituitary—releasing hormones secreted into a capillary network in the pituitary stalk travel directly to a second capillary bed in the anterior pituitary. Therefore, the anterior pituitary receives releasing hormones not diluted into the general circulation. This speeds up the responsiveness of the hypothalamic–anterior pituitary system and reduces the amount of releasing hormone that must be produced.

Hypophyseal (anterior pituitary) hormones are *trophic*—they stimulate the growth as well as the activity of their target glands. In most cases, the target gland produces one or more hormones in response to pituitary stimulation. There are five major classes of anterior pituitary hormones:

1. *Gonadotrophins*, which regulate the gonads (ovaries and testes).
2. *Prolactin*, which regulates the mammary glands (breasts).
3. *Adrenocorticotrophin* (or adrenocorticotrophic hormone, ACTH), which regulates the *cortex* of the adrenal glands.
4. *Thyroid-stimulating hormone* (TSH), which regulates the thyroid gland.
5. *Growth hormone* (GH), which regulates growth and metabolism of tissues in general.

A distinct panel of hypothalamic releasing hormones regulates each hypophyseal hormone. We discuss each of these pituitary hormones and their target glands below.

Remember This! **The posterior pituitary gland does not synthesize the hormones it secretes.**

Gonadotrophins Regulate the Gonads

The anterior pituitary produces *gonadotrophins* (follicle-stimulating hormone, FSH; and luteinizing hormone, LH) that act on the gonads. Gonadotrophins regulate the production of gametes (sperm or ova) and the production of gonadal steroids (testosterone, estrogen, progesterone). We discuss gonadotropins in detail in ➡ Chapter 17.

Prolactin Regulates Milk Synthesis

Prolactin stimulates milk synthesis by the mammary glands in a lactating woman's breasts. The primary regulator, hypothalamic dopamine, inhibits prolactin secretion. Therefore, triggers for milk synthesis (such as the baby's suckling) stimulate prolactin secretion by *inhibiting* hypothalamic dopamine secretion. The roles of prolactin in men and nonlactating women are less clear but may involve fertility and sexual activity.

15.32 Which lobe of the pituitary gland synthesizes the hormone it secretes?

15.33 True or false: Blood leaving the anterior pituitary gland has passed through a single capillary bed.

15.34 True or false: The anterior pituitary responds to releasing hormones, but the hypothalamus secretes them.

15.35 Which anterior pituitary hormone regulates the gonads?

15.36 Name the target gland for ACTH.

The Adrenal Gland

The **adrenal glands** (*ad-* = "upon," *renal* = "kidney") are a pair of dome-shaped, thumb-size organs, one of which perches atop each kidney (Fig. 15.20). The adrenal is two glands in one—it has an outer cortex that secretes hormones in response to chemical signals in blood and an inner *medulla* governed by the autonomic nervous system.

The **adrenal medulla** is nervous tissue and is an extension of the autonomic nervous system. It is a nodular mass of sympathetic postganglionic neurons specialized to secrete *epinephrine* and *norepinephrine* (⬅ Chapter 8 and ⬅ Chapter 11) into blood on command of the sympathetic nervous system. These hormones, collectively called **adrenomedullary hormones,** enhance the "fight or flight" response to stress: they increase the rate and force of heart contraction; increase systemic blood pressure and

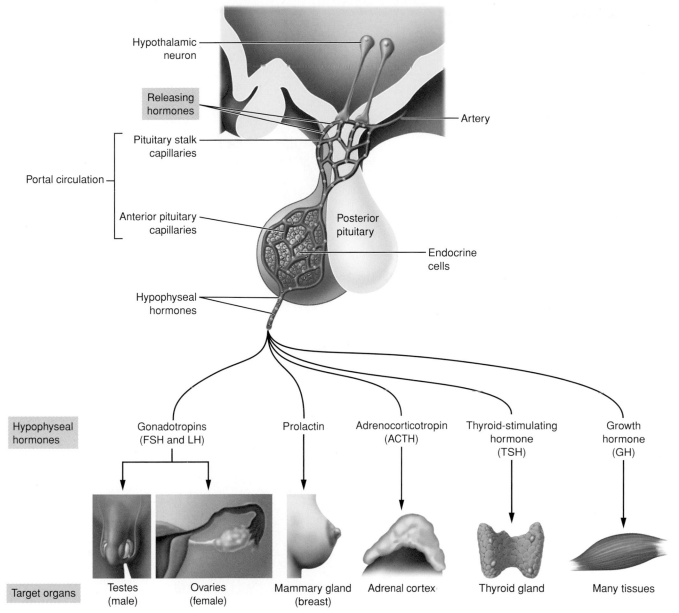

Figure 15.19. The anterior pituitary gland. This gland secretes its (hypophyseal) hormones under the influence of hypothalamic releasing hormones. The five classes of hypophyseal hormones regulate different target glands. *Which type of blood vessel supplies blood to the anterior pituitary?*

blood flow to heart, liver, and skeletal muscle; dilate airways; and increase blood levels of glucose and fatty acids.

The adrenal cortex secretes three types of steroid hormones, two of which are under ACTH control. These three hormones are collectively known as the **adrenocortical hormones:**

- **Glucocorticoids** are a family of hormones that regulate glucose and protein metabolism, the inflammatory reaction, and the immune system. ACTH from the anterior pituitary stimulates glucocorticoid secretion. The glucocorticoids are discussed in detail below.

- **Androgens** are a family of hormones that affect sexual characteristics, including sexual drive, the development of facial hair in men, and so on. They are also secreted in response to pituitary ACTH. Androgens are discussed in more detail in Chapter 17.

- **Mineralocorticoids** are a family of hormones, mainly *aldosterone*, that influence the kidney's handling of sodium and potassium. Mineralocorticoid secretion is largely independent of ACTH. Instead, plasma electrolytes and other hormones regulate their secretion, as discussed in ➡ Chapter 16.

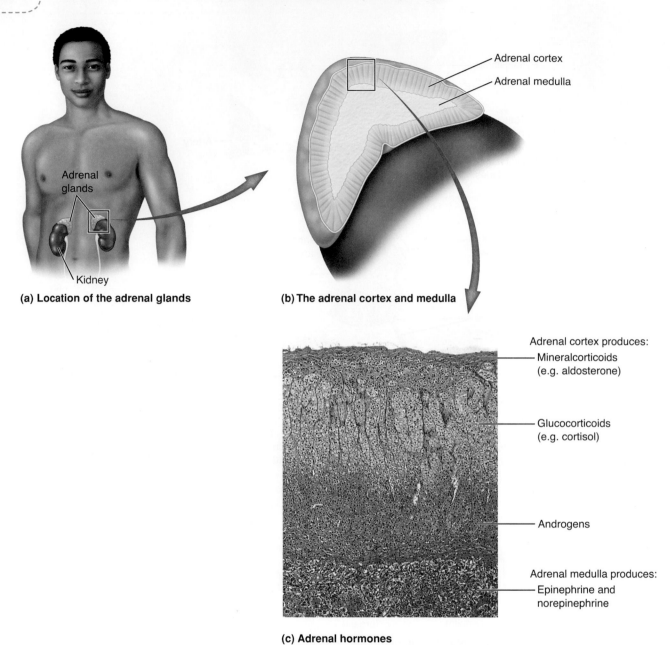

(a) Location of the adrenal glands

(b) The adrenal cortex and medulla

(c) Adrenal hormones

Figure 15.20. The adrenal glands. A. The adrenal glands sit on top of the kidneys. **B.** The adrenal cortex surrounds the inner adrenal medulla. **C.** The three regions of the adrenal cortex secrete three different types of steroid hormones. *Which hormone is secreted from the region closest to the medulla?*

CRH and ACTH Regulate Cortisol Secretion

The release of glucocorticoids, **cortisol** being chief among them, increases when the body is subjected to starvation, injury, or trauma (Fig. 15.21). Even the minor stressor of an overnight fast increases cortisol section. The hypothalamus secretes greater amounts of **corticotrophin-releasing hormone** (CRH) in response to stress (❶), which stimulates the pituitary gland to pro-

duce ACTH (❷), which in turn augments cortisol release from the adrenal gland (❸).

Cortisol Influences Glucose Metabolism and Inflammation

Cortisol cooperates with the sympathetic nervous system in mediating the stress response, but its effects are slower and longer lasting. In brief, cortisol discourages tissue building and encourages tissue breakdown,

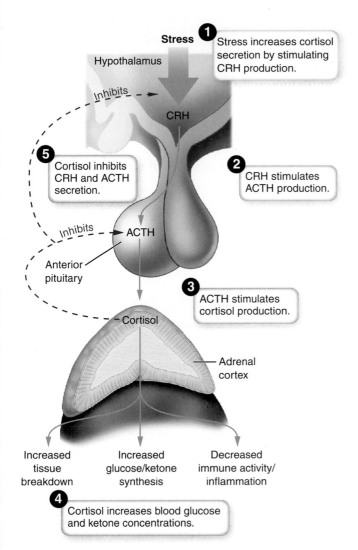

- Cortisol also inhibits immune system activity and suppresses the inflammatory reaction. While this action may appear counterproductive in coping with injury or disease, remember that excess immune activity causes harm. Cortisol thus promotes survival by preventing excess immune activity. This characteristic accounts for the widespread use of cortisol to suppress the immune system in autoimmune disease (rheumatoid arthritis, for example) or to suppress inflammation of other sorts (inflammation that narrows the bronchioles in asthma, for example).

In a classic homeostatic negative feedback loop, increased blood cortisol suppresses output of hypothalamic CRH and pituitary ACTH (**❺**). Note that neither adrenal androgens nor aldosterone inhibit CRH and ACTH production.

Case Note

15.16. Would a cortisol injection worsen or improve Santo's elevated blood glucose?

Excess Cortisol Causes Cushing's Syndrome

Excess cortisol production results in a constellation of clinical signs and symptoms known as **Cushing's syndrome** (Fig. 15.22). It can be prompted by any of three conditions:

- Infrequently, hyperplasia or a tumor of the pituitary gland secretes excess ACTH, which stimulates adrenocortical secretion of cortisol.
- More commonly, hyperplasia or tumor of the adrenal cortex itself secretes the extra cortisol.
- Most commonly, the cause is medical treatment with cortisol and related drugs, usually in patients with chronic inflammatory diseases such as rheumatoid arthritis. Steroid treatment must be carefully monitored and ideally only used for short periods because of the danger of developing Cushing's syndrome.

The effects of excess cortisol include:

- Obesity and abnormal fat deposition in face, shoulders and back—cortisol increases appetite and alters the production of other metabolic hormones.
- Easy bruising and skin striations (stripes)—cortisol stimulates connective tissue breakdown to generate amino acids.

Figure 15.21. Cortisol. Negative feedback usually prevents large variations in blood cortisol concentrations. Stress stimulates cortisol secretion by increasing CRH and ACTH production. Cortisol increases the availability of glucose and ketones in times of stress. *Does cortisol stimulate or inhibit ACTH secretion?*

ensuring that the body has enough available nutrients to fuel basic processes in times of fasting or illness. In brief (**❹**):

- Cortisol stimulates hepatic glucose synthesis (both from glycogen breakdown and from gluconeogenesis) and ketone synthesis.
- Cortisol increases the availability of raw materials for glucose and ketone synthesis by stimulating tissue breakdown. Proteins in muscle and connective tissue are catabolized into amino acids, and fats stored in adipose tissue are catabolized into glycerol and fatty acids.

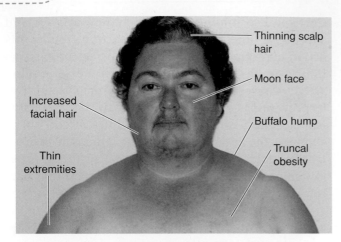

Figure 15.22. Hypercortisolism (Cushing's syndrome). Excess cortisol production causes many visible clinical signs, as shown by this woman with hypercortisolism. *What is the large fat pad on the upper back called?*

Labels in figure:
- Thinning scalp hair
- Moon face
- Buffalo hump
- Truncal obesity
- Increased facial hair
- Thin extremities

- Diabetes mellitus—cortisol decreases the ability of tissues to respond to insulin.
- Fluid retention (edema) and high blood pressure—very high levels of cortisol can activate mineralocorticoid receptors, which stimulates sodium (and thus water) retention.
- Frequent infections—cortisol suppresses the immune system.
- Excess body hair (hirsutism)—the adrenal cortex is usually also overproducing androgens.

Adrenocortical Insufficiency Is Called Addison's Disease

Adrenocortical hormone insufficiency is called **Addison's disease.** The most common cause is autoimmune destruction of the adrenal gland. Destruction of the medulla is inconsequential because the sympathetic nervous system compensates for its loss, but life cannot continue long without adrenocortical hormones. The loss of mineralocorticoids is most dangerous—failure to retain sodium due to loss of sodium in the urine lowers blood volume and blood pressure ← (Chapter 11) until vascular collapse and death occur. The loss of glucocorticoids adds to the insult by lowering blood glucose, which causes skeletal and cardiac muscle weakness and low cardiac output. Loss of androgens matters little in these circumstances. It is now settled history that President John F. Kennedy suffered from Addison's disease from the time he was a young man and required daily doses of adrenocortical hormones to remain alive.

15.37 Which adrenal hormone significantly inhibits ACTH secretion?

15.38 Which hormone stimulates protein breakdown—cortisol or glucagon?

15.39 Which hormone stimulates ketone synthesis—cortisol or glucagon?

15.40 Truncal obesity is a sign of excess cortisol secretion—true or false?

15.41 Adrenal androgen deficiency is the most serious problem with Addison's disease—true or false?

The Thyroid Gland

The **thyroid gland** is a butterfly-shaped, bilobed gland that secretes *thyroid hormones.* It lies immediately beneath the skin of the low anterior neck inferior to the larynx (Fig. 15.23A). A small bridge of thyroid tissue, the *isthmus,* joins the lower ends of the lobes. Several minute *parathyroid glands* are positioned on the posterior surface of thyroid (Fig. 15.23B). Thyroid tissue is fleshy, brownish-red, and composed of microscopic *follicles* (Fig. 15.23C), each of which consists of a single layer of follicular cells that enclose a viscous fluid called the *colloid.* As discussed further on, thyroid hormones are synthesized in the colloid, not within the follicular cells.

TRH and TSH Regulate the Production of Thyroid Hormone

The hypothalamus and anterior pituitary gland regulate the synthesis of thyroid hormones (Fig. 15.24). Thyroid gland activity increases in response to cold (especially in newborns) and decreases in response to stress, because cold stimulates the secretion of thyrotrophin-releasing hormone (TRH) from the hypothalamus, but stress inhibits it (**❶**). TRH then stimulates the release of thyroid-stimulating hormone (TSH) from the anterior pituitary (**❷**), and TSH stimulates thyroid growth and thyroid hormone synthesis (**❸**).

Thyroid Hormones Are Synthesized from Amino Acids

Thyroid hormones are not proteins or steroids—each thyroid hormone consists of two molecules of the amino acid *tyrosine,* to which three or four iodine atoms are

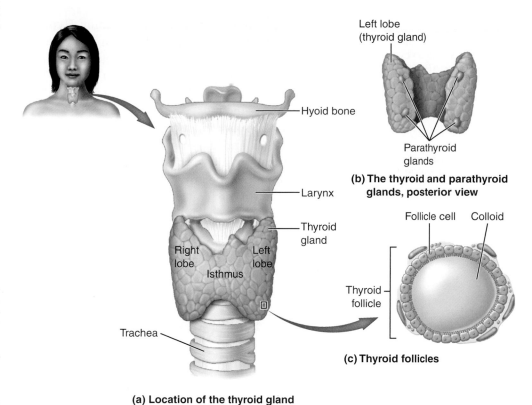

(b) **The thyroid and parathyroid glands, posterior view**

(c) **Thyroid follicles**

(a) **Location of the thyroid gland**

Figure 15.23. The thyroid gland. A. Anterior view. **B.** Posterior view and parathyroid glands. **C.** Each thyroid follicle consists of follicular cells surrounding a colloid-filled cavity. *Is the thyroid gland inferior or superior to the larynx?*

added. There are two major thyroid hormones, identical in every way but one: **thyroxine (T4)** has four iodine atoms; **triiodothyronine (T3)** has three.

Thyroid hormone synthesis is unique in that it occurs *extracellularly* in the colloid. The surrounding follicular cells provide the raw materials; they actively import iodine, and they assemble a large protein called *thyroglobulin,* which is the source of the tyrosines. Both the iodine and the thyroglobulin are secreted into the colloid, where enzymes link together nearby tyrosines within the thyroglobulin molecule and add iodines to create T3 and T4 (**❹**). Then, when thyroid hormone secretion is required, the newly synthesized thyroid hormones are freed from the thyroglobulin protein and secreted into the blood.

The thyroid gland produces much more T4 (90%) than T3 (10%), but T3 is the more potent form. Thyroid hormones circulate in blood bound to *thyroxine-binding globulin*, which transports them to tissues, where cells convert some T4 to more powerful T3 by removing an iodine (**❺**).

> ***Remember This!*** **Thyroid hormones are made in the colloid. Follicular cells provide the raw materials; iodine and tyrosines.**

Thyroid Hormones Increase Basal Metabolic Rate

Thyroid hormones increase the basal metabolic rate, sending cells into "overdrive" (Fig. 15.24, **❻**). Cells burn more nutrients and produce more ATP. As mentioned earlier, thyroid hormones actually *reduce* the efficiency of ATP generation so that a greater proportion of nutrient energy is converted into heat. Also, tissues become more sensitive to the effects of the sympathetic nervous system: the heart rate accelerates, cardiac contractions strengthen, and the respiratory rate increases.

Thyroid hormones are critical for normal brain development both prenatally and postnatally, and they support brain function even in adults. Moreover, normal fetal and postnatal growth and tissue development also require adequate amounts of thyroid hormones.

As you would expect, thyroid hormone levels are homeostatically stabilized by negative feedback (**❼**). Thyroid hormones influence both the hypothalamus and pituitary gland to reduce TRH and TSH secretion, thereby maintaining relatively constant circulating concentrations of thyroid hormones.

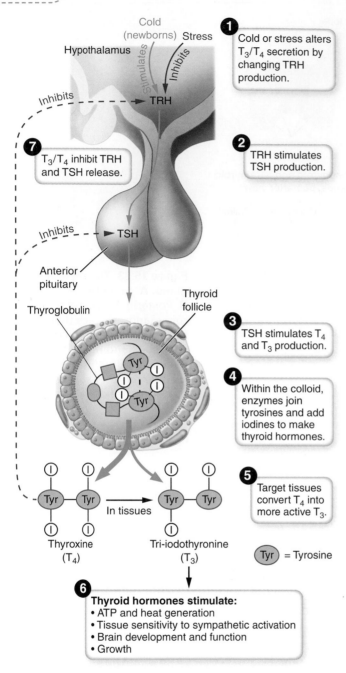

1 Cold or stress alters T_3/T_4 secretion by changing TRH production.

2 TRH stimulates TSH production.

7 T_3/T_4 inhibit TRH and TSH release.

3 TSH stimulates T_4 and T_3 production.

4 Within the colloid, enzymes join tyrosines and add iodines to make thyroid hormones.

5 Target tissues convert T_4 into more active T_3.

Thyroxine (T_4)

Tri-iodothyronine (T_3)

Tyr = Tyrosine

6 Thyroid hormones stimulate:
• ATP and heat generation
• Tissue sensitivity to sympathetic activation
• Brain development and function
• Growth

Figure 15.24. Thyroid hormones. Negative feedback usually prevents large variations in the blood's thyroid hormone concentrations. Cold and stress modify production by altering the secretion of TRH and TSH. Thyroid hormones increase the production of ATP and heat by body cells. *Name the amino acid and the inorganic element required for thyroid hormone synthesis.*

Excessive Thyroid Activity Causes Hyperthyroidism

The effect of thyroid hormones is particularly evident in cases of hormone excess (Fig. 15.25A, right side). Thyroid hormone excess, or **hyperthyroidism,** can reflect

TSH oversecretion from a pituitary tumor. Hyperthyroidism can also result from an autoimmune condition, in which the body synthesizes antibodies that massively stimulate thyroid gland activity, or from a tumor of the thyroid gland itself. However, overmedication with thyroid hormone is the most common cause.

In most patients with hyperthyroidism, the thyroid gland enlarges and may be visible as a bulge in the lower neck. An enlarged thyroid, no matter the cause, is called a *goiter.* Hyperthyroidism produces an excited neurologic state: nervousness, tremors of the hand, and "jumpy" or exaggerated reflexes. Patients are heat-intolerant, flinging off bedcovers when a bed partner is cold. Weight loss and sweating can be severe. Heart rate is rapid and muscle weakness and fatigue are common.

Insufficient Thyroid Activity Causes Hypothyroidism

Hypothyroidism, or thyroid insufficiency, can be due to low pituitary production of TSH, but most often there is some abnormality of the gland itself, usually inflammation (*thyroiditis*).

In hypothyroidism, the size of the gland varies according to the cause. The largest glands are those seen in patients who do not get sufficient dietary iodine and therefore cannot make enough thyroid hormone. In such cases the gland enlarges to compensate and the goiter may become very large (Fig. 15.25B). The symptoms are generally the opposite of hyperthyroidism—weight gain, cold intolerance, and impairment of the general activity of the nervous system: underactive reflexes, lethargy, depression, sleepiness, and fatigue (Fig. 15.25A, left side). Heart rate and blood pressure fall.

Hypothyroidism in fetuses and young children (*cretinism*) impairs growth and mental development. Hypothyroidism used to be quite common in inland populations that did not have access to the best nutritional iodine sources—seafood and vegetables grown in coastal environments. Lack of dietary iodine impairs the ability of the thyroid gland to produce thyroid hormones. The hypothalamus and pituitary try to correct the situation by producing a great amount of TSH, which stimulates growth of the gland. But alas, no iodine is available, so the gland grows ever larger and becomes a *goiter* (an enlarged thyroid). The result is *endemic goiter* (endemic = "occurring in a particular area"). Endemic goiter was so prevalent in early-twentieth-century inland North America that women wore a small fabric bands around their throats to hide the unsightly bulge. Goiter is no longer common in inland North America for two reasons:

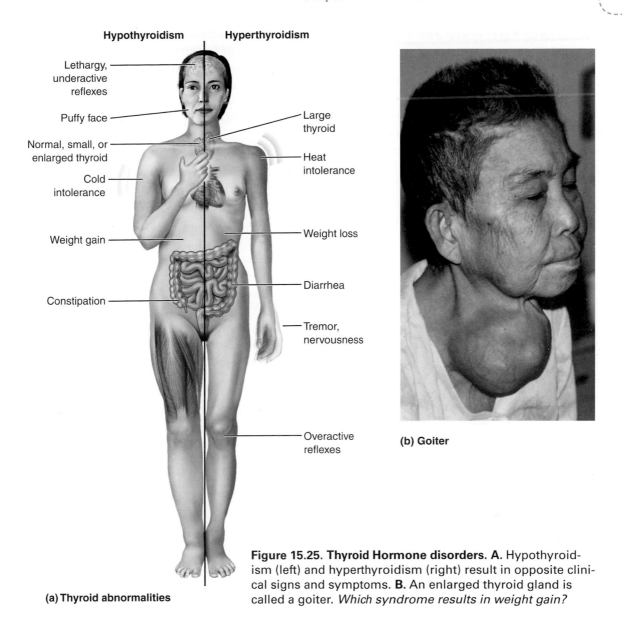

Hypothyroidism   **Hyperthyroidism**

Lethargy, underactive reflexes

Puffy face

Normal, small, or enlarged thyroid

Cold intolerance

Weight gain

Constipation

Large thyroid

Heat intolerance

Weight loss

Diarrhea

Tremor, nervousness

Overactive reflexes

(a) Thyroid abnormalities

(b) Goiter

Figure 15.25. Thyroid Hormone disorders. A. Hypothyroidism (left) and hyperthyroidism (right) result in opposite clinical signs and symptoms. **B.** An enlarged thyroid gland is called a goiter. *Which syndrome results in weight gain?*

table salt is now supplemented with iodine, and coastal vegetables and seafood are shipped everywhere.

15.42 Name the small tissue bridge that joins the two thyroid lobes.

15.43 What is the difference between thyroglobulin and thyroid binding globulin?

15.44 Which thyroid hormone variety is more potent—T3 or T4?

15.45 Would a hyperthyroid person be more comfortable in the Canadian winter or the Texas summer?

Case Note

15.17. A medical student examined Santos while his wife was out of the room. Would the student be more likely to suspect hypothyroidism or hyperthyroidism? Explain.

Growth Hormone

The pituitary hormones discussed so far—ACTH and TSH—stimulate the activity of other endocrine glands. **Growth hormone (GH),** in contrast, exerts its effects directly on target tissues, stimulating growth and tissue repair.

Hypothalamic Hormones Regulate GH Secretion

GH secretion varies during the day, over the course of a lifetime, and in response to stress (Fig. 15.26, ❶). It is greatest during deep sleep. When you were a teenager and your mother said you grew overnight, she was right. GH secretion peaks at puberty, inducing the pubertal growth spurt, and declines significantly in the elderly. Some of the symptoms of aging, including decreased muscle growth and skin thinning, can be attributed to this age-related reduction. Finally, stress (exercise, for example) stimulates GH secretion, presumably to facilitate tissue repair.

These alterations in GH secretion reflect the production of two hypothalamic hormones (❷):

- GH-releasing hormone (GHRH) stimulates GH secretion.
- **Somatostatin** is a hormone that inhibits GH secretion. Somatostatin was originally named and is often still referred to as *somatotropin-release inhibitory factor* Samatostatin (SRIF) because growth hormone used to be called *somatotropin*; however, we now know that somatostatin (SRIF) plays many roles unrelated to GH secretion.

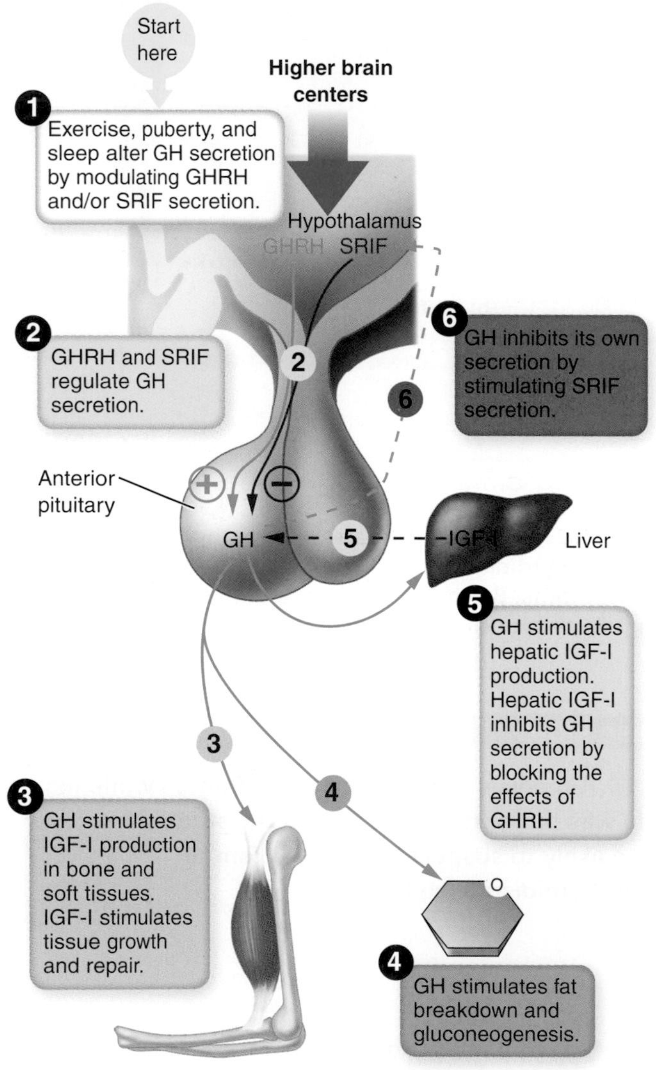

Case Note

15.18. Hypoglycemia stimulates GH secretion. If Santos does not take his insulin, what will happen to his GH secretion?

Growth Hormone Is Anabolic

GH stimulates tissue growth (anabolism) and modulates nutrient metabolism to do so. The *somatic* (growth-promoting) actions of GH are partially mediated by a second hormone, *insulin-like growth hormone-I (IGF-I)*. GH stimulates IGF-I production in bone, muscle, and other soft tissues, and IGF-I stimulates cell growth and cell division in these same tissues (Fig. 15.26, ❸). IGF-I is thus acting as a paracrine. Specifically:

- Before bone epiphyses close near age 20, GH and IGF-I stimulate linear bone growth. Increased height is the result.
- In adults and children, GH and IGF-I stimulate the growth of soft tissues such as muscle, cartilage, and skin. Added bulk is the result.

GH also stimulates fat breakdown and gluconeogenesis (❹), which raises blood concentrations of fatty acids and glucose. This ensures that cells have ample nutrients available to fuel growth. The metabolic effects of GH are completely independent of IGF-I, which tends to sensitize cells to the effects of insulin, resulting in *lower* concentrations of fatty acids and glucose.

GH secretion is modulated by two negative feedback loops. First, GH stimulates IGF-I production by the liver. Hepatic IGF-I reduces GH secretion by blocking the effects of the stimulatory hormone GHRH (❺). GH itself inhibits further GH secretion in a different manner, by enhancing the release of the inhibitory hormone SRIF (❻).

Figure 15.26. Growth hormone (GH). The secretion of GH is controlled by hypothalamic factors and negative feedback. GH stimulates growth in cooperation with IGF-I and independently increases the availability of glucose and fatty acids to fuel tissue growth and repair. *Which hormone stimulates GH secretion—IGF-I, SRIF, or GHRH?*

Abnormal GH Activity Affects Growth and Metabolism

Excess or deficient GH in children alters growth, but abnormal GH activity at any age disrupts metabolism (Fig. 15.27). In almost every instance of abnormal GH secretion, high or low, the problem lies in the pituitary, not the hypothalamus.

In children and young adults with open epiphyses, GH stimulates long bone growth and the growth of soft tissues. Thus, excessive GH secretion results in an unusually tall, massive person—hence the name *gigantism*. The metabolic effect is straightforward: there is excess gluconeogenesis and mobilization of fatty acids from adipose tissue, both of which are turned into glucose. High blood glucose and fatty acid concentration is the result.

In adults with closed epiphyses, long bones cannot lengthen, but excessive GH secretion can make them thicken. This effect is most notable in the bones of the hands and feet and in the membranous bones of the face and skull; hence the name of this disorder: *acromegaly* (Greek *akron* = "tip," in this case hand or foot; *mega* = "large"). The metabolic effect is the same as above.

In children, a deficiency of GH secretion reduces the linear growth of long bones. Small stature is the result. GH deficiency is one of the varied causes of *dwarfism*; there are many others. In adults, GH deficiency has no effect on bones. However, at any age the metabolic effect is decreased gluconeogenesis, resulting in hypoglycemia

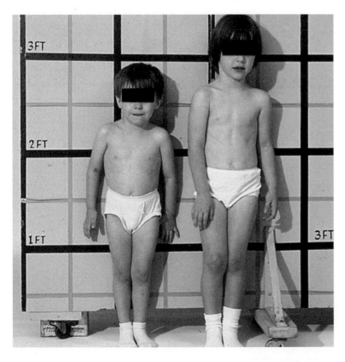

(a)

(b)

Figure 15.27. Growth hormone disorders. A. A 5.5-year-old boy with GH deficiency (left) is considerably shorter and chubbier than his fraternal twin sister. **B.** A man with gigantism alongside his identical twin. *Why does GH deficiency cause chubbiness?*

and reduced protein synthesis, which can lead to muscle weakness. The result is low blood glucose and weakness. GH is a legitimate treatment for GH-deficient children and adults, but it is also subject to abuse by athletes and other healthy individuals. See the nearby clinical snapshot titled *Growth Hormone: Is More Better?* for more information.

15.46   Growth hormone and insulin-like growth factor I exert opposite effects on growth: true or false?

15.47   Growth hormone and insulin-like growth factor I exert opposite effects on fat and carbohydrate metabolism: true or false?

15.48   Name a hypothalamic hormone that inhibits GH secretion.

15.49   If someone develops a GH-secreting tumor in adulthood, will they have gigantism or acromegaly?

Calcium Metabolism

Multiple hormones regulate calcium metabolism. Calcium metabolism, unlike nutrient metabolism, centers on one simple reaction:

$$Ca^{2+} + PO_4^{2-} \Leftrightarrow CaPO_4$$

$CaPO_4$ makes up the mineral portion of bones, whereas Ca^{2+} dissolves in fluids, such as blood, cytosol, and interstitial fluid. Recall from ← Chapter 6 that *osteoblasts* synthesize $CaPO_4$ from Ca^{2+} and PO_4^{2-} as they make new bone tissue. *Osteoclasts* separate $CaPO_4$ into Ca^{2+} plus PO_4^{2-} as they break down bone tissue (Fig. 6.10).

Ca^{2+} ions play many roles in the endocrine and nervous systems, controlling neurotransmitter release, muscle contraction, and some second messenger systems. All of these functions rely upon an extremely precise calcium concentration gradient between intracellular and extracellular fluids, so the blood calcium concentration must be closely regulated. The two most important calcium-regulating hormones are *parathyroid hormone (PTH)* and *calcitriol* (the active form of vitamin

D). Both hormones increase blood calcium concentration. A third hormone, *calcitonin,* is produced by the thyroid gland. It is a minor player in calcium homeostasis and is not mentioned further.

Vitamin D Increases Intestinal Calcium Absorption

Vitamin D is a family of cholesterol-like molecules that play a critical role in calcium metabolism. Many foods are fortified with vitamin D because, other than fatty fish, most foods contain little. What is more, upon exposure to sunlight, skin synthesizes vitamin D from a type of cholesterol. However, the liver and kidney must convert the ingested or synthesized vitamin D into its most active form, called 1,25-dihydroxyvitamin D3. This active form of vitamin D stimulates calcium absorption in the small intestine and thus increases blood calcium concentrations. As you would expect, low blood calcium levels stimulate vitamin D activation.

Vitamin D deficiency can cause disease. In growing children it is called *rickets;* in adults it is called *osteomalacia.* In each instance, not enough calcium can be absorbed from the diet and bone formation is compromised. Weak bones characterize both disorders. In adults the result is increased risk of fracture; in children, because bones are still growing, skeletal malformation is the main problem.

Parathyroid Hormone Increases Bone Breakdown and Calcium Retention

Like the pancreatic islets of Langerhans, the parathyroid glands are not modulated by the anterior pituitary. Normally there are four **parathyroid glands,** one each behind the upper and lower poles of the right and left lobes of the thyroid gland (see Fig. 15.23B). Occasionally there may be several extra ones in the nearby neck or upper chest. The parathyroid glands regulate calcium metabolism by means of their hormone, **parathyroid hormone (PTH).**

PTH release increases when the blood calcium concentration falls (Fig. 15.28). As you would expect, PTH increases the blood calcium concentration, using three different strategies:

1. PTH stimulates bone breakdown (by stimulating osteoclast activity) and inhibits bone deposition (by inhibiting osteoblast activity).
2. PTH increases calcium absorption from the intestines *indirectly,* by stimulating vitamin D activation. Remember that vitamin D stimulates calcium uptake from the intestine.

CLINICAL SNAPSHOT

Growth Hormone: Is More Better?

The effects of growth hormone read like a wish list for most of us—increased height, less fat, stronger muscles and better skin, and improved immune function; in short, overall enhanced physical attributes.

For many years we have known that GH-deficient children—say, those with a pituitary gland problem—benefit from GH injections. Numerous studies also suggest that lifelong GH therapy dramatically improves the quality of life of GH-deficient adults, whether or not they were treated with GH as children. But what about people with normal pituitary function and normal blood levels of GH? Can GH improve their quality of life? Let's look at three populations in which the use of GH is fairly common.

Older adults. In comparison with young adults, older adults are relatively GH-deficient. This normal age-related decline in pituitary GH production has been implicated in age-related changes such as thin, fragile skin, reduced muscle mass and strength, and reduced quality of life. Ever since some studies suggested that GH administration could remedy some of these problems, dietary supplements manufacturers have been promoting GH to older adults—and some seniors have been buying them. What these companies have not publicized is the fact that studies have also shown an increased risk of cancer and a higher overall death rate associated with GH treatment. Therefore the consensus of the scientific community is that the elderly should *not* invest in GH therapy.

Elite athletes. GH use is also common among elite athletes, who take GH injections hoping to build better and stronger muscles and gain a performance edge. Athletes participating in international competitions (such as the Olympics) cannot use GH because of potential strength enhancements resulting from GH supplementation. But even aside from the legal and ethical issues of using performance-enhancing drugs, GH supplementation is a poor strategy for athletes: scientific evidence suggests that it promotes muscle strength *only* in GH-deficient people, not in healthy athletes with normal pituitary function. Of greater concern are the adverse effects of GH when used by adults with normal pituitary function—they have more cancers and heart attacks than those who do not take GH.

Children with short stature. *Idiopathic short stature* is a condition in children who have normal blood levels of GH but who are unusually short compared with their peers. Some but not all studies show that GH therapy

GH use and abuse. GH therapy is a questionable choice for athletes and the elderly.

induces modest improvements in final adult height in these cases, especially if their parents are not short. Physicians postulate that extreme short stature impairs both the physical and psychological quality of life, and GH therapy should be provided to all children who are more than two2 standard deviations below the mean for height. Others are strongly opposed, asserting that short stature occurs by definition within any population, that it is not a disease, and that it should not be treated as such. They argue that children should not be subjected to drugs unnecessarily, and that the health care system should not have to be burdened with significant costs only to move a child closer to average on some purely physical parameter. Thus, in contrast to the use of GH in the first two populations, the jury is still out on this one.

3. PTH increases calcium retention by the kidney, reducing the amount of calcium lost in urine.

Since the actions of both vitamin D and PTH increase blood calcium concentrations, one would expect that low blood calcium levels would stimulate the production of both. This is indeed the case.

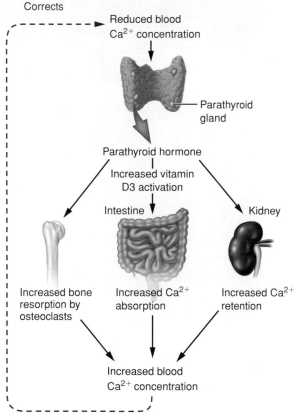

Figure 15.28. Calcium metabolism. PTH and vitamin D3 increase the calcium concentration in blood. *True or false: PTH increases the concentration of calcium in urine.*

Excess parathyroid activity, or **hyperparathyroidism,** pathologically elevates blood calcium levels at the expense of bone tissue. The excessive bone breakdown increases the risk of fractures, and high blood calcium spills into urine and precipitates the formation of kidney stones. Other effects include gastrointestinal upset as well as depression and malaise. PTH-secreting parathyroid tumors or overgrowth of the parathyroid glands are the most common causes. You can learn more about hyperparathyroidism by revisiting the case of Maggie H. in ◀ Chapter 6.

Hypoparathyroidism, or PTH deficiency, can be caused by autoimmune destruction of the parathyroid glands. It can also occur secondary to inadvertent surgical removal of the glands during surgical removal of the thyroid. The resulting hypocalcemia (low blood calcium concentration) can be fatal, because it disrupts the calcium concentration gradient and interferes with nerve signal transmission. Hypoparathyroidism is much rarer than hyperparathyroidism.

Pop Quiz

15.50 Would a drug that stimulates osteoclast activity promote bone formation or bone breakdown?

15.51 Which hormone(s) increase blood calcium concentrations: vitamin D3, parathyroid hormone, or both?

15.52 Describe the action of PTH at the kidney.

15.53 Does excessive production of PTH strengthen or weaken bones?

15. 54 Who is more likely to need vitamin D supplements, a darker-skinned Mexican immigrant living in Canada or a Canadian immigrant living in Mexico?

Word Parts		
Latin/Greek Word Parts	**English Equivalents**	**Examples**
cortico-	Cortex	Adrenocorticotropin: a substance stimulating the growth of the adrenal cortex
-crine	Secretion	Endocrine: secretion released within the body
endo-	Within	Endocrine: secretion released within the body
-gen, -genesis	Generates	Glycogen: a substance that generates glucose

Word Parts (continued)		
Latin/Greek Word Parts	**English Equivalents**	**Examples**
glyco-	Glucose	Glycolysis: glucose breakdown
hyper-	Excessive, above normal	Hyperthyroid: excessive thyroid function
hypo-	Deficient, below normal	Hypothyroid: deficient thyroid function
-lysis	Breakdown	Lipolysis: fat breakdown
neo-	New	Gluconeogenesis: generation of a new glucose molecule
para-	Alongside	Parathyroid glands: glands alongside the thyroid gland
troph/o	Nourishment, development, growth	Adrenocorticotropin: a substance stimulating the growth of the adrenal cortex

Chapter Challenge

CHAPTER RECALL

1. **Lactic acid generation permits cells to**
   a. regenerate NAD^+ supplies.
   b. generate ATP by the conversion of pyruvate into lactic acid.
   c. provide an alternate energy source for the brain.
   d. synthesize fatty acids.

2. **NADH**
   a. transfers a phosphate group to ADP to generate ATP.
   b. carries its hydrogen atom to the mitochondria, where it is used to generate ATP.
   c. is converted into oxoacetate by the citric acid cycle.
   d. Is used to generate acetyl-CoA.

3. **During the process of oxidative phosphorylation, high-energy electrons**
   a. combine with water.
   b. provide the energy for the active transport of hydrogen ions.
   c. interact with ATP synthase to provide the energy for ATP synthesis.
   d. combine with NAD+ and FAD+ to form NADH and FADH2, respectively.

4. **Put the stages of energy generation from a glucose molecule in the correct order.**
   a. Glycolysis, citric acid cycle, acetyl-CoA generation, oxidative phosphorylation
   b. Acetyl-CoA generation, glycolysis, citric acid cycle, oxidative phosphorylation
   c. Oxidative phosphorylation, glycolysis, citric acid cycle, acetyl-CoA generation,
   d. Glycolysis, acetyl-CoA generation, citric acid cycle, oxidative phosphorylation

5. **Which of the following is commonly synthesized from glucose?**
   a. Triglyceride
   b. Ketone bodies
   c. Amino acids
   d. All of the above

6. **The glycogen used to increase blood glucose levels is stored in**
   a. adipose tissue.
   b. muscle.
   c. liver.
   d. kidneys.

7. **The fasting state is characterized by**
   a. gluconeogenesis.
   b. glycogenesis.
   c. conversion of glucose into amino acids.
   d. conversion of glucose into fatty acids

8. **The basal metabolic rate can be measured when a person is**
   a. resting.
   b. eating.
   c. running on a treadmill.
   d. all of the above.

9. **Which of the following hormones stimulates appetite?**
   a. Leptin
   b. Ghrelin
   c. PYY
   d. CCK

10. **Ultraviolet or visible light waves transfer heat by**
    a. radiation.
    b. convection.
    c. conduction.
    d. evaporation.

11. **Insulin secretion is stimulated by**
    a. the presence of food in the intestine.
    b. increased fat stores.
    c. hypoglycemia.
    d. ketone bodies.

12. **An insulin-secreting tumor would result in**
    a. protein breakdown.
    b. hypoglycemia.
    c. fat breakdown.
    d. ketone synthesis.

13. **Glucagon directly stimulates**
    a. protein breakdown.
    b. ketone synthesis.
    c. hypoglycemia.
    d. protein synthesis.

14. **Untreated diabetes mellitus frequently causes**
    a. hypoglycemia.
    b. excess glycogenesis.
    c. excess gluconeogenesis.
    d. abnormal fat deposition in the trunk.

15. **The anterior pituitary gland**
    a. is nervous tissue.
    b. receives blood that has already passed through a capillary bed .
    c. secretes ADH.
    d. secretes releasing hormones.

Match each of the following hormones with its actions. Write the best answer in each blank. Each hormone can be used only once.

16. **Stimulates bone resorption**

17. **Directly stimulates gastrointestinal calcium uptake**

18. **Increases the basal metabolic rate**

19. **Inhibits inflammation**

20. **Stimulates bone and muscle growth and increases blood glucose concentrations**

21. **Stimulates sodium uptake in the kidney**

   a. Growth hormone
   b. Cortisol
   c. aldosterone
   d. T4 and T3
   e. PTH
   f. Vitamin D3

22. **Which hormone consists of two tyrosine molecules with two iodine atoms on each?**
    a. thyroxine (T4)
    b. parathyroid hormone (PTH)
    c. thyroid-stimulating hormone (TSH)
    d. triiodothyronine (T3)

23. **IGF-I production is stimulated by**
    a. parathyroid hormone (PTH).
    b. growth hormone (GH).
    c. insulin.
    d. glucagon.

CONCEPTUAL UNDERSTANDING

24. True or false: The amount of ATP a person produces in a given day is determined by the amount of food the person consumes. Explain your answer.

25. Vitamin D3 deficiency and PTH excess both result in weak bones, despite the fact that they have the same effect on blood calcium concentrations. Explain.

26. Which heat transfer method can cool the body but never heat it up?

APPLICATION

27. Bubba thinks that eating lots of protein will give him bigger muscles, because the extra amino acids will be "stored" as protein. Is he right?

28. Studies show that intravenous glucose does not induce the same insulin response as ingested glucose. Why?

29. Some professional athletes abuse glucocorticoids: when they are injured, they inject high doses of corticosteroids in order to continue to play in important games.
    a. How would the production of CRH and ACTH be affected?
    b. Predict some adverse effects that could result if the athlete continued this practice over a long time period and elevated circulating glucocorticoid concentrations.

You can find the answers to these questions on the student Web site at
http://thepoint.lww.com/McConnellandHull

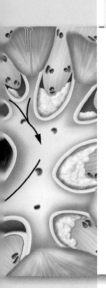

16

The Urinary System and Body Fluids

Major Themes

- Body sodium content determines body water content, and body water content determines blood pressure.

- The nephron, a vascular and epithelial structure, is the key functional unit of the kidney.

- The kidney filters a very large volume of fluid from blood and reabsorbs virtually all of it; the remainder is urine.

- The kidneys respond to changes in body fluid content and acidity by retaining or excreting more or less fluid and acid.

- The kidneys secrete hormones.

Chapter Objectives

8. Discuss the importance of (a) juxtamedullary nephrons and (b) ADH in the generation of concentrated urine.

Electrolyte and Water Balance 651

9. Describe the intervening steps between a drop in blood pressure and angiotensin II production.

10. Predict the changes in handling of kidney function and blood pressure and resulting from aldosterone deficiency, ADH deficiency, an angiotensin converting enzyme (ACE) inhibitor, or atrial natriuretic peptide excess. (Assume that no compensations occur.)

Fluid Imbalance: The Case of Santo G., Part 1 654

11. Use the case study to explain why diabetes mellitus disrupts fluid balance.

Acid–Base Balance 654

12. Use a chemical equation to show how buffers convert strong acids into weak acids.

13. Explain why, when we exercise, increased CO_2 production does not alter arterial pH.

14. Describe the body's response to fixed acid, such as the metabolic production of ketones.

Acid–Base Imbalance: The Case of Santo G., Part 2 659

15. Use the case study to explain why diabetes mellitus disrupts acid–base balance.

"He's having one of those acid attacks."

Note: Here we revisit the case of Santo G., whom we first met in Chapter 15. In this chapter we reveal previously undisclosed information about his case and discuss issues of electrolyte and fluid balance.

As you read through the following case study, assemble a list of the terms and concepts you must learn in order to understand it.

Clinical History: Santo G., an obese 47-year-old Hispanic man, required daily insulin injections for recently developed type 2 diabetes. His wife brought him to the emergency room early in the morning because he was confused, breathing rapidly, and having to urinate far more often than normal. "He's having one of those acid attacks," she said, "when everything gets out of balance." She further revealed that he had not been careful about his diet—the night before he had gorged on pizza and ice cream and had consumed 8 to 10 bottles of beer, becoming so inebriated that he went to sleep without taking his insulin.

Physical Examination and Other Data: Vital signs included temperature 39°C (102.2°F), heart rate 96 beats per minute (normal 72), and blood pressure 90/60 (normal 120/80). Respirations were noted to be "fast and hard," and his breath had an unusual "apple juice" odor. Santo was 5 ft 9 in. tall and weighed 285 lb. He was drowsy and confused about time and place.

Lab tests revealed abnormally high blood glucose and low (acidic) blood pH. His urine was also unusually acidic and contained large amounts of glucose and ketone bodies, neither of which should normally be present.

In addition to the above abnormalities, which we introduced in the last chapter, Santo had high blood osmolarity (indicating increased blood solutes), low blood bicarbonate, and low blood partial pressure of CO_2. His urine also had a very high specific gravity.

633

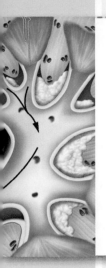

Clinical Course: The emergency room physician made a diagnosis of diabetic ketoacidosis and admitted Santo to the hospital. He was given insulin injections and large volumes of intravenous electrolyte solution containing bicarbonate. Further studies failed to find infection or another underlying cause for his fever.

Within 24 hours, Santo's blood pressure had returned to near normal. His breathing slowed to normal and he became fully awake and coherent. Blood glucose had decreased but was not yet in the normal range. Blood pH, bicarbonate and P_{CO_2} also had returned to near normal. Urinalysis revealed a stronger yellow color, less acidic pH, slightly positive glucose, and negative ketone bodies.

He was discharged on the third hospital day.

In Chapter 2, we discussed water as a chemical. We alluded to the earth as "the blue planet" because, alone among the planets of our solar system, earth has an abundance of liquid water. Water covers 70% of the surface of the globe, by far the majority of it being salty seawater. The archaeological record clearly indicates that life began in the sea; thus it should be no surprise that the most abundant substance in our body is water. Moreover, it is salty water, though not as salty as seawater.

This chapter is about body water, the solutes dissolved in it, and the role the kidneys play in maintaining a healthy volume of water and a healthy concentration of solutes.

"What is man, when you come to think upon him, but a minutely set, ingenious machine for turning, with infinite artfulness, the red wine of Shiraz into urine?"

Isak Dinesen (pseudonym of Karen Blixen), Danish writer, and author of *Out of Africa, 1938*

Body Fluid Compartments and Electrolytes

Water is the primary constituent of the body. In women, it comprises about 55% of body mass (weight); the remaining 45% is solids (Fig. 16.1). The proportions in males are 60% water and 40% solid. The difference is due to the greater amount of fat in women: fat contains little water.

In the study of body water, it is helpful to divide the body into spaces. These spaces are, for the most part, conceptual rather than anatomical; for example, the space within cells (*intracellular*) and the space outside of cells (*extracellular*); or the space within blood vessels (*intravascular*) and the space outside them (*extravascular*). In every instance, water is the most plentiful element in these spaces, so they are technically referred to as *fluid compartments*.

About 65% of body water is within cells in the **intracellular compartment;** that is, intracellular fluid (ICF) is the fluid of the cytosol. The remaining 35% is extracellular fluid (ECF) in the **extracellular compartment.**

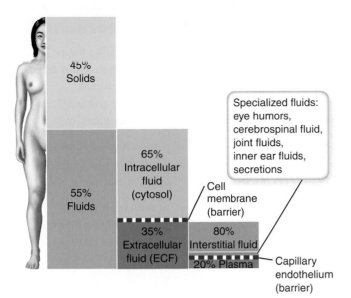

Figure 16.1. Body fluid compartments. The fluid content of the body can be divided into numerous compartments. This figure shows the approximate values for a lean female. *Name the two most abundant extracellular fluids.*

These two compartments are separated by a barrier, the cell membrane. Extracellular fluid includes blood plasma, interstitial fluid, and specialized fluids shown in Figure 16.1. Bone and other connective tissues contain most of the interstitial fluid. The vascular epithelium of blood vessels separates blood plasma (within the intravascular compartment) from the interstitial fluid.

instance, sodium chloride (NaCl, common table salt) dissociates in water into its component ions—Na^+ and Cl^-. Both the compound (NaCl) and its component ions (Na^+ and Cl^-) are considered to be electrolytes.

We gain salts through our diet. The most common dietary salt is NaCl. However, most humans consume far more than is necessary. Normally, we lose a modest amount of salt (mostly sodium) in urine and sweat. Little escapes in feces; however, patients with severe diarrhea can suffer fatal salt and water loss. Also, though the NaCl concentration in sweat is lower than that in blood, the sheer magnitude of sodium loss with heavy sweating can be a serious matter. This is the main reason that popular sport drinks contain a small amount of sodium and that people who sweat heavily for long periods are encouraged to take NaCl tablets with their drinking water.

Although most of our attention is devoted to NaCl, other important electrolytes in body fluids include potassium (K^+), bicarbonate (HCO_3^-), calcium (Ca^{2+}), and magnesium (Mg^{2+}). Since capillaries are permeable to these electrolytes, the ion composition of the interstitial fluid and plasma is roughly the same (Fig. 16.2). Normally, proteins cannot escape from capillaries, so proteins are largely absent from the interstitial fluid. Cell membranes, conversely, rigorously control ion flux, so the solute distribution differs enormously between intracellular and extracellular fluids. This electrolyte "disequilibrium" is at the heart of electrical communication, as discussed back

Case Note

16.1. Santo was very obese. Would his percentage of body water be higher or lower than that of a lean male?

Electrolytes Are Salts and Ions

Although body fluid compartments contains many solutes, the most important for our discussion here are **electrolytes**—salts that separate into ions when dissolved in water. Electrolytes were introduced in ← Chapter 2. Now, let us revisit the basic chemistry involved. An **ion** is an atom or molecule with a net positive or negative electric charge: **cations** (such as Na^+) are positively charged, whereas **anions** (such as Cl^- or HCO_3^-) are negatively charged. A **salt** is a compound of a cation (+) and an anion (−), so it is electrically neutral. Salts dissociate into their component ions in water. For

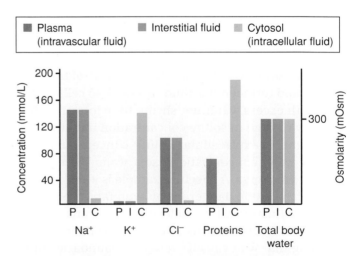

Figure 16.2. Electrolyte and water distribution. Ion concentrations differ between extracellular (plasma and interstitial fluid) and intracellular fluid compartments. Proteins are found only in plasma and cytosol. All cellular compartments are in osmotic equilibrium. *Which fluid compartment contains the highest concentration of potassium?*

in ◀ Chapter 4. Although they have other functions (for example, to carry electrical current), the relevance of electrolytes to this chapter is as follows:

- Electrolytes such as HCO_3^- (bicarbonate) help the body maintain *acid–base balance*, which we discuss at the end of this chapter.
- Electrolytes help determine the *osmolarity* of body fluids, by acting as *solutes*.

Electrolytes Determine Osmotic Gradients

The overall solute concentration, or **osmolarity,** of a solution is measured in *milliosmoles* per liter (mOsm/L). Remember that *water* concentration is low in fluids with high concentration of solute, and water moves down its gradient by *osmosis*. Thus, water moves from the solution with lower osmolarity (higher *water* concentration) to the solution with higher osmolarity (lower *water* concentration). That is, osmolarity is a measure of how much a solution "attracts" water.

If we were to add up all of the different electrolytes in each fluid compartments, the overall electrolyte concentration in all compartments would be virtually identical. To be specific, the osmolarity of all three compartments normally ranges from about 286 to 297 mOsm/L (Fig. 16.2, right side). This value is important, because intravenous fluids should be of the same osmolarity as that of blood and cells to avoid disruptions in fluid balance. A 0.9% (by weight) solution of NaCl (commonly called *normal saline*) equates to about 300 mOsm/L; it is thus the most commonly administered intravenous fluid. This *osmotic equilibrium* between the different compartments occurs because water passes freely between the three fluid compartments.

Under normal conditions, the osmolarity of the extracellular and intracellular fluids is equal, so cells neither swell with excess water nor shrink from loss of it. However, any change in solute concentration initiates water movement. If extracellular solute concentration falls (and the water concentration rises), water will enter the cell, and vice versa. A specific example is *water intoxication*, which can be dangerous or even fatal. This is perhaps best illustrated by a college hazing prank in which a student is forced to drink a very large volume of water. Recalling that water itself knows no boundaries in the body, brain cells absorb their share of the excess and enlarge. The net result is a swollen brain contained in a bony case that will not yield. Often the only adverse effect of such a prank is a bad headache, but convulsions or death can occur.

Case Notes

16.2. Santo was administered intravenous saline (NaCl solution) in the emergency room in order to rapidly increase his blood volume and blood pressure. What salt concentration (as a percent) would not disrupt Santo's osmotic equilibrium?

16.3. Santo's blood osmolarity was higher than normal. Does this mean that the solute concentration of his blood is abnormally high or abnormally low?

16.4. Santo's blood osmolarity was higher than normal. In the absence of other compensations, will the volume of body cells be increased or decreased?

Sodium and Water Balance Are Codependent

Body water is dynamic, constantly shifting between the intracellular fluid, interstitial fluid, and plasma to maintain osmotic equilibrium. *Sodium* (Na^+) is the major determinant of fluid shifts between one compartment and another. Recall that capillary walls are permeable to Na^+, which moves easily between plasma and interstitial fluid. However, the Na^+ concentration in the cytosol is low and Na^+ transport across cell membranes is strictly regulated. Intracellular *potassium* (K^+) provides the majority of osmotic pressure within the cell, just as albumin does for plasma. But it is Na^+ that fine-tunes fluid shifts between ECF and ICF.

Moreover, the overall body content of Na^+ is a major determinant of body water content. Think of the bloated feeling that follows the consumption of a big bag of salty popcorn and a jumbo soft drink. Water from the drink follows the sodium from the popcorn as it passes through intestinal cells into the vascular space, expanding blood volume. Some of this extra sodium and water will pass into the interstitial fluid, swelling tissues and producing "bloat." Interstitial fluid thus provides a buffer against large changes in blood volume.

But blood volume will still be increased somewhat, and blood volume (specifically *arterial* blood volume) is the major determinant of blood pressure. Thus, sodium balance and blood pressure are codependent—increased sodium retention increases blood pressure, and vice versa. This relationship is so absolute that our body uses body sodium content as an indicator of blood pressure. Our popcorn eater will not have a significant rise in blood pressure, because in the short term the cardiovascular system will respond homeostatically to avoid the increase and the kidneys will excrete the sodium load to restore electrolyte and water balance over the longer term. The corrective mechanisms restoring homeostasis are discussed later in this chapter.

The relationship between electrolyte balance and blood pressure has important clinical implications. Any pathological alteration in electrolyte balance will alter blood pressure. Chronic abnormally high blood pressure is **hypertension** (◀ Chapter 11). Hypertension accelerates the development of atherosclerosis, which has a marked ill effect on the heart and blood vessels and is associated with heart attacks and strokes. Although some persons are more susceptible than others to the effect of sodium on blood pressure, a high intake of dietary NaCl strongly correlates with increased blood pressure. Treatment of hypertension usually entails limiting dietary sodium intake and administering drugs (*diuretics*) to increase renal sodium excretion. On the bright side, we can also manipulate sodium balance to correct blood pressure problems.

Case Note

16.5. Santo's blood pressure is low. Would an increase in his blood volume improve or worsen the situation?

16.1   Name the two major extracellular fluids.

16.2   Despite differences in individual ion concentrations, the overall electrolyte concentration is roughly the same between intracellular and extracellular fluids. True or false?

16.3   Which fluid compartment contains the largest water volume—the intracellular compartment or the extracellular compartment?

16.4   What is the difference between a salt and an ion?

16.5   In the absence of compensation, what effect would a high-sodium diet have on blood pressure?

Overview of the Urinary System

The urinary system is involved in three interrelated types of homeostasis:

1. Electrolyte balance: Electrolyte input must match electrolyte output.

2. Water balance: Fluid output must be balanced with fluid input.

3. Acid–base balance: The pH of body fluids must remain relatively constant.

Other body systems also participate in these balancing acts, but in this chapter we concentrate on the renal contributions.

The kidneys, the major organs of the urinary system, perform these balancing acts by creating urine, into which they deliver water, electrolytes, and other solutes according to the needs of the body. For example, if the body needs more sodium, urine gets less of it; or, if the body has too much water, the kidneys eliminate it by producing more urine.

Urine Is a Waste Fluid

Because it is a waste product, **urine** gets about as much respect as feces. But the foul smell of poorly sanitized urinals derives from bacterial contamination and the products of bacterial metabolism, not from the normal constituents of urine. Most urine is clean enough to drink, or at least to taste, as ancient physicians regularly did to diagnose diabetes mellitus. Modern physicians use test strips to gain valuable diagnostic clues from urine. Read about urinalysis in the nearby clinical snapshot, titled *Liquid Gold*.

In addition to its role as a diagnostic fluid, urine across the ages has been used for almost every conceivable purpose—ancient civilizations used it as an antiseptic; Scots soaked wool in it to prevent shrinking; and others fermented and dried it to collect potassium nitrate (saltpeter) crystals, which they used to make explosives.

The waste products in urine mainly derive from protein metabolism, including urea, uric acid, creatinine, and ammonia. Urine gets its yellow color from *urobilin*, a waste product of hemoglobin metabolism. But urine is not merely metabolic waste; it also contains excess water, acid, and electrolytes, which the kidneys collect as they filter blood. Because 20% of cardiac output passes through the kidneys, our entire blood supply is filtered many times each hour to adjust blood osmolarity, pH, and electrolyte concentration.

As shown in Figure 16.3, after formation in the kidneys, urine travels down delivery tubes called *ureters* into the urinary bladder, which holds urine until it is ready for release. Together the ureters and bladder are referred to as the *collecting system*. Urine flows from the bladder through the urethra out of the body. Collectively, these organs constitute the **urinary system.** Each is discussed in more detail next.

CLINICAL SNAPSHOT

Liquid Gold

Your very life hangs on your kidneys' ability to make urine. Since it signals a healthy urinary system, urine is the diagnostic equivalent of liquid gold. The average person produces about 1 to 2 L of urine per day, of which about 99% is water. Because urine is easily obtained and inexpensive to analyze, **urinalysis** is one of the most common of all medical tests.

Color is important. For example, red urine may indicate bleeding somewhere in the urinary tract (or too many red beets in an earlier meal). Dark brownish-orange urine can be due to the presence of bilirubin (◀ Chapter 15) spilling from blood into urine in cases of jaundice, most often due to liver disease. However, the most common cause of abnormal color is medication—large amounts of some vitamins can produce strikingly orange urine.

Normal urine should be clear. Cloudy urine is usually caused by a precipitate, often crystals of some kind, but it may also be caused by white blood cells as a result of infection.

Given that metabolic waste is acidic, it should be no surprise that normal urine is slightly acidic: about pH 6.

It is the job of the kidney to eliminate most of the waste from protein metabolism, so large amounts of urea and creatinine are present. And because the kidney helps regulate electrolyte balance, urine contains variable amounts of sodium, potassium, calcium, and phosphate.

Especially important, however, are things that should not be present. Chief among these are bacteria and cells, especially red or white blood cells (RBCs or WBCs). WBCs may indicate infection. RBCs can mean anything from vaginal contamination in a menstruating female to kidney disease to cancer of the urinary tract. Because the glomerular barrier is designed to block protein from crossing into tubular fluid, normal urine should not contain a detectable amount of protein; nor should it contain glucose.

Urinalysis is usually performed by an automated device or, in an office or clinic, manually by the "dipstick" technique.

— Glucose
— Bilirubin
— Ketones
— Specific gravity
— Blood
— pH
— Protein

Urinalysis. The dipstick method evaluates the concentration of different substances in urine.

A dipstick is a flexible strip of plastic; on it are small pads impregnated with chemical detectors that change color to indicate either the presence of a certain substance or some other characteristic of urine. For example, the glucose strip is white. If glucose is not present, the pad remains white. If glucose is present, the pad turns blue. So, if the pad turns light blue, a small amount of glucose is present; dark blue, a lot.

Urine specific gravity is a measure of the concentration of dissolved solutes in urine. Increased specific gravity may indicate dehydration or a high concentration of glucose or proteins. Decreased specific gravity may indicate excessive fluid intake, as with alcohol abuse, or that kidney disease has limited the ability of the renal tubules to concentrate urine.

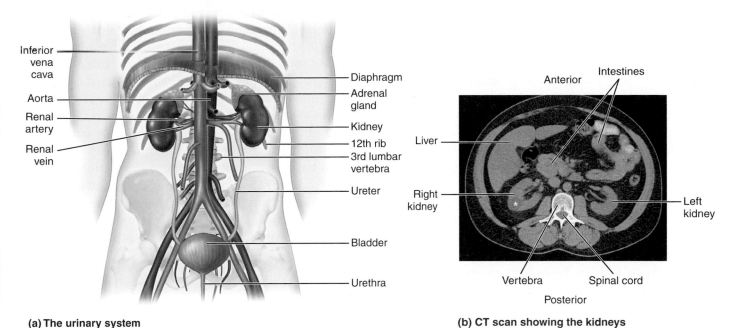

(a) The urinary system

(b) CT scan showing the kidneys

Figure 16.3. The urinary system. A. The urinary system consists of the kidneys, ureters, urinary bladder, and urethra. **B.** This computed tomography image shows a transverse section of the abdomen. *Identify the structures that carry urine to and from the bladder.*

Case Notes

16.6. Santo is too confused to provide a urine sample voluntarily, so a catheter is inserted into the bladder to extract the urine. Which tube does the catheter pass through—the ureter or the urethra?

16.7. If Santo were in ancient Greece, would a sip of Santo's urine be a useful diagnostic test?

The Kidneys Produce Urine

The **kidneys** are paired, fist-sized, bean-shaped organs (Fig. 16.3A). They rest near the vertebral column at a level between the twelfth thoracic vertebra and the third lumbar. The left sits slightly higher than the right, and the sheltering wing of the twelfth rib guards each one. However, the inferior ends peek out beneath the ribs and are vulnerable to trauma; hence the origin of the phrase "kidney punch" in boxing lingo. The right and left renal arteries supply arterial blood, and the corresponding renal veins collect the venous blood. The most posterior of the abdominal organs, they do not lie within the peritoneal cavity but posterior to it, in the retroperitoneum (Fig. 16.3B).

The ureter, blood vessels, and nerves enter and exit at an indentation in the medial side of each kidney (Figure 16.4). The upper end of each ureter flares out to

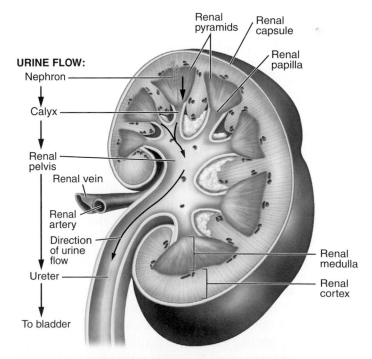

Figure 16.4. The kidney. Each kidney consists of an outer cortex and an inner medulla. Urine made by the renal tubules empties into the renal pelvis. *Are the renal pyramids found in the renal cortex or the renal medulla?*

form a funnel shape as it attaches to the kidney at the hilum. This wide attachment is often considered part of the kidney and is called the **renal pelvis.**

Internally, each kidney is composed of a pale outer **cortex** (Latin = "bark"), a rim of tissue about 1 cm thick, and a darker inner **medulla** (Fig. 16.4). The medulla is formed of several cone-shaped **renal pyramids,** which have their tips (the *papillae*) pointed toward the hilum. *Nephrons* are the kidneys' functional units, parts of which form the cortex and other parts the medulla. They are complex structures of blood vessels and epithelial tubules that form and collect urine. Their anatomic detail and how they produce urine is discussed in the next section of this chapter.

Urine formed by nephrons drains out the papilla into cup-shaped spaces called **calyces** (singular: *calyx*, Latin = "husk"). Urine flows from the calyces into the renal pelvis, which funnels urine into the upper end of a **ureter,** one of two long, thin, flexible muscular tubes that connect to the bladder.

The Ureters Transport Urine to the Bladder

The wall of the ureter is composed of three layers (Fig. 16.5A). The inner layer, the **mucosa,** is a lining of *transitional epithelial cells* (◀ Chapter 3), which have a capacity to stretch (to transition) from plump to flat to accommodate expansion of the tube. The middle layer, the **muscularis,** consists of smooth muscle—an inner longitudinal layer and an outer circular one. The outer layer, the **adventitia,** is a loose layer of fibrous tissue.

The ureters are embedded in a blanket of retroperitoneal fat and travel in the retroperitoneum on either side of the vertebral column. After crossing the posterior brim of the bony pelvis, they turn forward and travel in the bladder wall for a short distance before opening onto the bladder lumen. The ureters enter into the bladder floor; therefore, as the bladder fills, it expands upward and away from the **internal ureteral orifices**. The right and left ureteral orifices and the center line point where bladder and urethra join form a triangle in the floor of the bladder, which is called the *trigone*.

A flap of bladder wall tissue at each internal ureteral orifice creates a one-way valve that prevents urine from refluxing back into the ureter from the bladder. The valve is open when the bladder is relaxed, enabling the bladder to fill. Rising pressure during urination pushes the valve closed, preventing urine from reentering the ureter.

Some of the force conveying urine to the bladder is provided by gravity; however, the ureters actively massage urine downward by peristaltic waves in response to signals from the autonomic nervous system. Patients with spinal cord injury and paralysis lose some of this autonomic signaling. As a result, their urine may stagnate in the renal pelvis, ureters, and bladder which invites infection, formation of urinary stones (*urinary calculi*), and kidney disease. However, spinal cord injury is an uncommon cause of urinary stones. Most urinary stones are formed of calcium salts and occur in patients who do not have an underlying disease causing the stones to form; however, for unknown reasons, most of them have high urine calcium concentrations. Because a person is likely to form stones again and again, they are sometimes referred to as "stone formers" to distinguish them from others who form stones secondary to a known disease. You can see an example of a particularly large urinary stone in Figure 16.5C.

The Bladder Holds Urine Until It Is Released

Just as the stomach is a "holding tank" for food and the colon a holding tank for feces, the **urinary bladder** is a holding tank for urine. Protected behind the bony barrier of the symphysis pubis, the bladder is anterior to the rectum in males (Fig. 16.5B) or the vagina in females (◀ Fig. 17.7). The peritoneal membrane covers the superior surface of the bladder.

Like the ureters, the bladder wall is composed of three layers: an inner mucosal layer of transitional epithelium, a middle muscular layer, and an outer adventitia of fibrous tissue (the dome is covered by peritoneum). The muscular wall of the bladder is called the **detrusor muscle** (Latin *detrus* = "thrust down"). Like the muscular layer of the ureters, it is innervated by autonomic nerve fibers.

An empty bladder is shaped somewhat like a collapsed basketball: the dome falls downward into the cupped shape of the inferior side. In this condition, the mucosa is thrown into folds (rugae), and its transitional epithelial cells assume a plump, rounded shape. As it fills, the bladder pushes upward into the pelvis. This arrangement allows the bladder to fill easily and without stretching. In an average male, the maximum capacity of the bladder is about 700 to 800 mL; female capacity is slightly less because the uterus allows less room for bladder expansion, especially during pregnancy.

Urine Passes from the Bladder through the Urethra to the Environment

The **urethra** is a narrow tube leading from the neck of the bladder to the environment (Fig. 16.5B). The internal

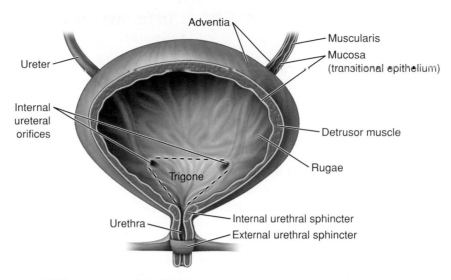

(a) The ureters and the bladder

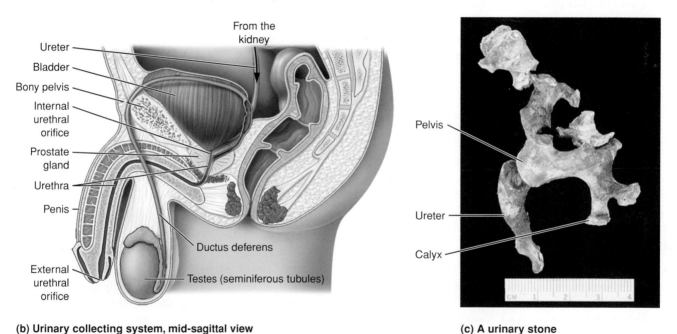

(b) Urinary collecting system, mid-sagittal view

(c) A urinary stone

Figure 16.5. The bladder and ureters. A. The bladder and ureters contain a muscular layer sandwiched between an outer covering and the inner mucosa. **B.** This midsagittal view shows the ureters, bladder, and urethra of a male. **C.** This large urinary stone completely filled the upper ureter, renal pelvis, and calyces. *What are the mucosal folds called?*

opening, where it is attached to the bladder, is the *internal urethral orifice*. The external opening is the *external urethral orifice*. In females the urethra lies immediately beneath the symphysis pubis; it is short, about 4 cm, embedded in the anterior wall of the vagina, and opens immediately anterior to the vaginal orifice (see ◀ (Fig. 17.6). In males it is longer, about 17 cm. It runs from the neck of the bladder through the center of the prostate gland, then through the center of the penis until it opens onto skin at the tip of the penis. Females are more prone to bladder infections than males because vaginal

mucosa is more hospitable to microbes than the skin of the penis, and the short urethra of females makes it easier for bacteria to gain access to the bladder.

Two rings of muscle encircle the internal urethral orifice (Fig. 16.5A). The upstream ring is composed of smooth muscle and is called the **internal urethral sphincter.** It is innervated by the autonomic nervous system and is not under voluntary control. Downstream about 3 to 4 cm is a second ring of muscle, the **external urethral sphincter,** which is composed of *skeletal* muscle under conscious control of the cerebral cortex.

Urination Is the Release of Urine from the Bladder

Urination, also called *micturition*, is release of urine from the bladder. It may be voluntary or involuntary.

Initial filling of the bladder does not stretch its wall—the collapsed dome is merely pushed upward. However, as the volume reaches about 300 mL, the wall must stretch to accommodate more volume. At lesser volumes, the brain is unaware of the volume; but as larger volumes accumulate, stretch receptors in the bladder wall begin sending signals to the cerebral cortex, which are interpreted as an awareness of bladder filling. As volume continues to grow and further stretching occurs, stretch receptors relay signals to the spinal cord, triggering an autonomic reflex, the **micturition reflex.** This reflex relays parasympathetic signals between spinal cord and bladder, causing contraction of the detrusor muscle and relaxation of the internal (involuntary smooth muscle) urethral sphincter. The cerebral cortex receives sensory input from the stretched bladder wall, which it interprets as the urge to urinate. Voluntary control of urination depends upon cortical control of the external urinary sphincter, which is composed of skeletal muscle. Relaxation of the external sphincter causes urination.

Early in life we learn to voluntarily overcome the inhibitory reflex signal that relaxes the external urethral sphincter. We can thus determine when urination occurs; but at very large volumes or in unconscious states, voluntary signals cannot restrain the reflex and involuntary urination occurs.

Case Note

16.8. Although he is confused, Santo is still able to maintain bladder control despite the presence of a significant volume of urine in his bladder. Which urethral sphincter is contracted and which is relaxed?

16.6 If we compared the kidney to a melon, which layer would be the melon rind—the cortex or the medulla?

16.7 What is the difference between the renal pyramids and the renal calyces?

16.8 Which part of the collecting system is actually part of the kidney?

16.9 Can urine move from the kidney to the bladder if you are standing on your head?

16.10 Which three points form the borders of the trigone?

16.11 The muscle layer of the bladder is called the *muscularis.* True or false?

16.12 Is the internal urethral sphincter contracted or relaxed if the detrusor muscle is contracting?

The Production of Urine

Urine formation requires precise exchanges between the tubular and vascular structures of the kidneys. So we begin our discussion with a closer look at the nephrons and associated blood vessels.

A Nephron Contains a Glomerulus and a Renal Tubule

The **nephron** is the functional unit of the kidney. It consists of an initial filtering unit, a tuft of capillaries called the **glomerulus,** and its associated epithelial tubule, the **renal tubule** (Fig. 16.6A). There are two types of nephrons: the shorter *cortical nephrons* lie nearer the surface of the cortex, whereas longer *juxtamedullary nephrons* lie more inward, near the medulla. The proximal tip of each tubule is the **glomerular capsule,** an enlarged portion of the tubule that envelops the glomerulus. The glomerulus is pushed into the glomerular capsule like a fingertip poked into the tip of a long, thin balloon. Fluid filters from the capillary fingertip into the balloon and flows down the tubule.

Each renal tubule is organized into three distinct sections (Fig. 16.6A), which from proximal to distal are:

- The **proximal tubule**
- The **nephron loop** (*loop of Henle*)
- The **distal tubule**

The proximal tubule of the nephron is quite twisty and entirely located in the cortex. Next, the nephron plunges downward into the medulla as a long, straight tube, the *descending limb* of the nephron loop. The loops of cortical nephrons are shorter and do not descend as deeply into the medulla as the loops of juxtamedullary nephrons. The loop then makes a hairpin turn and goes straight back up as the *ascending limb*. It reenters the cortex and becomes the distal tubule, which then empties along with other tubules into a **collecting duct.** Collecting ducts extend from the cortex back down into the

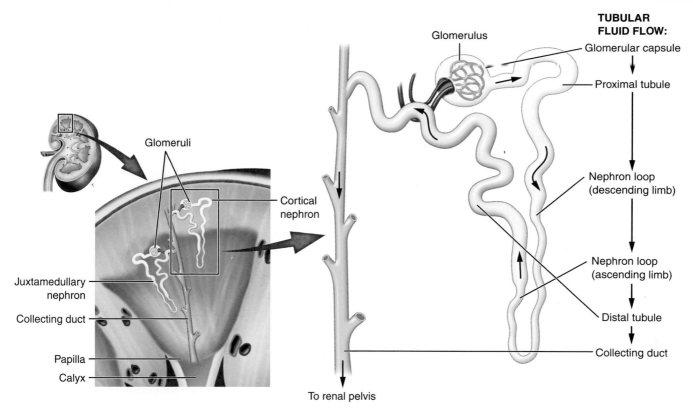

TUBULAR FLUID FLOW:

Glomerulus
Glomerular capsule
↓
Proximal tubule
↓
Nephron loop (descending limb)
↓
Nephron loop (ascending limb)
↓
Distal tubule
↓
Collecting duct

Glomeruli

Cortical nephron

Juxtamedullary nephron

Collecting duct

Papilla

Calyx

To renal pelvis

(a) A cortical nephron

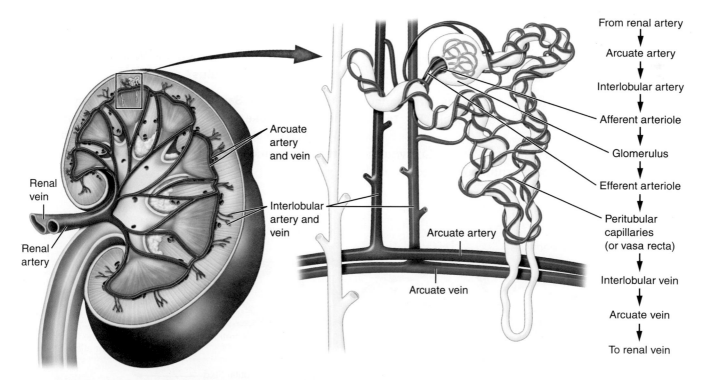

From renal artery
↓
Arcuate artery
↓
Interlobular artery
↓
Afferent arteriole
↓
Glomerulus
↓
Efferent arteriole
↓
Peritubular capillaries (or vasa recta)
↓
Interlobular vein
↓
Arcuate vein
↓
To renal vein

Arcuate artery and vein

Interlobular artery and vein

Renal vein

Renal artery

Arcuate artery

Arcuate vein

(b) Blood supply to a cortical nephron

Figure 16.6. The nephron and associated blood vessels. A. The nephron consists of the renal tubule and the glomerulus. Tubular fluid is formed in the glomerular capsule and passes through the renal tubule. **B.** The renal circulation sends blood through two separate capillary beds: the glomerulus and the peritubular capillaries/vasa recta. *Which blood vessel receives blood from the arcuate artery?*

medulla and empty through the tips of the pyramids into the calyces and the renal pelvis.

The kidneys receive about 20% of cardiac output—upward of 1 L of blood per minute—through the right and left **renal arteries,** each of which is a direct branch of the aorta and delivers blood to the glomeruli (Figure 16.6B). Each renal artery divides into successively smaller arteries that eventually branch into the **arcuate arteries** at the corticomedullary boundary. Small **interlobular arteries** branch off the arcuate artery, each supplying numerous **afferent arterioles.** Each afferent arteriole branches and curls to form the capillaries of the glomerulus. Blood flows out from the glomerulus through the **efferent arteriole** and subsequently into the **peritubular capillaries,** which surround nephrons. In juxtamedullary nephrons, the peritubular capillaries embrace the nephron loop in a ladderlike network called the *vasa recta.* After flowing around the nephrons, blood collects in the interlobular veins, which drain into the arcuate vein and eventually the renal vein. The right and left renal veins drain into the inferior vena cava.

> ***Remember This!*** You can remember that the afferent arteriole comes before the efferent arteriole because A comes before E.

Urine Production Is a Four-Step Process

The production of urine involves four types of exchanges between the renal tubule and blood vessels (Fig. 16.7):

1. In **glomerular filtration,** about 20% of fluid volume passing through the glomerulus is filtered through the glomerular capsule to become the **glomerular filtrate.** Glomerular filtrate contains water and many plasma solutes but no blood cells or protein, both of which are too large to pass through the filtration sieve. After the glomerular filtrate leaves the glomerular capsule and is acted upon by tubular epithelium, it is referred to as the **tubular fluid.**

2. **Tubular reabsorption** is the movement of substances from tubular fluid back into blood. Water and many other substances move from all portions of the tubule into peritubular capillaries.

3. **Tubular secretion** is the opposite of tubular reabsorption; substances move from peritubular blood into tubular fluid.

4. **Urine concentration** is a specific water reabsorption process that occurs in the collecting duct and nephron loop. The amount of water reabsorption determines the urine concentration.

> ***Remember This!*** Only tubular secretion moves solutes from blood to tubular fluid. All other processes move water and solutes from tubular fluid into blood.

Case Notes

16.9. Because of Santo's low blood pressure, less blood arrives at his glomerulus. Which vessel carries blood to the glomerulus?

16.10. Some of Santo's problems are caused by filtered glucose that does not return to the blood. Which process returns filtered substances to blood—reabsorption or secretion?

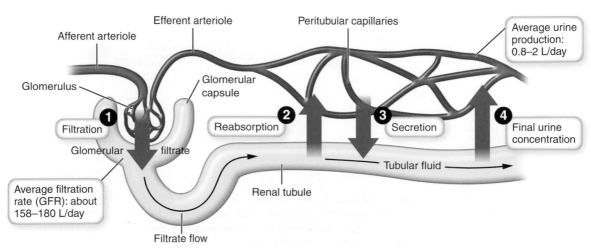

Afferent arteriole

Efferent arteriole

Peritubular capillaries

Average urine production: 0.8–2 L/day

Glomerulus

Glomerular capsule

1 Filtration

Glomerular filtrate

2 Reabsorption

3 Secretion

4 Final urine concentration

Tubular fluid

Average filtration rate (GFR): about 158–180 L/day

Renal tubule

Filtrate flow

Figure 16.7. Renal processes. Urine formation requires four types of exchanges between blood and tubular fluid. *Which process or processes move substances from the blood to the renal tubule?*

The Glomerulus Filters Blood to Create the Glomerular Filtrate

As noted earlier, each glomerulus can be conceived of as a finger of capillaries pressed deeply into one side of an inflated balloon (the blind sac end of a tubule) (Fig. 16.8A). The remaining space within the balloon is the lumen of the glomerular capsule and is called the **glomerular space** (*Bowman's space*). The cells of the balloon wall that touch the glomerular capillary fist are called *podocytes* because they extend footlike processes that reach out to touch the outer, basement membrane of glomerular capillaries. Podocyte processes are set close to one another, leaving tiny gaps (*filtration slits*) between

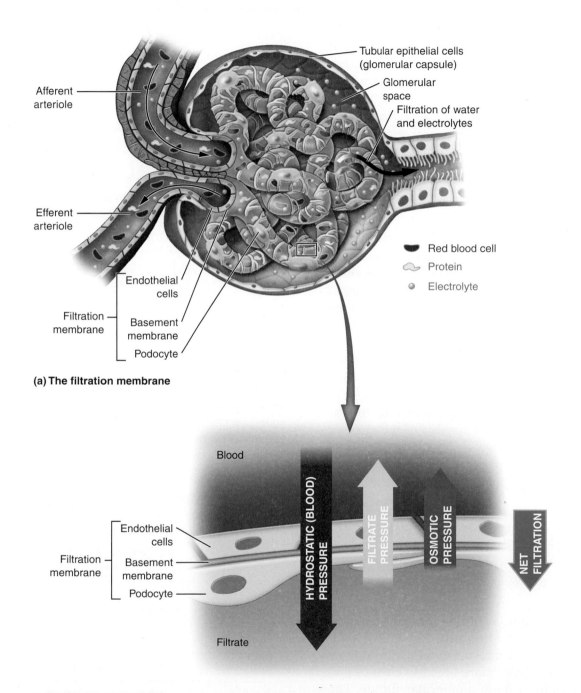

(a) The filtration membrane

(b) Formation of the filtrate

Figure 16.8. Filtration. A. The capillary endothelium, basement membrane, and podocytes of the glomerular capsule act like filter paper, keeping cells and protein in the blood and letting water and small solutes through. **B.** The hydrostatic pressure of blood (blood pressure) favors filtration; the hydrostatic pressure of the filtrate and the osmotic pressure of blood oppose filtration. *Which of the following substances would you expect to find in the glomerular capsule: large proteins, blood cells, or sodium?*

them. Together, podocytes and glomerular capillaries form the three-layer **filtration membrane,** which—from the vascular space to the glomerular capsule—consists of (a) capillary endothelial cell. (b) capillary basement membrane. and (c) podocyte.

The glomerular filtrate is formed as blood pressure forces fluid across the filtration membrane, from the vascular space into the glomerular capsule. Just as a coffee filter retains the coffee grounds, the filtration membrane retains proteins and blood cells because they are too large to pass through the filter. Everything else in blood—water, ions, nutrients, waste products, and other small molecules—passes into the glomerular filtrate.

The balance of forces determines how much glomerular filtrate is formed (Fig. 16.8B). The hydrostatic pressure of blood (that is, blood pressure) creates *filtration pressure*, which is the sole force pressing fluid from blood into the glomerular capsule. This is much like the hydrostatic pressure exerted as gravity pushes water through a coffee filter (red arrow). Two forces oppose filtration pressure. One is the hydrostatic pressure of fluid already in the glomerular capsule (yellow arrow). Another is the osmotic pressure gradient that exists between glomerular blood and glomerular filtrate (blue arrow). Glomerular blood contains a high concentration of solute—proteins—which are not in the filtrate, so the osmotic pressure gradient favors water movement back into the glomerulus. Nevertheless, the blood's hydrostatic pressure is greater than the sum of the other two forces combined, so *net filtration* (green arrow) occurs; that is, fluid moves from blood to tubule.

Glomerular blood pressure varies according to systemic blood pressure and the diameter of the afferent arteriole. Dilation of the afferent (upstream) arteriole increases blood flow and pressure in the glomerulus; constriction has the opposite effect.

> *Remember This!* Compare the forces governing glomerular filtration with those governing fluid movement across the capillary wall. You will see that they are virtually the same.

The volume of glomerular filtrate formed each minute is the **glomerular filtration rate (GFR).** The GFR averages about 100 to 125 mL/min, depending on body weight; larger people have more blood to cleanse and regulate and thus form more glomerular filtrate. The GFR determines filtrate volume, and higher filtrate volumes tend to increase urine volume. Changes in GFR thus affect how much fluid is retained in blood, and this, in turn, helps control blood pressure.

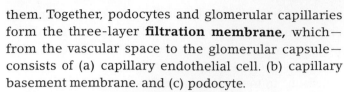

Case Note

16.11. Santo's urinalysis revealed the presence of protein. This finding indicates that proteins are passing through the filtration membrane into tubular fluid probably because of damage to the glomerulus. How will these filtrate proteins affect the osmotic gradient between the glomerulus and the filtrate, and what will happen to his GFR?

Reabsorption Occurs in the Proximal Tubule

The composition of urine is very different from the composition of filtrate, because the nephron exchanges water and dissolved substances between the tubule and the peritubular capillaries.

Tubular reabsorption is a heavy-duty task because a very large volume of glomerular filtrate must be reduced to a small volume of urine. Most of the work is done by the proximal tubule, which uses a combination of active and passive transport mechanisms to get most of the filtrate components back into blood (see Table 16.1). Here is how it works (Fig. 16.9):

1. Sodium ions are transported from the filtrate to the blood using *primary active transport* and the Na^+/K^+-ATPase protein. Chloride (Cl^-) ions follow the positively charged Na^+ ions.
2. The sodium gradient formed by step 1 is used to bring glucose, phosphate, and amino acids across, using *secondary active transport* ◀ (Chapter 3).
3. As solutes move from tubular fluid into blood, water follows along by osmosis.

Table 16.1	Renal Processing of Selected Substances		
	Percent of Filtered Solute Reabsorbed (%)		
Nephron Segment	**Sodium (and chloride)**	**Water**	**Glucose**
Proximal tubule	65%	65%	100%
Nephron loop	25%	10%	0
Distal tubule	5%	0	0
Collecting duct	4%–5% (regulated)	5%–24% (regulated)	0

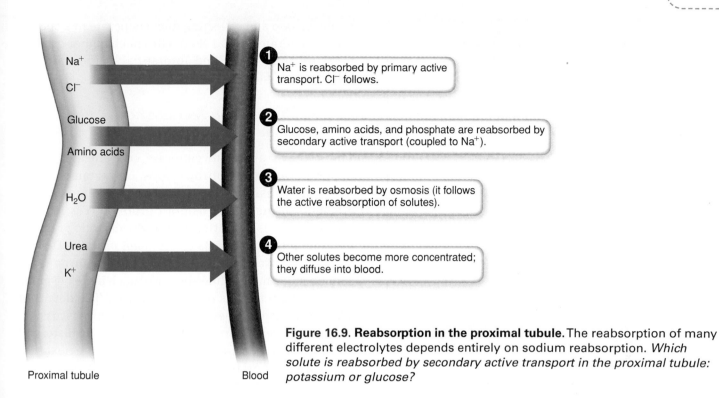

Figure 16.9. Reabsorption in the proximal tubule. The reabsorption of many different electrolytes depends entirely on sodium reabsorption. *Which solute is reabsorbed by secondary active transport in the proximal tubule: potassium or glucose?*

4. All of the remaining molecules (such as urea, calcium, and potassium) suddenly find themselves in a smaller filtrate volume—they are more *concentrated* than they were previously. This new concentration gradient drives urea and potassium from the filtrate into the blood.

Most filtered bicarbonate is also reabsorbed, through a specialized mechanism discussed further on.

Tubular Secretion Occurs throughout the Nephron

Although it is less common, tubular *secretion* also occurs. For instance, the proximal tubules secrete creatinine (from breakdown of muscle), hydrogen ions, and drugs from peritubular blood into the tubular fluid to be excreted in urine. As we will see, the proximal tubular secretion of hydrogen ions is important for acid–base balance. Other parts of the nephron also secrete hydrogen.

Regulated Reabsorption and Secretion Occur in the Distal Tubule and Collecting Duct

The distal tubule and collecting duct transfer relatively small amounts of water and solutes between the tubular fluid and the peritubular blood. They are nevertheless very important, because this is where the kidney regulates water and electrolyte balance. Water balance is discussed further on. Below we discuss electrolyte balance by focusing on the reciprocal secretion and reabsorption relationships of two electrolyte pairs: Na/K and Ca/phosphate. Under the influence of hormones, the first member of each pair is *reabsorbed* from tubular fluid into peritubular blood, and the second member is *secreted* from peritubular blood into tubular fluid.

Aldosterone Regulates Sodium Reabsorption and Potassium Secretion

Recall from ◀ Chapter 15 that adrenal *mineralocorticoids*, aldosterone chief among them, regulate sodium balance via their action on the renal tubule. Aldosterone increases renal sodium reabsorption by increasing the number of sodium channels and sodium-potassium pumps in cells of the cortical collecting duct (Fig. 16.10). The Na-K pump transports potassium in the opposite direction of sodium. Therefore, by increasing sodium *reabsorption*, aldosterone also increases potassium *secretion*—blood sodium rises as blood potassium falls, and vice versa.

Aldosterone secretion increases homeostatically as (a) blood pressure falls or (b) as blood potassium rises. Aldosterone's role in blood pressure and water balance is revisited further on.

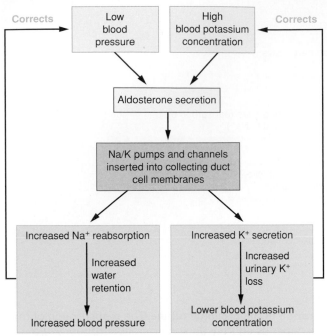

Figure 16.10. Aldosterone. Aldosterone secretion is regulated by negative feedback from blood pressure and potassium concentrations. *Bananas are a good source of potassium. How would the consumption of many bananas alter aldosterone secretion?*

Case Note

16.12. In all likelihood, Santo's low blood pressure would stimulate aldosterone production. How would this affect the amount of potassium in Santo's urine?

Parathyroid Hormone Stimulates Calcium Retention and Phosphate Excretion

Recall that calcium and phosphate are both reabsorbed in the proximal tubule. ◀ Chapters 6 and 15 introduced hormones that regulate calcium and phosphate balance, *parathyroid hormone* (PTH) in particular. As part of its widespread efforts to increase blood calcium concentration, PTH stimulates calcium reabsorption in the distal tubule and reduces calcium loss in the urine. Coincidentally, PTH also *reduces* phosphate reabsorption in the proximal tubule, thereby increasing phosphate excretion.

Urine Concentration Occurs in the Nephron Loop and Collecting Duct

The proximal tubule reabsorbs water until the osmolarity of tubular fluid is equal to that of blood and interstitial tissue—it is incapable of creating urine more concentrated than peritubular fluid and blood. However, the nephron loop and collecting duct cooperate in a special way to generate tubular fluid with much higher osmotic pressure than blood and interstitial fluid, a process called *urine concentration*.

The Nephron Loop Establishes a Unique Osmotic Gradient

The kidney's ability to generate concentrated urine depends on the nephron loops of juxtamedullary nephrons. These long loops establish a unique osmotic gradient, the **medullary osmotic gradient** (Fig. 16.11A). This gradient is the difference in the osmotic pressure between *different regions of the renal medulla*. Deeper medullary regions toward the tips of the nephron loops contain much more solute and have much higher osmotic pressure than regions nearer the cortex. The details of this mechanism are beyond the scope of this discussion, but you can read more about it in the nearby history of science box titled *How History Is Made*. In the next section, we learn how this osmotic gradient determines the concentration of urine.

An interesting fact apropos of this discussion: desert rats have many juxtamedullary nephrons with long nephron loops and are able to establish a much stronger osmotic gradient than humans. As a result, they can conserve water by excreting highly concentrated urine.

Antidiuretic Hormone Regulates Water Reabsorption

The nephron loop establishes the medullary osmotic gradient, but it is the collecting duct that uses the gradient to concentrate urine. The orientation of the collecting duct is all-important. It runs alongside the nephron loop and drains into the renal pelvis at medullary papilla (Fig. 16.11A) . As the tubular fluid descends in the collecting duct, osmotic pressure in the surrounding tissue becomes higher, and progressively more and more water is drawn out of tubular fluid and delivered into blood (Fig. 16.11B). As a result, tubular fluid, and the urine it is to become, becomes more concentrated. Note that, in contrast to osmosis in the proximal tubule, this mechanism *is not sodium-dependent*. The medullary osmotic gradient, not solute reabsorption, is the driving force for water reabsorption.

> ***Remember This!*** In the collecting duct, water reabsorption is independent of solute movement.

Although the medullary osmotic gradient creates the *circumstances* by which urine concentration can occur, little water will actually move out of the collecting duct in the absence of **antidiuretic hormone (ADH),** a

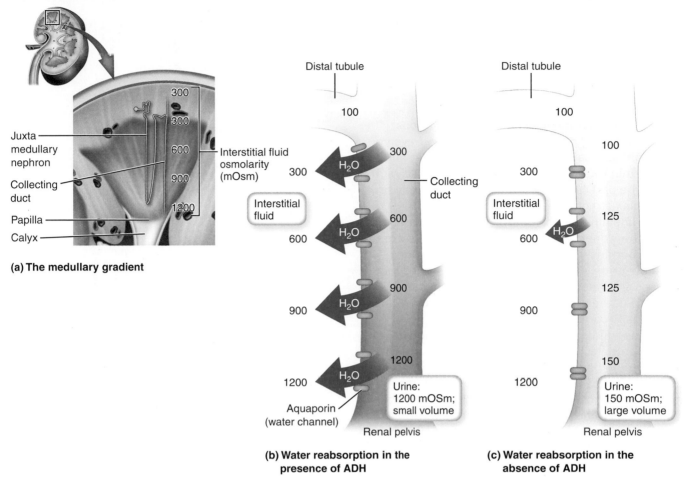

(a) The medullary gradient

Figure 16.11. Urine concentration. A. The nephron loop establishes an osmotic gradient in the renal medulla.
B. If ADH is present, water moves down the osmotic gradient, resulting in concentrated urine. **C.** If ADH is absent,
water cannot move down the osmotic gradient, and the urine remains dilute. *Where is the osmotic gradient the
strongest: by the distal tubule or by the renal pelvis?*

posterior pituitary hormone we met briefly in ◀ Chapter 15. As you may know, a *diuretic* is any substance that increases urine excretion; thus, ADH promotes water retention.

We think of osmosis as an unregulated process, and it usually is, because most cell membranes are permeable to water. However, collecting duct cells can be virtually *impermeable* to water without the influence of ADH. Here, osmosis will occur only if water channels, called **aquaporins,** are inserted into the cell membrane. ADH stimulates the insertion of aquaporins into the cell membranes of collecting duct cells, allowing osmosis to occur and urine to be concentrated (Fig. 16.11B). In the absence of ADH, very few aquaporins are present in the membrane, and most water remains in the tubular fluid to be excreted as large volumes of urine (Fig. 16.11C). The pathological absence of ADH is called *diabetes insipidus* (tasteless, high-volume urine). The volume of urine produced can be impressive: as much as 10 L per day.

As you would expect, a typical negative feedback loop operates and ADH secretion increases when blood osmolarity is high or blood pressure is low. By increasing water reabsorption from the renal tubule into blood, ADH decreases blood osmolarity back to the homeostatic norm. The reabsorbed water also expands blood volume, which increases blood pressure.

Case Notes

16.13. Santo's blood pressure is low, but his blood osmolarity is high. Predict the effect of these two perturbations on ADH synthesis.

16.14. Alcohol inhibits ADH production. Remember that Santo drank about 10 beers the night before. Would you expect his urine to be more concentrated or less concentrated than normal, simply as a result of his alcohol consumption?

THE HISTORY OF SCIENCE

How History Is Made

Nearby, we discuss the concentrations of various solutes in differing parts of the renal tubule. Have you ever caught yourself wondering, "How do they know—for sure—that the concentration of this or that is high or low at some particular place in these fantastically tiny and complex structures? Here is the answer.

By the late nineteenth century, simple microscopic study of the kidney had established the anatomy of the renal glomerulus and tubules. But little was known about how urine was formed. One camp argued that urine was a filtrate; the other argued that it was secreted. But they had no proof. We know now that both were partially correct.

It was not until the 1920s, with the invention of micropuncture technique, that our current understanding began to take shape. Almost impossibly thin glass tubules were handcrafted by experimenters and—using microscopes and mechanical manipulators capable of very fine adjustment—inserted into individual tubules of animal kidneys at various points. Samples of tubular fluid were obtained and analyzed, and renal physiology quickly began to advance.

Another technique helped to prove the medullary osmotic gradient by capitalizing on the fact that water with a high solute concentration has a lower freezing point than pure water. In 1951, experimenters froze animal kidneys and made ultrathin slices of renal medullary pyramids, which they laid out in order of their location from near the tip to those further toward

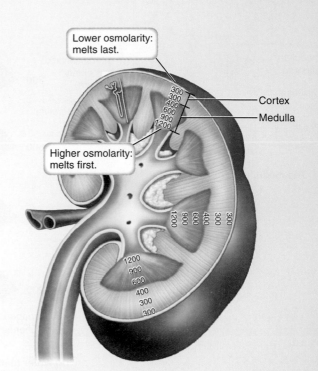

The medullary osmotic gradient.

the cortex. Then they raised the temperature very slowly. Slices from the tips of the renal pyramid melted first, whereas slices farther from the tip melted later, at higher temperatures. Voila! An osmotic gradient.

16.13 Name the two capillary beds found between the renal artery and the renal vein.

16.14 Which tubule portion drains directly in the collecting duct?

16.15 Which arteriole supplies blood to the peritubular capillaries?

16.16 If the kidneys filtered 200 L of plasma in a day, how much water would the kidneys return to blood: 198 L or 2 L?

16.17 The kidney eliminates many drugs by moving them from the peritubular blood to the tubular fluid. What is this process called?

16.18 From blood outward, name the three layers of the filtration membrane.

16.19 Which of the following substances is reabsorbed by secondary active transport: sodium, glucose, potassium?

16.20 Potassium is both reabsorbed and secreted by the renal tubule. Which tubule section reabsorbs potassium and which tubule section secretes potassium?

16.21 Which hormone stimulates sodium reabsorption: aldosterone or ADH?

16.22 Which hormone directly stimulates water reabsorption without stimulating solute movement: aldosterone or ADH?

16.23 Which tubule portion establishes the medullary osmotic gradient?

16.24 True or false: Urine concentration in the collecting duct relies on active sodium transport from the collecting duct to the peritubular capillaries.

Electrolyte and Water Balance

The kidney can *conserve* volume—by decreasing water loss—but cannot *replace* volume that has been lost. Lost volume can only be replaced by fluid gain, and the only natural way to gain fluids is to drink them. The urge to drink is regulated by the **thirst center** in the hypothalamus, which is especially sensitive to changes in blood osmolarity; some other stimuli also have an effect, including dry mouth, blood volume and blood pressure. Let us take a closer look.

Dehydration affects the water content of every compartment, including the intravascular compartment. With dehydration, blood osmolarity rises and blood volume and pressure fall ◀ (Chapter 11). The rise of blood osmolarity triggers osmoreceptors in the hypothalamus to signal the cerebral cortex, which perceives the signals as thirst. If you recall that blood volume is affected by body water and that low blood pressure may be a sign of low body water, it should not surprise you to learn that low blood pressure can also stimulate thirst (through mechanisms to be discussed later). Our perception of thirst prompts us to drink. Fluid intake quickly dilutes our blood and promotes increased blood volume and pressure, both of which tend to suppress thirst center signals, causing the sensation of thirst to disappear. Although a sip of water may temporarily decrease thirstiness, the sensation does not disappear until blood osmolarity returns to normal.

Case Note

16.15. Santo's blood osmolarity was 310 mOsm/L (286 to 297 mOsm/L is normal). Do you think that Santo is thirsty?

Potassium Is Important for Non–Fluid Balance Activities

Contrary to sodium, *potassium* is not important in fluid shifts, but it is very important for other, non–fluid balance activities.

For example, recall from ◀ Chapter 4 that K^+ is the most abundant cation in intracellular fluid and is responsible for maintaining the resting membrane potential. K^+ is also important in acid–base balance, discussed further on.

As it is for sodium, the kidney is the main regulator of K^+. The most important factor in potassium regulation is the plasma K^+ concentration. As plasma K^+ rises, K^+ diffuses into the interstitial fluid and into the cytosol of renal tubular cells. This prompts them to secrete K^+ into tubular fluid for excretion in urine. Increased blood K^+ also stimulates the release of aldosterone, which increases K^+ secretion into tubular fluid.

Too much or too little plasma K^+ is dangerous because of its effect on membrane potential. Foremost, it affects cardiac membrane potential. High plasma K^+ may cause fatal cardiac arrhythmia, slow heart rate or death from cardiac arrest. Intravenous injection of concentrated K^+ is quickly lethal. Conversely, moderately low plasma K^+ causes muscle weakness, which may be dangerous if respiratory effort is affected. Markedly low plasma K^+ increases myocardial irritability and causes cardiac arrhythmias, which can also be fatal. In fact, low potassium is responsible for some of the deaths associated with eating disorders, especially bulimia or laxative purging—as patients purge themselves of food, they also purge themselves of K^+.

Renal Regulation of Fluid Balance and Blood Pressure

Body sodium content determines body water content, and body water content determines blood pressure. Renal regulation of sodium and water balance is really about regulating blood pressure.

In Chapter 11, we discussed short-term blood pressure regulation by the vasomotor center in the brainstem. This center increases blood pressure by increasing cardiac output (which increases arterial blood volume) and stimulating vasoconstriction. Blood pressure is reduced by the inverse reactions. In this chapter, we discuss the role of the kidneys in blood pressure regulation. In contrast to the short-term actions vasomotor center, the kidneys modulate blood pressure over the medium and long term (Fig. 16.12). They do so primarily by regulating electrolyte and water balance.

The Renin–Angiotensin–Aldosterone System Increases Blood Pressure

When blood pressure is low, a cluster of renal cells collectively called the **juxtaglomerular apparatus (JGA)** secretes into the blood an enzyme called **renin,** which

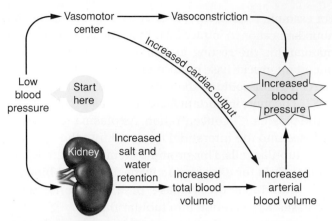

Figure 16.12. Blood pressure. The kidney and the vasomotor center regulate blood pressure by modulating arterial blood volume and vascular resistance. *Does changing cardiac output modify vascular resistance or arterial blood volume?*

activates a cascade of events that increases blood pressure. The JGA consists of two parts:

- Modified tubular epithelial cells of the distal tubule, called the **macula densa.** Macula densa cells respond to low Na^+ in the distal tubular fluid, which is an indicator of low filtration and thus low blood pressure (Fig. 16.13A).
- Modified afferent arteriolar smooth muscle cells, called **granular cells.** Granular cells secrete renin in response to signals from the macula densa. Granular cells are also directly activated by low blood pressure in the efferent arteriole.

As shown in Figure 16.13B, renin acts as an enzyme to convert *angiotensinogen*, a protein made by the liver, into *angiotensin I* (ATI). Then, ATI is converted into **angiotensin II (ATII)** by *angiotensin converting enzyme* (ACE), a protein made by pulmonary capillaries. ATII increases blood pressure by increasing peripheral resistance and blood volume in several ways.

1. It is a powerful vasoconstrictor, which increases peripheral resistance.
2. It stimulates aldosterone release from the adrenal cortex. Aldosterone increases renal sodium reabsorption, which causes water retention and increased blood volume.
3. It stimulates thirst, which increases water intake to expand blood volume.
4. It stimulates ADH release from the posterior pituitary, which decreases water loss in the urine. Note that ADH secretion is also directly stimulated by decreased blood pressure and increased plasma osmolarity.

Case Note

16.16. How will the kidney sense Santo's low blood pressure?

Atrial Natriuretic Peptide Lowers Blood Volume and Blood Pressure

All of the hormones discussed thus far—ADH, aldosterone, and angiotensin II—work together to increase arterial blood volume and therefore blood pressure. This redundancy makes sense: since low blood pressure lessens blood flow to the brain and vital organs, it is a much more imminent threat to life than high blood pressure. Only a single hormone opposes the "big three" and lowers blood pressure—**atrial natriuretic peptide (ANP;** Greek *nitron* = "salt crystals," and *ouresis* = "urination").

ANP is a hormone released by the heart's atria. High blood pressure stretches the atrial walls, which stimulates the release of ANP. It acts on renal tubules to reduce Na^+ absorption, and, by inhibiting ADH production from the posterior pituitary gland, also reduces water reabsorption. Increased loss of sodium (*natriuresis*) and water in the urine occurs and blood sodium and blood volume decline. Blood pressure returns to normal.

Case Note

16.17. Santo's blood volume is decreased and his osmolarity increased. How will these changes affect ANP, ATII, aldosterone, and ADH production?

16.25 Which brain region controls thirst?

16.26 Will high blood potassium levels stimulate potassium reabsorption or potassium secretion?

16.27 Where are granular cells found—the afferent arteriole or the distal tubule?

16.28 What will happen to renin production if the glomerular filtration rate decreases?

16.29 How does angiotensin II stimulate sodium retention?

16.30 If you were looking for a natural substance to act as a drug that would treat high blood pressure, which of the following we be a good candidate: ANP or ATII?

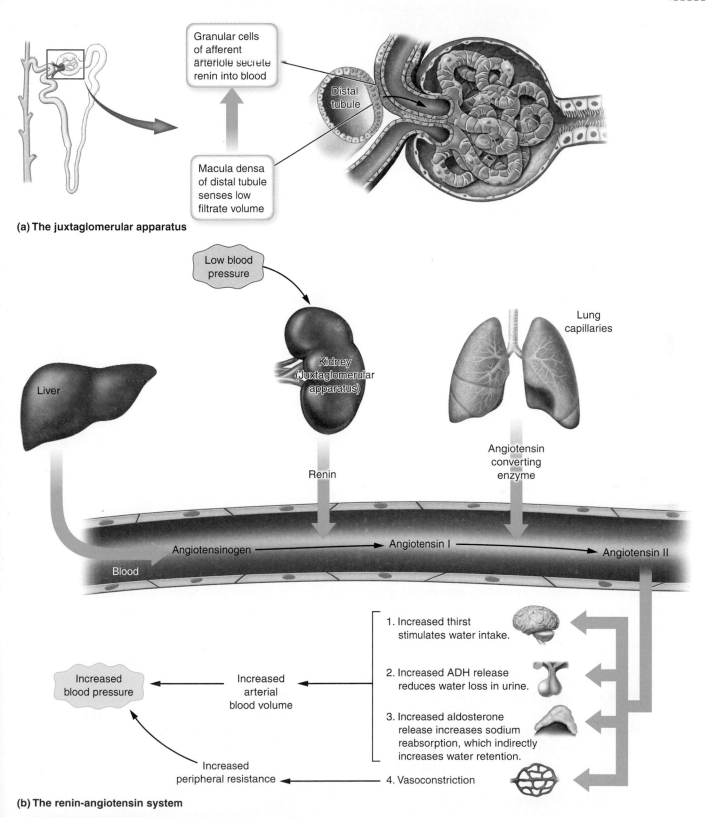

(a) The juxtaglomerular apparatus

Granular cells of afferent arteriole secrete renin into blood

Distal tubule

Macula densa of distal tubule senses low filtrate volume

Low blood pressure

Liver

Kidney (Juxtaglomerular apparatus)

Lung capillaries

Renin

Angiotensin converting enzyme

Angiotensinogen → Angiotensin I → Angiotensin II

Blood

1. Increased thirst stimulates water intake.

2. Increased ADH release reduces water loss in urine.

3. Increased aldosterone release increases sodium reabsorption, which indirectly increases water retention.

4. Vasoconstriction

Increased arterial blood volume

Increased blood pressure

Increased peripheral resistance

(b) The renin-angiotensin system

Figure 16.13. Renal blood pressure regulation. A. The granular cells of the juxtaglomerular apparatus secrete renin when filtrate volume and afferent arteriolar pressure are low. **B.** Renin initiates angiotensin II production. Angiotensin II acts at multiple sites to stimulate blood pressure. *Angiotensin II stimulates the production of an adrenocortical hormone. Which one?*

Case Discussion
The Case of Santo G.: Part 1

Those labouring with this Disease, [urinate] a great deal more than they drink, or take of any liquid aliment; and moreover they have always joyned with it continual thirst…. The Urine in all…was wonderfully sweet as it were imbued with Honey or Sugar….

Thomas Willis, 1679, writing of diabetes

When we first met Santo in Chapter 15, we focused on his fundamental metabolic defect: insufficient insulin action. When he failed to take his insulin, Santo's liver, muscle, and fat cells could not take up glucose, so glucose remained in his blood. The liver inadvertently worsened the situation, since it synthesized more glucose molecules from amino acids and converted fatty acids into an alternate energy source called *ketones*. Ketones can be used by the brain for energy, but they are weak acids and thus alter blood pH (see ◀ Chapter 2). As a result, Santo's blood glucose concentration rose very high and his blood pH became very acidic. The key lesson in ◀ Chapter 15 was that the administration of insulin allowed glucose to enter cells, where it could be metabolized. Although other short-term therapy was necessary, no treatment would have been successful if the underlying metabolic defect had not been corrected.

In this chapter we focus on two life-threatening problems that arose from Santo's lack of insulin:

● Water and solute balance problems associated with very high blood glucose concentration
● Problems of acid–base balance associated with ketone generation by the liver

In this part of the case study, we focus on water and solute balance. As shown in Figure 16.14:

1. Santo's high blood osmolarity (from the glucose and ketones) drew water out of his cells, brain cells for instance, resulting in shrunken cells that did not function normally. The effect on his brain accounts for his thirst, abnormal mental state, and abnormally high temperature.
2. Glucose passes freely into the glomerular filtrate and tubular fluid. Recall that the proximal tubules normally reabsorb all glucose from the glomerular filtrate. However, the glucose concentration in Santo's tubular fluid was so high that the proximal tubules could not

reabsorb it all. Glucose remained in the filtrate, increasing its osmotic pressure. As a result, water remained in the filtrate and was not reabsorbed into peritubular blood.
3. As a result of the increased volume of tubular fluid, he produced a high volume of watery, pale urine with high specific gravity (because of the glucose it contained). This robbed him of precious body water and caused his blood volume and blood pressure to fall.
4. Santo's low blood volume lowered his blood pressure. Although we do not have the relevant laboratory data, we can be sure that before his arrival at the hospital, Santo's low blood pressure triggered homeostatic reactions to increase blood volume, and with it blood pressure. It is certain that blood concentrations of ATII, aldosterone, and ADH were increased and ANP decreased. However, these compensations were unable to fully restore his blood pressure, which remained low upon admission.
5. The intravenous administration of normal saline increased body water and blood volume, which helped the underlying homeostatic reactions to boost blood pressure.

However, insulin administration was a critical intervention—it corrected the underlying metabolic defect.

Acid–Base Balance

As you learned in Chapter 2, an **acid** is a compound that releases hydrogen ions when dissolved in water. This increases the number of free hydrogen ions in the solution. Also recall that pH is a numerical expression of the degree of acidity or alkalinity of a solution: pH 7 is neutral, pH less than 7 is acidic, and pH greater than 7 is alkaline. The lower the pH, the more H^+ ions are in solution. For example, HCl is a **strong acid** because it completely dissociates into H^+ and Cl^- when dissolved in water (Fig. 16.15A). A **weak acid,** such as the acetic acid (HAc; Ac = acetate) in vinegar, holds onto its H^+ much more avidly, and is therefore less strongly acidic. Only a few HAc molecules dissociate into H^+ and Ac (Fig. 16.15B).

A **base** is a compound that decreases the number of free H^+ ions, usually by releasing hydroxide (OH^-) ions, which combine with H^+ to form H_2O (water). As the concentration of OH^- ions rises, the concentration of H^+ ions falls, and the solution becomes more basic. Normal arterial blood is slightly basic: its pH is homeostatically

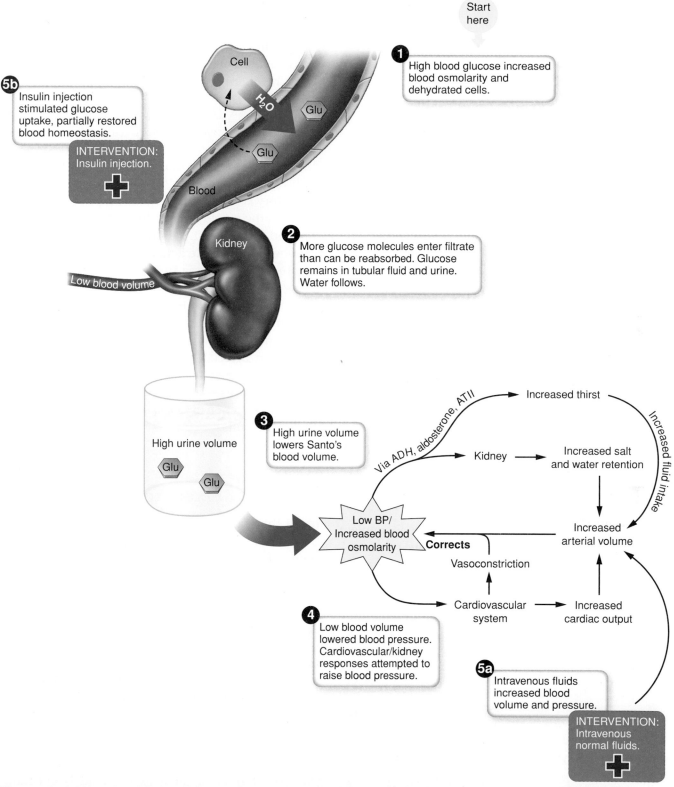

Figure 16.14. The case of Santo G: Part 1. Santo's high blood osmolarity and low blood pressure activated both cardiovascular and renal responses. He received two medical interventions: as shown in step 5a, intravenous fluids improved his blood pressure; as shown in step 5b, insulin injections addressed his insulin deficiency. Blood ketones, not illustrated, would have also contributed to high blood osmolarity and urine osmolarity. *Why did Santo's brain cells shrink, contributing to his mental confusion?*

(a) A strong acid (hydrochloric acid)   (b) A weak acid (acetic acid)

Figure 16.15. Strong and weak acids. A. HCl is a strong acid. It liberates all of its H$^+$ ions into water. **B.** Acetic acid (HAc) is a weak acid. It liberates only a few of its H$^+$ ions into water. *What is Ac?*

The Lungs Dispose of Carbon Dioxide

Carbon dioxide (CO_2) dissolved in water produces a weakly acidic solution by liberating H$^+$ ions in the following reaction:

$$CO_2 \quad + H_2O \Rightarrow \quad H_2CO_3 \quad \Rightarrow \quad HCO_3^- \Rightarrow H^+$$

Carbon dioxide + water $\Rightarrow$ carbonic acid $\Rightarrow$ bicarbonate + acid

This reaction occurs in blood as well as in cerebrospinal fluid. It rarely, however, causes acidosis, because the respiratory center of the medulla responds to increased acidity in the cerebrospinal fluid by increasing respiration and exhaling the extra CO_2. Carbon dioxide is described as a **volatile acid** because it can be exhaled in the breath. Alterations in tissue CO_2 production—say, during moderate exercise—are thus matched with respiration, and arterial pH stays relatively constant. But if lung function is significantly impaired, ventilation cannot adequately rid the body of CO_2. In such cases, it accumulates in arterial blood, and **respiratory acidosis** results. Conversely, if respiration is greater than CO_2 production requires—say, during voluntary hyperventilation—the reduction in CO_2 can raise pH and **respiratory alkalosis** results.

Acids Other Than Carbon Dioxide Are Called Fixed Acids

Recall that carbon dioxide is a volatile acid: it can be exhaled. Acids that cannot be exhaled are described as **fixed acids.** They include ingested amino acids (especially from meat) and ketones produced from fatty acid metabolism. Moreover, the reactions that generate and use ATP during intense exercise also release hydrogen ions that contribute to blood acidity. Abnormally high blood acidity resulting from fixed acids is called **metabolic acidosis.** Three systems compensate for, and eventually correct, an increase in blood acidity: buffers, the respiratory system, and the renal system (Fig. 16.16).

controlled in a very narrow range between 7.35 and 7.45. Abnormally low blood pH is **acidosis;** abnormally high is **alkalosis.**

The general principle of acid–base balance is simple: the kidneys excrete as much acid or base as we consume or generate metabolically. This principle is complicated somewhat by the ability of the respiratory system to exhale carbon dioxide, which, when dissolved in blood, acts as a weak acid. So before we discuss the kidneys' role in acid–base balance, let us return briefly to the respiratory system.

Case Notes

16.18. Santo's blood pH is 7.12. Does his blood contain more free hydrogen ions than normal blood, or fewer?

16.19. One of the ketone bodies detected in Santo's urine is called acetoacetic acid. Acetoacetic acid liberates only a small number of its H$^+$ ions. Is this ketone body a strong acid or a weak acid?

Buffer Systems Compensate for Fixed Acids

The first defense against excess acid is specialized substances called *buffers.* In physiology, a **buffer** is any substance that quickly acts to restrain a change in pH following the addition of an acid or base. All buffers in all body fluids bind newly introduced hydrogen ions and convert them into weak acids (Fig. 16.16 step 2). For

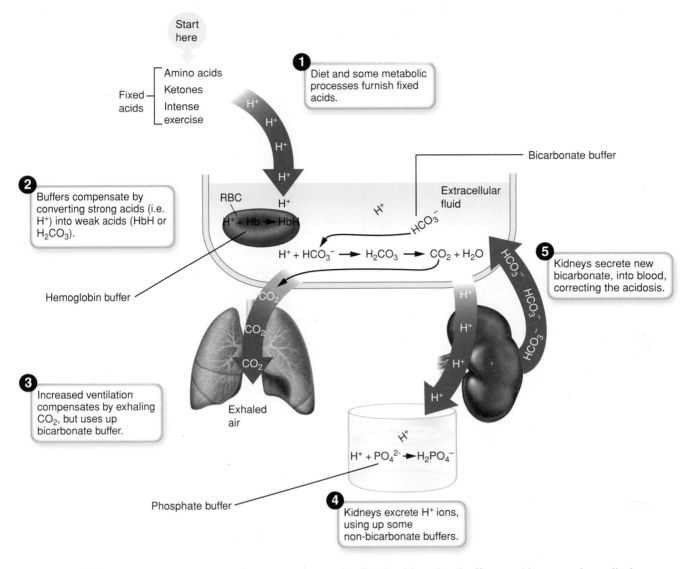

Figure 16.16. Metabolic acidosis. The body compensates for fixed acids using buffers and increased ventilation, but renal actions are required to correct the pH imbalance. *Where are new bicarbonate molecules generated?*

example, some hydrogen ions in urine are buffered by this reaction:

$$\underset{\text{Strong acid}}{H^+} \quad + \quad \underset{\text{buffer}}{HPO_4^{2-}} \quad \Rightarrow \quad \underset{\text{weak acid}}{H_2PO_4^-}$$

$H_2PO_4^-$ is a weak acid, so this reaction will decrease the acidity of the solution. This reaction is controlled by the concentration of the reagents, so the gain of a buffer drives the reaction to the right and decreases blood acidity. Conversely, the loss of a buffer drives the reaction to the left, liberating a hydrogen ion and increasing blood acidity. NH_3 (ammonia) also acts as a renal buffer, reacting with protons to form NH_4^+ (ammonium, a weak acid).

Hydrogen ions in blood cells are similarly buffered when they bind to hemoglobin, and hydrogen ions in other body cells bind to cellular proteins. But many hydrogen ions will find their way into extracellular fluid and blood plasma, where they meet with the most important buffer of all: the **bicarbonate buffering system.** It combines H^+ with bicarbonate (HCO_3^-) to generate the weak acid *carbonic acid,* as shown by this equation:

$$H^+ + \quad HCO_3^- \Rightarrow \quad H_2CO_3$$
Acid + bicarbonate ⇒ carbonic acid

All of the body's buffer systems work together to limit the impact of a change in acid concentration. However, chemical buffers are only a short-term solution, because they do not actually eliminate the added acid. Moreover, every buffering reaction uses up a buffer molecule, and buffer stores are not unlimited. The loss of buffers, such as the loss of bicarbonate that occurs with diarrhea, can cause metabolic acidosis as effectively as the gain of acid. Thus, buffers help *compensate* for the acid load but do not *correct* it.

Case Note

16.20. Santo was administered intravenous bicarbonate. Would this medical intervention have increased or decreased the production of carbonic acid, and what would have been the effect on his blood pH?

The Respiratory System Compensates for Fixed Acids

The respiratory system, like buffers, can compensate for metabolic acidosis (Fig. 16.16, step 3). The bicarbonate buffer reaction mentioned above can proceed to the production of CO_2, as shown below:

$$H^+ + HCO_3^- \Rightarrow H_2CO_3 \Rightarrow H_2O + CO_2$$

Acid + bicarbonate $\Rightarrow$ carbonic acid $\Rightarrow$ water + carbon dioxide

As you can see, the addition of H^+ from fixed acids increases blood CO_2. This, in turn, stimulates the rate and depth of breathing (Chapter 13), which eliminates the added CO_2. In other words, the body responds to metabolic acidosis by exhaling the newly generated CO_2. Although this response tends to resist acidosis, it results in the permanent loss of a bicarbonate molecule

and lessens reserve buffering capacity. In other words, for every molecule of CO_2 lost through the lungs, a molecule of HCO_3^- is removed from the body's pool of HCO_3^- buffer. Thus the respiratory response compensates for the acid load but cannot correct it because the fixed acids, which created the problem, have not been eliminated.

The Kidney Eliminates Fixed Acids

Buffers and the respiratory system essentially "hold the fort" in response to an acid load, preventing big changes in blood pH while the kidney acts to excrete acid in urine (Fig. 16.16, step 4) and generate new bicarbonate molecules (Figure 16.16, step 5).

Let us take a closer look at these two actions (Fig. 16.17):

1. Tubule cells in the kidney combine CO_2 and water to generate bicarbonate (HCO_3^-) and hydrogen ions (H^+).
2. The newly generated bicarbonate enters blood, where it is available to buffer hydrogen ions in plasma.
3. The H^+ is secreted into tubular fluid and passes into urine.

If the H^+ ions could be excreted "as is," the net result of these steps would be the loss of one H^+ and the gain of one buffer. However, excessively acidic urine could damage the urinary tract. Therefore most of these H^+ must be buffered, using HPO_4^{2-} or ammonia as buffers. Thus the loss of many H^+ ions is offset by the concomitant

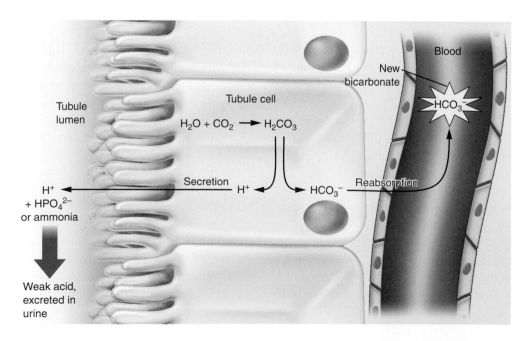

Figure 16.17. Renal response to acidity. The kidney responds to an increase in blood acidity by excreting acid urine and by secreting more bicarbonate into blood. *Does the bicarbonate synthesized by tubule cells enter blood or the tubular lumen?*

loss of buffers. Nevertheless, a patient with an unusually large acid load (such as Santo) may have a urine pH in the range of 4.0 to 5.0 instead of the normal average of 6.0, indicating that some hydrogen ions are lost unbuffered. Note that bicarbonate is *not* used to buffer urinary acid, because kidney-generated bicarbonate enters blood to replenish depleted bicarbonate stores.

> *Remember This!* **The most important renal mechanism to correct an acid load is the generation of new bicarbonate (buffer) molecules.**

Metabolic Alkalosis Is Relatively Rare

Metabolic alkalosis is much less common than metabolic acidosis. It results from the loss of acids, such as the loss of gastric acid following repeated vomiting, or the gain of bases, such as the ingestion of large amounts of antacids. In response to the rise in pH, three events occur:

1. Buffers tend to release their hydrogen ions. For instance, hemoglobin releases some hydrogen ions.
2. Respiration slows in order to retain more CO_2, which reacts to produce H^+ and bicarbonate.
3. Most importantly, the kidney excretes bicarbonate. As a result, hydrogen ions in blood remain unbuffered and pH decreases.

16.31 When citric acid is dissolved in water to keep cut fruit fresh, only a small proportion of the hydrogen ions dissociate from the citric acid molecules. Is citric acid a strong acid or a weak acid?

16.32 The ammonia buffer system involves two compounds: NH_3 (ammonia) and NH_4^+ (ammonium ion). Which compound is a buffer and which a weak acid?

16.33 What is the difference between respiratory acidosis and metabolic acidosis?

16.34 If five ketone molecules result in the production of five extra carbon dioxide molecules that are exhaled by the respiratory system, how many bicarbonate molecules will be lost?

16.35 If five molecules of buffered hydrogen are removed from the body by the kidneys, how many new bicarbonate molecules will be generated?

16.36 If a urine sample contains a high concentration of bicarbonate, which acid–base disorder would be suspected—acidosis or alkalosis?

Case Discussion
The Case of Santo G.: Part 2

In Chapter 15 we discussed Santo's fundamental metabolic defect, lack of insulin, as well as its consequences and its correction. Earlier in this chapter we covered his water and electrolyte problems. Now let us discuss his acid-base problems (Fig. 16.18).

1. Recall that Santo's liver responded to his cells' lack of glucose by generating ketones. Ketones are weak metabolic acids, so some ketone molecules gave up hydrogen ions, acidifying his blood. Thus, Santo was admitted to the emergency room in a state of *metabolic acidosis*. Some of the new hydrogen ions reacted with bicarbonate, producing carbon dioxide and depleting valuable bicarbonate buffer.
2. The respiratory center responded to the low pH (acidosis) by increasing the frequency and depth of respirations, which blew off CO_2. This *respiratory compensation* was so dramatic that it actually lowered blood CO_2 concentrations below normal, to 17 mmHg (normal 32 to 48), and kept blood pH from becoming even more acidic.
3. The kidneys responded by excreting free acid (both hydrogen ions and ketone molecules). The ketones and H^+ ions appeared in Santo's urine, and his urinary pH became quite acidic (4.1). At the same time the kidneys were generating bicarbonate to sustain the bicarbonate buffer reserve.
4. However, as shown by his low blood pH on admission, his kidneys and lungs together had not fully compensated for the acid overload, and medical intervention was required in the form of intravenous bicarbonate administration.

The intervention was successful: he responded quickly to the bicarbonate. All blood and urine abnormalities except urine protein began to return toward normal, his blood pressure improved, and his rapid breathing returned to normal. The proteinuria did not stop because it was likely due to long-term diabetic damage to the glomerular membrane, which allowed protein leakage into the glomerular filtrate. Filtered protein remained urine because renal tubules cannot resorb protein.

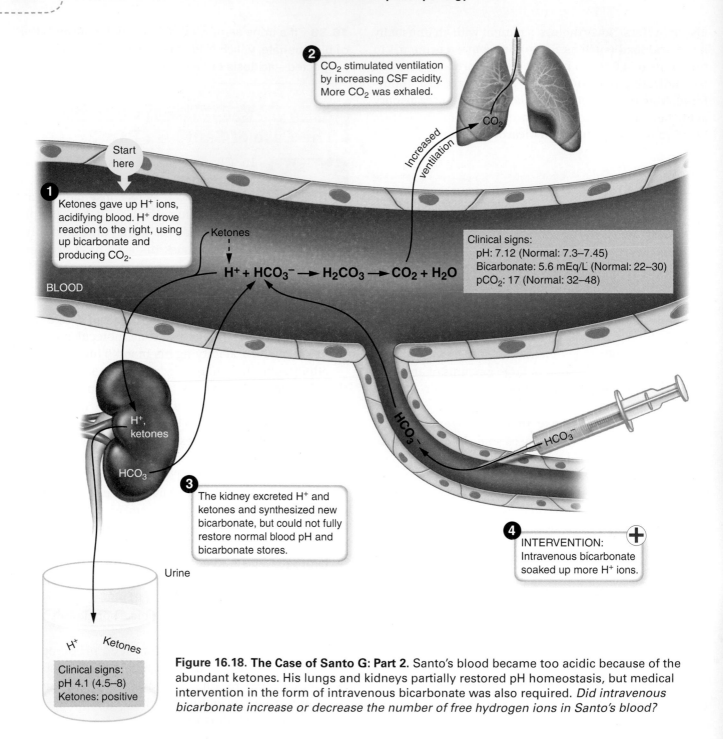

1 Ketones gave up H+ ions, acidifying blood. H+ drove reaction to the right, using up bicarbonate and producing CO2.

2 CO2 stimulated ventilation by increasing CSF acidity. More CO2 was exhaled.

Increased ventilation

Start here

Ketones

BLOOD

$$H^+ + HCO_3^- \longrightarrow H_2CO_3 \longrightarrow CO_2 + H_2O$$

Clinical signs:
pH: 7.12 (Normal: 7.3–7.45)
Bicarbonate: 5.6 mEq/L (Normal: 22–30)
pCO2: 17 (Normal: 32–48)

H+, ketones

HCO3

3 The kidney excreted H+ and ketones and synthesized new bicarbonate, but could not fully restore normal blood pH and bicarbonate stores.

Urine

4 INTERVENTION: Intravenous bicarbonate soaked up more H+ ions.

HCO3

H+   Ketones

Clinical signs:
pH 4.1 (4.5–8)
Ketones: positive

Figure 16.18. The Case of Santo G: Part 2. Santo's blood became too acidic because of the abundant ketones. His lungs and kidneys partially restored pH homeostasis, but medical intervention in the form of intravenous bicarbonate was also required. *Did intravenous bicarbonate increase or decrease the number of free hydrogen ions in Santo's blood?*

Word Parts

Word Part	English Equivalent	Example
cortic-, cortico-	Covering	Renal cortex: the outer, covering layer of the kidney
extra-	Outside	Extracellular: fluid outside cells
intra-	Inside	Intracellular: fluid inside cells
juxta-	Next to	Juxtaglomerular: next to the glomerulus
nephr/o	Kidney	Nephron: functional unit of the kidney
-o/sis	Condition	Acidosis: condition produced by low blood pH
papillo-	Nipple-like	Renal papillae: nipple-like extensions of the renal medulla into the renal pelvis
ren-/reno-	Kidney	Renal: referring to the kidney
-uria	Urine	Polyuria: large amount (poly-) of urine

Chapter Challenge

CHAPTER RECALL

1. **Which of the following is not an extracellular fluid?**
   a. Interstitial fluid
   b. Cytosol
   c. Blood plasma
   d. Both A and B

2. **Salts**
   a. have a positive charge.
   b. are acidic.
   c. are uncharged.
   d. are basic.

3. **The renal pelvis drains directly into the**
   a. urinary calyces.
   b. urethra.
   c. bladder.
   d. ureter.

4. **The ureters enter the bladder**
   a. on its superior surface.
   b. on its inferior surface.
   c. on its lateral surface.
   d. on none of the above; it varies from person to person.

5. **During urination**
   a. the detrusor muscle contracts.
   b. the internal urethral sphincter contracts.
   c. the external urethral sphincter contracts.
   d. all of the above.

6. **The kidneys are located**
   a. posterior to the peritoneum.
   b. superior to the diaphragm.
   c. inferior to the ribs.
   d. posterior to the vertebrae.

7. **In order, tubular fluid passes through the**
   a. collecting duct, proximal tubule, distal tubule, nephron loop.
   b. proximal tubule, distal tubule, nephron loop, collecting duct.
   c. proximal tubule, nephron loop, distal tubule, collecting duct.
   d. nephron loop, proximal tubule, collecting duct, distal tubule.

8. **In order, blood passes through the**
   a. afferent arteriole, glomerulus, efferent arteriole, peritubular capillaries.
   b. glomerulus, afferent arteriole, efferent arteriole, peritubular capillaries.
   c. afferent arteriole, peritubular capillaries, efferent arteriole, glomerulus.
   d. efferent arteriole, glomerulus, afferent arteriole, peritubular capillaries.

9. **A mystery solute can be detected in tubular fluid, but not in urine. The solute must be**
   a. secreted only.
   b. filtered only.
   c. filtered and reabsorbed.
   d. reabsorbed but not filtered.

10. **Which of the following problems would increase the glomerular filtration rate?**
    a. A blockage in a ureter, which increases the hydrostatic pressure of the filtrate.
    b. A decreased concentration of blood proteins, resulting from dietary protein. deficiency.
    c. Reduced blood pressure due to hemorrhage.
    d. All of the above.

11. **Glucose is reabsorbed in the proximal tubule by**
    a. primary active transport.
    b. secondary active transport.
    c. osmosis.
    d. facilitated diffusion.

12. **Aldosterone stimulates**
    a. potassium secretion and sodium reabsorption.
    b. calcium reabsorption and phosphate secretion.
    c. sodium secretion and potassium reabsorption.
    d. phosphate reabsorption and calcium secretion.

13. **The medullary osmotic gradient exists between**
    a. the glomerulus and the glomerular capsule.
    b. successive layers of the renal medulla.
    c. the peritubular capillaries and the nephron loop of cortical nephrons.
    d. the proximal tubule and the distal tubule.

14. **Antidiuretic hormone**
    a. stimulates sodium reabsorption in the collecting duct.
    b. increases the magnitude of the medullary osmotic gradient.
    c. increases the permeability of collecting duct's cell membranes to water.
    d. increases the volume of urine produced.

15. **The hypothalamic thirst center is stimulated by increased**
    a. blood pressure.
    b. blood water content.
    c. blood osmolarity.
    d. all of the above.

16. **Renin is produced by**
    a. granular cells.
    b. macula densa cells.
    c. glomerular cells.
    d. endothelial cells.

17. **Angiotensin II stimulates**
    a. aldosterone release.
    b. ADH release.
    c. thirst.
    d. all of the above.

18. **Atrial natriuretic peptide opposes the effects of**
    a. angiotensin II.
    b. aldosterone.
    c. ADH.
    d. all of the above.

19. **An example of a weak acid is**
    a. HCl.
    b. $NH4^+$.
    c. NaOH.
    d. bicarbonate (HCO_3^-).

20. **Urine does not usually contain**
    a. glucose.
    b. sodium.
    c. potassium.
    d. urea.

21. **The presence of acidic urine indicates that**
    a. the kidney is generating new bicarbonate molecules.
    b. the body is in a state of alkalosis.
    c. blood bicarbonate concentrations are high.
    d. all of the above.

CONCEPTUAL UNDERSTANDING

22. **Compare and contrast the following terms:**
    a. Urethra and ureter
    b. Glomerular capsule and glomerulus
    c. Afferent and efferent arteriole

23. **Renin is a chemical signal that circulates in blood, but it does not bind to a receptor. Explain how renin indirectly increases blood pressure.**

APPLICATION

24. **Ouabain is a poison that prevents the Na^+/K^+-ATPase from functioning. Ouabain also decreases the reabsorption of potassium in the proximal tubule, which occurs by facilitated diffusion. Explain how.**

25. **Young Evan is very angry because he is not allowed to whack his sister Lauren with a Tinkertoy. He holds his breath to punish his parents for their unfairness. What will happen to Evan's**
    a. blood carbon dioxide concentration.
    b. blood pH.
    c. blood bicarbonate concentration.

    Use a chemical equation to illustrate your answer.

You can find the answers to these questions on the student Web site at
http://thepoint.lww.com/McConnellandHull

17

The Reproductive System

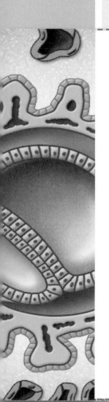

Major Themes

- Reproduction is the foremost biological purpose of life.

- Humans reproduce sexually: a male and a female each contribute genetic material.

- Sperm and ova (gametes) are produced by special organs— testes and ovaries (gonads)—devoted to the purpose.

- Both male and female genitalia comprise a system of tubes designed to bring sperm and ovum together.

- Sperm and ova each contain half the number of chromosomes found in all other (somatic) cells.

- Reproductive cycles are regulated by hypothalamic and anterior pituitary hormones.

- Conception, implantation, and embryonic development must progress perfectly to produce a normal infant.

- Pregnancy is a stress on the form and function of every body system.

Chapter Objectives

"We can't get pregnant."

As you read through the following case study, assemble a list of the terms and concepts you must learn in order to understand it.

Clinical history: Susan, age 28, and her husband Mark, age 31, went to see a fertility specialist because of their inability to conceive after 4 years of marriage.

"We can't get pregnant," Susan said. "I don't understand it. My periods are regular as clockwork, and we follow my temperature every day so we can have sex at exactly the right time, but it never works."

Each was also interviewed privately. In her interview, Susan confessed that during high school and college she suffered several episodes of what she called "tube infections." "I know now that those were sexually transmitted infections," she said. "I thought it was about the equivalent of a chest cold—a little penicillin and it was gone. But I know better now."

Mark's interview was unremarkable.

Physical examination and other data: Both were physically fit, with normal vital signs and unremarkable general laboratory tests.

Mark's genital examination was normal and he was able to produce a semen specimen. Semen volume, sperm count, sperm morphology and motility were normal.

Susan's examination, however, was not normal. During a manual examination of her pelvis, a slightly tender mass was detected in the region of her left ovary. Imaging studies confirmed the presence of a mass. In a special imaging study (hysterosalpingogram), dye visible by x-ray was injected into her uterus with enough pressure to force it through the uterine (fallopian) tubes. Both tubes were blocked a few centimeters from the uterus.

Clinical course: Exploratory surgery was performed on Susan. Surgeons found that her left fallopian tube and ovary had been destroyed by an old abscess and scar tissue. Repair was not possible and both tube and ovary were removed. On the right side, however, the tube, blocked by a narrow band of scar tissue, was otherwise normal. The scarred area was excised and the open ends brought together and joined. A postoperative study revealed that the tube was no longer blocked and appeared to be functional.

With further fertility advice and support, Susan became pregnant about 6 months later. The first two trimesters were uneventful. However, at the end of the 33rd week of pregnancy, she went into premature labor. Mark rushed her to the hospital, where a pelvic exam showed that her cervix had dilated (opened) to 4 cm. She was administered ibuprofen (an anti-inflammatory drug) and oxytocin blockers, and uterine contractions were quieted for a week. Then labor began again. Anticipating an inevitable premature birth, the obstetrician ordered administration of cortisol to accelerate fetal lung maturity. Susan's cervix continued to dilate, reaching 10 cm at the end of the 34th week, and she delivered a 2,155 g (4 lb, 12 oz) boy, who was named Toby. He was able to breathe on his own but had difficulties feeding. Susan used a breast pump to extract her milk, which was subsequently fed directly into Toby's stomach through a tube. After 7 days in neonatal intensive care, Toby was feeding well, and mother and child went home.

I don't know the question, but sex is definitely the answer.

Anonymous

The quotation above is funny because it slyly relies on a supreme biological truth—the primary goal of every species of life is to reproduce. So, from an evolutionary standpoint, sex is always the answer. Some species reproduce *asexually*; for example, bacteria simply halve themselves into two new organisms. However, our species, *Homo sapiens*, is perpetuated by *sexual reproduction*. **Sexual reproduction** requires that a species have two representatives, each of which supplies genetic material to create new offspring. In biology, the term **male** (from Latin *mas*) applies to the "giver" partner; the term **female** (from Latin *femina*) applies to the "receiver" partner.

That there can be many possible givers and receivers has a profoundly important genetic effect: it ensures *genetic diversity*, which in turn promotes good health in a population. Genetic *uniformity* is dangerous because it risks combining genetic defects which do not manifest as disease unless they are paired with a matching genetic defect. Matching defects are more likely to occur in close family members, which accounts for the social prohibition in most societies against procreation between close relatives.

We begin this chapter with a discussion of male reproductive anatomy and function, which is somewhat less complex than that of females.

Anatomy of the Male Reproductive System

The **male reproductive system** exists to manufacture sperm cells and deliver them to the female. **Sperm** (also called *spermatozoa*) are reproductive cells that carry the male's half of the genetic endowment. They are produced in the testes, the male *gonads* (reproductive glands). Sperm are stored, mature in, and conveyed to their destination by a ductal system, which is where we begin our exploration of male reproductive anatomy.

A System of Ducts Carries Sperm Outside the Body

The testes deliver immature sperm into the **epididymis,** a narrow tube about 20 ft (approximately 6 m) in length that is coiled into a 3- to 4-cm comma-shaped organ on the posterior aspect of each testis (Fig. 17.1). Lining the epididymis are tall columnar cells capped with large cilia that slowly sweep sperm through its length. During their 3-week journey through the epididymis, sperm gain the ability to swim. They remain in the epididymis until ejaculation, when they are forcefully ejected into the next ductal structure (the *ductus deferens*) by contraction of smooth muscle in the wall of the epididymis. Note that sperm are ejaculated from the epididymis, not from the testis.

The **ductus deferens** (or *vas deferens*; from Latin *vas* = "vessel" or "duct," and *deferens* = "carry away") receives sperm from the epididymis. It is a very firm muscular tube about 3 mm in diameter and 45 cm (18 in.) long, which runs from the epididymis superiorly. In the lower abdominal wall, the ductus loops over the anterior brim of the bony pelvis and descends into the soft tissues of the pelvis, traveling medially to the ureter. It then bends anteriorly to run along the underside of the bladder toward the prostate. Immediately before penetrating the prostate, it is joined by the duct of a *seminal vesicle* (a gland discussed below). The merged ducts form the **ejaculatory duct,** which then penetrates the prostate to connect with the proximal end of the urethra.

Like the epididymis, the ductus deferens is lined by tall, columnar ciliated epithelium. The wall of the ductus

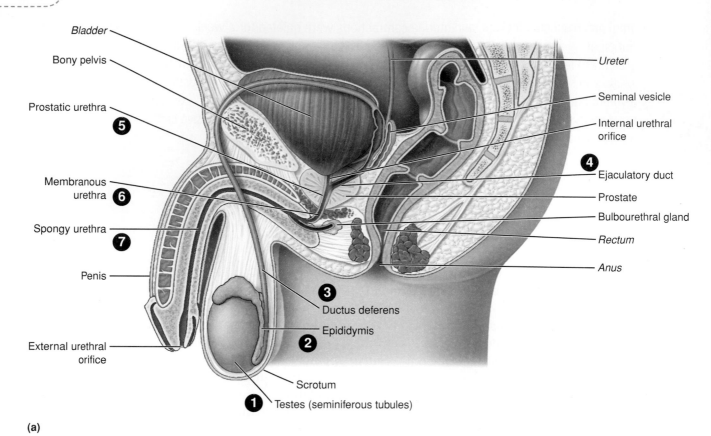

(a)

Figure 17.1. Male anatomy. This sagittal section illustrates the organs of the male reproductive system. The structures are numbered in the order in which they encounter sperm. *Into which duct does the bulbourethral gland drain—the ductus deferens or the urethra?*

is thick with bundles of smooth muscle, which contract rhythmically in very strong peristaltic waves to propel sperm and associated secretions into the urethra.

The **urethra** is the terminal male duct, which carries urine or semen to the outside. It originates at the *internal urethral orifice* at the neck of the bladder and terminates at the *external urethral orifice* on the tip of the penis. The most proximal segment is the *prostatic urethra*, which receives the ejaculatory duct. The next portion is the *membranous urethra*, which is embedded in a web of skeletal muscle fibers of the external urethral sphincter (◀ Chapter 16) and connective tissue known as the *urogenital diaphragm*. The distal segment is the *spongy (penile) urethra*, which is embedded in the spongy tissue of the penis. The prostatic urethra is lined by transitional epithelium, which gradually changes to pseudostratified columnar epithelium and, near the external opening, stratified squamous epithelium.

Case Note

17.1. When Mark was producing his semen sample, stored sperm left his epididymis. Name the next two tubes the sperm will encounter.

Accessory Sex Glands Secrete Fluid That Mixes with Sperm to Create Semen

Male accessory sex glands are the *seminal vesicles, prostate,* and *bulbourethral glands* (Fig. 17.1). Together they produce the bulk of semen.

Semen is a milky, slightly alkaline fluid, a mixture of sperm and accessory gland secretions that serves as a liquid transport and nutrient medium for the sperm. Semen forms when spermatozoa enter the ductus deferens and combine with accessory gland secretions. After delivery into the vagina, semen coagulates into a gel, which helps ensure that most of it remains close to the mouth of the cervix. Shortly, however, it liquefies, freeing sperm to begin their swim. What is more, the vagina is normally slightly acidic, and sperm do not swim well in acidic conditions; therefore the alkaline pH of semen neutralizes vaginal acidity and facilitates sperm motility.

The **seminal vesicles** are paired glands, each about the size of a little finger, which lie low on the postero-inferior surface of the bladder and anterior to the rectum. Like the epididymis, each is a coiled tube, but shorter, about 15 cm, and of greater diameter. They secrete **vesicular fluid**, a mildly alkaline fluid containing a mix of substances that nourish and empower sperm—fructose, vitamin C, and other substances. Vesicular fluid also contains an enzyme, **vesiculase**, which is responsible for semen's coagulation into a gel after it is expelled into the vagina. Vesicular fluid forms 60% of semen volume.

The **prostate gland** is a single gland about the size of a small plum (2 by 3 by 4 cm) that circles the proximal urethra like a doughnut immediately distal to the neck of the bladder (Fig. 17.1). It consists of numerous lobules of epithelial glands and ducts embedded in a dense stroma of fibrous tissue and smooth muscle. It secretes a mildly acidic, milky fluid containing a mixture of substances that nourish sperm, including citric acid, which sperm can use to generate ATP for energy. Included in prostatic secretions is **prostate-specific antigen (PSA),** the enzyme that slowly liquefies clotted semen and allows sperm to swim freely. Prostatic secretions account for about 25% to 35% of semen volume. For more information about PSA and prostate disease, see the nearby clinical snapshot titled *Prostate Disease*.

The **bulbourethral glands** are pea-sized glands, one on each side of the membranous urethra, that secrete a clear lubricant fluid into the urethra when a man becomes sexually excited. This fluid performs several functions. First, it is alkaline and neutralizes any urine, usually acidic, that remains in the urethra. Second, it makes its way down the length of the urethra, ensuring smooth flow of semen, and onto the head of the penis, providing lubrication to facilitate vaginal penetration.

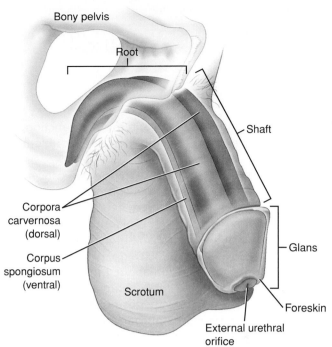

(a) The penis and scrotum

Case Notes

17.2. The liquefaction time (the time it takes for the semen gel to liquefy) was tested in Mark's semen sample. The activity of which enzyme is evaluated by this test?

17.3. Mark's ejaculate volume was 6 mL. About how much fluid did his seminal vesicles produce?

17.4. The pH of Mark's semen sample was normal: slightly alkaline, pH about 8. Would acidic semen indicate inadequate activity of the prostate gland or of the seminal vesicles?

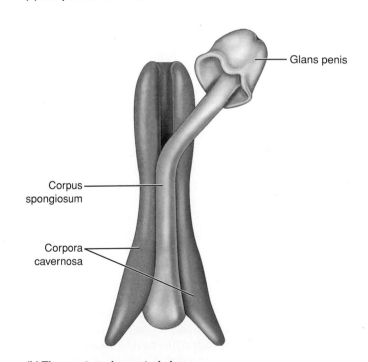

(b) The erect penis, ventral view

Figure 17.2. The penis. A. The shaft and the glans of the penis project outward from the lower torso, but the root remains within the pelvic cavity. **B.** A ventral view of the erect penis, with the corpus spongiosum displaced. *Does the foreskin (of an uncircumcised penis) cover the glans or the shaft?*

The Penis Is a Copulatory Organ

To have sex is to *copulate* (from Latin *copulare* = "to fasten together"), and the **penis** is the male copulatory organ.

It consists of three parts (Fig. 17.2A):

- The **root** is attached to the underside of the bony pelvis.
- The **shaft,** or body, is the long middle portion. It consists of three parallel cylindrical structures, two somewhat larger dorsolateral ones and a single smaller ventral one. Each dorsolateral cylinder is a

CLINICAL SNAPSHOT

Prostate Disease

Notwithstanding its small size and modest function, the prostate enjoys a king-size reputation as a troublemaker for men.

First and foremost, apart from minor skin cancers, *cancer of the prostate* is the most common malignancy of humans—even though women, half of the human species, do not have a prostate! This makes cancer of the prostate more common than lung, colon, or breast cancer. However, as malignancies go, prostate cancer is not very aggressive. Although about 1 in 6 men will develop it in their lifetimes, only about 1 in 30 will die of it. It grows slowly and metastasizes (spreads) late, so that even if a man gets prostate cancer, there is only a 10% chance that he will die of it. By contrast, over 90% of lung cancer patients die of their disease.

Principles that have long guided the diagnosis and treatment of prostate cancer are being reexamined after decades of settled agreement that early detection and treatment are key to survival. Although this is true for almost every other form of cancer, it is less so for prostate cancer because of the advanced age of most men who develop it and the sluggish nature of the malignancy—prostate cancer is much less likely to be fatal than other cancers. A very important consideration is that treatment (surgery, radiation) often affects sexual function and bladder control, serious problems that must be balanced against the benefit. Compounding the difficulty of the decision is that most prostate cancers occur in elderly men with limited life expectancy. A helpful tool in the debate about treatment of any disease is to calculate the number of patients who must be treated to save a life. In a perfect world it would be 1:1; that is, you save the life of every person you treat. For prostate cancer, you must treat about 50 to save one life, a ratio of 50:1. That means that about 49 men must suffer the serious consequences of treatment without gaining the intended life-saving benefit.

So, how is prostate cancer to be diagnosed? Malignant prostate tissue tends to be harder than normal prostate tissue, so insertion of a finger into the rectum to feel the prostate (digital rectal exam, or DRE) will detect some early cancers, although it misses many. Another diagnostic tool is measurement of the blood level of *prostate-specific antigen* (PSA), which increases with prostate cancer. But, as with DRE, the

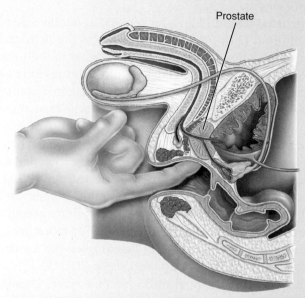

A digital rectal exam (DRE). Medical practitioners can check for prostate enlargement by this method.

results are often misleading: many times it is normal when cancer is present or high when cancer is absent. The entire problem is unsettled, so much so that the scientist who discovered PSA no longer recommends its use as a screening tool to diagnose early prostate cancer. A government panel now recommends PSA screening only for men 75 years of age or older. And an increasing number of physicians advocate no treatment for low-grade prostate cancer.

Another problem for some men is prostatic enlargement due to *benign prostatic hyperplasia*, a nonmalignant condition. Virtually every man over age 60 has an enlarged prostate, but in most it causes no problems. However, in about 10% of these the enlarged gland strangles the flow of urine, which slows the stream and prevents the bladder from emptying completely at urination. As a result, the bladder refills quickly and patients must urinate frequently, including getting up at night to urinate more than once (once is normal). Chronic prostatic obstruction of urine flow increases the risk of urinary infection and other complications.

corpus cavernosum (Latin *corpus* = "body," and *cavernosum* = "hollow") and is composed of a spongy mass of blood sinuses that can fill with blood to cause elongation and stiffening of the penis (*erection*, to be discussed further on). The single ventral cylinder is the **corpus spongiosum,** also composed of blood sinuses, which contains the urethra (Fig. 17.2B). Note that these directional terms refer to an erect penis pointing superiorly; the ventral surface is the *underside* of a nonerect penis, closest to the body.

● The **glans,** or head, is the enlarged distal portion of the corpus spongiosum (Fig. 17.2C). During an erection, the glans also stiffens and fills with blood. At the tip of the glans is the external urethral orifice.

The shaft of the penis is sheathed by an unstretchable fibrous fascia, which during erection limits expansion of the blood-filled penis, thereby ensuring adequate tension. The penis is covered by loose skin, which slides freely over the underlying fascia.

In its natural state, the penis has a collar of protective skin, the **foreskin** (or *prepuce*), which extends from the neck of the glans where it joins the shaft and folds distally to sheathe and cover the glans. According to social custom, the foreskin may or may not be cut away (*circumcision*; from Latin *circum* = "around," and *caedere* = "to cut"). The normal covering of the glans is a moist squamous mucosa. After circumcision, it is left unprotected and becomes the functional equivalent of skin.

Case Note

17.5. Mark and Susan decided against circumcision for their newborn boy. Which part of the penis does circumcision remove?

The Scrotum Contains the Testes

The **scrotum** is a pouch of skin inferior to the base of the penis that holds the testes (Fig. 17.2A). A midline septum divides the scrotum into right and left compartments.

Recall that the *ductus deferens* (vas deferens) travels superiorly from the scrotum in the lower, anterior abdominal wall before it loops over the anterior brim of the bony pelvis to descend. Between the scrotum and the pelvic brim it is bundled with nerves and blood vessels and sheathed with a thin layer of skeletal muscle fibers (the *cremaster muscle*). Collectively, this portion of the ductus deferens and associated tissues is called the **spermatic cord** (Fig. 17.3A).

Optimal temperature for sperm production is about 3°C lower than normal body temperature—approximately 34°C (93°F). For this reason, the testes hang outside the torso, away from core body temperature. Upon exposure to cold, the cremaster muscle fibers contract, bringing the testes nearer to the warmth of the body core. In warm conditions, the opposite occurs: the testicles hang lower, where they are better able to dissipate heat.

The **testes** (from Latin meaning "to testify," in the sense of "to prove," thus proof of maleness), or male gonads, are plum-sized ovoids that begin their embryonic existence high in the posterior abdominal wall and migrate into the scrotum by birth. Failure to migrate, *undescended testis (cryptorchidism)*, is an uncommon but serious problem: sterility is likely because the higher temperature in the abdomen suppresses sperm production, and undescended testes have a high risk for developing testicular malignancy.

A bilayered **tunica vaginalis** covers the testis, much like the bilayered pericardium covers the heart. In between the two layers is a potential space containing a small amount of slippery fluid, which allows each testis a limited degree of movement within the scrotal sac. Accumulation of fluid in this space is called *hydrocele* and can cause the scrotal contents to become alarmingly large, as if a tumor were present; however, the condition is usually innocuous.

A fibrous capsule called the **tunica albuginea** forms the outer surface of the testes. Septa extend from the tunica albuginea into the testis, dividing it into several hundred lobules. Coiled within each lobule are the sperm-producing **seminiferous tubules** (Fig. 17.3B). Interspersed in the supporting connective tissue between tubules are **Leydig cells** (*testicular interstitial cells*), which produce testosterone and other sex steroids. Lining the seminiferous tubules are two types of cells: (a) the *spermatogonia*, which are reproductive stem cells that give rise to sperm (see ◀ Chapter 3), and (b) support cells known as **Sertoli cells** (or *sustentacular cells*) that nourish maturing sperm. Sertoli cells are large, tall cells arrayed side by side; spermatogonia are smaller cells interspersed among them. Spermatogonia begin their maturation along the basement membrane of the seminiferous tubule. As they mature, they move toward the tubular lumen. Upon reaching the lumen, they are swept into the epididymis for storage and further maturation.

Case Note

17.6. Name the stem cells that gave rise to Mark's spermatozoa.

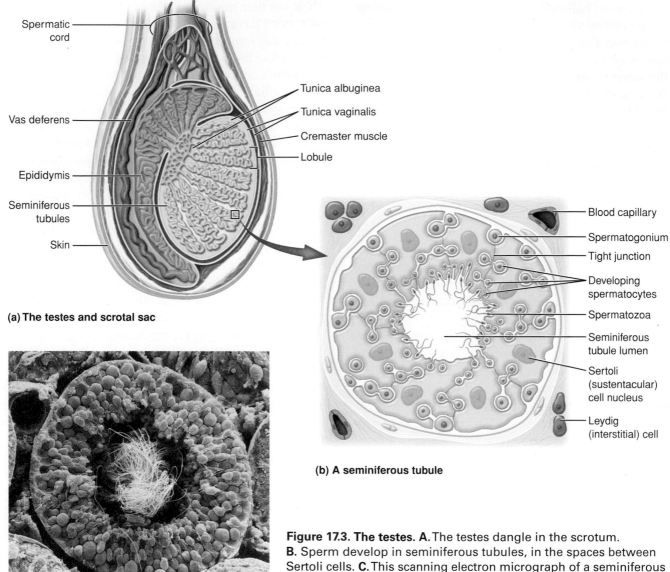

(a) The testes and scrotal sac

(b) A seminiferous tubule

(c) Cross-section of a seminiferous tubule

Figure 17.3. The testes. A. The testes dangle in the scrotum. **B.** Sperm develop in seminiferous tubules, in the spaces between Sertoli cells. **C.** This scanning electron micrograph of a seminiferous tubule shows spermatogonia at the tubule periphery and mature sperm with tails projecting into the lumen. *Name the steroid-producing cells of the testes.*

17.1   Name the three segments of the urethra, from the most proximal (to the bladder) to the most distal.

17.2   Indicate which of the following constituents of semen are nutrients and which are enzymes: vesiculase, citric acid, fructose, prostate-specific antigen.

17.3   Which part of the penis is dorsal—the corpus cavernosum or the corpus spongiosum?

17.4   What is the difference between spermatogonia and Sertoli cells?

Testicular Function

The testis performs two major functions. First is **spermatogenesis**—the production of sperm—which occurs in the seminiferous tubules. Second is production of the sex steroid **testosterone,** which is responsible for the development and maintenance of male secondary sex characteristics such as facial hair, heavier musculature, and deeper voice. Testosterone is produced by the Leydig (*interstitial*) cells.

Seminiferous Tubules Produce Sperm

Spermatogenesis produces sperm from spermatogonia. In doing so it halves the number of chromosomes from the normal 46 in all other body cells to 23. A reproductive cell whose chromosomal content has been halved is a **gamete.**

Recall from ◄ Chapter 3 that a gene is a segment of DNA that codes for a specific protein expressing a certain trait—eye color, for example—and that a chromosome is a distinct molecule of DNA containing many genes. Recall also that, with the exception of sex chromosomes (X, Y), all chromosomes occur as matching pairs, or *homologues*, one from each parent (although the sex chromosomes of a male are not identical, they contain a small homologous region and are thus considered homologues). Homologous chromosomes contain homologous (but not necessarily identical) genes. That is, you may have the brown-eyed gene from your mother and the blue-eyed gene from your father. Understanding homologous chromosomes and genes is important in understanding the details of spermatogenesis.

With the exception of ova (a female's egg cells) and sperm and their predecessors, *every* body cell contains 46 chromosomes—22 pairs (44) of nonsex (somatic) chromosomes (called *autosomes*), numbered 1 through 22, and one pair (2) of sex chromosomes. In females there are two copies of the X chromosome (designated XX) and in males one copy of the X and one copy of the Y chromosome (designated XY). The standard genetic notation for a genetically normal female is 46/XX; for a male it is 46/XY.

Notice that the spermatogonia from which sperm arise have 46 chromosomes. Because the 46 represent two sets of chromosomes, one from each parent, 46 is called the **diploid** set (Greek *diplous* = "double") and is conveniently designated as *2n* (n = 23). In producing a mature sperm, the number of chromosomes per cell must be halved from 46 to 23, a number called the **haploid** set (Greek *haplous* = "single") and designated *n*. The same process occurs in ova. Were this not the case, the uniting of sperm and ovum would double the number of chromosomes with each new generation—clearly an impossibility. The solution to this theoretical problem is **meiosis** (Greek *meion* = "less"), a form of cell division that halves the chromosome number from 46 in spermatogonia (and in ovarian stem cells) to 23 in gametes. Meiosis ensures that every normal sperm and ovum contains 22 nonsex chromosomes and one sex chromosome. The process of meiosis involves *two cycles of cell division*, which are called **meiosis I** and **meiosis II.**

Spermatogenesis begins when the stem cell, the **spermatogonium,** divides by normal *mitosis* into two daughter cells, each containing the normal complement of 46 chromosomes (Fig. 17.4, step 1). One of the daughter cells remains as a spermatogonium and the other differentiates into a **primary spermatocyte,** and each of these spermatocytes has a full diploid set of chromosomes.

In the next step, the primary spermatocyte duplicates the DNA in all 46 chromosomes; however, this new DNA does not separate from the old to form new chromosomes. Instead, all of the DNA remains together as a *replicated chromosome,* with the duplicates held together by a small protein disc called the *centromere*. The result is that a primary spermatocyte contains 46 chromosomes, each with twice the normal amount of DNA and two identical copies of each gene (step 2).

Next, the replicated chromosome homologues pair up along the cellular equator, one homologue on the northern side of the equator, the other on the southern side. And in these homologue pairings, the distribution across the cellular equator is random, so that there are some maternal and some paternal chromosomes on each side of the equator. For instance, your mother's chromosome number 16 may be on the north side of the equator and your dad's number 16 on the south, or vice versa. Then another unique thing occurs: the north and south chromosomes exchange *some* DNA before separating, a process called *crossing over*. As an example, imagine that homologues of replicated chromosome number 16 lined up across the cellular equator from one another. *Some* genes from your mother's and father's number 16 cross the equator to swap places; that is, a gene influencing eye color on mom's chromosome *may* change places with the matching eye color gene on dad's. As a result, your dad's eye color gene may end up in your mom's half of the DNA of your own sperm or ovum. This mixing of whole chromosomes and parts of chromosomes ensures genetic diversity, a point we touched on in the opening of this chapter when we discussed sexual versus nonsexual reproduction.

Finally, the replicated homologues move to the poles of their respective sides of the cellular equator (step 3). The cytoplasm then separates along the equator and cell division is complete. The new cells thus formed are called **secondary spermatocytes,** each with a haploid set of chromosomes. This brings us to the end of meiosis I: two cells, each with 23 *replicated* chromosomes (haploid, or *n*).

In *meiosis II,* the process is simpler: the 23 replicated chromosomes, each with enough DNA for two haploid cells, line up in the equator of the secondary spermatocyte. Then the replicated chromosomes finally split at the centromere: one half of the DNA moves to one pole of the cell, and the other half moves to the opposite pole. Following this, the cytoplasm divides through the equator and the second cell division is complete, the original diploid primary spermatocyte having produced four haploid spermatids (step 4).

In summary, meiosis accomplishes two important things. First, it halves the number of chromosomes in each mature sperm or ovum. Second, genetic variation is assured by the random alignment of homologous pairs on one side or another of the equator and the gene "trading" crossovers among paternal and maternal chromosomes; that is, brown eyes and big feet do not always have to be inherited together.

Finally, the spermatids mature by the process of *spermiogenesis* (Step 5). They lose almost all of their cyto-plasm and compact their DNA and mitochondria into a streamlined shape to become *spermatozoa* (sperm).

Each sperm has a head, a midpiece, and a tail (Fig. 17.4, bottom right). The head contains the nucleus, which stores the DNA; the midpiece is composed of mitochondria to provide energy; and the tail is a long flagellum that swishes

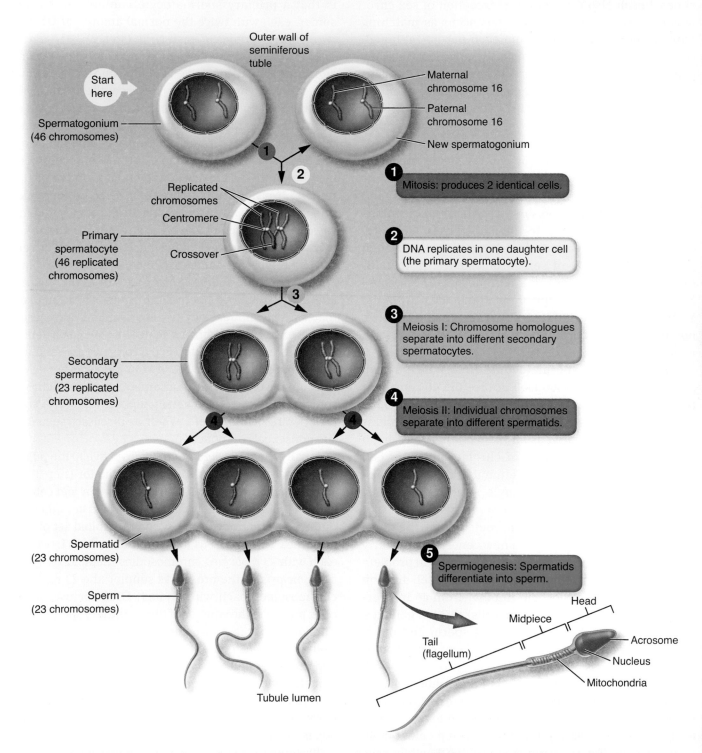

Figure 17.4. Spermatogenesis. The primary spermatocyte produces four spermatozoa by the process of meiosis. In the interests of space, only the fate of one pair of chromosomes is illustrated. *Which of the following cells contains replicated chromosomes: spermatogonia (nondividing), primary spermatocyte, secondary spermatocyte, spermatid?*

to and fro to propel the sperm forward like a snake. Sperm are streamlined to facilitate swimming and have virtually no cytoplasm to store nutrients for the arduous journey through the female reproductive tract. Instead, sperm mitochondria rely for nourishment on semen fructose and citrate and on nutrients they pick up along the way from the interior of the female genital tract.

Case Note

17.7. Will Mark's semen sample contain mostly haploid cells or mostly diploid cells?

Leydig Cells Produce Testosterone

The cells of the testis are influenced by a hormonal axis involving the hypothalamus and the anterior pituitary. The hypothalamus produces **gonadotropin-releasing hormone (GnRH),** which in turn stimulates production of two hormones from the anterior pituitary gland (Fig. 17.5):

● **Follicle-stimulating hormone (FSH)** is necessary for sperm production, but developing spermatocytes do not have FSH receptors. Instead, FSH acts indirectly by stimulating the production of sperm survival factors by Sertoli cells. These survival factors work

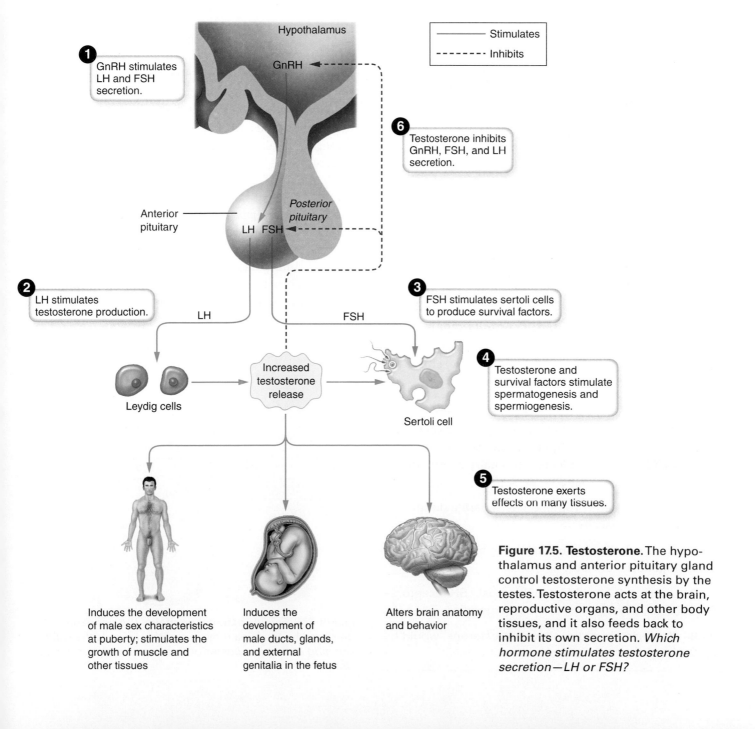

① GnRH stimulates LH and FSH secretion.

⑥ Testosterone inhibits GnRH, FSH, and LH secretion.

— Stimulates
----- Inhibits

Hypothalamus
GnRH

Anterior pituitary
LH   FSH
Posterior pituitary

② LH stimulates testosterone production.

③ FSH stimulates sertoli cells to produce survival factors.

LH

FSH

Leydig cells

Increased testosterone release

Sertoli cell

④ Testosterone and survival factors stimulate spermatogenesis and spermiogenesis.

⑤ Testosterone exerts effects on many tissues.

Induces the development of male sex characteristics at puberty; stimulates the growth of muscle and other tissues

Induces the development of male ducts, glands, and external genitalia in the fetus

Alters brain anatomy and behavior

Figure 17.5. Testosterone. The hypothalamus and anterior pituitary gland control testosterone synthesis by the testes. Testosterone acts at the brain, reproductive organs, and other body tissues, and it also feeds back to inhibit its own secretion. *Which hormone stimulates testosterone secretion—LH or FSH?*

together with testosterone (discussed shortly) to initiate and maintain spermatogenesis and spermiogenesis.

● **Luteinizing hormone (LH)** binds to Leydig cells and causes them to release testosterone.

LH and FSH were discovered first in women and are named for their ovarian effects.

> **Remember This!** Luteinizing hormone acts on <u>L</u>eydig cells.

A negative feedback loop regulates testicular hormone production. Testosterone and other gonadal factors act at the hypothalamus and pituitary to suppress the release of GnRH, LH, and FSH.

Testosterone stimulates the differentiation and growth of fetal internal and external genitalia. However, after sex organs develop in the fetus, hormone levels recede and do not surge again until puberty. As puberty nears in males, the hypothalamus becomes less sensitive to the suppressive feedback effect of testosterone, which allows more GnRH release. This results in a second surge of testosterone and development of specifically male characteristics known as *secondary sex characteristics*. The penis and testicles grow; genital and axillary hair appears; bones lengthen and become more robust; muscle mass increases; mature sperm appear in semen; skin thickens and becomes oilier; and the voice deepens as the larynx becomes larger and vocal cords lengthen.

Many of these masculinizing actions are mediated by testosterone derivatives produced within target cells. Behavior changes too, under the influence of testosterone. For instance, certain brain regions are larger in males than in females, and testosterone stimulates aggression and libido.

17.5 Which of the following cells are haploid? Spermatogonia, primary spermatocyte, secondary spermatocyte, spermatid

17.6 Do replicated chromosomes separate during meiosis I or meiosis II?

17.7 Does cell division occur during spermatogenesis, spermiogenesis, or both?

17.8 Which cells would have the most FSH receptors— Sertoli cells, Leydig cells, or spermatogonia?

17.9 If a male did not produce testosterone, would his GnRH levels be high, low, or normal?

Anatomy of the Female Reproductive System

The **female reproductive system** exists to produce ova (egg cells) and facilitate the conception, development, and birth of offspring. The ovaries are analogous to the male's testes in that they manufacture the ova—the reproductive cells that carry the female's half of the genetic endowment—and sex hormones. And like the male, the female has a system of ducts that guide ova and sperm to their union. We begin our discussion of the female reproductive system with the external genitalia.

The External Genitalia Form the Vulva

The female external genitalia is collectively known as the **vulva** (Fig. 17. 6).

The mons pubis (literally "pubic mountain") is a fatty mound of tissue covering the pubic bone, which in postpubertal females is covered by pubic hair. On each side it is continuous inferiorly with the **labia majora** ("large lips"), which are homologues of the scrotum. These are soft, rounded folds of skin that, like the mons pubis, are covered by genital hair in postpubertal women.

The **labia minora** ("small lips"), homologues of the ventral surface of the penis, are delicate folds of moist mucosa immediately medial and deep to the labia majora. They frame the *introitus* (Latin = "entryway"), the opening to the vagina, and act as movable flaps that

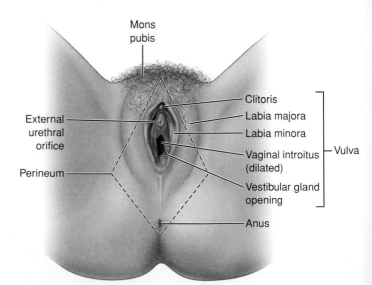

Figure 17.6. Female external genitalia. The perineum is bordered by the labia majora and the anus. *Is the vaginal opening anterior or posterior to the external urethral orifice?*

keep microbes out and moisture in but can easily be moved aside for intercourse. The vaginal mucosa itself has no glands. However, *vestibular glands*, homologues of the male bulbourethral glands, open immediately inside to the vaginal introitus, secreting mucus during sexual arousal and facilitating penile insertion. Vaginal lubrication also reflects the transudation of fluid across the vaginal wall.

Collectively, the area from the mons pubis to the anus is referred to as the **perineum.** In males, the perineum is the skin between the scrotum and the anus.

The **clitoris** (Greek = "little hill"), homologue of the penis, is a small (about 2 cm) elongated collection of highly sensitive erectile tissue located at the anterior juncture of the labia minora. The body of the clitoris, homologue to the penile shaft, is embedded beneath superior vaginal mucosa. The rounded, visible part is the *glans*, homologue of the glans (head) of the penis, the dorsal part of which is covered by a short *hood*, homologue of the male foreskin. During sexual arousal, the clitoris becomes engorged with blood, which increases its sensitivity.

The **external urethral orifice** opens within the labia minora immediately posterior to the clitoris.

Case Note

17.8. While Susan was giving birth, the skin posterior to the vaginal introitus tore slightly. What is the medical term describing the area from the mons pubis to the anus?

The Female Ductal System Guides Ova and Sperm to Their Union

The **female ductal system** consists of the vagina, the uterus, and the uterine (fallopian) tubes (Fig. 17.7).

The Vagina Is a Passageway

The **vagina** (Latin = "sheath") is a soft, thin-walled muscular tube about 8 to 10 cm (3 to 4 in.) that rests in the space between the bladder and the rectum (Fig. 17.7A). The urethra is anchored to its anterior wall by connective tissue. The vagina accepts the penis for intercourse, holds sperm after ejaculation, and serves as a passageway for menstrual flow from the uterus above and as an outlet for birth of a fetus.

The vagina has three layers: an outer, loose, fibrous layer; a middle layer of smooth muscle; and an inner mucosa lined by stratified squamous epithelium, which is typical of mucosae (the mouth, for example) that must

accommodate friction. The inner mucosal cells release glycogen, which vaginal bacteria convert into organic acids such as lactic acid. The resulting vaginal acidity protects against infection by pathogenic microbes. However, the alkaline pH of semen neutralizes vaginal acidity in the short term for safe passage of sperm.

The Uterus Incubates the Fetus

The **uterus** is a hollow, thick-walled organ about the size of four curled fingers (not an entire fist) that sits in the midline of the pelvis, anterior to the rectum and posterosuperior to the bladder (Fig. 17.7A). It shelters and nourishes the fertilized ovum until it becomes a full-grown fetus. At the end of pregnancy, the uterus contracts to force the fetus out of the body through the vagina.

The uterus is shaped something like an inverted pear, with the stem end fitted into the upper end of the vagina (Fig. 17.7A and B). The broad superior portion is the **body.** The narrow inferior end is the *cervix* (Latin = "neck"). It protrudes part way into the superior end of the vagina, creating around it a circular recess called the **vaginal fornix** (Latin = "domed chamber"). The cervix contains a central canal, the *cervical canal*, which opens into the vagina via a small central opening, the cervical **os** (Latin = "mouth"). The os is normally plugged with mucus as a barrier to contamination of the uterus by vaginal bacteria. Sperm, however, are specially equipped to penetrate cervical mucus. One cause of infertility is an inability of sperm to successfully penetrate cervical mucus.

The hollow interior of the uterus is the **endometrial cavity** (Latin *endo* = "within," and *metra* = "womb"), which connects below to the cervical canal and above to the uterine tubes (Fig. 17.7B). The lining of the uterus is a specialized tissue called the **endometrium.** During a woman's reproductive years—unless she is pregnant, taking birth control pills, or especially underweight—her endometrium is shed monthly by *menstruation*, a process which to be discussed in detail further on. The majority of the uterine wall is composed of a thick layer of smooth muscle, the **myometrium.** It produces the intense muscular contractions of childbirth as well as the less intense (but sometimes painful) contractions that help to expel the menstrual flow.

Case Note

17.9. Susan's exploratory surgery involved the passage of a small tube through the following structures: vagina, uterine body, cervical os, cervical canal, uterine tube, vaginal introitus. Put these structures in order, from the most external to the most internal.

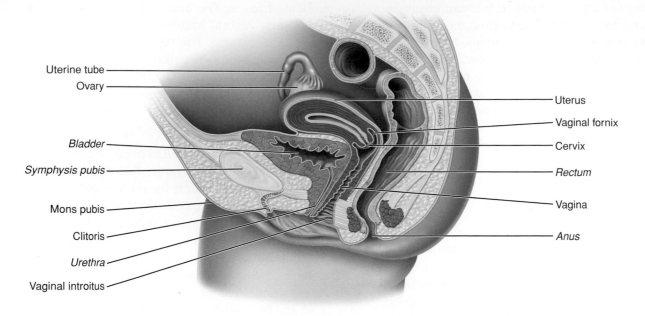

(a) Sagittal view

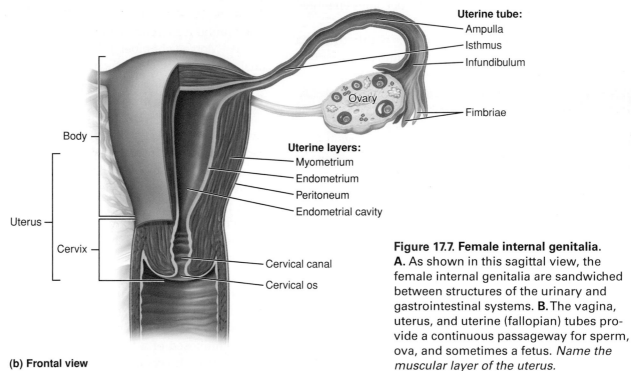

(b) Frontal view

Figure 17.7. Female internal genitalia.
A. As shown in this sagittal view, the female internal genitalia are sandwiched between structures of the urinary and gastrointestinal systems. **B.** The vagina, uterus, and uterine (fallopian) tubes provide a continuous passageway for sperm, ova, and sometimes a fetus. *Name the muscular layer of the uterus.*

The Uterine Tubes Convey Sperm and Ova

Females have two **uterine tubes** (also called *fallopian tubes* or *oviducts*), right and left. They extend laterally from the superior, lateral edges of the uterus and end very close to the ovaries but not attached to them. The narrow proximal third is called the **isthmus;** the somewhat wider distal two thirds is the **ampulla;** and the broad, funnel-shaped terminus is the **infundibulum.** Near each ovary, the mouth of the infundibulum hovers over the ovary ready to capture an ovum as soon as it is released. The distal edge of the infundibulum is fringed by delicate fingerlike projections called **fimbriae** (Latin = "fingers"), which move in wavelike unison to sweep an ejected ovum into the tube.

The wall of each uterine tube is formed of smooth muscle, which is covered with a surface layer of visceral peritoneum. Each tube is lined with a mixture of ciliated and nonciliated cells. Rhythmic contractions of smooth muscle and the coordinated beating of cilia move a captured ovum medially along the tube toward the uterus. Nonciliated cells secrete a fluid that keeps the ovum—and sperm, if present—nourished and moistened.

Fertilization of an ovum by a sperm usually occurs in the ampulla of the uterine tube. The fertilized ovum is then swept down the tube and into the uterus for implantation. However, sometimes implantation of a fertilized ovum can occur within the tube (most commonly) or on the ovary or elsewhere in the pelvis. This condition is known as an *ectopic pregnancy* (Greek *ektopos* = "off center," or "out of place") is discussed in more further on.

Infections and other conditions involving the uterine tubes can result in an inability to conceive. Many of these infections are sexually acquired (*sexually transmitted disease, or STD*). Infection can leave scars, kinks, or a complete blockage that inhibits the travel of sperm toward the ovum as well as the movement of the fertilized ovum toward the uterus, making both conception and implantation difficult. Surgical correction is sometimes possible.

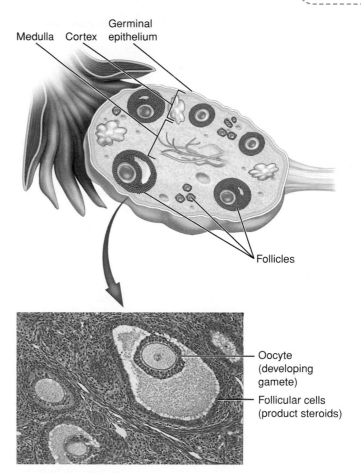

Figure 17.8. The ovary. A. The gametes and steroid-producing cells of the ovary are organized in follicles. **B.** This micrograph shows the oocyte within a mature follicle. *Are follicles found in the cortex or the medulla?*

Case Note

17.10. Look back at the case study. How did repeated STDs cause Susan's infertility?

The Ovaries Produce Ova and Female Hormones

The **ovaries** (Latin *ovum* = "egg") produce the female gametes and sex steroids. The female gametes are called **oocytes**; the mature gamete is called an **ovum.** Each ovary is about the size and shape of a dried apricot. There is one ovary on either side of the pelvis, very near the open, distal end of each uterine tube, a few centimeters from the uterus (Fig. 17.7B).

The bulk of the ovary consists of an inner medulla and an outer cortex, although there is no distinct separation between them (Fig 17.8A). The medulla consists of supporting connective tissue, blood vessels, and

nerves. The cortex contains **follicles** (Latin = "little bags") at various stages of development, each consisting of an oocyte and supporting *follicular cells* (Fig. 17.8B). Follicular cells protect and nourish the oocyte and also produce steroidal sex hormones: estrogen, progesterone, and related hormones.

17.10 Name the male homologue for each of the following: ovary, vestibular gland, labia majora, clitoris, clitoral hood.

17.11 Explain why the vaginal pH is acidic.

17.12 True or false: The uterine tubes connect to the cervical portion of the uterus.

17.13 Name two characteristics of the uterine tube wall that help propel ova down the tube.

The Female Reproductive Cycle

The years between puberty and menopause are a woman's *reproductive years*. During these years the anatomy and physiology of reproductive organs in nonpregnant women change dramatically each month. Predictable changes occur in the anatomy of the ovary and uterus, in the production of ova, and in the production of steroid sex hormones. Collectively these changes are referred to as the **female reproductive cycle.**

Each reproductive cycle is marked from the first day of menstruation (vaginal bleeding) and is usually about 28 days long (Fig. 17.9). However, cycles may be as short as 21 days or as long as 35 days or more. Once each cycle a mature gamete (ovum) is released from the ovary—an event called **ovulation.** Ovulation usually occurs about 14 days before the next menstrual cycle begins, whether or not the total cycle lasts 28 days. However, the day of ovulation is subject to considerable variation from one woman to the next and even from one month to the next in the same woman, a fact that makes achieving or avoiding pregnancy an imprecise science.

The phase of the cycle prior to ovulation is the **preovulatory phase;** this phase includes menstruation. The **postovulatory phase** spans the time period between ovulation and the beginning of the next reproductive cycle. Coordinated changes in the ovary and the uterus characterize these different phases. Although they are strictly dependent upon one another and constitute a harmonious whole, it is convenient to discuss ovarian and uterine changes separately.

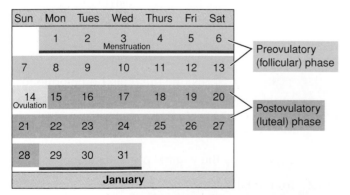

Figure 17.9. The reproductive cycle. Ovulation separates the preovulatory phase from the postovulatory phase. This figure shows a typical 28-day cycle, but in reality few cycles are "typical." *Does menstruation occur during the preovulatory or postovulatory phase?*

Case Note

17.11. To minimize discomfort, Susan's imaging study was performed just after her menstrual bleeding finished, when the cervix is easier to penetrate. Would Susan be in the preovulatory or postovulatory phase of her cycle?

The Ovarian Cycle

The **ovarian cycle** includes *oogenesis,* the production of an ovum (a process analogous to spermatogenesis) and *follicular development,* changes in the follicle that contains the ovum.

Oogenesis Produces Ova

Oogenesis is the production of mature female gametes from ovarian stem cells. It begins in the fetus as **oogonia,** diploid ($2n$) ovarian stem cells that are homologues of spermatogonia, undergo mitosis to produce more diploid oogonia (Fig. 17.10A). Recall that some of the offspring of spermatogonia remain as stem cells, even in adults. Males thus maintain their reproductive stem cells and can continue producing new sperm throughout life. In contrast, *all* of the oogonia differentiate into **primary oocytes** during fetal development by replicating their full set of 46 chromosomes and initiating meiosis. But about midway through meiosis I, these events suddenly halt. This state of suspended animation is described as *meiotic arrest.* These arrested primary oocytes and their halo of follicular cells are called **primary follicles.** Then, while still in the fetus, primary oocytes hibernate and are inactive until puberty.

Since all of the oogonia differentiate into primary oocytes, females stop producing new oocytes before they are even born. The fetal ovary contains perhaps 5 million primary oocytes contained within primary follicles. However, this number dwindles, so that at puberty only about 200,000 primary oocytes remain. As puberty arrives, a few follicles begin to develop. Their oocytes enlarge, and the rims of follicular cells multiply. During the rest of the woman's reproductive years, each cycle, one developing follicle is chosen to fully mature. How the particular oocyte is chosen remains a mystery. The chosen primary oocyte completes meiosis I and becomes a **secondary oocyte** by dividing into two haploid (n) cells with 23 replicated chromosomes each. It is a secondary oocyte that is released from the ovary to be fertilized, so it can also be called the ovum or egg. The follicle continues to enlarge and forms a tiny fluid-filled

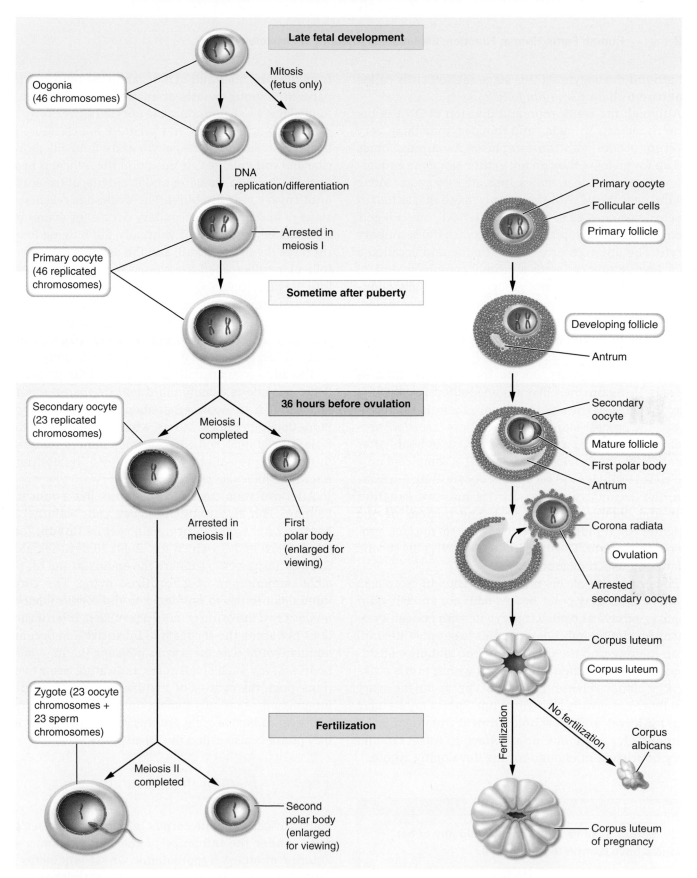

Late fetal development

Oogonia (46 chromosomes)

Mitosis (fetus only)

DNA replication/differentiation

Primary oocyte (46 replicated chromosomes)

Arrested in meiosis I

Sometime after puberty

Secondary oocyte (23 replicated chromosomes)

36 hours before ovulation

Meiosis I completed

Arrested in meiosis II

First polar body (enlarged for viewing)

Zygote (23 oocyte chromosomes + 23 sperm chromosomes)

Fertilization

Meiosis II completed

Second polar body (enlarged for viewing)

Primary oocyte
Follicular cells
Primary follicle

Developing follicle
Antrum

Secondary oocyte
Mature follicle
First polar body
Antrum

Corona radiata
Ovulation
Arrested secondary oocyte

Corpus luteum
Corpus luteum

Fertilization

No fertilization

Corpus albicans

Corpus luteum of pregnancy

(a) Oogenesis

(b) Follicular development

Figure 17.10. The ovarian cycle. A. Oogenesis. During a female's fetal development, all oogonia differentiate into primary oocytes, which cannot undergo mitosis. Primary oocytes complete meiosis I just before ovulation, and secondary oocytes complete meiosis II only if fertilized. Each meiotic division results in one gamete and one polar body. For simplicity's sake, the fate of only one pair of chromosomes is illustrated. The follicle surrounding each oocyte has been omitted for clarity. **B.** Follicular development. All follicles consist of follicular cells surrounding an oocyte. After oocyte release, the empty follicle differentiates into the corpus luteum, which degenerates into a corpus albicans if the oocyte is not fertilized. *Does a spermatozoon fertilize a primary oocyte or a secondary oocyte?*

sac around the oocyte. At this stage the follicle is called a **mature follicle** (*Graafian follicle*).

Although the replication and division of DNA is the same in meiosis in males and females, something very different occurs with the cytoplasm. As chromosomes line up for meiosis, they do not gather along the equatorial plane as they do in males. Instead, they gather along a plane near the edge of the cytoplasm, so that as half of the chromosomes go to each daughter cell, one cell gets almost all of the cytoplasm and becomes the secondary oocyte. The other receives almost none and is called a *polar body*, a tiny cell with a normal complement of 23 chromosomes but not enough cytoplasm or mitochondria to survive. Some polar bodies go on to complete meiosis II and divide into two even smaller polar bodies. Ultimately, however, all polar bodies degenerate and disappear.

Then the secondary oocyte immediately initiates meiosis II. Again the chromosomes align for division near the edge of the cell, but before meiosis II is completed, the process stops again. It is these "arrested" secondary oocytes that are ovulated, about 36 hours after the completion of meiosis I. Meiosis II is completed only if a sperm fertilizes the secondary oocyte. With fertilization, the secondary oocyte divides into two daughter cells and completes meiosis II. The nucleus of one daughter cell rapidly fuses with the sperm nucleus, producing the **zygote.** The other daughter cell is the second polar body, which degenerates.

The end result of this unequal division of cytoplasm is two or three tiny polar bodies with not enough cytoplasm to survive and one large zygote with enough cytoplasm, nutrients, and cell organelles to support life until it can migrate into the uterus, implant, and draw sustenance from the mother. The contrast with sperm is striking. Ejaculation releases millions of sperm, but the ovary usually releases only a single ovum. Sperm are slender, built for travel, and draw nourishment from their environment. Ova are bulky, do not have to travel far, and carry nourishment enough for the developing zygote.

Case Note

17.12. Before Susan's operation, would any of her gametes have completed meiosis II?

Ovulation Is the End Result of Follicular Development

It takes about 5 months for a follicle to fully mature to the point of ovulation. As the primary oocyte emerges

from the long hibernation it began in the fetus and prepares to continue meiosis, dramatic changes begin in the follicular cells surrounding the oocyte (Fig. 17.10B). The *primary* follicle evolves into a *mature* follicle. More follicular cells surround the oocyte, and a fluid-filled cavity (the *antrum)* enlarges. At this point the follicle is nearly an inch (2 cm) in diameter and is bulging at the surface of the ovary, ready to burst. The ovum has reached the stage of an "arrested" secondary oocyte (or ovum) with its attached polar body. Eventually the bulging follicle ruptures the ovarian wall, and the ovum and its halo of follicular cells (called the *corona radiata*) are expelled into the peritoneum. Ovulation may be accompanied by sharp, stabbing pain on one side of the abdomen, a symptom called *mittelschmerz*, which is German for "middle pain." The pain occurs in about 20% of women and can last for several minutes to several hours.

The adult ovary usually contains several mature follicles, but one is usually more mature than the others and bursts first. However, in about 5% to 10% of cycles, more than one follicle bursts, which may result in multiple conceptions. Follicular development is completed during the preovulatory phase, hence its alternative name, the *follicular phase.*

After ovulation, the follicle collapses like a punctured balloon. The remaining follicular cells enlarge and become bright yellow, forming a **corpus luteum** (literally, "yellow body"), which for about 10 days secretes a flood of hormones to prepare the endometrium for possible implantation of a fertilized ovum. The corpus luteum continues to enlarge into the *corpus luteum of pregnancy* if the ovum is indeed fertilized. If fertilization does not occur, the corpus luteum shrivels to become a nodular white scar, the *corpus albicans* (Latin = "white body"). Many such small scars, indicating prior ovulations, pock the ovaries of mature women. The corpus luteum is present only during the postovulatory phase of the reproductive cycle. For this reason, the postovulatory phase is also called the *luteal phase.*

The Follicle and Corpus Luteum Secrete Hormones

Follicular cells and the corpus luteum produce sex steroids, under the influence of the hypothalamus and anterior pituitary. Hypothalamic GnRH stimulates the production of FSH and LH from the anterior pituitary gland (Fig. 17.11, Step 1), and FSH and LH work together to control follicular development and ovarian steroidogenesis (step 2).

Follicular cells secrete **estrogen,** which derives its name from Latin *estrus*, the period of "heat" or sexual

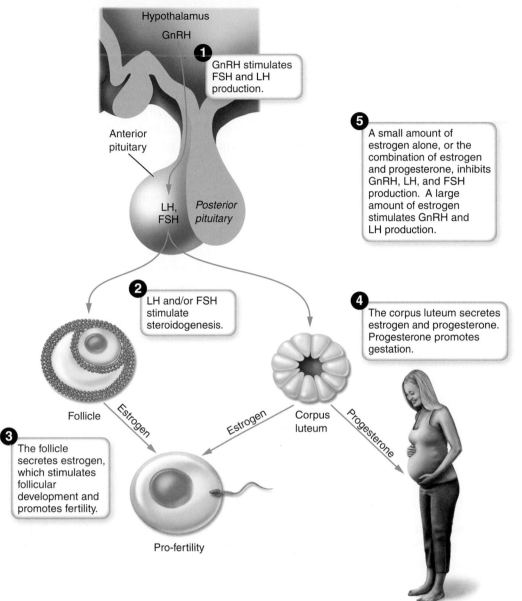

1 GnRH stimulates FSH and LH production.

Hypothalamus
GnRH

Anterior pituitary

LH, FSH

Posterior pituitary

5 A small amount of estrogen alone, or the combination of estrogen and progesterone, inhibits GnRH, LH, and FSH production. A large amount of estrogen stimulates GnRH and LH production.

2 LH and/or FSH stimulate steroidogenesis.

4 The corpus luteum secretes estrogen and progesterone. Progesterone promotes gestation.

Follicle

Estrogen

Estrogen

Corpus luteum

Progesterone

3 The follicle secretes estrogen, which stimulates follicular development and promotes fertility.

Pro-fertility

Pro-gestation

Figure 17.11. Ovarian steroids. The follicle secretes estrogen; the corpus luteum secretes both progesterone and estrogen. Estrogen also directly stimulates further estrogen production by the follicle. *Which hormone always inhibits LH production— estrogen or progesterone?*

receptivity in some mammals; and Latin *gen*, "to promote" (step 3). In human females, estrogen has a role in reproduction and in the health of numerous organs. Its activities include but are not limited to the promotion of follicular growth and steroid production; the growth and motility of the uterus, uterine tubes, and vagina; and maintenance of the health of breasts and external genitalia. At puberty, estrogen promotes a growth spurt owing to its effect on bone growth, and it plays a lifetime role in the maintenance of healthy bones.

The corpus luteum also secretes estrogen. But, unlike the follicle, it also secretes progesterone (step 4). The activity of **progesterone** is similarly implied by its name, which derives from Latin *pro* "for" and *gestare* "to carry

in the womb." That is, progesterone favors and supports pregnancy by promoting readiness of the endometrium to accept a fertilized oocyte and by promoting quietness and relaxation of the pregnant uterus.

> *Remember This!* **Estrogen promotes getting pregnant and progesterone promotes staying pregnant.**

The ovary also secretes small amounts of androgens. Combined with adrenal androgens, these hormones are responsible for the female sex drive and for the growth of body hair.

Recall from our earlier discussion that sperm and hormone production in the testis is regulated by a negative feedback loop, in which testosterone inhibits GnRH, FSH, and LH production. Although ovarian steroids also modulate GnRH and gonadotrophin secretion, the regulation of ovarian steroid production is complicated by two factors:

1. Low levels of estrogen alone, or the combination of estrogen and progesterone, inhibit GnRH and gonadotropin production. However, high levels of estrogen actually *stimulate* GnRH and LH production.
2. The number of steroid-producing cells varies from day to day. The larger the follicle or the corpus luteum, the more steroid hormones it produces.

As discussed further on, these complicating factors mean that steroid production depends much more on the *size* of the follicle or corpus luteum than on the amount of LH/FSH.

Case Note

17.13. Susan was tracking her temperature to pinpoint the moment when estrogen stimulates LH production. Would estrogen levels be relatively high or relatively low at this moment?

The Reproductive Cycle Is Regulated by the Hypothalamic–Pituitary Axis

Thus far we have discussed the ovarian production of gametes (ova) and steroids and how ovarian activity regulates and is regulated by the gonadotropin axis. Figure 17.12 shows how these events interact. Take a moment to orient yourself in this complex but critical figure. Begin at the top; note that the figure depicts a 28-day cycle and that ovulation at day 14 separates the preovulatory (follicular) phase (days 1 to 14) from the postovulatory (luteal) phase (days 15 to 28). Leaving the details of the uterine cycle at the bottom of the figure for now, let us begin at the left-hand side.

- *Day 1,* The developing follicle is very small, so it is secreting only small amounts of estrogen. Without estrogen's inhibitory effects, the pituitary secretes significant amounts of LH and FSH.
- *Days 2 to 11 (approximately).* The follicle grows under the influence of FSH and LH; as it grows, it secretes more estrogen (follicular cells connot produce progesterone). *Initially*, rising plasma estrogen exerts negative feedback, which reduces FSH and LH production. Estrogen is extremely effective at inhibiting

FSH production, which decreases during the follicular phase. Because estrogen cannot fully suppress LH production, LH levels continue to rise. However, as the follicle grows, it secretes more and more estrogen and a positive feedback loop becomes established. More follicular cells appear, which secrete more estrogen, which enlarges the follicle even more, which produces more estrogen, and so on. As we learned from ◄ Chapter 1, a positive feedback loop feeds on itself until it reaches a climax at an endpoint, which in this instance is ovulation about day 14.

- *About day 11.* Plasma estrogen levels reach the critical threshold and begin to *stimulate* the production of GnRH and LH. The resulting positive feedback loop sharply increases LH release—an event referred to as the *LH surge*. The increased GnRH production also stimulates a secondary, smaller rise in FSH production.
- *About day 14.* The surge of LH stimulates the following events related to ovulation:
  1. The secondary oocyte resumes meiosis (not shown on the figure).
  2. The follicle and the ovarian wall rupture, releasing the oocyte into the abdominopelvic cavity.
  3. The follicle stops producing estrogen; blood levels of estrogen rapidly decrease.
  4. The ruptured mature follicle converts into a corpus luteum.
- *Days 15 to 28 (approximately).* The follicle quickly becomes a corpus luteum and secretes both estrogen *and* progesterone, which together inhibit GnRH, LH, and FSH production, preventing the maturation of other follicles. After about 10 days, if no pregnancy occurs, the corpus luteum involutes into the corpus albicans, estrogen and progesterone levels fall, and GnRH is free to stimulate a new cycle of FSH and LH production.

Case Note

17.14. As part of Susan's fertility testing, her blood concentrations of several anterior pituitary hormones were monitored daily for a month. The test result noted a sudden dramatic increase in the blood concentration of one of these hormones, indicating that she was ovulating normally. Which hormone?

The Uterine Cycle Runs Parallel to the Ovarian Cycle

Above, we discussed the ovarian cycle and its interrelationship with the hypothalamic–pituitary axis. The **uterine**

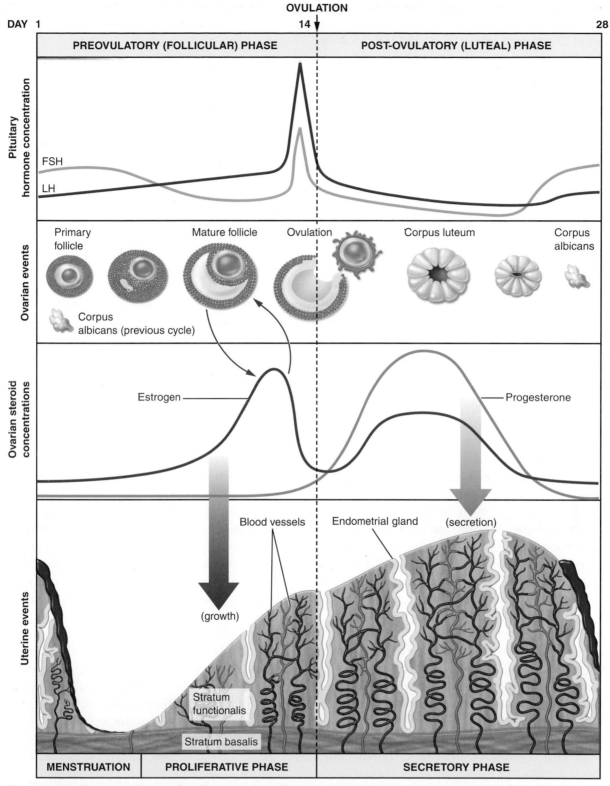

Figure 17.12. Reproductive cycles. The anterior pituitary gland secretes FSH and LH, which control the follicle and corpus luteum. The follicle and corpus luteum secrete ovarian steroids, which control the uterine cycle. *Why do FSH levels decrease during the follicular phase?*

cycle (or *menstrual cycle*) is a *consequence* of the changes occurring in the ovarian cycle. It is the monthly series of changes in the endometrium that render it thick and lush at exactly the right time to accept a fertilized ovum. The deepest layer, the thin **stratum basalis,** does not change over the uterine cycle. However, during a woman's reproductive years, the **stratum functionalis** lining the endometrial cavity thickens greatly as it readies for the possible arrival of a fertilized ovum. Because ovarian steroids control the growth and development of the stratum functionalis and ovarian steroids arise from the follicle, the uterine cycle is intimately related to follicular development. Below is the sequence of events, which have been idealized to a 28-day cycle (bottom of Fig. 17.12):

- *Days 1 to 5.* **Menstruation** (also *menses*; Latin *mensis* = "month"). Falling hormone levels at the end of the prior cycle lead to constriction of endometrial blood vessels and death of the endometrium; at this point menstrual bleeding begins. Day 1 is the first day of bleeding. The lush stratum functionalis layer of the endometrium, ready for pregnancy, dies in response to falling estrogen and progesterone support from the shrinking corpus luteum. The stratum functionalis flows out through the cervix and into the vagina as a mixture of blood and dead tissue.
- *Days 6 to 14.* **Proliferative phase.** In this phase a new growth (proliferation) of endometrium arises from the stratum basalis under the influence of rising levels of estrogen secreted by the growing follicle. Endometrial stroma thickens, blood vessels proliferate, and glands lengthen. Estrogen also thins the cervical mucus, facilitating the passage of sperm.
- *Days 15 to 28.* **Secretory phase.** The corpus luteum secretes rising levels of progesterone and estrogen as it grows. Progesterone stimulates endometrial glands to secrete a glycogen-rich fluid in anticipation of the nutrient needs of a fertilized ovum. Estrogen supports further stromal growth. Although the endometrium does not thicken further, the endometrial glands grow into wide, convoluted tubes. High blood levels of progesterone reconvert cervical mucus to a sticky plug to protect a potential pregnancy from vaginal bacteria. If pregnancy does not occur, menstruation starts and a new cycle begins.

Menstruation seems a heavy burden for females to bear in the service of reproduction: inconvenient and messy bleeding as well as the loss of blood and the vital iron that is carried away with it. On the other hand, endometrial infections are exceptionally rare because the flushing action of menstruation cleanses the uterus monthly. For insight into abnormal endometrial bleeding, see the nearby clinical snapshot titled *Abnormal Endometrial Bleeding.*

Ovulation and Menstruation Begin at Puberty and Cease at Menopause

During childhood, the ovaries release small amounts of estrogen, which inhibits hypothalamic release of GnRH. However, nearing puberty, the hypothalamus becomes less sensitive to estrogen and begins releasing bursts of GnRH. The pituitary follows by releasing FSH and LH. This continues until the amount of FSH and LH succeeds in completing a first full cycle at about age 12, whereupon the young woman has her first menstrual period, an event called **menarche** (Greek *men* = "month"; *arckhe* = "onset"). Ovulation usually does not occur initially, but after a few years the regular adult model of menstruation and ovulation is established.

Menopause is the normal, age-related cessation of ovulation and menstruation. This usually occurs around age 50, as the few remaining follicles in the ovary cease to respond to pituitary FSH and LH. In women of an appropriate age, it is signaled by missed or irregular periods and becomes "official" after 1 year with no menses. As ovarian cycles cease, *estrogen production falls.* This decline in estrogen prompts characteristic symptoms in postmenopausal women.

Some women have no unpleasant symptoms attributable to menopause, but many have hot flashes ("flushes") characterized by skin warmth, redness, and sweating. Hot flashes reflect peripheral vasodilation, and they vary in severity from a mild nuisance to nearly intolerable discomfort. Hot flashes are due to *declining* concentrations of estrogen, and they usually cease once estrogen levels reach their post menopausal set point. Oral estrogen tablets offer effective therapy but are not widely used because of a small but definite increase in the risk of certain types of cancer and vascular disease.

Most other postmenopausal symptoms can be traced to declining blood estrogen levels. These include depression, fatigue, irritability, and decline of sexual drive. Vaginal dryness and atrophy or fragility of the external genitalia also can occur.

As discussed in ◄ Chapter 6, estrogen is important in maintaining bone health. Postmenopausal women are therefore at increased risk of developing osteoporosis and experiencing a fracture. The relatively high blood estrogen levels present prior to menopause also offer women considerable protection against cardiovascular disease such as heart attack and stroke, which is much lower in premenopausal women than in men of comparable age. However, in postmenopausal women, cardiovascular disease begins to occur with increasing frequency.

CLINICAL SNAPSHOT

Abnormal Endometrial Bleeding

Menstrual problems—irregular periods, excessive bleeding, complete lack of menstruation, or pain—are the most common reason women seek gynecological care.

Variation in the expected time of menstruation is often unsettling to women, who find it inconvenient or wonder if it is indicative of pregnancy. One of the most common causes is failure to ovulate. When this occurs, the proliferative endometrium does not receive the expected flood of progesterone because no corpus luteum develops, and the endometrium sloughs away a few days after ovulation as an early period.

An abnormally heavy menstrual flow is also common. Sometimes heavy bleeding may accompany loss of an embryo in a woman who does not even realize she is pregnant. Alternatively, the patient may have a blood disorder that prevents normal blood clotting. Only rarely is the cause a cancerous or precancerous abnormality of the uterus. In any case, excessive menstrual bleeding is a common cause of anemia in women.

Menarche usually occurs before age 16. Failure to begin menstruation as a teenager is uncommon and is usually due to some underlying abnormality of the endocrine system, often the anterior pituitary, or a genetic problem. *Amenorrhea*, cessation of menses for 3 months or longer after a period of normal menstruation, can be due to severe stress or starvation and is also well known among female endurance athletes and women suffering from the eating disorder anorexia nervosa.

Dysmenorrhea, discomfort or pain during menstruation, is presumed to be due to uterine smooth muscle spasms. It is common, especially in young women, and usually does not indicate any underlying

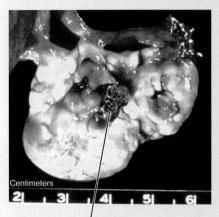

Centimeters

Endometriosis

Endometriosis. Deposits of endometrial tissue appear as bluish nodules on the ovary surface.

problem. It often improves with age and childbearing. Sometimes, however, dysmenorrhea occurs because of a condition called *endometriosis*, in which endometrial tissue grows outside of the uterus. These endometrial clumps can form on the surface of the ovaries, uterine tubes, or other abdominopelvic organs. Like normal endometrium, they grow and degenerate over the menstrual cycle in response to changing ovarian steroid levels. When abnormal endometrial deposits degenerate, the resulting blood and tissue remains within the abdominopelvic cavity, causing pain and inflammation. Endometriosis is also a leading cause of infertility in women.

17.14 Which of the following cells would not be found in an adult ovary—primary oocytes, oogonia, or secondary oocytes?

17.15 At what point in the ovarian cycle does meiosis I finish—the preovulatory phase or the postovulatory phase?

17.16 Which of the following contains a secondary oocyte—a primary follicle, a mature follicle, or the corpus luteum?

17.17 The corpus luteum produces both estrogen and progesterone. Will the combination of these two steroids stimulate or inhibit LH production?

17.18 Which ovarian steroid is produced only in the postovulatory phase? Explain.

17.19 Is the proliferative phase preovulatory or postovulatory?

17.20 Which hormone is proconception: estrogen or progesterone?

17.21 Which hormone acts directly on endometrial tissue to stimulate growth—LH or estrogen?

Sexual Behavior

Sexual behavior is a vast and provocative topic, which ranges from social custom and gender behavior to simple mechanical and biological facts, most of it beyond the purview of this discussion. Here, we discuss the sexual response and common sexually transmitted diseases.

The Sexual Response Includes Four Distinct Phases

The sexual response in both sexes can be divided into four phases: *arousal, plateau, orgasm,* and *resolution.* And despite anatomic differences, the physiology of the response is remarkably similar in men and women.

Arousal (sexual excitement) can be initiated or sustained by almost any stimulus: thought, sight, sound, touch, taste, or smell; stimuli which, by definition, are called *erotic* (Greek *eros* = "sexual love"). In males, arousal is primarily manifest by *erection,* an enlargement and stiffening of the normally flaccid penis. Parasympathetic nerve impulses stimulate vasodilation of penile arterioles, which flood penile blood sinuses with blood under high arterial pressure (a rare example of parasympathetic—"rest and digest"— control of arterioles). Sinus expansion blocks venous drainage, blood outflow is restricted, and penile pressure produces a firm erection.

In females engorgement and enlargement of the clitoris is analogous to penile erection. The woman's breasts and vaginal mucosa also engorge with blood, the nipples become erect, and the vestibular glands secrete lubricating fluid to ease penile insertion.

The *plateau* is a stable period during which continued erotic stimulation maintains arousal.

Orgasm is the sum of events surrounding the exquisitely pleasurable sensations that are the apex of the sexual experience, male or female.

In males, orgasm is almost always accompanied by ejaculation. **Ejaculation** (Latin *jacere* = "to throw") is propulsion of semen from the male ductal system. Immediately before semen is expelled, there occurs a short period, *emission,* during which the prostate, seminal vesicles, and ductus deferens squeeze a small amount of fluid into the urethra. To a certain point this process is consciously controllable, but as emission fluid continues to accumulate, a positive reinforcement loop emerges: as more fluid accumulates, more is secreted. This is accompanied by a mental sensation of inevitability as spinal cord reflexes assume control and ejacula-

tion cannot be stopped by mental effort. Ejaculation occurs as the positive reinforcement loops runs to its endpoint. Although erection, male or female, is mediated by parasympathetic nerve signals, emission and ejaculation are sympathetic—"fight or flight"—responses. Signals in sympathetic nerves produce rhythmic contraction of accessory glands and ducts, which expel semen. Somatic nerve signals cause muscles to contract at the base of the penis and induce rhythmic, thrusting contractions of pelvic, hip, and trunk muscles.

In females, orgasm is accompanied by increased muscular tension throughout the body, especially in the trunk and thighs. Blood pressure rises, and feelings of intense pleasure radiate through the genitals and sometimes to the lower back, hips, or other parts of the body. Uterine contractions, similar to the contractions of the male ductus deferens, also occur.

Resolution is the unwinding of these events. Ejaculation ends, excitement fades, arterioles in the base of the penis reassume their constricted state, blood drains from the penis, and within a few minutes the penis returns to its prearousal state. In females, clitoral erection fades in like manner. In males, a *refractory period* follows, which may be a few minutes or a few hours, during which erection and ejaculation is neurologically impossible. A refractory period does not occur in females; that is, most females are neurologically capable of having another orgasm right away and may have many in a single sexual encounter. However, in some women the clitoris becomes hypersensitive, so that for a short time another sexual experience would be uncomfortable.

Case Note

17.15. **When Mark and Susan conceived, they each experienced an orgasm. In the arousal stage of their episode, what part of the nervous system mediated their erections?**

Sexually Transmitted Diseases Can Have Lasting Health Effects

The sexual revolution is over and the microbes won.

P. J. O'Rourke (b. 1947), American humorist and political commentator, from *Give War a Chance*

Sexually transmitted diseases (STDs) are disease communicated by sexual contact. Some are transmitted *only* by sexual contact; infection by the bacterium *Neisseria gonorrhoeae*, a disease called *gonorrhea*, is an example. Others are diseases that can also be transmitted by non-sexual means—HIV/AIDS and some types of viral hepatitis may be transmitted by sex, needlestick, or blood transfusion (see the case study in ◀ Chapter 12).

Human papillomavirus (HPV) is the most common of all STDs—about 25% of the general population is infected, although most do not realize that they are. HPV is a family of viruses: some varieties cause benign skin warts (especially in children) and are transmitted by casual skin-to-skin contact, whereas others are passed by sexual contact. Two types of genital lesions occur. First is a warty growth on the labia or penis. However, in the cervix of infected women, HPV can cause precancerous changes (*cervical dysplasia*) or frankly malignant cancer. Most female genital HPV infections are silent and detection requires microscopic examination of cells scraped from the cervix (a *Pap smear*). It is important to know that dysplasia and cancer almost never occur as the result of a single infection—many infections are necessary and the risk rises with each reinfection. Vaccination is available to reduce the risk of genital HPV infection.

The bacteria *Chlamydia* and *Neisseria gonorrhoeae* are also very common but are easier to detect because they produce symptoms. In males, they usually cause painful inflammation of the penile urethra with a visible discharge of pus. In females, the uterine tubes are most often affected and the condition prompts lower abdominal pain. Repeated infections in males can produce scarring, which may interfere with the passage of urine. In females scarring and blockage of the uterine tubes can cause infertility owing to the inability of sperm to reach the ovulated egg. Antibiotics are usually effective but may not prevent scarring.

The common cold sore virus, *herpes simplex*, can also cause genital lesions. As with oral lesions, genital lesions appear as clusters of small, painful blisters. No cure is possible and most who become infected suffer from periodic recurrences for many years. Herpes simplex is very widespread, but its transmission can be easily prevented by using condoms. Genital herpes is only rarely associated with serious medical problems. However, pregnant women with an acute outbreak of infection near the time of birth are at risk of infecting their newborn infants as they pass through the birth canal. In such cases delivery by cesarean section is required.

Finally, *syphilis* is a disease caused by the corkscrew-shaped bacterium *Treponema pallidum*. Although relatively uncommon (less than 5% of STDs), syphilis holds a special place in medicine because, in antiquity, it was known to be incurable and to cause dementia and sometimes fatal vascular disease. Fresh infection is characterized by a solitary ulcer. In males, this usually forms on the head of the penis, whereas in females, it forms in the vagina or cervix where it cannot be seen. In any case, although the ulcer disappears in about a month, the bacteria spread silently throughout the body, where they may remain undetected for decades as they do their damage. In a pregnant woman, *Treponema* can cross the placenta and result in spontaneous abortion or stillbirth. Fortunately, syphilis is easily cured by a single injection of penicillin. But penicillin will not reverse damage already done.

Case Note

17.16. Susan's clinician mentioned that bacterial, not viral, infection was likely responsible for the scarring in her uterine tubes. Which of the following STDs is caused by bacteria—HPV, chlamydia, or herpes simplex?

17.22 Name the four phases of the sexual response.

17.23 Which STD could be treated with antibiotics—herpes simplex or gonorrhea? Explain your answer.

The Breasts

Mammals (archaic English *mamma* = "mother") are so named for their ability to nourish their young with milk. The **breast** is a modified sweat gland that produces milk for nourishing infants. The milk-producing glands in the breast are the **mammary glands.** Because males also have breasts—they are small and inactive but do have mammary glands—and because the breast plays no *primary* role in reproduction, female breasts are considered *secondary sexual characteristics*.

Mammary Glands Are Modified Sweat Glands

Each breast is covered by skin and lies anterior to the pectoralis major muscle of the anterior chest wall

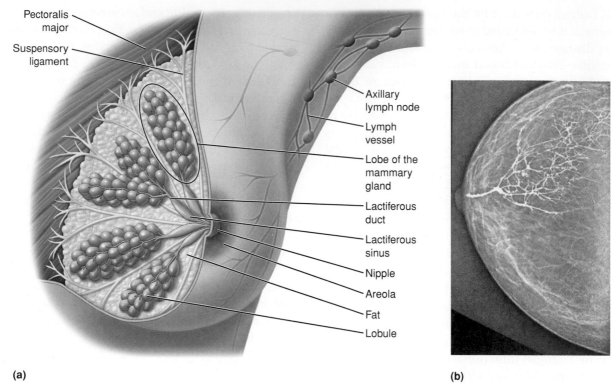

Pectoralis major

Suspensory ligament

Axillary lymph node

Lymph vessel

Lobe of the mammary gland

Lactiferous duct

Lactiferous sinus

Nipple

Areola

Fat

Lobule

(a)

(b)

Figure 17.13. The breasts. A. The breasts of pregnant or nursing women are packed with milk-producing glandular tissue. **B.** This x-ray image shows contrast material that was injected into a lactiferous duct at the nipple. It highlights the duct system of the breast. *What is the difference between the lactiferous sinus and the lactiferous duct?*

(Fig. 17.13A). The skin covering each contains a central pigmented circular area, the **areola** (Latin = "small space"), within which is a central projection, the **nipple.** The bulk of the breast is composed of fat, fibrous tissue, and mammary glands, which connect to the nipple by ducts. Mammary glands are small and inactive in non-pregnant females, but they enlarge during pregnancy to prepare for nourishing the newborn.

The glands of each breast are embedded in fat and divided into about 20 *lobes*, each of which consists of hundreds of smaller *lobules*, which in turn are composed of microscopic epithelial glands. Lobes are separated by thin bands of fibrous tissue called *suspensory ligaments*, which support and give shape to the breasts.

The *lactiferous ducts* draining the mammary glands merge and become progressively larger ducts as they near the nipple (Fig. 17.13A and B). Immediately beneath the areola they dilate to form *lactiferous sinuses* for milk storage. Ducts empty through numerous slitlike pores in the nipple. Specialized **myoepithelial cells** surround the glands and ducts. These cells contain contractile proteins that, when they contract, expel the contents of the gland and propel them along the duct.

The breast has a rich lymphatic network, most of which drain upward and outward to the axilla; lymphatics from the medial aspect of the breasts (between the nipples) drain into lymph nodes inside the chest on either side of the sternum (the internal mammary nodes).

Both fibrous tissue and fat are sensitive to estrogen and progesterone, which accounts for the growth of the breasts at puberty and for the engorgement and occasional tenderness of the breast during the menstrual cycle.

The most common abnormality of the female breast is *fibrocystic change*, a benign condition that consists of scarring (fibrosis), chronic inflammation, and cystic dilation of breast ducts, which may be painful and associated with alarming lumps. Because fibrocystic change occurs to some degree in most women and is therefore "normal," most physicians avoid referring to it as a "disease."

Breast tumors, like all tumors, are either benign or malignant. The most common tumor of the breast is a benign growth called *fibroadenoma*. Malignant tumors are less common; however, breast cancer is the most common cancer in women, affecting about 1 in every 8 women at some time in life. (Breast cancer is very rare in males.) Still, breast cancer is less lethal than lung cancer, which kills nearly twice as many women. Most breast cancers are malignant growths of the lining epithelium of breast ducts (about 80% of cases) or of mammary glands (10%). The single most important factor governing survival in breast cancer is whether or not it has spread to lymph nodes.

> **Case Note**
>
> **17.17.** Susan's son latched onto the pigmented area of the breast to feed. What is this area called?

Prolactin and Oxytocin Regulate Milk Production

Estrogen and progesterone prepare the breast for lactation, but two pituitary hormones are necessary for milk production. Prolactin, produced by the anterior pituitary gland, stimulates milk *synthesis* by the mammary gland. Oxytocin, synthesized by the hypothalamus and released from the posterior pituitary gland, stimulates milk *ejection*—oxytocin stimulates contraction of myoepithelial cells surrounding the glands and lining the milk ducts, ejecting milk through the nipple.

> *Remember This!* Milk production is stimulated by prolactin.

Mental stimuli (such as the sight of a baby) or physical stimuli (such as the baby suckling) stimulate both prolactin and oxytocin secretion, which increases milk production, which increases suckling. This positive feedback loop terminates when the baby detaches from the breast and ensures that a mother continues to produce milk as long as she is nursing her infant. Prolactin also has the added effect of suppressing GnRH production; therefore many lactating women do not ovulate or menstruate. The breasts begin producing milk about 3 or 4 days after birth and can continue production for a virtually unlimited period. In the interim between birth and milk production, the breasts produce a different fluid, called *colostrum,* which is rich in antibodies, proteins, and carbohydrates but low in difficult-to-digest fat.

Mature breast milk contains an easy-to-digest protein called *lactalbumin,* which (unlike cow's milk or soy milk proteins) rarely induces allergic reactions in the infant. It is rich in cholesterol and omega-3 fatty acids, which support neurological development. Human milk contains beneficial bacteria to help the infant establish its own intestinal bacterial flora, but it also contains antibodies and other substances to suppress the growth of harmful bacteria. Unlike formula, breast milk changes in composition to meet the needs of the infant as it grows. Amazingly, it also varies within the same feeding, beginning as thirst-quenching, watery foremilk and finishing with satiating, creamy hindmilk.

> **Case Note**
>
> **17.18.** Susan noticed that her uterus would contract when she breast-fed her baby. Can you think of a lactational hormone that also stimulates uterine contraction?

 17.24   Some people call the breasts the "mammary glands." Is this statement correct? Explain.

17.25   True or false: Oxytocin stimulates milk production, and prolactin stimulates milk ejection.

17.26   Which mammary gland product contains more fat—milk or colostrum?

Fertilization and Embryonic and Fetal Development

Pregnancy (*gestation*) begins with fertilization (conception) and ends with birth. This period of time, known as the **gestational period,** is normally 280 days (40 weeks) dated from the first day of the last menstrual period, which is easier to pinpoint than the date of conception. But because conception does not occur until approximately day 14, the actual gestation time is 2 weeks less than 280 days. Pregnancy is conveniently divided into three *trimesters* of about 3 months each. The developing human is called an **embryo** during the first 8 weeks and a **fetus** thereafter.

Pregnancy Begins with Fertilization

Pregnancy begins with **fertilization,** the union of oocyte and spermatozoa. The timing has to be just right—the oocyte is viable for only about 24 hours after ovulation, although the sperm can live for up to 5 days in the female reproductive tract. However, the newly ejaculated sperm must go through **capacitation,** a 7-hour maturation process, before they are capable of fertilization. To the extent that ovulation can be predicted, and it is a less than exact science; specialists advise couples who want children to have sex a day prior to ovulation, so that matured sperm are waiting in the oviduct for the oocyte.

The Optimal Fertilization Site Is the Uterine Tube

Semen is usually deposited into the vaginal *fornix* (Latin = "vaulted chamber"), the deepest portion of the vagina adjacent to the cervix, where it coagulates into a loose gel. A small number of sperm (perhaps less than 1%) swim out of the gel and up through the cervical mucus, using whiplike movements of the sperm tail. Mild uterine contractions aspirate this sperm-containing mucus into the endometrial cavity and assist the swimming sperm in reaching the uterine tube. Sperm can anchor themselves to the wall of the isthmus for several days to ensure complete capacitation, after which they release and swim distally into the ampulla.

Because the oocyte has no motile tail, its movement depends upon the actions of the uterine tube. The fimbriae bind to the corona radiata of the oocyte and sweep it into the uterine tube. Cilia lining the uterine tube propel the oocyte into the ampulla and the tide of sperm swimming to meet it.

The ampulla is the most common site for fertilization, though it can also occur in the pelvic space near the ovary (Fig. 17.14, day 0). Sperm are enticed to swim toward the oocyte by a chemical gradient of come-hither substances secreted by the oocyte.

On arrival at the oocyte, the capacitated sperm burrow through the remnant follicular cells of the corona radiata. One final barrier remains—the **zona pellucida,** a thick glycoprotein capsule that protects the ovum. The head of the sperm is especially fitted to penetrate the zona pellucida. Each sperm is covered by a cap, the **acrosome** (Greek *akron* = "tip" and *soma* = "body"), which contains digestive enzymes. Upon contact with the ovum, the acrosomes of hundreds of sperm release their enzymes, progressively dissolving the barrier. Eventually, the barrier is sufficiently degraded to allow proteins in the head of a single sperm to bind to receptors on the ovum's surface, which triggers fusion of the sperm and ovum cell membranes. Membrane fusion initiates a complex chemical and electrical reaction in the barrier that prevents any other sperm from entering.

Fertilization Restarts Meiosis II

Events move quickly once the cell membranes fuse—the ovum (secondary oocyte), which was suspended part way through meiosis II, completes its second meiotic division. The haploid nucleus of the resulting cell fuses with the haploid sperm nucleus to form the diploid zygote. The polar body produced by meiosis II degenerates.

The zygote begins to divide as it is slowly swept into the uterine cavity, exponentially increasing its cell number from 2 cells to 4, to 8, and so on. Each identical daughter cell, called a **blastomere** (*-blast* = "formative cell"), is capable of forming an entire human being.

About 4 to 5 days after fertilization, the embryonic cells begin to differentiate in a more specific way, forming a hollow structure called the **blastocyst** (Fig. 17.14, days 4 to 5).

The blastocyst has three components:

- The cyst wall, composed of *trophoblast* cells
- A nodule of cells on in the inner wall of the cyst, the *inner cell mass*
- A hollow cavity, the *blastocyst cavity*

Case Note

17.19. Temperature charting showed that Susan ovulated on day 15 of her reproductive cycle. Would a fertility specialist recommend that she and Mark have sex on day 14 or day 16 of her cycle? Explain why.

Trophoblasts Facilitate Implantation

The blastocyst is swept into the endometrial cavity and burrows into the lush secretory endometrium awaiting it, a process known as **implantation** (Fig. 17.14, days 6 to 9). The sticky trophoblast cells of the blastocyst wall adhere to the uterine lining and secrete proteolytic enzymes that form tunnels in the endometrium. Cords of trophoblast cells extend into these new crevices, pulling the rest of the blastocyst along behind. These trophoblast cords form the *chorion,* which evolves to form the fetal portion of the placenta. The chorion is discussed in greater detail further on.

Blastocyst secretions stimulate the endometrium to become more vascular and packed with nutrients. These newly succulent, plump endometrial cells form a tissue

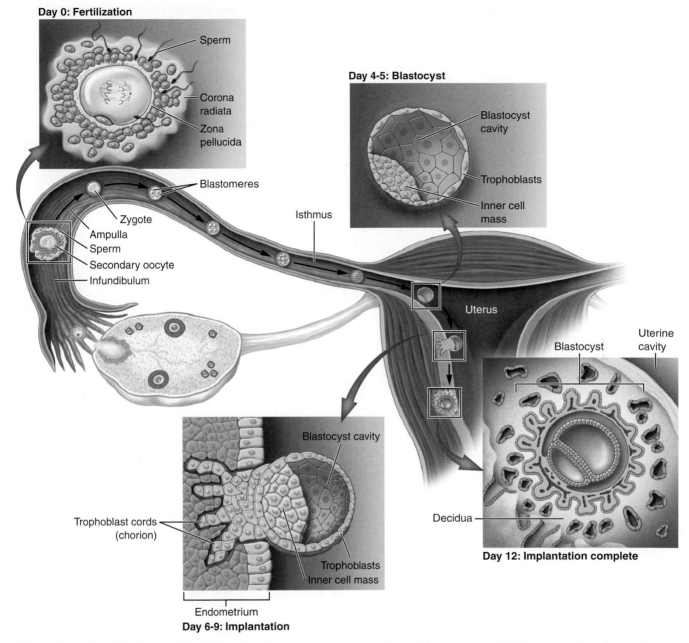

Day 0: Fertilization

Sperm

Corona radiata

Zona pellucida

Blastomeres

Zygote

Ampulla

Sperm

Secondary oocyte

Infundibulum

Isthmus

Uterus

Day 4-5: Blastocyst

Blastocyst cavity

Trophoblasts

Inner cell mass

Uterine cavity

Blastocyst

Decidua

Day 12: Implantation complete

Blastocyst cavity

Trophoblast cords (chorion)

Trophoblasts
Inner cell mass

Endometrium
Day 6-9: Implantation

Figure 17.14. Fertilization and implantation. The zygote develops into a blastocyst, which implants into the endometrium. *Which cells of the blastocyst contact the endometrium—trophoblasts or the inner cell mass?*

called **decidua,** which makes up the maternal part of the placenta. Eventually the blastocyst is fully enclosed within the maternal decidua (Fig. 7.14, day 12). Shortly after implantation, trophoblastic cells begin to secrete **human chorionic gonadotropin (hCG),** an analogue of pituitary LH, which stimulates the corpus luteum to maintain its output of estrogen and progesterone. Popular pregnancy tests rely on the detection of hCG in urine as an indication of pregnancy.

Case Note

17.20. About 22 days after her last menstrual period, Susan thought she was getting her period early, because she felt some pelvic discomfort and had some vaginal bleeding. The actions of which cell type is producing the bleeding—blastomeres, trophoblasts, or the inner cell mass?

The Inner Cell Mass Forms the Embryo and Several Embryonic Membranes

Some cells of the **inner cell mass** give rise to the embryo. Others follow a different developmental pathway and evolve into the membranes that surround and protect the embryo. Below we offer a brief summary of this intricate process.

Hypoblasts and Epiblasts Form the Yolk Sac, Amnion, and Embryo

By the time implantation is complete, the inner cell mass has differentiated into two different types of cells: **hypoblasts** on the side closest to the blastocyst cavity, and **epiblasts** on the side farthest from the cavity (Fig. 17.15A). Hypoblasts migrate around the wall of the blastocyst cavity and completely line its inner wall, forming

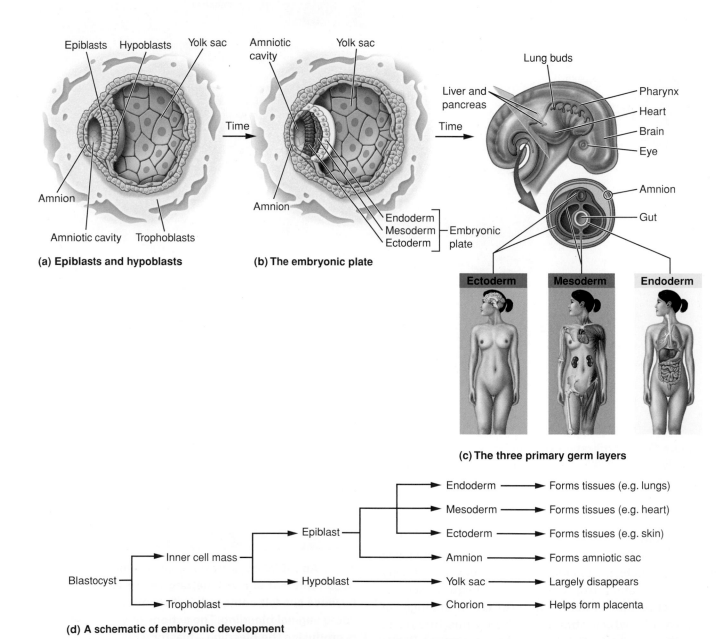

(a) Epiblasts and hypoblasts

(b) The embryonic plate

(c) The three primary germ layers

(d) A schematic of embryonic development

Figure 17.15. Development of the primary germ layers. A. The newly implanted blastocyst contains epiblasts and hypoblasts. **B.** Some epiblasts give rise to the three primary germ layers of the embryonic plate. **C.** The three primary germ layers give rise to distinct tissues. **D.** The derivation of fetal structures and membranes. *Is the brain of ectodermal or endodermal origin?*

a cavity called the *yolk sac* where the blastocyst cavity once was. During its existence, the yolk sac is the initial site of blood cell production, a task taken over for a while in the fetus by the liver and spleen before the bone marrow permanently assumes production after birth. Some yolk sac cells eventually form part of the intestinal wall. The yolk sac usually disappears by the ninth week.

A second cavity, called the **amniotic cavity,** forms within the epiblast cell mass. The epiblasts in contact with hypoblasts evolve into the embryo; those on the opposite side of the amniotic cavity form an embryonic membrane called the *amnion* or *amniotic membrane* (Fig. 17.15B). As discussed further on, the amnion eventually surrounds the entire embryo and forms the *amniotic sac* (see Fig. 7.17).

> **Remember This!** The embryo develops from a subpopulation of epiblasts. Hypoblasts do not contribute any cells to the embryo.

The Embryonic Plate Contains Three Germ Layers

Epiblasts take center stage in this next stage of development, displacing neighboring hypoblasts and forming the three-layer **embryonic plate** (Fig.17.15B). The displaced hypoblasts die by apoptosis. Cells of the epiblast migrate and differentiate to form the three primary **germ layers,** which each give rise to different tissues and organs (Fig. 17.15 B and C).

- Cells of the *endoderm* (endo = "inner") are destined to become the epithelial lining of the intestinal and respiratory tracts, the endocrine glands, and the liver and pancreas.
- Cells of the *mesoderm* (meso = "middle") are destined to become bone, muscle, blood cells, kidneys, gonads, heart and blood vessels, dermis, and connective tissues.
- Cells of the *ectoderm* (ecto = "outer") are destined to become epidermis, tooth enamel, and nervous tissue.

You can review the journey from blastocyst to germ layers in the flowchart of Figure 17.15D. It is worth noting that most organs contain elements of more than one layer. In skin, for instance, the epidermis is formed from ectoderm, the dermis from mesoderm.

Case Note

17.21. By the time Susan tells Mark that she is pregnant, the three germ layers have already developed. Which cell type produced these germ layers—epiblasts, hypoblasts, or both?

Organs Develop from the Germ Layers

In week 4, the embryonic plate folds into a cylindrical shape in which the germ layers develop into tissues and organs in a meticulously timed sequence. By the end of the embryonic period (week 8), the embryo has typically reached a length of about 3 cm (1.2 in.) and resembles a human being. Most of the major organs are formed, but they are not mature.

The first trimester is a vulnerable time for the embryo because so much complex development is occurring; it is the time during which most birth defects occur. During the first few weeks of embryonic development, the nervous system is particularly vulnerable to a deficiency of folate, one of the B vitamins. At this critical juncture the brain and spinal cord are evolving from an open trough into a hollow tube. Folate deficiency can retard these changes and leave the fetus with malformations called *neural tube defects* (NTDs), which can vary from innocuous to crippling paralysis or even death. In the United States folate-fortified breads and cereals have reduced the incidence of NTDs.

In addition, the first-trimester embryo/fetus is especially vulnerable to viruses. A cardinal example is the *rubella* virus, which causes a typically mild and transient illness called *German measles* in children and adults. In contrast, rubella infection in the first trimester can readily cross the placenta to cause blindness, deafness, heart defects, and other serious problems in the fetus.

Excessive alcohol intake is particularly hazardous for the fetus. Women who drink too much alcohol during pregnancy can give birth to infants with any of several alcohol-related disorders collectively known as *fetal alcohol spectrum disorders* (FASDs). The associated impairments, which can range from mild to severe, include facial abnormalities and cognitive deficits. How much alcohol is too much? Opinions vary, but most practitioners advise their pregnant patients to abstain entirely. Smoking, drug abuse, and even the use of certain prescription and over-the-counter medications can cause similar problems. A mother may not realize that her child is affected until later, when behavioral or learning problems appear.

The second trimester is a period of rapid growth and maturation, and the third trimester is a "polishing" stage during which final birth weight is achieved and organs complete their maturation. You can see the embryo and fetus at different developmental stages in Figure 17.16.

Particularly informative is the differentiation of the male and female reproductive organs. You can read about it in the accompanying box.

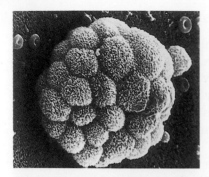

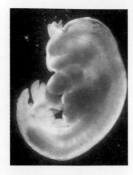

(a) Implantation blastocyst

(b) 32 day embryo

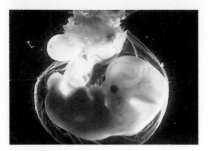

(c) 37 day embryo

(d) 20 week fetus

Figure 17.16. Embryos. A. An implanting blastocyst. **B.** A 32-day embryo. **C.** A 37-day embryo. **D.** A 12- to 15-week-old fetus. *Notice the tail in the 32-day embryo. Do you see this tail in the fetus?*

Case Notes

17.22. Because of her history of fertility problems, Susan had her first ultrasound 6 weeks after her last menstrual period. Did she see an embryo or a fetus?

17.23. Susan was very relieved the day she reached the second trimester, because most miscarriages occur during the first trimester. Has it been 12 weeks or 10 weeks since her fetus was conceived?

17.24. Toby was born before the third trimester was complete. Do any organs form during this trimester?

The Amniotic Sac Protects the Fetus

As the embryo grows, the **amnion** (also called the *amniotic membrane*) evolves into a tough, gray, semitransparent membrane that completely surrounds the embryo. The amniotic cavity of the blastocyst evolves into the **amniotic sac** (Fig. 17.17A). The amniotic sac slowly fills with watery *amniotic fluid* secreted by the amnion and grows to fill the entire endometrial cavity. Added to amniotic fluid is fetal urine and a small amount of fluid and cell debris issuing from the fetal bowel. Both are, of course, sterile. The fetus breathes amniotic fluid during

BASIC FORM, BASIC FUNCTION

"Is It a Boy or a Girl?"

Nowadays the question "Is it a boy or a girl?" is often the first question asked by a friend or relative learning of a new pregnancy. This occurs because modern imaging techniques can often allow visualization of the fetal external genitalia as early as the 10th week of pregnancy. But it is not widely known that "Is it a boy or a girl?" is sometimes the silent question of an obstetrician viewing the "ambiguous genitalia" of a third-trimester fetus on ultrasound or even of a newborn infant. That is, in rare instances, genitalia can form in a way that is neither clearly male nor clearly female because male and female genetalia are derived from the same embryologic structures.

Before we explain the development of the genitalia, it is important to understand the difference between *genotype* and *phenotype*. A human being's *genotype* is his or her genetic composition; phenotype is the *physical expression* of the genotype. For example, a person born

with two X chromosomes is forever a genotypic female and nothing can change that. However, a genotypic female can become a phenotypic male with surgery and testosterone therapy.

The initial indifferent configuration of the genitals consists of three structures—the urethral fold, genital tubercle, and labioscrotal swelling. The "default" developmental pathway results in female genitalia. That is, in the absence of testosterone, the genitalia develop into a phenotypic female: the tubercle develops into the clitoris, the urethral folds form the labia minora, and the labioscrotal swelling becomes the labia majora.

In contrast, male embryos have testes, which produce testosterone. A testosterone metabolite (called dihydrotestosterone, or DHT) stimulates the development of male external genitalia by binding to receptors in primitive genital tissue. The tubercle enlarges and lengthens to form the glans and shaft

"Is It a Boy or a Girl?" (Continued)

of the penis. Most of the urogenital folds fuse to form the spongy urethra, with only the most distal part remaining open as the urethral opening. Finally, the labioscrotal swelling fuses to form the scrotum.

Rarely, phenotype does not match genotype. How would the genitals develop in an XY fetus with defective testosterone receptors? The baby would be phenotypically female because female genitalia is the default pathway, and the infant would develop a clitoris, labia, and vagina instead of a penis, testes, and scrotum. Consider the alternative—an XX fetus exposed to high levels of testosterone during early development. Varying degrees of masculinization ensue, making it difficult to determine phenotypic sex. Some can have an organ that

appears to be a very large clitoris or a very small penis. Frequently, beneath this ambiguous structure, is an opening that might be a vagina, with swellings on each side that might be a divided scrotum or large labia.

In these instances, determining the infant's sex depends on genetic study of sex chromosomes and identification of gonads, either ovaries or testes. These cases require the utmost investigation and care because profound dilemmas are at hand. How should the child be raised? Is a genotypic XY male to be raised as female, and a genotypic XX female to be raised as a male, or are surgery and hormone therapy advisable?

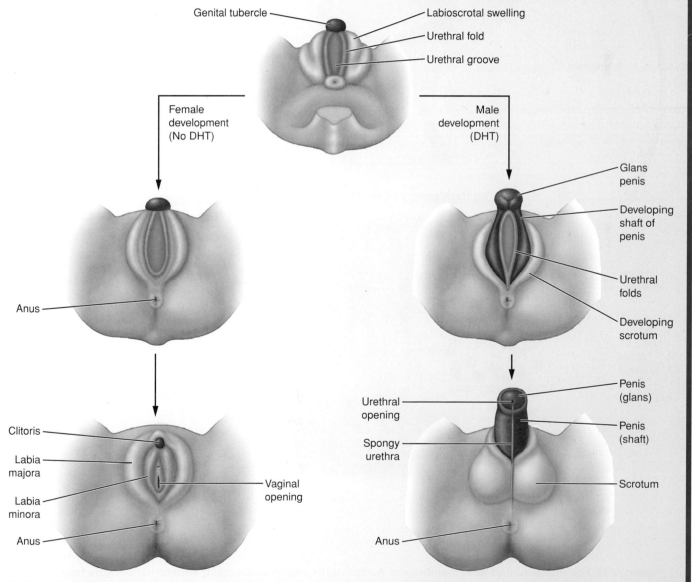

Embryonic development. The male and female external genitalia develop from the same initial structures.

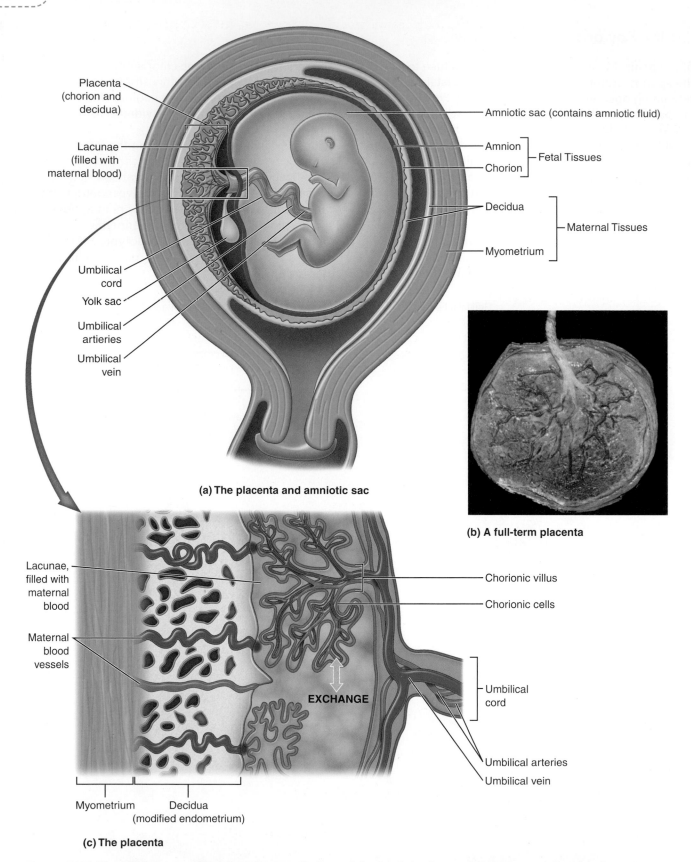

Placenta (chorion and decidua)

Amniotic sac (contains amniotic fluid)

Lacunae (filled with maternal blood)

Amnion — Fetal Tissues
Chorion

Decidua — Maternal Tissues
Myometrium

Umbilical cord
Yolk sac
Umbilical artieries
Umbilical vein

(a) The placenta and amniotic sac

(b) A full-term placenta

Lacunae, filled with maternal blood

Chorionic villus
Chorionic cells

Maternal blood vessels

EXCHANGE

Umbilical cord

Umbilical arteries
Umbilical vein

Myometrium   Decidua (modified endometrium)

(c) The placenta

Figure 17.17. The placenta and fetal membranes. A. Part of the fetal chorion and the maternal decidua form the placenta. **B.** This placenta from a full-term birth shows the blood vessels and the umbilical cord. **C.** Substances are exchanged between fetal chorionic blood vessels and the pool of maternal blood. *Which fetal membrane is closer to the fetus—the amnion or the chorion?*

its time in the uterus. The amniotic sac provides room for movement and growth and cushions the fetus against abdominal trauma to the mother.

The Placenta Nourishes the Fetus

Initially, the blastocyst nourishes itself from nutrients liberated by uterine cell digestion and also avails itself of endometrial secretions. However, these nutrient sources arrive by diffusion and cannot long support the enlarging mass of the embryo. Blood vessels are required, and the *placenta* and *umbilical cord* are the answer.

The **placenta** is a temporary organ of pregnancy that enables the transfer of nutrients and oxygen from mother to fetus and waste materials from fetus to mother. Blood travels between the embryo/fetus and the placenta via the **umbilical cord,** a ropelike structure covered by amniotic membrane that contains two arteries and one vein. At birth, the placenta forms a disc about 1 ft in diameter and 1 in. thick, weighing about 1 lb (Fig. 17.17B).

It is helpful to think of the placenta as the functional equivalent of the lungs, because the umbilical arteries carry oxygen-poor blood from the fetus to the placenta, where the blood picks up oxygen and drops off carbon dioxide (Fig. 17.17B). The single umbilical vein carries freshly oxygenated blood back to the fetus. Moreover, just as the pulmonary alveolar cells prevent air from entering blood but nevertheless allow the exchange of gases between air and blood, so the placenta keeps maternal and fetal blood from mixing but allows substances to diffuse back and forth between mother and fetus.

The placenta is composed of maternal and fetal tissue (Fig. 17-17C). The maternal portion is composed of part of the **decidua,** endometrium that is thickened and modified by the effect of progesterone. Part of the *chorion* forms the fetal portion. Recall from Figure 17.14 that the **chorion** consists of cords of trophoblasts that burrow into the maternal endometrium during implantation. These cords multiply and branch into innumerable **chorionic villi** to form the fetal layer of the placenta. Chorionic villi continue to secrete digestive enzymes that completely dissolve the surrounding endometrium, forming empty spaces (*lacunae)* between the villi, which fill with maternal blood. Fetal blood vessels grow into the chorionic villi, the ends of which float in the pool of maternal blood in the lacunae. Fetal blood in the vessels of the villi is separated from the pool of maternal blood in the endometrial lacunae by two exceedingly thin cell layers: the endothelial cells of fetal blood vessel walls and the chorionic cells covering the villi.

Nutrients, gases, and waste products and other small molecules are freely exchanged between mother and fetus; but large molecules, such as proteins and maternal blood cells, do not readily cross the membrane. Recall from Chapter 12 that, with the exception of identical twins, every person is immunologically alien to every other. Mixing of maternal and fetal blood could invoke serious or fatal antigen–antibody reactions in either mother or fetus.

> **Remember This!** Fetal blood vessels and maternal blood vessels are not continuous. Exchange between fetal and maternal blood occurs in the chorionic villi.

The placenta is well established by about 5 weeks of gestation and reaches its full potential about 2 months later. At this time the placenta takes over the important duty of hormone production from the corpus luteum. Placental hormones are discussed in greater detail further on.

Case Note

17.25. **Midway through labor, Susan's "waters broke" and a large fluid volume soaked the bedclothes. Which fluid was released?**

17.27   The enzyme-containing cap on the sperm head is called the _____.

17.28   How many chromosomes are in a zygote?

17.29   What is the difference between a zygote and a blastomere?

17.30   Which of the following secretes hCG—the decidua or the blastocyst?

17.31   Which blastocyst cavity forms from epiblasts and which from hypoblasts?

17.32   Are endometrial lacunae filled with maternal blood or fetal blood?

17.33   Which fetal structure helps form the placenta— the amnion or the chorion?

17.34   True or false: The hypoblast layer differentiates into the endoderm germ layer.

Pregnancy Changes Maternal Form and Function

Although pregnancy is a normal biological process, it is properly considered a medical condition (though not a disease!) because it can threaten health or even cause death. In any case, it is, like exercise, a physiological condition that causes material changes in physiology and anatomy. Many of these changes reflect the actions of placental hormones, while others result from physical constraints in the maternal abdomen.

Placental Hormones Modify Maternal Physiology

The placenta is an amazingly powerful endocrine organ, secreting large amounts of diverse hormones that support the pregnancy and prepare mother and fetus for childbirth. Some of these placental hormones (such as human chorionic gonadotrophin, hCG) are modified versions, or *analogues,* of anterior pituitary hormones. See Table 17.1 for a brief description of the actions of placental hormones.

Case Note

17.26. Susan noticed that some yoga stretches became easier when she was pregnant, because her joints had a greater range of movement. Which pregnancy hormone is responsible for this effect?

Pregnancy Affects Every Body System

As the fetus grows, the uterus enlarges to hold it, so that by the end of the third trimester, it fills almost the entire abdominal cavity, reaching the xyphoid process and pushing aside the liver, spleen, and intestines. These physical changes combine with hormonal fluctuations to cause many of the symptoms of pregnancy, as illustrated in Figure 17.18.

Although the average infant weighs about 7½ lb (3500 g), the recommended maternal weight gain is about 30 lb. The difference is accounted for by the placenta and amniotic fluid, the increased size of the uterus, enlargement of the breasts, increased maternal blood volume to service the uterus and fetus, and increased

Table 17.1 Placental Hormones

Name	Nonplacental Analogue	Actions
Human chorionic gonadotropin (hCG)	Pituitary luteinizing hormone (LH)	Maintains corpus luteum; stimulates progesterone and estrogen synthesis
Human chorionic somatomammotropin (hCS)	Pituitary growth hormone (GH) and prolactin (PRL)	Prepares mammary glands for lactation; spares glucose for fetus by reducing maternal glucose utilization; enhances maternal growth
Human chorionic thyrotropin (hCT)	Pituitary thyroid-stimulating hormone (TSH)	Stimulates thyroid hormone synthesis, which increases maternal metabolic rate
Relaxin	None	Increases joint laxity, especially in the pubic symphysis, to facilitate childbirth
Corticotropin-releasing hormone (CRH)	Hypothalamic CRH	Stimulates adrenal steroid production; important in fetal lung maturation and labor initiation
Progesterone	Ovarian progesterone	Maintains the endometrium, inhibits myometrial contractions, promotes breast development
Estrogen	Ovarian estrogen	Prepares the myometrium for labor; promotes breast development

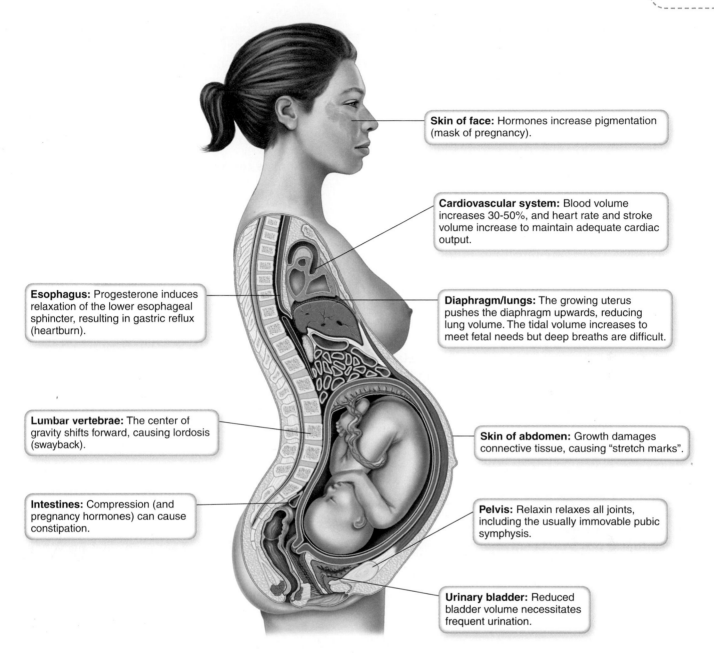

Skin of face: Hormones increase pigmentation (mask of pregnancy).

Cardiovascular system: Blood volume increases 30-50%, and heart rate and stroke volume increase to maintain adequate cardiac output.

Esophagus: Progesterone induces relaxation of the lower esophageal sphincter, resulting in gastric reflux (heartburn).

Diaphragm/lungs: The growing uterus pushes the diaphragm upwards, reducing lung volume. The tidal volume increases to meet fetal needs but deep breaths are difficult.

Lumbar vertebrae: The center of gravity shifts forward, causing lordosis (swayback).

Skin of abdomen: Growth damages connective tissue, causing "stretch marks".

Intestines: Compression (and pregnancy hormones) can cause constipation.

Pelvis: Relaxin relaxes all joints, including the usually immovable pubic symphysis.

Urinary bladder: Reduced bladder volume necessitates frequent urination.

Figure 17.18. Pregnancy. The signs and symptoms of pregnancy reflect hormonal changes and space requirements of the growing fetus. This figure shows the impact of a full-term fetus. *Despite the increasing energy needs of the fetus, pregnant women frequently choose to consume quite small meals. Why?*

maternal fat stores, the latter an adaptation to ensure enough nutrition for mother and fetus in lean times.

Much of the added weight is anterior, which forces pregnant women to adopt a swayback posture (lordosis) to maintain equilibrium. It also adds uncomfortable strain to the lower back.

Breasts, vagina, and labia become engorged with blood. Increased skin pigmentation is common. Areolae darken and the skin of the face, especially around the eyes, may darken, creating a "mask of pregnancy," which is called *chloasma*. In some women, small vascular growths (*spider angiomas*) occur on the face or chest. Bone-to-bone ligaments relax under the influence of the placental hormone *relaxin*, especially in the pelvis, facilitating passage of the fetus down the birth canal. All of these anatomical changes usually revert to prepregnant norms after birth.

Physiological changes are equally dramatic. Basal metabolic rate rises about 20%. Maternal blood volume increases 25% to 50%, and heart rate increases about

10%. Together this results in about a 20% to 30% increase in cardiac output. At the same time, vascular resistance falls, so that blood pressure usually rises only modestly. However, in 5% to 10% of pregnancies, abnormally high blood pressure can occur, especially in the third trimester (the last 13 weeks). This rare condition, *pregnancy-induced hypertension* (also *preeclampsia* or *toxemia of pregnancy*), can cause maternal seizures, kidney failure, or brain hemorrhage.

One of the most common serious complications of pregnancy is **ectopic pregnancy,** a condition in which the fertilized ovum implants in a site other than the endometrium. Various diseases can predispose to ectopic pregnancy, especially infections of the tube, which create a web of scars and kinks in the tube that impedes travel of the fertilized ovum to the uterus. Ectopic pregnancies rarely survive until term because they are not in the sheltering, nurturing environment of the uterus. Frequently they end with painful pelvic–abdominal hemorrhaging, which in some instances can be fatal.

Infection, especially viral infection, early in pregnancy may cause spontaneous abortion of the fetus (commonly called *miscarriage*) or a severe birth defect. Near time for delivery, infection can arise if the mother's amniotic membrane breaks, providing an entryway for vaginal bacteria to infect the uterus and fetus.

Case Note

17.27. Susan took antacids late in her pregnancy. Why was heartburn such a problem?

17.35 Name the placental hormone most similar to luteinizing hormone (LH).

17.36 When pregnant women develop high blood pressure, what is the condition called?

Parturition

Parturition (Latin = "giving birth") or *labor* is the process of giving birth. Normal birth occurs between 38 and 42 weeks of gestation. By definition, birth before 38 *full* weeks is *premature,* and birth after 42 *full* weeks is *postmature.* Premature infants are at risk for being slow learners, having cerebral palsy, and suffering from a

host of other conditions. Because fetal lungs are the last major organ to mature (apart from the brain, which does not mature fully until age 20 to 22), an especial hazard is *respiratory distress syndrome*, a potentially fatal respiratory condition (◀ Chapter 13).

Parturition Includes Three Stages

Parturition is conventionally divided into three stages:

- *Stage 1: Cervical Dilation.* The cervix dilates (opens) and thins (*effaces*) to permit the passage of the baby's head. Before labor begins, the cervix is "corked" by a mucus plug. Myometrial contractions slowly enlarge the cervical opening until it has a diameter of about 10 cm. Dilation can begin weeks before the baby is born in response to barely perceptible, painless contractions (*Braxton–Hicks contractions*). However, the contractions required to complete dilation are much stronger and are generally considered to be painful.
- *Stage 2: Expulsion of the Fetus.* Myometrial contractions, accompanied by maternal "pushing," expel the baby from the uterus.
- *Stage 3: Expulsion of the Placenta.* The placenta separates from the underlying layer of maternal endometrium, and further myometrial contractions expel it through the vagina.

Case Note

17.28. When Susan arrived at the hospital, her cervix was partially dilated. Which stage of labor was she in?

Maternal and Fetal Factors Initiate Parturition

Parturition begins when the uterus becomes more sensitive to the effects of oxytocin because more oxytocin receptors have accumulated in the uterine myometrium.

Once labor begins, contractions accelerate and strengthen until the baby is born. An oxytocin-related positive feedback loop controls the progress of events. You may remember this labor and birth feedback loop from ◀ Chapter 1, where it was used as an example of a positive feedback reaction. Oxytocin stimulates uterine contractions directly, but it also stimulates the production of prostaglandins, which themselves stimulate uterine contractions as well as other parturition-related events, such as relaxation of the cervix and rupture of the fetal membranes (Fig. 17.19). What is more, contractions push the baby's head against the cervix, dilating it

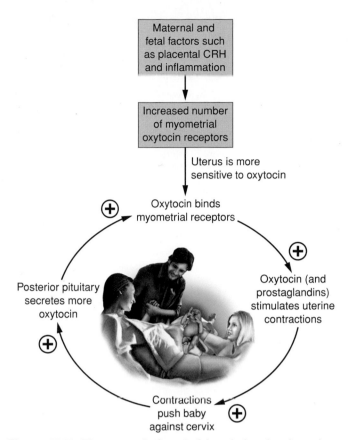

Figure 17.19. The control of parturition. Labor begins when there are enough oxytocin receptors in the uterine myometrium. *Name two chemical signals that stimulate uterine contractions.*

and sending signals to the hypothalamus to produce more oxytocin. The net result is that the more the uterus contracts, the more it is stimulated to contract, so that—as in all positive feedback loops—the process speeds to a conclusion.

If parturition begins either too soon or too late, the baby is likely to experience health problems. Although the factors responsible for preterm and delayed births are not fully elucidated, some studies implicate the production of placental corticotrophin-releasing hormone (CRH). Excess CRH is associated with preterm labor, inadequate CRH with labor failure. A second potential factor is infection. Research suggests that a normal inflammatory reaction within the uterus may increase the number of oxytocin receptors and initiate normal labor, whereas infection or other causes of inflammation prior to 36 weeks can trigger premature birth.

The Newborn Adapts to Extrauterine Life

Think for a moment of the challenge a newborn faces: to go from living under water, getting every one of life's

requirements from its mother, to living in air and apart from the mother. As daunting as it sounds, the newborn adapts remarkably quickly to extrauterine life. As the umbilical circulation is reduced and then ceases, accumulated carbon dioxide activates the newborn's respiratory center. The resulting first breath must fill tiny bronchioles and pry open collapsed alveoli. Without adequate pulmonary surfactant to ease expansion, respiratory distress syndrome occurs.

Prior to this first breath, the fetus has relied on its mother for oxygen, but with the first breath, it relies upon itself. The instantaneous transfer of a newborn's oxygen source from the placenta to the lungs depends on the design of the fetal circulation.

The essence of fetal circulation is this—*during intrauterine life, the placenta serves as lungs, kidneys, and intestines.* The fetal lungs are bypassed in two ways: (1) most blood entering the right atrium travels directly to the left side of the heart through a hole (the *foramen ovale*) in the atrial septum; therefore little blood enters the right ventricle to be pumped into the pulmonary artery and lungs; and (2) any blood that does find its way into the right ventricle and pulmonary artery is diverted into the aorta by the *ductus arteriosus* before it gets to the lungs (Fig. 17.20). A third modification, the *ductus venosus,* enables about 80% of the fetal blood supply to bypass the liver. This proportion increases in low-oxygen conditions, in order to maximize oxygen delivery to the all-important fetal heart and brain. With the first breath, this is what happens:

1. A cytokine released by the newly inflated lungs stimulates constriction of the ductus arteriosus, which suddenly directs a flood of blood into the lungs instead of allowing it to continue into the aorta.
2. Within several heartbeats, this blood from the lungs arrives back in the left atrium, thus raising left atrial pressure enough to equalize right and left atrial pressure.
3. This, in turn, causes blood to stop flowing through the foramen ovale. Even though it remains open for much of the first year of life, the pressure equalization ensures that there is negligible flow directly between the atria.

Persistence of blood flow across the atrial septum or through the ductus arteriosus in the neonate can cause serious health problems, as discussed in◄ Chapter 11.

All of the specialized fetal blood vessels—ductus arteriosus, umbilical arteries, and ductus venosus—wither and become ligaments within the first year of life.

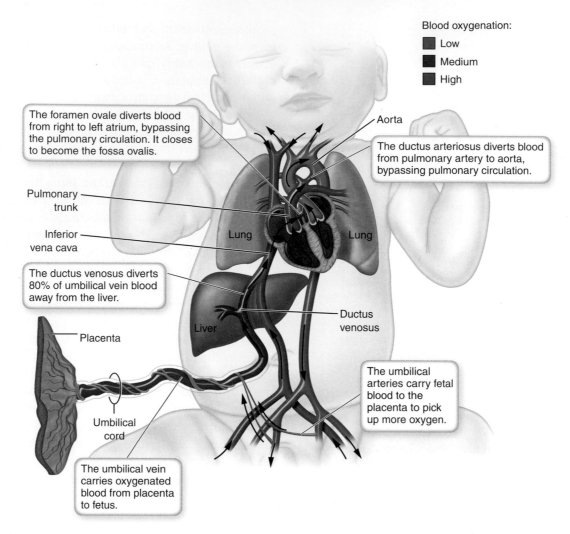

Figure 17.20. Fetal circulation. During fetal life, specialized vessels and openings shunt blood away from the nonfunctioning lungs and toward the placenta. *Which vessel shunts blood away from the pulmonary circulation—the ductus arteriosus or the ductus venosus?*

All of the other body systems come on line in short order, as the newborn begins to drink colostrum and produce urine and feces. The first feces are in the form of a sticky substance called *meconium*. We discuss the further maturation of body systems in Chapter 18.

17.37   During which stage of labor is the baby actually born?

17.38   True or false: Labor begins because estrogen receptor number in the uterus increases.

17.39   True or false: The fetal heart starts to beat only with the first breath.

17.40   Name the "hole" in the fetal atrial wall.

Case Discussion

Sexually Transmitted Disease and Infertility: The Case of Susan, Mark, and Toby

Let us return to our case.

Figure 17.21 outlines all of the steps required for conception and the birth of a healthy infant. A failure at any point of this complex, cooperative process means either no infant or an unhealthy one. Mark and Susan were subjected to a variety of tests to determine why they were failing to conceive.

The first test conducted was a semen analysis. The goal of this test was to see whether Mark could produce an adequate number of normal-appearing sperm with

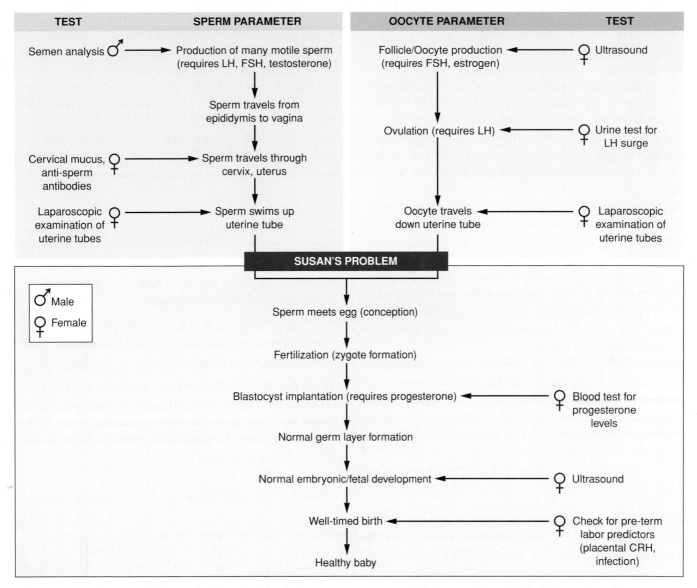

Figure 17.21. The case of Susan, Mark and Toby. Laboratory tests revealed the problem behind Susan's and Mark's infertility and the successful conception of Toby followed. *What does the laparoscopic examination of the uterine tubes evaluate?*

enough vigor to swim to the oocyte. The findings were entirely normal. If Mark's semen had been abnormal, blood tests would have been performed to see if he produced adequate amounts of LH, FSH, and testosterone, all of which are necessary for sperm production.

Next were tests of Susan's cervical mucus, which in some women can prevent passage of sperm into the uterus. For example, some women produce antisperm antibodies, which react with sperm and paralyze their ability to swim. Mark's sperm was mixed with Susan's cervical mucus and no such effect was found. In other women the cervical mucus is too thick for sperm to penetrate, but a check of Susan's mucus proved it to be normal.

The physician next ordered tests to determine whether Susan's ovaries were making follicles. Ultrasound of her ovaries revealed images proving that a normal number of developing follicles were present. The fact that Susan had regular periods suggested that follicles were releasing their oocytes, but this also needed to be checked, because some women have anovulatory cycles; they menstruate but do not ovulate. It is difficult to detect ovulation itself but relatively easy to detect the blood LH surge that precedes it. Blood tests proved that

Susan's LH levels increased dramatically about 10 days after her last menstrual period began, so, in all likelihood, she ovulated about 36 hours later.

The next step was to be sure the anatomical pathway from vagina to ovary was not obstructed. Susan's pelvic exam suggested a problem because a mass was found in the region of her left ovary. Dye visible on x-ray was injected under pressure into her cervix and was seen to travel into the uterine cavity and part of the way through her right fallopian tube. The left fallopian tube was not visible.

In the next stage, a surgeon made a small incision in Susan's lateral abdomen and inserted a narrow tube containing a tiny TV camera, a procedure called *laparoscopy* (Greek *lapara* = "flank"). Via the camera, the surgeon was able to examine Susan's abdominopelvic cavity and found that both uterine tubes were closed by scars from old infections. One tube was successfully repaired, but the other (along with its ovary) was not salvageable and was excised.

Postoperatively, Susan's fertility was certain to be lower than that of a normal woman, since only one ovary remained. Happily, however, within 6 months, she missed a menstrual period. She quickly bought a pregnancy test kit and tested her urine. The test was positive, indicating the presence of chorionic gonadotropin. An ultrasound image of her uterus confirmed that Mark's sperm had met up with one of Susan's oocytes, and fertilization and implantation had occurred normally.

But even at this happy stage, not all was guaranteed to proceed successfully. Some women, for example, do not produce enough of the progesterone necessary to maintain a rich endometrium and are prone to have an early spontaneous abortion. Fortunately, blood progesterone tests can detect this condition and the mother can be treated with supplemental progesterone. In Susan's case, however, blood progesterone was normal and a second ultrasound confirmed that the embryo was developing normally.

Unfortunately, however, at 33 weeks, Susan went into premature labor, which threatened to bring Toby into the world before his organ systems were mature enough to support life independently. Susan had additional tests to try to determine why she went into early labor, because one premature birth is a risk factor for future premature births. Infection prompts inflammation, and inflammatory signaling molecules (such as prostaglandins) can induce labor. Although no infection was found, there are other causes of inflammation; therefore, as a precautionary move, Susan was treated with ibuprofen, which inhibits prostaglandin production. As a second precautionary move, she was also given oxytocin blockers prevent uterine contractions.

Labor stopped for a week, during which blood tests revealed abnormally high levels of CRH, which may have initiated the early labor. Another precautionary move was to give Susan cortisone in an effort to force Toby's lungs toward mature production of pulmonary surfactant. That Toby did not have respiratory distress syndrome at birth suggests that this tactic was successful.

> ### Case Note
>
> **17.29.** **Explain why Susan received injections of cortisone, oxytocin blockers, and prostaglandin inhibitors (anti-inflammatories).**
>
> **17.30.** **In an attempt to understand why Susan began labor prematurely, her amniotic fluid was tested for infectious agents. Why?**
>
> **17.31.** **A section of Susan's placenta was sent to a research lab that investigates premature labor. They found that her placental tissue produced more CRH than was normal for a 32-week placenta. Why was this finding significant?**

Abortion and Contraception

The medical and lay definitions of abortion vary considerably. The public generally uses the term *miscarriage* to describe spontaneous early termination of pregnancy with death of the fetus and reserves the term *abortion* for *deliberate* termination of pregnancy. However, clinicians define **abortion** as *any* interruption of pregnancy—incidental, accidental, or intentional—with death of the fetus before the end of 22 full weeks of gestation or if the fetus weighs less than 500 g. *Miscarriage* is not a medical term.

A pregnancy that ends between 22 and 38 full weeks of gestation and with a living infant is known as a **premature birth.** A pregnancy that ends after 22 weeks with death of the fetus spontaneously in the uterus or during delivery is called a *stillbirth*.

Many early abortions go unnoticed. Perhaps half of all conceptions do not end with implantation. Even when implantation occurs, perhaps 50% or more abort without the woman even knowing she was pregnant, and 15% of *known* pregnancies end in spontaneous abortion. About half of spontaneous abortions are due to fatal genetic defects. Other causes include alcohol and drug abuse or a maternal infection that also infects the fetus (for example, rubella). Older mothers have more spontaneous abortions, and a woman who has had one is at increased risk for another.

Contraception Intends to Prevent Pregnancy

Contraception (Latin *contra* = "against," and *cept* = "catch") is any method intended to prevent conception or successful implantation of a fertilized ovum. Some methods prevent union of sperm and ovum; others prevent implantation after fertilization. Effective contraception requires dependability. The most dependable methods rely little on human behavior. Of course, some methods are better than others, but only one is 100% effective: total abstinence.

Because we have precisely defined contraception as prevention of fertilization or implantation, one method often considered a contraceptive is not included in the list below. *Mifepristone (RU-486)* is a drug that induces a very early chemical abortion. It blocks progesterone's ability to maintain a lush secretory endometrium and reverses the calming effect progesterone has on uterine smooth muscle. When administered with a small amount of prostaglandin, mifepristone causes the endometrium to begin to degenerate within 12 hours. Uterine contractions soon follow and expel the products of conception. Mifepristone is approved for use only under the care of a physician and only in the first 7 weeks of pregnancy.

Some Contraceptive Measures Prevent Fertilization

Methods to prevent fertilization can be a cooperative effort between partners or can interfere in other ways with the production or travel of either sperm or ova.

Behavioral methods include:

- *Coitus interruptus*, withdrawing the penis a moment before ejaculation, is too contrary to the visceral urges and reflexes involved to be a reliable method of contraception. Moreover, some sperm can enter the vagina prior to ejaculation.
- The *rhythm method*—abstinence during the few days just before and after ovulation—can be somewhat

(a) Male condoms   **(b) An intrauterine device (IUD)**

Figure 17.22. Contraception. A. Male condoms are made from latex, polyurethane, or natural materials. **B.** Intrauterine devices (IUDs) are implanted within the uterus. *Of the two types of contraceptives shown here, which type prevents sexually transmitted diseases?*

effective but requires monitoring ovulation by one method or another, and accurate determination of the date of ovulation is difficult to achieve with accuracy.

No method yet approved interferes with sperm *production*, but research is promising. Methods interfering with the *travel* of sperm include the following:

- *Spermicidal drugs* (drug-saturated sponge, gels, creams, etc.) can be introduced into the vagina prior to intercourse and kill sperm before they can reach the ovum.
- *Male condoms* are widely used and are effective in reducing sexually transmissible disease, especially when combined with spermicides. However, some males resist wearing them because they reduce sexual pleasure (Fig. 17.22A). Female condoms have been available since 1993, but their use is not widespread.
- *Vasectomy*, cutting out a section of both vas deferens and tying off the ends, blocks the passage of sperm into the distal portion of the ductal system. It is very effective, especially if semen analysis a few weeks after surgery proves than no sperm are present. Without further surgery to reattach the ends, vasectomy is permanent. Contrary to ill-informed concerns, vasectomy does not interfere with the pleasure of orgasm, the ability to have an erection, or male hormone production.
- *An intravaginal diaphragm* or *cervical cap* blocks sperm from entering the cervical canal. These devices are most often used in conjunction with spermicides.

Methods interfering with the *production* of ova are the most widely used and effective in developed nations.

- *"The pill."* Oral tablets containing a mixture of progesterone and estrogen are very effective and, when used consistently have a failure rate below 1%. Taken in cycles of 28 tablets, the first 21 tablets contain drug, the last 7 contain filler only. The steady state of blood hormones inhibits GnRH, LH, and FSH production by negative feedback. Follicular development and ovulation cease, and secretory endometrium does not develop, so menstrual flow is minimal.
- *Other hormonal methods.* A steady state of elevated blood progesterone can also be delivered by skin patch, subcutaneous implant, quarterly (every 3 months) injection, or a drug ring inserted into the vagina.

The only method interfering with the *travel* of ova is *tubal ligation*, a procedure similar to vasectomy: a portion of each uterine tube is removed in order to prevent meeting of ovum and sperm. Like vasectomy, this method is very effective and largely irreversible, especially if imaging studies done after surgery prove that the channel is sealed.

Some Contraceptive Measures Prevent Implantation

Rather than interfere with gamete production or travel, some methods aim to interfere with implantation.

- *Intrauterine devices (IUDs).* IUDs prevent implantation of a fertilized ovum (Fig. 17.22B). They are metal or plastic coils or similar devices saturated with progesterone that are inserted into the endometrial cavity. They are widely used, especially in developing nations, because they are cheap and effective. However, they are less popular in developed nations because their failure rate, though low, is higher than that of the pill, and they are sometimes associated with painful periods, uterine perforation, or infection.
- *"Morning-after pills" (MAPs).* MAPs contain estrogen and progesterone in concentrations higher than "the pill" and are designed to be taken within 3 days after unprotected intercourse. They work by scrambling natural hormonal signals in a way that prevents implantation of a fertilized ovum.

17.41 What is the difference between a spontaneous abortion and stillbirth?

17.42 Does vasectomy interfere with the production of sperm or with its transit?

17.43 Distinguish between mifepristone and the morning-after pill in terms of drug action and timing of use.

Word Parts

Latin/Greek Word Part	English Equivalent	Example
blast-, -blast, blasto-	Formative cell	Blastocyst: a fluid-containing sac that is the formative structure for the embryo
chori-, chorio-	Chorion (outer fetal membrane)	Human chorionic thyrotropin: a hormone produced by the chorion
cyst	Fluid-containing sac	Blastocyst: a fluid-containing sac that is the formative structure for the embryo
endo-	Inner	Endometrium: inner lining of the uterus
-genesis	Development, generation	Spermatogenesis: generation of sperm
lact-, lacti-, lacto-	Milk	Lactiferous duct: duct carrying milk

Word Parts (continued)

Latin/Greek Word Part	English Equivalent	Example
mamm-, mammo-	Breast	Human chorionic somatomammotropin: a chorionic hormone that stimulates breast growth
metr-, metro-	Uterus	Myometrium: muscular (myo-) layer of the uterus
oo, ov/o, ovul/o	Ovum, egg cell	Oocyte: cell that gives rise to an ovum
troph-, tropho-	Nourishment	Trophoblasts: cells that nourish the formative cells (the inner cell mass)

Chapter Challenge

CHAPTER RECALL

1. **The cremaster muscle is found in the**
   a. scrotum and spermatic cord.
   b. uterus.
   c. oviduct wall.
   d. bladder.

2. **Sperm are produced by the**
   a. seminal vesicles.
   b. prostate gland.
   c. seminiferous tubules.
   d. epididymis.

3. **Which of the following cells divides by mitosis?**
   a. Spermatogonia
   b. Primary spermatocytes
   c. Spermatids
   d. None of the above

4. **From anterior to posterior, the female perineal orifices are the**
   a. anus, urethra, vagina.
   b. vagina, urethra, anus.
   c. urethra, vagina, anus.
   d. urethra, anus, vagina.

5. **Luteinizing hormone stimulates**
   a. testosterone release.
   b. ovulation.
   c. progesterone release.
   d. all of the above.

6. **Which of the following is *not* part of the uterus?**
   a. Cervical os
   b. Introitus
   c. Myometrium
   d. Endometrium

7. **Which uterine layer is lost each month as a woman menstruates?**
   a. The myometrium
   b. The entire endometrium
   c. The stratum functionalis
   d. The stratum basalis

8. **The first polar body is a daughter cell of the**
   a. primary oocyte.
   b. secondary oocyte.
   c. ovum.
   d. oogonium.

9. A primary follicle contains
a. a secondary oocyte.
b. a large antrum.
c. an oogonium.
d. follicular cells.

10. Progesterone is produced
a. by the follicle.
b. during the preovulatory phase.
c. by the corpus luteum.
d. by all of the above.

11. The LH surge during the reproductive cycle stimulates
a. the resumption of meiosis II.
b. increased estrogen production from the follicle.
c. rupture of the ovarian wall.
d. all of the above.

12. Which stage of the sexual response occurs in males but not usually in females?
a. Arousal
b. Orgasm
c. Plateau
d. Refractory period

13. Breast milk is
a. not produced until several days after parturition.
b. ejected from a single opening in the middle of the nipple.
c. ejected as soon as the mammary glands synthesize it.
d. much lower in fat than colostrum.

14. Capacitation occurs
a. within the epididymis.
b. before the sperm contacts the ovum.
c. after the sperm penetrates the zona radiata.
d. after the nuclei of the sperm and ovum fuse.

15. Which of the following is a single cell that can develop into an entire organism?
a. Trophoblast
b. Blastocyst
c. Blastomere
d. A and C.

16. Trophoblasts differentiate to form
a. the chorion.
b. the amnion.
c. the decidua.
d. all of the above.

17. The fetal membrane contributing to the placenta is the
a. yolk sac membrane.
b. decidua.
c. amnion.
d. chorion.

18. A drug that inhibits oxytocin's action would impede
a. milk synthesis.
b. the initiation of labor.
c. development of the corpus luteum.
d. all of the above.

19. Pregnancy is commonly confirmed by measuring urinary levels of
a. human chorionic gonadotropin.
b. human chorionic thyrotropin.
c. progesterone and estrogen.
d. relaxin.

20. Which of the following birth control methods has the highest failure rate?
a. Coitus interruptus
b. Condoms
c. Birth control pills
d. Diaphragm

CONCEPTUAL UNDERSTANDING

21. Explain how a small follicle produces small amounts of estrogen but a large corpus luteum produces large amounts of estrogen and progesterone.

22. Compare and contrast the following:
a. The birth control pill and the morning-after pill
b. Vesiculase and prostate specific antigen
c. Amnion and chorion

23. Most body processes are controlled by negative feedback, but positive feedback regulates many female reproductive processes. Name and briefly describe three different positive feedback loops involved in female reproduction.

APPLICATION

24. Occasionally, men who cannot ejaculate but still produce viable spermatozoa still wish to produce children. A very fine needle is used to aspirate spermatozoa. Would it be better to extract the gametes from the epididymis or from the testis itself?

25. Ms. W. has had four spontaneous abortions (miscarriages) in 3 years. Blood tests revealed a deficiency in progesterone. Could this deficiency play a role in Ms. W.'s miscarriages?

You can find the answers to these questions on the student Web site at
http://thepoint.lww.com/McConnellandHull

18

Life

Major Themes

- In one manner or another every body system is dependent on every other body system.

- Not all body systems are mature at birth; some will not fully mature until near age 30.

- We age cell by cell.

- Decline of system function generally parallels age.

- Stress may have positive or negative effects on body function.

- Exercise is necessary for good health.

Chapter Objectives

Exercise 732

10. Describe the steps required to bring an oxygen molecule from the atmosphere to cells and use it to generate ATP.

11. Explain how different types of exercise training improve the functioning of different body systems.

Life and Death 735

12. List factors that accelerate and delay aging.

"She thought it was a wasp." A multigenerational family goes hiking.

As you read through the following case study, assemble a list of the terms and concepts you must learn in order to understand it. Note that, unlike other chapters, this chapter does not have a separate case discussion. Instead, the case is discussed throughout the narrative.

To celebrate his 71st birthday, Tom treated his children and grandchildren to a family vacation at a mountain dude ranch. After a picnic lunch, he decided to take his granddaughter Kate, 7, and her mother, Lea, 38, on a hike up to a nearby mountain meadow to see the wildflowers.

"There won't be any yellow jackets will there, Mommy?" Kate asked.

"No," Lea said. "Wasps don't live up this high in the mountains. To her dad she added, "A yellow jacket stung her last week."

As the trail led upward, Kate scampered ahead, while Tom and Lea lagged behind. Soon Tom became short of breath, his respirations deep and fast. He said to Lea, "Let's take a little break. There was a time I could have done this all day, but not any more."

Lea called for Kate, who frolicked back with a smile and no noticeable shortness of breath.

"Look at her, Dad. She doesn't even know it's 10,000 feet," Lea said. "I think of myself as fit, but even I'm a bit winded at this altitude."

After a quick gulp of water, Kate began tugging Lea ahead to the meadow. Soon they had disappeared over the next rise, leaving Tom, who had to take frequent stops to catch his breath, behind. Despite a regular exercise regimen incorporating weights, he was finding the climb difficult.

Upon reaching the meadow, Lea enjoyed the view while Kate picked some flowers. The placid scene was broken by a wild shriek from Kate, who began clawing at her hair and screaming about a wasp. Lea rushed to Kate's side and immediately saw a small yellow butterfly tangled in Kate's long, curly hair. While working to free the insect, she tried to explain to Kate that it was merely a butterfly, but Kate, still clutching a cluster of red wildflowers, was screaming and thrashing so frantically that it was several moments before her mother succeeded. Finally, the butterfly floated away and Lea took Kate in her arms and comforted her.

Meanwhile, Tom had heard the screams and broken into a run. He quickly found himself at his limit, but fearing that something terrible might have happened to Kate, he pushed himself until he finally reached the top.

Need to Know

Chapter 18 integrates information from many chapters. However, the following topics are the most relevant.

- Arrangement of electrons around the atomic nucleus ⬅ (Chapter 2)

- DNA, genes, and genetic mutations ⬅ (Chapter 3)

- How muscles generate ATP; determinants of muscle fatigue ⬅ (Chapter 7)

- Sympathetic nervous system ⬅ (Chapter 8)

- Determinants of cardiac output and blood pressure ⬅ (Chapter 11)

- Tidal volume, respiratory muscles ⬅ (Chapter 13)

- Adrenal gland, cortisol action, regulation of cortisol secretion, insulin action ⬅ (Chapter 15)

- Puberty ⬅ (Chapter 17)

"She's okay, Dad," Lea said. "Poor thing, a butterfly got caught in her hair. She thought it was a wasp. She got so scared I could feel her heart pounding as if it was about to leap out of her chest, and she was breathing harder than you are now."

Tom, still gasping for air, could only nod his head to signal that he understood. After a minute of rest he was able to say a few words of comfort to Kate, who had become calmer and was breathing easier, too.

Seeking to distract Kate, Tom initiated a game he regularly played with her, which depended on the fact that he was red-green colorblind. "What color are your flowers, Kate? Are they yellow?"

Kate, immensely pleased that she could outwit her grandfather at this simple task, immediately brightened and said, "No, Poppy, they're red! You're so silly."

Do you remember our overview of anatomy and physiology in ⬅ Chapter 1? There we thought of ourselves as looking at our subject from an airplane at 30,000 ft. Then, in our study of chemistry in ⬅ Chapter 2, we took a magnifying glass to study atoms and molecules, the building blocks of all matter, of all life. In the succeeding chapters we hiked through the hills and dales of anatomy and physiology, exploring cells and tissues and the various systems that compose the body. Put another way, we've been examining the trees, not the forest. Yet if there is truth in the observation that something can be "greater than the sum of its parts," then life itself surely must be one of those things.

So what is life? In the first paragraph of ⬅ Chapter 1 we defined it by saying, "When form and function combine to produce biologic activity, we call it *life*." But in retrospect, it is hard to settle for a definition as sterile as "biologic activity." It does not ring true, does it? You are an expert in life: after all, you live it, and you know there is more. So what is it?

It's a daunting question. Answers will vary from irreverent to clinical to transcendent. Our pages are limited, so we must be brief and focus on the sum of human form and function. And in doing so we hope, in this chapter, to give you a taste of life as it is lived—how our systems, integrated into a greater whole, work together in the sweep of time.

Age is an issue of mind over matter. If you don't mind, it doesn't matter.

Mark Twain, American humorist, writer and lecturer (1835–1910)

Genetics, Inheritance, and Life

Each of us is the sum of two factors: genetic gifts from our parents and environmental influences beginning in the womb. Here, we focus on genetics.

A **genome** is all of the genes that are common to a single species. The human genome consists of thousands of genes arranged in our 23 chromosome pairs (Fig. 18.1). Recall from the previous chapter that one of these pairs is the **sex chromosomes:** this pair in *females* consists of two X chromosomes and in *males* of one X sex chromosome and one Y sex chromosome, which is markedly different from the X. All other chromosomes are **autosomes;** that is, nonsex chromosomes. There are 22 autosome pairs, each consisting of two homologous chromosomes, one from each parent. The homologues that make up any given autosome pair contain, of course, the same genes; that is, genes that control the same trait.

Remember, each gene codes for a particular protein, and each protein contributes to one or more *traits*. A **trait** is a quality such as hair color or height that results from the expression of one or more genes. Several diseases—including cystic fibrosis, Huntington's disease, and rare forms of obesity—are controlled by a single gene and are described as **monogenic.** But many of our traits—facial appearance, for example—result from the influence of multiple genes and are said to be **polygenic.** The familiar observation to a mother about a newborn girl, "She looks just like you," reveals that,

deep in our cells, multiple genes are producing a composite effect.

When genes are normal, their effect fades into the everyday physiology of life. But when a gene or a combination of genes is abnormal, disease may arise. That some genetic diseases, such as cystic fibrosis (◀ Chapter 3), are inherited in a predictable fashion is proof of the specific and significant effect single genes can exert on our health and functioning. On the other hand, type 2 diabetes is encouraged by, but not strictly caused by, a combination of many genes, most of them as yet undiscovered.

Gene Pairs Control Genetic Traits

We've noted that homologous chromosomes contain homologous genes, one from your father and one from your mother, each devoted to the same trait. Geneticists use the term **allele** to identify a version of a gene as it occurs on one chromosome of a pair. The term is useful because the DNA of one allele may be "spelled" slightly differently from the DNA of another allele of the same gene. Again, with the exception of genes on the sex chromosomes, we can have two different alleles for each gene, one on each homologous chromosome.

To appreciate how each person's unique set of genes (or **genotype**) affects his or her physical and molecular characteristics (or **phenotype**), it is necessary to understand that some alleles are *dominant* and others are *recessive*. A **dominant** allele (indicated by uppercase italics, such as *C*) is one that expresses itself as a trait even when a recessive allele is present. That is to say, the

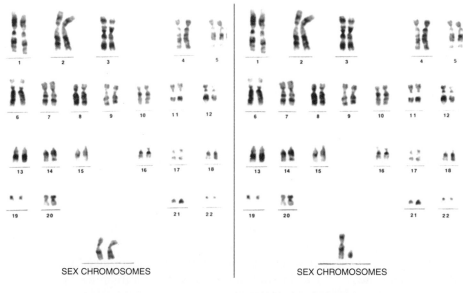

(a) Normal female karyotype

(b) Normal male karyotype

Figure 18.1. Human karyotypes. This photograph illustrates the chromosomes of a female (left side) and a male (right side). *Which gender has one long sex chromosome and one short sex chromosome—male or female?*

dominant allele has enough force of expression to overcome the effect of the other, different version of the same gene on the other chromosome of the pair. A **recessive** allele (indicated by lowercase italics) is one that does not express itself as a trait when a dominant allele is also present; that is, its effect (expression) is weak. For example, if the allele combination is *CC* or *Cc*, the trait will be whatever *C* says it will be. Only if the combination is *cc* will the *c* (recessive) trait be expressed.

As an example, let's return to Julia, the child with cystic fibrosis in our case study in ◄ Chapter 3. Cystic fibrosis results from a defect in the CFTR gene, which codes for a chloride ion channel. It is a common mistake to believe that this gene and the protein it encodes are present only in people with cystic fibrosis. In fact, everyone has the CFTR gene, even though few people have cystic fibrosis. The reason is this: the gene was first discovered in a patient with cystic fibrosis, and so it was named for the disease. There are different alleles, or versions, of this gene. One allele (*C*) is dominant and encodes a functionally *normal* transporter. A different allele (*c*) is recessive and is *abnormal*: it encodes a transporter that is not correctly targeted to the cell membrane and is therefore dysfunctional.

Each person has two alleles of the CFTR gene, one from each parent. There are three possible combinations of these two alleles, as shown in Figure 18.2A:

- Two normal alleles (*CC*, top left box): The individual inherits a *C* from each parent. This person makes normal transporters and does not have cystic fibrosis.
- One normal allele and one dysfunctional allele (*Cc*, top right and bottom left boxes): Although the individual inherits one dysfunctional allele (*c*), it is recessive; the dominant *C* enables the individual's cells to manufacture enough functional transporters for them to function effectively. This person does not have cystic fibrosis.
- Two dysfunctional alleles (*cc*, bottom right box): The individual inherits a *c* from each parent. Since this person's cells can make nothing but dysfunctional transporters, the person has cystic fibrosis. Because cystic fibrosis appears only in individuals with two recessive alleles, it is called an *autosomal recessive disease.*

We know that for the CFTR gene (located on chromosome 7), Julia's *phenotype* is "cystic fibrosis;" that is, she has the disease, which means that her *genotype* must be *cc*. We can describe this aspect of Julia's genotype as **homozygous** recessive (*homo* = "same"), because she has two copies of the same (recessive) allele. An individual with two dominant alleles is homozygous dominant for that trait;

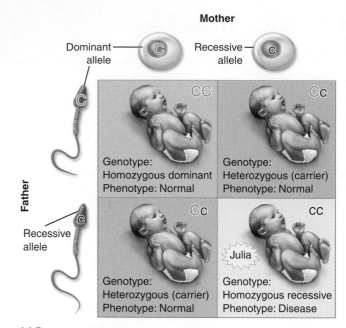

(a) Punnett square

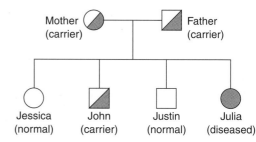

(b) Pedigree analysis

Figure 18.2. Genetic analysis. A. Punnett squares demonstrate all possible genotypes and phenotypes resulting from a particular match. **B.** Pedigrees map the inheritance of a particular genetic trait. *Which child is a homozygous dominant male—John or Justin?*

someone with one dominant and one recessive allele is **heterozygous** (hetero = "different") for that trait.

With a bit of further reasoning we can even predict the genotype of Julia's parents' CFTR gene. We know that neither has cystic fibrosis, so each must have at least one *C*. We also know that each of them must have a *c* allele, since Julia received two copies of *c*. The only possible answer is that both parents are *Cc*. Because both parents carry the defective allele but do not have the disease, both parents are **carriers.**

Most monogenic disorders are recessive and require two copies of a recessive allele to be expressed as disease, but a few are caused by dominant alleles; that is, only one copy of the allele is necessary for the disease to occur. *Huntington's disease,* for instance, is a fatal neurological disorder that, though genetic, does not appear until adulthood. The disease allele (*H*) is dominant, so only homozygous recessive (*hh*) individuals escape the disorder, and patients with *Hh* or *HH* have the disease.

There Are Several Ways to Illustrate Inheritance Patterns

Punnett squares (Fig. 18. 2A) are a useful way of depicting the likelihood of different phenotypes appearing in offspring. They are based on the simple principle that each parent contributes one allele, and that alleles combine randomly. The possible paternal alleles (from spermatozoa) are conventionally placed vertically and maternal alleles (from ova) horizontally, with the possible combinations in boxes. One of the four possible combinations in this example results in a child with cystic fibrosis, so the probability of two carriers having an afflicted child is one in four, or 25%. Whenever carrier parents have a child, there is also a 50% chance that the child will be a carrier (*Cc*), and a 25% chance that the child will be genetically normal (*CC*). However, note that these forecasts are of limited predictive value because probabilities are close to norms only in large sample sizes (e.g., 100). In Julia's family, for instance, genetic testing revealed that two of her siblings were homozygous dominant (*CC*) and only one sibling was a carrier (*Cc*).

Clinicians frequently assemble *pedigrees* to map the inheritance of a genetic disorder. A **pedigree** is a table of the generations arising originally from a single set of parents that shows the genotype and phenotype of each individual. As shown in (Fig. 18.2B), lines mark the relationships between individuals, and symbols indicate the gender and genotype of each individual. Julia (and all related females) are represented by circles, her father and other male relatives by squares. Julia's circle is solid black because she has the disorder in question. The symbols of her parents and one sibling are only half black, because they are heterozygotes and carry the trait. The two siblings without the disease allele are indicated by open, white symbols.

> *Remember This!* Heterozygotes for recessive traits are carriers but do not exhibit the trait. Heterozygotes for dominant traits exhibit the trait.

Some Single-Gene Defects Are Linked to Sex Chromosomes

Julia's cystic fibrosis arose due to a genetic mutation of a gene on chromosome 7, an autosome. Most genetic dis-

orders are caused by genes located on autosomes; however, some arise because of defects on sex chromosomes and are referred to as **sex-linked traits.** Almost all such disorders involve the X chromosome. This is because the Y chromosome contains few genes vital to life—after all, half of the population, women, do not have a Y chromosome, so no evolutionary survival value accrues to having essential genes on the Y. The few Y genes that exist are related to testicular development, and defects involving these genes can cause male infertility.

Referring back to our Case Study, Tom's red–green color blindness is caused by a defective gene on his X chromosome. It is inherited a bit differently from Julia's cystic fibrosis, as you can see by studying the pedigree in Figure 18.3A. This pedigree traces the inheritance of color blindness through three generations of Tom's family. Notice that females are either carriers of the trait (half-black circles) or normal (open circles). In females, color blindness behaves like an autosomal recessive disorder: they are heterozygous carriers of the trait. Males, conversely, are either color-blind (black squares) or normal (open squares) but *never* carriers.

Males have only one X chromosome, so it takes just one copy of the abnormal allele to manifest the disorder: no second X chromosome makes up for the defect with its normal allele. As you would expect, red-green color blindness is most common in males, afflicting up to 10% of the male population (and thereby frustrating the fashion sense of many female partners). The only way that a female can be red–green color-blind is to inherit the defective allele from both her mother and her father.

Sex-linked traits can also be analyzed by Punnett squares (Fig. 18.3B), but the identity of the sex chromosome (either X or Y) must be provided as well as the allelic variant. For this reason, the nomenclature for sex-linked traits indicates the sex chromosome. In our example, we abbreviate the dominant allele as X^R and the recessive (color blindness–causing) allele as X^r. The Y chromosome does not have the gene in question, so is simply abbreviated Y.

> ### Case Note
>
> **18.1. Look at Figure 18.3. What are Kate's and Lea's genotypes?**

Traits Are Determined by Genes and Environment

Each of the traits discussed above—cystic fibrosis, Huntington's disease, and color blindness—is entirely

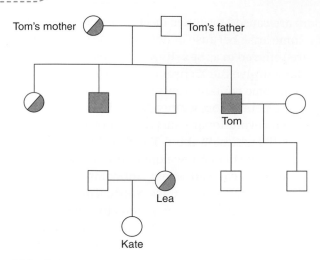

Tom's mother   Tom's father

Tom

Lea

Kate

(a) Pedigree analysis, sex-linked trait

Mother

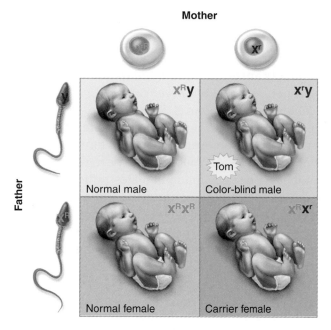

Father

	$X^R y$	$X^r y$
	Normal male	Tom Color-blind male
	$X^R X^R$	$X^R X^r$
	Normal female	Carrier female

(b) Punnett square, sex-linked trait

Figure 18.3. The inheritance of color blindness. A. A pedigree demonstrating the inheritance of red-green color blindness. **B.** A Punnett square representing the possible outcomes of mating between Tom's parents. *Is it possible for Tom's mother and father to produce a daughter with color blindness?*

determined by a specific abnormality of a single gene. That is, the environment has nothing to do with it. Julia will always have cystic fibrosis and Tom will always be color-blind, regardless of what Julia's mother ate during her pregnancy or how much alcohol Tom consumes with dinner. Most human traits are far more complex and reflect the actions of multiple interacting gene products as well as the influence of the environment.

Researchers attempting to distinguish between genetic and environmental effects often study identical (monozygotic) twins, which develop from a single zygote that split into two embryos. Monozygotic twins raised apart thus share identical genes but not an identical environment. Such twin studies have revealed, for instance, that obesity has a strong genetic component. Unfortunately, mice studies suggest that obesity is a polygenic trait: in mice, about 6,000 different genes contribute to the tendency to increased body weight. The polygenic nature of obesity makes it very difficult to determine exactly which genes are involved. And as strong as genetic influences are in obesity, environmental factors—home environment, socioeconomic status, food advertising, opportunities for physical activity, and so on—are thought to be significantly more important than genetics. Add these factors together and you can see just how daunting is the task to sort out what is genetic and what is environmental. For instance, the strongest predictor of childhood obesity is parental obesity. But is this relationship due more to genes, or to the similarity between the family members' diets and levels of physical activity? Both genetics and environment are involved.

Considering how difficult it is to determine factors influencing physical characteristics, do we have any hope of understanding behavior? That is, does DNA govern behavior, or is all behavior a reaction to the environment? Moreover, could your genes make you behave in some way that encourages obesity, cancer, migraine headaches, chronic fatigue syndrome, or any other condition? These are just some of the intriguing questions currently being studied by behavioral geneticists. Recently, after years of misuse in the service of racist agendas, behavioral genetics has become a respected academic discipline. Like their colleagues, behavioral geneticists often study identical twins raised apart. These careful twin studies have revealed some interesting similarities in the twins' behavior patterns. For instance, identical twins raised apart often have similar intellectual abilities and achievements, similar tastes in food and clothing, and similar levels of participation in churchgoing and other religious activities. On the other hand, these studies also show that the behaviors of identical twins differ in more ways than they correspond.

It seems safe, therefore, to liken our DNA to a genetic seed for which the environment provides soil, water, and sun. In one environment, a certain seed will sprout and flourish. In another, it will not thrive. Life offers countless examples of gifted young people who failed to fulfill their promise, and an equal number of people, supposedly made of plainer stuff, who soared to unexpected achievement. Important as it is, nothing in your DNA automatically destines you to live in a homeless shelter, just as nothing in the blueprint assures that you will sing at the White House.

> ### Case Note
>
> **18.2. Tom, Lea, and Kate all have a low body-mass index (BMI). Tom attributes their slim physiques to good food and lots of exercise. Is he telling the whole story?**

18.1 What is a polygenic trait?

18.2 True or false: Every person has a CFTR (cystic fibrosis) gene.

18.3 Which genotype is homozygous recessive—AA, Aa, or aa?

18.4 Is it possible to have a child with cystic fibrosis if one parent is a carrier but the other is not and does not have the disease?

18.5 The Huntington disease gene has been located on chromosome 4. Is Huntington disease autosomal or sex-linked?

18.6 What would be the genotype of a color-blind female?

18.7 What possible parental genotypes could result in a color-blind female?

Stages of Life

Most textbooks of anatomy and physiology, this one included, are based on the form and function of young adults—a stable, perfected stage of development. This has the added advantage of appealing to readers, most of whom are young adult students. However, there is much more to the anatomy and physiology of life. In this section we briefly consider anatomical and physiological distinctions of those younger and older than the perfected form we have been studying. We divide human development into convenient stages (Fig. 18.4):

- **Infancy** (Latin *infant* = "unable to speak") is the time between birth to 12 months (some prefer 18 or 24 months).

- **Childhood** spans the period between infancy and adolescence.
- **Adolescence** begins with the initiation of sexual maturity and ends with attainment of physical maturity.
- **Adulthood** begins with physical maturity, which is about age 20.
- **Older adulthood** is variously defined as beginning around age 60 to 70.

> ### Case Note
>
> **18.3. Kate has not yet developed breasts or other signs of sexual maturity. Is she a child or an adolescent?**

Infancy and Childhood Transform Needy Newborns into Independent Teenagers

Infancy is a period of splendid growth and maturation. Most babies triple their birth weight and double their length within the first year. Compared with other mammals, human infants are born in a remarkably immature and vulnerable state, one that requires long-term dependency and nurturing. For example, a newborn colt is able to stand within a short time after birth and quickly develops the ability to frolic alongside its mom in the pasture. The helpless human newborn, however, is capable of little more than feeding, eliminating, and crying and requires about a year to acquire the ability to walk just a few steps. However, by early toddlerhood, the child is walking, running, dancing, singing, and often even speaking a few words. These dramatic developmental changes reflect the maturation of every organ system.

The curvature of the spine is an easily observable example of the developmental changes at work. The newborn spine has a C-shape, fitting for the cramped space of the uterus. An 18-month old toddler has an S-shaped spine with a prominent lumbar curve and "pot belly." By the school-age years, the child's spine develops a more relaxed S-shape, which persists throughout life (see Fig. 6.28).

Infants are particularly susceptible to infectious disease, because their immune systems are immature and have not been programmed by interacting with potential environmental threats (◀ adaptive immunity, Chapter 12). For example, the average child catches six to ten colds per year, whereas most adults average only two to four. In the first few weeks after birth, infants are protected by some maternal antibodies that have crossed

Figure 18.4. The life stages of the authors. *Who has the worst haircut?*

the placenta; but after birth, these fade away. To a certain extent, they can be replaced by breast-feeding—mother's milk contains antibodies that, in the first 6 months of life, can be absorbed by the infant's intestine and pass into the bloodstream. Reflecting the fading reliance upon maternal antibodies, vaccinations are often begun as early as birth to 2 months of age in order to prevent life-threatening infectious illness.

The preschool and primary school years are marked by continued growth and development, albeit at a slower rate of change than during infancy and adolescence. Continued maturation of the neuromuscular system, for instance, is evidenced in the growing mastery of locomotion and fine motor skills. Body proportions continue to change as the legs grow faster than the trunk or the head (Fig. 18.5). The lymphatic and immune systems grow rapidly and peak about puberty. The thymus reaches maximum size near puberty and then involutes into anatomic insignificance in adults.

Case Notes

18.4. Is Kate's spine curved in a C-shape or an S-shape?

18.5. In proportion to height, who has a bigger head—Tom or Kate?

Adolescence Is the Transition to Sexual Maturity

Adolescence (Latin *adolescere* = "grow to maturity") is defined in different ways by various authors. Some equate this period with the teenage years (13 to 19), whereas others mark its beginning with the onset of puberty—which can occur as early as age 10—and its ending with legal adulthood, usually age 18. **Puberty** (Latin *pubes* = "adult") is a 24- to 36-month period of transition during which sexual maturity appears and sexual reproduction becomes possible. It is marked by

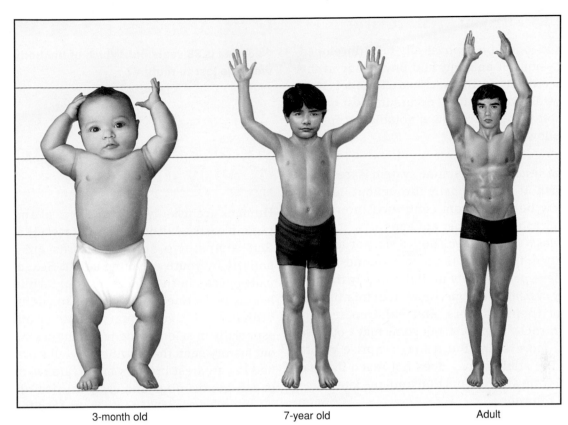

3-month old   7-year old   Adult

Figure 18.5. Body proportions. The head is much larger proportionally in the newborn than in the adult. *Who has the largest hands in proportion to the head—a baby or a 7-year-old?*

the development of secondary sex characteristics, such as axillary and pubic hair in boys and girls, breast development in girls, and the deepening of the voice in boys. In girls, menarche (the first menstrual period) is an unmistakable mark of its arrival. The age of onset of puberty varies considerably: it most often occurs around 10 to 12 years of age in girls and around 11 to 13 years of age in boys.

Puberty is accompanied by a growth spurt—a 20% to 25% gain of height. Indeed, one of the primary markers of the end of adolescence is the attainment of final adult height, which occurs roughly around age 18 (girls) or 21 (boys). In response to the increased load, the sacral bones begin to fuse into the sacrum, a process that is complete by the age of 25 to 33. The activity of sweat and sebaceous glands increases, which can result in acne.

The dramatic changes of adolescence—physical and emotional—are initiated and maintained by hormonal changes, as discussed in ◄ Chapter 17. Having the external biological trappings of adulthood—facial hair for boys, breasts for girls—adolescents consider themselves mature enough to do adult things. But they fail to realize that their frontal lobes—the seat of mature judgment—are not mature and will not be mature until they are in their twenties. As every parent knows, this leads to trouble.

For most adolescents, the teenage years are not associated with significant disease. However, eating disorders and body image disturbances often surface during this time, as both males and females attempt to conform to unrealistic media images of idealized bodies. In addition, for some teens, the promotion of safe sexual practices is critical to prevent pregnancy and sexually transmitted diseases. Unrestrained by mature frontal lobes and increasingly independent from their parents, teenagers frequently engage in reckless exploratory behaviors, including the abuse of alcohol and drugs. Motor vehicle accidents, drownings, and other trauma can follow: 80% of adolescent deaths are a result of injury, both accidental and deliberate. Finally, some cancers—such as leukemias, lymphomas, Hodgkin's disease, and bone cancers—are more common during adolescence than in childhood.

Adulthood Is Physical Maturity

An **adult** is someone who is physically fully developed, and our discussions of anatomy and physiology in this textbook concentrate on the early adult model. Near complete physiological maturity occurs in most organs in childhood or adolescence. Some notable exceptions include the following:

- *The immune system.* The immune system is fresh and naive at birth. It matures daily throughout life as it "learns" new behavior from continued interaction with new microbes and other antigens.
- *The musculoskeletal system.* Bones do not achieve peak density in females until the midtwenties; males do not achieve peak density until their early thirties.
- *The urinary system.* Everyone knows that infants cannot control their bladders, and children usually achieve control by age 3 or 4, but some may not do so for several more years. But it may surprise you to learn that the adult kidney does not reach its final anatomical form and full function until the late twenties or early thirties.

The examples above discuss physical maturity. Emotional maturity is another matter. Or is it? At birth, the brain is by far the least mature of organs, and it gains new form and new function—which is manifest by improved emotional maturity—year by year until in the early twenties, when physical maturity is complete.

But that is not the end of the story. Throughout life, the brain continues to evolve and adapt according to our life experiences. It is now clear that learning, something we do well for most of our adult lives, requires the brain to rewire itself in microscopic detail. These changes are subtle but nevertheless are genuine anatomical differences that accumulate as we age.

The behavior of the brain with some intracranial tumors and the recovery of some stroke patients offer examples that are easier to see. For example, some slow-growing brain tumors can, over the course of many years, press gently but firmly on the motor cortex that controls movement of a certain part of the body, say the right foot. But the process is so slow that as some neurons are damaged their function may be "remapped" to others. As a result, if the tumor is discovered and removed, the brain is left with an empty spot but no function has been lost. Similarly, some stroke patients lose their ability to speak (aphasia), but with speech therapy can regain lost function as the brain remaps the activity.

Case Note

18.6. Lea is 38 years old. Which of her body systems were the last to mature?

Aging and the Decline of Body Functions

Humans are usually at their physical and intellectual peak in early adulthood. It is not merely a coincidence that competitive athletic records are held almost entirely by youth or young adults. Sexual function in males peaks in the late teens, and fertility in females begins to decline before age 30. In addition, the most brilliant and revolutionary flights of human genius, especially in science and mathematics, have throughout history been the province of young brains: Einstein had his greatest insights in his late twenties, Newton in his late twenties and early thirties, and so on. But if you are over age 30, there is no need for dismay: our ability to act thoughtfully and with good judgment seems to improve well into advanced age. Although Alzheimer's disease and other forms of dementia rob some people of memory and cognition, many people find emotional fulfillment in the wisdom accumulated in a long life.

Aging is simply the process of growing older. It is not restricted to the elderly: even infants are aging. Still, after several decades have gone by, the physical and mental limitations associated with advancing years begin to interfere with form and function, reducing physical performance and the quality of life. These age-related deteriorations, which are technically known as **senescence,** are associated with:

- Reduced *physiological capacity.* That is, organs and systems no longer function optimally. Muscles contract with less force and ears hear less acutely.
- Diminished *adaptability.* Older adults cannot easily adjust to environmental changes (e.g., temperature). They succumb more easily to heat stroke and hypothermia.
- Increased disease *susceptibility.* Minor illnesses in adults (such as influenza) can be life-threatening in the elderly.

Many studies suggest that the simple measure of eating less may delay senescence. You can read about it in the nearby clinical snapshot, titled *Eat Less, Live Longer?*

CLINICAL SNAPSHOT

Eat Less, Live Longer?

To lengthen thy life, lessen thy meals.

Benjamin Franklin, *Poor Richard's Almanack* (1737)

Cosmetics, plastic surgery, elixirs, miracle lotions and potions, and countless other products and services—many with only the slimmest relation to science—form a multibillion-dollar industry and are hawked to a public obsessed with delaying the aging process. They want, and who can blame them, a quick fix to look and feel younger.

What if there were a prescription for extended youth that actually works *and* saves money? Benjamin Franklin proposed such a measure—eating less. Scientific research appears to support his thesis. When animals (rodents, primates, and invertebrates) are fed a balanced diet of 30% fewer calories than they would normally consume, they live up to 40% longer than animals fed a normal diet. They also have better health; age-related degeneration in all body systems is delayed (especially in the nervous system), the immune system is strengthened, and neoplasms develop less frequently. Although similar long-term feeding studies with humans are fraught with challenges, short-term studies suggest that individuals who either consume a low-calorie, nutrient-dense diet or who fast intermittently can obtain similar benefits. For, instance, blood pressure lowers, glucose tolerance improves, and blood lipid profiles show increased HDL and reduced LDL.

Theories abound attempting to explain the beneficial effects of caloric restriction. The reduction in adipose tissue may be important, since obesity is directly related to health problems and mortality rates. The *hormesis theory* suggests that chronic exposure to a low-level stressor (such as caloric restriction or even low-dose radiation) enhances the body's ability to deal with more intense stresses that have an ill effect on health. Caloric restriction also reduces free radical production and decreases cell damage.

Even if caloric restriction works in humans—is it worth it? It is commonly known that excess body weight (a body mass index [BMI] greater than 25) is associated with increased mortality. But it is less well known that low body weight (a BMI below 18.5) is also associated with increased mortality. Thus, caloric restriction would increase mortality if it drove BMI below 18.5. In addition, most people find a calorie-

Benjamin Franklin. Benjamin Franklin's abdominal girth suggests that he did not follow his own advice: "To lengthen thy life, lessen thy meals."

restricted diet highly challenging to adhere to in the long term. If you normally eat about 2,000 calories per day, you would have to drop that to about 1,400 calories. What is more, the food you did eat would have to be highly nutritious, providing all the necessary nutrients for the lower energy intake. That means you would have to say goodbye to pepperoni pizza and cookie-dough ice cream. Finally, the meticulous meal planning required would mean that you would be eating most of your meals at home. There go the parties, the barbecues, the restaurant meals. Could you do it—not just for a week or two but for a lifetime? Your answer probably depends on the relative value you place on years of life versus quality of life. So—if you would describe yourself as a chocoholic or a burger connoisseur, you'll probably want to pass on the calorie-restricted diet.

Case Note

18.7. Simply based on their age, who is at greater danger of heat stroke—Tom or Lea?

Senescence Affects Every Body System

The age-related decrease in quality of life and increase in death rate reflect changes in every body system. These are summarized in Figure 18.6.

Skin changes are the most obvious sign of aging. Gravity tugs relentlessly: buttocks and breasts droop and skin sags. Sun damage produces dark or light spots and skin cancer. Aging skin loses the subcutaneous fat and elastic fibers that keep it robust and unwrinkled. Too, aging skin is thinner, more easily wounded, and slower to heal. The sum is a bonanza for plastic surgeons and cosmetics sales.

Bone mineral density declines with age, more precipitously in women after menopause, and bones become more easily fractured and subject to abnormal spinal curvature. An abnormal forward spinal curve, *kyphosis*, is especially common. Bones must have weight-bearing stress to remain dense and strong. Recalling our case, Tom limits his loss of bone mineral density by regular weight-bearing exercise. Bones respond to the forces placed upon them by depositing more mineral and protein. Degenerative changes in the joints, described as *osteoarthritis,* also limit mobility.

Even with frequent exercise, muscle mass decreases with age. Tom's muscle mass is probably only about half of what it was when he was younger. Since muscle fiber number and size is a determinant of power development, his muscles have also lost strength. More subtle changes also impact muscle function; the proportion of type I fibers (slow twitch) increases at the expense of type II fibers (fast twitch). The loss of these faster, more powerful type II fibers results in more falls and accidents, because older individuals cannot respond with the fast, powerful movements required to recover balance.

Exacerbating this problem is declining functionality of the vestibular apparatus of the inner ear, which renders us unstable on our feet. Often, an older person with a cane or walker may not be using the device to make up for a sore knee or weak hip—his or her musculoskeletal system may be in good shape—the device serves to steady the person's balance. In order to offset these detrimental changes in sensation and response, Tom's trainer regularly includes exercises that require balancing, which strengthen trunk muscles important in maintaining balance.

Cardiovascular performance declines as we advance into adulthood. Studies suggest that the maximum heart rate that any individual can achieve during exercise is approximately 220 minus the person's age in years. That is, someone 30 years old can reach about 190 beats per minute, and someone 70 years old can reach about 150. In like manner, respiratory capacity declines: the average 70-year-old has 30% to 40% less vital capacity than at his or her physical peak in young adulthood.

The gastrointestinal system and liver are also affected by advancing age, but less noticeably so. Our ability to taste declines, too, but remember that most of the sense of taste is attributed to our sense of smell, which also fades with age. Indeed, some elderly people lose nearly all of their sense of smell. Of particular note is the decline in liver size and blood flow. Recall from ← Chapter 16 the central role of the liver in glucose metabolism. Reduced liver function impedes the ability of older adults to deal with low blood glucose levels. Moreover, the liver metabolizes and inactivates many drugs, including common anti-inflammatories such as ibuprofen. Prescribed dosages take this into account but are often based on the hepatic function of young adults. Reduced liver function means that medications can persist in the blood for a longer time in older adults, so starting dosages should, in some cases, be reduced.

With aging the kidneys gradually shrink, losing about 20% to 25% of their mass by age 75. The glomerular filtration rate, a key measure of renal function, declines as well, falling to about half its peak by late adulthood.

As discussed in ← Chapter 17, reproductive capacity declines too. Women cease ovulating at menopause, which usually occurs about age 50. In men, fertility does not cease so abruptly. After age 50, there is a slow decline in the number of sperm capable of fertilizing an ovum, but some viable sperm production may persist well into old age.

At several other points in this narrative we refer to the decline of mental function that to some degree affects every person living into their seventies and beyond. By age 80, the brain has lost about 5% to 15% of its peak mass, mainly because of the loss of neurons. Although it may not be evident in daily life, mental agility tests in older adults show a slowing in problem-solving ability; nerve conduction velocity slows, which slows reflexes; and voluntary motor movements slow.

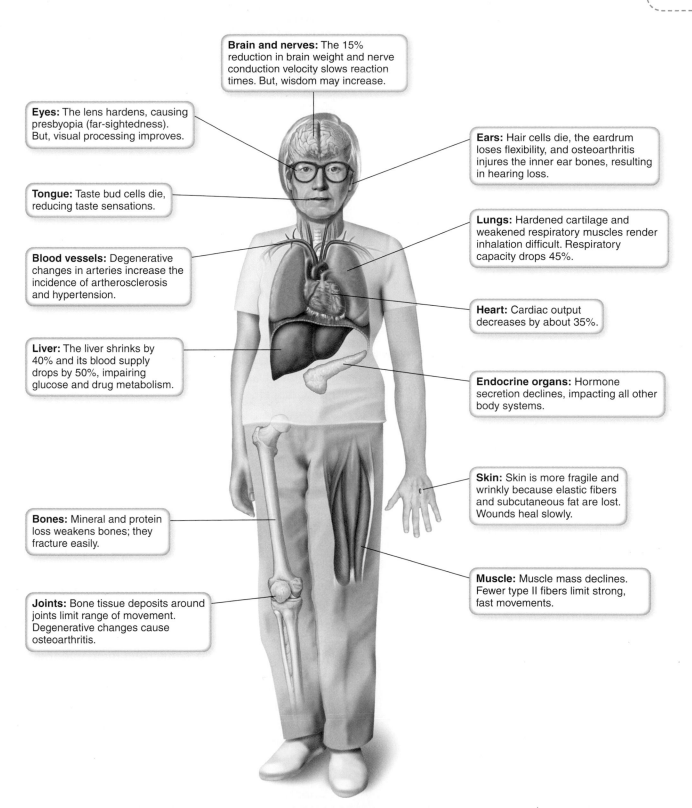

Brain and nerves: The 15% reduction in brain weight and nerve conduction velocity slows reaction times. But, wisdom may increase.

Eyes: The lens hardens, causing presbyopia (far-sightedness). But, visual processing improves.

Ears: Hair cells die, the eardrum loses flexibility, and osteoarthritis injures the inner ear bones, resulting in hearing loss.

Tongue: Taste bud cells die, reducing taste sensations.

Lungs: Hardened cartilage and weakened respiratory muscles render inhalation difficult. Respiratory capacity drops 45%.

Blood vessels: Degenerative changes in arteries increase the incidence of artherosclerosis and hypertension.

Heart: Cardiac output decreases by about 35%.

Liver: The liver shrinks by 40% and its blood supply drops by 50%, impairing glucose and drug metabolism.

Endocrine organs: Hormone secretion declines, impacting all other body systems.

Skin: Skin is more fragile and wrinkly because elastic fibers and subcutaneous fat are lost. Wounds heal slowly.

Bones: Mineral and protein loss weakens bones; they fracture easily.

Muscle: Muscle mass declines. Fewer type II fibers limit strong, fast movements.

Joints: Bone tissue deposits around joints limit range of movement. Degenerative changes cause osteoarthritis.

Figure 18.6. Senescence and body systems. Cell death (or lack of cell replacement) and impaired neural/endocrine communication negatively impact every body system. *How are bones affected by cellular senescence?*

18.8. Tom observed that a glass of wine has a greater effect on him than it would have when he was younger. Can any age-related changes in body systems account for this change?

Declining Endocrine Function Is Responsible for Some Senescence

Central to many age-related deteriorations is the decline of the endocrine system. Growth hormone secretion, for instance, can be as low in an elderly person as in a younger person who is symptomatic from GH deficiency. Decline in the production of sex steroid hormones is also important. In women, the decline in estrogen is important as the cause of postmenopausal hot flashes and in the increasing fragility of bones in older women. In men, the decline in testosterone production accounts for increasing abdominal girth due to belly fat deposits, declining sex drive, and loss of muscle mass and strength.

Some of the positive effect of hormones is due to their ability to stimulate cells to reproduce. And—given that cancer is an uncontrolled growth of new cells—some hormones, especially estrogen, can promote breast or uterine cancer in women. This is not to say that estrogen alone can cause cancer. However, it is indisputable that some cancers arise more readily with the aid of estrogen. For reasons not well understood, testosterone does not appear to promote the appearance of new prostate cancers in men but will accelerate the growth of existing prostate cancer and other cancers. Therefore the age-related decline of hormonal output in both women and men acts to some extent to limit the occurrence and growth of tumors.

18.9. Could endocrine decline partially account for Tom's loss of muscle mass? Explain.

Senescence Reflects Cellular Changes

An older adult's reduced function reflects cellular senescence. The senescence of hair cells in the inner ear, for instance, is one cause of hearing loss. There are two interrelated problems with aging tissues and cells: First, cells become damaged and no longer function normally. Second, cells lose the ability to replicate; therefore damaged cells are not replaced as readily.

Cells Accumulate Damage as They Age

Scientists now agree that aging is mainly the result of accumulating damage to the molecules that make up our cells—especially proteins, lipids, and nucleic acids (DNA and RNA). As molecules are damaged, cell function becomes less robust. As cell function declines, our tissues and organs do not perform as well, and soon our health begins to deteriorate. Moreover, DNA damage causes DNA mutations (changes in the DNA sequence), and mutations in key genes may result in cancer or impaired cellular function. But what causes the damage? Although pollutants and other environmental villains are obvious candidates, equally important are the adverse consequences of normal cellular processes and everyday necessities such as oxygen.

Reactive Oxygen Species Are the Dark Side of Oxygen

Throughout this textbook oxygen appears time and again as the most important molecule in the maintenance of life: without it, we die quickly. There is, however, the other side of the coin. The metabolic reactions necessary for life also produce **reactive oxygen species (ROS),** molecules containing an oxygen atom with an unpaired electron. An *unpaired electron* is an electron not paired with another in its own electron shell and that also is not part of a bond with another atom. Unpaired electrons are unstable and seek to pair with an electron in another atom. Examples include superoxide (O_2^-), the hydroxyl radical (OH^-), and hydrogen peroxide (H_2O_2).

ROS are a type of **free radical**—a general term describing any molecule or atom with an unpaired electron. Most free radicals act as *oxidants,* molecules that remove electrons from other molecules. In their quest for stability, they strip electrons from stable molecules (with paired electrons), converting them into free radicals. The newly radicalized molecules stop performing their normal function and instead set out to convert other stable molecules into free radicals.

ROS are a natural part of body chemistry and play a role in certain enzymatic reactions. In such reactions they are valuable but strictly controlled. However, if uncontrolled, they can cause considerable damage. For example, when they oxidize (strip electrons from) lipids in the cell membrane, they damage the membrane's ability to regulate the movement of substances into and out of the cell. This damage frequently kills the cell. ROS can also damage DNA, and damaged DNA is prone to copying errors that result in DNA mutations. Accumulated DNA mutations often promote the development of cancer.

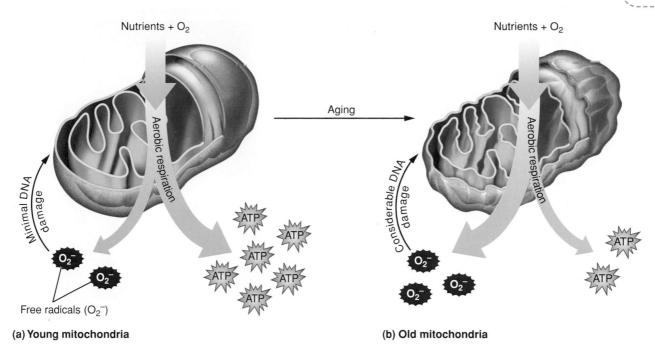

Figure 18.7. Mitochondria and reactive oxygen species (ROS). A. Young mitochondria produce a few ROS and many ATP molecules. ROS will damage mitochondrial (and nuclear) DNA if they are not eliminated, but they are in low concentration. **B.** Old mitochondria produce many ROS and relatively fewer ATP molecules. The ROS worsen the already existing DNA damage. *Are ROS generated during aerobic or anaerobic respiration?*

Not only does ROS damage accumulate as we age, it also appears to accelerate (Fig. 18.7). Remember that mitochondria contain their own DNA, which does not mix with nuclear DNA. Mutations accumulate in mitochondrial DNA as cell divisions occur. Older mitochondria, having divided many times, have many more mutations, are less efficient, and produce less ATP and more ROS. The consequence is a positive feedback loop: ROS damage to mitochondria produces more ROS, which causes even more damage. The result is a spiraling positive feedback loop that ends in cell death.

As we have seen, ROS are necessary for life, but they are also dangerous. Cells limit (but do not completely eliminate) ROS production and action by the use of **antioxidants.** Antioxidants include compounds such as vitamins C and E, which can give up electrons to an ROS without becoming free radicals themselves. Other antioxidants include enzymes that convert ROS into inactive compounds. For instance, one antioxidant enzyme system converts superoxide into hydrogen peroxide and then into water. Because there is so much evidence that free radicals can damage body systems, it is natural to think that increasing our antioxidant defenses can help us to live longer. For example, science has proven that fruit flies genetically engineered to clear away free radicals live 50% longer than normal fruit flies. *But do antioxidants delay aging and prevent human disease?* We do not know for sure. Whereas many studies support a benefit from antioxidant vitamins and minerals consumed naturally in whole foods, studies of antioxidant supplements (tablets, capsules, and so on) are inconclusive. In fact, some researchers even speculate that antioxidants taken in supplement form may act as prooxidants in some situations, promoting the formation of free radicals. The jury is still out on this question. Someday an antioxidant strategy might be developed that prolongs life and reduces the prevalence of chronic diseases, but currently no such magic elixir exists.

Case Note

18.10. Whose mitochondria would produce more ROS—Tom's or Lea's?

Telomere Length Limits Cell Replacement

Cells with damaged molecules frequently stop dividing. However, even undamaged old cells eventually lose their ability to reproduce, and the key to this behavior seems to be a structure at the end of each chromosome—the **telomere,** a region of DNA that loses a bit of its length with every cell division. Although much remains to be learned about telomeres, it appears that when a cell runs out of telomere, it "retires." After cells divide about 50 times, they quit the hard work of dividing and leave the task to other cells. Of course, organs with many retired cells do not function as well as others. Indirect proof of this point is that our organs shrink as we age. Compared with that of a healthy young adult, an old adult's brain is 5% to 15% smaller, the liver weighs 35% less, and the respiratory capacity of the lungs is about 50% less.

This built-in telomere limit may protect against cancer. Normal cells become cancerous by accumulating DNA mutations. As cancer cells divide and age, their accumulating mutations may make them become more aggressive. However, with each division, they also lose telomere length. Thus, they may be forced to retire before they accumulate enough mutations to become fully malignant or more aggressive.

ROS or poor DNA repair appears to accelerate telomere shortening. Thus, ROS not only damage cells but also insidiously limit their ability to produce new cells.

Case Note

18.11. Tom's chromosomes are minutely smaller than Kate's chromosomes. Why?

Abnormal Apoptosis Hastens Aging

Recall from Chapter 3 that cells have a natural life span, and after living out their natural life of a few days, a few months, or a human lifetime, they die by "natural suicide" in a carefully regulated, orderly process called **apoptosis.** Science now has evidence that some degenerative diseases, especially Alzheimer's disease and others affecting the nervous system, are caused by excessive apoptosis. So, in a sense, Alzheimer's can be considered abnormally rapid aging of the brain.

Progerias Are Disorders of Accelerated Aging

Proof that DNA and telomeres are important is provided by the *progerias,* a family of genetic diseases associated with premature aging. The most severe type, *Hutchinson–Gilford disease,* results from a defect in the *lamin* gene (Fig. 18.8). Lamin proteins direct DNA and RNA synthesis and help form the nuclear envelope. One of the many results of lamin mutations is that cells can no

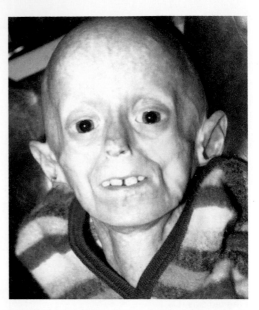

Figure 18.8. Progeria. This 10-year-old girl shows the typical signs of premature aging associated with progeria, such as hair loss; thin, wrinkled skin; and the loss of subcutaneous adipose tissue. *Some of these signs reflect accelerated programmed cell death. Is this process described as necrosis or apoptosis?*

longer readily reproduce themselves. Afflicted children begin to lose their hair as early as age 12 months and tend to develop the adult form of diabetes (type 2) during early childhood. Not surprisingly, they do not tend to develop cancers because, after all, cancer is an uncontrolled growth of new cells, and in progeria cells have a hard time reproducing. Most children with progeria die by the age of 13 as a result of cardiovascular disease.

In *Werner syndrome,* another variety of progeria, patients lack a protein that participates in DNA replication, telomere maintenance, and DNA repair. Without this protein, DNA replication frequently stalls and new cells are not produced to replace retiring ones. Although the cells of all progeria patients have difficulty dividing, their DNA develops mutations at a normal rate. But Werner syndrome patients also lack normal DNA repair, so they tend to develop cancers. People with Werner syndrome are also prone to cardiovascular disease and diabetes mellitus, and they usually die at about age 40.

18.8 When is the transition between infancy and childhood?

18.9 The fact that auto insurance rates are much lower for a 25-year-old than for an 18-year-old reflects the delayed maturation of which organ system?

18.10   What is the difference between senescence and aging?

18.11   What is the advantage of declining estrogen and growth hormone production in the elderly?

18.12   Many older adults wear hearing aids. Why does hearing acuity decrease as we age?

18.13   How do reactive oxygen species increase DNA mutations?

18.14   Do telomeres lengthen or shorten as we age?

Stress

Stress is difficult to define, but the origin of the word offers some insight: *stress* derives from Latin *strictus*, which means "drawn tight." Our modern understanding of stress begins with Hans Selye of McGill University (Canada) and his pioneering stress studies in the 1930s. Selye injected rats with various noxious substances, expecting to have the animals react one way to one substance and another way to another. But what he found was that the reaction was the same for every substance. Ultimately he came to the conclusion that all the substances equally induced a state of stress in the animals. He told reporters: "Everyone knows what stress is, but nobody really knows." Still, he did come up with a useful definition of stress as "the nonspecific response of the body to any demand for change."

Stress occurs in response to a **stressor,** which is unhelpfully defined as an agent that causes stress. Stressors can be physical, such as lifting more weight or running a longer distance. Deprivation can also be classified as a stressor: for instance, energy deprivation during dieting, fasting, or starving; oxygen deprivation when, like Tom, taking a high-altitude hike; deprivation of light during a long northern winter; or deprivation of caffeine or another drug in someone with an addiction, like Andy from our case in Chapter 4. In addition, most agents taken to excess can become stressors. For example, recall our case in Chapter 2—too much lemon juice was stressful (in this case, chemically toxic) to the student forced to drink it. Of course stressors can also be emotional, such as taking an important exam, moving to a new address, beginning or ending a relationship, or watching a scary movie.

Stress Involves Perception

Developing a more concise definition of either stress or stressors is extraordinarily difficult, because stress involves perception. That is, stressors differ between individuals and over time in the same individual. On a Saturday afternoon in the football stadium, what is stressful to some—the final score of the game—may bring relief and joy to others. Or consider Kate's stress response to the butterfly in her hair. She perceived a fluttering insect in her hair as a threat, based on her previous experience with a yellow jacket. Kate's stress was entirely due to her viewpoint: she thought the insect was going to sting her. A child with a different viewpoint may have been charmed to the point of giggling at having a butterfly in her hair. This example underlines the importance of *perception* in emotional stressors: it is often the perception of being in danger or out of control, not the reality of it, that produces stress.

Stressors are not always unpleasant. The Social Readjustment Rating Scale, developed by Thomas Holmes and Richard Rahe in the early 1970s, allocated varying numbers of points to different stressors. Marriage ranks quite highly on the list, as does "outstanding personal achievement." Even taking a vacation can be stressful.

Case Note

18.12.   Identify some of Lea's stressors as described in the case study.

The Stress Response Includes Three Phases

The stress response includes three phases: alarm, adaptation, and exhaustion. These are not immediately sequential events; rather, if alarm continues without adaptation, exhaustion will occur. Adaptation is a learned behavior that lessens with familiarity with the perceived threat.

Alarm Phase

The first response to a stressor is the *alarm phase* (Fig. 18.9). It has two components: sympathetic (autonomic) and endocrine.

The sympathetic nervous system reaction to a stressor is immediate and affects many body systems. When Kate's brain interpreted the fluttering insect as a threat, it immediately activated her sympathetic nervous system. Sympathetic nerves released norepinephrine and also stimulated epinephrine (and, to a lesser extent, norepinephrine) release from the adrenal medulla. As a result, her heart rate and cardiac output increased, which raised blood pressure to support the "fight or flight" reaction. For the same reason her bronchioles dilated to bring in more air; her blood glucose levels rose; and her pupils

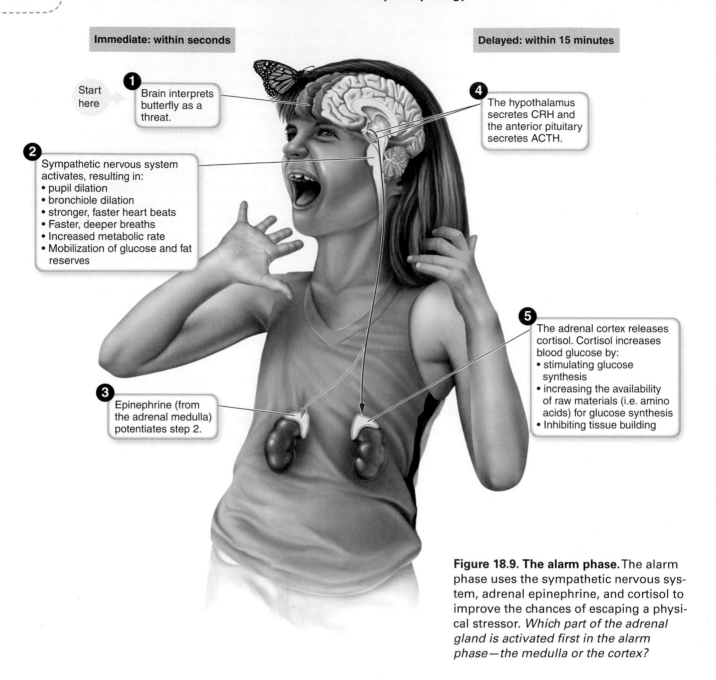

Immediate: within seconds

Delayed: within 15 minutes

Start here

1 Brain interprets butterfly as a threat.

2 Sympathetic nervous system activates, resulting in:
• pupil dilation
• bronchiole dilation
• stronger, faster heart beats
• Faster, deeper breaths
• Increased metabolic rate
• Mobilization of glucose and fat reserves

3 Epinephrine (from the adrenal medulla) potentiates step 2.

4 The hypothalamus secretes CRH and the anterior pituitary secretes ACTH.

5 The adrenal cortex releases cortisol. Cortisol increases blood glucose by:
• stimulating glucose synthesis
• increasing the availability of raw materials (i.e. amino acids) for glucose synthesis
• Inhibiting tissue building

Figure 18.9. The alarm phase. The alarm phase uses the sympathetic nervous system, adrenal epinephrine, and cortisol to improve the chances of escaping a physical stressor. *Which part of the adrenal gland is activated first in the alarm phase—the medulla or the cortex?*

dilated to gather more light. This coordinated response might have helped Kate fight or escape from the physical threat—which was real to her.

We noted above that the alarm phase has an endocrine component. However, endocrine reactions are slower than nervous system responses. Endocrine responses have an effect within several minutes after the stressor and peak at about 30 minutes. In the butterfly incident, Kate's hypothalamus secreted more corticotrophin-releasing hormone (CRH), which stimulated pituitary adrenocorticotropin (ACTH) production. ACTH traveled in blood to the adrenals to induce the adrenal cortex to synthesize and release a flood of

cortisol, which in turn increased blood glucose concentrations. This second, endocrine arm of the alarm phase is devoted to a longer, sustainable reaction to stress if the stressor persists. For example, it channels more of Kate's body resources toward intense muscular activities.

Case Note

18.13. Kate's heart rate was high when Tom reached her. Does increased heart rate reflect the actions of epinephrine or cortisol?

Adaptation Phase

As time passes and Kate matures and understands her butterfly encounter, she will probably become less alarmed by insects. To put it another way, she will enter the second phase of the stress response—*adaptation*. The sight of a yellow butterfly or a yellow jacket will not be quite so stressful and will no longer activate her sympathetic nervous system and her adrenal cortex. Insects are not likely to change, but Kate's perception is. We also adapt to physical stressors, as discussed further on. For instance, if Tom were to hike the same trail daily, the hike would no longer pose such a physical challenge.

Exhaustion Phase

If adaptation fails—if we do not adapt to the stressor or if the threat is ongoing and real—we enter the third phase: *exhaustion*. In this phase, cortisol secretion stays high for a time. Eventually, however, it ceases. The detrimental effects of stress become manifest, and death can result. Let us examine the effects of stress exhaustion more closely.

Stress Exhaustion Harms Cells and Systems

The stress response evolved as a means of coping with physical stressors, such as food deprivation and predators. Yet most modern stressors are psychological and frequently of long duration. When we do not adapt to these chronic stressors, we experience stress exhaustion, which is linked to three deleterious changes:

1. Chronic inflammation: Stress induces inflammation, which in turn damages tissues. Prior to menopause, estrogen offers some protection to women from stress-related vascular disease, which explains some of the higher rates of vascular disease in men under age 50. However, postmenopausal women, without the protective effects of estrogen on blood vessels, soon develop vascular disease at the same rate as men.
2. Accelerated cellular aging: Chronic stress may inhibit DNA repair, which increases the frequency of gene mutations and also accelerates telomere shortening. Cells age faster.
3. Increased oxidative stress: Stress increases ROS production, which further affects telomere length. Increased oxidative stress has, for instance, been linked to degenerative changes in the brain.

The steps linking stress to these three changes are not well understood but may partially reflect long-term

elevations in cortisol. The resulting chronically high blood glucose concentrations frequently induce insulin resistance. Cortisol coupled with insulin resistance leads to central obesity—that is, fat depositions in the abdominal cavity. To make matters worse, many people consume calorie-rich foods when they get stressed. Central fat stores release all sorts of hormones and inflammatory mediators that prompt or exacerbate disease.

Table 18.1 summarizes some of effects of stress exhaustion on different body systems. Of note is how many stress-related disorders have an autoimmune component.

Table 18.1	Diseases and Conditions Related to Stress
Body System	**Disease or Condition**
Gastrointestinal	Ulcers Irritable bowel syndrome Diarrhea or constipation
Skin	Eczema* Acne
Cardiovascular	Heart rhythm disturbances (arrhythmias) Coronary artery disease Hypertension Stroke
Respiratory	Asthma* Hay fever, other allergies* Frequent respiratory infections
Endocrine	Diabetes mellitus
Reproductive	Erectile dysfunction Infertility Amenorrhea (absence of menstrual periods)
Nervous	Migraines Fibromyalgia, other pain disorders Multiple sclerosis* Depression Insomnia or sleepiness Posttraumatic stress disorder Appetite changes Learning difficulties
Musculoskeletal	Backache, muscle spasms Rheumatoid arthritis*

*These disorders have or may have an autoimmune component.

18.15 True or false: Virtually anything can be a stressor.

18.16 During the alarm phase of the stress response, which hormone is secreted first—epinephrine or cortiosol?

18.17 Which substance increases ventilation—cortisol or epinephrine?

18.18 Name three autoimmune disorders that are also stress-related.

Exercise

Apart from the details in our discussion so far, it is clear from personal experience that our physical, and to a certain extent our mental, abilities begin to decline as soon as our organs are fully mature. Age exacts its inevitable toll, and stress adds to the burde—unhealthy stress, that is. Healthy stress is beneficial, and exercise is a special form of beneficial stress.

The field of exercise physiology has blossomed in importance in the last few decades. *Exercise physiology* is the study of the functional aspects of exercise. Slightly different is the field of *sports physiology*, which focuses on training for competitive excellence, and *sports medicine*, which focuses on the prevention and correction of sports-related injuries.

The main focus of exercise physiology is investigating the limits to which physiological systems can be stressed. A helpful way of quantifying physical stress is to measure the amount of energy expended. For instance, even the most severe illnesses increase metabolic rate by only about 100%, but highly trained endurance runners may increase their metabolic rate by 2,000%. Few of us are marathon runners, but what exercise physiologists learn by studying marathoners can help us understand lesser degrees of stress.

For most people, exercise is the most common challenge to normal homeostasis. Exercise requires that energy be converted into force. In the process, heat and metabolic waste (mainly acids and CO_2) are produced. This represents a homeostatic challenge because body temperature and blood pH must be modulated within a narrow range. Both must also remain at optimal levels for maximum performance. In addition, increased ventilation and cardiac output are required. When it all comes together perfectly, athletic performance is a symphony of power, endurance, speed, and flexibility that is the result of thousands of hours of practice sustained by a will to triumph.

In this section we examine how multiple body systems work together harmoniously to maximize exercise performance and how a problem with any one system can reduce performance. Critical in our discussion will be the **rate-limiting factor**—the weakest link in the chain that prevents the athlete from achieving greater speed or power. The concept of rate-limiting factors highlights the integrated nature of physiology—how every system is dependent on every other.

Oxygen Consumption Parallels Exercise Intensity

Remember that mitochondria, relying on oxygen, generate most of the ATP for exercising muscles. Energy production and oxygen consumption are thus proportional. A person's maximum capacity to consume oxygen (and thus generate energy) is his/her **$\dot{V}O_2$max,** which is expressed in liter of oxygen consumed per minute. Researchers typically measure $\dot{V}O_2$max by having the subject run on a treadmill at gradually increasing speed and incline. A device compares the oxygen content of inhaled and exhaled air in order to quantify the amount of oxygen extracted.

To analyze the possible factors limiting someone's $\dot{V}O_2$max, consider all of the necessary elements for oxygen's journey from the environment to its use in cellular metabolism (Fig. 18.10):

1. *Respiratory Muscles.* Respiratory muscles expand the thorax, creating a pressure gradient that draws air into the lungs. Both the depth and frequency of breathing determine how much air is available for gas exchange. So, to return to our case, Tom is breathing both deeply and rapidly.
2. *Atmospheric Oxygen.* Oxygen diffuses down its own partial pressure gradient, from the alveoli to pulmonary blood. For Tom, the high altitude of the mountain dude ranch may be interfering with this step. Atmospheric pressure, and thus the partial pressure of oxygen, are lower at the dude ranch than at Tom's home in Dallas. As a result, less oxygen diffuses from lungs to blood because the oxygen partial pressure gradient is lower.
3. *Hemoglobin.* Hemoglobin (in red blood cells) picks up about 98% of the oxygen that diffused into blood and carries it to the tissues. Neither Tom nor his offspring are anemic, so their hemoglobin stores are adequate

② Alveolar O_2 pressure helps determine diffusion into blood.

③ Hb concentration determines how much O_2 blood can carry.

① Respiratory muscles contract strongly and frequently to maximize ventilation, which maximizes alveolar oxygen partial pressure.

④ Heart rate and stroke volume determine cardiac output, which sends O_2 to tissues.

⑤ More capillaries increase O_2 supply to muscle.

⑥ Muscle cells with more mitochondria and enzymes can use more O_2 to generate more ATP.

Figure 18.10. Maximal oxygen consumption. The maximal level of exertion is determined by oxygen uptake, transport, or cellular use. *How could we increase the amount of oxygen supplied to an individual muscle (as compared with the other muscles)?*

for everyday life. Recall from ◀ Chapter 10 that low blood oxygen stimulates renal output of erythropoietin, which in turn stimulates red blood cell and hemoglobin production. Thus, if they stay for a while at the high-altitude ranch, their hemoglobin stores will increase to offset the problem described in step 2.

4. *Cardiac output.* The heart sends the oxygen-laden blood from the lungs to the periphery. Tom, Lea, and

Kate need a high cardiac output, so the sympathetic nervous system increases their heart rate and the strength of each cardiac muscle contraction. Cardiac output can increase 20-fold at maximum exercise intensity.

5. *Muscle Blood flow.* The amount of blood delivered to working muscles depends on blood flow. For example, patients with severe vascular disease and low blood flow to their legs learn that their legs tire long before their cardiovascular and respiratory reserves reach their limit.

6. *Mitochondrial capacity*. The amount of oxygen that muscles use to generate ATP depends on adequate supplies of mitochondria and metabolic enzymes. Muscle fiber types differ in this respect: type I (oxidative) muscle fibers have more mitochondria, while type II (glycolytic) fibers have less.

In the steep part of the hike, Tom was exercising all-out and felt as if he could not bring in enough air. This sensation indicates that Tom was working at his maximum intensity level—his maximum oxygen consumption, or $\dot{V}O_2$max.

In a healthy person, therefore, what determines $\dot{V}O_2$max? The amount of respiratory membrane for gas diffusion? The respiratory rate? Cardiac output? Blood flow? The amount of hemoglobin in blood? The number of mitochondria in muscle? In people with certain disorders, any one of these could be the limiting factor: the amount of pulmonary membrane in someone with emphysema, for example, or the hemoglobin of an anemic patient. But in a healthy person, which one is it?

Well, it is not the total surface area of respiratory membrane available for gas diffusion. Nor is it the amount of blood hemoglobin or the respiratory rate. The choke point is usually cardiac output—ventilation is typically at only 65% of maximum capacity when cardiac output is near 90%.

Case Note

18.14. **If Tom did the same hike at lower altitude, would he find it easier or harder? Explain.**

Specific Exercise Training Improves Exercise Capacity

Tom did not anticipate his difficulties with the hike, because he exercised regularly. However, recall from ← Chapter 7 that there are actually two types of exercise—resistance exercise, such as weight lifting, and endurance exercise, such as hiking or running. Recall that Tom's exercise regimen centered on weight lifting. Resistance exercise improves two aspects of muscle function. First, it increases muscle fiber size: Tom's muscles contain more contractile elements and can generate more power. Second, it improves anaerobic metabolism—the ability to generate ATP from glycolysis and creatine phosphate. Tom's muscle cells contain more

creatine phosphate and glycogen, so they can generate more ATP anaerobically.

However, if Tom wants to keep up with Lea on mountain hikes, he must specifically train for endurance exercise. Prevailing wisdom dictates that Tom should exercise at about 70% of his maximum heart rate (which equals about 50% to 55% of his $\dot{V}O_2$max) for at least 30 minutes three times weekly. As we noted earlier, a rough estimate of maximum heart rate is 220 minus age, which for Tom means about 150 beats per minute. So his 70% target for exercise is near 105 beats per minute. After about 6 weeks, he should see his exercise capacity improve. This training regimen addresses the limiting factor for most people—cardiac output (Fig. 18.11). Recall that two factors determine cardiac output: stroke volume and heart rate. Heart rate actually declines with endurance training, so it is stroke volume that improves. Endurance exercise increases maximum stroke volume, enabling the heart to provide more blood for the muscles' use with every stroke. It accomplishes this by increasing heart size and the number of cardiac muscle fibers.

Secondarily, endurance exercise also improves the aerobic capacity of working muscles—that is, how effectively they can generate ATP. Muscles that frequently participate in endurance exercise (such as the legs of a jogger) have more capillaries, so they receive a greater proportion of the blood supply. Moreover, trained muscles can generate ATP more efficiently from both carbohydrate and fats, because they have more enzymes and mitochondria.

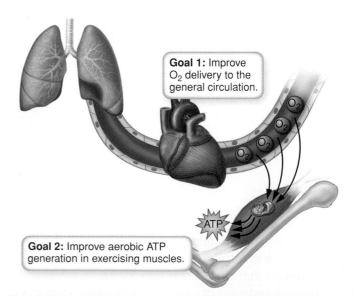

Goal 1: Improve O_2 delivery to the general circulation.

ATP

Goal 2: Improve aerobic ATP generation in exercising muscles.

Figure 18.11. Training increases maximal oxygen consumption. Endurance training improves oxygen delivery to working muscles and, secondarily, oxygen use by working muscles. *Does endurance training improve anaerobic or aerobic ATP generation?*

If Tom wants to simply improve his cardiovascular health, any endurance exercise will do—swimming, cycling, or running, for instance. However, improving his hiking performance requires an additional layer of specificity. Muscle blood supply and metabolism only improve in the specific muscles that are involved in the exercise. That is, swimming (which primarily relies on muscles of the trunk and upper limbs) would improve Tom's cardiac output and ventilatory capacity but would do nothing for his leg muscles. Thus, Tom needs to stick to exercises that call upon the large muscles of the lower limb and hip, such as hiking, jogging, stair climbing, or cross-country skiing.

> **Case Note**
>
> **18.15. Tom initiates an exercise regimen upon his return home. His friends suggest swimming to spare his aging joints. Will this form of exercise (which primarily involves the trunk and arm muscles) prepare him for next year's hike?**

Exercise Improves Health

Exercise is part of a healthy lifestyle at every age. If we don't "use it," we will indeed "lose it"; that is, we'll lose the capacity to perform even the modest activities of daily life. Healthy young adults take for granted the ability to run for a bus, carry a small child, dance at a party, or simply climb a flight of stairs, yet many adults lack the basic physical ability to perform these tasks comfortably. People who lead sedentary lives quickly begin to lose their physical capacity. Muscles lose mass and strength; joints stiffen; bones lose calcium and become easily fractured; blood flow becomes sluggish and venous thrombi are likely to form; urine stagnates in the collecting system; and skin develops bed sores. Even cognitive function begins to decline.

The remedy is exercise—and surprisingly little pays a big bonus. For example, the 1996 U.S. Surgeon General's report on physical activity stated that Americans need to accumulate only 30 minutes of physical activity on most days of the week to optimize their health. What's more, the exercise need not be performed all at once: a brisk 10-minute walk three times a day would fit the bill.

Regular exercise has many benefits, including:

- *Improved vascular function.* Regular exercise decreases the incidence of atherosclerosis in two ways. First, it improves lipid profiles, increasing the proportion of "good" high-density lipoprotein (HDL) cholesterol and reducing the proportion of "bad" low-density lipoprotein (LDL) cholesterol. It also decreases resting blood pressure.
- *Decreased body fat.* Endurance exercise can reduce body fat, and resistance exercise increases muscle mass. The improvement in lean body mass (that is, body mass minus fat mass) reduces the incidence (or severity) of obesity and many of the disorders associated with obesity, including type 2 diabetes, heart disease, stroke, sleep apnea, some cancers, infertility, and degenerative joint disease.
- *Enhanced psychological well-being.* Regular endurance or resistance training improves psychological well being, often to the same extent as antidepressant medication, and helps prevent stress exhaustion from psychological stressors.

18.19   Which of the following would be higher in a resting athlete than in a resting couch potato—stroke volume or heart rate?

18.20   True or false: For most people, the rate-limiting factor in exercise is the amount of blood pumped by the heart, not the amount of air brought into the lungs.

19.21   What is $\dot{V}O_2max$, and why do athletes determine their $\dot{V}O_2max$?

18.22   Which type of lipoprotein increases in response to regular exercise—LDL or HDL?

> **Case Note**
>
> **18.16. Since he began weight lifting, Tom's body weight has increased despite his consistent eating habits. Do you think weight lifting increased or decreased his risk of obesity-related disorders?**

Life and Death

It seems appropriate to end our discussion with death—the end of individual form and function. Our age at death defines our **life span;** anything that puts off death extends our **longevity;** that is, our life expectancy. To a certain extent, your death is foretold when you are a zygote: because of genetics—dogs live shorter lives than humans, while whales live much longer. However, within each species, and especially for humans, average life span varies: women on average live longer than men, and average life span varies by ethnicity and culture. And average life span is not fixed: evidence is clear that few Stone Age humans lived much beyond age 40. And in the twentieth century, improved health and nutrition increased average life span in most populations worldwide.

Every cell division shortens our telomeres, bringing us closer to the end. Some people are luckier in the genetic lottery than others—genes influence our susceptibility to autoimmune diseases and to cancer. They also influence how readily our cells repair DNA damage and how easily they replace themselves, both of which determine cell aging.

Death Is Necessary for Life

Our new Constitution is now established, and has an appearance that promises permanency; but in this world nothing can be said to be certain, except death and taxes.

Benjamin Franklin (1706–1790), letter to Jean-Baptiste LeRoy, November 13, 1789

The inevitability of death is critical to the continuance of life: without death there would not be enough room on the planet for new trees, bees, people, and countless other living organisms.

What is death and how is it defined? Scientifically, death is the cessation of metabolism. But determining exactly *when* death occurs is difficult. For example, even when the heart ceases to move blood through the vascular system and brain electrical activity has disappeared, some cells may continue to flicker with life for up to an hour or so. Among the general public the moment of death is usually considered to be cessation of heartbeat, but such a definition does not account for open-heart surgery, in which heart contractions are deliberately suspended and life maintained by artificial means while surgery is performed. And the legal definition of death is usually taken to be the absence of brain electrical activity for a prolonged period of time, which does not take into account the therapeutic, deathlike coma induced by medication for some types of brain or vascular surgery of the head and neck.

But what of dying itself? How do we die? Our penchant for orderliness has led some to propose that death proceeds in predictable ways, which can be studied and ranked: first this happens and then that. And in some instances predeath events do have a certain order, especially in patients suffering a long decline caused by some chronic condition such as cancer—our appetite declines, we grow listless and sleep more, our senses fail and we may become comatose, breathing becomes slower and irregular, blood pressure falls, and finally the heart stops and metabolism ceases. However, despite romanticized visions of an orderly, noble, peaceful ending, death is often messy, unpredictable, prolonged, and complicated. It is not how you die that counts but how you are remembered.

The prospect of imminent death affects the living, including the dying person. **Grief** is a deep sorrow over the approaching or actual loss of a loved one or a person's reaction upon learning that they are dying. Numerous theories about the experience of grieving have been developed. Among the most widely known is the Kübler-Ross model, which proposes five stages: denial, anger, bargaining, depression, and, finally, acceptance. While this model adds to our understanding, dying and grieving are not purely scientific, and Dr. Kübler-Ross herself cautioned that to expect an orderly, step-by-step process is to do a disservice to the individuality of people and circumstances.

Finally, death ends lives but not relationships—people are capable of holding for a lifetime a grudge against someone deceased, and most people conduct themselves in a manner that is true to established parental expectations long after that parent is dead.

Many Factors Contribute to a Longer, Happier Life

But how we live life is important too. Let us briefly review the key behaviors influencing longevity (Fig. 18.12):

- *Avoid unhealthy stress*. We must avoid unhealthy stress to the extent possible and develop effective coping mechanisms for reducing stress that is unavoidable. Stress exhaustion reduces longevity by accelerating cell aging and inducing or worsening many stress-related illnesses.
- *Do not smoke*. Smoking is the most pervasively destructive personal habit ever developed by humankind. We do not have room here to enumerate the many ways tobacco use destroys health. But to name a few, smoking promotes cancer (of many kinds, not just lung cancer), heart disease, high blood pressure, lung disease, and stroke.
- *Do not abuse alcohol and do not use illicit drugs*. Enough said.
- *Eat a diet rich in vitamins and minerals, and do not become overweight*. Good nutrition is essential to a

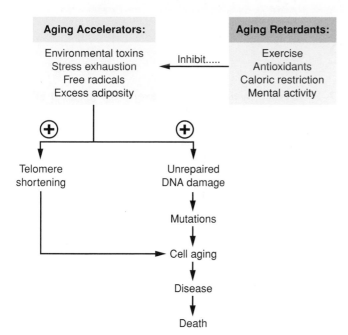

Figure 18.12. Death and aging. Death results from dysfunction (disease) in one or more body systems. One cause of disease is cell aging. All cells age, but some environmental and genetic factors hasten cell aging while other factors slow it down. *How do free radicals increase DNA mutations?*

healthy, happy life. Poor dietary choices—too many calories, too much salt, too much refined carbohydrate, or too much trans fat, for example—invite obesity and disease. And let us not forget that obesity itself, especially abdominal obesity, is a toxic, disease-causing condition that independently increases mortality risk.

● *Exercise regularly.* The appropriate physical stress of exercise reduces harmful stress, helps to prevent or reduce obesity, improves our cardiovascular and respiratory health and immune system, and promotes the peace of mind and feeling of well-being that comes from being fit.

● *Get enough sleep.* Research shows that lack of sleep disrupts every physiological system in the body, and there is no adapting to it. Most people need 7 to 9 hours, and the data indicate that disease risk begins to rise with fewer than 6 to 7 hours. Lack of sleep increases risk for certain types of cancer, heart attack, stroke, depression, and many other conditions.

We have focused primarily on the physical aspects of life, but life is a fusion of mind and matter, and mental health is equally important for a long and healthy life. So how is mental health maintained? For most it comes down to a matter of friends, family, faith, and lifelong learning. The support of friends and family has long been shown to improve health. Pets are also important. It has been shown that giving a pet to a person who has had a heart attack cuts in half the risk of having a second heart attack. Spiritual beliefs and practices also appear beneficial. The National Institutes of Health cites evidence that spiritual practices contribute to better health and longer life. The link is thought to involve improved immune and cardiovascular function. These effects are not limited to older adults: a recent national study of college and university students found that those who are spiritually inclined have better mental and emotional health, and a recent study of Americans of all ages found a positive association between reduced mortality risk and participation in spiritually oriented events. In addition, keeping the mind active with new learning—whether taking up a musical instrument, enrolling in a philosophy class, or challenging oneself to the crossword puzzle in the daily newspaper—can keep the mind sharp.

And finally, something must be said of happiness, the central element in most people's mental health. After studying people for a long time in many cultures, trying to tease out what it is that makes people happy, one scientist put it this way: "Happiness equals love."

Pop Quiz

18.23 Fill in the blank: Scientifically, death is the cessation of_____.

18.24 Exercise is a stressor. Does it increase or decrease longevity?

Chapter Challenge

CHAPTER RECALL

1. **Sickle cell anemia is equally common in men and women. Two unaffected parents can have a child with sickle cell anemia. This disease is most likely**
   a. autosomal dominant.
   b. sex-linked dominant.
   c. autosomal recessive.
   d. sex-linked recessive.

2. **In a pedigree, a heterozygotic female is represented by a**
   a. solid circle.
   b. half-filled circle.
   c. open square.
   d. half-filled square.

3. **As an adult ages from 40 to 60 years,**
   a. hormone production generally increases.
   b. the proportion of type II muscle fibers decreases.
   c. bones gain minerals, assuming the adult drinks enough milk.
   d. the sense of taste grows more acute.

4. **During the stress response, the adrenal cortex releases**
   a. cortisol.
   b. epinephrine.
   c. norepinephrine.
   d. all of the above.

5. **Ms. M. has been running 3 miles daily for 4 years. While the run was difficult at first, she now finds it very comfortable, and her pace has not changed for the past 2 years. Ms. M. is in which phase of the stress response?**
   a. Alarm
   b. Adaptation
   c. Exhaustion
   d. It is impossible to predict based on this information.

6. **Lea, aged 38, wants to train for endurance exercise. Her target heart rate is about**
   a. 127.
   b. 182.
   c. 220.
   d. 150.

7. **Resistance (weight) training exerts the greatest effect on**
   a. the quantity of glycogen and creatine phosphate stored in muscle cells.
   b. heart size.
   c. the number of capillaries in the working muscles.
   d. the quantity of mitochondrial enzymes in the working muscles.

8. **Frequent slow, long-distance bike rides would considerably improve**
   a. the number of capillaries in arm muscles.
   b. the number of mitochondria in leg muscles.
   c. the muscle fiber size in leg muscles.
   s. all of the above.

9. **Your favorite sporting event is the 5-km run. Which of the following would improve your 5-km running time?**
   a. Running a race at higher altitude, where the air is thinner
   b. Holding your breath as much as possible, to rest your respiratory muscles
   c. Increasing your heart rate without affecting your stroke volume
   d. Increasing your stroke volume without changing your heart rate

CONCEPTUAL UNDERSTANDING

10. **Explain how the $\dot{V}O_2$max is measured and why it is a measure of endurance exercise capacity.**

11. **Explain how DNA changes as we age, and describe the factor(s) causing these changes.**

APPLICATION

12. **Breast-fed babies often get their first respiratory illness at about 6 months of age. This increased susceptibility may reflect changes in their gastrointestinal systems. Explain.**

13. Endurance athletes may be at greater danger of ROS-induced aging than a sedentary person. Explain why.

14. There is a grain of truth in the saying, "Stress makes you old." Explain.

15. A common class of antihypertensive drugs, called beta agonists, substantially inhibit the sympathetic response. How would these drugs affect the stress response, assuming that they completely blocked the ability of tissues to respond to epinephrine/norepinephrine?

16. Refer to Figure 18.3A. Prepare a Punnett square for the offspring of Tom and his wife and answer the following questions:
    a. Is it possible for Tom to have a color-blind son? Why or why not?
    b. Is it possible for Tom to have a daughter who is not a carrier? Why or why not?

You can find the answers to these questions on the student Web site at
http://thepoint.lww.com/McConnellandHull

Glossary

abdominal cavity The superior portion of the abdominopelvic cavity, inferior to the diaphragm and superior to the pelvic bones.

abdominopelvic cavity The cavity that lies inferior to the diaphragm; the combined abdominal and pelvic cavities

ABO group A characterization of red blood cells according to their A or B antigen content

abortion Any interruption of pregnancy with death of the embryo or fetus before the 23rd week of pregnancy

accommodation In vision, the ability of the lens to vary its thickness in order to precisely focus light rays from objects on the retina

acid A compound that releases hydrogen ions when dissolved in water, which increases the number of free hydrogen ions

acidosis Abnormally low blood pH

actin A component protein of thin myofilaments in muscle

action potential A dramatic, transient, depolarizing change in membrane potential that travels the length of the cell in a self-propagating wave

activation In an adaptive immune reaction, the activation of an inactive lymphocyte by exposure to its target antigen

active immunity Immunity provided by one's own immune system

active transport The ATP-consuming (energy-consuming) transfer of a substance across the cell membrane, from a region of low concentration to high

acute Short term, more intense; the opposite of chronic

adaptation In sensory physiology, a decrease in the signal strength generated by the receptor during prolonged steady-state stimulation

adaptive immunity Acquired, targeted immune protection usually developed after birth, which requires education by an initial exposure to a particular pathogen

Addison disease A disease due to insufficient production of adrenocortical hormones

Adenosine triphosphate (ATP) A compound of adenosine, ribose, and three phosphate groups present in all living tissues, which stores energy in a phosphate bond; the molecule of intracellular energy transfer

adipose tissue Connective tissue containing large numbers of adipocytes (lipocytes) holding stored fat

adolescence The period of time from initiation of sexual maturity to the attainment of physical maturity, usually about age 20

adrenocortical hormone Any one of the hormones belonging to the three families of hormones produced by the adrenal cortex: glucocorticoids, mainly cortisol; androgens, mainly testosterone; and mineralocorticoids, mainly aldosterone

adrenomedullary hormone A hormone produced by the adrenal medulla, principally epinephrine and norepinephrine

adult A person who is physically fully developed

adulthood Physical and mental maturity

aerobic metabolism The generation of ATP (the production of energy) from nutrients by oxygen and mitochondrial respiration

agglutination Binding of one cell to another in clumps or chains

agonist A ligand that mimics or enhances the effect of an endogenous ligand

AIDS Acquired immunodeficiency syndrome; a constellation of clinical signs and symptoms (a syndrome) caused by defective immune function due to infection by the human immunodeficiency virus (HIV)

albumin The most abundant protein in blood

alkalosis Abnormally high blood pH

allele One of two or more alternative forms of a gene

allergen A substance capable of stimulating a hypersensitive (allergic) immune reaction

allergy An exaggerated (hypersensitive) immune reaction to certain substances (allergens) such as foods, pollens, metals or other substances in the external environment

alveolar ventilation rate The volume of fresh air reaching the alveoli each minute; the minute ventilation less an allowance for dead space

alveolus In the lung, a small air sac where gas exchange occurs

amino acids A small organic molecule containing both a carboxyl (–COOH) and an amino (–NH2) group that can be combined into proteins

amnion The inner fetal membrane that forms the amniotic sac

amniotic sac   The sac enclosing the fetus

anaerobic metabolism   The production of ATP (the production of energy) from creatine phosphate and glucose without reliance on oxygen and mitochondrial respiration

anatomy   Study of the form, structure and shape of body parts

androgen   Any one of a family of hormones that have masculinizing effects, especially testosterone

anemia   Too little hemoglobin in blood

anion   A negatively charged ion

antagonist   A ligand that blocks or diminishes the effect of an endogenous ligand. The opposite of agonist.

antibiotic   A small molecule capable of killing or halting the reproduction of bacteria

antibody   An immunoglobulin; an immune protein produced by the immune system in response to antigen stimulation; an anti-antigen

antibody-mediated immunity   The immunity mediated by B lymphocytes

antigen   Any substance, usually a protein, capable of inciting an immune reaction

antigen-presenting cell   A B lymphocyte, macrophage or dendritic cell that presents antigen to other cells of the immune system

antioxidant   A substance that inhibits the production or activity of reactive oxygen species

apocrine sweat gland   A skin gland in the groin and axillary regions; its salty secretion contains fatty acids and proteins, which bacteria metabolize, resulting in body odor

aponeurosis   A broad sheet of collagenous tissue similar to a tendon, which attaches bone to muscle

apoptosis   The normal, programmed death of aged cells

asthma   A condition in which expiration is obstructed by spasms of small airways

astrocyte   The most abundant glial cell of the central nervous system, which provides support, cohesion and homeostasis for neurons and can also act as a neuronal stem cell

atom   The smallest particle of an element that behaves like the element

atomic mass   The number of protons and neutrons in the nucleus of an atom

atomic number   The number of protons in the nucleus of an atom

atrial natriuretic peptide   A hormone released from the stretched atrium that acts on the renal tubule to reduce absorption of sodium into blood from tubular fluid

atrioventricular (AV) node   A cluster of cardiac myocytes with pacemaker potential located low in the interatrial septum, which conduct and delay slightly heartbeat action potentials before passing them downward into the ventricles

atrioventricular bundle   *The bundle of His*; A short bundle of cardiac conduction fibers that conducts heartbeat action potentials from the atria to the ventricles through the fibrous skeleton that electrically insulates the two

atrium   In the heart, one of the two low pressure upper chambers

autoimmune disease   Disease caused by immune attack upon self antigens

autonomic nervous system   That part of the peripheral nervous system involved in unconscious, involuntary (automatic) commands and responses

axon   A long cytoplasmic extension from a neuron cell body, which carries electrical signals from the neuron cell body to the axon terminal

axon terminal   The distal ending of an axon; frequently the site of neurotransmitter release.

B lymphocytes (B cells)   The cells of antibody-mediated adaptive immunity, which mature in the bone marrow

basal cell   An epithelial stem cell

basal cell   A cell in the deepest layer of the epidermis; a skin stem cell

basal metabolic rate   The rate of energy consumption at rest

base   A compound that decreases the number of free hydrogen ions, usually by releasing hydroxyl (OH–) ions

basement membrane   A thin sheet of extracellular material separating epithelium from underlying tissue

basophil   A granulocyte with dark purple (basic dye-loving) cytoplasmic granules

bile   A fluid secreted by the liver into bile ducts, which contains bile salts and metabolic waste

bile canaliculus   In the liver, a tiny channel running between hepatocytes that collects bile

bile salt   A cholesterol-related salt that emulsifies intestinal fat into small droplets for digestion

blood group   A characterization of red blood cells according to their antigen content

blood   The mixture of fluid and cells that circulates in the cardiovascular system

blood typing   The characterization of red blood cells according to their ABO and Rh groups

blood-brain barrier   A molecular barrier inherent in brain capillaries that limits the ability of substances in blood to enter brain tissue

bone   As a tissue: a complex, calcified tissue that forms the skeleton. As a structure: any anatomical object composed of bone tissue

Boyle's Law   The volume (V) and pressure (p) of a gas are inversely proportional; alternatively, the product of the pressure and volume of a gas is constant (k); $pV = k$.

bronchiole   A branch of the bronchial tree less than one millimeter in diameter

brush border enzyme   A digestive enzyme in the microvilli of small intestinal epithelial cells

buffer   Any substance that acts to restrain change of pH following the addition of acid or base

buffy coat   In centrifuged blood, a thin, tan layer of white blood cells and platelets at the interface between red blood cells and plasma

C-reactive protein   A protein synthesized by the liver in response to inflammation anywhere in the body, which attaches to pathogens and marks them as targets for phagocytosis

callus   A type of tissue formed in bone repair, which in its early *soft* stage contains granulation tissue, woven bone and cartilage, and later in its *hard* stage contains increasing amounts of bone tissue

capillary   The smallest and most numerous of all blood vessels, which carry blood through tissues

carcinoma   A malignancy of epithelial cells

cardiac conduction system   A branching network of specialized cardiac muscle fibers that conducts the electrical signals controlling myocardial contraction

cardiac cycle   The sequence of electrical and mechanical events from the beginning of one heartbeat to the beginning of the next

cardiac muscle   Heart muscle

cardiac output   The volume of blood ejected per minute by the left ventricle into the aorta

carrier   In genetics, a person or organism possessing a particular gene, especially a single recessive gene whose effect is masked by a dominant copy of the same gene

cartilage   As a tissue: firm, flexible, resilient connective tissue found in various forms in the ear, nose, respiratory tree and joints. As a structure: any anatomical object composed of cartilaginous tissue

cation   A positively charged ion

cell   The smallest structural and functional unit of an organism

cell cycle   The orderly sequence of events by which one cell reproduces into two

cell-mediated immunity   The immunity mediated by T lymphocytes

cellular respiration   The utilization of oxygen and production of carbon dioxide by body cells to generate ATP

central vein   The vein at the center of an hepatic lobule

centriole   A cylindrical organelle that organizes microtubules into an array for separating chromosomes in cell division

centromere   A small body that binds chromatids together during the process of cell division

centrosome   The region of cytoplasm near the nucleus that contains centrioles

cerumen   A waxy secretion of glands in the external ear canal

chemoreceptor   A sensory receptor activated by a chemical

chief cell   A gastric epithelial cell that produces *pepsinogen*, an inactive precursor to pepsin, a protein-digesting enzyme

childhood   The period between infancy and adolescence

cholecystokinin   A hormone secreted by intestina endocrine cells that stimulates bile release from the gallbladder and enzyme release from the pancreas

chordae tendineae   Fibrous cords that attach the edges of the atrioventricular valves to the papillary muscles of the ventricular wall

chorion   The outer fetal membrane, a portion of which forms the placenta

chromatin   The tangle of nuclear DNA visible in non-dividing cells

chromosome   An organized packet of DNA that contains genes. In humans, one of 23 pairs (46 total).

chronic   Long term, less intense; the opposite of acute

chylomicron   A lipid droplet wrapped in a coat of protein, which is assembled by intestinal epithelial cells from absorbed fat and channeled into lacteals for transport to blood

cilia   Hair-like projections of the cell membrane which move in unison to move microscopic material along the surface of a layer of cells

cirrhosis   A patterned scarring of the entire liver that is the final common pathway for many liver diseases

citric acid cycle   A circular series of chemical reactions that transfers nutrient energy to energy carriers (FADH and NADH2)

clonal expansion   In an adaptive immune reaction, the production of many identical copies of an original lymphocyte, each targeted at the same antigen

clot   A semisolid gel of fibrin, plasma and cellular elements of blood

coagulation factor   One of a family of proteins and other substances that regulate the formation of a clot

coagulation   The formation of a blood clot

cochlea   That part of the inner ear that senses sound waves

codon   In protein synthesis, a sequence of three DNA nucleotides that codes for a particular amino acid

collagen   The main structural protein of connective tissue; most abundant in the collagenous fibers of tendons and ligaments

columnar cell   A tall, thin, upright epithelial cell usually arrayed shoulder to shoulder with other columnar cells

common pathway   The final coagulation pathway, which when stimulated by either the tissue factor or contact activation pathway acts to cause the polymerization of fibrinogen into fibrin

compact bone   Normal, dense, smooth bone with an orderly microscopic structure and no grossly apparent spaces, which forms the outer (cortical) tissue of bones

complement system   Also, *complement*, a family of about 20 small proteins that participates in both innate and adaptive immunity.

compliance   The ease with which the volume of a space can be distended to accommodate increased content

compound   A substance containing at least two *different* elements linked by a chemical bond.

conducting zone   Respiratory air passages that convey air to the respiratory zone but do not participate in gas exchange

conduction   In energy transfer, the direct transfer of heat by contact

cone   In the retina, a light sensing cell responsible for color vision

connective tissue   Tissue that is not muscle, nerve or epithelium, which serves a structural or supportive role, and which contains a large amount of extracellular matrix.

consciousness   A quality of human brains that integrates sensory input into a composite picture of reality and one's place in it. This attribute of the brain binds sensory input into an internal picture of a "self" apart from the environment ("nonself"). This sense of self is made up of sensations, memories of the past, and expectations of the future, and has a concept of "time" based on immediate sensation and stored memory. Consciousness grants us an internal dialog with ourselves ("thinking") that is apparently uniquely human and endows us with the ability *to decide*.

contact activation pathway   One of the two initial branches of the coagulation process, which is initiated by plasma coming into contact with non-tissue substances outside the vascular space

contraception   Any method to prevent conception or successful implantation of a fertilized ovum

contractility   In cardiac and skeletal muscle, the strength of muscle cell contraction

convection   In energy transfer, the transfer of heat from an object to moving gas or liquid

convergence   In vision, the ability of the eyes to rotate nasally toward one another in order to remain focused on a closely approaching object, and to rotate temporally away from one another to focus on a receding object

coronary circulation   A system of arteries and veins that serve the heart

corpus luteum   A hormone-producing structure that develops from the follicle after ovulation

covalent bonds   A chemical bond formed between atoms sharing electrons

cranial cavity   The bony cavity formed by skull bones that contains the brain and associated structures

creatine phosphate   A molecule important in muscle energy production that contributes phosphate in the production of adenosine triphosphate (ATP)

cross-match   The trial mixing of donor and recipient blood to test for red cell agglutination

cuboidal cell   A boxy cell, usually part of an epithelium, especially in glands

Cushing syndrome   A constellation of signs and symptoms characteristic of a chronic excess of cortisol or related hormones

cytokine   One of a family of small proteins released by injured cells that regulate immune function by acting as messenger molecules as they travel between immune system cells

cytokinesis   The final act of cell division: division of the cytoplasm with the emergence of two new, independent cells

cytoplasm   All cell contents except the nucleus

cytoskeleton   A lacy network of protein filaments and tubules that provides a framework for cell structure and for intracellular transport

cytosol   The liquid part of the cytoplasm

cytotoxic T cell   A variety of T lymphocyte that attacks cells that are infected or cancerous

dead space   In pulmonary physiology, the space containing inspired air that does not participate in gas exchange; *anatomical dead space* is the air space of the *conducting zone*

decidua   The modified endometrium of pregnancy

dendrite   A short, branched cytoplasmic extension from a neuron cell body, which brings electrical signals from other cells or direct stimuli to the cell body.

dendritic cell   An antigen-presenting cell, present mainly in tissues (skin, respiratory and intestinal tracts) that come in contact with the environment.

depolarization   A reduction of the membrane potential

dermis   The thickest layer of skin located immediately beneath the epidermis, composed of connective tissue

diabetes mellitus   A disease characterized by abnormally high blood glucose due insufficient production or action of insulin

diapedesis   The crawling motion of white blood cells

diaphysis   The main shaft of a long bone

diastole   Myocardial relaxation and repolarization as the heart refills with blood

diffusion   The passive movement of substances from an area of high concentration to low

digestion   *Mechanical digestion* is the tearing and cutting of food into small pieces by the teeth and by the churning action of the stomach. *Chemical digestion* is the cleaving of large foodstuff molecules into smaller ones by the enzymes and other chemicals.

diploid   A cell having two complete sets of chromosomes

disease   An unhealthful state of form or function due to the effects of any kind of cellular damage

DNA   *Deoxyribonucleic acid.* A polymer containing deoxyribose nucleotides that is found in the cell nucleus. Contains the genetic code in the sequencing of its nucleotide bases.

eccrine sweat gland   A skin gland producing salty sweat that evaporates to decrease body temperature

ectopic pregnancy   Implantation of the fertilized ovum anywhere other than in the endometrium

effector   A structure, such as a muscle or gland, that carries out a command signal

ejaculation   Propulsion of semen from the male ductal system

ejection fraction   Stroke volume divided by end diastolic volume; the fraction of end diastolic volume (preload) ejected by left ventricular systole

elastance   The tendency of a distended space to recoil and return to its original dimension

elastin   A stretchy glycoprotein found in elastic fibers of some connective tissue

electrocardiogram   A graphical tracing of the grand sum of all myocardial voltage *changes*—greater or lesser than the moment before—caused by each heartbeat.

electrolyte   A salt that separates into ions when dissolved in water

element   A substance that cannot be separated into simpler substances by normal forces; e.g., oxygen.

embryo   In humans, an offspring during the first eight weeks of gestation

emphysema   A lung disease functionally characterized by expiratory obstruction and anatomically characterized by destruction of alveoli with loss of pulmonary membrane and coalescence of alveoli into large air sacs

end diastolic volume (EDV)   The maximum volume of blood in the left ventricle immediately prior to systole; also called *preload*

end systolic volume   The amount of blood remaining in the left ventricle after systole

endocardium   A thin layer of cells lining heart chambers and continuous with the endothelium of blood vessels

endochondral bone   Bone that in its formation passes through a cartilaginous phase

endocrine   Pertaining to glands that secrete their product into blood

endocrine glands   Discrete organs that secrete hormones into blood

endocytosis   The import of substances from the extracellular fluid or the cell membrane by invagination of the membrane to form a vesicle

endolymph   In the inner ear, the fluid within the membranous labyrinth

endoplasmic reticulum   An interconnected network of saccular membranes involved in protein synthesis, lipid synthesis, and the removal of toxic substances.

endosteum   A layer of bone-forming cells that lines the medullary cavity

endothelium   A single layer of flat cells that lines all blood vessels (also see *endocardium*)

energy   The capacity to do work.

enteric nervous system   A nerve network formed entirely of autonomic nerves in the wall of the gastrointestinal tract

enzyme   A specialized protein that facilitates a chemical reaction

eosinophil   A granulocyte with red (eosin dye-loving) cytoplasmic granules

ependymal cell   A specialized cell lining the CNS ventricles

epicardium   A thin layer of cells on the heart surface

epidermis   The surface layer of skin, composed of stratified squamous epithelium

epiphysis   The broadest part of a long bone at its end

epithelium   One or more layers of tightly packed cells that forms glands, covers body surfaces and lines hollow internal organs

equilibrium   In sensory physiology, the sense of balance

erythrocyte   A red blood cell

erythropoiesis   The production of new red blood cells

erythropoietin   A hormone secreted by the kidney that stimulated red blood cell production by the bone marrow

essential amino acid   One of nine amino acids necessary for life but which must be obtained from the diet

estrogen   A steroid hormone most abundant in females with multiple reproductive and non-reproductive roles

etiology   The cause of disease

eupnea   Quiet breathing; breathing in which the tidal volume is exchanged with each breath

evaporation   To turn from liquid into vapor

exocrine   Pertaining to glands that secrete their product into a duct

exocytosis   The export of substances out of the cell or into the cell membrane enabled by the fusion of a secretory vesicle with the cell membrane

expiration   Exhalation; the expelling of air from the lungs

external gas exchange   The absorption of oxygen from lung air into blood and the movement of carbon dioxide from blood into lung air

extracellular matrix   An acellular mixture of fibers and ground substance that characterizes different types of connective tissue

fat   A lipid that is solid at room temperature

fatty acid   A long chain of hydrogenated carbon atoms with an acidic molecule (the *carboxyl group*, COOH) at one end.

ferritin   An iron storage protein most abundant in bone marrow, liver, and muscle

fetus   In humans, an unborn offspring after the first eight weeks of gestation

fibrin   A long, thin strand of polymerized fibrinogen

fibrinogen   A plasma protein that polymerizes into long strands of fibrin to form a clot

fibroblast   A specialized connective tissue cell that synthesizes two types of connective tissue fibers: collagen and elastin

filtration membrane   Cells and basement membrane of the glomerulus and nephron capsule wall through which the glomerular filtrate passes from blood into the glomerular space

first heart sound (S1)   The sound made by simultaneous closure of the atrioventricular valves

fixed acid   An acid that cannot be exhaled as a gas and must be excreted by the kidneys; e.g., ketones

flagellum   A large, tail-like cell membrane extension of the spermatozoa that propels cell movement.

follicle   In the ovary, an oocyte and its casing of follicular cells

follicle stimulating hormone   An anterior pituitary hormone that in females stimulates oogenesis, maturation of ovarian follicles, and ovarian estrogen production; in males, indirectly promotes sperm production

Frank-Starling Law   The relationship between the initial fiber length (degree of stretch as indicated by end diastolic volume) of ventricular muscle fibers and their contractile strength

free radical   A general term describing any molecule or atom with an unpaired electron; an escaped anarchist.

frontal plane   Any vertical plane that divides structures into anterior and posterior parts.

G cell   An endocrine cell in the gastric mucosa of the pylorus that secretes the hormone gastrin

G-protein   An intracellular protein activated by a G-protein coupled receptor, which stimulates or inhibits the production of a second messenger

G-protein coupled receptor (GPCR)   A class of cell membrane receptors that uses G-proteins to alter second messenger synthesis when bound by an extracellular signal

gamete   A haploid germ cell able to unite with a gamete of the opposite sex to form a zygote

ganglion   In nervous tissue, a nodule of neuron cell bodies in the peripheral nervous system

gastric inhibitory peptide   A hormone secreted by intestinal mucosa that stimulates release of insulin by the pancreas

gene   A distinct segment of DNA devoted to the synthesis of a particular protein and therefore responsible for a particular trait

genome   All of the genes that are common to a single species

genotype   A set of genes unique to an individual

germ cell   A reproductive cell found only in the ovary or testis, which develops into either an ovum or sperm

germ layers   Three layers of tissue in the early embryo (endoderm, mesoderm, ectoderm), each of which gives rise to particular organs or types of tissue

gestational period   The length of pregnancy; in humans 280 days from the first day of the last menstrual period

ghrelin   A hormone synthesized by the stomach that stimulates appetite

gland   A collection of epithelial cells that synthesizes and secretes substances into blood or into a duct

glomerular capsule   The enlarged proximal portion of the renal tubule that envelops the glomerulus

glomerular filtrate   Water and solutes from blood that pass through the glomerular filtration membrane into the glomerular space

glomerular filtration rate   The volume of glomerular filtrate formed per unit of time

glomerular filtration   The passage of water and certain solutes through the glomerular filtration membrane into the glomerular space

glomerular space   The lumen of the glomerular capsule

glomerulus   A tuft of capillaries that is the filtering unit of a nephron

glucagon   A hormone secreted by pancreatic islets; its main action is to raise blood glucose levels by promoting hepatic glucose production

glucocorticoid   Any one of a family of adrenocortical hormones, mainly cortisol, with actions that modulate glucose and protein metabolism and the immune system

gluconeogenesis   The production of glucose from amino acids or other noncarbohydrate nutrients

glycogen   A glucose polymer (polysaccharide) used as a store of energy

glycogenesis   An anabolic reaction that converts glucose into glycogen

glycogenolysis   The breakdown of glycogen into glucose

glycolysis   A series of reactions in the cell cytosol that converts glucose into pyruvate and generates two ATPs

Golgi apparatus   A complex of folded membranes that processes and packages material from endoplasmic reticulum

Golgi tendon organ   A sensory organ in tendons that detects tension

gonadotropin-releasing hormone   A hypothalamic hormone that stimulates the anterior pituitary to release follicle stimulating hormone and luteinizing hormone

graded potential   A depolarizing or hyperpolarizing change in membrane potential that is proportional to the initiating stimulus and decays as it spreads

gradient   The difference in the quantity or concentration of a physical value between two areas

granulocyte   One of a family of white blood cells with large cytoplasmic granules

gray matter   In nervous tissue, a collection of neuron cell bodies

ground substance   An acellular fluid composed of water, minerals and small glycoproteins that is a component of the extracellular matrix of connective tissue

growth hormone   A pituitary hormone stimulates tissue growth and hepatic glucose production

gustation   The sense of taste

haploid   A cell having a single set of unpaired chromosomes

HDL   High density lipoprotein

healing   The natural repair of injury

heart block   A delay or complete blockage of transmission of cardiac action potentials from atria to ventricles

heat   A form of energy attributable to the motion of molecules in a substance

helper T cell   A variety of T lymphocyte that facilitates the activities of B lymphocytes and other T lymphocytes

hematocrit   The percent of blood volume occupied by red blood cells

hematopoiesis   The production of new blood cells

hemoglobin   1) an iron-bearing protein in red blood cells that binds oxygen for transport; 2) a measure of the weight of hemoglobin in a given volume of blood, typically grams per deciliter

hemoglobin saturation   The percent of hemoglobin bound to oxygen

hemolysis   Destruction of red blood cells

hemostasis   The body's collective system to prevent or stop hemorrhage

hepatic lobule   The basic organizational unit of the liver. Each surrounds a central vein and is bordered by portal triads

hepatic sinusoid   A large, leaky liver capillary between plates of hepatocytes, which carries a mixture of venous blood from the hepatic portal vein and arterial blood from the hepatic artery

hepatocyte   The principal functional cell of the liver

heterozygous   Having two different alleles of a particular gene

HIV   *Human immunodeficiency virus.* The virus that is the cause of acquired immunodeficiency syndrome (AIDS)

homeostasis   The body's collective communication and control effort to maintain internal conditions within a narrow, stable physiological range

homozygous   Having two identical alleles of a particular gene

horizontal plane   A transverse plane that divides structures into superior and inferior parts

hormone   A chemical signal that travels through blood to regulate the activity of distant cells.

human chorionic gonadotropin   A placental hormone that stimulates the corpus luteum to secrete estrogen and progesterone

hydrolysis   Chemical breakdown of a compound due to reaction with water

hydrophilic   Soluble in water

hydrophobic   Not soluble in water

hypercapnia   Above normal blood carbon dioxide concentration

hyperpolarization   An increase of the membrane potential

hyperthermia   In human physiology, a body temperature above 38°C or 100°F

hypoxia   Below normal blood oxygen concentration

iatrogenic   Caused by a physician

idiopathic   Of unknown cause (etiology)

immune   In physiology, the state of protection provided by the immune system against microbes or other external threats and against the internal threat of tumor development

immune surveillance   An activity of natural killer cells and cytotoxic T lymphocytes that identifies and kills cancer cells

immune system   A cellular defense system that protects against microbes and other threats

immunoglobulin   An antibody; a gamma globulin

infancy   The time between birth to 12 months

inflammation   The body's composite vascular and cellular response to injury

innate immunity   Inborn, general immune protection developed *in utero*, which requires no prior exposure to be activated and is directed broadly at any nonself antigens

inorganic   Not exclusive to living things; usually without carbon

inspiration   Inhalation; the intake of air into the lungs

insulin   A hormone secreted by pancreatic islets whose main action is to lower blood glucose levels by stimulating cell uptake and utilization of glucose

internal gas exchange   The transfer of oxygen from blood to body cells and the transfer of carbon dioxide from cells to blood

interneuron   A neuron that relays a signal from one neuron to another. Also, *association neuron*

interphase   The period of time between cell divisions.

interstitium   Supporting tissue and space between the functional cells of an organ

ion   An atom or molecule with a net positive or negative electric charge due to the gain or loss of an electron

isometric contraction   A muscle contraction that produces force but does not shorten the muscle

isotonic contraction   A muscle contraction that changes muscle length

isotope   Each of two or more forms of the same element (the same number of protons) but different numbers of neutrons

juxtaglomerular apparatus   A cluster of renal tubular and capillary cells near the glomerulus that is sensitive to blood pressure and secretes renin

keratinocyte   An epidermal cell containing keratin

Kupffer cell   A stationary macrophage in the liver

lacteal   A blunt-ended small intestinal lymph capillary devoted to the absorption of lipids

LDL   Low density lipoprotein

lesion   The structural abnormality associated with disease

leukemia   A malignancy of white blood cells in which malignant cells are present in blood

leukocyte   A white blood cell

leukopenia   An abnormally low number of leukocytes in blood

leukopoiesis   The production of new white blood cells

lifespan   The time period between birth and death

ligand   A chemical (such as a hormone) that binds to a larger molecule (such as a receptor)

lipid   A greasy, non-polar, hydrophobic compound composed mainly of carbon, hydrogen, and oxygen

lipogenesis   The synthesis of triglyceride from glycerol and free fatty acids

lipoprotein   A molecular complex formed of lipid and apoprotein that circulates in blood

longevity   Life expectancy beyond the norm for one's species

luteinizing hormone   An anterior pituitary hormone that in females triggers ovulation and ovarian progesterone production; in males it promotes testicular testosterone production

lymph node   A pea-sized organ in the network of lymphoid vessels composed of lymphocytes, macrophages, dendritic cells and supporting connective tissue

lymphatic system   A network of lymphoid vessels and lymphoid organs that has immune activity and other functions

lymphocyte   One of a family of nongranulocyte white blood cells; the principal cell of the immune system

lymphoma   A malignancy of lymphocytes or related immune cells in which malignant cells are not present in blood

lysosome   A membrane-bound packet of digestive enzymes that degrades internalized substances and worn-out organelles

macrophage   A tissue phagocyte that originated as a monocyte

major histocompatibility complex   A glycoprotein on the surface of every cell that binds and display antigens to the immune system

mechanoreceptor   A sensory receptor activated by mechanical activity

mediastinum   The space between the lungs, anterior to the spine, posterior to the sternum, superior to the diaphragm and inferior to the superior edge of the sternum, which contains the heart, great vessels, lower trachea, mainstem bronchi, thymus, and related structures

medullary cavity   In bones, the central core of spongy bone filled with marrow

medullary osmotic gradient   The gradual increase in osmotic pressure between the superficial and deep regions of the renal medulla

megakaryocyte   A bone marrow cell that produces platelets

meiosis   Germ cell division in which a diploid stem cell produces four haploid gametes

Meissner corpuscle   A sensory receptor activated by light touch, such as fluttering or stroking

melanin   A dark brown pigment that is produced by melanocytes and deposited in nearby tissue. Abundant in skin, hair, and eyes

melanocytes   A cell, most abundant in skin, that produces melanin

membrane potential   An electrical gradient that exists at the cell membrane, resulting from an excess of positive charges on one side and an excess of negative charges on the other; the voltage difference between the inside and outside of a cell membrane

membranous bone   Bone that in its formation begins as a fibrous membrane

memory cell   In adaptive immunity, an activated lymphocyte targeted to a particular antigen, which is held in reserve for quick response on reexposure to the antigen

menarche   A woman's first menstrual period

menopause   The normal age-related cessation of ovulation and menstruation

menstruation   The cyclical discharge of blood and dead tissue from the uterus during a woman's reproductive years

metabolism   The chemical reactions of a living organism necessary to sustain life

metaphysis   The flared, wider part of a long bone between the diaphysis and the epiphysis

micelle   An aggregation of monoglycerides, fatty acids, and small amounts of cholesterol surrounded by bile salts

microglia   Specialized macrophages of the central nervous system

microvilli   Short, hair-like projections of the cell membrane which facilitate absorption by increasing cell membrane surface area

micturition reflex   Autonomically modulated relaxation of the internal urinary sphincter triggered by a full urinary bladder

mineral   A *non-organic* element that can form a crystalline solid. Compounds composed of carbon, hydrogen, oxygen and nitrogen are not considered minerals because by definition they are organic.

mineralocorticoid   Any one of a family of adrenal hormones that influence renal sodium and potassium metabolism

minute ventilation   The total amount of air moved into and out of the lungs in one minute

mitochondrion   An organelle that houses the biochemical apparatus for converting nutrients into energy

mitosis   Division of the nucleus into two identical nuclei

molecule   Two or more atoms linked by a covalent bond

monocyte   A white blood cell that develops into a macrophage

monogenic   Trait attributable to a single gene

monomer   A molecule that can be bonded with other identical molecules to form a polymer

motor (efferent) neuron   A nerve cell that carries signal from the central nervous system to an effector organ such as a muscle or gland

mucosa   Mucous membrane

mucous membrane   *Mucosa.* A mucus-secreting epithelial tissue that lines body cavities exposed to the external environment (e.g., the respiratory tract)

murmur   An abnormal heart sound caused by abnormal, turbulent flow of blood in the heart or blood vessels

muscle spindle   Modified fibers in skeletal muscle that sense muscle length, which are important in proprioception and muscle reflexes

muscle tone   A state of subconscious isometric contraction that occurs relaxed skeletal muscle

mutation   A permanent change in DNA sequence resulting from DNA damage

myelin   A lipid that is the main component of the wrapping around the axons of selected neurons and which accelerates nerve impulse conduction

myelin   A wrapping around the axons of selected neurons that accelerates nerve impulse conduction.

myocardium Heart muscle

myofibril A bundle of myofilaments that runs the length of a skeletal muscle cell

myofilament A contractile element of muscle composed of thick or thin filaments

myoglobin An iron-containing compound in muscle, which stores oxygen

myosin A component protein of thick myofilaments in muscle

natural killer (NK) cells A lymphocyte of the innate immune system with an inborn capacity to kill virus-infected cells and cancer cells

necrosis Cell death due to injury or disease

negative feedback A process that decreases the magnitude of a change away from the set-point. The system's output reduces or reverses the original stimulus.

nephron The functional unit of the kidney, which consists of a glomerulus and its associated renal tubule

neuroglia Supporting cells of the nervous system

neurotransmitter A chemical signal that is stored in neurons, released in response to an electrical signal, and induces an electrical signal in an adjacent neuron.

neutropenia An abnormally low number of neutrophils in blood

neutrophil A granulocyte with neutral colored (tan) cytoplasmic granules

nociceptor A sensory receptor for pain

node of Ranvier The uncovered collar of cell membrane between successive sections of myelin sheath in myelinated neurons

nuclear envelope A two-layered membrane that surrounds the nuclear contents (nucleoplasm)

nucleic acid A polymer of nucleotides, such as deoxyribonucleic acid (DNA) or ribonucleic acid (RNA)

nucleolus A particularly dense nodule of nuclear chromatin visible in non-dividing (interphase) cells where ribosome assembly begins.

nucleotide The basic unit of DNA and RNA. Each is formed of one of two simple sugars (ribose or deoxyribose), one of five bases (five-carbon–ringed compounds: adenine, guanine, cytosine, thymine, or uracil), and one or more phosphate groups

nucleus In nervous tissue, a nodular collection of neuron cell bodies deep in the brain

nutrient Any chemical in food or drink that an organism needs to live or grow

odorant Any substance capable of stimulating olfactory cells

olfaction The sense of smell

oligodendrocyte A specialized glial cell that wraps CNS axons with a myelin sheath

oocyte A developmental stage in oogenesis. A primary oocyte has a diploid set of chromosomes. A secondary oocyte has a haploid set and is released from the ovary.

oogenesis The process of producing female gametes (ova)

oogonium The ovarian stem cell from which ova arise

opportunistic infection An infection by an organism that usually does not cause infection in people with normal immune systems, but common in those with immunodeficiency

opsonin A protein that attaches to a pathogen and marks it for phagocytosis

organ A self-contained structure devoted to a specific purpose or purposes

organ of Corti In the cochlea of the inner ear, a strip of tall receptor cells with cilia (hair cells) that detect sound waves

organ system A collection of organs devoted to a purpose or related purposes

organelle A specialized cellular subunit that performs a distinct function

organic Originating in a living thing; usually contain carbon

organism A complete life form

orgasm The apex of pleasurable sexual excitement, which in men is usually accompanied by ejaculation

osmolarity The number of particles of solute per liter of solvent (water)

osmosis The movement of water across a semipermeable membrane from an area of low osmolarity (high water concentration) to an area of higher osmolarity (lower water concentration)

ossicle Any of the three small bones of the middle ear

ossification The process of bone formation

osteoblasts Bone-forming cells

osteoclasts Normal, bone-dissolving cells which mobilize bone calcium and other constituents and return them to blood

osteocytes Bone cells that nourish and maintain mature bone tissue

osteoid A special form of collagen in bones that forms the initial structural network upon which bone tissue is built into bones

osteoporosis A disease of bone in which bones are weak and brittle due to lack of calcium

otolith organ The organ of the inner ear capable of sensing gravity and linear acceleration

ovum   A mature female gamete capable of union with a sperm to create a zygote; a secondary oocyte

oxidative phosphorylation   An oxygen-dependent metabolic pathway that uses the hydrogen ions of energy carriers (FADH and NADH2) to produce ATP

oxyhemoglobin   Oxygenated hemoglobin; hemoglobin with oxygen molecules attached

P wave   The electrocardiographic wave generated by atrial depolarization

pacemaker potential   An instability of the membrane potential of certain cardiac myocytes that gradually approaches the threshold potential, which results in paced firing of the action potentials that initiate each heartbeat

Pacinian corpuscle   A sensory pressure receptor in skin activated by vibration or sudden pressure

papillary dermis   The superficial, loose layer of dermis, which is formed into waffle-like ridges

papillary muscle   A nipple-like mound of cardiac muscle projecting from the interior wall of the ventricle to which chordae tendineae are attached from the edge of an atrioventricular valve

paracrine   A chemical signal that diffuses through extracellular fluid to alter the activity of nearby cells

parasympathetic   Referring to that part of the autonomic nervous system that stimulates the "rest and digest" reaction

parathyroid hormone   The hormone secreted by parathyroid glands, which acts to increase blood calcium

parietal   Pertaining to the wall of a body cavity. When applied to membranes, the membrane lining the wall of the cavity

parietal cell   A gastric epithelial cell that secretes *hydrochloric acid* and *intrinsic factor*

parietal pleura   The pleura that lines the chest cavity

parturition   The process of giving birth

passive immunity   Immunity provided by transfer of antibodies from another person or an animal

pathogen   An organism capable of causing disease

pathogenesis   The natural history and development of disease

pathology   The study of changes in body structure and function that occur as a result of disease

pathophysiology   The manner of abnormal function of disease

pedigree   A record of generations to map inheritance

pelvic cavity   The inferior portion of the abdominopelvic cavity, outlined by the pelvic bones.

peptide   A short chain of amino acids linked together by a bond (peptide bond) between the amino group of one and the peptide group of the other

pericardial cavity   The space between the two layers of the pericardium

pericardium   A sac in which the heart is contained

perilymph   In the inner ear, the fluid inside the bony labyrinth and outside the membranous labyrinth

periodic table   An array of the basic elements arranged in rows in order of their atomic number, so that elements with similar atomic structure (and therefore similar chemical properties) appear in vertical columns

periosteum   A tough fibrous membrane that covers bones

perisinusoidal space   In the liver, the space between an hepatic sinusoid and nearby liver cells, which connects to hepatic lymph vessels

peristalsis   A wave of circular muscle contraction that passes down the GI tract, propelling a food bolus ahead of it

peritoneal space   The space between the peritoneal membrane covering the surface of abdominopelvic organs and the peritoneal membrane lining the abdominopelvic cavity

peritoneum   The membrane lining the abdominopelvic cavity and covering many organs within

peroxisome   A membrane-bound packet of enzymes that metabolizes hydrogen peroxide and other harmful substances

phagocytosis   Ingestion of bacteria or other potentially harmful substances by mobile scavenger cells of the immune system

phenotype   The set of physical and molecular characteristics peculiar to particular genotype

phospholipid   A lipid containing a phosphate group, which is abundant in cell membranes

photoreceptor   A sensory receptor activated by light

phototransduction   In vision, the absorption of a photon and conversion of its energy into a chemical reaction; the initial step of light perception

physiology   Study of the function of body parts

placenta   A temporary organ of pregnancy that joins fetus to mother

plasma cell   A B lymphocyte that is actively producing antibodies

plasma   The liquid part of blood

platelets   A cellular element of blood; a fragment of megakaryocyte cytoplasm

pleura   One of two membranes that cover the lungs and line the chest cavity

pleura   A double-layered membrane that covers the surface of the lung and lines the interior of the thoracic cavity

pleural cavity   The space between the pleura that covers surface of the lung (the visceral pleura) and the pleura that lines the thoracic cavity (the parietal pleura)

pleural fluid   The fluid that occupies the pleural cavity

polycythemia   Too many red blood cells per unit of blood volume

polygenic   Trait attributable to more than one gene

polymer   A large molecule composed of many identical units (monomers)

portal triad   The trio of structures that mark the edge of an hepatic lobule: a bile duct, an hepatic artery arteriole and a portal venule

positive feedback   A process that increases the magnitude of a change away from the set-point. The system's output reinforces or increases the original stimulus.

preload   The maximum load of blood in the left ventricle immediately prior to systole; also called *end diastolic volume*

premature birth   Birth of a living infant after the 22nd week of pregnancy and before the end of the 38th week

primary oocyte   The offspring of an oogonium that undergoes meiosis I to produce one secondary oocyte and one polar body

primary spermatocyte   An offspring of spermatogonium that undergoes meiosis I to produce two secondary spermatocytes

progesterone   A steroid hormone that stimulates the uterus and endometrium to prepare for pregnancy

proprioception   The ability to sense the position of body parts relative to one another

prostaglandin   A lipid-derived chemical signal that usually acts as a paracrine

proteasome   A protein complex that degrades unneeded or faulty proteins

protein   A long chain of amino acids joined together by peptide bonds

puberty   The period of time during which adolescents gain sexual maturity and become capable of reproduction

Purkinje fibers   Small terminal fibers of the cardiac conduction system that convey action potentials into ventricular myocardium

QRS complex   The electrocardiographic wave generated by ventricular depolarization

radiation   The emission of energy as electromagnetic waves such a light

reactive oxygen species (ROS)   Highly reactive molecules containing an oxygen atom with an unpaired electron

receptor   A protein that changes the activity of the cell when bound by a specific chemical signal

red cell count   The number of red blood cells in a given volume of blood; typically one cubic millimeter (a microliter)

reflex   A rapid, involuntary response to a sensory signal

refraction   The bending of light waves

regulatory T cell   A variety of T lymphocyte that acts to suppress activation of the immune system and thereby maintain immune system homeostasis and self-tolerance; a suppressor T cell

releasing hormone   A hormone secreted by the hypothalamus that stimulates the anterior pituitary to synthesize and release a particular hormone

renal tubule   A tube of epithelial cells that concentrates and modulates tubular fluid to produce urine

repolarization   The return of membrane potential to its original state

respiratory center   A collection of neurons in the brainstem that initiates the respiratory cycle and modulates it in response to chemical or physical factors

respiratory membrane   The exposed surface of the alveolar wall through which gas exchange occurs

respiratory zone   That part of the respiratory system where gas exchange occurs

reticular dermis   The deep, dense layer of dermis, which contains most of the blood vessels, nerves, glands and other structures of the dermis

Rh group   A characterization of red blood cells according to their Rh D antigen content

rhodopsin   A molecule in the rods and cones of the retina that absorbs photons and initiates the process of light perception

ribosome   A cell organelle that synthesizes proteins

rigor mortis   Stiffening of skeletal muscle after death, which rigidifies the body for about 24 hours

RNA   *Ribonucleic acid*. A polymer containing ribose nucleotides that is found throughout the cell. The multiple types of RNA perform diverse roles in protein synthesis.

rod   In the retina, a light sensing cell that reacts in the same way to light of any color

Ruffini corpuscle   A sensory stretch receptor in skin that reacts to minute displacements

sagittal plane   Any vertical plane that divides structures into right and left parts

salt   An electrically neutral combination of a cation and an anion

saltatory conduction   The propagation of an action potential in myelinated neurons, in which the action potential jumps from one node of Ranvier to the next

sarcolemma   The cell membrane of a muscle cell

sarcomere   The basic structural unit of a myofibril in skeletal muscle

sarcoplasm   Muscle cell cytoplasm

sarcoplasmic reticulum   A network of fluid-filled tubules in muscle cells, which stores calcium

satellite cell   In skeletal muscle, a stem cell capable of differentiating into a muscle cell

saturated fat   A triglyceride, usually solid at room temperature, in which all of the carbon atoms in its fatty acids are joined by single bonds and which, therefore, have no capacity to bind additional hydrogen atoms

scarring   Fibrous repair; especially in association with severe tissue injury or in tissues with few stem cells

Schwann cell   A specialized cell in the peripheral nervous system similar to an oligodendrocyte in the central nervous system, which wraps PNS axons with a myelin sheath

sebaceous gland   A gland that secretes sebum into hair follicles of skin

sebum   An oily substance secreted by sebaceous glands in skin

second heart sound (S2)   The found made by simultaneous closure of the aortic and pulmonary valves

second messenger   An intracellular signal molecule synthesized in response to an initial extracellular signal molecule (the first messenger)

secondary oocyte   A mature ovum: the offspring of a primary oocyte that is released from ovary. If fertilized by a spermatozoon, it will complete meiosis II and fuse with the spermatozoa to create a zygote.

secondary spermatocyte   An offspring of a primary spermatocyte that undergoes meiosis II to produce two spermatids

secretin   A hormone secreted by duodenal mucosa that stimulates pancreatic bicarbonate production

self-tolerance   The ability of the immune system to recognize self antigens and not attack them

semen   A fluid mixture of sperm and accessory gland secretions

senescence   Age-related deterioration

sensation   Conscious awareness of a sensory signal

sensory (afferent) neuron   A cell carrying a nerve signal from sensory receptor to the central nervous system

sensory receptor   Specialized cells (in many cases modified neurons) or parts of cells that react to a specific type of stimulus, such as light or sound waves, and convert it into an electrical signal that the brain can integrate into a sensation.

serum   Plasma without coagulation factors; a fluid exuded by clotted blood

sex-linked traits   Genetic characteristics carried by sex chromosomes

sexual reproduction   A method of reproduction requiring two contributors to produce offspring

sign   A direct, measurable observation by an examiner

sinoatrial (SA) node   A cluster of cardiac myocytes in the right atrium with pacemaker potential that fire the regular action potentials that initiate each heartbeat.

skeletal muscle   Voluntary, striated muscle; usually attached to bone

smooth muscle   Involuntary, nonstriated muscle

solute   The minor component in a solution

solution   A mixture of solute and solvent

solvent   The major component in a solution

somatic cell   Any cell not a germ (reproductive) cell

somatic nervous system   That part of the nervous system involved in voluntary (conscious) commands and responses

sperm   (spermatozoon) The reproductive cell that carries the male's half of the genetic endowment

spermatid   An immature spermatozoon

spermatogenesis   The production of male gametes (sperm)

spermatogonium   The testicular stem cell from which sperm arise

spinal cavity   The bony cavity formed by vertebrae, which contains the spinal cord and related structures

spirometer   A device for measuring lung air volumes and rates of air flow

spongy bone   Normal bone containing multiple open spaces, which renders it lighter than compact bone

squamous cell   A flat, pancake-like cell; usually part of an epithelium

standard anatomic position   Standing, head upright, face forward, arms at sides, hands positioned with palms forward, feet forward and parallel

starch   A plant polysaccharide that is a good energy source

stem cell   An unspecialized (undifferentiated) cell that is capable of asymmetric division: the simultaneous production of replacement copy of itself and a more specialized cell

steroid   A class of lipid characterized by four interlocking carbon rings; includes cholesterol, some hormones, and some vitamins

stress   In biology, the response of a cell or system to demand for change

stressor In biology, any agent or situation that prompts demand for physical or mental change

stroke volume The volume of blood ejected by the left ventricle in one heartbeat

subcutaneous tissue The superficial layer of tissue beneath, and tightly bonded to, skin

sympathetic Referring to that part of the autonomic nervous system that stimulates the "fight-or-flight" reaction

symptom A complaint reported by a patient

synapse The site where an electrical signal passes from one cell to the next.

syndrome A distinctive collection of signs and symptoms

systole Myocardial muscle depolarization and contraction to eject blood from the heart

T cell antigen receptor A receptor in the T lymphocyte cell membrane that binds with antigen

T lymphocytes (T cells) The cells of cellular adaptive immunity, which mature in the thymus

T wave The electrocardiographic wave generated by ventricular repolarization

T-tubules A network of tubular extensions of sarcolemma into the sarcoplasm

tastant Any substance capable of stimulating taste cells

telomere A special length of DNA at the end of a chromosome, some of which is lost with each cell division, and which limits the number of cell divisions, after which the cell loses functionality or dies

tendon A thick, tough cord of collagenous tissue that attaches muscle to bone

testosterone A steroid hormone most abundant in males that promotes spermatogenesis and the development and maintenance of male secondary sex characteristics

thirst center A portion of the hypothalamus sensitive to blood osmolarity, which controls the urge to drink water

thoracic cavity The cavity that lies above the diaphragm; the chest cavity

thrombin A coagulation factor produced by conversion of prothrombin, which acts directly to stimulate the polymerization of fibrinogen into fibrin

thrombocytopenia Too few platelets per volume of blood

thrombopoiesis The production of new platelets

thrombopoietin A hormone that stimulates platelet production

thrombus An abnormal, pathologic, intravascular accumulation of cellular elements of blood

thymus A small lymphoid organ in the anterior chest where immature lymphocytes mature into T lymphocytes before migrating to other organs

thyroxine Also, *tetraiodothyronine or T4*. The thyroid hormone with four iodine molecules (also see triiodothyronine)

tissue A collection of similar or related cells

tissue factor A membrane protein found in extravascular tissues that initiates clotting when coagulation factor VII binds to it.

tissue factor pathway One of the two initial branches of the coagulation process, which is initiated by plasma coming into contact with tissue factor in the extravascular space

tonsils Masses of lymphoid tissue encircling the opening of the throat at the back of the mouth

toxin A substance of plant or animal origin that causes disease when present at low concentration in the body

tract In nervous tissue, a bundle of myelinated axons in the central nervous system

transcription The process of transferring the DNA code to RNA

transferrin A plasma protein that binds iron for transport

translation The process of translating the nucleotide sequence of messenger RNA (mRNA) into the amino acid sequence of a protein

transverse plane A horizontal plane that divides structures into superior and inferior parts

triglyceride A lipid composed of three fatty acids attached to glycerol

triiodothyronine Also *T3*. The thyroid hormone with three iodine molecules

tubular fluid The glomerular filtrate as it passes down the renal tubule

umbilical cord A flexible rope-like connection between fetus and placenta that contains blood vessels

unsaturated fat A triglyceride, usually liquid at room temperature, in which one or more fatty acids contains a double bond and which, therefore, can bind at least one additional hydrogen atom

uterine cycle (*menstrual cycle*) The monthly series of events in the endometrium from the beginning of one menstrual period to the next

vagal tone The steady parasympathetic action of the vagus nerve upon the heart, intestines and other vagus-innervated structures

vasoconstriction Narrowing of a blood vessel lumen due to contraction of smooth muscle cells in its wall

vasodilation Broadening of a blood vessel lumen due to relaxation of smooth muscle cells in its wall

ventricle   In the heart, one of the two high pressure lower chambers

vesicle   An intracellular, membrane-bound packet of material

vestibular apparatus   That part of the inner ear that senses motion and gravity

visceral   Pertaining to viscera or organs. When applied to membranes, the membrane covering the surface of the organ

visceral pleura   The pleura that covers the lungs

vitamin   A small organic molecule that is required for a particular metabolic reaction and must be obtained in the diet

vitamin D   A family of cholesterol-related molecules which promote increased intestinal absorption of calcium

V̇O₂ max   A person's maximal capacity to consume oxygen

volatile acid   An acid exhaled as a gas

white matter   In nervous tissue, collections of myelinated axons

woven bone   A type of bone with a haphazard microscopic structure that forms 1) temporarily in the process of repair of damaged bone tissue, or 2) as a normal stage of bone development in the formation of flat bones

Z-disc   The structure in skeletal muscle that divides one sarcomere from another

zygote   The diploid cell formed from the union of spermatozoon and ovum that develops into the embryo

Figure Credits

Chapter 1

Unnumbered Figure 1.2. 17th-century photo—oil on canvas 1632. The Koninklijk Kabinet van Schilderijen, Mauritshuis, The Hague.

Chapter 2

Unnumbered Figure 2.2. Reprinted from McArdle W, Katch F, Katch V. *Exercise Physiology: Energy, Nutrition, & Human Performance*, 6th ed. Baltimore: Lippincott Williams & Wilkins, 2007.

Chapter 3

Unnumbered figure. "The History of Science: Decoding the Rosetta Stone." Courtesy of Adam J. Zaner.

Figure 3.5A and B. Reprinted with permission from Cormack DH. *Essential Histology*, 2nd ed. Philadelphia: Lippincott Williams & Wilkins, 2001.

Figure 3.6. Reprinted with permission from Cormack DH. *Essential Histology*, 2nd ed. Philadelphia: Lippincott Williams & Wilkins, 2001.

Figure 3.7. Reprinted with permission from Cormack DH. *Essential Histology*, 2nd ed. Philadelphia: Lippincott Williams & Wilkins, 2001.

Figure 3.8 (right figure). Reprinted with permission from Cormack DH. *Essential Histology*, 2nd ed. Philadelphia: Lippincott Williams & Wilkins, 2001.

Figure 3.13 (left bottom). Reprinted with permission from McClatchey KD. *Clinical Laboratory Medicine*, 2nd Edition. Philadelphia: Lippincott Williams & Wilkins, 2002. (right bottom) Reprinted from Dean D, Herbener TE. *Cross-Sectional Human Anatomy*. Baltimore: Lippincott Williams & Wilkins, 2000.

Figure 3.26B. (SEMs courtesy of S. Erlandsen and P. Engelkirk.) Engelkirk PG, Burton GRW. *Burton's Microbiology for the Health Sciences*, 8th ed. Philadelphia: Lippincott Williams & Wilkins, 2007.

Figure 3.27. Reprinted with permission from Cormack DH. *Essential Histology*, 2nd ed. Philadelphia: Lippincott Williams & Wilkins, 2001.

Figure 3.30. (A, D, E, and F) Reprinted with permission from Cormack DH. *Essential Histology*, 2nd ed. Philadelphia: Lippincott Williams & Wilkins, 2001.
(B, C) Reprinted with permission from Mills SE, *Histology for Pathologists*, 3rd ed., Philadelphia: Lippincott Williams & Wilkins, 2007.
(G) Reprinted with permission from Gartner LP, Hiatt JL. *Color Atlas of Histology*, 3rd ed. Philadelphia: Lippincott Williams & Wilkins, 2000.
(H) Reprinted with permission from McClatchey KD. *Clinical Laboratory Medicine*, 2nd ed. Philadelphia: Lippincott Williams & Wilkins, 2002.

Chapter 5

Unnumbered Figure 5.1. Reprinted with permission from Goodheart HP. *Goodheart's Photoguide of Common Skin Disorders,* 2nd ed. Philadelphia: Lippincott Williams & Wilkins, 2003.

Unnumbered Figure 5.2A–D. Reprinted with permission from McConnell T. *The Nature of Disease*. Baltimore: Lippincott Williams & Wilkins, 2007.

Figure 5.5B. Reprinted with permission from Goodheart HP, MD. *Goodheart's Photoguide of Common Skin Disorders*, 2nd ed. Philadelphia: Lippincott Williams & Wilkins, 2003.

Figure 5.7. From Bickley LS, Szilagyi, P. Bates' *Guide to Physical Examination and History Taking*, 8th ed. Philadelphia: Lippincott Williams & Wilkins, 2003.

Figure 5.9B and C. Reprinted with permission from Cormack DH. *Essential Histology*, 2nd ed. Philadelphia: Lippincott Williams & Wilkins, 2001.

Unnumbered Figure 5.3A, B. (A) Reprinted with permission from McConnell T. *The Nature of Disease*. Baltimore: Lippincott Williams & Wilkins, 2007. (B) Reprinted from Smeltzer SC, Bare BG. *Textbook of Medical-Surgical Nursing*, 9th ed. Philadelphia: Lippincott Williams & Wilkins, 2000.

Table 5.2. (A) Reprinted with permission from Goodheart HP. *Goodheart's Photoguide of Common Skin Disorders*, 2nd ed. Philadelphia: Lippincott Williams & Wilkins, 2003. (B and C) From Fleisher GR, Ludwig S, Baskin MN. *Atlas of Pediatric Emergency Medicine*. Philadelphia: Lippincott Williams & Wilkins, 2004. (D) Image from Rubin E, Farber JL. *Pathology*, 3rd ed. Philadelphia: Lippincott Williams & Wilkins, 1999.

Figure 5.12 Epithelial membranes A–C: Reprinted with permission from Cormack DH. *Essential Histology*, 2nd ed. Philadelphia: Lippincott Williams & Wilkins, 2001.

Chapter 6

Figure 6.4C. Reprinted with permission from Rubin R, Strayer D. *Rubin's Pathology: Clinicopathologic Foundations of Medicine*, 5th ed. Philadelphia: Lippincott Williams & Wilkins, 2008.

Figure 6.8 A, B. Reprinted with permission from Rubin R, Strayer D. *Rubin's Pathology: Clinicopathologic Foundations of Medicine*, 5th ed. Philadelphia: Lippincott Williams & Wilkins, 2008.

Unnumbered Figure 6.1A–E. Reprinted with permission from McConnell T. *The Nature of Disease*. Baltimore: Lippincott Williams & Wilkins, 2007.

Unnumbered Figure 6.2B. Reprinted with permission from Rubin E, Farber JL. *Pathology*. 3rd ed. Philadelphia: Lippincott Williams & Wilkins, 1999 Philadelphia: Lippincott Williams & Wilkins, 2000.

Unnumbered Figure 6.3. Reprinted with permission from Daffner RH. *Clinical Radiology: The Essentials*, 3rd ed. Philadelphia: Lippincott Williams & Wilkins, 2007.

Chapter 7

Table 7.1. Reprinted with permission from Cormack DH. *Essential Histology*, 2nd ed. Philadelphia: Lippincott Williams & Wilkins, 2001.

Figure 7.10. Reprinted with permission from Rubin R, Strayer D. *Rubin's Pathology: Clinicopathologic Foundations of Medicine*, 5th ed. Philadelphia: Lippincott Williams & Wilkins, 2008.

Chapter 8

Figure 8.5B. Reprinted with permission from Gartner LP, Hiatt JL. *Color Atlas of Histology*, 3rd ed. Philadelphia: Lippincott Williams & Wilkins, 2000.

Chapter 10

Figure 10.2. (A) Reprinted with permission from McConnell T. *The Nature of Disease*. Baltimore: Lippincott Williams & Wilkins, 2007. (B) Reprinted with

permission from Gartner LP, Hiatt JL. *Color Atlas of Histology*, 3rd ed. Philadelphia: Lippincott Williams & Wilkins, 2000.

Figure 10.4A, B. Reprinted with permission from McConnell T. *The Nature of Disease*. Baltimore: Lippincott Williams & Wilkins, 2007.

Figure 10.5A. Reprinted with permission from Cohen BJ, Wood DL. *Memmler's The Human Body in Health and Disease*, 11th ed. Philadelphia: Lippincott Williams & Wilkins, 2009.

Figure 10.8B. Reprinted with permission from McKenzie SB. *Textbook of Hematology*, 2nd ed. Baltimore: Williams & Wilkins, 1996.

Chapter 11

Unnumbered Figure 11.1. Sigerist HE. (1965). *Große Ärzte*. München, Deutchland: J.F. Lehmans Verlag (5. auflage) (1. auflage 1958). plate 26 p 120.

Chapter 12

Figure 12.1. Bastion Castle © Copyright Civertan Grafikai Stúdió (Civertan Bt.), 1997–2006.

Figure 12.3. Virginia Commonwealth University Medical School, Department of Anatomy and Neuroscience, 2008.

Figure 12.20. Reprinted with permission from McConnell T. *The Nature of Disease*. Baltimore: Lippincott Williams & Wilkins, 2007.

Unnumbered Figure 12.2. Content Providers(s): CDC/James Hicks.

Chapter 13

Figure 13.6 C. Reprinted with permission from Cagle PT. *Color Atlas and Text of Pulmonary Pathology*. Philadelphia: Lippincott Williams & Wilkins, 2005.

Chapter 14

Figure 14.2. Copyright © 2008. For more information about The Healthy Eating Pyramid, please see The Nutrition Source, Department of Nutrition, Harvard School of Public Health, http://www.thenutritionsource.org, and Eat, Drink, and Be Healthy, by Walter C. Willett, M.D. and Patrick J. Skerrett (2005), Free Press/Simon & Schuster Inc.

Figure 14.18 B. Reprinted with permission from Gartner LP, Hiatt JL. *Color Atlas of Histology*, 3rd ed. Philadelphia: Lippincott Williams & Wilkins, 2000.

Figure 14.20 D. Reprinted with permission from Gartner LP, Hiatt JL. *Color Atlas of Histology*, 3rd ed. Philadelphia: Lippincott Williams & Wilkins, 2000.

Box Figure 3A,B. Reprinted with permission from McConnell, T. *The Nature of Disease*. Baltimore: Lippincott Williams & Wilkins, 2007.

Chapter 15

Figure 15.14B. Reprinted with permission from Cormack DH. *Essential Histology*, 2nd ed. Philadelphia: Lippincott Williams & Wilkins, 2001.

Figure 15.22. Reprinted with permission from Rubin E. *Essential Pathology*, 3rd ed. Philadelphia: Lippincott Williams & Wilkins, 2000.

Figure 15.25B. Reprinted with permission from Rubin E. *Essential Pathology*, 3rd ed. Philadelphia: Lippincott Williams & Wilkins, 2000.

Figure 15.27A, B. Reprinted with permission from Gagel R.F., McCutcheon I.E. [1999]. *Images in Clinical Medicine*. *New England Journal of Medicine* 340, 524. Copyright © 2003. Massachusetts Medical Society.

Chapter 16

Figure 16.3. Reprinted with permission from Daffner DH. *Clinical Radiology the Essentials,* 3rd ed. Philadelphia: Lippincott Williams & Wilkins, 2007.

Figure 16.5C. Reprinted with permission from McConnell T. *The Nature of Disease*. Baltimore: Lippincott Williams & Wilkins, 2007.

Chapter 17

Figure 17.3C: Reprinted with permission from Eroschenko VP, *di Fiore's Atlas of Histology with Functional Correlations*, 8th ed. Baltimore: Williams & Wilkins, 1995.

Figure 17.8. Courtesy of Dana Morse Bittus and BJ Cohen.

Figure 17.13B. Reprinted with permission from Mills SE. *Histology for Pathologists*, 3rd ed. Philadelphia: Lippincott Williams & Wilkins, 2007.

Figure 17.16C. Reprinted with permission from Pilliteri A. *Maternal and Child Health Nursing,* 4th ed. Philadelphia: Lippincott Williams & Wilkins, 2003.

Figure 17.22A–B. (A) Courtesy of Ansell Health Care, Inc., Personal Products Group and Carter-Wallace, Inc. Reprinted from Westheimer R, Lopater S. *Human Sexuality: A Psychosocial Perspective*. Baltimore: Lippincott Williams & Wilkins, 2002. (B) Courtesy of ALZA Pharmaceuticals, Palo Alto, CA.

Unnumbered Figure 17.1. Reprinted with permission from Sadler TW. *Langman's Medical Embryology*, 7th ed. Baltimore: Lippincott Williams & Wilkins, 1995.

Chapter 18

Figure 18.1. Reprinted with permission from Pillitteri A. *Maternal and Child Nursing*, 4th ed. Philadelphia: Lippincott, Williams & Wilkins, 2003.

Figure 18.8. Reprinted with permission from Rubin R, Strayer DS. *Rubin's Pathology: Clinicopathologic Foundations of Medicine*, 5th ed. Philadelphia: Lippincott Williams & Wilkins, 2008.

Index

Note: Page numbers followed by the letter f refer to figures; those followed the letter t refer to tables.